MUSCA DOMESTICA *LINNAEUS*

Balloon Wing in *Musca domestica*. A rare deformity, resulting from a malfunction in wing expansion. This photograph was first published as a cover feature of the *Annals of the Entomological Society of America*, Vol. 62, No. 5, 1969. Used by permission of the Society and of Dr. Robert P. Bodnaryk, of the Research Institute, Canadian Department of Agriculture, at Belleville, Ontario. Photo by Mr. T. Stovell, of the same Institute.

AN ANNOTATED BIBLIOGRAPHY OF *MUSCA DOMESTICA* LINNAEUS

by

Luther S. West Ph.D., D.Sc.

and

Oneita Beth Peters, M.A.

with the Technical Assistance of

Jane Elizabeth Phillips, M.A.

1973

DAWSONS OF PALL MALL
FOLKESTONE & LONDON

In conjunction with

NORTHERN MICHIGAN UNIVERSITY
MARQUETTE, MICHIGAN

First Published in 1973

DAWSONS OF PALL MALL
Cannon House
Folkestone, Kent, England

ISBN: 0 7129 0536 7

This publication was supported in part by
Grant LM 26,409 from the
National Library of Medicine
National Institutes of Health
United States Department of Health,
Education and Welfare.

Printed in Great Britain by C. H. Gee & Co. (Printers) Ltd., Leicester

TABLE OF CONTENTS

PREFACE

Not until recent years, has it become desirable or necessary to go to the expense of preparing and publishing "bibliographies" as independent contributions to the disciplines concerned. The great flood of research papers over the past quarter century has, however, definitely required more undertakings of this kind. The bibliography of the world's literature on schistosomiasis by Warren and Newill (1967) is an excellent example of a valuable aid to the parasitologist, likewise the bibliography entitled "Pathology in Invertebrates other than Insects" by Phyllis Johnson, 1968. Typical of bibliographies more limited in scope or in time covered is the list of references included in Scott's 1964 paper, "Human Myiasis in North America" (1952-1962 inclusive). An expecially valuable work because of the abstracts it contains, is Hocking's "Smell in Insects", published in 1960.

The present undertaking with relation to *Musca domestica* is an attempt to meet the needs of at least three groups of scientists who must acquire and maintain a fairly comprehensive acquaintance with the published literature which has a bearing on their work. These are: (1) Medically and public-health oriented investigators, usually associated with governmental agencies, philanthropic foundations, or medical schools (2) Agriculturally oriented personnel, normally employed by experiment stations, agricultural colleges and universities or veterinary schools (3) Graduate students, preparing for a professional career under the guidance of senior scholars, whose fundamental interest is likely to be in the more purely academic rather than applied aspects of biology and chemistry.

It is with such users in mind, that our annotations have been prepared. It should be pointed out, however, that an annotation is not an abstract. Consisting of from one to three sentences, the annotation seeks only to convey some insight into the content of the publication not set forth in the title. As compilers, we have therefore exercised personal judgement as to whether, in certain cases, an annotation is justified. Where the title is sufficiently informative, and especially if the publication is more of historical interest than of basic value to the modern investigator, the annotation is omitted.

We have, however, in many cases, indicated an informational source, where an abstract of the paper may be found. This is for the convenience of the user who may not have ready access to the original publication, but who has available one or more of the abstracting journals which cover the field of Sanitary Entomology. In the English language, *three* are especially important: *Biological Abstracts; Tropical Diseases Bulletin;* and the *Review of Applied Entomology*, Series B. In the pages which follow, these are indicated, respectively, by the abbreviations: BA, TDB and RAE—B.

The abbreviations used in this bibliography do not follow any *one* of the several check lists in general use. With occasional modifications, they are the abbreviations adopted in the senior author's volume "The Housefly, its Natural History, Medical Importance, and Control", published in 1951. Since the present publication is in fact, a continuation of compilation work begun in 1945, it has seemed desirable to disturb established techniques as little as possible.

Titles enclosed in brackets, [] are translations, usually into English, of titles published originally in another language.

Persons consulting this volume will note certain deviations from consistency when one Section is compared with another. There are several reasons for this. Section IV which parallels, in dates covered, the author's 1951 volume was prepared considerably in advance of other Sections and retains certain usages not followed in the preparation of Sections V, VI and VII. For example, English translations of titles originally in another language have the principal words capitalized in Section IV; rarely so in those which follow. Again, in citing abstracts, only the *page* of the abstracting journal is given in Section IV, as very few such journals numbered their abstracts in earlier days. An example would be "RAE-B 17:133, 1929." For later years, when the numbering of abstracts became more generally practiced, the page number is usually followed by an abstract number, placed in parenthesis. A typical entry would be "BA 50:4781 (50259), 1969." It was also decided, in preparing the later Sections, to indicate the month of issue, as many libraries postpone binding these journals into volumes, and the work of the investigator is often expedited by being able to call for a particular issue at the library desk. A complete citation of this sort reads "TBD 66:1185 (2309), Nov., 1969."

To re-edit the manuscript, in the interest of strict uniformity, would have delayed publication a great deal, to no good purpose, while further denying the availability of the Bibliography to a large number of persons who have long awaited its publication.

In the course of our labors, we have accumulated a considerable number of references of anonymous authorship. Because these are never reports of actual research, and fall almost altogether in the class of farmers' bulletins, sanitary education for the layman, or trade-journal articles, it has been decided not to include them in the present publication. On the other hand, important papers of the corporate authorship type (e.g. certain releases by the World Health Organization) are listed whenever their content is relevant to the purpose and objectives of our work.

A complete index in the usual meaning of the term is obviously not practicable for a publication of this type. It is possible, however, to give the user certain clues in searching out the literature most likely to have a bearing on the research in which he is engaged. The pages constituting item VIII in the Table of Contents are designed to give such clues under a number of subdivisions most likely to be sought. It is hoped that this feature will considerably enhance the general usefulness of the volume.

The following bibliographies include approximately 8000 entries, of which no less than 5800* are detailed references, listed either by surname of single author, or of the senior member of the team, if the authorship is multiple. The remaining entries identify joint authors, and consist of a cross reference to senior author and date, for each publication in which the team member had a part.

Scattered throughout the listing are references for which complete information was not available at the time of publication, but whose titles appeared too important to omit. It was felt that the users of this

* Actual listing terminates with the number 5720. Approximately 300 references, however, were added during the process of publication. This was done by utilizing the preceding number, followed by the letter "a". Thus Reference No. 642a would fall between 642 and 643, without disturbing the original numbering, or complicating the ongoing preparation of the Index.

bibliography would appreciate their inclusion, nevertheless. Should the publication of a supplement to the present work prove practicable during the next five years, such listings will be repeated with annotations and citation of available abstracts.

The compilers feel that great importance attaches to joint authorship, for two reasons: 1. Multiple authorship is becoming more frequent as "team" research replaces individual effort. 2. In many cases, the junior author is a future world authority who will continue the work of the senior author in a particular field when the latter is no longer active in publication.

ACKNOWLEDGEMENTS

Basic to all progress, has been the encouragement and assistance received from personnel connected with Northern Michigan University. Two former Directors of Research and Development, Dr. Roy Heath and Mr. J. Patrick Farrell, were most helpful in preparing proposals to the Department of Health, Education and Welfare, and in following through on details necessary to approval of the Grant which financed the final year of work.

Mr. James L. Carter, also of Research and Development, has labored constantly to further our enterprise in all possible ways, both during the operational period of the Grant and subsequently, with particular regard to arrangements for publication.

Dr. Roland S. Strolle, currently Director of that Office, also Dr. Thomas Griffith, then Dean of Arts and Science, gave the authors encouragement and support in various ways. We are especially grateful for the sustained interest of Dr. Jacob Vinocur, Vice President for Academic Affairs, through whose personal attention a substantial subsidy toward the cost of publication was provided by the University.

We are deeply indebted to Dr. Robert T. Wagner, Head of the University's Department of Physics, not only for the provision of excellent working space, but also for his willingness to share the services of his departmental secretary, Miss Rosemary Weza, whose skilful typing contributed significantly to the quality of the manuscript.

The Library staff, headed by Miss Helvi Walkonen, has responded promptly to all our requests and thus made our use of periodical literature a pleasurable experience. Besides Miss Mary McCarthy, former Periodicals Librarian, mention is due Mrs. Ted Rose, Mrs. Audrey Warren, and Miss Anne Rinnemaki, all of whom were expecially helpful in making available the various abstracting journals.

Mrs. Sally Bentley, officer in charge of funds for research grants, served the project with courtesy and efficiency. Her always cheerful cooperation was much appreciated.

Our profound thanks go to the Honorable Philip E. Ruppe, Member of Congress from the 11th District of Michigan, who went to great lengths to contact forty-one foreign embassies on our behalf. This was done so that we might obtain the mailing addresses of a large number of scientists with whom there was need to correspond. The prompt cooperation of most of the Embassies in either furnishing these adresses directly, or naming a suitable agency for contact, was beyond expectation and highly gratifying.

Our full-time secretary for the project, Mrs. Janice DeMarte, has been responsible, more than any other person, for keeping the work on schedule, and for assuring the availability of supplies as needed. Her many hours of arduous typing, sometimes at considerable sacrifice of time owed to her family has earned the lasting gratitude of the authors and associated academic personnel.

The painstaking work of our part-time typist, Miss Linda Hooper, has contributed greatly to the accuracy and acceptable appearance of the manuscript.

We are particularly grateful for the help of three persons who though busy at their own academic institutions, found time to give emergency service in praparing manuscript for publication. These are: Dr. David J. West, Assistant Professor of Biology at Lawrence University, Appleton, Wisconsin; Miss Susan A. Kamensky, of the North Central Forest Experiment Station, U.S. Forest Service, Marquette, Michigan; and Miss Lesley Ann West, of the Department of Zoology, at the University of Michigan, Ann Arbor.

The unique ability of our assistant in library technique, Mrs. Jane Philips, deserves special comment. Mrs. Phillips brought to her work a dedication to thoroughness and accuracy rarely found in auxiliary personnel. Her devotion to the project, even beyond the period of her original employment, has done much to keep errors at a minimum.

A word is in order concerning our choice of frontispiece, made available by Dr. R. P. Bodnaryk, to whom we are most grateful. The abnormality concerned has been observed to occur with a frequency somewhat less than $1{:}10^4$. In a personal communication to the senior author, Dr. Bodnaryk writes as follows: "Since balloon wing appears to be a malfunction in wing-expansion, it is tempting to speculate that an elucidation of the physical and chemical factors that govern normal wing expansion might lead to the design of agents that could disrupt this process—and hence to a novel method of fly control."

Such research falls obviously in the realm of biochemistry, biophysics, and physiology, an interdisciplinary front which is gaining in importance at a very rapid rate. An illustration depicting this anomaly seems not inappropriate as an introductory feature in a publication designed to serve modern biological research.

It is impossible to name here the large number of individuals and institutions that have contributed in various ways to the promotion of this work. Former students, graduate and undergraduate, have shared in the establishment and maintenance of files. Scientists in all parts of the world have contributed reprints of their work, have corresponded concerning best methods to pursue, and in many other ways have encouraged us in our undertaking. To all of these we express our hearty thanks. Without the continuing good will of colleagues, both at home and abroad, it would have been difficult, if not impossible to complete the work within the time available.

Luther S. West
Oneita Beth Peters

Section I.

Taxonomic Scope

All references to the genus *Musca*, so far as confirmed, are here included. We have taken special pains to cover the literature on the face fly, *Musca autumnalis* DeGeer, as it is the most likely of all forms to be confused with *Musca domestica* Linnaeus by farmers and dairymen. Outside the genus *Musca*, the only species which has received similar attention is the lesser housefly, *Fannia canicularis* Linnaeus. Any significant reference to the genus *Fannia* has therefore been included. Also, other genera are occasionally represented, as when an important contribution to muscoid biology rests primarily on experimental work with such groups as *Calliphora*, *Phaenicia* or some related genus. In the genetic field, it is inevitable that current findings for *Musca domestica* are compared by the investigator with earlier knowledge pertaining to *Drosophila*. Several titles bear this out. When, however, only the genus *Drosophila* is concerned, (even though the work be very recent), if there is no mention of or reference to the genus *Musca*, such titles have been omitted.

Section II.

Historical Periods

The following references are grouped into three historical periods, as indicated by dates of publication. The first of these concerns the literature up to and including 1949.* Approximately forty per cent of all references dealing directly or indirectly with the common housefly fall in this group. Since only about one-third of these were cited in the 1951 volume indicated below, a somewhat more complete listing seems justified at this point, even though the knowledge represented is, in many cases, chieffly of historical interest.

The second period (1950-1959) saw great emphasis on resistance to insecticides, particularly to the chlorinated hydrocarbons. Experiments dealt with such problems as the advantages and disadvantages of rotational use of chemicals versus "shot-gun" applications. The constant search for new insecticides, particularly of hitherto untested chemical groups, went on apace. Various "strains" of flies were developed and standardized in numerous laboratories throughout the world, for use in experiemental work. A considerable number of synergists were produced to increase the effectiveness of insecticidal substances by reducing the resistance of the fly. There was a growing realization that genetic differences were probably of prime importance in strain resistance and that selection under field conditions might account for most of the high resistance to modern insecticides.

Organophosphorus and carbamate insecticides were brought forward in impressive numbers only to find, in many instances, that cross-resistance already existed from exposure to their predecessors. Disillusionment with chemical control methods had, however, one wholesome effect: a return to the realization that basic sanitation is of fundamental importance in the control of insect vectors of disease, a principle temporarily lost sight of during the period when it was felt that near-perfect insecticides had been discovered.

The third period (1960-1969) is characterized by serious attention to complex biochemical phenomena, such as the metabolism of insecticides in the insects' body, the relation of cholesterol to these processes and the many organic substances, especially enzymes, in-

* The last year covered by "The Housefly, Its Natural History, Medical Importance and Control." Cornell University Press. 1951.

volved in these and other physiological phenomena. Current investigations, therefore, have taken on a strongly academic character. It is now appreciated, more than ever before, that basic research is fundamental to any significant improvement in economic practice. Investigations have been encouraged in such areas as histological and cellular research by use of the electron microscope; chromosomal loci of specific genes; and factors affecting gametogenesis. A practical area, closely related to the last involves the use of sterilization procedures, with subsequent release of sterilized insects among the wild population. Both X-ray and chemical means have been employed. The latter shows a greater range of possible uses. First developed to combat screw-worm flies, these techniques have since been widely tested on house flies and have accomplished appreciable reductions in reproduction rates in specific areas.

Another practical area, now being revived, is the manipulation of predators, parasites and especially microorganisms, such as certain bacteria, to affect adversely the natural reproduction of flies. *Bacillus thuringiensis* has been utilized in numerous studies. Among hymenopterous species, the pteromalid parasite, *Muscidifurax raptor*, is of real importance in many parts of the world.

In recent years, a significant number of research papers have been concerned with "Integrated Control Procedures", by which sanitary practices may be combined with the use of natural enemies and/or insecticides or repellents to reduce fly populations.

As this goes to the publisher, current research reports deal with control of *Musca sorbens* in Egypt by use of chemosterilants; insecticide metabolism by house fly enzymes; control of face fly larvae in manure by inclusion of insecticides in the diet of cattle; and with the development of juvenile hormone analogs as synergists for pyrethrins against house flies.

In conclusion, it should be pointed out that *Musca domestica*, long reared in laboratories in connection with the study of agricultural and public health problems, has in modern times achieved a place as one of the most frequently-employed experimental animals for academic studies. The numerous papers of Rockstein and associates (1960-1969), concerned chiefly with the physiology of aging, illustrate this very well, as do the studies of Greenberg (1954-1969) in gnotobiology, tissue culture, and germ-free research.

Section III.

Importance of Technique

Throughout the entire period covered by this bibliography, an impressive number of publications reflect the desire of investigators to share their technical experience with students and colleagues. Methods of maintaining dipterous colonies, devices for separating and/or counting eggs, larvae, pupae and adults, techniques of marking specimens for studies on dispersion (including the use of radioactive substances), preparation of baits containing chemical sterilants, and scores of other procedures which a particular author has invented or improved, are made the *raison d'etre* for technical papers or scientific notes. Always important, of course, are the constant refinements in controlled procedure for testing the comparative efficiency of available or potential insecticides. Your compilers have therefore thought it wise to separate from each of the foregoing groups of papers, those which deal purely with matters of technique, and combine these in a single, special list. This follows the list of references for the 1960-1969 period, and precedes the Index Portion.

Section IV.

References from Early Times through 1949

The authors' files contain approximately 3,300 references for this period, of which nearly one-third are general in nature, deal with many more taxonomic groups than the genus *Musca*, or are so non-technical as to be of little significance in the light of modern knowledge. To conserve space and at the same time to bring the cost of publication within reasonable limits, *for this portion alone*, we have endeavored to be selective. The following list, made up of approximately 1500 chosen titles, will, it is believed, serve adequately to acquaint the modern investigator with sufficient background information to constitute an appropriate foundation for advanced research.

LISTS OF REFERENCES
ALPHABETICALLY ARRANGED BY AUTHOR

Section IV.

References from Early Times through 1949

1. **Acharya, C. N.** and **Krishna Rao, K. S. 1945** Control of Fly-Breeding in Compost Heaps. Indian Med. Gaz. 80:272-273.
   Use of air-proof cloth assures asphyxiation of larvae, pupae and emerging adults, besides preventing additional oviposition. ABSTRACTS: BA 20:223, 1946; RAE-B 35:1, 1947.
2. **Acharya, C. N.** and **Krishna Rao, K. S. 1945** Experiments on the Control of Fly-Breeding in Compost Trenches. Indian J. Agric. Sci. 15:318-327.
   Use of tarred cloth over night-soil trenches filled in with earth and cattle dung. Some danger of pupation around edges. ABSTRACTS: BA 21:1808, 1947; RAE-B 37:142, 1949.
3. **Ackert, J. E. 1918** On the Life Cycle of the Fowl Cestode, *Davainea cesticillus* (Molin). J. Parasit. 5:41-43.
   Experimental proof that *M. domestica* can transmit this parasite.
4. **Ackert, J. E. 1919** On the Life History of *Davainea tetragona* (Molin), a Fowl Tapeworm. J. Parasit. 6:28-34.
   Chickens fed on flies which had consumed onchospheres (and/or parasite eggs) developed tape worms.

   Ackert, J. E., joint author. *See* Case, A. A., et al., 1939; Reid, W. M., et al., 1937.
5. **Acree, F., Jacobson, M.** and **Haller, H. L. 1945** An Amide possessing Inseciticidal Properties from Roots of *Erigeron affinis* D-C. J. Inorg. Chem. 10(3):236-242.
   An unsaturated isobutylamide of the same order of toxicity to *M. domestica* as pyrethrins. Proposed name, "affinin". ABSTRACT: RAE-B 34:191, 1946.
6. **Adams, C. F. 1941** A Preliminary List of Muscoid Flies in Missouri. Proc. Missouri Acad. Sci. 6:73-74.
   M. domestica one of 61 species falling in 45 genera.
7. **Aders, W. M. 1916** Entomology in Relation to Public Health and Medicine. Zanzibar Protectorate Med. and Publ. Health Repts. for 1916:47-49.
   M. domestica and other non-blood suckers scramble for last drop when blood-suckers finish feeding. ABSTRACT: RAE-B 6:46, 1918.
8. **Ainsworth, R. B. 1909** The House-Fly as a Disease Carrier. J. Roy. Army Med. Corps. 12:485-498.
   Studied infantile diarrhoea in Poona and Kirkee, India. Yearly curve shows close relationship with flies.
9. **Aksinin, J. S. 1929** [The Output of Free Ammonia by House-Fly Larvae (*Musca domestica* L.).] Plant Protection 6:379-382 (In Russian).
   ABSTRACT: RAE-B 18:276, 1930.

In the annotations accompanying these citations, all species of the genus *Musca* are referred to as *M. domestica*, *M. autumnalis*, etc. Any other genus beginning with the letter *M* is always spelled out (e.g. *Macrocheles muscaedomesticae*).

10. Aldridge, A. R. 1904 The Spread of the Infection of Enteric Fever by Flies. J. Roy. Army Med. Corps. 3:649-651.
Refers to *M. domestica determinata* Walker, which occurs at certain seasons in enormous numbers.

11. Aldridge, A. R. 1907 House-flies as Carrier of Enteric Fever Infection. J. Roy. Army Med. Corps. 9:558-571.
Greatest incidence of typhoid was among mounted troops, whose camps, because of horse manure, produced greatest number of flies.

12. Allen, T. C., Dicke, R. J. and **Harris, H. H. 1944** Sabadilla (*Schoenocaulon* spp.) with Reference to its Toxicity of Houseflies. J. Econ. Ent. 37:400-408.
Seed contains complex group of alkaloids (veratrine). Kerosene extracts effective against houseflies, but causes sneezing in man. ABSTRACT: RAE-B 33:66, 1945.

Allen, T. C., joint author. *See* Ikawa, M., et al., 1945; Kido, G. S., et al., 1947; Ball, H. J., et al., 1949.

13. Allnut, E. B. 1926 Some Experiences in the Control of Fly Breeding. J. Roy. Army Med. Corps. 47:105-120.
Recommendations for the handling of horse manure in Bermuda. ABSTRACT: RAE-B 14:187, 1926.

14. Alsterlund, J. 1946 Rothane D-3 in the Pest Control Field. Pests 14(5):10, Kansas City, Mo.
Is DDD (dichlordiphenyldichlorethane). Nearly complete kill of *M. domestica* by concentration of 0·5 per cent. ABSTRACT: RAE-B 37:50, 1949.

15. Alston, A. M. 1920 The Life History and Habits of two Parasites of Blowflies. Proc. Zool. Soc. London for 1920, (Part 3):195-243.
One of these, *Nasonia brevicornis* Ashm. is primarily a parasite of *M. domestica.* ABSTRACT: RAE-B 8:218, 1920.

16. Anderson, J. F. 1908 The Differentiation of Outbreaks of Typhoid Fever due to Infection by Water, Milk and Contact. Medical Record 74 (Nov. 28):909; also in Am. J. Publ. Health 19:251-259.
Flies of particular importance in relation to milk.

Anderson, J. F., joint author. *See* Lumsden, L. L., et al., 1911.

Anderson, T. G., joint author. *See* Kuhns, D. M., et al., 1944.

Anderson, W. H., joint author. *See* Cory, E. N., et al., 1936.

Ando, T., joint author. *See* Shinoda, O., et al., 1935.

17. Andre, C. 1908 Flies as Agents in the Dissemination of Koch's Bacillus. (Paper presented before Anti-Tuberculosis Congress, Washington, D.C.)

18. Andrews, H. W. 1925 Flies and Disease. Proc. S. London Ent. and Nat. Hist. Soc. 1924-25:45-62.
Part III deals with spread of disease by filth-feeding species. Is a compilation, not a research report.

19. Andrews, J. M. and **Simmons, S. W. 1948** Developments in the Use of the Newer Organic Insecticides of Public Health Importance. Am. J. Publ. Health 38:613-631.
Notes development of resistance to DDT by *M. domestica.* Evidence of toxicity to man. Names other currently useful compounds. ABSTRACT: RAE-B 22:2600, 1948.

20. Annand, P. N. 1944 Tests Conducted by the Bureau of Entomology and Plant Quarantine to Appraise the Usefulness of DDT as an

Insecticide. J. Econ. Ent. 37:125-159.
An introduction by the author, plus 49 short notes by others, 7 of which concern houseflies.

21. **Anthony, A. L. 1920** A Simple Form of Fly-proof Latrine as used in West Africa. J. Roy. Army Med. Corps 34:141-143.
Recommends that pit have minimum depth of 15 feet. ABSTRACT: RAE-B 8:91, 1920.

Anthony, M. V., joint author. *See* Marcovitch, S., et al., 1931.

22. **Ara, F.** and **Marengo, U. 1932** Sull 'importanza della mosca nella diffusione della febbre tifoide. Nota I. Boll. Soc. Ital. Biol. Sper. 7:150-154.
Forty-four per cent of flies in rooms of typhoid patients (and with access to infected feces) showed infection. ABSTRACT: RAE-B 20:230, 1932.

de Arêa Leano, A.-E., joint author. *See* de Marqarinos Torres, C. B., et. al., 1923.

23. **Arizumi, S. 1934** [On the Potential Transmission of *Bacillus leprae* by Certain Insects.] J. Med. Ass. Formosa 33(349):634-661. (In Japanese, with English summary.)
M. domestica and *Lucilia caesar* most likely carriers. Flies from hospital wards frequently showed *B. leprae* in gut.

24. **Ark, P. A.** and **Thomas, H. E. 1936** Persistence of *Erwinia amylovora* in Certain Insects. Phytopath. 26:375-381.

25. **Armstrong, D. B. 1914** Flies and Diarrhoeal Disease. Publication No. 79, New York Ass. for Improving the Conditions of the Poor. Bur. Publ. Health and Hyg., Dept. of Local Welfare, New York. 29 pp. 3 pls.
Incidence of such diseases fell off as a result of anti-fly measures carried out in a congested area in the Italian quarter. Remained high in a similar area in which nothing was done.

26. **Armstrong, D. B. 1914** The House-fly and Diarrhoeal Disease among Children. J. Am. Med. Ass. 62:200-201.
An abridged version of the foregoing report.

27. **Armstrong, H. E. 1908** House-flies and Disease; and the Duties of Sanitary Authorities in Relation Thereto. Special Rept. by Med. Officer of Health for Newcastle-upon-Tyne.

Arthur, J. M., joint author. *See* Harvill, E. K., et al., 1943; Synerholm, M. E., et al., 1945.

28. **Atkeson, F. W., Shaw, O. A., Smith, R. C.** and **Borgman, A. R. 1943** Some Investigations of Fly Control in Dairy Barns. J. Dairy Sci. 26:219-232.
Covers several aspects. Difficult to dispense sprays made with oils of higher viscosity than 50 seconds Saybolt. ABSTRACT: RAE-B 32:106. 1944.

29. **Atkins, W. R. G. 1921** Note on the Chemotropism of the House Fly. Ann. Appl. Biol. 8:216-217.
Includes observations on formalin, beer, crude and amyl alcohol, fuel oil constituents, methyl and other acetates. Objective: to attract insects to poison baits. ABSTRACT: RAE-B 10:35, 1922.

30. **Atkinson, E. L. 1916** The Fly Pest in Gallipoli. J. Roy. Naval Med. Serv. 2:147-152.
Treats with disposal of manure, care of latrines, poison baits. ABSTRACT: RAE-B 4:91, 1916.

31. **Attimonelli, R. 1940** Reperti parasitologici nell' acqua di lavaggio delle mosche. [Parasites found by washing flies in water.] Pathologica 32(581):111-112.
Obtained flagellates, ciliates and rhizopods; also a larva of *Acarus reflexus*. ABSTRACT: BA 21:1020, 1947.

Attimonelli, R., joint author. *See* Sangiorgi, G., et al., 1940.

32. **Austen, E. E. 1904** The House-fly and Certain Allied Species as Disseminators of Enteric Fever among Troops in the Field. J. Roy. Army Med. Corps 2:651-668.
Includes reference to very great numbers of flies at meal time near Para, in tropical Brazil; also to great concentration of maggots in an abandoned latrine.

33. **Austen, E. E. 1909** Notes on Flies Examined during 1908. Report Local Gov't. Bd. on Publ. Health and Med. Subjects, N. S. No. 5. Prelim. Repts. on Flies as Carriers of Infection, pp. 3-4.
How to distinguish the more important species of flies found in houses.

34. **Austen, E. E. 1910** A new Indian Species of *Musca*. Ann. Mag. Nat. Hist. (8)5:114-117.
Musca pattoni sp. n. Breeds in cow dung. Sucks drops of blood oozing from puncture spot after blood-sucker leaves.

35. **Austen, E. E. 1911** Memorandum on the Result of Examination of Flies, etc., from Postwick Village and Refuse Deposit. Repts. Local Gov't. Bd. on Publ. Health and Med. Subjects, N.S. No. 53. Further Repts. (No. 4) on Flies as Carriers of Infection, pp. 11-12.
Includes note on the occurrence of the lesser house-fly at Leeds. Appears earlier than *M. domestica.*

35a. **Austen, E. E. 1912** British Flies which cause Myiasis in Man. Repts. Local Gov't. Bd. Publ. Health and Med. Subjects. N.S. No. 66; Further Reports (No. 5) on Flies as Carriers of Infection, pp. 5-15.
Includes *M. domestica, Fannia canicularis* and *F. scalaris.*

36. **Austen, E. E. 1913** The House-Fly as a Danger to Health, Its Life-History and How to Deal with It. Brit. Mus. (Nat. Hist.), Econ. Ser. No. 1, 11 pp, 2 pls., 3 figs.
A good, popular-type presentation with instructions for control. ABSTRACT: RAE-B 1:66, 1913.

37. **Austen, E. E. 1914** Do House-Flies Hibernate? Entomologist (London) 47:69-70.
A request that specimens taken in attics, stables, barns, etc. be sent in for identification.

38. **Austen, E. E. 1920** The House-Fly, its Life-History and Practical Measures for its Suppression. Brit. Mus. (Nat. Hist.), Econ. Ser. No. 1a, 52 pp., 2 pls., 7 figs.
A revision of his 1913 bulletin, plus information gained through experience with troops in World War I.

39. **Austen, E. E. 1926** The House-Fly, its Life-History, Importance as a Disease Carrier and Practical Measures for its Suppression. Brit. Mus. (Nat. Hist.) Econ. Ser. No. 1a, Ed. 2, 68 pp., illus.
Mentions connection with *Habronema* in horses and possible relation to anthrax, hog cholera, swamp fever of horses, tapeworms in poultry. ABSTRACT: RAE-B 14:101, 1926.

40. **Austen, E. E. 1928** The House-Fly, its Life-History, Importance as a Disease Carrier and Practical Measures for its Suppression. Brit.

Mus. (Nat. Hist.) Econ. Ser. No. 1a, Ed. 3, 71 pp., 2 pls., 7 figs.
Refers to use of calcium cyanide, borax and methods of burying manure. ABSTRACT: RAE-B 16:229, 1928.

41. **Austen, E. E. 1939** The House-Fly as a Danger to Health. Its Life-History and How to Deal with It. Brit. Mus. (Nat. Hist.) Econ. Ser. No. 1, Ed. 4, 25 pp., 11 figs.
An up-dating of the content of similar releases (1913, 1920, 1926, 1928). ABSTRACT: TDB 37:380, 1940.

42. **Austin, T. A.** and **Mayne, L. C. 1935** A Review of the Incidence of Amoebiasis in Zomba, with Special Reference to European Cases. Ann. Med. Rept. Nyasaland for 1934, pp. 73-82.
Extract of a report by W. A. Lamborn, 1935, which see.

43. **Awati, P. R. 1915** Studies in Flies, I. Chaetotaxy and Pilotaxy of Muscidae and the Range of their Variability in the same Species. Indian J. Med. Res. 3:135-148.
Cephalic chaetotaxy of little use for species recognition in the genus *Musca*. Thoracic bristles important in distinguishing genera. ABSTRACT: RAE-B 3:211, 1915.

44. **Awati, P. R. 1916** Studies in Flies, II. Contributions to the study of Specific Differences in the genus *Musca*. Indian J. Med. Res. 3:510-529.
Goes into morphology of male genitalia in great detail. Disagrees with Hewitt (1914) on many points. ABSTRACT: RAE-B 4:74, 1916.

45. **Awati, R. R. 1916** Studies in Flies, II. Contribution to the Study of Specific Differences in the genus *Musca*. 2. Structures other than Genitalia. Indian J. Med. Res. 4:123-139.
Many characters reviewed, including variations in the fifth sternite of the female, also alar spines. Includes a diagnostic key. ABSTRACT: RAE-B 4:172, 1916.

46. **Awati, P. R. 1917** Studies in Flies, III. Classification of the Genus *Musca* and Description of the Indian Species. Indian J. Med. Res. 5:160-191.
Gives descriptions of most species keyed out in his 1916 paper. ABSTRACT: RAE-B 6:1, 1918.

47. **Awati, P. R. 1920** A Note on the Genitalia of Portchinsky's Species *M. corvina* (*vivipara*) and *M. corvina* (*ovipara*). Indian J. Med. Res. 8:89-92.
Two forms are not seasonal varieties, but separate species (*corvina* F. and *autumnalis* DeG.) as shown by genitalia. ABSTRACT: RAE-B 9:105, 1921.

48. **Awati, P. R. 1920** Bionomics of Houseflies: I, Outdoor Feeding Habits of Houseflies with Special Reference to *Musca promiscua* (*angustifrons*?). Indian J. Med. Res. 7:548-552.
Species of *Musca* divided into two groups: (1) hematophagous forms, usually associated with cattle, (2) those more commonly associated with man. ABSTRACT: RAE-B 8:159, 1920.

49. **Awati, P. R. 1920** Bionomics of Houseflies: II, Attraction of House-flies to Different Colours. Indian J. Med. Res. 7:553-559.
Yellow had greatest attraction; red and violet, the least. No sex difference noted. ABSTRACT: RAE-B 8:159, 1920.

50. **Awati, P. R. 1920** Bionomics of House-Flies: IV, Some Notes on the Life History of *Musca*. Indian J. Med. Res. 8:80-88.
Musca divaricata and *M. promiscua* compared. First prefers goat's dung, the latter human feces. ABSTRACT: RAE-B 9:103, 1921.

51. **Awati, P. R.** and **Swaminath, C. S. 1920** Bionomics of Houseflies: III, A Preliminary Note on Attraction of Houseflies to Certain Fermenting and Putrefying Substances. Indian J. Med. Res. 7:560-567.
Odors of ammonia, hydrogen sulphide, phosphorous compounds, etc. believed necessary before flies approach, though food may be satisfactory otherwise. ABSTRACT: RAE-B 8:160, 1920.

52. **Aylott, W. R. 1896** Do Flies spread Tuberculosis? Virginia Med. Semimonthly, June 1896. *Also in* Am. Monthly Micr. J., Aug. 1896.

53. **Aziz, M. 1943** A Simple Device for Destroying Adult Mosquitoes, House-flies and other Household Pests by Use of a Flame Thrower. Trans. Roy. Soc. Trop. Med. Hyg. 36:364-365.
Flit gun with burner to ignite the spray. For use only in situations safe from fire. ABSTRACT: RAE-B 31:232, 1943.

54. **Baber, E. 1918** A Method of Trapping Fly Larvae in Manure Heaps. Lancet I (4935):471.
Larvae, driven from manure by heat of fermentation migrate to sheet metal gutter with overhanging lip. ABSTRACT: RAE-B 7:5, 1919.

55. **Baber, E. 1925** Fly Control by Means of the Fly-larval-trap Manure Enclosure. J. Roy, Army Med. Corps 45:443-452.
Concrete or brick platform with marginal trench. Manure enclosed in wire fence. Various methods for disposing of trapped larvae and liquid waste. ABSTRACT: RAE-B 14:25, 1926.

56. **Bacot, A. W. 1911** On the Persistence of Bacilli in the Gut of an Insect during Metamorphosis. Trans. Ent. Soc. London, for 1911: Part II, pp. 497-500.
Bacillus pyocyaneus fed to *M. domestica.*

57. **Bacot, A. W. 1911** The Persistence of *Bacillus pyocyaneus* in Pupae and Imagines of *Musca domestica* raised from Larvae Experimentally Infected with the Bacillus. Parasitology 4:68-73.
Avoided possibility of secondary contamination by sterilizing the pupae and placing them in clean sand, in clean tubes.

58. **Bacot, A. 1916** The Improvement of Fly-spraying Fluids and the Control of Experimental Trials. Brit. Med. J. for 1916 (2919):801.
Concerned with problem of knock-down which fails to kill. ABSTRACT: RAE-B 5:24, 1917.

59. **Bahr, P. H. 1914** A Study of Epidemic Dysentery in the Fiji Islands. Brit. Med. J. for Feb. 7, 1914:294-296.
No persistence of organisms in gut of *M. domestica* beyond the fifth day. ABSTRACT: RAE-B 2:66, 1914.

60. **Bahr, P. H. 1915** Notes on Yaws in Ceylon, with Special Reference to its Distribution in that Island and its Tertiary Manifestation. Ann. Trop. Med. Parasit. Ser. T.M. 8:675-680.
Doubts relation to *M. domestica*, which is more abundant in elevated planting districts, where yaws is unknown. ABSTRACT: RAE-B 3:76, 1915.

61. **Baker, W. C., Scudder, H. I.** and **Guy, E. L. 1947** The Control of House Flies by DDT Sprays. Publ. Health Repts. 62:597-612.
Control effective in dairies for three months, in restaurants for four, about garbage cans for three to five weeks. ABSTRACTS: BA 22:720, 1948; RAE-B 38, 1950.

Baker, W. C., joint author. *See* Quarterman, K. D., et al., 1949.
Balch, R. E., joint author. *See* Twinn, E. R., et al., 1946.

61a. Baldwin, E. and **Needham, D. M. 1934** The Phosphorus Distribution in Resting Fly Muscle. J. Physiol. 80:221-237.
Arginine phosphoric acid is the phosphagen of invertebrates. Checked by these authors for certain Diptera.

62. Balfour, A. 1916 The Medical Entomology of Salonica. Publ. by Wellcome Bur. Sci. Res., London. 25 pp., 31 figs.
Gives appropriate attention to *M. domestica.* ABSTRACT: RAE-B 4:179, 1916.

63. Balfour, A. 1916 Fly-Trap for Camps, Hospital Precincts, and Trench Areas. J. Roy. Army Med. Corps. 27:61-72.
Useful in Egypt and other countries where conditions make difficult destruction of flies in breeding places. ABSTRACT: RAE-B 4:142, 1916.

64. Ball, H. J. and **Allen, T. C. 1949** Insecticidal Tests of Some New Organic Phosphates. J. Econ. Ent. 42:394-396.
Four compounds tested against house flies and other species. ABSTRACT: RAE-B 38:89, 1950.

65. Ball, S. C. 1918 Migrations of Insects to Rebecca Shoal Light-Station and the Tortugas Islands, with Special Reference to Mosquitoes and Flies. Pap. Dept. Marine Biol. Carnegie Inst. Washington 12:193-212.
Collected specimens which he considered to have been blown from Cuba (90 miles) and from Florida (105 miles).

Ball, W. H., joint author. *See* Fales, J. H., et al., 1948; Yeomans, A. H., et al., 1949.

65a. Bancroft, E. 1769 An Essay on the Natural History of Guiana in South America. London.
Bancroft corresponded with his brother July 8-Nov. 15, 1766. These letters were collected and published under the above title. Pages 385-386 cited by Gudger, E. W. 1910. (Reference No. 474.) Concerns flies which visit the lesions of yaws.

Bancroft, M. J., joint author. *See* Johnston, T. H., et al., 1920, 1921.

66. Bang, F. B. and **Glaser, R. W. 1943** The Persistence of Poliomyelitis Virus in Flies. Am. J. Hyg. 37:320-324.
Theiler's virus recovered from *M. domestica* up to 12 days after infection by oral route. Never found in flies reared from infected larvae. ABSTRACT: RAE-B 32:42, 1944.

Bankowski, R. A., joint author. *See* Furman, D. P., et al., 1949.

67. Banks, N. 1912 The Structure of Certain Dipterous Larvae with Particular Reference to those in Human Foods. Bull. 22, Tech. Ser., Bur. Ent., U.S. Dept. Agric. 44 pp., 8 pls.
Includes *M. domestica* and other muscoid species.

Barachini, B., joint author. *See* Bettini, S., et al., 1948.

68. Baranov, N. 1939 [Dung Heaps as Breeding Places of Flies in the Village of Mratzlin]. Vet. Arhiv. 9:280-287. Zagreb. (In Serbian, with German summary.)
Dung in sheds yielded *M. domestica, Fannia* and *Stomoxys.* Dung in open largely without maggots due to feeding by barn yard fowls. ABSTRACT: 27:168, 1939.

69. Barber, G. W. 1919 A Note on the Migration of Larvae of the House Fly. J. Econ. Ent. 12:466.
Larvae observed to pupate from 1 to 150 feet away from manure in which they had developed. ABSTRACT: RAE-B 8:35, 1920.

70. Barber, G. W. 1948 The Lethal Lines of the House Fly. J. Econ. Ent. 41:292-295.
Two non-viable types, isolated by selection, designated as "Lethal" and "Flat". Recognized by pupal characters. ABSTRACT: RAE-B 37:187, 1950.

71. Barber, G. W. and **Schmitt, J. B. 1949** Houseflies Resistant to DDT Residual Sprays. Bull. N.J. Agric. Exp. Sta. No. 742.
See following reference.
ABSTRACT: RAE-B 38:87, 1950.

72. Barber, G. W. and **Schmitt, J. B. 1949** Further Studies on Resistance to DDT in the Housefly. J. Econ. Ent. 42:287-292.
Ellenville (NAIDM) strain and Laboratory strains tested against "Technical DDT" and pp'isomer of DDT. ABSTRACT: RAE-B 38:87, 1950.

73. Barber, G. W. and **Schmitt, J. B. 1949** A Line of Houseflies Resistant to Methoxychlor. J. Econ. Ent. 42:844-845.
No evidence of resistance to DDT, but with methoxychlor, NAIDM line showed less kill than Wilson line. ABSTRACT: RAE-B 38:134, 1950.

74. Barber, G. W., Starnes, O. and **Starnes, E. B. 1948** Resistance of House Flies to Insecticides. Soap and Sanit. Chem. 24(11):120-121, 143.
ABSTRACT: RAE-B 38:203, 1950.
The four preceding references relate to the discovery and measurement of resistance to Gesarol. This period was one of disillusionment regarding the "miracle insecticides".

Barber, G. W., joint author. *See* Granett, P., et al., 1949; Hansens, E. J., et al., 1948.

Barbieri, V., joint author. *See* Patrissi, T., et al., 1949.

75. Barlow, J. 1912 The House Fly or Typhoid Fly. Bull. Ext. Dept. Rhode Island State Coll. No. 3, 12 pp.
Suggests wire traps over dust bins and garbage cans. ABSTRACT: RAE-B 1:67, 1913.

Barnes, Sarah, joint author. *See* Busvine, J. R., et al., 1947.

Barrett, J. P., joint author. *See* Ralston, A. W., et al., 1941.

76. Barthel, W. F., Gersdorff, W. A., LaForge, F. B. and **Graham, J. J. T. 1946** Evaluating Pyrethrum Extract. Soap and Sanit. Chem. 22(3):129, 131.
Commercially prepared concentrate showed 27 per cent loss of toxicity at end of 8 months storage at room temperature. ABSTRACT: RAE-B 37:63, 1949.

Barthell, W. F., joint author. *See* Gersdorff, W. A., et al., 1946.

77. Baxter, G. R. 1940 The House-fly, Public Enemy No. 1. Agric. J. Fiji 11(3):66-70. *Continued as* The Control of the Fly Nuisance. T.c. (4):96-98.
Cow dung principal breeding medium in Fiji for *M. domestica*. ABSTRACT: RAE-B 29:124, 1941.

78. Bayon, H. 1915 Leprosy: A Perspective of the Results of Experimental Study of the Disease. Ann. Trop. Med. Parasit. 9:1-90.
M. domestica can take up enormous numbers of Hansen's bacilli from sores containing these germs. ABSTRACT: RAE-B 120, 1915.

79. Beacher, J. H. and **Parker, W. L. 1948** Residual Toxicity. Chlorinated Camphene Compared to DDT for Toxic Residual Effects on Various Surfaces and in Paints against the House Fly. Soap and Sanit. Chem. 24(6):139, 141, 143, 163.
(Chlorinated camphene = toxaphene.) ABSTRACT: RAE-B 38:182, 1950.

Beacher, J. H., joint author. *See* Parker, W. L., et al., 1947.

80. **Beal, W. P. 1915** Epidemic of Verminous Enteritis among Sheep from Senegal. Annual Report for 1914. London. 18 pp., 3 pls., 9 tabs., 1 map.
Habronema muscae transmitted by housefly. ABSTRACT: RAE-B 4:12, 1916.

Beaulieu, A. A., joint author. *See* Munro, H. A. U., et al., 1947.

81. **Becker, E. R. 1923** Observations on the Morphology and Life-history of *Herpetomonas muscae-domesticae* in North American Muscoid Flies. J. Parasit. 9:199-213.
Found only in adult flies. Attempts to infect larvae failed. ABSTRACT: RAE-B 11:178, 1923.

82. **Becker, E. R. 1923** Transmission Experiments on the Specificity of *Herpetomonas muscae-domesticae* in Muscoid Flies. J. Parasit. 10:25-34.
Parasites from *M. domestica* and muscoids of 4 other genera mutually infective. ABSTRACT: RAE-B 12:26, 1924.

83. **Becker, R. 1910** Zur Kenntnis der Mundteile und des Kopfes des Dipterenlarven. [Toward Knowledge of the Mouth-parts and Head of Dipterous Larvae.] Zool. Jahrb. (Anat.) 29:281-314.
An important contribution, well illustrated.

84. **Bedford, G. A. H. 1929** Anoplura (Siphunculata and Mallophaga) from South African Hosts. 15th Rept. Vet. Ser. S. Afr. 1:501-547.
Immature specimen of *Linognathus* attached to *Musca interrupta* Walker. ABSTRACT: RAE-B 18:180, 1930.

85. **Beresoff, W. F. 1914** Die Schlafenden Fliegen also Infektionsträger. [Hibernating Flies as Transmitters of Infection.] Centralblt. Bakt. Parasit. Infekt., I Abt. Orig. 74:244-250.
Experiments indicate that both hibernating flies and their dead bodies can carry infection. ABSTRACT: RAE-B 2:174, 1914.

86. **Berlese, A. 1902** L'accoppiamento della mosca domestica. [Copulation in the domestic fly.] Rev. Patolog. Vegetale 9:345-357.

87. **Berlese, A. 1912** La distruzione della mosca domestica. [Destroying the domestic fly.] Redia 8:462-470.
Used arsenic spray bait, applied every 10 days. ABSTRACT: RAE-B 68-69, 1913.

88. **Berlese, A. 1926** La mosca domestica. Metodo di lotta pratico ed efficace. [The domestic fly. Practical and effective methods of control.] Riv. Agric. 31(9):131-133.
Poison bait spray. Various dilutions for different situations. ABSTRACT: RAE-B 14:72, 1926.

89. **Berlese, A. 1927** La distruzione della mosca domestica. [Elimination of the domestic fly.] Redia 16:1-11. Florence.
ABSTRACT: RAE-B 16:128, 1928.

90. **Bernstein, J. 1910** Summary of Literature Relating to the Bionomics of the Parasitic Fungus of Flies; *Empusa muscae* Cohn, with Special Reference to the Economic Aspect. Repts. Local Gov't Bd. on Publ. Health and Med. Subjects, N.S. No. 40, Further Repts. (No. 3) on Flies as Carriers of Infection, pp. 41-45.

91. **Bernstein, J. M. 1914** Repts. Local Gov't. Bd. on Publ. Health and Med. Subjects; N.S. No. 102. ii The Destruction of Flies by Means of Bacterial Cultures. pp. 27-31.
Flies fed spores of *Mucor* died as though infected with *Empusa*. ABSTRACT: RAE-B 3:90, 1915.

Berry, L. F., joint author. *See* Freeborn, S. B., *et al.*, 1935.

91a. Bertarelle, E. 1910 Verbreitung des Typhus durch die Fliegen. [Spread of Typhoid Fever by Flies.] Zentr. Bakt. Parasit. Infekt. I. Orig., 53:486-495.
Typhoid fever was frequently referred to as "typhus" in publications of this date.

Bessler, C., joint author. *See* Patrissi, T., 1949.

92. Bettini, S. 1948 Contributo allo studio della resistenza all'azione del DDT nelle mosche domestiche. [Contribution to the study of resistance to the action of DDT against the domestic fly.] Rend. Ist. Sup. Sanit. 11:1131-1136.
Demonstrates a true physiological resistance independent of the particular thickness of the cuticle.

93. Bettini, S. and **Barachini, B. 1948** Primi resultati della lotta con l'Octa-Klor e il Gammaesano contro le mosche domestiche resistenti al DDT. [First Results of the effort with Octa-Klor and Gammexane against Resistant House Flies.] Rend. Ist. Sup. Sanit. 11:841-848.
Flies showing high resistance to DDT all died on contact with Octa-Klor or Gammexane.

94. Bevan, L. E. W. 1926 Southern Rhodesia: Report of the Director of Veterinary Research for the year 1925. Fol., 9 pp. Salisbury.
Musca humilis and *M. domestica* among the species reared from larvae feeding on decomposing carcasses. Abstract: RAE-B 14:117, 1926.

95. Beyer, G. E. 1920 Supplementary Report on Disease-carrying Flies in Public Markets. Quart. Bull. La. State Bd. Health. 11:102-107.

96. Beyer, G. E. 1925 The Bacteriology of Market-flies of New Orleans. Quart. Bull. La. State Bd. Health. 16:110-116.

97. Beyer, H. G. 1910 The Dissemination of Disease by the Fly. New York Med. J. 91:677-685.
Similar to many papers published about this date, partly with the purpose of educating the medical profession and others.

98. Bezzi, M. 1921 *Musca inferior* Stein, Type of a New Genus of Philaematomyine Flies (Diptera). Ann. Trop. Med. Parasit. 14:333-340.
Erects new genus, *Ptilolepis*. Abstract: RAE-B 9:76, 1921.

99. Bezzi, M. 1923 Les males de *Musca albina* Wied. et de *Musca lucidula* Loew (Dipt.). Bull. Soc. Roy. Ent. Egypte for 1922 (15th year): 108-118.
Keys out males and females of several species.

Bieberdorf, G. A., joint author. *See* Fenton, F. A., et al., 1936.

100. Bigot, J. M. F. 1887 Diptera nouveaux provenant de l'Amerique du Nord. [New Diptera from North America.] Bull. Soc. Ent. Fr. 7(6):180-182.
Brief diagnostic descriptions. List includes *Musca atrifrons* and *M. flavipennis*.

101. Bishopp, F. C. 1930 Flytraps and their Operation. U.S. Dept. Agric. Farmer's Bull. 734, revised. 14 pp.
Gives formulae for preparing fly baits. Abstract: RAE-B 18:276, 1930.

102. Bishopp, F. C. 1939 Housefly Control. U.S. Dept. Agric. Leaflet No. 182, 6 pp.

103. Bishopp, F. C. 1946 Present Position of DDT in the Control of Insects of Medical Importance. Pests 14(8):14-28.
Slow knock-down, but surfaces remain lethal from 2 to 8 months. Abstract: BA 21:493, 1947.

104. Bishopp, F. C. 1946 Present Position of DDT in the Control of Insects of Medical Importance. Am. J. Publ. Health 36:593-606.
Production reached 3 million lbs. per month in the U.S. toward the end of World War II. ABSTRACT: BA 20:2256, 1946.

105. Bishopp, F. C. 1946 New Insecticides. A Progress Report on Results being Obtained in Preliminary Test Studies. Agric. Chem. 1(6): 19-22, 39-40.
Benzene hexachloride and chlorinated camphene "3956". ABSTRACT: BA 21:1537, 1947.

106. Bishopp, F. C., Dove, W. E. and **Parman, D. C. 1915** Notes on Certain Points of Economic Importance in the Biology of the House Fly. J. Econ. Ent. 8:54-71.
Touches on copulation, preoviposition period, egg-laying, developmental stages, breeding media, longevity, hibernation. ABSTRACT: RAE-B 3:92, 1915.

107. Bishopp, F. C. and **Laake, E. W. 1919** The Dispersion of Flies by Flight. J. Econ. Ent. 12:210-211.
Marked specimens of *M. domestica* collected 13 miles from place of liberation. ABSTRACT: RAE-B 7:121, 1919.

108. Bishopp, F. C. and **Laake, E. W. 1921** Disperson of Flies by Flight. J. Agric. Res. 21:729-766.
M. domestica found 6 miles from point of release in less than 24 hours. ABSTRACT: RAE-B 9:194, 1921.

109. Bishopp, F. C., Roark, R. C., Parman, D. C. and **Laake, E. W. 1925** Repellents and Larvicides for the Screw Worm and other Flies. J. Econ. Ent. 8:776-778.
Object is to combine a spray of high toxicity with strong repellent in carriers not irritating to animals. ABSTRACT: RAE-B 14:26, 1926.

Bishopp, F. C., joint author. *See* Parman, D. C., et al., 1927; Howard, L. O., et al., 1926.

Bishop, M. B., joint author. *See* Paul, J. R., et al., 1941; Power, M.E., et al., 1943.

Blackith, R. E., joint author. *See* Page, A. B. P., et al., 1949.

110. Blair, D. M. 1945 Baited Fly Traps. Rhod. Agric. J. 42:488-492. *Also as* Bull. Minist. Agric. S. Rhodesia, No. 1328. 6 pp.
Bait in relative darkness. Flies, after feeding, are attracted by light above into trapping chamber. Captured 4,500 flies in one hour.

111. Blair, K. G. 1942 How Does a Housefly Alight on a Ceiling? Ent. Mon. Mag. (London) 78:40-41.
Theoretical suggestion, not supported by photography. CITED: Ent. News 53:118, 1943.

Blakeslee, E. B., joint author. *See* Bruce, W. G., et al., 1946.

Blanc, G., joint author. *See* Nicolle, C., et al., 1919.

112. Blanchard, R. 1915 La lutte contre la mouche. Bull. No. 5 Ligue Sanitaire Française. 63 pp.
Informative publication, well illustrated.

113. Blaxter, K. L. 1945 Control of Flies in Cowsheds (Experiments with DDT). Agriculture (Grt. Brit.) 52(3):112-115.
Effective only in screened quarters, because of reinfestation. ABSTRACTS: RAE-B 34:62, 1946; BA 20:215, 1946.

114. Blickle, R. L., Capelle, A. and **Morse, W. J. 1948** Insecticide Resistant Houseflies. Soap and Sanit. Chem. 24(8):139, 141, 149.
Laboratory strain contaminated with BHC required larger doses of all insecticides, for control. ABSTRACT: RAE-B 38:183, 1950.

115. Bleisner, E. R. and **Brown, S. T. 1945** Eye Appeal is not Sanitation. Wisc. State Bd. of Health Quar. Bull. 7(16):16-18.
Emphasizes importance of what goes on behind the scenes in public eating and drinking places.

116. Blijdorp, P. A. 1933 Rapport inzake het onderzock der vliegenplaag op de stortplaats van het Haagsche stadviel te Wijster, uitgebracht aan de V.A.M. einde Augustus 1932. Versl. Pl. Ziektenk. Dienst. No. 71, 16 pp.
M. domestica predominated over *Lucilia* and *Calliphora* in dump to which refuse was brought by rail and left until composted. ABSTRACT: RAE-B 21:214, 1933.

117. Block, S. S. 1946 Tests of Screening Effectiveness against Insects. Science 104(2706):447.
M. domestica contained by 18 warp (length), 14 filler (width). More economical to manufacture than 16, 16.

118. Block, S. S. 1946 Insect Tests of Wire Screening Effectiveness. Am. J. Publ. Health 36:1279-1286.
M. domestica unable to penetrate 18 × 18, 16 × 16, 14 × 14, 18 × 16, 18 × 14, 18 × 12. ABSTRACT: BA 21:741, 1947.

119. Block, S. S. 1948 Insecticidal Surface Coatings. I. II. Soap and Sanit. Chem. 24(2-3):138-141, 151, 153, 171.
Used flies and cockroaches to test oil films, modified resins, waxes, etc., impregnated with various insecticides. ABSTRACTS: RAE-B 38:67, 1950; BA 23:2867, 1949.

120. Block, S. S. 1948 Residual Toxicity Tests on Insecticidal Protective Coatings. Soap and Sanit. Chem. 24(4):155, 157, 159, 161, 207, 213. Reprinted as Tech. Paper No. 19. Florida Engin. and Industr. Exp. Sta., Coll. of Engin., Univ. of Fla, Gainesville. 6 pp.
Houseflies and other insects used in testing numerous formulae and methods of application. Coatings containing DDT tend to become more toxic, with time. ABSTRACT: RAE-B 38:109, 1950.

121. Bodenheimer, F. S. 1924 Ueber die Voraussage der Generationenzahl von Insekten. II. Die Temperaturentwicklungskurve bei medizinisch wichtigen Insekten. [On Prediction of the Generation Frequency of Insects. II. The Temperature-affected developmental curve of Medically-important Insects.] Centrabl. Bakt. Parasit. Infekt. Ite Abt. Orig. 93:474-480. Jena.
Curves show duration of development at various temperatures for a number of medically important species, including *M. domestica.* ABSTRACT: RAE-B 13:33, 1925. See 1925 paper also.

122. Bodenheimer, F. S. 1925 On Predicting the Development Cycles of Insects. I. *Ceratitis capitata* Wied. Bull. Soc. Roy. Ent. Egypte for 1924 (2-4):149-157.
Makes use of Blunck's correction, as applied to curves published in preceding paper.

123. Bodenheimer, F. S. 1931 Erfahrung über die Biologie der Hausfliege (*Musca domestica* L.) in Palästina. [Information on the Biologie of the Housefly (*Musca domestica* L.) in Palestine]. Zeit. Angew. Ent. 18:492-504.
Elucidates ecological influences. Winter and summer conditions compared. ABSTRACT: RAE-B 20:17, 1932.

124. Bodkin, G. E. 1917 Report of the Economic Biologist, Brit. Guiana Dept. Sci. and Agric., for 1916. 14 pp.
Lime-peel refuse great source of *M. domestica.* Intestinal disorders

increase and decrease with rise and fall of fly population. ABSTRACT: RAE-B 6:178, 1918.

125. Bodkin, G. E. 1923 Government Economic Biologist Rept., Dept. Sci. and Agric., Brit. Guiana, 121, Appx. iii, pp. 36-40.
Large numbers of larvae and pupae of *M. domestica* in mule and cow pens. ABSTRACT: RAE-B 11:139, 1923.

126. Bogdanow, E. A. 1903 Zehn Generationen der Fliegen (*Musca domestica*) in veränderten Lebensbedingungen. [Ten Generations of Flies (*Musca domestica*) under varied conditions essential for life.] Allg. Zeit. f. Ent. 8:265-267.

Bohart, G. E., joint author. *See* Hall, D. G., et al., 1948

Bohart, R. M., joint author. *See* Travis, B. V., et al., 1946.

127. Booker, W. H. 1912 A Working Plan for an Anti-fly and Anti-mosquito Campaign. Bull. North Carolina State Bd. Health 27:126-133.

128. Booker, W. H. 1912 Essential Facts about Flies, and a Suggested Anti-fly Ordinance. Bull. North Carolina State Bd. Health 27: 133-141.
The two preceding papers are typical of the effort being made during this period to (1) abate the fly nuisance, (2) devise ordinances to prevent recurrence.

Borgmann, A. R., joint author. *See* Atkeson, F. W., et al., 1943.

129. Borre, P. de 1873 Note on Parasites of *Musca domestica.* Nature 8:263.
Very brief. Parasites not identified.

130. Borre, P. de 1873 *Musca domestica* with *Chelifer panzeri* Koch Attached Parasitically. Sitzunsber. Zool. Ges. Wien. 23:36.
Clarifies foregoing paper.

131. Bouché, P. F. 1834 Naturgeschichte der Insekten besonders in Hinsicht ihrer ersten Zustande als Larven und Puppen. [The Natural History of Insects especially in regard to their early Existence as Larvae and Pupae.] Berlin. 216 pp., 10 pls. *M. domestica,* pp. 65-66, pl. 5, figs. 20-24.

Bourne, A. J., joint author. *See* Shaw, F. R., et al., 1946.

132. Bovington, H. H. S. and **Coyne, F. P. 1944** Trichloraceto-nitrile as a Fumigant. Ann. Appl. Biol. 31:255-259.
M. domestica all died with sufficient exposure, but fumigant was better for bed bugs than for houseflies. ABSTRACT: RAE-B 33:150, 1945.

Bovington, H. H. S., joint author. *See* Grove, J. F., et al., 1947.

Bowery, T. G., joint author. *See* Granett, P., et al., 1949.

133. Boyd, J. E. M. 1922 The Botany and Natural History of Dyke-land near Sandwich, Kent, as far as they concern Medical Entomology. J. Roy. Army Med. Corps. 38(1-2): 41-47; 117-130.
M. domestica heads list of flies. Instructions for use of formalin (indoors), sodium arsenite or ammonium fluoride (for manure heaps). ABSTRACT: RAE-B 10:87, 1922.

134. Boye, G. and **Guyot, R. 1919** Contribution à la Lutte contre les Mouches. [Contribution to the War against Flies.] Bull. Acad. Méd. Paris. 81(3):80-84.
Potassium permanganate with formalin against larvae. Various poison baits against adults. ABSTRACT: RAE-B 7:67, 1919.

Bracey, P., joint author. *See* David, W. A. L., et al., 1947.

Bradley, B., joint author. *See* Cleland, J. B., et al., 1919.

Bragin, E. A., joint author. *See* Derbeneva-Ukhova, V. P., et al., 1943.

135. Brain, C. K. 1912 *Stomoxys calcitrans* Linn. Part I. Ann. Ent. Soc. Am. 5:421-430.
Figures of *M. domestica*, *Fannia canicularis*, *Stomoxys calcitrans*; all stages. Abstract: RAE-B 1:22, 1913.

136. Brain, C. K. 1918 Storage of Manure and Fly Suppression at Durban Remount Depot. J. Econ. Ent. 11:339-341.
East Africa. Describes method in detail. Sprays and poison baits for flies in stalls. Abstract: RAE-B 6:185, 1918.

137. Brauer, F. 1893 Vorarbeiten zu einer Monographie der Muscaria schizometopa (excl. Anthomyidae) besprochen. [Presentation of Preliminary Studies toward a monograph of the Myodaria Schizophora (excl. Anthomyidae).] Verh. Ges. Wien 43:447-525.
One of the author's important taxonomic contributions.

138. Brazil, V. 1926 A defeza contra a mosca. [A Defense against the Fly.] Mem. Inst. Butantan 3:189-203. São Paulo. (In Portuguese, with French summary.)
Mixing dry earth with animal manure prevents feeding movements of larvae. Abstract: RAE-B 15:115, 1917.

139. Brefeld, O. 1871 Untersuchungen über die Entwicklung der *Empusa muscae* und *E. radicans*. [Researches on the Development of *Empusa muscae* and *E. radicans*.] Abh. d. Naturf. Gesellsch. Halle. 12:1-50, pls. 1-4.
Fungus parasites of houseflies.

140. Brescia L., La Mer, V. K., Wilson, I. B., Rowell, J. C. and **Hodges, K. C. 1946** Relative Toxicity of DDT Aerosols to Mosquitoes and *Musca domestica*. Insect Balance. Ent. News 57:180-183.
M. domestica adult 4-6 times more resistant than salt marsh mosquito. Abstracts: BA 21:1004, 1947; RAE-B 37:16, 1949.

141. Brèthes, J. 1915 Sur la *Prospalangia platensis* (n. gen. n. sp.) (Hymen.) et sa Biologie. [Concerning *Prospalangia platensis* (n. gen. n. sp.) (Hymen.) and its Biology.] Anales Soc. Cien. Argentina (Buenos Aires.) 79(5-6):314, 320.
A chalcid, parasitic on pupae of *M. domestica*, *Stomoxys calcitrans* and other flies. Abstract: RAE-B 8:61, 1920.

142. Brett, C. H. and **Fenton, F. A. 1946** DDT as a Residual Insecticide for Fly Control in Barns. J. Econ. Ent. 39:397-398.
First application 5 per cent; second (6-8 weeks later) 2·5-3 per cent. Abstract: RAE-B 35:191, 1947.

143. Brewster, E. T. 1909 The Fly, the Disease of the House. McClure's Magazine. 33:564-568. Sept.
Proposes to make use of tropisms for ridding houses of flies.

144. Bridwell, J. C. 1918 Certain Aspects of Medical and Sanitary Entomology in Hawaii. Trans. Med. Soc. Hawaii for 1916-1917. (Honolulu.) pp 27-32.
Conditions sanitary in residential areas, but flies swarm in the country. Discussion of breeding media, etc. Abstract: RAE-B 6:164, 1918.

145. Britton, W. E. 1906 The Common House-fly (*M. domestica*) in its Relation to Public Health. Yale Med. J. 12:750-757.
Written to be read by medical personnel.

146. Britton, W. E. 1912 The House-fly and its Relation to Typhoid Fever. Proc. Sixth Conf. Health Officers of Connecticut. Hartford. 18 pp.
Reflects the thinking and concern of that period.

147. Britton, W. E. 1912 The Rôle of the House-fly and Certain other Insects in the Spread of Human Disease. Pop. Sci. Monthly, July 1912, pp. 36-49.

A commendable effort in public health education for the lay-public.

148. Britton, W. E. 1912 The House-fly as a Disease-carrier and How Controlled. (Leaflet) Publ. by Connecticut State Bd. of Health. 12 pp., 5 figs.

Includes instructions on practical procedure.

149. Brown, A. W. A., Robinson, D. B. W., Hurtig, H. and **Wenner, B. J. 1948** Toxicity of Selected Organic Compounds to Insects, Part I. Tests for General Toxicity on Larvae of *Musca*, *Tribolium* and *Ephestia* and Adults of *Sitophilus*. Canad. J. Res. (D) 26:177-187.

Tests run on 127 synthetic compounds. Most highly toxic were gammexane and chlordane. ABSTRACT: RAE-B 37:96, 1949.

150. Brown, F. M. 1927 Descriptions of New Bacteria Found in Insects. Am. Mus. Novit. No. 251, 11 pp.

Neisseria lucilarum, a pathogen of *M. domestica*, found in *Lucilia sericata*. ABSTRACT: RAE-B 15:117, 1927.

Brown, S. T., joint author. *See* Bleisner, E. R., et al., 1945.

151. Bru, M. 1946 Nouveaux emplois insecticides de'l'hexachlorocyclohexane. [New Insecticidal Uses of Benzene Hexachloride.] Compt. Rend. Agric. Fr. 32:771-772. Short introduction by **P. Vayssière.**

A preparation in which the gamma isomer predominated. Used in paint. *M. domestica* died within 12 hours after contact. ABSTRACT: RAE-B 37:34, 1949.

152. Bruce, W. G. 1936 Seasonal Appearance and Relative Abundance of Flies Caught in a Baited Trap at Fargo. Circ. N. Dak. Agric. Expt. Sta. No. 60:9-12.

Phormia regina most abundant May-June; *M. domestica*, Jul.-Oct. ABSTRACTS: RAE-B 25:217, 1937.

153. Bruce, W. G. 1939 Some observations on Insect Edaphology. J. Kans. Ent. Soc. 12(3):91-93.

Involved 1600 pupae of *M. domestica*. Adults failed to emerge from sand with water content above 16·1 per cent. ABSTRACT: RAE-B 28:16, 1940.

154. Bruce, W. G. and **Blakeslee, E. B. 1946** DDT to Control Insect Pests affecting Livestock. J. Econ. Ent. 39:367-374.

Emulsified solution or water suspension of 2·5 per cent DDT sprayed on painted surface controlled *M. domestica* for whole season if fly sanitation was good. ABSTRACT: RAE-B 35:189, 1947.

155. Bruce, W. G. and **Knipling, E. F. 1936** Seasonal Appearance and Relative Abundance of Flies attracted to Bait Traps. Iowa State Coll. J. of Sci. 10:361-366.

M. domestica made up 42·28 per cent of catch in 1933; 40·84 per cent in 1934.

156. Bruce, W. N. 1948 How to Control Flies. Pests 16(6):16-17.

ABSTRACT: BA 23:1367, 1949.

157. Bruce, W. N. 1949 Characteristics of Residual Insecticides Toxic to the House Fly. Bull. Illinois Nat. Hist. Surv. 25(Art. 1):1-32.

Greatest degradation of DDT on white-washed and concrete surfaces, except indoors where porous surfaces remained toxic longer than hard ones. ABSTRACT: TDB 49:559, 1952.

158. Buchanan, R. M. 1913 *Empusa muscae* as a Carrier of Bacterial Infection from the House-fly. Brit. Med. J. Nov. 22, 1913. 18 pp., 21 figs.

159. Buchanan, W. J. 1897 Cholera Diffusion and Flies. Indian Med. Gaz. for 1897, 86-87.

160. Bucher, G. E., Cameron, J. W. M. and **Wilkes, A. 1948** Studies on the Housefly (*Musca domestica* L.). II. The Effects of Low Temperatures on Laboratory-reared Puparia. Canad J. Res. (D) 26: 26-56. III. The Effects of Age, Temperature, and Light on the Feeding of Adults. Tc., pp. 57-61.

II. Survival of pupae was decreased by lowering the temperature of storage and by increasing the duration of the storage period, or both. All exposures were above 1°C (33·8°F).

III. Flies begin to feed at a temperature (and age) at which they first become active. Males feed before females. Some light is required. Flies die quickly if restricted as to liquid food.

See Wilkes, A., et al., for Part I. ABSTRACT: RAE-B 37:88, 1949.

Bucher, G. E., joint author. *See* Wilkes, A., et al., 1948.

161. Buchmann, W. 1929 Untersuchungen über die physiologische Wirkung von Pyrethrum-Insektenpulver auf Fliegenlarven. [Investigations concerning the Physiological Action of Pyrethrum Powder on Fly Larvae.] Zeit. Desinfekt. 21(3):72-76. (Aufgabe A.)

Regards pyrethrins as nerve poisons. ABSTRACT: RAE-B 17:179, 1929.

162. Buchmann, W. 1929 Weitere Untersuchungen über die Wirkung von Pyrethrumpulvern auf die Muscidenbrut. [Further Investigations concerning the Effect of Pyrethrum Powder on the Fly Hatch.] Zeit. Desinfekt. 21(1-2):14-16, 37-39.

Larvae of *M. domestica* and *Stomoxys* killed by pyrethrum powder, mixed with stable manure. ABSTRACT: RAE-B 18:58, 1930.

163. Buchman, W. 1930 Versuche zur quantitativen Wertbestimmung von Pyrethrum-Insect Pulvern durch den physiologischen Tierversuch. [Research on the Quantitative Evaluation of Pyrethrum Insect Powder through Physiological Animal Experimentation.] Zeit. Desinfekt. 23(5):414-418.

There is a gradual effect on the nervous system, leading up to total paralysis. *M. domestica* among the species studied. ABSTRACT: RAE:B 18:252, 1930.

164. Buck, J. E. 1915 Fly Baits. Alabama Agric. Expt. Sta., Auburn. Circ. No. 32. 6 pp.

Recommends glass or porcelain containers, since chemicals reacting with tin or zinc may repel flies. ABSTRACT: RAE-B 3:165, 1915.

Buckner, A. J., joint author. *See* Fay, R. W., et al., 1947, 1948.

165. Buichkov-Oreshnikov, V. A. 1934 [On the Microflora of the Flies of Some of the Camps and on the Role possibly played by them in the Distribution of Intestinal Diseases.] Trav. Acad. Milit. Méd. Armée rouge URSS 1:393-400. (In Russian).

Nine species of intestinal-type bacteria recovered. Fly gut flora fairly constant; surface contamination dependent on conditions. ABSTRACT: RAE-B 23:14, 1935.

Buichkov, V. A. joint author. *See* Pavlovskii, E. N., et al., 1932.

166. Bull, L. B. 1919 A Contribution to the Study of Habronemiasis: A Clinical, Pathological and Experimental Investigation of a Granulomatous Condition of the Horse. Habronemic Granuloma. Trans. Roy. Soc. S. Australia. 43:85-141.

M. domestica believed to act as intermediate host. Larvae penetrate wherever a lesion exists. ABSTRACT: RAE-B 8:51, 1920.

167. **Bull, L. B. 1922.** Habronemic Conjunctivitis in Man Producing a "Bung" Eye. Med. J. Australia. 9th Year, ii, No. 18:499-500.
Suggests that larva of *Habronema* escaped into eye of a child from *M. domestica* or *M. vetustissima*, seeking moisture. ABSTRACT: RAE-B 11:21, 1923.

168. **Burgess, A. F.** and **Sweetman, H. L. 1949** The Residual Property of DDT as Influenced by Temperature and Moisture. J. Econ. Ent. 42:420-423.
Tested against houseflies, screens kept at high temperature (37 degrees C) with high moisture (60-75 per cent R.H.) lost toxicity more quickly than those kept at 23°C and 25-40 per cent R.H. ABSTRACT: RAE-B 38:97, 1950.

169. **Burnett, J. E. 1918** Method of Combating Flies. Qtrly. Bull. Mich. Agric. Expt. Sta. I, No. 1:18-19.
Poison bait, repellent spray (on cattle), traps. ABSTRACT: RAE-B 7:37, 1919.

170. **Bushnell, L. D.** and **Hinshaw, W. R. 1924** Prevention and Control of Poultry Diseases. Kansas Agric. Expt. Sta. Circ. 106. 75 pp., 25 figs.
Houseflies and stable flies necessary intermediate hosts of tapeworms. ABSTRACT: RAE-B 13:21, 1925.

171. **Busvine, J. R.** and **Barnes, S. 1947** Observations on Mortality among Insects Exposed to Dry Insecticidal Films. Bull. Ent. Res. 38: 81-90.
M. domestica very susceptible to gammexane, pyrethrins, and DDT. ABSTRACT: RAE-B 147, 1947.

172. **Bütschli, O. 1888** Bemerkungen über die Entwicklungs-geschichte von *Musca*. [Observations on the Developmental History of *Musca*.] Morph. Jahrb. 14:170-174.

173. **Buxton, P. A. 1920** The Importance of the House-Fly as a Carrier of *E. histolytica*. Brit. Med. J. No. 3083:142-144.
In lower Mesopotamia the house-fly is a major factor in the spread of intestinal disorders. ABSTRACT: RAE-B 8:69, 1920.

174. **Byer, H. C. 1910** Dissemination of Disease by the Fly. New York Med. J., April 2, 1910.

Cabral, J., joint author. *See* DeMello, F., et al., 1926.

175. **Cameron, J. W. M. 1938** The Reactions of the Housefly *Musca domestica* Linn. to Light of Different Wave-lengths. Canad. J. Res. (D) 16:307-342.
Housefly much more strongly stimulated by ultra violet light of wave-length 3656Å than by any other part of spectrum tested. Range of test: 3022Å to 5780Å.

176. **Cameron, J. W. M. 1939** Reactions of House-flies to Light of Different Wave Lengths. Nature (London) 143(3614):208.
Is essentially an abstract of his paper published in the Canadian Journal of Research, 1938.

Cameron, J. W. M., joint author. *See* Bucher, G. E., et al., 1948; Wilkes, A., et al., 1948.

Campbell, A., joint author. *See* Cleland, J. B., et al., 1919.

177. **Campbell, C. 1901** House-flies and Disease. Brit. Med. J. for 1901, Second Vol., p. 980.

178. **Campbell, F. L., Sullivan, W. N.** and **Jones, H. A. 1934** Derris in Fly Sprays. Kerosene Extracts of Derris Root as House Fly Sprays. Part I. Methods and Results of Laboratory Tests of Extracts of Derris and Cubé Roots. Soap. 10(3):81-83, 85, 87, 103, 105, 107.
Methoxyl content may be better chemical index of insecticidal value than rotenone. ABSTRACT: RAE-B 22:131, 1934.

179. Campbell, F. L., Sullivan, W. N. and **Jones, H. A. 1934** Derris in Fly Sprays. Kerosene Extracts of Derris Root as House Fly Sprays. Part II. Comparative Tests of Derris and Pyrethrum. Soap 10(4): 83, 85, 103, 105.
Extract from 5 gm. derris root almost as effective as that from 20 gm. pyrethrum flowers. ABSTRACT: RAE-B 23:13, 1935.

180. Candura, G. S. 1939 La mosca domestica. Vita, danni, lotta e osservazioni nella Venezia Tridentina. [The Domestic Fly. Its life, harm done, struggle (for control), from observations in Venezia Tridentina.] Pubbl. prat. Fitopat. Igiene No. 1. 16 pp., 15 figs.
ABSTRACT: RAE-B 28:56, 1940.

Capelle, A., joint author. *See* Blickle, R. L., et al., 1948.

181. Carment, A. G. 1922 Report on Experiment of Fly Breeding from Stable Manure with a Short Account of the Finding of a Parasite. Agric. Circ., Fiji Dept. Agric., No. 3:1-5.
A hymenopterous parasite of pupae may be responsible for recent decrease in numbers of *M. domestica*. ABSTRACT: RAE-B 10:169, 1922.

182. Carr, E. C. 1927 Flies and their Eradication. U.S. Nav. Med. Bull. 25:528-542.
Incineration of garbage and manure. Sprays. Fly traps. ABSTRACTS: RAE-B 16-80, 1928; Hocking, B. 1960 Smell in Insects, p. 187 (5. 75).

183. Carter, H. J. 1861 On a Bi-sexual Nematoid Worm which Infests the Common House-fly (*Musca domestica*) in Bombay. Ann. Mag. Nat. Hist. ser. (3) Vol. 7:29-33.
Carter gave them the name *Filaria muscae*. One-third of the flies contained from 2 to 20 worms, chiefly concentrated in the proboscis.

184. Case, A. A. and **Ackert, J. E. 1939** Intermediate Hosts of Chicken Tape-worms found in Kansas. Trans. Kans. Acad. Sci. 42:437-442.
M. domestica one of many insect hosts of *Choanotaenia infundibulum*. ABSTRACT: RAE-B 29:162, 1941.

Casey, A. E., joint author. *See* Paul, J. R., et al., 1941.

185. Castellani, A. 1907 Experimental Investigation on *Framboesia tropica* (Yaws). J. Hygiene 7:558-569.
Concludes that under certain conditions yaws may be conveyed by flies.

Castro, M. P., joint author. *See* Pereira, C., et al., 1945.

Cattani, J., joint author. *See* Tizzoni, G., et al., 1886.

186. Chang, K. 1943 Domestic Flies as Mechanical Carriers of Certain Human Intestinal Parasites in Chengtu. J. West China Border Res. Soc. (B) 14:92-98.
Reports on dissection of flies for intestinal parasites. ABSTRACT: RAE-B 32:18, 1944.

187. Chapman, R. K. 1944 An Interesting Occurrence of *Musca domestica* L. Larvae in Infant Bedding. Canad. Ent. 76:230-232.
Maggots developed in mattress filling which had been saturated with urine. Confirmed by experiment. ABSTRACTS: BA 19:1270, 1945; RAE-B 34:17, 1946.

188. Charrier, H. 1927 Note préliminaire sur les mouches de la région de Tanger. Bull. Soc. Path. Exot. 20:619-622.
Eighteen Moroccan species of Muscidae. *M. domestica* abundant and important. ABSTRACT: RAE-B 16:57, 1928.

189. Chebotarevich, N. D. 1934 [Breeding Places of Flies and their Control under Conditions of the Camps of the Central Asiatic Military

District.] Trav. Acad. Milit. Méd. Armée Rouge. URSS 1:411-418. (In Russian.)
Human excreta and kitchen refuse chief breeding places. Methods of control. ABSTRACT: RAE-B 23:14, 1935.

Chilingarova, S. V., joint author. *See* Kalandadze, L. P., et al., 1940, 1942.

190. Chisholm, R. D., Nelson, R. H. and **Fleck, E. E. 1949** The Toxicity of DDT Deposits as Influenced by Sunlight. J. Econ. Ent. 42: 154-155.
Mortality of surface for *M. domestica* almost negligible after 32 hours. (Glass plates treated with emulsions.) ABSTRACT: RAE-B 38:56, 1950.

191. Chmelicek, J. F., 1899 My Observations on the Typhoid Epidemic in Southern Camps and its Treatment. New York Med. J. 70: 193-198.
Latrine pits were about 40 feet from the entrance of the kitchen tent. Number of flies—countless.

192. Cholodkovsky, N. A. 1909 Zur Kenntniss des weiblichen Geschlechtsapparatus der Musciden. [Toward a knowledge of the female sex apparatus of Muscidae.] Zeit. Wiss. Insektenbiol. 5:333-337.
One of the good, early contributions in this area. Well illustrated.

193. Clapham, P. A. 1939 On Flies as Intermediate Hosts of *Syngamous trachea*. J. Helminth. 17:61-64.
Maggots take up eggs; become flies which are sluggish, easy for birds to catch. ABSTRACT: RAE-B 28:95, 1940.

Clare, S., joint author. *See* Gaines, J. C., et al., 1937.

Clark, P. F., joint author. *See* Flexner, S., et al., 1911; Howard, C. W., et al., 1912.

194. Clarke, J. L. 1947 Ground Spraying of Entire Towns with DDT. Proc. Ann. Meeting New Jersey Mosq. Exterm. Assoc. 34:127-131.
Accomplished 90-95 per cent reduction in fly population in 6 weeks. ABSTRACT: BA 24:243, 1950.

195. Cleare, L. D. 1918 The House-fly. How it Lives, How it Spreads Disease and How to Destroy it. J. Brit. Guiana Bd. Agric. Georgetown. 11(2):13-27.
Practical résumé of up-to-date information. ABSTRACT: RAE-B 6:209, 1918.

196. Cleland, J. B. 1914 Prevalence of Domestic Flies in Lower Hawkesbury River District, New South Wales. Third Rept. Gov't. Bur. Microbiol. for 1912. Sydney. pp. 155-160.
Fannia canicularis numerous in Sept., when *M. domestica* is not. Latter reaches its peak in June. ABSTRACT: RAE-B 3:207, 1915.

197. Cleland, J. B., Campbell, A. W. and **Bradley, B. 1919** The Australian Epidemics of an Acute Polio-encephalo-myelitis. Rept. Dir. Gen. Publ. Health, New South Wales for 1917. Sydney. pp. 173-174.
Both *M. domestica* and *M. vetustissima* suspected as carriers. ABSTRACT: RAE-B 7:187, 1919.

198. Coates, B. H. 1863 Medical Note on the More Familiar Flies. Trans. Coll. Phys. Phila. N.S. 3:348.

199. Cobb, J. O. 1905 Is the Common House-fly a Factor in the Spread of Tuberculosis? Am. Med. for 1905. No. 9, pp. 475-477.
Believed so; via digestive tract.

200. Cobb, N. A. 1910 The House Fly. National Geographic Magazine. 21:371-380 (May).
Very informational for its time. Excellent photographs.

201. Cobb, N. A. 1910 Notes on the Distances Flies can travel. National Geographic Magazine. 21:380-383 (May).
A study of the wing muscles of *Sarcophaga.*

202. Cochrane, E. W. W. 1912 A Small Epidemic of Typhoid Fever in Connection with Specifically Infected Flies. J. Roy. Army Med. Corps. 18:271-276.
At St. George's, Bermuda. Fly collected in latrine near fatal case yielded washings positive for *Bacillus typhosus.*

203. Coffee, J. H. 1948 Emergency Fly Control Operations at Wilmington, Delaware. Proc. Ann. Meeting New Jersey Mosq. Exterm. Assoc. 35:163-173.
DDT by weekly air and ground sprays. Reduced fly population index 67 per cent in three days. ABSTRACT: BA 24:1628, 1950.

204. Cohen, M. 1946 Experiments with DDT Smokes. Ann. Appl. Biol. 33:125-126.
Burning of paper impregnated with DDT. Quick knock-down of flies, which die several hours later. ABSTRACT: RAE-B 35:103, 1947.

205. Cohn, F. 1855 *Empusa muscae* und die Krankheit der Stubenfliegen. [*Empusa muscae* and the Disease of Houseflies.] Nova Acti Acad. Caes. Leop. Carol. Germ. Nat. Cur. 25:301-360.

206. Cohn, M. 1898 Fliegeneier in den Entleerungen eines Säuglings. [Fly ova in the Bowel Movements of an Infant.] Dtsch. Med. Wochenschr. Jahrg. 24:191-193.

Colas-Belcour, J., joint author. *See* Roubaud, E., et al., 1936.

Colberg, W., joint author. *See* Munro, J. A., et al., 1947.

Cole, E. L., joint author. *See* Fay, R. W., et al., 1947.

207. Collet, —. 1925 Un procédé de destruction des moches adultes. [A Procedure for Destroying Adult Flies.] Bull. Soc. Medico-Chirurgicale. 1924.
Spray materials and techniques. ABSTRACTS: Bull. écon. Indochine, 28. Renseignementes April 1925, pp. 190-191. Hanoi; RAE-B 13:164, 1925.

Collet, R., joint author. *See* Tomlinson, T. G., et al., 1949.

Collins, D. L., joint author. *See* Glasgow, R. E., et al., 1946.

208. Connal, A. 1922 Medical Entomology (Nigeria). Ann. Rept. Med. Res. Inst. for 1920, pp. 17-22. Lagos.
Cysts of *Entamoeba* found in *M. domestica* and *Lucilla* spp. ABSTRACT: RAE-B 11:44, 1923.

Connola, D. P., joint author. *See* Granett, P., et al., 1949.

209. Conradi, A. F. 1913 Controlling Flies. Circ. No. 23, South Carolina Agric. Expt. Sta., Clemson Coll., S.C. 14 pp., 5 figs.

210. Converse, G. M. 1910 Amoebiasis. Bull. California State Bd. Health. October, 1910.
Possibility of flies feeding on infected stools, then flying to food of others.

211. Cook, F. C. and **Hutchison, R. H. 1916** Experiments during 1915 in the Destruction of Fly Larvae in Horse Manure. U.S. Dept. Agric. Bull. No. 408, 22 pp.
A summary of three years' work. Advocates use of borax or hellebore. ABSTRACT: RAE-B 5:8, 1917.

212. Cook, F. C., Hutchison, R. H. and **Scales, F. M. 1914** Experiments in the Destruction of Fly Larvae in Horse Manure. U.S. Dept. Agric. Bull. No. 118, 26 pp.
Background research on insecticides, stressing effect on larvae, effect on bacteria, chemical effect on manure. ABSTRACT: RAE-B 2:178, 1914.

213. Cook, F. C., Hutchison, R. H. and Scales, F. M. 1915 Further Experiments in the Destruction of Fly Larvae in Horse Manure. U.S. Dept. Agric. Bull. No. 245, 22 pp.
Deals principally with use of borax, which can endanger vegetation if used excessively on manure to be used as fertilizer. ABSTRACT: RAE-B 3:192, 1915.

Cook, F. C., joint author. *See* Parman, D. C., et al., 1927.

214. Cook, S. S. 1934 Insect Control. U.S. Nav. Med. Bull. 32(2):229-247.
Bionomics of mosquitoes, houseflies, etc. ABSTRACT: RAE-B 22:227, 1934.

215. Cook, W. C. 1926 The Effectiveness of Certain Paraffin Derivatives in Attracting Flies. J. Agric. Res. 32:347-358.
Iso—or branched—chain compounds relatively more attractive than their normal isomers. ABSTRACT: RAE-B 14:101, 1926.

216. Cook-Young, A. W. 1914 The Prevalence of Flies in Delhi and their Reduction. Proc. Third All-India Sanitary Conference. Lucknow. Jan. 19-27, 1914. II., pp. 141-147. (Suppl. to Ind. J. Med. Res.).
Reports bad outbreak of flies 10 days after spring house-cleaning festival. Bred in rubbish thrown out. ABSTRACT: RAE-B 2:191, 1914.

217. Cooke, O. B. 1946 Fly Control with DDT. Misc. Insect Notes. 46th Rept. Connecticut State Ent. Bull. 515, 111 pp.
Brief mention of *M. domestica.*

218. Copeman, S. M. 1913 Hibernation of House-flies (Preliminary Note). Repts. Local Gov't. Bd. Publ. Health and Med. Subjects, N.S. No. 85; Further Repts. (No. 6) on Flies as Carriers of Infection, pp. 14-19.

219. Copeman, S. M. 1916 Prevention of Fly-Breeding in Horse-manure. Lancet. I (No. 4841):1182-1184.
Advocates close packing on concrete or on ground treated with mineral wood-preserving oil. ABSTRACT: RAE-B 4:141, 1916.

220. Copeman, S. M. and **Austen, E. E. 1914** Do Flies Hibernate? Repts. Local Gov't. Bd. Publ. Health and Med. Subjects. N.S. No. 102, pp. 6-26.
Many flies taken in all parts of England in active state during winter months. ABSTRACT: RAE-B 3:88, 1915.

221. Copeman, S. M., Howlett, F. M. and **Merriman, G. 1911** An Experimental Investigation on the Range of Flight of Flies. Repts. Local Gov't. Bd. Publ. Health and Med. Subjects, N.S. No. 53; Further Repts. (No. 4) on Flies as Carriers of Infection, pp. 1-10.
Flies marked with yellow powder. Recovered at various distances. Both rural and urban experiments were carried out.

222. Corfield, W. F. 1919 Some Experiments upon the Control of Fly-breeding Areas in Camps. J. Roy. Army Med. Corps. 33:415-418.
Deterrents useless in latrines or refuse pits; merely drive larvae to edges. ABSTRACT: RAE-B 8:20, 1920.

Corl, C. S., joint author. *See* Gnadiger, C. B., et al., 1932.

223. Cornwall, J. W. and **Patton, W. S. 1914** Some Observations on the Salivary Secretion of the Common Blood-sucking Insects and Ticks. Indian J. Med. Res. 2:569-593.
Several species of *Musca* involved. ABSTRACT: RAE-B 3:42, 1915.

223a. Cory, E. N. 1917 The Protection of Dairy Cattle from Flies. J. Econ. Ent. 10:111-114.
A spray of 6 per cent emulsion of pine tar creosote was used, with excellent results. ABSTRACT: Hocking, B. 1960 Smell in Insects, p. 190 (5.93).

223b. Cory, E. N. 1918 The Control of House Flies by the Maggot Trap. Maryland State Coll. Agric. Expt. Sta. College Park. Bull. No. 213:103-126.
Close packing of manure, etc. Average percentage of maggot destruction, 95·8. ABSTRACT: RAE-B 7:10, 1919.

223c. Cory, E. N. 1928 The Protection of Cattle from Flies. Bull. Maryland Agric. Exp. Sta. No. 298, pp. 195-196.
To repel flies with pine oil for any considerable length of time, a 4 or 5 per cent emulsion must be used. ABSTRACT: Hocking, B. 1960 Smell in Insects, p. 190 (5.94).

224. Cory, E. N., Harns, H. G. and **Anderson, W. H. 1936** Dusts for Control of Flies on Cattle. J. Econ. Ent. 29:331-335.
Pyrethrum and derris dusts act as toxicants, not as repellents. ABSTRACT: RAE-B 24:198, 1936.

225. Cory, E. N. and **Langford, G. S. 1947** Fly Control in Dairy Barns and on Livestock by Cooperative Spray Services. J. Econ. Ent. 40:425-426.
Participating farmers agreed that protection was more complete and cost less than individual effort. ABSTRACT: RAE-B 37:80, 1949.

226. Cotte, J. and **Tian, A. 1924** Observations sur le métabolisme de la mouche domestique (*Musca domestica* L.) [Observations on the Metabolism of the Domestic Fly (*Musca domestica* L.)]. Soc. Biol., Reunion plénière 1924. pp. 193-196.
The effects of nutrition, temperature and other factors on basal metabolism.

227. Cotterel, G. S. 1920 The Life-history and Habits of the Yellow Dung Fly (*Scatophaga stercoraria*); a Possible Blow-fly Check. Proc. Zool. Soc. (London) for 1920, pt. 4, pp. 629-647.
Dung fly larvae known to devour large numbers of *Musca*. ABSTRACT: RAE-B 9:57, 1921.

228. Cotterel, G. S. 1940 Preliminary Investigations on the Fly Population of Stable Manure Heaps and Measures for the Prevention of Breeding. Pap. Third W. Afr. Agric. Conf. Nigeria, June 1938. Sect. Gold Coast. Lagos (1940) Vol. 1, pp 118-125.
Larvae and pupae not found deeper than 3 inches from surface of heap, below which temperature exceeds level of tolerance (50°C = 122°F). ABSTRACTS: RAE-B 29:46, 1941; TDB 38, 418, 1941.

229. Cowan, J. and **Mackie, F. J. 1919** A Note upon the Modes of Infection in Bacillary Dysentery. J. Roy. Army Med. Corps 32:209-214.
Flies play a considerable part, by contamination of food and drink. ABSTRACT: RAE-B 7:101, 1919.

230. Cox, G. L., Lewis, F. C. and **Glynn, E. F. 1912** The Number and Varieties of Bacteria Carried by the Common House Fly in Sanitary and Insanitary City Areas. J. Hygiene 12:290-319.
Bears close relation to the habits of people and the condition of the streets.

Coyne, F. P., joint author. *See* Bovington, H. H. S., et al., 1944.

231. Cragg, F. W. 1913 Studies on the Mouth Parts and Sucking Apparatus of the Blood-sucking Diptera. Sci. Mem. Med. and Sanit. Depts. Gov't. of India 60:1-56.
Includes *M. nebulo* and others. ABSTRACT: RAE-B 1:229, 1913.

232. Cragg, F. W. 1920 The Maggot Trap: A Means for the Safe Disposal of Horse Manure and Similar Refuse. Indian J. Med. Res. Spec. Indian Science Congress No., 1920, pp. 18-21.
More than 800,000 maggots caught in one week in one trap. This was 99 per cent of the larvae present. ABSTRACT: RAE-B 9:125, 1921.

Cragg, F. W., joint author. *See* Patton, W. S., et al., 1913.

233. Craig, C. F. 1917 The Occurrence of Endamebic Dysentery in the Troops Serving in the El Paso District from July, 1916 to December, 1916. Military Surgeon, Washington, D.C. 40(3-4): 286-302; 423-434.
Reports 156 cases. Flies considered chief cause of distribution. ABSTRACT: RAE-B 5:127, 1917.

Cram, E. B., joint author. *See* Schwarz, B., et al., 1925.

234. Crampton, G. C. 1941 The Terminal Abdominal Structures of Male Diptera. Psyche 48:79-94.
Basic to an understanding of the muscoids. *M. domestica* included in discussion.

235. Cristol, S. J., Haller, H. L. and **Lindquist, A. W. 1946** Toxicity of DDT Isomers to Some Insects Affecting Man. Science 102(2702): 343-344.
Certain isomers not effective against house flies. ABSTRACTS: RAE-B 37:33, 1949; BA 21:1004, 1947.

236. Crumb, S. E. and **Lyon, S. C. 1917** The Effect of Certain Chemicals upon Oviposition in the House-fly (*Musca domestica* L.). J. Econ. Ent. 10:532-536.
In nature, chief incitant is a product of fermentation, possibly $C0_2$, which gave 82·8 per cent higher stimulation to oviposition than ammonia. ABSTRACT: RAE-B 6:67, 1918.

237. Crumb, S. E. and **Lyon, S. C. 1921** Further Observations upon Oviposition in the House Fly. J. Econ. Ent. 14:461-465.
Thirteen additional substances tested. Sodium Carbonate most attractive. ABSTRACT: RAE-B 10:83, 1922.

Cuocolo, R., joint author. *See* De Mello, M. J., et al., 1943.

238. Curran, C. H. 1928 Diptera of the American Museum Congo Expedition. II. Bull. Am. Mus. Nat. Hist. 57:359-361.
Lists *M. ventrosa*, *M. spectanda*, *M. humilis* and *M. sp.*

239. Currie, D. H. 1910 Mosquitoes in Relation to the Transmission of Leprosy. Flies in Relation to the Transmission of Leprosy. U.S. Publ. Health Bull. 39. 42 pp.
Flies, having fed on leprous patients contain bacilli in intestinal tract and in feces for several days.

240. Cuthbertson, A. 1933 The Habits and Life Histories of Some Diptera in Southern Rhodesia. Proc. Rhod. Sci. Ass. 32:81-111.
Treats of several species of *Musca.* Compares mouth parts, feeding habits, etc.

241. Cuthbertson, A. 1934 Biological Notes on some Diptera in Southern Rhodesia. Proc. Rhod. Sci. Ass. 33:32-50.
Both sexes of *M. fasciata* seek exudations from cuts and sores of cattle and from biting sites of blood-suckers. ABSTRACT: RAE-B 22:185, 1934.

242. Cuthbertson, A. 1938 The Breeding Habits and Economic Significance of Some Common Muscoidean Flies (Diptera) in Southern Rhodesia. Proc. Rhod. Sci. Ass. 36:53-57.
Includes information on *M. lusoria*, *M. xanthomelas*, *M. interrupta*, *M. crassirostris*, *M. sorbens*, and *M. vetustissima*. ABSTRACT: RAE-B 27:249, 1939.

243. Dahm, P. A. and **Kearns, C. W. 1941** The Toxicity of Alkyl Secondary Amines to the Housefly. J. Econ. Ent. 34-462-466.
Studied 22 alkyl secondary amines. These differ considerably as to toxicity and rate of action, according to molecular structure.

Dahm, R. G., joint author. *See* Olson, T. A., et al., 1945, 1946.

244. Dakshinamurty, S. 1948 The Common House-fly, *Musca domestica* L. and its Behaviour to Temperature and Humidity. Bull. Ent. Res. 39:339-357.
M. domestica chooses the lower humidity on any gradient between 20 and 100 per cent at a constant temperature of 25°C. At a constant humidity, fly chooses 30°C in preference to either 20°C or 40°C. ABSTRACTS: RAE-B 37:57, 1949; BA 23:3137, 1949.

245. Dale, J. 1922 Flies on a Sanitary Site and Typhoid in a Boys' Home. Med. J. Australia. Sydney. 9th year. i, No. 25:694-695.
Great abundance of *M. domestica.* Twenty-two cases of enteric fever notified. ABSTRACT: RAE-B 10:198, 1922.

246. Darling, S. T. 1912 Experimental Infection of the Mule with *Trypanosoma hippicum* by means of *Musca domestica.* J. Exp. Med. 15:365-366.
Flies visit one open lesion after another (cuts, scratches, galls). Organism remains alive within the fly's proboscis.

247. Darling, S. T. 1913 The Part Played by Flies and Other Insects in the Spread of Infectious Diseases in the Tropics, with Special Reference to Ants and to the Transmission of *Tr. hippicum* by *Musca domestica.* Trans. 15th Internat. Congr. Hyg. and Demog. 5(4):182-185.
Trypanosoma hippicum, cause of murrina, available to flies in open sores.

248. David, W. A. L. 1945 Insecticidal Sprays and Flying Insects. Nature. 155:204.
Greatest number of droplets impinge upon the wings. Flies then clean wing by legs (which are permeable) and clean legs by proboscis which gives insecticide opportunity to act as a stomach poison. ABSTRACT: TDB 42:499, 1945.

249. David, W. A. L. 1946 The Quantity and Distribution of Sprays Collected by Insects Flying through Insecticidal Mists. Ann. Appl. Biol. 33:133-141.
Insecticide was colored with Soudan III. Absorbed through wings or legs color appeared later in Malpighian tubes and was then excreted. ABSTRACT: BA 20:2002, 1946.

250. David, W. A. L. and **Bracey, P. 1947** Factors Influencing the Interaction of Insecticidal Mists and Flying Insects. Part IV. Some Experiments with Adjuvants. Bull. Ent. Res. 37:393-398.
Refers to isobutyl undecylenamide, also sesame oil, as used to enhance efficiency of pyrethrum sprays against houseflies.

251. Davidson, J. 1918 Some Practical Methods Adopted for the Control of Flies in the Egyptian Campaign. Bull. Ent. Res. 8:297-309.
Musca bred in horse manure, camp and cook-house refuse and latrine trenches. Several control methods used. ABSTRACT: RAE-B 6:100, 1918.

252. Davidson, J. 1942 Flies, Fleas and Lice. Med. J. Australia for 1942. pp. 111-116. Sydney.
M. domestica in houses, living huts, mess huts, kitchens. *M. vetustissima* more in open situations. ABSTRACT: RAE-B 31:206, 1943.

253. Davidson, J. 1944 On the Relationship between Temperature and Rate of Development of Insects at Constant Temperatures. J. Animal Ecol. 13:26-38.
Develops an equation which applies to *M. domestica* and other species. ABSTRACT: RAE-B 33:99, 1945.

254. DeAlmeida, J. L. 1933 Noveaux agents de transmission de la Berne (*Dermatobia hominis* L. junior 1781) au Brésil. [New Agents of Transmission for Berne (*Dermatobia hominis* L. junior 1781) in Brazil.] Compt. Rend. Soc. Biol. 113:1274-1275.
Lists *M. domestica* and a species of *Promusca* as 2 of 9 forms demonstrated to be vectors. ABSTRACT: TDB 31:64, 1934.

Dearden, W. F., joint author. *See* Plowman, C. T., et al., 1915.

255. DeBach, P. 1939 A Hormone which Induces Pupation in the Common House Fly, *Musca domestica.* Ann. Ent. Soc. Am. 32:743-746.
Mature larvae ligatured anterior to segment 5 (Hewitt), pupated only posterior to ligament. If ligatured after critical period (15-25 hours before pupation), both ends pupate.

256. DeBach, P. 1943 The Effect of Low Storage Temperature on Reproduction in Certain Parasitic Hymenoptera. Pan Pacific Ent. 19(3): 112-119.
Refers to 2 pupal parasites of *M. domestica. Mormoniella vitripennis* and *Muscidifurax raptor* survived low storage temperatures very well. ABSTRACT: RAE-B 32:64, 1944.

257. De Bach, P. 1944 Environmental Contamination by an Insect Parasite and the Effect on Host Selection. Ann. Ent. Soc. Am. 37:70-74.
Female *Mormoniella,* while seeking host pupae to parasitize, so affects environment that other females leave the area alone and no further parasitization occurs. ABSTRACT: RAE-B 33:82, 1945.

Decker, G. C., joint author. *See* Kearns, C. W., et al., 1949.

257a. D'Emmerez de Charmoy, D. and **Moutia, A. 1925** Hints on the General Treatment of Insect Pests in Mauritius. Dept. Agric. Mauritius, Gen. Ser. Bull. 31. 16 pp. Port Louis.
Gives many formulae, with directions for their use. List includes poison baits for houseflies, and fly repellents for protection of cattle. ABSTRACT: Hocking, B. 1960 Smell in Insects, p. 192 (5.110).

258. DeGeer, C. 1776 Mémoires pour servir à l' Histoire des Insectes. [Memoirs Contributing to a History of Insects.] Vol. 6, pp. 71-78. Stockholm.
M. domestica treated on pages indicated.

Dekester, M. joint author. *See* Jausion, H., et al., 1923.

259. De la Paz, G. C. 1938 The Bacterial Flora of Flies Caught in Foodstores in the City of Manilla. Mon. Bull. Bur. Health Phillip. I. 18:393-412.
Practically all flies were *Musca.* Many disease organisms found. ABSTRACT: RAE-B 27:164, 1939.

260. De la Paz, G. C. 1938 The Fly Problem in Relation to Refuse Disposal in Manila. Mon. Bull. Bur. Health Phillip. I. 18:521-539.
Suggests that dumping be replaced by garbage reduction through use of zymothermic cells. ABSTRACT: RAE-B 27:164, 1939.

261. De la Paz, G. C. 1938 The Breeding of Flies in Garbage and their Control. Mon. Bull. Bur. Health Philipp. I. 18:515-519.
M. domestica vicina, M. sorbens and *M. nebulo* head the list, in order of abundance. ABSTRACT: RAE-B 27:164, 1939.

262. De la Paz, G. C. 1939 Our Common Houseflies, their Importance as Disease Transmitters, and their Eradication. Mon. Bull. Bur. Health Philipp. I. 19:219-230.
An educational type publication. Lists several species. *M. domestica vicina* of prime importance. ABSTRACT: RAE-B 28-80, 1940.

263. Delbue, A. 1940 Un esperimento di lotta contro le mosche a Valeggio sul Mincio. [An Experiment in the Control of Flies at Valeggio on the Mincio.] Ann. Igiene 50(2):70-78.
A successful 2-year campaign employing a variety of control practices. ABSTRACT: RAE-B 28:133, 1940.

Delisle, R., joint author. *See* Monro, H. A. U., et al., 1947.

264. Del Ponte, E. 1938 Les especies argentinas del género *Cochliomyia* T. T. (Dipt. Musc.) [The Argentine Species of the Genus *Cochliomyia* T. T.]. Rev. Ent. 8(3-4):273-281.
Considers *Musca laniara* Wied. a synonym of *Cochliomyia macellaria*. ABSTRACT: RAE-B 27:2, Jan. 1939.

265. DeMello, F. and **Cabral, J. 1926** Les Insectes sont ils susceptibles de transmettre la lepre? [Are Insects Capable of Transmitting Leprosy?] Bull. Soc. Path. Exot. 19:774-777.
Infection found in gut of 4 out of 10 specimens of *Musca bezzii* Patton and Craig.

266. DeMello, F. and **Jacques, J. E. 1919** Note sur l'existence de l' *Herpetomonas muscae-domesticae* à l'Inde portugaise. [Note on the existence of *Herpetomonas muscae-domesticae* in Portuguese India.] Bol. Ger. Med. e Farmacia, Nova-Goa. 5(5):194-195.
Infests less than 1 per cent of the house flies in Goa. ABSTRACT: RAE-B 9:129, 1921.

267. DeMello, M. J. and **Cuocola, R. 1943** Técnica para o xenodiagnóstico da habronemose gástrica dos equídeos. [Technique for the Xenodiagnosis of Gastric Habronemiasis of Equines.] Arq. Inst. Biol. 14:217-226. São Paulo. (English Summary.)
Uses either *M. domestica* or *Stomoxys calcitrans*, according to the parasite concerned. Infected flies liberate larvae into normal horse serum, when provided. ABSTRACT: RAE-B 33:103, 1945.

268. DeMello, M. J. and **Cuocolo, R. 1943** Alguns aspetos das relações da *Habronema muscae* (Carter, 1861) com a mosca domestica. [Certain Aspects of the Relationship of *Habronema muscae* to the Domestic Fly.] Arq. Inst. Biol. 14:227-234. São Paulo. (English Summary.)
Maggots become infected between the first and third day, probably by ingestion. Parasites form cyst-like bodies. ABSTRACT: RAE-B 33:105, 1945.

269. DeMello, M. J. and **Pereira, C. 1946** Determinismo da evasão das larvas de *Habronema* sp. da tromba da Mosca domestica. [Proof of the emergence of the larvae of *Habronema* from the proboscis of the domestic fly.] Arq. Inst. Biol. 17:259-266. São Paulo. (English Summary.)
ABSTRACT: RAE-B 36:155, 1948.

270. Deonier, C. C. 1938 Effects of Some Common Poisons in Sucrose Solutions on the Chemoreceptors of the Housefly, *Musca domestica*. J. Econ. Ent. 31:742-745.
Chemoreceptors on tarsal segments. Extension of proboscis is positive response; withdrawal, negative. ABSTRACT: RAE-B 27:102, 1939.

271. Deonier, C. C. and **Richardson, C. H. 1935** The Tarsal Chemoreceptor Response of the Housefly, *Musca domestica* L. to Sucrose and Levulose. Ann. Ent. Soc. Am. 28:467-474.
Levulose less effective than sucrose in stimulating response. Average of 10 per cent of fly population did not respond, even to strong solution of sucrose. ABSTRACT: RAE-B 24:64, 1936.

271a. Deonier, C. C. and **Richardson, C. H. 1939** Effects of Toxic Compounds on the Gustatory Chemoreceptors in Certain Diptera. Iowa State Coll. J. Sci. 14:22-23.
Tarsal chemoreceptors of *M. domestica* showed no olfactory response. Houseflies sensitive to very small amount of 1 M sucrose solutions. Certain chemical substances, e.g. $HgCl_2$, made such solutions repellent. Repellence decreased with age. ABSTRACT: Hocking, B. 1960 Smell in Insects, p. 10 (1.94).

272. Deoras, P. J. and **Jandu, A. S. 1943** The House-fly and its Control. Indian Fmg. 4:565-568. Delhi.
Five species of *Musca* found in and about houses, with *M. nebulo* the most common. ABSTRACTS: RAE-B 33:9, 1945; BA 20:1539, 1946.

273. Derbeneva-Ukhova, V. P. 1935 [Die Einwirkung der Imaginalen Ernährungsbedingungen auf die Entwicklung der Ovarien von *Musca domestica* L.] Med. Parasit. 4:394-403. (In Russian, with German summary.)
Virgin females kept at constant temperature and relative humidity were dissected daily and first follicle of egg tube measured. Effects of various diets compared. ABSTRACT: RAE-B 24:72, 1936.

274. Derbeneva-Ukhova, V. P. 1935 [Ueber die Zahl der Generationen bei *Musca domestica* L.] Med. Parasit. 4:404-407. (In Russian.)
Four generations in Moscow between May 20 and Sept. 30. Development accelerated by increase of temperature with decrease in R.H. ABSTRACT: RAE-B 24:72, 1936.

275. Derbeneva-Ukhova, V. P. 1937 [Die Oekologie der Larven von *Musca domestica* unter naturlichen Bedingungen.] Med. Parasit. 6:408-417. (In Russian, with German summary.)
Three times as many larvae in pig dung as in horse dung, or refuse. Much information of ecological nature. ABSTRACT: RAE-B 26:167, 1938.

276. Derbeneva-Ukhova, V. P. 1940 [Sur l'écologie des mouches de fumier à Kabarda.] Med. Parasit. 9:323-339. (In Russian.)
M. domestica absent in human feces probably because of predatory action of the larvae of *Fannia scalaris* and *Ophyra leucostoma.* Remarks on other species of *Musca.*

277. Derbeneva-Ukhova, V. P. 1940 [Influence de la temperature sur les larves de *Musca domestica* L.] Med. Parasit. 9:521-524.
Useful paper. Contains 2 graphs. ABSTRACTS: TDB 40:499, 1943; RAE-B 31:27, 1943.

278. Derbeneva-Ukhova, V. P. 1940 [Adaptation des larves de *Musca domestica* L. à des hautes températures.] Med. Parasit 9:525-527. (In Russian.)
ABSTRACTS: TDB 40:499, 1943; RAE-B 31:27, 1943.

279. Derbeneva-Ukhova, V. P. 1942 [The Fly-Maggots as Components of the Dung Biocenoses.] Med. Parasit. 11(3):79-86. (In Russian.)
Treats of food habits of *M. larvipara*, *M. domestica* and *Muscina stabulans.* ABSTRACT: RAE-B 32:52, 1944.

280. Derbeneva-Ukhova, V. P. 1942 [On the Ecology of *Musca domestica* and *Muscina stabulans* Flln.] Med. Parasit. 10:534-543. (In Russian.)
In Province of Archangel, *M. domestica* chiefly in calf dung; *Muscina* most in human feces not in latrines. ABSTRACT: RAE-B 31:223, 1943.

281. Derbeneva-Ukhova, V. P. 1942 [On the Development of Ovaries and on the imaginal Nutrition in Dung-flies.] Med. Parasit. 11(4): 85-97. (In Russian.)
Treats of *M. domestica* and 14 other species. Recognizes 5 groups, based on feeding habits. ABSTRACTS: BA 21:237, 1947; RAE-B 32:118, 1944.

282. **Derbeneva-Ukhova, V. P. 1943** [The Influence of Soil Humidity and Compactness on the Emergence of the Imago of *M. domestica.*] Med. Parasit. 12(3):72-76.
Loose soil or sand not an obstacle to emergence. High moisture content can be fatal to larvae. ABSTRACT: BA 19:2231, 1945.

283. **Derbeneva-Ukhova, V. P. 1947** [The Application of DDT Preparations against Flies.] Med. Parasit. 16(1):16-28.
The usual successful control, if applied to non-resistant strains. ABSTRACT: RAE-B 36:126, 1948.

284. **Derbeneva-Ukhova, V. P., Bragin, E. A.** and **Skvortsov, A. A. 1943** [Problems of Research on Flies as Vectors of Infection.] Med. Parasit. 12(5):39-44.
Identifies 6 categories of research. ABSTRACT: BA 19:2514, 1945.

285. **Derbeneva-Ukhova, V. P.** and **Kuzina, O. S. 1938** [Quelques observations au cours de l'essai de lutte contre les mouches aux constructions nouvelles.] Med. Parasit. 7:399-405. (In Russian, with French summary.)
Numbers of *M. domestica* in buildings in inverse proportion to distance from breeding places. Fly trap baits compared for efficiency. ABSTRACT: RAE-B 27:69, 1939.

286. **De Salles, J. F.** and **Hathaway, C. R. 1944** Nota sôbre a infestação de *Musca domestica* Linneu, 1758 per um Ficomiceto do genero *Empusa.* [Note on an infestation of *Musca domestica* Linnaeus, 1758, by a Phycomycete of the genus *Empusa.*] Mem. Inst. Oswaldo Cruz 41:95-99.
A histological study of tissues invaded by the parasite.

287. **Descazeaux** and **Morel. 1933** Diagnostic biologique (xenodiagnostic) des habronémoses gastriques du cheval. [Xenodiagnosis of gastric habronemoses in the horse.] Bull Soc. Path. Exot. 26:1010-1014.
Place eggs of *Musca* and/or *Stomoxys* on excreta of suspected animals. Five to six days after flies emerge, dissect proboscis for nematodes. ABSTRACT: RAE-B 22:14, 1934.

Descazeaux, J., joint author. *See* Roubaud, E., et al., 1921, 1922, 1923.

288. **De Souza-Araujo, H. C. 1943** Verificação da infeccão de moscas da familia Tachinidae pela *Empusa* Cohn 1885. Essas moscas, sugando ulceras lepróticas, se infestaram com o bacilo de Hansen. [Verification of the Infection of Flies of the Family Tachinidae by *Empusa* Cohn. 1885. Flies, feeding on leprous ulcers, infect themselves with Hansen's bacillus.] Mem. Inst. Oswaldo Cruz. 39 (Fasc. 2):201-203. (English summary.)
ABSTRACT: RAE-B 35:123, 1947.

289. **De Souza-Lopes, H. 1938** Sur quelques diptères porteurs d'oeufs de la *Dermatobia hominis* L., J. au Brézil. [On certain Dipterous Carriers of the eggs of *Dermatobia hominis* L. Collected in Brazil.] (Dipt.-Oestridae.) Compt. Rend. Soc. Biol. 129(27):427-428. Paris.
Ova found on 2 species of *Fannia* and 1 of *Morellia.* ABSTRACT: RAE-B 27:110, 1939.

290. **De Stefani, T. 1921** Importanza dell' Entomologia applicata nell' Economia sociale. Entomologia legale e dei Cadaveri. [Importance of the application of entomology in social economics. Legal entomology in regard to cadavers.] Allevamenti, (Palermo) II, No. 5,

May, pp. 131-133.
M. domestica appears among the first insects to attack the body. ABSTRACT: RAE-B 9:121, 1921.

291. **Dethier, V. G. 1947** Chemical Insect Attractants and Repellants. Philadelphia, The Blakiston Company. pp. xv + 289. 69 figs. Houseflies treated on pp. 29, 118, 152, 153, 160, 164, 165, 187, 191, 223.

292. **Devoe, A. 1945** House Flies. The American Mercury, October, 1945, pp. 473-477.
One of a series of popular articles on Natural History published in American Mercury under the Section Title "Down to Earth".

293. **Dews, S. K. and Morrill, A. W. 1946** DDT for Insect Control at Army Installations in the Fourth Service Command. J. Econ. Ent. 39:347-355.
DDT fly control became standard practice at all posts of this command. Necessary to treat at source (barns, bull pens, etc. within 500 ft.).

294. **Dhondy, B. S. 1940** Automatic Flyproof Latrine Seat. Indian Med. Gaz. 75:466-468.
Based on mobile foot rests, connected to levers. Release of foot pressure closes lid. ABSTRACT: TDB 38:234, 1941.

Dicke, R. J., joint author. *See* Allen, T. C., et al., 1944; Ikawa, M., et. al., 1945.

295. **Dickenson, G. K. 1907** The House-fly and its Connection with Disease Dissemination. Med. Rec. 71:134-139.
Prevailing concepts of that period.

296. **Diguet, L. 1915** The "Mosquero" or Spider's Nest used in Mexico as a Fly Trap. Bull. Soc. Nat. Acclimat. 62(6):170-171. Paris.
Nest may attain 40 cu. ft. in bulk. Spider measures one-twelfth of an inch. ABSTRACT: RAE-B 3:151, 1915.

297. **Diguet, L. 1915** Nouvelles observations sur le mosquero ou nid d'araignés sociales employé come piège à mouches dans certaines localités du Mexique. [New Observations on the "mosquero" or nest of the social spiders, used as a trap for flies in certain localities of Mexico.] Bull. Soc. Nat. Acclimat. 62(8):240-249.
Coenotele gregalis is nest maker. A Drassid spider, *Poecilochroa*, and a Clavicorn beetle *Corticaria*, are commensals. Cleaning duties may be performed by ants. ABSTRACT: RAE-B 4:9, 1916.

Dnyansagar, V. R., joint author. *See* Joshi, K. G., et al., 1945, 1947.

298. **Doane, R. W. 1911** An Annotated List of the Literature on Insects and Diseases for the year 1910. J. Econ. Ent. 4:386-398.

299. **Doane, R. W. 1912** An Annotated List of the Literature on Insects and Diseases for the year 1911. J. Econ. Ent. 5:268-285.

300. **Doane, R. W. 1914** Disease-bearing Insects in Samoa. Bull. Ent. Res. 4:265-269.
M. domestica probably connected with typhoid, framboesia and eye infections. ABSTRACT: RAE-B 2:83, 1914.

Dodd, W. L., joint author. *See* Orton, S. F., et al., 1910.

301. **Dolinskaya, T. Y., Klechetova, A. M.** and **Tregubov, A. M. 1946** [The Use of Chloride of Lime for the Control of Flies.] Med. Parasit. 15(6):81-82. (In Russian.)
Serves for emergency control of larvae, and as a repellent against adult flies. ABSTRACT: RAE-B 36:82, 1948.

302. Dönhoff. 1872 Beiträge zur Physiologie, I. Ueber das Verhalten Kaltblütiger Thiere gegen Frosttemperatur. [Contribution to Physiology, I. On the Behaviour of Cold-blooded Animals in the presence of Freezing Temperature.] Arch. Anat. und Phys. und Wiss. Med. von Reichert und Du Bois-Reymond., pp 724-727.
Flies killed at temperatures not many degrees below the freezing point of water.

303. Donovan, H. L. 1886 Can Flies Carry Cholera? Indian Med. J. 21:318.
A typical example of speculative writing for that period.

304. Douglas, J. R. 1947 Control of Flies on Cattle with DDT. Univ. of California, Div. of Entom. and Parasit., Davis. March 1947. 2 pp. mimeogr.
Suspension sprays more efficient than emulsions, because they weather more slowly.

305. Douglas, J. R. 1947 Control of Flies in Farm Building. Univ. of California, Div. of Entom. and Parasit., Davis. May, 1947. 2 pp. mimeogr.
Deposit of 300 mg. of DDT per sq. foot reduced fly population 94 per cent and was still effective after 100 days.

306. Dove, W. E. 1916 Some Notes Concerning Overwintering of the House Fly, *Musca domestica*, at Dallas, Texas. J. Econ. Ent. 9:528-538.
Adults become inactive at 45°F or below. Most favorable temperature for survival is under 60°F. Record longevity, 91 days. ABSTRACT: RAE-B 5:29, 1917.

307. Dove, W. E. 1947 Piperonyl Butoxide, a New and Safe Insecticide for Household and Field. Am. J. Trop. Med. 27:339-345.
M. domestica used in several types of tests. No hazard to warm blooded animals. Predicts use in combination with pyrethrins. ABSTRACT: BA 21:2321, 1947.

Dove, W. E., joint author. *See* Simmons, S. W., et al., 1942, 1945; Stoddard, R. B., et al., 1949; Bishopp, F. C., et al., 1915.

Dow, R. P., joint author. *See* Maier, P. P., et al., 1949.

Down, H. A., joint author. *See* Harris, A. H., et al., 1946.

308. Drbohlav, J. J. 1925 Studies on the Relation of Insect Herpetomonad and Crithidial Flagellates to Leishmaniasis. Parts I, II, III. Am. J. Hyg. 5:580-621.
Herpetomonas as found in *Musca, Fannia* and *Lucilia,* believed to represent but one species, *H. muscae-domesticae*. ABSTRACT: RAE-B 13:173, 1925.

309. Drbohlav, J. J. 1926 The Cultivation of *Herpetomonas muscarum* (Leidy 1866) Kent 1881, from *Lucilia sericata*. J. Parasit. 12: 183-190.
Parasites limited to lumen of alimentary tract. Species believed to be a synonym of *H. muscae-domesticae*. ABSTRACT: RAE-B 14:191, 1926.

310. Dresden, D. and **Krijgsman, B. J. 1948** Experiments on the Physiological Action of Contact Insecticides. Bull. Ent. Res. 38:575-578.
Ability of insecticide to penetrate the cuticle is of great importance. Order of toxicity changes when poisons are injected.

311. Dresner, E. 1949 Culture and Use of Entomogenous Fungi for the Control of Insect Pests. Contr. Boyce Thompson Inst. 15:319-335.
M. domestica susceptible to paralysis caused by fungus when it is in liquid medium. ABSTRACT: RAE-B 39:207, 1951.

312. Du Chanois. 1947 Toxicity of Gamma-benzene Hexachloride to Preimaginal Stages of the Housefly. J. Econ. Ent. 40:749-751.
Non toxic to ova, but strikingly toxic to second and third stage larvae and pupae. ABSTRACT: RAE-B 22:986, 1948.

313. Dudgeon, L. S. 1919 The Dysenteries: Bacillary and Amoebic. Brit. Med. J. (London) No. 3041:448-451. Apr. 12.
Disease most prevalent in Macedonia when flies were most numerous. Two peaks, spring and fall. ABSTRACT: RAE-B 7:102, 1919.

Dugas, A. L., joint author. *See* Manis, H. C., et al., 1942.

314. Duncan, J. T. 1926 On a Bactericidal Principle present in the Alimentary Canal of Insects and Arachnids. Parasitology 18:238-252.
Found in 8 species, including *M. domestica.* Bactericidal action greater at 37°C (98·6°F) than at room temperature. ABSTRACT: RAE-B 14:173, 1926.

315. Dunkerly, J. S. 1911 On Some Stages in the Life-history of *Leptomonas muscae-domesticae*, with Remarks on the Relationship of the Flagellate Parasites of Insects. Quart. J. Micr. Sci. 56:645-655.
Oval, cyst-like bodies in the rectum of flies pass out with the feces and are taken up by the proboscides of other flies.

316. Dunn, L. H. 1923 Observations on the Oviposition of the House Fly, *Musca domestica* L., in Panama. Bull. Ent. Res. 13:301-305.
Female remains fertilized for life after one successful pairing. Earliest oviposition 2¼ days after emergence. ABSTRACT: RAE-B 11:51, 1923.

317. Dunne, A. B. 1902 Typhoid Fever in South Africa; its Cause and Prevention. Brit. Med. J., March 8, 1902. p. 622.
Need for fly control emphasized.

Du Plessis, S., joint author. *See* Wahl, R. O., et al., 1923.

Duran-Borda, G., joint author. *See* Proto-Gomez, Y., et al., 1888-89.

318. Du Toit, R. and **Nieschulz, O. 1933** *Musca crassirostris*, a Blood-sucking Fly, new to South Africa. J. South Afric. Vet. Med. Ass. 4(2):97-98. Johannesburg.
Attacks horses on veldt and in kraals. ABSTRACT: RAE-B 22:23, 1934.

Du Toit, R., joint author. *See* Nieschulz, O., et al., 1933.

319. Dutton, W. F. 1909 Insect Carriers of Typhoid Fever. J. Am. Med. Ass. 53:1248-1252.

320. Dwyer, J. M. 1943 Some Notes on Fly Control. Med. J. Australia 25:501-503.
M. vetustissima, the bush fly, does not enter buildings like *M. domestica.* Both breed in human feces. ABSTRACT: Bull. War Med. 4(10):614, 1944.

321. Eagleson, C. 1938 Resistance of *Stomoxys calcitrans* (L.) to Laboratory Application of Pyrethrum Spray. J. Econ. Ent. 31:778.
Gives table comparing this species with *M. domestica.* ABSTRACT: RAE-B 27:105, 1939.

322. Eagleson, C. 1942 Effect of Temperature on Recovery of Houseflies. Soap. 18:115-117, 141.
Pyrethrum, also lethane, produce complete torpor in a dose far below the lethal one. Temperature affects the recovery rate. ABSTRACT: RAE-B 31:33, 1943.

323. Eagleson, C. 1942 Sesame in Insecticides. Soap. 18:125, 127.
Causes much lower concentration of pyrethrum to be lethal for flies. Does not harm coats of cattle. ABSTRACT: RAE-B 31:148, 1943.

Eastwood, T. joint author. *See* McDonnell, R. P., et al., 1917.

324. Eddy, C. O. 1927 Cyanogas Calcium Cyanide for House Fly Fumigation in Certain Types of Buildings. J. Econ. Ent. 20:270-281.
Successful in large rooms, barns, stables and stores. One-half ounce to 1000 cu ft. lethal for flies. ABSTRACT: RAE-B 15:147, 1927.

325. Eddy, C. O. 1927 House Fly Fumigation in Certain Types of Buildings. S. Carolina Agric. Exp. Sta. Bull. 237. 14 pp.
Size and shape of room affect results. Night fumigation best. Dust cloud superior to film method. ABSTRACT: RAE-B 15:174, 1927.

326. Eddy, C. O. 1929 House Fly Fumigation Experiments with Calcium Cyanide. S. Carolina Agric. Exp. Sta. Bull. 256. 48 pp.
Reports various trials under controlled conditions. ABSTRACT: RAE-B 17:179, 1929.

327. Edwards, F. W. 1916 On the Correct Names of some Common British Diptera. Ent. Mon. Mag. 52:59-63.
M. autumnalis DeG. (1776) the correct name for *M. corvina* Fab. 1781.

328. Egorov, P. I. 1940 [Essai d'utilisation de la fumigation du sol comme moyen de la destruction des agglomérations de nymphes de mouches.] Med. Parasit. 9:528-530. (In Russian.)
Used 1·2 ounces of paradichlorobenzene per square yard. Covered with paper after treatment. Pupae of *M. domestica* dead within three days. ABSTRACTS: TDB 40:501, 1943; RAE-B 31:28, 1943.

329. Elmore, J. C. and **Richardson, C. H. 1936** Toxic action of Formaldehyde on the Adult House Fly, *Musca domestica* L. J. Econ. Ent. 29:426-433.
Amount of poison consumed determined by weighing fly before and after feeding. Median lethal dose 2·24 mg. per gm. body weight. ABSTRACT: RAE-B 24:199, 1936.

330. Eltringham, H. 1916 Some Experiments on the House Fly in Relation to the Farm Manure Heap. J. Agric. Sci. Cambridge. 7:443-457.
Compares breeding habits and behavior of *M. domestica* and *M. autumnalis*. ABSTRACT: RAE-B 4:108, 1916.

331. Esten, W. M. and **Mason, C. J. 1908** Sources of Bacteria in Milk. Storrs Agric. Exp. Sta. Bull. 51:65-109.
Number of bacteria on the fly range from 550 to 6,600,000. Considered chief source of intestinal disease in children as summer progresses.

332. Evans, A. T. 1916 Some Observations on the Breeding Habits of the Common House-Fly, *Musca domestica*, Linnaeus. J. Econ. Ent. 9:354-362.
Females oviposit in manure because it is alkaline. Very dilute acid, added to manure prevented development. Flies visited acid garbage for feeding but not to oviposit. ABSTRACT: RAE-B 4:147, 1916.

333. Ewing, H. E. 1913 A New Parasite of the House-Fly (*Acarina, Gamasoidea*). Ent. News 24:452-456.
Macrocheles muscae, sp. n. from Ithaca, N.Y. and Corvallis, Oregon. ABSTRACT: RAE-B 2:6, 1914.

334. Ewing, H. E. and **Hartzell, A. 1918** The Chigger-Mites Affecting Man and Domestic Animals. J. Econ. Ent. 11:256-264.
Thrombidium muscarum Riley is confined to a single host, *M. domestica*. Never attaches to man. ABSTRACT: RAE-B 6:148, 1918.

335. Ewing, H. E. jr. 1942 The Relation of Flies (*Musca domestica* Linnaeus) to the Transmission of Bovine Mastitis. Am J. Vet. Res. 3:295-299.
Streptococcus agalactae can be carried by flies which contact external surface of teat sphincters. ABSTRACT: RAE-B 31:121, 1943.

336. Eyles, E. D. 1945 Landing on a Ceiling. New York Times. August 12, 1945.
Took high-speed motion pictures (1500 per sec.). Flies do a "half-roll".

337. Face, L. la 1949 La mosca domestica. [House Flies (*Musca domestica*).] Rev. Sanidad é Hig. Publ. (Madrid) 23:832-887. (A Review.)

338. Faichnie, N. 1909 Fly-borne Enteric Fever; the Source of Infection. J. Roy. Army Med. Corps 13:580-584.
Author reared 4000 flies from one-sixth of a cubic foot of ground from a latrine; 500 from a single dropping of human excrement.

339. Faichnie, N. 1909 *Bacillus typhosus* in Flies. J. Roy. Army Med. Corps 13:672-675.
The two foregoing papers stress the importance of flies bred from enteric excreta. They are carriers for the remainder of their lives.

340. Fairchild, David and **Marian. 1914** A book of Monsters. National Geographic Magazine 26:89-98 (July, 1914).
Dragon fly shown devouring muscoid fly. Can consume 40 in an hour.

341. Fairchild, H. E. Hoffman, R. A., Lindquist, A. W. and **Mote, D. C. 1949** A Comparison of the Chlorinated Hydrocarbon Insecticides for Control of the Sheep Tick. J. Econ. Ent. 42:410:414.
M. domestica used as the test insect. ABSTRACT: RAE-B 38:94, 1950.

Fairchild, Marian, joint author. *See* Fairchild, David and Marian, 1914.

342. Fales, J. H. and **Goodhue, L. D. 1944** Aerosols vs Oil Spray Insecticides. A Study of Comparative Efficiency. Soap. 20(7):107-108.
Aerosol superior when sprayed directly on flies, but requires a little more time for knock-down. ABSTRACT: RAE-B 33:97, 1945.

343. Fales, J. H., McGovran, E. R. and **Goodhue, L. D. 1946** Aerosol Toxicity. Effect of the Nonvolatile Content of a DDT Aerosol on Mortality of House-flies. Soap and Sanit. Chem. 22(6):157-158.
Greatest mortality of house flies when nonvolatile content is about 15 per cent. ABSTRACT: RAE-B 37:101, 1949.

344. Fales, J. H., McGovran, E. R. and **Fulton, R. A. 1947** Gamma-Benzene Hexachloride in the Liquified-gas Aerosol. J. Econ. Ent. 40:754.
Knockdown quicker than with DDT. Caused no corrosion of metal with standard tests. ABSTRACT: RAE-B 37:127, 1949.

345. Fales, J. H., Nelson, R. H., Ball, W. H. and **Fulton, R. A. 1948** Storage of Aerosols. Soap and Sanit. Chem. 24(11):127, 129.
Pyrethrum-DDT aerosols toxic up to 2½ years in storage. ABSTRACT: RAE-B 38:203, 1950.

Fales, J. H. joint author. *See* Goodhue, L. D., et al., 1942, 1945; McGovran, E. R., et al., 1943, 1946; Nelson, R. H., et al., 1949; Sullivan, W. N., et al., 1940.

346. Fantham, H. B. 1922 Some Parasitic Protozoa found in South Africa. V. S. Afr. J. Sci. 19:332-339.
Found *Herpetomonas muscae-domesticae* in only 4 flies (*M. domestica*) of 286 examined in 2 years. ABSTRACT: RAE-B 11:129, 1923.

347. Fantham, H. B. and **Robertson, K. G. 1927** Some Parasitic Protozoa found in South Africa. X. S. Afr. J. Sci. 24:441-449.
M. domestica infected in laboratory with *Herpetomonas luciliae* from gut content of *Lucilia sericata.* ABSTRACT: RAE-B 16:106, 1928.

348. Farrar, R. 1905 Reports of Medical Inspectors of the Local Govnt. Board, No. 216, p. 9.
Possible carriage of enteric fever by flies in Yorkshire.

349. Fay, R. W. 1939 A Control for the Larvae of Houseflies in Manure Piles. J. Econ. Ent. 32:851-854.
Advocates tall piling in ricks, covering, and trapping of larvae in ditch. ABSTRACT: RAE-B 28:173, 1940.

350. Fay, R. W., Buckner, A. J. and **Simmons, S. W. 1948** Laboratory Evaluation of DDT Residual Effectiveness against Houseflies, *Musca domestica*. Am J. Trop. Med. 28:877-888.
Female specimens less susceptible than males. Flies tested 3 days after emergence more resistant than any other age group. ABSTRACTS: RAE-B 40:37, 1952; BA 23:1749, 1949.

350a. Fay, R. W., Cole, E. L. and **Buckner, A. J. 1947** Comparative Residual Effectiveness of Organic Insecticides against House Flies and Malaria Mosquitoes. J. Econ. Ent. 40(5):635-640. Erratum T.c. No. 6, p. 909.
Five insecticides applied as xylene- or kerosene-based sprays on test panels. Most of them lost toxicity with time; gradually for mosquitoes, suddenly for houseflies. ABSTRACTS: RAE-B 37:125, 1949; BA 22:994, 1948.

351. Feldman-Muhsam, B. 1944 A Note on the Conditions of Pupation of *Musca domestica vicina* (Diptera) in Palestine, and its Application. Proc. Roy. Ent. Soc. London. Ser. A. Gen. Ent. 19:139-140.
Drying of surface of manure pile inhibits pre-pupal migration; may delay or prevent development. ABSTRACTS: BA 19:2243, 1945; RAE-B 33:178, 1945; TDB 42:413, 1945.

352. Feldman-Muhsam, B. 1944 Studies on the Ecology of the Levant House Fly (*Musca domestica vicina Macq.*). Bull. Ent. Res. 35: 53-67.
Though eggs are very sensitive to prolonged cold, they are more resistant than other stages to intense cold for a short time. ABSTRACTS: BA 19:193, 1945; TDB 41:496, 1944; RAE-B 32:134, 1944.

353. Feldman-Muhsam, B. and **Feldman-Muhsam, H. V. 1946** Life Tables for *Musca vicina* and *Calliphora erythrocephala*. Proc. Zool. Soc. London. 115(3/4):296-305.
Compares stages with corresponding periods in the life of man. ABSTRACT: BA 21:506, 1947.

354. Felt, E. P. 1909 The Economic Status of the House-fly. J. Econ. Ent. 2:39-44. (A review.)

355. Felt, E. P. 1909 The Typhoid or House-fly and Disease. *In* 24th Rept. of State Entomologist. *In* N.Y. State. Mus. Bull. No. 455.
Gives excellent bibliography.

356. Felt, E. P. 1910 Control of Flies and Household Insects. Bull. No. 465, New York State Educ. Dept. 53 pp., 34 fig.
Ten pages devoted to *M. domestica.*

357. Felt, E. P. 1910 Methods of Controlling the House-fly and thus Preventing the Dissemination of Disease. New York Med. J. 91:685-687 (April 2).

358. Felt, E. P. 1910 Typhoid or House-fly. Twenty-fifth Rept. of State Entomologist, Bull. No. 475, New York State Educ. Dept., pp. 12-17. *See also* J. Econ. Ent. 3:24-26.
Flies do not breed freely in darkness, but females will crawl into dark crevices to lay their eggs.

358a. Felt, E. P. 1913 Notes for the Year. Twenty-eighth Rept. State Entomologist on Injurious and other Insects of the State of New York. Museum Bull. No. 165, N.Y. State Museum, p. 93.
Records a case of myiasis caused by larvae of *M. domestica.* These had been swallowed with canned sardines which were left exposed to flies.

359. Feng, L.-C. 1933 Some Parasites of Mosquitoes and Flies found in China. Lingnan Sci. J. 12, Suppl. 23-31. Canton.
M. domestica frequently infested with a red mite in Amoy, South China. ABSTRACT: RAE-B 21:218, 1933.

360. Fenton, F. A. and **Bieberdorf, G. A. 1936** Fly Control on A. & M. Farms, Stillwater, Okla. J. Econ. Ent. 29:1003-1008.
Manure either removed within 24 hours or treated with hellebore or borax. Also used baited traps. ABSTRACT: RAE-B 25:55, 1937.

Fenton, F. A., joint author. *See* Brett, C. H., et al., 1946.

Fernier, L., joint author. *See* Parisot, J., et al., 1934.

361. Ferrière, C. 1920 Insectes et Epidémies. Rev. Internat. de la Croix-Rouge, Geneva. 2(2):149-173.
Flies (one of 4 categories) associated with typhoid, dysentery, cholera, diphtheria, ophthalmia, infantile paralysis, tuberculosis, leprosy. ABSTRACT: RAE-B 8:102, 1920.

362. Ferrière, C. 1933 Note sur les parasites de *Lyperosia exigua* de Meij. Rev. Suisse Zool. 40(34):637-643.
The encyrtid species *Tachinaephagus giraulti* has also been bred from *Musca.* ABSTRACT: RAE-B 23:90, 1935.

363. Fey, K. Y. 1922 [How to Control the House-flies.] vi + 166 pp. Shanghai Commercial Press. (In Chinese.)
Six species that occur in Southeast China are described in detail. ABSTRACT: RAE-B 13:19, 1925.

364. Ficker, M. 1903 Typhus und Fliegen. Arch. f. Hygiene 46:274-283.
In that period typhoid was referred to as "typhus".

365. Filippini, A. 1921 L'acido cianidrico nella disinfestazione. [Cyanic acid in disinfestation.] Ann. d'Igiene. Rome 31:419-427.
M. domestica (like *Lucilia* and *Sarcophaga*) killed in 2 minutes by use of 5 gms. of sodium cyanide per cu. meter of space. ABSTRACT: RAE-B 9:191, 1921.

366. Firth, R. H. and **Horrocks, W. H. 1902** An Inquiry into the Influence of Soil, Fabrics, and Flies in the Dissemination of Enteric Infection. Brit. Med. J. 2:936-943.
Infected mixture of soil and honey placed in cage containing flies, sterile agar plates, and dishes containing sterile broth. *Bacillus typhosus* recovered from both media.

367. Fisher, H. C. 1920 Report of the Health Department of the Panama Canal for the Calendar Year 1919. Mount Hope, C.Z. 134 pp., 20 pls.
Larvae in stables systematically killed with borax every 3 weeks. ABSTRACT: RAE-B 9:108, 1921.

368. Fitch, C. P. 1918 Animal Parasites affecting Equines. J. Am. Vet. Med. Ass. 53(NS6):312-330.
M. domestica among larval forms found occasionally in wounds on horses and mules. ABSTRACT: RAE-B 6:169, 1918.

368a. Fitch, J. B. and **Lush, R. H. 1926** A Study of the Use of Fly Repellents for the Control of Flies on Dairy Cattle. Kansas Agric. Exp. Sta. Bienn. Rept. for 1926:99-101.

368b. Fitch, J. B. and **Lush, R. H. 1928** A Study of the Use of Fly Repellents for Dairy Cattle. Kansas Agric. Exp. Sta. Bienn. Rept. for 1928:95.

Fleck, E. E., joint author. *See* Chisholm, R. D., et al., 1949.

369. Fletcher, J. 1901 Practical Entomology. Canad. Ent. 33:84. A review of "Insect Fauna of Human Excrement" by Howard, 1900.

370. Fletcher, T. B. 1917 Report of the Imperial Pathological Entomologist. Sci. Repts. Agric. Res. Inst., Pusa, for 1916-17. pp. 91-102.
Feeding and breeding habits of several species of *Musca*. ABSTRACT: RAE-B 6:64, 1918.

371. Flexner, S. and **Clark, P. F. 1911** Contamination of the House-fly with Poliomyelitis Virus. J. Am. Med. Ass. 56:1717-1718.
Contaminated flies harbored the virus for at least 48 hours.

372. Flint, W. P. 1922 The Control of Household Insects. Illinois Agric. Exp. Sta. Circ. 257. 24 pp., 15 figs.
Includes *M. domestica*. ABSTRACT: RAE-B 11:65, 1923.

373. Flu, P. C. 1911 Studien über die im Darm der Stubenfliege, *Musca domestica*, vorkommenden protozoären Gebilde. [Studies on the Protozoan Complex occurring in the Intestine of the Housefly, *Musca domestica*.] Centralbl. f. Bakt. 57:522-534.

374. Flu, P. C. 1916 Vliegen en Amoebendysenterie. [Flies and Amoebic Dysentery.] Geneesk. Tijdschr. v. Nederl.-Indië (Batavia) 56: 928-939.
Flies not as important as drinking water, in Java. ABSTRACT: RAE-B 5:66, 1917.

Fluno, H. J., joint author. *See* Jones, H. A., et al., 1946.

Folger, A. H., joint author. *See* Freeborn, S. B., et al., 1925, 1928.

(da) Fonseca, O., joint author. *See* (de) Margarinos-Torres, C. B., et. al., 1923.

Ford, J. H., joint author. *See* Kilgore, L. B., et al., 1942.

375. Foreman, F. W. and **Graham-Smith, G. S. 1917** Investigations on the Prevention of Nuisances Arising from Flies and Putrefaction. J. Hygiene (Cambridge) 16(2):109-226.
M. domestica among the most prevalent flies in Gallipoli. Suggests daily spraying of manure with creosote. ABSTRACT: RAE-B 6:25, 1918.

376. Forman, R. H. 1906 Indian Enteric and Latrines. J. Roy. Army Med. Corps. 7:304-305.

377. Fraenkel, G. 1939 The Function of the Halteres of Flies. Proc. Zool. Soc. London. Ser. A—General and Experimental. 109:69-78.
Worked largely with muscoids other than *M. domestica*. Refutes von Buddenbrock's theory (1919).

378. Fraenkel, G. and **Pringle, J. W. S. 1938** Halteres of Flies as Gyroscopic Organs of Equilibrium. Nature 141:919-920.
Experiments provide 6-point argument for this concept.

379. Franca, C. 1921 An Early Portuguese Contribution to Tropical Medicine. Trans. Roy. Soc. Trop. Med. Hyg. 15(1-2): 57-60.
Gabriel Soares de Souza, in 1587 described feeding of flies on lesions of yaws and subsequent contact with abrasions of healthy persons. Considered this a means of transmission. ABSTRACT: RAE-B 9:155, 1921.

380. Franchini, G. 1923 Inoculation de flagellés d'insectes dans le latex des *Euphorbes*. Bull. Soc. Path. Exot. 16:646-650.
Two species of *Euphorbia* failed to become diseased when inoculated with flagellates from digestive tract of *M. domestica*. ABSTRACT: RAE-B 12:35, 1924.

381. Franchini, G. and **Mantovani, M. 1915** Infection expérimentale du rat et de la souris par *Herpetomonas muscae-domesticae*. Bull. Soc. Path. Exot. 8:109-111.
Light infection can be produced in rats and mice by this and other flagellates from various insects. ABSTRACT: RAE-B 3:111, 1915.

382. Francis, C. F. 1893 Cholera Caused by a Fly? Brit. Med. J. for 1893, part 2, p. 65.

Francis, T., joint author. *See* Rendtorff, R. C., et al., 1943.

383. Franklin, G. D. 1906 Some Observations on the Breeding Ground of the Common House-fly. Indian Med. Gaz. 41:349-350.

384. Freeborn, S. B. 1945 Investigations with DDT in California, 1944. Use Against Household Insects or Insects of Medical Importance. *In* A Preliminary Report, prepared under the Direction of the Division of Entomology and Parasitology, Calif. Agr. Exp. Sta., Berkeley, pp. 3-4.

Reviews methods of application against *M. domestica, Stomoxys calcitrans* and other species. ABSTRACT: RAE-B 35:57, 1947.

385. Freeborn, S. B. and **Berry, L. J. 1935** Color Preferences of the House Fly, *Musca domestica* L. J. Econ. Ent. 28:913-916.

Flies prefer dark colors and rough surfaces. ABSTRACTS: RAE-B 24-61, 1936; TDB 33:637, 1936.

386. Freeborn, S. B., Regan, W. M. and **Folger, A. H. 1925** The Relation of Flies and Fly Sprays to Milk Production. J. Econ. Ent. 18: 779-790.

M. domestica second after *Stomoxys* in causing milk loss. Sprays containing oils also cause loss. ABSTRACT: RAE-B 14:27, 1926.

387. Freeborn, S. B., Regan, W. M. and **Folger, A. H. 1928** The Relation of Flies and Fly Sprays to Milk Production. J. Econ. Ent. 21: 494-501.

Oil sprays, especially, cause rise in temperature and respiratory rate in cows, with resulting milk loss. ABSTRACT: RAE-B 16:227, 1928.

Freeman, A. E., joint author. *See* Levy, E. C., et al., 1908.

388. Frew, J. G. H. 1929 Report on the Tsetse Fly Survey of Sierra Leone, Sept. 1927. 8 vo. 16 pp. Freetown.

Injury to cattle by *Musca* (*Biomyia*) *tempestata* Bezzi feeding at puncture made by proboscis of a blood-sucker. Causes sores. ABSTRACT: RAE-B 17:234, 1929.

389. Froggatt, W. W. 1910 The House-fly and the Disease it Spreads. Agric. Gaz. N.S. Wales, March 1910. pp. 243-250. (Also issued separately as Misc. Publ. No. 1311.)

390. Froggatt, W. W. 1917 Policeman Flies. Fossorial Wasps that Catch Flies. Agric. Gaz. N.S. Wales. 28:667-669.

Especially predaceous on *M. autumnalis* (*corvina*). ABSTRACT: RAE-B 6:68, 1918.

391. Froggatt, W. W. 1921 Sheep-maggot Flies and their Parasites. Agric. Gaz. New S. Wales. 32:725-731; 807-813.

Parasites include the chalcid, *Nasonia brevicornis*, which is primarily a parasite of *M. domestica*. ABSTRACT: RAE-B 10:56, 1922.

392. Frost, W. and **Vorhees, C. T. 1908** The House-fly Nuisance. Country Life in America. May issue.

393. Frye, W. W. and **Meleney, H. E. 1932** Investigations of *Endamoeba histolytica* and other Intestinal Protozoa in Tenessee. IV. A Study of Flies, Rats, Mice and some Domestic Animals as Possible Carriers of the Intestinal Protozoa of Man in a Rural Community. Am. J. Hyg. 16:729-749. Baltimore.

Of 7,948 flies examined, 6 contained characteristic cysts. Flies important when incidence of infection is high, and deposition of human feces promiscuous. ABSTRACT: RAE-B 21:35, 1933.

394. Fullaway, D. T. 1917 Description of a New Species of *Spalangia*. Proc. Hawaiian Ent. Soc. 3:292-294.
S. philippinensis n. sp., introduced from the Philippines in 1914. Has been bred from *M. domestica* and other Muscidae. ABSTRACT: RAE-B 5:137, 1917.

395. Fuller, M. E. 1932. The Larvae of the Australian Sheep Blowflies. Proc. Linn. Soc. New S. Wales 57:77-91.
M. ventrosa Wied. (*hilli* J. & B.) considered a tertiary invader. ABSTRACT: RAE-B 20:163, 1932.

395a. Fulton, P. W. 1889 Dispersion of Spores of Fungi by Agency of Insects, with Special Reference to the Phalloidei. Annals of Botany 3:207-238.
Makes reference to *Musca domestica* as one of the species involved.

Fulton, R. A., joint author. *See* Fales, J. H., et al., 1947, 1948; Nelson, R. H., et al., 1949.

396. Furman, D. P. and **Bankowski, R. A. 1949** Absorption of Benzene Hexachloride in Poultry. J. Econ. Ent. 42:980-982.
Tissues from birds protected against *Dermanyssus gallinae* by BHC yielded extracts variously toxic to *M. domestica.* ABSTRACT: RAE-B 38:154, 1950.

397. Fuschini, C. 1915 Ancora sulla distruzione dell mosche. [Baits for the Destruction of Flies.] Rib. Vitic. Enol. Agrar. (Conegliano) 21:425-426.
Commercial formalin, bread and milk. ABSTRACT: RAE-B 4:14, 1916.

398. Füsthy, O. 1937 [Observations Concerning the Development and Biological Characteristics of the Housefly.] Különlenyomat a Népegeszségügy. 1937. évi 11. számából. (In Hungarian.)
Shortest duration of total development 7¼ days at 37°C.

399. Gabbi, U. 1917 Dissenteria amebica. Malaria e Malattie dei Paesi Caldi. (Rome) 8(5-6):218-240.
M. domestica and *Calliphora erythrocephala* among the agents spreading this disease. ABSTRACT: RAE-B 6:52, 1918.

400. Gahan, J. B., Gilbert, I. H., Peffly, R. L. and **Wilson, H. G. 1948** Comparative Toxicity of Four Chlorinated Organic Compounds to Mosquitoes, House Flies and Cockroaches. J. Econ. Ent. 41:797-802.
BHC approximately 6 times as toxic to *M. domestica* as DDT. ABSTRACTS: RAE-B 38:3, 1950; BA 23:3129, 1949.

Gahan, J. B., joint author. *See* King, W. V., et al., 1949; Wilson, H. G., et al., 1948; Peffly, R. L., et al., 1949.

401. Gaines, J. C., Clare, S. and **Richardson, C. H. 1937** Weight of Adult Housefly and Effect of Sublethal Dose of Sodium Arsenite upon it. J. Econ. Ent. 30:363-366.
No significant change, but paper gives valuable statistical data on normal weight in both sexes. ABSTRACT: RAE-B 25:205, 1937.

402. Galaine, C. and **Houlbert, C. 1916** Pour chaser les mouches de nos habitations. [To Eliminate Flies from our Homes.] Compt. Rond. Acad. Sci. Paris. 163(5):132-135.
Suggests use of blue glass in windows to discourage entrance through doors! ABSTRACT: RAE-B 4:156, 1916.

403. Galli-Valerio, B. 1910 L'état actuel de nos connaissances sur le rôle des mouches dans la dissemination des maladies parasitaires, et sur les moyen de lutte à employer contre elles. [The True State of our Knowledge concerning the Role of Flies in the Spread of Parasitic Diseases and the Methods of Attack to Employ against Them.] Centralbl. f. Bakt. Abt. 1, Vol. 54, pp. 193-209.

404. Ganon, J. 1909 Cholera en vliegen. [Cholera and flies.] Geneesk. Tijdschr. V. Nederl. Indië. 48(2):227-233.
ABSTRACT: J. Trop. Med. and Hyg. 12(10):158, 1909.

Gardner, C. H., joint author. *See* Stiles, C. W., et al., 1910.

405. Gater, B. A. R. 1929 Annual Report of the Division of Entomology for the Year 1928. Ann. Rept. Med. Dept. F.M.S. 1928, pp. 69-74. Kuala Lumpur.
Common housefly of Malaya probably *M. bakeri* Patton. Breeds in material too moist for *M. domestica.*

406. Gauch, A. 1930 Une substance qui protège efficacement les plaies contre la visite des mouches. [A Material which Effectively Protects Sores against the Visits of Flies.] Rev. Vet. 82:638-639. Toulouse.
Formula of a repellent ointment for use with farm animals. ABSTRACT: RAE-B 19:65, 1931.

407. Gayon, J. P. 1903. A Note concerning the Transmission of Pathogenic Fungus by Flies and Mosquitoes. Publ. Health (U.S.A.) 28: 116-117.
Cultivated several species of molds from flies which he caught and dropped into nutrient gelatine.

408. Generali, G. 1886 Una larva di nematode della mosca commune. [Nematode larva in the common fly.] Atti Soc. d. Nat. di Modena, Rendic., Ser. (3), Vol. 2, pp. 88-89.
Called the worm *Nematodum sp.* (Is probably identical with *F. muscae* Carter.)

409. Gerhard, W. P. 1909 Bibliography on Flies and Mosquitoes as Carriers of Disease. Ent. News. 20:84-89; Additional Bibliography ——. Ent. News 20:207-211.

410. Gerhard, W. P. 1911 Flies and Mosquitoes as Carriers of Disease. Reprinted from Country Gentleman, Albany, N.Y. 14 pp.
A popular type discussion with educational value.

411. Gersdorff, W. A. 1946 DDT against House Flies. Soap and Sanit. Chem. 22(3):126-127.
Para-para isomer found 53 times as toxic as *ortho*-isomer. ABSTRACTS: BA 21:1005, 1947; RAE-B 37:63, 1949.

412. Gersdorff, W. A. 1947 Toxicity to House Flies of the Pyrethrins and Cinerins and Derivatives in Relation to Chemical Structure. J. Econ. Ent. 40:878-882.
Both pyrethrin II and cinerin II were about 25 per cent as toxic as corresponding Type I compounds. ABSTRACTS: RAE-B 37:155, 1949; BA 22: 1954, 1948.

413. Gersdorff, W. A. 1949 Toxicity to Houseflies of New Synthetic Pyrethroid. Soap and Sanit. Chem. 25(11):129, 131, 139.
Two isomeric forms of this compound tested by turntable method. ABSTRACT: RAE-B 39:25, 1951.

414. Gersdorff, W. A. 1949 Toxicity to House Flies of Synthetic Compounds of the Pyrethrin Type in Relation to Chemical Structure. J. Econ. Ent. 42:532-536.
Seven substituted cyclopentenolones tested. All had high knock-down effect. ABSTRACT: RAE-B 38:103, 1950.

415. Gersdorff, W. A. and **Barthell, W. F. 1946** Determination of Pyrethrins Deterioration. Soap and Sanit. Chem. 22(10):155, 157.
Much loss of toxicity at 40°C, considerable at room temperature. No loss in dark at 2°C. ABSTRACT: RAE-B 37:106, 1949.

416. Gersdorff, W. A. and **Gertler, S. I. 1944** Pyrethrum Synergists. Toxicity to Houseflies of certain N-substituted Piperonylamides and Benzamides combined with Pyrethrins in Oil Base Insect Sprays. Soap and Sanit. Chem. 20(2):123, 125.
In certain concentrations, had greater toxic effect on houseflies than corresponding preparations of pyrethrins. ABSTRACT: RAE-B 32:187, 1944.

417. Gersdorff, W. A. and **McGovran, E. R. 1944** Laboratory Tests on Houseflies with DDT in Contact Sprays. J. Econ. Ent. 37:137.
Very effective when used in deodorized kerosene, alone, or in conjunction with other insecticidal substances. ABSTRACT: RAE-B 32:204, 1944.

418. Gersdorff, W. A. and **McGovran, E. R. 1945** Insecticide Toxicity Studies. Experimental Results on the Comparative Toxicity of Benzene Hexachloride, DDT and Pyrethrum. Soap and Sanit. Chem. 21(11):117, 121.
For 50 per cent mortality, mean concentration for BHC was 0·0888 mg. per ml.; for DDT 0·788 mg.; for pyrethrins 1·61 mg. ABSTRACT: RAE-B 35:38, 1947.

419. Gersdorff, W. A. and **Nelson, R. H. 1948** Toxicity to House Flies of three Phosphorus Acid Esters. J. Econ. Ent. 41:333-334.
Parathion the most toxic. All three more effective than standard pyrethrum extract. ABSTRACT: RAE-B 37:190, 1949.

420. Gersdorff, W. A. and **Schechter, M. S. 1948** Space Spray Toxicity. Toxicity to House Flies of Space Sprays Containing Certain Diaryl Esters and Pyrethrum Extract. Soap and Sanit. Chem. 24(6):155, 157.
As tested by turn-table method, seven compounds showed strong synergistic properties. ABSTRACT: RAE-B 38:182, 1950.

Gersdorff, W. A., joint author. *See* Barthel, W. F., et al., 1946; McGovran, E. R., et al., 1945; Nelson, R. H., et al., 1949.

421. Gertler, S. I. 1949 Insect Repellent, U.S. Patent 2, 469, 228; application Feb. 18, 1946. Issued May 3, 1949.
Benzyl ether in zinc oxide. ABSTRACT: BA 23:2595, 1949.

Gertler, S. J., joint author. *See* Gersdorff, W. A., et al., 1944; Nelson, R. H., et al., 1949.

Ghosal, S. C., joint author. *See* Lal, R. B., et al., 1939.

Giannotti, O., joint author. *See* Lepage, H. S., et al., 1944.

422. Giard, A. 1879 Deux espèces d'Entomophthora nouveaux pour la flore française et la présence de la forme *Tarichium* sur une Muscide. [Two species of Entomophthora new to the flora of France and the occurrence of the type *Tarichium* on a Muscid.] Bull. Scient. du Department du Nord, Ser. 2, second year, No. 11, pp. 353-363.
Describes formation of conidiospores and their distribution.

Gilbert, J. H., joint author. *See* Gahan, J. B., et al., 1948.

423. Gill, C. A. and **Lal, R. B. 1931** The Epidemiology of Cholera with Special Reference to Transmission. Indian J. Med. Res. 18: 1255-1297.
Vibrios may disappear from fly's alimentary tract the second day after ingestion, only to reappear about the fifth day.

424. Gillogly, L. R. 1949 Interference of Sulphur with Bioassay Dip Test. J. Econ. Ent. 42:983-984.
M. domestica used routinely in these tests. ABSTRACTS: RAE-B 38:164, 372, 1950.

425. Gilmour, D. 1946 The Toxicity to Houseflies of Paints Containing DDT. J. Coun. Sci. Industr. Res. Australia 19(3):225-232.
Five types of paint were used. DDT of various known concentrations was ground up with the paint during its preparation. ABSTRACTS: BA 22:721, 1948; RAE-B 35:132, 1947.

426. Girault, A. A. and **Sanders, G. E. 1909** The Chalcidoid Parasites of the Common House or Typhoid Fly (*Musca domestica*, L.) and its Allies. Psyche 16:119—132, (to be continued).
Nasonia brevicornis Ashmead described as new record from Illinois.

427. Girault, A. A. and **Sanders, G. E. 1910** The Chalcidoid Parasites of the Common House or Typhoid Fly (*Musca domestica* Linn.) and its Allies. Psyche 17:9-28.
An account of the biology of the parasite.

428. Glaser, R. W. 1922 *Herpetomonas muscae-domesticae*, its Behavior and Effect in Laboratory Animals. J. Parasit. 8(3):99-108.
One of three species of flagellates found in digestive tract of *M. domestica*. ABSTRACT: RAE-B 10:162, 1922.

429. Glaser, R. W. 1923 The Effect of Food on Longevity and Reproduction in Flies. J. Exp. Zool. 38:383-412.
Carbohydrates important for longevity; protein for oviposition. ABSTRACT: RAE-B 12:105, 1924.

430. Glaser, R. W. 1924 A Bacterial Disease of Adult House Flies. Am. J. Hyg. 4:411-415.
Staphylococcus muscae sp. n. Male flies more susceptible. ABSTRACT: RAE-B 12:180, 1924.

431. Glaser, R. W. 1924 The Relation of Microorganisms to the Development and Longevity of Flies. Am. J. Trop. Med. 4:85-107.
Studied *M. domestica* and 3 other species. Bacteria supply needful food materials, which must be added if media are sterilized by heat. ABSTRACT: RAE-B 12:64, 1924.

432. Glaser, R. W. 1926 Further Experiments on a Bacterial Disease of Adult Flies with a Revision of the Etiological Agent. Ann. Ent. Soc. Am. 19:193-198.
Virulence of *Staphylococcus muscae* not enhanced by successive passage through flies. Larvae carry organism but do not suffer disease. RAE-B 14:179, 1926.

433. Glaser, R. W. 1926 The Isolation and Cultivation of *Herpetomonas muscae-domesticae*. Am. J. Trop. Med. 6:205-219.
Succeeded in obtaining cultures free from bacteria. ABSTRACT: RAE-B 14:133, 1926.

Glaser, R. W., joint author. *See* Bang, F. B., et al., 1943.

434. Glasgow, R. D. and **Collins, D. L. 1946** The Thermal Aerosol Fog Generator for Large Scale Applications of DDT and other Insecticides. J. Econ. Ent. 39:227-235.
Best particle size, 10 to 50 micra, according to the pest concerned. ABSTRACT: BA 21:485, 1947.

435. Gleichen, F. W. von 1766 Histoire de la mouche commune de nos appartemens. Donnée au Public par Jean C. Keller. Nuremberg.

34 pp., 4 pls., color. (Was first issued in 1764 under the title "Geschichte der Gemeinen Stubenfliege".)
The most comprehensive and detailed of all early accounts. Contains four excellent color plates. (Available edition bears date 1790.)

Glynn, E. F., joint author. *See* Cox, G. L., et al., 1912.

436. Gnadinger, C. B. and **Corl, C. S. 1932** The Relative Toxicity of the Pyrethrins and Rotenone as Fly Spray Ingredients. J. Econ. Ent. 25:1237-1240.
Oil solution of rotenone considerably less toxic to *M. domestica* than pyrethrin solutions of same concentrations. ABSTRACT: RAE-B 21:35, 1933.

Gnedina, M., joint author. *See* Podyapolskaya, V., et al., 1934.

Goddin, A. H., joint author. *See* Hansens, E. J., et al., 1949.

437. Goidanich, A. 1938 Esperimenti di lotta contro le larve della mosca domestica con l'impiego di calciocianamide sul letame. [Experiments Concerning Measures against Larvae of the Domestic Fly by use of Calcium Cyanamide on Manure.] Rass. Faun. 5, pt. 1, Reprint 33 pp., 14 figs., 18 graphs.
This compound not effective against *M. domestica.* ABSTRACT: RAE-B 26:224, 1938.

438. Goidanich, A. 1938 Si può combattere la mosca domestica con la calciocianamide? [Is it Possible to Combat the Domestic Fly with Calcium Cyanamide?] Italia Agric. 75(4):259-265.
Gives an account of his negative experiments. Disagrees with Grandori, 1938. ABSTRACT: RAE-B 27:143, 1939.

439. Göldi, E. A. 1917 Darmkanal und Rüssel der Stubenfliege vom sanitarischen Standpunkte aus. [Alimenary Canal and Proboscis of house flies from the Sanitary Standpoint.] Mitt. Schweiz. Ent. Ges. Berne. 12(9-10):418-431.
Discusses probable relation of flies to several diseases. ABSTRACT: RAE-B 7:28, 1919.

440. Golding, F. D. 1946 A New Method of Trapping Flies. Bull. Ent. Res. 37:143-154.
An adhesive prepared from plant materials. Used in Nigeria. ABSTRACT: RAE-B 34:161, 1946.

441. Gomeza-Ozamiz, J. M. 1945 El descubrimiento del nuevo insecticida 666. [Information concerning the New Insecticide 666.] Ion 5(53): 745-750. Madrid.
Is benzene hexachloride. *Gamma* isomer is most toxic, with *alpha* and *beta* following, in that order. ABSTRACT: RAE-B 36:120, 1948.

442. Gomilevsky, V. 1914 [The Question of Controlling Flies.] Progressive Fruit-growing and Market-gardening. Petrograd. 38:1175-1176.
Two formulae, based on plant materials; one to be used as a bait, one as a repellent. ABSTRACT: RAE-B 3:22, 1915.

443. Goodhue, L. D. 1942 Insecticidal Aerosol Production Spraying Solutions in Liquified Gases. Industr. Engin. Chem. (Industr. Ed.) 34:1456-1459.
Formula of choice produces 90 lbs. pressure per sq. inch at room temperature. Controls house flies. ABSTRACT: RAE-B 31:167, 1943.

444. Goodhue, L. D. 1944 Insecticidal Aerosols. J. Econ. Ent. 37:338-341.
Pyrethrum about twice as toxic to flies in aerosol form as in spray, but knock-down rate is slower. ABSTRACT: RAE-B 33:64, 1945.

444a. Goodhue, L. D. 1946 Aerosols and Their Application. J. Econ. Ent. 39(4):506-509.
Discusses types of liquified gas aerosols; their use in greenhouses, agricultural areas and households.

445. Goodhue, L. D., Fales, J. H. and **McGovran, E. R. 1945** Dispersants for Aerosols. Soap and Sanit. Chem. 21(4):123, 125, 127.
Need of substitute for Freon 12 (pledged to military use). Methyl chloride is toxic to warm bloods, but is excellent for use out-of-doors. ABSTRACT: RAE-B 34:150, 1946.

446. Goodhue, L. D. and **Sullivan, W. N. 1940** Toxicities to the Housefly of Smoke from Derris and Pyrethrum. J. Econ. Ent. 33:329-332.
Based on doses giving 50 per cent kill, derris was ten times as toxic as pyrethrum. ABSTRACT: RAE-B 29:44, 1941.

446a. Goodhue, L. D., Sullivan, W. N. and **Fales, J. H. 1942** The Effects of some Organic Halides on the Housefly. J. Econ. Ent. 35:533-536.
Produced aerosols by spraying insecticides in solution on hot plate. Tested 39 compounds. ABSTRACT: RAE-B 31:93, 1943.

Goodhue, L. D., joint author. *See* Fales, J. H., et al., 1944, 1946; Haller, H. L., et al., 1942; McGovran, E. R., et al., 1943, 1946; Sullivan, W. N., et al., 1939, 1940, 1941.

Gorham, R. P., joint author. *See* Kelsall, A., et al., 1926.

447. Görnitz, K. 1947 Erfahrungen bei der Rapslanzkäfer- und Fliegenbekämpfung mit Gesarol. [Experiences in Cabbage Beetle- and Fly Control with Gesarol.] Festschr. Appel 1947, pp. 41-44. Berlin-Dahlem, Biol. Zent. Anst. Land n. Forstw. 1947.
In one room, at least, 5 per cent DDT was still giving protection at the end of a year. ABSTRACT: RAE-B 37:216, 1949.

448. Gorodetzskii, A. S. and **Sukhova, M. N. 1936** [Nouveaux poisons pour la destruction des stades préimaginales de mouches.] Med. Parasit. 5(3):303-323. (In Russian, with French summary.)
Residues from aniline industry made breeding media poisonous to *M. domestica* larvae for 1 month. ABSTRACT: RAE-B 25:109, 1937.

449. Gorodetskii, A. S. and **Sukhova, M. N. 1936** [Les dépôts de fumier et les caisses à ordures comme pièges pour les larves des mouches.] Med. Parasit. 5(3):324-328. (In Russian, with French summary.)
Manure stored in covered boxes, with bottoms of wire netting. Larvae migrate for pupation and fall on ground poisoned with industrial waste. ABSTRACT: RAE-B 25:110, 1937.

450. Gosio, B. 1925 Ueber die Verbreitung der Bubonenpesterreger durch Insektenlarven. [On the Spread of Bubonic Plague by Insect Larvae]. Arch. Schiffs. u Trop. Hyg. 29. Beiheft 1:134-139.
M. domestica larvae (and others) complete development with numerous plague bacilli in gut and feces. Bacilli fewer in pupae, but multiply again in the fly. ABSTRACT: RAE-B 13:182, 1925.

451. Gothard, N. J. 1944 Another View of Aerosol Efficiency. Soap and Sanit. Chem. 20(7):113, 115.
Sprays give higher knock-down and kill, when amount of pyrethrins is the same. ABSTRACT: RAE-B 33:98, 1945.

452. Gourdon, G. 1929 La capture et la destruction des insectes par rayons ultraviolets. [The Capture and Destruction of Insects by Ultraviolet Rays.] Recherches Inventions 10:245-250.
Suggests that the ozone produced by thunder storms causes coma in flies.

Graham, J. J. T., joint author. *See* Barthel, W. F., et al., 1946.

453. Graham-Smith, G. S. 1909 Preliminary Note on Examinations of Flies for the Presence of Colon Bacilli. Repts. Local Gov't. Bd.

on Health and Med. Subjects. N.S. No. 16; Further Prelim. Repts. on Flies as Carriers of Infection, pp. 9-13.
Thirty-four per cent of flies captured in London and Cambridge carried (externally, internally, or both) bacteria of the colon group.

454. Graham-Smith, G. S. 1910 Observations on the Ways in which Artificially Infected Flies (*Musca domestica*) Carry and Distribute Pathogenic and Other Bacteria. Repts. Local Gov't. Bd. on Publ. Health and Med. Subjects. N.S. No. 40; Further Repts. (No. 3) on Flies as Carriers of Infection, pp. 1-41.
Capacity of the crop of *M. domestica* varies between 0·003 and 0·002 c.c.

455. Graham-Smith, G. S. 1911 Further Observations on the Ways in which Artificially Infected Flies (*Musca domestica* and *Calliphora erythrocephala*) Carry and Distribute Pathogenic and other Bacteria. Repts. Local Gov't. Bd. on Publ. Health and Med. Subjects. N.S. No. 53; Further Repts. (No. 4):31-48.
Gives an excellent account of the general behavior of flies during feeding.

456. Graham-Smith, G. S. 1912 An Investigation of the Incidence of the Micro-organisms Known as Non-lactose Fermenters in Flies in Normal Surroundings and in Surroundings Associated with Epidemic Diarrhoea. *In* Ann. Rept. Local Gov't. Bd.; Suppl. Rept. Med. Officer 1911-1912, pp. 304-329.
Certain that flies infected with Morgan's bacillus can contaminate materials on which they feed or over which they walk.

457. Graham-Smith, G. S. 1912 An Investigation into the Possibility of Pathogenic Micro-organisms being taken up by the Larva and Subsequently Distributed by the Fly. *In* 41st Ann. Rept. Local Gov't. Bd., Suppl. Rept. Med. Officer, 1911-1912, pp. 330-335.
Spores of *B. anthracis* persist in a large proportion of adult *M. domestica.* A few non-spore formers survive at times.

458. Graham-Smith, G. S. 1912 House-flies. Bedrock. No. 2:205-223.

459. Graham-Smith, G. S. 1913 Further Observations on Non-lactose-fermenting Bacilli in Flies, and the Sources from which they are Derived with Special Reference to Morgan's bacillus. Repts. Local Gov't. Bd. on Publ. Health and Med. Subjects. N.S. No. 85; Further Rept. (No. 6) on Flies as Carriers of Infection, pp. 43-46.
Refers to *Proteus morgani.*

460. Graham-Smith, G. S. 1914 Flies in Relation to Disease. Non-blood-sucking Flies. Cambridge University Press. xvi + 389 pp., illus.
NOTICE: RAE-B 2:19, 1914. An excellent volume, full of detailed information.

461. Graham-Smith, G. S. 1916 Observations on the Habits and Parasites of Common Flies. Parasitology 8:440-544.
Much information on behavior in relation to environment. ABSTRACT: RAE-B 4:143-147, 1916.

462. Graham-Smith, G. S. 1918 Hibernation of Flies in a Lincolnshire House. Parasitology 11:81-82.
M. autumnalis (*corvina*) most numerous and troublesome. ABSTRACT: RAE-B 7:71, 1919.

463. Graham-Smith, G. S. 1919 Further Observations on the Habits and Parasites of Common Flies. Parasitology 11:347-384.
Empusa mentioned as attacking *M. corvina* and several other species. ABSTRACT: RAE-B 8:22, 1920.

464. Graham-Smith, G. S. 1929 The Relation of the Decline in the Number of Horse-drawn Vehicles and Consequently of the Urban Breeding Grounds of Flies, to the Fall in the Summer Diarrhoea Death Rate. J. Hygiene 29(2):132-138.

Graham-Smith, G. S., joint author. *See* Forman, F. W., et al., 1917.

465. Grandori, R. 1938 L'azione disinfestante della calciocianamide contro la mosca domestica sperimentalmente dimonstrata. [Disinfecting Action of Calcium Cyanamid against the Domestic Fly Demonstrated Experimentally.] Boll. Zool. Agr. Bachic. 8:233-250. Turin.

Poison liberated when cyanamid comes in contact with water. Advocates use with manure, also in poison baits. ABSTRACT: RAE-B 27:84, 1939.

466. Granett, P., Haynes, H. L., Connola, D. P., Bowery, T. G. and **Barber, G. W. 1949** Two Butoxypolypropylene Glycol Compounds as Fly Repellents for Livestock. J. Econ. Ent. 42:281-286.

Tested against *M. domestica* in laboratory; *Stomoxys*, *Siphona* and *Tabanus* in field. Very low toxicity to mammals.

467. Grassi, B. 1883 Les méfaits des mouches. [The "Misdeeds" of the Fly.] Arch. Ital. de Biol. 4:205-228.

First to demonstrate that flies have the ability to ingest worm eggs, specifically *Taenia solium*.

468. Grassi, B. and **Rovelli, G. 1889** Embryologische Forschungen an Cestoden. [Embryological Investigations on Cestodes.] Centralblt. Bakt. Parasit. 5:370-377, 401-410.

Background work on Cestode life histories. *See* reference which follows.

469. Grassi, B. and **Rovelli, G. 1892** Ricerche embriologiche sui Cestodi. Atti. Accad. Gionia di Sci. Nat. (Catania). 4:1-108.

Larvae closely resembling those of *Choanotaenia infundibuliformis* (chicken tapeworm) found in body of *M. domestica.*

Green, A. A., joint author. *See* Higgins, A. E. H., et al., 1949; Parkin, E. A., et al., 1944, 1945, 1947.

Greenwood, W., joint author. *See* Veitch, R., et al., 1924.

470. Griffith, A. 1908 The Life-history of House-flies. Public Health 21: 122-127.

Succeeded in keeping a male fly alive for 16 weeks.

471. Griffith, F. 1907 Description of a House-fly Parasite. Med. Brief (St. Louis). 35:59-63.

472. Griffiths, J. T. Jr. 1946 DDT used to control Flies in Manila. J. Econ. Ent. 39:750-755.

Applied $\frac{1}{2}$ lb. per acre in diesel oil by airplane. ABSTRACTS: RAE-B 36:87, 1948; BA 21:1809, 1947.

Grindley, J., joint author. *See* Tomlinson, T. G., et al., 1949.

473. Grove, J. F. and **Bovingdon, H. H. S. 1947** Thiocyanate Insecticides. The Relation between Knock-down Activity and Chemical Constitution. Ann. Appl. Biol. 34:113-126.

The most effective ones proved too irritant for use as domestic fly sprays. ABSTRACT: RAE-B 37:49, 1949.

474. Gudger, E. W. 1910 An Early Note on Flies as Transmitters of Disease. Science. N.S. 31:31-32.

Refers to Edward Bancroft's "Essay on the Natural History of Guiana in South America". London 1769. Concerns yaws and flies which visit sores.

475. Gudger, E. W. 1910 A Second Early Note on the Transmission of Yaws by Flies. Science 32:632-633.
Refers to Koster's "Travels in Brazil in the Years from 1809 to 1815". Published in Philadelphia, 1817. Concerns flies which may or may not have been *M. domestica*. Yaws referred to as *bohas*.

476. Guilhon, J. 1946 Proprietés Insecticides des Isomères d l'Hexachlorocyclohexane. Compt. Rend. Agric. Fr. 32(18):754-760. Paris.
As tested against *M. domestica* (and other species) gamma isomer of BHC is the most toxic of the four isomers. ABSTRACT: RAE-B 37:33, 1949.

477. Günther, G. 1931 Die Wirkung verschiedener Antiparasitica auf Fliegen-larven. [The Action of Various Parasiticides on Fly Larvae.] Wien. Tierärztl. Mschr. 17:813-816. Vienna.
Creolin found of value, but it is neither odorless, nor cheap. ABSTRACT: RAE-B 19:203, 1931.

478. Güssow, H. T. 1913 *Empusa muscae* and the Extermination of the House-fly. Repts. Local Gov't. Bd. on Publ. Health and Med. Subjects, N.S. No. 85; Further Rept. (No. 6) on Flies as Carriers of Infection, pp. 10-14, 1 pl.

479. Güssow, H. T. 1917 *Empusa muscae* versus *Musca domestica* L. Ann. Appl. Biol. (London) 3(4):150-158.
Dying off of flies in autumn cannot be ascribed to this. Use in natural control doubtful. ABSTRACT: RAE-B 5:116, 1917.

480. Gutberlet, J. E. 1919 On the Life History of the Chicken Cestode *Hymenolepis carioca* (Magalhaes). J. Parasit. 6:35-38.
Is carried by *Stomoxys calcitrans*, much as *Choanotaenia* is transmitted by *M. domestica*.

Guy, E. L., joint author. *See* Baker, W. C., et al., 1947.

481. Guyénot, E. 1907 L'apparel digestif et la digestion de quelques larves des mouches. [The digestive apparatus and digestive process in the larvae of certain flies.] Bull. Sci. de la Fr. et de la Belguique. 41:353-359.
One of the early contributions to insect physiology.

Haas, B., joint author. *See* Rathay, E., et al., 1883.

482. Hadjinicolaou, J. 1948 Toxicity of DDT and "Gammexane" Sprays to House Flies (*Musca domestica* L.). Archives of Hygiene 6(7-12), 7 pp.
Order of effectiveness for DDT is: (1) wettable powder, (2) emulsion, (3) solution in oils. For gammexane, order is reversed (but differences are less).

483. Haecker, V. 1916 Zur Fliegenplage in Wohnungen und Lazaretten. [On the Fly Problem in Dwellings and Hospitals., Zeit. Angew. Ent. (Berlin). 3:204-209.
Recommends shutting all windows just before the sun reaches them. Keep so until again in shade. ABSTRACT: RAE-B 8:120, 1920.

484. Hafez, H. C. 1939 Some Ecological Observations on the Insect-fauna of Dung. Bull. Soc. Fouad 1er Ent. 23:241-287.
Refers particularly to *M. domestica vicina* Macq. An important paper. ABSTRACT: RAE-B 28:210, 1940.

485. Hafez, M. 1941 Investigations into the Problem of Fly Control in Egypt. Bull. Soc. Fouad 1er Ent. 25:99-144. Cairo.
An important assemblage of facts pertaining to sanitary problems of that country. ABSTRACTS: TDB 40:91, 1943; RAE-B 30:136, 1942.

486. Hafez, M. 1941 A Study of the Biology of the Egyptian Common House-fly: *Musca vicina* Macq. (Diptera: Muscidae). Bull. Soc. Fouad 1er Ent. 25:163-189. Cairo.
This is *M. domestica vicina* of many authors. ABSTRACTS: TDB 40:93, 1943; RAE-B 30:138, 1942.

487. Hafez, M. 1947 A Preliminary Note on the Influence of Low Temperature on the Numbers of *Musca domestica vicina* Macq. during the Winter. Ann. Mag. Nat. Hist. Ser. 11, Vol. 14, 810-812. (Published 1948.)
Low winter population attributed to (1) retardation of development, (2) females fail to oviposit below 15°C, (3) Larvae and pupae die from excess moisture.

487a. Hafez, M. 1947 Further Additions to the Dipterous Fauna of Dung in Egypt, with some Biological Observations. Bull. Soc. Fouad 1er Ent. 31:307-312.
List includes *Fannia canicularis* L. and *Musca* (*Eumusca*) *xanthomelas* Wied.

488. Hall, D. G. and **Bohart, G. E. 1948** The Sarcophagidae of Guam (Diptera). Proc. Ent. Soc. Wash. 50:127-135.
Includes reference to control of *M. sorbens* by DDT from the air. ABSTRACT: RAE-B 39:39, 1951.

489. Haller, H. L., LaForge, F. B. and **Sullivan, W. N. 1942** Some Compounds Related to Sesamin: Their Structure and their Synergistic Effect with Pyrethrum Insecticides. J. Org. Chem. 7(2):185-188. (Cited by Eagleson, 1942.)

490. Haller, H. L., LaForge, F. B. and **Sullivan, W. N. 1942** Effect of Sesamin and Related Compounds on the Insecticidal Action of Pyrethrum on House Flies. J. Econ. Ent. 35:247-248.
Sesamin, added to pyrethrum in refined kerosene (10 per cent acetone) raised mortality in 24 hours from 20 (on scale) to 85. ABSTRACT: RAE-B 31:8, 1943.

491. Haller, H. L., McGovran, E. R., Goodhue, L. D. and **Sullivan, W. R. 1942** The Synergistic Action of Sesamin with Pyrethrum Insecticides. J. Org. Chem. 7:183-184. (Cited by Eagleson, 1942.)

492. Haller, H. L. and **Sullivan, W. N. 1938** Toxicity of Hydrogenated Pyrethrins I and II to the Housefly (*Musca domestica* L.). J. Econ. Ent. 31:276-277.
Concentrates in which Pyrethrin I predominates more toxic than those in which Pyrethrin II predominates. ABSTRACT: RAE-B 26:208, 567, 1938.

Haller, joint author. *See* Acree, F., et al., 1945; Cristol, S. J., et al., 1945, 1946; Sullivan, W. N., et al., 1938, 1939, 1943; Jacobson, M., et al., 1947; Schechter, M. S., et al., 1946, 1947.

493. Hamer, W. H. 1908 Nuisance from Flies. Report by the Medical Officer ——, No. 1, 138 pp. 1-10, 2 figs., 3 diagrams.
Printed for London County Council (Public Health Committee). Deals with extent to which fly nuisance in London is produced by accumulation of offensive matter. Nine-tenths of the flies caught were *M. domestica*.

494. Hamer, W. H. 1908 Nuisance from Flies. Rept. Med. Officer of Health (Further report by Dr. Hamer). No. 1, 207, pp. 1-6, 4 diagrams.
Seventeen per cent of flies caught were *F. canicularis*; in another situation 24 per cent.

495. Hamer, W. H. 1908 The Breeding of Flies Summarized. Am. Med. No. 3, p. 431.

496. Hamer, W. H. 1910 Flies and Vermin. Rept. Med. Officer of Health (Reports by Dr. Hamer). Rept. Publ. Health Comm., London County Council for 1909. Appendix No. 4, pp. 1-9, 5 charts.
Gives diagram illustrating the seasonal prevalence of the six principal genera of flies caught in houses.

497. Hamilton, A. 1903 The Fly as a Carrier of Typhoid; an Inquiry into the Part Played by the Common House Fly in the Recent Epidemic of Typhoid Fever in Chicago. J. Am. Med. Ass. 40: 576-583.
Only 40 per cent of houses had sanitary plumbing. Discharges from patients often left exposed to flies in privies or yards.

498. Hamilton, A. 1904 The Common House-fly as a Carrier of Typhoid Fever. J. Am. Med. Ass. 42:1034.

499. Hamilton, A. 1906 The Role of the House-fly and other Insects in the Spread of Infectious Diseases. Illinois Med. J., Springfield. 9:583-587.

500. Hamilton, A. 1907 Isolation of Tubercle Bacillus from Flies Caught in a Privy. J. Am. Med. Ass., Aug. 28th, 1907. (See also Brit. Med. J. for 1903, p. 149.)

501. Hammer, O. 1941-1942 Biological and Ecological Investigations on Flies Associated with Pasturing Cattle and their Excrement. Vidensk. Medd. Dansk. Naturh. Forens. 105: 141-393 (1942). Reprint 257 pp., 50 figs. Copenhagen. B. Lunos. (15th November, 1941). (Danish summary.)
M. autumnalis Deg. and *M. tempestiva* Fall. regarded as facultative blood suckers. ABSTRACT: RAE-B 31:175, 1943.

Hammer, O., joint author. *See* Thomsen, M., et al., 1936.

502. Hanawalt, E. M. 1945 The Trend in Fly Control. Mich. State Coll. Vet. 5(4):163-167.
Various formulations of DDT gave residual control in barns, autopsy room, etc.

503. Hansens, E. J. and **Goddin, A. H. 1949** Reaction of Certain Fly Strains to DDT and Methoxychlor Deposits. J. Econ. Ent. 42: 843-844.
Gives percentage of kill for each of several formulations, with four recognized laboratory strains. ABSTRACT: RAE-B 38:134, 1950.

504. Hansens, E. J., Schmitt, J. B. and **Barber, G. W. 1948** Resistance of House Flies to Residual Applications of DDT in New Jersey. J. Econ. Ent. 41:802-803.
Flies from treated barns, quite resistant to DDT, were killed by deposits of chlordan, toxaphene, or parathion. ABSTRACT: RAE-B 38:4, 1950.

505. Hanson, H. G. 1946 DDT in the General Health Program. Am. J. Publ. Health 36:653-656.
Particularly effective against the common house fly. Used as solutions, suspensions, emulsions and dusts. ABSTRACT: BA 20:2257, 1946.

506. Hardy, G. H. 1934 Twelfth Annual Report of the Walter and Eliza Hall Fellow in Economic Biology. Fol., 17 pp. multigraph. Qd. Univ. Brisbane.
Deals chiefly with blow flies, but larvae in matted wool on newly blown sheep were found to be *M. domestica*. (Flies attracted by feces.) ABSTRACT: RAE-B 22:154, 1934.

507. Hardy, G. H. 1935 The Positions assumed by Copulating Diptera. Ann. Mag. Nat. Hist. (10) 16:419-434. (See also Hardy, 1944.)

508. Hardy, G. H. 1936 Notes on Australian Muscoidea II. Subfamily Muscinae. Proc. Roy. Soc. Queensland. 48:22-29.

509. Hardy, G. H. 1944 The Copulation and Terminal Segments of Diptera. Proc. Roy. Ent. Soc. London. Ser. A. 19(4-6):52-65.
Recognizes (1) superimposed, (2) opposed and (3) diverted positions. Evolution of terminal segments. ABSTRACT: BA:19:427, 1945.

510. Hardy, G. H. (Undated) The Book of the Fly. A Nature Study of the House-fly and its Kin, the Fly Plague and a Cure. New York, Rebman Company. vi + 124 pp., illus.

Hardy, G. H., joint author. *See* Johnston, T. H., et al., 1923.

511. Hargreaves, E. 1923 Entomological Notes from Taranto (Italy) with References to Faenza during 1917 and 1918. Bull. Ent. Res. 14:213-219.
M. domestica found breeding in enormous numbers in the crust of septic tanks. ABSTRACT: RAE-B 11:219, 1923.

Harns, H. G., joint author. *See* Cory, E. N., et al., 1936.

512. Harris, A. H. and **Down, H. A. 1946** Studies of the Dissemination of Cysts and Ova of Human Intestinal Parasites by Flies in Various Localities on Guam. Am. J. Trop. Med. 26:789-900.
Cysts or ova of ten human parasites identified microscopically from fly specks.

Harris, H. H., joint author. *See* Allen, T. C., et al., 1944.

513. Harris, W. H. 1903 The Dentition of the Diptera. J. Quekett Micro. Club. (2)3:389-398, pl. 19.

513a. Harrison, R. A. 1949 Toxicity to Houseflies of a Flat Oil Paint Containing DDT. New Zealand J. Sci. Tech., Sec. B, 31(2):24-30. (Cited by abstracting journals as published in 1950.)
Paint films scarcely toxic until crystals form, but insects walking on surface will cause this. Paint containing 5 per cent DDT remained toxic for 385 days. Films with 2 per cent DDT also effective (but not films with 1 per cent). ABSTRACTS: BA 26:1794, Jul., 1952; RAE-B 40:14, Jan., 1952.

514. Harsham, A. 1946 Debunking a Color Theory. Food Indust. 18(6): 851, 984.
Foil-wrapped packages attract more flies than color-wrapped ones. ABSTRACT: BA 20:2084, 1946.

515. Hartzell, A. 1945 Histological Effects of Certain Sprays and Activators on the Nerves and Muscles of the Housefly. Contr. Boyce Thompson Inst. 13(9):443-454.
Poison affects one tissue component; activator another. When used together, lower concentration suffices than when either is used alone. ABSTRACT: RAE-B 34:61, 1946.

516. Hartzell, A. 1949 Effectiveness of N-(2-ethylhexyl) bicyclo (2.2.1)-5-heptene-2, 3-dicarboximide on Houseflies. Contr. Boyce Thompson Inst. 15(7):337-339.
Small amounts very effective in combination with standard insecticides. Compound also known as Van Dyk 264 = Octacide 264. ABSTRACTS: TDB 49:1005, 1952; RAE-B 40:53, 1952; BA 23:2868, 1949.

517. Hartzell, A. and **Scudder, H. L. 1942** Histological Effects of Pyrethrum and an Activator on the Central Nervous System of the Housefly. J. Econ. Ent. 35:428-433.
Pyrethrum causes widespread clumping of chromatin in cell nuclei. Activator causes chromatolysis, particularly in fat cells.

518. Hartzell, A. and **Strong, M. 1944** Histological Effects of Piperine on the Central Nervous System of the Housefly. Contr. Boyce Thompson Inst. 13(5):253-257.
Alkaloid in dried fruit of *Piper nigrum* causes destruction of fiber tracts and vacuolisation of nerve tissue in the brain. Is more toxic to *M. domestica* than pyrethrum. ABSTRACT: RAE-B 33:25, 1945.

519. Hartzell, A. and **Wexler, E. 1946** Histological Effects of Sesamin on the Brain and Muscles of the Housefly. Contr. Boyce Thompson Inst. 14:123-126.
Larger nerve cells showed marked vacuolation. Muscle fibers showed accentuation of both nodes and Krause's membrane. ABSTRACT: RAE-B 37:159, 1949.

520. Hartzell, A. and **Wilcoxon, F. 1936** Relative Toxicity of Pyrethrins I and II to Insects. Contr. Boyce Thompson Inst. 8(3):183-188.
In kerosene solution, no perceptible difference. In aqueous sprays (emulsions of varying stability) physical condition of toxic factor determines result. This may explain the divergent results of various authors. ABSTRACT: RAE-B 25:13, 1937.

Hartzell, A., joint author. *See* Ewing, H. E., et al., 1918; Harvill, E. K., et al., 1943; Wilcoxon, F., et al., 1939; Prill, E. A., et al., 1946, 1947; Synerholm, M. E., et al., 1945.

521. Harvey, W. C. and **Hill, H. 1947** Insect Pests. 2nd Ed. xi + 347 pp. 2 pls., 4 figs. London, H. K. Lewis & Co., Ltd.
M. domestica given considerable space. REVIEW: RAE-B 35:155, 1947.

522. Harvill, E. K. and **Arthur, J. M. 1943** Toxicity of Organic Compounds to Houseflies. Contr. Boyce Thompson Inst. 13(2):79-86.
Allyl phenols possess insecticidal properties; more so, as the number of nuclear allyl groups is increased.

523. Harvill, E. K., Hartzell, A. and **Arthur, J. M. 1943** Toxicity of Piperine Solutions to Houseflies. Contr. Boyce Thompson Inst. 13(2):87-92.
Is more toxic than pyrethrum to houseflies. At concentration of 0·1 per cent killed 75 per cent of flies exposed. (Peet-Grady Method.) Is a good synergist. ABSTRACT: RAE-B 32:73, 1944.

524. Hase, A. 1916 Ein Beitrag zur Fliegenplage. [A Contribution pertaining to the Fly Nuisance.] Zeit. Angew. Ent. (Berlin). 3:117-123.
Village conditions in Poland conducive to heavy breeding of *M. domestica* and other flies. (Stables part of dwelling house.) ABSTRACT: RAE-B 8:119, 1920.

525. Hase, A. 1935 Ueber Wärmeentwicklung in Massenzuchten von Insekten sowie über ein einfaches Verfahren, Stubenfliegen daurend zu züchten. [On Temperature Production in Mass Culture of Insects as in Ordinary Procedures for Continuous Breeding of Houseflies.] Zool. Anz. 112(11-12):291-298.
Maximum temperature achieved, 42°C (107·6°F). Insects were very flourishing. ABSTRACT: RAE-B 24:215, 1936.

526. Haseman, L. 1917 The House-Fly and its Control. Missouri Agric. Exten. Serv. Circ. No. 16, 11 pp., 4 figs.
Nine to ten generations per year in Missouri. ABSTRACT: RAE-B 5: 137, 1917.

527. Hatch, E. 1911 The House-fly as a Carrier of Disease. Ann. Am. Acad. Pol. Soc. Sci., March 1911, pp. 412-423.

528. Hatch, E. 1913 Report of the Chairman of the Fly-fighting Committee of the American Civic Assn., Ann. Convention, Baltimore,

Md., 20th Nov., 1912. Separately issued, 16 pp.

Hathaway, C. R., joint author. *See* DeSalles, et al., 1944.

Haushalter, M., joint author. *See* Spielman, M., et al., 1887.

529. Haydon, L. G. 1922 Memorandum on the Disposal of Animal Manure and Garbage in Relation to Fly-breeding and the Prevention of Enteric Fever and other Intestinal Diseases. S. Afric. Med. Rec. (Cape Town). 20(12):230-232.
Used enclosure of wire mesh on brick or cement platforms. Larvae, driven out by heat of fermentation, are trapped in surrounding gutter. ABSTRACT: RAE-B 10:178, 1922.

530. Hayes, W. P. and **Liu, Y. S. 1947** Tarsal Chemoreceptors of the Housefly and their Possible Relation to DDT Toxicity. Ann. Ent. Soc. Am. 40:401-416.
Chemoreceptive sensilla occur only on tarsal segments 2 to 5. Are always lateral in position. ABSTRACT: RAE-B 38:11, 1950.

Haynes, H. L., joint author. *See* Granett, P., et al., 1949.

531. Hayward, E. H. 1904 The Fly as a Carrier of Tuberculosis Infection. New York Med. J. 80:643-644.
Bacilli remain virulent after passing through the digestive tract of the fly.

532. Headlee, T. J. 1914 Fly Control. Rept. New Jersey Agric. Exp. Sta. for 1913. New Brunswick. pp. 698-718.
Found iron sulphate and carbon bisulphide effective larvicides. ABSTRACT: RAE-B 3:51, 1915.

533. Heilbrunn, L. V. 1945 An Outline of General Physiology. Philadelphia and London. W. B. Saunders Co. xii + 748 pp. illus.
In table on page 628 gives maximum length of life for *M. domestica* as 76 days.

534. Hepburn, G. A. 1943 Sheep Blowfly Research V. Carcasses as Sources of Blowflies. Onderstepoort J. Vet. Sci. 18(1-2):59-72. Pretoria.
Lucilia and *Chrysomyia* predominate at first. *Musca* spp. come when carcass has been broken down, exposing content of digestive organs. ABSTRACT: RAE-B 33:8, 1945.

535. Hepworth, J. 1854 On the Structure of the Foot of the Fly. Quart. J. Micr. Sci. 2:158-160.

Herman, F. A., joint author. *See* Twinn, C. R., et al., 1927.

536. Herms, W. B. 1909 Essentials of House-fly Control. Bull. Berkeley (Cal.) Bd. of Health, 29th June. 9 pp.

537. Herms, W. B. 1909 The Berkeley House-fly Campaign. California J. Technol. 14(2), 11 pp., 3 figs.

538. Herms, W. B. 1909 The House-fly Problem. Pacific Slope Ass. Econ. Entomologists. Bull. Card No. 1.

539. Herms, W. B. 1910 Fight the Fly, —Why, —When, —Where, —How? Bull. Berkeley (Cal.) Bd. of Health.

540. Herms, W. B. 1910 How to Control the Common House-fly. Monthly Bull. Calif. State Bd. of Health, Vol. 5, May, pp. 269-277, 5 figs.
The five preceding references are more or less popular and educational in nature. They report very little original research.

541. Herms, W. B. 1911 The Housefly in its Relation to the Public Health. Univ. Calif. Agric. Exp. Sta. Bull. 215, pp. 513-548, 16 figs.
Is extensive and complete for its day. Quotes sanitary ordinances.

542. Herms, W. B. 1916 Flies: their Habits and Control. California State Bd. of Health. Sacramento. Spec. Bull. No. 20, 19 pp., 6 figs.
Discusses stable construction, use of manure bins, etc., ABSTRACT: RAE-B 5:162, 1917.

543. Herold, W. 1922 Beobachtungen an zwei Feinden der Stubenfliege: *Mellinus arvensis*, L. und *Vespa germanica*, Fabr. [Observations on Two Enemies of Houseflies ——.] Zeit. Angew. Ent. (Berlin). 8(2):459.
These two wasps recorded as preying on *M. domestica*. ABSTRACT: RAE:B 10:195, 1922.

544. Hervieux. 1904 [Report on Carriage of Smallpox by Flies.] Read to Academy of Medicine, Paris, June 5, 1904. Lancet pt. 1, p. 1761, June 16, 1904.
Distribution of disease attributed to direction of prevailing wind which was observed to distribute flies and mosquitoes.

545. Hesse, E. 1912 The Parasitic Fungus of the House-fly. Reprinted from Shrewsbury Chronicle, 29 Nov., 1912.

546. Hesse, E. 1913 A Parasitic Mould of the House Fly. Brit. Med. J. Jan. 4, 1913, p. 41.
Cultured *Empusa muscae* and killed *M. domestica* (also *Stomoxys* and *Fannia*) by feeding them on culture. Believed that spores germinate in crop. ABSTRACT: RAE-B 1:11, 1913.

547. *Hewitt, C. G. 1906 A Preliminary Account of the Life-history of the Common House-fly (*Musca domestica* L.). Mem. and Proc. Manchester Lit. Phil. Soc. 51, pt. 1, 4 pp.

548. Hewitt, C. G. 1907 The Proboscis of the Housefly. Brit. Med. J. Nov. 23, 1907, p. 1558.

549. Hewitt, C. G. 1907 On the Bionomics of Certain Calyptrate Muscidae and their Economic Significance with Special Reference to Flies Inhabiting Houses. J. Econ. Biol. 2:79-88.

550. Hewitt, C. G. 1907 House-flies. Ann. Rept. and Trans. Manchester Micro. Soc. for 1907, pp. 82-91, 1 pl.

551. Hewitt, C. G. 1907-08 The Structure, Development and Bionomics of the House-fly, *Musca domestica*, Linn.: Part I. The Anatomy of the Fly. Quart. J. Micr. Sci. 51:395-448, pls. 22-26. Part II. The Breeding Habits, Development and the Anatomy of the Larva. Ibid. 52:495-545.

552. Hewitt, C. G. 1908 The Biology of House-flies in Relation to Public Health. J. Roy. Inst. Publ. Health. 16:596-608, 3 figs.

553. Hewitt, C. G. 1909 The Structure, Development and Bionomics of the House-fly, *Musca domestica*, Linn.: Part III. The Bionomics, Allies, Parasites and the Relations of *M. domestica* to Disease. Quart. J. Micr. Sci. 54:347-414, 1 pl., 1 fig.

554. Hewitt, C. G. 1910 The House-fly, *Musca domestica*. A Study of its Structure, Development, Bionomics and Economy. 195 pp., 10 pls., 1 fig. (Manchester, The University Press.)

555. Hewitt, C. G. 1910 House-flies and Disease. Nature 84:73-75, 3 figs.

556. Hewitt, C. G. 1910 House-flies and the Public Health. Ottawa Naturalist 24:31-35.

557. Hewitt, C. G. 1910 House-flies and their Allies. Fortieth Ann. Rept., Ent. Soc. Ontario, pp. 30-36, 4 figs.

* Most of the references by C. G. Hewitt are not accompanied by annotations. This is because practically all of the scientific information contained therein is recapitulated in his classical volume, published in 1914. It is felt important to list most of Hewitt's papers, nevertheless, as a tribute to the historical impact of his work, and to the high quality of this author's scholarly achievements during his regrettably brief career.

558. Hewitt, C. G. 1911 The House-fly in Relation to Public Health. Publ. Health J. (Canada) 2:259-261.

559. Hewitt, C. G. 1911 The House-fly. Canad. Ent. 43:294-295.
A review of the 1910 volume of that title by L. O. Howard. Criticizes the absence of an index.

560. Hewitt, C. G. 1912 House-flies and How they Spread Disease. Cambridge University Press. Cambridge, Eng. 122 pp., 19 figs. (*In* Cambridge Manuals of Science and Literature.) Reprinted essentially without change in 1914.
Is a popular type presentation, similar to "The House-Fly, Disease Carrier" by L. O. Howard, 1910, but not as large.

561. Hewitt, C. G. 1912 *Fannia* (*Homalomyia*) *canicularis* Linn. and *F. scalaris* Fab. Parasitology 5:161-174, illus.

562. Hewitt, C. G. 1912 Observations on the Range of Flight of Flies. Repts. Local Gov't. Bd. on Publ. Health and Med. Subjects. N.S. No. 60. Further Repts. (No. 5) on Flies as Carriers of Infection, pp. 1-5, with map.

563. Hewitt, C. G. 1914 The Housefly, *Musca domestica* Linn. Structure, Habits, Development, Relation to Disease and Control. Cambridge University Press. xv + 382 pp. 104 illus.
A comprehensive, scientific work, based chiefly on the author's previously published technical papers. Contains an extensive and valuable bibliography. REVIEW: RAE-B 3:174, 1915.

564. Hewitt, C. G. 1914 The Predaceous Habits of *Scatophaga.* Canad. Ent. 46:2-3.
Adults seen capturing *M. domestica* and other species. Has piercing proboscis and sucks out body fluids. ABSTRACT: RAE-B 2:56, 1914.

565. Hewitt, C. G. 1914 Further Observations on Breeding Habit and Control of the House-fly, *Musca domestica.* J. Econ. Ent. 7:281-289.
Larvae leave manure pile and pupate in sand; up to 2 feet from heap, and down to 9 inches below surface. ABSTRACT: RAE-B 2:159, 1914.

566. Hewitt, C. G. 1915 Notes on the Pupation of the House-Fly (*Musca domestica*) and its Mode of Overwintering. Canad. Ent. 47(3): 73-78.
Pupae found out as far as 4 feet from heap. Greatest depth, 24 inches. Gives 4 ways by which flies pass the winter in northern latitudes. ABSTRACT: RAE-B 3:125-126, 1915.

567. Hewitt, C. G. 1915 House-Fly Control. Agric. Gaz. Canada (Ottawa). 2(5):418-421.
Urges control work early in season. Opposes "storing" of manure unless unavoidable. ABSTRACT: RAE-B 3-133, 1915.

568. Hewlett, H. T. 1903 Insects as Carriers of Disease. Med. Press and Circ., N.S. 76:439-442. London.

569. Hewlett, H. T. 1905 The Etiology of Epidemic Diarrhoea. J. Prevent. Med. 13:496-507.

Hicks, J. B., joint author. *See* Samuelson, J., et al., 1860.

570. Hickson, S. J. 1905 A Parasite of the House-fly. Nature, Oct. 26.
If *Chernes* (pseudoscorpion) feeds on mites attached to a fly, it should be considered a *friend* of the fly, not a foe.

571. Hiestand, W. A. 1932 Progressive Paralysis of the Nervous System of House Flies by Formaldehyde and Anesthetics. Proc. Indiana Acad. Sci. (for 1961) 41:433-437.
Begins at posterior extremity and ends with mouth parts and antennae. ABSTRACT: RAE-B 21:11, 1933.

572. Higgins, A. E. H. and **Green, A. A. 1949** A Combined Hand- or Power-operated Sprayer for Fly and Mosquito Control. Ann. Appl. Biol. 36:383-391.
Highly efficient for both knock-down and kill. ABSTRACTS: TDB 47:163, 1950; RAE-B 38:68, 1950; BA 24:1055, 1950.

573. Hill, G. F. 1918 Relationship of Insects to Parasitic Diseases in Stock. Proc. Roy. Soc. Victoria. Melbourne 31(1):11-107. 7 pls.
Embryos of *Habronema muscae* passed in feces of horse are taken up by larvae of *M. domestica*, after a period of preliminary development. ABSTRACT: RAE-B 7:118, 1919.

574. Hill, G. F. 1921 *Musca domestica* as a "Bush Fly" in Australia. Ann. Trop. Med. Parasit. 15(1):93-94.
Recorded on freshly skinned buffaloes that had been shot in scrub country, 3-6 miles from nearest habitation. ABSTRACT: RAE-B 9:118-119, 1921.

Hill, H., joint author. *See* Harvey, W. C., et al., 1947.

575. Hindle, E. 1913 Note on the Colour Preference of Flies. Repts. Local Gov't. Bd. on Publ. Health and Med. Subjects, N.S. No. 85; Further Repts. (No. 6) on Flies as Carriers of Infection, pp. 41-43.
Used colored cardboard. Concluded that flies do not display marked color preferences.

576. Hindle, E. 1914 The Flight of the House Fly. Proc. Cambridge Phil. Soc. 17(4):310-313.
Fifty observation stations for recording the distribution of 25,000 flies released. Flies tend to travel against or across wind. Many modifying factors. ABSTRACT: RAE-B 2:38-39, 1914.

577. Hindle, E. and **Merriman, G. 1914** The Range of Flight of *Musca domestica*. J. Hygiene. 14:23-25.
Direction influenced by odors. Flies in open country travel farther than those in town. Morning-released flies travel farther than those set free in the afternoon.

Hindle, E., joint author. *See* Nuttall, G. H. F., et al., 1913.

Hinshaw, W. R., joint author. *See* Bushnell, L. D., et al., 1924.

Hixon, E., joint author. *See* Muma, M. H., et al., 1949.

578. Ho, Ch'i. 1938 The Significance of the Female Terminalia of House Flies as a Grouping Character. Ann. Trop. Med. Parasit. 32: 287-312.
Flies in the genus *Musca* (broad sense) can be divided into 3 groups on the basis of oviposition habit and corresponding differences in structure. ABSTRACTS: RAE-B 27:65, 1939; TDB 36:426, 1939.

579. Hoare, C. A. 1924 A Note on the Specific Name of the Herpetomonad of the House Fly. Trans. Roy. Soc. Trop. Med. Hyg. 17:403-406.
Establishes the use of *H. muscarum*. (Followed by Steinhaus, 1967.)

580. Hobson, R. P. 1936 Sheep Blow-fly Investigations. III. Observations on the Chemotropism of *Lucilia sericata* Mg. Ann. Appl. Biol. 23: 845-851.
Various aliphatic alcohols, acids and esters attract *M. domestica*. ABSTRACT: RAE-B 25:53, 1937.

581. Hodge, C. F. 1910 A Practical Point in the Study of the Typhoid or Filth Fly. Nature Study Review, 6:195-199.

582. Hodge, C. F. 1911 A Plan to Exterminate the Typhoid or Filth-Disease Fly. LaFollette's Weekly Magazine, Madison, Wisconsin. Vol. 3, No. 15 (April 15), pp. 7-8.

583. **Hodge, C. F. 1911** How you can Make your Home, Town, or City, Flyless. Nature and Culture, Cincinatti, Ohio. Vol. 3, Nos. 2-3 (Jul.-Aug.), pp. 9-23.

584. **Hodge, C. F. 1911** Exterminating the Fly. California Outlook, Sept. 30.

585. **Hodge, C. F. 1913** The Distance House-Flies, Blue-Bottles and Stable Flies may Travel over Water. Science 38:512-513.
Flies found in abundance on cribs up to 6 miles out in Lake Erie. Perhaps blown by the wind. ABSTRACT: RAE-B 1:233, 1913.

586. **Hodge, C. F. 1913** A New Fly Trap. J. Econ. Ent. 6:110-112, 1 pl.
In stable cellar window. Cow inside is only bait. Caught many mosquitoes and biting flies as well.

Hodges, K. C., joint author. *See* Brescia, F., et al., 1946.

587. **Hofman, E. 1888** Ueber die Verbreitung der Tuberculose durch Stubenfliegen. [On the Spread of Tuberculosis by House-flies.] Correspondenzbl. d. ärztl. Kreis. und Bezirksvereine im Königr. Sachsen. 44:130-135.

588. **Hoffman, C. H.** and **Surber, E. W. 1949** Effects of Feeding DDT-sprayed Insects to Fresh-water Fish. Special Scientific Report. Fisheries No. 3, 9 pp. mimeo. U.S. Dept. Int., Fish and Wildlife Service, Washington.
Used larvae and adults of *M. domestica* (also certain other species) for testing. Effects on fish variable, erratic. Small mouth bass much more sensitive than gold fish.

589. **Hoffman, R. A.** and **Lindquist, A. 1949** Effect of Temperature on Knock-down and Mortality of House Flies Exposed to Residues of Several Chlorinated Hydrocarbon Insecticides. J. Econ. Ent. 42:891-893.
DDT and Methoxychlor gave faster knock-down of flies at 70°F than at 90°. Also caused greater mortality at the lower temperature. Reverse was true for heptachlor, parathion, chlordane, dieldrin and toxaphene. ABSTRACTS: BA 25:864, 1951; RAE-B 38:153, 1950.

590. **Hoffman, R. A.** and **Lindquist, A. W. 1949** Fumigating Properties of Several New Insecticides. J. Econ. Ent. 42:436-438.
Tested DDT, several isomers of BHC, chlordan, chlorinated camphene, parathion, TDE and the fluorine analog of DDT, using *M. domestica*. Materials applied at 100 mg. and 10 mg. per sq. ft. ABSTRACT: RAE-B 38:98, 1950.

Hoffman, R. A., joint author. *See* Fairchild, H. E., et al., 1949.

591. **Hollick, F. S. J. 1941** The Flight of the Dipterous Fly *Muscina stabulans*. Philos. Trans. Roy. Soc. Lond. (B) 230:357-390.
Diagrams of wing path for female *M. domestica* also included for comparison. Used three different conditions as to rate of air flow.

592. **Holmgren, N. 1904** Zur Morphologie des Insektenkopfes: II. Einiges über die Reduktion des Kopfes der Dipteren-larven. [On the Morphology of the Insect Head: II. Sundry (remarks) regarding the Reduction of the Head in Dipterous Larvae.] Zool. Anz. 27: 343-355, 12 figs.
Floor of pharynx traversed by eight grooves, which are separated by supporting structures called "T-ribs".

593. **Honeij, J. A.** and **Parker, R. R. 1914** Leprosy: Flies in Relation to the Transmission of the Disease. J. Med. Res. 30:127-130.
Organism survives several days in the fly, after a meal of contaminated material.

Horrocks, W. H., joint author. *See* Firth, R. H., et al., 1902.

594. Horstmann, D. M. 1945 The Role of Flies in the Epidemiology of Poliomyelitis. J. Bact. 50(2):236.
Food exposed to flies in nature (during an epidemic) found infective when fed to chimpanzees. ABSTRACT: BA 20:366, 1946.

Horstman, D. M., joint author. *See* Ward, R., et al., 1945.

Hoskins, W. M., joint author. *See* Wieting, J. O. G., et al., 1939.

595. Hough, G. de N. 1898 The Muscidae Collected by Dr. A. Donaldson Smith in Somaliland. Proc. Acad. Nat. Sci. Philadelphia 50: 165-187.
M. domestica L., *M. corvina* Fabr. and *M. biseta* sp. nov.

Houlbert, C., joint author. *See* Galaine, C., et al. 1916.

596. Houston, W. M. 1913 Formalin against Flies. Indian Med. Gaz. Feb. 1913, p. 84.
Formalin, milk and water sprinkled about in tiny pools. Used in jail kitchen at Rajkot. ABSTRACT: RAE-B 1:73, 1913.

597. Howard, C. W. 1913 The House Fly. Minnesota Insect Life. 2(3): 9-10.
Popular and educational.

598. Howard, C. W. 1914 Control of Flies in Rural Districts. Office of State Entomologist, Minnesota. Circ. No. 33, 12 pp., 2 figs.
Urges immediate spreading of manure and incineration of garbage. ABSTRACT: RAE-B 3:27, 1915.

599. Howard, C. W. 1914 Some Suggestions in Fly Control. Fifteenth Rept. State Entomologist of Minnesota, pp. 57-60.

600. Howard, C. W. 1917 Hibernation of the House-fly in Minnesota. J. Econ. Ent. 10:464-468.
Believes Minnesota winters not favorable for survival of any stage except adult, and that only when temperature and food supply permit. ABSTRACT: RAE-B 6:9, 1918.

601. Howard, C. W. 1917 What the House-fly Costs? Minnesota Insect Life. (St. Paul.) 4(2), 8 pp.
An attempt to estimate monetary loss through disease, also cost of screening, fly paper, poisons, nursing care and medical treatment of the sick, etc. ABSTRACT: RAE-B 5:114, 1917.

602. Howard, C. W. 1917 A Fly Control Exhibit. J. Econ. Ent. 10: 411-412, 1 pl.
Model exhibit called "The Flyless Farm", for use at State Fairs. ABSTRACT: RAE-B 5:168, 1917.

603. Howard, C. W. and **Clark, P. F. 1912** Experiments on Insect Transmission of the Virus of Poliomyelitis. J. Exp. Med. 16:850-859.
M. domestica can carry the virus in an active state for several days on body surface and for several hours in the alimentary tract. ABSTRACT: RAE-B 1:18, 1913.

604. Howard, L. O. 1896 House Flies (*In* The Principal Household Insects of the United States, by L. O. Howard and C. L. Marlatt). U.S. Dept. Agric., Div. of Ent., Bull. No. 4, N.S., pp. 43-47.
Larvae are often destroyed by predatory beetles and adults by the house centipede, *Scutigera forceps*.

605. Howard, L. O. 1898 Further Notes on the House-fly. (*In* Some Miscellaneous Results of the Work of the Division of Entomology.) U.S. Dept. Agric., Div. of Ent., Bull. No. 10, N.S., pp. 63-65.
Kerosene superior to lime for application to manure to prevent flies.

606. Howard, L. O. 1900 A Contribution to the Study of the Insect Fauna of Human Excrement (with Special Reference to the Spread of Typhoid Fever by Flies). Proc. Wash. Acad. Sci. 2:541-604.
M. domestica constitutes 98·8 per cent of the whole number of insects captured in houses throughout the whole country (circum 1900).

607. Howard, L. O. 1901 On Some Diptera Bred from Cow Manure. Canad. Ent. 33:42-44.
Includes *M. domestica*, which is one of 9 species that breed in human excrement, as well.

608. Howard, L. O. 1901 Flies and Typhoid. Pop. Sci. Monthly, January 1901, pp. 249-256.

609. Howard, L. O. 1901 The Carriage of Disease by Flies. U.S. Dept. Agric. Div. of Ent., Bull. 30 N.S., pp. 39-45.

610. Howard, L. O. 1906 House Flies. U.S. Dept. Agric., Div. of Ent. Circ. No. 71.
A revision of Circular No. 35, 1898.

611. Howard, L. O. 1909 Economic Loss to the People of the United States through Insects that Carry Disease. National Geographic Magazine 20 (Aug. 1909), pp. 735-749.
Summarizes the diseases which the housefly is believed to carry.

612. Howard, L. O. 1909 Economic Loss to the People of the United States through Insects that carry Disease. U.S. Dept. Agric., Bur. of Ent., Bull. No. 78, 40 pp.
Discusses *M. domestica*, pp. 23-36. Proposes the common name be changed to "typhoid fly".

613. Howard, L. O. 1911 The House Fly, Disease Carrier. An Account of its Dangerous Activities and the Means of Destroying It. New York, Frederick A. Stokes Company. xix + 312 pp., col. front, 39 figs., 21 cm.
Though written in popular style, contains much scientific information. Would have been more useful, if indexed.

614. Howard, L. O. 1911 House-flies. U.S. Dept. Agric., Farmers Bull. No. 459, 16 pp., 9 figs.

615. Howard, L. O. 1911 Flies as Carriers of Infection. Science, N.S. 34: 24-25 (Review).

616. Howard, L. O. 1916 Report of the Entomologist. U.S. Dept. Agric., Bur. of Ent., Aug. 1916, pp. 8-10.
In the latitude of Dallas, Texas, *M. domestica* passes the winter in the larval or pupal stage. ABSTRACT: RAE-B 5:58, 1917.

617. Howard, L. O. 1920 Report of the Entomologist. U.S. Dept. Agric., Bur. Ent., pp. 28-31.
Both *M. domestica* and *Stomoxys calcitrans* implicated in the spread of hog cholera. ABSTRACT: RAE-B 9:50, 1921.

618. Howard, L. O. 1920 The Housefly—Carrier of Disease. Ohio State Dept. Health, Columbus, May 1920, 16 pp. 11 figs.
Brief, useful, in popular style. ABSTRACT: RAE-B 9:71, 1921.

619. Howard, L. O. and **Bishopp, F. C. 1926** The House Fly and How to Suppress It. U.S. Dept. Agr. Farmers' Bull. No. 1409, 16 pp., illus.
A revision of similar, previous publications. *See also* U.S. Dept. Agric. Leaflet No. 182.

620. Howard, L. O. and **Hutchison, R. H. 1917** The House-Fly. U.S. Dept. Agric. Farmers' Bull. No. 851, 23 pp., 15 figs.
Gives all methods of control. Lists natural enemies of the fly, including the common house centipede, *Scutigera forceps*. ABSTRACT: RAE-B 6:103, 1918.

621. Howard, L. O. and **Marlatt, C. L. 1896** [Housefly Information.] U.S. Dept. Agric., Div. of Ent., Bull. No. 4:43-47.
A general account, with methods of control for that time.

622. Howard, L. O. and **Pierce, W. D. 1921** The Non-bloodsucking Diptera. *In* The Practice of Medicine in the Tropics, by W. Byam and R. G. Archibald. Vol. 1:420-448. London.

Howat, C. H., joint author. *See* Lamborn, W. A., et al., 1936.

623. Howell, D. E. 1941 The Use of Arsenicals for the Control of Housefly Larvae. Proc. Oklahoma Acad. Sci. (for 1941). 22:68-72. Edmond.
One-tenth per cent solution of sodium arsenite, used at less than 2 lb. per ton of manure, superior to borax or hellebore. Does not seem to harm crops. ABSTRACT: RAE-B 31:136, 1943.

Howlett, F. M., joint author. *See* Copeman, S. M., et al., 1911.

624. Hoyer, D., Zarkoski von Schmidt, S. and **Weed, A. 1936** Dosage-mortality Curve of Pyrethrum Sprays on the House Fly *Musca domestica* L. J. Econ. Ent. 29:598-600.
Used Peet-Grady method for testing. Pyrethrin content of spray determined by method of H. A. Seil. ABSTRACT: RAE-B 24:253, 1936.

Huan, I. C., joint author. *See* Yao, H. Y., et al., 1929.

625. Huang-Chen. 1945 [Analysis of the Fly's Feeding Habits and the Discrimination of its Breeding Places.] Fukien Acad. Inst. Zool. and Bot. Res. Bull. 1:59-75. (In Chinese, with English summary.)
M. domestica next important menace to health after *Chrysomyia megacephala*. ABSTRACT: BA 25:1146, 1951.

Huddleson, I. F., joint author. *See* Ruhland, H. H., et al., 1941.

626. Hughens, H. V. 1919 A Useful and Inexpensive Fly-trap. U.S. Nav. Med. Bull. 13(1):80-82.
Square trap. Gauze bottom raised to a ridge, roof-wise. One quarter inch holes along ridge.

Huie, D., joint author. *See* Yao, H. Y., et al., 1929.

627. Hunt, W. T. 1944 Relative Effectiveness of DDT and Rotenone against House Flies. J. Econ. Ent. 37:136.
Rotenone considered to be 2½ to 4 times as effective as DDT. Used tunnel method for testing. ABSTRACT: RAE-B 32:203, 1944.

628. Hunter, W. 1907 [Occurrence of Plague Bacilli in Alimentary Tract of Flies that had fed on Infected Material in Hong Kong.] Quoted in article on "The Danger of the Common Fly". Nursing Times, 28th Sept., 1907, p. 842.

Hurtig, H., joint author. *See* Brown, A. W. A., et al., 1948.

629. Husain, M. T. 1927 Yolk Formation in Some Arthropods. Nature 119:817.
In *Musca*, nucleolus divides at very early stage. Nucleoli begin to fragment, and fragments escape into cytoplasm. Fuchsinophil yolk arises in masses throughout the cytoplasm. No fatty yolk in this form.

Hussain, M., joint author. *See* Ross, W. C., et al., 1924.

630. Hutchinson, W. 1908 The Story of the Fly that Does not Wipe its Feet. Saturday Even. Post, 7 March, 1908.

631. Hutchinson, W. 1911 How Doth the Little Busy Fly? Country Life (U.S.A.) 20:31-33. Aug. 15.

632. Hutchison, R. H. 1914 The Migratory Habit of House Fly Larvae as Indicating a Favorable Remedial Measure. An Account of Progress. U.S. Dept. Agric. Bull. No. 14, 11 pp.
Larvae about to pupate seek drier regions, avoid light and high tempe-

ratures. Will leave manure if it is kept moist. Maggot trap plausible. ABSTRACT: RAE-B 2:72, 141.

633. Hutchison, R. H. 1915 A Maggot Trap in Practical Use; an Experiment in House-Fly Control. U.S. Dept. Agric. Bull. No. 200, 15 pp., 4 figs., 3 pls.
Manure on wooden platform with 1 inch cracks. Larvae fall into water in a flat concrete basin, 20 × 10, with 4 inch rim. Ninety-eight per cent of all larvae destroyed. ABSTRACT: RAE-B 3:134, 1915.

634. Hutchison, R. H. 1916 Notes on the Preoviposition Period of the House Fly, *Musca domestica*, L. U.S. Dept. Agric. Bull. No. 345, 13 pp.
Varies between 2½ and 23 days. Shortest in summer; longest in autumn. Pairing takes place from 1st to 47th day. Maximum longevity, 70 days. ABSTRACT: RAE-B 4:62, 1916.

635. Hutchison, R. H. 1918 Overwintering of the House Fly. J. Agr. Res. 13(3):149-169.
Continued breeding more widespread than generally thought, especially in cities. But flies do not survive well in heated buildings. ABSTRACT: RAE-B 6:158, 1918.

Hutchison, R. H., joint author. *See* Cook, F. C., et al., 1914, 1915, 1916; Howard, L. O., et al., 1917.

636. Ikawa, M., Dicke, R. J., Allen, T. C. and **Link, K. P. 1945** The Principal Alkaloids of Sabadilla Seed and their Toxicity to *Musca domestica* L. J. Biol. Chem. 159(2):517-524.
Kerosene extracts of *Schoenocaulon* sp. contains a combination of alkaloids (veratrine) toxic to flies. Three fractions were obtained. All gave quick and complete knock-downs. ABSTRACT: RAE-B 34:163, 1946.

637. Illingworth, J. F. 1923 Insect Fauna of Hen Manure. Proc. Hawaiian Ent. Soc. 5(2):270-273.
M. domestica one of eight species of larvae present. Ants of the genus *Pheidole* feed on both eggs and larvae. Newly emerged flies heavily beset by mites. ABSTRACT: RAE-B 11:195, 1923.

638. Illingworth, J. F. 1923 House Flies. Proc. Hawaiian Ent. Soc. 5(2): 275-276.
Questions occurrence of true *M. domestica* in Hawaii. Common housefly believed to be *M. flavinervis* Thomsen, or perhaps a hybrid. *Fannia canicularis* common. ABSTRACT: RAE-B 11:195, 1923.

639. Illingworth, J. F. 1923 Insects attracted to Carrion in Hawaii. Proc. Hawaiian Ent. Soc. 5(2):280-281.
Species which passes for *M. domestica* attracted by odor of decay.

640. Illingworth, J. F. 1926 The Common Muscoid Flies Occurring about Sweet-shops in Yokohama, Japan. Proc. Hawaiian Ent. Soc. 6(2): 260-261.
Used traps. *M. convexifrons* Thoms. appears in June list. *M. domestica* recorded in small numbers last week in July. ABSTRACT: RAE-B 14:182, 1926.

641. Illingworth, J. F. 1930 Pineapple Cannery Waste and its Insect Problems. Proc. Hawaiian Ent. Soc. 7(3):462-465.
Musca bred in enormous numbers in fermented scum on pools of wash water containing fruit fiber. Controlled by running arsenic solution into flume carrying cannery waste. ABSTRACT: RAE-B 19:169, 1931.

642. Ingle, L. 1943 An Apparatus for Testing Chemotropic Responses of Flying Insects. J. Econ. Ent. 36:108-110.
Used blue light to attract *M. domestica* (or *Stomoxys*) in order to record flies' response to substance on *one* of a group of screens. Various attractants, repellents and odiferous materials were tested. ABSTRACT: RAE-B 31:199, 1943.

643. **Isaac, P. V. 1944** Prevention of House-fly Breeding. Indian Fmg. 5(2):61-62.
Fresh dung and garbage left in pit for 7 days while flies laid eggs. Then covered with soil, tamped, and wetted down. Repeat twice. Maggots are trapped. ABSTRACT: BA 21:494, 1947.

644. **Iwanoff, X. 1934** Ueber Sommerwunden beim Rinde. [On Summer Sores in Cattle.] Arch. Tierheilk. 67:261-270.
Sores affecting cattle in Bulgaria relate to parasitic nematodes, carried by *M. domestica* and *Stomoxys calcitrans*. ABSTRACT: RAE-B 23:276, 1935.

645. **Jack, R. W. 1916** Home-made Fly Papers. Dept. Agric. Rhodesia. Bull. 249, 4 pp.
Gives formula for preparation which remains moist in Rhodesian climate. May be stored in closed tin for months. ABSTRACT: RAE-B 5:52, 1917.

646. **Jack, R. W. 1931** Report of the Chief Entomologist for the Year 1930. Rept. Secy. Dept. Agric. S. Rhodesia, 1930, pp. 65-73.
Several species of *Musca* feed on exudations from sores of cattle, at wounds, and at bites of true bloodsuckers. Breed mostly in fresh cow dung around cattle kraals. ABSTRACT: RAE-B 19:179, 1931.

647. **Jack, R. W.** and **Williams, W. L. 1937** The Effect of Temperature on the Reaction of *Glossina morsitans* Westw. to Light. A Preliminary Note. Bull Ent. Res. 28:499-503.
M. domestica vicina has greater ability to find comfortable temperature by trial and error than *Glossina*. ABSTRACT: RAE-B 26:29, 1938.

648. **Jackson, A. C.** and **Lefroy, N. M. 1917** Some Fly Poisons for Outdoor and Hospital Use. Bull. Ent. Res. 7:327-335.
Formaldehyde too volatile for use outdoors in a hot climate. ABSTRACT: RAE-B 5:146, 1917.

649. **Jackson, D. D. 1907** Pollution of New York Harbour as a Menace to Health by the Dissemination of Intestinal Disease through the Agency of the Common House-fly. A Report to the Committee on Pollution of the Merchants' Association of New York. 22 pp., 2 maps, 3 charts, 3 figs.
Points out that the fly and intestinal disease are more important in New York than mosquito-borne diseases. Report transmitted to Gov. Hughes.

650. **Jackson, D. D. 1908** Conveyance of Disease by Flies Summarized. Boston Med. and Surg. J. for 1908, p. 451.
Found over 1,000,000 bacteria on each fly caught at a swill barrel.

651. **Jackson, D. D. 1910** The Disease-Carrying House-fly. Review of Reviews (U.S.A.) July, 1910.

652. **Jacobson, M.** and **Haller, H. L. 1947** The Insecticidal Component of *Eugenia haitiensis* Identified as 1, 8-Cinerol. J. Am. Chem. Soc. 69:709.
Tested on *M. domestica*. Another species, *E. buxifolia* grows in Porto Rico. Natives use product as insecticidal spray (eucalyptol). ABSTRACT: RAE-B 38:173, 1950.

Jacobson, M., joint author. *See* Acree, F., et al., 1945.

Jacques, J. E., joint author. *See* DeMello, F., et al., 1919.

653. **James, H. C. 1928** On the Life-histories and Economic Status of Certain Cynipid Parasites of Dipterous Larvae, with Descriptions of Some New Larval Forms. Ann. Appl. Biol. 15:287-316.
Deals with several parasites of larvae and pupae. *M. domestica* one of the common hosts. ABSTRACT: RAE-B 16:198, 1928.

654. James, J. F. 1935 A Simple Fly Trap. J. Roy. Army Med. Corps 65:400-401, 2 figs.
ABSTRACT: RAE-B 24:64, 1936.

654a. Jamieson, H. C. 1938 The House Fly as a Cause of Nasal Allergy. J. Allergy. 9:273-274. 1938.
Cited by Shulman, S., 1967.

Jandu, A. S., joint author. *See* Deoras, P. J., et al., 1943.

655. Janes, R. L. 1949 A Program for House Fly Control. Michigan State Coll. Mimeo. Publ. 6068, 5 pp.
A clear, concise and useful pamphlet.

656. Janet, C. 1907 Sur l'origine du tissu adipeux imaginal, pendant la nymphose chez les Muscides (Dipt.). [On the Origin of the Adult Fat Body during the Pupal Stage in Muscidae.] Bull Soc. Ent. Fr. for 1907, pp. 350-351.
States that the imaginal fat body arises during development from mesodermal leucocytes.

657. Janisch, E. 1928 Die Lebens- und Entwicklungsdauer der Insekten als Temperaturfunktion. [Duration of Life and Development in Relation to Temperature.] Zeit. Wiss. Zool. 132.
(*See* Below.)

658. Janisch, E. 1933
(*See* Section VII, Publications on Research Techniques.)

659. Janisch, E. and **Maercks, H. 1933** Ueber die Berechnung der Kettenlinie als Ausdruck für die Temperaturabhängigkeit von Lebenserscheinungen. [On Computation of the Catenary Curve as an Expression for the Dependence of Life Phenomena on Temperature.] Arb. biol. Reichsanst. Dahlem 20.
Formula for symmetrical catenary curve: $y = \frac{m}{2}(a^{w-x} + a^{-w+x})$
where y is the time, m is shortest duration of development, w is the optimum temperature, x is the registered temperature, and a is a constant.

660. Jausion, H. and **Dekester, M. 1923** Sur la transmission comparée des kystes d'*Entamoeba dysenteriae* et de *Giardia intestinalis* par les mouches. [On the Comparative Transmission of the Cysts of *Entamoeba dysenteriae* and of *Giardia intestinalis* by Flies.] Arch. Inst. Pasteur Afr. Nord. 3(2):154-155; also Arch. Inst. Pasteur Algérie 1:154-155.
Small, resistant cysts of *Giardia* better adapted to carriage by flies than cysts of *Entamoeba*. ABSTRACT: RAE-B 11:173, 1923.

Jennings, E., joint author. *See* Lingard, A., et al., 1906.

Jensen, J. A., joint author. *See* Quarterman, K. D., et al., 1949.

661. Jepson, F. P. 1915 Report of the Entomologist. Dept. Agric., Fiji. Ann. Rept. for the year 1914. Suva. pp. 17-27.
M. domestica diminished greatly in province of Bua. Due largely, it is believed, to the predatory wasp, *Polistes habreus,* which feeds on both adults and larvae. ABSTRACT: RAE-B 4:41, 1916.

662. Jepson, F. P. 1926 House Flies and their Connection with Manuring Operations in Ceylon. Ceylon Dept. Agric. Bull. 74, 16 pp., 7 pls.
M. yerburyi Patton and *M. nebulo* F. were bred from artificial manure. Flies oviposited after blood meal. Both species invade bungalows. ABSTRACT: RAE-B 14:187, 1926.

663. Jepson, J. P. 1909 Some Observations on the Breeding of *Musca domestica* during the Winter Months. Repts. Local Gov't. Bd. on Publ. Health and Med. Subjects, N.S. No. 3. Prelim. Repts. on Flies as Carriers of Infection, pp. 5-8.
Reports flies more or less active in college sculleries throughout the winter months. These become the progenitors of "summer's millions".

664. Jepson, J. P. 1909 The Breeding of the Common House-fly (*Musca domestica*) during the Winter Months. J. Econ. Biol. 4:78-82.
Probable that flies are to be found in winter only in isolated colonies in certain warm places. An opportunity for extermination?

664a. Jettmar, H. M. 1927 Beiträge zum Studium der Pest unter den Insekten. II Mitteilung. [Contributions to the study of Plague among Insects.] Zeit. Hyg. Infekt. 107(3-4):498-509.
Flies (*M. domestica*, also Sarcophagidae) fed "plague material" and maintained at room temperature, lived over a week. Conclusion: the mortality of such flies, in comparison with that of the controls, was not significantly increased.

664b. Jettmar, H. M. 1940 Some Experiments on the Resistance of the Larvae of the Latrine Fly, *Chrysomyia megacephala* against Chemicals. Chinese Med. J. 57:74-85.
In Hanchung, S. Chensi, in 1938, *C. megacephala* was very abundant in July; *M. domestica*, not so until mid-August. Larvae of *C. megacephala* are best destroyed by pouring boiling water on the surface of the latrine.

665. Johannsen, O. A. 1911 The Typhoid Fly and its Allies. Maine Agric. Expt. Sta. Bull. 401, pp. 1-7.

666. Johnston, A. N. 1949 Studies on the Action of DDT on Anopheline Mosquitoes and House-flies. Bull. Ent. Res. 40:447-452.
Complete mortality among flies confined in tubes where treated flies had been kept for 2 hours. Duration of exposure needed to acquire a lethal dose, also time required to kill, very much shorter for flies than for mosquitoes. ABSTRACT: RAE-B 38:27, 1950.

667. Johnston, T. H. 1913 Notes on Some Entozoa. Proc. Roy. Soc. Queensland, 24:63-91, 5 pls.
Larvae of nematode, *Habronema muscae* found in *Stomoxys calcitrans*, *M. domestica* and common cattle fly, *M. vetustissima*. Last named frequents eyes of cattle. ABSTRACT: RAE-B 1:165, 1913.

668. Johnston, T. H. 1920 Flies as Transmitters of Certain Worm Parasites of Horses. Science & Industry (Melbourne). 2:369-372.
Many species found to harbor *Habronema*. Heavily infested flies do not live long. Accounts for low percentage of parasitism among flies captured in nature. ABSTRACT: RAE-B 8:188, 1920.

669. Johnston, T. H. 1920 Notes on Certain Queensland "Bush Flies". Trans. Australian Med. Congress, 11th Session. Brisbane, Queensland, 21-28 August, 1920, pp. 265-272.
M. convexifrons (*fergusoni*), *M. pumila* (*vetustissima*), *M. terraereginae* and *M. ventrosa* (*hilli*) all infest eyes, nose, mouth and invade any abrasion or sore (man and animals). ABSTRACT: RAE-B 10:168, 1922.

670. Johnston, T. H. 1921 The Sheep Maggot Fly Problem in Queensland. Queensland Agric. J., Brisbane, 15(6):244-248, 1 pl.
Mentions *Pachycrepoideus dubius* (*Hymenoptera*) as destroying house flies in North Queensland. ABSTRACT: RAE-B 9:156, 1921.

671. Johnston, T. H. 1922 Some Facts regarding the Biology of the House Fly. Med. J. Australia, 9th year, ii (18):494-499.
Notes on *M. domestica* and *Sarcophaga* spp. ABSTRACT: RAE-B 11:21, 1923.

672. Johnston, T. H. 1925 Remarks on the Common "Bushflies" of Australia. Health 3(4):110-113.
M. pumila (*vetustissima*) and *M. convexifrons* (*fergusoni*) readily settle on humans and may distribute enteric diseases. Are found in town streets as well as in the bush. ABSTRACT: RAE-B 13:164, 1915.

673. Johnston, T. H. and **Bancroft, M. J. 1920** The Life Histories of *Musca australis* Macq. and *M. vetustissima* Walker. Proc. Roy. Soc. Queensland. Brisbane. 31(12):181-203.
M. lusoria Wied., *australis* Macq. and *fergusoni* J. & B. are synonyms. This species is larviparous. *M. humilis* Wied. is identical with *vetustissima* Walker; is oviparous. ABSTRACT: RAE-B 10:14, 1922.

674. Johnston, T. H. and **Bancroft, M. L. 1920** Notes on the Biology of some Queensland Flies. Mem. Queensland Mus. 7:30-43.
M. terraereginae sp. n. lays eggs in cow and horse dung. *M. hilli* sp. n. has similar habits. Paper gives duration of immature stages for both species. ABSTRACT: RAE-B 8:189, 1920.

675. Johnston, T. H. and **Bancroft, M. J. 1920** The Life History of *Habronema* in Relation to *Musca domestica* and Native Flies in Queensland. Proc. Roy. Soc. Queensland. Brisbane. 32(5):61-88.
At least 5 species of *Musca* are involved, also other muscoid genera. Larvae escape from fly's proboscis while latter is about animals mouth; are swallowed. Escape into eye, or wound, causes habronemic conjunctivitis or granuloma. ABSTRACT: RAE-B 10:15, 1922.

676. Johnston, T. H. and **Hardy, G. H. 1923** A Synonymic List of some Described Australian Calliphorine Flies. Proc. Roy. Soc. Queensland. 34(10):191-194.
Includes *M. australis* Boisd. and *M. dorsalis* Walker, the latter as a synonym of *Anastellorhina augur* F. ABSTRACT: RAE-B 11:43, 1923.

677. Johnston, T. H. and **Tiegs, O. W. 1923** Notes on the Biology of Some of the More Common Queensland Muscoid Flies. Proc. Roy. Soc. Queensland. 34(3):77-104.
Life cycle of *M. domestica* requires 7-8 days in summer, 11-15 in autumn, 12-16 in winter. Paper also gives data on several other species of *Musca*. ABSTRACT: RAE-B 11:23, 1923.

678. Jolly, G. G. 1923 An Automatic Fly-proof Latrine Seat. Indian Med. Gaz. 58(12):575-578, 5 figs.
Complies with Caste prejudices. May be portable or fixed. ABSTRACT: RAE-B 12:25, 1924.

679. Jones. 1906 The Common House-fly as the Cause of Disease. Maritime Med. News (Halifax, Canada) 18:285-294.

680. Jones, F. W. C. 1907 Notes on Enteric Fever Prevention in India. J. Roy. Army Med. Corps 8:22-34.

681. Jones, H. A. and **Fluno, H. J. 1946** DDT-Xylene Emulsions for Use against Insects affecting Man. J. Econ. Ent. 39:735-740.
Triton X-100 is best emulsifier. Can be diluted to concentrations ranging from 0·1 to 10 per cent, and can be used with distilled, hard, or sea water. ABSTRACT: RAE-B 36:86, 1948.

682. Jones, H. A. and **Smith, C. M. 1936** Derris and Cubé. Approximate Chemical Evaluation of their Toxicity. Soap 12:113, 115, 117.
Principle extracted from root of *Lonchocarpus*. Tested for toxicity against *M. domestica*. ABSTRACT: RAE-B 25:9, 1937.

683. Jones, H. A. and **Sullivan, W. N. 1942** *Tephrosia* Extract against House Flies. Soap and Sanit. Chem. 18(9):94-95.
T. virginiana may prove a substitute for derris and cubé. Rotenone is

not the only toxic constituent of these insecticides. ABSTRACT: RAE-B 31:98, 1943.

Jones, H. A., joint author. *See* Campbell, F. L., et al., 1934; Schroeder, H. O., et al., 1946, 1948; Lindquist, A. W., et al., 1944, 1945, 1946; Madden, A. H., et al., 1947; McAlister, L. C., Jr., et al., 1947.

684. Joshi, K. G. and **Dnyansagar, V. R. 1945** Some Observations on Fly Breeding in Compost Trenches. Indian Med. Gaz. 80(7):358-361.
Bangalore method of composting. Under monsoon conditions only 10-20 per cent of the pupae proved viable. ABSTRACTS: RAE-B 35:33, 1947; BA 20:1013, 1946.

685. Joshi, K. G. and **Dnyansagar, V. R. 1947** Use of Crude Oil and Bleaching Powder in Controlling Fly Breeding in Compost Trenches. Indian Med. Gaz. 82(2):92-96.
Very effective during the monsoon. Breeding controlled to the extent of 85-90 per cent. ABSTRACTS: RAE-B 38:159, 1950; BA 22:209, 1948.

686. Kaji, S. 1944 [On the Pupation-hormone of *Lucilia.*] Trans. Kansai Ent. Soc. 14:87-91, illus. (In Japanese.)
Represents pioneer work on Muscoid types. Antedates similar findings for *M. domestica.* ABSTRACT: BA 24:1337, 1950.

687. Kalandadze, L. P. and **Chilingarova, S. V. 1940** [Résultata des observations faites sur la mouche domestique, *Musca vicina* Macq.] Med. Parasit. 9(4):305-354. (In Russian.)
Life cycle completed in 8½ to 27 days, depending on time of year. Hibernation occurs chiefly, but not always, in the adult stage. ABSTRACT: RAE-B 31:126, 1943.

688. Kalandadze, L. P. and **Chilingarova, S. V. 1940** [Contributions à l'étude des mouches de Géorgie (principalement synanthropes).] Med. Parasit. 9(5):518-520. (In Russian.)
Of 32 species of flies observed in Georgia, *M. domestica vicina* was the most abundant. Typical *M. domestica* was not seen.

689. Kalandadze, L. P. and **Chilingarova, S. V. 1942** [On the Use of Sticky Fly Paper against Flies.] Med. Parasit. 10(5-6):569-572. (In Russian.)
Formula based on beeswax and castor oil lost all effectiveness after 4 days. ABSTRACT: RAE-B 31:226, 1943.

690. Kalandadze, L. P. and **Chilingarova, S. V. 1942** [The Rôle of Substrates in the Oviposition and Pre-imaginal Development of *Musca vicina* Macq.] Med. Parasit. 11(4):105-112.
Compares suitability of various types of dung. ABSTRACT: RAE-B 32:120, 1944.

691. Karsch, F. 1886 Dipteren von Pungo-Andongo-II. Die Cyclorhaphen. Ent. Nachr. 12:257-264.
Includes *M. pungoana* sp. nov., page 259.

Kastle, joint author. *See* Rosenau, M. J., et al., 1909.

692. Katagai, T. 1935 Seasonal Fluctuation of the Numbers of *Musca domestica* L. in the City of Taihoku. Tokyo Iji-Shimshi No. 2929, pp. 1218-1223.
(Taihoku is in Formosa.) Flies scarce January to April. Sudden increase in May, decrease in June, second peak in September. ABSTRACT: RAE-B 23:226, 1935.

Kawabada, K., joint author. *See* Sugai, C., et al., 1918.

693. Kearns, C. W., Weinman, C. J. and **Decker, G. C. 1949** Insecticidal Properties of Some New Chlorinated Organic Compounds. J.

Econ. Ent. 42:127-134.
Very complex substances as compared with DDT, BHC, and chlordane. Tested on *M. domestica* and cockroaches. ABSTRACT: RAE-B 38:55, 1950.

Kearns, C. W., joint author. *See* Dahm, P. A., et al., 1941; Picard, J. P., et al., 1949.

694. Kearns, H. G. H. 1942 The Control of Flies in Country and Town. Ann. Appl. Biol. 29:310-313. London.
M. domestica and other flies increased enormously by breeding in food stuffs among rubble, after bombing. Repellents and poison sprays used. ABSTRACT: RAE-B 31:65-66, 1943.

694a. Keiding, J. and **Van Deurs, H. 1949** DDT-Resistance in House-Flies in Denmark. Nature 163:964.
First report of this phenomenon in that country. Contains comments on resistance in United States and other parts of Europe.

695. Keilin, D. 1919 On the Life History and Larval Anatomy of *Melinda cognata* Meigen (Diptera, Calliphorinae) Parasitic in the Snail, *Helicella* (*Heliomanes*) *virgata* DaCosta, with an Account of the other Diptera living upon Molluscs. Parasitology (Cambridge) 11(3-4):430-455, 4 pls., 4 figs.
Of 50 molluscs collected, 9 yielded larvae of *M. domestica*. All were under size as compared with those reared on normal medium. ABSTRACT: RAE-B 8:24, 1920.

696. Keilin, M. D. 1912 Structure du pharynx en fonction du régime chez les larves de Diptères cylorhaphes. [Structure of the Pharynx in Relation to Diet among Larvae of Cyclorrhaphous Diptera.] Compt. Rend. Acad. Sci. (Paris). 155:1548-1550.
All larvae which are parasitic on animals and plants, or are carnivorous, or bloodsucking, have no ribs on the floor of the pharynx. Ribs are present in all *saprophagous* larvae.

Keister, W. S., joint author. *See* Stiles, C. W., et al., 1913.

Keller, J. C., collaborator. *See* Gleichen, F. W. von, 1766.

697. Kellers, H. C. 1911 A Sanitary Garbage Can Holder. U.S. Nav. Med. Bull. 5:45.

698. Kelsall, A., Spittal, J. P., Gorham, R. P. and **Walker, G. P. 1926** Derris as an Insecticide. 56th Ann. Rept. Ent. Soc. Ontario for 1925, pp. 24-40. Toronto.
Five lbs. to 100 gal. used as spray on *M. domestica*. Majority died within 24 hours, all within 48 hours. Dust less effective. ABSTRACT: RAE-B 15:17, 1927.

699. Kerr (Dr.) 1906 Some Prevalent Diseases in Morocco. Paper read before Glasgow Medico-chirurgical Society. Dec. 7, 1906.
Considered that epidemics of syphilis were augmented by flies which feed on open sores.

700. Kerr, R. W. 1948 The Effect of Starvation on the Susceptibility of Houseflies to Pyrethrum Sprays. Australian J. Sci. Res. (B) 1(1): 76-92, 5 figs.
ABSTRACTS: BA 24:243, 1950; RAE-B 37:85, 1949.

Keyworth, W. D., joint author. *See* Morrison, J., et al., 1916.

701. Kido, G. S. and **Allen, T. C. 1947** Colloidal DDT. Its Use in Insecticide Sprays. Agric. Chem. 2(6):21-23, 67, 69.
Particles 1 micron or more in size, suspended in water, plus a dispersing agent. Various concentrations tested against *M. domestica*. ABSTRACT: RAE-B 37:100, 1949.

702. Kilgore, L. B. 1939 Insect Repellents. A Study of the Comparative Repellency by the Sandwich-Bait Method using Confined House Flies. Soap and Sanit. Chem. 15(6):103, 105, 107, 109, 111, 123.
Tested materials soluble in alcohol. Bait was molasses, dried on blotting paper. Preparations were placed in stock cage containing about 2000 flies.

702a. Kilgore, L. B., Ford, J. H. and **Wolfe, W. C. 1942** Insecticidal Properties of 1, 3-Indandiones. Effect of Acyl Groups. Industr. Engin. Chem. (Industr. Ed.) 34(4):494-497. Easton, Penn.
Very toxic to house flies, but action not sufficiently rapid for use alone. Can be substituted for major portion of pyrethrum in more concentrated sprays. ABSTRACT: RAE-B 32:156, 1944.

703. King, H. N. 1918 Some Unusual Methods of Disposal of Excreta in Camps. Indian Med. Gaz. 53(2):74-75.
Describes (1) smoke latrine; (2) daily "burning out" of latrine trenches; (3) deep trenches, sprayed daily with sodium arsenite (where fuel is scarce). ABSTRACT: RAE-B 6:221, 1918.

704. King, W. V. 1943 Some Entomological Aspects of Troop Mobilization. J. Econ. Ent. 36(4):577-580.
Includes mechanical transmission of intestinal diseases by flies. ABSTRACT: RAE-B 32:77, 1944.

705. King, W. V. and **Gahan, J. B. 1949** Failure of DDT to Control House Flies. J. Econ. Ent. 42:405-409.
Reports from 7 localities in 5 States. Resistance to DDT much greater than to methoxychlor, chlordan, and benzene hexachloride. ABSTRACT: RAE-B 38:93, 1950.

706. Kisliuk, M. 1917 Some Winter Observations of Muscoid Flies. Ohio J. Sci. (Columbus). 17(8):285-294.
Greatest longevity at 45°F mean temperature was 44 days. All stages of *M. domestica* possible under rare, artificial conditions, where temperature and breeding media are suitable. ABSTRACT: RAE-B 5:127, 1917.

707. Klechetova, A. M. 1946 [Hexachlorethane as an Insecticide for the Destruction of the Larvae of Flies.] Med. Parasit. 15(6):77-81. Moscow. (In Russian.)

Klechetova, A. M., joint author. *See* Dolinskaya, T. Y., et al., 1946.

708. Klein, E. 1908 Flies as Carriers of the *Bacillus typhosus*. Brit. Med. J. 2(Oct. 17):1150-1151.

709. Kling, C. and **Levaditi, C. 1913** Etudes sur la Poliomyetite Aigue Epidemique. [Studies on Acute, Epidemic Poliomyelitis.] Ann. Inst. Pasteur (Paris). 27(9):718-749.
Consider the work of Flexner and Clark (1911) too artificial. Mechanical transmission of infantile paralysis by flies possible, but not proven. ABSTRACT: RAE-B 1:216, 1913.

710. Kneeland, S. 1879 *Musca domestica* (?) in the Philippines. Proc. Boston Soc. Nat. Hist. 20:121.
States that 25 years previous to 1878, there was not a housefly in the Philippines!

711. Knipling, E. F. 1946 DDT to Control Insects Affecting Man. J. Econ. Ent. 39:360-366.
Both space sprays and residual sprays may be employed against flies. DDT is also an important adjunct for pyrethrum in liquified gas aerosols.

712. Knipling, E. F. 1948 Insect Control Investigations of the Orlando, Fla. Laboratory during World War II. From the Smithsonian Report for 1948(3968):331-348, 6 pls.
One section deals with researches using flies. ABSTRACT: BA 25:1146, 1951.

713. Knipling, E. F. 1949 DDT Developments. Further Light on Recent DDT Failures against the House Fly. DDT Bans Discussed. Soap and Sanit. Chem. 25(7):107, 109, 111, 131.
ABSTRACT: RAE-B 39:7, 1951.

Knipling, E. F., joint author. *See* Bruce, W. G., et al., 1936; Lindquist, A. W., et al., 1945; Madden, A. H., et al., 1945, 1946.

714. Kobayashi, H. 1919 Chosen no Hai (Daiichi Hoboku). [Flies in Korea. Report I.] Chosen Igakukwai Zasshi. [Journal of the Korean Medical Society.] Seoul. No. 24, 29 pp. Apr. 1919.
Collected 400,000 flies in city of Seoul. Two peaks, one in June, one in September. Latrines exposed to sunshine produce more flies than those in shade. ABSTRACT: RAE-B 7:142, 1919.

715. Kobayashi, H. 1921 Overwintering of Flies. Jap. Med. World 1(3): 11-14.
Eggs least resistant to cold; adults most tolerant; larvae and pupae intermediate. Winter temperatures prevent breeding *out of doors.* ABSTRACT: RAE-B 9:180, 1921.

716. Kobayashi, H. 1922 On the Further Notes of the Overwintering of Flies. Jap. Med. World 2(7):1-4.

717. Kobayashi, H. 1922 Further Notes of the Overwintering of Flies. Jap. Med. World 2(7):193-196.
M. domestica hibernates in the adult stage, both males and females surviving. ABSTRACT: RAE-B 10:206, 1922.

718. Kobayashi, H. 1929 General Survey on the Seasonal Prevalence of the House-fly in Chosen. First Report: Researches during 1928. Acta Medicinalia in Keijo 12(2):59-65. Keijo Imperial Univ.
Collected in 18 localities. Flies active in winter in many places. A low point in mid summer, which brings highest temperature. ABSTRACT: RAE-B 18:44, 1930.

719. Kobayashi, H. 1930 Study on the Seasonal Prevalence of House Flies in Chosen (Korea). Trans. 7th Congr. Far East Ass. Trop. Med. 1927, iii pp. 186-195, 10 charts. Calcutta.
Life cycle requires 1 month in summer, 2-5 months in winter. Minimum pre-oviposition period, 3 days. *M. domestica* increases in houses earlier than *Muscina* or *Fannia*. ABSTRACT: RAE-B 18:136, 1930.

720. Kobayashi, H. 1934 The Influence of Foods on the Fecundity of *Musca domestica.* Keijo J. Med. 5(1):36-67. Chosen.
Saccharose or soluble starch necessary for longevity; protein or peptone for maturation of ovary. ABSTRACT: RAE-B 22:154, 1934.

721. Kobayashi, H. 1934 General Survey on the Seasonal Prevalence of the House-fly in Chosen. Second Report: Research during 1929. Keijo J. Med. 5(2):69-76, 20 charts.
Collections from 20 localities. ABSTRACT: RAE-B 22:222, 1934.

722. Kobayashi, H. 1935 The Influence of Temperature upon the Development of Larvae of *Musca domestica.* Trans. Dynam. Develop. 10: 385-395. Moscow. (Russian summary.)
At optimum temperature (25°-30°C = 77°-86°F), development required 7-12 days. Larvae reared on by-product of soy beans. ABSTRACT: RAE-B 24:3, 1936.

723. Kobayashi, H. 1940 [Diapause of *Fannia canicularis.*] Zool. Mag. 52(3):118-119. Tokyo. (In Japanese.)
Some eggs laid Oct.-Nov. produce adults in December; others in February, March or April. Indicates a diapause in the prepupal stage of certain individuals.

724. Kobayashi, H. 1940 [Passing Winter in Flies.] Rept. Jap. Ass. Adv. Sci. 15(2):233-236. Tokyo. (In Japanese.)

In Korea, with *M. domestica*, oviposition and larval development occur down to 15°-16°C (59°-60·8°F). ABSTRACT: RAE-B 29:33, 1941.

725. Kobayashi, H. and **Mizushima, H. 1937** The Relation between the Laboratory Temperature and the Development of Flies. Keijo J. Med. 8(1):19-39.

Uses the hyperbola: $y = a \div (x - \alpha)^b$, where x is laboratory temperature, y is length of development in days, α is threshold temperature, a (primary constant for *M. domestica*) is 126·795 and b (secondary constant for *M. domestica*) is 0·96921. ABSTRACT: RAE-B 25:245, 1937.

726. Kobayashi, S. 1909 Nipponsan kajo no hassei oyobi shusei ni tsuite. [The Metamorphosis and Habits of the Japanese House-fly.] Dobuts Z. Tokyo 21:335-341.

727. Kober, G. M. 1905 Report on the Prevalence of Typhoid in the District of Columbia. Report Health Officer, D. C., for year ending June 30, 1905. pp. 253-292.

728. Koch, R. 1883 Bericht über die Thätigkeit der deutschen Cholera-Kommission in Aegypten und Ostindien. [Report on the Action of the German Cholera-Commission in Egypt and the East Indies.] Wien. Med. Wochenschr. 52:1548-1551.

Recognized two forms of ophthalmia, probably carried by flies.

Kollros, J. J., joint author. *See* Savit, J., et al., 1946.

728a. Koster, H. 1817 Travels in Brazil in the Years from 1809 to 1815. Philadelphia.

Pages 235-236 of Volume II cited by Gudger, E. W. 1910. (Reference No. 475.) Concerns probable transmission of yaws by flies.

Kostich, D., joint author. *See* Simitch, T., et al., 1937.

729. Kowalevsky, A. 1887 Zur Embryonalentwicklung der Musciden. [On the Embryological Development of Muscidae.] Biol. Centralbl. 6: 49-54.

A good description of early stages, for that time. ABSTRACT: J. Roy. Micro. Soc. (Ser. 2)6:429:430.

730. Kowalevski (y), A. 1887 Beiträge zur Kenntniss der nachembryonalen Entwicklung der Musciden. [Contribution to the Knowledge of post-embryonic Development in Muscidae.] Zeit. Wiss. Zool. 45: 542-594, pls. 26-30.

Kowalik, R., joint author. *See* Nicewicz, N., et al., 1946.

731. Kozhanchikov, I. V. 1946 On the Lower Thermal Limit in the Development of Insects. Compt. Rend. Acad. Sci. URSS (N.S.) 51(3): 241-244. Moscow.

Used pupae of *M. domestica*. Concludes that there is a threshold, a little above 5°C, below which developmental processes cease completely. ABSTRACT: RAE-B 35:144-145, 1947.

732. Kozhanchikov, I. V. 1947 [Nutritional Value of Proteins in the growth of Blow-fly Larvae.] Rev. Ent. URSS 28(3-4): 57-63. Moscow. (In Russian.)

ABSTRACT: RAE-B 36:123, 1948.

733. Kraepelin, K. 1883 Zur Anatomie und Physiologie des Rüssels von *Musca*. [On the Anatomy and Physiology of the Proboscis of *Musca*.] Zeit. Wiss. Zool. 39:683-719.

Kraepelin's studies were chiefly carried out on species now referred to the genus *Calliphora*.

734. Kramer, S. D. 1915 The Effect of Temperature on the Life Cycle of *Musca domestica* and *Culex pipiens*. Science 41(1067):874-877.
For *M. domestica*, elevating temperature from 20° to 30° cut in half the time between egg and adult. ABSTRACT: RAE-B 3:139, 1915.

Kraneveld, F. C., joint author. *See* Nieschulz, O., et al., 1929.

Krijgsman, B. J., joint author. *See* Dresden, D., et al., 1948.

Krishna Rao, K. S., joint author. *See* Acharya, C. N., et al., 1945, 1946.

735. Krontowski, A. 1913 Zur Frage über die Typhus und Dysenterieverbreitung deuch Fliegen. [On the Question of the Distribution of Typhoid and Dysentery by Flies.] Centralbl. Bakt. Parasit. Infekt. (Jena). 68:586-590.
Infected larvae of *M. domestica* and others with *B. dysenteriae* and allowed them to pupate. Feces of resulting flies gave negative results. ABSTRACT: RAE-B 1:117-118, 1913.

736. Krusé, C. W. 1948 The Airplane Application of DDT for Emergency Control of Common Flies in the Urban Community. Publ. Health Repts. 63(48):1535-1550.
Used 0·5 lb. of DDT per acre, applied as a 30 per cent suspension in Velsicol AR-60. Released Material at 100 to 150 feet elevation. ABSTRACTS: BA 23:1750, 1949; RAE-B 39:107, 1951.

737. Kubo (Kametaro). 1920 [Common Species of Flies in Houses in Manchuria.] Tokyo Iji Shinshi, No. 2180, pp. 1085-1087. (In Japanese.)
In the town of Eiko, *M. domestica* account for 98·58 per cent of the flies (8 species) collected in September and October in food shops. ABSTRACTS: TDB 18:22, 1921; RAE-B 10:45, 1922.

Kudo, R., joint author. *See* Noguchi, H., et al., 1917.

738. Kuhn, P. 1922 Untersuchungen über die Fliegenplage in Deutschland. [Researches on the Fly Problem in Germany.] Centralbl. Bakt., Parasit., Infekt. (Jena). Ite Abt. Orig. 88(3):186-204.
In Alsace and Baden, *M. domestica* as abundant in dwellings as in pig sties. Is a pest from April or May until November or December, with the peak in September. *Empusa muscae* perhaps a controlling factor. ABSTRACT: RAE-B 10:160-161, 1922.

739. Kuhns, D. M. and **Anderson, T. G. 1944** A Fly-borne Bacillary Dysentery Epidemic in a Large Military Organization. Am. J. Publ. Health. 34:750-755.
Isolated *Shigella paradysenteriae* from nine lots of flies collected in military kitchens and latrines.

740. Kulagin, N. M. 1916 [*Musca domestica* L. On the Question of its Control.] [Friend of Nature.] Petrograd. 11(3-4 & 6-7), pp. 93-100 & 189, 194.
Given as one of several lectures on the control of epidemic diseases. ABSTRACT: RAE-B 5:15, 1917.

741. Kuzina, O. S. 1936 [Fertilität und präimeginale Mortalität bei *Musca domestica* L.] Med. Parasit. Moscow 5(3):329-339. (In Russian, with German summary.)
Insufficient nutrition for larvae reduces the weight of resulting pupae and the size and fertility of adult flies. ABSTRACT: RAE-B 25:111, 1937.

742. Kuzina, O. S. 1938 [The Choice of Egg-laying by the House Fly (*Musca domestica* L.).] Med. Parasit. 7(2):244-257. (In Russian, with English summary.)
With all manure equally fresh, 70·70 per cent of visits were to horse dung, 23·83 per cent to pig, 5·47 per cent to cow, but old pig dung (kept covered) was most attractive of all. ABSTRACT: RAE-B 26:242, 1938.

743. **Kuzina, O. S. 1940** [Rôle des organes sensitife chez la *M. domestica* L. dans la recherche du fumier et dans la ponte des oeufs.] Med. Parasit. 9(4):340-349. (In Russian.)
Flies directed to a suitable medium almost entirely by smell, but taste is a necessary stimulant for the deposition of eggs. ABSTRACT: RAE-B 31:125, 1943.

744. **Kuzina, O. S. 1942** [On the Gonotrophic Relationships in *Stomoxys calcitrans* L. and *Haematobia stimulans* L.] Med. Parasit. 11(3): 70-78, 4 figs.
A confirmation, for these species, of the eight stages of ovarian development, as previously worked out for *M. domestica.* Goes back to Christopher's work on the mosquito. ABSTRACT: RAE-B 32:52, 1944.

Kuzina, O. S., joint author. *See* Derbeneva-Ukhova, V. P., et al., 1938; Smirnov, E., et al., 1933.

745. **Kvasnikova, P. A. 1931** [Flies Observed in Human Dwellings and Outhouses in the Town of Tomsk.] Wiss. Ber. Biol. Fak. Tomsk. St.-Univ. 1(1):9-47. (In Russian.)
Think larvae of *Calliphora* perhaps predatory on those of *M. domestica.* Direct relation between abundance of the latter and number of cases of dysentery in Tomsk. ABSTRACT: RAE-B 19:259, 1931.

746. **Laake, E. W. 1949** Livestock Parasite Control Investigations and Demonstrations in Brazil. J. Econ. Ent. 42:276-280.
Includes work with DDT against *M. domestica.* ABSTRACT: RAE-B 38:85, 1950.

747. **Laake, E. W., Parman, D. C., Bishopp, F. C.** and **Roark, R. C. 1931** The Chemotropic Responses of the House Fly, the Green Bottle Flies and the Black Blowfly. U.S. Dept. Agric. Techn. Bull. 270. 10 pp.
Pine tar oils are the most effective repellents. Gives formula for protecting meat from flies in camps. ABSTRACT: TDB 31:60, 1934.

Laake, E. W., joint author. *See* Parman, D. C., et al., 1927; Bishopp, F. C., et al., 1919, 1921, 1925.

LaForge, F. B., joint author. *See* Barthel, W. F., et al., 1946; Haller, H. L., et al., 1942.

748. **Laing, J. 1935** On the Ptilinum of the Blow-fly (*Calliphora erthrocephala*). Quart. J. Micr. Sci. 77:497-521.
Good morphological work, applicable to all Muscoidea.

749. **Lal, R. B., Ghosal, S. C.** and **Mukherji, B. 1939** Investigations on the Variation of Vibrios in the House Fly. Indian J. Med. Res. 26(3): 597-609.
Relates to *Vibrio cholerae.* Change of form, in an endemic region, could lead to species becoming microscopically unrecognizable though still present. ABSTRACTS: TDB 36:896, 1939; RAE-B 27:168, 1939.

Lal, R. B., joint author. *See* Gill, C. A., et al., 1931.

750. **Lamb, C. G. 1922** The Geometry of Insect Pairing. Proc. Roy. Soc. (Ser. B) 94:1-11.
Uses Diptera as examples.

751. **Lamborn, W. A. 1935** Ann. Rept. of the Medical Entomologist for 1934. Ann. Med. Rept. Nyasaland for 1934:65-69. Zomba.
Raises question as to transmission of *Bacillus leprae* by *M. sorbens,* which feeds on sores. ABSTRACT: RAE-B 24:187, 1936.

752. **Lamborn, W. A. 1935** The Passage of Leprosy Bacilli through the Intestine of the Fly, *Musca sorbens* Wied. Trans. R. Soc. Trop.

Med. Hyg. 29(1):3-4.
Flies fed on leprosy sores showed acid-fast bacilli in vomit 5 days later and in feces on seventh or eighth day. ABSTRACT: RAE-B 23:226, 1935.

753. Lamborn, W. A. 1936 Ann. Rept. of the Medical Entomologist for 1935. Ann. Med. Rept. Nyasaland, for 1935. pp. 50-52. Zomba.
Concerns possible transmission of *Bacillus leprae* by *M. sorbens* Wied., also relation of flies to yaws (*Treponema pertenue*). ABSTRACT: RAE-B 25:17, 1937.

754. Lamborn, W. A. 1936 The Experimental Transmission to Man of *Treponema pertenue* by the Fly *Musca sorbens* Wd. J. Trop. Med. Hyg. 39(20):235-239.
Transmission takes place as a result of contamination of proboscis by regurgitation. *Bacillus leprae* and *Trypanosoma rhodesiense* under investigation. ABSTRACT: RAE-B 25:44, 1937.

755. Lamborn, W. A. 1937 The Hematophagous Fly, *Musca sorbens*, Wied. in Relation to the Transmission of Leprosy. J. Trop. Med. Hyg. 40(4):37-42.
Is non-biting, but is attracted to slightest break in skin surface. Alimentary tract of fly gives acid fast reaction of *Bacillus leprae*. ABSTRACT: RAE-B 25:155, 1937.

756. Lamborn, W. A. 1939 Ann. Rept. of the Medical Entomologist for 1938. Ann. Med. Sanit. Rept. Nyasaland for 1938:40-48. Zomba.
Musca sorbens Wd. allowed to feed on fresh tubercular sputum. Organisms passed up to 15 days, apparently unchanged. Similar results with dried sputum, up to 79 days old. ABSTRACT: RAE-B 28:107, 1940.

757. Lamborn, W. A. 1940 Ann. Rept. of the Medical Entomologist for 1939. Ann. Med. Sanit. Rept. Nyasaland for 1939:26-31. Zomba.
Gut content of flies infected with TB bacillus, injected into guinea pigs, caused death of the latter from generalized tuberculosis. ABSTRACT: RAE-B 28:251, 1940.

758. Lamborn, W. A. and **Howat, C. H. 1936** A Possible Reservoir Host of *Trypanosoma rhodesiense*. Brit. Med. J. No. 3935, June 6, pp. 1153-1155.
Dogs were infected by *Musca sorbens* which fed on blood of a patient and was then transferred to an incision on the ear of the dog. Animal died 48 days later. ABSTRACT: RAE-B 25:17, 1937.

Lamborn, W. A., joint author. *See* Thomson, J. G., et al., 1934.

759. Lambotte, U. 1905 Insectes et maládies infectieuses. [Insects and Infectious Diseases.] Ann. Soc. Méd.—Chir. Liège, 44:371-389.

La Mer, V. K., joint author. *See* Brescia, F., et al., 1946.

760. Langfield, M. 1905 The Rôle of Insects in the Transmission of Disease. Trained Nurse — —, New York, Vol. 35, pp. 195, 263 and 336.

Lengford, G. S., joint author. *See* Cory, E. N., et al., 1947.

761. Larsen, E. B. 1943 Problems of Heat Death and Heat Injury. Experiments on Some Species of Diptera. K. Danske Vidensk. Selskab. Biol. Medd. 19(3):1-52.
Studied *M. domestica* and four other muscoids. Eggs most sensitive to high temperatures; larvae intermediate; pupae least so. ABSTRACT: BA 21:545, 1947.

762. Larsen, E. B. 1943 The Influence of Humidity on Life and Development of Insects. Vidensk. Medd. Dansk. Naturh. Foren. Bd. 107: 127-184.
An elaboration of the work cited above. Very thorough and informative. At moderate temperatures and humdities, effect of desiccation on length

of life, mortality and length of development believed due solely to increased evaporation.

763. Larsen, E. B. and **Thomsen, M. 1940** The Influence of Temperature on the Development of some Species of Diptera. Vidensk. Medd. Dansk. Naturh. Foren. Bd. 104, pp. (reprint) 1-75.
Chiefly concerned with *M. domestica.* Preimaginal growth very rapid. Weight is multiplied 54 times in 4 days at 25C°. Weight loss in post-alimentary period due to water output by larva.

764. Latreille. 1795 [*Astoma parasiticum*, parasitic mite of houseflies.] Magazin Encyclopedique Vol. IV, p. 15.

765. Laüger, P., Martin, H. and **Müller, P. 1944** Ueber Konstitution und neuen synthetischen insektentötenden Stoffen. Helv. Chim. Acta 27:892-928.
An account of the several years of work which led up to the discovery of DDT as an insecticide.

766. Läuger, P., Pulver, R., Montigel, C., Wiesmann, R. and **Wild, H. 1946** Mechanism of Intoxication of DDT Insecticides in Insects and Warm-blooded Animals. Address delivered July 31, 1945. 24 pp., Geigy Co., Inc., New York.
Included here because of the historical importance of DDT.

767. Laurans, R. 1946 Note sur l'action toxique du sulfure de polychlorocyclane a l'egard des aphaniptères. [Note on the toxic action of polychlorobenzene (in talc) against "Siphonaptera".] Bull. Soc. Path. Exot. 39(7-8):295-299.
Killed both fleas and houseflies in Petri dishes. ABSTRACT: BA 21:2061, 1947.

Lavine, I., joint author. *See* Zimmerman, O. T., et al., 1946.

768. Lawrence, S. M. 1909 Dangerous Dipterous Larvae. Brit. Med. J. 1:88.

769. Lebailly, C. 1924 Le mouches ne jouent pas de rôle dans la dissémination de la fièvre aphteuse. [Flies do not play a rôle in the dissemination of foot and mouth disease.] Compt. Rend. Hebdom. Acad. Sci. 179(21):1225-1227.
Experiments decisively refuted the contention that either *M. domestica* or *Muscina stabulans* were carriers of the infection.

770. Leboeuf, A. 1912 Dissémination du bacille de Hansen par le mouche domestique. [Distribution of Hansen's bacillus by the domestic fly.] Bull. Soc. Path. Exot. 5(10):860-868.
Fly can take up enormous numbers of bacilli, which seem neither to degenerate not multiply in its alimentary tract. Many found in fresh fecal specks. ABSTRACT: RAE-B 1:36, 1913.

771. Leboeuf, A. 1914 La lèpre en Nouvelle Calédonie et dépendences. [Leprosy in New Caledonia and Dependencies.] Ann. Hyg. Med. Colon. 17(1):177-197.
Nineteen of thirty-six flies captured in Leper's infirmary yielded Hansen's bacillus. ABSTRACT: RAE-B 2:80, 1914.

772. Leclerc, M. 1946 Observations écologiques sur les mouches de nos habitations. [Ecological Observations on the Flies in our Houses.] Rev. Franç. Ent. 13(2):76-79.
In climate of Liége, Belgium, houses serve as hibernation places for flies emerging in the autumn. ABSTRACT: BA 21:756, 1947.

773. Leclercq, M. 1948 La Transmission de la poliomyélite par les insectes. [The Transmission of Poliomyelitis by Insects.] Rev. Méd. Liége 3(7):154-156; (8):197.
Is chiefly a review of the American literature supporting the thesis that

houseflies may act as vectors of polio virus. ABSTRACT: BA 24:244, 1950.

773a. Leclercq, M. 1948 Encore à propos de la transmission de la poliomyélite par les mouches. [More concerning the transmission of poliomyelitis by flies.] Rev. Méd. Liége. 3(11):279-281.
Includes illustrations of *M. domestica* (male and female), *Protophormia terrae-novae*, *Lucilia sericata*, and *Stomoxys calcitrans*.

773b. Leclercq, M. 1950 La transmission de la poliomyélite (paralysie infantile) par les mouches domestiques. [The transmission of poliomyelitis (infantile paralysis) by domestic flies.] Nat. Belges Bull. Mens. 31(1): 7-12.
A review of the subject. ABSTRACT: BA 26:1529, June 1952.

774. Ledingham, J. G. 1911 On the Survival of Specific Microorganisms in Pupae and Imagines of *Musca domestica* raised from Experimentally Infected Larvae. Experiments with *Bacillus typhosus*. J. Hygiene (Cambridge) 11(3):333-340.

775. Ledingham, J. C. G. 1920 Dysentery and Enteric Disease in Mesopotamia from the Laboratory Standpoint. J. Roy. Army Med. Corps. London 34(4):306-320.
Flies abundant in April-May, and again in November. Flies almost disappear during hot months and dysentery drops, but not proportionately. ABSTRACT: RAE-B 8:110, 1920.

Ledingham, K. (J?) C. G., joint author. *See* Morgan, H. de R., et al., 1909.

776. Lefroy, H. M. 1916 The Control of Flies and Vermin in Mesopotamia. Agric. J. India 11(4):323-331.
Claims that "almost every disease in this country is carried by flies or water". Recommends various sanitary procedures especially in relation to military problems. ABSTRACT: RAE-B 5:23, 1917.

777. Lefroy, H. M. 1919 Fly Sprays. Trans. Soc. Trop. Med. Hyg. 13(1): 1-9.
Discusses a formula based on pyrethrum which was made effective in hot climates by adding ½ to 2 per cent castor oil. ABSTRACT: RAE-B 7:161, 1919.

Lefroy, H. M., joint author. *See* Jackson, A. C., et al., 1917.

778. Léger, L. 1903 Sur quelques Cercomonadines nouvelles ou peu connues parasites de l'intestin des insectes. [Concerning Certain New or Little Known Cercomonad Parasites of the Intestine of Insects.] Arch. f. Protistenk. 2:180-189, 4 figs.
Flies among the hosts examined.

779. Leidy, J. 1871 Flies as a Means of Communicating Contagious Diseases. Proc. Acad. Nat. Sci. Philadelphia for 1871, p. 297.
Leidy had observed gangrene during American Civil War, and had also studied the fly's habit of exuding, from the proboscis, material recently taken in while feeding.

780. Leidy, J. 1874 On a Parasitic Worm of the House-fly. Proc. Acad. Nat. Sci. Philadelphia. 26:139-140.
Probably a species of *Habronema*.

781. Leĭkina, L. I. 1942 [The Rôle of Various Substrata in the Breeding of *Musca domestica*.] Med. Parasit. 11(1-2):82-86. (In Russian.)
Female flies much attracted to horse manure for oviposition, though it is the least favorable of several types for rapid, normal growth of larvae. ABSTRACT: RAE-B 32:5, 1944.

782. Leon, N. 1920 Quelques observations sur les Pédiculides. [Some observations on the Pediculidae.] J. Parasit. 6(3):144-147.
Relates to typhus in Rumania. Experimental proof that lice can be carried by flies, and later attach themselves to a human host. ABSTRACT: RAE-B 8:145, 1920.

783. Lepage, H. S. and **Giannotti, O. 1944** Experiências com o DDT. Biológico 10(11):353-366. São Paulo.
Residual coatings in petri dishes caused paralysis of flies in 20 minutes, followed almost immediately by death. ABSTRACT: RAE-B 34:170, 1946.

Levaditi, C., joint author. *See* Kling, C., et al., 1913.

784. Lever, R. J. A. W. 1933 Status of Economic Entomology in the British Solomon Islands. Bull. Ent. Res. 24(2):253-256.
M. vicina, unknown there 20 years before, now very common on plantations where there are horses and cattle. ABSTRACT: RAE-B 21:200, 1933.

785. Lever, R. J. A. W. 1934 Entomology and Agriculture in the British Solomon Islands. Trop. Agric. 11(2):36-37. Trinidad.
Lyperosia exigua de Meij., pest of cattle imported from New Britain *circum* 1923, makes wound punctures at which *M. vicina* later feeds on oozing blood. ABSTRACT: RAE-B 22:65, 1934.

786. Lever, R. J. A. W. 1938 Entomological Notes 3. A Javenese Beetle to Control Houseflies. Agric. J. Fiji 9(4):15, 18. Suva.
Hister chinensis, introduced to control *M. domestica vicina*, tunnels in dung in pursuit of maggots. ABSTRACT: RAE-B 27:155, 1939.

787. Lever, R. J. A. W. 1944 Entomological Notes. Agric. J. Fiji 15(2): 45-50. Suva.
Concerns *M. domestica vicina*. Pupae found in cow dung about twice as numerous as those in pig dung. (Horses rare in area studied.) ABSTRACT: RAE-B 33:104, 1945.

788. Lever, R. J. A. W. 1945 Entomological Notes. Agric. J. Fiji. 16(3): 88-90. Suva.
Pachycrepoideus dubius Ashmead (Hymenoptera) reared from pupae of *M. domestica vicina*. New record for Fiji. ABSTRACT: RAE-B 35:19, 1947.

789. Lever, R. J. A. W. 1946 Entomological Notes. Agric. J. Fiji 17(1): 9-15. Suva.
Reports first use of DDT in Fiji. Used 5 per cent solution in benzene, diluted with water. Window panes remained toxic for 10 weeks. ABSTRACT: RAE-B 36:139, 1948.

790. Levick, G. T. 1927 The House Fly. J. Dept. Agric. Vict. 25(11):669-672.
Stresses control of fly breeding, and preventing of access to human food. ABSTRACT: RAE-B 16:102, 1928.

791. Levy, E. C. and **Freeman, A. E. 1908** Certain Conclusions Concerning Typhoid Fever in the South, as Deduced from a Study of Typhoid Fever in Richmond, Va. Old Dominion J. Med. and Surg., Vol. 8. (Reprint, 39 pp., 3 maps, 3 charts.)
Flies considered part of the aetiology.

792. Levy, E. C. and **Tuck, W. T. 1913** The Maggot-trap, — A New Weapon in our Warfare against the Typhoid Fly. Am. J. Publ. Health, 3(7): 657-660.
Place manure in receptacles from which larvae, seeking drier situation for pupation, can escape through wire gauze sides and bottom, then fall into pans provided.

Lewis, F. C., joint author. *See* Cox, G. L., et al., 1912.

Lhéritier, A., joint author. *See* Sergent, Ed., et al., 1918.

793. Liddo, S. 1933 Decistamento delle amebe nel tubo digerente delle mosche. [Fragmentation of the Cyst of Ameba in the Digestive Tract of the Fly.] Boll. Accad. Pugliese Sci. 1933, No. 2, p. 109.
Cysts of the *limax* type, particularly. Reasons that fly gut could be a sustaining environment for survival of organism which would have perished by desiccation outside. ABSTRACTS: Boll. Inst. Sieroter. Milan. 12(10): 815, 1933; RAE-B 22:20, 1934.

794. Liebermann, A. 1925 Korrelation zwischen den antennalen Geruchsorganen und der Biologie der Musciden. [Correlation between the Antennal Sense Organs and the Biology of Muscidae.] Zeit. Morph. u. Oekol. Tiere 5(1):1-97, 19 figs.
A detailed account of the antennal sense organs and their function. ABSTRACT: RAE-B 14:24, 1926.

795. Ligue Sanitaire Francaise contre la Mouche et le Rat. 1915 La lutte contre la mouche. Bull. No. 5 (issued Aug. 25, 1915). Paris.
States that *M. domestica* (along with *Muscina* and *Calliphora*) oviposits on corpses soon after death. *Lucilia* and *Sarcophaga* wait for putrefaction. ABSTRACT: RAE-B 4:17-18, 1916.

796. Ligue Sanitaire Francaise. 1916 Circulars 3 (May 1), 4 (May 15) and 5 (June 1). Paris.
Educational pamphlets, all concerned with fly control. ABSTRACTS: RAE-B 4:17 and 179, 1916.

797. Lilly, J. H. 1931 A Preliminary Study of the Presence of Bacteria in the Blood of the House Fly, *Musca domestica*. Unpublished Thesis, University of Wisconsin. 30 pp.
One or more species normally present.

798. Lindquist, A. W. 1936 Parasites of the Horn Fly and other Flies Breeding in Dung. J. Econ. Ent. 29(6):1154-1158.
The Pteromalid, *Spalangia muscidarum* var. *stomoxysiae* Gir. readily attacks pupae of *M. domestica*. ABSTRACT: RAE-B 25:84, 1937.

799. Lindquist, A. W., Jones, H. A. and **Madden, A. H. 1946** DDT Residual-type Sprays as Affected by Light. J. Econ. Ent. 39:55-59.
Solutions, emulsions and suspensions applied to various surfaces, which were then exposed to both sunlight and ultra violet. Effectiveness of insecticide was reduced. ABSTRACTS: RAE-B 35:90, 1947; BA 20:1758, 1946.

800. Lindquist, A. W., Madden, A. H. and **Wilson, H. G. 1947** Pretreating House Flies with Synergists before Applying Pyrethrum Sprays. J. Econ. Ent. 40:426-427.
High knock-down when 0·1 per cent pyrethrins were applied 1, 2 and 4 hours after pretreatment. Knock-down near zero when procedure is reversed. ABSTRACT;: BA 22:994, 1948; RAE-B 37:81, 1949.

801. Lindquist, A. W., Madden, A. H., Wilson, H. G. and **Jones, H. A. 1944** The Effectiveness of DDT as a Residual Spray against Houseflies. J. Econ. Ent. 37:132-134.
Treats of different solvents on different surfaces, dosage, duration of exposure, effect of temperature on knock-down and related matters. ABSTRACT: RAE-B 32:200, 1944.

802. Lindquist, A. W., Madden, A. H., Wilson, H. G. and **Knipling, E. F. 1945** DDT as a Residual-type Treatment for Control of Houseflies. J. Econ. Ent. 38:257-261.
Tested in kerosene, in water paint and in other ways. Because kerosene leaves visible deposit on furniture and walls, should be restricted to barns, etc. ABSTRACTS: RAE-B 34:133, 1946; BA 19:2517, 1945.

803. Lindquist, A. W., Schroeder, H. O., and **Knipling, E. F. 1945** Concentrated Insecticides. Preliminary Studies of the Use of Concentrated Sprays against Houseflies and Mosquitoes. Soap and Sanit. Chem. 21(7):109, 111, 113, 119.

DDT at 96·8 mg. per 1000 cu. ft. equally effective with 1 per cent and 16 per cent solutions. Other data. ABSTRACT: RAE-B 34:168, 1946.

804. Lindquist, A. W., Travis, B. V., Madden, A. H. and **Jones, H. A. 1946** Aerosol Formulation: Laboratory Tests with Pyrethrum and DDT Aerosols against the Common Malaria Mosquito and the Housefly. Soap and Sanit. Chem. 22(5):135, 137, 139, 141, 143.

DDT requires an auxiliary solvent, such as Cyclohexanone. Good knock-down effect with 5 per cent DDT plus 0·4 per cent pyrethrins. ABSTRACTS: RAE-B 37:65, 1949; BA 21:2027, 1947.

805. Lindquist, A. W., Travis, B. V., Madden, A. H., Schroeder, H. O., and **Jones, H. A. 1945** DDT and Pyrethrum Aerosols to Control Mosquitoes and Houseflies under Semi-practical Conditions. J. Econ. Ent. 38:255-257.

Addition of motor oil produces desired effect with reduced concentration of DDT and cyclohexanone. ABSTRACTS: BA 19:2517, 1945; RAE-B 34:132, 1946.

806. Lindquist, A. W. and **Wilson, H. G. 1948** Development of a Strain of Houseflies Resistant to DDT. Science 107:276.

Selective breeding produced a resistant strain in 14 generations. ABSTRACT: BA 22:1955, 1948.

807. Lindquist, A. W., Wilson, H. G., Schroeder, H. O. and **Madden, A. H. 1945** Effect of Temperature on Knock-down and Kill of Houseflies Exposed to DDT. J. Econ. Ent. 38:261-268.

Temperatures may influence effectiveness of DDT against a given species, as in different parts of the world. *M. domestica*, however, is susceptible over a wide temperature range. ABSTRACTS: RAE-B 34, 134, 1946; BA 19:2517, 1945.

Lindquist, A. W., joint author. *See* Hoffman, R. A., et al., 1949; Fairchild, H. E., et al., 1949; Cristol, S. J., et al., 1946; McDuffie, W. C., et al., 1945, 1946; Madden, A. H., et al., 1945, 1946, 1947; Roth, A. R., et al., 1949; Schroeder, H. O., et al., 1945, 1946, 1948.

Lindsay, D. R., joint author. *See* Melnick, J. L., et al., 1947; Watt, J., et al., 1948.

808. Lingard, A. and **Jennings E. 1906** Some Flagellate Forms found in the Intestinal Tracts of Diptera and other Genera. London, Adlard and Son. 25 pp., 5 pls.

Repeats Prowazek's earlier error, claiming double origin for the flagellum. Gives seasonal data for India. Some confusion with *Crithidia*.

Link, K. P., joint author. *See* Ikawa, M., et al., 1945.

809. Linné, Carl von (Linnaeus). 1758 Systema Naturae per Regna Tria Naturae Secundum Classes, Ordines, Genera, etc. 10th Edition.

Included here because of its historical importance to taxonomy.

810. Linstow, von. 1875 Beobachtungen an neuen und bekannten Helminthen. [Observations on New and Known Helminths.] Arch. f. Naturgesch. (1875):183-207.

Described a small nematode which he named *Filaria stomoxeos*, from the head of *Stomoxys calcitrans*. (Was considered by later authors to be a *Habronema*.)

Liu, Yu-Su, joint author. *See* Hayes, W. P., et al., 1947.

811. **List, G. M.** and **Payne, M. G. 1947** Insecticidal Action of 1-Trichloro-2, 2-Bis (p-Bromophenyl) ethane (Colorado 9). Science 105(2720):182-183.
Made by condensing one molecule of chlorohydrate with 2 of bromobenzene. Kills flies at low spray level of 0·1 ml. of 5 per cent petroleum solution per cu. meter of air space.

812. **Lloyd, L. 1920** On the Reasons for the Variation in the Effects of Formaldehyde as a Poison for House-flies. Bull. Ent. Res. 11(1): 47-63.
Effect depends on freedom of exposed fluid from formic acid, and to a lesser extent, from methylamine. Gives preferred formula. ABSTRACT: RAE-B 8:175, 1920.

813. **Lochmann, R. 1899** Eine epidemische auftretende Krankeit der Stubenfliege verursacht durch *Empusa muscae*. [An Epidemically Occurring Disease of the Housefly, brought about by *Empusa muscae*.] Pharm. Reformer (Wien.). Vol. 4, p. 127.

814. **Lodge, O. C. 1916** Fly Investigation Reports IV. Some Enquiry in the Question of Baits and Poisons for Flies, being a Report on the Experimental Work Carried out During 1915. Proc. Zool. Soc. London for 1916, Pt. 3, pp. 481-518.
Gives 6 formulations for use as bait. Formalin considered the most effective poison. Mention of *Empusa*. ABSTRACT: RAE-B 4:169, 1916.

815. **Lodge, O. C. 1918** An Examination of the Sense-Reactions of Flies. Bull. Ent. Res. (London) 9(2):141-151.
Extensive experiments, including amputation of antennae, painting over the eyes, etc. No appreciable difference in response between the sexes. ABSTRACT: RAE-B 7:56, 1919.

816. **Loeb, J. 1890** Der Heliotropismus der Thiere und seine Uebereinstimmung mit dem Heliotropismus der Pflanzen. [The Heliotropisms of Animals and their Synchronization with the Heliotropisms of Plants.] Wurzburg. 118 pp., 6 figs.
Larvae of housefly manifest a strong, negative, heliotropism and move rapidly away from light.

Longwell, J. H., joint author. *See* Munro, J. A., et al., 1945.

817. **Lord, F. T. 1904** Flies and Tuberculosis. Boston Med. and Surg. J. 151:651-654.
Flies may ingest TB sputum and excrete TB bacilli, the virulence of which may last for at least 15 days.

818. **Lörincz, F.** and **Makara, G. 1935** Observations and Experiments on Fly Control and the Biology of the House Fly. League of Nations Health Org. C.H./Hyg. rur./E.H. 5, 13 pp. multigraph. Geneva.
Investigations in Hungary. Almost any decomposing material suitable for breeding. Nearly full sized flies bred from tobacco powder! ABSTRACT: RAE-B 24:45, 1936.

819. **Lörincz, F.** and **Makara, G. 1936** Investigations in the Fly Density in Hungary in the Years 1934 and 1935. Quart. Bull. Health. Org. League of Nations 5(2):219-227.
Typhoid curve follows fly curve with a lag of about 4 weeks, e.g. fly peak, Aug.-Oct.; typhoid peak, Sept.-Nov. ABSTRACT: RAE-B 24:279, 1936.

820. **Lörincz, F., Szappanos, G.** and **Makara, G. 1936** On Flies Visiting Human Faeces in Hungary. Quart. Bull. Health Org. League of Nations 5(2):228-236.
Of 4,567 specimens of *M. domestica* taken, 13·62 per cent were captured

on feces, 86·38 per cent on over-ripe or decomposing fruit. ABSTRACT: RAE-B 24:281, 1936.

Lotmar, R., joint author. *See* Wiesmann, R., et al., 1949.

821. Loughnan, W. F. M. 1930 Bourgalt's Cattle-fly Trap. J. Roy. Army Med. Corps 54(3):208-211.
A black bullock, preferably tailless is driven about and finally through the trap, which consists of brushes made of coconut leaves. These last 6 weeks or more. Effective against *Stomoxys* and *Musca*. ABSTRACT: RAE-B 18:156, 1930.

821a. Lowman, M. S., Gersdorff, W. A., and **Mitlin, N. 1954** Pyrethrum Flower Toxicants. Soap and Chem. Spec. 30(8):139, 141, 143, 145, 159.
Flowers fermented in closed containers for 4-6 days, six months or more before drying, retain more pyrethrins during storage than those simply dried at room temperature. Tested on *M. domestica*. ABSTRACT: RAE-B 43:167, Nov. 1954.

822. Lowne, B. T. 1884 On the Compound Vision and the Morphology of the Eye in Insects. Trans. Linn. Soc. (Zool.) Vol. 2, pt. 11.

823. Lubbock, J. 1871 The Fly in its Sanitary Aspect. Lancet. Vol. II, p. 270.
Summarizes Sir John Lubbock's speech in Parliament. Includes quotation from a current elementary school text-book: "The fly keeps the warm air pure and wholesome by its swift and zigzag flight."

824. Lumsden, L. L. and **Anderson, J. F. 1911** The Origins and Prevalence of Typhoid Fever in the District of Columbia (1909-1910). Publ. Health and Marine Hosp. Serv., Hyg. Lab. Bull. No. 78.
Evidence that flies, though not playing the major part, were of considerable importance in spreading typhoid in Washington.

825. Lumsden, L. L., Stiles, C. W. and **Freeman, A. W. 1917** Safe Disposal of Human Excreta at Unsewered Homes. U.S. Publ. Health Serv. Publ. Health Bull. No. 68, 28 pp.
Instructions and diagrams concerning best procedure under various conditions. Includes distribution of effluent by open-joint tile (Precedes the introduction of septic tanks.)

Lumsden, L. L., joint author. *See* Rosenau, M. J., et al., 1909; Stiles, C. W., et al., 1916.

Lush, R. H., joint author. *See* Fitch, J. B., et al., 1926, 1928.

Lyman, F. E., joint author. *See* Melnick, J. L., et al., 1947.

826. Lyon, H. 1915 Does the House-fly Hibernate as a Pupa? Psyche 22(4): 140-141.
Experiments with 37 lots of 100 pupae each showed that *M. domestica* cannot readily hibernate as a pupa, though it can and does emerge until the middle of winter. ABSTRACT: RAE-B 3:215-216, 1915.

Lyon, S. C., joint author. *See* Crumb, S. E., et al., 1917, 1921.

827. McAlister, L. C., jr., Jones, H. A. and **Moore, D. H. 1947** Piperonyl Butoxide with Pyrethrins in Wettable Powders to Control Certain Agricultural and Household Insects. J. Econ. Ent. 40:906-909.
A water suspension applied at the rate of 50 mg. of butoxide and 5 mg. of pyrethrins per sq. ft. of unfinished plywood, gave complete knockdown for houseflies in 30 min. or less for 10 weeks. ABSTRACT: RAE-B 37:155, 1949.

828. McDaniel, E. 1927 Flies Commonly Found in Dwellings. Circ. Mich. Agric. Expt. Sta. No. 104, 15 pp. E. Lansing.
Covers *M. domestica* and several other species of Diptera. ABSTRACT: RAE-B 16:81, 1928.

829. McDaniel, E. I. 1935 Flies and Mosquitoes Commonly Found about Michigan Homes. Michigan State College Circ. Bull. No. 144, Revised. 27 pp., illus.

830. McDaniel, E. I. 1942 Houseflies. Mich. State Coll. Ext. Bull. 239, 8 pp.
Lists 12 kinds of "houseflies", with recommendations for control.

831. McDaniel, E. I. 1945 Pest Control in Dairies. J. Milk Technol. 8(6): 338-341.
Directions for use of DDT preparations against flies, cockroaches, etc. ABSTRACT: BA 20:1013, 1946.

832. McDonnell, R. P. and **Eastwood, T. 1917** A Note on the Mode of Existence of Flies during Winter. J. Roy. Army Med. Corps. 29(1): 98-100.
Larvae, 3 feet down in old manure, also 2 feet down in mixture of earth and human feces. Pupae, 2 feet away, 1 inch under surface. Observations made in March. ABSTRACT: RAE-B 5:154, 1917.

833. MacDougall, R. S. 1906 Insects and Arachnids in Relation to the Spread of Disease. Pharm. J. (London) Ser. 4, vol. 22, p. 60.

834. McDuffie, W. C., Lindquist, A. W. and **Madden, A. H. 1946** Control of Fly Larvae in Simulated Pit Latrines and in Carcasses. J. Econ. Ent. 39:743-749.
O-dichlorobenzene and p-dichlorobenzene destroyed eggs and gave effective larval control. ABSTRACTS: RAE-B 38:86, 1948; BA 21:1810, 1947.

835. McDuffie, W. C., Lindquist, A. W. and **Wilson, H. G. 1945** Studies on the Control of Fly Larvae in Simulated Pit Latrines and in Carcasses. Comm. on Med. Res. O.S.R.D. Insect Control Comm. Rept. No. 20, 12 pp. mimeogr.
Preliminary to 1946 publication of McDuffie, Lindquist and Madden.

836. McDuffie, W. C., Madden, A. H. and **Lindquist, A. W. 1945** Further Studies on the Destruction of Fly Larvae in Animal Carcasses. Comm. on Med. Res. O.S.R.D. Insect Control Comm. Rept. No. 80, 12 pp. mimeogr.
M. domestica not important here, but reference included for completeness of the series.

837. McFarland, J. 1902 Relation of Insects to the Spread of Disease. Medicine, Vol. 8, 15 pp., 12 figs.

838. Macfie, J. W. S. 1917 The Identifications of Insects Collected at Accra during the year 1916, and other Entomological Notes. Rept. Accra Lab. for the year 1916. London, pp. 67-75, 3 pls., 3 figs.
Of all *M. domestica* collected from butcher's stalls, 42·5 per cent were heavily infected with *Herpetomonas muscae domesticae*. ABSTRACT: RAE-B 6:16, 1918.

839. McGovran, E. R. and **Fales, J. H. 1946** Toxicity of Aerosols. Rate of Movement through and Height of Suspension in a Toxic Aerosol Influences Mortality of Caged House Flies. Soap and Sanit. Chem. 22(9):127-129, 3 graphs.
Toxicity greater at lower levels, due to settling. ABSTRACT: RAE-B 37: 104, 1949.

840. McGovran, E. R., Fales, J. H. and **Goodhue, L. D. 1943** Testing Aerosols against Houseflies. Soap and Sanit. Chem. 19(9):99, 101, 103, 105, 107, 4 figs.
Describes standardized testing procedure. Includes instructive diagram of dispenser. ABSTRACT: RAE-B 32:99, 1944.

841. McGovran, E. R., Fales, J. H. and **Goodhue, L. D. 1946** New Formulations of Aerosols Dispersed by Liquefied Gases. J. Econ. Ent. 39: 216-219.
Tests on *M. domestica* and mosquitoes with low percentage of pyrethrins, plus various synergists and additives. ABSTRACT: RAE-B 35:127, 1947.

842. McGovran, E. R. and **Gersdorff, W. A. 1945** The Effect of Fly Food on Resistance to Insecticides Containing DDT or Pyrethrum. Soap and Sanit. Chem. 21(12):165, 169.
Results compared for 6 different food preparations. Susceptibility definitely affected by nutrition. ABSTRACT: RAE:B 36:188, 1948.

843. McGovran, E. R. and **Piquett, P. G. 1943** Toxicity of Thiourea and Phthalonitrile to Housefly Larvae. J. Econ. Ent. 36:936.
Each of these substances, added to breeding medium at the rate of 0·112 per cent by weight, killed all third stage larvae. ABSTRACT: RAE-B 23:155, 1944.

844. McGovran, E. R. and **Piquett, P. G. 1945** Toxicity of Benzene Hexachloride to Housefly Larvae. J. Econ. Ent. 38:719.
Effectiveness compared with that of BHC, DDT, Borax and Thiourea. ABSTRACT: RAE-B 35:55, 1947.

845. McGovran, E. R., Richardson, H. H. and **Piquett, P. G. 1944** Toxicity of DDT to Bedbugs, Cockroaches, the Mexican Bean Beetle and Housefly Larvae. J. Econ. Ent. 37:139-140.
In similar concentrations against fly larvae, thiourea was more toxic than borax; borax more toxic than DDT. ABSTRACT: RAE-B 32:206, 1944.

846. McGovran, E. R. and **Sullivan, W. N. 1942** Two Activators for Pyrethrins in Fly Sprays. J. Econ. Ent. 35:792.
Three per cent methylphenylnitrosoamine and five per cent 2, 4-diamylcyclohexanol gave increase in kill. ABSTRACT: RAE-B 31:102, 1943.

847. McGovran, E. R., Sullivan, W. N. and **Phillips, G. L. 1939** Resistance to Insecticides. The Effect of Knockdown and Light Doses on the Resistance of Houseflies to Pyrethrum Sprays. Soap and Sanit. Chem. 15(8):88-90.
M. domestica, previously paralyzed with ether, acetone or low-content pyrethrum-kerosene spray, is more resistant to high-content pyrethrum spray than normal flies. ABSTRACT: RAE-B 28:21, 1940.

McGovran, E. R., joint author. *See* Fales, J. H., et al., 1946, 1947; Goodhue, L. D., et al., 1945; Haller, H. L., et al., 1942; Gersdorff, W. A., et al., 1944, 1945.

848. MacGregor, M. E. 1917 A Summary of our Knowledge of Insect Vectors of Disease. Bull. Ent. Res. 8:155-163.
Diplococcus intracellularis (*Neisseria intracellularis*) is probably carried by flies.

849. McHardy, J. W. 1930 Report of Government Entomologist. Ann. Rept. Med. Lab. Dar es Salaam for 1928, pp. 41-52, 1 pl.
Musca can fly through a mesh one inch wide without first settling; never through mesh of ¾ inch. Such frames, coated with fly gum could keep flies from exposed food. ABSTRACT: RAE-B 18:201, 1930.

850. McKenzie, J. W. 1917 The House Fly. J. Dept. Agric. Victoria (Melbourne) 15(10): 628-631.
Concerns intestinal infections of infants; also ophthalmia. ABSTRACT: RAE-B 6:41, 1918.

851. Mackerras, I. M. 1929 The Seasonal Prevalence of House-flies in Sydney, New South Wales—Preliminary Report. Rept. Direct. Publ.

Health. N.S.W. for 1927, pp. 205-207.
Includes records of fly paper catches for 10 collecting centers. ABSTRACT: RAE-B 17:134, 1929.

Mackie, F. J., joint author. *See* Cowan, J., et al., 1919.

852. **Mackinnon, D. L. 1910** Herpetomonads from the Alimentary Tract of Certain Dung-flies. Parasitology 3:255-274.
Never found any larval infection in *M. domestica*; suggests that infection of the adult fly is fresh, and independent.

853. **Macloskie, G. 1880** The Proboscis of the House-fly. Am. Nat. 14:153-161.
Simple illustrations. First published account of the morphology of this organ.

854. **McMahon, J. P. 1935** Preliminary Notes on the Control of Flies. East African Med. J. 12(5):128-135.
Shallow burial of night soil produces many muscoid flies. Recommends naphthalene or paradichlorbenzene as repellents. ABSTRACT: TDB 32: 910, 1935.

McMiller, H., joint author. *See* Stiles, C. W., et al., 1911.

855. **McMurdo, H. B. 1927** Construction and Use of the Fly Trap Stand. Military Surgeon 60(4):423-424.
Sharp angle of stand to windward provides sheltered landing point for flies. Baited trap, so mounted, caught 4 times as many flies as trap without stand. ABSTRACT: RAE-B 15:174, 1927.

856. **Macrae, R. 1895** Flies and Cholera Diffusion. Indian Med. Gaz. 29: 407-412.
Experiments carried out in the gaol at Gaya. Flies had access to cholera stools, food of prisoners, and to boiled milk (which became infected).

857. **Madden, A. H., Lindquist, A. W.** and **Jones, H. A. 1947** Fly Larvicide Tests. Soap and Sanit. Chem. 23(3):141, 143.
Thirty different solvents were *not* made more larvicidal by 5 per cent DDT. Larvae also resistant to a number of harsh chemicals. Integument believed to be mechanically protective. ABSTRACT: RAE-B 37:187, 1949.

858. **Madden, A. H., Lindquist, A. W.** and **Knipling, E. F. 1945** DDT Treatment of Airplanes to Prevent Introduction of Noxious Insects. J. Econ. Ent. 38:252-254.
Treatment proved toxic to mosquitoes and to *M. domestica* for at least 6 weeks Most deposits on glass resistant to wiping with dry cloth. ABSTRACT: RAE-B 34:131, 1946.

859. **Madden, A. H., Schroeder, H. O., Knipling, E. F.** and **Lindquist, A. W. 1946** A Modified Aerosol Formula for Use against Mosquitoes and Houseflies. J. Econ. Ent. 39:620-623.
Motor oil as effective as sesame oil in pyrethrum aerosols. ABSTRACTS: RAE-B 36:46, 1948; BA 21:1810, 1947.

Madden, A. H., joint author. *See* McDuffie, W. C., et al., 1945, 1946; Lindquist, A. W., et al., 1944, 1945, 1946, 1947; Schroeder, H. O., et al., 1945, 1946.

860. **Maddox, R. L. 1885** Experiments in Feeding some Insects with the Curved or *Comma* Bacillus, and also with Another Bacillus (*B. subtilis*). J. Roy. Micro. Soc. ser. 2, Vol. 5, pp. 602-607, 941-952.
Found cholera organism in feces of *Eristalis tenax* and *Calliphora vomitoria.*

Maercks, H., joint author. *See* Janisch, E., et al., 1933.

861. **Maheux, G. 1926** Household Insects. —18th Ann. Rept. Quebec Soc. Prot. Plants, 1925-1926, pp. 96-112.
Includes material on *M. domestica.* ABSTRACT: RAE-B 15:191, 1927.

862. Maier, P. P. and **Dow, R. P. 1949** Diarrheal Disease Control Studies. II Conical Net for Collecting Flies. Publ. Health Repts. 64(19): 604-607.
Illustrations show method of construction and manner of operation. ABSTRACTS: RAE-B 39:133, 1951; BA 24:741, 1950.

863. Makara, G. 1935 [The Breeding Places of *Musca domestica* in Hungary and the Fly Control.] Rep. Hung. Agric. Exp. Sta. 38(5-6):286-291. (In Magyar, with German, French and English summaries.)
Pig dung is the preferred breeding medium in that country. ABSTRACT: RAE-B 24:132, 1936.

Makara, G., joint author. *See* Lorincz, F., et al., 1935, 1936.

864. Malloch, J. R. 1928 Fauna Sumatrensis (Beitrag No. 56) Family Muscidae (Dipt.) Ent. Mitt. 17:310-316. (In English.)
Keys to genera, subgenera and species. Considers *M. vicina* a form of *M. domestica.*

865. Malloch, J. R. 1932 Muscidae of the Marquesas Islands. Bull. Bishop Mus. No. 98:193-203, 1 fig. Honolulu.
Considers *M. vicina* Macq. to be at most a variety of *M. domestica* confined to tropical regions of the old world. ABSTRACT: RAE-B 22:7, 1934.

866. Mally, C. W. 1915 The House-fly. Reprinted from the Farmers' Weekly, Bloemfontein, 13th, 20th and 27th Oct. 1915, 9 pp.
In South Africa, *M. domestica* reaches maturity 11 days after oviposition. Twelve generations per year, under favorable conditions. ABSTRACT: RAE-B 4:80, 1916.

867. Mally, C. W. 1915 Notes on the Use of Poisoned Bait for Controlling The House-Fly, *Musca domestica* L. S.Afr. J. Sci. (Cape Town). 11(9):321-328.
Placed aresenite of sodium bait on branches of trees with firm foliage. Material is slow to drop and eliminates most of the danger to children and animals.

868. Malmgren, B. 1935 Réapparition de la tularémie en Suède au cours de l'annee 1934. [Reappearance of tularemia in Sweden during the year 1934.] Bull. Off. Int. Hyg. Publ. 27(11):2184-2191, 1 pl. 1 map. Paris.
Outbreak in Sweden. *M. domestica* suspect, but found negative for *Bacterium tularense.* Ticks believed responsible. ABSTRACT: RAE-B 24:99, 1936.

869. Manewaring, W. H. 1903 Flies as Carrier of Bacteria. J. Appl. Micro. 6:2402. Rochester, N.Y.

870. Manis, H. C., Dugas, H. L. and **Fox. I. 1942** Toxicity of Paradichlorobenzene to Third Stage Larvae of the Housefly. J. Econ. Ent. 35: 662-664.
Concentration of 1234·6 mg. per 100 grams of treated food caused 100 per cent mortality. Toxic action believed due to combined action as fumigant and stomach poison. ABSTRACTS: RAE-B 31:99, 1943; BA 17:890 (10484), 1943.

Mann, H. D., joint author. *See* Carter, R. H., et al., 1949.

871. Manson-Bahr, P. H. 1919 Bacillary Dysentery. Trans. Soc. Trop. Med. Hyg. 13:64-72.
Fly must feed on blood and mucus of stool almost immediately if it is to carry infection. Can carry 4 days (gives opportunity for dispersion). ABSTRACT: RAE-B 8:75, 1920.

Mantovani, M., joint author. *See* Franchini, G., et al., 1915.

872. **Manwaring, W. H. 1945** Poliomyelitis from Fly-contaminated Food. Calif. and West. Med. 63(2):59.
Infections successfully transmitted to monkeys. ABSTRACT: BA 19:2517, 1945.

873. **March, R. B.** and **Metcalf, R. L. 1949** Development of Resistance to Organic Insecticides other than DDT by Houseflies. J. Econ. Ent. 42:990.
Various intensities of resistance reported for BCH, dieldrin and parathion, with two strains of flies. ABSTRACT: RAE-B 38:155, 1950.

874. **Marchionatto, J. B. 1945** Nota sobre algunos hongos entomogenos. [A Note on Certain Entomogenous Fungi.] Inst. Sanid. Veg. (A) 1(8), 11 pp., 1 col. pl., 3 figs. Buenos Aires.
Aspergillus parasiticus on *M. domestica* from Entre Riós and *Empusa americana* on *Lucilia caesar* from Buenos Aires. ABSTRACT: RAE-B 34: 169, 1946.

875. **Marchoux, E. 1916** Transmission de la lèpre par les mouches (*Musca domestica*). [Transmission of Leprosy by the Fly (*Musca domestica*).] Ann. Inst. Pasteur 30:61-68.
Considered probable that infection can take place by deposition of fresh fly excrement on wounds. ABSTRACT: RAE-B 4:85, 1916.

876. **Marcovitch, S. 1929** Sodium Fluosilicate as a House-fly Poison. J. Econ. Ent. 22(3):602.
Dishes containing saturated solution placed near windows. No other water available. Flies drank readily and died within 30 minutes. ABSTRACT: RAE-B 17:209, 1929.

877. **Marcovitch, S.** and **Anthony, M. V. 1931** A Preliminary Report on the Effectiveness of Sodium Fluosilicate as compared with Borax in Controlling the House Fly (*Musca domestica* Linne). J. Econ. Ent. 24:490-497.
Silicate superior to borax as a larvicide. Neither will destroy eggs. Recommends sprinkling manure with saturated solution of silicate each day. ABSTRACT: RAE-B 19:181, 1931.

Marengo, U., joint author. *See* Ara, F., et al., 1932.

878. **Marett, P. J. 1915** Sanitation in War. J. Roy. Army Med. Corps. 24(4):359-366.
Advocates burning as much manure as possible. Remainder hauled to depression, covered with earth and planted with grass and other seeds. ABSTRACT: RAE-B 3:163, 1915.

879. **Marett, P. J. 1915** Fly Prevention Measures. J. Roy. Army Med. Corps. 25(4):456-460.
Recommends combination of trapping, poison baits, incineration and hand gathering of egg clusters. ABSTRACT: RAE-B 4:20, 1916.

880. **Margarinos-Torres, C. 1925** Déterminisme de la libération spontanée des larves d'*Habronema muscae* (Carter 1861) par la trompe de la mouche domestique: importance de l'hématotropisme. [Determination of the Spontaneous Liberation of the Larvae of *Habronema muscae* (Carter 1861) by the Proboscis of the Domestic Fly: Importance of (positive) Haematotropism.] Compt. Rend. Soc. Biol. 93(20):33-35.
Horse blood, even if cold, will cause ejection of larvae, if the humidity is right. Infected fly lives on. ABSTRACT: RAE-B 13:130, 1925.

881. **Margarinos-Torres, C. 1925** L'hémalotropisme des larvae mûres d'*Habronema muscae* (Carter 1861). [Hematotropism of the Mature

Larvae of *Habronema muscae* (Carter 1861).] Compt. Rend. Soc. Biol. 93(20):38-39.

Ejection of larvae stimulated by horse blood only; not by blood of man, guinea pig or rabbit. ABSTRACT: RAE-B 13:131, 1925.

882. (de) Margarinos-Torres, C. -B., da Fonseca, O. and **de Arêo Leao, A.-E. 1923** Sur la "Esponja" (Habronémose cutanée des Equidés). Du parasitisme des mouches par l'*Habronema muscae* (Carter). [Concerning "Esponja" (Cutaneous Habronemosis of Horses). Parasitism of the fly by *Habronema muscae* (Carter).] Compt. Rend. Soc. Biol. 89(27):767-768.

Of 89 horses and donkeys examined in Rio, 7·8 per cent were infected. Of 164 adult *M. domestica*, 18·9 per cent had larvae in their heads. ABSTRACT: RAE-B 12:2, 1924.

Marlatt, C. L., joint author. *See* Howard, L. O., et al., 1896.

883. Marpman, G. 1884 Die Verbreitungen von Spaltpilzen dur Fliegen. [The Distribution of Bacteria by Flies.] Arch. f. Hygiene 2:560-563.

Colonies of bacteria grown from infected flies.

884. Marpmann, G. 1897 Bacteriologische Mitteilungen III. Ueber den Zusammenhang von pathogenen Bakterien mit Fliegen. [Bacteriological Notices III. On the Close Association of Pathogenic Bacteria with Flies.] Centralbl. (as Zentralbl.) Bakt., Parasit., Infekt. 1(22): 122-132.

Fed flies on cultures of *Bacillus septicus* (*Bacterium agrigenum*) and inoculated contents of flies into mice 12 hours later. The majority died.

885. Martin, A. W. 1903 Flies in Relation to Typhoid Fever and Summer Diarrhoea. Publ. Health (London). 15:652-653.

One of the early contributions in this area of interest.

886. Martin, C. J. 1913 Horace Dobell Lectures on Insect Porters of Bacterial Infections, delivered before the Royal College of Physicians. Brit. Med. J., Jan. 4, 1913, pp. 1-8; Jan. 11, 1913, pp. 59-68.

The first of the two lectures deals with the relation of house flies to typhoid, infantile diarrhoea and other conditions. ABSTRACT: RAE-B 1:12, 1913.

887. Martin, H. and **Wain, R. L. 1944** DDT: Its Properties and Possible Uses in Horticultural Pest Control. J. Roy. Horticultural Soc. 69: 366-369.

House flies killed by contact with glass surface treated with DDT at the rate of 10^{-5} micrograms per square centimeter.

Martin, H., joint author. *See* Lauger, P., et al., 1944.

888. Martinez, M. 1928 La mosca domestica en la ciudad de Mexico. [The Domestic Fly in the City of Mexico.] Mem. Soc. "Ant. Alzate" 48(7-12):391-418, 8 pls.

A somewhat popular and educational account of the fly's life history in that locality. ABSTRACT: RAE-B 17:180, 1929.

Mason, C. J., joint author. *See* Esten, W. M., et al., 1908.

889. Matthysse, J. G. 1945 Observations on Housefly Overwintering. J. Econ. Ent. 38(4):493-494.

M. domestica bred throughout the winter in a dairy barn in northern New York. Temperature always above 0°C. ABSTRACTS: RAE-B 34: 174, 1946; BA 20:828, 1946.

890. Mayer, K. and **Strenzke, K. 1948** Versuche mit Gix zur Bekämpfung von Fliegen und Mücken. [Experiments with "Gix" for the Control of Flies and Midges.] Schädlingsk. Berlin 21(5):74-76.

Is a preparation of fluoro-DDT, used chiefly in the form of an emulsified solution. ABSTRACTS: RAE-B 39:93-94, 1951; TDB 48:1052, 1951.

891. **Mayne, B. 1929** The Nature of the "Black Spores" Associated with the Malaria Parasite in the Mosquito and their Relationship to the Tracheal System. Ind. J. Med. Res. 17(1):109-134, 4 pls.
Included here because these spores also occur in the genus *Musca*. They react to the chitin test, and are believed to be chitinous thickenings of tracheal tubes. ABSTRACT: RAE-B 17:254, 1929.

Mayne, L. E., joint author. *See* Austin, T. A., et al., 1935.

892. **Mays, T. J. 1905** The Fly and Tuberculosis. New York Med. J. and Phila. Med. J. 82:437-438.
Questions the conclusions of J. O. Cobb, 1905.

893. **Mégnin, J. P. 1874** Du transport et de l'inoculation du virus charbonneux et autres par les mouches. [On the Conveyance and Inoculation of the Virus of anthrax and other (diseases) by Flies.] Compt. Rend. de l'Acad. des Sci., Paris. 69:1338-1340.

894. **Mégnin, J. P. 1875** Memoire sur la question du transport et de l'inoculation du virus par les mouches. [Memoir on the Question of the Conveyance and Inoculation of Virus by Flies.] J. de l'Anat et de Physiol., Paris. 11:121-133; *also in* J. de Méd. Vétér. Mil. Paris. 12:461-475.

895. **Meijere, J. C. H. de. 1902** Ueber die Prothorakalstigmen der Dipterenpuppen. [On the Prothoracic Spiracles of Dipterous Pupae.] Zool. Jahrb. (Anat.) 16:623-692.

896. **Meijere, J. C. H. de. 1916** Beitrage zur Kenntniss der Dipteren-Larven und Puppen. [Contribution to (our) Knowledge of Dipterous Larvae and Pupae.] Zool. Jahrb. (Syst.) 40:177-322.

Meleney, H. E., joint author. *See* Frye, W. W., et al., 1932.

897. **Mellanby, K. 1934** The Influence of Starvation on the Thermal Deathpoint of Insects. J. Exp. Biol. 11(1):48-53.
Author worked with *Culex* and *Pediculus*, but also discusses published figures on *Musca*. Several factors involved. ABSTRACT: TDB 31:740, 1934.

898. **Mellor, J. E. M. 1919** Observations on the Habits of Certain Flies, especially of those Breeding in Manure. Ann. Appl. Biol. Cambridge 6(1):53-88.
Manure which has finished fermenting does not attract house-flies; nor will it nourish their larvae. ABSTRACTS: RAE-B 7:192, 1919; Hocking, B. 1960 Smell in Insects, p. 216 (5.284).

899. **Melnick, J. L. 1949** Isolation of Poliomyelitis Virus from single Species of Flies during an Urban Epidemic. Am. J. Hyg. 49:8-16.
M. domestica found positive, though constituting only 4 per cent of the fly population trapped. ABSTRACTS: RAE-B 40:68, 1952; BA 23:1751, 1949.

900. **Melnick, J. L.** and **Penner, L. R. 1947** Experimental Infection of Flies with Human Poliomyelitis Virus. Proc. Soc. Exp. Biol. and Med. 65(2):342-346.
Used *Phormia regina*. Virus recovered up to two weeks later, from body; for three weeks from feces. ABSTRACTS: RAE-B 38:10, 1950; BA 21:2556, 1947.

901. **Melnick, J. L., Ward, R., Lindsay, D. R.** and **Lyman, F. E. 1947** Fly Abatement Studies in Urban Poliomyelitis Epidemics during 1945. Publ. Health Repts. 62(25):910-922.
Measures were instituted after epidemic peak. ABSTRACTS: BA 22:915, 1948; RAE-B 38:210, 1950.

Melnick, J. L., joint author. *See* Trask, J. D., et al., 1943; Ward, R., et al., 1945; Paul, J. R., et al., 1941; Power, M. E., et al, 1943, 1945.

902. Melvin, R. 1932 Physiological Studies on the Effect of Flies and Fly Sprays on Cattle. J. Econ. Ent. 25:1151-1164.
M. domestica had very little effect on cow's temperature; *Stomoxys* a great deal more. ABSTRACT: RAE-B 21:33, 1933.

903. Melvin, R. 1934 Incubation Period of Eggs of Certain Muscoid Flies at Different Constant Temperatures. Ann. Ent. Soc. Am. 27:406-410.
Both high and low temperatures prolong the incubation period. Range studied was from 15°C to 42·8°C. ABSTRACT: RAE-B 22:255, 1934.

904. Meng, Ch'ing-hua and **Winfield, G. F. 1938** Studies on the Control of Fecal-borne Diseases in North China. V. A Preliminary Study of the Density, Species Make Up, and Breeding Habits of the House Frequenting Fly Population of Tsinan, Shantung, China. Chinese Med. J. for 1938. Suppl. No. 2, pp. 463-486.
M. vicina breeds in all the animal manures as well as in human feces. Is second most common fly in houses (after *Chrysomyia megacephala*). ABSTRACTS: RAE-B 26:145, 1938; TDB 35:917, 1938.

905. Meng, Ch'ing-hua and **Winfield, G. F. 1941** Studies on the Control of Fecal-borne Disease in North China. XIII. An Approach to the Quantitative Study of the House-frequenting Fly Population. A. The Estimation of Trapping Rates. Peking Nat. Hist. Bull. 15(4): 317-331. XIV B. The Characteristics of an Urban Fly Population. T. c. pp. 333-351.
Concerns *M. domestica vicina* Macq. and *Chrysomyia megacephala* F.

906. Mercier, L. 1925 Diptères "buveurs du sang" et Diptères "succeurs de suer". [Dipterous Drinkers of Blood and Dipterous Imbibers of Sweat.] Compt. Rend. Soc. Biol. 92(3):135-136.
M. vitripennis and *M. autumnalis* feed on skin exudations, but readily ingest blood from wounds left by true blood-feeders (or of any other cause). ABSTRACT: RAE-B 13:50, 1925.

906a. Mercurialis. 1577 De Pestilentia. Venice. Cited by Abel, 1899.
Believed that flies carry the virus of plague from active cases to the food of healthy people.

907. Merilatt, L. A. 1920 Fly Repellent. Am. J. Vet. Med., (Chicago) 15(7): 333.
For use by veterinarians during surgical operations and in treatment of sick animals. ABSTRACT: RAE-B 8:157, 1920.

908. Merlin, A. A. C. E. 1897 The Foot of the Housefly. J. Quekett Micro. Club. (2) Vol. 6, p. 348.
One of the early studies of foot morphology.

909. Merlin, A. A. C. E. 1905 Supplementary Note on the Foot of the House-fly. J. Quekett Micro. Club. (2) Vol. 9, pp. 167-168.
An amplification of the above.

Merriman, J., joint author. *See* Copeman, S. M., et al., 1911; Hindle, E., et al., 1914.

Metcalf, R. L., joint author. *See* March, R. B., et al., 1949.

910. Metelkin, A. 1935 [The Rôle of Flies in the Spread of Coccidiosis in Animals and Men.] Med. Parasit. & Parasitic Dis. Moscow. 4(1-2): 75-82. (In Russian, with English summary.)
M. domestica one of 6 species used. All can ingest oocysts which are viable in fly's discharges until the latter dry up. ABSTRACTS: RAE-B 23:164, 1935; TDB 32:660, 1935.

911. Michal, K. 1931 Die Beziehung der Populationsdichte zum Lebensoptimum und Einfluss der Lebensoptimums auf das Zahlenverhältnis der Geschlechter bei Mehlwurm und Stubenfliege. [The Relation of Population Density to Life Duration and the Influence of Life Span on the Numerical Relation of the Sexes, in Meal Worms and House Flies.] Biol. Generalis 7(3):631-646. Vienna.
ABSTRACT: RAE-B 20:180, 1932.

912. Michelbacher, A. E., Smith, R. F. and **Smith, G. L. 1945** Control of Flies in a Dairy Barn with DDT. *In* Investigations with DDT in California, 1944. A Preliminary Report, prepared under the Direction of the Division of Entomology and Parasitology, Calif. Agr. Expt. Sta., Berkeley, pp. 8-9.
Very good control for 3 weeks; at seven weeks some supplementary spraying required. ABSTRACT: RAE-B 35:57, 1947.

913. Michelbacher, A. E. and **Smith, G. L. 1946** Control of Flies in Dairy Barns. *In* Circular 365 (Investigations with DDT and other New Insecticides in 1945). Div. of Ent. and Parasit., Calif. Agr. Expt. Sta., Berkeley, pp. 94-97.
Three, properly timed spray applications of DDT gave good control for period of one year.

914. Michigan Department of Health. 1937 The Collection and Disposal of Garbage. Engineering Bull. No. 6. Lansing. 40 pp., illus.
Very complete for its time. Well illustrated and instructive.

915. Michigan Department of Health. 1941 Sewerage and Sewage Disposal. Engineering Bull. No. 11. Lansing, 58 pp., illus.
Excellent for its time. Very well illustrated.

916. Miller, A. C. and **Simanton, W. A. 1938** Biological Factors in Peet-Grady Results. Soap 14(5):103, 105, 107, 109, 111, 113.
Truer evaluation would be obtained if based on an average from tests on several cultures, rather than one. ABSTRACT: RAE-B 26:244, 1938.

Miller, A. C., joint author. *See* Simanton, W. A., et al., 1937, 1938.

917. Miller, D. 1932 The Biology and Economic Status of New Zealand. Muscidae and Calliphoridae. Part I. Historical Review. Bull. Ent. Res. 23(4):469-476.
M. domestica believed to have reached New Zealand prior to 1856, by way of Australian cattle ships. ABSTRACT: RAE-B 21:71, 1933.

918. Miller, D. F. 1929 Determining the Effects of Change in Temperature upon the Locomotor Movements of Fly Larvae. J. Exp. Zool. 52: 293-313.
Studied *Lucilia sericata.* Below 10°C (50°F) locomotor activities are sharply reduced. Slow increase up to 30°C, slow decrease above 42°C.

919. Miller, L. L. 1948 Sanitary Land-Fill Provides for Modern Garbage Disposal. Michigan Publ. Health 36(7):125, 128, 136.
An informative presentation of this method of waste disposal.

920. Milliken, F. B. 1911 Another Breeding-Place for the House-fly. J. Econ. Ent. 4:275.
Larvae developed in alfalfa ensilage, in good feeding condition.

921. Minett, E. P. 1911 The Question of Flies as Leprosy Carriers. J. London School Trop. Med. 1:31-35.
An informative discussion.

Min(n)ett, E. P., joint author. *See* Wise, K. S., et al., 1912.

Minett, F. C., joint author. *See* Sen, S. K., et al., 1944.

921a. Mironov, V. S. 1940 [The use of *Acorus calamus* for insecticidal and repellent preparations.] Med. Parasit. (Moscow) 9:409-410.
A powder from the rhizomes of the "sweet rush" killed adult *M. domestica* and *Anopheles maculipennis* within 40 minutes; was slightly repellent to certain ticks. ABSTRACT: Hocking, B. 1960 Smell in Insects, p. 217 (5.291).

922. Mitchell, P. C. 1915 Practical Advice on the Fly Question. Zool. Soc. London. 7 pp.
Recommends sodium arsenite on municipal dumps and covering manure heaps with oil-soaked earth. ABSTRACT: RAE-B 3:206, 1915.

923. Mitscherlich, E. 1943 Die Uebertragung der Kerato-Conjunctivitis infectiosa durch Fliegen und die Tenazität von *Rickettsia conjunctivae* in der Aussenweet. [The Transference of Infectious Kerato-conjunctivitis by Flies and the Persistence of *Rickettsia conjunctivae* in the Environment. Dtsch. Tropenmend. Zeit. 47(3):57-64. Leipzig.
M. domestica and *Stomoxys calcitrans* transmitted infection when placed on eyes of calves within 8 hours after becoming themselves infected. ABSTRACT: RAE-B 32:55, 1944.

924. Mitzmain, M. B. 1912 The Role of *Stomoxys calcitrans* in the Transmission of *Trypanosoma evansi* Philipp. J. Sci., Sect. B., (Philipp. J. Trop. Med.) 7:475-518, 5 pls.
States that surra organisms have been found in the mouth parts and stomachs of house flies. ABSTRACT: TDB 2:130-133, 1913.

925. Mitzmain, M. B. 1916 A Digest of the Insect Transmission of Disease in the Orient with especial Reference to the Experimental Conveyance of *Trypanosoma evansi.* New Orleans Med. & Surg. J. 69(6) 416-424.
Positive transmission by *M. domestica*, when flies fed at bites of *Stomoxys* and had access to wounds of healthy animals. ABSTRACT: RAE-B 5:60, 1917.

Mizushima, H., joint author. *See* Kobayashi, H., et al., 1937.

926. Mohler, J. R. 1919 Report of the Chief of the Bureau of Animal In dustry. U.S. Dept. Agric. Washington, D.C. Sept. 29, 1919, 63 pp.
M. domestica may harbor virus of hog cholera for several days and possibly infect animals by feeding on eyes or fresh wounds. ABSTRACT: RAE-B 8:99, 1920.

927. Moniez, R. 1874 Apropos des publications récentes sur le faux parasitisme des Chernétides sur différent arthropodes. [Concerning Recent Publications on the False Parasitism of Chelifers on Different Arthropods.] Rev. Biol. du Nord de la France. 6:47-54.
Holds the view that *Chernes* uses *M. domestica* for dispersal only. Is usually attached to the legs of the fly.

928. Monro, H. A. U., Beaulieu, A. A. and **Delisle, R. 1947** DDT Residues. Their Toxicity to Houseflies on various Surfaces and Materials. Soap and Sanit. Chem. 23(8):123, 125, 127, 129, 143, 145.
Summarizes two-year results at Montreal. ABSTRACT: RAE-B 38:33, 1950.

929. Montfils, A. J. 1776 D'une maladie fréquente connue en Bourgogne sous le nom de Puce maligne. [Concerning a disease frequently reported in Bourgogne under the name of "puce maligne".] J. de Médecine 45:500.
Considered flies possible distributors of anthrax.

Montigel, C., joint author. *See* Lauger, P., et al., 1946.

Moore, D. H., joint author. *See* McAlister, L. C., Jr., et al., 1947.

930. **Moore, W. 1893** Diseases Probably Caused by Flies. Brit. Med. J. for June 3, 1893. Pt. 1, p. 1154. Also in Med. Magaz. Jul., 1893.
One of many articles published on this subject in the latter part of the nineteenth century.

931. **Moore, W. 1917** Toxicity of Various Benzene Derivatives to Insects. J. Agric. Res. Washington, D.C. 9(11):371-381.
M. domestica succumbs more readily to compounds with low boiling point than *Lucilia sericata*. Reverse is true with compounds of high boiling point. ABSTRACT: RAE-B 5:132, 1917.

932. **Moore, W. 1917** Volatility of Organic Compounds as an Index of the Toxicity of their Vapours to Insects. J. Agric. Res. Washington, D.C. 10(7):365-371.
In tests with *M. domestica*, general rule is that the less volatile the chemical the more toxic it is. But compounds with boiling points 225°-250°C are usually so slightly volatile, that they kill insects only after very long exposures. ABSTRACT: RAE-B 5:174, 1917

933. **Moore, W. 1934** Esters as Repellents. J. New York Ent. Soc. 42(2): 185-192.
Tests on *M. domestica* confirmed the principle, already established for mosquitoes, that terpene alcohols, as a class, are more effective than terpene hydrocarbons. In each instance, the ester is superior to the corresponding alcohol. ABSTRACTS: RAE-B 22:165, 1934; Hocking, B. 1960 Smell in Insects, p. 138 (3.96).

934. **Moorehead, S.** and **Weiser, H. H. 1946** The Survival of Staphylococci Food Poisoning Strain in the Gut and Excreta of the House Fly. J. Milk Technol. 9(5):253-259.
M. domestica may serve as reservoir host and may initiate or augment an outbreak by spreading organisms from contaminated handlers or dirty equipment to foodstuffs. ABSTRACT: BA 21:151, 1947.

Morel, joint author. *See* Descazeaux, et al., 1933.

935. **Morgan, B. B. 1942** The Viability of *Trichomonas foetus* (Protozoa) in the House Fly (*Musca domestica*). Proc. Helminth. Soc. Wash. 9(1): 17-20.
Concludes that *M. domestica* may transmit *T. foetus* to cattle by defecating or regurgitating on the genitalia. ABSTRACT: RAE-B 32:101, 1944.

936. **Morgan, H. de R. 1906** Upon the Bacteriology of the Summer Diarrhoea of Infants. Brit. Med. J. Apr. 21, 1906. (12 pp.) and July 6, 1906 (11 pp.)
Isolated a bacillus which he designated "No. 1" (Morgan's bacillus of subsequent authors). Belongs to the non-lactose fermenting group.

937. **Morgan, H. de R.** and **Ledingham, J. E. G. 1909** The Bacteriology of Summer Diarrhoea. Proc. Roy. Soc. Med. 2:133-158. (Separate pagination in reprint.)
Further information regarding Morgan's No. 1 bacillus. Is *Proteus morganii* of modern literature.

938. **Morison, J. 1915** The Causes of Monsoon Diarrhoea and Dysentery in Poona. Second Report. Indian J. Med. Res. 2(4):950-976.
Flies not the main cause of epidemics in Poona (1912-1914). They are nevertheless a factor in the spread of infection under various circumstances. ABSTRACT: RAE-B 3:127, 1915.

939. **Morison, J.** and **Keyworth, W. D. 1916** Flies and their Relation to Epidemic Diarrhoea and Dysentery in Poona. Indian J. Med. Res. (Calcutta.) 3(4):619-627.
Close correlation between number of flies and cases of diarrhoea June 10-

Aug. 12. Polluted water supply chief cause of the epidemic, however. ABSTRACT: RAE:B 4:124, 1916.

940. **Monard. 1917** La lutte contre les mouches. [The War against Flies.] Progrès Agricole, Amiens. 31(1533):264.
Various formulae given, using formalin, sulphur or cresyl. ABSTRACT: RAE-B 5:121, 1917.

941. **Morrill, A. W. 1914** Some American Insects and Arachnids Concerned in the Transmission of Disease. Arizona Med. J. (Phoenix). (Reprint, 12 pp.)
Average number of bacteria on bodies of 414 house flies found to be 1¼ million; for those from swill barrels, 4 million. ABSTRACT: RAE-B 2:137, 1914.

942. **Morrill, A. W. 1914** Experiments with House-fly Baits and Poisons. J. Econ. Ent. 7:268-273.
Value of sticky fly paper enhanced by placing small amount of attractive bait at center of sheet. ABSTRACT: RAE-B 2:159, 1914.

Morrill, A. W., joint author. *See* Dews, S. C., et al., 1946.

943. **Morris, H. 1919** Anthrax: Transmission of Infection by Non-biting Flies. Louisiana Agric. Expt. Sta., Baton Rouge. Bull. 168, 12 pp.
M. domestica and other muscoids capable of carrying infection to wounds on healthy animals after feeding on anthrax-infected flesh. ABSTRACT: RAE-B 10:26, 1922.

944. **Morris, H. 1920** Some Carriers of Anthrax Infection. J. Am. Vet. Med. Ass. Washington, D.C. 56:606-608.
Anthrax spores do not form in *unopened* animal carcasses. Flies bred in *mutilated* carcasses, however, will carry infection on their bodies. Argentine ant also a carrier. ABSTRACT: RAE-B 8:105, 1920.

945. **Morrison, F. O. 1948** Know the Flies to Control Them. Pests 16(6): 9-14.
ABSTRACT: BA 23:1368, 1949.

Morse, W. J., joint author. *See* Blickle, R. L., et al., 1948.

946. **Mosna, E. 1949** Octa Klor, gammaesano e toxaphene usati contro le mosche DDT resistenti. [The Use of Octa-Klor, gammexane and toxaphene against DDT-resistant flies.] Estratto dai Rendiconti dell' Istituto Sup. di Sanit. 12 (Parte VII-VIII-IX):467-489.
Does not recommend toxaphene for practical use against DDT-resistant flies.

Mote, D. C., joint author. *See* Fairchild, H. E., et al., 1949.

Moussatche, J., joint author. *See* Oliveira, S. J. de, et al., 1947.

947. **Moursi, A. A. 1946** The Effect of Temperature on the Sex Ratio of Parasitic Hymenoptera. Bull Soc. Fouad Ier Ent. 30:21-37. Cairo. The Effect of Temperature on Development and Reproduction of *Mormoniella vitripennis* (Walker) (Hymenoptera:Chalcidoidea-Pteromalidae) T. c., pp. 39-61.
This is a pupal parasite of *M. domestica* and other muscoids. *Nasonia brevicornis* is a synonym. ABSTRACT: RAE-B 36:202, 1948.

948. **Moutia, A. 1926** La mouche domestique — *Musca domestica* (Linn.) Rev. Agric. Ile Maurice, No. 26, pp. 66-68. Mauritius.
A popular account of *M. domestica* and *Stomoxys nigra*, with recommendations for control. ABSTRACT: RAE-B 15:117, 1927.

Muhsam, H. V., joint author. *See* Feldman-Muhsam, B., et al., 1946.

Muirden, M. J., joint author. *See* Tomlinson, T. G., et al., 1949.

Mukherjee, P. K., joint author. *See* Roy, D. N., et al., 1937.

Mukherjee, S. P., joint author. *See* Roy, D. N., et al., 1940.

Mukherji, B., joint author. *See* Lal, R. B., et al., 1939.

Müller, P., joint author. *See* Lauger, P., et al., 1944.

949. **Muma, M. H.** and **Hixson, E. 1949** Effects of Weather, Sanitation and Chlorinated Chemical Residues on House and Stable Fly Populations on Nebraska Farms. J. Econ. Ent. 42:231-238.
DDT, methoxychlor, DDD, toxaphene and chlordane all more or less equally effective against *M. domestica* under comparable conditions. ABSTRACT: RAE-B 38:82, 1950.

950. **Munro, J. A. 1936** Fly Trapping and its Application to Human Welfare Circ. N. Dak. Agric. Expt. Sta. No. 60, pp. 3-8. Fargo.
Popular type account of fly bionomics, with instructions for control. ABSTRACT: RAE-B 25:217, 1937.

951. **Munro, J. A., Post, R. L.** and **Colberg, W. 1947** The New Insecticides in Fly Control. Bimonthly Bull. N. Dak. Agric. Expt. Sta. 9(5): 123-128. Fargo.
A report on the use of various DDT preparations and BHC. ABSTRACT: RAE-B 37:47, 1949.

952. **Munro, J. A., Redman, K.** and **Longwell, J. H. 1945** Effectiveness of DDT against Flies in Livestock Barns. Bimonthly Bull. N. Dak. Agric. Expt. Sta. 7(5):21-23. Fargo.
Semi-emulsifying spray in pig barn produced residual effect from Aug. 10 to end of season. ABSTRACT: RAE-B 34:125, 1946.

Munson, S. C., joint author. *See* Yeagar, J. F., et al., 1945, 1949.

Murasawa, I., joint author. *See* Takano, T., et al., 1947.

953. **Murphy, W. 1949** ACS at San Francisco. Science 109:387. (Concerns 115th national meeting of the American Chemical Society, Mar. 28-April 1, 1949.)
Mention made of Compound 497, said to be more effective against cockroaches and houseflies than DDT.

954. **Murray, A. 1877** Economic Entomology. Aptera. London, p. 129.
Larval mites parasitic on houseflies. Not as prevalent in England as in North America.

955. **Murray, C. A. 1937** A Statistical Analysis of Fly Mortality Data. Soap 13(8):88-99, 101, 103, 105.
Far greater susceptibility among males. Unfed flies more susceptible in both sexes. ABSTRACT: RAE-B 27:22, 1939.

956. **Nabokov, V. A.** and **Tiburskaya, N. A. 1935** [On the Use of Pyrethrum in Insect Control.] Med. Parasit. 4(6):486-491. (In Russian.)
M. domestica more resistant than mosquitoes to "Flicide" (benzene extract of pyrethrum plus phenyl and methyl salicylates). ABSTRACT: RAE-B 24:139, 1936.

957. **Nageotte, J. 1943** Le principe d'inertie dans la physiologie sensorielle: Etude sur la balancier des Diptères. [The Principle of Inertia in Sensory Physiology: Study on the Haltere of Diptera.] Arch. Zool. Exp. et Gen. 83(3):99-111.
Compares activity of the haltere to that of a gyroscope and ball governor. ABSTRACT: BA 20:2269, 1946.

958. **Najera Angulo, L. 1947** La lucha contra las moscas. [The War against Flies.] Madrid, Dir. Gen. Sanid. 204 pp., 22 pls., 14 figs.
A volume of 5 chapters. Proper disposal of manure most important method of control for Spain. ABSTRACT: RAE-B 37:53, 1949.

959. **Nash, J. C. T. 1903** The Seasonal Incidence of Typhoid Fever and of Diarrhoea. Trans. Epidemiol. Soc. (London) N.S., vol. 22, pp. 110-138.
See collective annotation at the end of this series of publications by Nash.

960. Nash, J. C. T. 1903 The Etiology of Summer Diarrhoea. Lancet, p. 330.

961. Nash, J. C. T. 1904 Some Points in the Prevention of Epidemic Diarrhoea. Lancet, p. 892.

962. Nash, J. C. T. 1905 The Waste of Infant Life. J. Roy. Sanit. Inst. 26: 494-498.

963. Nash, J. C. T. 1906 The Prevention of Summer or Epidemic Diarrhoea. The Practitioner, 12 pp.

964. Nash, J. C. T. 1906 Special Report on Epidemic Diarrhoea, Borough of Southend-on-Sea. 16 pp., 1 chart.

965. Nash, J. C. T. 1906 Second Report on Epidemic Diarrhoea, Borough of Southend-on-Sea. 28 pp.

966. Nash, J. C. T. 1906 Annual Report of the Medical Officer of Health, Borough of Southend-on-Sea.

967. Nash, J. C. T. 1908 A Note on the Bacterial Contamination of Milk, as Illustrating the Connection between Flies and Epidemic Diarrhoea. Lancet, Pt. 2, pp. 1668-1669.

968. Nash, J. C. T. 1908 Flies as a Nuisance and Flies as a Dangerous Nuisance. Lancet, pp. 131-132.

969. Nash, J. C. T. 1908-09 House-flies as Carriers of Disease. Norwich Rep. Mus. Assn. Vol. 2, pp. 14-19.

970. Nash, J. C. T. 1909 House-flies as Carriers of Disease. J. Hygiene 9:141-169.

Nash's papers were perhaps more educational than scientific in their content. They reiterate the concept that flies carry infective material from filth to human food supplies, and suggest a correlation between cool weather, diminished fly production and termination of infant deaths from enteritis. Experiments are reported showing that milk exposed to flies soon contains many more microorganisms than milk protected from them. Various breeding media are indicated as productive of flies, and appropriate sanitary procedure recommended. For this particular period in the history of sanitary entomology, Nash's publications were unquestionably important in moulding public opinion.

Needham, D. M., joint author. *See* Baldwin, E., et al., 1934.

971. Nelson, R. H., Fulton, R. A., Fales, J. H. and **Yeomans, A. H. 1949** Low Pressure Aerosols — Their Efficiency in Relation to Formulation and Particle Size. Soap and Sanit. Chem. 25(1):120-121, 123, 125, 166.

Best formulations appear to be those containing 10 or 15 per cent of nonvolatile constituents. ABSTRACT: RAE-B 38:206, 1950.

972. Nelson, R. H., Gersdorff, W. A. and **Gertler, S. I. 1949** Toxicity to House Flies of DDT and two DDT Analogs. J. Econ. Ent. 42: 158-159.

Knockdown by bromine and fluorine analogs lower than by DDT at the same concentration. ABSTRACT: RAE-B 38:57, 1950.

Nelson, R. H., joint author. *See* Gersdorff, W. A., et al., 1948; Fales, J., et al., 1948; Chisholm, R. D., et al., 1949.

973. Newsholme, A. 1903 Annual Report on Health of Brighton, p. 21.

Deals with infection of milk by flies.

974. Newsholme, A. 1906 Domestic Infection in Relation to Epidemic Diarrhoea. J. Hygiene 6:139-148.

A rather full treatment of the problem.

975. Newsholme, A. 1909-10 A Report on Infant and Child Mortality, being a Supplement to the Report of the Medical Health Officer in the

39th Ann. Rept. of the Local Gov't. Bd. 1909-1910. (Separate, 110 pp.)
Mortality highest where, under urban conditions of life, filthy privies are permitted and where streets are not paved.

976. Newsholme, A. 1910 Enteric Fever in Durham. (Partial quotation of Newsholme's Report.) Times, Weekly Edition, Aug. 5, 1910.
Durham has highest death rate from enteric fever of any County in England or Wales. Filthy domestic arrangements believed responsible.

977. Newstead, R. 1907 Preliminary Report on the Habits, Life-Cycle and Breeding Places of the Common House-fly (*Musca domestica* Linn.) as Observed in Liverpool, with Suggestions as to the Best Means of Checking its Increase. Liverpool, 23 pp., 14 figs.

978. Newstead, R. 1909 Second Interim Report on the House-fly as Observed in the City of Liverpool. 4 pp.

979. Newstead, R. 1908 On the Habits, Life-cycle and Breeding-places of the Common House-fly. Ann. Trop. Med. Parasit. 1(4):507-520.
Though bearing an earlier imprint date than the item just preceeding, this is the most important paper of the three. It sums up the content of the others and is considered a reliable work of reference.

980. Nicewicz, N., Nicewicz, W. and **Kowalik, R. 1946** Opis drobnoustrojów stwierdzonych droga analizy bakteriologicznej w przewodzie pokarmowym plusky domowej, muchy domowej i karaczana wschodniego. [Bacteriological Analysis of Microorganisms in the Alimentary Tracts of the Bedbug, House fly and Cockroach.] Ann. Univ. Mariae Curie-Sklodowska Sect. C. Biol. 1(2):35-38.
Lists 15 species each, from *Cimex*, *Musca* and *Periplaneta*. ABSTRACT: BA 22:721, 1948.

Nicewicz, W., joint author. *See* Nicewicz, N., et al., 1946.

981. Nicholas, G. E. 1873 The Fly in its Sanitary Aspect. Lancet, Vol. 2, p. 724.
A shrewd and intelligent interpretation of the relation of flies to cholera, as observed in Malta, on Naval Vessels and in England.

982. Nicholls, L. 1912 The Transmission of Pathogenic Microorganisms by Flies in St. Lucia. Bull. Ent. Res. 3:81-88.
Considers the majority of cases of yaws in the West Indies to be caused by inoculation of surface injuries by the fly, *Oscinis pallipes*.

983. Nicol, G. 1946 Parasites of the Horse. J. Dept. Agric. Victoria 44(2): 53-56. Melbourne.
Refers to *Habronema muscae* and *H. megastoma* as transmitted by *M. domestica*; *H. microstoma* by *Stomoxys*.

984. Nicoll, W. 1911 On the Varieties of *Bacillus coli* associated with the House-fly (*Musca domestica*). J. Hygiene 11(3):381-389.
Flies may carry at least 27 varieties of *B. coli*. These appear to be derived about equally from excremental and from other sources.

985. Nicoll, W. 1911 On the Part Played by Flies in the Dispersal of the Eggs of Parasitic Worms. Rept. to Local Gov't. Bd. of Publ. Health and Med. Subjects. N.S. 53:13-30.
Flies cannot ingest particles of larger size than 0·045 mm., but will take up and pass in feces ova that are smaller, e.g. *Taenia solium*, *Oxyuris*, *Trichocephalus*.

986. Nicoll, W. 1917 Flies and Bacillary Enteritis. Brit. Med. J., (London) No. 2948, pp. 870-872.
M. domestica though not the exclusive carrier of any human disease, is perhaps the most effective single distributor of enteric infections. ABSTRACT: RAE-B 5:126, 1917.

987. Nicoll, W. 1917 Flies and Typhoid. J. Hygiene, Cambridge. 15(4):505: 526.
Fly is mechanical carrier, not a host. *B. typhosus* does not survive long in fly's intestinal tract. ABSTRACT: RAE-B 5:149, 1917.

988. Nicolle, C. 1922 État de nos connaissances d'ordre expérimental sur le trachome. [The State of our Knowledge of an Experimental Nature concerning Trachoma.] Bull. Inst. Pasteur 19:881-894.
An up-dating of the paper listed below.

989. Nicolle, C., Cuenod, A. and **Blanc, G. 1919** Transmission of Trachoma by Flies. Presse Med. Dec. 20, 1919.
Flies can transmit disease for at least 24 hours after contact with infected eyes, or with bandages up to 6 hours after removal from an active case. ABSTRACTS: Ann. d'Igiene. Rome 31(1):66; RAE-B 9:94, 1921.

990. Nieschulz, O. 1933 Ueber die Bestimmung der Vorzugstemperatur von Insekten (besonders von Fliegen und Mücken). [On the Determination of Temperature Preference on the part of Insects (especially Flies and Midges).] Zool. Anz. 103(1-2):21-29.
Used long metal tank, one end kept at 50°C, the other in an ice box. Glass wall permitted observation of where insect came to rest. ABSTRACT: TDB 31:48, 1934.

991. Nieschulz, O. 1933 Ueber die Temperaturebegrenzug der Aktivitätsstufen von *Stomoxys calcitrans.* [Concerning the Temperature Limitation on Degree of Activity in *Stomoxys* . . .] Zeit. Parasitenk. 6:220-242.
One of a series of experiments from which Nieschultz concluded that each species has its own optimum temperature and range of normal activity.

992. Nieschulz, O. 1928 Enkele multouroverbrengingsproeven met Tabaniden, Musciden en Muskieten. [Anthrax Transmission Experiments with Tabanidae, Muscidae and Mosquitoes.] Veeartsenijk Meded. No. 67, pp. 1-23. Buitenzorg.
Direct transmission to guinea pigs succeeded in 1 of 6 experiments with *M. inferior* Stein. Transmission capacity of tabanids and muscids higher than for surra. ABSTRACT: RAE-B 17:230, 1929.

993. Nieschulz, O. 1933 Some Remarks about the Role of True Bloodsucking *Musca* Species as Transmitters of Diseases. Ann. Trop. Med. Parasit. 27(2):213-214.
Reviews experiments by which *M. crassirostris* has been proved capable of transmitting surra, anthrax and haemorrhagic septicaemia of buffaloes. ABSTRACT: RAE-B 21:203, 1933.

994. Nieschulz, O. 1934 Uber die Vorzugstemperatur von *Stomoxys calcitrans.* [On the Preferred Temperature of *Stomoxys calcitrans.*] Zeit. Angew. Ent. 21:224-228.
One of a series of papers in which *Stomoxys*, *Musca* and *Fannia* are compared.

995. Nieschulz, O. 1935 Ueber die Temperaturabhängigkeit der Aktivität und die Vorzugstemperater von *Musca domestica* und *Fannia canicularis.* [Concerning the Dependence of Activity on Temperature and the Preferred Temperature of *Musca domestica* . . . etc.] Zool. Anz. 110(9-10):225-233. Leipzig.
Activity for *M. domestica* began at 6·7°C. Optimum for females, 33·1°C; for males, 34·2°C. Heat paralysis began at 44·6°C; complete at 46·5°C. ABSTRACT: RAE-B 23:222, 1935.

996. **Nieschulz, O.** and **Du Toit, R. 1933** Ueber die Temperaturabhangigkeit der Aktivität und die Vorzugstemperatur von *Stomoxys calcitrans, Musca vicina* und *M. crassirostris* in Sudafrika. [Concerning the Dependence of Activity on Temperature, and the Preferred Temperature of *Stomoxys calcitrans, Musca vicina* and *M. crassirostris* in South Africa.] Zentralblt. Bakt. (2)89(8-12):244-249. Jena.
Each species showed its own, peculiar characteristics. ABSTRACT: RAE-B 22:22, 1934.

997. **Nieschulz, O.** and **Kraneveld, F. C. 1929** Experimentelle Untersuchungen über die Uebertragung der Büffleseuche durch Insekten. [Experimental Investigations on the Transmission of Buffalo-sickness by Insects.] Zentralblt. Bakt. (1) Orig. 113:(5-6):403-417.
Haemorrhagic septicaemia of buffaloes transmitted from rabbit to rabbit by tabanids, muscoids and mosquitoes. ABSTRACT: RAE-B 17:228, 1929.

Nieschulz, O., joint author. *See* DuToit, R., et al., 1933.

998. **Niewenglowski, G. H. 1913** La Transmission des Maladies par les Mouches. [The Transmission of Disease by Flies.] Naturaliste Canadien 40:33-38.
An informative article, typical of that time.

999. **Nijkamp, J. A.** and **Swellengrebel, N. H. 1934** Proeven over anophelesvernietiging met pyrethrum praeparaten. [Tests on *Anopheles* Control with Pyrethrum Preparations.] Ned. Tijdschr. Geneesk. 78(21): 2327-2338, 2 pls. (In Dutch, with German and English summaries.)
M. domestica used as a test species. Pyrethrum somewhat superior to powder from *Artemisia maritima*, a plant native to Holland. ABSTRACT: RAE-B 22, 197, 1934.

1000. **Nikol'skii, A. L. 1945** [Rational Preparation of Kizyak as a Means of Controlling Flies.] Med. Parsit. 14(2):70-72. Moscow. (In Russian.)
Sun dried bricks of cow dung, used as fuel, will not breed flies if less than 8 cm. thick, and if turned repeatedly while drying. ABSTRACT: RAE-B 34:182, 1946.

1001. **Nitzulescu, V. 1924** Sur l'ingestion du bleu méthylene par *Stomoxys calcitrans*. Compt. Rend. Soc. Biol 90(2):155-156.
Both this species and *M. domestica* died after ingesting water containing a concentrated solution of methylene blue. ABSTRACT: RAE-B 12:53. 1924.

1002. **Niven, J. 1904-1912** Annual Reports on the Health of the City of Manchester.

1003. **Niven. J. 1910** The House-fly in Relation to Summer Diarrhoea and Enteric Fever. Proc. Roy. Soc. Med. April. 1910. (Reprint, 83 pp.)
Of 8553 flies collected in Manchester, 8196 were *M. domestica.* Niven's studies, epidemiological in nature, made a strong impression on health authorities of that period. Important guide lines are set forth for the guidance of future workers.

1004. **Noack, W. 1901.** Beiträge zur Entwicklungsgeschichte der Musciden. [Contributions to the Developmental History of Muscidae.] Zeit. Wiss. Zool. 70:1-56, 10 figs., 5 pls.
Important for its time. Contains a good bibliography.

1005. **Noc, —. 1920** Rôle des mouches en pathologie intestinal. Prophylaxie. [The Role of Flies in Intestinal Pathology. Control.] Bull. Soc. Med. — Chirurg. Francaise d l'Ouest-Afr. (Dakar) 10(10):280-283.
Records a great increase in the number of flies in connection with an epidemic of bacillary and amoebic dysentery. ABSTRACT: RAE-B 10:155, 1922.

1006. Noel, P. 1913 La Guerre aux mouche. [The War against Flies.] Bull. du Laboratoire Régional d'Entomologie Agricole, Roven. 1913:4-5.
Flies attracted to meat traps, produced larvae which were fed to fishes in a fish culture establishment. ABSTRACT: RAE-B 1:30-31, 1913.

1007. Noel, P. 1914 La Destruction des mouches. [The Destruction of Flies.] Bull. Trim. Lab. Rég. d'Entom. Agric. Seine — Infér; Roven. Jan.-Mar. 1914, pp. 12-14.
Two types of traps described, with formulas for bait. ABSTRACT: RAE-B 2:68, 1914.

1008. Noguchi, H. 1926 The Differentiation of Herpetomonads and Leishmanias by Biological Tests. Science 63(1637):503-504.
H. muscidarum, sp. n., from house fly ferments 14 carbohydrates. *H. media*, sp. n., from *Calliphora* ferments 7. ABSTRACT: RAE-B 14: 133, 1926.

1009. Noguchi, H. and **Kudo, R. 1917** The Relation of Mosquitoes and Flies to the Epidemiology of Acute Poliomyelitis. J. Exp. Med. 26(1): 49-57.
Found no trace of polio virus in filtrate of emulsion made from *M. domestica* reared on infective material. ABSTRACT: RAE-B 6:171, 1918.

1010. Noguchi, H. and **Tilden, E. B. 1926** Cultivation of Herpetomonads from Insects and Plants. J. Exp. Med. 44(3):307-325.
One strain isolated from *M. domestica.*

1011. Noguchi, H. 1926 Differentiation of the Organisms by Serological Reactions and Fermentation Tests. J. Exp. Med. 44(3):327-337.
Discusses differentiation of *Herpetomonas* natural to the insect, and developmental stages of *Leishmania.* ABSTRACT of the two foregoing papers: RAE-B 16:17, 1928.

Norwood, V. H., joint author. *See* Robinson, W., et al., 1933, 1934.

1012. Novy, F. G. 1912 Disease Carriers. Science 36(914):1-9.

1013. Nuttal, G. H. F. 1897 Zur aufklärung der Rolle, welche Insekten bei der Verbreitung der Pest spielen — Ueber die Empfindlichkeit verschiedener Thiere für dieselbe. [Toward Elucidation of the Role which Insects play in the Dissemination of Plague — On the Susceptibility of Different Animals for the Same.] Centralbl. f. Bakt. 22: 87-97.
(*See* Annotation following this series of papers by Nuttal.)

1014. Nuttall, G. H. F. 1899 On the Role of Insects, Arachnids, and Myriapods in the Spread of Bacterial Diseases in Man and Animals: a Critical and Historical Study. John Hopkins Hospital Reports, Vol. 8, pp. 1-155, 3 pls.

1015. Nuttall, G. H. F. 1899 Die Rolle der Insekten Arachniden (Ixodes) und Myriapoden als Träger bei der Verbreitung von durch Bakterien und thierische Parasiten verursachten Krankheiten des Menschen und der Thiere. [The Rôle of Insects, Arachnids (Ixodes) and Myriapods as Carriers, in the Spread of Diseases of Men and Animals, Caused by Bacteria and Animal Parasites.] Hyg. Rundschau, Vol. 9 (72 pp. reprint).

1016. Nuttall, G. H. F. 1899 The Part Played by Insects, Arachnids and Myriapods in the Propagation of Infective Diseases of Man and Animals. Brit. Med. J., 9 Sept. 1899 (4 pp. reprint).

1017. Nuttall, G. H. F. 1899 The Rôle of Insects, Arachnids and Myriapods in the Propagation of Infective Diseases of Man and Animals. J. Trop. Med. Nov. 1899, pp. 107-110.

1018. **Nuttall, G. H. F. 1907** Insects as Carriers of Disease. Reprint from "Bericht über den XIV Intern. Kongress f. Hyg. u. Demographie. Berlin."
Collective Annotation: Nuttall was perhaps the leading world authority on medical entomology at the turn of the century. He also appears to have taken seriously his role as a health educator, and readily accepted invitations to speak on the subject or to prepare articles for publication in professional journals. This accounts for the fact that several of his papers bear essentially the same title. Nor do they differ greatly as to content, save that while some contain much explanatory detail, others are much condensed, as conditions of publication might require. Collectively, they constitute an important chapter in the history of Sanitary Entomology.

1019. **Nuttall, G. H. F., Hindle, E.** and **Merriman, G. 1913** The Range of Flight of *Musca domestica.* Repts. Local Gov't Bd. on Publ. Health and Med. Subjects. N.S. No. 85. Further Repts. (No. 6) on Flies as Carriers of Infection, pp. 20-41, 11 charts.
Seems to depend considerably on how far the fly must travel to find food and/or breeding medium.

1020. **Nuttall, G. H. F.** and **Jepson, F. P. 1909** The Part Played by *Musca domestica* and Allied (non-biting) Flies in the Spread of Infective Diseases. A Summary of our Present Knowledge. Rept. to Local Gov't. Bd. on Publ. Health and Med. Subjects. N.S. No. 16:13-41.
The authors stress particularly typhoid and cholera. They state: "In potential possibilities, the dropping of one fly may, in certain circumstances, weigh in the balance as against buckets of water or milk."

O'Connor, F. W., joint author. *See* Wenyon, C. M., et al., 1917.

1021. **Odlum, W. H. 1908** Are Flies the Cause of Enteric Fever? J. Roy. Army Med. Corps. 10:528-530.

1022. **Okamoto, H. 1924** The Insect Fauna of Quelpart Island (Saishiu-to). Bull. Agric. Expt. Sta. Gov't. Gen. Chosen. 1(2), pp. i-iv and 47-233, 4 pls., 1 map. Korea.
List includes *M. domestica* and *M. autumnalis* De Geer (=*corvina* F.)
ABSTRACT: RAE-B 12:155, 1924.

1023. **Okulov, V. 1947** [The Results of Tests of the New Soviet Insecticide, Pantachlorin Paste.] Med. Parasit. 16(1):33-35. Moscow. (In Russian.)
Contained 40 per cent DDT. Applied as a spray. Residual effect less satisfactory against *M. domestica* than most DDT preparations. ABSTRACT: RAE-B 36:128, 1948.

1024. **Olenev, N. O. 1936** [Notes sur la parasitologie en Carélie.] Med. Parasitol. 5(6):957. Moscow. (In Russian.)
M. domestica L., *M. tempestiva* Fall., and *M. autumnalis* DeG. listed as annoying species. ABSTRACT: RAE-B 25:200, 1937.

1025. **Olive, E. W. 1906** Cytological Studies on the Entomophthoreae. i. The Morphology and Development of *Empusa.* Bot. Gaz. 41:192, 2 pls.
This is the well-known organism which causes the death of numerous houseflies.

1026. **Oliveira, S. J. de** and **Moussatche, I. 1947** Acão do DDT (Dichlorodifenil-tricloretana) sôbre larvas e pupas de *Musca domestica* Linneu. [The Action of DDT . . . on Larvae and Pupae of *Musca domestica* Linnaeus.] Rev. Brasil Biol. 7(1):67-72. (German Summary.)
At concentrations used, from 75 to 99 per cent of the immature forms were either killed or developed abnormally. ABSTRACTS: RAE-B 36:157, 1948; BA 21:2549, 1947.

1027. Olsen, A. B. 1912 Only a Flyspeck. Good Health, Vol. 10, No. 7, pp. 195-200, 3 figs.
A semi-popular type discussion, typical of this period.

1028. Olson, T. A. and **Dahms, R. G. 1945** Control of Housefly Breeding in Partly Digested Sewage Sludge. J. Econ. Ent. 38:602-604.
Records the results from use of several combinations of DDT, also borax and rotenone. ABSTRACTS: RAE-B 35:26, 1947; BA 20:828, 1946.

1029. Olson, T. A. and **Dahms, R. G. 1946** Control of Housefly Breeding in Partly Digested Sewage Sludge. Public Works 77:5, 24.
Reports rotenone and DDT emulsions as of little larvicidal value. Obtained 100 per cent kill with borax and with kerosene. ABSTRACT: BA 20:1758, 1946.

Ono, W., joint author. *See* Takano, T., et al., 1947.

1030. Ono, Z., 1939 [On Insects of Hygienic Importance in Shinkyo and its Environs, Manchuria.] Ent. World 7(59):4-11. Tokyo. (In Japanese.)
M. domestica first appear in April, become abundant in mid-June, decrease in November. ABSTRACT: RAE-B 27:108, 1939.

1031. Oosthuizen, M. J. and **Smit, B. 1936** Electrocuting Insects. Fmg. in S. Africa 11:103-104, 121. (Reprint No. 27, 3 pp., 3 figs.) Pretoria.
Electrically charged screens (of more than one type) found satisfactory in control of *M. domestica.* ABSTRACT: RAE-B 24:214, 1936.

1032. Orton, S. F. and **Dodd, W. L. 1910** Experiments on Transmission of Bacteria by Flies with Special Relation to an Epidemic of Bacillary Dysentery at the Worcester State Hospital, Masschusetts. Boston Med. Surg. Jour. 163:863-868.
Organisms recovered from flies several days after insects were contaminated. The flies had by this time invaded other rooms of the hospital.

1033. Osten-Sacken, C. R. 1887 On M. Portchinski's Publication on the Larvae of Muscidae, including a Detailed Abstract of his last Paper; Comparative Biology of the Necrophagous and Coprophagous Larvae. Berlin Ent. Zeit. 31:17-28.
Contains much information concerning the morphology and behavior of mature and immature stages of *M. domestica* and *M. corvina.*

1034. Ostrolenk, M. and **Welch, H. 1942** The House Fly as a Vector of Food Poisoning Organisms in Food Producing Establishments. Am. J. Publ. Health 32(5):487-494.
Flies fed on food infected with *Salmonella enteritidis* may infect other flies as well as food, water and miscellaneous surfaces. Organisms can live in the fly for at least 4 weeks (=life of the fly). ABSTRACT: BA 17:2025, 1943.

1035. Ostrolenk, M. and **Welch, H. 1942** The Common Housefly (*Musca domestica*) as a Source of Pollution in Food Establishments. Food Res. 7:192-200.
Gives photos of fly cages and control procedures. Good bibliography.

1036. Otway, A. L. 1926 A Method of Excreta Disposal in the Tropics which Entirely Prevents Fly Dissemination. J. Roy. Army Med. Corps. 46(1):14-22.
Employs a large pit, covered with timber, brushwood and beaten earth, tarred or treated with heavy oil. Wire cone trap, placed near one end (where light is allowed to enter) captured 250,000 flies in 5 days. ABSTRACT: RAE-B 14:26, 1926.

1037. **Packard, A. S. 1874** On the Transformation of the Common House-fly with Notes on Allied Forms. Proc. Boston Soc. Nat. Hist. 16: 136-150, 1 pl.
A fairly complete account of the fly's anatomy and life history. Shortest time for complete development, 10 days.

1038. **Packchanian, A. A. 1944** Malaria Thick Films Contaminated with Excretions of Flies Containing Flagellates (Herpetomonas). Am. J. Trop. Med. 24(2):141-143.
Blood contaminated with *H. muscae domesticae* may be falsely determined as positive for *Trypanosoma*, if technician is inexperienced. Blood films should be protected.

1039. **Page, A. B. P., Stringer, A.** and **Blackith, R. E. 1949** Bioassay Systems for the Pyrethrins. I. Waterbase Sprays against *Aedes aegypti* and other Flying Insects. Ann. Appl. Biol 36(2):225-243.
M. domestica used as a test insect. ABSTRACT: RAE-B 38:34, 1950.

1040. **Page, G. P. 1919** Experiments with Insectox (A Substance for the Destruction of Flies). J. Roy. Naval Med. Ser. London 5(4):432-436.
As applied, killed all insect life in a room after 30 minute exposure. ABSTRACT: RAE-B 7:192, 1919.

1041. **Paillot, A. 1933** L'infection chez les insectes. [Infection in Insects.] Trévoux, G. Patissier. 535 pp.
Figures the mycelium and conidiophores of the fungus *Empusa muscae*.

1042. **Paine, J. H. 1912** The House-fly in Relation to City Garbage. Psyche. 19:156-159.
Of flies issuing from garbage during the summer, 22 per cent were *Musca domestica*.

1043. **Palmer, J. W. 1910** The Relation the House-fly Bears to Typhoid and other Infectious Diseases. Atlantic J. Rec. of Med. for August.

1044. **Paraf, J. 1920** The Spread of Bacillary Dysentery by Flies. Rev. d'Hyg. et de Police Sanitaire. (For 1920), p. 24.
Flies captured near latrines, in wards, and at table, during epidemic at Vinewil, carried dysentery bacillus. ABSTRACTS: Ann. d'Igiene, Rome 31(1):66, 1921; RAE-B 9:94, 1921.

1045. **Paraf, J. 1920** Etude expérimentale du rôle des mouches dans la propagation de la dysenterie bacillaire. [Experimental Study of the rôle of flies in the propagation of bacillary dysentery.] Rev. Hyg. 62:241-244.
Bacillus remained in fly's intestine five days after initial infection.

1046. **Paramonov, S. J. 1934** Dipterenlarven zur biologischen Behandlung von Osteomyelitis und Gasbrand. [Dipterous Larvae in the Biological Management of Osteomyelitis and Gas Gangrene.] Zeit. Wiss. Insektenbiol. 27(5-6):82-85.
Considers larvae of *M. domestica* safer for use in maggot therapy than larvae of *Calliphora*. ABSTRACT: RAE-B 23:57, 1938.

1047. **Paramonov, S. Y. 1934** Dipterenlarven als Mittel gegen die Gangräne, Osteomyelitis u. s. w. [Dipterous Larvae as a Remedy against Gas Gangrene, Osteomyelitis and Similar Conditions.] J. Cycle Bio-Zool. Acad. Sci. Ukr. No. 3(7):73-83. Kiev. (In Ukrainian, with German Summary.)
A more extensive paper than the preceding. Records cases saved from almost certain amputation.

1048. **Parant, G. 1905** Un procédé de destruction des mouches. Bull. Soc. Autun, Vol. 17, pp. 118-124.

1049. Parenti, G. 1932 La citta senza mosche. [The Fly-free Town.] 11 pp., 9 figs., Montecatini.
Montecatini Terme, Italy, rendered free from flies by spraying all possible breeding places with arsenical bait and placing bait in animal quarters. ABSTRACT: RAE-B 20:208, 1923.

1050. Parenti, G. 1934 A proposito di un esteso esperimento di lotta contro le mosche. [On the Design of an Extensive Experiment concerning the War against Flies.] Boll. Soc. Ital. Biol. Sper. 9(5):374-378. Milan.
At Rimini, use of 3,300 gal. of spray (beet molasses, sodium arsenite, water) failed to control *M. domestica*. Gives reasons. ABSTRACT: RAE-B 22:155, 1934.

1051. Parisot, J. and **Fernier, L. 1934** The Best Methods of Treating Manure-heaps to Prevent the Hatching of Flies. Quart. Bull. Health Org. League of Nations. 3 (Extract No. 1), 31 pp., 5 pls., 7 figs.
Suggests receptacles for watering manure without contamination of soil. Rise in temperature kills eggs and larvae. ABSTRACT: RAE-B 22:154, 1934.

1052. Parker, H. L. and **Thompson, W. R. 1928** Contribution à la biologie des chalcidiens entomophages. [Contribution to the Biology of Entomophagous Chalcids.] Ann. Soc. Ent. Fr. 97:425-465.
Mormoniella vitripennis and *Spalangia* sp. reared from pupae of several muscid genera. ABSTRACT: RAE-B 17:101, 1929.

1053. Parker, J. B. 1917 A Revision of the Bembecine Wasps of America North of Mexico. Proc. U.S. Nat. Mus., Washington 52:1-55.
M. domestica one of 11 species of flies found in nests of *Bembex spinolae*. ABSTRACT: RAE-B 5:74, 1917.

1054. Parker, K. G. 1936 Fire Blight: Overwintering, Dissemination and Control of the Pathogen. Cornell Agric. Expt. Sta. Mem. 193.
Flies believed important in transmission of infection from oozing cankers to blossoms and from flower to flower.

1055. Parker, R. R. 1914 Summary of "Report to the Montana State Board of Entomology concerning Fly Investigations Conducted in the Yellowstone Valley during the Summer of 1914". First Bienn. Rept. Montana State Bd. Ent. 1913-14. Helena, pp. 35-50.
Ninety per cent of the flies captured on garbage were house flies. ABSTRACT: RAE-B 3:61, 1915.

1056. Parker, R. R. 1915 New Evidence Concerning the Dispersal of the House Fly. Bull. Dept. Publ. Health Montana 9(7-8):3-7.
Concludes that flies from a given breeding place may spread over a city of at least 5 square miles. ABSTRACT: RAE-B 4:78, 1916.

1057. Parker, R. R. 1916 The House-fly in Relation to Public Health in Montana. Article 1. Some Facts Concerning its Habits. Bull. Dept. Publ. Health. Helena 9(9-10):6-11.
One thousand pounds of horse manure may contain 450,000 maggots. ABSTRACT: RAE-B 5:71, 1917.

1058. Parker, R. R. 1916 The House-fly in Relation to Public Health in Montana. Article 2. The House-fly as a Disease Carrier. Bull. Dept. Publ. Health, Helena 9(11):5-11.
Of flies captured in dwellings, 75 per cent were found infected with intestinal bacteria. ABSTRACT: RAE-B 5:71, 1917.

1059. Parker, R. R. 1916 Dispersion of *Musca domestica* Linn. under City Conditions in Montana. J. Econ. Ent. 9(3):325-354.
Marked flies recaptured at 78 stations located from 50 to 3,500 yards from 4 points of release. ABSTRACT: RAE-B 4:147, 1916.

1060. Parker, R. R. 1916 The House Fly and the Control of Flies. Second Bienn. Rept., Montana State Bd. Ent. 1915-1916. Bozeman, pp. 57-66.
Recommends incinceration of garbage, for cities; maggot traps for the country. ABSTRACT: RAE-B 5:81, 1917.

1061. Parker, R. R. 1917 Seasonable Abundance of Flies in Montana. Ent. News 28(6):278-282.
M. domestica reaches a peak during the first 3 weeks in August, decreases abruptly thereafter, but many seek houses, to escape cold. ABSTRACT: RAE-B 5:125, 1917.

1062. Parker, R. R. 1918 Data concerning Flies that Frequent Privy Vaults in Montana. (Diptera.) Ent. News 29(4):143-146.
In all species, females predominated. Concludes that any bait which will attract females primarily is best. Example, beer and oatmeal. ABSTRACT: RAE-B 6:116, 1918.

Parker, R. R., joint author. *See* Honey, J. A., et al., 1914.

1063. Parker, W. L. and **Beacher, J. H. 1947** Toxaphene, a Chlorinated Hydrocarbon with Insecticidal Properties. Bull. Del. Agric. Expt. Sta. No. 26 (Technical No. 36) 27 pp., 4 figs., Newark.
A residual insecticide with slow knock-down. Gives high kills, however, at low concentrations. ABSTRACT: RAE-B 36:198.

Parker, W. L., joint author. *See* Beacher, J. H., et al., 1948.

1064. Parkes, L. C. 1911 The Common House-fly. J. Roy. Sanit. Inst., May, 1911.

1065. Parkin, E. A. and **Green, A. A. 1944** Activation of Pyrethrins by Sesame Oil. Nature, London 154:16-17.
One of the early uses of synergists. Activation for kill does not depend on reduction of the rate of knock-down.

1066. Parkin, E. A. and **Green, A. A. 1945** The Toxicity of DDT to the Housefly, *Musca domestica* L. Bull. Ent. Res. 36(2):149-162.
DDT acts more slowly when dissolved in industrial methylated spirit than in kerosene. Full description of action on flies. ABSTRACTS: RAE-B 34:3, 1946; BA 20:1007, 1946.

1067. Parkin, E. A. and **Green, A. A. 1945** Residual Films of DDT. Nature 155(3944), p. 668.
Concerns changes in toxicity of DDT on treated surfaces with passage of time; crystal formation, etc. ABSTRACT: RAE-B 34:122, 1946.

1068. Parkin, E. A. and **Green, A. A. 1947** DDT Residual Films. The Persistence and Toxicity of Deposits from Kerosene Solutions on Wall Board. Bull. Ent. Res. 38(2):311-325.
Absorbent surfaces require as high a concentration as is consistent with reasonable coverage, to avoid much DDT being carried into the substratum by the solvent. ABSTRACT: RAE-B 36:17-18, 1948.

1069. Parman, D. C. 1920 Observations on the Effect of Storm Phenomena on Insect Activity. J. Econ. Ent. 13(4):339-343.
M. domestica (and other species), when the barometer is falling, become nervously active, then pass into a coma. ABSTRACT: RAE-B 8:186, 1920.

1070. Parman, D. C., Bishopp, F. C., Laake, E. W., Cook, F. C. and **Roark, R. C. 1927** Chemotrophic Tests with the Screw-worm Fly. U.S. Dept. Agric. Bull. 1472. 32 pp. Washington.
Repellents tested on *Cochliomyia macellaria*, with comparative tests on *Phormia*, *Lucilia*, and *M. domestica*.

Parman, D. C., joint author. *See* Bishopp, F. C., et al., 1915, 1925.

1071. Parrott, P. J. 1900 To Rid the House of Flies. Bull. No. 99, Kansas State Expt. Sta.
Includes design for a fly trap to be used in windows.

1072. Parrott, P. J. 1927 Progress Report on Light Traps for Insect Control. 12 pp. New York, N.Y. Empire State Gas & Electric Assoc.
Light used to attract flies to electric traps in dairy barn. Very effective. ABSTRACT: RAE-B 16:81, 1928.

1073. Patrissi, T., Barbieri, V., and **Bessler, C. 1949** Resistenza di *Musca domestica* al DDT ed impeigo di altri insetticidi. [Resistance of *Musca domestica* to DDT and the Use of other Insecticides.] Riv. di Malariol. 28(3):1-17. (English summary.)
Various explanations suggested. Report on experiments with Octa-Klor.

Patton, R. L., joint author. *See* Sarkaria, D. S., et al., 1949.

1074. Patton, W. S. 1908 *Herpetomonas lygaei.* Arch. f. Protistenk. 13:1-18, 1 pl.
"Double flagellum" state interpreted as the beginning of longitudinal fission. Applies also to *H. muscae-domesticae.*

1075. Patton, W. S. 1909 A Critical Review of our Present Knowledge of the Haemoflagellates and Allied Forms. Parasitology 2:91-139.

1076. Patton, W. S. 1919 Note on the Etiology of Oriental Sore in Mesopotamia. Bull. Soc. Path. Exot., Paris 12(8):500-504.
Ophthalmia carried by *M. domestica determinata* Walk., which also carries bacillary dysentery. *M. humilis* Wied. likewise incriminated. ABSTRACT: RAE-B 8:15, 1920.

1077. Patton, W. S. 1920 Some Notes on the Arthropods of Medical and Veterinary Importance in Mesopotamia and their Relation to Disease. Part II. Mesopotamian House Flies and their Allies. Indian J. Med. Res. 7(4):751-777, 4 pls., 6 figs.
Concerns differences in habit of *M. determinata, M. humilis, M. mesopotamiensis, M. tempestiva* and *M. vitripennis.* Control measures. ABSTRACT: RAE-B 9:15, 1921.

1078. Patton, W. S. 1920 Some Notes on the Arthropods of Medical and Veterinary Importance in Mesopotamia, and on their Relation to Disease. Part III. The Bot Flies of Mespotamia. Indian J. Med. Res. 8(1):1-16, 2 pls. 2 figs.
Includes synonymy of *M. humilis* Wied., considered one of the most important house flies of the East. ABSTRACT: RAE-B 9:102, 1921.

1079. Patton, W. S. 1921 Studies on the Flagellates of the Genera *Herpetomonas, Crithidia* and *Rhynchoidomonas.* No. 7. Some Miscellaneous Notes on Insect Flagellates. Indian J. Med. Res. 9(2):230-239, 3 pls.
H. muscae-domesticae Burnett is parasitic in *M. nebulo, M. humilis, Fannia, Borborus, Drosophila, Lucilia* and other forms. ABSTRACT: RAE-B 9:208, 1921.

1080. Patton, W. S. 1922 Notes on the Species of the Genus *Musca* Linnaeus. Part I. Bull. Ent. Res. 12(4):411-426.
Divides genus into two groups: 1, House-fly group; 2, Wild species group. ABSTRACT: RAE-B 10:103-104, 1922.

1081. Patton, W. S. 1922 New Indian Species of the Genus *Musca.* Indian J. Med. Res. 10(1):69-77.
M. prashadi, incerta, senior-whitei, villeneuvei described as new. ABSTRACT: RAE-B 10:196, 1922.

1082. Patton, W. S. 1923 Some Philippine Species of the Genus *Musca.* Philipp. J. Sci. 23:309-322.
Recognizes three groups: 1, Non-biting, occasionally haematophagous;

2, Non-biting, normally haematophagous; 3, True biting, blood sucking. ABSTRACT: RAE-B 12:7, 1924.

1083. Patton, W. S. 1923 A New Oriental Species of the Genus *Musca* with a Note on the Occurrence of *Musca dasyops* Stein, in China and a Revised List of the Oriental Species of the Genus *Musca* Linnaeus. Philipp. J. Sci. 23(4):323-335.

Corrects certain taxonomic errors and indicates synonymy. Eight species names involved. ABSTRACT: RAE-B 12:7, 1924.

1084. Patton, W. S. 1925 Diptera of Medical and Veterinary Importance. Philipp. J. Sci. 27:177-200, 397-411.

Lists 45 species of *Musca* known to himself, with synonyms and notes. ABSTRACT: RAE-B 13:162, 165, 1925.

1085. Patton, W. S. 1926 The Ethiopian Species of the Genus *Musca* Linnaeus. Rec. Indian Mus. 28(1):29-52, 4 pls.

A practical guide to identification of species, with keys to both sexes. ABSTRACT: RAE-B 15:12, 1927.

1086. Patton, W. S. 1926 Blood-sucking Arthropods of Medical and Veterinary Importance in China. China Med. J. 40(6-7):543-553, 603, 612.

Mentions that *M. domestica* occurs in North China, but believes that an Oriental species takes its place in the south. ABSTRACT: RAE-B 14: 190, 1926.

1087. Patton, W. S. 1932 Studies on the Higher Diptera of Medical and Veterinary Importance. A Revision of the Genus *Musca* based on a Comparative Study of the Male Terminalia. I. The Natural Grouping of the Species and their Relationship to Each Other. Ann. Trop. Med. Parasit. 26(3):347-405.

Utilizes the paramere, phallosome, anal cerci and lateral processes of the fifth ventrite. Divides the genus into three groups. ABSTRACTS: RAE-B 20:282, 1932; TDB 30:309, 1933.

1088. Patton, W. S. 1933 Studies on the Higher Diptera of Medical and Veterinary Importance. A Revision of the Genera of the Tribe Muscini, Subfamily Muscinae, based on a comparative Study of the Male Terminalia. I. The Genus *Musca* Linnaeus. Ann. Trop. Med. Parasit. 27(1):135-156.

The blood feeding species of *Musca* cannot penetrate the skin. Teeth of proboscis can merely remove a scab or scratch and tear the tissue. ABSTRACTS: RAE-B 21:144, 1933; TDB 30:614, 1933.

1089. Patton, W. S. 1933 Studies on the Higher Diptera of Veterinary and Medical Importance. A Revision of the Species of the Genus *Musca*, based on a Comparative Study of the Male Terminalia. II. A Practical Guide to the Palaearctic Species. Ann. Trop. Med. Parasit. 27:327-345; 397-430.

Useful keys to both sexes of 16 palaearctic species, founded on external characters. Terminology differs somewhat from other authors. ABSTRACTS: RAE-B 21:286, 1933; TDB 31:744, 1934.

1090. Patton, W. S. 1933 A New Haematophagous Species of *Musca* from Malaya. Ann. Trop Med. Parasit. 27(3):477-480, 2 figs.

Concerns *M. greeni* sp. nov. ABSTRACT: RAE-B 21:286, 1933.

1091. Patton, W. S. 1934 The Blood-sucking Species of the Genus *Musca* and the Evolution of the Blood-drawing Proboscis in the Genus. Proc. 5th Pacific Sci. Congress. Canada. 1933. 5:3361-3366.

Much morphological information. ABSTRACT: RAE-B 23:208, 1935.

1092. Patton, W. S. 1936 Studies on the Higher Diptera of Medical and Veterinary Importance. A Revision of the Species of the Genus

Musca, based on a Comparative Study of the Male Terminalia. III. A Practical Guide to the Ethiopian Species. Ann. Trop. Med. Parasit. 30:469-490.
Gives figures of larval spiracles and adult male genitalia. Gives separate keys to males and females. ABSTRACT: RAE-B 25:128, 1937.

1093. Patton, W. S. 1937 Studies on the Higher Diptera of Medical and Veterinary Importance. A Revision of the Species of the Genus *Musca*, based on a Comparative Study of the Male Terminalia. IV. A Practical Guide to the Oriental Special. Ann. Trop. Med. Parasit. 31:127-140; 195-213. Similar to Paper III (Ethiopian).
ABSTRACT: RAE-B 25:192 and 256, 1937.

1094. Patton, W. S. 1937 Male Genitalia of *Musca dasyops*. Ann. Trop. Med. Parasit. 31:209.

1095. Patton, W. S. and **Cragg, F. W. 1939** On Certain Haematophagous Species of the Genus *Musca* with Descriptions of Two New Species. Indian J. Med. Res. 1:11-25, 5 pls.
M. gibsoni, sp. nov., South India and *M. bezzi*, sp. nov., Nilgiri Hills. Remarks on feeding habits of the group. ABSTRACT: RAE-B 1:209, 1913.

1096. Patton, W. S. and **Senior-White, R. 1924** The Oriental Species of the Genus *Musca* Linnaeus. Rec. Indian Mus. 26(6):553-577, 5 pls.
Redescribes all species of *Musca* known from the Oriental Region, including *M. fletcheri* sp. nov. from India. ABSTRACT: RAE-B 13:46, 1925.

Patton, W. S., joint author. *See* Cornwall, J. W., et al., 1914.

1097. Paul, J. R., Trask, J. D., Bishop, M. B., Melnick, J. L. and **Casey, A. E. 1941** The Detection of Poliomyelitis Virus in Flies. Science 94:395-396.
From a summer camp in Connecticut and from Jasper, Alabama, flies were collected from which inoculation into monkeys caused polio. Several species of flies were involved.

Paul, J. R., joint author. *See* Trask, J. D., et al., 1943.

1098. Pauli, M. E. 1927 Die Entwicklung geschnürter und centrifugierte Eier von *Calliphora erythrocephala* und *Musca domestica*. [The Development of Ligatured and Centrifuged Eggs of *Calliphora erythrocephala* and *Musca domestica*.] Zeit. Wiss. Zool. 129:483-540.

1099. Pavlovski, E. N. 1923 [Flies.] 100 pp., 48 figs., 1 graph. Moscow. National Kommisariat Health Protect.
First of five parts devoted chiefly to *M. domestica*, its structure, life history, etc. ABSTRACT: RAE-B 11:66, 1923.

1100. Pavloskii, E. N., Stein, A. K. and **Buichkov, V. A. 1932** Experimentelle Untersuchungen über die Wirkung einiger Verdaungssekrete von *Musca domestica* auf die Hautdecken des Menschen. [Experimental Investigations on the Action of Some Digestive Secretions of *Musca domestica* on the Skin Surface of Humans.] Mag. Paras. Inst. Zoon Acad. Sci. URSS 3:131-147, 7 figs. (In Russian, with German summary.)
Saliva of the fly contains thermolabile elements which have an irritating effect on human skin. ABSTRACT: RAE-B 23:53, 1935.

Payne, M. G., joint author. *See* List, G. M., et al., 1947.

1101. Peairs, L. M. 1927 Some Phases of the Relation of Temperature to the Development of Insects. Coll. Agric., West Virginia Univ. Agric.

Expt. Sta. Bull. 208.
Studied several species, including *M. domestica.* Curves do not agree with those of Larsen and Thomsen, 1940.

1102. Pearsall, D. E. and **Wallace, P. P. 1946** Insecticidal Cords. A Convenient Method of Generating Insecticidal Smoke by Burning Cords Impregnated with DDT or other Toxic Materials. Soap and Sanit. Chem. 22(10):139, 141, 143, 161, 163.
Used rotenone, pyrethrum, nicotine, and DDT. At concentration used, 1 foot of cord per 50-60 cubic feet gave complete kill. ABSTRACT: RAE-B 37:105, 1949.

1103. Pearson, A. M. 1935 The Role of Pine Oil in Cattle Fly Sprays. Bull. Del. Agric. Expt. Sta. No. 196, 63 pp., 7 figs., 39 graphs. Newark.
Pine oil increases toxicity of pyrethrum and improves repellency. ABSTRACT: RAE-B 24:304, 1936.

1104. Pearson, A. M. and **Richardson, C. H. 1933** The Relative Toxicity of Trisodium Arsenite and Arsenious Acids to the House Fly, *Musca domestica* L. J. Econ. Ent. 26:486-493.
Median lethal dose for either solution was 0·14 mg. arsenic per gram body weight. Repellency varied with pH. ABSTRACT: RAE-B 21:175, 1933.

1105. Pease, H. D. 1910 Relation of Flies to the Transmission of Infectious Disease. Long Island Med. J. Dec. 1910.

1106. Pech, M. J. 1913 Note sur l'etiologie de la fièvere typhoide. La Caducée, 15 February, 1913, p. 48.
Ammoniacal fermentation is an energetic sterilizer and *may* lessen spread of typhoid by flies which settle in military latrines. ABSTRACT: RAE-B 1:50, 1913.

1107. Peet, C. H. and **Grady, A. G. 1928** Studies in Insecticidal Activity (4 parts). J. Econ. Ent. 21:612-625. Geneva, N.Y.
Describes standard testing procedure, using flies. Aim is to hold all factors constant, except the biological element. ABSTRACT: RAE-B 26:255, 1928. Note: There have been several revisions of this technique, over the years.

1108. Peffly, R. L. and **Gahan, J. B. 1949** Residual Toxicity of DDT Analogs and related chlorinated Hydrocarbons to House Flies and Mosquitoes. J. Econ. Ent. 42:113-116.
Tested 21 analogs and 19 other new chlorinated hydrocarbons. ABSTRACT: RAE-B 38:53, 1950.

Peffly, R. L., joint author. *See* Gahan, J. B., et al., 1948; Sundararaman, S., et al., 1949.

Penner, L. R., joint author. *See* Melnick, J. L., et al., 1947.

1109. Peppler, H. J. 1944 Usefulness of Microorganisms in Studying Dispersion of Flies. Bull. U.S. Army Med. Dept. No. 75, pp. 121-123. Washington.
Flies sprayed with yeast culture at breeding places were recovered in a bivouac area 3 miles away. ABSTRACT: RAE-B 33:70, 1945.

1110. Pereira, C. 1947 A luta contra as moscas. [Regarding the Struggle against flies.] Biológico (São Paulo) 13(2):25-43, 1 pl., 4 figs.
A compilation of information on many aspects of the fly problem, including control procedures. ABSTRACTS: RAE-B 36:158, 1948; BA 22:209, 1948.

1111. Pereira, C. and **de Castro, M. P. 1945** Contribuicão para o conhecimento da espécie tipo de *Macrocheles* Latr. (Acarina): *M. muscaedomesticae* (Scopoli, 1772) emend. [Contribution to our knowledge of the type species of *Macrocheles* Latr. (Acarina): *M. muscaedomesticae* (Scopoli, 1772) Correction.] Arq. Inst. Biol. 16:153-186. (São Paulo.)

Considered *Macrocheles muscae* Ewing and *Holostaspis badius* Koch to be synonyms of *M. muscaedomesticae*. Mites feed on fly eggs and use adults for transportation. ABSTRACTS: BA 22:219, 1948; RAE-B 36:24, 1948.

Pereira, C., joint author. *See* DeMello, M. J., et al., 1946.

1112. Perje, A. M. 1948 Studies on Spermatogenesis in *Musca domestica* L. Hereditas 34:209-232.

First contribution to such knowledge for the housefly.

1113. Pestico, J. F. 1918 La mosca domestica. Rev. Agric. Bogotá 4(2): 98-101, 1 fig.

Describes possible control measures relative to an anti-fly campaign about to be undertaken in Medéllin, Colombia. ABSTRACT: RAE-B 6:161, 1918.

1114. Peterson, A. 1918 House-fly Investigations. Rept. Dept. Ent for 1917, New Jersey Agric. Expt. Sta., New Brunswick, pp. 479-484.

Flies controlled at Beach Haven, N.J. through manure disposal, trapping and use of poison baits. ABSTRACT: RAE-B 8:29, 1920.

1115. Peterson, A. 1945 Some Insect Infants. Sci. Monthly 60(6):426-442, 59 figs.

Includes larva of *M. domestica.*

1116. Petrishcheva, P. A. 1932 Zur Biologie der Hausfliege in den Bedingungen der Stadt Samara. [Concerning the Biology of Houseflies under Conditions (found) in the State of Samara.] Mag. Paras. Inst. Zool. Acad. Sci. (URSS) 3:161-182. (In Russian.)

In Samara, on the Volga, *M. domestica* has 6 generations a year, averaging 4 weeks each. Fertile females hibernate. ABSTRACT: RAE-B 23:53, 1935.

1117. Petrova, E. F. 1942 [Synanthropic Flies of Alma-Ata.] Med. Parasit. 11(1-2):86-89, 1 graph. (In Russian.)

Lists 19 species, of which *M. domestica* is the most prevalent. *Phormia azurea* dominates in slaughter houses. ABSTRACTS: RAE-B 32:5, 1944 BA 20:655, 1946.

1118. Petrova, E. F. 1944 [On the Question of the Method of Applying Arsenic for the Control of Flies.] Izv. Kazakhsk. Fil. Akad. Nauk SSSR (Ser. Zool.) 3:119-122. Alma-Ata. (In Russian.)

Miofanin (5 per cent sugar plus 1·0-2·5 per cent sodium arsenite) applied to inner sides of window panes, 0·03 fluid oz. per sq. yd. Lower concentrations not effective. ABSTRACT: RAE-B 35:159, 1947.

1119. Phalen, J. M. 1917 U.S. Army Methods of Disposal of Camp Refuse. Amer. J. Publ. Health 7(5):481-484.

Concerns garbage, animal manure, human excreta, night urine, etc. ABSTRACT: RAE-B 5:129, 1917.

1120. Phelps, E. B. 1916 Fly Poisons. Studies on Sodium Salicylate, a new Muscicide, and on the Use of Formaldehyde. U.S. Publ. Health Repts. 31(44):3033-3035.

Gives formulae and methods of use in the household. ABSTRACT: RAE-B 5:9, 1917.

1121. Phelps, E. B. and **Stevenson, A. F. 1916** Experimental Studies with Muscicides and other Fly-destroying Agencies. U.S. Publ. Health Serv. Hyg. Lab. Bull. 108.

Found sodium fluoride effective, but dangerous to use.

1122. Philip, C. B. 1937 The Transmission of Disease by Flies. Publ. Health Repts. Suppl. No. 29, 22 pp.
Includes photo of a fly's foot, magnified.

Phillips, G. L. joint author. *See* McGovran, E. R., et al., 1939; Sullivan, W. N., et al., 1938.

1123. Piana, G. P. 1896 Osservazioni sul *Dispharagus nasutus* Rud. dei polli e sulle Nematoelmintiche delle mosche e dei porcellioni. [Observations on *Dispharagus nasutus* of Poultry and on a Roundworm of Flies and of Pigs.] Atti Soc. Ital. Sci. Nat. 36:239-262, 21 figs.
Worm from proboscis of *M. domestica*. Both male and female genital organs in the same individual.

1124. Picard, J. P. and **Kearns, C. W. 1949** Analysis of the Essential Structural Features of DDT by a Study of the Toxicity of Closely Related Compounds to Roaches and to Houseflies. Canad. J. Res. (D) 27(2):59-67.
Toxicity affected by the "whole shape" of the molecule. ABSTRACT: RAE-B 38:41, 1950.

1125. Pickard-Cambridge, O. 1892 On the British Species of False-Scorpions. Proc. Dorset Nat. Hist. and Antiq. Field Club. 13:199-231, 3 pls.
Describes *Chernes nodosus*, frequently found attached to legs of houseflies, thus achieving dispersal.

1126. Piedrola Gil, G. and **Valdes Garcia, A. 1949** Un nuevo método de lucha totalitaria contra la mosca adulta en les medios rural y urbano. [A New Method of Total War against the Adult Fly for Rural and Urban Benefit.] Med. Colonial (Madrid) 13:83-131.
Concerns DDT and benzene hexachloride as used in Spain. ABSTRACT: BA 24:244, 1950.

Pierce, W. D., joint author. *See* Howard, L. O., et al., 1921.

1127. Pierpont, R. L. 1939 Terpene Esters in Pyrethrum and Rotenone Fly Sprays. Bull. Del. Agric. Expt. Sta. No. 217, 59 pp., 15 figs. Newark.
Recommends inclusion of ethylene ether of pinene in fly sprays. ABSTRACT: RAE-B 28:180, 1940.

1128. Pierpont, R. L. 1945 Terpin Diacetate as an Activator for Pyrethrum. J. Econ. Ent. 38:123-124.
Peet-Grady test showed toxicity always increased by addition of this compound. ABSTRACT: RAE-B 34:83, 1946.

1129. Pierpont, R. L. 1945 Development of a Terpene Thiocyano Ester (Thanite) as a Fly Spray Concentrate. Bull. Del. Agric. Expt. Sta. No. 253, 58 pp., 14 figs.
Led to various formulations, all termed "thanite". Gives more complete knockdown than pyrethrum, but action is a little slower. ABSTRACT: RAE-B 36:165, 1948.

1130. Pinkus, H. 1913 The Life-History and Habits of *Spalangia muscidarum* Richardson; a parasite of the Stable fly. Psyche 20(5):148-158, 1 pl., 1 fig.
This parasite has also been bred from *M. domestica*. Adult deposits 1 egg in each host. ABSTRACT: RAE-B 2:22-23, 1914.

1131. Pinto, C. 1939 Disseminacão da malaria pela aviação; biologia do *Anopheles gambiae* e outros anofelineos do Brasil. [Dissemination of Malaria at Airport; Biology of *Anopheles gambiae* and other Anophelines in Brazil.] Mem. Inst. Oswaldo Cruz 34(3):293-430, 35 figs., 61 pls.
Fourth stage larvae of *A. gambiae* reared on bread crumbs and fragments of the thorax of flies (*M. domestica*). ABSTRACT: RAE-B 28:169, 1940.

1132. Pipkin, A. C. 1942 Filth Flies as Transmitters of *Endamoeba histolytica.* Proc. Soc. Exp. Biol. and Med. 49(1):46-48. New York.
M. domestica and other muscoids studied as to survival of trophozoites and cysts on body surface and in various organs of the flies. ABSTRACT: RAE-B 30:114, 1942.

1133. Pipkin, A. C. 1943 Experimental Studies on the Role of Filth Flies in the Transmission of Certain Helminthic and Protozoan Infections of Man. Abstr. Theses Tulane Univ. 44(1):9-13, 1943.
Small helminth ova ingested, larger ova carried only on exterior of fly's body. Further data on *Endamoeba histolytica.* ABSTRACT: BA 20: 1756, 1946.

1134. Pipkin, A. C. 1949 Experimental Studies on the Role of Filth Flies in the Transmission of *Endamoeba histolytica.* Am. J. Hyg. 49(3): 255-275.
Transmission by way of the alimentary tract important. Viability of parasite not reduced by storage in fly's crop. ABSTRACTS: RAE-B 40:97, 1952; BA 23:2596, 1949.

Piquett, P. G., joint author. *See* McGovran, E. R., et al., 1943, 1944, 1945.

1135. Pivovarov, V. M. 1939 Méthode de lutte contre les mouches au Moyen des pondres. [Technique of the War against Flies by Means of Dusting.] Med. Parasit. 8(3):362-363. Moscow. (In Russian.)
Powdered sugar plus 5-6 parts sodium arsenite or arsenic trioxide. At 0·5 gm. per 10 sq. ft., killed most flies in 2-3 hours. ABSTRACT: RAE-B 28:40, 1940.

1136. Place, F. E. 1916 The Flies that Defile. The Register, Adelaide 81(21 & 39):6, 1916.
Discusses relation of *M. domestica, Fannia scalaris* and blow flies to intestinal and other infections. ABSTRACT: RAE-B 5:49 1917.

1137. Plowman, C. F. and **Dearden, W. F. 1915** Fighting the Fly Peril. A Popular and Practical Handbook. London, T. Fisher Unwin, Ltd. 127 pp.
Title conveys content. Has a clever introduction by A. E. Shipley, Master of Christ's College, Cambridge.

1138. Podyapolskaya, V. and **Gnedina, M. 1934** [Sur le rôle des mouches dans l'épidémiologie des helminthoses.] Med. Parasit 3(2):179-185. Moscow. (In Russian, with French summary.)
Average of 3 helminth eggs found in specks deposited by 34 flies in 12 hours. (Human feces available.) ABSTRACTS: RAE-B 22:196, 1934; TDB 32:232, 1935.

1139. Pokrovskii, S. N. and **Zima, G. G. 1938** Mouches comme transporteurs des oeufs des helminthes dans les conditions naturelles. [Flies as Carriers of the Eggs of Helminths under Natural Conditions.] Med. Parasit. 7(2):262-264. (In Russian.)
Dissected nearly 3000 flies, chiefly *M. domestica.* Helminth ova found in 11. ABSTRACT: RAE-B 26:244, 1938.

1140. Poore, C. V. 1901 Flies and the Science of Scavenging. Lancet, pt. 1, pp. 1389-1391, May 18, 1901.

1141. Portchinsky, I. A. 1892 Biology des Mouches Coprophagues et Nécrophagues. Hor. Soc. Ent. Ross. 26:63-131.
An extensive paper, well illustrated.

1142. Portchinsky, I. A. 1910 Recherches biologiques sur le *Stomoxys calcitrans* L. et biologie comparée des mouches coprohagues. Publica-

tion of the Entomological Bureau, Russian Dept. of Land Administr. and Agriculture. Vol. 8, No. 8, 91 pp., 97 figs. (In Russian.)
Latter portion deals with *M. domestica* and allies.

1143. Portchinsky, I. A. 1911, 1913 *Hydrotaea dentipes* F.; Sa biologie et la destruction par ses larves de celles des *Musca domestica* L. [*Hydrotaea dentipes* F.; Its biology and the destruction by its larvae of the larvae of *Musca domestica*.] Mem. Bur. Ent. Russian Dept. of Land Administr. and Agriculture. Vol. 9, No. 5, 30 pp., 25 figs.
Larvae become predaceous in third stage and devour larvae of at least 3 genera of muscoids. Are long-lived in the larval state. ABSTRACT: RAE-B 1:149, 1913.

1144. Portchinsky, I. A. 1913 *Muscina stabulans* Fall., Mouche nuisible à l'homme et à son ménage, en état larvaire destructeuse des larves de *Musca domestica*. [*Muscina stabulans* Fall., Pest Fly of Man and his Environment, in the Larval State a Destroyer of the Larvae of *M. domestica*.] Publications of the Entomological Bureau, Russian Dept. of Land Administr. and Agriculture, Vol. 10, No. 1, 39 pp., 32 figs.
Larvae of *Muscina*, after completion of the second stage, follow and attack larvae of *M. domestica* and exterminate them. ABSTRACT: RAE-B 1:108-110, 1913.

1145. Porter, A. 1909 The Life-cycle of *Herpetomonas jaculum* (Léger), Parasitic in the Alimentary Tract of *Nepa cinerea*. Parasitology 2:367-391, 1 pl.
Confirms Patton's (1908-1909) account of the life history of *H. muscae-domesticae*.

1146. Porter, A. 1911 The Structure and Life-history of *Crithidia pulicis* n. sp., Parasitic in the Alimentary Tract of the Human Flea, *Pulex irritans*. Parasitology 4:237-254, 1 pl.
Corrects faulty generic interpretation of Dunkerly (1911) in regard to a parasite of *M. domestica*.

Post, R. L., joint author. *See* Munro, J. A., et al., 1947.

1147. Potgieter, J. T. 1945 Blow-flies and Flies in Towns. Fmg. in S. Africa, 1945. Reprint No. 1, 4 pp., 1 fig. Pretoria.
Includes usual measures to control *M. domestica*. Recommends poison bait, based on sodium arsenite. ABSTRACT: RAE-B 35:66, 1947.

1148. Pouillaude, I. 1913-1914 Les Mouches communes. Insecta Rennes iii. Nos. 34-36, Oct.-Dec. 1913, pp. 410-412, 444-448, 479-482; iv, Nos, 37-41, Jan.-May 1914, pp. 27-34, 73-75, 99-108, 146-148, 173-180, 25 figs.
Contains much information, all compiled. Lists a bibliography of 50 works. ABSTRACT: RAE-B 2:134, 1914.

1149. Power, M. E. and **Melnick, J. L. 1945** A Three-year Survey of the Fly Population in New Haven during Epidemic and Non-epidemic Years for Poliomyelitis. Yale J. Biol. and Med. 18(1):55-69.
The green bottle fly (*Phaenicia sericata*) proved dominant. *M. domestica* did not exceed 10 per cent of any single collection. ABSTRACT: BA 20: 829, 1946.

1150. Power, M. E., Melnick, J. L. and **Bishop, M. B. 1943** A Study of the 1942 Fly Population of New Haven. Yale J. Biol. and Med. 15(5): 693-705.
A non-epidemic year for poliomyelitis. Four or more genera of muscoids common. Bait preferences compared. ABSTRACT: RAE-B 32:174, 1944.

1151. Pratt, F. C. 1912 Insects Bred from Cow Manure. Canad. Ent. 44: 180-184.
M. domestica well represented. (A posthumous paper, compiled from notes by W. D. Hunter.)

1152. Prill, E. A., Hartzell, A. and **Arthur, J. M. 1946** Insecticidal Thio Esthers Derived from Safrole, Isosafrole, and other Aryl Olefins. Contr. Boyce Thompson Inst. 14(3):127-150.
These products, in general, have a synergistic effect against house flies, when combined with pyrethrins. ABSTRACT: RAE-B 37:159, 1949.

1153. Prill, E. A., Hartzell, A. and **Arthur, J. M. 1947** Some Cyclic Acetals Containing the 3,4-methylenedioxyphenyl Radical and their Insecticidal Effectiveness against Houseflies. Contr. Boyce Thompson Inst. 14:397-403.
Tested 17 compounds, ranging from low to high activity. ABSTRACT: RAE-B 38:147, 1950.

1154. Prill, E. A., Synerholm, M. E. and **Hartzell, A. 1946** Some Compounds Related to the Insecticide "DDT" and their Effectiveness against Mosquito Larvae and Houseflies. Contr. Boyce Thompson Inst 14(6):341-353.
Tested 31 analogs of DDT, about half of which were new. ABSTRACTS: RAE-B 38:110, 1950; BA 21:732, 1947.

Pringle, J. W. S., joint author. *See* Fraenkel, G., et al., 1938.

1155. Proto-Gomez, Y. and **Durán-Borda, G. 1888-1889** Sobre la causa de la meurte de los moscas en Bogotá. [On the Cause of Death of Flies in Bogotá.] Rev. Med. Bogotá 12:65-74.

1156. Prowazek, S. 1904 Die Entwicklung von *Herpetomonas*, einem mit den Trypanosomen verwandten Flagellaten. [The Development of *Herpetomonas*, one of the Trypanosome-related Flagellates.] Arb. Kaiserl. Gesundheitsamte. 20:440-452, 7 figs.
Describes in great detail, successive stages in the life cycle of *H. muscae-domesticae.*

1157. Prowazek, S. 1913 Notiz zur *Herpetomonas*-Morphologie sowie Bemerkung zu der Arbeit von Wenyon. [Memorandum concerning *Herpetomonas*-Morphology as a Comment on the Work of Wenyon.] Arch. f. Protistenk. 31:37-38.
Corrects what he considers to be misrepresentations of statements in his 1904 publication. ABSTRACT: TDB 2:465, 1913.

1158. Pruthi, H. S. 1946 Studies on House-fly and other Diptera. Abridg. Sci. Rept. Agric. Res. Inst. New Delhi 1941-1944, pp. 69-70.
Six species of *Musca* are found about Delhi, but only *M. nebulo* and *M. domestica vicina* are abundant. Control measures. ABSTRACT: RAE-B 37:5, 1949.

Pulver, R., joint author. *See* Läuger, P., et al., 1946.

1159. Purdy, J. S. 1910 Flies and Fleas as Factors in Disease. J. Roy. Sanit. Inst., Trans. Vol. 30, pp. 496-503.

1160. Puri, I. M. 1943 The House-frequenting Flies, their Relation to Disease and their Control. Health Bull. No. 31 Second Ed. 31 pp., illus. Gov't. of India Press, Simla.
Includes a key to the house-frequenting species of *Musca* which occur in India.

1161. Quarterman, K. D., Baker, W. C. and **Jensen, J. A. 1949** The Importance of Sanitation in Municipal Fly Control. Am. J. Trop. Med. 29:973-982.
Surveys indicate a continuing need for good sanitary practice regardless

of the success of newer insecticides. ABSTRACTS: RAE-B 40:164, 1952; BA 24:3570, 1950.

1162. Quill, R. H. 1900 Report of an Outbreak of Enteric Fever at Diyatalawa Camp, Ceylon, among the 2nd King's Royal Rifles. Army Med. Dept. Rept. Appendix 4, p. 425.
Outbreak at Boer camp preceded that among the troops. Flies very numerous; amounted to almost a plague. Camps were adjacent.

1163. Raabe, H. 1920 Studja nad Muchą domową. Przeglad Epidemjologiczny, Warsaw. 1(1):45-55. (In Polish, with French summary.)
Most adult flies die in the winter. Author believes that eggs, larvae, and pupae hibernate. ABSTRACT: RAE-B 8:193, 1920.

1164. Raimbert, A. 1869 Recherches expérimentales sur latr ansmission du charbon par les mouches. Compt. Rend. Acad. Sci., Paris, Vol. 69, pp. 805-812. *See also:* Paris Acad. Méd. Bull. 35:50, 215, 471, 1870 *and* Union Med. Paris 9:209, 350, 507, 709, 1870.
Gives experimental proof that house flies and meat flies are able to carry the anthrax bacillus which he found on their proboscides and legs.

1164a. Ralston, A. W. and **Barrett, J. P. 1941** Insect Repellent Activity of Fatty Acid Derivatives. Oil and Soap. 18:89-91.
Decyl, undecyl, undecenyl and dodecyl alcohols are highly repellent. Aliphatic nitriles containing 10-14 carbon atoms also have high repellency for flies. ABSTRACT: Hocking, B. 1960 Smell in Insects, p. 140 (3.111).

1165. Ramakrishna Iyer, T. V. 1935 The Housefly Nuisance and its Control with Maggot Traps. Madras Agric. J. 23(3):96-98, 2 pls.
Four types of traps used, all based on the principle that mature larva seeks drier habitat for pupation. ABSTRACT: RAE-B 23:182, 1935.

1166. Ramirez, R. 1898 The Diptera from a Hygienic Point of View. Public Health (U.S.A.) Vol. 24, pp. 257-259.

1167. Ramsbottom, J. 1914 Repts. to the Local Gov't. Bd. on Publ. Health and Med. Subjects. N.S. No. 102, iii, An Investigation of Mr. Hesse's Work on the Supposed Relationship of *Empusa muscae* and *Mucor racemosus*, pp. 31-32.
Empusa spore *never* gives rise to mycelium or fruit of the *racemosus* type, in spite of previous interpretations to that effect. ABSTRACT: RAE-B 3:90, 1915.

1168. Ransom, B. H. 1911 The Life-history of a Parasitic Nematode (*Habronema muscae*). Science, N.S. Vol. 34, pp. 690-692.
Adult worm is a parasite of the horse, the house-fly acting as a carrier of the larval form. Found that 28 per cent of the flies he collected were infected.

1169. Ransom, B. H. 1913 The Life History of *Habronema muscae* (Carter), a Parasite of the Horse Transmitted by the House Fly. U.S. Dept. Agric. Bur. Animal Ind. Bull. 163, pp. 1-36, 41 figs.
Embryos, thrown out with dejecta of the horse, enter the bodies of fly larvae which have hatched from eggs laid on the manure. ABSTRACT: RAE-B 1:223, 1913.

1169a. Rathay, E. and **Haas, B. 1883** Über *Phallus impudicus* (L.) und einige *Coprinus* Arten. [On *Phallus impudicus* and some forms of *Coprinus*.] Akad. Wiss. Vienna Math.—Natur-wissenschaft. 87:18-44.
M. domestica listed among species visiting *Phallus impudicus*, with a tabulation of the substances sought.

1170. Rancourt, M. 1945 Découverte récente d'un nouvel insecticide hexachlorocyclo-hexane. [Recent Discovery of a New Insecticide, Hexa-

chloro-hexane.] La Nature No. 3093:235-236. Paris.
Is benzene hexachloride, which can act as a stomach poison, a contact poison, or fumigant. Kills houseflies.

1171. **Réaumur, R. A. F. de. 1738** Mémoires pour servir à l'Histoire des insectes. Paris Vol. 4, p. 384 (*M. domestica*).
A classical work, important in the history of entomology.

Redman, K., joint author. *See* Munro, J. A., et al., 1945.

1172. **Reed, W. 1899** War Dept. Ann. Rept. for 1899, pp. 627-633. Washington.
Flies the cause of the outbreak of typhoid in the U.S. Army in 1899.

1173. **Reed, W. Vaughan, V. C.** and **Shakespeare, E. O. 1904** Report on the origin and spread of typhoid fever in U.S. military camp during the Spanish War of 1898. Vol. 1, text, 720 pp., Vol. 2, maps and charts.

Regan, W. M., joint author. *See* Freeborn, S. B., et al., 1925, 1928.

1174. **Reh, L. 1927** Ungewöhnliches Massen-Vorkommen von Fliegen in Häusern. [Unusual Mass Concentration of Flies in Houses.] Zeit. Desinfekt. u. Gesundheitswesen for 1927. No. 6, reprint, 4 pp. Königsbrück.
M. autumnalis DeG. occasionally invade houses in great numbers. Recommends fumigation with sulphur dioxide or HCN gas. ABSTRACT: RAE-B 15:174, 1927.

1175. **Reid, W. M.** and **Ackert, J. E. 1937** The Cysticercoid of *Choanotaenia infundibulum* (Block) and the Housefly as its Host. Trans. Am. Micr. Soc. 56(1):99-104.
Cysticercoids removed from body cavities of house flies. Good description.

1176. **Reingard, L. V.** and **Zabudko-Reingard, T. N. 1945** [The Common Mayweed (*Matricaria inodora*) as a Good but Little Studied Insecticide.] Med. Parasit. 14(3):92. (In Russian.)
Powdered, dried flowers as effective in laboratory tests as commercial pyrethrum. Used *M. autumnalis* as test species. ABSTRACT: RAE-B 34:208, 1946.

1177. **Reith, F. 1925** Die Entwicklung des Musca-Eies nach Ausschaltung verschiedener Eibereiche. [The Development of *Musca* Ova after the Removal of Different Egg Zones.] Zeit. Wiss. Zool. 126:181-238. 39 figs.
Contains a good bibliography.

1178. **Rendtorff, R. C.** and **Francis, T. 1943** Survival of Poliomyelitis Virus in the Common House Fly, *Musca domestica* L. J. Infect. Dis. 73(3): 198-205.
Virus found in feces and vomitus up to 6 hours after feeding.

1179. **Reum, W. 1914** Der Weisse Tod der *Musca domestica*. Societas Entomologica, Zurich. 29:13-14.

1180. **Riabov, M. A. 1943** [Arsenicals in the Control of the Larvae of the House Fly.] Med. Parasit. 12(5):53-66. (In Russian.)
Arsenic trioxide effective against larvae of *M. vicina*. ABSTRACT: BA 19:2518, 1945.

1181. **Richardson, C. H. 1913** An Undescribed Hymenopterous Parasite of the House-Fly. Psyche 20(1):38-39, 1 pl.
Spalangia muscidarum sp. nov., Family Pteromalidae, from pupal stage of host. ABSTRACT: RAE-B 1:94, 1913.

1182. Richardson, C. H. 1913 Studies on the Habits and Development of a Hymenopterous Parasite, *Spalangia muscidarum* Richardson. J. Morph. 24:513-549.
M. domestica is host to *S. muscidarum*, *S. nigra* and *S. hirta.* Parasite consumes blood plasma of the pupa. ABSTRACT: RAE-B 2:23-24, 1914.

1183. Richardson, C. H. 1915 Fly Control on the College Farm. Rept. Ent. Dept., New Jersey Agric. Coll. Expt. Sta., Paterson, for 1914, pp. 382-399.
Horse manure chief breeding medium, but cow, pig and chicken manure were also used. ABSTRACT: RAE-B 4:5, 1916.

1184. Richardson, C. H. 1916 The Response of the House Fly (*Musca domestica* L.) to Ammonia and other Substances. New Jersey Agric. Expt. Sta. Bull. No. 292, 19 pp.
Tested many substances. Results set forth in 16 tables.

1184a. Richardson, C. H. 1916 The Attraction of Diptera to Ammonia. Ann. Ent. Soc. Am. 9:408-413.
Species spending part of their lives in animal excrement are attracted to ammonia. Responses complicated by many factors, as shown by studies on *M. domestica.* ABSTRACT: Hocking, B. 1960 Smell in Insects, p. 106 (2.445).

1185. Richardson, C. H. 1916 A Chemotropic Response of the House-Fly (*Musca domestica* L.). Science 43(1113):613-616.
Positive results only with ammonium hydroxide and ammonium carbonate, especially the latter. ABSTRACT: RAE-B 4:106, 1916.

1186. Richardson, C. H. 1917 The Response of the House-fly to Certain Foods and their Fermentation Products. Rept. Dept. Ent. for 1916. New Jersey Agric. Expt. Sta. New Brunswick, pp. 511-519. Also published in J. Econ. Ent. 10(1):102-109.
No sugars very attractive (lactose the most). Gives formula for an effective poison bait. ABSTRACT: RAE-B 5:72, 1917.

1187. Richardson, C. H. 1917 The Domestic Flies of New Jersey. New Jersey Agric. Expt. Sta. Bull. No. 307, 28 pp., 18 figs.
M. domestica and others. Includes a key for their identification. ABSTRACT: RAE-B 5:145, 1917.

1188. Richardson, C. H. 1936 Flies as Household Pests in Iowa. Bull. Iowa Agric. Exp. Sta. No. 345, pp. 217-238, 13 figs.
Deals with recognition characters, biology, and control. ABSTRACT: RAE-B 24:247, 1936.

1189. Richardson, C. H. and **Richardson, E. H. 1922** Is the Housefly in its Natural Environment Attracted to Carbon Dioxide? J. Econ. Ent. 15:425-430.
Bran which volitalizes CO_2 alone will not induce oviposition. Decomposition products of ammonium carbonate do. ABSTRACT: RAE-B 11:31, 1923.

Richardson, C. H., joint author. *See* Elmore, J. C., et al., 1936; Deonier, C. C., et al., 1935; Gaines, J. C., et al., 1937; Pearson, A. M., et al., 1933.

Richardson, E. H., joint author. *See* Richardson, C. H., et al., 1922.

1190. Richardson, H. H. 1931 Research on Kerosene Extracts of Pyrethrum. J. Econ. Ent. 24:763-764.
Tests on house flies showed no deterioration in extracts kept for 12 months. Paralyzed flies sometimes recover, without loss of fertility. ABSTRACT: RAE-B 19:199, 1931.

1191. Richardson, H. H. 1931 The Pyrethrin I Content of Powders as an Index of Insecticidal Power. J. Econ. Ent. 24:1098-1106.
Power estimated by speed of paralytic action against *M. domestica*. Coefficient of correlation between the two factors—0·987. ABSTRACT: RAE-B 20:15, 1932.

1192. Richardson, H. H. 1932 Insecticidal Studies of Midcontinent Distillates as Bases of Pyrethrum Extracts. Industr. Engin. Chem. (Industr. Ed.) 24(12):1394-1397.
Distillates with heavier specific gravities (or higher boiling points) were most rapid in paralyzing action against *M. domestica*; also most toxic. ABSTRACT: RAE-B 22:10, 1934.

1192a. Richardson, H. H. 1933 Extractive Efficiency of Kerosene on Pyrethrum Powders of Varying Fineness. J. Econ. Ent. 26:252-259.
Time, in seconds, recorded for 50 per cent of the exposed flies to become paralyzed. (Mortality counts not a sufficiently sensitive indicator.) This author was the first to use the 50 per cent knockdown point for evaluation of insecticides.

1193. Richardson, H. H. 1933 A Note on the Insecticidal Efficiency of Kerosene Extracts of Derris alone, and in Combination with Kerosene Extracts of Pyrethrum. J. Econ. Ent. 26:914-915.
Addition of kerosene extract of quassia chips or hellebore did not appreciably increase toxicity. ABSTRACT: RAE-B 21:255, 1933.

1193a. Richardson, H. H. 1943 Toxicity of Derris, Nicotine, and other Insecticides to Eggs of the Housefly and the Angoumois Grain Moth. J. Econ. Ent. 36(5):729-731.
Against eggs 1½ to 5 hours old, suspensions of derris powder, derris extract or rotenone were very effective. Nicotine was effective only against very young eggs, 1 to 2 hours old.

Richardson, H. H., joint author. *See* McGovran, E. R., et al., 1944.

1194. Ridlon, J. R. 1911 An Investigation of the Prevalence of Typhoid Fever at Charlestown, W. Va. Publ. Health Repts., Publ. Health and Marine Hosp. Serv. (Washington) Vol. 26, pp. 1789-1799.
Most probable source of infection in 5 cases was from flies. Stresses sanitary practices.

1195. Riley, W. A. 1910 Earlier References to the Relation of Flies to Disease. Science 31:263-264.
Refers to Josiah Nott (1849), Kircher (1658), Mercurialis (1530-1607).

1196. Riley, W. A. 1920 How to Fight the Dangerous House Fly. Univ. Minn. Agric. Ext. Bull. 48, 8 pp.
Advocates bread and milk bait. Describes maggot trap. ABSTRACT: RAE-B 8:127, 1920.

1197. Ringdahl, O. 1929 Bestämningstabeller till Svenska muscidsläkten. I Muscinae. Ent. Tidskr. 50:8-13, 273.
Separates the Anthomyinae from the Muscinae, with a key to the genera of the latter group.

1198. Ringdahl, O. 1929 Übersicht der in Schweden gefundenen *Hylemyia*—arten mit posteroventraler Apikalborste an den Hinterschienen. [Review of the Swedish forms of *Hylemyia* with posterioventral Apical Bristle on the Hind Segment.] Ent. Tidschr. 50:268-273.
Proposes *Pseudomorellia*, nov. gen., for *Musca albolineata* of authors.

Roark, R. C., joint author. *See* Parman, D. C., et al., 1927; Bishopp, F. C., et al., 1925.

1199. Roberts, E. W. 1947 The Part Played by the Faeces and Vomit-drop in the Transmission of *Entamoeba histolytica* by *Musca domestica*. Ann. Trop. Med. Parasit. 41(1):129-142.
Discusses particle size in relation to feeding. Both vomitus and feces infective. ABSTRACTS: RAE-B 38:50, 1950; BA 22:456, 1948.

1200. Roberts, R. A. 1930 The Wintering Habits of Muscoid Flies in Iowa. Ann. Ent. Soc. Am. 23(4):784-792.
Trapping experiments, spring and fall. *M. domestica*, most abundant species in autumn, was taken in heated buildings all winter. ABSTRACT: RAE-B 19:94, 1931.

1201. Robertson, A. 1908 Flies as Carriers of Contagion in Yaws (*Framboesia tropica*). J. Trop. Med. Hyg. 11(14):213.
Concludes that *M. domestica* is capable of carrying the *virus* (broad sense) of yaws.

1202. Robertson, J. 1909 Report of the Medical Officer of Health of the City of Birmingham for the year 1909, 144 pp.
Of 24,562 flies collected in houses, 22,360 (91 per cent) were *M. domestica*.

1203. Robertson, J. 1917 Flies and Stable Litter. Public Health, London. 30(12):245-246.
No flies bred out of manure with fresh peat-litter; very *few* from sawdust and shavings; *many* from straw. ABSTRACT: RAE-B 5:185, 1917.

Robertson, K. G., joint author. *See* Fantham, H. B., et al., 1927.

1204. Robineau-Desvoidy, J. B. 1830 Essai sur les Myodaires. Mémoires des Savants etrangers de l'Academie des Sciences de Paris. Vol. II. Paris, 813 pp.
Included because of its importance as a classic work in muscoid taxonomy.

Robinson, D. B. W., joint author. *See* Brown, A. W. A., et al., 1948.

1205. Robinson, W. 1935 Stimulation of Healing in Nonhealing Wounds by Allantoin Occurring in Maggot Secretions and of Wide Biological Distribution. J. Bone Joint Surg. 17:267-271.
One of the studies which came out of the successful use of maggot therapy for osteomyelitis.

1206. Robinson, W. 1937 The Healing Properties of Allantoin and Urea Discovered through the Use of Maggots in Human Wounds. Smithsonian Report for 1937, pages 451-461. Washington.
A popular review of the subject, with 20 references.

1207. Robinson, W. 1940 Ammonium Bicarbonate Secreted by Surgical Maggots Stimulates Healing in Purulent Wounds. Am. J. Surg. (N.S.) 47(1):111-155.
Applied as a sterile solution, on gauze, shows healing properties like those of allantoin. ABSTRACT: RAE-B 30:36, 1942.

1208. Robinson, W. 1940 Ammonia as a Cell Proliferant and its Spontaneous Production from Urea by the Enzyme Urease. Am. J. Surg. 49: 319-325.
Shows that chemical therapy of wounds, without use of maggots is feasible (and more sanitary).

1209. Robinson, W. and **Norwood, V. H. 1933** The Role of Surgical Maggots in the Disinfection of Osteomyelitis and other Infected Wounds. J. Bone Joint Surg. 15:409-412.
This, and the four papers listed above, were basic in maggot therapy research in the United States.

1210. **Robinson, W.** and **Norwood, V. H. 1934** Destruction of Phygenic Bacteria in the Alimentary Tract of Surgical Maggots Implanted in Infected Wounds. J. Lab. Clin. Med. 19:581-586.
Used *Lucilia sericata*. Bacteria, in their course through the maggot's alimentary tract, become reduced in number and finally disappear. Stomach is site of greatest mortality.

1211. **Robson, M. H. 1879** On the Development of the House Fly and its Parasite. Sci. Gossip 15:7 et seq.

1212. **Rödel, H. 1886** Über das vitale Temperaturminimum wirbelloser Tiere. Zeit. Naturw. 59:183-214.
Flies are killed at temperatures not many degrees below the freezing point of water.

Rodova, R. A., joint author. *See* Vainshtein, B. A., et al., 1940.

Rogers, E. E., joint author. *See* Yeomans, A. H., et al., 1949.

1213. **Rogers, L. 1905** The Conditions Affecting the Development of Flagellated Organisms from Leishman Bodies and their Bearing on the Probable Mode of Infection. Lancet, June 3, 1909, pp. 1484-1487.
Points out the similarity of *Herpetomonas muscae-domestica* to the organism of Kala-azar.

1214. **Romanov, A. N. 1940** L'écologie des mouches synanthropes du Tadjikistan méridional. [The ecology of Synanthropic Flies in Southern Tadjikistan.] Med. Parasit. 9(4):355-362.
Concerns the development, in cow dung, of *M. larvipara* and *M. tempestiva.* ABSTRACT: RAE-B 31:126, 1943.

1215. **Root, F. M. 1921** Experiments on the Carriage of Intestinal Protozoa of Man by Flies. Am. J. Hyg. 1(2):131-153.
Fly feces only dangerous to human beings if deposited on moist or liquid foods. Early hours after fly's feeding most important. ABSTRACT: RAE-B 9:126, 1921.

1216. **Rosenau, M. J., Lumsden** and **Kastle. 1909** Report No. 3 on Origin and Prevalence of Typhoid Fever in the District of Columbia. Bull. No. 52, Hyg. Lab. U.S. Publ. Health and Marine Hosp. Serv. (Washington), p. 30.

1217. **Rosenbusch, F. 1910** Ueber eine neue Encystierung bei *Crithidia muscae-domesticae.* [On a new cyst stage in *Crithidia muscae-domesticae.*] Centralblt. f. Bakt. 1 Abt. Orig. Vol. 33, pp. 387-393, 1 pl.
Gives dimensions of flagellate and post-flagellate stages.

1218. **Ross, E. H. 1913** The Reduction of Domestic Flies. London. John Murray. 103 pp., illus.
A good argument for civic effort against flies. Reproduces Cobb's 1910 photographs.

1219. **Ross, E. H. 1914** House-Flies and Disease. J. Roy. Soc. Arts., London 62 (Nos. 3200, 3201, 3202), pp. 388-397, 423, 442.
Dust bins thought to be principal source of flies in London. Recommends daily collection and burning. ABSTRACT: RAE-B 2:92-93, 1914.

1220. **Ross, I. C. 1929** Observations on the Hydatid Parasite (*Echinococcus granulosus*) and the Control of Hydatid Disease in Australia. Bull. Council Sci. Ind. Res. Australia. 40, 63 pp., 4 figs.
Flies ingest eggs of parasite and may carry them from feces of dogs to human food. ABSTRACT: RAE-B 17:133, 1929.

1221. **Ross, J. N. M. 1916** Medical Impressions of the Gallipoli Campaign from a Battalion Medical Officer's Standpoint. J. Roy. Naval Med.

Serv., London 2(3):313-324, 1 chart.
Infection undoubtedly transmitted by flies to food consumed. ABSTRACT: RAE-B 4:143, 1916.

1222. Ross, T. S. 1916 Flies in a Jail. Indian Med. Gaz., Calcutta 51(4): 133-134.
Flies bred in latrine trenches, where only 5 inches of earth covering was used. Larvae migrated toward surface for pupation. ABSTRACT: RAE-B 4:92, 1916.

1223. Ross, T. S. and **Hussain, M. 1924** On the Life History of *Herpetomonas muscae-domesticae*. A Preliminary Note. Indian Med. Gaz. 59(12): 614-615.
This is a tissue parasite, not an intestinal one. Half of 7000 flies examined were infected. ABSTRACT: RAE-B 13:32, 1925.

1224. Rottrup, S. 1922 Stuefluens Bekaempelse. Almanak 1922 Kjøbenhavns Observ., 8 pp. Copenhagen.
Pig sties are favorite breeding places for houseflies in Denmark. Oviposit on fresh feces. ABSTRACT: RAE-B 12:10, 1924.

1225. Roth, A. R. and **Lindquist, A. W. 1949** Comparative Effectiveness of DDT, Methoxychlor, and Dichlorodiphenyl Dichlorethane Residues against House Flies and *Aedes* Floodwater Mosquitoes. J. Econ. Ent. 42:871-873.
At varying doses and exposure times, DDT proved approximately 10 times as toxic as methoxychlor, from which many flies recovered. ABSTRACT: RAE-B 38:150, 1950.

1226. Roubaud, E. 1911 Sur la biologie et la viviparite poecilogoniques de la mouche des bestiaux (*Musca corvina* Fab.) en Afrique tropicale. [On the Biology and Special Viviparity of the Fly of Livestock (*Musca corvina* Fab.) in Tropical Africa.] Compt. Rend Acad. Sci Paris 152:158-160.
Is viviparous in tropical Africa (also in Crimea) but sometimes extrudes a single, large egg containing a larva about to hatch. A geographical race or subspecies? (Is oviparous in Northern Russia.)

1227. Roubaud, E. 1915 Etudes biologiques sur la mouche domestique, methode biothermique de destruction des oeufs dans le tas de fumier. [Biological Studies on the Domestic Fly. Biothermic Method of Destroying the Eggs in the Dung Hill.] Compt. Rend. Soc. Biol. (Paris) 78(18):615-616.
Eggs cannot survive a temperature above 115°F. Recommends digging hole in old heap and putting fresh manure in, to ferment and raise temperature. ABSTRACT: RAE-B 4:22-23, 1916.

1228. Roubaud, E. 1915 Production et auto-destruction par le fumier de cheval des mouches domestiques. [Production and Self-destruction of Domestic Flies by Horse Manure.] Compt. Rend. Acad. Sci. Paris. 161(11):325-327.
Manure from one horse can produce 40,000-50,000 flies per month. Covering fresh manure with layer of "hot" manure (20 cm.) will kill all larvae and eggs. "Méthode biothermique." ABSTRACT: RAE-B 3:197 1915.

1229. Roubaud, E. 1918 Le rôle des mouches dans la dispersion des amibes dysentériques et autres protozoaires intestinaux. [The Role of Flies in the Dispersion of the Amebae of Dysentery and other Intestinal Protozoa.] Bull. Soc. Path. Exot. 11:166-171.
Excreta of flies must be deposited directly into a liquid or upon moist food if cysts are to reach human hosts with enough vitality to survive. ABSTRACT: RAE-B 6:112, 1918.

1230. Roubaud, E. 1921 Fécondité et longévité de la mouche domestique. Compt. Rend. Acad. Sci. (Paris). 173(22):1126-1128.
Calculates that 4000 trillions of descendents may be derived from a single fertilized female between May 1 and Sept. 30. ABSTRACT: RAE-B 10:51, 1922.

1231. Roubaud, E. 1921 La fermentation du tas de fumier au service de la basse-cour dans la lutte contre les mouches. [Fermentation in the Manure Pile of Service to the Poultry Yard in the War against Flies.] Bull. Mus. Natnl. Hist. Nat. (Paris) for 1921, No. 1, pp. 48-52.
Enclosing heap at sides forces all larvae to come to the top. Fowls destroy great numbers. ABSTRACT: RAE-B 9:212, 1921.

1232. Roubaud, E. 1922 Recherches sur la fécondité et la longévité de la mouche domestique. Ann .Inst. Pasteur 36:765-783.
Adults, as a rule do not hibernate, but continue to reproduce throughout the winter in warm rooms or stables. ABSTRACT: RAE-B 11:27, 1923.

1233. Roubaud, E. 1922 Etudes sur le sommeil d'hiver pré-imaginal des Muscides. [Studies on the pre-imaginal winter sleep of Muscidae.] Bull. Biol. France & Belg. 56(4):455-544.
Distinguishes between "homodynamic" species, such as *M. domestica* (no diapause) and "heterodynamic" forms such as *Lucilia caesar*, in which development is abruptly suspended without regard to temperature. ABSTRACT: RAE-B 11:55, 1923.

1234. Roubaud, E. 1922 Sommeil d'hiver cédant à l'hiver chez les larves et nymphes de Muscides. [Hibernation in Relation to Winter among Larvae and Pupae of Muscidae.] Compt. Rend. Acad. Sci., Paris, 174(14):964-966.
M. domestica, also *Stomoxys* and *Drosophila*, suspend activities *only* because of cold; with *Lucilia*, *Sarcophaga*, et al., a period of inertia is essential for development. ABSTRACT: RAE-B 10:135, 1922.

1235. Roubaud, E. 1927 Sur l'hibernation de quelques mouches communes. Bull. Soc. Ent. France for 1927 (2):24-25.
M. autumnalis DeG (*corvina* F.) and *Pollenia rudis* are heterodynamic; *Muscina stabulans*, homodynamic, as is *Calliphora erythrocephala*. ABSTRACT: RAE-B 15:86, 1927.

1236. Roubaud, E. 1936 The Biothermic Method of Fly Destruction and the Ease with which it can be Adapted to Rural Conditions. Quart. Bull. Health Org. League of Nations 5(2):214-218.
Advocates placing fresh manure in hollow scooped out at top of fermenting pile. Eggs will be destroyed by heat. ABSTRACT: RAE-B 24:279, 1936.

1237. Roubaud, E. and **Colas-Belcour, J. 1936** Observations biologiques sur les glossines . . . Bull. Soc. Path. Exot. 29(6):691-696.
Reared the Eulophid, *Syntomosphyrum glossinae* Wtstn. through one generation on the pupae of *M. domestica*. ABSTRACT: RAE-B 24:238, 1936.

1238. Roubaud, E. and **Descazeaux, J. 1921** Contribution a l'histoire de la mouche domestique comme agent vecteur des habronémoses d'equidés. Cycle évolutif at parasitisme de l'*Habronema megastoma* (Rudolphi 1819) chez la mouche. [Contribution to the Story of the Domestic Fly as the Vector of Habronemoses. Life Cycle and Parasitism of *Habronema megastoma* (Rudolphi 1819) in the Fly.] Bull. Soc. Path. Exot. (Paris), 14(8):471-506.
Infection of the horse produced by flies alighting on wounds, on nostrils or on lips. Hence cutaneous, pulmonary or gastric habronemosis. ABSTRACT: RAE-B 10:12-13, 1922.

1239. **Robaud, E.** and **Descazeaux, J. 1922** Evolution de l'*Habronema muscae* Carter chez la mouche domestique et de l'*H. microstomum* Schneider chez la stomoxe. (Note préliminaire.) Bull. Soc. Path. Exot. (Paris), 15(7):572-574.
H. megastomum is a parasite of the insect's malpighian tubes, and causes a tumor there. *H. muscae* and *H. microstomum* are parasites of adipose tissue. ABSTRACT: RAE-B 10:219-220, 1922.

1240. **Roubaud, E.** and **Descazeaux, J. 1922** Deuxième contribution à l'etude des mouches dans leur rapport avec l'evolution des Habronèmes d'equides. [Second Contribution to a Study of Flies in their Relation to the Development of Habronemas in Horses.] Bull. Soc. Path. Exot. 15(10):978-1001.
Third stage larvae of *Habronema*, on leaving the fly's proboscis is unable to pierce the skin of the vertebrate host. Must be deposited at a favorable point. ABSTRACT: RAE-B 11:57, 1923.

1241. **Roubaud, E.** and **Descazeaux, J. 1923** Sur un agent bactérien pathogéne pour les mouches communes. *Bacterium delendae-muscae* n. sp. [On a Bacterial Agent Pathogenic to Common Flies . . .] Compt. Rend. Acad. Sci., Paris, 177(16): 716-717.
Infection contracted in the larval stage. Death occurs in *M. domestica* at the end of pupation. Value in control? ABSTRACT: RAE-B 11:209, 1923.

1242. **Roubaud, E.** and **Treillard, M. 1935** Un coccobacille pathogène pour les mouches tsétsés. [A coccus bacterium pathogenic for the tsetse fly.] Compt. Rend. Acad. Sci. (Paris), 201(4):304-306.
Bacterium mathisi n. sp. In 3-18 hours, killed *Glossina morsitans*, *Lucilia sericata*, *Musca domestica*, *Sarcophaga carnaria* and other species. ABSTRACT: RAE-B 23:242, 1935.

1243. **Roubaud, E.** and **Veillon, R. 1922** Recherches sur l'attraction des mouches communes par les substances de fermentation et de putréfaction. [Researches on the Attraction for Common Flies of Fermentive and Putrefactive Substances.] Ann. Inst. Pasteur 36(11): 725-764.
Species differ from one another, sexes also. *M. domestica* oviposits in response to a complex mixture of gaseous emanations produced at a certain stage of decomposition. ABSTRACT: RAE-B 11:25, 1923.

Rovelli, G., joint author. *See* Grassi, B., et al., 1889, 1892.

Rowell, J. C., joint author. *See* Brescia, F., et al., 1946.

1244. **Roy, D. N. 1938** Number of Eggs of Common House-frequenting Flies of Calcutta. Indian J. Med. Res. 26:531-533.
Counted mature eggs in ovaries of gravid females. *M. domestica vicina* averaged 97; *M. nebulo*, 75; *Chrysomyia megacephala*, 122. ABSTRACT: RAE-B 27:109, 1939.

1245. **Roy, D. N. 1946** Entomology, Medical and Veterinary. xii + 358 pp., 162 figs. Calcutta. Saraswaty Library.
More than half of this volume is devoted to Diptera. REVIEW: RAE-B 34:188, 1946.

1246. **Roy, D. N.** and **Mukherjee, P. K. 1937** *Allantonema muscae* sp. nov., A New Parasitic Nematode of the Family Rhabditidae from the Haemocoele of *Musca* (*domestica*) *vicina*. Ann. Trop. Med. Parasit. 31(4):449-451, 1 fig. Liverpool.
Describes only female, larvae, and ova. Becomes an addition to Hall's 1929 list. May prove a synonym of *A. mirabile*. ABSTRACT: RAE-B 26:80, 1938.

1247. Roy, D. N. and **Mukherjee, P. K. 1937** *Allantonema stricklandi* sp. nov., a parasitic Nematode of Houseflies *Musca* (*domestica*) *vicina*. Ann. Trop. Med. Parasit. 31(4):453-456, 1 fig.

The second species of this genus to be described by these authors, from *M. vicina*. Gives 9 points of difference between the two. Abstract: RAE-B 26:80, 1938.

1248. Roy, D. N. and **Siddons, L. B. 1939** A List of Hymenoptera of Superfamily Chalcidoidea Parasites of Calyptrate Muscoidea. Rec. Indian Mus. 41(3):223-224. Delhi.

Spalangia sp. from several species, including *M. vicina*. *Dirhinus pachycerus* from *M. inferior*. Abstract: RAE-B 28:104, 1940.

1249. Roy, D. N., Siddons, L. B. and **Mukherjee, S. P. 1940** The Bionomics of *Dirhinus pachycerus* Masi (Hymenoptera:Chalcidoidea) a Pupal Parasite of Muscoid Flies. Indian J. Ent. 2(2):229-240, 9 figs. New Delhi.

Attacks various species of *Musca* and *Sarcophaga*. Ignores *Stomoxys* and *Drosophila*. Abstract: RAE-B 29:169, 1941.

Roy, D. N., joint author. *See* Siddons, L. B., et al., 1940.

1250. Ruata, G. 1937 La lotta contro le mosche in Italia. Ann. Igiène 47(4): 180-190.

Summarizes 10 years of health regulation. Describes use of sweetened arsenical spray in Montecatini and elsewhere. Abstract: RAE-B 25: 206, 1937.

1251. Ruata, G. 1941 La lotta contro le mosche in Italia. Ann. Igiène 51(2): 111-113. Rome.

Formula for sweetened arsenical bait spray. Is hygroscopic; does not stain. Abstract: RAE-B 29:170, 1941.

1252. Ruhland, H. H. and **Huddleson, I. F. 1941** The Role of Our Species of Cockroach and Several Species of Flies in the Dissemination of *Brucella*. Am. J. Vet. Res. 2:371-372.

Flies can pick up *B. abortus* from infected placenta or fetus, and carry infection for 4 days or longer.

1253. Russo, C. 1930 Recherches expérimentales sur l'épidémiogenèse de la peste bubonique par les insects. [Experimental Investigations on the Epidemiology of Bubonic Plague due to Insects.] Bull. Off. Int. Hyg. Publ. 22(11):2108-2120, Paris.

Carcasses of infected rats fed on (in order) by *M. domestica*, *Calliphora vomitoria*, *Sarcophaga carnaria*, *Lucilia caesar*. Abstract: RAE-B 19: 86, 1931.

1254. Sabin, A. B. and **Ward, R. 1941** Flies as Carriers of Poliomyelitis Virus in Urban Epidemics. Science 94(2451):590-591.

Virus recovered from flies trapped in urban areas close to houses in which disease had occurred. Abstract: RAE-B 30:93, 1942.

1255. Sabin, A. B. and **Ward, R. 1942** Insects and Epidemiology of Poliomyelitis. Science 95(2464):300-301.

Disease considered primarily of the intestinal tract, secondarily of the nervous system. Incidence rises in summer, flies being of some importance. Abstract: RAE-B 30:115, 1942.

1256. Saccá, G. 1947 Sull'esistenza di mosche domestiche resistente al DDT. [On the occurrence of a domestic fly resistant to DDT.] Riv. di Parrasit. 8(2-3):127-129.

Proposes the name *M. domestica*, var. *tiberina* for resistant flies. Abstracts: BA 22:1007, 1948; RAE-B 38-129, 1950.

1257. Saccá, G. 1948 Osservazioni su di una popolazione di mosche dapo tratamento con DDT & Okta-Klor. [Observations on a Population of Flies after Treatment with DDT and Okta-Klor.] Estratto Dai Rendiconti Dell Ist., Sup. Sanit. 11(6):1354-1361.
Identifies 16 species; *Lyperosia irritans*, *Stomoxys calcitrans* and *Musca domestica* the most prevalent, in that order.

1258. Saceghem, R. van. 1918 Cause étiologique et traitement de la dermite granuleuse. [Causative Agent and Treatment of Granular Dermatitis.] Bull. Soc. Path. Exot. 11(7):575-578.
Pertains to *Habronema muscae*, carried by *M. domestica* and others. Advocates "heat" method of manure disposal. ABSTRACT: RAE-B 6: 202, 1918.

1259. Saceghem, R. van. 1919 Cause étiologique et traitement de la dermite granuleuse. Ann. Méd. Vét. Brussels 64(5-6):151-154.
M. domestica can be infected only during the larval period. Gives prophylactic treatment for horses, by use of arsenic. ABSTRACT: RAE-B 7:159, 1919.

1260. Samuelson, J. and **Hicks, J. B. 1860** The Earthworm and the Common House-fly. Humble Creatures. Pt. 1, 79 pp., 8 pls. London. House-fly, pp. 26-79, pls. 3-8.
Of historical interest. Account is very superficial and contains much that is inaccurate.

1261. Sanders, D. A. 1940 *Musca domestica*, a Vector of Bovine Mastitis (Preliminary Report). J. Am. Vet. Med. Ass. 97(761):120-122 Chicago.
Having gained access to infected milk, flies feed at the orifices of the teats of healthy cows. ABSTRACT: RAE-B 28:203, 1940.

1262. Sanders, G. E. 1942 Housefly Control in Relation to Poliomyelitis. Pests 10(3):22-26.
A review of various methods of fly control, including information on primary and tertiary parasites. ABSTRACT: BA 17:527 (5959), 1943.

Sanders, G. E., joint author. *See* Girault, A. A., et al., 1909, 1910.

1263. Sandilands, J. E. 1906 Epidemic Diarrhoea and the Bacterial Content of Food. J. Hygiene 8:77-92.
Believes that house-flies convey these diseases from the excrement of infected infants.

1264. Sandwith, F. M. 1911 The Danger of the House-fly. Clinical Journal, Vol. 39, No. 4, Nov. 1, 1911.

1265. Sangiori, G. and **Attimonelli, R. 1940** Saggi sul valore pratico di un nuova mezzo moschicida. [Notes on the Practical Value of a New Septic Muscicide.] Ann. Igiene 50(2):67-69.
Whey and arsenic, diluated with water to make a bait spray. If it dries, moistening restores both attractiveness and toxicity. ABSTRACT: RAE-B 28:133, 1940.

1266. Sangree, E. B. 1899 Flies and Typhoid Fever. New York Med. Record, Vol. 55, pp. 88-89, 4 figs.
Recovered various bacteria, including anthrax, from the tracks of flies on culture plates.

1267. Sarkaria, D. S. and **Patton, R. L. 1949** Histological and Morphological Factors in the Penetration of DDT through the Pulvilli of Several Insect Species. Trans. Am. Ent. Soc. 75:71-82, 2 pls.
Microscopic anatomy of the pulvilli and tarsal segments. Walking fly brings pulvillus into contact with surface at all times.

1268. Sarkissian, S. M. 1941 Mutations in *Musca domestica.* Drosophila Information Service 15:19.
Believed to be the first attempt to make formal study of the genetics of this species.

1269. Sasaki, C. 1926 [Flies Frequenting Houses.] Dobutsugaku Zasshi (Zool. Mag.) 38(447):34-36. Tokyo. (In Japanese.)
A popular account of *M. domestica, Fannia canicularis, Calliphora erythrocephala,* all common in Japan. ABSTRACT: RAE-B 14:101, 1926.

1270. Saunders, P. T. 1914 Notes on some Parasites of Live Stock in the West Indies. West Indian Bull., Barbados, 14(2):132-138.
Records *M. domestica* from St. Kitts, Antigua and St. Vincent. ABSTRACT: RAE-B 2:175, 1914.

1271. Saunders, W. H. 1916 Fly Investigations Reports. I. Some Observations on the Life-History of the Blow-Fly and of the House-Fly, made from August to September, 1915, for the Zoological Society of London. Proc. Zool. Soc. London for 1916. Part iii, pp. 461-463.
Life cycle of *M. domestica,* up to emergence of adult, 9-14 days (at 100·4°F); 23-25 days (at 40°- 60°F). ABSTRACT: RAE-B 4:167, 1916.

1272. Saunders, W. H. 1916 Fly Investigations Reports. II. Trials for Catching, Repelling and Exterminating Flies in Houses, Made during the Year 1915 for the Zoological Society of London. Proc. Zool. Soc. London for 1916. Part iii, pp. 465-468.
ABSTRACT: RAE-B 4:167, 1916.

1273. Saunders, W. H. 1916 Fly Investigations Reports. III. Investigations into Stable Manure to check the Breeding of House-Flies, made during the Year 1915 for the Zoological Society of London. Proc. Zool. Soc. London for 1916. Part 3, pp. 469-479.
Two successful methods: (1) Surface dressing of manure with green tar oil or neutral blast furnace oil and soil. (2) Tetrachlorethane, miscible or pure. Both harmless to plants. (For Report IV, see Lodge, O. C. 1916.

1274. Savit, J., Kolros, J. J. and **Tobias, J. M. 1946** Measured Dose of Gamma Hexachlorocyclohexane (γ666) Required to Kill Flies and Cockroaches, and a Comparison with DDT. Proc. Soc. Exp. Biol. and Med. 62(1):44-48.
Lethal Dose-50 is approximately 0·4 mg. per kg. for newly emerged *M. domestica.* For older adults, 0·8 mg. per kg. ABSTRACT: BA 20:2004, 1946.

1275. Sawtchenko, J. G. 1892 Le rôle des mouches dans la propagation de l'épidémie cholérique. [The Role of Flies in the Propagation of an Epidemic of Cholera.] Vratch, St. Petersburg. (Reviewed in Ann. de l'Institut. Pasteur, Vol. 7, p. 222, 1892.)
Flies fed on cholera spirillum were disinfected externally. Cultures were then made from the dissected-out alimentary canals.

Scales, F. M., joint author. *See* Cook, F. C., et al., 1914, 1915.

1276. Scharff, J. W. 1940 Composting. Journal of the Malay Branch of the British Medical Association. Vol. 4, No. 1, June, 1940.
Composting manures with green vegetation generates tremendous heat and destroys any eggs, larvae, or pupae that may be present. This method used effectively by Scharff in Malaya.

Schechter, M. S., joint author. *See* Gersdorff, W. A., et al, 1948; Sullivan, W. N., et al., 1943.

1277. **Schilling, C. 1907** Die Uebertragung von Krankheiten durch Insekten und ihre Bekämpfung. [The Transmission of Diseases by Insects and their Control.] Gesundh. Ingenieur. 30:300-303.
Includes information regarding flies.

Schmitt, J. B., joint author. *See* Barber, G. W., et al., 1949; Hansens, E. J. et al., 1948.

1278. **Schroeder, H. O., Jones, H. A.** and **Lindquist, A. W. 1948** Certain Compounds Containing the Methlenedioxyphenyl Group as Synergists for Pyrethrum to Control Flies and Mosquitoes. J. Econ. Ent. 41:890-894.
Tested 30 compounds on laboratory reared insects. About 10 proved good synergists against houseflies. ABSTRACT: RAE-B 38:28, 1950.

1279. **Schroeder, H. O., Madden, A. H., Lindquist, A. W.** and **Jones, H. A. 1946** Concentrated Sprays tested for Uutility in Control of Flies and Mosquitoes. Soap and Sanit. Chem. 22(4):145, 147, 153.
Chief finding was that the least effective spray was quite as effective as the aerosol. ABSTRACT: RAE-B 37:64, 1949.

1280. **Schroeder, H. O., Madden, A. H., Wilson, H. G.** and **Lindquist, A. W. 1945** Residual Action of DDT Aerosols against House Flies. J. Econ. Ent. 38:277-278.
Though intended primarily to kill flying insects quickly by contact, repeated use of aerosols will eventually leave a deposit which becomes a valuable adjunct in contol. ABSTRACTS: BA 19:2518, 1945; RAE-B 34: 135, 1946.

Schroeder, H. O., joint author. *See* Madden, A. H., et al., 1946; Lindquist, A. W., et al., 1945.

1281. **Schuckmann, W. von. 1923** Ueber Mittel zur Fliegenbekämpfung. [On Method(s) of Fly Control.] Zeit. Angew. Ent. 9(11):81-104.
Covers use of borax, slaked lime, iron sulphate, etc. States that "heat method" of treating manure satisfies all requirements. ABSTRACT: RAE-B 11:73, 1923.

1282. **Schuckmann, W. von. 1927** Die Fleigenplage und ihre Bekämpfung. [The Fly Plague and its Control.] 36 pp., 11 figs., 1 pl. Berlin. Julius Springer.
A popular account of the medical importance of flies, illustrated in color. (Issued by the German Imperial Health Office.) ABSTRACT: RAE-B 15:236, 1927.

1283. **Schuckmann, W. von. 1927** Zur Fliegen-und Mückenbekämpfung. [On Fly and Mosquito Control.] Zeit. Angew. Ent. 12(2):332-339.
Recommends a fumigant giving off $S0_2$ for use in buildings. Discusses proper mesh, for screening. ABSTRACT: RAE-B 15:13, 1927.

1283a. **Schwardt, H. H. 1945** DDT—its Possibilities and Limitations. J. Milk Technol. 8(6):356-359.
Proper manure disposal essential in any case. Combining DDT with white-wash reduces its efficiency. ABSTRACT: BA 20:1013, 1946.

1284. **Schwardt, H. H. 1946** DDT Spells Death to Flies in Dairy Barns. Farm Research 12(3), July 1946, 2 pp.
Properly applied, two DDT sprayings yearly were found sufficient for fly control.

1285. **Schwartz, B.** and **Cram, E. B. 1925** Horse Parasites Collected in the Philippine Islands. Philipp. J. Sci. 27(4):495-505.
Concerns the three fly-borne nematodes, *Habronema megastoma*, *H. microstoma* and *H. muscae*. ABSTRACT: RAE-B 13:180, 1925.

1286. Schwetz, J. 1927 Notes sur les trypanosomiasis animales du Haut-Katanga. [Notes on the Trypanosomiasis of Upper Katanga.] Ann. Soc. Belge. Méd. Trop. 7(2):135-145.
T. congolense is possibly transmitted by *M. spectanda* Wied. which is found on both healthy and affected animals. ABSTRACT: RAE-B 16:36, 1928.

1287. Scott, H. H. 1915 An Investigation into the Causes of the Prevalence of Enteric Fever in Kingston, Jamaica; with Special Reference to the Question of Unrecognized Carriers. Ann. Trop. Med. and Parasit. 9(2):239-284, 10 charts.
Flies are especially numerous during "mango season" which begins about May. Number of enteric fever cases increases from this time on. ABSTRACT: RAE-B 3:152, 1915.

1288. Scott, J. 1909 The Dangerous House-fly. Indian Publ. Health 5:292-298.

1289. Scott, J. R. 1917 Studies upon the Common House-fly (*Musca domestica* Linn.). I. A General Study of the Bacteriology of the House-fly in the District of Columbia. J. Med. Res. 37(32, N.S.):101-119

1290. Scott, J. R. 1917 Studies upon the Common House-fly (*Musca domestica* Linn.). II. Isolation of *B. cuniculicida*, a Hitherto Unreported Isolation. J. Med. Res. 37:121-124.
The two forgoing papers are in fact, one study. Titles are self-explanatory.

1291. Scott, J. W. 1924 The Experimental Transmission of Swamp Fever, or Infectious Anemia by Means of Secretions. Wyoming Agric. Expt. Sta. Bull. 138, pp. 19-62, 11 pls., 3 figs. Laramie.
Eye and nasal secretions overflow in affected horses. *M. domestica* feeds on these, and may possibly infect eyes of healthy animals. (Principal transmission is by biting flies.) ABSTRACT: RAE-B 13:86, 1925.

1292. Scudder, H. I. 1947 A New Technique for Sampling the Density of Housefly Populations. Publ. Health Repts. 62(19):681-686. (Reprint No. 2785.)
Introduces the "fly grill", consisting of parallel wooden strips. Three foot grill, slats $\frac{3}{4}$ inch apart, for outdoor use. Eighteen inch grill, slats $\frac{1}{4}$ inch apart, for use in restaurants, etc. ABSTRACT: RAE-B 38:158, 1950.

1293. Scudder, H. I. 1949 Some Principles of Fly Control for the Sanitarian. Am. J. Trop. Med. 29(4):609-623.
Long-distance migrations of house flies and blow flies identified as the overflow of heavy breeding areas, out of which flies are forced by competition with each other. ABSTRACT: BA 24:1062, 1950.

Scudder, H. I., joint author. *See* Baker, W. C., et al., 1947.

Scudder, H. L., joint author. *See* Hartzell, A., et al., 1942.

1294. Séguy, E. 1937. Diptera, Family Muscidae. *In* Wytsman. Genera Insectorum, fasc. 205, pp. 1-604, pls. 1-9.
A basic, taxonomic reference.

1295. Séguy, E. Sur les Caractères Communs aux Muscides et aux Oestrides Gastricoles. *His* Encyclopédie Entomologique, Sér. B, 2, Diptera. T.9:1-21.

1296. Séguy, W. 1929 Etudes sur les Diptères à larves commensals ou parasites des oiseaux de l'Europe occidentale. [Studies on the Diptera with Larvae Commensal or Parasitic on the Birds of Western Europe.] Encyc. Ent. Sér. B, II. Dipt. V, pp. 63-82, 27 figs.
Includes data on the Pteromalid, *Mormoniella vitripennis*, Wlk, recorded from *M. domestica* and other genera. ABSTRACT: RAE-B 18:161, 1930.

1297. Sen, P. 1938 A Note on the Overwintering of the Housefly, *Musca domestica*. Indian J. Med. Res. 26:535-536.
Concludes that *M. domestica* does not hibernate in the Indian Hills, where its presence is due to more or less continuous breeding, as temperature etc. permit. ABSTRACT: RAE-B 27:109, 1939.

1298. Sen, S. K. 1926 Possibility of Fly Transmission of Rinderpest. Rept. Imp. Inst. Vet. Res. Muktesar, for 1924-25, p. 36, Calcutta.
Rather elaborate experiments with virulent blood, nasal discharges and fresh feces from cattle. Concludes that *M. domestica* is not a natural transmitter. ABSTRACT: RAE-B 14:180, 1926.

1299. Sen, S. K. 1926 Experiments on the Transmission of Rinderpest by Means of Insects. Mem. Dept. Agric. India Ent. Ser. ix, No. 5, pp. 59-185.
Obtains transmission through *M. domestica* only when bodies of flies fed on infective material were inoculated into susceptible bulls. Technical note: Reared larvae of *Aedes* mosquitoes on dead flies! ABSTRACT: RAE-B 15:43, 1927.

1300. Sen, S. K. and **Minett, F. C. 1944** Experiments on the Transmission of Anthrax through Flies. Indian J. Vet. Sci. and An. Hus. 14(3): 149-158.
Both *M. domestica* and *Calliphora erythrocephala*, having fed at diseased carcasses, transmitted the disease when placed in contact with the cauterized skin of goats. ABSTRACTS: RAE-B 35:17, 1947; BA 20:607, 1946.

1301. Senior-White, R. 1945 Some Notes on the Life-History of *Musca planiceps* Wied. Indian J. Vet. Sci. 14(2):123-125.
Species is blood-sucking and larviparous. Breeds in cow dung, along with *M. conducens* Wlk. Is Zoophilic. ABSTRACT: RAE-B 34:102, 1946.

Senior-White, R., joint author. *See* Patton, W. S., et al., 1924.

1302. Serebrovsky, A. S. 1941 Mutations in *Musca domestica.* Drosophila Inform. Serv. 15:19.
Together with Sarkissian (1941), this author lists a total of 9 house fly mutants which either arose spontaneously or were induced by X-ray.

1303. Sergent, Ed. and **Lheritier, A. 1918** Fósse à fumier sans mouches. [Manure Pits without Flies.] Rev. d'Hyg. et de Police Sanitaire, Sept.-Oct. 1918, p. 553.
Utilized twin pits of concrete, on pillars, so cart may be placed beneath. Fourteen days to fill, 10 more days to ferment, which produces a larvicidal temperature of 201°F (94°C). ABSTRACTS: RAE-B 8:77-78, 1920; Bull. Off. Internat. Hyg. Publique, Paris 11(4):430-431, 1919.

1304. Sergent, Ed. and **Sergent, Et. 1934** Fly-free Manure Heaps. Quart. Bull. Health Org. League of Nations. 3(2):299-303.
For towns, concrete chambers on pillars; for country, modified Baber traps. ABSTRACT: RAE-B 23:12, 1935.

Sergent, Et., joint author. *See* Sergent, Ed., et al., 1934.

1305. Seriziat. 1875 Etudes sur l'oasis de Biskra. Paris. 1875.
Asserts that flies carry "Bouton de Biskra". (Believed to be the same as "Oriental Sore", caused by *Leishmania tropica.*)

1306. Shaffi, M. 1936 A Simple, Cheap and Effective Fly-trap. Agric. and Live-stock. (India) 6(1):60-62, 2 figs.
A preparation using mustard oil cake attracted 2000 flies, chiefly *M. domestica*, in 2 hours. ABSTRACT: RAE-B 24:104, 1936.

Shakespeare, E. O., joint author. *See* Reed, W., et al., 1904.

1307. Shane, M. S. 1948 Effect of DDT Spray on Reservoir Biological Balance. J. Am. Water Works Ass. 40(3):333-336.
The city of Wilmington, Delaware, was sprayed repeatedly from the air for fly control. Microcrustacea in an open reservoir were killed, allowing an increase of the diatom *Synedra* to nearly 10,000 per milliliter! Abstract: BA 22:1771, 1948.

1308. Sharpe, W. S. 1900 Influence of Dust and Flies in the Contamination of Food and the Dissemination of Disease. Lancet, pt. 1, June 2, 1900.

1309. Shaw, F. R. and **Bourne, A. J. 1946** The Abundance and Control of Flies on a College Campus. J. Econ. Ent. 39:543-544.
Amherts, Massachusetts. Used electric traps, other traps, insecticides. Stored manure covered with black canvas. Abstract: BA 21:746, 1947.

Shaw, O. A., joint author. *See* Atkeson, F. W., et al., 1943.

1310. Sherrick, J. L. 1940 What Standard for Stock Sprays? Soap and Sanit. Chem. 16(9):92-97, 111.
Repellency more important than knock-down. Many light oil sprays scorch or injure hide. Abstract: RAE-B 29:31, 1941.

1311. Shingareva, N. I. 1926 [Observations and Control of Hibernating Mosquitoes in Sanatoria.] Russ. J. Trop. Med. 1926 (4):23-25. (In Russian.)
M. domestica found hibernating in the same places as *Anopheles maculipennis;* chiefly cellars or unoccupied rooms of even temperature, average humidity, no draft and little light. Abstract: RAE-B 14:203, 1926.

1312. Shinoda, O. and **Ando, T. 1935** [Diurnal Rhythm of Flies.] Bot. and Zool. 3(1):117-121. (In Japanese.)
M. domestica commonest from Aug.-Sept. on. Greatest numbers collected at about 28°C (82·4°F). With *Lucilia*, preferred temperature is 30°C (86°F). Abstract: RAE-B 23:125, 1935.

1313. Shipley, A. E. 1905 Infinite Torment of Flies. Camb. Univ. Press (Printed privately), 23 pp.

1314. Shipley, A. E. 1907 The Danger of Flies. Science Progress, Vol. 1, pp. 723-729. April.

1315. Shipley, A. E. 1908 Infinite Torment of Flies, and the Danger of Flies. *In* Pearls and Parasites. London, John Murray, pp. 155-173 and 174-182.

1316. Shircore, T. O. 1916 A Note on some Helminthic Diseases, with Special Reference to the House-fly as a Natural Carrier of the Ova. Parasitology 8(3):239-243.
About 10 per cent of flies captured contained ova of *Trichocephalus dispar*, *Taenia saginata* or *Ancylostoma duodenale*. Abstract: RAE-B 4:54, 1916.

1317. Shoemaker, E. M. and **Waggoner, A. 1903** Flies as Carriers of Bacteria. School Science, April, 1903.

1318. Shooter, R. A. and **Waterworth, P. M. 1944** A Note on the Transmissibility of Haemolytic Streptococcal Infection by Flies. Brit. Med. J. No. 4337, pp. 247-248.
One-third of the flies caught in wards where there were such infections, gave cultures of haemolytic streptococcus when allowed to walk on culture medium. Abstract: RAE-B 32:105, 1944.

1319. Shope, R. E. 1927 Bacteriophage Isolated from the Common House Fly (*Musca domestica*). J. Exp. Med. 45:1037-1044.
Salt solution extract of house flies contained phage active against *Eberthella typhosa, Salmonella pyratyphi, Escherichia coli,* and *Staphylococcus muscae.*

1320. Showalter, W. J. 1914 Redeeming the Tropics. National Geographic Magazine 25:344-364. March.
Bath of sterile water brought 100,000 germs from the body of a single fly.

1321. Shuzo, Asami. 1933 Propagation de la lèpre par les insectes. [Transmission of Leprosy by Insects.] La Lepro No. 1. Published in Abstract form in Bull. Off. Int. Hyg. Publ. 25(11):2006, Paris.
Of 447 flies collected in wards with advanced cases, 118 contained leprosy bacilli. *M. domestica, Calliphora lata* and *Lucilia caprina* were concerned. ABSTRACT: RAE-B 22:19, 1934.

1322. Sibthorpe, E. H. 1896 Cholera and Flies. Brit. Med. J. Sept. 1896, p. 700.
Flies not wholly bad. Larvae serve as scavengers, consume cadavers and excrement and thus may help to abate the disease!

1323. Siddons, L. B. and **Roy, D. N. 1940** The Early Stages of *Musca inferior* Stein. Indian J. Med. Res. 27(3):819-822, 1 pl.
Describes ovum, 3rd stage larva, puparium. Adult is a bloodsucker, feeding on cattle. Breeds in isolated pats of cow dung.

Siddons, L. B., joint author. *See* Roy, D. N., et al., 1939.

1324. Sievers, A. F. and **Sullivan, W. N. 1939** Toxicity of *Tephrosia.* A Study of the Toxicity of *Tephrosia virginiana* Roots prepared by Several Methods. Soap and Sanit. Chem. 15(9):111, 113.
Is a possible source of rotenone. All tests made on *M. domestica* by the Campbell-Sullivan turntable method. ABSTRACT: RAE-B 28:55, 1940.

1325. Sieyro, L. 1942 Die Hausfliege (*Musca domestica*) als Ueberträger von *Entamoeba histolytica* und anderen Darmprotozoen. [The Housefly (*Musca domestica*) as Transmitter of *Entamoeba histolytica* and other Intestinal Protozoa.] Dtsch. Tropenmed. Zeit. 46(14):361-372.
Cysts pass through fly in from 1 minute to 24 hours. Those which stain with eosin usually considered dead; those not taking stain, alive. Author doubts reliability of eosin test, however. ABSTRACT: TDB 40: 698, 1943.

1326. Siler, J. F. 1931 Report of the Health Department of the Panama Canal for the Calendar Year 1930. 8 vo., 126 pp. Balboa Heights, C.Z.
M. domestica and *Stomoxys calcitrans* most numerous in April, May and June. Due to good sanitary practices, Colon, Panama City, and the Canal Zone are relatively free from flies. ABSTRACT: RAE-B 20:109, 1932.

1327. Simanton, W. A. and **Miller, A. C. 1937** Housefly Age as a Factor in Susceptibility to Pyrethrum Sprays. J. Econ. Ent. 30:917-921.
Flies used for testing should be 5 days old. Very young flies are paralyzed readily, but are much less easily killed than older individuals. ABSTRACT: RAE-B 26:86, 1938.

1328. Simanton, W. A. and **Miller, A. C. 1938** Greater Speed and Accuracy with Modified Peet-Grady Method. Soap and Sanit. Chem. 14(5): 115, 117.
By starting with 3 battery jar cultures at the same time, their method enables one operator to evaluate 12-18 samples a week, instead of 5, by older methods. ABSTRACT: RAE-B 26:245, 1938.

Simanton, W. A., joint author. *See* Miller, A. C., et al., 1938.

1329. **Simitch, T.** and **Kostitch, D. 1937** Rôle de la mouche domestique dans la propagation du *Trichomonas intestinalis* chez l'homme. [The Role of the Domestic Fly in the Transmission of *Trichomonas intestinalis* among Humans.] Ann. Parasit. Hum. Comp. 15(4):324-325. Paris.
M. domestica can carry this parasite, mechanically, from human or animal feces to moist or liquid food, by feet or proboscis. *T. intestinalis* can survive in the intestine of the fly for 8 hours. ABSTRACT: RAE-B 25:277, 1937.

1330. **Simmonds, H. W. 1922** Entomological Notes. Agric. Circ. Dept. Agric. Fiji. (Suva) 3(4):24.
A hymenopterous parasite of *M. domestica*, identified as a species of *Spalangia*, is believed to have been introduced from Queensland or Hawaii. ABSTRACT: RAE-B 10:226, 1922.

1331. **Simmonds, H. W. 1925** House Fly Pest and its Control in Fiji. Agric. Circ. Dept. Agric. Fiji 5(2):85-86.
Considers the ant. *Pheidole megacephala*, an important agent of Natural Control of flies in Fiji. ABSTRACT: 13:119, 1925.

1332. **Simmonds, H. W. 1928** The House Fly Problem in Fiji. Agric. J. Dept. Agric. Fiji 1(2):12-13. Suva.
A species (perhaps not *M. domestica*) breeds largely in cow droppings, in open fields. ABSTRACT: RAE-B 16:253, 1928.

1333. **Simmonds, H. W. 1929** Introduction of *Spalangia cameroni*, Parasite of the Housefly, into Fiji. Agric. J. Fiji 2(1):35. Suva.
A parasite of *Lyperosia* in Hawaii. Will develop in pupae of housefly and fleshflies. ABSTRACT: RAE-B 17:209, 1929.

1334. **Simmonds, H. W. 1929** Introduction of Natural Enemies against the House Fly in Fiji. Agric. J. Fiji 2(2):46.
Besides *Spalangia cameroni*, the dung-burying beetle, *Copris incertus* var. *prociduus* was imported from Hawaii. Expected to break up cow droppings, and thus make the dung unsuitable for maggots. ABSTRACT: RAE-B 17:235, 1929.

1335. **Simmonds, H. W. 1929** Experiments with House Flies in Fiji. Agric. J. Fiji 2(2):46-47.
Flies abundant in the wet (summer) season, but very few in the dry (winter) period. Eggs are laid, but larvae fail to mature. ABSTRACT: RAE-B 17:235, 1929.

1336. **Simmonds, H. W. 1932** The House-Fly Problem in Fiji. Fiji Ann. Med. & Health Rept. for the year 1932. pp. 46-53.
M. vicina (narrow fronted housefly) is abundant and troublesome in the Rewa Valley and elsewhere. Eggs laid in depressions in crust of cow dung less than 24 hours old. ABSTRACT: TDB 31:358, 1934.

1337. **Simmonds, H. W. 1940** Investigations with a View to the Biological Control of Houseflies in Fiji. Trop. Agric. 17(10):197-199. Trinidad. *See also* Agric. J. Fiji 11(1):21. Suva.
Reviews effects of the introduced ant (*Pheidole*), the two parasites (*Spalangia* and *Eucoila*), the dung beetle (*Copris*), the Java beetle (*Hister*), et. al. ABSTRACT: RAE-B 29:57, 1941.

1338. **Simmonds, M. 1892** Fliegen und Choleraübertragen. [Flies and Cholera Transmission.] Dtsch. Med. Wochenschr. No. 41, p. 931.
One of several early demonstrations that flies may take up the vibrios.

1339. **Simmons, P. 1923** A House-fly Plague in the American Expeditionary Force. J. Econ. Ent. 16:357-363.
In southwest France, severe dysentery among the troops coincided with the presence of enormous numbers of flies. Advocates "Commissioned Entomologists" for the armed forces. ABSTRACT: RAE-B 11:171, 1923.

1340. Simmons, S. W. and **Dove, W. E. 1942** Waste Celery as a Breeding Medium for the Stable fly or "Dog Fly" with Suggestions for Control. J. Econ. Ent. 35:709-715.
Many *M. domestica* found associated with *Stomoxys* in this medium. ABSTRACT: RAE-B 31:100, 1943.

1341. Simmons, S. W. and **Dove, W. E. 1945** Experimental Use of Gas Condensate for the Prevention of Fly Breeding. J. Econ. Ent. 38: 23-25.
Is a biproduct from the manufacture of cooking gas. Twenty per cent in water or ten per cent in fuel oil kills larvae of *Musca* and *Stomoxys* in celery waste. ABSTRACT: RAE-B 34:80, 1946.

1342. Simmons, S. W. and **Wright, M. 1944** The Use of DDT in the Treatment of Manure for Fly Control. J. Econ. Ent. 37:135.
Gives procedure for use of DDT emulsions against larvae of *M. domestica*. ABSTRACT: RAE-B 32:203, 1944.

Simmons, S. W., joint author. *See* Fay, R. W., et al., 1948; Andrews, J. M., et al., 1948.

1343. Simpson, W. J. R. 1918 The Sanitary Aspects of Warfare in South-Eastern Europe. Trans. Soc. Trop. Med. Hyg. (London). 12(1):1-12, 2 pls.
Important because it offers suggestions for the reorganization of sanitary services in the army for the tropics and unhealthy regions. ABSTRACT: RAE-B 7:53, 1919.

1344. Skinner, H. 1909 The Relation of House-flies to the Spread of Disease. New Orleans Med. & Surg. J. 61:950-959.
One of the more influential papers about this time.

1345. Skinner H. 1913 How Does the House-fly Pass the Winter? Ent. News 24(7):303-304.
Concludes that houseflies pass the winter as pupae. ABSTRACT: RAE-B 1:146, 1913. (This is disputed by Hewitt, 1914, p. 85. Footnote.)

1346. Skinner, H. 1915 How Does the House-fly Pass the Winter? (Dipt.). Ent. News 26(6):263-264.
Holds that both sexes emerge from pupae in late winter and that these are responsible for large summer broods. (Philadelphia.) ABSTRACT: RAE-B 3:149, 1915.

1347. Skinner, H. 1917 Insects and War. Ent. News 28(7):330-331.
Recalls Spanish American War, in which only 454 Americans were killed, but 5,277 died of disease, mostly typhoid fever, carried by flies. ABSTRACT: RAE-B 5:145, 1917.

Skvortsov, A. A., joint author. *See* Derbeneva-Ukhova, V. P., et al., 1943.

1348. Slater, J. W. 1881 On Diptera as Spreaders of Disease. J. of Science, London, Ser. 3, pp. 533-539.

1349. Smirnov, E. S. 1937 Résultats sommaires du travail du laboratoire dans sa lutte contre les mouches. [Summarized Results of Laboratory Work in the War against Flies.] Med. Parasit. 6(6):872-879. (In Russian.)
M. domestica found active by night as well as by day; were taken in traps, in warm rooms, in total darkness. ABSTRACT: RAE-B 26:240, 1938.

1350. Smirnov, E. S. 1940 Le problème des Mouches à Tadjikistane. Med. Parasit. 9(5):515-517. Moscow. (In Russian.)
M. domestica vicina most abundant; *M. sorbens* next (especially in markets and food displays). Several other muscoids mentioned. ABSTRACT: RAE-B 31:26, 1943.

1351. Smirnov, E. S. 1942 [Mechanical Control of the Larvae and Pupae of Synanthropic Flies.] Med. Parasit. 11(4):97-105. (In Russian.)
M. domestica adults failed to reach the surface when pupae were buried under 20 inches of compressed clay. ABSTRACT: RAE-B 32:119, 1944.

1352. Smirnov, E. and Kuzina, O. 1933 Experimentalökologische Studien an Fliegenparasiten. [Ecological Experimental Studies on Fly Parasites.] I. Teil. Zool. J. 12(4):96-108. (In Russian, with German Summary.)
Reared 40 *Mormoniella vitripennis* from 1 host pupa of *Calliphora*; 33 from 1 pupa of *Musca*. ABSTRACT: RAE-B 22:216, 1934.

1353. Smirnov, E. and Wladimirow, M. 1934 Studien über Vermehrungsfähigkeit der Ptermalide *Mormoniella vitripennis* Wlk. [Studies on the Reproductive Capacity of the Pteromalid, *Mormoniella vitripennis* Wlk.] Zeit. Wiss. Zool. 145(4):507-522.
M. domestica among several hosts studied. *Calliphora* preferred. Parasite cannot locate host pupa 1 mm. under sand. ABSTRACT: RAE-B 22: 256, 1934.

Smirnov, E. S., joint author. *See* Vladimirova, M. S., et al., 1938; Wladimirow, M., et al., 1934.

1354. Smit, B. 1926 Sheep Blow-fly Control. Fly-traps: Their Construction and Operation. J. Dept. Agric. Union S. Africa 12(2):132-134.
Used a modification of the Wahl-DuPlessis trap, which was very effective. For *M. domestica*, a bait using curds of sour milk and cheap brown sugar served best. ABSTRACT: RAE-B 14:73, 1926.

1355. Smit, B. 1938 The Control of Household Insects in South Africa. Bull. Dept. Agric. S. Africa. No. 192, 52 pp., 20 figs. Pretoria.
M. domestica given appropriate attention. ABSTRACT: RAE-B 28:21, 1940.

1356. Smit, B. 1945 The New DDT Insecticide. Fmg. in S. Africa 20(231): 337-340, 356. Pretoria.
Window panes, sprayed on August 30, killed all flies on contact until washed, Nov. 7. *M. domestica* was the most abundant of the species killed. ABSTRACT: RAE-B 35:66, 1947.

1357. Smit, B. 1945 Fly-traps: Their Construction and Operation. Bull. Dept. Agric. S. Africa No. 262, 10 pp., 5 figs. Pretoria.
This is an up-dating of the author's 1926 publication. ABSTRACT: RAE-B 36:176, 1948.

Smit, B., joint author. *See* Oosthuizen, M. J., et al., 1936.

Smith, C. M., joint author. *See* Jones, H. A., et al., 1936.

1358. Smith, F. 1903 Municipal Sewage. J. Trop. Med. 6:285-291, 304, 308, 330-334, 353-355, 381-383.
Fly may be considered a "dysentery inoculator" in military camps with open latrines. Example from South Africa.

1359. Smith, F. 1907 House-flies and their Ways at Benares. J. Roy. Army Med. Corps. 9:150-155 and 447.
Mentions *M. domestica determinata* Wlk., also *M. enteniata* Bigot. Notes that higher temperatures shorten the life history of *M. domestica* and related forms.

Smith, G. L., joint author. *See* Michelbacher, A. E., et al., 1945, 1946.

1360. Smith, J. L. 1904 An Investigation into the Conditions Affecting the Occurrence of Typhoid Fever in Belfast. J. Hygiene 4:407-433.

1360a. Smith, K. M. 1919 A Comparative Study of Certain Sense Organs in the Antennae and Palpi of Diptera. Proc. Zool. Soc. London. 89: 31-69.
Gives information on such organs in 27 Families. *M. domestica* among

the examples cited. Abstract: Hocking, B. 1960 Smell in Insects, p. 49 (1.481).

Smith, R. C., joint author. *See* Atkeson, F. W., et al., 1943.

Smith, R. F., joint author. *See* Michelbacher, A. E., et al., 1945.

1361. Smith, R. I. 1911 Formalin for Poisoning House-flies Proves very Attractive when Used with Sweet Milk. J. Econ. Ent. 4:417-419.
Advocates 1 oz. of 40 per cent formaldehyde to 16 oz. of fresh milk (or half milk, half water).

1362. Smith, T. 1908 The House-fly as an Agent in Dissemination of Infectious Diseases. Am. J. Publ. Hyg. Aug. 1908, pp. 312-313.
Relates habits of flies to dangerous sources of contamination.

1363. Smith, T. 1909 The House-fly at the Bar. Merchants' Assn., New York, pp. 1-48.
A compilation of letters from various authorities and quotations from the publications of various authors.

1364. Smith, W. W. 1948 Fly Control in a Small City by Use of DDT-Oil Mist. J. Econ. Ent. 41:828-829.
Reports a successful campaign carried out in Hattiesburg, Mississippi, using a 5 per cent DDT-oil solution. Abstract: RAE-B 38:5, 1950.

1365. Smith, W. W. 1948 Reductions of Fly Indices in the Business District of a Small City by the Use of DDT residual Sprays. J. Econ. Ent. 41(5):829-830.
Refers to procedures carried out at Columbus, Mississippi. Used 5 per cent DDT-water emulsions. Abstract: RAE-B 38:6, 1950.

1366. Soberón y Parra, G. 1946 Algo acerca de la transmisión del mal del pinto. [Something Pertaining to the Transmission of the Disease (called) Pinto.] Aliis Vivere 3(14). Mexico, D.F. Reprinted in Kuba 2(2):40-43. Havana.
Experimented with *Hippelates* and *M. domestica*. Both retained the spirochaete, *Treponema caratea* and regurgitated the organism in vomit drops. Abstract: RAE-B 36:25, 1948.

1367. Spaar, E. E. 1925 Some Observations on the Common Endemic Fevers of Ceylon. J. Trop. Med. Hyg. 28(19):349-352.
Seasonal incidence of typhoid coincided with the breeding peak of *M. domestica*. Problem reduced when sewage of Colombo became water-borne. Abstract: RAE-B 13:172, 1925.

1368. Speyer, E. R. 1920 Notes on Chemotropism in the House-Fly. Ann. Appl. Biol., Cambridge. 7(1):124-140.
Experimented with decomposition products of banana and chemically related substances. Over-ripe banana alone, extremely attractive to flies. Abstract: RAE-B 8:215, 1920.

1369. Spielman, M. and **Haushalter, M. 1887** Dissémination du bacille de la tuberculose par la mouche. Compt. Rend. Acad. Sci. 105:352-353.
Believed to be the first published statement to the effect that houseflies which have fed on tubercular sputum may serve as carriers.

Spittall, J. P. 1926, joint author. *See* Kelsall, A., et al., 1926.

1370. Stage, H. H. 1945 Use of DDT in Control of Flies on Cattle and around Farm Buildings. U.S. Dept. Agric. Bur. of Ent. Mimeographed Document E-675, 6 pp.
Contains directions for various methods of application.

1371. Stage, H. H. 1947 DDT to Control Insects affecting Man and Animals in a Tropical Village. J. Econ. Ent. 40:759-761.
Was effective against houseflies and other pests in a jungle village in Surinam. Abstract: BA 23:2022, 1949.

1372. **Stallman, G. P. 1912** Ants Destroying the Larvae of Flies. Military Surgeon 31(3):325-326.
Red ants predatory on fly larvae in Arizona.
Starnes, E. B., joint author. *See* Barber, G. W., et al., 1948.
Starnes, O., joint author. *See* Barber, G. W., et al., 1948.
Stein, A. K., joint author. *See* Pavlovskii, E. N., et al., 1932.

1373. **Stein, F. R. 1878** Der Organisms des Infusionsthiere. 154 pp., 24 pls. Leipzig, W. Engelmann.
Was first to use the specific name *Cercomonas muscae domesticae*. Species later transferred to the genus *Herpetomonas*, by Kent (1880-1882).

1374. **Stein, P. 1909** Neue Javanische Anthomyiden. Tijdschr. Ent. 52:205-271.
Paper records occurrence of *M. augustifrons* in Java; also descriptions of new species: *M. pollinosa*, *M. nigrithorax*, and *M. inferior*.

1375. **Stein, P. 1910** Diptera Anthomyidae, mit den Gattungen *Rhinia* und *Idiella*. Trans. Linn. Soc. London (2):14:149-163.
Includes listing of *M. niveisquama* Thoms. (China, Manila, Malacca), also *M. fasciata* sp. nov. (Seychelles).

1376. **Steiner, G. 1942** Zur quantitativen Analyse tierischer Geruchsreaktionen. [On the quantitative analysis of animal olfactory reactions.] Naturwissenschaften 30(41-42):647-648.
Distinguishes (a) positive food reaction, (b) positive oviposition reaction, peculiar to females, (c) negative protective reaction (has highest threshhold).

1377. **Steiner, G. 1945** Fallenversuche zur Kennzeichnung des Verhaltens von Schmeissfliegen gegenüber verschiedenen Merkmalen ihrer Umgebung. [Trapping Studies to Determine the Behavior of Blowflies in Response to Various Properties of their Surroundings.] 2 + 37 pp., 14 figs. Mult. Darmstadt, Zool. Inst. Tech. Hochsch., 1945.
Worked with *Phormia regina*. Odors at first strongly repellent became less so, as flies became accustomed to them. ABSTRACT: RAE-B 36:25, 1948.

1378. **Steinhaus, E. A. 1942** Catalogue of Bacteria Associated Extracellularly with Insects and Ticks. Burgess Publishing Co., Minneapolis. 206 pp.

1379. **Steinhaus, E. A. 1946** Insect Microbiology: An Account of the Microbes Associated with Insects and Ticks with Special Reference to the Biologic Relationships Involved. Ithaca, Comstock Publ. Co. x + 763 pp. *See also* Steinhaus, 1967.

1380. **Steinhaus, E. A. 1949** Principles of Insect Pathology. New York, McGraw-Hill. First Edition.
Housefly is mentioned in an important way on 18 different pages.

1381. **Stephens, J. W. W. 1905** Transmission of Disease by Insects. Bart's Hosp. J. 12:131-134.

1382. **Stephens, J. W. W. 1905** Two Cases of Intestinal Myiasis. Thompson Yates and Johnstone Laboratories Report. Vol. 6, part 1, pp. 119-121.
Two larvae, passed from rectum, considered to be *Fannia canicularis* and *M. corvina*.

1383. **Sternberg, G. M. 1899** Sanitary Lessons of the War. Philad. Med. J., June 10th and 17th, 1899.
Aware of danger from flies, the writer states, "The surface of the faecal matter should be covered with fresh earth, or quick-lime or ashes three times a day".

1384. Sterngold, E. Y. 1937 [Flies and Intestinal Infections.] Social Sci. Tech. 5(9):93-94. Tashkent. (In Russian.)
Bacillus dysenteriae and other species survive in the fly up to 12 days. Are passed in the feces. ABSTRACT: RAE-B 26:174, 1938.

1385. Sterzinger, O. 1929 Verhalten und Umstimmungen der Stubenfliege bei Gewitter. [Behavior and Changed Actions of Houseflies in Thunderstorms.] Zeit Psychol. 109:229-230.

Stevenson, A. F., joint author. *See* Phelps, E. B., et al., 1916.

1386. Stewart, W. 1944 On the Viability and Transmission of Dysentery Bacilli by Flies in North Africa. J. Roy. Army Med. Corps. 83(1): 42-46.
Observations in Algiers. Flies can carry bacilli up to 11 or 12 days. ABSTRACT: TDB 42:34, 1945.

1387. Stiles, C. W. 1901 Insects as Disseminators of Disease. Virginia Med. Semi-monthly. Vol. 6, pp. 53-58. May 10.

1388. Stiles, C. W. 1910 The Sanitary Privy; its Purpose and Construction. Publ. Health Bull. No. 37, U.S. Publ. Health and Marine Hosp. Service, 24 pp., 12 figs.

1389. Stiles, C. W. 1913 Contamination of Food Supplies. Publ. Health Repts. U.S. Publ. Health Serv. Vol. 28, pp. 290-291.
The three foregoing references deal with different aspects of disease transmission by flies. Some stress is laid on the possibility of flies distributing intestinal protozoa, such as *Entamoeba coli, Lamblia duodenale*, and *Trichomonas intestinalis*.

1390. Stiles, W. C. 1917 Notice to the Zoological Profession of a Possible Suspension of the International Rules of Zoological Nomenclature in the Cases of *Musca* Linnaeus 1758, and *Calliphora* Desvoidy 1830 (Dipt.). Ent News 28:231.
Strict application of rules would require *Musca* to take either *M. caesar* or *M. vomitoria* as type. Problem should be resolved.

1391. Stiles, C. W. 1923 *Musca* Linnaeus, 1758 and *Calliphora* Desvoidy 1830. Science 57:176.
Notice of vote to have been taken May 1, 1918, by International Commission on Zoological Nomenclature, to establish *M. domestica* as type of genus *Musca* and to validate the genus *Calliphora* Desvoidy 1930, with *C. vomitoria* as type. Due to the intervention of World War I, voting was delayed until early in 1924. Explains publication date (1923).

1392. Stiles, C. W. and **Gardner, C. H. 1910** Further Observations on the Disposal of Excreta. U.S. Publ. Health Repts. 25:1825-1830.
Fly-blown excreta buried beneath 48 inches of clean sterilized sand. Adult *M. domestica* emerged nevertheless. An undetermined species issued through 72 inches!

1393. Stiles, C. W. and **Keister, W. S. 1913** Flies as Carriers of *Lamblia* spores. The Contamination of Food with Human Excreta. U.S. Publ. Health Repts. 28:2530-2534. (In Reprint, pp. 3-7.)
Evidence that flies commonly act as carriers of such spores (cysts) is not very conclusive.

1394. Stiles, C. W. and **Lumsden, L. L. 1916** The Sanitary Privy. U.S. Dept. Agric. Farmer's Bull. 463, 32 pp.
Farms may have sanitary privy while churches and rural schools have not. Family with high standards may thus be exposed to disease.

1395. Stiles, C. W. and **Miller, H. M. 1911** The Ability of Fly Larvae to Crawl through Sand. Publ. Health Repts., Publ. Health and Marine Hosp. Serv., 26:1277.
Migratory powers of the larvae responsible for emergence of flies pro-

duced by ova buried 4 to 6 feet under sand. Pupation is close to surface.

Stiles, C. W., joint author. *See* Lumsden, L. L., et al., 1917.

1396. Stoddard, R. B. 1939 What of Moribund Kill? Soap and Sanit. Chem. 15(10):93, 95, 97, Oct.
Peet-Grady test was devised for evaluating pyrethrum, which paralyzes rapidly. Use with other insecticides requires modified interpretation. ABSTRACT: RAE-B 28:79, 1940.

1397. Stoddard, R. B. and **Dove, W. E. 1949** Cinerin I Homolog Tested. A Preliminary Evaluation of the Insecticidal Effectiveness of the Completely Synthetic Allyl Homolog of Cinerin I. Soap and Sanit. Chem. 25(10):118-121, 161.
This is allethrin, which has a high degree of effectiveness against many species. ABSTRACT: RAE-B 39:24, 1951.

1398. Stratton, C. H. 1907 The Prevention of Enteric Fever in India. J. Roy. Army Med. Corps. 8:224.
In Meerut, India, fever at low ebb during the monsoon, recurs with the reappearance of dust and flies.

1399. Strauss, P. 1922 Sur la destruction des mouches domestique. Circulaire du Ministre de l'Hygiene d l'Assistance et de la Prevoyance sociales. J. d'Agric. Prat. Paris, 37(18):361-362.
Educational in nature. Emphasises importance of destroying flies. ABSTRACT: RAE-B 10:224, 1922.

1400. Strauss, P. 1922 Comment détruire les mosches. [How to Exterminate Flies.] Vie Agric. et Rur., Paris 20(25):432-433.
Advocates spreading manure in thin layers so poultry can pick out fly larvae. Also treats of traps, insecticides, fumigation, etc.

Strenzke, K., joint author. *See* Mayer, K., et al., 1948.

1401. Strickland, E. H. 1945 A Method for Permanently Reducing the Number of Blowflies in Screened Houses. Bull. Brooklyn Ent. Soc. 40(2): 59-60.
Flies are positively phototactic, and when walking, negatively geotactic. Will escape through pencil-hole in upper right hand corner of screen. ABSTRACTS: RAE-B 34:120, 1946; BA 20:224, 1946.

Stringer, A., joint author. *See* Page, A. B. P., et al., 1949.

Strong, M., joint author. *See* Hartzell, A., et al., 1944.

1402. Sugai, C. and **Kawabada, K. 1918** Leprosy and Tubercle Bacilli, Viability of, in Alimentary Tract of the Fish and Fly. Igaku Chuo Zasshi [Central J. of Med. Sci.] No. 271, Feb. 1918.
Flies fed on suspension of TB bacilli, retained organisms in alimentary tract. Preparations made from feces and abdominal contents 5 hours later, yielded bacilli pathogenic to guinea pigs. ABSTRACTS: RAE-B 8:138, 1920; China Med. J. 34(2):170, 1920.

Sukhova, M. N., joint author. *See* Gorodetzkii, A. S., et al., 1936.

1403. Sullivan, W. N., Goodhue, L. D. and **Fales, J. H. 1940** Insecticide Dispersion. A New Method of Dispersing Pyrethrum and Rotenone in Air. Soap and Sanit. Chem. 16(6):121, 123, 125. June.
Safrol (solvent), sprayed on heated surface, gives an aerosol, which remains in air several hours. Toxic action accelerated as compared with spray. ABSTRACT: RAE-B 28:240, 1940.

1404. Sullivan, W. N., Goodhue, L. D. and **Haller, H. L. 1939** Rotenone Series Compounds. A Study of Toxicity to the Housefly (*Musca domestica* L.) of Optically Active and Inactive Compounds of the Rotenone Series. Soap and Sanit. Chem. 15(7):107, 109, 111, 113. ABSTRACT: RAE-B 28:16, 1940.

1405. **Sullivan, W. N., Haller, H. L., McGovran, E. R.** and **Phillips, G. L. 1938** Knockdown in Fly Sprays. Comparison of Toxicities of Pyrethrins I and II as Determined by Method for Knockdown and Mortality. Soap and Sanit. Chem. 14(9):101, 103, 105. Sept.
Used turn-table method, with glass covers to permit photography. Difference between knockdown and mortality found significant. ABSTRACT: RAE-B 27:24, 1939.

1406. **Sullivan, W. N., McGovran, E. R.** and **Goodhue, L. D. 1941** Fumigating Action of a Mixture of Orthodichlorobenzene and Naphthalene applied by a New Method. J. Econ. Ent. 34:79-80.
Material sprayed on surface heated to 375°C (707°F). Volitalizes rapidly. *M. domestica* exposed 24 hours in 216 cu. ft. chamber. All died. ABSTRACT: RAE-B 29:174, 1941.

1407. **Sullivan, W. N., Schechter, M. S.** and **Haller, H. L. 1943** Insecticidal Tests with *Phellodendron amurense* Extractive and Several of its Fractions. J. Econ. Ent. 36:937-938.
A petroleum-ether extract of the fruit of this species (also that of *P. lavallei*) proved ineffective in fly sprays. ABSTRACT: RAE-B 32:155, 1944.

Sullivan, W. N., joint author. *See* Haller, H. L., et al., 1938, 1942; McGovran, E. R., et al., 1939, 1942; Jones, H. A., et al., 1942; Goodhue, L. D., et al., 1940, 1942; Campbell, F. L., et al., 1934; Sievers, A. F., et al., 1939.

1408. **Sundararaman, S.** and **Peffly, R. L. 1949** Effectiveness against Flies and Mosquitoes of DDT Applications to Clay, Palm and Straw Surfaces. J. Nat. Mal. Soc. 8:267-269.
Studied effects of various formulations against *M. domestica* and *Anopheles quadrimaculatus*. On all surfaces, suspensions were more effective than solutions or emulsions. ABSTRACTS: RAE-B 41:53, 1953; BA 24: 2657, 1950.

Surber, E. W., joint author. *See* Hoffman, C. H., et al., 1949.

1409. **Surface, H. A. 1915** To Keep Down House Flies. Zool. Press. Bull., Div. of Zool., Penn. Dept. Agric., Harrisburg. No. 313, April 26.
Ground phosphate rock, scattered over manure heaps, destroys larvae of flies, also increases value of manure. ABSTRACT: RAE-B 3:118, 1915.

Swaminath, C. S., joint author. *See* Awati, P. R., et al., 1920.

1410. **Sweet, E. A. 1916** The Transmission of Disease by Flies. Supplement No. 29 to the Publ. Health Repts. Second Ed., 20 pp., illus.
A popular type pamphlet, educational and informative.

1411. **Sweetman, H. L. 1936** The Biological Control of Insects. Comstock Publ. Co., Ithaca, N.Y. 461 pp.
Gives considerable attention to the parasites and infective organisms to which *M. domestica* is host.

1412. **Sweetman, H. L. 1946** DDT as a Spot Treatment for Flies. J. Econ. Ent. 39:380-381.
DDT sprays are sometimes temporarily repellent, due to oil carriers, but this is a minor factor. Applications need be made only in spots where flies congregate, as where fly specks are thick. ABSTRACTS: RAE-B 35:190, 1947; BA 21:730, 1947.

1413. **Sweetman, H. L. 1947** Comparative Effectiveness of DDT and DDD for Control of Flies. J. Econ. Ent. 40:565-566.
Author reports excellent control with both insecticides. (DDD is dichlordiphenyldichlorethane.) ABSTRACT: RAE-B 37:115, 1949.

Sweetman, H. L., joint author. *See* Burgess, A. F., et al., 1949.

Swellengrebel, N. H., joint author. *See* Nijkamp, J. A., et al., 1934.

1414. Sydenham, T. 1666 Sydenham's Works. Syd. Soc. Ed., Vol. 1, p. 271.
Held that if swarms of flies were abundant during the summer, the autumn would be unhealthy.

1415. Synerholm, M. E. and **Hartzell, A. 1945** Some Compounds Containing the 3, 4-Methylenedioxyphenyl Group and Their Toxicities toward Houseflies. Contrib. Boyce Thompson Inst. 14(2):79-90.
Piperine, also several closely related amides and esters of piperic acid, shown to possess exceptional toxicity when used alone, as well as with pyrethrins. ABSTRACT: RAE-B 35:186, 1947.

1416. Synerholm, M. E., Hartzell, A. and **Arthur, J. M. 1945** Derivatives of Piperic Acid and their Toxicity toward Houseflies. Contr. Boyce Thompson Inst. 13(9):433-442.
Various amides and esters of piperic acid tested by Peet-Grady method. Most are highly toxic, though piperic acid itself is not. ABSTRACT: RAE-B 34:61, 1946.

Synerholm, M. E., joint author. *See* Prill, E. A., et al., 1946.

Szappanos, G., joint author. *See* Lörincz, F., et al., 1936.

1417. Taggart, R. S. 1946 DDT. Cornell Vet. 36(2):159-169.
Effective against houseflies and stableflies when applied as a spray for residual effect. ABSTRACT: BA 21:485, 1947.

Takacs, W. S., joint author. *See* Toomey, J. A., et al., 1941.

1418. Takano, T., Ueda, M., Murasawa, I. and **Ono, M. 1947** [On the Mosquitocide Incense made of Pyrethrum mixed with Benzophenone.] Botyu Kagaku [Sci. Insect Control.] 7:11-15. (In Japanese.)
Tested against *M. domestica.* Benzophenone has a synergistic effect. ABSTRACT: BA 24:736, 1950.

1419. Tao, S. M. 1927 A Comparative Study of the Early Larval Stages of Some Common Flies. Am. J. Hyg. 7(6):735-761, 7 pls.
Keys to first and second stage larvae of a number of genera of Muscoids. ABSTRACT: RAE-B 16:62, 1928.

1420. Taschenberg, E. L. 1880 Praktische Insektenkunde, Part IV; (*M. domestica*, pp. 102-107, fig. 27).
Gives a somewhat popular account of habits and life history. Incorporates the work of Gleichen and Bouché.

1421. Tattersfield, F. and **Hobson, R. P. 1929** Pyrethrins I and II. (Second Paper.) Their Estimation in Pyrethrum (*Chrysanthemum cinerariaefolium*). J. Agric. Sci. 19(2):443-437.
Gives a short method for the evaluation of pyrethrum by a determination of pyrethrin I.

1422. Taylor, J. F. 1919 The Rôle of The Fly as a Carrier of Bacillary Dysentery in the Salonica Command. Med. Research Committee Nat. Health Insurance. London. Spec. Rept. Ser. 40, pp. 68-83, 2 charts, 1 fig.
Incidence of dysentery corresponds to prevalence of flies. Experiments with Flexner and Shiga bacilli showed flies capable of carrying both. ABSTRACT: RAE-B 8:4, 1920.

1423. Taylor, T. 1883 *Musca domestica* as a Carrier of Contagion. Proc. Am. Ass. Adv. Sci. Vol. 31, p. 528.
Includes report of a rare observation; a minute thread worm emerging from the fly's ruptured proboscis.

1424. Tebbutt, H. 1913 On the Influence of the Metamorphosis of *Musca domestica* upon Bacteria Administered in the Larval Stage. J. Hygiene 12:516-526.
Process appears to destroy most bacteria present in the larvae. Only a remote possibility of the fly retaining the organism.

1425. Teichmann, E. 1918 Die Bekämpfung der Fliegenplage. [The Conquest of the Fly Plague.] Zeit. Angew. Ent. (Berlin), 4(3):347-365.
Recommends use of HCN gas (1·2096 grams of cyanide per cubic metre). ABSTRACT: RAE-B 8:121, 1920.

1426. Telford, H. S. 1945 DDT Toxicity. Soap and Sanit. Chem. 21(12):161, 163, 167, 169.
Houseflies were killed by feeding on milk from a goat that had received 1 gram of DDT per 8-9 pounds of body weight. ABSTRACT: RAE-B 36: 188, 1948.

1427. Terry, C. E. 1912 Extermination of the House-fly in Cities; its Necessity and Possibility. Am. J. Publ. Health 2:14-22.

1428. Terry, C. E. 1913 Fly-borne Typhoid Fever and its Control in Jacksonville (Florida, U.S.A.) Publ. Health Repts. U.S. Public Health Serv. Vol. 28, pp. 68-73.
The two foregoing papers set forth the urban fly problem and how it may be handled. Typhoid was greatly reduced in Jacksonville by compulsory fly-proofing of privies.

Teterovskaia, T. D., joint author. *See* Zimin, L. S., et al. 1943.

1429. Thaxter, R. 1888 The Entomopthoreae of the United States. Mem. Boston Soc. Nat. Hist. Vol. 4, pp. 133-201, pls. 14-21.
Records *Empusa sphaero-sperma* (Fres.) Thaxter and *E. americana* Thaxter as attacking houseflies.

1430. Theobald, F. V. 1904 Swarms of Flies Bred in House-refuse. Second Rept. on Economic Zool. Brit. Mus. Nat. Hist., pp. 125-126.

1431. Theobald, F. V. 1905 Flies in Distribution of Disease. Nursing Times. London, Vol. 1, p. 461.

1432. Theobald, F. V. 1905 The House-fly. Rept. on Economic Zoology, S.E. Agric. Coll., Wye, pp. 109-111, 2 figs.

1433. Theobald, F. V. 1905 Flies and Ticks as Agents in the Distribution of Disease. Proc. Ass. Econ. Biol. Vol. 1, pt. 1, pp. 17-26.

1434. Theobald, F. V. 1907 The House-fly Annoyance. Rept. on Economic Zool., S.E. Agric. Coll., Wye, pp. 141-143, 1 fig.

1435. Theobald, F. V. 1911 House-flies, their Destruction and Prevention. Rept. on Economic Zoology, S.E. Agric. Coll., Wye, pp. 133-137, 1 fig.
The six foregoing papers are essentially educational and practical. They are aimed at arousing interest and stimulating improved sanitary practice. Lessons learned in the Spanish-American War and the Boer War, led to a large output of this type of publication during the first decade of the twentieth century.

Thomas, H. E., joint author. *See* Ark, P. A., et al., 1936.

1436. Thompson, R. L. 1915 Some Household Insects. Dept. Agric., Salisbury. Rhodesia. Bull. No. 214, 11 pp.
Gives duration range for all stages in life cycle of *M. domestica* in Rhodesia. Also covers larvicides, poison baits, etc. ABSTRACT: RAE-B 3:164, 1915.

Thompson, W. R., joint author. *See* Parker, H. L., et al., 1928.

1437. Thomsen, Ellen. 1938 Über die Kreislauf in Flügel der Musciden, mit besonderer Berücksichtigung der akzessorischen pulsierenden Organe. [On the Circulation in the Wings of Muscidae, with special Reference to the Accessory Pulsation Organs.] Zeit. Morph. und Ökol. Tiere. 34(3):416-438, 12 figs.
Gives several illustrations pertaining to *Musca domestica.*

1438. Thomsen, E. and **Thomsen, M. 1937** Ueber das Thermopräferendum der Larven einiger, Fliegenarten. Zeit. Vergl. Physiol. 24(3):343-380.
Introduced larvae into manure-filled trough of uniform humidity, one end heated, the other cooled. *M. domestica* larvae chose portion 30°-37°C for feeding; later migrated to lower temperature for pupation. ABSTRACTS: RAE-B 25:244, 1937; TDB 35:143-144, 1938.

1439. Thomsen, M. 1934 Fly Control in Denmark. Quart. Bull. Hlth. Org. League of Nations No. 3, pp. 304-324, 13 pls. (mostly in color).
M. domestica preferred pig manure over horse and used cow dung rarely. Calf dung attractive because of milk content. Various recommendations for control. ABSTRACT: RAE-B 22:230, 1934.

1440. Thomsen, M. 1935 The Problem of Fly Control. League of Nations Hlth. Org., C. H./Hyg. rur/E. H. 4. 17 pp. multigraph.
Discusses practicality of covering pig dung (favorite medium) with a layer of cow dung, seaweed, peat moss, and/or a tarpaulin, to discourage oviposition. ABSTRACT: RAE-B 24:43, 1936.

1441. Thomsen, M. 1935 A Comparative Study of the Development of the Stomoxydinae (especially *Haematobia stimulans* Meigen) with Remarks on other Coprophagus Muscids. Proc. Zool. Soc. for 1935, pt. 3, pp. 531-550, 8 pls.
Cephalo-pharyngeal skeleton of larva compared with that of *M. domestica* and other species. ABSTRACT: RAE-B 23:242, 1935.

1442. Thomsen, M. 1936 Fluerne og deres Bekaempelse. [Flies and their Control.] 23 pp., 11 figs. Aarhus, Dansk. Mejerinforen. Faellesorganis.
Deals largely with *M. domestica*. Stresses need for covering pig dung to prevent breeding. ABSTRACT: RAE-B 25:282, 1937.

1443. Thomsen, M. 1938 Stuefluen (*Musca domestica* og Stikfluen (*Stomoxys calcitrans*) . . . Beretn. Vet. og Landbohøjsk Forsøgslab. No. 176, 352 pp., 107 figs., 13 pls.
An important publication. Two hundred thirty-seven pages deal with *M. domestica*. Summarizes research in Denmark from 1932 to 1936. ABSTRACT: RAE-B 26:112, 1938.

1444. Thomsen, M. and **Hammer, O. 1936** The Breeding Media of some Common Flies. Bull. Ent. Res. 27(4):559-587, 4 figs.
Denmark produced 4·8 million tons of pig manure in 1934. Retains its capacity to support fly breeding much longer than horse or cow dung. ABSTRACT: RAE-B 25:92, 1937.

Thomsen, M., joint author. *See* Larsen, E. B., et al., 1940; Thomsen, E., et al., 1937.

1445. Thomson, F. W. 1912 The House-fly as a Carrier of Typhoid Infection. J. Trop. Med. Hyg. 15:273-277.
Ingestion of typhoid organisms has no bad effect on the health of the flies. They carry living germs internally for 24 hours, and on their exterior, 6 hours.

1446. Thomson, J. G. and **Lamborn, W. A. 1934** Mechanical Transmission of Trypanosomiasis, Leishmaniasis, and Yaws through the Agency of Non-biting Haematophagous Flies. (Preliminary Note on Experiments.) Brit. Med. J., Vol. 2 (3845):506-509.
Flies readily engorge on blood, serum, and serous exudates from ulcers, sores and on secretions from eyes, nose and mouth. Species have preferred hosts, e.g. in Nyasaland, *M. spectanda* prefers man. ABSTRACT: RAE-B 22:214, 1934.

1447. Thomson, R. C. M. 1947 Notes on the Breeding Habits and Early Stages of some Muscids associated with Cattle in Assam. Proc. Roy. Ent. Soc. (London) (A) 22(10-12):89-100.
Larvae of *Anaclysta flexa* Wied. could not be reared on dung alone. Fed on full grown larvae of *Musca*. ABSTRACT: RAE-B 38:156, 1950.

Tian, A., joint author. *See* Cotte, J., et al., 1924.

Tiburskaya, N. A., joint author. *See* Nabokov, V. A., et al., 1935.

Tiegs, O. W., joint author. *See* Johnston, T. H., et al., 1923.

Tilden, E. B., joint author. *See* Noguchi, H., et al., 1926.

Tischer, L. A., joint author. *See* Toomey, J. A., et al., 1941.

1448. Tischler, N. 1931 Reproductivity of Flies Exposed to Pyrethrum Sprays. J. Econ. Ent. 24(2):558.
Flies which recovered from pyrethrum paralysis will feed and reproduce normally, despite claims of certain manufacturers to the contrary. ABSTRACT: RAE-B 19:181, 1931.

1449. Tischler, N. 1931 A Satisfactory Nutriment for Adult Houseflies. J. Econ. Ent. 24(2):559.
Found bread soaked in milk entirely adequate.

1450. Tischchenko, O. D. 1929 [On the Question of the Hibernation of the House-fly (*Musca domestica*).] Sanitarno-Entomol. Bynl. i, No. 2, pp. 10-14. Kharkov. (In Ukrainian.)
Enter hibernation quarters within 90 yards of last breeding site. Favor warmer rooms and remain active. Author believes there is no diapause in *M. domestica*. ABSTRACT: RAE-B 18:116, 1930.

1451. Tizzoni, G. and **Cattani, J. 1886** Untersuchungen über Cholera. Centralbl. f. d. Med. Wissensch. (Berlin). 24:269-271.
Epidemic in Bologna. Flies shown to be capable of carrying the "comma bacillus" on their feet.

Tobias, J. M., joint author. *See* Savit, J., et al., 1946.

1452. Tomlinson, T. G., Grindley, J., Collett, R. and **Muirden, M. J. 1949** Control of Flies Breeding in Percolating Filters. 2. J. and Proc. Inst. Sewage Purification (London). 2:127.
M. domestica is not involved here. The species concerned fall in *Psychoda*, *Anisopus*, and *Hypogastrura*. Because of the importance of the problem, it seems desirable to include at least one reference in this field. ABSTRACT: BA 25:2520, 1951.

1453. Tonnelier, A. C. 1918 Las Moscas, *Musca domestica* L. Métodos de Destrucción. Anales Soc. Rural Argentina. 52(3):170-174.
Makes conventional recommendations. No fresh information. ABSTRACT: RAE-B 180, 1918.

1454. Toomey, J. A., Takacs, W. S. and **Tischer, L. A. 1941** Poliomyelitis Virus from Flies. Proc. Soc. Exp. Biol. Med. 48(3):637-639.
Flies trapped near stream carrying raw sewage from Cleveland were used to prepare a suspension which was then injected into monkeys. Animals developed poliomyelitis. ABSTRACT: RAE-B 31:17, 1943.

1455. Tooth, H. H. 1900 Enteric Fever in the Army in South Africa. Brit. Med. J., Nov. 10, 1900 (Reprint, 5 pp.).

1456. Tooth, H. H. 1901 Some Personal Experiences of the Epidemic of Enteric Fever among the Troops in South Africa, in the Orange River Colony. Trans. Clin. Soc. Vol. 34 (Reprint, 64 pp.).
Flies increased with the coming of the army, due to large amounts of waste for them to breed in. Illness ceased abruptly with the coming of frost which killed most of the flies. Waste disposal seen to be as important as a pure water supply.

1457. Torrey, J. C. 1912 Numbers and Types of Bacteria Carried by City Flies. J. Infect. Dis. 10:166-177.

Flies taken in early June were free from fecal bacteria, but those collected in July and August carried "several millions each". Bacteria in fly's intestine 8·6 times as numerous as those on surface. *B. paratyphosus*, Type A was isolated 3 times, in pure culture.

1458. Townsend, C. H. T. 1911 Review of Work by Pantel and Portchinski on Reproductive and Early Stage Characters of Muscoid Flies. Proc. Ent. Soc. Wash. 13:151-170.

Northern form of *Eumusca corvina* deposits large eggs, from which come only *two* larval stages. Crimean form deposits *third* stage larvae. Species considered to have an "incubating uterus".

1459. Townsend, C. H. T. 1915 Correction of the Misuse of the Generic Name *Musca*, with Descriptions of Two New Genera. J. Wash. Acad. Sci. 5:433-436.

Would erect new genus, *Promusca*, for *Musca domestica* L. Holds that type of genus *Musca* is *vomitoria* (*Calliphora vomitoria* of current usage). Townsends' pronouncements led to much controversy, and ultimately to special action. *See* Stiles, C. W. 1917 and 1923.

1460. Townsend, C. H. T. 1916 On Australian Muscoidea, with Descriptions of New Forms. Insec. Inscit. Menstruus 4(4-6):44-45.

Eumusca australis Macq. (figured by Froggat as *M. autumnalis*) (= *corvina*) is recorded from New South Wales. Abstract: RAE-B 4:159, 1916.

1461. Townsend, C. H. T. 1921 Some New Muscoid Genera, Ancient and Recent. Insec. Inscit. Menstruus 9(7-9):132-134.

Proposes *Awatia*, n.g. for *M. indica* Awati. Differs from others of the *Philaematomyia* group by being larviparous. Abstract: RAE-B 9:207, 1921.

1462. Trask, J. D. and **Paul, J. R. 1943** The Detection of Poliomyelitis Virus in Flies Collected during Epidemics of Poliomyelitis. II. Clinical Circumstances under which Flies were Collected. J. Exp. Med. 77(6):545-556.

Positive flies found only in lots collected within ten days of the onset of a local case. Areas 15-20 acres in extent used for sampling. Abstract: RAE-B 33:110, 1945.

1463. Trask, J. D., Paul, J. R. and **Melnick, J. L. 1943** The Detection of Poliomyelitis Virus in Flies Collected during Epidemics of Poliomyelitis. I. Methods, Results and Types of Flies Involved. J. Exp. Med. 77(6):531-544.

Positive results obtained with the monkey species, *Macacus cynomolgus*; negative with *Cercopithecus* and *Rhesus*. *M. domestica* much fewer in number than *Phormia* and *Lucilia*, in the collections made.

Trask, J. D., joint author. *See* Paul, J. R., et al., 1941.

1464. Travis, B. V. and **Bohart, R. M. 1946** DDT to Control Maggots in Latrines. J. Econ. Ent. 39:740-742.

Field trials made in Okinawa. One quart of 4 to 5 per cent solution in Diesel oil or kerosene per seat hole, once a week, controlled flies in pits built according to specifications. Abstracts: RAE-B 36:86; BA 21:1810, 1947.

Travis, B. V., joint author. *See* Lindquist, A. W., et al., 1945, 1946.

Tregubov, A. N., joint author. *See* Dolinskaya, T. Y., et al., 1946.

Treillard, M., joint author. *See* Roubaud, E., et al., 1935.

1465. Trofinov, G. K. 1942 [On the Fauna of the Synanthropic Flies of Azerbaijan.] Med. Parasit. 10(5-6):561-562. (In Russian.)
M. sorbens Wied. taken on human feces and cow dung in 6 localities, chiefly in southern Azerbaijan. ABSTRACT: RAE-B 31:224, 1943.

1466. Tsuzuki, J. 1904 Bericht über meine epidemiologischen Beobactungen und Forschungen während der Choleraepidemie in Nordchina im Jahre 1902 etc. [Report on my Observations and Researches during the Cholera Epidemic in North China in the Year 1902, etc.] Arch. f. Schiffs. u. Tropen-Hyg. 8:71-81.
Flies, taken in a cholera house were incubated in bouillon, from which plate cultures were made. Concludes that flies are able, under natural conditions to carry the cholera spirillum.

Tuck, W. T., joint author. *See* Levy, E. C., et al., 1913.

1467. Tucker, E. S. 1913 Formaldehyde Gas not Effective upon Flies. Biol. Papers, Kansas Acad. Sci. 25 (1912):53.
Used tightly closed room 12 ft. × 14 ft. × 8 ft. 8 ins. Concluded that formaldehyde is an insecticide only in very concentrated form, with long exposure. ABSTRACT: RAE-B 2:57, 1914.

1468. Tullgren, A. 1918 Våra Insekter såson Sjukdomsspridare. [Our Insects as Carriers of Disease.] Stockholm, 30 pp., 24 figs.
A popular treatise, with due emphasis on the importance of *M. domestica.* ABSTRACT: RAE-B 7:157, 1919.

1469. Tully, E. J. 1945 Sanitary Garbage Disposal. Wisc. State Board of Health. Quart. Bull. 7(16):3-7. (Oct.-Dec.)
A good discussion. Includes instructions for both platform feeding and ground feeding of hogs so as to insure minimum fly production. Three photographs.

1470. Tulpius, N. 1672 Observationes medicae, Vol. 2, pp. 173-174.
Twenty-one small larvae passed from patient's urethra. Figure seems to be larva of *Fannia canicularis.*

1471. Twinn, C. R. 1947 Livestock Insect Control with DDT. Dom. of Canada Dept. of Agric. Div. of Entomol. Processed Publication No. 65, 12 pp.
Gives effective formulations for use against flies and other livestock pests.

1472. Twinn, C. R. and **Balch, R. E. 1946** Insect Control in Lumber Camps with DDT. Div. of Entomol. Processed Publ. No. 38, 8 pp. Ottawa, Canada.
Formulations given for use against bedbugs, lice, fleas, cockroaches, flies, mosquitoes, etc. ABSTRACT: RAE-B 35:41, 1947.

1473. Twinn, C. R. and **Balch, R. E. 1946** How to Use DDT in Lumber Camps. Div. of Emtomol. Processed Publ. No. 39, 3 pp. Ottawa, Canada.
Is a brief condensation of Publication No. 38. ABSTRACT: RAE-B 35: 41, 1947.

1474. Twinn, C. R. and **Herman, F. A. 1927** A Cheap and Effective Fly Spray. Sci. Agric. (Ottawa) 8(7):441-445.
Any solution of pyrethrum in kerosene stronger than 1 per cent brings down flies. Ten per cent gives best mortality, but 6 per cent is satisfactory, for all practical purposes. ABSTRACT: RAE-B 16:137, 1928.

Ueda, M., joint author. *See* Takano, T., et al., 1947.

1475. Uganda Medical Department, Corporate Author. 1940 Annual Report for 1939. Med. 8 vo, 56 pp. Entebbe.
Of the flies found frequenting ulcers and wounds, 95 per cent were

M. sorbens Wied., a species which breeds in human feces. ABSTRACT: RAE-B 28:250, 1940.

1476. **Underwood, W. L. 1903** House-fly as Carrier of Disease. Boston.

1477. **Uribe, C. 1926** A New Invertebrate Host of *Trypanosoma cruzi* Chagas. J. Parasit. 12(4):213-215, 1 fig.
Concerns the Reduviid bug, *Apiomeris pilipes. M. domestica* proved acceptable food for this species in the laboratory. ABSTRACT: RAE-B 14:191, 1926.

1478. **Uvarov, B. P. 1931** Insects and Climate. Trans. Ent. Soc. London. 79: 1-247.
As quoted by Wigglesworth: "Houseflies (*M. domestica*) in a cold room with a single Bunsen burner will collect in a circle at a fixed distance around the source of heat".

1479. **Vacher, F. 1909** Report of County Medical Officer of Health on "Some recent Investigations regarding the propagation of Disease by Flies". Cheshire.

1480. **Vaillard, (Dr.) 1913** Au sujet des measures à prendre contre les mouches. [On the Topic of Measures to Take against Flies.] Bull. Mens. Office Internat. d'Hyg. Publique, for 1913, pp. 1313-1336. Paris.
Flies can transmit microbes 74 hours after infection. Can also carry eggs of parasitic worms (*Oxyuris*, *Trichocephalus*, *Taenia echinococcus*). ABSTRACT: RAE-B 1:191, 1913.

1481. **Vaillard, (Dr.) 1914** Pour lutter contre les mouches. [To Strive against Flies.] Vie Agric. et Rur. (Paris) 3(14):373-378, 3 figs.
Various control measures outlined. Mentions *Bembex* and *Empusa* as natural enemies. ABSTRACT: RAE-B 2:111, 1914.

1482. **Vainshtein, B. A.** and **Rodova, R. A. 1940** Les lieux de développement des mouches de fumier dans les conditions du Tadjikistan Montagneux. [The Places of Development of Flies from Manure under Conditions in Mountainous Tadjikistan.] Med. Parasit. (4):364-368. (In Russian.)
No cows or sheep. Horse dung dries rapidly. Garbage quickly disposed of. Various species of *Musca* use pig or human feces. ABSTRACT: RAE-B 31:126, 1943.

Valdes Garcia, A., joint author. *See* Piedrola Gil, G., et al., 1949.

1483. **Vandenberg, S. R. 1930** Report of the Entomologist. Rept. Guam Agric. Expt. Sta. for 1928, pp. 23-31. Washington.
Spalangia sp., parasite of the pupa of *M. domestica*, recently introduced into Guam from Hawaii. Requires 15-17 days for complete life cycle (egg to adult). ABSTRACT: RAE-B 19:89, 1931.

1484. **Vandenberg, S. R. 1931** Report of the Entomologist. Rept. Gaum Agric. Expt. Sta. for 1929, pp. 16-17. Washington.
Gives improved laboratory methods for rearing *Spalangia*, pupal parasite of *Musca* and *Stomoxys*. Pine shavings better medium than earth; eliminates fungus infection. ABSTRACT: RAE-B 19:219, 1931.

1485. **Vandenberg, S. R. 1931** Report of the Entomologist. Rept. Guam Agric. Expt. Sta. for 1930, pp. 23-25. Washington.
Spalangia well established and spreading. No longer necessary to breed and release these parasites. ABSTRACT: RAE-B 20:107, 1932.

Van Deurs, H., joint author. *See* Keiding, J., et al., 1949.

1485a. **Van Leeuwen, E. R. 1944** Residual Effect of DDT against Houseflies. J. Econ. Ent. 37:134.
All tests favorable for complete control. (Resistance of extreme nature was unknown at this time.) ABSTRACT: RAE-B 32:201, 1944.

1486. **Vanskaya, R. A. 1942** [The Use of Ammonium Carbonate for the Control of *Musca domestica* L.] Med. Parasit 10(5-6):562-567. (In Russian.)
Ammonium carbonate attracts females about to oviposit. Might be used to attract them to a medium not suitable for larval development. ABSTRACT: RAE-B 31:225, 1943.

1487. **Vanskaya, R. A. 1942** Hibernation of *Musca domestica* L. Med. Parasit. 11(3):87-90. (In Russian.)
Believes that housefly hibernates in any stage except the adult, in Russia. ABSTRACTS: BA 20:1541, 1946; RAE-B 32:53, 1944.

1488. **Varley, C. 1853** Microscopical Observations on a Malady Affecting the Common House-Fly. Trans. Micr. Soc. (London). 3:55-57, 1 pl.

1489. **Vaughan, V. C. 1899** Some Remarks on Typhoid Fever Among our Soldiers during the Late War with Spain. Am. J. Med. Sci. 118: 10-24.
Observed that flies, crawling over the food of soldiers, frequently carried on their legs and bodies, particles of lime acquired while feeding or ovipositing in latrines.

Vaughan, V. C., joint author. *See* Reed, W., et al., 1904.

1490. **Veeder, M. A. 1898** The Spread of Typhoid and Dysenteric Diseases by Flies. Public Health (U.S.A.) 24:260-262.

1491. **Veeder, M. A. 1898** Flies as Spreaders of Sickness in Camps. Med. Record 54(12):429-430.
In the two foregoing references, the author gives a very strong argument, incriminating flies. Most camps became "stale" in a few weeks, and all water within reach unfit for use. Flies numerous on feces in trenches.

1492. **Veeder, M. A. 1899** The Relative Importance of Flies and Water Supply in Spreading Disease. Med. Record 55:10-12.
Anticipates Reed, Vaughan and Shakespeare (1904) in regarding flies, not water, as the chief cause of the Spanish War epidemic.

1493. **Veeder, M. A. 1902** Typhoid Fever from Sources other than Water Supply. Med. Record, 62:121-124.
Stresses again the importance of filth-feeding flies.

Veillon, R., joint author. *See* Roubaud, E., et al., 1922.

1494. **Veitch, R.** and **Greenwood, W. 1924** The Food Plants or Hosts of some Fijian insects. Part II. Proc. Linn. Soc. New S. Wales, 49, pt. 2, No. 196, pp. 153-161.
Larvae of *M. domestica* feed on rotten seeds of *Vigna catiang* Walp., a plant of the family Leguminosæ. ABSTRACT: RAE-B 12:153, 1924.

1495. **Velbinger, H. H. 1947** Wirkung und Anwendung des Phenothiazins als Antiparasiticum. Neue Untersuchungen zur Feststellung der vermiziden und insektiziden Wirksamkeit. [Action and Use of Phenothiazine as a Paracide. New Researches toward the Confirmation of its Vermicidal and Insecticidal Efficiency.] Dtsch. Tierärztl. Wochenschr. 54(17-18):130-133. Hanover.
Is a cerebrospinal contact poison, with about the same range of effectiveness as rotenone. Used as a dust, it is less effective against *M. domestica* than against fleas and lice. ABSTRACT: RAE-B 38:28, 1950.

1496. **Venables, E. P. 1914** A Note on the Food Habits of adult Tenthredinidae. Canad. Ent. 46(4):121.
Captive specimens of *Tenthredo variegatus* fed voraciously on house flies, by making a wound in the body through which contents were extracted. ABSTRACT: RAE-B 2:129, 1914.

1497. Vignon, P. 1901 Recherches de Cytologie générale sur les Epithéliums, l'appareil pariétal protecteur ou moteur; le rôle de la co-ordination biologique. [Researches concerning the General Cytology of the Epithelium, the Parietal Protective or Motile Apparatus; its Role in Biological Coordination.] Arch. Zool. Exp. et Gén. 9:371-720, pls. 15-18.
A study of the peritrophic (growing) membrane, which is probably represented in *M. domestica* by the mucous intima of the ventriculus.

1498. Villeneuve, J. 1913 Notes synonymiques. Ent. Zeit. Wien. 32:128.
Believes *Plaxemyia beckeri* Schnabl. identical with *Musca speculifera* Bezzi.

1499. Villeneuve, J. 1922 Description d'especes nouvelles du genre *Musca*. Ann. Sci. Nat. (Zool.) Paris. (10)5:335-336.

1500. Villeneuve, J. 1936 Description d'une nouvelle espèce africaine du genre *Musca*. [Description of a New African Species of the Genus Musca.] Bull. Ann. Soc. Ent. Belg. 76:414.
Musca lasiopa sp. nov.

1501. Vladimirova, M. S. and **Smirnov, E. S. 1938** Concurrence vitale dans une population homogène de *Musca domestica* L. de *Phormia groenlandica* et entre ces deux espèces. [Vital Competition in a Homogenous Population of *Musca domestica*, of *Phormia groenlandica* and between the Two Species.] Med. Parasit. 7(5):755-777, 12 graphs. (In Russian, with French summary.)
M. domestica can survive in larger numbers under difficult conditions than *Phormia terraenovae* (*groenlandica*), and can survive in small numbers under extreme conditions, where *Phormia* cannot. In competition with each other, only *M. domestica* survived. ABSTRACT: RAE-B 27:175, 1939.

1502. Vogler, C. H. 1900 Weitere Beiträge zur Kenntnis von Dipteren Larven. Illustr. Zeit. f. Ent. 5:273-276; 289-292, 8 figs.
Includes pharyngeal apparatus, tracheal system and segmentation of *Homalomyia* (=*Fannia*) *scalaris*.

1503. Volkenberg, H. L. van. 1932 Report of the Parasitologist. Rept. Porto Rico Agr. Expt. Sta. for 1931, pp. 24-27. Mayaquez.
M. domestica less abundant in the country than *Stomoxys calcitrans*. ABSTRACT: RAE-B 20:268, 1932.

Vorhees, C. T., joint author. *See* Frost, W., et al., 1908.

Waggoner, A., joint author. *See* Shoemaker, E. M., et al., 1903.

1504. Wahl, B. 1914 Ueber die Kopfbildung Cyclorapher Dipterenlarven und die postembryonale Entwicklung des Fliegenkopfes. [On the Head Structure of the Larvae of Cyclorrhaphous Diptera and the Postembryonic Development of the Fly's Head.] Arb. Zool. Inst. Wien. Vol. 20, 114 pp. 20 figs., 3 pls.

1505. Wahl, R. O. and **du Plessis, S. 1923** Combating Sheep-maggot Flies. J. Dept. Agric. Union S. Africa 7(5):428-432.
Directions for constructing traps. Half the flies caught were *M. domestica*, as compared with 23 per cent, sheep-maggot flies. ABSTRACT: RAE-B 12:20, 1924.

Wain, R. L., joint author. *See* Martin H., et al., 1944.

1506. Walker, F. 1856 Insecta Saundersiana, i, Diptera, p. 345.
M. domestica determinata described as new, from the East Indies.

1507. Walker, F. 1864 Catalogue of the Dipterous Insects Collected in Waigon, Mysol and North Ceram by Mr. A. R. Wallace, with

Descriptions of New Species. J. Proc. Linn. Soc. London. 7:202-238.
Includes *M. sordidissima*, sp. nov. and *M. arcioides*, sp. nov.

Walker, G. P., joint author. *See* Kelsall, A., et al., 1926.

1508. Wallace, D. H. 1927 Some Aspects of House Fly Control. The Viability of the Newly "Hatched" Fly under Adverse Conditions. Health Bull. Dept. Publ. Health, Victoria No. 11, pp. 338-341. Melbourne.
Burial of manure cannot be relied on to destroy larvae and pupae already present. *M. domestica* can emerge through 7 inches of sandy soil. ABSTRACT: RAE-B 17:22, 1929.

Wallace, P. P. joint author. *See* Pearsall, D. E., et al., 1946.

1509. Wang, L. N. 1926 [House-flies, especially *Musca domestica*.] 8 vo., 78 pp., 15 figs. Shanghai, Commercial Press, Ltd. (In Chinese.)
ABSTRACT: RAE-B 16:48, 1928.

1510. Wanhill, C. F. 1909 An Investigation into the Causes of the Prevalence of Enteric Fever among the Troops Stationed in Bermuda, giving Details of the Measures Adopted to Combat the Disease, and Showing the Results of these Measures, during the Years 1904-1906. J. Roy. Army Med. Corps. Vol. 12, pp. 28-45.
From 1893 to 1902 Bermuda had the highest enteric fever rate among all British troops. Through fly control, the disease was almost eradicated.

1511. Ward, R., Melnick, J. L. and **Horstman, D. M. 1945** Poliomyelitis Virus in Fly-contaminated Food Collected at an Epidemic. Science 101(2628):491-493.
Food fed to young chimpanzees, free from infection. Both showed a rise in temperature. Excreta remained infected 14 to 21 days. Infection considered "subclinical". ABSTRACT: TDB 33:126, 1945.

Ward, R., joint author. *See* Melnick, J. L., et al., 1947; Sabin, A. B., et al., 1941, 1942.

1512. Ware, F. 1924 A Case of Habronemiasis in England. J. Comp. Path. and Therap. 37(3):160-162.
Reviews the supposed insect vectors: *Stomoxys* conveys *H. microstoma*; *M. domestica* is carrier of *M. megastoma* and *M. muscae*. ABSTRACT: RAE-B 12:185, 1924.

1513. Washburn, F. L. 1910 The Typhoid Fly on the Minnesota Iron Range. Thirteenth Report of the State Entomologist of Minnesota, for 1909-1910, pp. 135-141, 6 figs.
Open privies and almost no use of screens contributed to much fly-borne typhoid and dysentery. Food habits of certain groups a factor, also.

1514. Washburn, F. L. 1912 The Minnesota Fly Trap. J. Econ. Ent. 5:400-402.
Two photos and one diagram to show construction.

1515. Washburn, F. L. 1914 A Model Fly Ordinance. Minnesota Insect Life 2(10):2.
A good example. Consists of four sections.

1516. Waterhouse, D. F. 1947 An Examination of the Peet-Grady Method for the Evaluation of Household Fly Sprays. Australian Counc. Sci. Industr. Res. Bull. 216:1-24.
A discussion of the "large group" method in terms of the future use of DDT. ABSTRACT: BA 23:246, 1949.

1517. Waterhouse, D. F. 1948 The Effect of Colour on the Numbers of Houseflies Resting on Painted Surfaces. Australian J. Sci. Res. (B) 1(1):67-75. Melbourne.
Red most commonly preferred, with dusky blue a second choice. Least used were white or sky blue. ABSTRACTS: BA 23:2873, 1949; RAE-B 37:85, 1949.

1518. Waterston, J. 1916 On the Occurrence of *Stenomalus muscarum* (Linn.) in Company with Hibernating Flies. Scottish Naturalist. (Edinburgh), No. 54, June 1916, pp. 140-142.
This chalcid has been reported bred from puparia of *Musca*. Eggs believed to be deposited on muscid larvae. ABSTRACT: RAE-B 4:121, 1916.

1519. Waterston, J. 1918 Notes on some Blood-sucking and other Arthropods (except Culicidae) collected in Macedonia in 1917. Bull. Ent. Res. (London) 9(2):153-155.
M. domestica serious in October and November. Used traps, poison baits, sprays. ABSTRACT: RAE-B 7:57, 1919.

Waterworth, P. M., joint author. *See* Shooter, R. A., et al., 1944.

1520. Watt, J. and **Lindsay, D. R. 1948** Diarrheal Disease Control Studies. I. Effect of Fly Control in a High Morbidity Area. Reprint No. 2890 Public Health Repts. 63(41):1319-1334.
Removal of breeding sites considered to be a more basic form of control than use of insecticides. (Lower Rio Grande valley.) ABSTRACTS: RAE-B 39:86, 1951; BA 23:1751, 1949.

1521. Wayson, N. E. 1914 Plague and Plague-like Disease. A Report on their Transmission by *Stomoxys calcitrans* and *Musca domestica*. Publ. Health Repts. 29(51):3390-3393. Washington.
Bacterium tularense experimentally transmitted from infected viscera of dead animal to conjunctivae of guinea pigs by house flies. ABSTRACT: RAE-B 3:46, 1915.

1522. Weddell, J. A. 1934 The House Fly. Queensland Agric. J. 41(1):43-47. (Also issued as Ent. Leaflet Dept. Agric. Stk. Qd. No. 28, 6 pp. Brisbane.)
A brief, popular account of the bionomics and control of *M. domestica*. ABSTRACT: RAE-B 22:96, 1934.

Weed, A., joint author. *See* Hoyer, D., et al., 1936.

Weinman, C. J., joint author. *See* Kearns, C. W., et al., 1949.

Weiser, H. H., joint author. *See* Moorehead, S., et al., 1946.

1523. Welander. 1896 [Gonorrhoel Ophthalmia Conveyed by Flies.] Wien. Klin. Wochenschr. No. 52.
Flies found bearing living gonococci upon their feet three hours after they became contaminated.

1524. Welch, E. V. 1939 Insects Found on Aircraft at Miami, Fla. in 1938. Publ. Health Repts. 54(14):561-566.
Among a considerable list of insects, *M. domestica*, living and dead, was the most prevalent species throughout the year. ABSTRACT: RAE-B 27:215, 1939.

Welch, H., joint author. *See* Ostrolenk, M., et al., 1942.

1525. Welden, L. 1946 Shoo Fly. The American Girl, June 1946, p. 30.
Four paragraphs of interesting, popular information.

1526. Wellington, W. G. 1944 Barotaxis in Diptera and its Possible Significance to Economic Entomology. Nature (London) 154(3917):671-672.
Slight decreases in pressure cause increase in flight activity. Behavior becomes normal again at still lower pressure. ABSTRACT: BA 19:1950 1945.

1527. Wellington, W. G. 1945 Conditions Governing the Distribution of Insects in the Free Atmosphere. Canad. Ent. 77:7-15, 21-28, 44-49.
First publication of the suggestion that insects exhibit a *positive* response to decreased pressure.

1528. Wellington, W. G. 1946 The Effects of Variations in Atmospheric Pressure upon Insects. Canad. J. Res. Sect. D., Zool. Sci. 24:51-70.
Gives barometric equivalents: One millibar = 1,000 dynes of force per sq. cm. = a force able to support 0·75 mm. (0·0295 inches) of mercury.

1529. Wellington, W. G. 1946 Some Reactions of Muscoid Diptera to Changes in Atmospheric Pressure. Canad. J. Res. Sect. D, Zool. Sci. 24(4):105-117.
Antennal aristae sensitive to slight fluctuations in pressure; may be considered "external baroreceptors". ABSTRACT: BA 21:155, 1947.

1530. Wells, R. W. 1931 Some Observations on Electrified Screens and Traps. J. Econ. Ent. 24(6):1242-1247.
M. domestica constituted 69 per cent of flies killed by electric screens. Where electric current is not available, cylindrical baited traps work well. ABSTRACT: RAE-B 20:56, 1932.

1531. Wenyon, C. M. 1911 Oriental Sore in Bagdad, together with Observations on a Gregarine in *Stegomyia fasciata*, the Haemogregarine of Dogs, and the Flagellates of House-flies. Parasitology 4:273-344.
Includes life-history of *Herpetomonas muscae-domesticæ*.

1532. Wenyon, C. M. 1913 Observations on *Herpetomonas muscae-domesticae* and some allied Flagellates with Special Reference to the Structure of their Nuclei. Arch. f. Protistenk. 31:1-36, 3 pls., 6 figs, 1 diagram.
Gives a careful account of the cytology of the flagellate. ABSTRACT: TDB 2(9):463-465, 1913.

1533. Wenyon, C. M. and **O'Connor, F. W. 1917** The Carriage of Cysts of *Entamoeba histolytica* and other Intestinal Protozoa and Eggs of Parasitic Worms by House-flies with some Notes on the Resistance of Cysts to Disinfectants and other Agents. J. Roy. Army Med. Corps 28(5):522-527.
M. domestica and other muscoids distribute cysts by way of fecal specks (not by regurgitation). ABSTRACT: RAE-B 5:117, 1917.

1534. Wenyon, C. M. and **O'Connor, F. W. 1917** An Inquiry into some Problems affecting the Spread and Incidence of Intestinal Protozoal Infections of British Troops and Natives in Egypt . . . Part IV Experimental Work with the Human Intestinal Protozoa, their Carriage by House-flies and the Resistance of their Cysts to Disinfectant and other Agents. J. Roy. Army Med. Corps 28(6):686-698.
Flies take up cysts while feeding on feces and may retain them in the gut for 42 hours. Pass cysts while feeding on human food. ABSTRACT: RAE-B 5:151, 1917.

1535. Wenyon, C. M. and **O'Connor, F. W. 1917** Human Intestinal Protozoa in the Near East. Wellcome Bureau of Sci. Res. London. 218 pp.
Includes discussion of dipterous vectors.

1536. Werner, H. 1908 Ueber eine eigeisselige Flagellatenform im Darm der Stubenfliege. [Concerning a single whip-lash Flagellate in the Intestine of the House-fly.] Arch. f. Protistenk. 13:19-22, 2 pls.
Crithidia muscae-domesticae Werner. Occurred in alimentary tract of 4 out of 82 flies examined.

West, A. S. Jr., joint author. *See* Wilkes, A., et al., 1948.

1537. West, L. S. 1945 The Order Diptera; Myiasis. *In* Manual of Tropical Medicine by T. T. Mackie, G. W. Hunter and C. B. Worth. Philadelphia and London. W. B. Saunders. pp. 569-646.

1538. Westcott, S. 1913 Flies and Disease in the British Army. J. State Med. 21:480-488.

Wexler, E., joint author. *See* Hartzell, A., et al., 1946.

1539. Wheeler, W. M. 1891 The Embryology of a Common Fly. Psyche 6: 97-99.
Is a review and criticism of Graber's paper on the muscid genera *Calliphora* and *Lucilia*.

1540. Wheeler, W. M. 1910 Ants. Their Structure, Development and Behavior. New York. Columbia University Press. Biol. Ser. No. 9, 633 pp., 286 figs.
Ants frequently destroy *M. domestica* in egg, larval and adult stages.

1541. Wherry, W. B. 1908 Notes on Rat Leprosy and on the Fate of Human and Rat Lepra Bacilli in Flies. Publ. Health Repts. Publ. Health and Marine Hosp. Serv. (Washington). Vol. 23, No. 42, 8 pp. (Reprint.)
M. domestica, also *Calliphora vomitoria* and *Lucilia caesar* take up great numbers of lepra bacilli from the carcase of a leper rat and deposit them with their feces.

1542. Wherry, W. B. 1908 Further Notes on the Rat Leprosy and on the Fate of the Human and the Rat Leper Bacillus in Flies. J. Infect. Dis. Vol. 5, No. 5.
Flies found clear of the bacilli 48 hours after becoming contaminated. Larvae hatching in leprous carcases become heavily infected, but if they are transferred to non-infected meat, usually transform into non-infected flies.

1543. Whipple, G. C. 1908 Typhoid Fever, its Causation, Transmission and Prevention. New York, Wiley and Sons; London, Chapman and Hall.

1544. White, T. C. 1874-1877 On *Empusa muscae*. J. Quekett Micro. Club. (London), Vol. 4, pp. 211-213.
Fungus causes death of house-flies.

1545. Whitehead, F. E. 1944 The Effect of Various Diluents on the Toxicity of Derris Dust to House Flies. (*Musca domestica* L.) Proc. Oklahoma Acad. Sci. 24:27-28.
Several mixtures proved more toxic than 3 parts talc, 1 part derris, the combination usually used. ABSTRACT: RAE-B 33:48, 1945.

1546. Whitmire, H. E. 1939 Test Methods for Recording Moribund Kill, Soap and Sanit. Chem. 15(11):99, 101, 103, 123, 2 figs.
Describes two modifications of Peet-Grady test which permit its use while allowing for moribund kill. ABSTRACT: RAE-B 28:79, 1940.

1547. Wiesmann, R. 1943 Eine neue Methode der Bekämpfung der Fliegenplagen in Ställen. [A New Method for Combatting the Fly Nuisance in Stables.] Anz. Schädlingsk. 19(1):5-8. Berlin.
A very early use of Gesarol (DDT). Gives time-tables for primary paralysis, secondary paralysis, and death for *M. domestica* and *Stomoxys calcitrans*. ABSTRACT: RAE-B 32:175, 1944.

1548. Wiesmann, R. 1947 Untersuchungen über das physiologische Verhalten von *Musca domestica* L. verschiedener Provenienzen. [Investigations into the Physiological Relationship of *Musca domestica* L. of Different Derivation.] Mitt. Schweiz. Ent. Ges. 20(5):484-504. Berne.
Arnäs and Basle strains, which differ as to structure of tarsi and pulvilli, differ also regarding susceptibility to DDT, tolerance of marginal temperatures and ability to recover from narcosis. ABSTRACT: RAE-B 37: 216, 1949.

1549. Wiesmann, R. and **Lotmar, R. 1949** Beobachtungen und Untersuchungen über den Wirkungsbereich des neuen Repellent "Kik-Geigy". [Observations and Researches on the Effective Range of

the New Repellent, "Kik-Geigy".] Acta Tropica 6:292-349. Basle.
Repellency proved moderate to poor against *M. domestica* and *Stomoxys calcitrans*. ABSTRACT: TDB 47(3):279.

Wiesmann, R., joint author. *See* Lauger, P., et al., 1946.

1550. Wieting, J. O. G. and **Hoskins, W. M. 1939** The Olfactory Responses of Flies in a new Type of Insect Olfactometer. II. Responses of the Housefly to Ammonia, Carbon Dioxide and Ethyl Alcohol. J. Econ. Ent. 32:24:29.
Ethyl alcohol attracts at very low concentrations, repels above 0·05 per cent; CO^2 exercises no effect up to 2 per cent; Ammonia attracts at 0·012 per cent, repels at 0·03 per cent. ABSTRACT: RAE-B 27:166, 1939.

1551. Wilcox, E. V. 1908 Fighting the House-fly. Country Life in America. May, 1908.

1552. Wilcoxon, Frank, Hartzell, A. and **Wilcoxon, Fredericka. 1939** Insecticidal Properties of Male Fern (*Aspidium felix-mas* [L.] Sw.). Contr. Boyce Thompson Inst. 11(1):1-4.
Tested on *M. domestica* by Peet-Grady method. Spray combining filicin with pyrethrins almost as effective as official test insecticide. ABSTRACT: RAE-B 28:187, 1940.

Wilcoxon, Frank, joint author. *See* Hartzell, A., et al., 1936.

Wilcoxon, Fredericka, joint author. *See* Wilcoxon, Frank, et al., 1939.

Wild, H., joint author. *See* Lauger, P., et al., 1946.

1553. Wilhelmi, J. 1919 Zur Biologie der klienen Stubenfliege, *Fannia canicularis* L. [On the Biology of the Little Housefly, *Fannia cancularis* L.] Zeit. Angew. Ent. (Berlin) 5(2):261-266.
Compares life history with that of *M. domestica*. Is less attracted to food than common housefly. ABSTRACT: RAE-B 8:122, 1920.

1554. Wilhelmi, J. 1920 Zur Ueberwinterung von Musciden. [On the Overwintering of Muscidae.] Zeit. Angew. Ent. (Berlin) 6(2):296-301.
M. domestica hibernates as a mobile adult, probably also in pre-imaginal stage. ABSTRACT: RAE-B 9:28-29, 1921.

1555. Wilhelmi, J. 1920 Versuche zur Bekämpfung der in Kot, Mist und anderen organischen Abfallstoffen lebenden Muscidenbrut, insbesondere der gemeinen Stechfliege mit Kalisalzen und anderen Chemikalien. [Experiments on the Control of Muscoid Larvae, especially the Common Biting Flies, living in Dung, Trash and other Organic By-products, by means of Potassium Salts and other Chemicals.] Mitt. Landesanstalt Wasserhygiene, Berlin-Dahlem. No. 25, p. 190.
Worked with *M. domestica*, *Stomoxys calcitrans* and *Lyperosia irritans*. Used slaked lime, borax, gas lime and other chemicals. ABSTRACTS: Biedermann's Centralbl., Leipsic 1(2) pp. 71-73, 1921; RAE-B 9:89, 1921.

1556. Wilkerson. 1904 Flies as Carriers of Disease. Mobile Med. and Surg. J. Vol. 4, pp. 125-141.

1557. Wilkes, A., Bucher, G. E., Cameron, J. W. Mac.B, and **West, A. S. Jr. 1948** Studies on the Housefly (*Musca domestica* L.) I. The Biology and Large Scale Production of Laboratory Populations. Canad. J. Res. (D) 26(1):8-25. Ottawa.
By using eggs from genetically-selected stock of known age, and rearing in a medium of which the temperature was kept constant, production of flies was increased, and maintained at a constantly uniform rate. For Parts II and III, *see* Bucher, G. E., et al., 1948. ABSTRACT: RAE-B 37:88, 1949.

Wilkes, A., joint author. *See* Bucher, G. E., et al., 1948.

Williams, W. L., joint author. *See* Jack, R. W., et al., 1937.

1558. Willcocks, F. C. 1917 Notes on Some Insects found in Egypt of Medical and Veterinary Interest. Bull. Soc. Ent. d'Egypte (Cairo), 10(3): 79-90.
Lists *M. domestica* and *M. autumnalis* (*corvina*). Fly problem very great in Egyptian cities. ABSTRACT: RAE-B 6:239, 1918.

1559. Williams, F. X. 1943 Mosquitoes and some Other Noxious Flies that occur in New Caledonia. Hawaii Plant Rec. 47(4):205-222. Honolulu.
List includes *M. domestica vicina* Macq. ABSTRACT: RAE-B 32:148, 1944.

1560. Wilson, H. G. and **Gahan, J. B. 1948** Susceptibility of DDT-resistant Houseflies to Other Insecticidal Sprays. Science 107:276-277.
Flies proved somewhat resistant to five different compounds. Seemed to indicate that selection against DDT had produced an "unusually strong stock of flies". ABSTRACT: BA 22: 1955, 1948.

Wilson, H. G., joint author. *See* McDuffie, W. C., et al., 1945; Gahan, J. B., et al., 1948; Schroeder, H. O., et al., 1945; Lindquist, A. W., et al., 1944, 1945, 1947, 1948.

Wilson, I. B., joint author. *See* Brescia, F., et al., 1946.

Winfield, G. F., joint author. *See* Meng (Ch'ing-hua), et al., 1938, 1941.

1561. Winter, G. 1881 Zewi neue Entomophthoren. Bot. Centralbl. Vol. 5, p. 62.
States that he found resting spores in specimens of *M. domestica* occurring indoors; also conidia of *E. muscae*.

1562. Wise, S. K. and **Minnett, E. P. 1912** Experiments with Crude Carbolic Acid as a Larvicide in British Guiana. Ann. Trop. Med. Parasit. 6:327-330.

1563. Wladimirov, M. and **Smirnov, E. 1934** Ueber das Verhalten der Schlupfwespe *Mormoniella vitripennis* Wlk. zu verschiedenen Fliegenarten. [On the Relation of the Ichneumon-fly *Mormoniella vitripennis* Wlk. to Various Fly Species.] Zool. Anz. 107(3-4):85-89.
This chalcid prefers puparia in the following order: *Calliphora erythrocephala*, *Lucilia caesar*, *Phormia terrae-novae* (*groenlandica*), *Musca domestica*. ABSTRACT: RAE-B 22:166, 1934.

Wladimirov (*or* **Wladimirow**), **M.,** joint author. *See* Smirnov, E., et al., 1934.

Wolfe, W. C., joint author. *See* Kilgore, L. B., et al., 1942.

1564. Wollman, E. 1911 Sur l'elevage des mouches steriles. Contribution à la connaissance du rôle des microbes dans les voies digestive. [On the Rearing of Sterile Flies. Contribution to our Knowledge of the Role of Microorganisms in the Digestive Canal.] Ann. Inst. Pasteur 25:79-88.
Believed to be the first successful maintenance of fly cultures on media free from microbes.

1565. Wollman, E. 1921 Le rôle des mouches dans le transport de germes pathogènes etudié par le technique des elevages aseptique. [The Role of Flies in the Transmission of Pathogenic Germs, Studied by the Technique of Aseptic Rearing.] Compt. Rend. Hebdom. Acad. Sci. Paris. 72(5):298-301. Also published in Ann. Inst. Pasteur 35(7):431-449.
Fly larva, infected with typhoid or tubercle bacilli, remains so in pupa stage; but adult is free of such orgasnisms unless it becomes contaminated from the exterior of the puparium. ABSTRACT: RAE-B 9:71, 1921.

1566. Wollman, E. 1922 Biologie de la mouche domestique et des larves de mouches à viande, en élevages aseptiques. [Biology of the Domestic Fly and Fly Larvae on Meat, in Aseptic Rearing.] Ann. Inst. Pasteur 36(11):784-788.
Larvae can absorb only liquid food. Proteolytic action of bacteria may unite with ferments secreted by larvae, to liquefy solid matter. ABSTRACT: RAE-B 11:27, 1923.

1567. Wollman, E. 1927 Le rôle des mouches dans le transport de quelques germes importants por la pathologie tunisienne. [The Role of the Fly in the Transmission of Certain Germs Important in Tunisian Pathology.] Arch. Inst. Pasteur Tunis. 16(4):347-364.
Infected aseptically reared flies with organisms of conjunctivitis, plague, and contagious abortion. ABSTRACT: RAE-B 16:103, 1928.

1568. Woodcock, H. M. 1918 Note on the Epidemiology of Dysentery. J. Roy. Army Med. Corps. 30(1):110-111.
Considers water more important than flies in the spread of amoebic dysentery. Bacillary dysentery on the other hand, occurs under hot, dry conditions, and is more likely to be conveyed by flies. ABSTRACT: RAE-B 6:85, 1918.

1569. Woodcock, H. M. 1919 Note on the Epidemiology of Amoebic Dysentery. J. Roy. Army Med. Corps. 32(3):231-235.
Reiterates his opinion that flies are not a very great factor in the spread of this disease. ABSTRACT: RAE-B 7:102, 1919.

1570. Woodhouse. 1910 Notes on the Causation and Prevention of Enteric Fever in India. J. Roy. Army Med. Corps. Vol. 10, p. 616.

1571. Wright, M. 1945 Dragonflies Predaceous on the Stablefly *Stomoxys calcitrans* (L). Florida Entom. 28(2):31-32.
Seize flies while they are swarming about livestock, or resting on walls. Included here because of general interest in "Natural Control" of undesirable species. ABSTRACT: RAE-B 35:39, 1947.

Wright, M., joint author. *See* Simmons, S. W., et al., 1944.

1572. Yao, H. Y., Huan, I. C. and **Huie, D. 1929** The Relation of Flies, Beverages and Well Water to Gastro-intestinal Diseases in Peiping. Nat. Med. J. China 15:410-418.
Flies from slums carried (externally) an average of 3,683,000 bacteria; those from cleanest district, an average of 1,941,000. Insides of flies harbored 8 to 10 times as many bacteria as the outsides.

1573. Yeager, J. F. and **Munson, S. C. 1945** Physiological Evidence of a Site of Action of DDT in an Insect. Science 102(2647):305-307.
Nicotine affects ganglia; DDT the nerves, somewhere along their length. This work, done on cockroaches, became a basis for subsequent research on flies. ABSTRACT: RAE-B 34:3, 1946.

1574. Yeager, J. F. and **Munson, S. C. 1949** Relationship between Knock-down and Survival Time for DDT-poisoned Flies and Roaches. J. Econ. Ent. 42:874-877.
Concentration-survival time curves for *Musca domestica* differ from those of the cockroach, by exhibiting no inflections. May be described by the same equation, nevertheless. ABSTRACT: RAE-B 38:151, 1950.

1575. Yeomans, A. H., Rogers, E. E. and **Ball, W. H. 1949** Deposition of Aerosol Particles. J. Econ. Ent. 42:591-596.
In tests with *M. domestica*, optimum size particle for fly in motion at 4 miles per hour was approximately 22·4 micra. ABSTRACT: RAE-B 38:116, 1950.

Yeomans, A. H., joint author. *See* Nelson, R. H., et al., 1949.

1576. Yersin 1894 La peste bubonique à Hongkong. Ann. Inst. Pasteur 8: 622-667.
Virulent plague bacilli recovered from dead flies in laboratory where autopsies were performed on animals dead of plague.

Zabud'ko-Reingard, T. N., joint author. *See* Reingard, L. V., et al., 1945.

1577. Zanini, E. 1930 El'*Holostaspis badius* (Koch) parassita della mosca domestica. [*Holostaspis badius* Koch, Parasite of the Domestic Fly.] Boll. Lab. Zool. Agrar. Bachic, Milano 1(1928-29):59-73. Milan.
M. domestica found heavily infested with the gamasid mite, *H. badius*, in Venetia and in Lombardy. ABSTRACT: RAE-B 19:216, 1931.

Zarkoski von Schmidt, S., joint author. *See* Hoyer, D., et la., 1936.

1578. Zetek, J. 1914 Dispersal of *Musca domestica* Linne. Ann. Ent. Soc. Am. 7:70-72.
Marked flies found their way into dwelling a half a mile away and 150 feet lower elevation than manure which served as the breeding place. ABSTRACT: RAE-B 2:101, 1914.

Zima, G. G., joint author. *See* Pokrovskii, S. N., et al., 1938.

1579. Zimin, L. S. 1939-1941 [A Survey of the Synanthropic Diptera of Tadzhikistan. Conf. on Parasitological Problems. Summaries of Reports.] Izd. Akad. Nauk SSSR, pp. 35-38. Moscow. (In Russian.)
Collections from houses, hospitals, restaurants, markets, abbattoirs, and animal quarters, at 3 hour intervals yielded 120 species in 23 families. Included *Fannia scalaris*, *M. domestica vicina*, *M. sorbens*. ABSTRACT: RAE-B 34:111, 1946.

1580. Zimin, L. S. 1939-1941 [An Experiment in the Control of Flies under the Conditions of Southern Tadzhikistan. Conf. on Parasit. Problems. Summaries of Reports.] Izd. Akad. Nauk. SSSR, pp. 38-39. Moscow. (In Russian.)
By use of larvicides, applied to breeding places every 3-5 days, fly populations in four villages were reduced 76-96 per cent. ABSTRACT: RAE-B 34:111, 1946.

1581. Zimin, L. S. 1948 [Key to the Third-instar Larvae of the Synanthropic Flies of Tadzhikistan.] Opred. Faune SSSR No. 28, 115 pp., 61 figs. Izd. Zool. Inst. Akad. Nauk SSSR. Mowcow. (In Russian.)
Largely Muscidae, Calliphoridae and Sarcophagidae. ABSTRACT: RAE-B 38:28, 1950.

1582. Zimin, L. S. and Teterovskaia, T. D. 1943 [Seasonal Variation in Numbers of the House Fly, *Musca vicina*.] Med. Parasit. 12(5): 44-53, 1 fig.
Regards *M. vicina* as a southern variety of *M. domestica*. ABSTRACT: BA 19:2519, 1945.

1583. Zimmerman, O. T. and **Lavine, I. 1946** DDT-Killer of Killers. Dover, N. H. Industrial Research Service xi + 180 pp., illus.
Contains many photos pertaining to the practical use of this insecticide. ABSTRACT: RAE-B 36:6, 1948.

1584. Zumpf, F. 1949 Hausfliegen als Ueberträger von Seuchen, insbesondere der spinalen Kinderlähmung. [Houseflies as Carriers of Contagious Disease, especially Infantile Paralysis.] Anz. Schädlingsk. 22(11): 161-163.
ABSTRACT: BA 24:2937, 1950.

1585. Zwick, K. G. 1914 Massnahmen gegen die Uebertragung von Infektionskrankheiten durch die Hausfliegen. [Precautionary Measures against the Transmission of Infectious Diseases by Houseflies.] Schweizerischen Rundschau für Med. Vol. 14. (Reprint, pp. 1-16), No. 13.

Section V.

References from 1950 to 1959 Inclusive

1586. Abedi, Z. H. 1958 Inheritance of Aldrin Resistance in the Indian House Fly, *Musca domestica nebulo* F. Bull. Ent. Res. 49(4):637-642.
Offspring from reciprocal crosses, when tested by topical application, indicate that the resistance is not due to a sex-linked multifactorial heredity. ABSTRACTS: BA 35(23):5793(68565), Dec., 1960; RAE-B 47:21, Feb., 1959.

1587. Abraham, E. V. and **Muthukrishnan, T. S. 1955** A Note on the Control of the Housefly in Farmyards. Madras Agric. Jour. 42(9):394-397.
BHC proved to be effective as a spray against oviposition, also as a larvicide; it was also somewhat effective against adult houseflies. ABSTRACT: BA 30:2740(27302), Sept., 1956.

1588. Acree, F., jr., Davis, P. L., Spear, S. F., LaBrecque, G. C. and **Wilson, H. G. 1959** Nature of the Attractant in Sucrose fed on by House Flies. J. Econ. Ent. 52(5):981-985.
Attraction is related to a relative humidity gradient generated by the moisture from other flies feeding. ABSTRACTS: BA 35:771(8837), Feb., 1960; RAE-B 49:9, Jan., 1961; TDB 57:304, Mar., 1960.

Adam J. P., joint author. *See* Jaiyou, C., et al., 1950.

1589. Adolphi (?). 1958 Examination of a Pyrethrum Synergist. Preliminary information. Pyrethrum Post 4(4):3-5.
Results of tests indicate that S421 (octachlorodipropylether) seems to be suitable as a synergist for pyrethrum sprays used indoors. ABSTRACT: RAE-B 48:133-134, Aug., 1960.

1590. Afifi, S. E. D. 1956 Reproductive Potential Life Span and Weight of House Flies, *Musca domestica* L. Surviving Initial Exposure to an Insecticide. Diss. Absts. 16:521.

1591. Afifi, S. E. D. and **Knutson, H. 1956** Reproductive Potential, Longevity, and Weight of House Flies which Survived one Insecticidal Treatment. J. Econ. Ent. 49:310-313.
No significant correlations between the length of life and average weight or between length of life and number of progeny. ABSTRACTS: BA 30:3296(32933), Nov., 1956; RAE-B 45:112-113, July, 1957.

Afifi, S. E. D., joint author. *See* Knutson, H., et al., 1958; Rai, L., et al., 1956.

1592. Ahmed, M. K., Casida, J. E. and **Nichols, R. E. 1958** Bovine Metabolism of Organophosphorus Insecticides. Significance of Rumen Fluid with particular reference to Parathion. J. Agric. Fd. Chem. 6(10):741-746.
The toxicological significance of the hydrolysis of many organophosphate insecticides by bovine rumen fluid is considered in relation to the toxicity of various derivatives. ABSTRACT: RAE-B 47:87, June, 1959.

1593. Akatov, V. A. 1955 [Duration of Survival of Trichomonads in *Musca domestica.*] Veterinariya 32(8):84. (In Russian.)
Trichomonads could survive in the flies up to eight hours after their ingestion and so could possibly be transmitted by them. ABSTRACT: RAE-B 45:29, Feb., 1957.

Alessandrini, M., joint author. *See* Mosna, E., et al., 1954.

1594. d'Alessandro, G. and **Mariani, M. 1953** Osservazioni sulla eredità dei caratteri "resistenza" e "sensibilita" agli insetticidi clorunati in

Musca domestica L. [Observations on the heredity of resistance and sensitivity of *M. domestica* to chlorinated insecticides.] Boll. Soc. Ital. Biol. Sper. 29(4):687-689.
See following reference.

1595. d'Alessandro, G. and **Mariani, M. 1954** Osservazioni sulla eredità dei caratteri "resistenza" e "sensibilita" al DDT in *Musca domestica* L. [The Inheritance of Resistance and Sensitivity to DDT in *Musca domestica* L.] Riv. di Parrasit. 15(2):85-94. (English summary.)
Results are inconsistent with a single-gene theory of DDT resistance, which character appears to be due to multiple genes. Authors conclude that resistance is polygenically determined. ABSTRACTS: RAE-B 44:203, Dec., 1956; TDB 51:1201-1202, Nov., 1954.

1596. d'Alessandro, G., Mariani, M. and **Gagliani, M. 1951** La resistenza al DDT della *M. domestica* si attenuata nel corso delle generazione allevate in assenza dell' insetticida? [Is resistance to DDT in *M. domestica* reduced in Corsica during a generation reared in the absence of the insecticide?] Boll. Soc. Ital. Biol. Sper. 27(12):1746-1747.
Authors conclude that a pure, resistant strain, composed of homogeneous individuals with a high order of resistance to chlorinated insecticides, maintains unaltered that characteristic for many generations in the absence of the insecticide.

1597. d'Alessandro, G., Mariani, M. and **Gagliani, M. 1952** Ancora sulla resistenza della *Musca domestica* al DDT e ad altri insetticidi clorurati. Selezione di ceppi forniti di caratteristiche differenti da unica popolazione catturata in natura. [Resistance of *Musca domestica* to DDT and other Chlorinated Insecticides: Selection of Strains having Different Characteristics from a Single Population Captured in Nature.] Riv. di Parassit. 13:169-175.
Resistance to chlordane and DDT extends also to other chlorinated insecticides. ABSTRACT: TDB 49:1004, Oct., 1952.

1598. Alexander, B. H., Barker, R. J. and **Babers, F. H. 1958** The Phosphatase Activity of Susceptible and Resistant House Flies and German Cockroaches. J. Econ. Ent. 51(2):211-213.
Differences in acid and alkaline phosphatase activity believed due to strain variations rather than to resistance. ABSTRACTS: RAE-B 47:105, July, 1959; TDB 55:1057, Sept., 1958.

Alexander, B. H., joint author. *See* Barker, R. J., et al., 1958; Piquette, P. G., et al., 1957, 1958.

1599. Alexander, C. C. 1955 Diazinon, a New Fly Killer. Pest Control 23(6): 14-17.
This organic phosphorus compound was effective against houseflies and other insects resistant to DDT. Gave 5-8 week residual control, under prevailing conditions. ABSTRACT: BA 30:306(3059), Jan., 1956.

Alexander, C. C., joint author. *See* Harris, H. J., et al., 1950.

Allred, A. M., joint author. *See* Sumerford, W. T., et al., 1951.

Anders, R. S., joint author. *See* Gahan, J. B., et al., 1953.

1600. Anderson, A. D. and **Patton, R. L. 1953** Efficiency of Sulfhydril Compounds in the Detoxication of Arsenite in two Species of Insects. J. Econ. Ent. 46:423-426.
Methionine, cystine and cysteine, reduced glutathione and BAL were fed to or injected into adult houseflies and mealworm larvae, and the effect of this treatment upon the toxicity of sodium metarsenite was determined.

1601. Anderson, A. D. 1955 Determination of Housefly Succinic Dehydrogenase with Triphenyltetrazolium and Neotetrazolium Chloride. Science 122:694.
The action of cyanide and malonate on neotetrazolium reduction is in agreement with the hypothesis that this tetrazolium compound is reduced directly by succinic dehydrogenase, while the results with TPTZ may indicate the mediation of a cyanide-sensitive hydrogen carrier.

1602. Anderson, A. D., March, R. B. and **Metcalf, R. L. 1954** Inhibition of the Succinoxidase System of Susceptible and Resistant House Flies by DDT and Related Compounds. Ann. Ent. Soc. Am. 47:595-602.
The inhibition of succinoxidase and its components by DDT is not a primary factor in the mode of action of DDT and the DDT-synergists or in the mechanism of DDT-resistance in the house-fly.

1603. Andrews, J. M. 1952 The Importance of Household Insecticides in Public Health. Proc. 38th Mid-year Meeting of Chem. Spec. Mfg. Assoc. pp. 102-106.
Includes appropriate remarks on fly control.

Arakawa, X. Y., joint author. *See* Hall, I. M., et al., 1959.

1604. Arthur, B. W. and **Casida, J. E. 1958** Biological Activity of Several 0,0-Dialkyl Alpha-Acyloxy-ethyl Phosphonates. J. Agric. Fd. Chem. 6(5):360-365.
Toxicity and anti-cholinesterase activity compared. Initial site of *in vivo* hydrolysis is major factor in the selective toxicity of these compounds. ABSTRACT: RAE-B 47(4):64, Apr., 1959.

1605. Arvy, L. 1954 Données sur la leucopoièse chez *Musca domestica* L. [Data on leukopoiesis in *Musca domestica* L.] Proc. Roy. Ent. Soc. London. Ser. A. 29(1/3):39-41.
Leucopoietic organ observed microscopically in larvae of 3 hours, 6 hours, 21 hours (illustrated), 30 hours and 45 hours. Generalizations.

1606. Ascher, K. R. S. 1955 Insect-Resistance to Dieldrin. A Survey of the Literature. Riv. di Parassit. 16:31-40.
Refers to a very high level of resistance, encountered in field housefly tests. ABSTRACT: RAE-B 45:152, Sept., 1957.

1607. Ascher, K. R. 6 [**S591.d** Housefly breeing in rural areas of Israel and methods suggested for its reduction.] *Tavrua* (*Sanitation*) 9:21-28. (In Hebrew.)

1608. Ascher, K. R. S. 1956 Some Remarks on the Problem of Insecticide Resistance. *Tavrua* (*Sanitation*) 9:29-34.
Special English issue on the occasion of World Health Day.

1609. Ascher, K. R. S. 1956 [A report on fly breeding in rural areas of Israel based on a survey in 94 localities.] Medical Research Laboratories of the Israel Medical Corps. 27 pp. (In Hebrew.)
Mimeographed document.

1610. Ascher, K. R. S. 1957 Prevention of Oviposition in the Housefly through Tarsal Contact Agents. Science 125(3254):938-939.
Gives graphical formulae of 2 organic compounds which, when applied topically in acetone or by tarsal contact, inhibit oviposition. Insemination and ovarian development are normal but there is a retention of the eggs. ABSTRACTS: RAE-B 47:66, 1959; TDB 54:1138, Sept., 1957.

1611. Ascher, K. R. S. 1957 Enhanced Susceptibility of a DDT-Resistant Strain of Houseflies towards Alkyl and Aryl Bromo-Acetates. Riv. di Parassit. 18(3):185-197.
Residual insecticidal activity found in esters of bromoacetic acid such

as cetyl bromoacetate (CBA). ABSTRACTS: RAE-B 48:57, Mar., 1960; TDB 55:462, Apr., 1958.

1612. Ascher, K. R. S., 1957 Houseflies in Israel I. A. Resistance Survey in Rural Areas. Riv. di Parassit 18(2):113-122. Rome. (Summary in Italian.)

1613. Ascher, K. R. S. 1958 A Common Information Source for all Resistance Problems in Biology is Urgently Needed in view of the Paucity of Effective Countermeasures to Insecticide-resistance. Indian J. Malariol. 12(4):615-625.
Discusses the failure of synergists to provide a practical answer to the problems of resistance and the use of tarsal contact agents that prevent oviposition, of desiccating and sticking agents and of abrasive dusts. Suggests that information on resistance in every field of biology should be collated in abstract form. ABSTRACT: RAE-B 49(10):232, Oct., 1961.

1614. Ascher, K. R. S. 1958 Enhanced Knock-down in Various Laboratory Reared Resistant Housefly Strains Subjected to Continuous Contact with Cetylbromoacetate. Zeit. Naturforsch. 13b(2):138-139.
Cetyl bromoacetate was the only substance capable of enhancing knock-down in extremely DDT-resistant and polyresistant flies. (Swiss strain K.) Continuous tarsal contact was necessary. ABSTRACT: BA 33:2256(27777), May, 1959.

1615. Ascher, K. R. S. 1958 The Attraction of the Levant Housefly, *Musca vicina* Macq. to Natural Breeding Media. Acta Tropica 15(1):1-14. (Summaries in French and German.)
Pig dung was most attractive to adults, including the males. A solution of ammonium carbonate also attracted both sexes; 5 per cent solution of yeast was attractive only for females. ABSTRACTS: RAE-B 49:66, March, 1961; TDB 55:1056, Sept., 1958.

1616. Ascher, K. R. S. 1958 Insecticidal Properties of N-substituted Fluoro-Acetamides. Riv. di Parassit. 19(3):229-231.
Gives data on the knockdown of females of *Musca domestica vicina* Macq. in a highly DDT-resistant strain by 6 N-substituted fluoroacetamides. Compounds tested are rather weak contact poisons for resistant house-flies. ABSTRACT: RAE-B 50:9, Jan., 1962.

1617. Ascher, K. R. S. 1958 Preferential Knockdown Action of Cetyl Bromoacetate for Certain Laboratory-reared Resistant Strains of Houseflies. Bull. W. H. O. 18(4):675-677. Geneva.
Cetyl bromoacetate was less toxic for the two normal, susceptible strains. It was almost equally toxic to all strains showing some degree of resistance, or developed under selection pressure, whether DDT resistant or not. ABSTRACTS: RAE-B 49:96, May, 1961; TDB 55:1179, Oct., 1958.

1618. Ascher, K. R. S. 1958 Chemicals Affecting the Preimaginal Stages of the Housefly. VII. The Contact Toxicity of Some Alkyl Bromo- and Chloroacetates for Third Stage Larvae. Riv. di Parassit. (Rome) 19(2):139-155. 2 figs. 28 refs. (Summary in Italian.)
Rapidly penetrating toxic substances applied at adequate concentrations kill a high percentage of larvae; those that penetrate more slowly allow partial pupation or even the formation of abnormal pupae. ABSTRACTS: RAE-B 50:6, June, 1962; TDB 56:670, June, 1959.

1619. Ascher, K. R. S. 1958 Aufsuchung der durch Resistenz induzierten, erhöhten Empfindlichkeit-eine neue Versuchsrichtung zur Bekämpfung der Resistenz gegen Insekticide. [Location of a resistance-induced, enhanced susceptibility—a new method of research in the

combating of resistance towards insecticides.] Arzneimettel-Forschung 8:771-773.
A new experimental pathway to combat insecticide resistance in insects is described, consisting in an endeavour to locate "resistance-induced enhanced susceptibility" to specific agents, in the resistant insects. The above term, abbreviated. R.I.E.S. is proposed for this property in a strain resistant to one insecticide but abnormally susceptible to another. ABSTRACT: RAE-B 49:61, Mar., 1961.

1620. Ascher, K. R. S. 1959 "Di-(p-chlorophenyl) Compounds" and Oviposition in the Housefly. Riv. di Parassit. 20(2):143-144. 9 refs. (Summary in Italian.)
None of the compounds tested showed any oviposition-inhibiting activity. ABSTRACT: RAE-B 50:194, Sept., 1962.

1621. Ascher, K. R. S. and **Kocher, C. 1954** Housefly Baits Containing Diazinon. I. Laboratory Experiments. J. R. Geigy, S. A., Basle.
Mimeographed report.

1622. Ascher, K. R. S. and **Kocher, C. 1954** Synergizing Diazinon in Kerosene Sprays. J. R. Geigy, S.A., Basle.
Mimeographed report.

1623. Ascher, K. R. S. and **Kocher, C. 1954** Gesarol liquid (GL) as Synergist in Diazinon Kerosene Sprays. J. R. Geigy, S. A., Basle.
Mimeographed report.

1624. Ascher, K. R. S. and **Kocher, C. 1954** Synergizing Diazinon with Chlorobenzilate and Related Substances. J. R. Geigy, S. A., Basle.
Mimeographed report.

1625. Ascher, K. R. S. and **Kocher, C. 1954** Enhanced Susceptibility of a Highly Resistant Strain of Houseflies to Ingestion of Potassium Bromide. Experientia (Basel) 10(11):465-467.
A DDT resistant strain was more susceptible than a normal one, showing faster knockdown by orally administered KBr and other inorganic bromides. First discovery of negatively correlated compounds. ABSTRACT: RAE-B 44:15, Jan., 1956.

1626. Ascher, K. R. S. and **Levinson, Z. H. 1953** Chemicals Affecting the Preimaginal Stages of the Housefly. I. A Review of the Literature. Riv. di Parassit. (Rome) 14(4):235-259.
Review mentions practically all the more important contributions to the subject in recent years up to that time. Review is rather uncritical, but is useful in providing a guide to the difficulties of controlling housefly larvae, especially with the development of insecticide resistance. ABSTRACTS: RAE-B 44:203, Dec., 1956; TDB 51:653, June, 1954.

1627. Ascher, K. R. S. and **Levinson, Z. H. 1954** Chemicals Affecting the Preimaginal Stages of the Housefly. III. Contact Toxicity for Third Stage Larvae of some Chlorinated Hydrocarbons deposited on Absorbent Surfaces. Riv. di Parassit. (Rome) 15(1):57-61.
Gamma-BHC and decachlorobutane were effective contact poisons at certain concentrations. DDT, DDD, chlordane, toxaphene and dieldren were ineffective. Method of evaluation is described. ABSTRACTS: RAE-B 4:203. Dec., 1956; TDB 51:1116, Oct., 1954.

1628. Ascher, K. R. S. and **Levinson, Z. H. 1956** The Influence of Protein Addition to the Larval Diet on Oviposition of the Housefly. Riv. di Parassit. (Rome) 17(4):217-222.
Protein reserves laid down in the larval state cannot be mobilized for egg formation. Flies feeding on dead flies may lead to errors in interpreting oviposition studies. ABSTRACT: TDB 54:886, July, 1957.

1629. **Ascher, K. R. S.** and **Levinson, Z. H. 1956** Toxicity of Decachlorobutane to Adult Houseflies. Pesticides Abstract and News Summary, p. 141.
From Ascher's personal list of publications.

1630. **Ascher, K. R. S., Levinson, Z. H., Silverman, P. H.** and **Tahori, A. S. 1954** Chemicals Affecting the Preimaginal Stages of the Housefly. II. The Perchloroparaffins. Riv. di Parassit. (Rome) 15(1):45-55.
Using *Musca vicina*, authors employed several new testing methods. Mixed with media, decachlorobutane was the most effective of the larvicides tested. ABSTRACTS: RAE-B 44:203, Dec., 1956; TDB 51:1116, Oct., 1954.

1631. **Ascher, K. R. S., Reuter, S.** and **Levinson, Z. 1951** The Insecticidal Properties of DDT as Related to its Physical State in Residue. Laboratory Experiments on Glass Surfaces. Advances in Insecticide-Research. (Jerusalem) 1:18 pp.
First unit concerns benzyl benzoate as adjuvant to DDT; second treats of DDT-by-products as adjuvant to p-p̍ DDT. Various formulations compared. ABSTRACTS: TDB 48:1155, Dec., 1951; Chem. Abstracts 46: 1698, 1952.

1632. **Ascher, K. R. S.** and **Roch, D. 1957** Houseflies in Israel—1. A Resistance Survey in Rural Areas. Riv. di Parassit. 18(2):113-122.
Resistance was calculated from the ratio of the 50 per cent lethal deposit rate to that obtained with a susceptible laboratory colony. Results showed varying degrees of resistance to DDT and gamma BHC in different parts of Israel, with no evident relation to geographical location. ABSTRACT: TDB 54:1253, Oct., 1957.

Ascher, K. R. S., joint author. *See* Kocher, C., et al., 1954; Levinson, Z. H., et al., 1954; Gratz, N. G., et al., 1957; Reuter, S., et al., 1956.

Ashnafi, S., joint author. *See* Lambremont, E. N., et al., 1959.

Atkeson, F. W., joint author. *See* Dahm, P. A., et al., 1950.

Ault, A. K., joint author. *See* Incho, H. H., et al., 1954.

Austin, H. C., joint author. *See* Burns, E. C., et al., 1959.

Avery, B. W., joint author. *See* Cheng, T.-H., et al., 1957.

1633. **Babers, F. H.** and **Mitlin, N. 1955** The Toxicology of Homologs and Derivatives of Bayer L 13/59. J. Econ. Ent. 48:430-431.
Toxicity and cholinesterase inhibitory action of the dehydrochlorinated products were much greater than for L 13/59 but generally decreased with lengthening of carbon chain connected by oxygen to the phosphorus atom. ABSTRACTS: BA 30:1153(11754), Apr., 1956; RAE-B 44:103, July, 1956.

1634. **Babers, F. H., Mitlin, N.** and **Shortino, T. J. 1956** The Fate of Radiophosphorus Ingested by House Flies and German Cockroaches. J. Econ. Ent. 49:820-822.
Half life in housefly was 8 days. Fed P^{32} for six days; houseflies did not oviposit. If fed for a shorter time, oviposition and fertility were inhibited. Normalcy returned with normal diet. Larvae reared on P^{32} compound developed normally. ABSTRACT: RAE-B 46:23, Feb., 1958.

1635. **Babers, F. H.** and **Pratt, J. J. 1950** Studies on the Resistance of Insects to Insecticides. I. Cholinesterase in House Flies (*Musca domestica* L.) resistant to DDT. Physiol. Zool. 23(1):58-63.
Cholinesterase activity in heads of resistant houseflies was lower in first 5 days than in normal flies. Males show more cholinesterase activity than do females. ABSTRACTS: RAE-B 40:75, May, 1952; TDB 49:1004, Oct., 1952.

1636. **Babers, F. H.** and **Pratt, J. J. 1951** A Comparison of the Cholinesterase in the Heads of the House Fly, the Cockroach, and the Honey Bee. Physiol. Zool. 24(2):127-131.
Cholinesterase in the heads of insects inhibited by excess ACH but not by excess MECH or TA. In the housefly one enzyme apparently splits both the acetylcholine and the methyl derivatives.

1637. **Babers, F. H.** and **Pratt, J. J. 1953** Resistance of Insects to Insecticides. The Metabolism of Injected DDT. J. Econ. Ent. 46-977-982.
DDT metabolizes to DDE in susceptible and resistant houseflies. Much still to be learned regarding resistance to insecticides. ABSTRACTS: BA 28:1927, Aug., 1954; RAE-B 42:133, Sept., 1954; TDB 51:852, Aug., 1954.

1638. **Babers, F. H., Pratt, J. J.** and **Williams, M. 1953** Some Biological Variations between Strains of Resistant and Susceptible House Flies. J. Econ. Ent. 46:914-915.
Flies grown under crowded conditions were smaller and weaker than those under normal conditions. However, smaller insects from resistant colony grown under crowded conditions were more resistant than susceptible flies raised normally. ABSTRACTS: BA 28:1686, July, 1954; RAE-B 42:76, May, 1954; TDB 51:438, Apr., 1954.

1639. **Babers, F. H.** and **Roan, C. C. 1954** Distribution of Radioactive Phosphorus in Susceptible and Resistant House Flies. J. Econ. Ent. 47:973-975.
Housefly larvae utilize inorganic phosphorus to synthesize phosphates of different chemical nature. Except in lipoid fraction radioactive phosphorus decreased rapidly. ABSTRACTS: BA 29:2259, Sept., 1955; RAE-B 43:188, Dec., 1955.

Babers, F. H., joint author. *See* Alexander, B. H., et al., 1958; Mitlin, N., et al., 1955, 1956; Pratt, J. J., et al., 1950, 1953.

1640. **Badanov, M. I. 1955** [Resistance of *Musca domestica* to DDT.] Med. Parazit. (Moskva) 24(2):170-174. (In Russian.) Referat. Zhur., Biol., 1956, No. 27656 (Translation).
In a town where DDT had been used for 6 years to combat flies, a resistant population of *Musca domestica* arose. In 1950 experiments, death rate was 93 to 100 per cent; in 1953, 5 to 10 per cent. ABSTRACT: BA 30:3588(35771), Dec., 1956.

1641. **Badanov, M. I. 1956** [Resistance of Summer Populations of Southern House Fly to DDT preparations.] Vopr. Kraevoi Patol. Akad. Nauk UzSSR 1956(8):151-153; Referat. Zhur., Biol., 1958, No. 5422 (Translation).
A comparison of effective activity of DDT applied in 2 per cent dust, 3 per cent aqueous emulsions and 2·1 per cent acetone solutions showed an advantage for the dust. ABSTRACT: BA 35:2918(33178), June, 1960.

Bailey, S. F., joint author. *See* Wall, W. J., et al., 1952.

1642. **Baker, F. D.** and **Panetsky, D. 1958** Studies on the Enzyme Lipase in the Housefly. Arch. Biochem. N.Y. 77(2):329-335.

Baker, F. S., joint author. *See* Guthrie, F. E., et al., 1954.

Baker, G. J., joint author. *See* Campau, E. J., et al., 1953.

1643. **Baker, W. C.** and **Schoof, H. F. 1955** Prevention and Control of Fly Breeding in Animal Carcasses. J. Econ. Ent. 48:181-183.
Dog carcasses were treated with emulsion applications of dieldrin, endrin, aldrin, malathion, Diazinon, BHC and chlordane to evaluate efficacy of each in preventing blow fly infestation. Organic phosphorus compounds produced a more rapid kill of mature larvae than chlorinated hydrocarbon insecticides. ABSTRACTS: BA 30:306, Jan., 1956; RAE-B 44:55, Apr., 1956; TDB 42:939, Sept., 1955.

Baker, W. C., joint author. *See* Maier, P. B., et al., 1952.

1644. Bakry, G. 1955 Preliminary Investigations of the Seasonal Incidence of *Musca sorbens* Wied. in Relation to Acute Ophthalmias in Egypt J. Egypt Med. Ass. 38(9):507-510.

Bami, H. L., joint author. *See* Sharma, M. I. F., et al., 1957.

1645. Baranyovits, F. 1951 Fly Reactions to Insecticidal Deposits: A New Test Technique. [Correspondence.] Nature 168(4283):960-961.
Flies remain on surface having BHC deposit until a lethal dose is acquired. On a DDT deposit flies tend to leave quickly and may not be killed if no additional contact is made. ABSTRACT: TDB 49:452, Apr., 1952.

1646. Barbesgaard, P. and **Keiding, J. 1955** Crossing Experiments with Insecticide-resistant House Flies. Vidensk. Medd. Dansk. Naturh. Foren. 117:84-116.
Crossings among (1) a strain resistant to DDT, BHC and chlordane, (2) a strain resistant to DDT only, and (3) a susceptible strain indicate that no cytoplasmic or sex-linked factors are involved, also that several genes are responsible for BHC-resistance and perhaps for chlordane-resistance. A single recessive gene may be responsible for DDT-resistance, at least in the multiresistant strain. ABSTRACT: RAE-B 45:213, Dec., 1957.

1647. Barker, R. J. 1953 Effect of Temperature on Absorption and Detoxification of DDT in *Musca domestica* L. 165 pp. Dissertation Publ. 5219, Univ. of Illinois.
The effects are determined of 8 different temperatures on absorption, degradation, and mortality of survivors in 3 strains of houseflies, which had received 3 dosages of DDT, applied topically. ABSTRACT: BA 28: 948(9704), Apr., 1954.

1648. Barker, R. J. 1957 The Absorption and Degradation of DDT at Varied Dosages by the House Fly. J. Econ. Ent. 50:748-750.
The amount of degraded DDT increases with higher dosages varying from 84 to 99 per cent of the absorbed DDT, but shows no trend dependent on dosage. Only indistinguishable differences in absorption are shown by survivors of resistant strains. ABSTRACTS: BA 32:1491(17705), May, 1958 (as *Baker*, R. J.); RAE-B 46:141, Sept., 1958; TDB 55:593, May, 1958.

1649. Barker, R. J. 1957 DDT Absorption and Degradation in House Flies of Varied Age. J. Econ. Ent. 50:499-500.
Age does not change total DDT degradation in treated flies; however, in older flies DDE (the dehydrochlorination product) does appear in increased amounts. ABSTRACTS: BA 31:3589(39824), Dec., 1957; TDB 54:1372, Nov., 1957.

1650. Barker, R. J. 1957 Some Effects of Temperature on Adult House Flies Treated with DDT. J. Econ. Ent. 50:446-450.
Untreated house flies tolerate temperatures from 5° to 40°C without adverse effects. Effects of DDT are manifested earlier and at temperatures of over 30° survival is lower; DDT degradation increases with higher temperatures. ABSTRACTS: BA 31:3589(39823), Dec., 1957; RAE-B 46: 139, Sept., 1958.

1651. Baker, R. J. and **Alexander, B. H. 1958** Acid and Alkaline Phosphatases in House Flies of Different Ages. Ann. Ent. Soc. Am. 51(3): 255-257.
Acid phosphatase activity is highest in eggs and puparia and greater in females than males. Alkaline phosphatase activity is highest in 2-day-old larvae.

1652. **Barker, R. J.** and **Edmunds, L. N., Jr. 1958** Toxicity of DDT in Oil and in Acetone to Adult DDT-resistant House Flies. J. Econ. Ent 51(6):914-915.
Tests were on 2-to-3 day old adult flies of LDD strain. Table shows effect on males and females, with limits for fiducial probability of 95 per cent. ABSTRACTS: BA 33:2256(27778), May, 1959; RAE-B 48:60, Mar., 1960.

1653. **Barker, R. J.** and **Rawhy, Abd-El-Rahman. 1957** Toxicity of DDT in Acetone and in Oil to Adult House Flies. J. Econ. Ent. 50:105.
Flies absorb more DDT from an acetone solution than from an oil solution, but the difference seems insufficient to explain the great difference in mortality in normal flies. In resistant flies more toxicity results from DDT in oil. ABSTRACTS: BA 31:2997(33093), Oct., 1957; RAE-B 46:49, Nov., 1958; TDB 54:764, June, 1957.

Barker, R. J., joint author. *See* Alexander, B. H., et al., 1958.

1654. **Barlow, F.** and **Hadaway, A. B. 1952** Some Factors affecting the Availability of Contact Insecticides. Bull. Ent. Res. 43:91-100.
Addition of lanoline (5 per cent) in kerosene solutions of DDT increased its effectiveness toward *Musca domestica* but only for three weeks. ABSTRACT: RAE-B 40:109, July, 1952.

Barlow, F., joint author. *See* Hadaway, A., et al., 1952, 1957, 1958.

Barlow, J. S., joint author. *See* House, H. L., et al., 1958.

Barnes, D., joint author. *See* Ramirez Genel, M., et al., 1953.

1655. **Barnhart, C. S.** and **Chadwick, L. E. 1953** A "Fly Factor" in Attractant Studies. Science 117:104-105.
Bait visited by flies seems more attractive to other flies than similar baits not previously visited. ABSTRACTS: BA 27:2930, Nov., 1953; TDB 50:573, June, 1953.

Baroody, A. M., joint author. *See* Mitlin, N., et al., 1958.

1656. **Barrett, W. L. 1952** Control of House Flies with Methoxychlor in Texas Dairy Barns. J. Econ. Ent. 45:90-93.
Methoxychlor provides 5-7 weeks control of houseflies in unscreened dairies around which good sanitation is practiced. If methoxychlor deposits are masked with fly excrement in favored places, maximum benefit is lost. ABSTRACTS: RAE-B 40:94, June, 1952; TDB 49:809, Aug., 1952.

1657. **Barsa, M. C.** and **Ludwig, D. 1959** Effects of DDT on the Respiratory Enzymes of the Mealworm, *Tenebrio molotor* Linnaeus, and of the House Fly, *Musca domestica* Linnaeus. Ann. Ent. Soc. Am. 52(2): 179-182.
The amount of inhibition caused by 10^{-3} M DDT increased with a decrease in enzyme concentration both with cytochrome oxidase and succinic dehydrogenase of mealworm and house fly homogenates; hence, DDT appears to be irreversible in its action.

Barsa, M. C., joint author. *See* Ludwig, D., et al., 1959.

Barthel, W. F., joint author. *See* Mitlin, N., et al., 1956; Piquette, P. G., et al., 1957, 1958.

Barthez, W. J., joint author. *See* Beroza, M., et al., 1957.

Bartley, C. E., joint author. *See* Hansens, E. J., et al., 1953.

Batte, E. G., joint author. *See* Thompson, R. K., et al., 1953.

1658. **Beard, R. L. 1958** Laboratory Studies on House Fly Populations. I. A Continuous Rearing System. II. Characteristics of Closed Populations Reared Continuously. Bull. Conn. Agr. Exp. Sta. 619, 12 pp.
As an introduction to a series of laboratory studies on the long range effects of insecticides on house fly populations, this bulletin describes

rearing techniques and the characteristics of untreated populations maintained for 20 generations. ABSTRACT: RAE-B 51:120, June, 1963.

Bekman, A. M., joint author. *See* Derbeneva-Ukhova, V. P., et al., 1959.

Benjamini, E., joint author. *See* Menn, J. J., et al., 1957.

Benetti, M. P., joint author. *See* Sacca, G., et al., 1957, 1958.

1659. Berezantsev, I. U. A. 1959 Tennyi kolpak dlia ucheta gnusa. [A dark cap for diptera count.] Med. Parazit. (Moskva) 28(1):97-99.

1660. Bergmann, E. D. and **Levinson, Z. H. 1954** Steroid Requirements of Housefly Larvae. Nature (London) 173:211-212.
Specificity of steroids as growth factors for larvae of *M. vicina* Macq. It was concluded that the saturated ketone acts as an antimetabolite. The larvae of *M. vicina* are incapable of dehydrogenating the saturated sterol ring system.

1660a. Bergmann, E. D. and **Levinson, Z. H. 1958** Fate of Beta-sitosterol in Housefly Larvae. Nature (London) 182(4637):723-724.
M. vicina Macq. convert dietary sitosterol into a sterol nutritionally equivalent to cholesterol or 7-dehyrocholesterol in *Dermestes* larvae.

1661. Bergmann, E. D., Levinson, Z. H. and **Mechoulam, R. 1958** The Toxicity of "Veratrum" and "Solanum" Alkaloids to Housefly Larvae. J. Insect. Physiol. 2(3):162-177.
The toxicity of ester alkaloids depends on the intactness of their molecules, their constituent alkamines being non-toxic. The more ester groups contained, the more toxicity is present. ABSTRACT: RAE-B 49: 60, Mar., 1961.

1662. Bergmann, E. D., Moses, P. and **Neeman, M. 1957** Studies on Organic Fluorine Compounds. VIII. N.-substituted Fluoroacetamides as Insecticides and Rodenticides. J. Sci. Fd. Agr. 8(7):400-404.
Results are given regarding compounds tested for toxicity on 3rd instar larvae of *Musca domestica vicinia* Macq. of a strain slightly resistant to DDT, on females (2-3 days old) of a strain of *M. domestica vicina* highly resistant to DDT, and on male white rats. Possible mode of action of the toxicants discussed. ABSTRACT: RAE-B 46:99, June, 1958.

Bergmann, E. D., joint author. *See* Levinson, Z. H., et al., 1957, 1959.

Berlin, F. D., joint author. *See* Schroeder, H. O., et al., 1950.

1663. Berni, A., Capirchio, F. and **Jerace, E. 1959** Ricerche sperimentali con un nuovo larvicida nella lotta contro le mosche. [Experimental Studies with a New Larvicide in the Control of Flies.] Nuovi Ann. Igiene Microb. 10(1):62-67.
Laboratory and field experiments confirm that "Fly flakes" (active ingredients: o, o dimethyl-dithiophosphate or diethyl-mercapto-succinate) are a practical, economic and effective agent for controlling larval and adult houseflies. No special apparatus required; may be sprinkled by hand. ABSTRACT: TDB 57:191, Feb., 1960.

1664. Beroza, M. 1954 Pyrethrum Synergists in Sesame Oil. Sesamolin, a Potent Synergist. J. Am. Oil Chem. Soc. 31:302-305.
Sesamolin and sesamin, two fractions of sesame oil account for practically all of the synergistic activity of the oil. Studies on the chemical properties of sesamolin, about 5 times as active as sesamin, have been made and a chemical formula proposed. ABSTRACT: RAE-B 44:97, July, 1956.

1665. Beroza, M. 1955 The Structure of Sesamolin and its Stereochemical Relationship to Sesamin, Asarinin and Pinoresinal. J. Am. Chem. Soc. 77:3332-3334.
Not a fly study. Of interest because of the similarity of these chemicals to certain insecticides and/or synergists. ABSTRACT: RAE-B 44:97, July, 1956.

1666. Beroza, M. 1956 Sesamolin and Related compounds as Synergists for Pyrethrum. Soap and Chem. Spec. 32:128, 133, 134.
Presence of the methylenedioxyphenyl group is necessary for the synergistic activity.

1667. Beroza, M. 1956 Insecticide Synergists. Determination of Methylenedioxyphenyl-Containing Synergists Used in Analysis of Fly Sprays. Agric. Food Chem. 4(1):53-56.
The chromotrophic-sulfuric acid method proved applicable in the estimation of synergists and in overcoming interferences encountered with several constituents.

1668. Beroza, M. 1956 Insecticide Synergists. 3,4-Methylenedioxyphenoxy Compounds as Synergists for Natural and Synthetic Pyrethrins. Agric. Food Chem. 4(1):49-53.
Sixty-six new compounds containing the methylenedioxyphenoxy group were tested for synergistic properties. Almost all ethers, acetals, and sulfonates showed synergism. Urethans showed less, and esters of carboxylic acids showed none.

1669. Beroza, M. 1957 A Review of Some Fundamental Studies on Pyrethrum Synergists in Sesame Oil. Pyrethrum Post. January. (Unpaged reprint.)
A review of the methylenedioxyphenyl group containing sesamin and sesamolin, most effective synergists with pyrethrum against the housefly.

1670. Beroza, M. and **Barthel, W. F. 1957** Chemical Structure and Activity of Pyrethrin and Allethrin Synergists for Control of the Housefly. J. Agric. Food Chem. 5(11):855-859.
Few compounds not containing 3, 4-methylenedioxyphenyl group were effective synergists. More than 200 methylenedioxyphenyl compounds were investigated. ABSTRACT: BA 32:2678(32087), Sept., 1958.

Beroza, M., joint author. *See* Fales, J. H., et al., 1956, 1957; Gersdorff, W. A., et al., 1954, 1956, 1957.

1671. Bettini, S. 1958 Osservazioni sul grado di sensibilita di *Musca domestica* L. verso il malathion nella zona di Chianciano dopo quattro anni di trattamento. [Observations on the degree of susceptibility of *Musca domestica* to malathion in the Chianciano area after 4 years of treatment.] Riv. di Parassit. 19(1):73-78. *Also:* Rend. Ist. Sup. Sanit. (Roma) 21(9):787-793.
Simple tests on the F_1 progeny showed no increase in resistance in the treated zone. May be due to (i) small extent of the treated zone, (ii) insecticide was used only against adult flies, (iii) simultaneous measures were taken to reduce breeding sites for flies. ABSTRACT: TDB 55:950, Aug., 1958.

1672. Bettini, S. and **Boccacci, M. 1952** Sostanze alogenate alchilanti come insetticidi di contatto per le mosche resistenti. [Halogenated alkylating substances as contact insecticides against resistant flies.] Rend. Ist. Sup. Sanit. 15(8):609-611.
A series of monohalogenated acetic acids and their esters and salts are active against roaches and flies resistant to DDT, chlordan, etc., Death is preceded by paralysis of muscles in contraction. ABSTRACT: BA 27: 1932(20750), July, 1953.

1673. Bettini, S. and **Boccacci, M. 1952** Sostanze alogenate alchilanti come insetticide di contatto per le mosche resistenti. Riv. di Parassit. 13: 165-167.
See preceding reference.

1674. Bettini, S. and **Boccacci, M. 1954** Localizzazione e distribuzione della succinodeidrogenasi nel muscolo di insetti. [Localization and distribution of succinic dehydrogenase in the muscle of insects.] Rend. Ist. Sup. Sanit. 17:188-191. (English summary.)
Musca domestica among species studied. Enzyme concentration in muscle varies in the same insect in proportion to the kinetic activity and pigmentation of the muscle. ABSTRACT: TDB 51:1199, Nov., 1954.

1675. Bettini, S. and **Boccacci, M. 1955** Azione tossica degli acidi iodo e cloroacetico sugli insetti. Inibizione della triosofosfato deidrogenasi. Ric. di Parassit. (Rome) 16(1):13-29. (English summary.)
(See Bettini, S. and Boccacci, M. 1956.) ABSTRACTS: RAE-B 45:139, Sept., 1957; TDB 52:1155, Nov., 1955.

1676. Bettini, S. and **Boccacci, M. 1956** Azione tossica degli acidi iodo-e cloroacetico sugli insetti. Inibizione della triosofosfato deidrogenasi. [Toxic action of iodo- and chloracetic acid on insects through inhibition of triosephosphate dehydrogenase.] Rend Ist. Sup. Sanit. 19:1086-1106.
Reports experiments on *Periplaneta* and 2 strains of *Musca domestica*. Failed to increase tolerance of *M. domestica* toward chloroacetic acid by 30 generations of selection.

1677. Bettini, S., Boccacci, M. and **Natalizi, G. 1958** Behavior of Various Susceptible and Resistant Strains of *Musca domestica* L. Towards Cetyl and Myristyl Bromo- and Chloro-acetates. Indian J. Malariol. 12(4):447-452.
Speed of action and LD50 values were determined at 24 hours for 6 strains (2 susceptible, 4 resistant). For the insecticide-resistant strains, differences in susceptibility are due to characteristics of the strains.

1678. Bettini, S., Boccacci, M. and **Natalizi, G. 1958** A Comparative Study on the Speed of Action of some Halogen-Containing Thiol Alkylating Agents on Resistant House Flies. J. Econ. Ent. 51(6):880-882. 4 figs.
Cetyl and myristyl bromoacetates act faster than corresponding chloroacetates in inhibiting SH-containing enzymes, which is their basic mode of action in affecting carbohydrate metabolism. ABSTRACTS: RAE-B 48(3):57, Mar., 1960; TDB 56(5):576, May, 1959.

Bettini, S., joint author. *See* Boccacci, M., et al., 1956, 1957, 1959.

Bhatia, M. L., joint author. *See* Sharma, M. I. D., et al., 1957.

1679. Bigelow, R. S. and **LeRoux, E. J. 1954** Distinct Morphological Differences between DDT-resistant and non-DDT-resistant Strains of the House Fly, *Musca domestica* L. Canad. Ent. 86(2):78-86.
Morphological differences found, two of which were associated genetically with factors producing DDT-resistance. A discriminant linear function in which differences can be combined was developed and can be used as an accurate method of detecting DDT resistance in wild house fly populations. ABSTRACTS: RAE-B 43:163, Nov., 1955; TDB 53: 506, Apr., 1956.

1680. Bishopp, F. C. and **Henderson, L. S. 1950** House Fly Control. Leaflet No. 182, U.S. Dept. Agr. 8 pp., illus.
Emphasizes space sprays and residual sprays.

Bishopp, F. C., joint author. *See* Knipling, E. F., et al., 1952.

1681. Black, D. M. and **Mazur, S. 1958** Impregnated Cords in the Control of House Flies. Canad. J. Publ. Health 49(5):248-253.
Use of cords impregnated with parathion or diazinon (Kilpatrick and Schoof's method) proved very efficient in the control of flies under difficult conditions in British Colombia. One application of cords

appeared to be effective over an entire fly season in B.C. Method is safe, efficient and less costly than spraying and/or fogging.

Blakeslee, E. B., joint author. *See* Byers, G. W., et al., 1956; Edmunds, L. R., et al., 1959; Wheeler, C. M., et al., 1958.

1682. Blázquez, J. 1959 The Residual Insecticidal Effect of Bayer Compound S-1752 (Bayer 29 493) in Laboratory Tests. J. Econ. Ent. 52(6): 1096-1099.
Chemical, physical and biological properties of Bayer S-1752 are described. The duration of persistence of S-1752 depends upon the absorptivity of the treated surface. No difference in efficacy of S-1752 was found against susceptible or DDT-resistant house flies. ABSTRACT: RAE-B 49:30, Feb., 1961.

1683. Block, R. J., LeStrange, R. and **Zweig, G. 1952** Paper Chromatography. Academic Press, Inc. New York 195 pp.
Included because of the increasing importance of paper chromatography in physiological studies of Diptera.

1684. Blois, F. 1956 La lotta contro le mosche e provvedimenti da adottare nell'ambiente. [The war against the fly and measures to be taken in the environment.] Ann. Med. Nav. (Roma) 61(3):303-319.

1685. Blum, M. S. 1958 The Toxicities of Fluorinated Cyclohexanes to House Flies. J. Econ. Ent. 51(3):413-414.
None of the compounds evaluated exhibited appreciable activity, either as insecticides or synergists. ABSTRACT: BA 32:3513(42234), Dec., 1958.

1686. Blum, M. S. and **Bower, F. A. 1957** The Evaluation of Triethyl Tin Hydroxide and its Esters as Insecticides. J. Econ. Ent 50(1):84-86.
Triethyl tin hydroxide and its esters were toxic to both susceptible and DDT-resistant strains of house flies, but all were slightly more toxic to the susceptible strain. Compounds tested produced quick paralysis. ABSTRACT: RAE-B 46:47, Mar., 1958.

1687. Blum, M. S. and **Kearns, C. W. 1956** The Effect of Pyrethrum Activators on the Toxicity of Sabadilla to House Flies. J. Econ. Ent 49:283.
Toxicity of sabadilla is increased by six pyrethrum synergists, sulfoxide and piperonyl butoxide being the most effective. ABSTRACT: RAE-B 45: 81, May, 1957.

1688. Blum, M. S. and **Kearns, C. W. 1956** The Synergistic Activity of Piperonyl Butoxide Applied at Different Intervals to Pyrethrum-Treated House Flies. J. Econ. Ent. 49:496-497.
High ratio of synergist to insecticide produces substantial kills even if the synergist is applied 8 hours after the insecticide. At similar concentrations, injected applications of the synergist are more efficient than topical applications when applied to pyrethrum-treated flies. ABSTRACT: RAE-B 45:143, Sept., 1957.

1689. Blum, M. S. and **Pratt, J. J., Jr. 1958** Relationships between Structure and Synergistic Activity of some Structural Analogs of DDT. J. Econ. Ent. 51(5):647-650.
Twenty-eight compounds related to DDT were evaluated as DDT synergists. Synergistic activity was found only with the parachloroanalog. ABSTRACT: RAE-B 48:22-23, Jan., 1960.

1690. Blum, M. S., Pratt, J. J., Jr. and **Bornstein, J. 1959** Fluorinated Analogs of DDT as Toxicants and DDT Synergists. J. Econ. Ent. 52(4):626.
Maximum synergistic activity in those derivatives containing p-chloro- or bromo-groups, the fluoro-being inferior, and not significantly toxic to houseflies. ABSTRACTS: BA 34:1056, Nov., 1959; RAE-B 48:191-192, Nov., 1960.

1691. Boccacci, M. and Bettini, S. 1956 Co-Enzyme A Content in the American Cockroach (*Periplaneta americana* L.) and the Housefly (*Musca domestica* L.). Experientia (Basel) 12:432-433. (Italian summary.)
In *M. domestica* practically equal values were obtained by estimation of CoA in the whole insect and in the thorax alone.

1692. Boccacci, M. and **Bettini, S. 1956** Potere insetticida per contatto dell'acido cloroacetico e di alcuni suoi esteri per le mosche "resistenti". [Chloroacetic acid and some of its esters as contact-insecticides for resistant flies.] Rend. Ist. Sup. Sanit. 19:1237-1246. (English summary.)
Compounds tested were shown to possess an insecticidal power of the same order as that of the well-known organic insecticides. Maximum residual action observed was less than a week. Some compounds of this class are lachrymatory. ABSTRACT: TDB 54:1022, Aug., 1957.

1693. Boccacci, M. and **Bettini, S. 1957** Contenuto in coenzima A nella blatta (*Periplaneta americana* L.) e nella mosca domestica (*Musca domestica* L.) [Coenzyme A content in the Cockroach *Periplaneta americana* L. and in the House-fly *Musca domestica* L.] Rend Ist. Sup. Sanit. (Roma) 20(7-8):689-691. (English summary.)

1694. Boccacci, M. and **Bettini, S. 1959** Toxicity of Benzenethiol and its Derivatives on *Musca domestica* (DDT-resistant strain) and on *Periplaneta americana*. Experientia 15(1):19. (Italian summary.)
Benzenethiol shows a toxicity superior to that of its chloro- and bromoacetates. The only derivative to maintain equal toxicity was phenyl disulfide; thioesters are much less toxic. ABSTRACT: BA 33:2582(31697), May, 1959.

Boccacci, M., joint author. *See* Bettini, S., et al., 1952, 1954, 1955, 1956, 1958.

Bodenstein, O. F., joint author. *See* Fales, J. H., et al., 1951, 1954. 1955, 1956, 1957, 1958.

Bogart, R., joint author. *See* Johnston, E. F., et al., 1954.

1695. Bøggild, O. and **Keiding, J. 1958** Competition in House-Fly Larvae. Experiments involving a DDT-resistant and a Susceptible Strain, Oikos 9(1):1-25.
Mortality caused by competition is proportional to the square of the density. At high density the resistant strain survived better than the susceptible one; its larvae also developed faster. ABSTRACTS: BA 33:622, Feb., 1959; RAE-B 50:118, May, 1962.

Bogue, M. D., joint author. *See* Maier, P. B., et al., 1952; Kilpatrick, J. W., et al., 1956.

1696. Bohart, G. E. and **Gressitt, J. L. 1951** Filth-inhabiting Flies of Guam. Bull. Bishop Mus. 204, 152 pp., illus.
Chrysomyia megacephala and *Musca sorbens* are most important species. *Megaselia scalaris* is abundant indoors. Full synonomy given for *M. sorbens* and *M. vicina*. About 100 species of flies are treated, with keys to all stages. ABSTRACT: BA 26:1816, July, 1952.

1697. Bohart, R. M. 1957 Five years of Fly Control on an Agricultural University Campus. Calif. Vector Views 4(4):24-27.
Reviews the extensive management of the fly problem on the Davis, California campus covering removal of manure from the various barns, insecticide spraying and use of baits. Byproducts of the program include fly research involving insecticides, population movements and sanitation measures.

1698. **Bohart, R. M., Davies, C. S., Deal, A. S., Furman, D. P., March, R. B., Smith, A. C.** and **Swift, J. E. 1958** Insecticides for Control of Flies on Poultry Ranches, Dairies, Homes and Food Processing Plants. Calif. Vector Views 5(6):35-41.
Discusses sanitation and management practices, chemical control by baits, sprays, cords and larvicides, safety precautions in control measures. Very complete tables are given showing type of treatment, formulation amounts, etc.

1699. **Bolwig, N. 1952** Hunger-Reaction of Flies (*Musca*) and the Functions of their Stomatogastric System. Nature 169:197-198.
Drinking and sucking movements occur independently of the hunger or thirst condition of fly; actual hunger-reaction is caused by the effect of exhausted haemolymph on the Central Nervous System. ABSTRACT: Hocking, B. 1960 Smell in Insects, p. 62 (2.44).

Bonner, F. L., joint author. *See* Burns, E. C., et al., 1959.

Borash, A. J., joint author. *See* Haynes, H. L., et al., 1954.

Borgman, A. R., joint author. *See* Kitselman, C. H., et al., 1950.

1700. **Born, D. E. 1954** Mold Control in Fly Rearing Media. J. Econ. Ent. 47:367.
Prevented growth of mold by adding a one-inch layer of sand on top of the medium, 48 hours after the introduction of eggs. Addition of the sand less than 48 hours after the introduction of eggs inhibits hatching. ABSTRACT: RAE-B 43:26, Feb., 1955.

1701. **Born, D. E.** and **Davidson, R. H. 1955** The Effect of Pyrethrins in Combination with Chlorinated Hydrocarbons on Resistant and Non-resistant House Flies. J. Econ. Ent. 48:413-414.
All compounds tested gave increased mortality when combined with pyrethrins and tested against DDT-resistant flies. ABSTRACTS: BA 30: 1153(11756), Apr., 1956; RAE-B 44:102, Jul., 1956.

Bornstein, J., joint author. *See* Blum, M. S., et al., 1959.

1702. **Bosch, H. M. 1953** Global Opportunity for a Better Environment. Am. J. Publ. Health 43(6)Part II:20-25.
American methods for dealing with foreign health problems are not necessarily satisfactory. Permanent solution of the sanitation needs of any country will be one that utilizes local material and local labor and which does not deviate from the established cultural patterns. ABSTRACT: BA 28:1138(11557), May, 1954.

Boush, G. M., joint author. *See* DeLong, D. M., et al., 1952.

1703. **Bovingdon, H. H. S. 1958** An Apparatus for Screening Compounds for Repellency to Flies and Mosquitoes. Ann. Appl. Biol. 46(1):47-54.
Paper includes detailed determinations concerning repellency of butoxy-polypropylene glycol for *M. domestica.* See also listing in Section VII—Publications on Research Techniques. ABSTRACTS: TDB 55:1055, Sept., 1958; Hocking, B. 1960 Smell in Insects, p. 129 (3.15).

Bower, J. A., joint author. *See* Blum, M. S., et al., 1957.

1704. **Bowness, J. M.** and **Wolken, J. J. 1959** A Light-Sensitive Yellow Pigment from the Housefly. J. Gen. Physiol. 42:779-792.
A yellow photo sensitive pigment (absorption peak of 437 mμ), of unknown chemical nature, was isolated from *M. domestica.* It was not believed, at that time, to contain retinal. *See* Walker, Bowness and Scheer, 1960.

1705. **Bradbury, F. R. 1957** Absorption and Metabolism of BHC in Susceptible and Resistant Houseflies. J. Sci. Fd. Agric. 8(2):90-96.
The part played by metabolism in making an insect resistant to poisoning is discussed in terms of the effect of the action of concentration and

time. It is concluded that detoxication by metabolism is essential for the complete recovery of an insect following absorption of the insecticide. ABSTRACT: RAE-B 46:52, Mar., 1958.

1706. **Bradbury, F. R., Campbell, A.** and **O'Carroll, F. M. 1958** The Waxes and Lipoids of Resistant Houseflies and Mosquitoes. Indian J. Malariol. 12(4):547-564.
The weight of epicuticular waxes and the lipoids were not found to differ between BHC-resistant and susceptible strains. Five strains of *M. domestica* were used. ABSTRACT: RAE-B 49:230-231, Oct., 1961.

1707. **Bradbury, F. R., Nield, P.** and **Newman, J. F. 1953** Amount of Gamma-Benzene Hexachloride picked up by Resistant Houseflies bred on a Medium Containing Benzene Hexachloride. [Correspondence.] Nature 172:1052.
An "inside" and "outside" pick-up was recorded. Resistance to gamma-benzene hexachloride is not due to penetration resistance of the insects. A detoxification mechanism is unlikely in resistant flies since they carry in their tissues from the larval stage enough insecticide to kill normal flies in a few minutes. ABSTRACT: TDB 51:321, Mar., 1954.

1708. **Bradbury, F. R.** and **Standen, H. 1955** The Fate of γ-Benzene Hexachloride in Normal and Resistant Houseflies. I. J. Sci. Fd. Agric. 6(2):90-99.
Normal and resistant houseflies were exposed to γBHC labelled with ^{14}C. After exposure, fractions of γBHC in the inside and outside tissues were separately removed, by extraction with carbon tetrachloride, before and after the flies were ground. ABSTRACT: RAE-B 43:187, Dec., 1955.

1709. **Bradbury, F. R.** and **Standen, H. 1956** The Fate of γ-Benzene Hexachloride in Normal and Resistant Houseflies. II. J. Sci. Fd. Agric. 7(6):389-396.
The combined effect of reduced absorption and increased decomposition was to reduce the amount of γBHC in Resistant flies to one-quarter that in normal houseflies four hours after exposure for 15 minutes. ABSTRACT: RAE-B 45:30, Feb., 1957.

1710. **Bradbury, F. R.** and **Standen, H. 1958** The Fate of γ-Benzene Hexachloride in Resistant and Susceptible Houseflies. III. J. Sci. Fd. Agric. 9:203-212.
Authors propose, with some reservations, that PHC (pentachlorocyclohexene) is an intermediate in the metabolism of BHC by the Uruguayan flies. Prefer to regard it, and the accompanying small amount of TCB (trichlorobenzene), as the results of a secondary dehydrochlorination process. ABSTRACT: RAE-B 48:98, May, 1960.

1711. **Bradbury, F. R.** and **Standen, H. 1959** Metabolism of Benzene Hexachloride by Resistant Houseflies, *Musca domestica*. Nature (London) 183(4666):983-984.
Metabolism of benzene hexachloride to produce water soluble compounds involves the formation of a C-S bond. ABSTRACT: RAE-B 50:14, Jan., 1962.

Brannon, C. C., joint author. *See* Farrar, M. D., et al., 1953.

Braums, W., joint author. *See* Heinz, J. J., et al., 1955.

1712. **Bridges, P. M. 1957** Absorption and Metabolism of [^{14}C] Allethrin by the Adult Housefly, *Musca domestica* L. Biochem. J. 66:316-320.
^{14}C allethrin was either incubated with enzyme extracts or homogenates or injected into adult houseflies. After a "metabolism" period, any unchanged allethrin and its metabolites were extracted, resolved by paper chromatography and determined radiometrically. ABSTRACTS: RAE-B 46:100, June, 1958; TDB 54:1138, Sept., 1957.

Bridges, P. M., joint author. *See* Winteringham, F. P., et al., 1955.

1713. **Bridges, R. G. 1959** Pentachlorocyclohexane as a Possible Intermediate Metabolite of Benzene Hexachloride in Houseflies. Nature 184 (4695) Supp. 17:1337-1338.

Suggests that monodehydrochlorination is not the first step in any important procedure for the metabolism of γBHC but rather that there is a direct substitution of the chlorine atoms. The detection of γPHC may indicate dehydrochlorination as an alternative (though secondarily important) pathway. ABSTRACTS: BA 35:1726(19826), Apr., 1960; RAE-B 50:52, Mar., 1962; TDB 57:540, May, 1960.

1714. **Bridges, R. G.** and **Cox, J. T. 1959** Resistance of Houseflies to γ-Benzene Hexachloride and Dieldrin. Nature 184:1740-1741.

Recognized 3 levels of resistance: (1) not related to enhanced metabolism of the insecticide, (2) perhaps due to enhanced metabolism, (3) associated with adult dietary fat. The last 2 differences occur only when the flies are under dieldrin pressure.

1715. **British Museum (Natural History) Economic Series No. 1. 1959** The Housefly as a Danger to Health. Its Life History and How to Deal with It. 7th Edition, 17 pp., 11 figs., London.

1716. **Brookes, V. J.** and **Fraenkel, G. 1958** The Nutrition of the Larva of the Housefly, *Musca domestica* L. Physiol. Zool. 13(3):208-223.

Describes a medium on which larvae will grow under aseptic conditions. Larvae require biotin, folic acid, nicotinic acid, pantothenic acid, pyridoxine, riboflavin and thiamin.

1717. **Brown, A. W. A. 1951** Insect Control by Chemicals. Illus. John Wiley and Sons, Inc., New York City.

ABSTRACTS: BA 26:2330, Sept., 1952; RAE-B 40:121, Aug., 1952; TDB 49:1092, Nov., 1952.

1718. **Brown, A. W. A. 1952** Insect Control by Chemicals. Soap and Sanit. Chem. 28(7):135-137.

Also appears in Proc. 38th Mid-year Meeting of the Chem. Spec. Mfg. Assoc. 1952, pp. 86-88.

1719. **Brown, A. W. A. 1953** The Control of Insects by Chemicals. Canad. J. Publ. Health 44:1-8.

A general review. ABSTRACT: BA 27-1676, June, 1953.

1720 **Brown, A. W. A. 1956** DDT-dehydrochlorinase Activity in Resistant Houseflies and Mosquitoes. Bull. W. H. O. 14(4):807-812.

Activity of homogenates and acetone powders was obtained from a DDT-resistant and a Dilan-resistant housefly strain. Similar preparations from mosquito larvae of a DDT resistant strain showed no DDE production under test conditions although DDE production could be demonstrated in living larvae. ABSTRACTS: RAE-B 46:163, Oct., 1958; TDB 54:97, Jan., 1957.

1721. **Brown, A. W. A. 1958** Insecticide Resistance in Arthropods. W.H.O. Monograph Series No. 38, 240 pp., 29 figs., 28 pp., refs. Geneva: Palais des Nations.

Brings together all that is known about resistance to insecticides in arthropods of medical or veterinary importance, during the past decade. Work divided into 4 chapters, the third (pp. 104-186) concerns *Musca domestica* L. ABSTRACTS: RAE-B 46:185, Dec., 1958; TDB 56:247, Feb., 1959.

1722. **Brown, A. W. A.** and **Perry, A. S. 1956** Dehydrochlorination of DDT by Resistant Houseflies and Mosquitoes. Nature 178(4529): 368-369.

Values are given from assessment of DDT dehydrochlorinase content in two strains of resistant flies. ABSTRACT: TDB 53:1492, Dec., 1956.

1723. **Brown, A. W. A. (Chairman), et al., 1951** Symposium on the Development of Resistance of Insects to Insecticides. 81st Ann. Rept. Ent. Soc. Ontario. 1950 81(1950):33-50. 79 refs.

1. Introduction. A. W. A. Brown. Outlines history of resistance.
2. Resistance of Houseflies to D.D.T. H. H. Schwart. Outlines history.
3. The Relation of Space Fumigants to the Problem of Insect Resistance to Insecticides. H. A. U. Monro. Paper includes: Short history of resistance to fumigant types of insecticides; development of resistance to quarantine fumigants; fumigant hazards.
4. Morphological and Habit Changes Correlated with Resistance to Insecticidal Treatment. F. O. Morrison. Brings together references reporting habit or morphological differences in resistant insect forms. References may be classified into: (a) those which refer to changes in habit; (b) those which refer to observed morphological differences; and (c) references to tolerance, a result of increase in general vigour.
5. Selection for DDT Tolerance in *Macrocentrus ancylivorus*. D. P. Pielou.
6. The Development of Resistance of Insects to Insecticides. H. Martin. Short statement.
7. Concluding Remarks. A. W. A. Brown.

Brown, A. W. A., joint author. *See* Roy, D. N., et al., 1954.

1724. **Brown, V. E. 1953** Synopsis of Medical Entomology. viii + 219 pp. Illus. Milwaukee, Wisconsin. Privately printed.

1725. **Brown, W. E. 1954** Big Stinky. The Regenerative Fly Trap. 18 pp. mimeogr. Copyright by author.
Explains design and operation of apparatus manufactured by Dioptran Company of Milwaukee, Wisconsin.

1726. **Bruce, W. N. 1950** Current Report on Housefly Resistance to Insecticides. Reprinted from Pest Control of April, 1950. 3 pp.
Summarises available control procedures, with recommendations for combined use. Would give "sanitation" first place.

1727. **Bruce, W. N. 1952** Insecticides and Flies. *In* Insects. The Yearbook of Agriculture. U.S. Dept. Agr., Gov't. Printing Office, Washington, D.C., pp. 320-327.
Excellent discussion citing many important studies. For example, the use of the micro-syringe and spray chamber in fly research.

1728. **Bruce, W. N. 1953** Laboratory and Field Evaluation of Factors affecting the Performance of Fly Repellents. Diss. Absts. 13:1309-1310.
A new technique using gravimetric measurements rather than an optimentric method was used in tests conducted to evaluate repellent effectiveness against *Stomoxys* and *Tabanus* on cattle. The weights of lactose pellets consumed by houseflies *Musca domestica* L. were used as criteria for measuring percentage repellency of test materials.

1729. **Bruce, W. N. 1954** A Return to Fly Control. Pest Control 22(5):19-20.
A discussion of the role of aerosols, baits and new residuals in the control of DDT-resistant flies. ABSTRACT: BA 29:207(2170), Jan., 1955.

1730. **Bruce, W. N. 1956** Are Fly Repellents Practical? Pest Control 24(4): 14, 58.
Tested 125 materials as residual repellents for all sorts of uses. All were of low toxicity to mammals. ABSTRACT: BA 30:3011(30033), Oct., 1956.

1731. **Bruce, W. N. 1957** Evaluation of Synergized Pyrethrins on Various Strains of House Flies. J. Econ. Ent. 50:828-829.
LD-50 values reveal a substantial difference among strains in susceptibility to synergized pyrethrins. Insecticides having combined synergists gave better results. Best single synergist tested was sesoxane.

1732. **Bruce, W. N.** and **Decker, G. C. 1950** House Fly Tolerance for Insecticides. Soap and Sanit. Chem. 26:122-125, 145-147.
Intensive selection and inbreeding resulting from exposure of adults and larvae produced strains showing at least some degree of tolerance for all the chemicals tested. ABSTRACT: RAE-B 39:39, Mar., 1951.

Bruce, W. N., joint author. *See* Decker, G. C., et al., 1951, 1952; Lichtwardt, E. T., et al., 1955; Sternburg, J., et al., 1950; Varzandeh, M., et al., 1954; Zingrone, L. D., et al., 1959.

1733. **Brunetto, A.** and **Cocino-Frumento, L. 1959** Osservazioni sulla biologia di *Musca domestica* L. Inincrocio e fertilità. [Observations on the biology of *Musca domestica.* Inbreeding and fecundity.] Rapporto Sessi. Symposia Genetica 6:1-17.

1734. **Buck, J. B., Keister, M. L.** and **Posner, I. 1952** Physiological Effects of DDT on *Phormia* Larvae. Ann. Ent. Soc. Am. 45:369-384.
DDT enhances oxygen uptake, water loss under dry conditions, water gain under wet conditions. It must be introduced into blood or the tracheal system. (Included here as a contribution to muscoid biology.) ABSTRACT: RAE-B 41:49, Apr., 1953.

Buckner, A. J., joint author. *See* Perry, A. S., et al., 1953, 1958, 1959.

1735. **Buei, K. 1958** Effects of Population Density on the Development of the Common House Fly Larvae, *Musca domestica vicina* Macq. Ecological Studies of the Flies of Medical Importance. I. Botyu-Kagaku 23: 173-176. (In Japanese with English summary.)
A population density of 100 considered optimum for the procurement of many healthy individuals of uniform size.

1736. **Buei, K. 1959** Seasonal Prevalence of Flies in Osaka. Ecological Studies of the Flies of Medical Importance. III. Botyu-Kagaku 24:47-54. (In Japanese with English summary.)
Flies were collected from 1953 through 1955 and the species numbers and sex ratios tabulated. Reasons suggested for the influence of temperature and seasonal change on the numbers of the species collected.

1737. **Buei, K. 1959** The Effect of Temperature on Fecundity of the Common Housefly, *Musca domestica vicina* Macq. Ecological Studies of the Flies of Medical Importance. IV. Botyu-Kagaku 24:78-82. (In Japanese with English summary.)
Rate of reproduction was maximal at 29°C; decreased at 33°C and above. Females deposited 73·4 per cent of their eggs within the first 10 days of their reproductive period. Mating observed at 15°C, but eggs were not obtained. ABSTRACT: RAE-B 49:112, June, 1961.

1738. **Bull. World Health Organization.** *See* World Health Organization. No. 2788a.

Bunden, G. S., joint author. *See* Cole, M. M., et al., 1959.

1739. **Burkhardt, C. C.** and **Dahm, P. A. 1953** An Urban Fly Control Campaign. J. Econ. Ent. 46:409-414.
The most abundant fly species in two cities during the period of this experiment was the house fly, with the greenbottle fly *Phaenicia* (*Lucilia*) second. In Manhatten, Kansas, improved sanitation and DDT application resulted in 70 per cent control. ABSTRACTS: RAE-B 42:3, Jan., 1954; TDB 50:1091, Nov., 1953.

1740. **Burnett, G. F.** and **Thompson, B. W. 1956** Aircraft Application of Insecticides in East Africa. X. An Investigation of the Behaviour of Coarse Aerosol Clouds in Woodland. Bull. Ent. Res. 47:495-524.

Wild-caught *Musca* (*Eumusca*) *lusoria* held in cages proved more susceptible to DDT in oil than *Glossina palpalis fuscipes*. Effects of wind, density of vegetation and open spaces are discussed. ABSTRACT: TDB 54:243, Feb., 1957.

1741. **Burns, E. C., Tower, B. A., Bonner, F. L.** and **Austin, H. C. 1959** Feeding Polybor 3 for Fly Control under Caged Layers. J. Econ. Ent. 52(3):446-448.
Effective larval control of the housefly was obtained at the higher rates of feeding, but boron residues were found in eggs and tissues at all levels. ABSTRACT: RAE-B 48:165, Sept., 1966.

Burns, E. C., joint author. *See* Harris, W. S., et al., 1959.

1742. **Busvine, J. R. 1951** Mechanism of Resistance to Insecticide in Houseflies. Nature 168:193-195.
Dehydrochlorination considered the predominant (but not the only) mechanism for DDT resistance in the "Sardinian" strain. An additional defense mechanism is postulated. This appears to predominate in the "Italian" strain. ABSTRACT: TDB 49:89, Jan., 1952.

1743. **Busvine, J. R. 1952** Symposium on Insecticides. IV. The Newer Insecticides in Relation to Pests of Medical Importance. Trans. Roy. Soc. Trop. Med. Hyg. 46:245-252.
Gives historical review of the discovery of modern insecticides. Concludes with general remarks on chlorinated insecticides, organo-phosphorus compounds and pyrethrum synergists. ABSTRACT: RAE-B 42:82, June, 1954.

1744. **Busvine, J. R. 1953** Forms of Insecticide-Resistance in Houseflies and Body Lice. Nature 171:118-119.
A resistance to BHC and chlordane was shown to be independent of DDT tolerance. Supports author's 1951 suggestion that there is a DDT resistant factor here not concerned with dehydrochlorination. ABSTRACTS: RAE-B 41:158, Oct., 1953; TDB 50:657, Jul., 1953.

1745. **Busvine, J. R. 1953** Laboratory Investigations with Insecticide-resistant Houseflies. *In* Trans. IXth Inter. Cong. Ent. Amsterdam. Vol. II. Symposia, pp. 335-339.
Laboratory work done with 2 strains of houseflies—one from Italy after a DDT-spraying campaign and one from Sardinia, after DDT and chlordane had been used. Shows that tolerance of insecticides is due to a heritable factor. Table summarizes results of various workers who have examined the theory that the DDT-resistance of a fly depends on its ability to metabolise DDT to DDE. ABSTRACT: RAE-B 41:122, Aug., 1953.

1746. **Busvine, J. R. 1954** Houseflies Resistant to a Group of Chlorinated Hydrocarbon Insecticides. Nature, London 174:783-785.
A similar molecular structure found in chlordane, aldrin, dieldrin, endrin, isodrin and in the γform (isomer) of BHC, accounts for the high insecticidal activity of these compounds. ABSTRACTS: RAE-B 44:111, July, 1956; TDB 52:222, Feb., 1955.

1747. **Busvine, J. R. 1956** Resistance of Insects to Insecticides. The Occurrence and Status of Insecticide Resistant Strains. Chemistry and Industry for 1956; pp. 1190-1194.
Summarizes the cases of insecticide resistance reported in insects since 1940. Discusses various aspects of the problem and emphasizes the caution to be observed in deducing resistance solely from field observations on an insects' abundance. ABSTRACT: TDB 54:505, Apr., 1957.

1748. **Busvine, J. R. 1957** Inheritance of Insecticide Resistance in the Housefly. Bull. W. H. O. 16(1):205-206.
A short note discussing the inconsistencies and conflicting results of

research on the mode of inheritance of DDT-resistance. ABSTRACT: TDB 54:1022, Aug., 1957.

1749. **Busvine, J. R. 1957** Insecticide-Resistant Strains of Insects of Public Health Importance. Trans. Roy. Soc. Trop. Med. Hyg. 51:11-31.
Housefly one of 7 important examples in the Hexapoda. Treats of factors which favor development of resistance. Program to oppose resistance: (1) prevention, (2) early detection, (3) overcoming existing resistance. ABSTRACTS: BA 31:2413(26103), Aug., 1957; RAE-B 47:71, May, 1959; TDB 54:767, June, 1957.

1750. **Busvine, J. R. 1959** Patterns of Insecticide Resistance to Organo-Phosphorous Compounds in Strains of Houseflies, from Various Sources. Ent. Exp. Appl. 2(1):58-67.
In 3 resistant strains (originating in the field) there was evidence of two distinct types of defense mechanism. Similar forms showed minor differences not attributable to chance. Concludes that resistance to organophosphorus compounds is by no means a simple phenomenon. ABSTRACTS: RAE-B 48:161, Sept., 1960; TDB 56:894, Aug., 1959.

1751. **Busvine, J. R.** and **Harrison, C. M. 1953** Tests for Insecticide-Resistance in Lice, Mosquitoes and House-flies. Bull. Ent. Res. 44:729-738.
Employed *Musca domestica vicina* (caught in market sheds), also *Musca domestica domestica* and *Musca sorbens* (caught in full sunlight). New techniques for rearing *M. domestica vicina* are described. ABSTRACTS: RAE-B 42:26, Feb., 1954; TDB 51:515, May, 1954.

1752. **Busvine, J. R.** and **Khan, N. H. 1955** Inheritance of BHC-Resistance in the Housefly. Trans. Roy. Soc. Trop. Med. Hyg. 49(5):455-459.
A preliminary investigation dealing with gamma BHC. Samples of the population were exposed to various concentrations of insecticide and the percentage kill recorded. ABSTRACTS: RAE-B 45:51, Apr., 1957; TDB 53:678, May, 1956.

1753. **Busvine, J. R.** and **Nash, R. 1953** The Potency and Persistence of Some New Synthetic Insecticides. Bull. Ent. Res. 44:371-376.
Pyrolan was quite as toxic to two DDT-resistant strains of *M. domestica* as to normally susceptible ones. ABSTRACT: RAE-B 41:156, Oct., 1953.

Butt, B. A., joint author. *See* Mitlin, N., et al., 1957.

Butts, J. S., joint author. *See* Hoffman, R. A., et al., 1951, 1952; Lindquist, A. W., et al., 1951; Yates, W. W., et al., 1952.

1754. **Buxton, P. A. 1952** Symposium on Insecticides (arranged by P. A. Buxton). Trans. Roy. Soc. Trop. Med. Hyg. 46:213-274. (Seven papers by a number of authors.)
Paper No. 1, "The Place of Insecticides in Tropical Medicine, An Introduction", covers pages 216-226. Author cites a number of important, modern works of reference. ABSTRACT: TDB 49:1006, Oct., 1952.

1755. **Buzzati-Traverso, A. A.** and **Rechnitzer, A. B. 1953** Paper Partition Chromatography in Taxonomic Studies. Science 117:58-59.
This reference deals with fishes but was included because of increasing use of chromatography in taxonomic studies of insects.

1756. **Byers, G. W., Wheeler, G. M.** and **Blakeslee, T. E. 1956** A Study of Insecticide Resistance in House Flies of Japan and Okinawa. J. Econ. Ent. 49:556-557.
Four strains studied were moderately DDT-resistant and highly susceptible to dieldrin and malathion. Two strains, one exposed to lindane, and the other to BHC for less than a year were 10 times as resistant to lindane as those not exposed. ABSTRACTS: RAE-B 45-146, Sept., 1957; TDB 54:239, Feb., 1957.

Cable, R., joint author. *See* Fletcher, O. K., et al., 1956.

1757. Cabrera, B. D. and **Rozeboom, L. E. 1956** The Flies of the Genus *Musca* in the Philippines. Philipp. J. Sci. 85(4):425-449.
Includes key to known Philippine forms. Special notes describe male, female, distribution and habits for each species. ABSTRACT: BA 32: 905(10663), Mar., 1958.

Caiman, R., joint author. *See* Jaiyou, C., et al., 1950.

1758. Calhoun, E. L., Dodge, H. R. and **Fay, R. W. 1956** Description and Rearing of Various Stages of *Dendrophaonia scabra* (Giglio-Tos) (Diptera: Muscidae). Ann. Ent. Soc. Am. 49:49-54.
Cow manure was the preferred oviposition medium. Larval stage consisted of 3 instars which could be distinguished by anatomical differences. The minimum life cycle (at 80°F and 70 per cent relative humidity) was approximately 29 days. Included here as a contribution to muscoid biology.

Calsetta, D. R., joint author. *See* Starr, D. F., et al., 1955.

Campbell, A., joint author. *See* Bradbury, F. R., et al., 1958.

Capirchio, F., joint author. *See* Berni, A., et al., 1959.

1759. Casida, J. E., Augustinsson, K.-B. and **Johnsson, G. 1960** *See* Section VI, Ref. No. 3203. (Has been erroneously cited as a 1959 publication).

Casida, J. E., joint author. *See* Ahmed, M. K., et al., 1958; Arthur, B. W., et al., 1958; Gatterdam, P. E., et al, 1959; Krueger, H. R., et al., 1957; Mengle, D. C., et al., 1958; Plapp, F. W. Jr., et al., 1958.

Cass, J. S., joint author. *See* Hayes, W. J., et al., 1951.

Catella, F., joint author. *See* Melis, R., et al., 1956, 1957.

Catts, E. P., joint author. *See* Furman, D. P., et al., 1959.

1760. Chadwick, L. E. 1952 The Current Status of Physiological Studies on DDT Resistance. Am J. Trop Med. Hyg. 1:404-411, 45 refs.
Panel Papers on Resistance of Insects to Insecticides No. V. An excellent review of existing knowledge on the subject. ABSTRACTS: BA 27:1325, May, 1953; RAE-B 42:55, Apr., 1954; TDB 49:1161, Dec., 1952.

1761. Chadwick, L. E. 1954 Recent Advances in Basic Studies on Insect Physiology in Relation to Mechanisms of Resistance to Insecticides. *In* First Inter. Symp. on Control of Insect Vectors of Disease. Ist. Sup. Sanit., pp. 219-234.
Treats of biochemical degradation of DDT, new knowledge of its mode of action and resistance to an organic phosphate.

1762. Chadwick, L. E. 1957 Progress in Physiological Studies of Insecticide Resistance. Bull. W. H. O. 16(6):1203-1218.
Toxic effects of DDT believed not due to inhibition of any vital enzyme system. Resistant strains metabolize DDT to DDE. Enzyme is DDT-dehydro-chlorinase. Lindane and related hydrocarbons have CNS, not peripheral nervous system as site of action.

1763. Chadwick, L. E. 1957 Temperature Dependence of Cholinesterase Activity. *In* Johnson, F. H. (Edit.) *The Influence of Temperature on Biological Systems*. Waverley Press, Inc., Baltimore, Md., pp. 45-59.
See next reference.

1764. Chadwick, L. W. and **Lovell, J. B. 1956 (1958)** The Effect of Temperature on the Activity of Fly Head Cholinesterase. Proc. Tenth Internat. Congr. Ent. 2:19-28.
At pH 8·0, ChE activity was directly proportional to temperatures between -1·6°C and 30°C. Rates of inactivation of ChE (by two methods) were also measured.

1765. **Chadwick, L. E., Lovell, J. B.** and **Egner, V. E. 1953** The Effect of Various Suspension Media on the Activity of Cholinesterase from Flies. Biol. Bull. 104:323-333.

1766. **Chadwick, L. E., Lovell, J. B** and **Egner, V. E. 1954** The Relationship between pH and the Activity of Cholinesterase from Flies. Biol. Bull. 106(2):139-148.
Activity of enzyme measured at 25·0°C (with Ach. Br0.D15 *M* as substrate), as a function of the pH of the assay medium (three were used). Optimum pH was between 8·0 and 9·0.

Chadwick, L. E., joint author. *See* Barnhart, C. S., et al., 1953.

1767. **Chamberlain, R. W. 1950** An Investigation on the Action of Piperonyl Butoxide with Pyrethrum. Am. J. Hyg. 52:153-183.
Lends support to the hypothesis that piperonyl butoxide activates the pyrethrum esters by inhibiting lipase, an enzyme that can detoxify them by hydrolysis. ABSTRACTS: RAE-B 41:61, Apr., 1953; TDB 48:209, Feb., 1951.

1768. **Chang, J. T.** and **Wang, M. Y. 1958** Nutritional Requirements of the Common Housefly, *Musca domestica vicina* Macq. Nature 181 (4608):566.
Casein hydrochloride hydrolysate failed to support growth, due to lack of tryptophan. Pancreatic hydrolysate of casein will, as tryptophan is not here destroyed during hydrolysis. ABSTRACT: TDB 55:704, June, 1958.

Chang, J. T., joint author. *See* Kun, K. Y., et al., 1958.

1769. **Chatterjee, A. K.** and **Mitra, R. D. 1954** Experiments on the Induced Resistance to Benzene Hexachloride in the House Fly, *Musca nebulo* F. Zschr. Tropenmed. und Parasit. 5:113-115. (German summary.)
Did not develop resistance to *gamma* BHC in 5 successive generations, completed in 12 weeks. Dosage: the standard concentration of 10 mg. *gamma* BHC per sq. ft.

1770. **Chefurka, W. 1954** Oxidative Metabolism of Carbohydrates in Insects. I. Glycolysis in the Housefly *Musca domestica* L. Enzymologia. (The Hague) 17(2):73-89.
Glycolysis is of the phosphorylative type. All of the intermediates of the Embden-Meyerhoff System were metabolized. ABSTRACT: TDB 52: 696, July, 1955.

1771. **Chefurka, W. 1957** Oxidative Metabolism of Carbohydrates in Insects. II. Glucose-6-phosphate Dehydrogenase and 6-phosphogluconate Dehydrogenase in the housefly *Musca domestica* L. Enzymologia (Den Haag) 18(4):209-227.
Both are found exclusively in the soluble fraction of the flight muscles. Activity of both is dependent upon the presence of active sulphydryl groups and is strongly inhibited by certain sulfur fungicides.

1772. **Chefurka, W. 1958** Oxidative Metabolism of Carbohydrates in Insects. III. Hexose Monophosphate Oxidative Cycle in the Housefly, *Musca domestica* L. Canad. J. Biochem. & Physiol. 36(1):83-102.
Deals with the individual reaction patterns of the hexose monophosphate oxidative pathway. Five principal steps of the cycle are given. ABSTRACT: TDB 55:593, May, 1958.

1773. **Chefurka, W.** and **Smallman, B. N. 1955** Identity of the Acetylcholine-like Substance in the Housefly. Nature 175(4465):946-947.
Authors used a chemical method, chromatography and electrophoresis to examine the pharmacologically active substance extracted from the fly. Results show, consistently, that this substance is identical with acetylcholine.

1774. **Chefurka, W.** and **Smallman, B. N. 1956** The Occurrence of Acetylcholine in the Housefly, *Musca domestica* L. Canad. J. Biochem. Physiol. 34:731-742.
A continuation of the researches reported in 1955.

Chen, H.-H., joint author. *See* Liu, S.-Y., et al., 1957.

1775. **Cheng, T.-H., Patterson, R. E., Avery, B. W.** and **Vanderberg, J. P.** An Electric- Eye-Controlled Sprayer for Application of Insecticides to Livestock. Pennsylvania Agric. Exp. Sta. Bull. 626:1-14.
Describes system of nozzles. Devised for biting flies, but not limited to these. Several insecticides tested. No adverse effect on cattle during 3 months testing period.

Chih-Hy Kuan, joint author. *See* Luh, P.-L., et al., 1950.

Chikamoto, T., joint author. *See* Katsuda, Y., et al., 1958.

1776. **Chow, C. Y.** and **Thevasagayam, E. S. 1953** A Simple and Effective Device for Housefly Control with Insecticides. Bull. W.H.O. 8: 491-495.
A portable wooden frame which can be treated with insecticide and placed anywhere in a fly-frequented room. ABSTRACT: TDB 50:1092, Nov., 1953.

1777. **Chow, C. Y., Thevasagayam, E. S.** and **Tharumarajah, K. 1954** Insects of Public Health Importance in Ceylon. Rev. Ecuatoriana de Entom. y Parasit. 2(1/2):105-150.
Gives up to date information, based on the literature and observations, on insect-borne diseases in Ceylon and the bionomics and control of the vectors. A bibliography of over 200 articles on insects, ticks and mites of public health importance in Ceylon, published up to the end of 1953, is appended. ABSTRACT: TDB 53:1058, Aug., 1956.

Chuvakhina, Z. F., joint author. *See* Smirnov, E. S., et al., 1956.

1778. **Clari, L. 1954** Prove di demuscazione delle stalle con l'Emmaton 50. [Tests of Emmaton 50 against Flies in Animal Quarters.] Ann. Sper. Agric. (Rome) (N.S.) 8(5):Suppl. pp. lv-lvii. (English summary.)
Emmaton 50 is a proprietary emulsion concentrate containing 50 per cent malathion. Results indicate that spraying with 2-2·5 per cent Emmaton 50 would give control for 20 days. ABSTRACTS: RAE-B 43-80, May, 1955; TDB 52:1028, Oct., 1955.

1779. **Clark, C. C. 1950** Insecticidal Compositions Containing Primary Polyhalophenylethlamine. U.S. Patent 2,504,803. Issued April 18, 1950.
Solutions killed houseflies and had a high knockdown value. ABSTRACT: BA 24:2930, 1950.

Clark, P. H., joint author. *See* Cole, M. M., et al., 1958.

1780. **Clifford, P. A. 1957** Pesticide Residues in Fluid Market Milk. Publ. Health Repts. 72(8):729-734.
More than 60 per cent of 801 market milk samples collected in a countrywide survey (1955) contained residues of chlorinated organic pesticides. Fly bioassay procedure used. ABSTRACT: RAE-B 47:191, Dec., 1959.

Cocino-Freemento, L., joint author. *See* Brunetto, A., et al., 1959.

1781. **Coffey, J. H. 1951** Location and Community Fly Control. Pest Control 19(5):18, 20, 36.
Four main fly breeding sources: domestic animal shelters, industrial wastes, inadequate refuse handling, and insanitary privies. Sanitation of the environment plus chemical application can result in nearly complete elimination of flies. Chemicals alone are not sufficient. ABSTRACT: BA 26:230(2342), Jan., 1952.

Coffey, J. H., joint author. *See* Schoof, H. F., et al., 1951.

Cohen, N. W., joint author. *See* Hoffman, R. A., et al., 1954.

1782. Cohen, S. and **Tahori, A. S. 1957** Mode of Action of Di-(p-chlorophenyl)-(trifluoromethyl)-carbinol, as a Synergist to DDT against DDT-resistant Houseflies. J. Agric. Fd. Chem. 5(7):519-523.
This compound is a very active synergist, and is rapidly absorbed through the fly cuticle. It inhibits the dehydrochlorination of DDT by DDT dehydrochlorinase. Gives optimum ratio of synergist to DDT for maximum effect. ABSTRACT: BA 31:3583(39761), Dec., 1957.

Cohen, S., joint author. *See* Reuter, S., et al., 1956; Tahori, A. S., et al., 1958.

1783. Cole, M. M. and **Wilson, H. G. 1958** Effect of a Combination of Unrelated Synergists on the Toxicity of Pyrethrins to Body Lice and House Flies. J. Econ. Ent. 51(5):742-743.
Experiment involved a combination of IN-930 (N-isobutyl undecenamide) with piperonyl butoxide. Combination was no more effective than piperonyl butoxide alone, but was better than IN-930 by itself.

1784. Coluzzi, A. and **Raffaele, G. 1951** Indagini sull'azione residua del DDT e del chlordane sulla mosca domestica e sugli anofeli. [Inquiry into the residual action of DDT and chlordane on the house fly and certain anophelines.] Riv. di Malariol. 30(3):113-136.
Laboratory experiments, tests on wild-caught flies, and field tests were employed with both insecticides. Conclusions and recommendations given. ABSTRACTS: BA 26:3197, Dec., 1952; BA 26:473, Feb., 1952; TDB 48:1154, Dec., 1951.

1785. Cooke, F. C. 1950 [Fly proof Composting in a Prison Camp in Singapore during Japanese Regime.] World Crops 2(1):24-25. (Personal letter of Dr. J. W. Scharff.)
Describes special modifications and methods as used to prevent fly breeding in the compost heaps.

1786. Corbo, S. 1951 La mosca domestica principale responsabile della mortalita' infantile per malattie gastroenteriche. [The Housefly principally responsible for infantile mortality due to gastrointestinal diseases.] Riv. di Parassit. 12(1):37-45.
Mortality curve drops considerably when houseflies are eliminated. Whenever and wherever resistance to insecticides prevents fly control, mortality rate remains high.

Cotter, G. D., joint author. *See* DuBois, K. P., et al., 1955.

1787. Cotty, V. 1956 Respiratory Metabolism of Prepupae and Pupae of the House Fly, *Musca domestica* L. and of their Homogenates. Reprint No. 790, Contr. Boyce Thompson Inst. 18(6):253-262.
Oxygen consumption measured during metamorphosis follows a U-shaped curve characteristic of other hexapods; that of pupal homogenates does not.

1788. Cotty, V. F. and **Henry, S. M. 1958** Glutathione Metabolism in DDT-resistant and -susceptible House Flies. Contr. Boyce Thompson Inst. 19(5):393-401, 3 figs., 4 refs. Yonkers, New York.
Glutathione concentrations measured in prepuape, pupae, and newly emerged susceptible and resistant flies. Newly emerged adult had a higher level of glutathione than any other stage. Injected cysteine appeared to be incorporated into glutathione more rapidly by susceptible than by resistant flies. ABSTRACT: RAE-B 48:205, Dec., 1960.

1789. Cotty, V. F., Henry, S. M. and **Hilchey, J. D. 1958** The Sulfur Metabolism of Insects. III. The Metabolism of Cystine, Methionine, Taurine and Sulfate by the House Fly, *Musca domestica* L. Contr. Boyce Thompson Inst. 19:379-392.
Extracts, hydrolyzates and feces analyzed by chromatography and auto-

radiography for radioactive metabolites. Results indicate an active transulfuration system in houseflies.

Cotty, V. F., joint author. *See* Hilchey, J. D., et al., 1957.

Couch, M. D., joint author. *See* Gilbert, I. H., et al., 1953.

1790. Coutinho, J. O., Taunay, A. and **Lima, L. 1957** Importance of *Musca domestica* as a Vector of Organisms Pathogenic to Humans. Rev. Inst. Adolfo Lutz 17(1):5-23.
Report observations on the finding of intestinal parasites and bacteria of human origin in flies. Suggest that *M. domestica*, under normal conditions, is not an important agent in the transmission of human enteric infections. ABSTRACT: BA 35:772(8840), Feb., 1960.

1791. Cova Garcia, P. 1957 Las moscas problema de salud publica y organizacion del servicio de aseo urbano y domiciliario en la ciudad de Valencia. [The fly problem in public health and organization of sanitary service to the community and individual in the city of Valencia.] Salud Publica 11(10):755-792.
Gives data based on survey-type studies of flies, their breeding sites, seasonal abundance, relation to the spread of intestinal disorders, etc. Many tables and graphs.

1792. Cova Garcia, P. and **Suàrez, O. M. 1953** Estudio sobre la morfología de la mosca doméstica. [Study of the morphology of the domestic fly.] Rev. Sanidad y Asistencia Social (Caracas) 18(5-6):887-919. (English summary.)
Reviews earlier opinions of the status of the various forms of the *Musca domestica* complex. Obtained specimens from Spain, U.S.A., Italy, Greece, Egypt and locally from Maracay and Rancho Grande. Records a series of measurements of different parts of the head, thorax and wings of males and females. ABSTRACT: TDB 53:1185, Sept., 1956.

1793. Cova Garcia, P. and **Sutil, O. E. 1959** La propreté domestique et l'emploi des insecticides phosphorés dans la lutte contre les mouches. [Domestic sanitation and the use of phosphorus insecticides in the war against flies.] Bull. W. H. O. 20(5):849-859.
Use of organo-phosphorus insecticides in stables, cow-sheds, etc. greatly reduced the population of *M. domestica.* No effect on species which do not come indoors. ABSTRACTS: RAE-B 49:235, Oct., 1961; TDB 57:85, Jan., 1960.

Cox, J. T., joint author. *See* Bridges, R. G., et al., 1959.

Crowell, H. H., joint author. *See* Morrison, H. E., et al., 1950.

Crowell, R. L., joint author. *See* Resnick, S. L., et al., 1951.

Crutkhow, C., joint author. *See* Schulz, K. H., et al., 1957.

1794. Cunningham, H. B. and **Eden, W. G. 1955** Toxicity of Several Insecticides to House Fly Larvae. J. Econ. Ent. 48:109-110.
LD-50 concentrations of endrin, aldrin, dieldrin, chlordane and DDT used against 3rd stage larvae, in larval medium. Emergence of adults used as a measure of toxicity. Endrin proved the most toxic. ABSTRACTS: BA 29:2483, Oct., 1955; RAE-B 44:25, Feb., 1956; TDB 52:586, June, 1955.

1795. Cunningham, H. B., Little, C. D., Edgar, S. A. and **Eden, W. G. 1955** Species and Relative Abundance of Flies Collected from Chicken Manure in Alabama. J. Econ. Ent. 48:620.
M. domestica constituted 50·198 per cent of all specimens taken. Forty-five species of Diptera were represented. ABSTRACTS: BA 30:1153 (11756B), Apr., 1956; RAE-B 44:174, Nov., 1956.

Cutkomp, L. K., joint author. *See* Hunter, P. E., et al., 1958, 1959.

1796. **Cwilich, R.** and **Mer, G. G. 1954** Determination of the Age of the Housefly *Musca domestica vicina* Macq. by the Persistence of Larval Fat Body Cells in the Imago. Riv. di Parassit. 15:357-359. (Italian summary.)
Dissection of the 5th and following abdominal segments of adults of *M. domestica vicina* of both sexes 1-4 days after emergence showed that the number of larval fat-body cells decreased with age. ABSTRACT: RAE-B 45:62, Apr., 1957.

Cwilich, R., joint author. *See* Mer, G. G., et al., 1957.

1797. **Dahm, P. A., Fountaine, F. C.** and **Pankaskie, J. E. 1950** The Experimental Feeding of Parathion to Dairy Cows. Science 112:254-255.
No parathion found in milk by analytical or bioassay method, using houseflies. No detectable taste. No effect on cow's health.

1798. **Dahm, P. A., Fountaine, F. C., Pankaskie, J. E., Smith, R. C.** and **Atkeson, F. W. 1950** The Effects of Feeding Parathion to Dairy Cows. J. Dairy Sci. 33(10):747-757.
Biological assays of the milk, using adult houseflies (*M. domestica*) gave no mortality, thus confirming the negative results of previous analytical findings.

1799. **Dahm, P. A., Gurland, J., Hibbs, E. T., Orgell, W. H., Pfaeffle, W. O.** and **Lee, I. 1959** Field Sampling of Alfalfa for the Estimation of Guthion Residues. J. Econ. Ent. 52(5):791-798.
Water emulsion sprayed on field at the rate of 1 pound of Guthion per acre. Sampling techniques described. Both chemical analyses and biological assays were performed. For the latter, adult female houseflies (*M. domestica*) were used.

1800. **Dahm, P. A.** and **Kearns, C. W. 1951** A Study of Certain Metabolic Intermediates in the Normal and DDT-poisoned House Fly Adult. Ann. Ent. Soc. Am. 44:573-580.
A distribution method was employed for studying some of the phosphorylated metabolic intermediates occurring in homogenized aggregates of whole normal and whole DDT-poisoned adult house flies. Significant differences found between fed normals and normals held six hours without food and water. ABSTRACT: BA26:2330, Sept., 1952.

1801. **Dahm, P. A.** and **Raun, E. S. 1955** Fly Control on Farms with Several Organic Thiophosphate Insecticides. J. Econ. Ent. 48:317-322.
House flies and stable flies together constituted nearly all (99 per cent) of the Diptera observed. Diazinon most effective against *Stomoxys*; Diazinon, Penthion-80, and Pirazinon all killed houseflies effectively. ABSTRACT: RAE-B 44:71, May, 1956.

1802. **Dahm, P. A., Gurland, J., Lee, I.** and **Berlin, J. 1961** *See* Section VI, Ref. No. 3287.

Dahm, P. A., joint author. *See* Burkhardt, C. C., et al., 1953; Lindquist, D. A., et al., 1957; Teotia, T. P. S., et al., 1951; Stansbury, R. E., et al., 1951; Kitselman, C. H., et al., 1950.

Davey, T. H., joint author. *See* Muirhead-Thomson, R. C., et al., 1952,

1803. **Davidovici, S., Levinson, Z.** and **Reuther, S. 1950** The Toxicity of DDT-Lanoline Residues to Flies and Mosquitoes. W. H. O./Mal/37:1-4. March 20. (Processed.)
The addition of lanoline and parent substances to DDT kerosene solutions enhances their toxicity, even to resistant strains. Effect is of 3 weeks duration.

Davidovici, S., joint author. *See* Mer, G. G., et al., 1950.

1804. **Davidson, G. 1955** The Principles and Practice of the Use of Residual Contact Insecticides for the Control of Insects of Medical Importance. J. Trop. Med. Hyg. 58(3/4):49-56; 73-80.
Explains relation between physical form of the insecticides and their toxicity under various conditions. Indicates most efficient and economical way to use them (as of that date). ABSTRACT: TDB 53:114, Jan., 1956.

Davidson, G., joint author. *See* Wilson-Jones, K., et al., 1958.

Davidson, R. H., joint author. *See* Born, D. E., et al., 1955.

1805. **Davies, L. 1950** The Hatching Mechanism of Muscid Eggs (Diptera). J. Exp. Biol. 27:437-445.
(The 7 species studied did not include *M. domestica.*) Increasing humidity aids the process by an effect on the chorion. Ovum changes shape and chorion is put under strain, making it easier for the larva to effect a rupture. ABSTRACT: RAE-B 40:203, Dec., 1952.

1806. **Davies, M., Keiding, J.** and **Von Hofsten, C. G. 1958** Resistance to Pyrethrins and to Pyrethrins-Piperonyl Butoxide in a Wild Strain of *Musca domestica* L. in Sweden. [Correspondence.] Nature 182 (4652):1816-1817.
Farm near Stockholm. Resistance to certain hydrocarbons, also to parathion, had taken place. P-P-B then controlled flies for about a year. Swedish strain compared with Laboratory strain. ABSTRACTS: RAE-B 49:297, Dec., 1961; TDB 56:670, June, 1959.

Davies, M., joint author. *See* Goodwin-Bailey, K. F., et al., 1954.

Davis, C. S., joint author. *See* Bohart, R. M., et al., 1958.

Davis, P. L., joint author. *See* Acree, F., Jr., et al., 1959.

Deal, A. S., joint author. *See* Bohart, R. M., et al., 1958.

1807. **Decker, G. C. 1955** Fly Control on Livestock. Does it Pay? Soap and Chem. Spec. (June, July). Reprint, 6 pp.
A review of methods for abating the fly nuisance. Relation to milk and meat production. Use of "safe" methods definitely contributes to profit from the enterprise.

1808. **Decker, G. C.** and **Bruce, W. N. 1951** Where are We Going With Fly Resistance? Soap and Sanit. Chem. 27(6):139, 141, 143, 159.
Concerns sanitation, mechanical devices, space sprays. Mixtures containing several lasting-type insecticides should be avoided. Larvicides should not be closely related to insecticides leaving toxic deposits to control adults. ABSTRACT: RAE-B 40:25, Feb., 1952.

1809. **Decker, G. C.** and **Bruce, W. N. 1952** House Fly Resistance to Chemicals. Am. J. Trop. Med. Hyg. 1:395-403. Panel Papers on Resistance of Insects to Insecticides. No. IV.
A review of resistance to insecticides, with emphasis on understanding the variables encountered in resistance investigations. ABSTRACTS: BA 27:1685, June, 1953; RAE-B 42:54, Apr., 1954; TDB 49:1159, Dec., 1952.

Decker, G. C., joint author. *See* Bruce, W. N., et al., 1950; Lichtwardt, E. T., et al., 1955; Zingrone, L. D., et al., 1959; Varzandeh, M., et al., 1954.

1810. **DeFoliart, G. R. 1956** Fly Control in Wyoming Barns. J. Econ. Ent. 49:341-344.
Residual sprays of diazinon, methoxychlor and malathion, at appropriate strength (with or without sugar) were effective in that order. ASTRACTS: BA 31:284(2864A), Jan., 1957; RAE-B 45:113, July, 1957.

1811. **Delaplane, W. K., Jr. 1959** The Effects of Sublethal Dosages of DDT upon the Insecticidal Tolerance and Biotic Potential of a Labora-

tory Susceptible Strain of the Housefly, *Musca domestica* L (Diptera, Muscidae). Diss. Absts. 19(10):2637-2638.
Positive correlation between acquisition of resistance, and total oviposition, increased length of the larval stage and the mean length of adult female life. . . . Resistant flies will therefore greatly outnumber susceptible flies, other factors remaining equal.

1812. **DeLong, D. M.** and **Boush, G. M. 1952** Is the Housefly being Replaced by other Diptera as the Major Insect Pest of Food Markets? Ohio J. Sci. 52:217-218.
Three-year observation showed *Phaenicia sericata* 66·1 per cent; *Musca domestica* 28·9 per cent; *Phormia regina* 5·0 per cent. Should be taken into account in formulating fly control methods for restaurants and food markets. Abstracts: BA 26:3197, Dec., 1952; RAE-B 42:63, Apr., 1954; TDB 51:1000, Sept., 1954.

1813. **DeLong, D. M., Boush, G. M.** and **Lea, A. O. 1952** The Comparative Residual Toxicity of Fog Applications of Certain Organic Insecticides to House Flies. J. Econ. Ent. 45:323-325.
Achieved from 4 weeks to 3 months control in buildings, depending on the nature of the surface, uniformity of coverage, and accumulation of foreign matter. Abstracts: RAE-B 40:130, Aug., 1952; TDB 49:1003, Oct., 1952.

1814. **Deoras, P. J.** and **Ranade, D. R. 1953** Studies on Muscidae. I. Collection of and some Observations on the Muscid Flies of Poona City. J. Univ. Bombay, Sect. B, Biol. Sci. 22(3):1-17.
Nine species of *Musca* were found in the city, with *M. domestica nebulo* and *M. domestica vicina* predominating. Fly population was greatest in dairies, with highest count in the rainy season. Abstracts: BA 30:900 (9182), Mar., 1956; TDB 51:652, June, 1954.

1815. **Deoras, P. J.** and **Ranade, D. R. 1957** Life History of the Common Indian Housefly *Musca domestica nebulo* Fabricius. Indian J. Ent. 19:15-22.
Large numbers bred at B. K. Rajderkar Insect Breeding Station, Poona. Paper treats of external morphology of all stages, also of life history and environmental influences. Abstract: BA 33:629, Feb., 1959.

1816. **Derbeneva-Ukhova, V. P. 1957** [Results of research on synantropic flies and development of fly control in the U.S.S.R.] Med. Parazit. i Parazitarn. (Moskva) 26(5):567-574. (In Russian.)

1817. **Derbeneva-Ukhova, V. P.** and **Lineva, V. A. 1959** Opyt primeneniia khlorofosa v bor'be s populiatsiei *Musca domestica* L., ustoichivoi k khlorixovannym uglevodorodam. [Result of the applications of chlorophos in control of *Musca domestica* L. resistant to chlorinated hydrocarbons.] Med. Parazit. (Moskva) 28(1):44-53. (English summary.)
During the first year of use, trichlorphon (chlorophos) proved effective as a contact poison against populations of *M. domestica* that had become resistant to DDT. No symptoms of resistance were observed. Abstracts: BA 45:309, Jan. 1, 1964; RAE-B 50:101, May, 1962.

1818. **Derbeneva-Ukhova, V. P., Timoshkov, V. V.** and **Bekman, A. M. 1959** [Effect of refuse dump organizations on flies' multiplication.] Gig. I Sanit. (Moskva) 24(4):33-37. (In Russian.)

1819. **Derbeneva-Ukhova, V. P.** and **Vuslaev, M. A. 1953** Kvoprosu o merakh bor'by s mukhami perenoschikami infektsii. Sovet Med. 17(7): 38-42.
Gives measures for control of infection-carrying flies.

1820. **Desai, R. M.** and **Killy, B. A. 1958** Some aspects of nitrogen metabolism in the fat body of the larva of *Calliphora erythrocephala*. Arch. Int. Physiol. Biochim. 66(2):248-259.
Included here because of close similarity in the physiological processes of *Calliphora* and *Musca*. Intermediary nitrogen metabolism similar to that in mammalian tissues.

DeSouza, S. H., joint author. *See* Ricciardi, I., et al., 1952.

1821. **Dethier, V. G. 1954** Olfactory Responses of Blowflies to Aliphatic Aldehydes. J. Gen. Physiol. 37:743-751.
Studied *Phormia regina*. Refers to stimulating effect of iso-valeraldehyde on *M. domestica*. Suggests that relation of olfactory stimulations at equal thermodynamic activities by homologous aliphatic compounds (at least for homologues of intermediate chain length) may be of rather general application in olfaction.

1822. **Dethier, V. G. 1955** The Tarsal Chemoreceptors of the Housefly. Proc. Roy. Ent. Soc. London, Ser. A., Pts. 7/9:87-90.
Double-walled hairs, innervated by bipolar neurons, are found on all legs, but most abundantly on those of the prothorax. Identified as tarsal chemoreceptors by direct individual and group stimulation with sucrose. ABSTRACT: TDB 53:378, Mar., 1956.

1823. **Dethier, V. G. 1955** The Physiology and Histology of the Contact Chemoreceptors of the Blowfly. Quart. Rev. Biol. 30(4):348-371.
An informative discussion on *Phormia regina*, *Protophormia terrae-novae*, *M. domestica*, *Calliphora erythrocephala*, *C. vomitoria* and *Drosophila melanogaster*, all of which have contact chemoreceptors on legs and labellum.

1824. **Dethier, V. G. 1955** Mode of Action of Sugar-baited Fly Traps. J. Econ. Ent. 48:235-239.
Concerns *Phormia regina* and *M. domestica*. A reaction between sugar and exudates from the flies produces the attractive situation. ABSTRACTS: BA 30:307, Jan., 1956; RAE-B 44:67, May, 1956.

1825. **Dethier, V. G., Hackley, B. E.** and **Wagner-Jauregg, T. 1952** Attraction of Flies by Iso-Valeraldehyde. Science 115:141-142.
Of several hundred natural products and extracts thereof, the most attractive for *M. domestica* was a 10 per cent aqueous solution of malt extract. It's odor resembles that of iso-valeraldehyde. Comment on probable importance of naturally occurring aldehydes.

Dewey, J. E., joint author. *See* Pimentel, D., et al., 1950, 1951, 1953, 1954.

1826. **Dicke, R. J. 1953** Barn Flies and their Control. Univ. Wisconsin Ext. Ser., Madison, Circular 452. 8 pp.
Concerns the development, recognition and control of the housefly, stable fly and horn fly. Very good photographs.

1827. **Dicke, R. J. 1953** Sanitation Management in Housefly Control. Soap and Sanit. Chem. (July) Reprint, 2 pp.
Concludes that most available insecticides would be adequate and effective if used in combination with proper sanitation procedures.

1828. **Dicke, R. J. 1954** Fly Control on Dairy Farms. Resistance to Insecticides Less Important as Fly Control Factor than Poor Sanitation. Soap and Sanit. Chem. 30(5):177-179, 183.
Example: Resistance to methoxychlor not chief reason for fly control failure on farms. Adequate sanitation management is mandatory. ABSTRACTS: RAE-B 43:11, Jan., 1955; TDB 51:1117, Oct., 1954.

1829. Dicke, R. J., Lugthart, G. J. and **Jones, R. H. 1955** Control of maggots in Turkey Dung with Malathion. J. Econ. Ent. 48:342-343.
Malathion used in surface application, is a very effective larvicide for housefly maggot control. No toxic effects on turkey poults were observed. ABSTRACTS: BA 30:307, Jan., 1956; RAE-B 44:75, May, 1956.

1830. Dicke, R. J., Moore, G. D. and **Hilsenhoff, W. L. 1952** House Fly Response to Volatilized Chlorinated Hydrocarbon Insecticides. J. Econ. Ent. 45:722-725; 46:184.
The olfactory effect of 12 insecticides or diluents or other adjuncts indicate that all of these have an influence on the effectiveness of insecticidal residues. ABSTRACTS: RAE-B 41:43, Mar., 1953; TDB 50:769, Aug., 1953; TDB 50:258, Mar., 1953.

1831. Dicke, R. J. and **Paul, J. J. 1951** Space Spray Combinations of Chlorinated Insecticides. J. Econ. Ent. 44:896-898.
DDT, BHC, toxaphene, chlordane, aldrin and dieldrin in various combinations, tested against houseflies in kerosene solution. Most combinations synergistic; a few antagonistic. No significant synergism in combinations of DDT with dieldrin. ABSTRACTS: BA 26:2066, Aug., 1952; RAE-B 40:58, Apr., 1952.

Dicke, R. J., joint author. *See* Kohls, R. E., et al., 1958.

Dickinson, B. E., joint author. *See* Moore, D. H., et al., 1954.

1832. Digby, P. S. B. 1958 Flight Activity in the Blowfly, *Calliphora erythrocephala*, in Relation to Wind Speed, with Special Reference to Adaptation. J. Exp. Biol. 35(4):776-795.
Wind can cause an increase or decrease of activity according to the range of wind speed. Included because of probable application to *Musca* spp. ABSTRACT: BA 33:1668(20626), Apr., 1959.

Dobson, R. C., joint author. *See* Matthew, D. L., et al., 1959.

1833. Dodge, H. R. 1953 Identifying Common Flies. Publ. Health Repts. 68(3):345-350.
Especially written for personnel engaged in survey and control operations in which a large number of specimens is involved, and the principal objective is the measurement of fly densities. ABSTRACT: BA 27:2939, Nov., 1953.

Dodge, H. R., joint author. *See* Calhoun, E. L., et al., 1956; Schoof, H. F., et al., 1956.

1834. Döhring, E. 1955 Über Fliegenplage in Kleintierställen, Einsatz chemischer Bekämpfungmittel und Resistenz. [On the fly nuisance in small animal stalls, the introduction of chemical means of combat and resistance.] Anz. Schädlingsk. 28(9):131-134.
Reports experiments with DDT and lindane. Urges proper sanitary care of stalls, summer and winter.

Dolmatova, A. V., joint author. *See* Morozova, V. P., et al., 1957.

Doucet, J., joint author. *See* Binson, G., et al., 1956.

1835. Dove, W. E. 1957 Push Button Control for Dairy Flies. Agric. Chem. 12(12):43-44; 13:113.
Discusses types of devices for dispensing insecticides in barns (spray aerosols, treadle sprayer, push-button controlled spray guns). Safety to animals insured by Dept. of Agriculture insisting that formulators bring their formulas within the limits set by the Miller Bill. ABSTRACT: BA 35:2470(28000), May, 1960.

Dove, W. E., joint author. *See* Moore, D. H., et al., 1954.

Dow, R. P., joint author. *See* Melnick, J. L., et al., 1953.

1836. **Dowling, M. A. 1955** Porton Low-Pressure Aerosol Dispenser; Report on a Series of Experiments to Confirm the Efficiency of the Dispenser against Different Test Insects. J. Roy. Army Med. Corps. 101:335-342.

Dreiss, J. M., joint author. *See* Eddy, G. W., et al., 1954.

1837. **Dremova, V. P. 1956** [Results of campaign against fly imago in Samarkanda.] Med. Parazit. i Parazitarn. (Bolezni) 25(4):364-367. Moskva. (In Russian, with English summary.)

1838. **Dremova, V. P. 1957** [On the factors determining the distribution of *Musca Domestica* L. in the rooms and the exchange of population between rooms and the street.] Zool. Zhur. 36(4):561-568. (In Russian, with English summary.)

Flies tend to show a preference for elevated temperature and lower humidity. Applies to both vertical and horizontal distribution. The higher the temperature, the less concentrated the fly population, generally. ABSTRACT: RAE-B 47:146-147, Oct., 1959.

1839. **Dremova, V. P. 1957** Osmovnye napravleniya borby s sinantropnymie mukhami. [Basic instructions for control of synantropic flies.] Med. Zhur. Uzbekistana (1):64-66.

Fly oviposition behaviour is discussed and sanitary measures are given for the prevention of oviposition by flies and control of larvae and adults. ABSTRACT: BA 35:2918(33182), June 15, 1960.

1840. **Dremova, V. P. 1958** [Control of the larvae of synanthropic flies in uncanalized lavatories by means of benzene hexachloride.] Med. Parazit. (Moskva) 27(2):227-228.

Drew, W. A., joint author. *See* Guyer, G. E., et al., 1956.

1841. **Dubinskaia, G. I. 1957** [Results of application of thiodiphenylamine (phenthiozine) and of kerosene in control of fly larvae during 1955-1956.] Zhur. Mikrobiol., Epidemiol., i Immunobiol. 28(8):82-84.

1842. **DuBois, K. P.** and **Cotter, G. J. 1955** Studies on the Toxicity and Mechanism of Action of Dipterex. A.M.A. Arch. Industr., Hlth. 11:53-60.

Tests on mammals. Rapid detoxification accounts for the low toxicity and short duration thereof in mammalia. Its toxic effect (in any organism) is due to its anti-cholinesterase action. ABSTRACT: RAE-B 45:16, Jan., 1957.

Duca, E., joint author. *See* Duca, M., et al., 1958.

1843. **Duca, M., Duca, E., Tomescu, E.** and **Dana, C. 1958** Cercetări asupra rolului mustei de casă in transmiterea infectiei cu virus Coxsackie. [Investigations on the part played by *Musca domestica* in the transmission of infection with Coxsackie virus.] Stud. Cercet. Inframicrobiol. 9(1):31-41. 20 refs. Bucharest. (Summary in French and Russian.)

The coxsackie virus may be excreted by the fly 16 days after an infective meal and can survive in the fly excreta 37 days at room temperature, therefore, control of flies and protection of food is high recommended. ABSTRACT: RAE-B 48:117, July, 1960.

DuChanois, F. R., joint author. *See* Sullivan, W. N., et al., 1958.

Dupuy, R., joint author. *See* Gaud, J., et al., 1955.

Durant, R. C., joint author. *See* McGuire, C. D., et al., 1957.

Eastwood, R. E., joint author. *See* Matheny, R. W., et al., 1957.

1844. **Eddy, G. W., McGregor, W. S., Hopkins, D. E., Dreiss, J. M.** and **Radeleff, R. D. 1954** Effects on Some Insects of the Blood and Manure of Cattle fed Certain Chlorinated Hydrocarbon Insecticides. J. Econ. Ent. 47:35-38.

Blood and excreta showed some toxicity for *Stomoxys*, but all tests with

M. domestica were negative. ABSTRACTS: BA 28:2700, Nov., 1954; RAE-B 42:187, Dec., 1954.

Eden, W. G., joint author. *See* Cunningham, H. B., et al., 1955; Oliver, A. D., et al., 1955; Marsh, M. W., et al., 1955.

Edgar, S. A., joint author. *See* Cunningham, H. B., et al., 1955.

1845. Edmunds, L. R. and **Blakeslee, T. E. 1959** Intensive Limited Area Control of Potential Anthropod Disease Vectors. J. Econ. Ent. 52(6):1050-1053.
Control of arthropod vectors in military manuevers must be early and effective. ABSTRACTS: BA 35:1969(22470), Apr., 1960; TDB 57:662, June, 1960.

1846. Edwards, R. L. 1954 The Host-finding and Oviposition Behaviour of *Mormoniella vitripennis* (Walker, Hym., Pteromalidae), a Parasite of Muscoid Flies. Behaviour 7:88-112.
Puparia of houseflies and blowflies have no attractive odor of their own, but if contaminated with larval food will elicit a turning response in female *Mormoniella* when they detect the odor. ABSTRACT: Hocking, B. 1960 Smell in Insects, p. 74 (2.145).

Eenne, E. J., joint author. *See* Wells, A. L., et al., 1958.

Egner, V. E., joint author. *See* Chadwick, L. E., et al., 1954.

1847. Eichler, W. 1959 Suggestions on Fly Control. Dtsch. Gesundh. 14: 1154-1170.

1848. Eldefrawi, M. E., Miskus, R. and **Hoskins, W. M. 1959** Resistance to Sevin by DDT- and Parathion-Resistant Houseflies and Sesoxane as Sevin Synergist. Science 129(3353):898-899.
Sesamex (Sesoxane) greatly increases the effectiveness of Sevin, a low toxicity insecticide to which *Musca domestica* L. has considerable resistance. ABSTRACT: RAE-B 49:166, Aug., 1961.

El-Moursy, A. A., joint author. *See* Hafez, M., et al., 1959.

Emmons, J., joint author. *See* Melnick, J. L., et al., 1954.

Erofeeva, T. V., joint author. *See* Sukhova, M. N., et al., 1959; Vashkov, V. I., et al., 1958.

1849. Esther, H. 1951 Über das Auftreten "DDT-resistenter" Fliegen (*Musca domestica* L.) im Jahre 1950. [Appearance of DDT resistant Houseflies in Germany in 1950.] Deut. Gesundheits-wesen 34:967-970.
ABSTRACT: TDB 48:1155, Dec., 1951.

1850. Evans, W. A. L. 1956 Studies on the Digestive Enzymes of the Blow-fly *Calliphora erythrocephala*. I. The Carbohydrases. Exptl. Parasitol. 5(2):191-206.
Identification of blow-fly enzymes. ABSTRACT: BA 30:3395(33859), Dec., 1956.

Facetti, D., joint author. *See* Grandori, R., et al., 1951, 1952.

1851. Fairchild, H. E. 1958 "Sesoxane" as a Synergist for Methoxychlor. Soap and Chem. Spec. 34(1):82, 84, 151; 4 refs. New York, N.Y.
Gives no effect alone, but has synergistic action on a wide variety of insects, including *M. domestica* for which it imparts greater kill and increased knockdown. ABSTRACT: RAE-B 48:175-176, Oct., 1960.

1852. Fales, J. H. and **Bodenstein, O. F. 1958** Aerosols vs. House Flies. Tests of Aerosols against Native Strains of House Flies add Further Evidence that a more Realistic View must be Taken. Soap and Chem. Spec. 34(12):115-117, 121.
Recent marketing of preparations that do not depend on a chlorinated hydrocarbon as the main killing agent, is considered a start in the right direction. ABSTRACT: TDB 56:671, June, 1959.

1853. Fales, J. H., Bodenstein, O. F. and **Beroza, M. 1956** Effectiveness Against House Flies of Some 3,4-Methylenedioxyphenoxy Compounds as Synergists for Pyrethrins and Allethrin. J. Econ. Ent. 49:419-420.
Of 8 compounds tested, 4 proved equal to or better than piperonyl butoxide as a synergist. ABSTRACTS: BA 31:284(2865A), Jan., 1957; RAE-B 45:118, July, 1957.

1854. Fales, J. H., Bodenstein, O. F. and **Beroza, M. 1957** New Pyrethrum Synergist. Evaluation of 3,4-methylenedioxyphenyl Acetal as Synergist for Pyrethrins and Allethrin against House Flies, Mosquitoes, Roaches, Japanese Beetles. Soap and Chem. Spec. 33(2):79-82.
The Acetal, identical to sesoxane, was tested against houseflies. Increasing the synergist increased the spray's effectiveness. In most tests ENT-20871 caused greater knockdown and kill than piperonyl butoxide or sulfoxide. ABSTRACTS: RAE-B 45:147, Sept., 1957; TDB 54:887, July, 1957.

1855. Fales, J. H., Bodenstein, O. F. and **Nelson, R. H. 1954** The Synergistic Action of Sulfoxide in Insecticide Sprays and Aerosols. J. Econ. Ent. 47:27-29.
May be applied in sprays, similar to the use of piperonyl butoxide. Is *more* effective in aerosols (with pyrethrum and DDT). ABSTRACTS: BA 28:2703, Nov., 1954; RAE-B 42:186, Dec., 1954.

1856. Fales, J. H., Bodenstein, O. F., Nelson, R. H. and **Fulton, R. A. 1951** Effect of Storage on Allethrin Formulations. J. Econ. Ent. 44:991-992.
Was checked at 6, 10, and 15 months. There was no loss of effectiveness against *M. domestica.* ABSTRACT: RAE-B 40:61, Apr., 1952.

1857. Fales, J. H., Bodenstein, O. F. and **Piquett, P. G. 1955** Tests with Furethrin Sprays and Aerosols against House Flies, Mosquitoes, and Cockroaches. J. Econ. Ent. 48:49-51.
Similar in action to allethrin. Slightly less kill, but better knock-down. ABSTRACTS: BA 29:2483, Oct., 1955; RAE-B 44:22, Feb., 1956; TDB 59:621, June, 1962.

1858. Fales, J. H., Fulton, R. A. and **Bodenstein, O. F. 1956** Low-pressure Insecticidal Space Sprays. Aerosol Age 1(8):35-38.
A variety of tests against houseflies, comparing the effect of aerosols and of low pressure space sprays containing the same ingredients in various proportions. ABSTRACT: RAE-B 46:31, Feb., 1958.

1859. Fales, J. H., Nelson, R. H., Fulton, R. A. and **Bodenstein, O. F. 1951** Insecticidal Effectiveness of Sprays and Aerosols containing Allethrin. J. Econ. Ent. 44:23-28.
The synthetic allyl homologue of cinerin I compared favorably with natural pyrethrins in kerosene-base sprays against *M. domestica.* Was equally satisfactory in aerosols (against house flies) but less so with mosquitoes and cockroaches. ABSTRACTS: RAE-B 39:134, 1951; TDB 48:1158, Dec., 1951.

1860. Fales, J. H., Nelson, R. H., Fulton, R. A. and **Bodenstein, O. F. 1951** Insecticidal Effectiveness of Aerosols and Sprays Containing Esters of Synthetic Cyclopentenolones with Natural Chrysanthemum Monocarboxylic Acid. J. Econ. Ent. 44(2):250-253.
Various formulations evaluated against houseflies, mosquitoes and cockroaches.

1861. Fales, J. H., Nelson, R. H., Fulton, R. A. and **Bodenstein, O. F. 1951** Tests with Two Insecticidal Aerosols for Use on Aircraft. J. Econ. Ent. 44:621.
Formula G-651 suggested as a substitute for G-382, to avoid use of

cyclohexanone. ABSTRACTS: BA 26:1522, June, 1952; RAE-B 39:186, Nov., 1951.

1862. Fan, Tze-Teh and **Shi, Teh-Chi. 1959** The Breeding Habits of the Common Flies in Shanghai District. Acta Ent. Sinica 9(4):342-365. (In Chinese.)
A general survey of breeding places in 1954-1955 revealed that the 15 most common species were all proven vectors of disease. List includes *M. domestica vicina* and *M. sorbens*. ABSTRACTS: BA 35:5512(64862), Nov., 1960; TDB 57:83-84, Jan., 1960.

1863. Farrar, M. D. and **Brannon, C. C. 1953** Fly Control on Dairy Farms. J. Econ. Ent. 46:172.
Concerns chiefly TEPP (tetraethyl pyrophosphate). ABSTRACTS: RAE-B 41:145, Sept., 1953; TDB 50:864, Sept., 1953.

1864. Fatakhov, I. U. M. 1953 Sites of Incubation of Flies in Garbage from Various Industries. Gig. I Sanit. (Moskva) 8:54.

1865. Fatakhov, I. U. M. 1955 Razvitie mukh v otkhodakh proizvodstva. [Development of Flies in Industrial Wastes.] Med. Parazit. (Moskva) 24:179-180; Referat Zhur., Biol., 1956, No. 27639. (Translation.)
Musca domestica vicina was found in wastes from industries such as leather goods factories or silk reeling mills, also from food manufacturing establishments except dairies. *Muscina stabulans* developed mainly in plant wastes from the silk mills. Other species found were more restricted. ABSTRACT: BA 30:3588(35775), Dec., 1956.

Faure, P., joint author. *See* Gaud, J., et al., 1950, 1951, 1954, 1956.

1866. Fay, R. W. and **Kilpatrick, J. W. 1958** Insecticides for Control of Adult Diptera. Ann. Rev. Ent. 3:401-420.
Considers a number of insecticides which either produce death or restrict the activities of adult flies.

1867. Fay, R. W., Kilpatrick, J. W. and **Morris, G. C., Jr. 1958** Malathion Resistance Studies on the House Fly. J. Econ. Ent. 51(4):452-453.
A strain of *M. domestica* L. physiologically resistant to malathion was also shown to be behavioristically resistant to malathion baits. ABSTRACT: TDB 56:109, Jan., 1959.

1868. Fay, R. W. and **Lindquist, D. A. 1954** Laboratory Studies on Factors Influencing the Efficiency of Insecticide Impregnated Cords for House Fly Control. J. Econ. Ent. 47:975-980.
The attractiveness of the cord increases with its diameter. Red or black cords favored, also those previously used by flies. Of several insecticides, tested, parathion-xylene solution proved most effective. ABSTRACTS: BA 29:2259, Sept., 1955; RAE-B 43:189, Dec., 1955; TDB 52:489, May, 1955.

1868a. Fay, R. W., Stenberg, R. L. and **Ernst, A. H. 1952** A Device for Recording Insecticicidal Knockdown of House Flies and Evaluating Residual Deposits. J. Econ. Ent. 45:288-293.
ABSTRACTS: RAE-B 40:129, Aug., 1952; TDB 49:1083, Nov., 1952. (See also SECTION VII, TECHNIQUES.)

Fay, R. W., joint author. *See* Calhoun, E. L., et al., 1956; Jensen, J. A., et al., 1951; Lindquist, D. A., et al., 1956; McCawley, R. H. Jr., et al., 1955; Perry, A. S., et al., 1953; Pimentel, D., et al., 1955; Sumerford, W. T., et al., 1951.

Fedder, M. L., joint author. *See* Vashkov, V. I., et al., 1958.

Fedorova, K. G., joint author. *See* Shura-Bura, B. L., et al., 1956, 1958.

1869. Fehn, C. F. 1958 House Fly Control at Scout Camps with Insecticide-Impregnated Cord. Calif. Vector Views 5(9):62-63.
Commercially prepared, parathion or diazinon-impregnated cords were used effectively for fly control at 1957 National Boy Scout Jamboree.

Feng, L. C., joint author. *See* Li, H. H., et al., 1951.

Ferguson, F. F., joint author. *See* Hayes, W. J., et al., 1951.

1870. Filipponi, A. 1955 Sulla natura dell' associazione tra *Macrocheles muscaedomesticae* e *Musca domestica.* [Nature of the association between *Macrocheles muscadomesticae* and *Musca domestica* L.] Rend. Ist. Sup. Sanit. (Roma) 18(11):1129-1152. *Also* published in Riv. Parassit. 16:83-102.
Adult and immature mites attack all stages but the puparia of the housefly, killing mostly first stage larvae and eggs. ABSTRACT: TDB 53:680, May, 1956.

1871. Finney, D. J. 1952 Probit Analysis: A Statistical Treatment of the Sigmoid Response Curve, Second Ed. xvi + 318 pages. New York-London, Cambridge University Press.
Included here because of the extensive use of probit analysis in the preparation of research reports.

Fischer, R. L., joint author. *See* Guyer, G. E., et al., 1956.

Fish, F. W., joint author. *See* Champlain, R. A., et al., 1954; Lambremont, E. N., et al., 1959.

1872. Fish, W. A. 1952 Embryology of *Phaenicia sericata* (Meigen) (Diptera: Calliphoridae). Part IV. The Inner Layer and Mesenteron Rudiments. Ann. Ent. Soc. Am. 45:1-22.
No mention of *M. domestica,* but much of the material presented is believed to hold true for most muscoid Diptera. Is very well illustrated. ABSTRACT: BA 26:2360, Sept., 1952.

1873. Fisher, E. H. 1955 A Dairy-barn Fogging Method for Fly Control. J. Econ. Ent. 48:330-331.
Compressed air fog application before milking provided effective fly control in barns and for the cattle following their release. Piperonyl butoxide plus pyrethrins used in formulation. ABSTRACTS: BA 30:307 (3078), Jan., 1956; RAE-B 44:72, May, 1956.

1874. Fisher, R. W. 1952 The Importance of the Locus of Application on the Effectiveness of DDT for the House Fly, *Musca domestica* L. (Diptera: Muscidae). Canad. J. Zool. 30:254-266.
Effectiveness increased (1) as loci of application approached body or head, (2) as the area of intact integument around the locus was increased. ABSTRACTS: RAE-B 41:208, Dec., 1953; TDB 50:69, Jan., 1953.

Fisher, R. W., joint author. *See* O'Brien, R. D., et al., 1958.

1875. Fisk, F. W. 1958 Toxicity of BHC in Milk to House Flies as Related to Butter Fat Content of the Milk. J. Econ. Ent. 51(4):560-561.
Controlled feeding and residue studies showed that increased butter fat in milk reduces the toxicity of lindane ingested by flies. This is because there is more solvent butter fat for a given amount of lindane, resulting in a lower percentage of lindane in the fat. Less lindane will therefore diffuse into the fly's haemocoele and nerve tissue. ABSTRACT: TDB 56: 109, Jan., 1959.

1876. Fletcher, O. K., Jr., Major, J. and **Cable, R. 1956** Studies on Fly Breeding in Sanitary Pit Privies in South Georgia. Am. J. Trop. Med. Hyg. 5(3):562-572.
Of 44,597 insects taken in 600 or more trap collections, only 137 (3/10 per cent) were *M. domestica.* Of 4,000,000 larvae taken from pits, none

were *Musca*. Darkness repels them. Suggests that properly-constructed pit privies need not necessarily be fly-proofed. ABSTRACTS: BA 31:284 (2865B), Jan., 1957; RAE-B 46:130, Aug., 1958; TDB 53:1391, Nov., 1956.

1877. Floch, H. 1954 Quelques remarques au sujet des "resistances" de divers insectes (notamment *Anopheles darlingi*) au DDT et autres insecticides. [Some remarks on the subject of "resistances" of various insects (notably *Anopheles darlingi*) to DDT and other insecticides.] Bull. Soc. Path. Exot. 47:555-560.

Notes that *M. domestica* is still conspicuously present in French Guiana. ABSTRACT: BA 30:3303(32998), Nov., 1956.

1878. Fluno, J. A. 1955 Insecticidal Phosphorus Compounds. Soap and Chem. Spec. 31(11):151-154, 203.

Tabulates results of tests against various insects of Bayer 13/59, DDVP and other phosphorus compounds. Were applied against houseflies as contact sprays, sugar baits, and residual films. ABSTRACT: RAE-B 45: 89, June, 1957.

1879. Forgash, A. J. and **Hansens, E. J. 1959** Cross Resistance in a Diazinon-Resistant Strain of *Musca domestica* (L). J. Econ. Ent. 52(4):733.

Most contemporary insecticides proved ineffective against Diazinon-resistant flies. ABSTRACT: RAE-B 48(11):196-197, Nov., 1960.

1880. Foster, A. C. 1951 Some Plant Responses to Certain Insecticides in the Soil. Circ. No. 862. U.S. Dept. Agric.

Concerns DDT, BHC, Chlordane, Toxaphene and Parathion. Heavy annual use of the first two, at least, presents the danger of reducing the productivity of soils in comparatively few years.

Fountaine, F. C., joint author. *See* Dahm, P. A., et al., 1950.

Fowler, K. S., joint author. *See* Lewis, S. E., et al., 1956.

Fraenkel, G., joint author. *See* Rasso, S. C., et al., 1954; Brookes, V. J., et al., 1958; Galun, R., et al., 1957.

Freeman, S. K., joint author. *See* Gersdorff, W. A., et al., 1959.

Frings, H., joint author. *See* Knipe, F. W., et al., 1953.

1881. Frontali, N. 1955 Attitivà della colinesterasi in ceppi sensibili e in ceppi resistenti al DDT di *Musca domestica* L. [Activity of cholinesterases in strains of *Musca domestica* L. susceptible to and resistant to DDT.] Riv. di Parassit. 16:241-252.

DDT had no inhibitory action on cholinesterase when applied to head homogenates. Variability within strains eliminated the idea of differences between strains. ABSTRACTS: TDB 54:239, Feb., 1957; TDB 53:678, May, 1956.

1882. Frontali, N. 1955 Attività della colinesterasi in ceppi sensibili e in ceppi resistenti al DDT di *Musca domestica* L. [Activity of cholinesterase in a strain of *Musca domestica* L. susceptible (to) and in a strain resistant to DDT.] Boll. Soc. Ital. Biol. Sper. 31(9-10):1312-1314.

A briefer presentation of the content of the preceding reference.

1883. Frontali, N. 1956 Attività della colinesterasi in ceppi sensibili e in ceppi resistenti al DDT di *Musca domestica* L. [Activity of cholinesterase in a strain of *Musca domestica* L. susceptible (to) and in a strain resistant to DDT.] Rend. Ist. Sup. Sanit. (Roma) 19:555-568.

Similar in content to the preceding reference.

1884. Frontali, N. 1957 Ichetoacidi in mosche domestiche intossicate con tetraetilpiroforaethyl. Boll. Soc. Ital. Biol. Sper. 33(10-11):1511-1512.

1885. Frontali, N. 1958 Acetylcholine Synthesis in the Housefly Head. J. Insect Physiol. 1(4):319-326.
The significance of acetylcholine in the nervous transmission of insects. Substance synthesized behaves chemically and chromatographically as acetylcholine.

1886. Frontali, N. 1959 La sintese de acetilcolina nella testa di mosca domestica. [The synthesis of acetylcholine in the head of *Musca domestica.*] Rend. Ist. Sup. Sanit. (Rome) 22(1):94-103.
See preceding reference.

Fryer, H. C., joint author. *See* Rai, L., et al., 1956.

1887. Fujito, S., Buei, K., Endo, T., Takumi, T., Natsume, T. and **Inaba, Y. 1959** [Field test of organic phosphorus insecticide.] Jap. J. Sanit. Zool. 10(3):197-201. (In Japanese, with English summary.)
Used oil solution, emulsion and dust to apply malathion. Fly-ribbons and fly-grills were employed for evaluation. In cesspools once-dusting controlled larvae until the next "throw" of feces.

1888. Fukuto, T. R. and **Metcalf, R. L. 1959** Insecticidal Activity of the Enantiomorphs of 0-Ethyl S-2-(Ethylthio)-ethyl Ethylphosphonothiolate. J. Econ. Ent. 52(4):739-740.
Dextro, *levo* and racemic forms of this substance were prepared, and examined for insecticidal properties. The *levo* isomer was about 10 times more toxic to certain insects (including the housefly) than the *dextro*, and activated fly-brain cholinesterase 11 times faster. ABSTRACT: RAE-B 48(11):197, Nov., 1960.

1889. Fukuto, T. R., Metcalfe, R. L. and **Winton, M. 1959** Alkylphosphonic Acid Esters as Insecticides. J. Econ. Ent. 52(6):1121-1127.
Testing of a large number of alkyl p-nitrophenyl alkyphosphonates and their sulphur analogues show them extremely effective against *M. domestica*. Toxicity is largely a function of the reactivity of the molecule. ABSTRACTS: BA 35:2230(25366), May, 1960; RAE-B 49:31-32, Feb., 1961.

Fukuto, T. R., joint author. *See* Metcalf, R. L., et al., 1958, 1959.

1890. Fullmer, O. H. and **Hoskins, W. M. 1951** Effects of DDT upon the Respiration of Susceptible and Resistant Houseflies. J. Econ. Ent. 44:858-870.
Addition of piperonyl cyclonene to DDT causes the respiratory curves for resistant flies to resemble those for susceptible ones, but the maxima reached are lower. At 35°C treated and untreated flies of both strains have the same total respiration, and DDT appears to have less influence than environment in causing death. ABSTRACTS: BA 26:2066, Aug., 1952; RAE-B 40:57, Apr., 1952; TDB 49:561, May, 1952.

1891. Fulton, R. A., Gelardo, R. P. and **Sullivan, W. N. 1952** Relative Efficiency of Methods of Applying Lindane in Enclosed Spaces. J. Econ. Ent. 45:540-541.
Four methods of applying lindane (suspension spray; lindane screen with fan; vaporizer; carbon dioxide-propelled solution) were all effective against insects susceptible to low concentrations of vapor. Table shows results against house flies. Screens treated with lindane should be protected with air filters to prevent an accumulation of dust, which hampers insecticidal activity. ABSTRACT: RAE-B 40:176, Nov., 1952.

1892. Fulton, R. A., Nelson, R. H. and **Smith, F. F. 1950** The Toxicity of Lindane Vapor to Insects. J. Econ. Ent. 43:223-224.
M. domestica one of four species tested. Housefly is suceptible to minute traces of this insecticide, which shows fumigating action when applied in an aerosol, or as a spray containing a water-dispersible powder. ABSTRACT: RAE-B 38:209, 1950.

1893. **Fulton, R. A., Sullivan, W. N.** and **Mangan, G. F. 1953** Effectiveness of Lindane Vaporizers. J. Econ. Ent. 46:639-641.
In ventilated rooms, even with air conditioning, flies could be controlled after 3 hours. Concentrations ranged from 0·11 microgram lindane per liter in a closed room to 0·044 microgram in a room with 3 windows and a door, open. ABSTRACTS: RAE-B 42:37, Mar., 1954; BA 28:955, Apr., 1954.

Fulton, R. A., joint author. *See* Fales, J. H., et al., 1951, 1956; McBride, O. C., et al., 1950.

1894. **Furman, D. P.** and **Bankowski, R. A. 1950** Absorption of Benzene Hexachloride in Poultry. J. Econ. Ent. 42:980-982.
This substance is absorbed by poultry tissues following treatment of roosts, but with continuous exposure, the tissues lose toxicity within 3 to 6 weeks after application.

1895. **Furman, D. P., Young, R. D.** and **Catts, E. P. 1959** *Hermetia illucens* (Linnaeus) as a Factor in the Natural Control of *Musca domestica* Linnaeus. J. Econ. Ent. 52(5):917-921.
As *Hermetia illucens* larvae increased in number, housefly larvae decreased (in mixed culture). *H. illucens* larvae were not predaceous though they ate dead flies. Introduction of *H. illucens* larvae into poultry manure prevented housefly breeding. ABSTRACTS: BA 35:772(8842), Feb., 1960; TDB 57:761, July, 1960.

Furman, D. P., joint author. *See* Bohart, R. M., et al., 1958.

Furmanska, W., joint author. *See* Mer, G. G., et al., 1953.

1896. **Fusco, R., Losco, G.** and **Peri, C. A. 1959** Esteri fosforici con funzione azo, azossi e azometinica. [Phosphoric esters with azoazoxy- and azomethine functions.] Gazz. Chim. Ital. 89(5/6):1282-1297.
Several compounds of three new series of phosphoric esters of phenols have been synthesized, showing strong fly killing properties and a low toxicity to mammals. ABSTRACT: BA 35:1033(11735), Feb., 1960.

Gagliani, M., joint author. *See* d'Alessandro, G., et al., 1951, 1952.

1897. **Gahan, J. B. 1953** Residual Toxicity of Some New Insecticides to House Flies under Conditions Prevalent in Egypt. J. Egyptian Pub. Health Ass. 28:181-196.
Dieldrin, chlordane, benzene hexachloride, methoxychlor and aldrin all proved promising in lab. tests for use on dried, clay surfaces of Egyptian houses. Practical tests made in small villages. All seemed better than DDT. ABSTRACT: TDB 51:998, Sept., 1954.

1898. **Gahan, J. B. 1953** Studies with Chlordane Sprays to Control House Flies in Egyptian Villages. J. Egyptian Pub' Health Ass. 28:197-210.
In village of Moneeb chlordane was used as a combination residual spray and larvicide. In small village (Ayoub) chlordane was sprayed on floors alone. Village of Ashtor served as untreated control. Results presented as tables and graphs. ABSTRACT: TDB 51:999, Sept., 1954.

1899. **Gahan, J. B., Anders, R. S., Highland, H.** and **Wilson, H. G. 1953** Baits for the Control of Resistant Flies. J. Econ. Ent. 46:965-969.
TEPP, sodium fluoroacetate, sodium arsenate, and sodium arsenite were the most effective of 15 water-soluble toxicants. Sodium arsenate gave good results with the least labor. Is probably the safest, also. ABSTRACTS: BA 28:2194, Sept., 1954; RAE-B 42:132, Sept., 1954; TDB 51:851, Aug., 1954.

1900. **Gahan, J. B.** and **Weir, J. M. 1950** [U.S. Naval Medical Research Unit No. 3, Cairo, Egypt.] Houseflies Resistant to Benzene Hexachloride. Science, 111:651-652.
Had given temporary benefit against DDT-resistant flies in village of

Quarantil. Controlled tests with *M. domestica vicina* proved the existence of a BHC-resistant strain, beyond a doubt.

1901. **Gahan, J. B., Wilson, H. G.** and **McDuffie, W. C. 1954** Organic Phosphorus Compounds as Toxicants in House Fly Baits. J. Econ. Ent. 47:335-340.
Malathion, Diazinon and Bayer $\frac{L13}{59}$ were evaluated against DDT resistant flies. Diazinon and $\frac{L13}{59}$ were more toxic than malathion. $\frac{L13}{59}$ and malathion were less acceptable to flies than others. ABSTRACTS: BA 29:207, Jan., 1955; RAE-B 43:25, Feb., 1955; TDB 51:1202, Nov., 1954.

1902. **Gahan, J. B., Wilson, H. G., Keller, J. C.** and **Smith, C. 1957** Organic Phosphorus Insecticides as Residual Sprays for the Control of House Flies. J. Econ. Ent. 50:789-793.
Promising residual sprays include Dow E-14, diazinon, Bayer $\frac{21}{199}$ and EPN. Diazinon and malathion were somewhat more effective under Nebraska conditions than in Florida. ABSTRACTS: BA 32:1796(21343), June, 1958; TDB 55:832, July, 1958.

Gahan, J. B., joint author. *See* Wilson, H. G., et al., 1953, 1957, 1959.

Gaines, J. C., joint author. *See* Robertson, R. S. Jr., et al., 1951.

1903. **Galbiati, F. 1956** Il "Dition", nuovo insetticida per uso civile. [Dition, new Insecticide for Civil Use.] Riv. di Malariol. 35:59-71.
Dition is a synthesized phosphoric ester, persistent because it is the least volatile of those tested and the most stable. It may be applied as a spray or as a paste. ABSTRACT: TDB 54:240, Feb., 1957.

1903a. **Galun, R.** and **Fraenkel, G. 1957** Physiological Effects of Carbohydrates in the Nutrition of a Mosquito, *Aedes aegypti* and Two Flies, *Sarcophaga bullata* and *Musca domestica*. J. Cell. and Comp. Physiol. 50(1):1-23.
Adult insects possess the carbohydrases necessary for hydrolysis of most of the utilized oligo- and poly-saccharides and glycosides. Larvae of *M. domestica* possess only invertase and acquire the others during the pupal stage.

Gamal-Eddin, F. M., joint author. *See* Hafez, M., et al., 1959.

1904. **Gasser, R. 1953** Über ein neues Insektizid mit beitem Wirkungsspektrum. [Concerning a new insecticide with broad spectrum effect.] Zeit. Naturforsch. 8b:225-232.
Refers to the phosphorus insecticide, Diazinon, by Swiss Geigy Co. Seems promising against houseflies resistant to chlorinated hydrocarbons. ABSTRACT: TDB 51:746, Jul., 1954.

1905. **Gatterdam, P. E., Casida, J. E.** and **Stoutamire, D. W. 1959** Relation of Structure to Stability, Antiesterase Activity and Toxicity with Substituted-Vinyl Phosphate Insecticides. J. Econ. Ent. 52:270-276.
Musca domestica was used in toxicity tests of these compounds, many of which proved highly toxic. ABSTRACT: RAE-B 48:122, July, 1960.

1906. **Gaud, J., Dupuy, R.** and **N'Haili, A. 1955** Mouches capturees dans l'agglomeration urbanede Rabat-Sali. [Flies captured in the urban area of Rabat, Morocco.] Bull. Inst. Hyg. Maroc. 15(3-4):209-219.
Most of the thousands of flies trapped were either *M. domestica vicina* or *M. sorbens*. ABSTRACT: TDB 55:109, Jan., 1958.

1907. Gaud, J. and **Faure, P. 1951** Effet de la lutte antimouches sur l'incidence des maladies oculaires dans le sud marocain. [Effect of the war against flies on the incidence of eye disease on Southern Morocco.] Bull. Soc. Path. Exot. 47(7/8):446-448.
Use of BHC reduced flies 99 per cent. Ocular disease in children dropped from 88 per cent to 36 per cent. Other species of insects were reduced also. ABSTRACT: BA 26:1527, June, 1952.

1908. Gaud, J., Laurent, J. and **Faure, P. 1954** Biologie de *Musca sorbens* et rôle vecteur probable de cette espèce en pathologie humaine au Maroc. [The biology of *Musca sorbens* and the probable vector role of this species in human pathology in Morocco.] Bull. Soc. Path. Exot. (Paris) 47:97-101.
Life history and behavior of *M. sorbens*, also of *M. domestica vicina* and *M. domestica cuthbertsoni*. All are called "houseflies". *M. sorbens* the more important vector, if numbers are equal. ABSTRACT: RAE-B 43: 178, Nov., 1955.

1909. Gaud, J., Laurent, J. and **Faure, P. 1956** Arthropodes vecteurs possibles de maladies au Maroc. [Possible arthropod vectors of illness in Morocco.] Maroc. Med. 35:1259-1266.
Insects of medical importance are reviewed. A table is given of the several species of flies. ABSTRACT: TDB 54:625, May, 1957.

1910. Gaud, J., Maurice, A., Faure, P. and **Lalu, P. 1950** Expériences de lutte contre les mouches au Maroc. [Experiences in the campaign against flies in Morocco.] Bull. Inst. Hyg. Maroc. 10(1/2):55-71.
Treats of aerial spraying (BHC), insecticidal mists from ground vehicles, weekly spraying of local sites, etc. The readiness with which flies develop resistance, indicates need for improved sanitary conditions in rural communities. ABSTRACT: TDB 49:729, Jul., 1952.

1911. Gebel'skii, S. G. 1954 [Effect of dose of hexachloro-cyclohexane and of duration of exposure on intoxication of housefly.] Med. Parazit. (Moskva) 1:58-60. (In Russian.)

Gelardo, R. P., joint author. *See* Fulton, R. A., et al., 1952.

Gentbe, S. I., joint author. *See* Gersdorff, W. A., et al., 1950, 1952.

1912. Georgopoulos, G. 1951 [On the effects of insecticide chlordane (Octa-Clor) on the house fly (*Musca domestica* L.).] Arch. Hyg. Athens 4(1-4):108-119. (In Greek, with English summary.)
Gave good control in first season of use. ABSTRACT: TDB 49:653, June, 1952.

1913. Gerberich, J. B. 1952 The Housefly (*Musca domestica* Linn.) as a Vector of *Salmonella pullorum* (Retteger) Bergy, the Agent of White Diarrhea of Chickens. Ohio J. Sci. 52:287-290.
S. pullorum, when ingested by housefly larvae, survived through metamorphosis. Established in chicken, by housefly vectorship, was later recovered from chicken's feces. ABSTRACT: RAE-B 42:81, June, 1954.

1914. Gerolt, P. 1957 Improved Persistence of Dieldrin Deposits on Sorptive Mud Surface. Nature 180:394-395.
A wettable powder, produced by grinding solidified melts of dieldrin plus powdered coumarone resin, when applied to such surfaces, greatly prolonged the duration of the residual effect. Tested by exposing houseflies 30 sec. to the residues. ABSTRACT: TDB 54:1479, Dec., 1957.

1915. Gerolt, P. 1959 "Vapour Toxicity" of Solid Insecticides. Nature 183: 1121-1122.
Experiments utilized *Aedes*, *Drosophila* and *M. domestica*. Flies were exposed to dry residue of hexane solution (various dilutions) of over-

night vapor deposit. Results were compared with those simultaneously obtained by exposure to known amounts of insecticide.

1916. **Gersdorff, W. A. 1956** Effect of Acetone on Toxicity in Pyrethrum and Allethrin Space Sprays. J. Econ. Ent. 49:849-851.
When calculated by usual procedure, pyrethrins in mixtures containing acetone were more toxic to *M. domestica* than pyrethrins in kerosene. But due to the volatile properties of acetone, valid comparisons should use standards with same proportion of acetone to kerosene or allow for differences in proportion. ABSTRACT: RAE-B 46:23, Feb., 1958.

1917. **Gersdorff, W. A., Freeman, S. K.** and **Piquett, P. G. 1959** Some Barthrin Isomers and their Toxicity to Houseflies in Space Sprays. J. Agric. Fd. Chem. 7(8):548-550; 2 graphs, 18 refs.
Retention of toxicity in the spray mixture was considered good. Knockdown values were excellent at highly toxic concentrations, but proved much inferior to allethrin at comparable concentrations. ABSTRACT: RAE-B 48:175, Oct., 1960.

1918. **Gersdorff, W. A.** and **Mitlin, N. 1950** Relative Toxicity to House Flies of Tetra-n-Propyl Dithionopyrophosphate. J. Econ. Ent. 43:562.
Proved about 3 times as toxic as pyrethrins. Further tests desirable. ABSTRACT: RAE-B 39:64, Apr., 1951.

1919. **Gersdorff, W. A.** and **Mitlin, N. 1951** Relative Toxicity of Allethrin Analogs to House Flies. J. Econ. Ent. 44:70-73.
Compounds tested ranged in toxicity from one-fifth to one-and-a-half times that of pyrethrins. ABSTRACTS: RAE-B 39:136, 1951; TDB 48:932, Oct., 1951.

1920. **Gersdorff, W. A.** and **Mitlin, N. 1951** Joint Toxic Action against House Flies in Mixtures of Parathion and its Methyl Homolog. J. Econ. Ent. 44:474-476.
Homolog is 68 per cent as toxic as parathion. Toxicity of the various mixtures used fell at 72, 88, and 91 per cent. ABSTRACTS: BA 26:1522, June, 1952; RAE-B 39:182, Nov., 1951; TDB 48:1155, Dec., 1951.

1921. **Gersdorff, W. A.** and **Mitlin, N. 1952** Relative Toxicity to House Flies of Scabrin and its Mixtures with some Pyrethrum Synergists. J. Econ. Ent. 45:519-523.
Deals with extractives of the root of *Heliopsis scabra*, which are nearly as toxic as pyrethrins. Their chemical nature. Report of experiments. ABSTRACT: RAE-B 40:175, Nov., 1952.

1922. **Gersdorff, W. A.** and **Mitlin, N. 1952.** A Bioassay of Some Stereoisomeric Constituents of Allethrin. J. Wash. Acad. Sci. 42:313-318.
An evaluation of their relative toxicity and that of their mixtures when applied as contact insecticides in refined kerosene on the Campbell turntable. *M. domestica*, the test insect. ABSTRACTS: RAE-B 41:125, Aug., 1953; BA 27:1032, Apr., 1953.

1923. **Gersdorff, W. A.** and **Mitlin, N. 1953** Relative Toxicity of Furethrin, its Natural Acid Isomer, and Pyrethrins to House Flies. J. Econ. Ent. 45:849-850.
Furethrin proved as toxic as natural pyrethrins, and one-third as toxic as allethrin. A d-trano acid isomer was twice as toxic as furethrin. Both had high knock-down value. ABSTRACTS: RAE-B 41:74, May, 1953; TDB 50:352, Apr., 1953.

1924. **Gersdorff, W. A.** and **Mitlin, N. 1953** The Relative Toxicity to House Flies of the Methyl and Ethyl Analogs of Allethrin. J. Econ. Ent. 46:945-948.
In corresponding esters a hydrogenation of an alkyl side chain increases toxicity. ABSTRACTS: BA 28:1927, Aug., 1954; RAE-B 42:131, Sept., 1954.

1925. **Gersdorff, W. A.** and **Mitlin, N. 1953** Effect of Molecular Configuration on Relative Toxicity to House Flies as Demonstrated with the Four *Trans* Isomers of Allethrin. J. Econ. Ent. 46:999-1003.
Molecular configuration is of great importance in regard to the insecticidal action of compound, e.g., the d-allethrolene d-trans chrysanthemumic acid ester is 150 times as toxic as the least toxic λ-λ isomer. ABSTRACTS: BA 28:1928, Aug., 1954; RAE-B 42:135, Sept., 1954.

1926. **Gersdorff, W. A.** and **Mitlin, N. 1954** The Relative Toxicity of Some Aryl Analogs of Allethrin to House Flies. J. Econ. Ent. 47:888-890.
Five analogs differed from allethrin in that the side chains of the keto alcoholic components were aryl groups instead of the allyl group. All had considerable knock-down value, but were less toxic than allethrin and pyrethrins. ABSTRACTS: BA 29:2005, Aug., 1955; RAE-B 43:157, Oct., 1955; TDB 52:303, Mar., 1955.

1927. **Gersdorff, W. A., Mitlin, N.** and **Beroza, M. 1954** Comparative Effects of Sesamolin, Sesamin, and Sesamol in Pyrethrum and Allethrin Mixtures as House Fly Sprays. J. Econ. Ent. 47:839-842.
Sesamol and a certain sterol had no synergistic effect with pyrethrins. Sesamolin was more effective as a synergist with pyrethrins than sesamin. Both synergize allethrin but slightly. ABSTRACTS: BA 29:2005, Aug., 1955;; RAE-B 43:157, Oct., 1955; TDB 52:303, Mar., 1955.

1928. **Gersdorff, W. A., Mitlin, N.** and **Gertler, S. I. 1950** Preliminary Tests with certain N Substituted p-Bromobenzene Sulfonamides and p-Toluenesulfonamides for Synergistic Action in Pyrethrum Fly Sprays. U.S. Dept. of Agr. E-805, 1-4. 5 pp. mimeo.
Toxicity tables show pyrethrins equivalents.

1929. **Gersdorff, W. A., Mitlin, N.** and **Gertler, S. I. 1952** Preliminary Tests with Certain N-Substituted Piperonylamides for Synergistic Action in Allethrin Fly Sprays. U.S. Dept. of Agr. E-848, 1-6. Processed.
Compounds were tested in oil sprays, alone, and in mixtures with allethrin.

1930. **Gersdorff, W. A., Mitlin, N.** and **Nelson, R. H. 1952** The Relative Toxicity to House Flies of Endrin and Isodrin in Kerosene Sprays. U.S. Dept. of Agr. (Agr. Res. Admin. Bur. of Ent. and Plant Quarantine): E-843, 1-4. Processed.
Endrin proved 67 per cent as toxic as aldrin; isodrin, 61 per cent.

1931. **Gersdorff, W. A., Mitlin, N.** and **Nelson, R. H. 1954** New House Fly Insecticides. Soap and Sanit. Chem. 30:133, 159.
The organic phosphorus compounds are compared. Diazinon proves to be very effective. ABSTRACTS: RAE-B 42:109, Jul., 1954; TDB 51:746, July, 1954.

1932. **Gersdorff, W. A., Mitlin, N.** and **Nelson, R. H. 1955** The Comparative Effect of a Sulfone in Pyrethrum and Allethrin Mixtures as House Fly Sprays. J. Econ. Ent. 48:9-11.
Sulfone synergized both insecticides; the pyrethrins much more powerfully than the allethrin. Sulfone alone causes little knock down or mortality. ABSTRACTS: BA 29:2717, Nov., 1955; RAE-B 44:19, Feb., 1956; TDB 52:842, Aug., 1955.

1933. **Gersdorff, W. A., Mitlin, N.** and **Piquett, P. G. 1957** The Relative Toxicity, Synergistic Activity, and Knockdown Effectiveness of Mixtures of Piperonyl Butoxide with Allethrin and Its Trans Fraction and Isomers as House Fly Sprays. J. Econ. Ent. 50:150-156.
Synergism with respect to mortality occurred in all the mixtures. ABSTRACTS: RAE-B 46:79, May, 1958; BA 31:2710(29723), Sept., 1957; TDB 54:1138, Sept., 1957.

1934. **Gersdorff, W. A., Nelson, R. H.** and **Mitlin, N. 1950** Chlorinated Hydrocarbons. The relative Toxicity of Heptachlor, Aldrin and Dieldrin to House Flies when Applied as Space Sprays using Campbell Turntable Method. Soap and Sanit. Chem. 26(4):137, 139.
Results agreed with those of previous workers. None of these insecticides caused an appreciable knockdown. ABSTRACT: RAE-B 39:85, May, 1951.

1935. **Gersdorff, W. A., Nelson, R. H.** and **Mitlin, N. 1952** The Relative Effect of Several Pyrethrum Synergists in Fly Sprays Containing Allethrin. J. Econ. Ent. 44(6):921-927.
Experimental results given in toxicity tables. Knockdown is slightly slower with allethrin than with natural pyrethrins. ABSTRACTS: BA 26: 2067, Aug., 1952; RAE-B 40:59, Apr., 1952; TDB 49:563, May, 1952.

1936. **Gersdorff, W. A., Nelson, R. H.** and **Mitlin, N. 1952** Sulfoxide and 3,4-Methylenedioxybenzyl *N*-Propyl Ether in Pyrethrum and Allethrin Mixtures as House Fly Sprays. J. Econ. Ent. 45:905-908.
Sulfoxide the better synergist. Pyrethrum responded better than allethrin to both synergists; gave knock-down in 20 minutes. ABSTRACTS: RAE-B 41:101, July, 1953; TDB 50:574, June, 1953.

1937. **Gersdorff, W. A.** and **Piquett, P. G. 1955** A Comparison of Cyclethrin, Allethrin, Pyrethrins and Mixtures of Piperonyl Butoxide or Sulfoxide with them in House Fly Sprays. J. Econ. Ent. 48:407-409.
Cyclethrin has less toxic effect than allethrin, more than pyrethrins. It is synergized more strongly than allethrin, but not so strongly as pyrethrins. When synergized, pyrethrins are more effective than allethrin in knockdown. ABSTRACTS: BA 30:1154(11762), Apr., 1956; RAE-B 44: 101, July, 1956; TDB 53:111, Jan., 1956.

1938. **Gersdorff, W. A.** and **Piquett, P. G. 1957** Comparative Effects of Piperettine in Pyrethrum and Allethrin Mixtures as House Fly Sprays. J. Econ. Ent. 50:164-166.
One of the best commercial synergists with pyrethrins; less effective with allethrin. Knockdown complete, in sprays, but none at all when used alone. ABSTRACTS: BA 31:2710(29724), Sept., 1957; RAE-B 46:80, May, 1958; TDB 54:1139, Sept., 1957.

1939. **Gersdorff, W. A.** and **Piquett, P. G. 1958** Effect of Molecular Configuration on Relative Toxicity to House Flies as Demonstrated with the Four *cis* Isomers of Allethrin. J. Econ. Ent. 51(2):181-184.
Effect of a change in each component simultaneously from the λ to the *d* form was equal to the product of the separate effects. All stereoisomers caused complete or nearly complete knockdown at concentrations used. ABSTRACT: RAE-B 47:103, July, 1959.

1940. **Gersdorff, W. A.** and **Piquett, P. G. 1958** Relative Toxicity and Intensity of Synergism of Mixtures of Piperonyl Butoxide with the *cis* isomers of Allethrin as House Fly Sprays. J. Econ. Ent. 51(5):675-678.
Mixtures of piperonyl butoxide with *trans* allethrin, *cis* allethrin, and allethrin were tested, comparatively, against *M. domestica.* Synergism was demonstrated in all three. All sprays caused more than 90 per cent knockdown.

1941. **Gersdorff, W. A.** and **Piquett, P. G. 1958** The Toxicity of Certain Alkobenzyl Esters of Chrysanthemumic Acid in Comparison with Allethrin and Pyrethrins as House Fly Sprays. J. Econ. Ent. 51(6): 810-812.
The *m*-form of these esters proved far more toxic than the *o*- and *p*-derivatives. All these compounds have inferior knockdown value. ABSTRACT: BA 33:2257(27791), May, 1959.

1942. Gersdorff, W. A. and **Piquett, P. G. 1958** Relative Toxicity to House Flies of the *cis*- and *trans*-Fractions of Allethrin and the Intensity of Synergism of their Mixtures with Piperonyl Butoxide. J. Econ. Ent. 51:76-78.

These two mixtures, also allethrin-piperonyl butoxide mixture, are nearly 3 times as toxic as pyrethroids alone; both fractions are synergized to the same degree as allethrin. ABSTRACT: BA 32:1791(21280), June, 1958.

1943. Gersdorff, W. A. and **Piquett, P. G. 1959** Relative Toxicity, Knockdown Effectiveness and Intensity of Synergism of Mixtures of Piperonyl Butoxide with Barthrin and Some of Its Isomers. J. Econ. Ent. 52(6):1168-1171.

DC-50 estimates indicate that *barthrin* is about 6 per cent as effective as allethrin in knockdown action. Its value, relative to allethrin, as a knockdown agent against houseflies is therefore about one-fifth to one-sixth its value as a lethal agent. ABSTRACTS: BA 35:2230(25368), May, 1960; RAE-B 49:32, Feb., 1961.

1944. Gersdorff, W. A. and **Piquett, P. G. 1959** Toxicity of Some Alkyl Benzyl Esters of Chrysanthemumic Acid and Two Related Esters in Comparison with Allethrin and Pyrethrins as House Fly Sprays. J. Econ. Ent. 52(3):521.

Twelve related synthetic compounds proved to have knockdown ability inferior to that of allethrin or pyrethrins. Excellent at highly toxic concentrations, however. ABSTRACTS: RAE-B 48:168, Sept., 1960; RAE-B 48:54, Mar., 1960.

1945. Gersdorff, W. A. and **Piquett, P. G. 1959** Toxicity of the Piperonyl Ester of Chrysanthemumic Acid and some of its Derivatives in Comparison with Allethrin and Pyrethrins as House Fly Sprays. J. Econ. Ent. 52(1):85-88.

Eighteen related synthetic compounds had excellent knockdown values at high concentration, but were nevertheless inferior to allethrin and pyrethrins. Barthrin was one of the more toxic compounds tested. ABSTRACTS: BA 33:2894-2895(35372), June, 1959; RAE-B 48:96, May, 1960.

1946. Gersdorff, W. A., Piquett, P. G. and **Beroza, M. 1956** Comparative Synergistic Effects of Synthetic 3, 4-Methylenedioxyphenoxy Compounds in Pyrethrum and Allethrin Fly Sprays. J. Agric. Fd. Chem. 4:858-862.

Forty-three compounds were shown to be synergistic with each insecticide. Eighteen of these rated so high as to achieve six times the expected effectiveness of pyrethrum alone or three times that expected for allethrin alone. ABSTRACT: RAE-B 45:214, Dec., 1957.

1947. Gersdorff, W. A., Piquett, P. G. and **Beroza, M.** 1957. Comparative Effects of the Optical Forms of Epiasarinin, Asarinin, and Sesamin in Pyrethrum Mixtures as House Fly Sprays. J. Econ. Ent. 50:409-411.

At proportions tested, three forms of sesamin proved equally effective synergists with pyrethrins. Asarinin and epiasarinin somewhat less so. Isomers alone caused no knockdown; with pyrethrins knockdown was complete. ABSTRACTS: BA 31:3583(39765), Dec., 1957; RAE-B 46:137, Sept., 1958; TDB 54:1477, Dec., 1957.

1948. Gersdorff, W. A., Piquett, P. G. and **Beroza, M. 1957** The Intensity of Synergism of Some Methylenedioxyphenyl Acetals with Pyrethrins or Allethrin in House Fly Sprays. J. Econ. Ent. 50:406-409.

Determination was of mixtures in oil sprays, containing ten times as much synergist as insecticide. Knockdown was high, for all mixtures. ABSTRACTS: BA 31:3583(39764), Dec., 1957; RAE-B 46:137, Sept., 1958; TDB 54:1477, Dec., 1957.

1949. **Gersdorff, W. A., Piquett, P. G.** and **Mitlin, N. 1956** The Crystalline Isomer of Allethrin as a Standard of Comparison for Fly Sprays. J. Econ. Ent. 49:450-453.
Considered as a possible replacement for pyrethrins as standard, but proved to be no better biometrically than several samples of purified allethrin. ABSTRACT: RAE-B 45:141, Sept., 1957.

1950. **Gersdorff, W. A., Piquett, P. G.** and **Nelson, R. H. 1955** The Relative Toxicity to House Flies of Pirazinon, Am. Cyanamid 4124 and Malathion in Comparison with Parathion and Pyrethrins. J. Econ. Ent. 48:680-681.
The three compounds were, respectively, 0·36, 0·60 and 0·062 as toxic as parathion and 12, 20 and 2 times as toxic as pyrethrins. Caused only negligible knockdown of flies in 25 minutes. ABSTRACTS: BA 30:2075 (20850), July, 1956; RAE-B 45:4, Jan., 1957; TDB 53:679, May, 1956.

Gersdorff, W. A., joint author. *See* LaForge, et al., 1952; Lowman, M. S., et al., 1954; Mitlin, N., et al., 1951, 1953, 1954; Nelson, R. H., et al., 1950; Piquett, P. G., et al., 1958, 1959.

Gertler, S. I., joint author. *See* Mitlin, N., et al., 1951.

1951. **Gharpure, P. B.** and **Perti, S. L. 1957** Comparative Efficacy and Economics of Pyrethrins and Mixtures of Pyrethrins and Piperonyl Butoxide as Space Sprays against Flies and Mosquitoes. J. Sci. and Indust. Res. 16A(10):469-471.
Pyrethrins are more economical than mixtures of pyrethrins and piperonyl butoxide, but the mixtures effect comparable levels of high mortality. ABSTRACT: TDB 55:336, Mar., 1958.

1952. **Gibson, A. W., 1955** Fly Control Program on the Illinois State Fairground. Pest Control 23(4):14-16, 54.
The fly problem was met with the use of a malathion spray and bait pyrethrum space sprays, plus strict sanitary measures begun two weeks prior to the fair and continued through its duration. ABSTRACT: BA 30:308(3079), Jan., 1956.

1953. **Gilbert, I. H., Couch, M. D.** and **McDuffie, W. C. 1953** Development of Resistance to Insecticides in Natural Populations of House Flies. J. Econ. Ent. 46(1):48-50.
Studied change of resistance in natural populations of flies during 1950-1951. General conclusion: Resistance increases as the season progresses. ABSTRACTS: RAE-B 41:137, Sept., 1953; TDB 50:861, Sept., 1953.

1954. **Gil Collado, J. 1956** Problemas sanitarios derivados de la resistencia de los insectos a los insecticidas. [Sanitary problems arising out of the resistance of insects to insecticides.] An. Acad. Farm. (Madrid) 22:158-166.

1955. **Gill, G. D. 1955** Filth Flies of Central Alaska. J. Econ. Ent. 48:648-653.
M. domestica not recorded in this survey, though *Muscina assimilis* and *Muscina stabulans* were; also 6 species of *Fannia*. Total list impressive. ABSTRACTS: BA 30:2075(20851), July, 1956; RAE-B 45:2, Jan., 1957.

Glazunova, A. Y. joint author. *See* Shura-Bura, B. L., et al., 1956, 1958.

1956. **Gliwic, W. 1950** Zastosowanie kilku preparató DDT krajwey produkciji oraz kilku ptynnych substancji chemuznych do walki z muchami. [The Application of Several Preparations of DDT of Native Pro-

duction and Several Liquid Chemical Substances for the Combat of Flies.] Proc. 2nd Meeting Polish Parasitol. Soc. (1950):104.
The substances used were on the whole effective. ABSTRACT: BA 26: 2342(26402), Sept., 1952.

Goette, M. B., joint author. *See* Sumerford, W. T., et al., 1951.

1957. Goodhue, L. D. 1953 Some Observations on the Use of Isoparaffin Oils as Carriers for Agricultural Chemicals. J. Econ. Ent. 46:986-988.
Isoparaffin oils such as soltrol 140 and 180 are good carriers of insecticides since they spread out well and also because they are nontoxic when applied to the skin of mammals. ABSTRACT: RAE-B 42:134 Sept., 1954.

1958. Goodhue, L. D. and **Stansbury, R. E. 1953** Some New Fly Repellents from Laboratory Screening Tests. J. Econ. Ent. 46:982-985.
A testing of 500 chemicals. Four gave high repellent action against house flies. These were then tested against *Stomoxys.* Most effective compound was di-n-propyl isocincho-meronate. ABSTRACTS: BA 28: 1928, Aug., 1954; RAE-B 42:132, Sept., 1954.

Goodhue, L. D., joint author. *See* Howell, D. E., et al., 1955.

1959. Goodman, J. G. 1958 Effects of Feeding Boron to Hens to Prevent Flies. Progr. Rep. Ser. Ala. Agric. Exp. Sta. 71:2.
Boron, fed in small quantities as Polybor-3, to laying hens greatly reduced fly development in the manure. Feeding for 14 weeks at the rate of 1-2 ounces per 100 pounds of feed, does not cause boron to appear in the eggs, or affect hatchability. Hatchability *was reduced* by feeding for 22 weeks. ABSTRACT: RAE-B 49:188, Aug., 1961.

1960. Goodwin, W. J. and **Gressette, F. R. 1956** Residual House Fly Control in Dairy and Beef Barns in South Carolina. J. Econ. Ent. 49:622-624.
Diazinon, chlorthion, Am. Cyanamid 4124, malrin and Bayer 21/199 gave satisfactory control for 4 weeks or longer. Chlorthion was best in beef barns where sanitation was poor. ABSTRACTS: BA 31:931(9322), Mar., 1957; RAE-B 45:174, Nov., 1957.

1961. Goodwin, W. J. and **Schwardt, H. H. 1953** House Fly Control in New York State Dairy Barns. J. Econ. Ent. 46:299-301.
For a two-year period (1951-1952), four methods of application were used to apply 9 compounds in 172 dairy barns. None gave effective control with overall sprays or spot sprays. Strip methods gave control for 2 years, if dieldrin had not been used before. ABSTRACTS: RAE-B 41:175, Nov., 1953; TDB 50:863, Sept., 1953.

1962. Goodwin-Bailey, K. F. and **Davies, M. 1954** Occurrence of Dieldrin-Resistance in Wild *Musca domestica* L. in England. [Correspondence.] Nature, 173:216-217. January 30.
Wild flies, exposed to DDT or *gamma* BHC over 4 years of control work, proved resistant to these insecticides; also to dieldrin, which had not been used. ABSTRACT: TDB 51:654, June, 1954.

1963. Gorbov, V. A. and **Lineva, V. A. 1958** [Sanitary control of inhabited areas in prevention of infections and infestations.] Gig. I Sanit. (Moscow) 23(1):47-50. (In Russian.)

Gordon, R. M., joint author. *See* Muirhead-Thomson, R. C., et al., 1952.

Govind, R., joint author. *See* Mukerjea, T. D., et al., 1958.

Graham, H. M., joint author. *See* Webb, J. E., et al., 1956.

Grainger, M. M., joint author. *See* McCawley, R. H., Jr., et al., 1955.

1964. **Grandori, L.** and **Facetti, D. 1952** Ulteriore sperimentazione sugli effeti dell'etiluretano sulla mosca domestica adulta. [Further experiments on the effect of ethyl urethane on adult domestic flies.] Boll. Zool. Agr. Bachic. 18(1):21-28. (Summaries in French and English.)
Injection of ethyl urethane into flies causes no irreversible narcotic effect, though an eventual toxic action is not excluded. ABSTRACTS: RAE-B 42:192, Dec., 1954; TDB 52:587, June, 1955.

1965. **Grandori, L., Reali, G.** and **Facetti, D. 1952** Effetti narcotici dell' etiluretano sulla mosca domestica adulta. [Narcotic effects of urethane on adult domestic flies.] Boll. Zool. Agr. Bachic. 18(1):3-10. (Summaries in French and English.)
Physical, pharmacological and biochemical properties of ethylurethane are reviewed. The effect on flies is probably due to penetration of the respiratory system by vapors from the sublimation of the compound. ABSTRACTS: RAE-B 42:192, Dec., 1954; TDB 52:587, June, 1955.

Grandori, L., joint author. *See* Grandori, R., et al., 1951.

1966. **Grandori, R., Grandori, L.** and **Facetti, D. 1951** Effetti tossici selettivi della carbacolina su alcune specie di insetti. [Selective toxic effects of carbacholine on some species of insects.] Boll. Zool. Agr. Bachic. 17(2):123-129. Turin.
Physical and chemical properties of carbacholine are given. It acts as a selective insecticide after the manner of urethane 19258 (Dimetan), and is as toxic to resistant flies as to normal ones. ABSTRACT: RAE-B 42:130, Sept., 1954.

1967. **Granett, P.** and **Shea, W. D. 1955** Laboratory Tests of Diazinon-butoxy Polypropylene Glycol Residues. J. Econ. Ent. 48:487-488.
Applications of butoxy polypropylene glycol had no residual toxicity under the conditions of these tests. Under altered conditions, perhaps. ABSTRACT: BA 30:1154(11763), Apr., 1956.

Granett, P., joint author. *See* Helrich, K., et al., 1958; Hansens, E. J., et al., 1955.

1968. **Gratwick, M. 1957** The Uptake of DDT andot her Lipophilic Particles by Blowflies Walking over Deposits. Bull. Ent. Res. 48(4):733-740. London.
Used *Phormia* (*Protophormia*) *terrae-novae* RD as the test species. It is probable that these observations would apply equally well to *M. domestica*. ABSTRACT: RAE-B 46:28, Feb., 1958.

1969. **Gratz, N. G., Rosen, P.** and **Ascher, K. R. S. 1957** Houseflies in Israel —II. A Preliminary Survey of Species breeding in Rural Areas. Riv. di Parassit. 18(2):123-127. April.
Flies collected at breeding sites were mainly *Musca domestica*. A total of 13 species was recorded. ABSTRACT: TDB 54:1253, Oct., 1957.

Gratz, N. G., joint author. *See* Rosen, P., et al., 1959.

1970. **Gray, H. E. 1952** Insects Observed at Sea. J. Econ. Ent. 45:1081.
Large numbers of *M. domestica* appeared on Naval vessel 22 miles from Virginia shore for 2 successive days at 5:00-6:00 P.M. Clean ship, no breeding aboard, no ships in the area. Conclusion: Insects have more flight stamina than is usually acknowledged. ABSTRACTS: RAE-A 41: 211, July, 1953; RAE-B 41:107, July, 1953; TDB 50:573, June, 1953.

Gray, T. M., joint author. *See* Knutson, H., et al., 1958.

1971. **Green, A. A. 1951** Blowflies in Slaughterhouses. J. Roy. Sanit. Inst. 71:138-146. (Reprint, pp. 1-6).
Concerns *Lucilia*, *Calliphora* and *Phormia*. For flies bred on premises, improved sanitation; for flies attracted to the site by slaughtering conditions, chemical control. ABSTRACT: BA 27:1686, June, 1953.

1972. Green, A. A. 1951 The Control of Blowflies Infesting Slaughter-Houses I. Field Observations on the Habits of Blowflies. Ann. Appl. Biol. 38(2):475-494.
Covers all aspects of fly life-history against the background of slaughter house conditions. Recommendations for prevention and/or control. ABSTRACT: TDB 51:1000, Sept., 1954.

1973. Green, A. A. 1952 The Blowfly Problem. J. Roy. Sanit. Inst. 72(5): 621-626.
Touches on blowflies as vectors of disease, their breeding places, control and responsibility for control. Concerted action by all authorities desirable.

1974. Green, A. A. 1953 The Control of Blowflies Infesting Slaughter-Houses. II. Large Scale Experiments. Ann. Appl. Biol. 40:705-716.
Larvicide treatment was impractical but daily surface treatment with DDT reduced the fly number. A method for its evaluation in refuse and the surrounding area was devised. ABSTRACTS: BA 28:1455(14881) June, 1954; RAE-B 42:195, Dec., 1954; TDB 51:1000, Sept., 1954.

1975. Green, A. A. and **Kane, J. 1954** The Control of Blowflies Infesting Slaughter-Houses. III. Large-Scale Experiments at a Domestic-Refuse Depot. Ann. Appl. Biol. 41:165-173.
DDT and BHC were used for the control of blowfly larvae from depot refuse railcars. A trap was devised to test the effectiveness and overall control, which amounted to 99 per cent. ABSTRACTS: BA 28:2194(22289), Sept., 1954; RAE-B 43:37, Mar., 1955; TDB 51:1000, Sept., 1954.

Green, N., joint author. *See* LaForge, F. B., et al., 1952; Matsui, M., et al., 1952.

1976. Greenberg, B. 1955 Fecundity and Cold Survival of the House Fly. J. Econ. Ent. 48:654-657.
Fecundity: Paired house flies isolated from others compared well with controls as to fecundity and longevity. Cold Survival: Larvae, pupae, and adults were exposed to various modifications of temperature. Data on mortality. ABSTRACTS: BA 30:2075(20853), July, 1956; RAE-B 45:3, Jan., 1957; TDB 53:816, June, 1956.

1977. Greenberg, B. 1959 Persistence of Bacteria in the Developmental Stages of the Housefly. I. Survival of Enteric Pathogens in the Normal and Aseptically Reared Host. Am. J. Trop. Med. Hyg. 8(4):405-411.
Utilized *Salmonella typhi*, *flexneri*, *paratyphi* and *enteritidis*. Each behaved differently, due, perhaps to interspecific competition in the larval medium. ABSTRACT: BA 34:1061(11828), Nov., 1959.

1978. Greenberg, B. 1960 Persistence of Bacteria in the Developmental Stages of the Housefly. II. Quantitative Study of the Host Contaminant Relationship in Flies Breeding under Natural Conditions. Am. J. Trop. Med. Hyg. 8(4):412-416.
Mature maggots support populations of 10^7 bacteria. This falls through prepupal and pupal stages to an average of 10^2 or 10^3 (Range is from complete sterility to 10^6). ABSTRACT: BA 34:1061(11829), Nov., 1959.

1979. Greenberg, B. 1959 Persistence of Bacteria in the Developmental Stages of the Housefly. III. Quantitative Distribution in Pre-pupae and Pupae. Am. J. Trop. Med. Hyg. 8:613-617.
Concludes, from starvation experiments, that the more than 90 per cent decline in bacterial count in the prepupa is a result of the cessation of feeding, while elimination still goes on. ABSTRACTS: BA 35:1269(14463), Mar., 1960; RAE-B 49:196, Sept., 1961.

1980. **Greenberg, B. 1959** Persistence of Bacteria in the Developmental Stages of the Housefly. IV. Infectivity of the Newly Emerged Adult. Am. J. Trop. Med. Hyg. 8:618-622.
Found few bacteria on surface, in feces, or in alimentary tract. Epidemiological significance discussed. ABSTRACT: BA 35:1269(14464), Mar., 1960.

1981. **Greenberg, B. 1959** House Fly Nutrition: I. Quantitative Study of the Protein and Sugar Requirements of Males and Females. J. Cell. and Comp. Physiol. 53:169-177.
Males and females consume approximately the same amount of sucrose, but for ovipositing females the protein requirement is 2-3 times that of a male or virgin female. Ratio of sucrose to protein consumed is 7:1 for egg-layers; 16:1 for others.

1982. **Greenberg, B.** and **Paretsky, D. 1955** Proteolytic Enzymes in the House Fly, *Musca domestica* (L). Ann. Ent. Soc. Am. 48:46-50.
Pepsin or a pepsin-like enzyme has been demonstrated in homogenates of whole maggots in all three instars, as well as in dissected guts of third instar maggots of *Musca domestica* L.

Greenberg, H., joint author. *See* Incho, H. H., et al., 1952.

1983. **Gregor, F. 1958** Versuch einer Klassifikation der Synanthropen Fliegen (Diptera). [Attempt to classify synanthropic flies (Diptera).] J. Hyg., Epidemiol., Microbiol. and Immunol. 2(2):205-216.
There are two types of flies, as recognized in Soviet literature: the eusynanthropic forms, such as *Musca domestica*, appearing in human environment; and the hemisynanthropic forms living in "natural surroundings", facultatively contacting man. ABSTRACT: BA 35:3998 (46047), Aug. 15, 1960.

Gregor, F., joint author. *See* Povolny, D., et al., 1958.

Gressette, F. R., joint author. *See* Goodwin, W. J., et al., 1956

Gressitt, J. L., joint author. *See* Bohart, G. E., et al., 1951.

1984. **Gross, H.** and **Preuss, U. 1951-1953** Infektionsversuche an Fliegen mit darmpathogenen Keimen. [Infection research on flies with intestinal pathogens.] Mitteilung I. Zbl. Bact., Parasit., Infek. und Hyg. 156: 371-377. Mitteilung II, idem. 160:526-529.
Found *M. domestica*, *Lucilia sericata*, *Calliphora erythrocephala*, and *Sarcophaga haemorrhoidalis* capable of transmitting infectious intestinal diseases. *L. sericata*, surprisingly, stood on a par with *M. domestica* as a potential vector. Recommends planned control against this species.

Grudtsina, M. V., joint author. *See* Sychevskaya, V. I., et al., 1959.

Gubev, E., joint author. *See* Veselinov, V., et al., 1954, 1956.

1985. **Gudoshchikova, V. I. 1959** Osennie nabliudeniia za sutochnoi migratsiei komnatnykh mukh. [Fall observations on daily migration of house flies.] Med. Parazit. (Moskva) 28(1):57-60.
ABSTRACT: BA 36:3480(42795), July 1, 1961.

Gudoshchikova-krasil' nikova, V. I., joint author. *See* Polovodova U. P., et al., 1956.

Guest, H. R., joint author. *See* Haynes, H. L., et al., 1954.

Gurland, J., joint author. *See* Dahm, P. A., et al., 1959.

1986. **Gustafson, C. G. Jr., Lies, T. A.** and **Wagner-Jauregg, T. 1953** Action of Some Aliphatic Thiocyanates against DDT-resistant Strains of the Housefly. J. Econ. Ent. 46:620-624.
The Thiocyanidrins have low insecticidal activity. Esterification of the hydroxyl group of thiocyanidrins with lower aliphatic acids produces substances with marked knockdown and toxicity for normal and resistant

flies when tested in contact insecticides. ABSTRACTS: BA 28:949(9709), Apr., 1954; RAE-B 42:36, Mar., 1954; TDB 51:321, Mar., 1954.

1987. Guthrie, F. E. and **Baker, F. S. 1954** Malathion and Diazinon for Control of House Flies in North Florida Cattle Barns. Florida Ent. 37:13-17.
In a heavy infestation 1 per cent Diazinon gave control for 6 weeks, malathion for 2 weeks. ABSTRACT: BA 28:2195(22290), Sept., 1954.

1988. Guyer, G. E., King, H. L., Fischer, R. L. and **Drew, W. A. 1956** The Emergence of Flies Reared from Grass Silage in Michigan. J. Econ. Ent. 49:619-622.
Species composition and ecological requirements of the insect populations associated with grass silage were investigated. Twenty species of flies were identified. ABSTRACTS: BA 31:931(9323), Mar., 1957; RAE-B 45:174, Nov., 1957.

1989. Guyer, G. E., King, H. L., Smiley, E. S. and **Ralston, N. P. 1954** Chemical Fly Control. Quart. Bull. Mich. Agr. Expt. Sta. 37:254-263.
In 10 dairy barns and a composting unit, good control was obtained by using all available types of chemical control, plus observance of adequate sanitary practices. ABSTRACT: BA 30:1154(11764), Apr., 1956.

Guyer, G. E., joint author. *See* King, H., et al., 1953.

Guzer, G. E., joint author. *See* Wells, A. L., et al., 1958.

Hackley, B. E., joint author. *See* Dethier, V. G., et al., 1952.

1990. Hadaway, A. B. 1956 Cumulative Effect of Sublethal Doses of Insecticides on Houseflies. Nature 178(4525):149-150. July.
Flies from a *normal* strain can eliminate, metabolize or store a *proportion* of the absorbed insecticide. Relation to resistance requires investigation. ABSTRACTS: RAE-B 45:51, Apr., 1957; TDB 53:1393, Nov., 1956.

1991. Hadaway, A. B. and **Barlow, F. 1952** Symposium on Insecticides. III. Some Physical Factors Affecting the Efficiency of Insecticides. Trans. Roy. Soc. Trop. Med. Hyg. 46:236-242.
Treats of larvicides, aerosols, sprays, and residual deposits of oil solutions, emulsions and wettable powders.

1992. Hadaway, A. B. and **Barlow, F. 1957** The Toxicity of Three Organic Phosphorus Insecticides to Houseflies and Mosquitoes. Bull. W. H. O. 16(4):870-873. Geneva.
Studied diazinon, chlorthion and malathion. Relative toxicity to insects varied from one species to another. In order of decreasing toxicity to adult female *M. domestica*: diazinon > chlorthion > malathion. ABSTRACT: TDB 55:337, Mar., 1958.

1993. Hadaway, A. B. and **Barlow, F. 1957** The Influence of Temperature and Humidity upon the Action of Insecticides. I. During the Post-Treatment Period. Ann. Trop. Med. Parasit. 51(2):187-193. June.
M. domestica (and mosquitoes) stored after treatment at various temperatures and humidities. As temperature rose from 20° to 30°C, toxicity of DDT *decreased*; that of dieldrin and diazinon *increased*. No significant effect on toxicity of gamma BHC. Humidity not important. ABSTRACTS: RAE-B 46:193, Dec., 1958; TDB 55:107, Jan., 1958.

1994. Hadaway, A. B. and **Barlow, F. 1957** The Influence of Temperature and Humidity upon the Action of Insecticides. II. Temperature during Pre-Treatment Period. Ann. Trop. Med. Parasit. 51(2):194-200. June.
M. domestica (and mosquitoes) held at 25°C and 30°C. Those kept at higher temperatures produced eggs a day earlier. Pre-treatment temperature appears to have no effect on the susceptibility of insects to insecticides. ABSTRACTS: RAE-B 46:193, Dec., 1958; TDB 55:108, Jan., 1958.

1995. **Hadaway, A. B.** and **Barlow, F. 1958** Some Aspects of the Effect of the Solvent on the Toxicity of Solutions of Insecticide. Ann. Appl. Biol. 46(2):133-148. London.
Tested adult female *Aedes aegypti* and adult female *M. domestica* with solvents having similar physical properties. Insecticides were more effective in parafinnic hydrocarbons than in aliphatic long chain esters; least effective in aromatic esters. ABSTRACT: RAE-B 47:67, May, 1959.

Hadaway, A. B., joint author. *See* Barlow, F., et al., 1952.

1996. **Hadjinicolaou, J.** and **Hansens, E. J. 1953** Chlorinated Hydrocarbon Insecticides to Control House Fly Larvae. J. Econ. Ent. 46:34-37.
Dieldrin, aldrin, chlordan and heptachlor were effective against normal larvae when added to media in extremely small amounts. In the field, larvicides *alone* were not effective. Larvae from resistant adults were resistant to same substances. ABSTRACTS: RAE-B 41:135, Sept., 1953; TDB 50:863, Sept., 1953.

1997. **Hafez, M. 1950** On the Behaviour and Sensory Physiology of the House-Fly Larva, *Musca domestica* L. I. Feeding Stage. Parasitology 40(3-4):215-236.
Studied reactions of house-fly larvae to various combinations of humidity, temperature, odor, light. Discussion of the orientation mechanisms involved. ABSTRACTS: RAE-B 40:208, Dec., 1952; TDB 48:294, Mar., 1951.

1998. **Hafez, M. 1953** On the Behaviour and Sensory Physiology of the House-fly Larva, *Musca domestica* L. II. Prepupating stage. J. Exp. Zool. 124(2):199-225.
Studied reactions of the pupating larva to humidity, temperature, odor. Used a circular arena divided into 2 halves. Results not essentially different from those obtained with feeding larvae. ABSTRACTS: BA 28:1111, May, 1954; RAE-B 43:129, Sept., 1955; TDB 51:995, Sept., 1954.

1999. **Hafez, M. 1954** (*see* **Hafez, 1955**). House Fly Problem in Egypt. J. Egyptian Pub. Health Ass. 24(4):93-102.
Treats of biology, methods of control, sanitation and resistance.

2000. **Hafez, M. 1955** (*see* **Hafez, 1954**). House Fly Problem in Egypt; Resumé of a Lecture given at Namru, Cairo. Med. J. Egypt. Armed Forces 1(1):30-37.
Approximately equivalent in content to the preceding reference.

2001. **Hafez, M. 1958** On the Susceptibility of the Egyptian House-fly, *Musca domestica vicina* Macq. to Several Insecticides. (Diptera.) Bull. Soc. Ent. Egypte 42:75-81.
In all cases the female was normally less suceptible than the male to the six insecticides tested. The species shows high tolerance to DDT, great susceptibility to lindane and dieldrin. ABSTRACT: BA 36:7685 (82593), Dec. 1, 1961.

2002. **Hafez, M.** and **Attia, M. A. 1958** On the Developmental Stages of *Musca sorbens* Wied. with Special Reference to Larval Behaviour (Diptera: Muscidae). Bull. Soc. Ent. Egypte 42:123-161.
Orientation mechanisms by which larvae react to changes in temperature, humidity and odors. Treats of klinotaxis, klinokinesis, orthokinesis. ABSTRACTS: BA 36:8122(86820), Dec. 15, 1961; RAE-B 48:78-80, April, 1960; TDB 57:958, Sept., 1960.

2003. **Hafez, M.** and **Attia, M. A. 1958** Studies on the Ecology of *Musca sorbens* Wied. in Egypt (Diptera: Muscidae). Bull. Soc. Ent. Egypte 42:83-121.
Biological and ecological observations. Calls attention to close resem-

blance of adult *M. sorbens* to *M. domestica*. ABSTRACTS: BA 36:7685 (82594), Dec. 1, 1961; RAE-B 48:78-80, April, 1960; TDB 57:958, Sept., 1960.

2004. Hafez, M. and **Attia, M. A. 1958** The Relation of *Musca sorbens* Wied. to Eye Diseases in Egypt (Diptera: Muscidae). Bull. Soc. Ent. Egypte 42:275-283.
Much evidence to link *M. sorbens* with ophthalmiasis. Attraction of fly to diseased eyes believed due to an olfactory response rather than to a humidity reaction. ABSTRACTS: BA 36:7685(82595), Dec. 1, 1961; RAE-B 48:78-80, April, 1960; TDB 57:959, Sept., 1960.

2005. Hafez, M. and **Gamal-Eddin, F. M. 1959** Ecological Studies on *Stomoxys calcitrans* L. and *sitiens*. Rond. in Egypt, with Suggestions on Their Control. Bull Soc. End. Egypte 43:245-283.
Similarities and differences in the two species, with reference to adult habits, breeding, longevity, fecundity, sex ratios, control, etc. Though not concerned with *M. domestica*, this reference and the next are included because of the senior author's extensive work on muscoids of medical importance.

2006. Hafez, M. and **Gamal-Eddin, F. M. 1959** On the Feeding Habits of *Stomoxys calcitrans* L. and *sitiens* Rond., with Special Reference to their Biting Cycle in Nature. Bull. Soc. Ent. Egypte 43:291-301.
Both species prefer donkey blood; after that (for *sitiens*): camel, horse, buffalo, cow, sheep, goat. *S. sitiens*, in general, displays greater biting activity than *calcitrans*.

Haggag, G., joint author. *See* Hafez, M., et al., 1959.

2007. Hagley, E. A. C. and **Morrison, F. O.** The Synergistic Action of Certain Chemicals used in Combination with DDT. Proc. 10th Int. Congr. Ent., pp. 175-185.
Four methylene dioxyphenyl compounds were tested by topical application for their synergistic effect. The p-chlorophenyl compounds were significantly effective in increasing the lethal action of DDT. ABSTRACT: RAE-B 47:83, June, 1959.

Hagley, E. A. C., joint author. *See* Morrison, F. O., et al., 1959.

2008. Haines, T. W. 1953 Breeding Media of Common Flies. I. In Urban Areas. Am. J. Trop. Med. Hyg. 2(5):933-940.
M. domestica was produced by a wider range of breeding media, for the period of observation (1 year) than any other species, by a wide margin. ABSTRACTS: BA 28:1455, June, 1954; RAE-B 43:46, Mar., 1955; TDB 51:437, Apr., 1954.

2009. Haines, T. W. Breeding Media of Common Flies. II. In Rural Areas. Am. J. Trop. Med. Hyg. 4:1125-1130.
M. domestica most diverse species in use of breeding material. 87·1 per cent of flies from all breeding materials were produced from animal excrement and pen litter. ABSTRACTS: BA 30:2075(20855), July, 1956; RAE-B 45:165, Oct., 1957; TDB 53:936, July, 1956.

2010. Hall, I. M. and **Arakawa, K. Y. 1959** The Susceptibility of the Housefly *Musca domestica* Linnaeus, to *Bacillus thuringiensis* var. *thuringiensis* Berliner. J. Insect Pathol. 1(4):351-355.
Proved only slightly susceptible. Heavy doses did kill maggots of several strains. Believed due to toxic effect of crystalline inclusion bodies in the infective material. ABSTRACTS: BA 35:1961(22404), Apr., 1960; RAE-B 49:3, Jan., 1961.

2011. Hammen, C. S. 1956 Nutrition of *Musca domestica* in Single-Pair Culture. Ann. Ent. Soc. Am. 49:365-368.
A satisfactory semi-synthetic medium supplying all nutritional require-

ments for larvae and adults can be in a single bottle, needing no manipulation except the addition of wood shavings for the pupation site. ABSTRACT: BA 31:943(9460), Mar., 1957.

2012. Hammen, C. S. 1957 A Growth-promoting Effect of Cholesterol in the Diet of Larvae of the House Fly, *Musca domestica* L. Ann. Ent. Soc. Am. 50(2):125-127.
In the semi-synthetic medium, the addition of pure cholesterol increased adult size by 13 per cent. ABSTRACT: BA 31:2211(23676), Aug., 1957.

2013. Hampton, U. M. 1952 Reproduction in the Housefly (*Musca domestica* L.). Proc. Roy. Entom. Soc. London. Ser. A. 27(4/6):29-32.
Virgin flies lay fewer batches, fewer eggs, have a longer pre-oviposition period, lay less frequently than fertilized flies. Life span is prolonged when fertilization is delayed until late in life. ABSTRACT: TDB 50:257, Mar., 1953.

2014. Hanec., W. 1956 A Study of the Environmental Factors affecting the Dispersion of House Flies (*Musca domestica* L.) in a Dairy Community near Fort Whyte, Manitoba. Canad. Ent. 88(6):270-272. Ottawa.
Flies tagged with radioactive phosphorus in aqueous solution, sweetened with sucrose. Adults orient to wind-borne odors from farmyards. This assists in migration. ABSTRACT: RAE-B 46:8, Jan., 1958.

2015. Hanec, W. 1956 Investigations Concerning Overwintering of House Flies in Manitoba. Canad. Ent. 88:516-519.
Flies breed continually during winter in heated barns. In unheated barns they are able to survive through the winter. Studies on cold resistance of houseflies indicate that the pupa is the most resistant stage. ABSTRACTS: BA 31:931(9324), Mar., 1957; RAE-B 46:51, Mar., 1958.

2016. Hansens, E. J. 1950 House Fly Control in Dairy Barns. J. Econ. Ent. 43(6):852-858.
Several insecticides tested. Methoxychlor and lindane proved most satisfactory, and were recommended for use the following year. ABSTRACTS: RAE-B 39:120, July, 1951; TDB 48:590, 1951.

2017. Hansens, E. J. 1953 Failure of Residual Insecticides to Control House Flies. J. Econ. Ent. 46:246-248.
Seven strains of flies tested. Each was resistant to the insecticide used in the barn from which the strain was collected. Most were resistant to all 5 insecticides under investigation. ABSTRACTS: RAE-B 41:171, Nov., 1953; TDB 50:863, Sept., 1953.

2018. Hansens, E. J. 1956 Control of House Flies in Dairy Barns with Special Reference to Diazinon. J. Econ. Ent. 49:27-32.
Various materials tested as residual insecticides and as wet and dry baits. Diazinon was most effective compound tested. Did not appear in cow's milk. ABSTRACTS: BA 30:2413(24207), Aug., 1956; RAE-B 45:35, Mar., 1957; TDB 53:937, July, 1956.

2019. Hansens, E. J. 1958 House Fly Resistance to Diazinon. J. Econ. Ent. 51(4):497-499.
Resistance was proportional to the amount of Diazinon used in the barns during the several years preceding. Most resistant strain reproduced in the laboratory for 4 generations without loss of resistance. ABSTRACT: BA 33:622, Feb., 1959.

2020. Hansens, E. J. and **Bartley, C. E. 1953** Three New Insecticides for Housefly Control in Barns. J. Econ. Ent. 46:372-374.
Pyrolan and Pyramat not sufficiently lasting for economical use in dairy barns. Diazinon gave excellent control for a whole season (72 days) in

a horse barn and for 3 weeks in a dairy barn against non-resistant flies. ABSTRACTS: BA 28:955, Apr., 1954; RAE-B 41:178, Nov., 1953; TDB 50:864, Sept., 1953.

2021. Hansens, E. J., Granett, P. and **O'Connor, C. T. 1955** Fly Control in Dairy Barns in 1954. J. Econ. Ent. 48:306-310.
Residual sprays proved more effective than baits. American Cyanamid 4124 gave best results. Chlorthion, methoxychlor, malathion, and lindane were less effective. Diazinon, Bayer 13/69 and chlorthion in baits gave kills, but were not lasting. ABSTRACT: RAE-B 44:73, May, 1956.

2022. Hansens, E. J. and **Scott, R. 1955** Diazinon and Pirazinon in Fly Control. J. Econ. Ent. 48:337-338.
During 1954, these insecticides were used in 21 New Jersey barns, where they gave equally good results. Control ranged from 40-50 days to the entire season. ABSTRACT: RAE-B 44:73, May, 1956.

Hansens, E. J., joint author. *See* Forgash, A. J., et al., 1959; Hadjinicolaou, J., et al., 1953; Harris, H. J., et al., 1950; Helrich, K., et al., 1958.

2022a. Harada, F. 1953 On the Fly as a Carrier of Hookworm Larvae. Med. Biol. 29:28-30.

2023. Harada, F. 1954 Investigations of Hookworm Larvae. IV. On the Fly as a Carrier of Infective Larvae. Yokohama M. Bull. 5(4):284-286.

Harding, W. C., joint author. *See* Langford, G. S., et al., 1954.

Hargett, L. T., joint author. *See* Turner, E. C. Jr., et al., 1958.

2024. Harris, H. H., Hansens, E. J. and **Alexander, C. C. 1950** Determination of DDT in Milk Produced in Barns Sprayed with DDT Insecticides. Agric. Chem. 5(1):51-52.
Four dairy barns were sprayed with 5 per cent DDT in clay base as 50 per cent DDT wettable powder. Seven of 28 milk samples showed photometric evidence of containing 0·03 to 0·05 p.p.m. of DDT. Four of these seven samples came from barns with least sanitary milk-handling. ABSTRACT: RAE-B 39:8, 1951.

2025. Harris, W. G. and **Burns, E. C. 1959** Phosphate Resistance in a Field Strain of House Flies. Am. J. Trop. Med. Hyg. 8(5):580-582.
During 1958, following use of malathion, collections were made of resistant flies. The larval period was longer, and pupation and emergence extended over a greater period than characterizes normal flies. ABSTRACTS: BA 35:514(5998), Jan., 1960; TDB 57:192, Feb., 1960; RAE-B 49:174, Aug., 1961.

Harrison, A., joint author. *See* Winteringham, F. P. W., et al., 1951, 1955, 1956, 1959.

2026. Harrison, C. M. 1950 DDT-resistant House-flies. Ann. Appl. Biol. 37: 306-309.
Italian strain made resistant to DDT by injection. Resistance clearly not due to tarsal structure, as in strain reported by Wiesmann. ABSTRACT: RAE-B 38:158, 1950.

2027. Harrison, C. M. 1951 Inheritance of Resistant to DDT in the Housefly, *Musca domestica* L. (Correspondence). Nature 167:855-856.
Appears to rest on one Mendelian factor. Considers susceptibility to paralysis as partially dominant. ABSTRACT: TDB 48:930, Oct., 1951.

2028. Harrison, C. M. 1952 DDT Resistance in an Italian Strain of *Musca domestica* L. Bull. Ent. Res. 42:761-768.
Two strains named for localities (TP, resistant, from field, and R, non-resistant, in Rome). Older flies more susceptible to DDT than young individuals. Males more susceptible than females. Small flies more

susceptible than large. Need to standardize for age, size, and sex, for testing purposes. ABSTRACTS: RAE-B 40:80, May, 1952; TDB 49:560, May, 1952.

2029. Harrison, C. M. 1952 The Resistance of Insects to Insecticides. Trans. Roy. Soc. Trop. Med. Hyg. 46:255-261.
Resistance in flies seems to be inherited, but genetic factors as yet unrecognized. Refers to pertinent literature. ABSTRACTS: BA 26:3193, Dec., 1952; RAE-B 42:82-83, 1954.

2030. Harrison, C. M. 1953 DDT-Resistance and its Inheritance in the House Fly. J. Econ. Ent. 46:528-530.
Work on resistance to DDT and other insecticides in strains of housefly from Sardinia and Italy is reviewed. Theories of resistance inheritance are discussed. ABSTRACTS: RAE-B 42:7, Jan., 1954; TDB 50:1092, Nov., 1953.

Harrison, C. M., joint author. *See* Busvine, J. R., et al., 1953.

2031. Harrison, R. A. 1950 Toxicity to Houseflies of a Flat Oil Paint Containing DDT.
See Section IV, Reference 513a, for complete citation. Item relisted here because certain abstracting journals assign it a 1950 publication date.

2032. Harrison, R. A. 1951 Tests with some Insecticides on DDT-resistant Houseflies. New Zealand J. Sci. Tech. 32(B) (6):5-11. Wellington.
DDT-resistant flies were highly resistant to Rhothane. With benzene hexachloride, chlordane, and compound 497, mortalities were sufficiently high to suggest that they may be substituted for DDT. ABSTRACT: RAE-B 40:116, July, 1952.

2033. Harrison, R. A. 1951 The Occurrence in New Zealand of Houseflies Resistant to DDT. New Zealand J. Sci. Tech. 32(B) (5):40-43.
Six lines of flies were collected from DDT-treated premises in Auckland, and bred in the laboratory. Knockdown was slower and mortality less than with laboratory strain, when flies were placed on DDT residue. ABSTRACTS: RAE-B 40:92, June, 1952; TDB 49:1083, Nov., 1952.

2034. Harrison, R. A. 1951 Further Studies on DDT-Resistant Houseflies in New Zealand. New Zealand J. Sci. Tech. 33(B):92-95.
A line of DDT-resistant flies, bred without exposure to DDT for 19 generations, still showed high resistance; required an exposure 18 times that needed with non-resistant flies to produce the same kill. ABSTRACTS: RAE-B 40:149, Sept., 1952; TDB 50:161, Feb., 1953.

2035. Harrison, R. A. 1951 A Note on the Toxicity to Houseflies of DDT Residues. New Zealand J. Sci. Tech. 33(B) (2):96-98.
Good control achieved for non-resistant flies with 200 mg./sq. ft. on enamel for 1353 days; on oil bound water paint for 1316 days; on water paint for 909 days and on old oil paint for 324 days. Residue of 800 mg./sq. ft. on unpainted wood gave good control for 1399 days. ABSTRACTS: RAE-B 40:149, Sept., 1952; TDB 50:162, Feb. 1953.

2036. Harrison, R. A. 1952 Tests with Benzene Hexachloride Residues against Houseflies. New Zealand J. Sci. Tech. 34(B):134-138.
Gamma isomer of BHC gives satisfactory knockdown, but its effective life is short, due to dispersal of the residue by evaporation. Initial strength and type of surface also important. ABSTRACTS: RAE-B 42:50, Apr., 1954; BA 28:2195, Sept., 1954.

2037. Harrison, R. A. 1953 Toxicity of Pyrethrum to Houseflies. New Zealand J. Sci. Tech. 35(B) (1):22-29.
DDT-resistant and DDT-susceptible flies not significantly different for knockdown and mortality when exposed to pyrethrum. Males were more susceptible to both knockdown and mortality from pyrethrum than were females. ABSTRACT: RAE-B 42:171, Nov., 1954.

2038. **Harrison, R. A. 1953** Toxicity to Houseflies of Paints containing DDT. New Zealand J. Sci. Tech. 35(B) (1):43-44.
Flat oil paint containing 2·5 per cent and 5·0 per cent DDT produced 95 per cent and 100 per cent mortality respectively after 1262 days. DDT-resistant flies were *not* controlled. ABSTRACT: RAE-B 42:171, Nov., 1954.

Harrison, R. A., II, joint author. *See* Satchell, G. H., et al., 1953; Smith, A. G., et al., 1951.

2039. **Hart, S. A. 1957** Chicken House Fly Control Through Manure Handling. Calif. Vector Views 4(11):67-68.
Various methods discussed. Emphasis on building construction, minimum number of man hours required, storage and/or disposal. No "best method" recommended.

2040. **Hawley, J. E., Penner, L. R., Wedberg, S. E.** and **Kulp, W. L. 1951** The Role of the House Fly, *Musca domestica* in the Multiplication of Certain Enteric Bacteria. Am. J. Trop. Med. 31:572-582.
Precise methods are described for handling and feeding of flies and the collection and analysis of their excrement for bacterial content. Small numbers of bacteria fed to flies are not passed in their feces. Flies fed 1000 or more of several bacterial species excreted more than they were fed. ABSTRACTS: BA 26:1241, May, 1952; RAE-B 41:187, Nov., 1953.

2041. **Haws, L. D. 1959** Insecticide Treatment of Manure for Control of Houseflies with Notes on the Control of Cattle Lice and the Sheep Tick. Dissertation Absts. 19(9):2201.

Hayden, D. L., joint author. *See* Sullivan, W. N., et al., 1958.

2042. **Hayes, W. J., Jr., Ferguson, F. F.** and **Cass, J. S. 1951** The Toxicology of Dieldrin and its Bearing on Field Use of the Compound. Am. J. Trop. Med. 31:519-522.
Contamination with 100 c.c. or more of 25 per cent solution may be dangerous to man if not washed off promptly. Can be safely used in public health programs by properly trained personnel, especially in sprays, out of doors. ABSTRACT: RAE-B 41:186, Nov., 1953.

2043. **Haynes, H. L., Guest, H. R., Stansbury, H. A., Sousa, A. A.** and **Borash, A. J. 1954** Cyclethrin, a new Insecticide of the Pyrethrins-type. Contr. Boyce Thompson Inst. 18(1):1-16.
Similar to natural pyrethrins in many respects, including toxicity to mammals. Is synergized by pyrethrin synergists more readily than allethrin. ABSTRACTS: BA 29:2000, Aug., 1955; RAE-B 43:56, Apr., 1955.

2044. **Haynes, H. L., Lambreck, J. A.** and **Moorefield, H. H. 1957** Insecticidal Properties and Characteristics of 1- Napthyl N-Methylcarbamate. Contr. Boyce Thompson Inst. 18(11):507-513. Yonkers, N.Y.
Designated as "Sevin". Has broad insecticidal properties and a low order of mammalian toxicity. Has some residual effect. Its primary action, as an insecticide, appears to be of the "anticholinesterase" type. ABSTRACT: RAE-B 45:156, Oct., 1957.

2045. **Heinz, H. J.** and **Brauns, W. 1955** The Ability of Flies to Transmit Ova of *Echinococcus granulosus* to Human Foods. S. Afr. J. Med. Sci. 20 (3-4):131-132.
Used *Sarcophaga tibialis* Macq.; *M. domestica* also mentioned. Few ova cling to fly's exterior, but many are ingested and passed through fly's alimentary tract. Ova, found in milk and other food, proved viable, and produced cysts in the livers of rabbits.

Hellyer, G. C., joint author. *See* Winteringham, F. P., et al., 1953, 1954, 1955.

2046. **Helrich, K., Hansens, E. J.** and **Granett, P. 1958** Residues in Milk from Dairy Cattle treated with Methoxychlor for Fly Control. J. Agric. Fd. Chem. 6(4):281-283.
Milk analyzed at various intervals after spraying or dusting. Methoxychlor present in minute, but detectable amounts. Concentration diminishes rapidly with successive samplings. ABSTRACT: RAE-B 47:64, April, 1959.

Henderson, L. S., joint author. *See* Bishopp, F. C., et al., 1950.

2047. **Henry, S. J. 1958** Fly control that Works in Supermarkets. Pest Control 26(4):9, 11, 71.
Effective fly control depends upon sanitation, prevention of fly entry and chemical control. A 1 per cent diazinon emulsion was effective for a 3-year period. ABSTRACT: BA 33:289(3696), Jan., 1959.

Henry, S. M., joint author. *See* Cotty, V. F., et al., 1958; Hilchey, J. D., et al., 1957.

2048. **Herget, C. M. 1950** Reaction Time of the Common Housefly (*Musca domestica*). Science 112:62.
Fly's support jerked away by a bullet striking a plate. Room temperature about 75°F. First wing beat 21 msec. later. Extension of legs 1 msec. before first wing stroke. Leg motion executed 3 times, about 1 msec. apart.

2049. **Hess, A. D. 1952** The Significance of Insecticide Resistance in Vector Control Programs. Panel Papers on Resistance of Insects to Insecticides No. II. Am. J. Trop. Med. Hyg. 1:371-388, 95 refs.
Recognizes 3 types of resistance: 1. Physiological; 2. Morphological; 3. Behaviouristic. The status of insecticide resistance in 18 countries and areas is presented and several generalizations having implications concerning control programs are set forth. Among these are: rates of resistance development; types of resistance; operational practices; effects on life-history and vigor. ABSTRACT: TDB 49:1159, Dec., 1952.

2050. **Hess, A. D. 1953** Current Status of Insecticide Resistance in Insects of Public Health Importance. Am. J. Trop Med. Hyg. 2:311-317.
Resistance confirmed by experiment for *M. domestica*, *M. nebulo*, many mosquitoes, cockroaches, and lice. ABSTRACTS: BA 27:3151, Dec., 1953; RAE-B 42:139, Sept., 1954.

2051. **Hewlett, P. S. 1958** Interpretation of Dosage-Mortality Data for DDT-Resistant Houseflies. Ann. Appl. Biol. 46(1):37-46.
Slopes of the lines relating probit mortality to the Log. dose of DDT are generally lower for resistant than for susceptible strains. Two possible reasons discussed. ABSTRACTS: BA 32:2682(32120), Sept., 1958; TDB 55:832, July, 1958.

Hewlett, P. S., joint author. *See* Wilson-Jones, K., et al., 1958.

Hibbs, E. T., joint author. *See* Dahm, P. A., et al., 1959.

Highland, H. A., joint author. *See* Gahan, J. B., et al., 1953.

2052. **Hilchey, J. D., Cotty, V. F.** and **Henry, S. M. 1957** The Sulfur Metabolism of Insects. II. The Metabolism of Cystine-S^{35} by the Housefly, *Musca domestica* L. Contrib. Boyce Thompson Inst. 19:189-200.
In both male and female flies, conversion is to taurine and sulphate, the taurine being partly retained and partly excreted. All the sulphate is excreted in from 24 to 48 hours.

Hilchey, J. D., joint author. *See* Cotty, V. F., et al., 1958.

Hilsenhoff, W. L., joint author. *See* Dicke, R. J., et al., 1952.

Hirakoso, S., joint author. *See* Ogata, K., et al., 1957.

Hiroyoshi, T., joint author. *See* Sokal, R. R., et al., 1959.

2053. **Hirsh, J. 1959** The Meaning of the Fly for Medicine. Military Med. 124:733-736.
Reviews the fly's life history and medical importance. Lists 12 bacterial and viral diseases, 3 protozoan and spirochaetal diseases and 13 parasitic diseases, where vectorship may be involved.

2054. **Hocking, B.** and **Lindsay, I. S. 1958** Reactions to Insects of the Olfactory Stimuli from the Components of an Insecticidal Spray. Bull. Ent. Res. 49(4):675-683, 4 figures.
Studies 2 strains of *M. domestica* and 5 other dipterous species. Two types of equipment used. One (the newer) can be used in the field, and gives rapid and reproducible evaluations. DDT-resistant flies were less sensitive than normal flies to repellent components. ABSTRACTS: TDB 56:578, May, 1959; Hocking, B. 1960 Smell in Insects, p. 134 (3.53).

2055. **Hodge, A. J. 1955** Studies on the Structure of Muscle. III. Phase contrast and Electron Microscopy of Dipteran Flight Muscle. J. Bioph. Biochem. Cytol. 1(4):361-380.

2056. **Hodgson, E. S.** and **Roeder, K. D. 1956** Electrophysiological Studies of Arthropod Chemoreception. J. Cell and Comp. Physiol. 48: 51-75.
Used blow flies to study impulses from labellar chemosensory neurons. Observed responses of primary chemoreceptor cells to sugars, salts, acids and alcohols. Chemoreceptor neurons show sensitivity to mechanical stimulation and temperature changes.

2057. **Hodgson, E. S.** and **Smyth, T., Jr. 1955** Localization of Some Insect Sense Organs by Use of DDT. Ann. Ent. Soc. Am. 48:507-511.
Internal movement of physiologically active DDT studied in *Sarcophaga bullata* Parker using both behavioral and electrophysiological methods.

2058. **Hoffman, R. A. 1954** Observations on Fly Control by Sanitation. J. Econ. Ent. 47:194.
Very good control obtained by strict sanitation, with only minimal amounts of insecticides. In the dry climate of California, manure was dried. Barns were swept but never washed. ABSTRACTS: BA 28:2703, Nov., 1954; RAE-B 42:192, Dec., 1954.

2059. **Hoffman, R. A.** and **Cohen, N. W. 1954** House Fly Control with Residual Sprays of Organic Phosphorus Insecticides. J. Econ. Ent. 47: 701-703.
Piperonyl butoxide increased the mortality of flies and the longevity of residues, but not significantly. Diazinon sulfoxide gave promise of usefulness. ABSTRACTS: BA 29:966, Apr., 1955; RAE-B 43:118, Aug., 1955; TDB 52:223, Feb., 1955.

2060. **Hoffman, R. A., Hopkins, T. L.** and **Lindquist, A. W. 1954** Tests with Pyrethrum Synergists Combined with Some Organic Phosphorus Compounds Against DDT-Resistant Flies. J. Econ. Ent. 47:72-76.
Seven organic phosphorus compounds, combined in various ways with 19 candidate synergists were laboratory tested. Initial effectiveness, duration of effectiveness and effect of various chemical formulations were tabulated. ABSTRACTS: BA 28:2700, Nov., 1954; RAE-B 42:187, Dec., 1954; TDB 51:1312, Dec., 1954.

2061. **Hoffman, R. A.** and **Lindquist, A. W. 1952** Absorption and Metabolism of DDT, Toxaphene, and Chlordane by Resistant House Flies as Determined by Bioassay. J. Econ. Ent. 45:233-235.
Bioassay method indicated that resistant flies metabolized 4·0 μgm. of toxaphene, 3·5 μgm. of DDT, 59 μgm. of chlordane. ABSTRACT: RAE-B 40:128, Aug., 1952; TDB 49:910, Dec., 1952.

2062. **Hoffman, R. A., Lindquist, A. W.** and **Butts, J. S. 1951** Studies on Treatment of Flies with Radioactive Phosphorus. J. Econ. Ent. 44: 471-473.
Flies fed sugar solutions of radioactive phosphorus *or* reared in a medium containing it. There was some effect on fertility and large doses resulted in a carry-over into the following generation. Fertility adversely affected in feeding tests, but not in rearing tests. Feeding adults is the more efficient method of tagging with P^{32}. ABSTRACTS: BA 26:1522, June, 1952; RAE-B 39:181, Nov., 1951; TDB 49:88, Jan., 1952.

2063. **Hoffman, R. A.** and **Monroe, R. E. 1956** Control of House Fly Larvae in Poultry Droppings. J. Econ. Ent. 49:704-705.
Emulsion sprays of malathion, chlorthion, American Cyanamid 4124, Diazinon and dieldrin; also water solutions of Bayer L 13/59 were applied to manure. Diazinon and Bayer L 13/59 proved superior as larvicides. ABSTRACTS: BA 31:930(9325), Mar., 1957; RAE-B 45:178, Nov., 1957.

2064. **Hoffman, R. A.** and **Monroe, R. E.** Further Tests on the Control of Fly Larvae in Poultry and Cattle Manure. J. Econ. Ent. 50:515.
Dipterex gave fair control when applied to poultry manure. With cattle manure, poor penetration was the chief barrier to any effective larvicide application. ABSTRACTS: BA 32:285(3253), Jan., 1958; RAE-B 46:142, Sept., 1958.

2065. **Hoffman, R. A., Roth, A. R.** and **Lindquist, A. W. 1951** Effect on House Flies of Intermittent Exposures to Small Amounts of DDT Residues. J. Econ. Ent. 44:734-736.
Mortality decreased as the time between exposures was lengthened. Flies believed to detoxify themselves. ABSTRACTS: BA 26:1794, July, 1952; RAE-B 40:44, Mar., 1952; TDB 49:452, Apr., 1952.

2066. **Hoffman, R. A., Roth, A. R., Lindquist, A. W.** and **Butts, J. S. 1952** Absorption of DDT in Houseflies over an Extended Period. Science 115:312-313.
Reports radioassays. Time of storage of flies makes a great difference as to amount of radioactive DDT absorbed. Rate of absorption about the same in dead as in living flies. Delay of 9 days causes absorption to increase 62 per cent.

Hoffman, R. A., joint author. *See* Hopkins, T. L., et al., 1955; Lindquist, A. W., et al., 1951.

Holdaway, F. G., joint author. *See* Tanada, Y., et al., 1950.

Holt, C. J., joint author. *See* Rendtorff, R. C., et al., 1954.

2067. **Holway, R. T., Mitchell, W. A.** and **Salah, A. A. 1951** Studies on the Seasonal Prevalence and Dispersal of the Egyptian Housefly. Pt. I. The Adult Flies. Ann. Ent. Soc. Am. 44:381-398. Pt. II. The Larvae and their Breeding Areas. Ann. Ent. Soc. Am. 44:489-510.
Concerns *M. domestica vicina.* Primary peak in June; secondary peak in September. Flies concentrate in warmer sites when air is below 20°C; in cooler places when air is above 30°C. Move indoors when outdoor temperature exceeds 25°; outdoors when outdoor temperature falls below this. Odors of food a factor also. ABSTRACTS: BA 26:699, Mar., 1952; *also* 26:2342, Sept., 1952; RAE-B 40:165, Oct., 1952; *also* 40: 204, Dec., 1952; TDB 49:729, July, 1952; *also* 50:656, July, 1953.

Hopkins, D. E., joint author. *See* Eddy, G. W., et al., 1954.

2068. **Hopkins, T. L.** and **Hoffman, R. A. 1955** Effectiveness of Dilan and Certain Candidate Synergists Against DDT-resistant House Flies. J. Econ. Ent. 48:146-147.
Effectiveness of Dilan increased by the addition of piperonyl butoxide in

separate or in combined treatments. ABSTRACTS: BA 30:308, Jan., 1956; RAE-B 44:52, Apr., 1956; TDB 52:939, Sept., 1955.

2069. Hopkins, T. L. and **Robbins, W. E. 1957** The Absorption, Metabolism and Excretion of C^{14}-Labelled Allethrin by House Flies. J. Econ. Ent. 50:684-687.
Almost all of the absorbed dose of allethrin was metabolized in 24 hours. No qualitative differences in metabolism: (allethrin alone, or with piperonyl butoxide). ABSTRACTS: BA 32:891(10495), Mar., 1958; RAE-B 46:171, Nov., 1958; TDB 55:460, Apr., 1958.

Hopkins, T. L., joint author. *See* Hoffman, R. A., et al., 1954.

2070. Hori, K. 1950 Morphological Studies on Muscoid Flies of Medical Importance in Japan. 1. Notes on the Common House-fly of Japan and Adjacent Territory. Oyo-Dobutsugaku-Zasshi 16(1/2):59-63. (In Japanese with English summary.)
Concerns *M. domestica vicina*. ABSTRACT: BA 27:2939, Nov., 1953.

2071. Hori, K. 1952 On Some Flies of Medical Importance Obtained from Korea and Adjacent Districts. Oyo-Dobutsugaku-Zasshi 17(1/2): 77-82. (In Japanese with English summary.)
Records 48 species falling in 19 genera and 6 families. ABSTRACT: BA 27:2939, Nov., 1953.

2072. Hornstein, I. and **Sullivan, W. N. 1953** The Role of Chlorinated Polyphenyls in Improving Lindane Residues. J. Econ. Ent. 46:937-940.
Chlorinated terphenyl film-forming materials prolong the residual effectiveness of volatile insecticides (such as lindane) by lowering the vapor pressure. ABSTRACT: RAE-B 42:131, Sept., 1954.

2073. Hornstein, I., Sullivan, W. N. and **Murphy, R. 1958** Fumigation Properties of Dow ET-57. J. Econ. Ent. 51(3):408-409.
DOW ET-57 has a vapor pressure which makes air concentration so low as to *not* be a health hazard. But increasing temperature raises this (and also the effectiveness).

2074. Hornstein, I., Sullivan, W. N. and **Tsao, C. 1955** Residual Effectiveness of Mixtures of Organic Phosphorus Insecticides with Chlorinated Terphenyls. J. Econ. Ent. 48:482-483.
Addition of chlorinated terphenyls to volatile organic phosphorus insecticides prolonged the insecticidal effectiveness of the residues. ABSTRACT: RAE-B 44:105, July, 1956.

2075. Hornstein, I., Sullivan, W. N., Tsao, Ching-hsi, and **Yeomans, A. H. 1954** The Persistence of Lindane-Chlorinated Terphenyl Residues on Outdoor Foliage. J. Econ. Ent. 47:332-335.
Terphenyl residues gave longer protection than lindane alone. Best to use an unstable emulsion or an emulsifying agent which readily decomposes after application. ABSTRACTS: RAE-A 43:51, Feb., 1955; RAE-B 43:25, Feb., 1955.

Hornstein, I., joint author. *See* Tsao, Ching-Hsi, et al., 1953, 1954; Sullivan, W. N., et al., 1953, 1955.

Horvath, J., joint author. *See* Vostal, Z., et al., 1958.

Hoskins, W. M., joint author. *See* Eldefrawi, M. E., et al., 1959; Fullmer, O. H., et al., 1951; McKenzie, R. E., et al., 1954; Menn, J. J., et al., 1957; Tahori, A. S., et al., 1953; Perry, A. S., et al., 1950, 1951.

2076. House, H. L. and **Barlow, J. S. 1958** Vitamin Requirements of the House Fly, *Musca domestica* L. (Diptera: Muscidae). Ann. Ent. Soc. Am. 51(3):299-302.
Seven vitamins considered a dietary requirement. Omission of riboflavin,

calcium pantothenate, nicotinic acid, or biotin causes high first instar mortality; of thiamine, third instar mortality, of choline or pyridoxine, high adult mortality.

2077. Howell, D. E. 1952 Fly Control in Oklahoma. Bull. Oklahoma Agric. Expt. Sta. B-385:1-20.
Stresses prevention of breeding. ABSTRACT: BA 27:2939, Nov., 1953.

2078. Howell, D. E. and **Fenton, F. A. 1951** The Repellency of a Pyrethrum-Thiocyanate Oil Spray to Flies Attacking Cattle. J. Econ. Ent. 37(5):677-680.
An oil base cattle spray, containing pyrethrum and thiocyanate was particularly effective against horn fly.

2079. Howell, D. E. and **Goodhue, L. D. 1955** Fly Repellents. Soap and Chem. Spec. 31(10):181, 185-189, 221. October.
(Five page reprint bears Title: "Laboratory and Field Tests on Fly Repellents".) Wettable powder formulations are superior to other types of sprays. R-326 an outstanding repellent. R-11 and R-44 strongly repellent to stable flies. Either is synergistic to R-236. ABSTRACT: TDB 53:255, Feb., 1956.

2080. Hoyt, C. P. 1953 Deformed Abdominal Tergites in *Musca domestica* Linnaeus. Pan Pacific Ent. 29:208-210.
Several deformed flies discovered in an in-bred strain. Environmental factors clearly not responsible. ABSTRACT: BA 29:714, Mar., 1955.

Hughes, J. C., joint author. *See* Lewis, C. T., et al., 1957.

2081. Hunter, P. E., Cutkomp, L. K. and **Kolkaila, A. M. 1958** Reproduction in DDT- and Diazinon-Treated House Flies. J. Econ. Ent. 51(5): 579-582.
A reduction in population was apparent in measurements of fertility, fecundity and survival rate of offspring. DDT-treated females also had a shorter life-span. ABSTRACT: RAE-B 48:17-18, Jan., 1960.

2082. Hunter, P. E., Cutkomp, L. K. and **Kolkaila, A. M. 1959** Reproduction Following Insecticidal Treatment in Two Resistant Strains of House Flies. J. Econ. Ent. 52(4):765.
For two additional strains of DDT-resistant flies, there was a reduction in all phases of reproduction, except that in one strain, egg-fertility was normal. ABSTRACTS: RAE-B 48:198-199, Nov., 1960; TDB 57:307, Mar., 1960.

Hunter, P. E., joint author. *See* Sokal, R. R., et al., 1955.

2083. Hurlbut, H. S. 1950 The Recovery of Poliomyelitis Virus after Parenteral Introduction into Cockroaches and Houseflies. J. Infect. Dis. 86(1):103-104.
Virus of mouse-adapted Lansing strain was inoculated into the haemocoel. Emulsions of houseflies surviving 12 days were inoculated intracerebrally into mice, which developed typical paralysis. ABSTRACT: BA 24:1569, 1950.

2084. Iaguzhinskaia, L. V. 1953 Reaktsiia na svet mukh, *Musca domestica* L. otravlennykh DDT. [Reaction to light of *Musca domestica* L. poisoned with DDT.] Med. Parazit. (Moskva) 3:242-246.

2085. Ihndris, R. W. and **Sullivan, W. N. 1958** Laboratory Fumigation Tests of Organic Compounds. J. Econ. Ent. 51(5):638-639.
Tested 78 compounds. Five halogenated phosphorus compounds and three halogenated hydrocarbons caused complete knockdown of houseflies in less than 2 hours. ABSTRACT: BA 33:1266(15742), Mar., 1959.

2086. Ikeda, Y. 1958 Insect Repellents and Attractants. II. On the Estimation of Efficiencies of Pyrethrins and Allethrin as Fly Repellent. Botyu Kagaku 23(1):33-36, 9 refs. (Summary in Japanese.)
Filter papers were impregnated with benzene solutions of test com-

pounds; lactose pellet place in center and *M. domestica vicina* exposed to papers. Amount of feeding used to measure repellency of test compound. ABSTRACT: RAE-B 49(4):77, Apr., 1961.

2087. Ikeda, Y. 1959 Behaviour of Housefly to Certain Organic Phosphorus Insecticides. Botyu-Kagaku 24(1):22-25, 5 refs. Kyoto. (Summary in Japanese.)
Filter papers impregnated with solutions of phosphorus insecticides, dried, and then moistened with sugar solutions. Exposed to *M. domestica vicina*. Results given. DDVP showed no repellent effect. ABSTRACT: RAE-B 49:112, June, 1961.

2088. Ikeshoji, T. and **Suzuki, T. 1959** Relationship Between Insecticide Resistance in House Flies and the Recovery of Knock-down. Jap. J. Exptl. Med. 29(5):481-491.
Tests with *M. domestica vicina*. No recovery noticed when dieldrin, diazinon or DDVP were applied topically to dorsum of thorax. Flies treated with DDT, lindane or lethane showed marked recovery. Remarks on detoxication of lindane. ABSTRACTS: BA 35:3993(46007), Aug. 15, 1960; RAE-B 50:286, Dec., 1962; TDB 57:960, Sept., 1960.

Ikeshoji, T., joint author. *See* Suzuki, T., et al., 1958.

2089. Iliyas, M. 1957 Observations on Fly Densities in Karachi. Medicus 13(4):159-169. Map.
Describes grill method technique of counting flies. Discusses fly counts on 4 different attractants. Lists prevalent types of flies. ABSTRACT: BA 33:1604-1605, Mar., 1959.

2090. Incho, H. H. and **Ault, A. K. 1954** The Toxicity to House Flies of Allethrin Analogs in Combination with Piperonyl Butoxide Analogs. J. Econ. Ent. 47:664-672.
The analogs furfuryl and thenyl are equal to or more effective than allethrin. The benzyl analog showed a lower activity than the other two. ABSTRACTS: BA 29:966, Apr., 1955; RAE-B 43:117, Aug., 1955; TDB 52:224, Feb., 1955.

2091. Incho, H. H. and **Greenberg, H. 1952** Synergistic Effect of Piperonyl Butoxide with the Active Principles of Pyrethrum and with Allethrolone Esters of Chrysanthemum Acids. J. Econ. Ent. 45:794-795.
As tested against houseflies by the turntable method, cinerins exhibited more synergism in combination with piperonyl butoxide than pyrethrins. Influence on insecticidal effectiveness is brought about by optical and geometric isomerism in the acid portion of the molecule. ABSTRACTS: RAE-B 41:71, May, 1953; TDB 50:352, Apr., 1953.

Incho, H. H., joint author. *See* Jones, H. A., et al., 1950.

2092. Ingram, R. L. 1955 Water Loss from Insects Treated with Pyrethrum. Ann. Ent. Soc. Am. 48:481-485.
Water loss not the direct cause of death; can be accounted for in terms of neurotoxic action of pyrethrum on a hypothetical regulatory mechanism that involves nervously controlled secretory activity on the part of the epidermal cells.

2093. Ingram, R. L., Larsen, J. R., Jr. and **Pippen, W. F. 1956** Biological and Bacteriological Studies on *Escherichia coli* in the Housefly, *Musca domestica*. Am. J. Trop Med. Hyg. 5:820-830.
E. coli was tagged with P^{32} and fed to flies. Fairly good correlation between radioactivity and bacterial counts, for several days. Slightly over 50 per cent of the ingested bacteria were recovered. Apparently no multiplication in alimentary tract. ABSTRACT: RAE-B 46:173, Nov., 1959.

Ionova, A. I., joint author. *See* Karavashkova, A. I., et al., 1957.

Isaias, J. M., joint author. *See* Salgado, S., et al., 1958.

2094. **Istituto Superiore di Sanita. 1956** Selected Scientific Papers. Vol. I, Pt. 1, 233 pp. Rome: Fondazione Emanuele Paterno, Viale Regina Elena, 299. Oxford: Blackwell Scientific Publications, 24-25 Broad Street.
A valuable collection of research papers, many of them treating with *M. domestica* in one aspect or another. ABSTRACT: TDB 54:504, Apr., 1957.

Ivanova, E. V., joint author. *See* Shura-Bura, B. L., et al., 1956, 1958.

2095. **Jacobson, M. 1952** Constituents of *Heliopsis* species. II. Synthesis of Compounds related to Scabrin. J. Am. Chem. Soc. 74:3423-3424.
Compounds were not effective against houseflies, even at high concentrations. ABSTRACT: RAE-B 41:125, Aug., 1953.

2096. **Jarman, R. T. 1959** The Deposition of Airborne Droplets on Dead Houseflies (*Musca domestica* L.). Bull Ent. Res. 50(2):327-332.
Droplets were dyed. In controlled wind (1 meter per sec.), collecting efficiency of a fly varied from 70 per cent (droplet diam. 27 micra) to about 200 per cent (droplet diam. 75 micra). Optimum diameter for deposition on flies in woodland, 20-40 micra. ABSTRACTS: BA 36:324 (2819), Jan. 1, 1961; RAE-B 47:151-152, Oct., 1959.

2097. **Jaujou, C., Caiman, R.** and **Adam, J. P. 1950** Apparition en Corse de mouches domestiques résistantes au DDT. [Appearance in Corsica of domestic flies resistant to DDT.] Bull. Soc. Path. Exot. 43:490-496.
Resistance appeared during second annual anti-anopheline campaign. Mortality of 50 per cent among resistant flies achieved 47 hours after 1 hour on a treated wall. Occurs with normal flies within 10 hours after same exposure. ABSTRACTS: TDB 48:208, Feb., 1951; BA 25:1706(18738), June, 1951.

2098. **Jenkins, D. W. 1956** A qué distancia vuelan las moscas. [To what distance do flies travel.] Medicina (Mexico) 36(745), April 10. Suppl. pp. 52-55.

2099. **Jenkins, D. W. 1956** A quelle distance une mouche peut-elle voler? [What Distance can a Fly Fly?] Naturaliste Canadien 83(5):95-102.
Some results presented on dispersion of flies and mosquitoes. Mainly a summary of the use of radioactive isotopes in marking insecticides, insects and protozoa in research on disease. ABSTRACT: BA 33:623, Feb., 1959.

Jensen, J. A., joint author. *See* Schoof, H. F., et al., 1952.

Jerace, E., joint author. *See* Bernie, A., et al., 1959.

2100. **Jettmar, H. M. 1956** Ueber die erworbene Resistenz medizinisch wichtiger Gliederfüssler gegen moderne Insektizide. [Concerning the acquired resistance of medically important arthropods to modern insecticides.] Wien. Klin. Wochenschr. 68:714-718.
Reviews an impressive amount of literature, including many items involving houseflies. Recognizes physiological, morphological, and behavioral resistance; distinguishes types of insecticides and recommends areas for research.

2101. **Jezioranska, A. 1954** [The Distribution of Domestic Fly in Warsaw area in 1949-1950.] Acta Parasitol. Polonica 2(1/6):1-15. (In Polish, with English summary.)
About 95-96 per cent of flies were *Muscidae*; *Musca domestica* was the most abundant. Other statistics given. ABSTRACT: BA 29:1743(17567), July, 1955.

2102. **Johnson, W. T., Langford, G. S.** and **Lall, B. S. 1956** Tests with Organic Phosphorus Insecticides for Fly Control. J. Econ. Ent 49:77-80.
Efficient and safe when used in sweet baits were: diazinon, Am. Cyanamid 4124, chlorthion, pirazinon and Bayer 13/59. For wall sprays: malathion, diazinon, Am. cyanamid 4124. ABSTRACTS: BA 30:2413(24208), Aug., 1956; RAE-B 45:39, Mar., 1957; TDB 53:937, July, 1956.

Johnson, W. T., joint author. *See* Langford, G. S., et al., 1954.

2103. **Johnston, E. F., Bogart, R.** and **Lindquist, A. W. 1954** The Resistance to DDT by Houseflies. Some Genetic and Environmental Factors. J. Heredity 45:177-182.
Purpose: To determine the mode of inheritance of resistance to DDT and some of the factors affecting resistance. Concludes that factor responsible for DDT resistance is particulate. ABSTRACT: TDB 52:101, Jan., 1955.

2104. **Jones, H. A., Schroeder, H. O.** and **Incho, H. H. 1950** Allethrin with Synergists. Soap and Sanit. Chem. 26:109:133-139.
Allethrin-synergist mixtures were compared with pyrethrin-synergist mixtures. Results (LO_{50} basis): allethrin-synergist combinations varied from being equal to being about half as effective against houseflies as pyrethrin mixtures. ABSTRACT: TDB 48:211, Feb., 1951.

Jones, H. A., joint author. *See* Schroeder, H. O., et al., 1957.

Jones, R. H., joint author. *See* Dicke, R. J., et al., 1955.

2105. **Jong, C. de 1959** Mens en Dier. 8. Zijn vliegen altijd schadelijk? [Men and Animals. VIII. Are Flies Always Harmful?] Ned. Milit. Genesk. T. 12:333-337.

2106. **Judd, W. W. 1955** Mites (Anoetidae) Fungi (*Empusa* spp.) and *Pollinia* of Milkweed (*Asclepias syriaca*) Transported by Calyptrate Flies. Canad. Ent. 87(8):366-369.
Found on *M. domestica* were mites (deutonymphs of *Bonomoia*) and fungi (*Empusa muscae*).

2107. **Judd, W. W. 1956** Results of a Survey of Calyptrate Flies of Medical Importance Conducted at London, Ontario During 1953. Am. Midland Nat. 56(2):388-405.
Four families, including 20 species made up 142,975 flies counted, of which 5,898 were *M. domestica.* Its peak of abundance fell on Oct. 23. Most were caught in traps near manure heaps, stables, and animal remains. ABSTRACT: BA 31:2998(33099), Oct., 1957.

2108. **Judd, W. W. 1958** Studies of the Byron Bog in Southwestern Ontario. VIII. Seasonal Distribution of Filth Flies. Am. Midland Nat. 60(1): 186-195.
Four families (including 19 species) represented among 20,848 flies trapped and counted between May 15 and Nov. 15, 1956. Of these 126 were *M. domestica.* Peak of abundance, July 25. ABSTRACT: BA 33: 2557-2258(27794), May, 1959.

2109. **Kalmykov, E. S. 1959** Effektivnost′ khlorofosa v bor'be s mukhami v gorode. [Effectiveness of chlorophos in eradication of flies in cities.] Med. Parazit. (Moskva) 28(1):53-56.
Spraying of outdoor sanitary installations and surrounding surfaces (fences, etc.) with chlorophos was very effective in controlling flies. *M. domestica* was the dominant species. ABSTRACT: BA 44:1579(21278), Dec. 1, 1963.

Kaluszyner, A., joint author. *See* Reuther, S., et al., 1956; Tahori, A. S., et al., 1958.

Kane, J., joint author. *See* Green, A. A., et al., 1954.

2110. Kano, R. 1951 Notes on the Flies of Medical Importance in Japan. IV. Flies of Hachijo Area. Jap. J. Exptl. Med. 21:223-227.
Lists *M. vicina* and *M. convexifrons.*

2111. Kao, C. M., Li, S. C., Kow, Y. F., Wei, P. H. and **Sun, T. S. 1958** A Preliminary Study of the Species and Seasonal Prevalence of the Fly Population of Poating. Acta. Ent. Sinica 8(4):335-339.
Lists *M. vicina* and *M. stabulans* both of which reach a peak in September. ABSTRACT: BA 36:2356(28831), May 1, 1961.

2112. Karavashkova, A. I., Ryk-Bogdaniko, M. G. and **Ionova, A. I. 1957** [Result of application of insecticide mixture of DDT in controlling flies.] Gig. I Sanit. 22(6):87-88. Moskva. (In Russian.)

2113. Katsuda, Y., Chikamotor, T. and **Nagasawa, S. 1958** Relationships between Stereoisomerism and Biological Activity of Pyrethroids. III. The Toxicity of Allethronyl Homochrysanthemates and the Related Compounds. Bull. Agric. Chem. Soc. Japan 22(6):393-398.
Tests were made on *M. domestica.* Allethronyl homochrysanthemate most highly toxic, especially the dextrorotatory form. Further elongation of ester linkage reduced toxicity. All lactones tested were non-toxic. ABSTRACT: BA 34:254(2827), Sept., 1959.

Kearns, C. W., joint author. *See* Lovell, J. B., et al., 1959; Blum, M. S., et al., 1956; Dahm, P. A., et al., 1951; Mitlin, S. S., et al., 1957; Sternburg, J., et al., 1950, 1952, 1953, 1956.

2114. Keiding, J. 1953 Development of Resistance in the Field and Studies of Inheritance. Repr. from Transactions of the IXth International Congress of Entomology. 2:340-345. Amsterdam.
Reports various degrees of resistance to DDT, BHC, Chlordane, Toxaphene. Experiments tend to show dominance of susceptibility. ABSTRACT: RAE-B 41:123, Aug., 1953.

2115. Keiding, J. 1956 Resistance to Organic Phosphorus Insecticides of the Housefly. Science 123(3209):1173-1174.
Reports significant field resistance on the part of houseflies in Denmark to Bayer 21/199 (Resitox). Is an organic thionophosphate. ABSTRACTS: BA 31:1528(15505), May, 1957; TDB 53:1394, Nov., 1956.

2116. Keiding, J. 1958 The Relation between Resistance to Insecticides and the Effect of Cetyl Bromoacetate in Danish Strains of *Musca domestica* L. Indian J. Malariol. 12:453-468.
Used 2 susceptible strains and 8 strains from farms, resistant either to chlorinated hydrocarbons alone, or to these plus organophosphorus compounds. A positive correlation between resistance to insecticides and tolerance to knock-down action of C B A. (Disagrees with Ascher, 1958.) ABSTRACT: RAE-B 49:230, Oct., 1961.

2117. Keiding, J. 1969 House-fly Control and Resistance to Insecticides on Danish Farms. Ann. Appl. Biol. 47(3):612-618.
Discusses development of resistance between 1945 and 1958 to parathion, diazinon, Resitox, Dipterex and thiourea. Compares organophosphorous compounds with chlorinated hydrocarbons. Doubts possibility of finding 2 "negatively correlated" insecticides that can be alternated indefinitely. ABSTRACTS: BA 35:1726(19833), April, 1960; RAE-B 48:137-138, Aug., 1960; TDB 57:424, April, 1960.

Keiding, J., joint author. *See* Boggild, O., et al., 1958; Barbesgoard, P., et al., 1955; Davies, M., et al., 1958.

Keister, M. L., joint author. *See* Buck, J. B., et al., 1952.

2118. Keller, J. C. 1955 Poison Bait for the Control of House Flies on Military Reservations. J. Econ. Ent. 48:528-529.
Baits consisting of 1 or 2 per cent chlorthion in granulated sugar gave excellent control of houseflies. ABSTRACTS: BA 30:1486(15143), May, 1956; RAE-B 44:169, Nov., 1956; TDB 53:380, Mar., 1956.

2119. Keller, J. C. and **Wilson, H. G. 1955** Granular Baits for the Control of House Flies. J. Econ. Ent. 48:642-643.
Baits with insoluble granular carriers were devised to prevent dissolving, soaking away, etc., Sand and cornmeal worked well with chlorthion or malathion. Diazinon preparations did not last long enough to be effective. ABSTRACTS: BA 30:2076(20858), July, 1956; RAE-B 45:2, Jan., 1957; TDB 53:679, May, 1956.

2120. Keller, J. C., Wilson, H. G. and **Smith, C. N. 1955** Poison Baits for the Control of Blow Flies and House Flies. J. Econ. Ent. 48:563-565.
Chlorthion sugar bait gave excellent control around military camps at concentrations averaging 1-2 per cent insecticide in 200 grams of bait. ABSTRACTS: BA 30:1486(15144), May, 1956; RAE-B 44:170, Nov., 1956; TDB 53:379, Mar., 1956.

2121. Keller, J. C., Wilson, H. G. and **Smith, C. N. 1956** Bait Stations for the Control of House Flies. J. Econ. Ent. 49:751-752.
Permanent bait stations, using Bayer L 13/59, proved satisfactory. Unit consisted of tongue-depressor, coated with toxicant. Various baits were effective from 28 to 98 days. ABSTRACTS: BA 31:1232(12389), Apr., 1957; RAE-B 46:20, Feb., 1958; TDB 54:504, Apr., 1957.

Keller, J. C., joint author. *See* Gahan, J. B., et al., 1957.

2122. Kenaga, E. E. 1957 Some Biological, Chemical and Physical Properties of Sulfuryl Fluoride as an Insecticidal Fumigant. J. Econ. Ent. 50(1):1-6.
Doses effective on *M. domestica* adults and pupae, per cubic foot of space, were 0·96 oz. and 1·36 oz., respectively. Also tested on *Periplaneta americana* (L) and *Blattella germanica* (L). ABSTRACT: RAE-B 46:42, March, 1958.

2123. Kerr, R. W. 1951 Adjuvants for Pyrethrins in Fly Sprays. Bull. Commonwealth Sci. and Indust. Res. Organ. Australia. 261. 63 pp., illus.
Tested 32 materials from Australian plants, using houseflies. Five or six were effective synergists. ABSTRACTS: BA 26:2067, Aug., 1952; RAE-B 41:165, Oct., 1953; TDB 49:564, May, 1952.

2124. Kerr, R. W., Venables, D. G., Roulston, W. J. and **Schnitzerling, H. J. 1957** Specific DDT-Resistance in Houseflies. Nature 180(4595): 1132-1133.
Worked with 3 strains. "D" made resistant by use of DDT in larval medium; "L" made resistant by breeding only from late-emerging flies (Pimentel, et al., 1951); U- unselected. Males, in all cases, less resistant than females.

Khan, N. H., joint author. *See* Busvine, J. R., et al., 1955.

Khudadov, G. D., joint author. *See* Vashkov, V. I., et al., 1958.

2125. Kilgore, L. B. 1950 N-substituted Alkenyloxyacetamides. U.S. Patent 2,504,427; Appln. August 19, 1947. Issd. April 18, 1950.
Several acetamides are repellent to house flies and mosquitoes. They are also insecticidal. ABSTRACT: BA 24:2931, 1950.

Killy, B. A., joint author. *See* Desai, R. M., et al., 1958.

2126. Kilpatrick, J. W. 1955 The Control of Rural Fly Populations in South-Eastern Georgia with Parathion-impregnated Cords. Am J. Trop. Med. Hyg. 4(4):758-761.
Impregnated cords substantially reduced fly populations at a cost comparable to that of a standard DDT residual application. ABSTRACT: BA 30:1155(11766), Apr., 1956.

2127. Kilpatrick, J. W. and **Bogue, M. D. 1956** Adult Fly Production from Garbage Can Sites and Privy Pits in the Lower Rio Grande Valley. Am. J. Trop. Med. Hyg. 5(2):331-339.
Resistance to Dieldrin had become so great that flies were more numerous in treated than in untreated sites. ABSTRACTS: BA 31:620(6248), Feb., 1957; RAE-B 46:57, Apr., 1958; TDB 53:1284, Oct., 1956.

2128. Kilpatrick, J. W. and **Quarterman, K. D. 1952** Field Studies on the Resting Habits of Flies in Relation to Chemical Control. II. In Rural Areas. Am. J. Trop. Med. Hyg. 1(6):1026-1031.
Flies tend to concentrate at night. Selective application of space sprays after dark might result in a more economical expenditure of time and materials. ABSTRACTS: BA 27:1687, June, 1953; RAE-B 42:122, Aug., 1954; TDB 50:350, Apr., 1953.

2129. Kilpatrick, J. W. and **Schoof, H. F. 1954** House Fly Control in Dairies Near Savannah, Georgia, with Residual Applications of CS=708, NPD and Malathion. J. Econ. Ent. 47:999-1001.
These 3 insecticides tested against resistant house flies. CS-708, applied in various ways, gave longest lasting results. Excellent control for 8 weeks. ABSTRACTS: BA 29:2260, Sept., 1955; RAE-B 43:191, Dec., 1955.

2130. Kilpatrick, J. W. and **Schoof, H. F. 1955** DDVP as a Toxicant in Poison Baits for House Fly Control. J. Econ. Ent. 48:623-624.
This organophosphorus toxicant is effective at concentrations far below a number of others, but requires frequent application to insure low fly densities. ABSTRACTS: BA 30:1486(15145), May, 1956; RAE-B 44:175, Nov., 1956; TDB 53:379, Mar., 1956.

2131. Kilpatrick, J. W. and **Schoof, H. F. 1956** Fly Production in Treated and Untreated Privies. Publ. Health Repts. 71(8):787-796.
For reasons as yet unknown, treatment with dieldrin, aldrin, BHC, or chlordane increases the breeding of *M. domestica* in Savannah, Georgia. ABSTRACTS: BA 31:1821(19028), June, 1957; RAE-B 46:161, Oct., 1958; TDB 54:98, Jan., 1957.

2132. Kilpatrick, J. W. and **Schoof, H. F. 1956** The Use of Insecticide Treated Cords for Housefly Control. Publ. Health Repts. 71(2):144-150.
Cotton cords treated with parathion-xylene solution gave seasonal control of housefly populations in dairy barns with very little contamination of the air, and no significant harm to man. ABSTRACTS: RAE-B 44:126, Aug., 1956; TDB 53:816, June, 1956.

2133. Kilpatrick, J. W. and **Schoof, H. F. 1957** House Fly Control in Dairy Barns with Residual Treatments of Phosphorus Compounds. J. Econ. Ent. 50(1):36-39.
Six organophosphorus compounds were field-tested during 1954-1955 for the control of multi-resistant housefly populations. Locale more influential than formulations in variability of results. ABSTRACTS: BA 31:2998 (33101), Oct., 1957; RAE-B 46:43, Mar., 1958; TDB 54:765, June, 1957.

2134. Kilpatrick, J. W. and **Schoof, H. F. 1957** Fly Production Studies in Urban, Suburban, and Rural Privies in Southeastern Georgia. Am. J. Trop. Med. Hyg. 6(1):171-179.
Of the 97 species of Diptera collected, *M. domestica* was never very

important as to numbers. Maximum prevalence at any one time, 6 per cent. ABSTRACTS: BA 31:2998(33100), Oct., 1957; TDB 54:1137, September, 1957.

2135. Kilpatrick, J. W. and **Schoof, H. F. 1958** House Fly Control Studies in Chatham County, Georgia. J. Econ. Ent. 51(6):908-910.
A comparative study and evaluation of the effectiveness of chlorinated hydrocarbons, organophosphorus compounds and of insecticide impregnated cords. ABSTRACTS: BA 33:2258(27795), May, 1959; RAE-B 48:59, Mar., 1960.

2136. Kilpatrick, J. W. and **Schoof, H. F. 1958** A Field Strain of Malathion-Resistant House Flies. J. Econ. Ent. 51(1):18-19.
Responses of a strain of houseflies to malathion baits reveal both a physiologic and behavioristic resistance, with the latter probably the more important for loss of control. Said to be first reported resistance to this insecticide in the United States. ABSTRACTS: BA 32:1796(21344), June, 1958; TDB 55:706, June, 1958.

2137. Kilpatrick, J. W. and **Schoof, H. F. 1959** The Effectiveness of Ronnel as a Cord Impregnant for Housc Fly Control. J. Econ. Ent. 52(4):779-780.
This organophosphorus compound, of low mammalian toxicity, was found more effective as a cord impregnant than as a residual insecticide. ABSTRACT: RAE-B 48:201, Nov., 1960.

2138. Kilpatrick, J. W. and **Schoof, H. F. 1959** A Semiautomatic Liquid Fly Bait Dispenser. J. Econ. Ent. 52(4):775:776.
Use of a semiautomatic chicken-watering device for dispensing fly bait provided treatment for a 6-week period. ABSTRACT: RAE-B 48:200, Nov., 1960.

2139. Kilpatrick, J. W. and **Schoof, H. F. 1959** Interrelationship of Water and *Hermetia illucens* Breeding to *Musca domestica* Production in Human Excrement. Am. J. Trop. Med. Hyg. 8(5):597-602.
Houseflies cannot breed in semiliquid conditions produced by larvae of *Hermetia*. Use of insecticides in privies reduces *Hermetia*, thus altering conditions and permitting insecticide-resistant houseflies to flourish. ABSTRACTS: BA 35:515(6000), Jan. 15, 1960; RAE-B 49:176, Aug., 1961; TDB 57:305, Mar., 1960.

Kilpatrick, J. W., joint author. *See* Maier, P. B., et al., 1952; Fay, R. W., et al., 1958; Ogden, L. J., et al., 1958; Quarterman, K. D., et al., 1954; Schoof, H. F., et al., 1957, 1958, 1959.

2140. King, H. L., Guyer, G. and **Ralston, N. P. 1953** Fly Control in Dairy Barns. Mich. State Coll. Agric. Expt. Sta. Quart. Bull. 36:179-186.
Residual wall sprays of malathion or diazinon satisfactory to control *M. domestica* and *Stomoxys calcitrans* in conventional dairy barns. Loose-housing barns required direct spraying of animals. In milking parlors, bags sprinkled with malathion-bait mixture gave excellent control. ABSTRACT: BA 28:1935, Aug., 1954.

King, H. L., joint author. *See* Guyer, G. E., et al., 1954, 1956.

2141. King, W. V. 1950 DDT-resistant House Flies and Mosquitoes. J. Econ. Ent. 43(4):527-532.
Against DDT-resistant flies, most consistently effective insecticide was chlordane emulsion. Dieldrin emulsion was next in effectiveness. ABSTRACTS: BA 25:1409, May, 1951; RAE-B 39:61, Apr., 1951; TDB 48:413, Apr., 1951.

2142. King, W. V. 1954 Chemicals Evaluated as Insecticides and Repellents at Orlando, Fla. Agric. Handb. U.S. Dept. Agric. No. 69. 397 pp.

Washington, D.C.
Reports relative effectiveness of many substances tested against *M. domestica* as residual deposits and as space sprays. ABSTRACT: Hocking, B. 1960 Smell in Insects, p. 135 (3.68).

2143. **Kirchberg, E. 1951** Untersuchungen über die Fliegen-fauna menschlicher Fäkalien. [Researches on the fly-fauna of human excrement.] Zeit. Hyg. Zool. 39(5/6):129-139.
Predominant species were *Lucilia sericata* (53·5 per cent); *Sarcophaga* spp. (12·7 per cent); *Muscina stabulans* (12·1 per cent). Only one female *M. domestica* in 2,346 total; less than 0·1 per cent. Negligible catch of *M. domestica* discussed. ABSTRACT: TDB 48:1152, Dec., 1951.

2144. **Kirchberg, E. (von) 1957** Über einige Musciden von hygienischer Bedeutung. [Concerning several Muscidae of sanitary importance.] Verhandl. Deutsch. Ges. Angew. Ent. 14:36-42.
Contains a practical key to more common species and groups. Author considers *M. domestica* to have been somewhat overemphasized among medically important forms.

Kirk, R. L., joint author. *See* Maelzer, D. A., et al., 1953.

2145. **Kitselman, C. H., Dahm, P. A.** and **Borgmann, A. R. 1950** Toxicologic Studies of Aldrin (Compound 118) on Large Animals. Am. J. Vet. Res. 11(41):378-381.
Several samples from treated alfalfa were analyzed for aldrin residues by biological assays, using houseflies. Not more than 8 parts per million were found at any one time.

Kivichenko, A. T., joint author. *See* Morozova, V. P., et al., 1957.

Klechedov, G. D., joint author. *See* Vashkov, V. I., et al., 1958.

Klechetova, A. M., joint author. *See* Vashkov, V. I., et al., 1958.

2146. **Klimmer, O. R. 1955** Experimentelle Untersuchungen über die Toxikologie insecticider chlorierter Kohlenwasserstoffe. [Experimental studies on the Toxicology of insecticidal chlorinated hydrocarbons.] Arch. Exp. Pathol. Pharmakol. (Berlin) 227(2):183-195.
Treats of tests with *M. domestica*, also toxic effects on white rats, including histological changes in nervous tissue.

Klock, J. W., joint author. *See* Pimentel, D., et al., 1954.

2147. **Knapp, F. W.** and **Knutson, H. 1958** Reproductive Potential and Longevity of Two Relatively Isolated Field Populations of Insecticide-Susceptible House Flies. J. Econ. Ent. 51(1):43-45.
Ellsworth (Kansas) strain produced 24·3 adult offspring per female parent as compared with 49·8 produced by the Wilmore (Kansas) population. Difference due primarily to relatively high survival of Wilmore ova, laid mostly during second and third day of adult life, also during the 10th-18th day period. ABSTRACTS: BA 32:1791(21283), June, 1968; RAE-B 47:28, Feb., 1959; TDB 55:704, June, 1958.

2148. **Knapp, F. W., Terhaar, C. J.** and **Roan, C. C. 1958** Dow ET-57 as a Fly Larvicide. J. Econ. Ent. 51(3):361-362.
Two larvicides tested under caged hens. Dow ET-57 gave control of 78 per cent to 99 per cent (end of first and second weeks); malathion, a corresponding 0 per cent to 67 per cent. ABSTRACT: TDB 55:1289, Nov., 1958.

2149. **Knipe, F. W.** and **Frings, H. 1953** Oviposition by House Flies. J. Econ. Ent. 46:155-156.
Location of eggs related to length of ovipositor which can be extended up to 12 mm. ABSTRACTS: RAE-B 41:142, Sept., 1953; TDB 50:861, 1953.

2150. **Knipling, E. B.** and **Sullivan, W. N. 1957** Insect Mortality at Low Temperatures. J. Econ. Ent. 50(3):368-369.
Housefly showed 100 per cent mortality at −15°C for 60-minute exposure or −20°C for 30-minute exposure. Other aircraft "hitchhikers" also studied. ABSTRACTS: BA 31:3583(39768), Dec., 1957; TDB 54:1475, Dec., 1957; RAE-B 46:93, June, 1958.

2151. **Knipling, E. B.** and **Sullivan, W. N. 1958** The Thermal Death Points of Several Species of Insects. J. Econ. Ent. 51(3):344-346.
Tested 14 species of insects, including *M. domestica.* All were dead when exposed 15 minutes at 60°C. Has quarantine significance. Insects may be killed in uninsulated aircraft parked in the sun.

2152. **Knipling, E. F. 1952** Statements on the Insecticide-Resistance Problem in the United States—Comments of Dr. Knipling at Royal Sanitary Institute of Health Congress at Margate, England—April 22-25, 1952. To be published in J. Roy. Sanit. Institute.
Sets forth problems in control caused by resistance to various insecticides. Possibilities in procedure.

2153. **Knipling, E. F.** and **Bishopp, F. C. 1952** Progress in Medical Entomology in the United States. Trans. IXth Internatl. Congr. Ent. 1: 929-934.
A resumé of American research. Due attention is given to synanthropic flies. ABSTRACT: BA 30:885(8988), Mar., 1956.

2154. **Knipling, E. F.** and **McDuffie, W. C. 1957** Controlling Flies on Dairy Cattle and in Dairy Barns. Bull. W. H. O. 16(4):865-870. Geneva.
Discusses house-flies as well as blood-sucking species. Covers use of insecticides, also liquid and dry baits. ABSTRACT: TDB 55:112, Jan., 1958.

2155. **Knutson, H. 1959** Changes in Reproductive Potential in House Flies in Response to Dieldrin. Misc. Publ. Ent. Soc. Am. 1(1):27-32. Washington, D.C.
Discusses various biological characteristics which may be found to differ in resistant and normal strains. Survey covers behavior differences, morphological changes and alterations in reproductive potential. ABSTRACTS: RAE-B 48:211, Dec., 1960; TDB 58:149, Jan., 1961.

2156. **Knutson, H., Afifi, S. E. D.** and **Gray, T. M. 1958** Reproductive Potential in Relatively Isolated House Fly Populations following Repeated Insecticide Applications. J. Kans. Ent. Soc. 31(4):257-266.
Major, innate differences between field and lab.-reared populations, both insecticide-susceptible and insecticide-resistant. Concerns survival of various life-history stages, also over-wintering. Control measures, certain in the laboratory may not be so in the field. ABSTRACTS: BA 33:2895 (35375), June, 1959; RAE-B 48:189-190, Nov., 1960.

Knutson, H., joint author. *See* Afifi, S. E. D., et al., 1956; Knapp, F. W., et al., 1957, 1958; Ouye, M. T., et al., 1957.

Kobayashi, H., joint author. *See* Tsukamoto, M., et al., 1957.

2157. **Kobayashi, J. 1958** Studies on Habronemiasis in Kanto District, Japan. I. Epidemiological Studies on Habronemiasis. II. Experimental Transmission of *Habronema microstoma* by Various Species of Flies. III. The Vectors of *Habronema microstoma* among Flies in Natural Environment. O-Chanomigu Medical Magazine. 6(7):54-77 (868-891).
H. microstoma (temperate zone form) predominated in horses examined, Eggs and larvae transmitted to 16 species of flies. *Stomoxys calcitrans. Lyperosia exigua* and *Musca hervei* positive. *Stomoxys* believed best transmitter in nature.

2158. Kocher, C. and **Ascher, K. R. S. 1954** Topical Application of Organic Solvents to Houseflies. Riv. di Parassit. 15:103-109.
Of 47 substances tested, 14 (among them petroleum ether, benzene, and cyclohexane) were found non-toxic at a rate of 0·0004 ml. per fly. Solvents were applied to ventral abdomen of houseflies, kept at 23°C. ABSTRACT: RAE-B 44:203, Dec., 1956.

2159. Kocher, C., Roth, W. and **Treboux, J. 1953** Die Bekämpfung resistenter Stubenfliegen (*Musca domestica* L.) mit Diazinon. [The combatting of resistant houseflies (*Musca domestica* L.) with diazinon.] Anz. Schädlingsk. 26(5):65-71.
Flies resistant to DDT and other insecticides died within 15 minutes after treatment with diazinon spray powders. Neither livestock nor personnel were harmed. ABSTRACTS: BA 27:3151, Dec., 1953; RAE-B 42:193, Dec., 1954.

2160. Kocher, C. and **Treboux, J. 1957** Über die insektizide Wirkung von Diazinonbelägen in Abhängigkeit der Bruhenkonzentration. [The insecticidal effect of Diazinon deposits in relation to spray concentration.] Anz. Schädlingsk. 30(7):104-107. Berlin.
Population of *M. domestica* resistant to DDT exposed for 2 hours to deposits of diazinon. Effectiveness increased in general as the quantity of spray in which the insecticide was applied, decreased. Believed due to smaller quantity of more concentrated spray being deposited in discontinuous droplets. ABSTRACT: RAE-B 47:36, Mar., 1959.

Kocher, C., joint author. *See* Ascher, K. R. S., et al., 1954; Wiesmann, R., et al., 1951.

2161. Kohls, R. E., Todd, A. C. and **Dicke, R. J. 1957** Development of the House-fly *Musca domestica* L. when Exposed to Phenothiazine in Artificial Cultures. Veterinary Medicine 52:108-110.
Phenothiazine affected larval development and adult emergence in direct proportion to its concentration, but possible effectiveness as a cattle feed additive is doubtful.

2162. Kohls, R. E., Todd, A. C. and **Dicke, R. J. 1958** Size and Rate of Development of House Fly Larvae in Artificial Cultures Containing Phenothiazine. J. Parasit. 44(5):522.
Larvae of flies reared in cultures containing 2 g. of 2μ particle size were much smaller than check larvae; larvae in cultures containing 4 g. were still smaller. ABSTRACT: BA 33:1922(23806), April, 1959.

Kolkaila, A. M., joint author. *See* Hunter, P. E., et al., 1958, 1959.

2163. Konecky, M. S. and **Mitlin, N. 1955** Chemical Impairment of Development in House Flies. J. Econ. Ent. 48:219-220.
Studies to measure *indirect* effect of chemical substances rather than direct kill. The use of mitotic poisons and anti-metabolites may be a fruitful approach. ABSTRACTS: BA 30:308, Jan., 1956; RAE-B 44:57, Apr., 1956; TDB 52:938, Sept., 1955.

Konecky, M. S., joint author. *See* Mitlin, N., et al., 1954, 1955.

Koshi, T., joint author. *See* Damodar, P., et al., 1962.

Koshimizu, N., joint author. *See* Ogata, K., et al., 1960.

Kow, Y. T., joint author. *See* Kao, C. M., et al., 1958.

2164. Krastin, N. I. 1952 [The Decipherment of the Cycle of Development of the Nematode *Thelazia skrjabini* Erschow, 1928, a Parasite of the Eyes of Cattle.] Dokl. Akad. Nauk, SSSR (N.S.) 82:829-831. Moscow. (In Russian.)
Musca spp. serve as intermediate hosts for *Thelazia*. Viviparous *M.*

convexifrons for *T. rhodesi* and *T. gulosa*; oviparous *M. amica* for *T. skrjabini*. Oviparous *M. autumnalis* and viviparous *M. larvipara*, replace above species in the European part of the Soviet Union. ABSTRACT: RAE-B 42:97, July, 1954.

Krishnamurthy, B. S., joint author. *See* Pal, R., et al., 1953, 1954; Sharma, M. I. D., et al., 1957, 1958.

2165. Krueger, H. R. and **Casida, J. E. 1957** Toxicity of Fifteen Organophosphorus Insecticides to Several Insect Species and to Rats. J. Econ. Ent. 50(3):356-358.

Includes tables giving LD 50's of organophosphorus insecticides for 6 insect species and rats, and of paraoxon for 22 additional species of insects. Paraoxon was the most toxic to flies. ABSTRACT: RAE-B 46:91, June, 1958.

2166. Krueger, H. R. and **Casida, J. E. 1961** Hydrolysis of Certain Organophosphate Insecticides by House Fly Enzymes. J. Econ. Ent. 54(2): 239-244.

Presence of manganese ion enhanced *in vitro* enzyme activity of house fly homogenates. More than one housefly esterase acts on the phosphates. ABSTRACTS: RAE-B 50:107, May, 1962; TDB 58:962, Aug., 1961.

2167. Krueger, H. R. and **O'Brien, R. D. 1959** Relationship Between Metabolism and Differential Toxicity of Malathion in Insects and Mice. J. Econ. Ent. 52(6):1063-1067.

Eleven metabolites found in *Blattella germanica*, *Periplaneta americana* and *M. domestica*, and seven in mice. Most were identified. Degradation of malathion more extensive in mice than insects. Malaoxon production is correspondingly lower. ABSTRACTS: BA 35:2231(25375), May, 1960; RAE-B 49:29, Feb., 1961.

Krystanek, E., joint author. *See* Kulesza, J., et al., 1956.

2168. Kuenen, D. J. 1958 Competition in Laboratory Cultures of *Musca domestica*. Proc. 10th Int. Congr. Ent. Section on Behaviour, including Social Insects. Section on Ecology. pp. 767-774. Ottawa.

Records made of number and weight of puparia formed from various concentrations of eggs of 100 to 3,000 per liter container of culture media. Cultures gave 80-95 per cent hatching. ABSTRACT: RAE-B 47:97, July, 1959.

Kuklina, N. P., joint author. *See* Sukhova, M. N., et al., 1959.

2169. Kulesza, J. 1956 Z zagadnień aktywnósci kontaktowych środków dezynseckcyjnych. Przemysl Chemiczny. 12:581-586.

Tests on *M. domestica* of the effect when other compounds or insecticides are added to DDT. Addition of substances facilitating crystallization exerts a positive influence on the activity of the product.

2170. Kulesza, J. and **Krystanek, E. 1956** Przyczynek do selektywnego tepienia muchy domowej w środowiskach wiejskach. [A contribution to the selective destruction of house flies in the rural areas.] Roczniki Pánstwowego Zaktadu Hig. (Warsaw) 7:543-553. (English summary.)

Flies found to avoid feeding places that contain small amounts of p-dichlorbenzene. HCH and DDT sprays utilized. Conclusion drawn to reduce use of contact poisons by utilizing photo- and thermotropism of flies. ABSTRACT: TDB 54:503, Apr., 1957.

Kulp, W. L., joint author. *See* Hawley, J. E., et al., 1951.

2171. Kun, K. Y., Sun, Y. C., Ma, P. Y. and **Chang, J. T. 1958** Studies on House Flies Resistance to DDT. I. The Development of Resistance to DDT and BHC. Acta. Ent. Sinica 8(1):57-66. (In Chinese, with

English summary.)
Resistance to DDT increased gradually first 20 generations and more rapidly thereafter. BHC resistance increased gradually for 8 generations, more rapidly for next seven, then fluctuated. ABSTRACTS: BA 35:5233 (61146), Nov. 1, 1960; RAE-B 47:191, Dec., 1959.

2172. Laake, E. W., Howell, D. E., Dahm, P. A., Stone, P. C. and **Cuff, R. L. 1950** Relative Effectiveness of Various Insecticides for Control of House Flies in Dairy Barns and Horn Flies on Cattle. J. Econ. Ent. 43(6):858-861.
Testing done at three state colleges on effectiveness of four insecticides in current use. Sanitation appeared to be a major, controlling factor in density of flies around barns. ABSTRACTS: RAE-B 39:121, July, 1951; TDB 48:590, 1951.

2173. LaBrecque, G. C., Bowman, M. C., Acree, F., Jr. and **Smith, C. N. 1959** Absorption of DDT and its Absorption and Conversion to DDE by House Flies Exposed to Residues. J. Econ. Ent. 52(3): 528-529.
Different colonies of flies picked up DDT at different rates. A colony selected with malathion converted DDT to DDE in relatively short exposures. ABSTRACTS: BA 33:3945(47165), Aug., 1959; RAE-B 48:168, Sept., 1960.

2174. LaBrecque, G. C. and **Wilson, H. G. 1957** House Fly Resistance to Organophosphorus Compounds. Agric. Chem. 12(9):46-47, 147, 149. Baltimore, Maryland.
Flies displayed marked (many-fold) resistance to four contact sprays and two baits. ABSTRACTS: BA 35:2470(28009), May, 1960; RAE-B 47:85, June, 1959.

2175. LaBrecque, G. C. and **Wilson, H. G. 1958** Resistance of Houseflies (*Musca domestica* L.) in the United States to Organophosphorus Insecticides. Indian J. Malariol. 12:423-426.
Resistance now found very general. Deep concern regarding fly control in immediate future, especially in Florida poultry houses.

2176. LaBrecque, G. C. and **Wilson, H. G. 1959** Laboratory Tests with Sixty-five Compounds as Repellents against House Flies. Florida Ent. 42(4):175-177.
Three compounds proved effective for more than 90 days. ABSTRACTS: BA 35:1726(19835), April, 1960; RAE-B 49:193, Sept., 1961.

2177. LaBrecque, G. C., Wilson, H. G., Bowman, M. C. and **Gahan, J. B. 1959** Studies on the Development of Resistance to DDT and Malathion in House Flies. Florida Ent. 42(2):69-71.
Tables show the effectiveness of these two insecticides used individually, alternately, and in combination. ABSTRACTS: BA 35:1726(19834), April, 1960; RAE-B 49:97, May, 1961.

2178. LaBrecque, G. C., Wilson, H. G. and **Gahan, J. B. 1958** Synergized Pyrethrins and Allethrin Baits for the Control of Resistant House Flies. J. Econ. Ent. 51(6):798-800.
Insecticides were effective in sugar baits when synergized by piperonyl butoxide in ratio of 1:10. ABSTRACTS: BA 33:2258(27797), May, 1959; RAE-B 48:52, Mar., 1960.

2179. LaBrecque, G. C., Wilson, H. G. and **Gahan, J. B. 1958** Resistance of House Flies in Florida to Organophosphorus Insecticides. J. Econ. Ent. 51(5):616-617.
Reports a great increase in resistance, especially to Dipterex, since 1954 and 1956 surveys.

2180. LaBrecque, G. C., Wilson, H. G. and **Smith, C. N. 1959** Effectiveness of Two Carbamates against DDT and Malathion-resistant House Flies. J. Econ. Ent. 52(1):178-179.
Methyl carbamate and pyrolan show promise but may not be effective in some areas; may not long remain so in others. ABSTRACTS: BA 33: 3222(39233), July, 1959; RAE-B 48:97, May, 1960.

LaBrecque, G. C., joint author. *See* Cole, M. M., et al., 1959; Wilson, H. G., et al., 1958, 1959; Peffly, R. L., et al., 1956; Schmidt, C. H., et. al., 1959.

2181. LaFace, L. 1952 Sul comportamento ereditario della resistenza ad alcuni insetticidi in *Musca domestica.* [On the hereditary behavior of resistance of certain insecticides in *Musca domestica.*] (a) Riv. di Parassit. 13:57-60. Jan. (b) Rend. Ist. Sup. Sanit. 15:376-380.
The degree of resistance greatly varies after selection with the insecticide over many generations. There is no evidence of simple Mendelian inheritance in successive generations. ABSTRACTS: TDB 49:1004, Oct., 1952; BA 27:792, Apr., 1953.

2182. La Face, L. 1956 Contributo alle ricerche sui fattori determinanti la resistenza al DDT e al chlordane in *Musca domestica.* [Contribution to research on the factors determining resistance to DDT and chlordane in *Musca domestica.*] Rend. Ist. Sup. Sanit. (Roma) 19(4-5):363-397.
DDT-resistant flies crossed with chlordane-resistant. Descendants were less resistant to both. Resistance appears to depend on specific factors.

2183. LaFace, L. 1956 La mosca domestica; el enemigo en nuestra propia mesa. [The domestic fly; the enemy on our very table.] Medicina; Revista Mexicana 36:164-167.

2184. LaForge, F. B., Gersdorff, W. A., Green, N. and **Schechter, M. S. 1952** Allethrin-type Esters of Cyclopropane-carboxylic Acids and their Relative Toxicities to House Flies. J. Org. Chem. 17(3):381-389.
The most toxic ester, that of dl-dihydrochrysanthemum monocarboxylic acid was equal in toxicity to natural pyrethrins. ABSTRACT: RAE-B 41: 124, Aug., 1953.

LaForge, F. B., joint author. *See* Matsui, M., et al., 1952; Schechter, M. S., et al., 1951.

2185. Laird, M. 1959 Protozoa, including an Entozoan of Guinea Pigs from House Flies. Canad. J. Zool. 37:467-468.
Of 179 *M. domestica* examined, 7 harbored *Octosporea muscae-domesticae*, 2 *Herpetomonas muscarum*, 2 *Retortomonas caviae*; the last probably ingested from moist, fresh feces of guinea pigs. ABSTRACT: BA 35:2700 (30652), June, 1960.

2186. Lal, B. N. and **Srivastava, S. B. 1950** Control of Fly Breeding in Composting. I. Effect of Chemicals on Fly Breeding. Indian J. Agric. Sci. 20(2):239-250.
Lime, borax, wood ash, DDT, gamexane and bleaching powder were tested for effectiveness in prevention of fly breeding in manure. Borax was the most effective if layered with earth over the refuse. ABSTRACT: TDB 49:88, Jan., 1952.

Lall, S. B., joint author. *See* Johnson, W. T., et al., 1956.

Lalu, P., joint author. *See* Gaud, J., et al., 1950.

Lambreck, J. A., joint author. *See* Haynes, H. L., et al., 1957.

2187. **Lambremont, E. N., Fish, F. W.** and **Ashrafi, S. 1959** Pepsin-like Enzyme in Larvae of Stable Flies. Science 129(3361):1484-1485.
The activity of the enzyme is detectable in homogenates of whole larvae and of isolated mid-guts. Optimum pH for activity, 2·4.

2188. **Langenbuch, R. 1953** Über den Einfluss des Lösungemittels auf die insektizide Wirkung des Lindans. [Concerning the influence of solvents on the insecticidal action of lindane.] Zeit. Pfl. Krankh. 60(4):167-181. Ludwigsburg.
Musca domestica was used as one of the test insects. The part played by volatile and nonvolatile solvents is discussed. Lindane solution ability of the solvent important. ABSTRACT: RAE-B 42:169, Nov., 1954.

2189. **Langford, G. S. 1955** Sweetened Bait Used in Fly Control. World's Poultry Sci. J. 11(1):17-18.
A bait using 2 ozs. of 50 per cent malathion emulsions, 2 lbs. sugar and 2 gallons of water applied as a stock barn spray was most effective in controlling flies. ABSTRACT: BA 29:2717, Nov., 1955.

2190. **Langford, G. S., Johnson, W. T.** and **Harding, W. C. 1954** Bait Studies for Fly Control. J. Econ. Ent. 47:438-441.
Of 27 chemicals evaluated for fly control, for urban and rural usage, three phosphorus base chemicals showed promise: malathion, diazinone and Bayer's L 13/59. ABSTRACTS: BA 29:445(4609), Feb., 1955; RAE-B 43:102, July, 1955; TDB 51:1311, Dec., 1954.

Langford, G. S., joint author. *See* Johnson, W. T., et al., 1956

Lareg, E. P., joint author. *See* Davidow, B., et al., 1955.

Lariukhin, M. A., joint author. *See* Nabokov, V. A., et al., 1956.

Larsen, J. R., Jr., joint author. *See* Ingram, R. L., et al., 1956.

Lauderdale, R. W., joint author. *See* Morrison, H. E., et al., 1950.

Laurent, J., joint author. *See* Gaud, J., et al., 1954, 1956.

Lea, A. O., joint author. *See* DeLong, D. M., et al., 1952.

Lee, I., joint author. *See* Dahm, P. A., et al., 1959.

Leibman, A. L., joint author. *See* Morozova, V. P., et al., 1955, 157.

2191. **Le Roux, E. J.** and **Morrison, F. O. 1953** Dosage of DDT Needed to Kill a House Fly. J. Econ. Ent. 46:1109-1110.
Absorption and distribution of 10-20 per cent of the applied dosage produced the full effect. Locus of application did *not* influence rate of absorption; did influence effectiveness. ABSTRACTS: BA 28:2195, Sept., 1954; RAE-B 42:139, Sept., 1954; TDB 51:742, July, 1954.

2192. **Le Roux, E. J.** and **Morrison, F. O. 1954** The Adsorption, Distribution and Site of Action of DDT in DDT-Resistant and DDT-Susceptible House Flies Using Carbon^{14} Labelled DDT. J. Econ. Ent. 47:1058-1066.
Haemolymph distributes DDT from point of application to head of fly, the site of action of the insecticide. The closer the point of application to the head, the lower the observed LD 50. ABSTRACTS: BA 29:2260, Sept., 1955; RAE-B 43:192, Dec., 1955; TDB 52:488, May, 1955.

Le Roux, E. J., joint author. *See* Bigelow, R. S., et al., 1954; Morrison, F. O., et al., 1954.

2193. **Levenbook, L.** and **Williams, C. M. 1956** Mitochondria in the Flight Muscles of Insects. III. Mitochondrial Cytochrome *c* in Relation to the Aging and Wing Beat Frequency of Flies. J. Gen. Physiol. 39:497-512.
Concerns *Phormia regina*. During growth the number of sarcosomes remains constant, but the mass increases. Cytochrome components maintain constant ratio to one another and to sarcosomal weight. ABSTRACT: TDB 53:1062, Aug., 1956.

2194. **Leviev, P. I. A. 1955** Kvoprosu ob ustoichvosti *Musca domestica vicina* k preparatu DDT. [Resistance of *Musca domestica vicina* to DDT preparation.] Med. Parazit. (Moskva) 24(2):175-179.
As resistance to DDT developed after the 2nd and 3rd year of application, males flies developed more resistance; however, with age, susceptibility increased. ABSTRACT: BA 31:2125(22746), July, 1957. Cites pp. 160-170, in error.

2195. **Leviev, P. I. A. 1957** [The problem of house fly resistance to BHC preparations.] Med. Parazitol. i Parazitarn. (Bolezni) 26(1):22-52. (In Russian.)
Resistance appeared in the 3rd generation. That of 3-4 day old flies was greater than that of other ages, up to 8 days. ABSTRACT: BA 35: 2918(33185), June 15, 1960.

2196. **Levinson, Z. H. 1955** Nutritional Requirements of Insects. Riv. di Parassit. 16:1-48.
Reports recent work on sterol metabolism. Treats also of water, salts, protein carbohydrates, lipids and certain accessory nutritional factors.

2197. **Levinson, Z. H. 1955** Chemicals Affecting the Preimaginal Stages of the Housefly. V. Vapour Toxicity of the Dichlorobenzenes to Housefly Pupae. Riv. di Parassit. (Rome) 16:253-255.
At saturation concentration, order of toxicity of the dichlorobenzene isomers to house fly pupae was: ODB > MDB > PDB. ABSTRACT: TDB 53:679, May, 1956.

2198. **Levinson, Z. H. 1956** Chemicals Affecting the Preimaginal Stages of the Housefly. VI. Further Tests with Highly Chlorinated Aliphatic and Alicyclic Compounds. Riv. di Parassit. (Rome) 17(1):51-57.
An evaluation of these substances. Of those screened, the less volatile were the least toxic. Hexachlorobutadiene was very effective; more so if applied as a vapour. ABSTRACT: TDB 53:1285, Oct., 1956.

2199. **Levinson, Z. H.** and **Ascher, K. R. S. 1954** Chemicals Affecting the Preimaginal Stages of the Housefly. IV. The Fatty Acids. Riv. di Parassit. 15(2):111-119.
Several types tested for toxicity to larvae of *M. vicina* Macq. Caproic, caprylic and capric acid proved to be rapidly acting poisons. ABSTRACT: RAE-B 44:203, Dec., 1956.

2200. **Levinson, Z. H.** and **Bergmann, E. D. 1957** Steroid Utilization and Fatty Acid Synthesis by the Larva of the Housefly *Musca vicina* Macq. Biochem. J. 65(2):254-260.
Synthesis of body fat is from dietary protein. No carbohydrate or fatty acids required. Larval growth is proportional, up to a limit, to the concentration of dietary cholesterol. Utilization of steroids is dependent upon steroid structure. ABSTRACT: TDB 54:628, May, 1957.

2201. **Levinson, Z. H.** and **Bergmann, E. D. 1959** Vitamin Deficiencies in the Housefly produced by Antivitamins. J. Insect Physiol. (London) 3(3):293-305.
Larval growth severely hampered by any 1 of 10 antivitamins; somewhat hampered by 2 others. Five, in the diet of adult flies, impeded oviposition. Larval exposure caused pathological abnormalities in larvae, pupae, and adults. ABSTRACT: TDB 57:424, April, 1960.

2202. **Levinson, Z. H.** and **Silverman, P. H. 1954** Studies on the Lipids of *Musca vicina* (Macq.) during Growth and Metamorphosis. Biochem. J. (London) 58:294-297.
A systematic analysis of lipid at various stages of development to determine relation of sterols and other lipids to these stages. Moisture and

lipid content important. ABSTRACT: TDB 52:1024, Oct., 1955.

Levinson, Z. H., joint author. *See* Ascher, K. R. S., et al., 1951, 1953, 1954, 1956; Bergmann, E. D., et al., 1954, 1958; Silverman, P. H., et al., 1954; Davidovici, S., et al., 1950.

2203. Levkovich, E. N. and **Sukhova, M. N. 1957** [Duration of retention and excretion of poliomyelitis virus by synantrophic flies and its relation to dissemination and prevention of poliomyelitis.] Med. Parazit. i Parazitarn. (Bolezni) 26(3):343-347. (In Russian.)
Polio virus survived up to 10 days in fly intestine. ABSTRACT: BA 35: 3118(35478), July, 1960.

2204. Lewallen, L. L. 1950 Bristle Density of the Fifth Abdominal Sternite of Two House Fly Strains. Pan Pacific Ent. 26(3):138.
Hyman strain (from Illinois) has a higher average number of bristles per sternite than a strain from California.

2205. Lewallen, L. L. 1952 Laboratory Studies of the False Stable Fly. J. Econ. Ent. 45:515-517.
Life cycle of *Muscina stabulans* (Fallén) given as 20-25 days, in laboratory, using CSMA rearing method. This species not resistant to DDT, lindane, or methoxychlor. ABSTRACT: RAE-B 40:174, Nov., 1952.

2206. Lewallen, L. L. 1954 Biological and Toxicological Studies of the Little House Fly. J. Econ. Ent. 47:1137-1141.
Proposes a laboratory method for rearing *Fannia canicularis* (L.). Shows more tolerance than housefly to DDT, lindane, methoxychlor; more resistance than house-fly to dieldrin and pyrethrins. Effective organophosphorus insecticides are chlorothion, diazinon and malathion. ABSTRACT: BA 29:2260, Sept., 1955; RAE-B 43:193, Dec., 1955.

2207. Lewallen, L. L. 1958 Paper Chromatographic Analysis of Certain Protein Constituents from the House Fly, *Musca domestica* Linnaeus. (Diptera, Muscidae.) Ann. Ent. Soc. Am. 51:167-173.
Details technique for 1-way and 2-way chromatograms. Constituents found believed to be simple peptides.

Lewallen, L. L., joint author. *See* March, R. B., et al., 1950, 1952, 1956.

2208. Lewis, C. T. 1954 Contact Chemoreceptors of Blowfly Tarsi. Nature (London) 173(4394):130-131.
A preliminary report. Structural characteristics compared in *Phormia*, *Lucilia* and *Calliphora*.

2209. Lewis, C. T. 1954 Studies Concerning the Uptake of Contact Insecticides. I. The Anatomy of the Tarsi of Certain Diptera of Medical Importance. Bull. Ent. Res. 45:711-722.
M. domestica compared with *Phormia terraenovae* and *Glossina palpalis*. Each chemoreceptor is a differentiated, hollow seta with an extremely thin frontal membrane of cuticle, through which a lipoid-soluble insectiside might penetrate to sensory neurocytes. ABSTRACTS: RAE-B 43:1 Jan., 1955; TDB 52:412, Apr., 1955; Hocking, B. 1960 Smell in Insects, p. 243 (6.113).

2210. Lewis, C. T. and **Hughes, J. C. 1957** Studies concerning the Uptake of Contact Insecticides. II. The Contamination of Flies exposed to particulate Deposits. Bull. Ent. Res. 48(4):755-768. London.
Concerns *Phormia terraenovae*. Presence of an oil hinders the uptake of particles. "Cleaning movements" of insects believed to facilitate toxic action by transfer of particles. ABSTRACT: RAE-B 46:27, Feb., 1958.

2211. Lewis, D. J. 1954 Muscidae of Medical Interest in the Anglo-Egyptian Sudan. Bull. Ent. Res. 45:783-796.
Gives breeding sites of various species of *Musca*, of which 13 species

and 1 subspecies are found in the Sudan. Other genera of Muscidae also treated. ABSTRACT: TDB 52:486, May, 1955.

2212. Lewis, D. J. 1956 The Medical Entomology of the Tonkolili Valley, Sierra Leone. Ann. Trop. Med. Parasit. 50:299-313.
Many groups listed. Houseflies likely to become numerous as mining operations bring in laborers. A new species of *Musca* found at Nunkegoro. ABSTRACT: TDB 54:235, Feb., 1957.

2213. Lewis, S. E. and **Fowler, K. S. 1956** Effect of Diisopropylphosphorofluoridate on the Acetylcholine Content of Flies. Nature 178 (4539): 919-920.
Amount of acetylcholine slowly increases after treatment with D——. Decreases in head; slight increase in thorax; great increase in abdomen, mostly in content of hind gut. Used *M. domestica* and *Calliphora erythrocephala*. ABSTRACT: TDB 54:239, Feb., 1957.

2214. Lewis, S. E. and **Smallman, B. N. 1956** The Estimation of Acetylcholine in Insects. J. Physiol. 134:241-256.
Treats primarily of *Calliphora erythrocephala*, using fresh, frozen, and boiled heads. The last-named method, having proved the most valid, was then used with *Lucilia sericata*, *Periplaneta americana*, *Tenebrio molitor* and *M. domestica*.

2215. Li, H. H. and **Feng, L. C. 1951** Morphological Studies of the Common House Fly, *Musca vicina* in China. Peking Nat. Hist. Bull. 19:278-284 (pts. 2/3, Dec.-Mar.).
Is perhaps the only housefly as far north as Peking. All specimens examined conformed to James' formula. (Width of male frons, compared with width of third antennal segment less than three: not *domestica*.) Ratio averaged 2·13. All believed to be *vicina*. ABSTRACTS: BA 26:488, Feb., 1952; TDB 49:324, Mar., 1952.

Li, S. C., joint author. *See* Kao, C. M., et al., 1958.

2216. Lichtwardt, E. T. 1956 Genetics of DDT-Resistance of an Inbred Line of Houseflies. J. Heredity. 47(1):11-16.
Confirms suggestion of Maelzer and Kirk (1953) regarding a single, dominant gene (R) for high resistance to DDT. "R" factor does not seem to reduce fertility. ABSTRACTS: BA 31:979(9794), Apr., 1957; TDB 53:1063, Aug., 1956.

2217. Lichtwardt, E. T., Bruce, W. N. and **Decker, G. C. 1955** Notes on the Inbreeding of House Flies. J. Econ. Ent. 48:301-303.
Two, phenotypically homogeneous strains of inbred flies were developed in the laboratory; one susceptible to DDT, the other resistant. ABSTRACTS: BA 30:308, Jan., 1956; RAE-B 44:71, May, 1956; TDB 52: 1242, Dec., 1955.

2218. Lichtwardt, E. T., Luce, W. M., Decker, G. C. and **Bruce, W. N. 1955** A Genetic Test of DDT Resistance in Field House Flies. Ann. Ent. Soc. Am. 48(4):205-210.
Mass crosses revealed only one autosomal dominant gene for high resistance to DDT. Believed to be an allele of the "R" factor.

Lieberman, H. M., joint author. *See* Rockstein, M., et al., 1958, 1959.

Lien, Jik-Ching, joint author. *See* Liu, Su-Yung, et al., 1957.

Lilly, C. H., joint author. *See* Marke, D. J. B., et al., 1951.

Lima, L., joint author. *See* Coutinho, J. O., et al., 1957.

2219. Lin, S. and **Richards, A. G. 1955** Oxygen Consumption of Resistant and Susceptible House Flies. J. Econ. Ent. 48:627-628.
As judged by temperature kinetic determinations, there was no indication of a qualitative difference in respiratory enzyme complement controlling temperature effect in three strains studied. ABSTRACTS: RAE-B 44:175, Nov., 1956; TDB 53:379, Mar., 1956.

2220. Lin, S. and **Richards, A. G. 1956** A Comparison of Two Digestive Enzymes in the House Fly and American Cockroach. Ann. Ent. Soc. Am. 49:239-241.
Levels of activity of invertase and proteinase are similar. In houseflies invertase is present in both salivary glands and midgut, but proteinase is absent from salivary glands.

2221. Linam, J. H. and **Rees, D. M. 1956/1957** Flies Attracted to Animal Carcasses in Carey and Vicinity, Blaine County, Idaho. Proc. Utah Acad. Sci. Arts and Letters 34:57-58.
Included here because of negative results. No specimens of *Musca* were collected.

2222. Lindquist, A. W. 1952 Radioactive Materials in Entomological Research. J. Econ. Ent. 45:264-270.
Recommends use of radioisotopes in studies on biology and control of noxious forms. Cites previous work with *M. domestica.* ABSTRACT: RAE-B 40:128, Aug., 1952.

2223. Lindquist, A. W. 1954 Flies Attracted to Decomposing Liver in Lake County, California. Pan Pacific Ent. 30:147-152.
M. domestica was captured each month from April through November. Totalled 87 per cent of all flies taken. ABSTRACT: BA 29:2717, Nov., 1955.

2224. Lindquist, A. W. 1957 Effectiveness of Organo-Phosphorus Insecticides against Houseflies and Mosquitoes. Bull. W.H.O. 16(1):33-39.
At time of report, these insecticides were considered effective substitutes for DDT and other chlorinated hydrocarbon preparations. Residual applications were not as long-lasting as those of DDT. ABSTRACT: TDB 54:1021, Aug., 1957.

2225. Lindquist, A. W., Roth, A. R., Hoffman, R. A. and **Butts, J. S. 1951** The Distribution of Radioactive DDT in House Flies. J. Econ. Ent. 44:931-934.
Flies treated topically with radioactive DDT absorbed 26-34 per cent in internal organs, the remainder in the cuticle. Exposure to residual DDT showed similar distribution but only one-third to one-quarter as much total quantity. ABSTRACTS: BA 26:2067, Aug., 1952; RAE-B 40:60, Apr., 1952; TDB 49:561, May, 1952.

2226. Lindquist, A. W., Roth, A. R., Yates, W. W., Hoffman, R. A. and **Butts, J. S. 1951** Use of Radioactive Tracers in Studies of Penetration and Metabolism of DDT in House Flies. J. Econ. Ent. 44: 167-172.
With susceptible flies approximately 31-71 per cent of the DDT that penetrated the cuticle was metabolized to a non-toxic product. The rate of metabolism was similar to that which occurs in resistant flies. ABSTRACTS: RAE-B 39:153, Sept., 1951; TDB 48:931, Oct., 1951.

2227. Lindquist, A. W., Yates, W. W., Hoffman, R. A. and **Butts, J. S. 1951** Studies of the Flight Habits of Three Species of Flies Tagged with Radioactive Phosphorus. J. Econ. Ent. 44:397-400.
M. domestica recovered 12 miles from release point. P^{32} considered superior to color marking for flight studies. ABSTRACT: RAE-B 39:168, 1951.

Lindquist, A. W., joint author. *See* Hoffman, R. A., et al., 1951, 1952, 1954; Johnston, E. F., et al., 1954; Yates, W. W., et al., 1950, 1952; Roth, A. R., et al., 1953.

2228. Lindquist, D. A. and **Dahm, P. A. 1957** Some Chemical and Biological Experiments with Thiodan. J. Econ. Ent. 50:483-486.
Ultraviolet and infrared spectra were obtained of Thiodan, its two isomers, and Thiodan alcohol. LD_{50} values, using female house flies of

a non-resistant strain, were determined for Thiodan (technical and purified), also for the two isomers. ABSTRACT: RAE-B 46:140, Sept., 1958.

2229. **Lindquist, D. A., Fairchild, M. L., Dahm, P. A.** and **Gurland, J. 1959** Thiodan Residues on Corn Plants. J. Econ. Ent. 52(1):102-106.
Extracts of treated corn plants were examined by 3 analytical methods, to estimate total toxicity of the compounds and to distinguish between thiodin and thiodin alcohol. One method was "housefly bioassay."

2230. **Lindquist, D. A.** and **Fay, R. W. 1956** Laboratory Comparison of Eight Organic Phosphorus Insecticides as Larvicides against Non-resistant House Flies. J. Econ. Ent. 49:463-465.
The compounds had the following decreasing larvicidal effect: diazinon, EPN, parathion, Bayer 21/199, NPD, demeton, Bayer L 13/59 and malathion. The greatest effect was on small and medium larvae, with no effect on the pupae. ABSTRACTS: BA 31:620(6249), Feb., 1957; RAE-B 45:141, Sept., 1957; TDB 54:239, Feb., 1957.

Lindquist, D. A., joint author. *See* Fay, R. W., et al., 1954; McCawley, R. H., Jr., et al., 1955.

2231. **Lindsay, D. R.** and **Scudder, H. I. 1956** Non-biting Flies and Disease. Ann. Rev. of Ent. 1:323-346.
Excellent review. Cites 187 references from 1921 through June of 1955. Flies considered "indicator organisms", reflecting, by their abundance, the sanitary level of the community.

2232. **Lindsay, D. R., Stewart, W. H.** and **Watt, J. 1953** Effect of Fly Control on Diarrheal Disease in an Area of Moderate Morbidity. Publ. Health Repts. 68:361-367. Washington, D.C.
During effective fly control, in an area of moderate morbidity from diarrheal disease, the prevailing rate of such morbidity (including *Shigella* infections), was significantly lowered. ABSTRACTS: BA 27:2486, Sept., 1953; RAE-B 41:126, Aug., 1953.

Lindsay, I. S., joint author. *See* Hocking, B., et al., 1958.

2233. **Lineva, V. A. 1953** O fiziologicheskom vozraste samok komnatnoi mukhi *Musca domestica* L. (Diptera: Muscidae). [Physiological age of the female of the common fly, *M. domestica*.] Ent. Obozrenie 33:161-173.
ABSTRACTS: BA 29:604 Mar., 1955; RAE-B 43:106, July, 1955.

2234. **Lineva, V. A. 1955** Izmeneniya v oogeneze komnatnoi mukhi (*Musca domestica* L.) pod deistviem DDT. [Changes in oogenesis in the house fly (*Musca domestica* L.) under the influence of DDT.] Zool. Zhur. 34(6):1320-1325. Referat: Zhur., Biol., 1956, No. 63667.
Prolonged exposure through several generations to sublethal doses of DDT causes chronic and metatoxic poisoning, manifested in disturbances of oogenesis. ABSTRACT: BA 32:891(10497), Mar., 1958.

2235. **Lineva, V. A. 1956** [Hatching of flies in manure.] Gig. I. San. (Moskva) 21(4):41-43. (In Russian.)

Lineva, V. A., joint author. *See* Derbeneva-Ukhova, V. P., et al., 1959; Gorbov, V. A., et al., 1958; Zolotanen, E. Kh., et al., 1957.

Lipke, H., joint author. *See* Miyake, S. S., et al., 1957.

2235a. **List, G. M. 1952** Persistence of DDT on a Treated Surface as Shown by House Fly Knockdown and Kill. J. Econ. Ent 45:127-129.
Deposits of DDT insufficient for satisfactory control may be important in selecting resistant individuals of many species. Observations are given for a single treatment given in 1944, the effect of which persisted 7 years later. ABSTRACT: RAE-B 40:95, June, 1952.

Little, C. D., joint author. *See* Cunningham, H. B., et al., 1955.

2236. Liu, S. Y. 1958 A Summary of Recent Insecticidal Tests on some Insects of Medical Importance in Taiwan. Bull. W.H.O. 18(4):623-649. Geneva.

Investigations in 1956-57 for DDT and BHC resistance among insects showed that in *Musca domestica vicina* a 35 and 20 fold resistance was present in two strains compared with a susceptible strain of typical *Musca domestica* from England. ABSTRACT: RAE-B 49:93, May, 1961.

2237. Liu, S. Y., Chen, H. H. and **Lien, J. C. 1957** A Brief Study of the Bionomics of Fly Breeding in Kellung City, Taiwan. J. Formosan Med. Ass. 56(9/10):417-425.

Musca domestica vicina, *Musca sorbens*, *Chrysomyia megacephala* and *Chrysomyia rufifacies* were the most common flies visiting homes. The importance of *Chrysomyia* as a possible vector of excremental disease in the orient is emphasized. ABSTRACT: TDB 55:950, Aug., 1958.

2238. Loddo, B. 1954 Sull'importanza della colinesterasi sulla sensibilità della mosca domestica al DDT. [On the importance of cholinesterases in the suceptibility of the house fly to DDT.] Rass. Med. Sarda 5:167-174.

2239. Loddo, B. 1956 Attivita di vari insetticidi del commercio sulla *Musca domestica*. [Activity of various commercial insecticides on *Musca domestica*.] Ann. San. Pubb. (Roma) 17(3):695-738.

2240. Loddo, B. 1956 Sull' attività di alcuni prodotti insetticidi su colture larvali di *M. domestica*. [On the activity of some insecticidal products on larval culture of *M. domestica*.] Arch. Ital. Sci. Farm. 6(1):45-50.

2241. Loddo, B. and **Piras, L. 1956** Studio della popolazione di mosche di dieci paesi della Sardegna. [A study of fly population in ten villages of Sardinia.] Riv. Ital. Igiene (Pisa) 16:343-356.

Data on population changes was similar from all sites. The overall picture was one of a peak in early spring and another in late autumn correlating with times of high wind velocities. ABSTRACT: TDB 54:763, June, 1957.

2242. Loddo, B. and **Piras, L. 1957** Successivi studi sulla popolazione di mosche di dieci paesi della Sardegna. [Continued studies on fly population in ten villages of Sardinia.] Riv. Ital. Igiene 17(5-6): 271-281.

Using the same techniques as in the 1955-1956 study fly populations were found low at times of high wind velocities, in contrast to the previous findings. High levels of absolute humidity seemed to check the abundance of adult flies. ABSTRACT: TDB 55:335, Mar., 1958.

Loddo, B., joint author. *See* Piras, L., et al., 1957.

2243. Logan, J. A. 1953 The Sardinian Project: an Experiment in the Eradication of an indigenous Malarious Vector. Amer. J. Hyg. Monogr. Ser. No. 20 xxix + 415 pp. Baltimore, Maryland. Johns Hopkins Press.

Includes a chapter on fly control, effectiveness of which had a bearing on co-operation of population in malaria program. ABSTRACT: RAE-B 42:111, July, 1954.

2244. Lord, K. A. and **Solly, S. R. B. 1956** The Rate of Disappearance of Paraoxon from two Strains of Houseflies. Chem. and Ind. for 1956, pp. 1352-1353.

Paraoxon decreased rapidly and similarly in a resistant and susceptible strain. Resistance could develop from differences in localized rates of penetration or detoxification of poison. ABSTRACT: RAE-B 46:76, May, 1958.

2245. Losco, G. and **Peri, C. A. 1959** Esteri fosforici di umbelliferoni e loro proprietà insetticide. [Phosphoric esters of the Umbelliferae and their insecticidal properties.] Gazz. Chim. Ital. 89(5/6):1298-1314.
Compounds synthesized from umbelliferones are more or less effective against houseflies, even those resisting chlorinated insecticides. ABSTRACT: BA 35:1034(11739), Feb., 1960.

Losco, G., joint author. *See* Fusco, R., et al., 1959; Speroni, G., et al., 1951, 1953, 19?? (undated reprint).

Loveday, P. M., joint author. *See* Winteringham, F. P. W., et al., 1951, 1953.

2246. Lovell, J. B. and **Kearns, C. W. 1959** Inheritance of DDT-Dehydrochlorinase in the House Fly. J. Econ. Ent. 52(5):931-935.
Pattern of inheritance of resistance and of DDT'*ase* activity indicates that resistance to DDT may be governed by a single, partially dominant gene. ABSTRACTS: RAE-B 49:7, Jan., 1961; TDB 57:306, Mar., 1960.

Lovell, J. B., joint author. *See* Chadwick, L. E., et al., 1954, 1956.

2247. Lozano, M. A. 1954 Repercusiones sanitarias de la resistencia de los insectos a los insecticidas. [Sanitary repercussions from the resistance of insects to insecticides.] Medicina (Mexico) 34 (Suppl.):67-70; *also in* Rev. San. (Madrid) 27:623-628.

Luce, W. M., joint author. *See* Lichtwardt, E. T., et al., 1955.

2248. Ludwig, D. and **Barsa, M. C. 1959** Activities of Respiratory Enzymes During the Metamorphosis of the Housefly, *Musca domestica* (L.). J. New York Ent. Soc. 67(3/4):151-156.
Treats of succinic, malic, total alpha-glycerophosphate, also alcohol, lactic, and glutamic dehydrogenases, plus others. Activity curve for succinic dehy. is U-shaped during metamorphosis (as is the respiratory curve). ABSTRACT: BA 35:1422(16307), Mar., 1960.

Ludwig, D., joint author. *See* Barsa, M. C., et al., 1959.

2249. Lugthart, G. J. 1959 Biology and Control of House Fly Larvae. Diss. Absts. 19(8):1882-1883.

Lugthart, G. J., joint author. *See* Dicke, R. J., et al., 1955.

2250. Luh, P.-L. and **Chih, H.-K. 1950** A Note on the House-frequenting Calliphorids of Peking (Diptera: Calliphoridae). Peking Nat. Hist. Bull. 18(3):165-170.
A preliminary report of a study made in 1948. Lists five species of house-frequenting Calliphorids, the most common being *Calliphora megacephala* and *Lucilia sericata*. ABSTRACT: BA 25:299(3186), 1951.

2251. Ma, C. Y. 1959 A Preliminary Report on the Synanthropic Calyptrate Flies of North-east China. Acta. Ent. Sinica 9(2):154-160.
Collections made during 1956-1957 report 68 species belonging to 26 genera in 5 families. Of these 6 are new records for China. ABSTRACT: BA 35:5519(64965), Nov., 1960.

Ma, P. Y., joint author. *See* Kun, K. Y., et al., 1958.

2252. McBride, O. C., Sullivan, W. N. and **Fulton, R. A. 1950** Treatment of Airplanes to Prevent the Transportation of Insects. J. Econ. Ent. 43:66-70.
Used both residual insecticides and aerosols. Houseflies and mosquitoes very susceptible to treatments employed. ABSTRACT: RAE-B 38:187, 1950.

2253. McCauley, R. H., Jr., Grainger, M. M., Lindquist, D. A. and **Fay, R. W. 1955** Laboratory Comparison of Some Insecticides as Larvicides against Non-resistant House-Flies. J. Econ. Ent. 48:269-273.
Seven chlorinated hydrocarbons tested as larvicides in spray applications. Effectiveness determined by numbers of pupae and of defective

and normal adults in treated cultures as compared with check cultures. Data in tabular form. ABSTRACTS: BA 30:308, Jan., 1956; RAE-B 44: 69, May, 1956; TDB 62:1242, Dec., 1955.

2254. McCoy, C. E. 1958 Suppression of House Fly Breeding in Tomato Cannery Waste Lagoons by Means of Polyethylene Cover. J. Econ. Ent. 51(3):411:412.

Results very favorable. Considering probable health benefits, cost is minor. ABSTRACT: BA 32:3518(42298), Dec., 1958.

2255. McCullock, R. N. 1955 DDT-resistance in House Flies. Experience at Roseworthy Agricultural College. J. Dep. Agric. S. Aust. 59(5): 200-202.

Attempt to offset break-down in fly control with DDT by daily thin-spreading of manure to insure quick drying. Gave good reduction of fly numbers for a time, then failed. ABSTRACT: RAE-B 45:136, Aug., 1957.

McDuffie, W. C., joint author. *See* Knipling, E. F., et al., 1957; Gilbert, I. H., et al., 1953; Gahan, J. B., et al., 1954; Wilson, H. G., et al., 1953.

McGregor, W. S., joint author. *See* Eddy, G. W., et al., 1954.

2256. McGuire, C. D. and **Durant, R. C. 1957** The Role of Flies in the Transmission of Eye Disease in Egypt. Am. J. Trop. Med. Hyg. 6(3):569-575.

M. domestica vicina and *M. sorbens*, collected from faces of Egyptian children, bore bacteria similar to those found in both normal and diseased eyes. Most common organisms isolated were hemolytic micrococci. ABSTRACT: BA 32:510(5843), Feb., 1958.

2257. McKenzie, R. E. and **Hoskins, W. M. 1954** Correlation Between the Length of the Larval Period of *Musca domestica* L. and Resistance of Adult Flies to Insecticides. J. Econ. Ent. 47:984-992.

Two substrains selected for pupation time showed that those having the longer pupation became more resistant during 15 generations. ABSTRACTS: BA 29:2260(22683), Sept., 1955; RAE-B 43:190, Dec., 1955; TDB 52: 487, May, 1955.

2258. McLeod, W. S. 1957 Biological Testing of Household Insecticides. Proc. Ent. Soc. Manitoba. 13:11-18.

2259. McMahon, J. 1957 A Note on House Fly Resistance to Residual Insecticides. E. African Med. J. 34(9):507-508.

Resistance to dieldrin, gammexane and DDT in the Nandi Reserve and Kisumu Township was reported, sometimes from areas never before tested. Resistance against DDT was observed only where it had been used, but resistance to gammexane occurred if dieldrin only had previously been applied. ABSTRACT: TDB 55:110, Jan., 1958.

2260. McNamee, R. W. 1950 Allethrin Symposium. General Nature of Allethrin. Soap and Sanit. Chem. 26:106.

Stresses allyl cinerin, a compound analogous to the pyrethrins. ABSTRACT: TDB 48:211, Feb., 1951.

2261. Madwar, S. and **Zahar, A. R. 1951** Preliminary Studies on Houseflies in Egypt. Bull. W. H. O. (Geneva) 3:621-636.

Scudder grid used. Reliance has been placed chiefly on chemical control, but neither DDT, BHC nor chlordane were effective in heavily infested areas. Several reasons for this, including resistant strains. ABSTRACT: TDB 48:929, Oct., 1951.

2262. Madwar, S. and **Zahar, A. R. 1953** Some Ecological Observations on Houseflies in Egypt. Bull. W. H. O. (Geneva) 8:513-519.

Two fly peaks are recognizable, spring and fall. Flies decline in the summer. Relates to higher temperature, not to humidity. ABSTRACT: TDB 50:1090, Nov., 1953.

2263. **Maelzer, D. A.** and **Kirk, R. L. 1953** A Preliminary Study of the Genetics of DDT Resistance in Houseflies. Australian J. Biol. Sci. 6 244-256.
The highly resistant "Illinois" strain is perhaps heterozygous, being composed of "weak" and "strong" individuals with respect to resistance to DDT. Each type was crossed with non-resistant "Canberra" strain and results analyzed. ABSTRACTS: BA 28:266, Feb., 1954; RAE-B 42: 108, July, 1954; TDB 51:320, Mar., 1954.

2264. **Maier, P. B., Baker, W. C., Bogue, M. D., Kilpatrick, J. W.** and **Quarterman, K. D. 1952** Field Studies on the Resting Habits of Flies in Relation to Chemical Control. I. In Urban Areas. Am. J. Trop. Med. Hyg. 1:1020-1025.
Diurnal and nocturnal observations of fly resting places were made near Pharr, Texas, and Savannah, Georgia. It is suggested that spraying the night resting places might result in more efficient control. ABSTRACTS: RAE-B 42:122, Aug., 1954; TDB 50:350, Apr., 1953.

2265. **Mail, G. A.** and **Schoof, H. F. 1954** Overwintering Habits of Domestic Flies at Charleston, West Virginia. Ann. Ent. Soc. Am. 47:668-676.
Treats of 8 species including *M. domestica.* The latter bred in cat excrement and garbage sludge in basements of heated buildings in mid-winter. ABSTRACT: BA 30:309, Jan., 1956.

Mail, G. A., joint author. *See* Schoof, H. F., et al., 1953, 1954.

Major, J., joint author. *See* Fletcher, O. K., et al., 1956.

2266. **Makhneva, A. N. 1957** [Data on the duration of the insecticidal action of DDT dust on flies in the conditions of Kantsky Payon Frunze Oblast.] Sov. Zdravookhr. Kirgizii 1:52-53.
Treatment of surfaces with hexachlorane is more effective than treatment with DDT but the effectiveness is dependent upon the season of the year and the character of the surface treated. ABSTRACT: BA 35:4675 (54320), Oct. 1, 1960.

2267. **Makhover, M. V. 1959** Tsitokhimiya oogeneza Kommatnoi mukhi. [The cytochemistry of oogenesis in the housefly.] Tr. Leningradsk. Sanit.-Gig. Med. Inst. 43:69-75. (English summary.) Referat. Zhur., Biol., 1960, No. 38385.
The presence of DNA and RNA in the oogonia and oocytes of the housefly and the significance of the disappearance of DNA at the onset of vitellogenesis are discussed cytochemically. ABSTRACT: BA 45:5435 (66801), Aug. 1, 1964.

2268. **Mallis, A. 1954** Handbook of Pest Control. The Behavior, Life History, and Control of Household Pests. 2nd Ed. 1068 pp., 231 figs. New York. MacNair-Dorland Co.
That portion dealing with *M. domestica* is considerably expanded over the previous edition. ABSTRACT: RAE-B 42:147, Oct., 1954.

2269. **Mallis, A., Miller, A. C.** and **Sharpless, R. V. 1952** Effectiveness against House Flies of Six Pyrethrum Synergists Alone and in Combination with Piperonyl Butoxide. J. Econ. Ent. 45:341-343.
Combinations showed largely only additive effects. There was no "hyper-synergism" of importance. Piperonyl butoxide, sulfoxide, and piperonyl cyclonene (on a weight basis) were the most efficient synergists for pyrethrum. ABSTRACT: RAE-B 40:131, Aug., 1952.

Mallis, A., joint author. *See* Miller, A. C., et al., 1952.

Mangan, G. F., Jr., joint author. *See* Fulton, R. A., et al., 1953.

2270. Manville, R. H. 1952 The Principles of Taxonomy. Turtox News 30(1): 12:16; (2):50-52.
An excellent, brief presentation of the principles concerned, using *M. domestica* as the prime example.

2271. March, R. B. 1959 Resistance to Organophosphorus Insecticides. Misc. Publ. Ent. Soc. Amer. 1(1):13-19, 15 refs. Washington, D.C.
Unlike resistance to chlorinated hydrocarbons, resistance to organophosphorus insecticides is relatively unstable, decreasing initially at a rapid rate, on removal of selection pressure. Such resistance is not primarily related to high cholinesterase activity. ABSTRACT: RAE-B 48: 210, Dec., 1960.

2272. March, R. B. and **Lewallen, L. L. 1950** A Comparison of DDT-resistant and Non-resistant House Flies. J. Econ. Ent. 43:721-722.
Thickness of cuticula and general vigor not concerned in the 300-fold increase in resistance of the "Bellflower" strain. Nor do dimensions of tarsal segments differ significantly between "Laboratory" strain and "Bellflower". ABSTRACT: RAE-B 39:84, May, 1951.

2273. March, R. B. and **Lewallen, L. L. 1956** Paper Chromatography of Fresh Tissue Extracts of Suceptible and Insecticide-resistant Strains of the House Fly, *Musca domestica* L. Ann. Ent. Soc. Am. 49: 571-575.
Does not distinguish between susceptible and resistant flies. Taxonomic value of chromatography also questioned. ABSTRACTS: RAE-B 46:114, July, 1958; TDB 55:1288, Nov., 1958.

2274. March, R. B. and **Mercalf, R. L. 1950** Insecticide-resistant Flies. Soap and Sanit. Chem. 26(7):121, 123, 125, 139.
Failure of DDT to control flies in California first noted in 1948. Paper deals with two resistant strains. Discussion and suggestions. ABSTRACT: RAE-B 40:14, Jan., 1952.

2275. March, R. B. and **Mercalf, R. L. 1952** Insecticide Research for the Control of Resistant Houseflies. Pest Control 20(4):12-18.
Advocates: (1) Search for new insecticides against which resistance will not develop; (2) development of synergists; (3) better use of present insecticides to avoid building of resistance. ABSTRACT: BA 26:2330, Sept., 1952.

2276. March, R. B., Metcalf, R. L. and **Lewallen, L. L. 1952** Synergists for DDT against Insecticide-Resistant House Flies. J. Econ. Ent. 45: 851:860.
Of ninety-six compounds tested, sixteen showed some promise. Might be more profitable to experiment with specific synergists for methoxychlor or lindane, against which resistance in the field is not as great as against DDT. ABSTRACTS: RAE-B 41:74, May, 1953; TDB 50:463, May, 1953.

March, R. B., joint author. *See* Anderson, A. D., et al., 1954; Metcalf, R. L., et al., 1953, 1956, 1959; Bohart, R. M., et al., 1958.

2277. Mariani, M. 1953 Comportamento di mosche DDT-sensibili e DDT-resistenti di fronte al malathion. [Behavior of DDT-susceptible and DDT-resistant flies, exposed to malathion.] Boll. Soc. Ital. Sper. 29:1774-1776.
Flies of both resistant and susceptible strains given topical exposure. Mortality rates were 20 per cent and 94 per cent, respectively.

2278. Mariani, M. and **Oddo, F. 1959** Sull'impiego di un nuovo insetticida organo-fosforico nella lotta contro la *Musca domestica*. [The use of a new organophosphorus insecticide in the campaign against the

house-fly.] Riv. di Parassit. Rome 20(2):125-140. (English summary.)
Results are given of laboratory and field experiments in Sicily during 1957-1958 with deposits from sprays of an organophosphorus insecticide, Dition. Emulsifiable pastes containing Dition, DDT, with or without dimethoate, spreaders and diluents, were rapidly effective against *Musca domestica.* ABSTRACTS: RAE-B 50:193, Sept., 1962; TDB 57:307, Mar., 1960.

2279. Mariani, M. and **Smiraglia, B. 1954** Anatomia comparate e fisiologia dei pretarsi di alcuni ditteri calipterati. [Comparative anatomy and physiology of the pretarsi of some calyptrate Diptera.] Boll. Lab. di Zoologia Gen. e Agr. "Filippo Silvestri" Anno 1954, p. 69.
In vivo observations on the chemoreceptors of the tarsi, and on tarsal reaction to contact with various substances. (*Musca domestica.*) Mechanism of absorption and the limits on cuticular diffusion for different insecticides.

Mariani, M., joint author. *See* d'Alesandro, G., et al., 1951, 1952, 1953, 1954.

2280. Marke, D. J. B. and **Lilly, C. H. 1951** Smoke Generators for the Dispersion of Pesticides. J. Sci. Fd. Agric. 2(2):56-65. London.
M. domestica used as the test insect. ABSTRACT: RAE-B 40:74, May, 1952.

2281. Marsh, M. W. and **Eden, W. G. 1955** Relative Toxicity of Six Insecticides to Two Strains of the House Fly. J. Econ. Ent. 48:610-611.
LD-50s were not significantly different for the "Auburn" (more resistant) strain and the "Orlando" (less resistant) strain. Exception: *Diazinon,* which proved decidedly more toxic to "Auburn" flies. ABSTRACT: RAE-B 44:174, Nov., 1956.

2282. Marshall, D. R. 1955 The Bioassay and Control of Insecticide-resistant Flies. I. A review of the Literature on the Chemical Control of Flies in Municipal and Urban Areas. R. San. Inst. J. (London) 75:489-495.
The control of flies by insecticides in refuse tips is reviewed. ABSTRACT: TDB 53:256, Feb., 1956.

2283. Marshall, D. R. 1955 The Bioassay and Control of Insecticide-resistant Flies. II. The Investigation and Bioassay of a Resistant Strain of *M. domestica* Breeding on a Refuse Tip. R. Soc. Promot. Health J. (London) 75:633:639.
A simple method is described for the measuring of resistance by confining flies in a flask coated with DDT. Results of its use in field tests are reported. ABSTRACT: TDB 53:256, Feb., 1956.

2284. Marshall, D. R. 1955 The Bioassay and Control of Insecticide-resistant Flies. III. The Control of DDT-resistant Flies with Organo-phosphorus Insecticides. R. Soc. Promot. Health J. (London) 75:741-744.
Tests for fly control were made with parathion, diazinon and malathion, three phosphorus insecticides. ABSTRACT: TDB 53:256, Feb., 1956.

2285. Maruichi, T. 1955 [On the relation between height of drying position of fish and number of attached flies.] Bull. Jap. Soc. Sci. Fish. 20(12):1089-1091. (In Japanese, with English summary.)
The attraction of flies to fish drying outdoors is greater in the afternoon than in the morning. A positive correlation between flies and atmospheric pressure is possible. ABSTRACT: BA 31:2998(33103), Oct., 1957.

2286. Matheny, R. W. 1958 Another Lesser House Fly Source. Calif. Vector Views. 5(12):80.
Fannia canicularis larvae collected from grass clippings adhering to the housing of rotary mowers.

2287. Matheny, R. W. and **Eastwood, R. E. 1957** Analysis of Fly Service Requests in Santa Clara County. Calif. Vector Views 4:42-43.
The flies most responsible for complaints and requests for services from the Vector Control Section of the Health Department are *Fannia canicularis* (43 per cent in 1955, 46 per cent in 1956) and *Musca domestica* (44 per cent in 1955, 34 per cent in 1956).

2288. Matheson, R. 1950 Medical Entomology. 2nd Edition. Ithaca, Comstock Publishing Company, xii + 612 pp., illus.
Houseflies and their allies treated in chapter XVI, pp. 459-493; Myiasis in chapter XVII, pp. 494-537.

Mathis, W., joint author. *See* Quarterman, K. D., et al., 1952, 1954.

2289. Matsui, M., LaForge, F. B., Green, N. Y. and **Schechter, M. S. 1952** Furethrin. J. Am. Chem. Soc. 74:2181-2182.
Is about equal in toxicity to pyrethrins. Production cost might be less than for allethrin.

Mattson, A. M., joint author. *See* Parry, A. S., et al., 1953, 1958.

Maurice, A., joint author. *See* Gaud, J., et al., 1950.

Maxon, M., joint author. *See* Metcalf, R. L., et al., 1956.

2290. Mayeux, H. S. 1954 Malathion for House Fly Control. Florida Ent. 37(2):97-104.
Malathion baits applied in a wide variety of ways were extremely effective, sometimes entirely eliminating the adult population of flies. ABSTRACT: BA 29:207(2179), Jan., 1955.

2291. Mayeux, H. S. 1954 Granular Insecticidal Baits against Flies. Florida Ent. 37:171-190.
Granular baits were prepared by coating the toxicant (1 per cent malathion) along with attractants, dispersants and other adjuvants on particles of shell, marble, quartz, expanded mica, etc. Baits could be spread without special equipment. ABSTRACT: BA 29:2717(27294), Nov., 1955.

Mazur, S., joint author. *See* Black, D. M., et al., 1958.

Mechoulam, R., joint author. *See* Bergmann, E. D., et al., 1958; Reuter, S., et al., 1956.

2292. Medvedeva, A. M. 1958 [Results of fly control in the Kirov region of Stalingrad during 1955-1956.] Med. Parazit. i Parazitarn. (Bolezni) 27(3):361. (In Russian.)

2293. Melis, R. 1957 Nuove applicazioni di lotta organezzata contro la mosca domestica nel biennio 1955-1956. [New applications of the organized effort against the domestic fly during the two-year period 1955-1956.] Igiene e San Pubbl. 13(9-10):548-549. Rome.
Fly campaigns were carried out in four areas, using a mixture of malathion in a molasses solution. Results were very effective resulting in a 90·5 per cent reduction in 1955 and a 96·5 per cent reduction in 1956. ABSTRACT: TDB 55:337, Mar., 1958.

2294. Melis, R. and **Catella, F. 1956** La lotta organizata contro la mosca domestica; prime applicazioni in alcuni centri turistici e di cura della Toscana. [The organized campaign against the domestic fly; first applications in some tourist centers and in the parish of Tuscany.] Riv. Ital. Igiene 16(3-4):136-169.
During field trials of 1953-1954, four to eight treatments with malathion in a beet molasses bait reduced the fly population by 83-87 per cent. ABSTRACT: TDB 54:240, Feb., 1957.

2295. Melis, R. and **Catella, F. 1957** Prove di lotta contro la mosca domestica mediante l'uso di malathion con o senza l'aggiunta di cloroderivati e unito a differenti sostanze attrattive. [Test of the war against the

domestic fly by the use of malathion with or without a chloro-derived adjuvant and an unrelated attractive substance.] Igiene e San Pubbl. 13(7-8):401-415. (Rome).
About 75-80 per cent control of houseflies was reached with baits of malathion in refined sugar or beet-sugar molasses, the refined sugars being very successful attractants. ABSTRACT: TDB 55:337, Mar., 1958.

2296. Melnick, J. L. and **Dow, R. P. 1953** Poliomyelitis and Coxsacki Viruses from Flies. Am. J. Hyg. 58:288-309.
Both types of virus were present in *M. domestica* but there was no evidence of a true, host-parasite relationship. ABSTRACT: RAE-B 43-87, June, 1955.

2297. Melnick, J. L., Emmons, J., Coffey, J. H. and **Schoof, H. 1954** Seasonal Distribution of Coxsackie Viruses in Urban Sewage and Flies. Am. J. Hyg. 59:164-184.
Virus found more regularly in sewage than in flies. No evidence that its presence in either is a link in the passage from an infected to a non-infected person. ABSTRACT: RAE-B 43:139, Sept., 1955.

2298. Melnick, J. L. and **Penner, L. R. 1952** The Survival of Poliomyelitis and Coxsackie Viruses Following their Ingestion by Flies. J. Exp. Med. 96:255-271.
Used *M. domestica*, *Phormia regina* and *Phaenicia sericata*. Polio virus from human sources detectable in flies between 5th and 17th day and in fly's excreta between 4th and 10th day. Flies from maggots fed virus were free from the agent. ABSTRACTS: RAE-B 43:58, Apr., 1955; BA 27:669, Mar., 1953.

2299. Meltzer, J. 1958 Unspecific Resistance Mechanisms in the House-fly, *Musca domestica* L. Indian J. Malariol. 12:579-588.
Unlikely that multi-resistance can be explained by specific detoxifying mechanisms. Unspecific mechanisms more probable. ABSTRACT: RAE-B 49:231, Oct., 1961.

2300. Meng, C. H. and **Winfield, G. F. 1950** Studies on the Control of Fecal-borne Diseases in North China. XV, XVI, XVII. An Approach to Quantitative Study of the House-frequenting Fly Population. C. The Characteristics of a Rural Fly Population. D. The Breeding Habits of the Common North China Flies. E. The Food Preferences of the Common North China Flies. Philipp. J. Sci. 79(1):67-85; 79(2):165-192; 79:431-442.
Reports density indices and species make-up of the population. Besides *M. vicina* and *M. sarkens*, includes data on *Chrysomyia*, *Sarcophaga*, *Lucilia*, and *Muscina*. ABSTRACT: BA 26:2343, Sept., 1952.

2301. Mengle, D. C. and **Casida, J. E. 1958** Inhibition and Recovery of Brain Cholinesterase Activity in House Flies Poisoned with Organophosphate and Carbamate Compounds. J. Econ. Ent. 51(6):750-757.
One sulfamate, seventeen organophosphate and three carbamate compounds were topically applied to female flies. Time of highest mortality rate related to time of maximum enzyme inhibition. ABSTRACTS: BA 33:2249(27707), May, 1959; RAE-B 48:50, Mar., 1960.

2302. Menn, J. J., Benjamini, E. and **Hoskins, W. M. 1957** The Effects of Temperature and Stage of Life Cycle upon the Toxicity and Metabolism of DDT in the Housefly. J. Econ. Ent. 50(1):67-74.
Increasing temperature after exposure to DDT favors survival if conditions are such that shortly after exposure, the rate of detoxification or excretion exceeds rate of intake. ABSTRACTS: BA 31:1814(18952), June, 1957; RAE-B 46:46, Mar., 1958; TDB 54:1252, Oct., 1957.

2303. Mer, G. G. 1953 Daytime Distribution of DDT-Resistant Houseflies Inside DDT-Sprayed Buildings. Bull. W. H. O. 8:521-526.
Windows, doors and furniture more frequented than walls and ceilings (which had received the spray). ABSTRACT: TDB 50:1093, Nov., 1953.

2303a. Mer, G. G. 1954 Observations on the Behavior and Control of House-Flies in a Rural Area in Israel. First Internat. Symposium on the Control of Insect Vectors of Disease. Istituto Superiore di Sanita. Rome. 1954. pp. 201-218.
Recommends research on methods of storing manure and refuse, so as to limit reproduction of flies without impairing the value of such material for agricultural use.

2304. Mer, G. G. and **Cwilich, R. 1957** Observations on the Behavior and Control of Houseflies in a Rural Area, in Israel. II. A Laboratory and Field Study of Diazinon (0,0-diethyl-0-2 isopropyl 1-4-methyl-pyrimidil-6-thiophosphate) as a Larvicide. Riv. di Parassit. 17(4): 209-216. Rome.
Flies developing in a breeding medium having a low Diazinon concentration develop resistance in four generations. Therefore, the field use of diazinon as a larvicide should be limited to selected breeding sites so that the selection of resistant strains may be considerably reduced. ABSTRACT: TDB 54:888, July, 1957.

2305. Mer, G. G. and **Cwilich, R. 1957** Observations on the Behavior and Control of Houseflies in a Rural Area in Israel. III. On the Use of Diazinon Dust as Fly Larvicide. Riv. di Parassit. 18(1):35-42. Rome.
Treatments of manure with diazinon dust was a far less expensive method of fly control in materials and labor than the spraying of pyrethrum aerosol. ABSTRACT: TDB 54:1021, August, 1957.

2306. Mer, G. G. and **Davidovici, S. 1950** A Method for Estimating the Toxic Effect of Contact Insecticides on Mosquitoes and House-flies. Parasitology 40:87-92.
Method is based on a comparison of the death-rate and length of survival of insects after a very short and exactly timed contact with a surface sprayed with a known concentration of the insecticide. ABSTRACT: RAE-B 40:181, Nov., 1952.

2307. Mer, G. G. and **Furmanska, W. 1953** The Effect of the Fat Content in the Fly Food on the Resistant to DDT. Riv. di Parassit 14(1): 49-54.
The addition of fat to the diet renders it necessary to double the exposure to a DDT residue to obtain a 50 per cent knockdown. ABSTRACTS: RAE-B 42:31, Feb., 1954; TDB 51:229, Feb., 1954.

Mer, G. G., joint author. *See* Cwilich, R., et al., 1954.

2308. Metcalf, R. L. 1955 Physiological Basis for Insect Resistance to Insecticides. Physiol. Rev. 35:197-232.
Much emphasis on *M. domestica.* (The abstract indicated below is excellent.) ABSTRACT: TDB 52:843, Aug., 1955.

2309. Metcalf, R. L., Fukuto, T. R. and **March, R. B. 1958** Mechanisms of Action of Anticholinesterase Insecticides. Proc. 10th Int. Cong. Ent., pp. 13-18.
The effect of phosphorus and carbamate insecticides upon cholinesterase enzymes is due to simple bimolecular reactions resulting, in the case of the phosphorus compounds, in phosphorylation of the enzymes, and in the carbamates, in blocking of the active site by a stable competition. ABSTRACT: RAE-B 47:82, June, 1959.

2310. Metcalf, R. L., Fukuto, T. R. and **March, R. B. 1959** Toxic action of Dipterex and DDVP to the House Fly. J. Econ. Ent. 52(1):44-49.
DDVP formation is necessary for *in vitro* cholinesterase inhibition and is responsible for *in vivo* toxic action of Dipterex (which is degraded to DDVP). ABSTRACTS: BA 33:2889(35319), June, 1959; RAE-B 48:93, May, 1960.

2311. Metcalf, R. L. and **March, R. B. 1953** Further Studies on the Mode of Action of Organic Thionophosphate Insecticides. Ann. Ent. Soc. Am. 46:63-74.
Experiments show that conversion of thionophosphate insecticides such as parathion and methylparathion to anticholinesterases, presumably the oxygen analogues, occurs in the presence of mammalian and insect tissues under appropriate conditions. ABSTRACTS: RAE-A 43:37, Feb., 1955; TDB 52:698, July, 1955.

2312. Metcalf, R. L. and **March, R. B. 1953** The Isomerization of Organic Thionophosphate Insecticides. J. Econ. Ent. 46:288-294.
The thionophosphate insecticides dealt with were parathion, methyl-parathion and isopropyl-parathion, malathion and EPN. Their isomerization was by exchange of the double-bonded sulphur on the phosphorus atom for double-bonded oxygen. Concerns the effect of isomerization on anticholinesterase activity, hydrolysis rates and toxicity and decomposition rates of residues. ABSTRACT: RAE-B 41:173, Nov., 1953.

2313. Metcalf, R. L., Maxon, M., Fukuto, T. R. and **March, R. B. 1956** Aromatic Esterase in Insects. Ann. Ent. Soc. Am. 49:274-279.
Experiments with selective inhibitors paraoxon and prostigmine show that tissues of several insects, including *Musca domestica*, contain three types of esterases, a specific cholinesterase (CHE), an aliphatie esterase (AliE) and an aromatic esterase (ArE).

Metcalf, R. L., joint author. *See* Anderson, A. D., et al., 1954; Fukuto, T. R., et al., 1959; March, R. B., et al., 1950, 1952; Rogoff, W. M., et al., 1951.

2314. Micks, D. W. and **Singh, K. R. 1958** Infrared Spectra of Acetone Extracts of Susceptible and Insecticide-resistant Strains of House Flies. Texas Repts. Biol. Med. 16(3):355-362.
Certain strains of susceptible and resistant flies may be differentiated by subtle, reproducible differences in infrared spectra obtained from acetone extracts. Not linked with resistance *per se*. ABSTRACT: RAE-B 48:136-137, Aug., 1960.

2315. Milani, R. 1954 The Genetics of the House Fly, Preliminary Note. Atti IX Intern. Congr. Genetics, Caryologia. Suppl., pp. 791-796.
Lists and describes 21 mutant conditions, most of them recessive, autosomal. Treats also of gynandromorphs and inbreeding.

2316. Milani, R. 1954-1956 Comportamento mendeliano della resistenza alla azione abbattente del DDT e correlazione tra abbattimento e mortalita in *Musca domestica* L. [Mendelian behavior of resistance to the knock-down action of DDT and correlation between knock-down and mortality in *Musca domestica* L.] (a) Riv. di Parassit. (Rome) 15(4):513-542. (English summary.) (b) Rend. Ist Sup. Sanit. 19:1107-1143., Publ. 1956; (c) Selec. Sci. Papers, Ist. Sup. Santia 1:176-212. (English translation.)
ABSTRACTS: RAE-B 45:63, Apr., 1957; TDB 52:1025, Oct., 1955.

2317. Milani, R. 1955 Comportamento ereditario dei caratteri knock-down-resistance (kdr) e plexus (plx) in *Musca domestica* L. (nota preliminare). [Hereditary behavior of the characters knock-down-resistance (kdr) and plexus (plx) in *Musca domestica* L. (preliminary note).] (a) Rend. Ist. Sup. Sanit. (Roma) 18(10):889-897. (b) Convegno Genetica, Suppl. "Ric. Scient." 25:66-73. (Summaries in English, German, and French.)

Three characters common in a natural population in Latina Province. Inheritance of *kdr* is monofactorial, of *plx*, problematical. ABSTRACT: TDB 53:379, Mar., 1956.

2318. Milani, R. 1955 Considerazioni genetiche su alcuni aspetti della biologia dell a mosca domestica e osservazioni sulla fecondità e sulla fertilità. [Genetic considerations of certain aspects of the biology of the domestic fly and observations on fertility and prolificacy.] Riv. di Parassit. 16:257-270. (English summary.)

Genetic system probably tends to keep species in a heterozygous condition. Crowding of larvae influences both fecundity and fertility.

2319. Milani, R. 1956 Recenti sviluppi delle ricerche genetiche sulla mosca domestica. [Recent developments in genetic research on the domestic fly.] Boll. Zoologia. 23:749-764.

Linkage detected for 3 genes, including *kdr* (DDT-resistance).

2320. Milani, R. 1957 Errori genetici in ricerche sulla ereditarieta della resistenza agli insetticidi. [Genetic errors in research on the inheritance of resistance to insecticides.] Riv. di Parassit. 18(2):129-132.

ABSTRACT: TDB 54:1255, Oct., 1957.

2321. Milani, R. 1957 Ricerche genetiche sulla resistenza degli insetti alla azione delle sostanze tossiche. [Genetic research on the resistance of insects to the action of toxic substances.] Rend. Ist. Sup. Sanit. (Roma) 20(7-8):713-772.

Low fertility and a prolonged period of development were repeatedly found in resistant strains of flies. Genes responsible for resistance in laboratory strains exist also in natural populations.

2322. Milani, R. 1958 The Genetics of the House Fly (*Musca domestica*): New Mutants, Linkage Groups and Abnormal Segregations. Proc. Xth Congr. Genet. Montreal 2:189.

Several mutant types found in *M. domestica*. Linked genes: brown body (bwb); divergent wings (dv); knock-down resistance (kdr). Crossover very low in male; high in female.

2323. Milani, R. 1959 Differenze osservate tra ceppi di *Musca domestica* L. nelle variazione di riposta ai trattamenti tossicologici determinate da condizioni sperimentali. [Differences observed between strains of *M. domestica* in variations of response to toxicological treatments determined by experimental conditions.] Riv. di Parassit. 20(2):111-124.

Effect of temperature on response to treatment differs with strain, e.g.: without effect; acts similarly in both sexes; acts differently according to sex when dose is high. ABSTRACTS: RAE-B 50:193, Sept., 1962; TDB 57:305, Mar., 1960.

2324. Milani, R. 1959 Observations on Intra-specific Differentiation, Genetic Variability, Sex-limited Inheritance, DDT-resistance and Aspects of Sexual Behaviour in *Musca domestica* L. Symposia Genetica et Biologica Italica 9:1-16.

Suggests that genotype of the house fly is loaded with recessive genes, unfavourable when homozygous. Specific modifying genes, also outbreeding, may impart protection.

2325. Milani, R. and **Franco, M. D. 1959** Fertilità, rapporto sessi e segregazione del mutante *bwb* in incroci tra ceppi geograficamente

separati di *Musca domestica* L. [Fecundity, sex ratio and segregation of the mutant *bwb* in crosses between geographically distinct strains of *Musca domestica* L.] Symposia Genetica 6:249-268.
Fertility the same in all pairs involving strain OR females. OR males gave offspring only when mated to females of their own strain. Some strain crosses produced gymandromorphs.

2326. Milani, R. and **Franco, M. G. 1959** Comportamento ereditario della resistenza al DDT in incroci tra il ceppo Orlando—R. e ceppi *kdr* e *kdr+* di *Musca domestica* L. [Hereditary behavior of resistance to DDT in crosses between the Orlando-R strain and the strains *kdr* and *kdr+* of *Musca domestica* L.] Symposia Genetica 6:269-303.
Resistance factor of OR strain and of *kdr* strain have similar phenotypic effect, but are not allelic. Former should be known as *kdr-o*.

2327. Milani, R. and **Rivosecchi, L. 1955** Malformazioni e mutazioni di *Musca domestica* L. di interesse per la conoscenze dei segmenti terminali maschili dei Ditteri. [Malformations and mutations of *Musca domestica* L. of interest towards the knowledge of male terminal segments of Diptera.] (a) Boll. Zoologia 22(2):341-372. (b) Rent Ist. Sup. Sanit. 20:600-634. 1957.
Rotation of hypopygium can stop at any position between 0° and 360°. Direction of rotation is determined by a single gene. Anticlockwise rotation is termed "countercoiled". ABSTRACT: BA 39:2018(25458), Sept. 15, 1962.

2328. Milani, R. and **Rivosecchi, L. 1955** Osservazioni sugli scleriti dei segmenti terminali di individui sessual menti anormali di *M. domestica* L. [Observations on the sclerites of the terminal segments of sexually abnormal individuals of *Musca domestica* L.] (a) Boll. Zoologia 22(2):373-390. (b) Rend. Ist. Sup. Sanit. 20:635-654. 1957.
Developmental rotation may be incomplete when small female parts are present in mainly male structures. Reverse may cause torsions or flexions. ABSTRACT: BA 38:930(12027), May 1, 1962.

2329. Milani, R. and **Travaglino, A. 1957** Ricerche genetiche sulla resistenza al DDT in *Musca domestica* concatenazione del gene *kdr* (knock-down resistance) con due mutanti morfologigi. [Genetic research on resistance to DDT in *Musca domestica* linking the gene *kdr* (knock-down resistance) with two morphological mutations.] Riv. di Parassit. 18(3):199-202.
Tables show segregation and recombination of genes *bwb* (brown body), *dv* (divergent) and *kdr*. ABSTRACT: TDB 55:461, Apr., 1958.

2330. Milani, R. and **Travaglino, A. 1959** Esperimenti di incrocio tra due ceppi DDT-resistenti di *Musca domestica* L. di origine diversa. [Experiments with crosses between two DDT-resistant strains of *Musca domestica* of diverse origin.] Symposia Genetica 6:213-248.
If OR parents are males, females are less variable and more susceptible than when male parent is of strain *kdr*. Two genes for resistance probably *not* allelic.

2331. Miller, A. C., Mallis, A. and **Sharpless, R. V. 1952** Aerosol Insecticides, their Evaluation against House Flies, Cockroaches. Soap and Sanit. Chem. 28(2):151, 153, 181; (3)143, 145, 147, 149.
Tests took account of dosage, test temperature, height and manner of discharge, delivery rate of dispensers, size of test populations, type of mortality determination. ABSTRACTS: RAE-B 40:97, June, 1952; TDB 49:808, Aug., 1952.

2332. **Miller, A. C., Pellegrini, J. P., Pozefsky, A.** and **Tomlinson, J. R. 1952** Synergistic Action of Piperonyl Butoxide Fractions and Observations Refuting a Pyrethrins-butoxide Complex. J. Econ. Ent. 45: 94-97.
Experiments disclosed no evidence for the concept of a *molecular complex*. ABSTRACT: RAE-B 40:94, June, 1952.

Miller, A. C., joint author. *See* Mallis, A., et al., 1952; Pellegrini, J. P., et al., 1952.

Miskus, R., joint author. *See* Eldefrawi, M. E., et al., 1959.

2333. **Missiroli, A. 1950** The Control of Domestic Insects in Italy. Am. J. Trop. Med. Hyg. 30:773-783.
DDT-resistant strains of *M. domestica* and *Culex pipiens autogenicus* were easily killed with chlordane. ABSTRACTS: BA 25:873, 1951; RAE-B 41:60, Apr., 1953.

2334. **Missiroli, A. 1951** Resistenza agli insetticidi di alcune razze di *Musca domestica*. [Resistance to insecticides on the part of some races of *Musca domestica*.] Riv. di Parassit. (Rome) 12(1):5-25. (English summary.)
Author has selected out three races, one susceptible to all insecticides, one resistant to DDT, one resistant to chlordane and several others. ABSTRACT: TDB 49:89, Jan., 1952.

Mitchell, W. A., joint author. *See* Holway, R. T., et al., 1951.

2335. **Mitlin, N. 1956** Inhibition of Development in the House Fly by 3, 4-Methylenedioxyphenyl Compounds. J. Econ. Ent. 49:683-684.
Compounds are both useful as synergists and toxic in their own right. Resistant as well as normally susceptible flies are affected. ABSTRACTS: RAE-B 45:177, Nov., 1957; TDB 54:240, Feb., 1957; BA 31:922(9254), Mar., 1957.

2336. **Mitlin, N.** and **Babers, F. H. 1955** Relative Toxicity of Topically Applied Allethrin and Pyrethrins to House Flies and Cockroaches. J. Econ. Ent. 48-747-748.
Toxicity of allethrin found to be equal to that of pyrethrins in two strains of flies to which it was administered. ABSTRACTS: BA 30:2076 (20863), July, 1956; RAE-B 45:6, Jan., 1957; TDB 53:677, May, 1956.

2337. **Mitlin, N.** and **Babers, F. H. 1956** The Action of Radiostrontium in the House Fly and the German Cockroach. J. Econ. Ent. 49:714-715.
Continuous feeding inhibited oviposition. Shorter periods caused marked decrease in egg viability. Ova absorb and retain strontium. ABSTRACT: RAE-B 45:180, Nov., 1957.

2338. **Mitlin, N., Babers, F. H.** and **Barthel, W. F. 1956** The Relative Toxicity and Mode of Action of some Chlorinated Phosphates. J. Econ. Ent. 49:544-546.
Five esters of 1, 2, 2, 2- tetrachloroethyl phosphoric acid were tested on houseflies for cholinesterase inhibition and toxicity. Toxicity tended to decrease with increase in the carbon chain of the alkyl group. ABSTRACT: RAE-B 45:144, Sept., 1957.

2339. **Mitlin, N.** and **Baroody, A. M. 1958** Use of the Housefly as Screening Agent for Tumor-inhibiting Agents. Cancer Research 18(6):708-710.
Tested 26 growth-inhibiting agents of diverse biological systems. Ovarian growth used as a criterion. Fifteen agents inhibited development.

2340. **Mitlin, N.** and **Baroody, A. M. 1958** The Effect of Some Biologically Active Compounds on Growth of House Fly Ovaries. J. Econ. Ent. 51(3):384-385.
Ovarian growth was completely inhibited by coumarin, 1-phenyl-2-

thiourea, piperonyl butoxide, p-quinone and thiourea. ABSTRACT: TDB 55:1289, Nov., 1958.

2341. Mitlin, N., Butt, B. A. and **Shortino, T. J. 1957** Effect of Mitotic Poisons on House Fly Oviposition. Physiol. Zool. 30(2):133-136.
Aminopterin, a nitrogen mustard, and colchicine prevented ovarian development and caused sterility in females. Feeding mitotic poisons did not affect fertility in males. ABSTRACT: RAE-B 46:117, Aug., 1958.

2342. Mitlin, N. and **Gersdorff, W. A. 1953** The Stabilization of Scabrin Oil Sprays. J. Econ. Ent. 46:698-699.
Sprays were tested on houseflies by the Campbell turntable method. Phenol S (a mixture of isopropyl cresols) and hydroquinone stabilize scabrin in kerosene sprays for $1\frac{1}{2}$ years. ABSTRACTS: BA 28:949, Apr., 1954; RAE-B 42:41, Mar., 1954.

2343. Mitlin, N., Gersdorff, W. A. and **Nelson, R. H. 1953** Toxicity to the House Fly of two Chlorinated Turpentines in Space Sprays. J. Econ. Ent. 45:1083.
Proved about one-sixth as toxic as toxaphene; one-third as toxic as pyrethrins. ABSTRACT: RAE-B 41:107, July, 1953.

2344. Mitlin, N., Gersdorff, W. A. and **Nelson, R. H. 1954** Strobane-relative Toxicity against House Flies. Soap and Sanit. Chem. 30(3):157, 189.
A chlorinated mixture of α-pinene isomers was found to be about 0·8 as toxic as toxaphene. Caused only negligible knock-down at concentrations used. ABSTRACT: RAE-B 42:130, Sept., 1954.

2345. Mitlin, N. and **Gertler, S. I. 1951** Preliminary Tests with Certain N-Substituted Benzamides for Synergistic Action in Pyrethrum Fly Sprays. U.S. Dept. of Agr. E-828 1-3, November. Mimeographed document.
Thirty-two different preparations comparatively evaluated.

2346. Mitlin, N. Gertler, S. I. and **Gersdorff, W. A. 1953** Preliminary Tests with Certain N-substituted 2, 4-dinitroanilines and 0-chlorobenz-. amides for Synergistic Action in Allethrin Fly Sprays. U.S. Dept, Agric., Agric. Res. Admin., Bur. Ent. and Plant Quarantine. E-869 pp. 1-8, November. Mimeographed document.
Intensity of synergism not very great for any material tested. Best raised toxicity to twice that of allethrin alone.

2347. Mitlin, N. Green, W. A., Gersdorff, W. A. and **Schechter, M. S. 1953** Screening Tests against House Flies with some Derivatives of Chrysanthemumic Acid. U.S. Dept. Agric., Agric. Res. Admin., Bur. Ent. and Plant Quarantine. E-865, pp. 1-5. October. Mimeographed document.
At concentrations tested, little toxicity was shown.

2348. Mitlin, N. and **Konecky, M. S. 1955** The Inhibition of Development in the House Fly by Piperonyl Butoxide. J. Econ. Ent. 48:93-94.
Added in various concentrations to CSMA larval medium, caused inhibition of development. Length of larval life directly proportional to concentration; percentage of adult emergence inversely proportional. Greater effect on DDT-resistant strain than on normal flies. ABSTRACTS: BA 29:2484, Oct., 1955; RAE-B 44:23, Feb., 1956; TDB 52:696, July, 1955.

2349. Mitlin, N., Konecky, M. S. and **Piquett, P. G. 1954** The Effect of a Folic Acid Antagonist on the House Fly. J. Econ. Ent. 47:932-933.
Sodium salt of aminopterin, a structural analog of folic acid, was added to CSMA larval medium. Concentrations of 0·1 and 0·05 per cent prevented adult emergence. ABSTRACTS: BA 29:2005, Aug., 1955; RAE-B 43:157, Oct., 1955; TDB 52:303, Mar., 1955.

2350. **Mitlin, N., Nelson, R. H.** and **Gersdorff, W. A. 1951** Toxicity to House Flies of TDE Ethyl Analog and Heptachlor. Soap and Sanit. Chem. 27(11):139, 143.
Ethyl analog about one-fifth as toxic as DDT. Heptachlor about 26 times as toxic as pyrethrins. ABSTRACT: RAE-B 40:53, Apr., 1952; TDB 49:453, Apr., 1952.

Mitlin, N., joint author. *See* Gersdorff, W. A., et al., 1950, 1951, 1952, 1953, 1954, 1955, 1957; Konecky, M. S., et al., 1955; Babers, F. H., et al., 1955, 1956; Lowman, M. S., et al., 1954; Nelson, R. H., et al., 1950.

2351. **Mitra, R. D. 1953** Relation of Larval Diet to the Toxicity of Insecticide on the House Fly *Musca nebulo* F. Sci. Cult. 18(11):551-553.
M. nebulo bred in a natural medium, like cow dung, possesses a higher resistance to DDT in the adult stage. ABSTRACT: BA 28:956(9773), Apr., 1954.

Mitra, R. D., joint author. *See* Chatterjee, A. K., et al., 1954.

Mitriukova, M. S., joint author. *See* Shura-Bura, B. L., et al., 1956, 1958.

2352. **Miyake, S. S., Kearns, C. W.** and **Lipke, H. 1957** Distribution of DDT-Dehydrochlorinase in Various Tissues of DDT-Resistant House Flies. J. Econ. Ent. 50:359-360.
DDT-dehydrochlorinase activity was at a high level in the brain and fat body, at intermediate amounts in the cuticle, muscle and hemolymph and occurred in small amounts or not at all in the ovary and intestinal tract. ABSTRACTS: BA 32:891(10498), Mar., 1958; RAE-B 46:91, June, 1958; TDB 55:111, Jan., 1958.

Modiane, A., joint author. *See* Neeman, M., et al., 1956.

2353. **Mohieldin, M. S. 1959** On the Study of Two Newly Isolated Organisms (Enterobacteriaceae) from *Musca vicina*. J. Egypt. Med. Ass. 42: 528-539.

2354. **Monchy, H. M. De 1957** A Simple Control of Fly Breeding Places in Compulsory Communities. Docum. Med. Geogr. Trop. (Amsterdam) 9(4):331-333.
Recommends pouring large quantities of boiling water on breeding places for five consecutive days.

2355. **Monroe, R. E. 1959** Role of Cholesterol in House Fly Reproduction. Nature 184 (suppl. 19):1513.
A cholesterol-deficient diet (1) prevents egg hatch; (2) inhibits larval development in a medium containing sterols. ABSTRACT: BA 35:1516 (17487), March, 1960.

2356. **Monroe, R. E.** and **Robbins, W. E. 1959.** Studies of the Mode of Action of Synergized Bayer 21/199 and its Corresponding Phosphate in the House Fly. J. Econ. Ent. 52(4):643.
Applied topically, the two were isotoxic. Were 2·8 times as toxic with piperonyl butoxide. Reversibility of cholinesterase inhibition occurred in flies treated with Bayer 21/199. ABSTRACT: RAE-B 48:193, Nov., 1960.

Monroe, R. E., joint author. *See* Hoffman, R. A., et al., 1956, 1957.

2357. **Moore, D. H., Dove, W. E.** and **Dickinson, B. C. 1954** Fly Control on Livestock. Agric. Chem. 9(8):31-34, 109, 111, 113, 115, 117. Baltimore.
Gives effectiveness of various formulations of piperonyl butoxide with pyrethrins and/or allethrin. Houseflies *not* the major concern. ABSTRACT: RAE-B 43:12, Jan., 1955.

Moore, G. D., joint author. *See* Dicke, R. J., et al., 1952.

2358. Moore, J. B. 1950 Relative Toxicity to Insects of Natural Pyrethrins and Synthetic Allyl Analog of Cinerin I. J. Econ. Ent. 43:207-213.
M. domestica the chief test insect used. ABSTRACT: RAE-B 38:208, 1950.

2359. Moore, S., Toczydlowski, A. H. and **Sweetman, H. L. 1951** Fly Control Experiments in Massachusetts in 1950. J. Econ. Ent. 44:731-733.
No apparent resistance to several insecticides including methoxychlor, lindane and DDT. Tests included use of "continuous vaporization of insecticide" (a new method of application). ABSTRACTS: BA 26:1804, July, 1952; RAE-B 40:44, Mar., 1952; TDB 49:199, Feb., 1952.

Moore, S., joint author. *See* Sweetman, H. L., et al., 1951.

2360. Moorefield, H. H. 1956 Purification of DDT-Dehydrochlorinase from Resistant House Flies. Contr. Boyce Thompson Inst. 18:303-310.
Steps include: acetone powdering of adults; purification with ammonium sulphate; dialysis, adsorption with activated charcoal; salt concentration of resultant protein solution. There are probably only 4 major protein components. ABSTRACTS: BA 31:1814(18953), June, 1957; RAE-B 45:29, Feb., 1957.

2361. Moorefield, H. H. 1958 Synergism of the Carbamate Insecticides. Contr. Boyce Thompson Inst. 19(6):501-507.
Both aromatic and heterocyclic carbamates may be synergized with pyrethrins adjuvants. Isolan and pyrolan were synergized about 10-fold. Combinations effective against house flies. ABSTRACT: BA 33:1605, Mar., 1959.

2362. Moorefield, H. H. 1958 The Origin of DDT Resistance in the House Fly. Contr. Boyce Thompson Inst. 19(5):403-409. Yonkers, N.Y.
The detoxication enzyme, DDT-dehydrochlorinase has been demonstrated in larvae never exposed to DDT. Genetic origin of resistance appears to be substantiated. ABSTRACT: RAE-B 48:205, Dec., 1960.

2363. Moorefield, H. H. and **Kearns, C. W. 1955** The Mechanism of Action of Certain Synergists for DDT against Resistant House Flies. J. Econ. Ent. 48:403-406.
The selective action of DDT alone cannot result in the maximum potential level of resistance, but by the use of DDT and synergist in combination, selection proceeds further, resulting in a fly strain having a high titer of DDT-dehydrochlorinase. ABSTRACTS: BA 30:1146(11690), April, 1956; RAE-B 44:100, July, 1956; TDB 53:110, Jan., 1956.

2364. Moorefield, H. H. and **Kearns, C. W. 1957** Levels of DDT-dehydrochlorinase During Metamorphosis of the Resistant House Fly. J. Econ. Ent. 50(1):11-13.
The detoxification enzyme, DDT-dehydrochlorinase, increases during the larval stage in resistant houseflies. About 50 per cent reduction in the total activity occurs at pupation and the resultant level is maintained through pupal and adult life. ABSTRACTS: BA 31:1815(18954), June, 1957; RAE-B 46:42, Mar., 1958; TDB 54:764, June, 1957.

2365. Moorefield, H. H. and **Tefft, E. R. 1959** Evaluation of 2-(3, 5-dichloro-2-biphenylyloxy)-triethylamine as an Insecticide Adjuvant. Cont. Boyce Thompson Inst. 20(4):293-298.
Synergism of pyrethrins and the carbamates not a correlative phenomenon. Specific carbamate synergists are feasible.

Moorefield, H. H., joint author. *See* Haynes, H. L., et al., 1957.

2366. Morellini, M. and **Saccà, G. 1953** Alcuni aspetti del meccanismo di diffusione del b. tubercolare da parte di *M. domestica* (studio sperimentale). [Some aspects of the mechanism of distribution of the tubercle bacillus by *M. domestica* (experimental study).] Rend. Ist. Sup. Sanit. (Roma) 16:267-285.

Flies are particularly attracted to, and nourish themselves on patient's sputum. This results in a diarrhea of the fly; but fly vomitus is more important in the spread of tuberculosis than fly feces.

2367. Morozova, V. P., Dolmatova, A. V., Kirichenko, A. G., Leibman, A. L. and **Taguzov, T. U. 1957** [Organization of fly control in Yalta.] Med. Parazit. i Parazitarn. (Bolenzi) 26(1):17-21. (In Russian.)

2368. Morozova, V. P., Leibman, A. L. and **Kirichenko, A. T. 1955** Oprimenenii zelenogo masla v bòrbe s lichinbcami mukh. [The use of green oil in combating fly larvae.] Med. Parasit. (Moscow) for 1955 (3):266.

Green oil from petroleum is useful for fly control if used on refuse and trash. Manure so treated is not fit for fertilizer. ABSTRACT: BA 31:2125 (22750), July, 1957.

2369. Morrill, A. W., Jr. 1953 Army Insect Control Operations in the Far East. J. Econ. Ent. 46:270-276.

Musca, *Chrysomyia* and *Lucilia* were troublesome. Tend to alight on floor or food, avoiding residual insecticides on walls. Flies ride in great numbers in Army vehicles. Policy was to replace dumps with sanitary land fills. ABSTRACTS: RAE-B 41:172, Nov., 1953; TDB 50:1083, Nov., 1953.

Morrill, A. W., joint author. *See* Wesley, C., Jr., et al., 1956.

Morris, G. C., Jr., joint author. *See* Fay, R. W., et al., 1958.

2370. Morrison, F. O. 1957 The Ratio of the Width to the Length of the Second Abdominal Sternite of the House Fly. J. Econ. Ent. 50: 554-556.

Large ratios tend to go with resistance to DDT; small ratios with susceptibility. Correlation not perfect. Cultures with a known common origin had ratios with a narrow range. ABSTRACTS: BA 32:891(10499), Mar., 1958; RAE-B 46:167, Nov., 1958.

2371. Morrison, F. O. and **Hagley, E. A. 1959** The Effect of Three p-chlorophenyl Compounds on the Biology of the House Fly. J. Econ. Ent. 52(1):61-62.

These synergizing compounds were used to moisten larval substrate. Surviving individuals showed normal growth and development, but were more susceptible to contact with DDT. Pupation was retarded, but fecundity, viability and longevity were normal. ABSTRACTS: BA 33:2896 (35382), June, 1959; RAE-B 48:94, May, 1960.

2372. Morrison, F. O. and **LeRoux, E. J. 1954** House Fly Head as Site of Lethal Action of DDT. Canad. J. Agr. Sci. 34:316-318.

DDT may be translocated in fly's haemolymph. Fly's head is a critical region for lethal action. ABSTRACTS: BA 29:701, Mar., 1955; RAE-B 43:188, Dec., 1955; TDB 53:507, Apr., 1956.

Morrison, F. O., joint author. *See* Hagley, E. A. C., et al., 1958; Campau, E. J., et al., 1953; LeRoux, E. J., et al., 1953, 1954.

2373. Morrison, H. E., Lauderdale, R. W., Crowell, H. H. and **Mote, D. C. 1950** Space Spraying for Fly Control in Dairy Barns. J. Econ. Ent. 43(6):846-850.

DDT, chlordane and pyrenone were found satisfactory and economically feasible. Pyrenone has some advantage in that milk contamination is unlikely. ABSTRACTS: RAE-B 39:119, July, 1951; TDB 48:590, 1951.

2374. **Mosebach, E. 1951** Unterschiedliche Empfindlichkeit von Männchen und Weibchen der Stubenfliege. [Susceptibility (to DDT and BHC) of male and female houseflies.] Zeit. Hyg. Zool. 39(5/6):139-145.
Males knocked down more quickly than females, and a smaller proportion recover. This should be taken into account in evaluating effects of a control program against a native population, e.g., Were the sexes out of balance prior to the campaign?

Moses, P., joint author. *See* Bergmann, E. D., et al., 1957.

2375. **Mosna, E. 1950** Il controllo con Octa-Klor delle mosche domestiche DDT resistenti. [Control of DDT-resistant house flies with octa-klor.] Riv. di Parassit. 11(1):27-35. (English summary.)
Residual action decreases sharply after 4 months, but some insecticidal property persists for about a year.

2376. **Mosna, E. 1951** Sulla resistenza delle mosche domestiche al chlordane. [On the resistance of domestic flies to chlordane.] Riv. di Parassit. (Rome) 12(1):27-36. *Also* (under same title) Estratto Rend. Ist. Sup. Sanit. 14:563-673.
The possibility of successful house fly control by continued use of chlorinated residual insecticides seems very remote. ABSTRACT: TDB 49:90, Jan., 1952.

2377. **Mosna, E.** and **Alessandrini, M. 1954** L'impiego degli esteri fosforic. nella lotta contro la mosca domestica nella provincia di Latina. [The employment of phosphoric esters in the war against the housefly in the province of Latina.] Riv. di Parassit. Rome 15:543-556. (English summary.)
From 1952-1954 tests on fly control were carried out using deposits of diazinon and malathion as sprays or suitably placed sprayed objects. Although successful, the cost of control was almost double that when chlordane was used. ABSTRACTS: RAE-B 45:64, Apr., 1957; TDB 52: 699, July, 1955.

Mote, D. C., joint author. *See* Morrison, H. E., et al., 1950; Roth, A. R., et al., 1950.

2378. **Motta, L. C. Da 1956** Morte aos insectos transmissores de doencas, o perigo dos insectos e em especial o perigo das moscas; a luta contra as moscas. [Death from insect vectors of disease, the danger from insects and especially the danger from flies; a campaign against flies.] Gazeta Médica Portuguesa (Lisboa) 9:325-332.

2378a. **Mukerjea, T. D.** and **Govind, R. 1958** Studies on Indigenous Insecticidal Plants: Part II—*Annona squamosa.* J. Sci and Indust. Res. 17(1):9-15.
Ether extract of this species is one-seventh as toxic as DDT to *M. nebulo.* Residual toxicity lasts 8 days. Treatment with petroleum ether increases toxicity 6 times.

Murphy, R., joint author. *See* Hornstein, I., et al., 1958.

Muthukrishnan, T. S., joint author. *See* Abraham, E. V., et al., 1955.

2379. **Nabokov, V. A., Lariukhin, M. A.** and **Nikiforova, A. V. 1956** [Results of application of chlorophos and of diazinone in control of flies resistant to chlorinated hydrocarbons.] Med. Parazit. i Parazitarn. (Bolezni) 25:256-258.

2380. **Nabokov, V. A., Lariukhin, M. A.** and **Zhukova, L. I. 1956** [Insecticide effect of diasinone on *Musca domestica* L. larvae and nymphs resistant to DDT; preliminary communication.] Zhur. Mikrobiol. (Moskva) 27(1):83-87.

2381. Nagasawa, S. 1952 [Studies on the Biological Assay of Insecticides. XVIII. On the Knockdown Effect of Some Insecticidal Powders to Adults of the Common Housefly (*Musca domestica* L.) with Special Reference to the Sexual Difference of Susceptibility.] Oyo-Kontyo 8(1):29-33. (In Japanese, with English summary.)

Females are more susceptible than males to all insecticidal powders. ABSTRACT: BA 28:949, Apr., 1954.

Nagasawa, S., joint author. *See* Katsuda, Y., et al., 1958.

2382. Nakajyo, E. 1955 Some Histological Observations on Fatal Flies (*Musca domestica* Linné) Contacted with γ-BHC. J. Osaka City Med. Center 4:55-58, 97. (In Japanese.)

Observations of nervous bulb, cells of muscle and gastrointestinal tract made from time of γ-BHC application to 24 hours following death. ABSTRACT: TDB 53:379, Mar., 1956.

Nascimento, L. P., joint author. *See* Ricciardi, I., et al., 1952.

2383. Nash, K. B. 1950 Biological Tests of Allethrin without a Synergist. Soap and Sanit. Chem. 26(9):127, 129.

Reports results of tests from 7 laboratories. In aerosols, allethrin gave better knockdown and kill than pyrethrins. ABSTRACT: RAE-B 40:15, Jan., 1952.

2384. Nash, R. 1954 Studies on the Synergistic Effect of Piperonyl Butoxide and Isobutylundecyleneamide on Pyrethrins and Allethrin. Ann. Appl. Biol. 41:652-663.

Piperonyl butoxide is the more powerful synergist, increasing the potency of pyrethrin 5 times and allethrin 4 times to *M. domestica.* (Pyrethrins are basically twice as toxic as allethrin to house flies.) ABSTRACTS: RAE-B 43:86, June, 1955; TDB 52:412, Apr., 1955.

Nash, R., joint author. *See* Busvine, J. R., et al., 1953.

Natalizi, G., joint author. *See* Bettini, S., et al., 1958.

2385. National Academy of Sciences 1952 Conference on Insecticide Resistance and Insect Physiology. December 8-9, 1951, Univ. of Cincinnati. Nat. Acad. Sci. Nat. Res. Council, Washington D.C. Publ. 219, 99 pp.

Contributions from 23 scientists. Very frequent reference to *M. domestica,* both with regard to control problems and as a test species.

2386. Neeman, M. and **Modiane, A. 1956** Studies in Selective Toxicity, I. Syntheses of N-Alkylbenzene-sulfonailides. J. Org. Chem. 21(6): 667-670.

Preparation, properties and biological activity described. Compounds not toxic alone but some were active synergists for DDT against *M. domestica vicina* Macq. ABSTRACT: RAE-B 45:89, June, 1957.

2387. Neeman, M., Modiane, A. and **Shor, Y. 1956** Studies in Selective Toxicity. II. Substituted Phenyl Benzoates and Benzenesulfonates. J. Org. Chem. 21(6):671-672.

Compounds synthesized and tested for toxicity on eggs and larvae of. *M. domestica vicina* Macq. No significant toxicity observed. ABSTRACT: RAE-B 45:90, June, 1957.

Neeman, M., joint author. *See* Bergmann, E. D., et al., 1957.

2388. Nelson, R. H., Gersdorff, W. A. and **Mitlin, N. 1950** Relative Toxicity to House Flies of DDT, Methoxychlor, and Dichlorodiphenyl Dichloroethane. J. Econ. Ent. 43:393-394.

Methoxychlor about three-fifths and DDD (=TDE) about one-quarter as toxic as DDT. Order of knockdown: (1) Methoxychlor; (2) DDD (TDE); (3) DDT. ABSTRACT: RAE-B 39:6, 1951.

Nelson, R. H., joint author. *See* Gersdorff, W. A., et al., 1950, 1951, 1952, 1954, 1955; Fales, J. H., et al., 1951, 1954; Fulton, R. A., et. al., 1950; Mitlin, N., et al., 1951, 1953, 1954.

Nestervodskaia, E. M., joint author. *See* Zinchenko, V. S., et al., 1956.

2389. Newman, J. F. 1953 Resistance of Flies to Different Insecticides. *In* Trans. IXth Internat. Congress Entom. Amsterdam. August 17-24, 1951. Vol. II. Symposia, pp. 331-334.

Populations of *M. domestica* were tested for distribution of resistance, particularly to BHC. Resistant strains found to have as many susceptible individuals as "normal" strains, but some individuals 5 times as resistant. Results with inbred resistant strains reported. ABSTRACT: RAE-B 41:121, Aug., 1953.

Newman, J. F., joint author. *See* Bradbury, F. R., et al., 1953.

Newson, H. D., joint author. *See* Wheeler, C. M., et al., 1958.

N'Haili, A., joint author. *See* Gaud, J., et al., 1955.

Nichols, R. E., joint author. *See* Ahmed, M. K., et al., 1958.

2390. Nicholson, A. J. 1954 Experimental Demonstrations of Balance in Populations. Nature (London) 173(4410):862-863.

Worked with *Lucilia cuprina.* "Low population" condition induced rapid generation of new individuals; "high population" condition suppressed reproduction. *Density-induced reaction* holds population in balance in its environment.

Nield, P., joint author. *See* Bradbury, F. R., et al., 1953.

Nikiforova, A. V., joint author. *See* Nabokov, V. A., et al., 1956.

2391. Niko, M. and **Ogata, K. 1958** [Seasonal occurrence of the flies of public health importance (that) emerged (from fowl droppings) on a poultry ranch in Horishima Prefecture, Japan.] Jap. J. Sanit. Zool. 9:51-55. Tokyo. (In Japanese, with English summary.)

Fifteen species collected in one year, chiefly between May and September. Among them were *M. vicina, Fannia canicularis* and other medically-important forms.

Norton, L. B., joint author. *See* Pimentel, D., et al., 1950, 1951.

2392. Norton, R. J. 1953 Inheritance of DDT Tolerance in the House Fly. Contr. Boyce Thompson Inst. 17(2):105-126.

Relative tolerance to 5 insecticides was determined in the immature life stages for 5 more or less resistant strains and 1 non-tolerant laboratory strain. Employed crosses and back-crosses. ABSTRACTS: BA 27:2212, Aug., 1953; RAE-B 42:94, June, 1954.

Oana, C., joint author. *See* Duca, M., et al., 1958.

2393. O'Brien, R. D. 1956 The Inhibition of Cholinesterase and Succinoxidase by Malathion and its Isomer. J. Econ. Ent. 49:484-490.

In vitro results support the view that malathion kills by an anti-cholinesterase action, though *in vivo* findings do not. Succinoxidase inhibition apparently not cause of death in *M. domestica.* ABSTRACT: RAE-B 45: 142, Sept., 1957.

2394. O'Brien, R. D. 1957 The Effect of Malathion and its Isomer on Carbohydrate Metabolism of the Mouse, Cockroach, and House Fly. J. Econ. Ent. 50(1):79-84.

Unlikely that insecticidal action has any connection with small interference effect on citrate oxidizing system of the tricarboxylic acid cycle pathways of carbohydrate metabolism. ABSTRACTS: BA 31:1815(18957), June, 1957; RAE-B 46:47, Mar., 1958.

2395. **O'Brien, R. D., Thorn, G. D.** and **Fisher, R. W. 1958** New Organophosphate Insecticides Developed on Rational Principles. J. Econ Ent. 51(5):714-718.
Thionosphosphates containing carboxyester group all show selective toxicity. Metabolism of these compounds by mouse liver, cockroach and houseflies discussed. ABSTRACT: RAE-B 48:25-26, Jan., 1960.

2396. **O'Brien, R. D.** and **Wolfe, L. S. 1959** The Metabolism of Co-Ral (Bayer 21/199) by Tissues of the House Fly, Cattle Grub, Ox, Rat and Mouse. J. Econ. Ent. 52(4):692-695.
In the housefly, which is susceptible to Co-Ral, there is an activating but no degrading system. ABSTRACT: RAE-B 48:195, Nov., 1960.

O'Brien, R. D., joint author. *See* Krueger, H. R., et al., 1959.

O'Carroll, F. M., joint author. *See* Bradbury, F. R., et al., 1958.

O'Connor, C. T., Jr., joint author. *See* Hansens, E. J., et al., 1955.

Oddo, F., joint author. *See* Mariani, M., et al., 1959.

Ogaki, M., joint author. *See* Tsukamoto, M., et al., 1957.

2397. **Ogata, K. 1959** [Studies on the behavior of *Musca domestica vicina* and *Fannia canicularis* in winter season, as observed at inside and outside of a residence in Kawasaki City.] Jap. J. Sanit. Zool. 10(4):251-257. Tokyo. (In Japanese, with English summary.)
Stresses diurnal rhythm as dependent on changes in temperature and light intensity. *M. d. vicina* tends to be found indoors in morning, evening and at night; prefers a sunny, outdoor situation in the daytime. ABSTRACT: RAE-B 51:270, Dec., 1963.

2398. **Ogata, K., Suzuki, T., Osada, Y.** and **Hirakoso, S. 1957** [Some notes on the habits of early stages of *Fannia canicularis* L. in the northern part of Japan.] Jap. J. Sanit. Zool. 8(4):198-205. (In Japanese, with English summary.)
Most important habitat is "steepers" for preparation of *takywan* (pickled garden radishes). Contained rice bran and salt, used as the breeding medium. ABSTRACT: RAE-B 51:270, Dec., 1953.

Ogata, K., joint author. *See* Niko, M., et al., 1958.

2399. **Ogden, L. J.** and **Kilpatrick, J. W. 1958** Control of *Fannia canicularis* L. in Utah Dairy Barns. J. Econ. Ent. 51(5):611-612.
Parathion-Diazinon-impregnated cords controlled *F. canicularis* populations satisfactory. Malathion-DDVP poison baits did not. *M. domestica* populations too small to be evaluated.

2400. **O'Keefe, W. B.** and **Schorsch, C. B. 1954** Studies on Some Enteric Bacteria Associated with Flies (Diptera: Calliphoridae, Muscidae, Anthomyidae). Med. Technicians Bull. 5(1):15-17.

2401. **Oliver, A. D.** and **Eden, W. G. 1955** Toxicity of Several Insecticides to Two Strains of the House Fly. J. Econ. Ent. 48:111.
Resistance of native houseflies to DDT, lindane, endrin, TEPP and malathion was determined by topical application. Only in regard to DDT did the two strains differ. ABSTRACTS: BA 29:2484, Oct., 1955; RAE-B 44:25, Feb., 1956; TDB 52:586, June, 1955.

Onuchin, A. M., joint author. *See* Shura-Bura, B. L., et al., 1956.

2402. **Oppenoorth, F. J. 1954** Metabolism of Gamma-Benzene Hexachloride in Susceptible and Resistant Houseflies. [Correspondence.] Nature 173:1000-1001.
Insecticide injected into thorax in an emulsion of peanut oil. No BHC found in excreta. Rapid break-down in resistant strains (was metabolized). Slow in susceptibles, then heavy symptoms, death. ABSTRACT: TDB 51:1006, Sept., 1954.

2403. **Oppenoorth, F. J. 1955** Differences between Rates of Metabolism of Benzene Hexachloride in Resistant and Susceptible Houseflies. Nature 175:124-125.
Alpha and gamma isomers of BHC injected in peanut oil. Assumes that all isomers are metabolized by the same mechanism. Interpretation and hypotheses. ABSTRACTS: BA 29:2001, Aug., 1955; TDB 52:488, May, 1955.

2404. **Oppenoorth, F. J. 1956** Resistance to *gamma* Hexachlorocyclohexane in *Musca domestica* L. Arch. Néerland. Zool. 12(1):1-62.
Deals with physiological causes of resistance. Three processes studied: Absorption, detoxication, excretion. ABSTRACT: BA 32:891(10502), Mar., 1958.

2405. **Oppenoorth, F. J. 1958** A Mechanism of Resistance to Parathion in *Musca domestica* (L). Nature 181(4606):425-426.
Assumed that parathion is oxidized in flies to paraoxon. Some relation exists between resistance to *gamma* BHC and to parathion, though they are very different chemically. ABSTRACT: TDB 55:705, June, 1958.

2406. **Oppenoorth, F. J. 1958** Resistentie van *Musca domestica* tegen fosfor insekticiden Overdruk uit Mededelingen van de Landbouwhogeschool en de opzoekingsstations van de Staat te Gent. 1958. Deel 33(3-4)-709-714.
Susceptibility to several organophosphorus insecticides studied in 3 strains of flies already resistant to parathion or diazinone. More than one mechanism of resistance is indicated.

2407. **Oppenoorth, F. J. 1959** Resistance Patterns to Various Organophosphorus Insecticides in some Strains of Houseflies. Ent. Exp. Appl. 2(3):216-223. Amsterdam. (German summary.)
Each of 6 resistant strains manifested its own pattern of resistance when tested against 13 organophosphorus compounds. A large number of resistance mechanisms are probably involved. ABSTRACT: RAE-B 48: 212, Dec., 1960.

2408. **Oppenoorth, F. J. 1959** Genetics of Resistance to Organophosphorus Compounds and Low Ali-esterese Activity in the Housefly. Ent. Exp. Appl. 2(4):304-319. (German summary.)
Believed that there are at least 4 different genes for low aliesterase activity, plus alleles. One of these gives resistance to malathion, another, to diazinon. ABSTRACTS: BA 35:2512(28461), June, 1960; TDB 57:761, July, 1960.

Orgell, W. H., joint author. *See* Dahm, P. A., et al., 1959.

Osada, Y., joint author. *See* Ogata, K. S., et al., 1957.

2409. **Osmun, J. V. 1955 [1956]** Development of Resistance in Insects to Insecticides. *In*: The Past, Present and Future use of Insecticides. Proc. Indiana Acad. Sci. 65:139-144.
Points out that housefly resistance to DDT dates from 1947. ABSTRACT: BA 35:3993(46010), Aug. 15, 1960.

2410. **Ostashev, S. N. 1958** Sud'ba bakterii pri metamorfoze komnatnoi mukhi. [The fate of bacteria during the metamorphosis of the house fly.] Byul. Nauchn.—Tekh. Inform. Vses. Nauchn.—Issled. Inst. Vet. Sanit. 3:18-19; Referat. Zhur., Biol., 1960, No. 43889.
ABSTRACT: BA 45:4339(53529), June 15, 1964.

2411. **Ostachev, S. N. 1958** [Transfer of viruses by flies.] Byul. Nauchn.—Tekh. Inform. Vses. Nauchn.—Issled. Inst. Vet. Sanit. 3:16-17, 1958. Referat. Zhur., Biol., 1960, No. 50962 (Translation).
The housefly can transfer the virus of Asiatic fowl pest. ABSTRACT: BA 45:3884(47836), June 1, 1964.

2412. Ouye, M. T. 1959 Effects of Malathion and Piperonyl Butoxide Combinations on the Rate of Oxygen Consumption of House Flies, *Musca domestica* L. Diss. Absts. 20, Pt. 1, No. 3, p. 1056.
0_2 consumption of KUN(susceptible) females treated with 1·08 mg. of malathion per fly was increased 2·6 times normal. In CAL(resistant) females increase was 2·4 times normal. In both strains lowering the dosage lowered the degree of stimulation. ABSTRACTS: BA 35:1034 (11742), Feb. 15, 1960; BA 36:6873(74149), Nov. 1, 1961.

2413. Ouye, M. T. and **Knutson, H. 1957** Reproductive Potential, Longevity, and Weight of House Flies Following Treatments of Larvae with Malathion. J. Econ. Ent. 50:490-493.
Parent flies surviving 3 treatments in larval media produced 12 per cent more potential adult progeny than un-treated. Progeny of treated adults weighed less. ABSTRACTS: BA 32:1491(17708), May, 1958; RAE-B 46: 140, Sept., 1958; TDB 55:112, Jan., 1958.

2414. Owen, B. L. 1959 Investigations on the Development of Resistance to Chemicals by the House Fly, *Musca domestica* L. Diss. Absts. 20, Pt. 1, No. 3, p. 1107.
Studied cross tolerance responses of Orlando Laboratory strain. This strain showed low tolerance for DDT and TDE, a very high tolerance for dieldrin and a fairly low tolerance for endrin. ABSTRACT: BA 35: 1034(11743), Feb. 15, 1960.

Owen, D. F., joint author. *See* Parmenter, L., et al., 1954.

2415. Ozerov, A. S. 1958 [The synanthropic flies in Vologda.] Med. Parazit., (Moskva) 27(1):103-104.

2416. Packchanian, A. 1954 Altitude Tolerance of Normal and Infected Insects. J. Econ. Ent. 47:230-238.
Special apparatus provided the low temperatures and low pressures of higher altitudes. Studied 5 species of arthropods, including *M. domestica. Herpetomonas muscae domesticae* lived on in bodies of flies killed by extreme conditions. ABSTRACT: TDB 51:1303, Dec., 1954.

2417. Packchanian, A. 1955 The Effect of Simulated Altitude on Certain Insect Eggs. Texas Repts. Biol. Med. 13:454-469.
Material studied included pupae of *M. domestica.* None hatched after 15 minutes exposure to −30°C with 314 mm Hg. (equiv. 22,500 ft. above sea level) or to −44°C, with 225 mm Hg. (equiv. 30,000 ft). Relation to high-altitude flying of airplanes. ABSTRACT: TDB 53:375, Mar., 1956.

2418. Packchanian, A. 1957 The Isolation and Cultivation of Hemoflagellates in Pure Culture from Six Species of Insects. Texas Repts. Biol. Med. 15(3):399-410.
Haemoflagellates were isolated, and cultured, bacteria free, from 6 species, including *M. domestica.* Stock cultures maintained *in vitro* on blood agar slants for four to five years. Those from *Musca* were of leptomonad type. ABSTRACT: BA 32:286(3268), Jan., 1958.

2419. Pal, R. 1950 The Wetting of Insect Cuticle. Bull. Ent. Res. 41:121-239.
Thirty species of insects used. Larvae of *Musca* proved both lipophilic and hydrophilic. ABSTRACT: BA 27:1032, Apr., 1953.

2420. Pal, R. 1951 DDT Resistant Strain of *Musca nebulo.* Trans. Roy. Soc. Trop. Med. Hyg. 45(1):125-126.
There was only slight resistance after 3 generations and no intensification for the next 9. ABSTRACTS: RAE-B 41:94, June, 1953; TDB 49:90, Jan., 1952.

2421. Pal, R. 1952 The Relationship between the Physical Forms of DDT and their Biological Effectiveness on Solid Surfaces. Indian J.

Malariol. 6(3):239-250.
If insecticide exists as supersaturated droplets, several types of crystal formation may take place. Physical state influences effectiveness. House-flies used to test effectiveness of crystal types on glass panels. Abstract: TDB 51:121, Jan., 1954.

2422. Pal, R. 1952 Dieldrin for Malaria Control. Indian J. Malariol. 6:325-330.
Reviews use of this insecticide in controlling mosquitoes and house flies. Abstract: BA 28:1687, July, 1954.

2423. Pal, R. and **Sharma, M. I. D. 1952** Rapid Loss of Biological Effectiveness of DDT Applied to Mud Surfaces. Ind. J. Malariol. 6(3):251-263.
With flies especially, a very rapid decline in effectiveness took place within a week or so. Loss not considered to be due wholly to decomposition of DDT. Abstract: TDB 51:232, Feb., 1954.

2424. Pal, R. and **Sharma, M. I. D. 1953** Laboratory Studies on the Development of a Resistant Strain of *Musca nebulo*. *In* Trans. IXth Internat. Congress Entom. Amsterdam. August 17-24, 1951. Volume II. Symposia, pp. 346-350.
After selection for 24 generations, species was not more than twice as resistant, by knock-down, as normal flies. Abstract: RAE-B 41:124, Aug., 1953.

2425. Pal, R., Sharma, M. I. D. and **Krishnamurthy, B. S. 1952** Studies on the Development of Resistant Strains of House-flies and Mosquitoes. Ind. J. Malariol. 6(3):303-316.
Continued selection caused *M. nebulo* to become 13 to 29 times as resistant to kill (though only 2 times as resistant to knock-down). Strain lost resistance rapidly when not exposed to DDT. Abstracts: BA 28:2240, Oct., 1954; TDB 51:319, Mar., 1954.

2426. Pal, R., Sharma, M. I. D. and **Krishnamurthy, B. S. 1954** The Relationship between the Lipoid Solubility of DDT and its Contact Toxicity to Houseflies and Mosquitoes. Indian J. Malariol. 8(1):29-31.
Cuticular lipoid extracts prepared from *Anopheles stephensi* and *M. domestica* var. *nebulo* progressively dissolved fine DDT crystals. *Culex fatigans* lipoid extract did not. Possible explanation of different susceptibilities. Abstract: TDB 52:102, Jan., 1955.

Panetsky, D., joint author. *See* Baker, F. D., et al., 1958.

Pankaskie, J. E., joint author. *See* Dahm, P. A., et al., 1950.

Paretsky, D., joint author. *See* Greenberg, B., et al., 1955.

2427. Parmenter, L. and **Owen, D. F. 1954** The Swift, *Apus apus* L. as a Predator of Flies. J. Soc. Brit. Ent. 5(1):27-33.
Food balls show that Muscidae constitute only 1 per cent of the bulk. Large size and low hovering habits of muscids make them unlikely prey. Abstract: BA 30:322, Jan., 1956.

2428. Patel, G. A. and **Patel, N. G. 1956** Laboratory Tests on the Residual Effect of DDT Water Dispersible Powder against *Musca nebulo* Fabricius. Indian J. Ent. 18:10-28. New Delhi.
Median knockdown time decreased with increased deposit up to 15-25 mg. DDT per sq. ft. but further increases in deposit appeared to reduce speed of action. Abstracts: BA 32:1193(14133), April, 1958; RAE-B 45:155, Oct., 1957.

Patel, N. G., joint author. *See* Patel, G. A., et al., 1956.

2429. **Paterson, H. E. 1956** Status of the Two Forms of Housefly Occurring in South Africa. Nature 178:928-929.
Makes use of ratio of frons (narrowest width) to head (broadest width); also inferior forceps of male genitalia. *M. curviforceps* (*vicina* of authors) is mainly domestic; *M. cuthbertsoni* prefers the fields. *M. domestica*, s. str., is interfertile with *curviforceps*.

Patterson, R. E., joint author. *See* Cheng, T.-H., et al., 1957.

2430. **Patterson, R. S. 1957** On the Causes of Broken Wings of the House Fly. J. Econ. Ent. 50:104-105.
This condition in laboratory flies is confined largely to males and is produced chiefly by thrusts, on the part of females with their metathoracic legs in resisting attempts at mating. ABSTRACTS: BA 31:2431 (26336), Aug., 1957; RAE-B 46:49, Mar., 1958.

Patton, R. L., joint author. *See* Anderson, A. D., et al., 1953.

Paul, J. J., joint author. *See* Dicke, R. J., et al., 1951.

2431. **Paulini, E.** and **Ricciardi, I. 1952** Estudos sôbre a ação inseticida dos vapôres de BHC, DDT e Chlordane. [Studies on the insecticidal action of the vapors of BHC, DDT and Chlordane.] Rev. Brasil Malariol. e Doenças Trop. 4:375-384.
By means of a special apparatus, specimens of *M. domestica* were exposed to air saturated with insecticide. ABSTRACT: BA 28:451, Feb., 1954.

2432. **Paulini, E.** and **Ricciardi, I. 1954** Investigações sôbre o modo de acão dos inseticidas DDT e BHC sôbre *Musca domestica* L. [Investigations on the mode of action of the insecticides DDT and BHC on *Musca domestica* L.]. Rev. Brasil Malariol. e Doenças Trop. 6: 115-122.
Anesthetized flies did not die on contact with DDT, and very few with BHC; suggesting that intoxication was caused by vapors penetrating respiratory system and not by contact. ABSTRACT: BA 79:702, Mar., 1955.

Paulini, E., joint author. *See* Ricciardi, I., et al., 1955, 1956.

2433. **Pavillard, E. R.** and **Wright, E. A. 1957** An Antibiotic from Maggots. Nature (London) 180(4592):916-917.
Larvae of black blow-fly (*Phormia terrae-novae*) produce a substance effective against haemolytic streptococci, pneumococcus Type I and some others. Activity optimum at pH 7-8.

Peel, R. D., joint author. *See* Waldrop, R. H., et al., 1958.

2434. **Peffly, R. L. 1953** A Summary of Recent Studies on House Flies in Egypt. J. Egyptian Pub. Health Ass. 28:55-74.
Treats of surveys in native villages, biological studies and control methods. *M. domestica vicina* most prevalent of the 11 forms of *Musca* found in lower Egypt. ABSTRACT: TDB 51:996, Sept., 1954.

2435. **Peffly, R. L. 1953** Crossing and Sexual Isolation of Egyptian Forms of *Musca domestica* (Diptera: Muscidae). Evolution. 7(1):65-75.
Characteristics of subspecies *vicina* appear to be dominant to those of subspecies *domestica*. *M. cuthbertsoni* (short bristles) and *M. vicina* (long bristles) not always easy to separate, in nature.

2436. **Peffly, R. L. 1953** Effects of Some Environmental Factors on the Biology of *Musca domestica vicina* Macq. J. Egyptian Pub. Health Ass. 28:157-166.
Experiments concerned influence of temperature, humidity and illumination on oviposition, duration of life-history stages, size of offsprings, emergence from puparia, etc. ABSTRACT: TDB 51:996, Sept., 1954.

2437. Peffly, R. L. 1953 The Relative Importance[of] Different Fly-Breeding Materials in an Egyptian Village. J. Egyptian Pub. Health Ass. 28: 167-180.
Human stools and fuel cakes produce 70 per cent of the larvae; latrine dumps, animal quarters and compost piles, 30 per cent. *M. vicina* favors horse dung for oviposition. ABSTRACT: TDB 51:997, Sept., 1954.

2438. Peffly, R. L. 1953 Fly Control in Egypt. Research Project No. NM 005083.06.04. U.S. Naval Medical Research Unit No. 3 (Cairo, Egypt), 9 pp.
See next listed reference of same title.

2439. Peffly, R. L. 1954 Fly Control in Egypt. First International Symposium on the Control of Insect Vectors of Disease. Ist. Sup. Sanit., pp. 331-343.
Various formulations and methods of application evaluated under rural conditions. Observations on resistance. Plans for new studies.

2440. Peffly, R. L. and **Labrecque, G. C. 1956** Marking and Trapping Studies on Dispersal and Abundance of Egyptian House Flies. J. Econ. Ent. 49:214-217.
Phenophthalein used to mark 44,000 flies (95 per cent *M. domestica vicina*, 5 per cent *M. sorbens sorbens*, a few *Sarcophaga*). Recovered 7·15 per cent of the marked flies from traps within 1½ miles of release point. ABSTRACTS: BA 30:3011(30045), Oct., 1956; RAE-B 45:79, May, 1957; TDB 53:1393, Nov., 1956.

2441. Peffly, R. L. and **Shawarby, A. A. 1956** The Loss and Redevelopment of Insecticide Resistance in Egyptian House Flies. Am J. Trop Med. Hyg. 5(1):183-189.
Concerns chlorinated hydrocarbon insecticides. Other types of insecticide or technique should be used. ABSTRACTS: BA 30:3303(33002), Nov., 1956; RAE-B 46:13, Jan., 1958; TDB 53:1063, Aug., 1956.

2442. Pellegrini, J. P., Miller, A. C. and **Sharpless, R. V. 1952** Biosynthesis of Radioactive Pyrethrins using $C^{14}0_2$. J. Econ. Ent. 45:532-536.
Involves growing pyrethrum plants in an atmosphere containing $C^{14}0_2$ with a view to tracing fate of pyrethrins in insect's body. ABSTRACT: RAE-B 40:176, Nov., 1952.

2443. Pellegrini, J. P. Jr., Miller, A. C. and **Sharpless, R. V. 1952** Biosynthetic Preparation of C^{14}-labelled Pyrethrins. Proc. 38th Mid-Year Meeting of the Chem. Spec. Manu. Assoc., Inc. 3 pp.
Gives technique of producing "hot" pyrethrins. Cost, about $18,000,000 per pound!

Pellegrini, J. P., joint author. *See* Miller, A. C., et al., 1952.

Penner, L. R., joint author. *See* Hawley, J. E., et al., 1951; Melnick, J. L., et al., 1952.

Peri, C. A., joint author. *See* Losco, G., et al., 1959; Fusco, R., et al., 1959; Speroni, G., et al., 1951, 1953.

2444. Perry, A. S. 1956 (1958) Factors Associated with DDT-resistance in the House Fly *Musca domestica* L. Proc. 10th Internat. Cong. Ent. 2:157-172.
Classifies protective mechanisms as morphologic, physiologic and behavioristic. A strain may possess its own combination of resistance characters, perhaps unique.

2445. Perry, A. S. and **Buckner, A. J. 1959** The Metabolic Fate of Prolan in a Dilan-resistant Strain of House Flies. J. Econ. Ent. 52(5):997 1002.
Females flies treated topically. At 24 hours, 42·5 per cent of dosage had been absorbed; at 48 hours, 62·8 per cent. Two toxic compounds (meta-

bolites) one acid, one neutral, in fly's excreta. ABSTRACTS: BA 35:764 (8762), Feb., 1960; RAE-B 49:10, Jan., 1961; TDB 57:307, March, 1960.

2446. **Perry, A. S., Fay, R. W.** and **Bucker, A. J. 1953** Dehydrochlorination as a Measure of DDT-resistant in House Flies. J. Econ. Ent. 46: 972-976.
Involves colorimetric determinations of DDT and DDE after applying measured dosage of insecticide. Good correlation between dehydrochlorination rate and survival of flies. ABSTRACTS: BA 28:2196, Sept., 1954; RAE-B 42:132, Sept., 1954; TDB 51:851, Aug., 1954.

2447. **Perry, A. S.** and **Hoskins, W. M. 1950** The Detoxification of DDT by Resistant Houseflies and Inhibition of this Process by Piperonyl Cyclonene. Science 111:600.
Conversion of DDT to DDE largely prevented when piperonyl cyclonene is present. Other decomposition products (besides DDE) are formed.

2448. **Perry, A. S.** and **Hoskins, W. M. 1951** Detoxification of DDT as a Factor in the Resistance of Houseflies. J. Econ. Ent. 44:850-857.
Some DDT is detoxified otherwise than by degradation to DDE. Search for an unknown metabolite. Also, resistant flies *can* contain lethal amounts of DDT, probably stored in non-vital tissue. ABSTRACTS: BA 25:2068, Aug., 1952; RAE-B 40:56, Apr., 1952; TDB 49:561, May, 1952.

2449. **Perry, A. S.** and **Hoskins, W. M. 1951** Synergistic action with DDT Toward Resistant Houseflies. J. Econ. Ent. 44:839-850.
Piperonyl cyclonene has slight synergistic effect with DDT in DDT-susceptible strains; strongly increases mortality in DDT-resistant strains. Results are similar, though less marked with TDE and with methoxychlor. ABSTRACTS: BA 26:2067, Aug., 1952; RAE-B 40:55, Apr., 1952; TDB 49:561, May, 1952.

2450. **Perry, A. S., Jensen, J. A.** and **Pearce, G. W. 1955** Colorimetric and Radiometric Determinations of DDT and its Metabolites in Resistant Houseflies. Agric. Food Chem. 3(12):1008-1011.
Only significant product of DDT metabolism is DDE. Both are found in ether-soluble portion of fly's excreta. Ratio DDE/DDT increases with longer time intervals.

2451. **Perry, A. S., Mattson, A. M.** and **Buckner, A. J. 1953** The Mechanism of Synergistic Action of DMC with DDT against Resistant House Flies. Repr. from Biol. Bull. 104(3):426-438.
A small amount markedly enhances the effectiveness of DDT, when the two are applied together, but DMC has no insecticidal properties alone. Synergist competes with insecticide for mechanism of detoxification. ABSTRACTS: BA 27:2931, Nov., 1953; RAE-B 43:28, Feb., 1955; TDB 52:697, July, 1955.

2452. **Perry, A. S., Mattson, A. M.** and **Buckner, A. J. 1958** The Metabolism of Heptachlor by Resistant and Susceptible House Flies. J. Econ. Ent. 51(3):346-351.
Metabolite of heptachlor shown to be heptachlor epoxide in both resistant and susceptible flies. Is as toxic as heptachlor. Detected by chlorimetric analysis and infrared spectroscopy.

2453. **Perry, A. S.** and **Sacktor, B. 1955** Detoxification of DDT in Relation to Cytochrome Oxidase Activity in Resistant and Susceptible House Flies. Ann. Ent. Soc. Am. 48:329-323.
Resistant strains convert absorbed DDT to DDE. Absorption rate of DDT, also cytochrome oxidase activity vary between strains. No direct relationship between c.o. activity and degradation of DDT (all strains).

Perry, A. S., joint author. *See* Brown, A. W., et al., 1956.

Perti, S. L., joint author. *See* Gharpure, P. B., et al., 1957.

2454. **Perttunen, V. 1950** Experiments on the Humidity Reactions of Some Tsetse Fly Species. (Dipt. Muscidae.) Ann. Ent. Fenn. 16(2):41-44.
Tsetse flies avoid relative humidity below 35 per cent, in contrast to *M. domestica* which chooses the lower humidities, even 20-40 per cent. ABSTRACT: Hocking, B. 1960 Smell in Insects, p. 100 (2.395).

2455. **Peters, R. F. 1957** The Future of Health Department Activities in Fly Control. Calif. Vector Views 4(11):69-70.
An appraisal of current domestic fly problems in California as they affect changing ways of life and how they must be handled.

Petrova, T. A., joint author. *See* Sychevskaya, V. I., et al., 1958.

Pfaeffle, W. O., joint author. *See* Dahm, P. A., et al., 1959.

2456. **de Pietri-Tonelli, P. 1950** Contributo alla conoscenza della resistenza al DDT della *Musca domestica* L. [Contribution to our knowledge of resistance to DDT on the part of *Musca domestica* L.] Redia 35: 407-440.
Describes isolation, in laboratory, of a strain completely insensitive to the action of DDT, descended from a group of flies possessing a first stage resistance to this insecticide. ABSTRACT: RAE-B 41:81, June, 1953.

2457. **Pimentel, D. 1955** Relationship of Ants to Fly Control in Puerto Rico. J. Econ. Ent. 48:288-30.
Ants destroyed 91 per cent of a potential fly population between egg and adult stage. Most important species was the fire ant, *Solenopsis geminata*. ABSTRACTS: RAE-B 44:20, Feb., 1956; TDB 52:588, June, 1955.

2458. **Pimentel, D.** and **Dewey, J. E. 1950** Laboratory Tests with House Flies and House Fly Larvae Resistant to DDT. J. Econ. Ent. 43:105.
Field-collected adults found 7 times more resistant than laboratory flies. Larvae of the resistant strain showed 46 per cent mortality after controlled exposure; larvae of laboratory strain, 84 per cent. Difference found statistically significant. ABSTRACT: RAE-B 38:187, 1950.

2459. **Pimentel, D., Dewey, J. E.** and **Schwardt, H. H. 1951** An Increase in the Duration of the Life Cycle of DDT-resistant Strains of the House Fly. J. Econ. Ent. 44:477-481.
Larval periods found significantly longer than normal in four resistant strains. Length of egg stage or pupal stage not affected. ABSTRACTS: BA 26:1523, June, 1952; RAE-B 39:182, Nov., 1951; TDB 49:90, Jan., 1952.

2460. **Pimentel, D.** and **Fay, R. W. 1955** Dispersion of Radioactivity Tagged *Drosophila* from Pit Privies. J. Econ. Ent. 48:19-22.
Included here because of similarity of habits to those of *Fannia* and *Musca*. Majority of *Drosophila* found in houses had frequented privies. ABSTRACT: BA 29:2718, Nov., 1955.

2461. **Pimentel, D.** and **Klock, J. W. 1954** Disinsectization of Aircraft by Residual Deposits of Insecticides. Am. J. Trop. Med. Hyg. 3:191-194.
Lindane residues used to protect passenger cabins for up to 3 weeks. Baggage compartments (with heavier dosage) were protected for 5 weeks. ABSTRACT: TDB 51:649, June, 1954.

2462. **Pimentel, D., Schwardt, H. H.** and **Dewey, J. E. 1953** Development and Loss of Insecticide Resistance in the House Fly. J. Econ. Ent. 46:295-298.
Build-up of resistance to DDT and lindane is rapid; to parathion much slower. Loss of resistance to DDT required freedom from exposure for 10-20 generations. ABSTRACTS: RAE-B 41:174, Nov., 1953; TDB 50: 862, Sept., 1953.

2463. Pimentel, D., Schwardt, H. H. and **Dewey, J. E 1954** The Inheritance of DDT-Resistance in the House Fly. Ann. Ent. Soc. Am. 47(1): 208-213.
Rigid selection and inbreeding failed to produce a homogeneous population. Female flies influenced progeny more than males, but resistance was not sex-linked. ABSTRACTS: BA 28:2944, Dec., 1954; RAE-B 44:126, Aug., 1956.

2464. Pimentel, D., Schwardt, H. H. and **Norton, L. B. 1950** House Fly Control in Dairy Barns. J. Econ. Ent. 43:510-515.
Gives formulations for lindane, chlordane, toxaphene and DDT. Ob, tained four weeks or more of control. ABSTRACTS: RAE-B 39:58, Apr. 1951; TDB 48:413, Apr., 1951.

2465. Pimentel, D., Schwardt, H. H. and **Norton, L. B. 1951** New Methods of House Fly Control in Dairy Barns. Soap and Sanit. Chem. 27(1): 102-105, 112A-112C, 141.
Reports success of various formulations used. Period of control could be increased 2 or 3 times by good barn sanitation. ABSTRACT: TDB 48: 680, 1951.

2466. Pino, F. 1955 El empleo de los ésteres fosfóricos en el control de la mosca doméstica. [The employment of phosphoric esters in the control of the domestic fly.] Bol. Chileno Parasit. 10(3):55-57.

Pippen, W. F., joint author. *See* Ingram, R. L., et al., 1956.

2467. Piquett, P. G., Alexander, B. H. and **Barthel, W. F. 1958** A Study of certain Acetals as Synergists for Pyrethrins or Allethrin in House Fly Sprays. J. Econ. Ent. 51(1):39-40.
Of 17 compounds tested, 9, containing 3,4-methylenedioxyphenyl were found effective. Physical properties of all compounds given. ABSTRACTS: BA 32:1791(21284), June, 1958; RAE-B 47:28, Feb., 1959.

2468. Piquett, P. G. and **Gersdorff, W. A. 1958** Screening Tests with Some Synthetic Compounds for Toxicity to House Flies when Applied as Space Sprays. J. Econ. Ent. 51(6):791-793.
Eighteen compounds, esters of chrysanthemumic acid or a closely related cyclopropanecarboxylic acid were found sufficiently toxic to warrant further study. ABSTRACTS: BA 33:2258(27802), May, 1959; RAE-B 49:7, Jan., 1961.

2469. Piquett, P. G. and **Gersdorff, W. A. 1959** Screening Tests in 1957 with Synthetic Compounds for Toxicity to House Flies when Applied in Space Sprays. J. Econ. Ent. 52(5):954-955.
Eleven compounds found sufficiently toxic to warrant further study. All but one were esters of *dl-cis-trans*-chrysanthemumic acid, phosphoric acid, or a substituted phosphonic acid. ABSTRACTS: BA 35:773(8850), Feb., 1960; RAE-B 48:54, March, 1960; TDB 57:308, March, 1960.

Piquett, P. G., joint author. *See* Fales, J. H., et al., 1955; Gersdorff, W. A., et al., 1955, 1956, 1957, 1958, 1959; Mitlin, N., et al., 1954.

2470. Piras, L. and **Loddo, B. 1957** Esperimenti di lotta antimosche in comuni rurali. [Experiments in the war against flies in rural communities.] Nuovi Ann. Igiene Microb. 8(4):329-337.
Most efficacious products found to be (1) those containing phosphoric esters and DDT, and (2) a DDT synergized preparation. ABSTRACT: TDB 55:112, Jan., 1958.

Piras, L., joint author. *See* Loddo, B., et al., 1956, 1957.

2471. Plapp, F. W. and **Casida, J. E. 1958** Bovine Metabolism of Organophosphorus Insecticides. Metabolic Fate of 0, 0-Dimethyl 0-(2, 4, 5-trichlorophenyl) Phosphorothioate in Rats and a Cow. J. Agric.

Fd. Chem. 6(9):662-667.
Includes assay of trolene for toxicity to flies following topical application. LD50 values determined. Hydrolysis products show primary site of hydrolysis to be at the phorphorus-oxygen-phenyl bond. ABSTRACT: RAE-B 47:87, June, 1959.

2472. **Platkowska, W.** and **Skierska, B. 1953** Sklad jakościowy oraz ilósciowe wahania sezonowe muchówek na terenie Gdánska w roku 1951. [Qualitative and quantitative seasonal variation of flies in Gdánsk in 1951.] Bull. Inst. Marine. Trop. Med. Gdánsk. 5:237-254. (Russian and English summaries.)
Principal representative on fly-paper traps hung in dwellings was *M. domestica* (56·83 per cent). Maximum collection occurred in August for this species and was associated with high morbidity of diarrhea in children. ABSTRACTS: TDB 51:850, Aug., 1954; RAE-B 43:11, Jan., 1955.

2473. **Polovodova, V. P., Gudoshchikova-krasil'nikova, V. I., Tokmacheva, S. S. 1956** [Entomological bases in fly control.] Med. Parazit. i Parazitarn. (Bolezni) 25(4):358-363. Referat. Zhur., Biol., 1959, No. 52596 (translation).
ABSTRACT: BA 45:2113(26424), Mar. 15, 1964.

2474. **Ponghis, G. 1957** Quelques observations sur le rôle de la mouche dans la transmission des conjonctivities saisonnières dans le Sud-Marocain. [Some observations on the role of the fly in the transmission of seasonal conjunctivitis in South Morocco.] Bull. W. H. O. 16(5): 1013-1027. (English summary.)
Work in 24 villages confirms importance of flies. Rational fly control, by both insecticides and environmental sanitation, deemed essential.

2475. **Povolný, D.** and **Gregor, F. 1958** Versuch einer Klassifikation der synanthropen Fliegen (Diptera). [Attempt at a classification of the synanthropic flies.] J. Hyg. Epidem. (Praha) 2(2):205-216.

2476. **Porter, B. A. 1952** Insects are Harder to Kill. *In* Insects. The Year Book of Agriculture. U.S. Dept. Agric. pp. 317-320.
A good discussion of resistance. No particular emphasis on flies.

2477. **Pospíšil, J. 1958** Nékterē Problémy Čichu u Saprofilnīch Much. [Some problems of the smell of the saprophilic flies.] Acta Soc. Ent. Cechoslov. 55(4):316-334.
M. domestica one of six species studied by means of a "two-choice" olfactometer. Attractants and repellents not the same for all species.

2478. **Potemkin, V. I. 1958** Primenenie Khlorofosa v bor'be s mukhami v zhivotnovodcheskikh khozyaistvakh. [The use of khlorofos for the control of flies on cattle farms.] Veterinariya 35(6):51-52.
The toxicity of chlorophos well demonstrated in its effect on flies resistant to DDT. ABSTRACTS: BA 44:1580, Dec. 1, 1963; RAE-B 49:24, Jan., 1961.

Pozefsky, A., joint author. *See* Miller, A. C., et al., 1952.

2479. **Pratt, J. J.** and **Babers, F. H. 1950** Cross Tolerances in Resistant Houseflies. Science 112:141-142.
Cross-resistance of a strain of *M. domestica* to several compounds suggests that resistance to chemicals other than the one for which the strain was selected is due to increased vigor, and not to the functioning of a protective mechanism. ABSTRACT: TDB 48:207, Feb., 1951.

2480. **Pratt, J. J.** and **Babers, F. H. 1953** Sensitivity to DDT of Nerve Ganglia of Susceptible and Resistant House Flies. J. Econ. Ent. 46:700-702.
Ganglionic tissue of resistant flies is less sensitive to, and can recover

more rapidly from, DDT poisoning. Believed that dosage threshhold is probably higher for poisoning effect. ABSTRACTS: BA 28:949, Apr., 1954; RAE-B 42:41, Mar., 1954; TDB 51:517, May, 1954.

2481. Pratt, J. J. and **Babers, F. H. 1953** The Resistance of Insects to Insecticides. Some Differences Between Strains of House Flies. J. Econ. Ent. 46:864-869.
Treats of cholinesterase activity, oxygen consumption and mortality in 6 susceptible and 3 resistant strains. Occasional low activity of cholinesterase in resistant flies not considered a true strain difference. ABSTRACTS: BA 28:1683, July, 1954; RAE-B 42:74, May, 1954; TDB 51:516, May, 1954.

Pratt, J. J., joint author. *See* Babers, F. H., et al., 1950, 1953; Blum, M. S., et al., 1958, 1959.

Preuss, U., joint author. *See* Gross, H., et al., 1951-1953.

Price, G. M., joint author. *See* Lewis, S. E., et al., 1956.

2482. Prill, E. A. and **Smith, W. R. 1955** Comparisons of Ethylenedioxyphenyl and Methylenedioxyphenyl Compounds as Extenders for Pyrethrins. Contr. Boyce Thompson Inst. 18(4):187-192.
Many methylenedioxyphenyl compounds known to be effective synergists or extenders. Ethylenedioxyphenyl compounds found *inactive* in this respect. ABSTRACT: RAE-B 44:16, Jan., 1956.

2483. Quarterman, K. D., Kilpatrick, J. W. and **Mathis, W. 1954** Fly Dispersal in a Rural Area near Savannah, Georgia. J. Econ. Ent 47(3):413-419.
Dispersal proved to be at random over an area 8-10 miles in diameter. ABSTRACTS: BA 29:445, Feb., 1955; RAE-B 43:101, July, 1955; TDB 52:101, Jan., 1955.

2484. Quarterman, K. D. and **Mathis, W. 1952** Field Studies on the Use of Insecticides to Control Fly Breeding in Garbage Cans. Am. J. Trop. Med. Hyd. 1:1032-1037.
All four of the compounds tested were more effective against blow flies than house flies. ABSTRACTS: BA 27:1687, June, 1953; RAE-B 42:123, Aug., 1954; TDB 50:351, Apr., 1953.

2485. Quarterman, K. D., Mathis, W. and **Kilpatrick, J. W. 1954** Urban Fly Dispersal in the Area of Savannah, Georgia. J. Econ. Ent. 47(3):405-412.
Chemically- and dust-marked flies dispersed rapidly, continuously, and in all directions. Majority were captured within ½-mile of release point but one-third were taken over a mile away. ABSTRACTS: BA 29:445, Feb., 1955; RAE-B 43:100, July, 1955; TDB 52:100, Jan., 1955.

2486. Quarterman, K. D. and **Schoof, H. F. 1958** The Status of Insecticide Resistance in Arthropods of Public Health Importance in 1956. Am. J. Trop. Med. Hyg. 7(1):74-83.
M. domestica physiologically resistant in North and South America, Europe, and Pacific areas; *M. vicina* similarly resistant in Africa, Middle East and Far East.

2487. Quarterman, K. D. and **Sullivan, W. N. 1953** Disinfectization of Aircraft by Lindane Vapors from Filters in the Ventilating System. J. Econ. Ent. 46(4):715-716.
Carried out under both pressurized and non-pressurized conditions. Released susceptible flies tagged with 1-inch lengths of colored thread. Recovered 76 per cent. ABSTRACTS: RAE-B 42:42, Mar., 1954; TDB 51:234, Feb., 1954.

2488. **Quarterman, K. D.** and **Wright, J. W. 1956** Hay que terminar con las moscas. [There (can be) an end with flies.] Medicina (Mexico) 36(745) (Suppl.):49-52.

Quarterman, K. D., joint author. *See* Maier, P. B., et al., 1952; Kilpatrick, J. W., et al., 1952; Summerford, W. T., et al., 1951.

Quisenberry, J. H., joint author. *See* Tanada, Y., et al., 1950.

Rach, D., joint author. *See* Ascher, K. R. S., et al., 1957.

Radeleff, R. D., joint author. *See* Eddy, G. W., et al., 1954.

Raffaele, G., joint author. *See* Coluzzi, A., et al., 1951.

2489. **Rai, L., Afifi, S. E. D., Fryer, H. C.** and **Roan, C. C. 1956** The Effects of Different Temperatures and Piperonyl Butoxide on the Action of Malathion on Susceptible and DDT-Resistant Strains of House Flies. J. Econ. Ent. 49:307-310.

There were significant differences in the effects of the 2 treatments, the responses of the 2 strains and the effects of temperature at the different levels. ABSTRACTS: RAE-B 45:111, July, 1957; TDB 53:1395, Nov., 1956.

2490. **Rai, L.** and **Roan, C. C. 1956** Effects of Piperonyl Butoxide on the Anticholinesterase Activities of Some Organic Phosphorus Insecticides on House Fly and Purified Bovine Erythrocyte Cholinesterases. J. Econ. Ent. 49:591-595.

A significant decrease in cholinesterase inhibition was found whenever malathion was used in combination with piperonyl butoxide. ABSTRACT: RAE-B 45:172, Nov., 1957.

2491. **Rai, L.** and **Roan, C. C. 1959** Interactions of Topically Applied Piperonyl Butoxide and Malathion in Producing Lethal Action in House Flies. J. Econ. Ent. 52(2):218-220.

Four different experiments were used to study the nature of the antagonistic effects of piperonyl butoxide to the toxic action of malathion. ABSTRACT: TDB 56:892, August, 1959.

2492. **Rai, L.** and **Roan, C. C. 1963** Effects of Route of Administration to Houseflies on the Interactions of Malathion and Piperonyl Butoxide. J. Kans. Ent. Soc. 36(3):161-165.

Synergism or antagonism noted under various types of administration. ABSTRACTS: BA 45:7394(93359), Nov. 1, 1964; RAE-B 53:131, July, 1965.

Ralston, N. P., joint author. *See* Guyer, G. E., et al., 1954; King, H., et. al., 1953.

2493. **Ramirez Genel, M.** and **Barnes, D. 1953** Control de las moscas en los establos. [Control of flies in cattle barns.] Mexico. Sec. Agric. y Ganad. Oficina Estud. Foll. Divulg. 13:1-29.

ABSTRACT: BA 29:2718, Nov., 1955.

2494. **Ranade, D. R. 1956** Some Observations on the Range of Flight of the Common Indian Housefly, *Musca domestica nebulo*, Fabr. J. Animal Morph. Physiol. 3(2):104-108.

Range varied from 150 to 350 yards in congested areas; from 400 to 600 yards in thinly populated situations. ABSTRACTS: BA 32:905(10674), Mar., 1958; RAE-B 49:265, Dec., 1961.

2495. **Ranade, D. R. 1957** Effect of Temperature on the Longevity and Life-history of *Musca domestica nebulo* Fabr. Current Science, July, 1957. pp. 26, 213-214.

Temperature found inversely proportional to longevity. An increase in temperature quickens the rate of development.

2496. Ranade, D. R. 1959 A Study of the Cephalo-pharyngeal Skeleton of the Larva of *Musca domestica nebulo* Fabr. J. Animal Morph. Physiol. 6(1):10-15.
This structure is recognizably different in each instar.

Ranade, D. R., joint author. *See* Deoras, P. J., et al., 1953, 1957; Vishveshwaraiya, S., et al., 1951.

2497. Rao, A. V. K. M. 1957 The Effect of DDT in the Production of Tolerant and Sensitive Strains of the House Fly (*Musca domestica* L.) Diss. Absts. 17(6):1196-1197.
Determined the effect of increasing dosages of DDT applied in the larval medium. ABSTRACT: BA 32:1193(14134), April, 1958.

2498. Rasso, S. C. and **Fraenkel, G. 1954** The Food Requirements of the Adult Female Blow-fly, *Phormia regina* (Meigen) in Relation to Ovarian Development. Ann. Ent. Soc. Am. 47:636-645.
No ovarian development on a diet of sugar and water; in the absence of sugar, no eggs produced on a diet of protein. Proteins must be in soluble form. ABSTRACT: BA 30:82, Jan., 1956.

Raun, E. S., joint author. *See* Dahm, P. A., et al., 1955.

Rawhy, A-El-R., joint author. *See* Barker, R. J., et al., 1957.

Reali, G., joint author. *See* Grandori, L., et al., 1952.

Rechnitzer, A. B., joint author. *See* Buzzati-Traverso, A. A., et al., 1953.

Rees, D. M., joint author. *See* Linam, J. H., et al., 1956, 1957.

2499. Reid, J. A. 1953 Notes on House-Flies and Blowflies in Malaya. Bull. Inst. Med. Res. Fed. Malaya, N.S. 7:1-26.
Sets forth the author's observations and experiences, emphasizing the special problems which arise in fly control under Malayan conditions. ABSTRACT: TDB 51:652, June, 1954.

2500. Reiff, M. 1958 Über unspezifische Abwehrreaktionen bei polyvalent resistenten Fliegenstämmen. [On unspecific defense reactions in polyvalent resistant strains of flies.] Rev. Suisse Zool. 65(2):411-418.
In addition to a common basis for polyvalent resistance, it is believed that there is a protective mechanism for each insecticide, e.g. enzymatic degradation of DDT. ABSTRACT: BA 35:2231(25382), May, 1960.

Remmert, L. F., joint author. *See* Sacklin, J. A., et al., 1955.

2501. Rendtorff, R. E. and **Holt, C. J. 1954** The Experimental Transmission of Human Intestinal Protozoan Parasites. III. Attempts to Transmit *Endamoeba coli* and *Giardia lamblia* cysts by Flies. Am. J. Hyg. 60:320-326.
Flies exposed to cysts, then released in dining area. Are not efficient transmitters of cysts, but can be if conditions are suitable. ABSTRACT: RAE-B 44:41, March, 1956.

2502. Resnick, S. L. and **Crowell, R. L. 1951** Comparative Evaluation of Certain High Pressure Insecticidal Aerosols against *Musca domestica*. J. Nat. Malaria Soc. 10:248-256.
Used a modified Peet-Grady method to test G-382 and several other formulations. Lethane 384 gives promise as an effective ingredient in aerosols. ABSTRACTS: RAE-B 41:204, Dec., 1953; TDB 49:199, Feb. 1952.

2503. Reuter, S. and **Ascher, K. R. S. 1956** The Action of Di-(p-chlorophenyl) trichloromethylcarbinol (DTMC) and Di-(p-chlorophenyl) trifluoromethylcarbinol on Houseflies. Experientia (Basel) 12(8): 316-318. (French summary.)
Contact toxicity of DTMC to resistant flies is of short duration. Per-

sistence prolonged by DDT. Probably *not* a synergism. DTMC much more active against susceptible flies than the trifluoromethyl-carbinol. ABSTRACT: RAE-B 45:119, July, 1957.

2504. Reuter, S., Cohen, S., Mechoulam, R., Kaluszyner, A. and **Tahori, A. S. 1956** On the Mechanisms of DDT-Resistance. Riv. di Parassit. (Rome) 17(2):125-127.
A study of certain synergists, particularly carbinols (which themselves exert a certain toxic effect). In no case were DDT-resistant flies rendered fully susceptible. ABSTRACT: TDB 53:1394, Nov., 1956.

Reuter, S. (sometimes Reuther, S.), joint author. *See* Ascher, K. R. S., et al., 1951; Davidovici, S., et al., 1950.

2505. Ricciardi, I., Nascimento, L. P. and **DeSouza, S. H. 1952** Oservações sôbre um mutante de *Musca domestica* L., aparecido depois de contato com o isômero gama do hexaclorobenzeno, em prova de laboratorio. (Nota previa.) [Observations on a mutant of *Musca domestica* L., appearing after contact with the gamma isomer of benzenehexachloride, under test in the laboratory. (Preliminary Notc.)[Rev. Brasil. Malariol. e Doenças Trop. (Rio de Janiero) 4(4):427-431. (English summary.)
Descendents of flies so exposed presented "opened" wings, a character thereafter transmitted from generation to generation. ABSTRACT: TDB 51:229, Feb., 1954 (Questions findings).

2505a. Ricciardi, I. and **Paulini, E. 1955** Normas gerais para determinação da densidade de *Musca domestica* em uma localidade. [Standard methods for determining the density of *Musca domestica* in a given locality.] Rev. Brasil. Malariol. e Doenças Trop. 7(1):93-101.
Fly grill method used for observing adults. Larval density estimated by counting pupae and larvae in a sample taken by means of a specially constructed shallow pan.

2506. Ricciardi, I. and **Paulini, E. 1956** Experiência com um novo inseticida no combate à *Musca domestica.* [Experiments with a new insecticide for combatting the domestic fly.] Rev. Brasil. Malariol. e Doenças Trop. 8(3):493-498.
Tugon (Bayer) is furnished as impregnated cardboard discs active for 3-4 months. Poison is water soluble and effective if ingested, but discs have no special attraction for flies. ABSTRACT: BA 32:1796(21352), June, 1958.

Ricciardi, I., joint author. *See* Paulini, E., et al., 1952, 1954.

Richards, A. G., joint author. *See* Lin, S., et al., 1955, 1956.

Riehl, L. A., joint author. *See* Rodriquez, J. L., et al., 1959.

2507. Riemschneider, R. 1953 Literatur zur HCH- und Dien-Gruppe. [Literature on HCH and related groups.] Liste V. Zeit. Angew. Ent. 34: 405-460. Supplements list noticed in Rev. Appl. Ent. Ser. B. 40:207.
Lists 1,100 works in Supplement. ABSTRACT: RAE-B 42:169, Nov., 1954.

2508. Ringel, S. J. 1956 The Copper Content of Resistant and Susceptible House Flies. J. Econ. Ent. 49:569-570.
Reports a sex-difference, but no differences as to *strain.* ABSTRACT: RAE-B 45:147, Sept., 1957.

2509. Rivosecchi, L. 1955 Armatura genitale maschile e femminile di *Musca domestica.* [Armature of the male and female genitalia of *Musca domestica.*] Rend. 1st. Sup. Sanit. (Roma) 18(6):496-515.
Gives synonymy of nomenclature between male and female, based on comparative, embryological and teratological research. Tables compare interpretations of various authors.

2510. Rivosecchi, L. 1955 Dimorfismo sessuale in *Musca domestica.* [Sexual dimorphism in *Musca domestica.*] Rend. 1st. Sup. Sanit. (Roma) 18(9):826-848.
Forty-one characters analyzed in 50 individuals of each sex. Six tables. ABSTRACT: TDB 53:255, Feb., 1956.

2511. Rivosecchi, L. 1956 Osservazioni preliminari sull' inseminazione in *Musca domestica* L. [Preliminary observations on insemination in *Musca domestica* L.] Boll. Zoologia 23(2):735-748. *See* Reference 2513.

2512. Rivosecchi, L. 1958 Gli organi della riproducione in *Musca domestica.* [The organs of reproduction in *Musca domestica.*] Rend. 1st. Sup. Sanit. (Roma) 21(4/5):458-488.
A thorough morphological study, well illustrated. ABSTRACT: TDB 56: 109, Jan., 1959.

2513. Rivosecchi, L. 1958 Osservazioni preliminari sull'inseminazione in *Musca domestica.* [Preliminary observations on insemination in *Musca domestica.*] Rend. Ist. Sup. Sanit. (Roma) 21(3):267-277.
Direct introduction of sperm into the spermathecae does not appear to take place.

Rivosecchi, L., joint author. *See* Milani, R., et al., 1955, 1957; Sacca, G., et al., 1954, 1955, 1957, 1958.

2514. Rizzi, A. F. 1950 Sanitary Land Fill—A New, Economical Method of Garbage Disposal. Bimonthly Bull. Wisc. State Bd. of Hlth. 9(5): 14-15.
As of that date, cost of disposal was estimated as ranging from 35 cents to $1.00 per ton, depending on the local situation.

2515. Roadhouse, L. A. O. 1953 Laboratory Studies of DDT-resistant House Flies. (Diptera) in Canada. Canad. Ent. 85:340-346.
Wild flies caught in 1949, and cultured, averaged 20 times the resistance of normal laboratory flies; those collected in 1951, and cultured, averaged 2,000 times the resistance of the laboratory colony. ABSTRACTS: BA 28:957, Apr., 1954; RAE-B 43:36, Mar., 1955.

2516. Roan, C. C. and **Babers, F. H. 1954** Factors Affecting the Rate of Penetration of DDT. J. Econ. Ent. 47(5):798-800.
Data indicate limiting factors in increasing rate of penetration, e.g. optimum concentration per unit area beyond which there would be little or no effect over a relatively short period.

Roan, C. C., joint author. *See* Babers, F. H., et al., 1954; Knapp, F. W., et al., 1958; Zaharis, J. L., et al., 1959; Ware, G. W., et al., 1957; Rai, L., et al., 1956, 1959.

Robbins, W. E., joint author. *See* Hopkins, T. L., et al., 1957; Monroe, R. E., et al., 1959.

2517. Robertson, R. S., Jr., and **Gaines, J. C. 1951** The Toxicity of Certain Organic Insecticides to Different Strains of House Flies. J. Econ. Ent. 44:114-115.
DDT remained toxic on glass to a natural strain of flies for a period of 10 weeks. ABSTRACT: RAE-B 39:140, Aug., 1951; TDB 48:932, Oct., 1951.

2518. Rock, R. E. 1953 Yaws in Guam: Treatment and Environmental Considerations. Am. J. Trop. Med. Hyg. 2:74-78.
Fewest relapses among population in villages with best DDT control of flies and best trash clean up. ABSTRACT: BA 27:1589, June, 1953.

2519. Rockstein, M. 1956 Some Biochemical Aspects of Ageing in Insects. J. Gerontology 11(3):282-285.
Males of *M. domestica* showed consistently shorter average longevity (under 21 days) than did females. Males lose power of flight at end of second week.

2520. Rockstein, M. 1957 Longevity of male and female house flies. J. Gerontology 12(3):253-256.
Adult life of male averages 15 days (maximum 40). Females average 20-30 days (maximum 45-60, depending on nutrition).

2521. Rockstein, M. 1958 Aging in Insects. Publ. Hlth. Repts. 73(12):1114-1115.
Aging factor in male houseflies considered relatively immutable. Longevity factor in female is labile, its expression being influenced by age of parents and by limited diet.

2522. Rockstein, M. 1959 The Biology of Ageing in Insects. [Repr. from] Civa Foundation Symp. on the Lifespan of Animals, pp. 247-264.
Used over 8,500 houseflies. Mean female longevity of 29 days enhanced by addition of whole milk to diet of sugar and water. Average male longevity (17 days) not so affected.

2523. Rockstein, M. and **Lieberman, H. M. 1958** Survival Curves for Male and Female House Flies (*Musca domestica* L.). Nature 181(4611): 787-788.
At 80°F and 45 per cent relative humidity, 50 per cent of the females still survived after 30 days, but only 5 per cent of the males. Half of the males were dead by the 16th day.

2524. Rockstein, M. and **Lieberman, H. M. 1959** A Life Table for the Common House Fly, *Musca domestica*. Gerontologia (Basel) 3:23-36.
Covers 10 generations. Mean length of life for females, 29 days; for males, 17 days.

2525. Rodriguez, J. L., Jr. and **Riehl, L. A. 1959** Results with Cockerels for House Fly Control in Poultry Droppings. J. Econ. Ent. 52(3):542-543.
Two week old cockerels released under wire cages. Practically no larvae or pupae in manure suggests that chicks quickly search out and eat larvae. ABSTRACT: RAE-B 48:170, Sept., 1960.

2526. Roeder, K. D. 1952 Insects as Experimental Material. Science 115 (2985):275-280.
Treats of flight, ganglia, nerve activity, sensory physiology, behavior; also practical considerations. Example: Resistance of flies to DDT and related insecticides. ABSTRACT: BA 26:2356, Sept., 1952.

2527. Roeder, K. D. (Editor). 1953 Insect Physiology. New York, J. Wiley & Sons. London, Chapman & Hall, Ltd.
Includes material on *M. domestica* and related forms. ABSTRACT: RAE-B 41:184, Sept., 1953.

Roeder, K. D., joint author. *See* Hodgson, E. S., et al., 1956.

2528. Rogoff, W. M. 1952 Some Observations on Insecticide Tolerance in Houseflies. Proc. S. Dakota Acad. Sci. 31:36-40.
Brookings, S. D. strain highly tolerant to DDT; also (in descending order) to methoxychlor, heptachlor, CS-645A, and lindane. ABSTRACT: BA 28:957, April, 1954.

2529. Rogoff, W. M. 1954 Fly Control on the Farm. South Dakota Farm and Home Research 5(3):52-55.
House flies, being non-bloodsucking, are best controlled by treatment of the premises, rather than the stock.

2530. **Rogoff, W. M. 1955** Farm Fly Control. Bull. 452. Agr. Expt. Sta. S. Dakota State College. (Brookings) 11 pp., illus.
Treats of stable fly, hornfly, and housefly. Stresses unsanitary habits of the last.

2531. **Rogoff, W. M.** and **Metcalf, R. L. 1951** Some Insecticidal Properties of Heptachlor. J. Econ. Ent. 44:910-918.
Heptachlor found equivalent to aldrin and dieldrin in toxicity to house flies, but is less toxic than *gamma* benzene hexachloride or parathion. ABSTRACT: RAE-B 40:59, Apr., 1952.

2532. **Rosen, P.** and **Gratz, N. S. 1959** Tests with Organo-Phosphorus Dry Sugar Baits against Houseflies in Israel. Bull. W. H. O. 20(5):841-847. (French summary.)
Malathion 2·5 per cent, diazinon 1 per cent, and Dipterex 1 per cent gave equally effective results. Chlorthion 1 per cent and malathion 1 per cent and 2 per cent were less effective. ABSTRACTS: RAE-B 49:235, Oct., 1961; TDB 57:191, Feb., 1960.

Rosen, P., joint author. *See* Gratz, N. G., et al., 1957.

Ross, E. S., joint author. *See* Sherman, M., et al., 1959.

2533. **Ross Institute Industrial Advisory Committee** [Wigglesworth, A., Chairman.] London School of Hygiene and Tropical Medicine. Information and Advisory Service, 1950. The Housefly and its Control. 20 pp., 12 figs., Nov. 5.
Useful pamphlet. Contains some misleading statements. Emphasizes prevention of fly breeding. ABSTRACT: TDB 49:558, May, 1952.

2534. **Roth, A. R., Lindquist, A. W.** and **Terriere, L. C. 1953** Effect of Temperature and the Activity of House Flies on Their Absorption of DDT. J. Econ. Ent. 46:127-130.
Experiments with radioactive DDT were made to compare amounts absorbed by DDT resistant flies at 70°F and 90°F. ABSTRACTS: RAE-B 41:140, Sept., 1953; TDB 50:861, Sept., 1953.

2535. **Roth, A. R.** and **Mote, D. C. 1950** DDT-resistant Flies in Oregon. J. Econ. Ent. 43(6):937.
First demonstration of resistance in Oregon. (Control was poor in 1949.) ABSTRACT: RAE-B 39:123, July, 1951.

Roth, A. R., joint author. *See* Lindquist, A. W., et al., 1951; Hoffman, R. A., et al., 1951, 1952; Kocker, C., et al., 1953.

Roulston, W. J., joint author. *See* Kerr, R. W., et al., 1957.

2536. **Roy, D. N.** and **Brown, A. W. A. 1954** Entomology (Medical and Veterinary) Including Insecticides and Insect and Rat Control. 2nd Edn. Calcutta, Excelsior Press.
Synanthropic flies given appropriate treatment. ABSTRACT: RAE-B 42:143, Sept., 1954.

Roys, C. C., joint author. *See* Smyth, T., Jr., et al., 1955.

Rozeboom, L. E., joint author. *See* Cabrera, D., et al., 1956.

Ryk-Bogdaniko, M. G., joint author. *See* Karavashkova, A. I., et al., 1957.

2537. **Ryke, P. A. J. 1958** South African Mites of the Superfamily Uropodoidea (Acarina). Proc. Zool. Soc. London 130(2):217-230.
Describes 6 species of mites from South Africa, 5 of which are new to science. Ref. included here because of possible ectoparasitic relation to adult flies.

2538. **Sabrosky, C. W. 1951** Taxonomic Problems in Egyptian Flies of the Genus *Musca*. Division of Insect Identification. Bur. of Ent. and

Plant Quarantine; U.S. Dept. Agr. pp. 1-34. (Mimeographed document.)
Presents data on *M. vicina*, *M. cuthbertsoni*, *M. sorbens* and 5 others.

2539. Sabrosky, C. W. 1952 House Flies in Egypt. Am. J. Trop Med. Hyg. 1(2):333-336.
Relates to the work of NAMRU-3, at Cairo. Treats of the 11 species and subspecies of *Musca* found in Lower Egypt. ABSTRACTS: BA 26: 2614, Oct., 1952; RAE-B 42:29, Feb., 1954; TDB 50:257, Mar., 1953.

2540. Sabrosky, C. W. 1956 *Musca autumnalis* in Upstate New York. Proc. Ent. Soc. Wash. 58(6):347.
Species was first recorded from Nova Scotia in 1952, then from Long Island in 1953. Here reported as found in nests of *Bembia pruinosa* by H. E. Evans in Oswego Co., New York. Nests also contained *M. domestica*. ABSTRACT: BA 31:1828(19115), June, 1957.*

2541. Sabrosky, C. W. 1959 *Musca autumnalis* De Geer in Virginia. Proc. Ent. Soc. Wash. 61:6.
Has been taken in houses with cluster flies (*Pollenia*) in Ithaca, N.Y. and Leesburg, Va. First new world record south of New York. ABSTRACT: BA 33:2592(31807), May, 1959.

2542. Sabrosky, C. W. 1959 Recognition of Species of *Musca*. Cooperative Econ. Insect Rept. 9(45):1 page, illustrated.
Gives distinguishing characters of *M. domestica* and *M. autumnalis*.

2543. Sabrosky, C. W. 1959 *Musca autumnalis* in the Central States. J. Econ. Ent. 52(5):1030-1031.
Recorded from houses in 2 localities in northern Virginia. In Ohio and Indiana it was found on cattle. ABSTRACT: RAE-B 49:13, Jan., 1961.

2544. Saccà, G. 1951 Esperienze d'incrocio fra *Musca domestica* L., *Musca vicina* Macq., *Musca nebulo* F. [Experiments in crossing *Musca domestica* L., *Musca vicina* Macq., *Musca nebulo* F.] Riv. di Parassit. (Rome) 12(1):47-52.
All crosses fertile; further report pending. ABSTRACT: TDB 48:839, Sept., 1951.

2545. Saccà, G. 1951 Esperienze d'incrocio fra *Musca domestica* L., *Musca vicina* Mq., *Musca nebulo* F. [Experiments in crossing *Musca domestica* L., *Musca vicina* Mq., *and M. nebulo* F.] Rend Ist. Sup. Sanit. 14:937-943.
M. domestica × *M. vicina*, 100 per cent successful; *M. domestica* × *M. nebulo*, 55 per cent successful; *M. vicina* × *M. nebulo*, 50 per cent successful. All offspring fertile for at least 2 generations. ABSTRACT: BA 26:2360, Sept., 1952.

2546. Saccà, G. 1952 Variabilità fenotipica in *Musca domestica* L. [Phenotypic variability in *Musca domestica* L.] A manuscript of 5 pages. Intended for publication in Arch. Zool. Ital. (Atti. del Convegno della Unione Zool. Ital., Milano, 25-29 Settembre).
Illustrates various types of abdominal pattern.

* In 1966, T. A. Smith and associates published the first instalment of a "Bibliography of *Musca autumnalis* in North America." (See Section VI.) A number of subsequent instalments have since appeared in California Vector Views, all duty noted in our list. Aware that this might eliminate the need for including references to *M. autumnalis* in the present work, the compilers conferred with various co-workers, all of whom advised retaining such references, on the grounds that this is the species most likely to be confused with *M. domestica* by farmers, ranchers, and lay-personnel. As this goes to print, no known references to *M. autumnalis* have been omitted.

2547. Saccà, G. 1952 Due mosche nuove per la faune d'Europa. *Musca sorbens* Wied. e *Limnophora tonitrui* Wied. in Sicilia (Diptera, Muscidae). [Two flies new to the European fauna: *Musca sorbens* Wied. and *Limnophora tonitrui* Wied., in Sicily.] Riv. di Parassit. (Roma) 13:177-180. (English summary.)
Stresses relation of *M. sorbens* to eye infections. ABSTRACT: TDB 49: 1003, Oct., 1952.

2548. Saccà, G. 1953 Contributo alla conoscenza tassonomica del "gruppo" *domestica* (Diptera, Muscidae). [Contribution to our taxonomic knowledge of the *domestica* complex (Diptera, Muscidae).] Riv. di Parassit. 14(2):97-114. (English summary.) Also in Rend. Ist. Sup. Sanit. 16:442-464.
Records changes in measurements after several generations of laboratory rearing. *M. cuthbertsoni* and *M. vicina* probably one species. ABSTRACTS: BA 29:975, Apr., 1955; RAE-B 42:50, Apr., 1954; TDB 50:1088, Nov., 1953.

2549. Saccà, G. 1953 Variabilità fenotipica in *Musca domestica* L. [Phenotypic variability in *Musca domestica* L.] Rend. Ist. Sup. Sanit. (Roma) 16:465-470.
With decreasing temperatures the adult fly shows a wider vertex, a thinned-out body and black pigmentation. These characters are non-transmissible. ABSTRACT: BA 29:975, Apr., 1955.

2550. Saccà, G. 1954 Sulla biologia invernale di *Musca domestica* L., in Italia centrale. [On the winter biology of *Musca domestica* L. in central Italy.] Rend. Ist. Sup. Sanit. (Roma) 17:44-54. (English summary.)
In laboratory experiments a small percentage of adults survived up to 84 days. In pre-imaginal stages, low temperature could retard development to a maximum of 90 days. ABSTRACTS: BA 30:885(8990), Mar., 1956; TDB 51:850, Aug., 1954.

2551. Saccà, G. 1955 Reperto di ginandromorfi in *Musca domestica* L. (Diptera: Muscidae). [Report of gynandromorphs in *Musca domestica*.] Rend. Ist. Sup. Sanit. (Roma) 18(5):380-383.
Observed in both field and laboratory strains. Maximum occurrence 3·7 per cent.

2552. Saccà, G. 1955 Richerche preliminari sui rapporti intercorrenti fra *Musca domestica cuthbertsoni* Patton e *Musca domestica vicina* Mq. [Preliminary investigation of interspecific mating of *Musca domestica cuthbertsoni* Patton and *Musca domestica vicina* Mq.] Rend. Ist. Sup. Sanit. (Roma) 18(5):384-405. (English summary.) *Also in* Boll. Zool. 21(2):481-502.
Though independent in nature, these subspecies show complete interfertility. ABSTRACT: TDB 52:1242, Dec., 1955.

2553. Saccà, G. 1956 On the Winter Biology of *Musca domestica* L. in Central Italy. Proc. XIV Int. Cong. Zool. Copenhagen 1953. pp. 389-394. Copenhagen, 1956.
Survives three ways: (1) By adult alternating between torpor and activity; (2) By slowed development of immature stages; (3) By active reproduction, if temperature is over 20°C.

2554. Saccà, G. 1956 Speciation in the House Fly. I. Recent Views on the Taxonomic Problem (Diptera, Muscidae, gen. *Musca*). Selected Scientific Papers from Istituto Sup. Sanit. Reprinted from Vol. 1(1):141-154. *Originally published as*: Ricerche sulla speciazione

nelle mosche domestiche. I. Reconti vedute sul problema tassonomica (Diptera, Muscidae, gen. *Musca*). Estratto Rend. Ist. Sup. Sanit. 19:1072-1083.
Considers *M. domestica*, *M. vicina*, and *M. nebulo* as "geographical forms". *M. cuthbertsoni* and *M. vicina* are sympatric and interfertile. ABSTRACT: TDB 54:502, Apr., 1957.

2555. Saccà, G. 1957 Ricerche sulla speciazione nelle mosche domestiche. II. Sulla' evoluzione spontanea in laboratorio di *Musca domestica cuthbertsoni* Patton. [Researches on the speciation in the domestic fly II. Spontaneous evolution in the laboratory of *Musca domestica cuthbertsoni* Patton.] Rend. Ist. Sup. Sanit. (Roma) 20:235-246.
In the laboratory *M. domestica cuthbertsoni* lost its differentiating characteristics and transformed into *vicina* in 16 generations. These forms are distinct in N. Africa, due to different biological requirements. ABSTRACT: TDB 54:1317, Nov., 1957.

2556. Saccà, G. 1957 Ricerche sulla speciazione nelle mosche domestiche. IV. Esperienze sull'isolamento sessuale fra le subspecie di *Musca domestica* L. [Researches on the speciation in the domestic fly. IV. Experiences on sexual isolation between the subspecies of *Musca domestica* L.] Rend Ist. Sup. Sanit. (Roma) 20(7-8):702-712.
Complete isolation (irreversible through 20 generations) found between *cuthbertsoni* males and *curviforceps* females. Between *curviforceps* males and *cuthbertsoni* females isolation is negligible. The two forms are sympatric. ABSTRACT: TDB 55:108, Jan., 1958.

2557. Saccà, G. 1957 La resistenza di *Musca domestica* L. agli esteri fosforici in provincia di Latina. [The resistance of *Musca domestica* L. to phosphorous esters in the Province of Latina.] Riv. di Parassit. 18(4):289-292. *See also* Saccà, 1959, same title.
ABSTRACT: TDB 55:462, Apr., 1958.

2558. Saccà, G. 1958 Ricerche sulla speciazione nelle mosche domestiche. VI. Ibridismo naturale e ibridismo sperimentale fra le subspecie di *Musca domestica* L. [Researches on the speciation of the domestic fly. VI. Natural and experimental hybrids of the subspecies of *Musca domestica* L.] Rend. Ist. Sup. Sanit. (Roma) 21(2):1170-1184.
The 3 known subspecies (*cuthbertsoni*, *curviforceps* and *domestica*) are all interfertile. Crosses show hybrid vigor.

2559. Saccà, G. 1959 La resistenza di *Musca domestica* L. agli esteri fosforici in provincia di Latina. Rend. Ist. Sup. Sanit. (Roma) 22(1):88-93.
Resistance against organic phosphorus compounds, especially diazinon, appeared after four years. In the field, reached 60 times normal. ABSTRACT: RAE-B 49:291, Dec., 1961.

2560. Saccà, G. 1959 Esperienze sulla resistenza agli esteri fosforici e resistenza crociata nella mosca domestica. [Experiences with resistance to phosphoric esters and cross resistance in *Musca domestica*.] Rend. Ist. Sup. Sanit. (Roma) 22(11/12):1173-1188. (English summary.)
Object of tests: To study cross resistance and find a technique which could be standardized and used for further studies on resistance. Results summarized and discussed. ABSTRACTS: BA 35:4948(57240), Oct., 1960; TDB 57:960, Sept., 1960.

2561. Saccà, G. and **Benetti, M. P. 1957-1958** Fattori ambientali e fisiologici della inseminazione in *Musca domestica* L. e nelle sue subspecie. [Environmental and physiological factors in insemination in *Musca domestica* L. and its subspecies.] (a) Riv. di Parassit. (Rome) 18(2):

103-112. (English summary.) (b) 1958 Rend. Ist. Sup. Sanit. (Roma). 21(2):110-119.

M. cuthbertsoni requires longer time to mature sexually, a certain light intensity for mating, ample space, and relatively high temperature. *M. domestica* can mate and be fertilized under unfavorable conditions. *M. curviforceps* requires intermediate conditions. ABSTRACTS: TDB 54: 1251, Oct., 1957; TDB 55:1056, Sept., 1958.

2562. Saccà, G. and **Rivosecchi, L. 1955** Research on Speciation in the Domestic Fly. III. Differential Diagnosis of *Musca domestica curviforceps* subsp. nova. Boll. Zool. 22(2):215-224.

See Sacca and Rivosecchi, 1957 (title in Italian.) ABSTRACT: BA 37: 2047(20454), Mar., 1962.

2563. Saccà, G. and **Rivosecchi, L. 1955** Una nuova sottospecie di *Musca domestica* L. della regione etiopica. [A new subspecies of *M. domestica* from the Ethiopian region.] Rend. Acad. Naz. Lincei, Class. Sci. Ser. VIII, Vol. XIX, fasc. 6:497-498.

M. domestica curviforceps. Distinctive characteristics of the terminalia described and illustrated.

2564. Saccà, G. and **Rivosecchi, L. 1957** Ricerche sulla speciazione nelle mosche domestiche. III. Diagnosi differenziale di *M. domestica curviforceps.* subsp. nova. [Researches on the speciation of the domestic fly. III. Differential diagnosis of *M. domestica curviforceps*, new subspecies.] Rend. Ist. Sup. Sanit. (Roma) 20(3):247-257.

Distinguished from other subspecies by shape of forceps and fifth sternite. ABSTRACTS: BA 32:592(6870), Feb., 1958; TDB 54:1372, Nov., 1957.

2565. Saccà, G. and **Rivosecchi, L. 1958** Ricerche sulla speciazione nelle mosche domestiche. V. L'areale di distribuzione delle subspecie di *Musca domestica* L. (Diptera: Muscidae). [Research on the speciation of the domestic fly. V. Territorial distribution of the subspecies of *Musca domestica* L.] Rend. Ist. Sup. Sanit. (Roma) 21(2):1149-1169.

A study based on 1,909 male flies collected in Africa and elsewhere. Subspecies *curviforceps*, *cuthbertsoni* and *domestica* considered "concentric". Subspecies *domestica* s. s., *vicina* and *nebulo* occupy a lower taxonomic category.

Saccà, G., joint author. *See* Morellini, M., et al., 1953.

2566. Sacklin, J. A., Terriere, L. C. and **Remmert, L. F. 1955** Effect of DDT on Enzymatic Oxidation and Phosphorylation. Science 122(3165): 377-378.

Apparently DDT inhibits some reaction involved in oxidation via the citric acid cycle; this occurs at concentrations far lower than those at which cytochrome oxidase seems to be affected.

2567. Sacktor, B. 1950 A Comparison of the Cytochrome Oxidase Activity of Two Strains of House Flies. J. Econ. Ent. 43(6):832-838.

Resistance to DDT may be explained by a greater cytochrome oxidase activity in resistant flies, though other factors are involved. ABSTRACTS: RAE-B 39:118, July, 1951; TDB 48:591, June, 1951.

2568. Sacktor, B. 1951 Some Aspects of Respiratory Metabolism during Metamorphosis of Normal and DDT-resistant House Flies, *Musca domestica* L. Biol. Bull. 100:229-243.

Possible mechanisms of resistance: by-pass of the cytochrome system, a difference in reversibility of enzyme-inhibitor complex, or a detoxification of the inhibitor. O_2 consumption curve during metamorphosis is of the same order of magnitude in both strains. ABSTRACT: RAE-B 40:135, Aug., 1952.

2569. Sacktor, B. 1953 Investigations on the Mitochondria of the Housefly, *Musca domestica* L. I. Adenosinetriphosphatases. J. Gen. Physiol. 36:371-387.
Distribution and nature of mitochondria at phase activity in isolated fractions of the housefly.

2570. Sacktor, B. 1953 Investigations on the Mitochondria of the Housefly *Musca domestic* L. II. Oxidative Enzymes with Special Reference to Malic Oxidase. Arch. Biochem. and Biophysics. 45(2):349-365.

2571. Sacktor, B. 1954 Investigations on the Mitochondria of the Housefly, *Musca domestica* L. III. Requirements for Oxidative Phosphorylation. J. Gen. Physiol. 37(3):343-359.
Mitochondria isolated from flight muscle can effect oxidative phosphorylation. An investigation of factors regulating this coupling.

2572. Sacktor, B. 1955 Cell Structure and the Metabolism of Insect Flight Muscle. J. Bioph. Biochem. Cytol. 1(1):29-46.

Sacktor, B., joint author. *See* Perry, A. S., et al., 1955.

Sadanand, A. V., joint author. *See* Sharma, M. I. D., et al., 1957

Salah, A. A., joint author. *See* Holway, R. T., et al., 1951.

2573. Salas, G. 1956 Flies, in Relation to Gastro-Enteric Infections. J. Philippine Med. Ass. 32:599-601.

2574. Salgado, S. and **Isaias, J. M. 1958** [A Study of Chemical Control of *Musca domestica* L., with four chlorinated insecticides, in stables near Texcoco, Mexico.] *In* Memoria del Primer Congreso Nacional de Entomologia y Fitopatologia. Escuela Nacional de Agricultura, Chapingo, Mexico, pp. 55-64. Illus.
Methoxychlor and lindane were most effective; required only 1 treatment per year to maintain population index of less than 10. ABSTRACT: BA 38:1887(24202), June 15, 1962.

2575. Sampson, W. W. 1954 The Control of Insect Flight in Food Processing Plants by Residual Insecticides. J. Econ. Ent. 47:87-93.
BHC, DDT and Methoxychlor gave complete flight control for 5 months at 20-minute length exposure; but residual insecticides not considered substitutes for basic sanitation. ABSTRACTS: BA 28:2704(27438), Nov., 1954; RAE-B 42:189, Dec., 1954.

2576. Sampson, W. W. 1956 Insecticides Used to Control House Fly Larvae. J. Econ. Ent. 49:74-77.
After field experiments with many insecticides, author lists those which proved: (1) highly effective; (2) nearly as effective; (3) low in effectiveness; (4) worthless. Various ecological factors are discussed. ABSTRACTS: BA 30:2414(24217), Aug., 1956; RAE-B 45:38, Mar., 1957.

Santi, R., joint author. *See* Speroni, G., et al.. 1953.

2577. Satchell, G. H. and **Harrison, R. A. 1953** II. Experimental Observations on the Possibility of Transmission of Yaws by Wound-feeding Diptera in Western Samoa. Trans. Roy. Soc. Trop. Med. Hyg. 47:148-153.
This follows an informative paper on the disease itself, by M. J. Marples and D. F. Bacon, idem., pp. 141-147.
M. domestica vicina and *M. sorbens* considered important potential vectors of *Treponema pertenue*. ABSTRACTS: BA 27:2940, Nov., 1953; RAE-B 42:197, Dec., 1954.

2578. **Savage, E. P.** and **Schoof, H. F. 1955** The Species Composition of Fly Populations at Several Types of Problem Sites in Urban Areas. Ann. Ent. Soc. Am. 48:251-257.

Phaenicia sericata dominated (42-70 per cent) on garbage dumps; *M. domestica* (95 per cent) on hog farms, about horse stables and on poultry ranches. Surveys included localities in Michigan, New York, Kansas and Arizona. ABSTRACT: BA 30:2076(20866), July, 1956.

Savage, E. P., joint author. *See* Schoof, H. F., et al., 1954, 1955, 1956.

2579. **Schaerffenberg, B.** and **Kupka, E. 1951** Untersuchungen über die geruchliche Orientierung blutsaugende Insekten. I. Über die Wirkung eines Blutduftstoffes auf *Stomoxys* und *Culex*.. [Investigatons concerning the olfactory orientation of blood-sucking insects. I. On the effect of blood odor on *Stomoxys* and *Culex*.] Ost. Zool. Zeit. 3:410-424.

Females of *Stomoxys calcitrans* and *M. domestica* were attracted to an extract of human blood.

2580. **Schaerffenberg, B.** and **Kupka, E. 1953** Orientie Rungsversuche an *Stomoxys calcitrans* und *Culex pipiens* mit eiven Blutduftstoff. [Orientation research on *Stomoxys calcitrans* and *Culex pipiens* with a blood-odor material.]Trans. IXth Int. Congr. Ent. 1:359-361.

Gives reactions of *Stomoxys*, *Culex* and *Musca* to an odorous fraction obtained from blood.

2581. **Schechter, M. S., LaForge, F. B., Zimmerli, A.** and **Thomas, J. M. 1951** Crystalline Allethrin Isomer. J. Am. Chem. Soc. 73:3541.

M. domestica the test insect. *Alpha*-dl-trans isomer less effective, *beta*-dl-trans isomer more effective, than allethrin. ABSTRACT: RAE-B 40:116, July, 1952.

Schechter, M. S., joint author. *See* LaForge, F. B., et al., 1952; Matsui, M., et al., 1952.

Schenck, S. L., joint author. *See* Summerford, W. T., et al., 1951.

2582. **Schmidt, C. H.** and **LaBrecque, G. C. 1959** Acceptability and Toxicity of Poisoned Baits to House Flies Resistance to Organophosphorus Insecticides. J. Econ. Ent. 52(2):345-346.

Malathion bait killed 40 per cent of Grothe (resistant) strain females, 85 per cent of Grothe males, 100 per cent of normal flies. All flies, of both strains killed by Dipterex bait. ABSTRACT: RAE-B 48:125, July, 1960.

2583. **Schneider, E. V. 1958** Baits Containing Organophosphorus Insecticides for Use against House-flies. Zhur. Mikrobiol., Epidemiol., I Immunobiol. 29(1/2):263-268.

Describes where and how to use chlorophos baits and gauze netting soaked with bait. ABSTRACT: BA 36:1565(18840), March 15, 1961.

Schnitzerling, H. J., joint author. *See* Kerr, R. W., et al., 1957.

Schonbrod, R. D., joint author. *See* Terriere, L. C., et al., 1955.

2584. **Schoof, H. F. 1951** Entomologic Appraisal of Fly Control Programs. Repr. from CDC Bulletin Fed. Sec. Agen., P.H.S. (CDC, Atlanta, Ga.) 10(2):25-28.

Gives rules: (1) Employ same technique for each appraisal; (2) Size of sample must be adequate for area; (3) Make observations at 2-week intervals or less; (4) Schedule successive surveys to minimize ecological effects.

2585. Schoof, H. F. 1955 Mechanism of Development of Insect Resistance to Insecticides. Office of Naval Research. Medical Project Reports. N.R. 120-328: pp. 1-46. Washington.

2586. Schoof, H. F. 1955 Survey and Appraisal Methods for Community Fly Control Programs. Pub. Hlth. Mono. 33 (Public Health Service Publication 443): 18 pp. Illustrated. U.S. Government Printing Office, Washington, D.C.
Discusses Scudder grill, reconnaissance, and fly trap methods. (The last is least reliable.) ABSTRACTS: RAE-B 44:127, Aug., 1956; TDB 53:936, July, 1956.

2587. Schoof, H. F. 1959 How Far do Flies Fly and what Effect does Flight Pattern have on their Control? Pest Control 27(4):16-24, 66.
Overall movement for *M. domestica* is from 1 to 2 miles; tend to congregate around feeding and/or breeding sites. ABSTRACT: BA 33:3223(39247), July, 1959.

2588. Schoof, H. F. and **Kilpatrick, J. W. 1957** House Fly Control with Parathion and Diazinon impregnated Cords in Dairy Barns and Dining Halls. J. Econ. Ent. 50(1):24-27.
Diazinon-xylene impregnated cords give better results than parathion-xylene impregnated cords, have a lower mammalian toxicity, are more expensive. ABSTRACTS: BA 31:2999(33107), Oct., 1957; RAE-B 46:43, Mar., 1958; TDB 54:765, June, 1957.

2589. Schoof, H. F. and **Kilpatrick, J. W. 1957** Organic Phosphorus Compounds for the Control of Resistant House Flies in Dairy Barns. J. Econ. Ent. 51(1):20-23.
Tests run on several insecticides including Diazinon, Dipterex, Dow-ET-IS, parathion, DDVP-Aroclor. ABSTRACTS: BA 32:1796(21353), June, 1958; TDB 55:705, June, 1958.

2590. Schoof, H. F. and **Kilpatrick, J. W. 1958** House Fly Resistance to Organophosphorous Compounds in Arizona and Georgia. J. Econ. Ent. 51(4):546.
Treats of 2 field strains of *M. domestica* resistant to malathion, Diazinon and parathion. ABSTRACT: TDB 56:109, Jan., 1959.

2591. School, H. F. and **Kilpatrick, J. W. 1959** Dilan Formulations for Adult House Fly Control. J. Econ. Ent. 52(4):776-777.
Only effective formulation is a Dilan:DDT:cotton seed oil preparation. ABSTRACT: RAE-B 48:200-201, Nov., 1960.

2592. Schoof, H. F. and **Mail, G. A. 1953** Dispersal Habits of *Phormia regina* in Charlestown, W. Va. J. Econ. Ent. 46:258-262.
Dyed and radioactive flies dispersed to 10 miles from point of release. Control operations should extend 3-4 miles beyond city boundaries. (Farther than for *M. domestica.*) ABSTRACT: RAE-B 41:172, Nov., 1953.

2593. Schoof, H. F., Mail, G. A. and **Savage, E. P. 1954** Fly Production Sources in Urban Communities. J. Econ. Ent. 47:245-253.
In the three cities studied, *M. domestica* showed greatest diversity in breeding habits of all species of flies, utilizing 11 of the 13 substrates examined. ABSTRACTS: BA 29:208, Jan., 1955; RAE-B 43:23, Feb., 1955; TDB 51:1200, Nov., 1954.

2594. Schoof, H. F. and **Savage, E. P. 1955** Comparative Studies of Urban Fly Populations in Arizona, Kansas, Michigan, New York, and West Virginia. Ann. Ent. Soc. Am. 48:1-12.
M. domestica predominated in the southwest; *Phaenicia* and *Phormia* in the northeast. In most areas, average density indices were lower in 1950 than in 1949.

2595. **Schoof, H. F., Savage, E. P.** and **Dodge, H. R. 1956** Comparative Studies of Urban Fly Populations in Arizona, Kansas, Michigan, New York and West Virginia. II. Seasonal abundance of Minor Species. Ann. Ent. Soc. Am. 49:59-66.
Nine genera involved, including *Fannia.* Data from 10 municipalities in the 5 States listed. ABSTRACT: BA 30:3029(30258), Oct., 1956.

2596. **Schoof, H. F.** and **Siverley, R. E. 1954** Urban Fly Dispersion Studies with Special Reference to Movement Pattern of *Musca domestica.* Am. J. Trop. Med. Hyg. 3:539-547.
Liberated 147,000 radioactive flies. Traps at secondary and tertiary sites. Results show that flies have an inherent instinct to wander, beyond the need to seek breeding or feeding places. ABSTRACTS: BA 29:445, Feb., 1955; RAE-B 43:175, Nov., 1955; TDB 51:1310, Dec., 1954.

2597. **Schoof, H. F.** and **Siverly, R. E. 1954** Privies as a Source of Fly Production in an Urban Area. Am. J. Trop. Med. Hyg. 3:930-935.
Untreated privies showed 3·6 per cent *M. domestica,* 59 per cent *Muscina*; chemically treated privies 31·9 per cent *M. domestica,* 20·7 per cent *Muscina.* Reasons for reversal not clear. ABSTRACTS: RAE-B 44:30, Feb., 1956; TDB 52:221, Feb., 1955.

2598. **Schoof, H. F.** and **Siverley, R. E. 1954** Multiple Release Studies on the Dispersion of *Musca domestica* at Phoenix, Arizona. J. Econ. Ent. 47:830-838.
Flies infiltrated into city from ½ mile beyond periphery; *not* from 1·8 miles or more. Control program for 1 mile out should be adequate. ABSTRACTS: BA 29:2006, Aug., 1955; RAE-B 43:156, Oct., 1955; TDB 52:411, Apr., 1955.

2599. **Schoof, H. F., Siverly, R. E.** and **Coffey, J. H. 1951** Dieldrin as a Chemical Control Material on Community Fly Control Programs. J. Econ. Ent. 44:803-807.
At first successful with DDT-resistant flies, then failure. (Effective for 2 years in certain Texas areas.) ABSTRACTS: RAE-B 40:44, Mar., 1952; TDB 49-324, Mar., 1952; BA 26:1805, July, 1952.

2600. **Schoof, H. F., Siverly, R. E.** and **Jensen, J. A. 1952** House Fly Dispersion Studies in Metropolitan Areas. J. Econ. Ent. 45:675-683.
Two studies (June and September) of marked flies recaptured. 88 per cent and 81 per cent were found in the 1 mile zone, but some flies reached 4 miles in 72 hours. Flies will re-enter the city if it contains "Attractivity" sites. ABSTRACTS: RAE-B 41:42, Mar., 1953; TDB 50:257, Mar., 1953.

Schoof, H. F., joint author. *See* Kilpatrick, J. W., et al., 1954, 1955, 1956, 1957, 1958, 1959; Baker, W. C., et al., 1955; Mail, G. A., et al., 1954; Waldrop, R. H., et al., 1958; Welch, S. F., et al., 1953; Quarterman, K. D., et al., 1958; Savage, E. P., et al., 1955.

Schorsch, C. B., joint author. *See* O'Keefe, W. B., et al., 1954.

2601. **Schroeder, H. O. 1950** Piperonyl Butoxide in Low Pressure Aerosol Insecticides. Soap and Sanit. Chem. 26:145, 147, 148A, 148C.
Combinations of piperonyl butoxide and pyrethrins when used alone or with DDT permit a marked reduction in the amount of pyrethrins required for both knockdown and kill of house flies. Results in tabular form. ABSTRACT: RAE-B 39:172, 1951.

2602. **Schroeder, H. O.** and **Berlin, F. D. 1950** Allethrin in Aerosols. Soap and Sanit. Chem., June, 1950. Reprint, 3 pp.
Allethrin with piperonyl butoxide about half as effective as equivalent amount of pyrethrins with this synergist. Allethrin plus pyrethrins plus piperonyl butoxide gives only an "additive" effect.

2603. Schroeder, H. O. and **Jones, H. A. 1957** Household Aerosol Insecticides. How do theyFare Against House Flies Resistant to Chlorinated Pesticides? Soap and Chem. Spec. 33(12):115, 117, 121, 123, 265.
Official test Aerosol and commercial preparations gave poor results against resistant flies, but pyrethrum plus piperonyl butoxide was satisfactory. ABSTRACT: RAE-B 46:164, Oct., 1958.

Schroeder, H. O., joint author. *See* Jones, H. A., et al., 1950.

2604. Schulz, K. H. and **Crutkhow, C. 1957** A Demonstration Project of Fly Control using "Chlordane" in a Monsoon Area. Thailand. J. Trop. Med. Hyg. 60(6):141-145.
Residual insecticide spray used only on premises where food-stuffs were handled, stored, or sold. ABSTRACT: TDB 54:1139, Sept., 1957. (Lists a third author, C. Sitachitta.)

Schwardt, H. H., joint author. *See* Goodwin, W. J., et al., 1953; Pimentel, D., et al., 1950, 1951, 1953, 1954.

2605. Schwenke, W. 1958 Zur Ernährungsweise der *Muscina*-Larven (Diptera: Muscidae). [On the feeding habits of *Muscina* larvae (Diptera: Muscidae).] Beitr. Ent. 8(1/2):8-22. (English and Russian summaries.)
Larvae are phyto- or zoo-necrophageous. After second stage can live carnivorously on larvae of other Diptera. ABSTRACT: BA 36:6886(74367), Nov. 1, 1961.

Scott, R. H., joint author. *See* Hansens, E. J., et al., 1955.

Scrivani, P., joint author. *See* Speroni, G., et al., 19??.

Scudder, H. I., joint author. *See* Lindsay, D. R., et al., 1956.

2606. Seago, J. M. 1954 A New Species of *Fannia* from North America (Diptera, Muscidae). J. Kans. Ent. Soc. 26:141-142.
Fannia enotahensis is described as new, from Georgia. ABSTRACT: BA 29:456, Feb., 1955.

2607. Sedee, D. J. 1954 Qualitative Amino Acid Requirements of Larvae of *Calliphora erythrocephala* (Meigen). Acta Physiol. Pharm. Neerl. 3:262-269.
Lists 10 amino acids as essential; 8 dispensible. Requirements are in good accord with those of *Drosophila, Attagenus* and honey bee.

2608. Sedee, P. D. J. W. 1953 Qualitative Vitamin Requirements for Growth of Larvae of *Calliphora erythrocephala* (Meig.). Experientia 9(4): 142-143. (German summary.)
Vitamins required for normal growth include thiamin, riboflavin, nicotinic acid, pyridoxin, pantothenic acid and choline. Folic acid and biotin also important. ABSTRACT: BA 28:544, Mar., 1954.

2609. Sen. P. 1959 Insecticide Resistance in Houseflies (*Musca domestica vicina*) of Calcutta. Bull. Calcutta Sch. Trop. Med. 7(1):12-13.
Exposure to 4 per cent DDT for 4 hours was required to reach LD50 for Calcutta houseflies. Flies still susceptible to dieldrin and to the organophosphate group. ABSTRACT: BA 35:4675(54324), Oct. 1, 1960.

2610. Sen, P. 1959 Studies on Insecticide Resistance in Insects of Public Health Importance in West Bengal, India. Indian J. Malariol. 13(1): 19-33.
Records current susceptibility of various medically-important forms to modern insecticides, chiefly chlorinated hydrocrabons. ABSTRACT: RAE-B 49:264, Dec., 1961; BA 40:1925(25281), Dec. 15, 1962.

Shaikov, A. D., joint author. *See* Shura-Bura, B. L., et al., 1956, 1958.

2611. Shanahan, G. J. 1959 Genetics of Dieldrin Resistance in *Lucilia cuprina* Wied. Nature 183(4674):1540-1541.
Rests on a partially dominant single gene. Crossed susceptible females with resistant male. F2 1:3. Back cross 1:1. ABSTRACT: BA 34:258(2876), Sept., 1959.

2612. Sharma, M. I. D. and **Bami, H. L. 1957** Comparative Efficacy of 50 per cent Dieldrin-Resin Dispersible Powder and 50 per cent Dieldrin Dispersible Powder. Indian J. Malariol. 11(2):169-172.
Test insects were *Culex fatigans* and *Musca nebulo*. Dieldrin-resin preparation produced the higher initial kill and its residual toxicity lasted for a longer period. ABSTRACT: TDB 55:336, Mar., 1958.

2613. Sharma, M. I. D., Bhatia, M. L., Krishnamurthy, B. S. and **Sadanand, A. V. 1957** Studies on the Intergrading Forms of *Musca vicinia*, Macquart and *M. nebulo*, Fabricius Females. Bull. Nat. Soc. India for Malaria and other Mosquito-Borne Dis. 5(5):244-245.
Authors point out that "morphologically", *M. vicina* and *M. nebulo* are not two distinct species. ABSTRACT: TDB 55:593, May, 1958.

2614. Sharma, M. I. D., Krishnamurthy, B. S. and **Singh, N. N. 1958** Note on the Susceptibility to DDT of Houseflies of Delhi and Rajasthan. Indian J. Malariol. 12(3):203-207.
Flies from a DDT-sprayed village (Delhi) proved more tolerant of DDT than those from an unsprayed village (Rajasthan). ABSTRACTS: RAE-B 49:202, Sept., 1961; TDB 56:778, July, 1959.

Sharma, M. I. D., joint author. *See* Pal, R., et al., 1952, 1953, 1954.

Sharpless, R. V., joint author. *See* Mallis, A., et al., 1952; Miller, A. C., et al., 1952; Pellegrini, J. P., et al., 1952.

Shawarby, A. A., joint author. *See* Peffly, R. L., et al., 1956.

Shea, W. D., joint author. *See* Granett, P., et al., 1955.

2615. Sherman, M. and **Ross, E. 1959** Toxicity to House Fly Larvae of Insecticides Administered as Single Oral Dosages to Chicks. J. Econ. Ent. 52(4):719-723.
Degree of fecal toxicity to larvae directly related to dosage of insecticide ingested by chick; inversely related to period of time after ingestion that feces were voided. ABSTRACT: RAE-B 48:196, Nov., 1960.

Shi, Teh-Chi, joint author. *See* Fan, Tze-Teh, et al., 1959.

2616. Shih, S. and **Wancheng, H. 1952** [The bacteriological examination of dish water and flies obtained from the restaurants in Taipeh Provincial Public Health Demonstration Districts.] J. Formosan Med. Ass. 51(9):402-405. (In Chinese, with English summary.)
Average total bacterial count on flies after 24-hour incubation at 37°C was 192,443 per fly. About 91·5 per cent were of the coli-aerogenes group. ABSTRACT: BA 28:650, Mar., 1954.

2617. Shimogama, M. 1959 [Studies on the control of flies. 3. Field experiments for the house fly control by residual sprays and insecticide impregnated cords.] Endemic Dis. Bull. Nagasaki Univ. 1(3):330-342.

2618. Shimogama, M. 1959 [Studies on the control of flies. 4. Control experiments of fly maggots in the privies by various larvacides.] Endemic Dis. Bull. Nagasaki Univ. 1(4):414-422. (English summary.)
A 1:1000 dilution of 17 per cent Diazinon emulsion most economic and effective in killing maggots and preventing appearance of young larvae. ABSTRACT: BA 36:5572(58213), Sept. 1, 1961.

Shirai, M., joint author. *See* Suzuki, T., et al., 1958, 1959.

Schnaider, E. V., joint author. *See* Vashkov, V. I., et al., 1959; Sukhova, M. N., et al., 1959.

Shor, Y., joint author. *See* Neeman, M., et al., 1956.

Shortino, T. J., joint author. *See* Babers, F. H., et al., 1956; Mitlin, N., et al., 1957.

2619. Shura-Bura, B. L. 1952 [An attempt to study the migration of house flies by using radioactive tracers.] Zool. Zhur. (USSR)31: 410-412. (In Russian.)

2620. Shura-Bura, B. L. 1952 Zagriaznenie fruktov sinantropnymi mukhami. [The contamination of fruits by synanthropic flies.] Ent. Obozrenie (Moscow) 32:117-125. (In Russian.)

Treats of *Sarcophaga*, *Lucilia*, *Muscina*, *Fannia*, as well as *M. domestica vicina*. Flies were found to contaminate grapes, pears, and apples with various forms of *Escherichia coli*. ABSTRACTS: BA 28:2440, Oct., 1954; RAE-B 43:57, Apr., 1955.

2621. Shura-Bura, B. L. 1955 [Result of investigation of flies from garbage dumps using the method of labelled substances.] Gig. I. Sanit. (Moskva) for 1955. No. 8 (August):12-15. (In Russian.)

2622. Shura-Bura, B. L. 1957 [Causes of insufficient activity of B.H.C. on flies in Sevastopol.] Med. Parazit. (Moskva) 26(1):26-31. (In Russian.)

2623. Shura-Bura, B. L., Ivanova, E. V., Onuchin, A. N., Glazunova, A. Ya. and Shaikov, A. D. 1956 [Ways of fly dispersion from places of mass breeding in Leningrad.] Rev. Ent. URSS 35(2):334-346. Moscow. (In Russian.)

Dispersion occurred in all directions, but inhabited districts were the most attractive. ABSTRACT: RAE-B 45:210, Dec., 1957.

2624. Shura-Bura, B. L., Shaikov, A. D., Ivanova, E. V., Glazunova, A. Ya., Mitriukova, M. S. and **Federova, K. G. 1956** [Migration of synanthropic flies to the cities from open fields.] Med. Parazit. i Parazitarn. (Bolezni) 25(4):368-372.

2625. Shura-Bura, B. L., Shaikov, A. D., Ivanova, E. V., Glazunova, A. Ya., Mitriukova, M. S. and **Fedorova, K. G. 1958** [The character of dispersion from the point of release in certain species of flies of medical importance.] Ent. Obozrenie. [Transl.] 37(2):282-290.

Main migration was towards points of special attraction (areas with unsanitary conditions). Main direction was towards Leningrad (15 km. north of release point). ABSTRACTS: BA 35:2700(30653), June, 1960; RAE-B 49:36-37, Feb., 1961.

Silverman, L., joint author. *See* Silverman, P. H., et al., 1953.

2626. Silverman, P. H. and **Levinson, Z. H. 1954** Lipid Requirements of the Larva of the Housefly *Musca vicina* (Macq.) Reared under Nonaseptic Conditions. Biochem. J. (London) 58:291-294.

Only lipid required is contained in sterol fraction of wheat-bran medium; identified as "sitosterol". Resists bacterial infection in larvae, is a factor in pupation, and promotes growth. ABSTRACT: TDB 52:1024, Oct., 1955.

2627. Silverman, P. H. and **Mer, G. G. 1952** Preliminary Report: Behavior of a DDT-Resistant Strain of Flies at an Agricultural Settlement in Israel. Riv. di Parassit. 13(1):123-128.

Resistance (so-called) is due at least in part to the flies migrating at night to resting places on nearby vegetation; thereby avoiding lethal doses of insecticides.

2628. Silverman, P. H. and **Silverman, L. 1953** Growth Measurements on *Musca vicina* (Macq.) reared with a Known Bacterial Flora. Riv. di Parassit. 14(2):89-95.
Microflora believed to break down medium for larval use and thus function in producing desired temperature. If too cool, development is retarded; if too warm, larvae are small. ABSTRACT: TDB 50:1090, Nov., 1953.

Silverman, P. H., joint author. *See* Levinson, A. H., et al., 1954.

2629. Singer, C. 1955 The First English Microscopist: Robert Hooke (1635-1703). Endeavor. 14:12-18.
Shows Hooke's illustration of the fly's foot, also compound eye (with reflection of laboratory windows in every facet).

Singh, K. R., joint author. *See* Micks, D. W., et al., 1958.

Singh, N. N., joint author. *See* Sharma, M. I. D., et al., 1958.

2630. Siverly, R. E. 1953 Natural Gas as a Blowfly Attractant. J. Econ. Ent. 46:156.
Flies attracted to leaks in natural gas lines, especially females. Eggs found in debris. Suggests that H_2S and mercaptan serve as attractants.

2631. Siverly, R. E. 1958 Effects of Chilling of Pupae on Subsequent Emergence of Resistant and Susceptible House Flies. J. Econ. Ent. 51(5): 666-668.
No significant difference between strains. Linear relationship between time intervals of chilling, delay of emergence, and numbers of emerged adults.

2632. Siverly, R. E. and **Schoof, H. F. 1955** Utilization of Various Production Media by Muscoid Flies in a Metropolitan Area. I. Adaptability of Different Flies for Infestation of Prevalent Media. Ann. Ent. Soc. Am. 48:258-262.
Twelve-month survey in Arizona. *M. domestica* predominated in a wide variety of substrates. ABSTRACT: BA 30:2077(20867), July, 1956.

2633. Siverly, R. E. and **Schoof, H. F. 1955** Utilization of Various Production Media by Muscoid Flies in a Metropolitan Area. II. Seasonal Influence on Degree and Extent of Fly Production. Ann. Ent. Soc. Am. 48:320-324.
Twelve-month survey indicates year-long breeding for *M. domestica.* ABSTRACT: BA 30:2077(20868), July, 1956.

2634. Siverly, R. E. and **Schoof, H. F. 1955** Utilization of Various Production Media by Muscoid Flies in a Metropolitan Area. III. Fly Production in Relation to City Block Environment. Ann Ent. Soc. Am. 48:325-329.
In Phoenix, Arizona, business blocks were least infested, residential blocks with poor garbage disposal, heavily so. Animal pens on some blocks added to the count. ABSTRACT: BA 30-2077(20869), July, 1956.

Siverly, R. E., joint author. *See* Schoof, H. F., et al., 1951, 1952, 1954.

Skierska, B., joint author. *See* Platkowska, W., et al., 1953.

2635. Smallman, B. N. 1958 The Physiological Basis for the Mode of Action of Organophosphorus Insecticides. Proc. 10th Int. Congr. Ent. pp. 5-12.
Examines the physiological basis for the hypothesis that organophosphorus compounds exert their insecticidal action by inhibition of cholinesterase. ABSTRACT: RAE-B 47:82, June, 1959.

Smallman, B. N., joint author. *See* Chefurka, W., et al., 1955, 1956; Lewis, S. E., et al., 1956.

Smiley, E. S., joint author. *See* Guyer, G. E., et al., 1954.

2636. Smirnov, E. S. and **Chuvakhina, Z. F. 1956** [Visual discrimination of number, size and form in *Musca domestica* L.] Zool. Zhur. 35:

560-571. Moscow. (In Russian, with English summary.)
Attraction to small pieces of black paper was studied in relation to size, number and shape of the pieces. Difference in attractiveness connected with a "gregarious reflex". ABSTRACTS: RAE-B 46:18, Feb., 1958; BA 32:2463(29509), Sept., 1958.

2637. Smith, A. C. 1954 Flies, Chickens and People. Calif. Vector Views. 1(6):1, 3, 5.
Considers fly problems on poultry ranches to be the result of ignorance or neglect.

2638. Smith, A. C. 1956 Fly prevention in Dairy Operations. Calif. Vector Views. 3(11):57, 59, 60.
Flies bred on dairy, poultry, or rabbit farms move into communities. Basic sanitation more important in control than chemicals.

2639. Smith, A. C. 1957 Does Fly Control Mean a Profit or Loss for PCO's. Pest Control, May.
Control people know breeding source of the species being attacked. Procedure is then most economical.

2640. Smith, A. C. 1957 What do you Know About Flies? Pest Control, Reprint, 5 pp.
A photo-quiz of 20 questions.

2641. Smith, A. C. 1958 Flies from Bulbs. Calif. Vector Views. 5(12):80.
M. domestica and *Muscina assimilis* collected from a soft spot on a dahlia bulb.

2642. Smith, A. C. 1958 Fly-cord Studies in California. Calif. Vector Views. 5(9):57-61.
Fly cords impregnated with parathion or diazinon were effective eight months after exposure in farm buildings.

2643. Smith, A. C. 1958 High Protein Feed Waste as a Source of Lesser House Fly. Calif. Vector Views. 5(9):63.
Fannia canicularis bred in piles of cattle feed concentrate (20 per cent protein) near feed storage area.

2644. Smith, A. C. 1959 Flies from Fish Meal. Calif. Vector Views. 6(1):87.
Muscina stabulans collected from flower pots fertilized with fish emulsion.

2645. Smith, A. C. 1959 Fly Control. Is Species Important? Pest Control, 27(4):32-38.
Paper identifies commonly found flies and gives their life cycles, with a view to making control practices more effective. ABSTRACT: BA 33: 3223(39248), July, 1959.

Smith, A. C., joint author. *See* Bohart, R. M., et al., 1958.

2646. Smith, A. G. and **Harrison, R. A. 1951** Notes on Laboratory Breeding of the Housefly (*Musca domestica* L.). New Zealand J. Sci. Tech. 33B(1):1-4.
DDT-resistant lines have a lower female percentage and a later emergence peak than non-resistant lines. ABSTRACTS: BA 26:3222, Dec., 1952; TDB 50:161, Feb., 1953.

Smith, A. L., joint author. *See* Travis, B. V., et al., 1951.

Smith, C. N., joint author. *See* Gahan, J. B., et al., 1957; Keller, J. C., et al., 1956, 1957; Wilson, H. G., et al., 1959.

2647. Smith, E. A. 1954 Fly control on Poultry Ranches in Santa Clara County. Calif. Vector Views. 1(6):1, 4, 5.
Treats of County Health Dept., citizens' committees, surveys, educational and practical assistance to farmers.

Smith, F. F., joint author. *See* Fulton, R. A., et al., 1950.

Smith, W. R., joint author. *See* Prill, E. A., et al., 1955.

2648. Smyth, T., Jr. and **Roys, C. C. 1955** Chemoreception in Insects and the Action of DDT. Biol. Bull. 108:66-76.
Proboscis extension reflex controlled by tactile stimuli as well as by chemoreceptor activity. Failure of DDT to affect this reflex due to some mechanism intrinsic in the nervous system. ABSTRACTS: RAE-B 44:185, Nov., 1956; Hocking, B. 1960 Smell in Insects, p. 50 (1.482).

Smyth, T., Jr., joint author. *See* Hodgson, E. S., et al., 1955.

2649. Sokal, R. R. and **Hiroyoshi, T. 1959** The Supposed Correlation between the Ratio X_5 and DDT-Resistance in House Flies. J. Econ. Ent. 52(6):1077-1080.
The X_5 ratio = width of 2nd abdominal sternite divided by its length. Measured on both sexes. No correlation demonstrated between X_5 and resistance. ABSTRACT: RAE-B 49:30, Feb., 1961.

2650. Sokal, R. R. and **Hunter, P. E. 1955** A Morphometric Analysis of DDT-resistant and Non-resistant House Fly Strains. Ann. Ent. Soc. Am. 48:499-507.
Sixteen morphological characters were measured; DDT-resistance is not correlated with any of these characters. Possible reasons for this are discussed.

Solly, S. R. B., joint author. *See* Lord, K. A., et al., 1956.

2651. Sömme, L. 1958 The Number of Stable Flies in Norwegian Barns, and their Resistance to DDT. J. Econ. Ent. 51(5):599-601.
Made regular observations on the number of house flies (*M. domestica*), stable flies (*S. calcitrans*) and *Fannia* sp. in barns of eastern Norway. Tested stable flies for DDT-resistance.

2652. Sömme, L. 1959 On the Numbers of Stable Flies and House Flies on Norwegian Farms. Norsk. Ent. Tidsskr. 11(1/2):7-15.
On farms near Oslo stable flies were most numerous in cow-houses; house flies in pig pens. *Fannia* and other Muscidae also recorded. ABSTRACTS: BA 36:764(9087), Feb., 1961; RAE-B 49:176, Aug., 1961.

2653. Sömme, L. 1959 Resistance to Chlorinated Insecticides in Norwegian House-flies. Nytt Mag. Zool. 8:56-60.
In 1957 only a few fly strains showed resistance to DDT, lindane, chlordane and dieldrin. In 1958 tendency was general. ABSTRACT: RAE-B 48:85, May, 1960.

Sousa, A. A., joint author. *See* Haynes, H. L., et al., 1954.

2654. Spear, P. J. and **Sweetman, H. L. 1952** Continuous Vaporization of Insecticides with special Reference to DDT. J. Econ. Ent. 45:869-873.
Insecticide vaporized by heat in closed spaces. Effective against flies if non-resistant. ABSTRACT: RAE-B 41:75, May, 1953.

Spear, S. F., joint author. *See* Acree, F., Jr., et al., 1959.

2655. Speroni, G. 1952 Impiego di sinergici nella lotta contro la mosca domestica resistente agli insetticidi di contatto. [Employment of synergists in the war against domestic flies resistant to contact insecticides.] Estr. Riv. Chim. e Industr. 34:391-403.
Experiments are recorded in which 115 compounds were tested as synergists with DDT. ABSTRACTS: RAE-B 43:27, Feb., 1955; TDB 52: 696, July, 1955.

2656. Speroni, G., Losco, G. and **Peri, C. A. 1951** Sugli esteri dell' acido tiocianacetico e sulle loro proprietà insetticide. [Concerning the esters of acetic acid and on their insecticidal properties.] Estr. Riv. Chim. e Industr. 33:209-215.
Compounds evaluated by tests with *M. domestica*.

2657. Speroni, G., Losco, G., Santi, R. and **Peri, C. A. 1953** Recherches sur des produits synergiques des insecticides chlorurés. [Researches on the synergical effects of chlorinated insecticides.] Chim. e Industr. 69(4):658-666.
Tables show mortality of flies from mixed applications of DDT, chlordane and methoxychlor.

2658. Speroni, G., Losco, G. and **Scrivani, P. 19??** Properties insecticides de products de la chloruration du camphene. [Insecticidal properties of products from the chlorination of camphene.] Extrait de "Chimie et Industrie", numéro spécial du XXIe Congrès Internationale de Chimie Industrielle. 4 pp.
Various compounds subjected to biological assay using *M. domestica.*

Spierska, B., joint author. *See* Piatkowska, W., et al., 1954.

Srivastava, S. B., joint author. *See* Lal, B. N., et al., 1950.

Standen, H., joint author. *See* Bradbury, F. R., et al., 1955, 1956, 1958, 1959.

2659. Standifer, L. N. 1955 Larvicides for Control of the House Fly. J. Econ. Ent. 48:731-733.
Twenty-five formulations of chlorinated hydrocarbon and phosphate insecticides tested against third stage larvae. Aldrin emulsion proved superior to all others. ABSTRACTS: BA 30:2077(20870), July, 1956; RAE-B 45:5, Jan., 1957; TDB 53:677, May, 1956.

Stansbury, H. A., joint author. *See* Haynes, H. L., et al., 1954.

2660. Stansbury, R. E. and **Dahm, P. A. 1951** The Effect of Alfalfa Dehydration Upon Residues of Aldrin, Chlordane, Parathion and Toxaphene. J. Econ. Ent. 44:45-51.
A uniform method was developed for biological assays using *M. domestica* as the test species.

Stansbury, R. E., joint author. *See* Goodhue, L. D., et al., 1953.

Starkweather, R., joint author. *See* McGuire, C. D., et al., 1957.

2661. Starr, D. F. and **Calsetta, D. R. 1955** Ryania as a Housefly Larvicide. Agric. Chem. (Baltimore). 9(11):50-53.
Effected 99·9 per cent mortality with newly hatched larvae; 48 per cent mortality with larvae 1-2 days old. No significant resistance in 50 generations. ABSTRACTS: BA 30:309, Jan., 1956; RAE-B 43:55, Apr., 1955; TDB 52:1027, Oct., 1955.

2662. Statens Skadedyrlaboratorium. 1951 Arsberetning 1949-1950, 41 pp., 10 figs. Springforbi. (In Danish, with English summaries.)
Reports investigations on resistance of *M. domestica* to insecticides in Denmark. Chlordan was effective at the time of this report. ABSTRACT: RAE-B 40:115, July, 1952.

2663. Statens Skadedyrlaboratorium. 1953 Arsberetning 1950-1951. [Government Pest Infestation Laboratory (Stored Products and Household Pests). Annual Report.] 45 pp. Springforbi. (In Danish, with English summaries.)
Continues studies on resistance in *M. domestica.* ABSTRACT: RAE-B 42: 93, June, 1954.

2664. Stegwee, D. 1959 Esterase Inhibition and Organophosphorus Poisoning in the House-fly. Nature 184(4694):1253-1254. (Suppl. 16.)
Inhibition of the aliphatic esterase is of minor importance compared to the inhibition of cholinesterase. ABSTRACT: BA 35:1513(17439), Mar., 1960.

Stelmach, Z., joint author. *See* Wells, A. L., et al., 1958.

2665. Sternburg, J. and **Kearns, C. W. 1950** Degradation of DDT by Resistant and Susceptible Strains of House Flies. Ann. Ent. Soc. Am. 43:444-458.
Both larvae and adults of resistant strains can degrade oral dosages to the dehydrochlorinated compound DDE. Accumulates mostly in the cuticle-hypodermis. ABSTRACTS: RAE-B 39:211, Dec., 1951; TDB 49 562, May, 1952.

2666. Sternburg, J. and **Kearns, C. W. 1952** The Presence of Toxins other than DDT in the Blood of DDT-Poisoned Roaches. Science 116: 144-147.
Blood is equally toxic, by injection, to normal and to DDT-resistant flies. ABSTRACT: TDB 49:1163, Dec., 1952.

2667. Sternburg, J. and **Kearns, C. W. 1956** Pentachlorocyclohexane, an Intermediate in the Metabolism of Lindane by House Flies. J. Econ. Ent. 49:548-552.
Both susceptible and lindane-tolerant flies metabolize lindane, the tolerant much faster. Production of pentachlorocyclohexane shown by spectrophotometric, chromatographic and colorimetric techniques. ABSTRACTS: RAE-B 45:145, Sept., 1957; TDB 54:362, Mar., 1957.

2668. Sternburg, J., Kearns, C. W. and **Bruce, W. N. 1950** Absorption and Metabolism of DDT by Resistant and Susceptible House Flies. J. Econ. Ent. 43:214-219.
Susceptible flies cannot metabolize DDT to form DDE or DDA; resistant flies do. Very little of either product is excreted. ABSTRACT: RAE-B 38:209, 1950.

2669. Sternburg, J., Vinson, E. B. and **Kearns, C. W. 1953** Enzymatic Dehydrochlorination of DDT by Resistant Flies. J. Econ. Ent. 46: 513-515.
Demonstrates the presence of an enzyme system in a DDT-resistant strain capable of rapidly dehydrochlorinating DDT to DDE. ABSTRACTS: RAE-B 42:6, Jan., 1954; TDB 50:1173, Dec., 1953.

2670. Steve, P. C. 1959 Parasites and Predators of *Fannia canicularis* (L.) and *Fannia scalaris* (F.). J. Econ. Ent. 52(3):530-531.
Observations establish the following new host records of parasitism and predation: *Macrocheles musca domestica* on *F. canicularis* and *Pachycrepoideus vindemmiae* on *F. canicularis* and *F. scalaris*. ABSTRACTS: BA 34:258(2877), Sept., 1959; RAE-B 48:169, Sept., 1960.

2671. Steve, P. C. 1960 Biology and Control of the Little House Fly, *Fannia canicularis*, in Massachusetts. J. Econ. Ent. 53(6):999-1004.
The use of parathion-Diazinon treated cords gave excellent seasonal control. ABSTRACTS: BA 36:1833(22200), April 1, 1961; RAE-B 49: 217, Oct., 1961.

2672. Stewart, C. D. 1951 The Ingenious Housefly. The Milwaukee Journal. Sunday, Sept. 23.
Educational article, for popular reading.

Stewart, W. H., joint author. *See* Lindsay, D. R., et al., 1953.

2673. Steyn, J. J. 1959 Use of Social Spiders against Gastro-intestinal Infections Spread by House Flies. S. Afr. Med. J. 33:730-731.
Suggests that social spiders (*Stegodyphus mimosarum*), be established for the control of houseflies. Appear to be completely harmless to man. ABSTRACTS: RAE-B 51:19, Feb., 1963; TDB 57:191, Feb., 1960.

2674. Steyskal, G. C. 1957 The Relative Abundance of Flies (Diptera) Collected at Human Feces. Zeit. Angew. Zool. 44(1):79-83.
Only 2 per cent of the flies captured on faeces were *M. domestica*. Flies

were captured on 8 days in July, August and September. ABSTRACT: TDB 54:886, July, 1957.

Stoutamire, D. W., joint author. *See* Gatterdam, P. E., et al., 1959.

Suarez, O. M., joint author. *See* Cova Garcia, P., et al., 1953.

2675. Suenaga, O. 1958 [Ecological studies of flies. I. On the amount of larvae of housefly and stablefly breeding in animal manure.] Nagasaki Med. J. 33:124-133. Suppl. 11. (In Japanese, with English summary.)

Examined the number of larvae breeding from 1 week's cow and pig manure, collected once a month for one year. ABSTRACT: BA 35:1270 (14471), Mar., 1960.

2676. Suenaga, O. 1959 [Ecological studies of flies. IV. On the flies breeding out from the carcasses of small animals.] Endemic Dis. Bull. Nagasaki Univ. 1(3):343-352. (In Japanese, with English summary.)

Musca domestica vicina bred from carcasses of frog, rat, snake and guinea pig which were exposed to flies in or near dwellings. ABSTRACT: BA 36:6485(70198), Oct., 1961.

2677. Sukhova, M. N. 1953 O znachenii bazarnoi mukhi *Musca sorbens* Wied. v epidemiologii ostrogo epidemicheskogo kon' 'iunktivita v Zapadnoi Turkmenii. Gig. I Sanit. (Moskva) 7:40-42.

Stresses significance of *M. sorbens* in the epidemiology of acute epidemic conjunctivitis in western Turkmenia.

2678. Sukhova, M. N., Shnaider, E. V., Erofeeva, T. V., Zlalkooskaia, E. V. and **Kuklina, N. P. 1959** [Comparative evaluation of the effectiveness of measures of combatting synanthropic flies with DDT, BHC preparation, chlorophos and further prospects in regard to extermination of these insects.] Zhur. Mikrobiol., Epidemiol., i Immunobiol. (Moskva) 30(6):66-73.

Sukhova, M. N., joint author. *See* Levkovich, E. N., et al., 1957.

2679. Sullivan, R. L. 1958 Sex Limitation of Several Loci in the Housefly. (Abstr.) Proc. 10th Intern. Congr. Genet. 2:282 (August 20-27, McGill University, Montreal.)

Mutant "brown body" restricted to female (certain strain). In *Pavia* strain, behaves as an autosomal recessive. Strains were crossed. Possible linkage of brown with "green eye color".

2680. Sullivan, R. L. 1961 Linkage and Sex Limitation of Several Loci in the Housefly. J. Heredity 52(6):282-286.

A translocation between a portion of the second chromosome carrying the loci and the Y chromosome, is postulated as causing the type of inheritance discussed. ABSTRACTS: BA 39:1361(17234), Sept., 1962; TDB 59:943, Sept., 1962.

2681. Sullivan, W. N. 1952 Tests with Lindane Vapor for Freeing Aircraft from Insects. J. Econ. Ent. 45:544-545.

Method of dispersal utilizes the air conditioning system. The vapor has the penetrating properties of a fumigant, leaves no visible deposit and is invisible to the passengers. ABSTRACT: RAE-B 40:177, Nov., 1952.

2682. Sullivan, W. N., DuChanois, F. R. and **Hayden, D. L. 1958** Insect Survival in Jet Aircraft. J. Econ. Ent. 51(2):239-241.

All insects tested (including *M. domestica*) were killed on a 3-hour flight at 40,000 feet; outside air temperature, −51°C (−59·8°F). ABSTRACT: RAE-B 47:106, July, 1959.

2683. **Sullivan, W. N.** and **Hornstein, I. 1953** Concentrations and Exposure Time of Lindane Vapor Required to Kill Insects. J. Econ. Ent. 46: 898-899.
Two minute exposure of *M. domestica* at 24°C gave 59 per cent mortality; at 30°C, 75 per cent. Ten minute exposure at 24°C gave 90 per cent mortality; at 30°C, 100 per cent. ABSTRACTS: RAE-B 42:75, May, 1954; TDB 51:516, May, 1954.

2684. **Sullivan, W. N., Hornstein, I., Yeomans, A. H.** and **Tsao, C.-H. 1955** Improved Deposits for Controlling Insects Outdoors. J. Econ. Ent. 48:153-155.
DDT deposits remain toxic for a longer period on foliage when applied in a methyl ethyl ketone solution. ABSTRACT: BA 30:309, Jan., 1956.

Sullivan, W. N., joint author. *See* McBride, O. C., et al., 1950; Ihndris, R. W., et al., 1958; Knipling, E. B., et al., 1957, 1958; Hornstein, I., et al., 1954, 1955, 1958; Fulton, R. A., et al., 1952, 1953; Tsao, C.-H., et al., 1953, 1954; Quarterman, K. D., et al., 1953.

2685. **Sumerford, W. T., Goette, M. B., Quarterman, K. D.** and **Schenck, S. L. 1951** The Potentiation of DDT against Resistant Houseflies by Several Structurally Related Compounds. Science 114:6-7.
First report on the synergistic activity of 11 DDT analogs. ABSTRACT: TDB 49:91, Jan., 1952.

2686. **Sumerford, W. T., Fay, R. W., Goette, M. B.** and **Allred, A. M. 1951** Promising DDT-Synergist Combinations for the Control of Resistant Flies. J. Nat. Malaria Soc. 10:345-349.
Tested 2,400 compounds. Seventeen proved equal to or better than DMC (1, 1-bis(p-chlorphenyl)-ethanol). ABSTRACTS: BA 26:2606, Oct., 1952; TDB 49:562, May, 1952; RAE-B 41:206, Dec., 1953.

Sun, T. S., joint author. *See* Kao, C. M., et al., 1958.

Sun, Y. C., joint author. *See* Kun, K. Y., et al., 1958.

Sutil, O. E., joint author. *See* Cova Garcia, P., et al., 1959.

2687. **Suzuki, T., Ikeshoji, T.** and **Shirai, M. 1958** Insecticide Resistance in Several Strains of House Flies in Japan. Jap. J. Exptl. Med. 28(6): 395-403.
Resistance of 5 strains of *M. vicina* Macq. to 8 insecticides was estimated by topical application. All were more or less resistant to p,p'-DDT; some to lindane and dieldrin. ABSTRACTS: BA 35:764(8766), Feb., 1960; RAE-B 50:99, May, 1962.

2688. **Suzuki, T.** and **Shirai, M. 1959** Comparative Effect of *gamma* 60 per cent BHC with Lindane to Resistant and Susceptible House Flies. Jap. J. Exptl. Med. 29(3):181-189.
Comparing LD50s after 24 hours, 60 per cent BHC was more effective than lindane on resistant flies; less effect with susceptible flies. Used 5 strains of *M. vicina* Macq. ABSTRACTS: BA 35:764-765(8767), Feb., 1960; RAE-B 50:285, Dec., 1962.

Suzuki, T., joint author. *See* Ikeshoji, T., et al., 1959; Ogata, K., et al., 1957.

2689. **Sweetman, H. L., Wells, L. F., Jr., Toczydlowski, A. H.** and **Moore, S. III. 1951** Mange and Fly Control with Lindane in a Piggery. J. Econ. Ent. 44(1):112.
Insecticide was vaporized continuously. No detrimental effects on pigs. ABSTRACT: RAE-B 39:139, Aug., 1951.

Sweetman, H. L., joint author. *See* Toczydlowski, A. H., et al., 1951; Spear, P. J., et al., 1952.

Swift, J. E., joint author. *See* Bohart, R. M., et al., 1958.

2690. Sychevskaya, V. I. 1954 [The Displacement of the temperature limits of activity of synantropic species of the genus *Fannia* R.-D. in seasonal and diurnal aspects.] Zool. Zhur. 33(3):637-643. (In Russian.)

2691. Sychevskaya, V. I. 1956 [Biology and ecology of synantropic *Fannia* in the environment of Samarkand. Report 1.] Tr. Uzbekist. Inst. Maliarii i Med. Parazitol. for 1956 (2):293-306.
Discusses species composition, comparative numbers of *Fannia* species and the quantitative relationship of different species at different stages during March-November. ABSTRACT: BA 35:2918(33186), June 15, 1960.

2692. Sychevskaya, V. I. 1957 [On the seasonal occurrence of a number of synantropic flies in various zones of Uzbekistan.] Zool. Zhur. 36(5): 719-728.
Housefly is one of the most abundant species in all areas; mountain, foothill, and desert. ABSTRACT: BA 35:4679(54361), Oct. 1, 1960.

2693. Sychevskaya, V. I. 1957 [Synanthropic flies in the environs of Belovodsko (northern Kirgizia).] Ent. Obozrenie 36(1):108-115.
Fifty-two species were found; *Musca domestica* is the most common. Duration of development is given for 33 species. ABSTRACT: BA 35: 2239(25473), May, 1960.

2694. Sychevskaya, V. I., Grudtsina, M. V. and **Vyrvikhovst, L. A. 1959** [The epidemiological significance of synanthropic flies in Bukhara.] Ent. Obozrenie 38(3):568-578.
Seasonal peak of dysentery associated with great abundance of *M. domestica vicina* Macq. (But control of all common synanthropic flies is held important.) ABSTRACT: BA 35:5239(61203), Nov. 1, 1960.

2695. Sychevskaya, V. I. and **Petrova, T. A. 1958** [The role of flies in spreading the eggs of helminths in Uzbekistan.] Zool. Zhur. 37(4):563-569. (In Russian, with English summary.) Referat. Zhur. Biol. 1959, No. 52590.
Among the flies caught in market places, *M. domestica vicina* was found to be contaminated with *Taenia* onchospheres. ABSTRACT: BA 45:310, Jan. 1, 1964 (Transl. of Summary).

Taguzov, T. V., joint author. *See* Morozova, V. P., et al., 1957

2696. Tahori, A. S. 1955 Diaryl-trifluoromethyl-carbinols as Synergists for DDT against DDT-resistant Houseflies. J. Econ. Ent. 48:638-642.
Tested 29 compounds. Bis-(p-chlorophenyl)-trifluoromethylcarbinol equal to the best synergists. Activity reduced by replacement, elimination, or change of configuration. ABSTRACTS: BA 30:2409(24178), Aug., 1956; RAE-B 45:1, Jan., 1957; TDB 53:677, May, 1956.

2697. Tahori, A. S., Cohen, S. and **Kaluszyner, A. 1958** DDT-Analogs as Synergists for DDT. Experientia (Birkhäuser Verlag, Basel, Schwiez.) 14:25-29.
Treats of analogs containing chlorine and/or fluorine in the methyl group and importance of the hydroxyl group in these compounds.

2698. Tahori, A. S. and **Hoskins, W. M. 1953** The Absorption, Distribution and Metabolism of DDT in DDT-resistant House Flies. J. Econ. Ent. 46:302-306, and 829-937.
Notes effect of temperature on absorption, detoxification and mortality. Discusses relative importance of the processes controlling resistance in the strains studied. ABSTRACTS: RAE-B 42:73, May, 1954; RAE-B 41: 176, Nov., 1953; TDB 51:653, June, 1954.

Tahori, A. S., joint author. *See* Cohen, S., et al., 1957; Reuter, S., et al., 1956; Zahavi, M., et al., 1964, 1965, 1968.

2699. **Tálice, R. V. 1952** Estudios sobre la biologia de las moscas domésticas; influencia de diversos factores sobre la captura mediante trampas con cebo de puesta. [Studies on the biology of domestic flies; influence of diverse factors on the capture by traps with place bait.] Anales Facult. Med. Montevideo 37:423-433.

Fish bait most attractive, but manures gave best catches of *M. domestica*. Gray and white traps attactive, blue traps least so. ABSTRACT: TDB 51: 319, Mar., 1954.

2700. **Tanada, Y., Holdaway, F. G.** and **Quisenberry, J. H. 1950** DDT to Control Flies Breeding in Poultry Manure. J. Econ. Ent. 43:30-36.

M. domestica predominant in Univ. of Hawaii chicken farm. Water emulsion spray most effective, suspension spray less so. Dust unsatisfactory. ABSTRACT: RAE-B 38:184, 1950.

Taunay, A., joint author. *See* Coutinho, J. O., et al., 1957.

2701. **Teotia, T. P. S.** and **Dahm, P. A. 1951** The Effect of Temperature, Humidity, and Weathering on the Residual Toxicities to the House Fly of Five Organic Insecticides. J. Econ. Ent. 43(6):864-876.

Concerns aldrin, chlordane, dieldrin, lindane and parathion. Average median lethal dosage values and slope of regression lines obtained by exposing three-day-old adult houseflies to residues for 48 hours. ABSTRACTS: RAE-B 39:122, July, 1951; TDB 48:762, Aug., 1951.

Terhaar, C. J., joint author. *See* Knapp, J. W., et al., 1958.

2702. **Terriere, L. C.** and **Schonbrod, R. D. 1955** The Excretion of a Radioactive Metabolite by House Flies Treated with Carbon^{14}. J. Econ. Ent. 48(6):736-739.

Given sublethal doses, flies excrete up to 88 per cent of the dose in the form of a water-soluble conjugate. Begins first day after treatment, and continues until absorbed dose is metabolized. ABSTRACTS: RAE-B 45:5, Jan., 1957; TDB 53:816, June, 1956.

Terriere, L. C., joint author. *See* Sacklin, J. A., et al., 1955; Roth, A. R., et al., 1953.

2702a. **Teschner, D. 1959** Hausfliegen als Fäkaliensbesucher im Stadtgebiet. [Houseflies as frequenters of fecal matter in a town district.] Zeit. Angew. Zool. (Berlin) 46:358-363.

Of the principal species taken, *M. domestica* constituted 60·4 per cent. Tables gives seasonal distribution, by months, for 22 species, as taken in the open and in dwellings.

Tharumarajah, K., joint author. *See* Chow, C. Y., et al., 1954.

Thevasgayam, E. S., joint author. *See* Chow, C. Y., et al., 1953, 1954.

2703. **Thevenard, P. 1953** Emploi de la radio-cinématographie pour la recherche biologique, et application de cette méthode â l'étude de la métamorphose chez la mouche. [Employment of radio-cinematography for biological research, and the application of this method to the study of metamorphosis in the fly.] Compt. Rend. Acad. Sci. 237:1791-1793.

Thomas, J. M., joint author. *See* Schechter, M. S., et al., 1951.

Thompson, B. W., joint author. *See* Burnett, G. F., et al., 1956.

2704. **Thompson, R. K., Whipp, A. A., Davis, D. L.** and **Batte, E. G. 1953** Fly Control with a New Bait Application Method. J. Econ. Ent. 46:404-409.

Bait contains 0·15 per cent lindane, 0·06 per cent TEPP. Applied with sprinkling can. ABSTRACTS: RAE-B 42:3, Jan., 1954; TDB 50:1090, Nov., 1953.

2705. **Thomsen, M. 1951** Weismann's Ring and Related Organs in Larvae of Diptera. Dan. Biol. Skr. 6(5):32 pp., xiv pls.
A detailed description of the anatomy and cytology of the ring gland in third stage larva of *Calliphora erythrocephala*. Special emphasis on R-cells. ABSTRACT: BA 25:3382, Dec., 1951.

2706. **Thomson, A. G. 1953** Recent British Investigations Show Promise of Improved Fly Control. Food in Canada 13(12):22, 26, 28.
Stresses good sanitary practice and community cooperation. ABSTRACT: BA 28:1936, Aug., 1954.

Thorn, G. D., joint author. *See* O'Brien, R. D., et al., 1958.

2707. **Tiel, N. van. 1952** Improvement of the Residual Toxicity of DDT Solutions by the Addition of Coumarone Resin. Bull. Ent. Res. 43:413-419.
Additive reduces crystal size and increases toxicity. ABSTRACTS: BA 27: 1933, July, 1953; RAE-B 40:148, Sept., 1952; TDB 49:1085, Nov., 1952.

Timoshkov, V. V., joint author. *See* Derbeneva-Ukhova, V. P., et al., 1959.

2708. **Tischler, W. 1950** Biozönotische Untersuchungen bei Hausfliegen. [Bio-ecological investigations with houseflies.] Zeit. Angew. Ent. 32(2): 195-207.
Six species of large flies are dominant in cities; their control may be affected by spraying lamps and windows with proprietary formulations of DDT and its flourine analog. ABSTRACT: BA 25:3112, Nov., 1951.

2709. **Tishchenko, T. A. 1958** [Survival of dysentery bacilli in *Musca domestica*.] Tr. Khar'kovsk. Med. Inst. 40:113-117. Referat. Zhur., Biol., 1959, No. 83034. Courtesy NSF(PL83-480), 1963.
First experiment: flies infected by feeding on suspension of dysentery bacilli in sweetened saline. Second experiment: flies fed on contaminated human feces. ABSTRACT: BA 45:3225(39942), May 1, 1964.

Toczydlowski, A. H., joint author. *See* Moore, S., et al., 1951; Sweetman, H. L., et al., 1951.

Todd, A. C., joint author. *See* Kohls, R. E., et al., 1958.

Tokmacheva, S. S., joint author. *See* Polovodova, U. P., et al., 1956.

Tomescu, E., joint author. *See* Duca, M., et al., 1958.

Tomlinson, J. R., joint author. *See* Miller, A. C., et al., 1952.

Tower, B. A., joint author. *See* Burns, E. C., et al., 1959.

Travaglino, A., joint author. *See* Milani, R., et al., 1957, 1959.

2710. **Travis, B. V.** and **Smith, A. L. 1951** Tests with Repellents against the Stable Fly. J. Amer. Vet. Med. Ass. 119:214-216.
Indalone and 3 other compounds effective. Reference included here because *Stomoxys calcitrans* is also known as the "biting house fly". (Is so designated in this paper.) ABSTRACT: TDB 49:201, Feb., 1952.

Treboux, J., joint author. *See* Kocher, C., et al., 1953, 1957.

Tsao, C. C., joint author. *See* Hornstein, I., et al., 1955, 1954.

2711. **Tsao, C.-H., Hornstein, I.** and **Sullivan, W. N. 1954** The Joint Action of Chlorinated Terphenyl with Lindane and with Allethrin. J. Econ. Ent. 47:796-798.
Lindane mixture (at strength used) was only 0·8 as toxic as lindane alone. Allethrin mixture (at strength used) was 1·3 to 1·4 times as toxic as allethrin alone. ABSTRACTS: BA 29:2006, Aug., 1955; RAE-B 43:154, Oct., 1955; TDB 52:304, Mar., 1955.

2712. **Tsao, C.-H., Sullivan, W. N.** and **Hornstein, I. 1953** A Comparison of Evaporation Rates and Toxicity to House Flies of Lindane and

Lindane-Chlorinated Polyphenyl Deposits. J. Econ. Ent. 46:882-894.

Chlorinated polyphenyl cuts down rate of lindane evaporation and prolongs its residual toxicity. (May have a synergistic effect also.) ABSTRACT BA 28:1684, July, 1954; RAE-B 42-74, May, 1954; TDB 51:438, Apr., 1954.

Tsao, C.-H., joint author. *See* Sullivan, W. N., et al., 1955.

2713. **Tsintsadze, G. G., Shnaider, E. V.** and **Vashkove, V. I. 1959** [Comparative evaluation of the insecticidal properties of methoxychlor and chlorophos aerosols.] Zhur. Mikrobiol., Epidemiol., i Immunobiol. (Moskva) 30(6):60-65.

Chlorophos aerosols insecticidal; flies destroyed within 60 min. at 0·1 g. activating substance per m^3. Methoxychlor aerosols considerably weaker. ABSTRACT: BA 35(23):5796(68602), Dec., 1960.

2714. **Tsukamotor, M., Ogaki, M.** and **Kobayashi, H. 1957** Malaria Control and the Development of DDT-Resistant Insects in Hikone City, Japan. Publ. by Inst. Insect Control, Kyoto Univ. for WHO. English Edition of paper published in Japan. J. Sanit. Zool. 8:118-122.

Pictures 3 mutant wing venations among progeny of DDT-resistant houseflies.

2715. **Turner, E. C., Jr.** and **Hargett, L. T. 1958** The Effect of Residual Barn Sprays on Control of Horn Flies. J. Econ. Ent. 51(4):559-560.

Diazinon, used primarily for housefly control, also controls hornflies. Residual applications to quarters eliminates direct treatment of dairy animals. ABSTRACT: RAE-B 47:170-171, Nov., 1959.

2716. **Ueno, H. 1959** Autoecological Investigations on the Common Housefly, *Musca domestica vicina,* Survived from the Insecticidal Treatment at the Larval Stage. Analysis of Ecological Factors in Biological Assay of Insecticide. III. Botyu-Kagaku 25:54. (English résumé.)

2717. **U.S. Dept. Agric. Yearbook. 1952** Insects. Washington, D.C.

Three articles deal with resistance to insecticides, particularly on the part of *M. domestica.* ABSTRACT: RAE-B 41:197, Dec., 1953.

2718. **U.S. Public Health Service. 1959** Report from the Communicable Disease Center. Public Health Pesticides for Mosquitoes, Flies, Fleas, Roaches, Bed bugs, Ticks, Rodents. Reprinted from *Pest Control,* March. 11 pp., 1 fig.

M. domestica and other Diptera reported on at length. Concerns resistance to malathion and parathion, including behavioristic resistance to malathion-molasses baits. ABSTRACT: TDB 56:991, Sept., 1959.

2719. **Van Asperen, K. 1958** The Mode of Action of an Organophosphorus Insecticide (DDVP). Some Experiments and a Theoretical Discussion. Ent. Exp. Appl. 1(2):130-137.

Cholinesterase activity studied *in vitro.* Toxic action of DDVP felt to be related to its inhibition of the action of cholinesterase. ABSTRACT: RAE-B 47:132, Sept., 1959.

2720. **Van Asperen, K. 1958** Mode of Action of Organophosphorus Insecticides. Nature 181:355-356.

Parathion and paraoxon inhibit ali-esterase activity (splitting methyl butyrate). Perhaps more important than cholinesterase. ABSTRACT: TDB 55:704, June, 1958.

2721. Van Asperen, K. 1959 Distribution and Substrate Specificity of Esterases in the Housefly, *Musca domestica* L. J. Insect Physiol. (London) 3(3):306-322.
Housefly homogenates perform high cholinesterase and ali-esterase activity, first in head, later in thorax and abdomen. Evidence of a third enzyme concerned. ABSTRACT: TDB 57:424, April, 1960.

2722. Van Asperen, K. and **Oppenoorth, F. J. 1959** Organophosphate Resistance and Esterase Activity in Houseflies. Ent. Exp. Appl. 2(1):48-57.
Ali-esterase activity lower in resistant strains, whereas cholinesterase activity is the same in resistant and susceptible flies. ABSTRACTS: RAE-B 48:161, Sept., 1960; TDB 56:778, July, 1959.

Vanderberg, J. P., joint author. *See* Cheng, T.-H., et al., 1957.

2723. Vanderplank, F. L. 1950 Air-Speed/Wing-Tip Speed Ratios of Insect Flight. [Correspondence.] Nature. May. pp. 806-807.
Work done on *Glossina palpalis*. Used stroboscope and flash photographs. Wings go through 120 cycles per sec. Maximum speed, 16 miles per hour. Contribution to muscoid biology. ABSTRACT: TDB 47:820, 1950.

2724. Vanskaia, R. A. 1957 [Number of *Musca domestica* L. as index of effectiveness of sanitary conditions and of sanitary preventive measures in large cities.] Gig. I. Sanit. (Moskva) 22(6):85-86. (In Russian.)

2725. Vanskaia, R. A. 1958 [Count of the most important species of synanthropic flies in Moscow in 1954.] Med. Parazit. (Moskva) 27(1): 101-103. (In Russian.)

Van Tiel, N. *See* Tiel, N. van, 1952.

2726. Van Vuren, J. P. J. 1950 Soil Fertility and Sewage. Publ. by Faber and Faber.
Numerous references to fly-breeding in connection with composting practices in South Africa.

2727. Vargas, L. 1959 Nota Sobre La Identificacion De Los Generos Mas Comunes De Moscas Caseras Mexicanas. [Note on the identification of the most common species of Mexican houseflies.] Rev. Inst. Salubr., Enferm. Trop. 19(2):173-176.
Two keys to Mexican houseflies, one graphic. ABSTRACT: BA 41:307 (4075), Jan. 1, 1963.

2728. Varzandeh, M., Bruce, W. N. and **Decker, G. C. 1954** Resistance to Insecticides as a Factor Influencing the Biotic Potential of the House Fly. J. Econ. Ent. 47:129-134.
Inheritance of factors associated with vigor (such as egg production, weight, longevity, hatchibility and survival) were found independent of factors associated with resistance. ABSTRACTS: BA 28:2700, Nov., 1954; RAE-B 42:189, Dec., 1954; TDB 51:1311, Dec., 1954.

2729. Vashchinskaia, N. V. 1956 [Phenology of two subspecies of domestic flies in the Armenian SSR.] Med. Parazit. (Moskva) 25(1):40-42. Referat. Zhur., Biol., 1956, No. 64051. (In Russian.)

2730. Vashchinskaya, N. V. 1957 [On synanthropic flies in the Armenian S.S.R.] Med. Parazit. i Parazitarn. (Bolezni) 26(4):463-470.
Dates given on seasonal occurrence of different species. Analysis of connection between disease rate of acute intestinal infections and coincident numbers of houseflies in different climatic zones. ABSTRACT: BA 35:2476(28056), May, 1960.

2731. **Vashkov, V. I., Fedder, M. L., Klechetova, A. M., Erofeeva, T. V.** and **Khudadov, G. D. 1958** [The resistance of domestic flies to DDT and hexachlorocyclohexane.] Gig. i Sanit. 23(4):28-32. Referat. Zhur., Biol., 1959, No. 52602. (In Russian, with English summary.)
Studied flies from Yalta where these 2 insecticides had been used 4 years, and from an area where they had not been used. No marked difference in resistance was noted. ABSTRACT: BA 44:952(12827), Nov. 1, 1963.

Vashkov, V. I., joint author. *See* Tsintsadze, G. G., et al., 1959.

Venables, D. G., joint author. *See* Kerr, R. W., et al., 1957.

2732. **Vendramini, R. 1955** La lotta contro le mosche. Nuovi orientamenti e nuovi aspetti. [The campaign against the fly. New outlook and fresh aspects.] Igiene e San. Pubbl. (Rome). 11(5-6):292-307 and (7-8):360-381. May-June and July-August. (English summary.)
Author calls for new anti-fly campaigns based on old and new methods. ABSTRACT: TDB 53:817, June, 1956.

2733. **Veselinov, V.** and **Gubev, E. 1954 (1956)** [Synanthropic flies as an epidemiologic factor in intestinal infections.] Nauchni Trudove na Visshiia Medit. Inst. "Vulko Chervenkov" (Sofia) 2(4):29-44.

Vinson, E. B., joint author. *See* Sternburg, J., et al., 1953.

2734. **Vishveshwaraiya, S.** and **Ranade, D. R. 1951** A Preliminary Account of Cytology of the House-fly *Musca nebulo* F. Current Science 20: 333. December.
In an inbred race of *M. nebulo*, male flies showed haploid condition (6 chromosomes) while females showed 12 (6 homologous pairs).

2735. **Vittal, M. 1958** Experiments on the Use of Insecticides in the Control of Fly Nuisance. Indian J. Ent. 20:1-6.
Organophosphorus insecticides had longer period of effectiveness than hydrocarbons. Worked with *M. domestica nebulo* Fabr. and *M. domestica vicina* Marq. ABSTRACTS: BA 35:515(6009), Jan., 1960; RAE-B 48:44, Feb., 1960.

2736. **Vockeroth, J. R. 1953** *Musca autumnalis* Deg. in North America. (Diptera: Muscidae.) Canad. Ent. 85:422-423.
Claims to be first report from North America. Includes key for separating *M. autumnalis* and *M. domestica*. ABSTRACTS: BA 28:1696, July, 1954; RAE-B 43:55, Apr., 1955.

Von Hofstein, C. G., joint author. *See* Davies, M., et al., 1958.

2737. **Vostál, Z.** and **Horváth, J. 1958** [Determination of DDT Resistance in Flies by a Test in Petri Dishes.] Ceskoslov. Hyg. (Prague) 3(7): 416:420. (English summary.)
Assay method used in determining resistance of the progeny of flies from 4 localities where annual spraying with DDT had been carried out. Values of 1·46 and 4·00 were obtained for resistance index, based on LT 50 of test and standard strains. ABSTRACT: TDB 56:893, Aug., 1959.

2738. **Vranchan, Z. E. 1959** [Effect of DDT and hexachlorane aerosols on house flies. Thesis of a report.] Tr. Vses. Nauchn.—Issled. Inst. Vet. Sanit. 14:8-9; Referat. Zhur., Biol., 1960, No. 50965. — — Courtesy NSF(PL 83-480), 1963.

Vuslaev, M. A., joint author. *See* Derbenheva-Ukhova, V. P., et al., 1953.

Vyrvikhovst, L. A., joint author. *See* Sychevskaya, V. I., et al., 1959.

Wagner-Jauregg, T., joint author. *See* Dethier, V. G., et al., 1952; Gustafson, C. G., Jr., et al., 1953.

2739. **Waldrop, R. H., Peel, R.** and **Schoof, H. F. 1958** Fly Reduction through Sanitary Improvements. Fly Movement Study Shows Benefits of Control Program. Reprinted from Modern Sanitation and Building Maintenance 10(9):3.
Flies tend to migrate toward filth. Urges sanitary effort at local community level. ABSTRACT: TDB 58:961, Aug., 1961.

2740. **Wall, W. J.** and **Bailey, S. F. 1952** Repellency of Cotton Tufts to Flies. J. Econ. Ent. 45:749-750.
Suspended ball repelled a significant number of flies when it was swinging, but not when it was stationary. ABSTRACT: RAE-B 41-47, Mar., 1953.

Wancheng, H., joint author. *See* Shih, S., et al., 1952.

2741. **Wang, K. C. 1959** [Report on the survey of flies in Foochow area.] Foochow Health Sta. China; Acta Ent. Sinica. 9(6):555-563. (English summary.)
Detailed investigations attempted concerning larval breeding places and habitats of adults. Possibility of *M. d. vicina* spreading dysentery. ABSTRACT: BA 35:5514(64891), Nov., 1960.

Wang, M. Y., joint author. *See* Chang, J. T., et al., 1958.

2742. **Ward, R. 1952** Poliomyelitis and Coxsackie Viruses from Egyptian Flies. Fed. Proc. 11(1):486. An Abstract.
Flies collected in or near Cairo. One yielded polio virus (in monkey), nine yielded Coxsackie-like viruses (in new born mice). ABSTRACT: BA 26:2549, Oct., 1952.

2743. **Ware, G. W.** and **Roan, C. C. 1957** The Interaction of Piperonyl Butoxide with Malathion and Five Analogs Applied Topically to Male House Flies. J. Econ. Ent. 50:825-827.
Type of chemical structure determines type of activity. Presence of phosphorus indicates *synergism*; of thiophosphoryl, *antagonism.*

2744. **Ware, G. W., Jr. 1957** The Relation of the Structure of a Series of Cholinesterase Inhibiting Insecticides to Synergism and Antagonism with Piperonyl Butoxide on the Housefly. *Musca domestica* L. Diss. Absts. 17(3):704-705.

2745. **Watanbe, S. 1959** [Studies on the oral organs of the flies of medical importance as to pathogenic microorganisms ingested.] Osaka City Med. Center 8(7):869-881. (English summary.)

Watt, J. T., joint author. *See* Lindsay, D. R., et al., 1953.

2746. **Wattal, B. L.** and **Mammen, M. L. 1959** A Preliminary Note on Colour Preference for Oviposition of *Musca domestica nebulo* Fab. Indian J. Malariol. 13(4):185-187.
Nine colors tested using 4-5 day old female flies; 40·4 per cent of all egss were deposited on black. ABSTRACT: BA 40:1926(25285), Dec. 15, 1962.

2747. **Webb, J. E., Jr. 1959** A Study of Insecticide Resistance in House Flies of Germany and France. J. Econ. Ent. 52(3):419.
Field-collected flies colonized in laboratory. All were resistant to DDT; some showed resistance to lindane and dieldrin; none were significantly resistant to malathion. ABSTRACTS: BA 34:258(2880), Sept., 1959; RAE-B 48:164, Sept., 1960.

2748. **Webb, J. E.** and **Graham, H. M. 1956** Observations on Some Filth Flies in the Vicinity of Fort Churchill, Manitoba, Canada, 1953-1954. J. Econ. Ent. 49:595-600.
Twenty-seven species recorded. *M. domestica* and *Calliphora* spp. visited

mess halls. ABSTRACTS:BA 31:933(9343), Mar., 1957; RAE-B 45:172, Nov., 1957; TDB 54:241, Feb., 1957.

Wedberg, S. E., joint author. *See* Hawley, J. E., et al., 1951.

Wei, P. H., joint author. *See* Kao, C. M., et al., 1958.

2749. Weiant, E. A. 1955 Electrophysiological and Behavioral Studies on DDT-sensitive and DDT-resistant House Flies. Ann. Ent. Soc. Am. 48:489-492.
Sensory nerves of resistant flies are less sensitive to the direct action of DDT.

Weidhaas, D. E., joint author. *See* Cole, M. M., et al., 1958.

Weir, J. M., joint author. *See* Gahan, J. B., et al., 1950.

2750. Weiser, J. 1958 [The biological action of some synthetic Pyrethrins.] Vest. Cesk. Spol. Zool. 22(4):353-370.
ABSTRACT: BA 35:5233(61152), Nov. 1, 1960.

2751. Welch, S. F. and **Schoof, H. F. 1953** The Reliability of "Visual Surveys" in Evaluating Fly Densities for Community Control Programs. Am. J. Trop. Med. Hyg. 2(6):1131-1136.
Using 2,850 attractant sites, grill estimates were compared with actual counts. Agreed within an accuracy of 69-89 per cent. ABSTRACTS: BA 28:1687, July, 1954; RAE-B 43:64, Apr., 1955; TDB 51:651, June, 1954.

2752. Wellmann, G. 1952 Experimentelle Untersuchungen über die Möglichkeit der Brucellenübertragung durch Insekten. [Experimental studies on the possibility of transmitting *Brucella* by insects.] Zentralbl. Bakt. Parasit. Infekt. Abt. I. Orig. 159(1/2):71-86. (Russian, French and English summaries.)
Used *M. domestica* to transfer infection to injured and uninjured skin, also eyes, of laboratory animals (guinea pigs, cows). ABSTRACT: BA 28: 699, March, 1954.

2753. Wellmann, G. 1953 Insekten als Brucelloseüberträger-Neue Experimentella Ergebnisse. [Insects as Brucellosis Carriers—Results of New Experiments.] Proceedings of the XVth International Veterinary Congress. Stockholm 1953. pp. 3-9.
Results of experiments are presented in which Bang's Disease was transmitted by *Musca domestica* and *Stomoxys calcitrans* from the placenta (containing *Brucella*) of an infected cow, to pregnant heifers, resulting in abortion or a latent infection in many of them.

2754. Wellmann, G. 1954 Die Übertragung der Schweinerotlaufinfektion durch die Stubenfliege (*Musca domestica*). Zentralbl. Bakt. Parasit. Infekt. I. Orig. 162:261-264.
Musca domestica, having access to an erysipelas-culture in one room, could fly into a stable and were able to transmit a clinical illness to pigs.

2755. Wells, A. L., Stelmach, Z., Guyer, G. E. and **Benne, E. J. 1958** A Study of Malathion for Controlling Flies in Dairy Barns and on Dairy Cattle. Quart. Bull. Mich. Agr. Exp. Sta. 40(4):786-795. East Lansing.
Satisfactory control was obtained with a 1 per cent spray applied as a residual treatment every 3 weeks to barn and cattle. Milk production, body weights and general activities of cattle were not affected. ABSTRACT: RAE-B 48:205, Dec., 1960.

Wells, L. F., Jr., joint author. *See* Sweetman, H. L., et al., 1951.

2756. Wesley, C., Jr. and **Morrill, A. W., Jr. 1956** Effects of Various Insecticide Solutions on Different Kinds of Insects Screens. Mosquito News 16:206-208.
Insecticides used had no deleterious effects on screens (copper, brass, bronze, galvanized and others). ABSTRACT: BA 31:614(6207), Feb., 1957.

2757. West, L. S. 1950 The Status of *Rhynchiodexia* (*Dinera*) *robusta* Curran, Together with a Consideration of Certain Cephalic and other Characters Useful in Muscoid Taxonomy. Papers Mich. Acad. Sci., Arts and Letters. 34(1948):109-117, plus 4 plates. (Published in 1950.)

Not concerned with *M. domestica*, but useful by reason of the illustrations of cephalic and alar characters.

2758. West, L. S. 1951 The Housefly. Its Natural History, Medical Importance and Control. xi + 584 pp., illus. Comstock Publ. Co., Ithaca, N.Y.

Summarizes all important research published up to and including the year 1949. ABSTRACTS: BA 26:699, Mar., 1952; RAE-B 40:20, Jan., 95 TDB 49:735, July, 1952.

2759. West, L. S. 1952 Recent Advances in our Knowledge of Housefly Biology. Proc. 38th Mid-Year Meeting of the Chemical Specialties Manufacturers Association, Boston, Massachusetts, June, 1952, pp. 97-101.

A summary of selected research reports published between 1949 and 1952.

2759a. West, L. S. 1953 Fly Control in the Eastern Mediterranean and Elsewhere. Report of a Survey and Study. W.H.O. Div. of Environ. Sanit. 139 pages. (Mimeographed Document.)

Summarizes the author's findings as Scientist-Consultant to the World Health Organization during a three-month period of full-time service.

2760. Wheeler, C. M., Newson, H. D. and **Blakeslee, T. E. 1958** Insecticide Resistance in Houseflies of Japan, Korea, and the Ryuku Islands. U.S. Armed Forces Med. J. 9(1):68-76.

Median effective dosage (M50) values determined for DDT, lindane, dieldrin, and malathion with 20 strains of flies. All but 2 were highly resistant to DDT. ABSTRACT: TDB 55:832, July, 1958.

Wheeler, C. M., joint author. *See* Byers, G. W., et al., 1956.

Whipp, A. A., joint author. *See* Thompson, R. K., et al., 1953.

2761. Wichmand, H. 1953 Control of Multi-Resistant Houseflies. Nature 172:758-759.

Parathion-treated gauze strips used on nearly 100,000 Danish farms. Remained effective after 6 months in room kept at 20°C. ABSTRACTS: BA 28:1688, July, 1954; TDB 51:322, Mar., 1954.

Wies, T. A., joint author. *See* Gustafson, C. G., Jr., et al., 1953.

2762. Wiesmann, R. 1957 Das Problem der Insektizidresistenz. [The problem of insecticide resistance.] Anz. Schädlingsk. 30(1):2-7.

2763. Wiesmann, R. 1957 Vergleichend histologische Untersuchungen an Normalsensiblen und gegen DDT-Substanz resistenten Stämmen von *Musca domestica* L. [Comparative histological investigations on normally susceptible and DDT-resistant strains of *Musca domestica* L.] J. Insect Physiol. 1(2):187-197.

Differences found in the lipoid and lipoprotein content of the epicuticle and the total cuticle; also the epidermal cells of the tarsi and thorax. ABSTRACT: TDB 55:111, Jan., 1958.

2764. Wiesmann, R. and **Kocher, C. 1951** Untersuchungen über ein neuses, gegen resistente *Musca domestica* L. wirksames Insektizid. [Investigations on a new insecticide effective against resistant *Musca domestica* L.] Zeit. Angew. Ent. 33(1/2):297-321.

Pyrolan (1-phenyl-3-methyl-pyrazolyl-(5)-dimethyl carbamate) found effective against DDT-resistant strains. No selective resistance developed in the laboratory. ABSTRACTS: BA 26:2331, Sept., 1952; RAE-B 41:69, May, 1953; TDB 50:1093, Nov., 1953.

2765. Wigglesworth, V. B. 1953 The Principles of Insect Physiology. 5th Ed. Dutton.

Gives due attention to the physiology of *Musca* and related forms.

Williams, C. M., joint author. *See* Levenbook, L., et al., 1956.

Williams, M., joint author. *See* Babers, F. H., et al., 1953.

2766. Williams, R. W. 1954 A Study of Filth Flies in New York City, 1953. J. Econ. Ent. 47:556-563.

M. domestica was taken only 8 times during July, August, and September and represented only 0·044 per cent of the catch. Housefly is becoming much less common in the *Eastern* U.S.! ABSTRACTS: BA 29:966, Apr., 1955; RAE-B 43:114, Aug., 1955.

2767. Williams, R. W. 1956 Studies on the Filth Flies at the University of Michigan Biological Station, Douglas Lake, Michigan—Summer 1954. Am. Midland Nat. 55:126-130.

At least 48 species captured. *M. domestica* found 61 per cent of the time near the kitchen but only 44 specimens were taken. Not as important a pest as several others. ABSTRACT: BA 31:1536(15581), May, 1957.

2768. Williams, R. W. 1958 A Study of the Summer Filth Fly Population of the Bermuda Islands. J. Parasit. 44(3):339-342.

Decomposing fish, as bait, more attractive to *M. domestica* in Bermuda than in the United States. Same 12 species of filth flies present. ABSTRACT: RAE-B 48:64, Mar., 1960.

2769. Williams, R. W. 1959 The Reported Biting and Filth Frequenting Arthropods of the Bermuda Islands Exclusive of the Ixodoidae and Araneida. Proc. Ent. Soc. Wash. 61(5):234-238.

Lists six species of Muscidae, including *Musca domestica* L. and *Fannia pusio* (Wied).

2770. Wilson, H. B. 1959 Control of Flies on Poultry Farms. J. Agric. (Victoria, Australia) 57(11):701, 703, 722.

Treats of sanitary practices, space sprays, residual insecticides, baits. ABSTRACT: BA 35:5514(64892), Nov. 15, 1960.

2771. Wilson, H. G. and **Gahan, J. B. 1957** Control of House Fly Larvae in Poultry Houses. J. Econ. Ent. 50:613-614.

Used water solution of Dipterex and emulsions of 7 other organic phosphorus compounds. Diazinon dust, 25 per cent, gave over 2 weeks control. Dusts are visible, insuring uniform coverage, and hasten drying of the medium. ABSTRACTS: BA 32:1194(14138), Apr., 1958; RAE-B 46: 169, Nov., 1958.

2772. Wilson, H. G., Gahan, J. B. and **McDuQe, W. C. 1953** Toxicity of Vapors of Four Chlorinated Hydrocarbon Insecticides to Resistant House Flies. J. Econ. Ent. 46:699-700.

Lindane, chlordane, dieldrin and aldrin tested on non-resistant flies, flies resistant only to DDT, and flies resistant to several chlorinated hydrocarbons. Satisfactory mortality required too long an exposure to make vapor method practical. ABSTRACTS: BA 28:950, Apr., 1954; RAE-B 42:41, Mar., 1954.

2773. Wilson, H. G. and **LaBrecque, G. C. 1958** Tests with Organophosphorus Compounds as House Fly Larvicides in Poultry Houses. Florida Ent. 41(1):5-7.

Diazinon, Dipterex, Dow ET-57, Trithion and malathion in kerosene gave good control for about a week. Diazinon was most persistent. ABSTRACT: RAE-B 47:146, Oct., 1959.

2774. **Wilson, H. G.** and **LaBrecque, G. C. 1959** Evaluation of Some Space Sprays for the Control of House Flies. J. Econ. Ent. 52(4):704.
Tests in barns with pyrethrins, 3 organophosphorus compounds, and 4 chrysanthemumates gave 90 per cent control in 10 minutes, but no control at 24 hours. Daily spraying gave only minor reductions. ABSTRACT: RAE-B 48:196, Nov., 1960.

2775. **Wilson, H. G., LaBrecque, G. C., Gahan, J. G.** and **Smith, C. N. 1959** Insect Resistance in Seven Orlando House Fly Colonies. J. Econ. Ent. 52:308-310.
First colony to be resistant to DDT was maintained (270th generation) in cages coated with DDT! Other colonies were variously resistant to lindane, dieldrin, Dipterex, Bayer 21/199, Malathion, Diazinon, parathion. ABSTRACT: RAE-B 48:122, July, 1960.

Wilson, H. G., joint author. *See* Gahan, J. B., et al., 1953, 1954, 1957; Cole, M. M., et al., 1958; Keller, J. C., et al., 1955, 1956.

2776. **Wilson-Jones, K., Davidson, G.** and **Hewlett, P. S. 1958** Dosage-mortality Curves for Houseflies Susceptible and Resistant to DDT. Nature (London) 182(4632):403-404.
Discusses Hewlett's (1958) work with DDT. Is followed by a short note, in reply, bearing signature of P. S. Hewlett. ABSTRACT: RAE-B 49:261, Dec., 1961.

Winfield, G. F., joint author. *See* Meng, C.-H., et al., 1950, 1951.

2777. **Wingo, C. W. 1954** House Fly Control with Diazinon. J. Econ. Ent. 47:632-635.
Preliminary tests indicated good control of larvae, also good knockdown of adults, with small amounts of insecticide. Field tests in heavily-infested area gave control for 35 days. Heat and sunlight reduce knockdown effect and subsequent toxicity. ABSTRACTS: BA 29:966, Apr., 1955; RAE-B 43:117, Aug., 1955; TDB 52:223, Feb., 1955.

2778. **Winteringham, F. P. W. 1956** Labelled Metabolic Pools for Studying Quantitatively the Biochemistry of Toxic Action. Int. J. Appl. Radiat. Isotopes 1:57-65.
Houseflies fed carrier-free $^{32}PO_4^{3-}$ Entire pool of phosphorylated intermediates of glycolysis, muscle contraction, etc., becomes uniformly labelled. Such intermediates may be extracted, separated on paper chromatograms, and determined by radiometric scanning. ABSTRACT: RAE-B 45:138, Sept., 1957.

2779. **Winteringham, F. P. W., Bridges, P. M.** and **Hellyer, G. C. 1955** Mode of Insecticidal Action Studied with Labelled Systems. Phosphorylated Compounds in the Muscle of the Adult Housefly, *Musca domestica* L. Biochem. J. 59(1):13-21.
Major fractions separated are tentatively identified as ATP, ADP, AMP, glucose 6-phosphate, orginine phosphoric acid, phosphoglycerate and free orthophosphate. ABSTRACTS: TDB 52:413, Apr., 1955; RAE-B 44: 185, Nov., 1956.

2780. **Winteringham, F. P. W.** and **Harrison, A. 1956** Study of Anticholinesterase Action in Insects by a Labelled Pool Technique. Nature 178(4524):81-83.
By injection of 2-^{14}C-acetate into adult houseflies, labelling was extended to non-phosphorus-containing metabolites. There is a temporary rise in acetylcholine in head and thoracic tissues, when fly is exposed to a phosphorus insecticide.

2781. Winteringham, F. P. W. and **Harrison, A. 1959** Mechanisms of Resistance of Adult Houseflies to the Insecticide Dieldrin. Nature 184 (4684):608-610.
Experiments suggest that this resistance is *not* due to lack of cuticular penetration, or to a gross difference in rates of excretion or metabolism of the insecticide. ABSTRACTS: BA 35:1034-1035(11751), Feb., 1960; TDB 57:306, Mar., 1960.

2782. Winteringham, F. P. W., Harrison, A. and **Bridges, P. M. 1955** Absorption and Metabolism of [^{14}C] Pyrethroids by the Adult Housefly *Musca domestica* L., in Vivo. Biochem. J. 61:359-367.
Natural mixture of pyrethroids, labelled biosynthetically with ^{14}C was resolved by means of reversed phase paper chromatography into chrysanthemic esters, pyrethric esters and unidentified, non-insecticidal impurities. ABSTRACT: RAE-B 44:186, Nov., 1956.

2783. Winteringham, F. P. W. and **Hellyer, G. C. 1954** Effects of Methylbromide, Ethylenedibromide, and Ethylenedichloride on the Phosphorus Metabolism of *Musca domestica* L. Proc. Biochem. Soc. Lond. 333rd Meeting, 15-16 October. Issued with Biochem. J. 58(4), pp. xlv-xlvi.
Spectacular depletion of ATP and Arg-P by MB suggests a rapid blocking of the phosphorylation of nucleotide acceptors, these possibly then being lost to alternative metabolic pathways.

2784. Winteringham, F. P. W., Loveday, P. M. and **Harrison, A. 1951** Resistance of Houseflies to DDT. Nature 167(4238):106-107. London.
Metabolism of insecticide in resistant flies probably enzymic in nature. Presence of piperonyl cyclonene believed to completely inhibit dehydrohalogenation. ABSTRACT: TDB 48:501-502, May, 1951.

2785. Winteringham, F. P. W., Loveday, P. M. and **Hellyer, G. C. 1953** Phosphorus Metabolism in the Housefly *Musca domestica*. Proc. Biochem. Soc. Lond. 321st Meeting, September. Issued with Biochem. J. 55(5), pp. xxxiii-xxxiv.
Autoradiographs have demonstrated a high concentration of labelled material in the gut wall, of composition the same as that found in muscle.

Winton, M. Y., joint author. *See* Fukuto, T. R., et al., 1959.

2786. Wolf, H. W. 1955 Housefly Breeding in Sewage Sludge. Sewage and Indust. Wastes 27(2):172-176.
Bioassays indicate that pH, total solids, volatile solids and moisture content have no effect on housefly numbers. Some relation between percentage of volatile solids in sewage content, and pupal size. ABSTRACT: BA 30:310, Jan., 1956.

Wolfe, L. S., joint author. *See* O'Brien, R. D., et al., 1959.

2787. Wolfinsohn, M. 1951 The Influence of the Water Content of the Medium on the Accumulation of Larvae and Pupae of the House-Fly. Bull. Res. Counc. of Israel 1:166-168.
Housefly larvae accumulated, before pupation, in the drier portion of a mixture of bran, straw and water in a sloping container. ABSTRACTS: RAE-B 41:199, Dec., 1953; TDB 51:651, June, 1954.

Wolken, J. J., joint author. *See* Bowness, J. M., et al., 1959.

2788. World Health Organization. 1952 Technical Report Ser. No. 54. Expert Committee on Insecticides. Fourth Report. Geneva, November 28-December 4, 1951. 99 pp., 12 figs.
Most specifications suggested at earlier meetings revised. Some progress made with definition and standardization of aerosols. ABSTRACT: TDB 50:577, June, 1953.

2788a. World Health Organization. 1953 Bull. W.H.O. (Geneva) 8(4):527-533. 14 refs. Techniques and Materials for the Disinsectization of Aircraft. (Prepared by Staff of Communicable Disease Center, Public Health Service, U.S. Dept. of Health, Educ. and Welfare. Savannah, Georgia.)
Treatments adequate to kill mosquitoes in a passenger plane destroy only a small proportion of houseflies. Higher doses might annoy passengers and crew. ABSTRACT: TDB 50:1100, Nov., 1953.

Wright, E. A., joint author. *See* Pavillard, E. R., et al., 1957.

Wright, J. W., joint author. *See* Quarterman, K. D., et al., 1956.

2789. Wylie, H. G. 1958 Factors that Affect Host Finding by *Nasonia vitripennis* (Walk.) (Hymenoptera:Pteromalidae). Canad. Ent. 90:597-608.
Experimental hosts: *M. domestica, Calliphora vicina* Desv. [=*erythrocephala* (Meig.)] and *Polietes lardaria* (F.). Lengthy summary of results. ABSTRACT: RAE-B 48:43, Feb., 1960.

2790. Yates, W. W. 1951 Ammonium Carbonate to Attract House Flies. J. Econ. Ent. 44:1004-1006.
Concentration of 20 per cent in baits attracts more females than males, more flies over 5 days old than under that age. Wet mixtures, also aged mixtures with some decomposing protein attract best. Ammonium carbonate decomposes rapidly above 80°F. ABSTRACTS: RAE-B 40:61, Apr., 1952; BA 26:2074, Aug., 1952.

2791. Yates, W. W. and **Lindquist, A. W. 1950** Exposure of House Flies to Residues of Certain Chemicals before Exposure to Residues of Pyrethrum. J. Econ. Ent. 43:653-655.
Tested 51 chemicals for synergistic action. Piperonyl butoxide, N-isobutyl-undecylenamide, and sesamin concentrate showed greatest synergistic properties. ABSTRACT: RAE-B 39:82, May, 1951.

2792. Yates, W. W., Lindquist, A. W. and **Butts, J. S. 1952** Further Studies of Dispersion of Flies Tagged with Radioactive Phosphoric Acid. J. Econ. Ent. 45:547-548.
Flies were fed radioactive phosphoric acid (20 microcuries per 1,000 insects). Results given in table. ABSTRACTS: RAE-B 40:177, Nov., 1952; TDB 49:1083, Nov., 1952.

Yates, W. W., joint author. *See* Lindquist, A. W., et al., 1951.

Yeomans, A. H., joint author. *See* Fulton, R. A., et al., 1963; Schechter, M. S., et al., 1960; Sullivan, W. N., et al., 1960; Hornstein, I., et al., 1954.

Young, R. D., joint author. *See* Furman, D. P., et al., 1959.

2793. Zagnebel'nyli, L. F. 1957 [Development of resistance in house flies to DDT preparations.] Med. Parazit. (Moskva) 26(1):31-33.

2794. Zaharis, J. L. and **Roan, C. C. 1959** A Unique Culture Difference Between two Strains of House-flies. J. Econ. Ent. 52(3):542.
Watering pads in fly cages develop a violet color under laboratory conditions (Resistant CAL strain). With a certain insecticide-susceptible strain, no violet color appears, but pads become butter-yellow after 4 days exposure to flies. ABSTRACT: BA 34:258(2881), Sept., 1959.

2795. Zhovtuii, I. F. 1951 The Number of Generations and the Duration of the Cycle of Development of *Musca domestica* L. under Conditions of Baraba (Western Siberia). Dokl. Akad. Nauk SSSR (N.S.) 80(3): 477-480. Moscow. (In Russian.)
These varied with temperature and type of breeding medium. ABSTRACTS: RAE-B 41:199, Dec., 1953; TDB 51:650, June, 1954.

2796. **Zhovtyi, I. F. 1954** [Relation of seasonal rises of morbidity of dysentery to phenology of *Musca domestica* L.]. Med. Parazit. (Moskva) 1:43-45. (In Russian.)

Seasonal incidence of dysentery correlated with numbers of flies. ABSTRACT: RAE-B 43:108, July, 1955.

Zhukova, L. I., joint author. *See* Nabokov, V. A., et al., 1956.

Zimmerli, A., joint author. *See* Schechter, M. S., et al., 1951.

2797. **Zinchenko, V. S.** and **Nestervodskaia, E. M. 1956** [Problem of the role of fly factor in seasonal outbreaks of dysentery.] Zhur. Mikrobiol., Epidemiol., i Immunobiol. (Moskva) 27(10):33. (In Russian.)

2798. **Zingrone, L. D., Bruce, W. N.** and **Decker, G. C. 1959** A Mating Study of the Female House Fly. J. Econ. Ent. 52(2):236.

Of 100 females that produced offspring, only 2 gave evidence (eye color of offspring) of mating *twice*. One fertilization normally lasts for the entire egg-laying period. ABSTRACTS: BA 33:3224(39257), July, 1959; RAE-B 48:120, July, 1960.

Zlalkooskaia, E. V., joint author. *See* Sukhova, M. N., et al., 1959.

2799. **Zolotarev, E. K. 1957** [New substances toxic to house flies.] Vestn. Mosk. Univ. Ser. Biol. Pochvoved., Geol., Geogr. [for] 1957 (1): 141-146.

Four intramolecular metallocyclic compounds caused 100 per cent destruction of flies in 24 hours in doses of 11 mg/m^2. ABSTRACT: BA 35: 2232(25393), May, 1960.

2800. **Zolotarev, E. Kh.** and **Lineva, V. A. 1957** [A substance for poisoning DDT-resistant flies.] Vestn. Mosk. Univ. Ser. Biol. Pochvoved., Geol., Geogr. [for] 1957 (1):147-152.

MGU-22, a substance with an intramolecular metallocyclical structure, is highly toxic to DDT-resistant houseflies. LD_{50} was one-tenth that of DDT. ABSTRACT: BA 35:2233(25394), May, 1960.

Section VI.
References from 1960 to 1969 Inclusive

Although this bibliography makes no pretense of completeness beyond the year 1969, a number of papers published in 1970, some of them quite important, had come to hand in time to be included in the manuscript, and appear in the list which follows. In the course of publication, it has also been possible to insert a few additional items of subsequent dates which, in the opinion of the authors contribute substantially to the usefulness of the present work.

2801. Abasa, R. O. 1967 An Apholate-resistant Strain of House Fly (*Musca domestica*), its Ovarian Development, Oögenesis and Resistance to other Chemosterilsants and to Insecticides. Diss. Absts. 28(5B): 1972-B.

First resistance appeared after 15th generation. At 20th generation, flies had greater oviposition capacity and more ovarioles, a low cross-resistance to metepa and a decreased resistance to lindane, diazinon, dimethoate and dimetilan.

2802. Abasa, R. O. 1968 An Apholate-Resistant Strain of the House Fly, *Musca domestica*. II. Ovarian Growth and Oögenesis. Ann. Ent. Soc. Am. 61(6):1351-1354.

Second portion of technical report based on thesis work cited above. ABSTRACTS: BA 50:3709(38933), Apr. 1, 1969; RAE-B 57:151(547), Aug., 1969; TDB 66:360(708), Apr., 1969.

2803. Abasa, R. O. and **Hansens, E. J. 1969** An Apholate-Resistant Strain of House Flies. I. Resistance to Other Chemosterilants and Insecticides. J. Econ. Ent. 62(2):334-338.

First portion of technical report based on thesis work cited above. Reduction of resistance to lindane and related insecticides after selection with apholate, due possibly to *cessation* of selection with these insecticides. ABSTRACTS: RAE-B 57:183(655), Sept., 1969; TDB 66:1338(2534), Dec., 1969.

2804. Abdallah, M. D. 1963 Interaction of Some Organophosphorus Compounds in Susceptible and Resistant Houseflies (*Musca domestica* L.). Mededel Landbouwhogeschool Wageningen 63(11):1-97. (Dutch summary.)

Tri-o-cresyl phosphate (TOCP) used to trace entry of parathion or paraoxon through cuticle, and to study the role of insect cuticle in resistance. ABSTRACTS: BA 46:1082(13666), Feb. 1, 1965; RAE-B 53:218, Nov., 1965.

2805. Abd El-Aziz, S., Metcalf, R. L. and **Fukuto, T. R. 1969** Physiological Factors Influencing the Toxicity of Carbamate Insecticides to Insects. J. Econ. Ent. 62(2):318-324.

Toxicity affected by: (1) sex differences in metabolism; (2) life stages; (3) age of adults. Decrease in effect directly related to enzyme activity. ABSTRACT: RAE-B 57:182-183(653), Sept., 1969.

2806. Abdel-Razig, M. E. 1967 Evaluation of Compounds of Cadmium and Related Metals, as Reproduction Inhibitors in Insects. Diss. Absts. 27(12B):4594B.

Concerns *M. domestica, Blatella, Periplaneta, Drosophila.*

2807. Abdullah, M. 1961 Inheritance of Dieldrin Resistance in the Housefly. J. Heredity 52(4):179-182.
Using single pairs and topical applications, concludes that resistance to Dieldrin is of polygenic origin. No evidence for a sex-linked factor. ABSTRACTS: BA 38:330(4589), Apr. 1, 1962; RAE-B 52:100, May, 1964; TDB 59:310, Mar., 1962.

Abezguaz, Iz, joint author. *See* Lineva, V. A., et al., 1960.

Abou-Gareeb, A. H., joint author. *See* Sinha, R., et al., 1967.

2808. Aboul-Nasr, A. 1960 Antagonistic Effect of Chloroform on Etherized *Musca domestica vicina* (Diptera:Muscidae). Bull. Soc. Ent. Egypte 44:157-165.
Access of chloroform to etherized flies decreases the anaesthetic effect; access of ether to chloroform stupified flies, increases the anaesthetic effect. ABSTRACT: BA 38:941(12263), May 1, 1962.

2809. Adams, T. S. 1967 The Relationship of Ovaries and the Corpus Allatum to Mating in the House Fly, *Musca domestica*. Amer. Zool. 7(4):724.
Ovariectomized females mated normally. Ovariectomized and allatectomized individuals (also those simply allatectomized) showed reduced mating activity. Relates to titre of juvenile hormone in the insect.

2810. Adams, T. S. 1970 Ovarian Regulation of the Corpus Allatum in the Housefly, *Musca domestica*. J. Insect Physiol. 16:349-360.
Corpus allatum area decreases before vitellogenesis, then increases (by stages 5 or 6 of ovarian development). A small corpus allatum releases juvenile hormone; a large one stores it. ABSTRACT: BA 51:8566(87349), Aug. 1, 1970.

2811. Adams, T. S. and **Hintz, A. M. 1969** Relationship of Age, Ovarian Development and the Corpus Allatum to Mating in the House-fly, *Musca domestica*. J. Insect Physiol. 15(2):201-215.
Juvenile hormone stimulates both ovarian maturation and mating. Intensity in mating increases as ovaries develop. Males prefer females in stages 6-10. ABSTRACTS: BA 50:8095(84267), Aug. 1, 1969; TDB 66:755(1504), July, 1969; RAE-B 58:89(353), March, 1970.

2812. Adams, T. S., Hintz, A. M. and **Pomonis, J. G. 1968** Oöstatic Hormone Production in Houseflies, *Musca domestica*, with Developing Ovaries. J. Insect Physiol. 14(7):983-993.
Suppresses the developmental rate of eggs in stages 2-4. Maintains cycle of egg production. ABSTRACT: BA 49:9771(108221), Oct. 15, 1968.

2812a. Adams, T. S., Johnson, J. D., Fatland, C. L. and **Olstad, G. 1970** Preparation of a Semipurified Extract of the Oöstatic Hormone and its Effect on Egg Maturation in the House Fly. Ann. Ent. Soc. Am. 63(6):1565-1569.
Injected oöstatic hormone retarded egg development by prolonging stage 4 of öogenesis. Extract was prepared from mature female flies.

2813. Adams, T. S. and **Nelson, D. R. 1968** Bioassay of Crude Extracts for the Factor that Prevents Second Matings in Female *Musca domestica*. Ann. Ent. Soc. Am. 61:112-116.
Extracts from various male tissues, also from newly mated females, injected into virgin females. Some of these caused 50 per cent inhibition of mating. (Test 3 days later.) ABSTRACTS: BA 49:5301(59487), June 1, 1968; RAE-B 56:159(568), Aug., 1968; TDB 66:360(709), Apr., 1969.

2814. Adams, T. S. and **Nelson, D. R. 1969** Effect of Corpus Allatum and Ovaries on Amount of Pupal and Adult Fat Body in the Housefly, *Musca domestica*. J. Insect Physiol. 15(10):1729-1747.
Pupal fat body decreases during stages 4-7 of oogensis; is gone by stage

10. Adult fat body increases during stages 2-7 and 9-10; decreases during stages 7-9. ABSTRACT: BA 51:3376(34919), Mar. 15, 1970.

Adams, T. S., joint author. *See* Nelson, D. R., et al., 1969.

2815. Adelung, D. and **Karlson, P. 1969** Eine verbesserte, sehr empfindliche Methode zur biologischen Auswertung des Insektenhormones ecdyson. [An improved, very sensitive method for the biological evaluation of the insect hormone ecdysone.] J. Insect Physiol. 15(8): 1301-1307.
M. domestica larvae 5 times more sensitive to moulting hormone than *Calliphora erythrocephala.* Only partly due to difference in size. ABSTRACT: BA 51:2796(29019), Mar. 1, 1970.

2816. Adkins, T. R., Jr. 1963 Absence of Residues in Milk after Barns were Sprayed with Dimethoate. J. Econ. Ent. 56:119.
Test followed use of 2 per cent water emulsion of dimethoate on walls. ABSTRACT: RAE-B 51:94, May, 1963.

2817. Adkins, T. R., Jr. and **Seawright, J. A. 1967** A Simplified Dusting Station to Control Face Flies and Horn Flies on Cattle. J. Econ. Ent. 60:864-868.
For polled or dehorned cattle. Ten pounds of dust maintained in each suspended double sack. ABSTRACTS: BA 48:11207(124478), Dec. 15, 1967; RAE-B 55:192(661), Oct., 1967.

Adkins, T. R., Jr., joint author. *See* Hair, J. A., et al., 1964, 1965; Seawright, J. A., et al., 1968; Poindexter, C. E., et al., 1970.

2818. Adolphi, H. 1963 Untersuchungen über das Insektizid Nankor. [Investigations on the insecticide fenchlorophos.] Z. Angew. Zool. 50(3): 273-295.
Effectiveness assessed in comparison with other chemicals on a number of insect species, including *M. domestica.* ABSTRACT: RAE-B 53:131, July, 1965.

2819. Adolphi, H. 1964 A Case for Organic Phosphorus Insecticides. Biokemia 3:14-19. Midland, Michigan.
An abridged version of the author's 1963 report. ABSTRACT: RAE-B 53:131, July, 1965.

2820. Afrikyan, E. K. 1963 Entomopatogennye svoistva baketerii i ikh practiche skoe z nachenie. [Entomopathogenic characteristics of bacteria and their practical significance.] Ozvest. Akad. Nauk Armyansk. SSR Biol. Nauki 16(1):23-28.
Cultures of the *Bacillus cereus* and *B. thuringiensis* group were highly effective in controlling *M. domestica.* Larvae show an avitaminosis. ABSTRACT: BA 45:2868(35560), Apr. 15, 1964.

Aganina, V. M., joint author. *See* Fomicheva, V. S., et al., 1967.

2821. Agarwal, H. C. 1961 An Unusual Sterol from Houseflies. J. Insect Physiol. 7(1):32-45.
Sterols from whole adult fly and pupal exuviae. Major one close to cholesterol; 3 others close to Δ^7-cholesterol, 7-dehydrocholesterol and methostenol. ABSTRACT: BA 37:1150(12092), Feb. 1, 1962.

2822. Agarwal, H. C. 1961 Studies on House Fly Sterols and Cholesterol-C^{14} Metabolism in House Flies and Cockroaches. Diss. Absts. 21 (Pt. 4, No. 10):2872-2873.
Same sterol (termed *muscasterol*) predominated in eggs, larvae, pupae, exuviae, and adults. Study of metabolism of cholesteryl-4-C^{14} acetate revealed no conversion of cholesterol to any minor sterol, probably not to the major sterol either. ABSTRACT: BA 37:2058(20618), Mar., 1962.

2823. **Agarwal, H. C.** and **Casida, J. E. 1960** Nature of Housefly Sterols. Biochem. Biophys. Res. Commun. 3:508-512.
Three sterols recovered from adults and eggs; four from larvae and pupae. Major fly sterol (muscasterol) varies from cholesterol only in side chain (isopropyl structure missing).

2824. **Agarwal, P. N., Chada, D. B., Dixit, R. S.** and **Perti, S. L. 1963** The Synergism of DDT by Synthetic Pine Oil. Indian J. Malariol. 17: 71-73. Delhi.
Pine oil increased toxic action of DDT against houseflies and mosquitoes (*Culex fatigans*). Used in space spray, or topically applied was especially effective against *C. fatigans*. ABSTRACT: RAE-B 54:39, Feb., 1966.

2825. **Agarwal, P. N.** and **Paul, C. F. 1963** Evaluation of Insecticidal Properties of Telodrin—a New Chlorinated Hydrocarbon Insecticide. Indian J. Ent. 25(3):183-187.
Poor knockdown effect as a space spray against *M. nebulo*. Topically applied, less toxic than dieldrin, but more toxic than dieldrin by film technique. ABSTRACTS: BA 46:1431(17878), Feb. 15, 1965; RAE-B 52:74, Apr., 1964.

Agarwal, P. N., joint author. *See* Damodar, P., et al., 1964; Paul, C. F., et al., 1963; Tiwari, B. K., et al., 1966; Srivastava, A. P., et al., 1965.

2826. **Agosin, M., Fine, B. C., Scaramelli, N., Ilivicky, J.** and **Aravena, L. 1966** The Effect of DDT on the Incorporation of Glucose and Glycine into Various Intermediates in DDT-resistant Strains of *Musca domestica* L. Comp. Biochem. Physiol. 19(2):339-349.
Cross-resistant strains have remarkable metabolic differences as to levels of NADP, relative participation of glycolysis and the pentose phosphate pathway in glucose utilization rates of synthesis and glutathione turnover. ABSTRACT: BA 48:1891(20518), Feb. 15, 1967.

Agosin, M., joint author. *See* Balazs, I., et al., 1968; Gil, L., et al., 1968; Litvak, S., et al., 1967, 1968.

Ahmad, D., joint author. *See* Kahn, N. H., et al., 1964, 1965.

2827. **Ahmad, M.** and **Shafiq, S. A. 1965** Metamorphosis of Fat Body in the Housefly, *Musca vicina*, Macquart (Diptera:Muscidae): I. The Fat Cells of the Fully Grown Larva. Pakistan J. Sci. Res. 17(1):31-34.
These cells were found to be rich in RNA, mitochondria and Golgi elements. ABSTRACT: BA 48:11241(124858), Dec. 15, 1967.

2828. **Ahmad, N.-ud-D. 1966** Ecology of the House-fly (*Musca domestica* Linn.). Pakistan J. Sci. Res. 18(3/4):98-101.
Incubation, larval and pupal periods change in duration with the temperature. Optimum condition for fly emergence: 35°C combined with 91 per cent relative humidity. ABSTRACT: RAE-B 57:215(780), Oct., 1969 (pages given: 98-100, 107).

2829. **Ahmad, N.-ud-D. 1967** The Development of the External Male Genitalia of the House Fly, *Musca domestica* Linn. Pakistan J. Sci. Res. 19(1):11-17.
Traces the origin of all distinctive structures. ABSTRACT: BA 50:527 (5512), Jan. 1, 1969.

2830. **Ahmad, S. 1970** Studies on Aliesterase, Lipase and Peptidase in Susceptible and Organophosphate-resistant Strains of Housefly (*Musca domestica* L.). Comp. Biochem. Physiol. 32(3):465-474.
"Low esterase" activity of resistant strains was found to be entirely due to aliesterase and not to lipase or peptidase. Hydrolysis of fats and oils by fly lipase was found to be very low. ABSTRACT: BA 51:11473(116519), Oct. 15, 1970.

2831. Ahmad, S. 1970 The Recovery of Esterases in Organophosphate-treated Housefly (*Musca domestica* L.). Comp. Biochem. Physiol. 33(3):579-586.
A rapid recovery of aliesterase and cholinesterase was observed (more pronounced for ChE). Short-phased recovery could be due to reversal of some of the inhibited enzyme. Resynthesis rate similar for both esterases. ABSTRACT: BA 51:10310(104858), Sept. 15, 1970.

Aikawa, T., joint author. *See* Eto, M., et al., 1968.

2832. Aircraft Disinsection. 1961 Eleventh report of the Expert Committee on Insecticides—Tech. Rep. ser. W. H. O. 206. 26 pp, 18 refs. Geneva.
Explains many special terms. Houseflies used for bioassay of various preparations. W. H. O. can furnish a suitable, susceptible strain. ABSTRACT: RAE-B 51:72, Apr., 1963.

Akai, H., joint author. *See* Kobayashi, M., et al., 1967.

Akamatsu, T., joint author. *See* Miyamoto, K., et al., 1965.

2833. Akmurzaev, T. A. 1962 Podstavka dlya prosmotra i opredeleniya lichinok mukh (sinatropnykh). [Basis for the survey and identification of fly larvae (synanthropic).] Med. Parazit. i Parazitarn. (Bolezni) 30(4):477. Referat. Zhur. Biol. 1962, No. 2K138.

2834. Akov, S. and **Borkovec, A. B. 1968** Metabolism of the Chemosterilant HEMPA by Carbamate-resistant Houseflies. Life Sci. Biochem. Gen. Molec. Biol. 7(22, pt. 2):1215-1218.
Microsomal preparations from carbamate-resistant flies metabolize HEMPA more rapidly than those from susceptible flies. No resistance is conferred to the sterilizing effect of HEMPA. ABSTRACT: BA 50:10828 (112094), Oct. 15, 1969.

2835. Akov, S., Oliver, J. E. and **Borkovec, A. B. 1968** N-demethylation of the Chemosterilant HEMPA by House Fly Microsomes. Life Sci. Biochem. Gen. Molec. Biol. 7(22, pt. 2):1207-1213.
HEMPA is demethylated to pentamethylphosphoric triamide, an ineffective chemosterilant. N-demethylation of many pesticides in insects is carried out by microsomal enzymes requiring oxygen and NADPH. ABSTRACT: BA 50:10828(112095), Oct. 15, 1969.

Akov, S., joint author. *See* Shargel, L., et al., 1969.

Alejandro, Ortega-Corona, joint author. *See* MacGregor-Loaeza, R., et al., 1965.

Alexander, B. H., joint author. *See* Gersdorff, W. A., et al., 1960; Piquette, P. G., et al., 1960.

Al-Hafidh, R., joint author. *See* Pimentel, D., et al., 1965.

2836. Allen, W. C. 1968 Insects and Other Arthropods of Importance during 1968 in Households and on Livestock in Ontario. Proc. Ent. Soc. Ont. 99:11-12. Map.
Imported face fly (*Musca autumnalis*) present in very large numbers in widely separated areas. ABSTRACT: BA 50:5863(61420), June 1, 1969.

Amerson, G. M., joint author. *See* Hays, S. B., et al., 1967.

2837. Amonkar, S. V., Kalle, G. P. and **Nair, K. 1967** Mechanism of Pathogenicity of *Pseudomonas* in the House Fly. J. Invertebr. Path. 9(2): 235-240.
Concerns *M. domestica nebulo*. Bacteria elaborate an exotoxin during growth. ABSTRACTS: BA 48:10187(114466), Nov. 15, 1967; RAE-B 56: 144(504), July, 1968.

2838. **Amonkar, S. V.** and **Nair, K. K. 1965** Pathogenicity of *Aspergillus flavus* Link to *Musca domestica nebulo* Fabricius. J. Invertebr. Path. 7:513-514.

Previous to this publication, *Metarrhizium anisopliae* was the only hypomycetous fungus reported pathogenic to *M. domestica*. ABSTRACT: BA 47:5535(64893), July 1, 1966; RAE-B 54:135, July, 1966.

Anderson, E. L., joint author. *See* Davies, D. M., et al., 1967.

2839. **Anderson, J. R. 1964** Methods for Distinguishing Nulliparous from Parous Flies and for Estimating the Ages of *Fannia canicularis* and some other Cyclorraphous Diptera. Ann. Ent. Soc. Am. 57:226-236.

M. domestica among the species studied. The number of functional ovarioles decreased, as the number of gonotrophic cycles mounted. ABSTRACTS: BA 45:5078(62192), July 15, 1964; RAE-B 52:193, Oct., 1964.

2840. **Anderson, J. R. 1964** The Behavior and Ecology of Various Flies (Dipetra) Associated with Poultry (Aves) Ranches in Northern California. Proc. Pap. Ann. Conf. Calif. Mosq. Contr. Assoc., Inc. 32:30-34.

Gives special attention to the ecological relations of adult flies (female) during the 5 to 10 days before they are able to lay their first eggs.

2841. **Anderson, J. R. 1965** A Preliminary Study of Integrated Fly Control on Northern California USA Poultry Ranches. Proc. Pap. Ann. Conf. Calif. Mosq. Contr. Assoc., Inc. 33:42-44.

Integrating of chemical with biological control appears promising for the future. Further investigation needed. ABSTRACT: RAE-B 56:203(736), Oct., 1968.

2842. **Anderson, J. R. 1966** Recent Developments in the Control of some Arthropods of Public Health and Veterinary Importance: Muscoid Flies. Bull. Ent. Soc. Am. 12(3):342-348.

Changing agricultural techniques affect and change the origin of various fly problems and how they can be approached. ABSTRACT: BA 48:8784 (98690), Oct. 1, 1967.

2843. **Anderson, J. R. 1966** Biological Interrelationships Between Feces and Flies. Repr. from Proc. Nat. Symp. on Management of Farm Animal Wastes. May 5, 6, 7. pp. 20, 21, 22, 23.

A challenging discussion of modern problems with suggestions for future practices.

2844. **Anderson, J. R., Deal, A. S., Legner, E. F., Loomis, E. C.** and **Swanson, M. H. 1968** Fly Control on Poultry Ranches. Univ. Calif. Agr. Ext. Serv. AXT-72 Revised.

Stresses manure management, water systems and general farm sanitation.

2845. **Anderson, J. R.** and **Poorbaugh, J. H. 1964** Biological Control Possibility for House Flies. Calif. Agric. 18(9):2-4. Berkeley.

Larvae of the black garbage fly *Ophyra leucostoma* (Wied.) kill and feed on maggots of the house fly and associated species. ABSTRACT: RAE-B 54:220, Nov., 1966.

2846. **Anderson, J. R.** and **Poorbaugh, J. H. 1964** Observations on the Ethology and Ecology of Various Diptera Associated with Northern California Poultry Ranches. J. Med. Ent. 1(2):131-147.

Sets forth methodological advantages of sticky tapes in survey work. Describes a rapid method of collecting them. ABSTRACT: RAE-B 54:7, Jan., 1966.

2847. **Anderson, J. R.** and **Poorbaugh, J. H. 1965** Refinements for Collecting and Processing Sticky Fly Tapes Used for Sampling Populations of Domestic Flies. J. Econ. Ent. 58:497-500.

Explains use of cellophane in handling field-exposed sticky tapes. With proper illumination, common genera and many species can be identified through the transparent sheet. ABSTRACTS: BA 47:1601(19380), Feb. 15, 1966; RAE-B 53:179, Sept., 1965.

2848. Anderson, J. R. and **Poorbaugh, J. H. 1968** The Face Fly, *Musca autumnalis* . . . A New Livestock Fly is now Moving Toward California. Calif. Agric. 22(3):4-6.
Species is already recorded from Washington and Oregon. Notes that immature stages develop *only* in fresh cattle droppings. ABSTRACT: BA 49:8199(91427), Sept. 1, 1968.

Anderson, J. R., joint author. *See* Loomis, E. C., et al., 1967; Poorbaugh, J. H., et al., 1968; Burger, J. F., et al., 1970; Peck, J. H., et al., 1969.

Anderson, R. F., joint author. *See* Shieh, T. R., et al., 1968.

Anderson, W. F., joint author. *See* Hansens, E. J., et al., 1970.

2849. Andrawes, N. R. 1967 Metabolism of 2-methyl-2-methyl-thio Propionaldehyde 0-Methyl Carbamoyl Oxime Temik Insectic in Mammals and Insects. Diss. Absts. 28(2B):724-B.
M. domestica, treated by topical application, metabolizes Temik to its sulfinyl and sulfonyl derivatives. Sulfinyls predominate during the first 4 hours.

2850. Andreev, V. P. and **Vranchan, Z. E. 1961** On the Ecology of *Musca domestica* L. in Animal Breeding Farms. Med. Parazit. (Mosk) 30: 64-66.

Aniya, K., joint author. *See* Kano, R., et al., 1964.

2851. Ansari, J. A. 1969 Studies on the Isolation and Genetic Nature of Specific Insecticide Resistance in Houseflies. Botyu Kagaku (Bull. Inst. Insect Contr.) 34(2):70-76. (Japanese summary.)
Used back-cross and selection pressure techniques. Dieldrin resistance appears due to multi-factorial inheritance; DDT resistance, to a major single-factor type. ABSTRACT: BA 51:3972(41001), April 15, 1970.

2852. Ansari, J. A. and **Riaz, N. 1965** Temperature Coefficient of House Fly Resistance to Insecticides. Botyu-Kagaku. (Bull. Inst. Insect Contr.) 30(4):105-108.
Susceptibility of *M. domestica nebulo* to dieldrin increased with a rise in temperature, whereas DDT and BHC had a negative temperature coefficient. Susceptible and resistant strains responded in the same way. ABSTRACT: RAE-B 56:53(173), Mar., 1968.

Ansari, J. A., joint author. *See* Kahn, N. H., et al., 1964, 1965.

2853. Anthony, D. W., Hooven, N. W. and **Bodenstein, O. 1961** Toxicity to Face Fly and House Fly Larvae of Feces from Insecticide-Fed Cattle. J. Econ. Ent. 50(3):406-408.
Co-Ral and Bayer 2208 completely inhibited development of face fly larvae and survival of house fly larvae in manure of cattle so treated. Ronnel did likewise if dose is large, but at 2·5 mg/kg. controlled face flies only. ABSTRACTS: BA 36:5560(58115), Sept. 1, 1961; RAE-B 50:125, June, 1962.

Aravena, L., joint author. *See* Agosin, M., et al., 1966.

2854. Arevad, K. 1965 On the Orientation of Houseflies to Various Surfaces. Ent. Exp. Appl. (Amsterdam) 8:175-188. (German summary.)
Light surfaces less visited than darker ones; rough surfaces preferred to smooth; metallic surfaces avoided. ABSTRACTS: BA 47:5097(59810), June 15, 1966; RAE-B 54:68, Apr., 1966; TDB 63:486, Apr., 1966.

2855. Arevad, K. 1966 Undersøgelser over fluernes ad faerd med henblik på nye bek aempelsesmetoder. [Investigations of housefly behaviour

in relation to new methods of fly control.] Ann. Rept. (1965) Govt. Pest Infest. Lab. Denmark, pp. 52-56.
State of nutrition and especially lack of water are factors in determining fly's reaction to a black plate.

2856. **Arevad, K. 1967** Stuefluens biologi og adfaerd. Undersøgelser med henblik på ud vikling af strate gisk flue bekaempelse. Laboratorie-undersøgelser over stuefluers adfaerd. [Housefly biology and behaviour. Investigations in relation to development of strategic fly control. Laboratory investigations of housefly behaviour.] Ann. Rept. (1966) Govt. Pest Infest. Lab. Denmark, pp. 49-50, 52-53.
Size and shape of dark areas important factors in attracting flies to such areas.

2857. **Arevad, K. 1967** Metode til best emmelse af stuefluers alder. [A Method for estimating the age of houseflies.] Ann. Rept. (1966) Govt. Pest Infest. Lab. Denmark, pp. 55-56.
Color stages of the parafrontals, with due consideration given to temperature, can be used to estimate early periods of adult life.

2858. **Arevad, K. 1968** Laboratorieundersøgelser over nykloekkede stuefluers ad faerd. [Laboratory investigations on the behaviour of newly emerged houseflies.] Ann. Rept. (1967) Govt. Pest Infest. Lab. Denmark, pp. 55-58.
Wingless stage lasts an average of 14 minutes at 22°-24°C. Wing unfolding lasts 4 minutes; cuticle hardening 15-30 minutes. Flies manifest negative geotaxis and tend to avoid light; when able to fly, prefer rough, dark horizontal surfaces, facing downward.

2859. **Arevad, K. 1968** The Orientation of Newly Hatched Houseflies. 13th Int. Congr. Ent. Moscow. Abstracts, p. 15.
Initial high motor activity followed by inactivity while cuticle hardens. Treats of factors that influence direction of next movement and choice of resting place.

2860. **Arevad, K. 1969** Stuefluers adfaerd. [Housefly behaviour.] Ann. Rept. (1968) Govt. Pest Infest. Lab. Lyngby, pp. 52-55.
Treats of factors influencing choice of resting place soon after emergence, such as temperature and texture of surface, moisture on surface, etc.

2861. **Arias, R. O. 1963** The *in vitro* Hydroxylation of Naphthalene-1-C^{14} by Housefly (*Musca domestica* L.) Microsomes. Diss. Absts. 24(5): 1817-1818.

2862. **Arias, R. O.** and **Terriere, L. C. 1962** The Hydroxylation of Naphthalene-1-C^{14} by House Fly Microsomes. J. Econ. Ent. 55:925-929.
Microsomes of older adults show greatest activity; those of larvae, the least. ABSTRACTS: BA 42:658(8140), Apr. 15, 1963; RAE-B 51:68, Mar., 1963.

2863. **Arnold, F. D. 1966** An Evaluation of the Housefly Problem in the Major Populated Counties in Utah, U.S.A. Diss. Absts. 27(4B): 1037-B.

2864. **Aronson, J. 1963** Observations on the Variations in Size of the A Region of Arthropod Muscle. J. Cell. Biol. 19(2):359-367.
Used *Drosophila* as type. Included here as a contribution to the histology of higher Diptera. ABSTRACT: BA 45:4003(49265), June 15, 1964.

Arroyo-Bornstein, J. A., joint author. *See* Varela, G., et al., 1964.

2865. **Arskii, V. G., Gadzhei, E. F., Zatsepin, N. I.** and **Iasinskii, A. V. 1961** [Role of flies in seasonal characteristics of dysentery.] Zhur. Mikrobiol. 32:27-32. (In Russian.) *Also* J. Microb. Epidem. Immunol. 32:1013-1020. (In English.)

Arthur, B. W., joint author. *See* Buttram, J. R., et al., 1961; Brady, U. E., et al., 1960, 1962, 1963; Camp, H. B., et al., 1967; Dorough, H. W., et al., 1961; Knowles, C. O., et al., 1966, 1967; Parish, J. C., et al., 1965; Vickery, D. S., et al., 1960.

Artigas, R. T., joint author. *See* Diaz-Ungria, C., et al., 1966.

Arurkar, S. K., joint author. *See* Knowles, C. O., et al., 1969.

Asada, S., joint author. *See* Buei, K., et al., 1963, 1965, 1966, 1968.

2866. Asahina, S. (Chairman). 1964 Symposium on Insecticide Resistance of Pest Insects. Jap. J. Med. Sci. Biol. 17:39-57. Tokyo.
Includes contributions by 5 authors. Concerns *M. domestica vicina* along with *Drosophila* and other genera. ABSTRACT: RAE-B 54:107, June, 1966.

2867. Asahina, S., and Others. 1962 [Critical tests regarding current residual-spray control against medical insects in Japan.] I. Jap. J. Sanit. Zool. 13(3):213-219.
DDT and BCH sprays used in experimental summer huts near Tokyo. Susceptible and resistant flies added periodically. Concerns chiefly, *M. domestica vicina*. ABSTRACT: RAE-B 53:37, Feb., 1965.

2868. Asahina, S., and Others. 1963 [Critical tests regarding current residual-spray control against medical insects in Japan.] II. Jap. J. Sanit. Zool. 14(1):48-52.
Tests made with fenchlorphos (Nankor), dichlorvos (DDVP), naled (Dibrom). Deposits of organophosphorus insecticides remain effective only 15 days if temperature goes over 30°C [86°F]. ABSTRACT: RAE-B 53:37, Feb., 1965.

2869. Asano, S. and **Nagasawa, S. 1963** [DDT resistance of the so-called "Takatsuki" strain of the common house fly shown by the knock-down effect.] Botyu-Kagaku 28(1):8-12. Kyoto. (In Japanese, with English summary.)
Susceptibility to DDT tested for 13 populations. Two methods: (1) dusting apparatus; (2) topical application. Dusting showed greater differences between populations. ABSTRACT: RAE-B 54:200, Oct. 1, 1966.

2870. Ascher, K. R. S. 1960 A Review of Resistance-induced Enhanced Susceptibility in Insects, with some Notes on Similar Phenomena (especially "collateral sensitivity") in Microorganisms. Arzneimittel-Forschung 10:450-461.
Deals partially with houseflies.

2871. Ascher, K. R. S. 1961 DDT-resistance-induced Enhanced Susceptibility towards Cetyl Fluoride (CF) and Cetyl Fluoroacetate (CFA)—a Preliminary Report. Proc. XIth Int. Cong. Ent., Vienna, 2:557; 3:239.
Deals partially with houseflies.

2872. Ascher, K. R. S. 1961 Houseflies in Israel. III. Some Observations at Breeding Sites in Rural Areas, and Considerations about the Influence of Israeli Manure Handling Methods on Housefly Breeding. Zeit. Angew. Ent. 48(2):115-162.
Paper has 5 parts. Treats extensively of country in general; settlements, rural areas, manure handling, fly breeding sites, findings of fly survey. Recommendations for reduction of fly breeding. ABSTRACTS: RAE-B 51:64, Mar., 1963; TDB 59:98, Jan., 1962.

2873. Ascher, K. R. S. 1962 G. G. Mer's Contribution to Insect Toxicology. Estr. Rivista di Malariol. 41(4-6):reprint, 10 pp.
A lecture delivered at the Rehovot Symposium on Insect Toxicology, dedicated to the memory of G. G. Mer. Mer suspected DDT-resistance as early as 1946.

2874. Ascher, K. R. S. 1962 Two New Research Approaches to the Resistance Problem, using the Housefly as Experimental Animal. J. Hyg., Epidemiol., Microbiol., and Immunol. (Praha) 6:256-264.
Both relate to chemical control: (1) concerning compounds with negative correlation to resistance; (2) concerning chemosterilants and oviposition inhibitors. ABSTRACT: BA 42:1240(15710), May 15, 1963.

2875. Ascher, K. R. S. 1963 N, N-Dialkyl-p-Halobenzene-sulfonamides: their Insecticidal Properties and Joint Action with DDT. Riv. di Parassit. Rome 24:45-66.
Reviews basic research in Israel from 1955-1956. Treats of WARF as an anti-resistant (used as a synergist with DDT). *M. domestica vicina* Macq., the experimental animal. ABSTRACTS: RAE-B 52:200, Nov., 1964; TDB 61-108, January, 1964.

2876. Ascher, K. R. S. 1964 A Review of Chemosterilants and Oviposition-inhibitors in Insects. Wld. Rev. Pest Control 3(1):7-27.
Deals partially with houseflies.

2877. Ascher, K. R. S. 1964 The Housefly and the Cube of Sugar. Proc. 12th Int. Cong. Ent., London, July. p. 294.
A brief summary of the lecture.

2878. Ascher, K. R. S. 1964 Oviposition-inhibiting Agents: a Screening for Simple Model Substances. Proc. 12th Int. Cong. of Ent. London, July. p. 514.
A brief summary of the lecture.

2879. Ascher, K. R. S. 1965 Oviposition Inhibiting Agents: a Screening for Simple Model Substances. Int. Pest Contr. 7(1):8-11.
Housefly fecundity affected by such substances as benzene, tetrahydrofuran, benzyl alcohol, m-xylohydroquinone. ABSTRACTS: RAE-B 54:247, Dec., 1966; TDB 64:1364, Dec., 1967.

2880. Ascher, K. R. S. 1967 The Effect of Fentins on the Fertility of the Male Housefly. Abstracta, 6th Int. Cong. Plant Prot. Vienna, September. pp. 93.
A brief summary of the lecture.

2881. Ascher, K. R. S. and **Avdat, N. 1966** Sterilizing the Male Housefly with M-xylohydroquinone. Part I. Int. Pest. Contr. 8(6):16-17, 20-21, 23, 25.
Treats of feeding time, concentration of the chemical substance, stability of the compound and its solutions. (One-day feeding with 0·2 per cent gave poor results.) ABSTRACTS: RAE-B 57:83(290), May, 1969; TDB 64:1364, Dec., 1967.

2882. Ascher, K. R. S. and **Avdat, N. 1967** Sterilizing the Male House-fly with M-xylohydroquinone. Part II. Int. Pest Contr. 9(2):8-9, 11-13.
Good results from feeding 0·5 per cent m-x. for one day only, to one-day-old males. ABSTRACT: TDB 64:1364, Dec., 1967.

2883. Ascher, K. R. S. and **Avdat, N. 1967** Stabilisation of Aqueous Solutions of the Insect Chemosterilant M-xylohydroquinone (*m*-XHQ) by Vitamin C. Experientia 23:679.
This substance is a rather efficient inhibitor of oviposition in female house flies, besides serving as a chemosterilant in males.

2884. **Ascher, K. R. S.** and **Bergmann, E. D. 1962** Enhanced Susceptibility to Cetyl Fluoride (CF) and Cetyl Fluoroacetate (CFA) in DDT-resistant Insects. Ent. Exp. Appl. 5(2):88-98. Amsterdam. (German summary.)
Negative correlation established for CF to DDT resistant housefly and anopheline larvae. Both CF and CFA are very poor larvicides. ABSTRACTS: BA 40:1296(17016), Nov. 15, 1962; RAE-B 51:118, June, 1963; TDB 60:76, Jan., 1963.

2885. **Ascher, K. R. S.** and **Borkovec, A. 1967** Insect Chemosterilants. Advances in Pest Control Research Vol. VII (Editor: R. L. Metcalf) Interscience Publishers, New York, London, Sydney 1966. x + 143-565. *From* World Review of Pest Control 6(3):90-95.
A review of the volume cited. Appropriate attention given to work with flies.

2886. **Ascher, K. R. S.** and **Hirsch, I. 1961** Inhibition of Oviposition in the Housefly by Ingestion of Acaricides (Ovicides). Riv. di Malariol. 40(4-6):139-145.
Outstanding in their inhibitory effect were: (1) chlorbenside (p-chlorobenzyl p-chlorophenyl sulphide); (2) Tedion (2, 4, 5, 4-tetrachlorodiphenyl sulphone). Recommended dosages of either, inhibited egg laying almost completely without shortening the female life span. ABSTRACTS: RAE-B 52:150, Aug., 1964; TDB 59:847, Aug., 1962.

2887. **Ascher, K. R. S.** and **Hirsch, I. 1963** The effect of M-xylohydroquinone on Oviposition in the Housefly. Ent. Exp. Appl. 6(4):337-338. Amsterdam.
Oviposition strongly inhibited at concentrations between 0·15 per cent and 0·1 per cent. No toxicity noted below 10·15 per cent. ABSTRACTS: RAE-B 52:140, Aug., 1964; TDB 61:972, Sept., 1964.

2888. **Ascher, K. R. S.** and **Hirsch, I. 1965** The Housefly and the Cube of Sugar. Wld. Rev. Pest Control 4(3):103-111. Cambridge.
Treats of experiments concerning the fly factor (attraction to sites previously visited by flies), herd instinct, attraction to certain colors. ABSTRACTS: BA 48:6503(72614), July 15, 1967; RAE-B 54:26, Feb., 1966; TDB 65:1062(2411), Aug., 1968.

2889. **Ascher, K. R. S., Meisner, J.** and **Moscowitz, J. 1967** Chemo-sterilizing and Anti-feeding Effects of Organo Tins in the House Fly. Abstr. Int. Pflanzenschutz-Kong. (Int. Cong. Plant Prot.) 30:93.

2890. **Ascher, K. R. S., Meisner, J.** and **Nissim, S. 1968** The Effect of Fentins on the Fertility of the Male Housefly. Wld. Rev. Pest Control. 7(2): 84-96.
Fertility of eggs from females mated to treated males diminished more and more after third day of laying. Believed due chiefly to slow poisoning of sperm in the female spermathecae. Sperm cells gradually lose motility and die. ABSTRACTS: RAE-B 57:24(103), Jan., 1969; TDB 66:880(1756), Aug., 1969; BA 51:5671(158194), May 15, 1970.

2891. **Ascher, K. R. S., Moscowitz, J.** and **Nissim, S. 1967** An Anti-feeding Effect of Triphenyltins on Housefly Larvae. Leaflet No. 51. Reprinted from Tin and its Uses. 73(8-9):2 pp.
Fentin acetate and fentin hydroxide in larval food resulted in reduced pupal weights due to reduced larval feeding. ABSTRACT: RAE-B 58:33 (137), Feb., 1970.

2892. **Ascher, K. R. S.** and **Moscowitz, J. 1968** Fentins (triphenyltins) Have an Anti-feedant Effect for Housefly Larvae. Int. Pest Contr. 10(3):10-13.
As above. These substances are toxic at higher concentrations.

2893. **Ascher, K. R. S.** and **Moscowitz, J. 1969** Pennsalt TD-5032, an Experimental Organotin Insecticide with Antifeedant Properties. Int. Pest Contr. 11(1).
LD_{50} and LD_{90} for this substance, topically applied, are 0·62 and 1·25 micrograms per fly, respectively.

2894. **Ascher, K. R. S.** and **Neri, I. 1961** Lipoid Content and Resistance in the Housefly *Musca domestica* L. Ent. Expt. Appl. 4(1):7-19. (German summary.)
No change was brought about in resistance (or susceptibility) level of resistant or normal adults by replacing wheat bran lipids with fats of resistant or normal pupae or with cholesterol in the breeding medium. ABSTRACTS: BA 36:5562(58136), Sept. 1, 1961; TDB 58:1191, Oct., 1961.

2895. **Ascher, K. R. S.** and **Nissim, S. 1964** Organotin Compounds and their Potential Use in Insect Control. Wld. Rev. Pest Contr. 3(4):188-211.
Deals partially with *M. domestica.*

2896. **Ascher, K. R. S.** and **Rones, G. 1965** Fluoride Detoxification in the House-fly by Other Halides. Experientia 21(7):404-405. (German summary.)
House flies are protected by chloride and, to a lesser extent, by iodide against the toxic effects of bromide. Those which survive sodium fluoride poisoning appear less resistant to DDT than those which succumb. ABSTRACTS: BA 48:1891(20519), Feb. 15, 1967; RAE-B 56:106 (371), May, 1968.

Ascher, K. R. S., joint author. *See* Levinson, Z. H., et al., 1954; Kocher, C., et al., 1954; Gratz, N. G., et al., 1957.

Ashiba, M., joint author. *See* Fujito, S., et al., 1963.

2897. **Ashrafi, S. H., Khan, M. A. Q.** and **Murtuza, S. M. 1963** Toxicity of Makrolin against Cockroaches, House Flies and Mosquito Larvae as Compared with other Chlorinated Insecticides. Pakistan J. Sci. and Ind. Res. 6(3):192-197.
Toxicity of Makrolin to houseflies is markedly less than that of Heptachlor, Aldrin, Chlordane or DDT.

2898. **Ashrafi, S. H., Murtuza, S. S. M.** and **Asmatullah, D. 1966** Pesticidal Activity of Different Chlorinated Petroleum against Housefly *Musca domestica.* Sci. Res. Quart. J. East Reg. Lab. Posir Pakistan Counc. Sci. Ind. Re. 3(3):186-189.

2899. **Ashrafi, S. H., Murtuza, S. M., Mohiuddin, S.** and **Ashmatullah, D. 1969** A Study of Joint Action of Insecticides with Petkolin against *Musca domestica* L. Pak. J. Zool. 11(2):221-228.
Of the insecticides tested, 5 per cent Dimecron produced maximum synergizing effect on toxicity of Petkolin. ABSTRACT: BA 51:9700(98707), Sept. 1, 1970.

Ashrafi, S. H., joint author. *See* Zuberi, R. I., et al., 1969.

Askerov, G. A., joint author. *See* Rumianstev, I. N., et al., 1961.

Asmatullah, D., joint author. *See* Ashrafi, S. H., et al., 1969.

Atanasiu, A., joint author. *See* Duport, M., et al., 1963; Lupasco, G., et. al., 1963.

Atanosova, N., joint author. *See* Sarbova, S., et al., 1968.

Ausat, A., joint author. *See* Damodar, P., et al., 1962.

Avdat, N., joint author. *See* Ascher, K. R. S., et al., 1966, 1967.

Avramova, S., joint author. *See* Serbova, S., et al., 1966, 1968 (as Sarbova).

2900. Axtell, R. C. 1961 New records of North American Macrochelidae (Acarina:Mesostigmata) and their Predation Rates on the House Fly. Ann. Ent. Soc. Am. 54(4):748.
Several species of Macrochelidae should be considered in any investigation of the role of mites as natural control agents of the house fly. ABSTRACT: BA 37:307(4098), Jan. 1, 1962.

2901. Axtell, R. C. 1961 Mites—Enemies of House Flies. Farm Research 27(4):4-5.
Eggs and first stage larvae are consumed by mites of the genera *Macrocheles* and *Glyptholaspis*.

2902. Axtell, R. C. 1963 Manure-inhabiting Macrochelidae (Acarina:Mesotigmata) Predaceous on the House Fly. *In* Naegele, J. A. (ed.), Advances in Acarology, Vol. 1, xii + 480 pp. Ithaca, New York, Cornell University Press. pp. 55-59.
Seven species of this family were collected from domestic animal manure. ABSTRACT: BA 46:3674(45389), May 15, 1965.

2903. Axtell, R. C. 1963 Effect of Macrochelidae (Acarina:Mesostigmata) on House Fly Production from Dairy Cattle Manure. J. Econ. Ent. 56:317-321.
Shows, by various approaches, that when mites are destroyed (by use of Kelthane), a much larger percentage of housefly ova develop into flies. Used both *Macrocheles muscaedomesticae* (Scopoli) and *Glyptholaspis confusa* (Foa). ABSTRACTS: BA 44:623(8366), Oct. 15, 1963; RAE-B 51:201, Sept., 1963; TDB 60:1084, Nov., 1963.

2904. Axtell, R. C. 1963 Macrochelidae (Acarina:Mesostigmata) Inhabiting Manure and their Effect on House Fly Production. Diss. Absts. 23(5):1831.
Identified seven species, of which four are common. Considers the latter important in natural control. ABSTRACT: BA 44:623(8365), Oct. 15, 1963.

2905. Axtell, R. C. 1963 Acarina Occurring in Domestic Animal Manure. Ann. Ent. Soc. Am. 56(5):628-633.
A study of 211 samples, taken from 92 different farms.

2906. Axtell, R. C. 1964 Phoretic Relationship of Some Common Manure-Inhabiting Macrochelidae (Acarina:Mesostigmata) to the House Fly. Ann. Ent. Soc. Am. 57(5):584-587.
Macrocheles muscaedomesticae and *M. subbadius* were commonly found attached to flies. Others rarely (not truly phoretic). ABSTRACTS: BA 45: 8412(106882), Dec. 15, 1964; TDB 62:68, Jan., 1965.

2907. Axtell, R. C. 1966 Comparative Toxicities of Insecticides to House Fly Larvae and *Macrocheles muscaedomesticae*, a Mite Predator of the House Fly. J. Econ. Ent. 59:1128-1130.
Studied 17 insecticides for selectivity. Several, including Kepone and Coumaphos, were more toxic to fly larvae than to mites. Advocates intensive efforts to identify selective chemicals. ABSTRACTS: BA 48:1419 (15369), Feb. 1, 1967; RAE-B 55:12(31), Jan., 1967; TDB 64:325, Mar., 1967.

2908. Axtell, R. C. 1967 Macrochelidae (Acarina:Mesostigmata) as Biological Control Agents for Synanthropic Flies. WHO/VBC/69.119. 17 pp. Processed. (pp. 401-416 of Proceedings, published 1969.)
Presented at Second Inter. Congr. of Acarology, Nottingham, England. July 19-25, 1967. Reports experience and conveys views of the author. (Not to be quoted or abstracted without an agreement with W.H.O.)

2909. Axtell, R. C. 1968 Integrated House Fly Control: Populations of Fly Larvae and Predaceous Mites, *Macrocheles muscaedomesticae*, in Poultry Manure after Larvicide Treatment. J. Econ. Ent. 61:245-249.
Points out that larviciding manure with non-selective insecticides is detrimental to mite predators. ABSTRACTS: RAE-B 56:141(490), July, 1968; BA 49:6245(69845), July 1, 1968; TDB 65:843(1874), June, 1968.

2910. Axtell, R. C. 1970 Integrated Fly-control Program for Caged-Poultry Houses. J. Econ. Ent. 63(2):400-405.
Early-season removal of manure, and fly control by insecticide bait stations. Five or six selective insecticide applications to certain surfaces. (*Fannia* more easily controlled than *M. domestica.*)

2910a. Axtell, R. C. 1970 Fly Control in Caged-Poultry Houses: Comparison of Larviciding and Integrated Control Programs. J. Econ. Ent. 63(6):1734-1737.
Integrated program based on selective adulticiding with RaVap at 2- to 5-week intervals gave control as satisfactory as weekly larviciding. Mites, predaceous on immature stages of housefly, not harmed by selective adulticiding.

2910b. Axtell, R. C. and **Edwards, T. D. 1970** *Hermetia illucens* Control in Poultry Manure by Larviciding. J. Econ. Ent. 63(6):1786-1787.
Determined the effectiveness of 5 larvicides. *Musca domestica* larvae were absent when *H. illucens* larvae were present; became abundant when *H. illucens* larvae were controlled.

Axtell, R. C., joint author. *See* Farish, D. J., et al., 1966; O'Donnell, A. E., et al., 1965; Willis, R. R., et al., 1967, 1968.

Ayers, E. L., joint author. *See* Steiner, L. T., et al., 1961.

2911. Ayre, G. L. 1966 Colony Size and Food Consumption of Three Species of *Formica*. Ent. Exp. Appl. 9(4):461-467. (German summary.)
All colonies were maintained on 50 per cent honey solution plus laboratory-reared housefly larvae. ABSTRACT: BA 48:5549(62039), June 15, 1967.

2912. Azab, A. K., Tawfik, M. F. S. and **Awadallah, K. T. 1964** Insect Enemies of Common Flies in Giza Region. Bull. Soc. Ent. Egypte 47(1963): 277-286.
Reports 23·1 per cent of *Musca* pupae parasitized by *Nasonia vitripennis* Walker. ABSTRACT: RAE-B 55:84(300), May, 1967.

Baba, Y., joint author. *See* Tsukamotor, M., et al., 1961.

2913. Bachmann, F. M. 1960 Über die insektizide Wirkung von DDVP (0,0-Dimethyl-2,2-dichlorvinyl-phosphat). [On the insecticidal effect of dichlorvos.] Anz. Schädlingsk. 33(3):41-44. Berlin.
Dichlorvos (DDVP) proved more toxic than diazinon, gamma-BHC (lindane), aldrin, parathion or DDT and much more rapid in action than DDT. ABSTRACT: RAE-B 49:192-193, Sept., 1961.

2914. Bagent, J. L. 1967 Investigations on the Possible Existence of Rhythms in Sensitivity to Methyl Parathion in 2 Species of Insects. Diss. Absts. 28(5B):1972-B-1973-B.
Treated houseflies by topical technique. No evidence of rhythm. Tolerance to methyl parathion increased with age.

2915. Bagirov, G. A. 1964 [On the duration of the contact and intestinal action of chlorophos bait on house flies.] Azerbaidzh. Med. Zh. 1:74-76.

Bagirov, G. A., joint author. *See* Trofimov, G. K., et al., 1961.

Bagirov, K. S., joint author. *See* Rumiantsev, I. N., et al., 1961.

2916. **Bailey, D. L., LaBrecque, G. C.** and **Bishop, P. M. 1967** Residual Sprays for the Control of House Flies, *Musca domestica*, in Dairy Barns. Florida Ent. 50(3):161-163.
Nine insecticides evaluated for use in barns, in which bio-degradable compounds, leaving no residue, are considered desirable. An emulsion of dimethoate was most effective. ABSTRACT: BA 49:1905(21440), Feb. 15, 1968.

2917. **Bailey, D. L., LaBrecque, G. C., Meifert, D. W.** and **Bishop, P. M. 1968** Insecticides in Dry Sugar Baits Against Two Strains of House Flies. J. Econ. Ent. 61(3):743-747.
Of 62 compounds tested, 29 were superior to a trichlorfon standard Resistant strain showed more than 2-fold cross-resistance to all compounds except naled. ABSTRACTS: BA 49:10244(113078), Nov. 1, 1968; RAE-B 56:219(801), Nov., 1968; TDB 65:1308(2966), Oct., 1968.

2918. **Bailey, D. L., LaBrecque, G. C.** and **Whitfield, T. L. 1970** Laboratory Evaluation of Insecticides as Contact Sprays Against Adult House Flies. J. Econ. Ent. 63(1):275-276.
Of 81 insecticides tested, six were more effective than standard ronnel against both a resistant and a susceptible strain. Sixteen others were more effective than ronnel against resistant flies, less effective against the susceptible strain. ABSTRACTS: RAE-B 58:233(954), July, 1970; BA 51:9098(92604), Aug. 15, 1970.

2918a. **Bailey, D. L., LaBrecque, G. C.** and **Whitfield, T. L. 1970** Insecticides in Dry Sugar Baits for Control of House Flies in Florida Dairy Barns. J. Econ. Ent. 63(6):2000-2001.
Dimethoate, fenthion, formothion, naled, ronnel and trichlorfon gave an average of better than 75 per cent control for 18 days.

2919. **Bailey, D. L., LaBrecque, G. C.** and **Whitfield, T. L. 1970** Insecticides Applied as Low-Volume and Conventional Sprays to Control Larvae of the House Fly in Poultry Houses. J. Econ. Ent. 63(3): 891-893.
With CO_2 as the propellent, CIBA C-9491 and Monsanto CP-5143 were the most effective. Control lasted 11 days, at concentrations used. ABSTRACTS: BA 51:13122(133521), Dec. 1, 1970; TDB 67:1293(2509), Oct., 1970.

2920. **Bailey, D. L., LaBrecque, G. C.** and **Whitfield, T. L. 1970** Resistance of House Flies (Diptera:Muscidae) to Dimethoate and Ronnel in Florida. Florida Ent. 53(1):1-5.
Compared to susceptible Orlando strain, flies from dairies were 4·9 to 21·2 fold more resistant to ronnel, and 3·4 to 31·0 fold more resistant to dimethoate. With flies from poultry farms, figures were 3·8 to 54·5 and 1·8 to 28·5. ABSTRACT: BA 51:8530(86934), Aug. 1, 1970.

2921. **Bailey, D. L., Meifert, D. W.** and **Bishop, P. M. 1968** Control of Houseflies in Poultry Houses with Larvicides. Florida Ent. 51(2): 107-111.
Of 19 larvicides tested, none were effective beyond the ninth day. Using diazinon, field strain was 75·6 times harder to kill at LC 50 level than susceptible laboratory strain, 120·5 times harder at LC 90. ABSTRACT: BA 49:8721(97066), Sept. 15, 1968.

Bailey, P. T., joint author. *See* Monro, J., et al., 1965.

Bajpai, V. N., joint author. *See* Tiwari, B. K., et al., 1966.

2922. **Baker, G. J. 1963** The "Dual Synergist System" of Piperonyl Butoxide and MGK 264. Pyrethrum Post 7(1):16-18.
MGK 264 not an effective synergist by itself, but mixed with piperonyl

butoxide, shows some advantage over p-b alone for resistant flies. (Is less expensive than piperonyl butoxide.) ABSTRACT: RAE-B 53:122, June, 1965.

Baker, J. T., joint author. *See* Fay, R. W., et al., 1963.

2923. Bakry, N., Metcalf, R. L. and **Fukuto, T. R. 1968** Organothiocyanates as Insecticides and Carbamate Synergists. J. Econ. Ent. 61(5):1303-1309.

Benzyl thiocyanates, benzyl isothiocyanates, phenyl thiocyanates, and alkyl thiocyanates tested as synergists and as insecticides. Both types of activity require the intact organothiocyanate molecule. ABSTRACTS: BA 50:4221(44417), Apr. 15, 1969; RAE-B 57:51(183), Mar., 1969; TDB 62:166(302), Feb., 1969.

Bakuniak, E., joint author. *See* Gwiazda, M., et al., 1967.

2924. Balabaskaran, S., Clark, A. G., Cundell, A. and **Smith, J. N. 1968** Inhibitors for DDT-dehydrochlorinase. Australian J. Pharm. 49 (584) Suppl. No. 66, pp. S 69-S 70.

A comparison of certain phthaleins, sulphonphthaleins, and other phenyl methane dyes with WARF as inhibitors. Some compounds, effective *in vitro* were ineffective *in vivo.*

2925. Balan, N. K. 1967 [Toxicity of Dipterex for *Musca domestica* L. strains of flies with different levels of sensitivity to insecticides.] Med. Parazit. i Parazitarn. (Bolez.) 36(6):744-748. (In Russian, with English summary.)

Toxicity to flies of weak solutions of chlorophos, used as an intestinal poison, depends on insect's previous contact with insecticides. Manner of exposure (contact or ingestion) makes a difference. ABSTRACTS: BA 50:1051(11013), Jan. 15, 1969; TDB 65:541(1179), Apr., 1968. [Vol. 45 should read Vol. 36.] RAE-B 58:179(744), June, 1970.

2926. Balan, N. K. 1968 Sravnitel'naya privlekatel' nost' i toksichnost' dlya *M. domestica* khlorofosa idiptereksa. [Comparative attracting power and toxicity of chlorophos and Dipterex for *Musca domestica.*] Med. Parazit. i Parazitarn. (Bolez.) 37(4):422-423. (English summary.)

Technical quality chlorophos, used as an intestinal poison, is more effective in fly control than *purified* preparations of Dipterex. ABSTRACT: BA 50:9196(95044), Sept. 1, 1969.

2927. Balan, N. K. 1969 Sravnitel'naya privlekatel' nost' dlya komnatnoi mukhi zhidkikh primanok raznykh tsvetov. [Comparative attractiveness of liquid baits of different color for house flies.] Med. Parazit. i Parazitarn. (Bolez.) 38(1):46-48. (English summary.)

M. domestica prefers baits of (1) red color, (2) orange. Other colors, including black are less attractive. Contrast of bait color and background color is of little importance. ABSTRACTS: BA 50:10281(106342), Oct. 1, 1969; TDB 66:633(1252), June, 1969.

2928. Balashov, Yu. S., Gutsevich, A. V., Derbeneva-Ukhova, V. P. and **Shipitsyna, N. K. 1968** [Medical and veterinary entomology in the Soviet Union.] Ént. Obozrenie 47(2):298-316. 52 refs. (Summary in English). *Also* (English translation) Ent. Rev. 47(2):175-185. New York.

Synanthropic flies receive proportionate attention. ABSTRACT: RAE-B 58:3(7), Jan., 1970.

2929. Balazs, I. 1967 Labelling Pattern of RNA in various strains of *Musca domestica.* 1. Incorporation of Carbon-14 Uracil and the Effect of DDT Insecticide. Arch. Biol. Med. Exp. 4(1/2):191. (Abstract.)

2930. **Balazs, I.** and **Agosin, M. 1968** The Effect of 1, 1, 1-trichloro-2, 2-bis (para-chlorophenyl) ethane on Ribonucleic Acid Metabolism in *Musca domestica* L. Biochim. Biophys. Acta 157:1-7.
Examines the effect of pretreatment with DDT on RNA metabolism in DDT-resistant houseflies. DDT increases the incorporation of [2-^{14}C] uracil into total RNA. ABSTRACT: BA 49:6774(75550), July 15, 1968.

2931. **Balazs, I.** and **Agosin, M. 1968** Isolation and Characterization of Ribonucleic Acid from *Musca domestica* L. Comp. Biochem. Physiol. 27(1):227-237.
Ribonucleic acid isolated from adult female houseflies was separated into three principal fractions, differing in base composition, as in rat liver and *Drosophila*. ABSTRACT: BA 50:5345(56147), May 15, 1969.

Balazs, I., joint author. *See* Gil, L., et al., 1968.

2932. **Ballard, R. C. 1966** A Study of Electrocardiograms from Intact Male and Female House Flies. Ann. Ent. Soc. Am. 59(4):619-627.
For both sexes, pattern consists of slow systolic wave with spikes before, or on it, or both. Patterns of heat potentials variable. Female systolic wave has greater output and duration, but lower frequency. ABSTRACT: BA 47:9941(115000), Dec. 1, 1966.

2933. **Ballard, R. C.** and **Hall, M. S. 1969** A Suction-Electrode Study of the House Fly Heart. Ann. Ent. Soc. Am. 62(2):375-379.
Electrode potentials variable. Show spikelike components associated with a slow wave. Variability can exist in one segment independent of frequency, amplitude and contour of potential in adjacent one. ABSTRACT: BA 50:8651(89858), Aug. 15, 1969.

2934. **Ballard, R. C.** and **Holcomb, B. 1965** An Investigation of the House Fly Heartbeat through Transection Experiments. Ann. Ent. Soc. Am. 58:608-611.
Demonstrated at least 11 regions in the female and 10 in the male that can beat for extended periods. Origin of the beat not clearly understood. ABSTRACT: BA 46:8913(109934), Dec. 15, 1965.

Ballard, R. C., joint author. *See* Krueger, H. S., et al., 1965; Yale, T. H., et al., 1966.

2935. **Ballschmiter, K.** and **Tölg, G. 1966** Metabolismus des Thiodans in Insekten. [Metabolism of Thiodan in insects.] Angew. Chem. Int. Edition-Engl. 5(8):730; Deutsch Ed. 78(16):775-776.
Metabolite of Thiodan in *M. domestica* compared with corresponding product in the migratory locust, *Pachytilus migratorius migratorioides*. Diagrams of structural formulae. ABSTRACT: BA 48:3263(35990), Apr. 1, 1967.

Balsbaugh, E. U., Jr., joint author. *See* Kantack, B. H., et al., 1967.

Bamberg, K. J., joint author. *See* Dann, O., et al., 1968.

2936. **Ban, T.** and **Nagata, C. 1966** The Electronic Structure of Carbamate Derivatives as the Inhibitors of Cholinesterase. Jap. J. Pharmacol. 16(1):32-38.
Establishes relationship between electronic structure and inhibitory potency of several derivatives against true (acetyl) cholinesterase from fly brain. ABSTRACT: BA 48:2360(25807), Mar. 1, 1967.

2937. **Barfield, S. 1966** Practical Fly Control for Horse Keeping on Hillside Residential Lots. J. Environ. Hlth. 28(4):283-287.

2938. **Barkai, A.** and **Lidror, R. 1966** The Susceptibility of Rural and Urban Strains of Houseflies in Israel to Several Insecticides. Riv. di Parassit. 27(4)-261-268.
Urban houseflies showed a higher resistance to malathion than rural

ones. Resistance to dipterex, diazinon, and lindane was at a lower level than to malathion. No cross resistance to dibrom. ABSTRACTS: BA 49: 2836(31709), Mar. 15, 1968; RAE-B 57:24(105), Jan. 1969; TDB 64:1038, Sept., 1967.

2939. Barker, R. J. 1960 Syntheses of the Aliphatic Deuterium Analogs of DDT and TDE and Their Toxicity and Degradation When Applied to Adult House Flies. J. Econ. Ent. 53(1):35-41.
Synthesized analogs used to measure isotope effect on absorption spectra, insecticidal action and chemical stability. Deuteration gave little change in toxicity or degradation. ABSTRACT: RAE-B 49:81, Apr., 1961.

2940. Barker, R. J. and **Edmunds, L. N., Jr. 1963** Comparing Resistance of House Flies to the Eight Stereoisomers of Allethrin. J. Econ. Ent. 56:152-155.
Toxicity involved both optical and geometric specificity. Dosage response slopes were not parallel for all isomers and showed no change correlated with resistance. ABSTRACTS: BA 43:1329(16481), Aug. 15, 1963; RAE-B 51:146, July, 1963; TDB 60:903, Sept., 1963.

Barker, R. J., joint author. *See* Goldsmith, T. H., et al., 1964; Cohen, C. F., et al., 1962, 1963.

Barlow, F., joint author. *See* Hadaway, A., et al., 1963.

2941. Barlow, J. S. 1966 Effects of Diet on the Composition of Body Fat in *Lucilia sericata* (Meigen). Nature 212(5069):1478-1479.
Essentially the same as *M. domestica* for the three fatty acids: palmitic, palmitoleic, and oleic. (See next reference.) ABSTRACT: BA 48:7434 (83210), Aug. 15, 1967.

2942. Barlow, J. S. 1966 Effects of Diet on the Composition of Body Fat in *Musca domestica* L. Canad. J. Zool. 44(5):775-779.
Concentration of lipids in the body was directly related to the amount of fatty acid in the diet. Fats had little or no effect on rate of larval growth and development. ABSTRACTS: RAE-B 56:34(107), Feb., 1968; BA 48:464(4935), Jan. 1, 1967.

2943. Barnes, J. R. and **Fellig, J. 1969** Synergism of Carbamate Insecticides by Phenyl 2-Propynyl Ethers. J. Econ. Ent. 62(1):86-89.
These ethers strongly synergize 11 carbamate insecticides against *M. domestica*; were 2 to 5 times more active than the methylenedioxphenyl synergists and have a broad biological spectrum of activity. ABSTRACT: BA 51:539(5386), Jan. 1, 1970.

Barnes, J. R., joint author. *See* Fellig, J., et al., 1970.

2944. Barnes, W. W. 1965 The Absorption and Metabolism of C^{14}-labelled Endosulfan in the Housefly [*Musca domestica*]. Diss. Absts. 25(8): 4871.

2945. Barnes, W. W. and **Ware, G. W. 1965** The Absorption and Metabolism of C^{14}-Labelled Endosulfan in the House Fly. J. Econ. Ent. 58(2): 286-291.
Gas and paper chromatography used to analyze external, internal, and fecal extracts in hexane from cyclodiene-resistant and cyclodiene-susceptible female flies to which toxicant had been topically applied. Figures include chromatograms. ABSTRACTS: BA 46:5144(63875), July 15, 1965; RAE-B 53:128, July, 1965; TDB 62:811, Aug., 1965.

Barsy, G., joint author. *See* Sztankay-Gulyas, M., et al., 1962.

2946. Barthel, W. F. 1961 Synthetic Pyrethroids. Advances in Pest Control Res. 4:33-74.
Synthetic and natural pyrethroids compared. Synergical qualities of different synthetic pyrethroids. A compilation from 143 references. ABSTRACT: BA 36:5562(58138), Sept. 1, 1961.

2947. **Barthel, W. F. 1964** Insecticidal Activity of Barthrin, Dimethrin, and Related Compounds. Wld. Rev. Pest Contr. 3(2):97-101.
These compounds, of low mammalian toxicity, were used to control larvae in feces and adult flies in areas around animals. Residue problems minimal.

2948. **Barton-Browne, L.** and **Kerr, R. W. 1967** The Response of the Labellar Taste Receptors of DDT-resistant and Non-resistant Houseflies (*Musca domestica*) to DDT. Ent. Exp. Appl. 10(3-4):337-346. (German summary.)
Two characteristics of DDT resistance: (1) a significantly higher threshhold to DDT, (2) ability to recover from DDT poisoning. Resistance of strain HR is fully expressed at the level of chemoreceptor hairs on the labella; is mainly due to characteristic (2), above. ABSTRACT: RAE-B 56:104(363), May, 1968.

2949. **Bar-Zeev, M. 1962** The effect of acceptable and unacceptable compounds on the orientation of houseflies and mosquitoes. XI. Int. Kong. für Entomologies, Wien, 17 bis 25 Aug. Band II. Sektion X. Medizinische und veterinärmedizinische Entomologie. Verh. XI Int. Kongr. Ent. 2:444-447.
Studied the effects on orientation of sweet, sour, salt and bitter stimuli, acting through the contact chemoreceptors on the tarsi of houseflies. ABSTRACT: RAE-B 51:245, Nov., 1963.

2950. **Basheir, S. E. 1967** Causes of Resistance to DDT in a Diazinon-selected and a DDT-selected strain of House Flies. Ent. Exp. Appl. 10(1):111-126. (German summary.)
Ability to convert DDT to DDE not necessarily the most important cause of DDT resistance in all strains. Selection with diazinon may result in resistance to DDT by more than one mechanism. ABSTRACT: BA 48:11241(124865), Dec. 15, 1967.

2951. **Basheir, El S.** and **Lord, K. A. 1965** DDT Tolerance in Diazinon-selected and DDT-selected Strains of Houseflies. Chem. and Ind., pp. 1598-1599. September 11, 1965.
DDT changed to non-toxic metabolites much faster in both resistant strains than in the susceptible; but diazinon-selected flies degrade DDE much faster and more completely than DDT-selected strain.

2952. **Basurmanova, O. K. 1966** [Ultrastructural changes in the synaptic zone of the I optic ganglion of insects under various functional conditions.] Biofizika 11(2):263-266. (In Russian.)
The retina and I and II optic ganglia from *Musca domestica* were prepared for electron microscope study from control, light stimulated and dark adapted animals. ABSTRACT: BA 48:11242(124866), Dec. 15, 1967.

2953. **Basurmanova, O. K., et al. 1968** [The effect of ultrasonic waves on brain cells in larvae.] Dokl. Akad. Nauk, SSSR 181:476-477. (In Russian.)

Basurmanova, O. K., joint author. *See* El' Piner, L. E., et al., 1965.

2954. **Bates, A. N., Hewlett, P. S.** and **Lloyd, C. J. 1965** Synergistic Effects of some Compounds Related to 2-diethylamino-ethyl 2, 2-diphenyl-n-pentanoate (SKF525A) on the Insecticidal Activity of Pyrethrins. J. Sci. Fd. Agric. 16(6):289-292. London.
Several compounds proved to have synergistic capacity but none approached piperonyl butoxide. ABSTRACTS: BA 48:4644(51874), May 15, 1967; RAE-B 54:45, Feb., 1966.

Bates, A. N., joint author. *See* Hewlett, P. S., et al., 1961.

2955. Bathtel, W. F. 1964 Insecticidal Activity of Barthrin, Dimethrin and Related Compounds. Wld. Rev. Pest Contr. 3(2):97-101.

Treats of synthetic pyrethroids; Concerns *Aedes aegypti, Anopheles quadrimaculatus, Musca autumnalis.*

2955a. Batth, S. S. and **Stalker, J. M. 1970** A Survey of Canadian Populations of the House Fly for Resistance to Insecticides. J. Econ. Ent. 63(6):1947-1950.

All collections showed high resistance to DDT, *gamma* BHC and malathion; most were moderately resistant to ronnel. No substantial resistance to dichlorvos.

Batova, B., joint author. *See* Havlik, B., et al., 1961.

2956. Bauer, J. W. and **Monroe, R. E. 1969** Metabolism of Cholesterol-4-C^{14} in Insecticide-Resistant and Susceptible Strains of Aseptically Reared House Flies, *Musca domestica.* Ann. Ent. Soc. Am. 62(5): 1021-1025.

Uptake was different for all six strains, but always higher for females than for males. No correlation of cholesterol content as between resistant and susceptible flies. ABSTRACTS: BA 51:4513(46494), Apr., 1970; RAE-B 58:193(801), June, 1970.

2957. Bay, D. E., Pitts, C. W. and **Ward, G. 1958** Oviposition and Development of the Face Fly in Feces of Six Species of Animals. J. Econ. Ent. 61(6):1733-1735.

Feces of swine, bison and cattle served as good substrates. Unless reconstructed to a comparable moisture content, feces of horse, deer, and sheep were unsatisfactory. ABSTRACTS: BA 50:4776(50206), May 1, 1969; RAE-B 57:89(310), May, 1969.

2958. Bay, D. E., Pitts, C. W. and **Ward, G. 1969** Influence of Moisture Content of Bovine Feces on Oviposition and Development of the Face Fly. J. Econ. Ent. 62(1):41-44.

Feces containing 35 per cent moisture were most attractive. Some preference for dung from different diets was also observed. ABSTRACTS: BA:563(5661), Jan. 1, 1970; RAE-B 57:117(416), July, 1969.

2958a. Bay, D. E., Pitts, C. W. and **Ward, G. 1970** Face Fly Larvae Development in Relation to Bovine Sex and Quantity of Feces. J. Econ. Ent. 63(6):1973.

A minimal amount of 2·0 g. feces per larva was necessary for maximal pupal weight and per cent adult emergence. Sexual origin of feces not significant.

Bay, E. C., joint author. *See* Legner, E. F., et al., 1965, 1966, 1967, 1970.

2959. Bazante, G. 1966 Un problème à éclaircir: celui de la tue-mouche: l'Amanite tue-mouche, bien ou mal nommée? (Suite). [A problem to be clarified: that of fly-bane: the Amanita fly bane, well or badly named?] Rev. Mycol. (Paris) 31(3):261-268.

Reviews history of *Amanita muscaria.* Was regarded as flykiller and used for this purpose. Various reports indicate toxicity may vary in different countries. ABSTRACT: BA 50:1551(16300), Feb. 1, 1969.

Beard, J. R., joint author. *See* Bridges, R. G., et al., 1962.

2960. Beard, R. L. 1960 Laboratory Studies on House Fly Populations. III. The Influence of Insecticides on Population Trends. Bull. Connecticut Agric. Expt. Sta. 631, 22 pp.

As a further report on studies of housefly populations reared in specialized laboratory environments, this bulletin considers insecticides as ecological factors affecting numbers. ABSTRACTS: BA 37:1584(15814), Feb. 15), 1962; RAE-B 51:120, June, 1963.

2961. **Beard, R. L. 1963** Fighting the Face Fly (Diptera). Biological Control is Far from Simple. Frontiers of Plant Science 16(1):4.
Mentions use of face fly parasites in current research and deals with problems of "domestication", e.g. flies adapted for laboratory rearing may differ genetically from the field population.

2962. **Beard, R. L. 1964** Pathogenic Stinging of House Fly Pupae by *Nasonia vitripennis* (Walker). J. Insect Path. 6(1):1-7.
Heavy mortality in house fly pupae results from repeated stinging by female wasps. If this occurs during first 24 hours of pupal life, the *Nasonia* offspring within also dies; if later, it can survive. ABSTRACTS: BA 46:4071(50352), June 1, 1965; RAE-B 52:116, June, 1964.

2963. **Beard, R. L. 1964** Parasites of Muscoid Flies. Bull. W. H. O. 31:491-493.
Comments on two parasites: *Nasonia vitripennis* and *Aphaereta pallipes*; their culture, host-parasite relations, etc. ABSTRACTS: BA 47:4219(49492), May 15, 1966; RAE-B 55:26(88), Feb., 1967; TDB 62:932. Sept., 1965.

2964. **Beard, R. L. 1965** Ovarian Suppression by DDT and Resistance in the House Fly (*Musca domestica* L.). Ent. Exp. Appl. 8(3):193-204. (German summary.)
Concerns chronically toxic action of DDT expressed in reduced fecundity in females fed sub-lethal amounts of the insecticide. A DDT-resistant fly may consume DDT contaminated food and show ovarian suppression. ABSTRACTS: BA 47:5099(59831), June 15, 1966; RAE-B 54:69, Apr., 1966.

2965. **Beard, R. L. 1965** Observations on House Flies in High-Ozone Environments. Ann. Ent. Soc. Am. 58:404-405.
Low concentrations of ozone are beneficial to flies, directly or indirectly. High concentrations can reduce fecundity. Duration of exposure also important. ABSTRACTS: BA 46:6268(77839), Sept. 1, 1965; RAE-B 53:195, Oct., 1965.

2966. **Beard, R. L. 1965** Competition between DDT-Resistant and Susceptible House Flies. J. Econ. Ent. 58:584.
Strains differ in regard to survival traits under competition, but this phenomenon need not be associated with resistance to an insecticide. ABSTRACTS: BA 47:3378(39871), Apr. 15, 1966; RAE-B 53:183, Sept., 1965; TDB 63:237, Feb., 1966.

2967. **Beard, R. L. 1970** Outbred, Inbred, and Hybrid House Flies as Test Insects. J. Econ. Ent. 63(1):229-236.
Comparison of many characteristics indicate much more variability than is desirable for a test insect. Hybrid flies usually not advantageous, except where vigor is the prime consideration. ABSTRACTS: BA 51:9141 (93109), Aug. 15, 1970; RAE-B 58:232(950), July, 1970.

2968. **Beard, R. L.** and **Walton, G. S. 1965** An *Aspergillus* Toxin Lethal to Larvae of the Housefly. J. Invertebr. Path. 7:522-523.
A water-soluble toxin produced by this fungus (*A. flavus*) acts as an insecticide on fly maggots; usually during the first instar, but is potent on second and third stage larvae also. ABSTRACTS: BA 47:5535(64895), July 1, 1966; RAE-B 54:135, July, 1966.

2969. **Becker, G. 1963** Magnetfeld-Orienterung von Dipteren. Naturwissenschaften 50(21):664.
M. domestica and others tend to orient in the direction of, or at right angles to the earth's magnetic field, especially when the weather is calm and sunny. ABSTRACT: BA 45:4447(54138), July 1, 1964.

2970. **Becker, G. 1965** Zur Magnetfeld-Orientierung von Dipteren. Zeit. Vergl. Physiol. 51:135-150.
Observed orientation of imagines of *M. domestica* in landing and resting.

In a natural magnetic field preference for a N/S or E/W direction could be verified.

2971. **Becker, G.** and **Speck, U. 1964** Untersuchungen über die Magnetfeld-Orientierung von Dipteren. [Investigations concerning the magnetic-field-orientation of Diptera.] Zeit. Vergl. Physiol. 49(4): 301-340.

Mechanoreceptors perceive geomagnetic field forces and the fly responds by adjusting to a distinct direction. ABSTRACT: BA 48:1891(20520), Feb. 15, 1967.

Beechey, R. B., joint author. *See* Donnellan, J. F., et al., 1969.

Behnke, C. N., joint author. *See* Yendol, W. G., et al., 1968.

2972. **Behrenz, W.** and **Böcker, E. 1965** Baygon, a New, Promising Public Health Insecticide of the Organic Carbamate Group. Pfl. Schutz-Nachr. Bayer 18(2):53-81.

Gives details of chemical and physical properties of arprocarb, the active ingredient of Baygon preparations. Tested against 18 species of insects, including flies. ABSTRACT: RAE-B 55:109(392), June, 1967.

2973. **Bell, D. D., Bowen, W. R., Deal, A. S.** and **Loomis, E. C. 1965** Diazinon Dust for Fly Control in Poultry Manure. Calif. Agric. 19(2): 8-9.

A dust made from Diazinon wettable powder and gypsum killed 4 species of domestic flies, but failed to control *M. domestica*, apparently because of resistance to this insecticide. ABSTRACTS: BA 47:1599(19357), Feb. 15, 1966; RAE-B 55-89(317), May, 1967.

2974. **Bell, D.** and **Daehnert, R. H. 1962** Control of House Flies on Poultry Ranches with Antiresistant/DDT. J. Econ. Ent. 55(5):817-819.

AR/DDT accomplished a rapid kill. If application had been accompanied by proper handling of manure, residual fly control would have extended beyond the 2-3 weeks of the study. ABSTRACTS: BA 41:955(12416), Feb. 1, 1963; RAE-B 51:45, Feb., 1963.

2975. **Bell, J. D. 1968** Patterns of Cross Resistance to Organophosphates and Carbamates in *Musca domestica*. Bull. Ent. Res. 58(1):137-151.

Shows 2 new types of resistance mechanisms. Demonstrates existence of at least 3 resistant mechanisms for carbamates. ABSTRACTS: BA 50:2673 (27893), Mar. 1, 1969; RAE-B 56:243(901), Dec., 1968; TDB 65:1484 (3295), Dec., 1968.

2976. **Bell, J. D. 1968** Genetical Investigations on a Strain of House-flies Resistant to Organophosphates and Carbamates. Bull. Ent. Res. 58(2):191-200.

Observed segregation of resistance to arprocarb in F_1, F_2 and out-cross flies. Believes 2 genes involved: (1) concerned with an esterase-destroying system, (2) concerned with the microsomal oxidative system. ABSTRACTS: RAE-B 57:31(128), Feb., 1969; BA 50:5382(56519), 1969; TDB 66:1338 (2533), Dec., 1969.

2977. **Bell, J. D.** and **Busvine, J. R. 1967** Synergism of Organophosphates in *Musca domestica* and *Chrysomya putoria*. Ent. Exp. Appl. 10(2): 263-269. (German summary.)

Eight organophosphates synergized five insecticides (especially malathion) more effectively against resistant than against normal strains. ABSTRACTS: BA 49:927(10404), Jan. 15, 1968; RAE-B 56:10(32), Jan., 1968; TDB 65:340(784), Mar., 1968.

Bellenger, G. J., joint author. *See* Rogers, P. A., et al., 1967.

Belozerskii, A. N., joint author. *See* Ermokhina, T. M., et al., 1966.

Beltz, A. D., joint author. *See* Rogoff, W. M., et al., 1964.

2978. Benavides Gomez, M. 1967 Control quimico de la mosca casera. [Chemical control of the housefly.] Cenicafe 18(1):13-19.
Dipterex-Nuvan (DDVP)-Sugar produced an average mortality of 95·12 per cent, and was residually potent for 10 weeks; Malathion-Nuvan-Sugar an average mortality of 93·69 per cent. Latter was residually potent for 8 weeks. ABSTRACT: BA 50:2099(22017), Feb. 15, 1969.

Benetti, M. P., joint author. *See* Sacca, G., et al., 1960.

2979. Benezet, H. J. and **Hansens, E. J. 1967** A Study of the Moisture Preference for Pupation of Several Strains of House Flies. J. Econ. Ent. 60:777-781.
Resistant strains selected drier areas for pupation. Statistically significant differences found in moisture preference between media with 20, 10, and 5 per cent moisture differences; also between strains when there were moisture differences of 5 and 10 per cent. ABSTRACTS: BA 49:1417(15888), Feb. 1, 1968; RAE-B 55:189(652), Oct., 1967; TDB 65:81(182), Jan., 1968.

Benezet, H. J., joint author. *See* Hansens, E. J., et al., 1967.

2980. Benschoter, C. A. 1967 Effect of Dietary Biotin on Reproduction of the House Fly. J. Econ. Ent. 60(5):1326-1328.
Continuous feeding on 2 per cent biotin in food rendered females completely sterile, but males were only slightly affected. A percentage of 0·25 reduced egg hatch by half. ABSTRACTS: BA 49:1399(15663), Feb. 1, 1968 (*as* Benschotier); RAE-B 56:64(211), Mar., 1968; TDB 65:538(1174), Apr., 1968.

2981. Benson, O. L. and **Wingo, C. W. 1963** Investigations of the Face Fly in Missouri. J. Econ. Ent. 56:251-258.
Treats of hibernation, dispersal, feeding, parasites, control (including baits, sprays and back-rubbers). Face fly first found in Missouri in 1960. ABSTRACTS: BA 44:622(8356), Oct. 15, 1963; RAE-B 51:199, Sept., 1963.

2982. Benson, R. L. and **Friedman, S. 1970** Allosteric Control of Glucosamine Phosphate Isomerase from the Adult Housefly and its Role in the Synthesis of Glucosamine 6-phosphate. J. Biol. Chem. 245: 2219-2228.
Identifies the enzyme catalyzing chitin production, also necessary components of the reaction. Work done with actively feeding young flies. ABSTRACT: BA 51:10895(110679), Oct. 1, 1970.

2983. Beran, F. 1960 Die DDT-Empfindlechkeit verschiedener österreichischer Herkünfte von *Musca domestica* L. [The susceptibility to DDT of different Austrian strains of *M. domestica*.] Bodenkultur Ausgabe A, 11(2):131-140. Vienna.
Used topical applications; results interpreted by probit analysis. Several native strains showed slight to medium resistance. Recommends rotation of insecticides. ABSTRACT: RAE-B 49:111, June, 1961.

2984. Berberet, R. C. and **Landers, P. L. 1969** Correlation of Cellular Activity in Weismann's Ring with Molting Cycles in Larvae of *Musca autumnalis*. Ann. Ent. Soc. Am. 62(2):446-447.
Cells of Weismann's ring showed definite increase in activity during intermolt period; increased activity correlated with release of molting hormone.

2984a. Bergamini, P. G., Palmas, G., Piantelli, F. and **Rigato, M. 1970** Absorption of 137Cs by *Musca domestica* and Consequent Environmental Contamination. Health Phys. 18(5):491-498.
Larvae, pupae and adults were tagged with 137Cs of different concentra-

tions and the specific activity studied throughout the organism's life cycle into the ensuing generation. ABSTRACT: BA 51:12051(122385), Nov. 1, 1970.

2985. **Berge, O. I.** and **Fisher, E. H. 1964** Mechanized Fly Control for the Dairy Herd: Circular 604. University of Wisconsin Extension Service, College of Agriculture, Madison. 8 pp.
Utilizes a compressed air fogger in self-treatment area for cattle. Excellent control.

2986. **Berger, R. S. 1962** The Enzymatic Dyhydrohalogenation of DDT and Several other Diarylhaloethanes by Houseflies. Diss. Absts. 22(7): 2169-2170.

2987. **Berger, R. S.** and **Young, R. G. 1962** Specificity of Diarylhaleothane-Dehydrohalogenase of Susceptible and DDT-Resistant House Flies. J. Econ. Ent. 55:533-536.
Determined rates of enzymatic dehydrohalogenation by fly homogenates. No evidence of enzymatic preference for either enantiomorph of a racemic mixture of an optically active substrate. ABSTRACTS: BA 40: 1633(21312), Dec. 1, 1962 (Authorship given as Berger alone); RAE-B 51:4, Jan., 1963; TDB 59:1212, Dec., 1962.

2988. **Bergh, S. G. Van den** and **Slater, E. C. 1960** The Respiratory Activity and Respiratory Control of Sarcosomes Isolated from the Thoracic Muscle of the Housefly. Biochim. Biophys. Acta 40:176-177.
P:O ratios and respiratory control measured in isolated sarcosomes. Respiratory rates measured in the presence of ADP were much higher than previously reported.

2989. **Bergh, S. G. Van den** and **Slater, E. C. 1962** The Respiratory Activity and Permeability of Housefly Sarcosomes. Biochem. J. 82:362-371.
Sarcosomes from thoracic muscle can oxidize OC glycerophosphate and pyruvate at rates comparable with those for the flying insect. High P:O ratios in the absence of serum albumin. ABSTRACT: BA 38:299(4119), Apr. 1, 1962.

2990. **Bergmann, E. D.** and **Levinson, Z. H. 1966** Utilization of Steroid Derivatives by Larvae of *Musca vicina* (Macq.) and *Dermestes maculatus* Deg. J. Insect. Physiol. 12(1):77-81.
Fifteen derivatives and isomers of cholesterol tested as to biological availability for *M. vicina* larvae. Authors discuss the effect of various changes in the cholesterol skeleton. ABSTRACT: BA 47:5100(59833), June 15, 1966.

Bergmann, E. D., joint author. *See* Ascher, K. R. S., et al., 1962.

Bergstrom, R. E., joint author. *See* Walsh, J. D., et al., 1968.

2991. **Bernardini, P. M., Vecchi, M. L.** and **Laudani, U. 1965** Localizzazione istochimica della citocromo-C ossidasi nelle pupe di *Musca domestica* L. [Histochemical localization of cytochrome-c oxidase in the pupa of *Muscia domestica* L.] Boll. Zoologia 30(2):759-764. Atti XXXIV Convegno dell'U.Z.I.
Activity is localized mainly in the thorax; slightly in the abdomen. In the head, activity is confined to the neuroglandular histoblasts.

Berndt, W. L., joint author. *See* Kantack, B. H., et al., 1967.

2992. **Bernhard, C. G. 1967** Light Transmission and its Regulation in the Compound Eye. Med. Biol. Illus. 17:100-107.
A review of recent electro-physiological, spectrophotometric and micro-wave model experiments which have helped to explain the transfer of incident light to the photo-sensitive parts of the receptor cells.

2993. **Beroza, M. 1963** (Transferred to Section VII, Technique.)

2994. Beroza, M. and **LaBrecque, G. C. 1967** Chemosterilant Activity of Oils, especially Oil of *Sterculia foetida*, in the House Fly. J. Econ. Ent. 60:196-199.
Sterilized females only. Flies on diets containing 2·5 and 5 per cent of the ingredient produced no eggs. Oil considered edible, therefore might be practical in areas where aziridine sterilants cannot be used. ABSTRACTS: BA 48:6011(67184), July 1, 1967; RAE-B 55:106(376), June, 1967; TDB 64:1140, Oct., 1967.

Beroza, M., joint author. *See* Bowman, M. C., et al., 1968; Miller, R. W., et al., 1970.

2995. Berreur, P. and **Fraenkel, G. 1969** Puparium Formation in Flies: Contraction to Puparium Induced by Ecdysone. Science 164(3884): 1182-1183.
Ecdysone controls both puparial contraction and tanning. Work done with *Calliphora erythrocephala*, but principle presumably applies to all muscoids. ABSTRACT: BA 50:11958(123650), Nov. 15, 1969.

Berry, I. L., joint author. *See* Goffman, R. A., et al., 1965.

2996. Berteau, P. E. and **Casida, J. E. 1969** Synthesis and Insecticidal Activity of some Pyrethroid-like Compounds Including Ones Lacking Cyclopropane or Ester Groupings. J. Agr. Fd. Chem. 17(5):931-938.
Compounds resemble structurally allethrin or chrysanthemumate. Some are similar to pyrethrum in their action. Refers to literature on flies. ABSTRACT: BA 51:2196(22727), Feb. 15, 1970.

2997. Bessonova, I. V., Vashkov, V. I., Volkov, Yu. P., Volkova, A. P., Zhuk, E. B., Zubova, G. M., Polischuk, L. A. and **Tsetlin, V. M. 1969** Relative Activity of Pyrethroids in Aerosol Form. J. Hyg. Epidemiol. Microbiol. Immunol. 13(1):56-63. (French, German, Spanish summaries.)
Concerns natural pyrethrins with and without piperonyl butoxide. Neopinamine showed greatest activity. Good against *M. domestica* and is non-toxic to mammals. ABSTRACT: BA 50:11924(123253), Nov. 15, 1969.

2998. Bessonova, I. V., Vashkov, V. I., Volkov, Yu. P., Volkova, A. P., Zhuk, E. B., Zubova, G. M. and **Tsetlin, V. M. 1968** [Use of dimetrin to control houseflies.] Gig. I. Sanit. 33(5):116-118.
Dimetrin not suitable as a substitute for pyrethrins for filling aerosol insecticide balloons. Dimetrin-octaclorodipropyl ether (1:10) and dimetrin-pyrethrin (7:3) recommended mixtures. ABSTRACT: BA 49:11334 (124376), Dec. 1, 1968.

2999. Bettini, S. 1965 Acquired Immune Response of the House Fly, *Musca domestica* (Linnaeus) to Injected Venom of the Spider *Latrodectus mactans tredecimguttatus* (Rossi). J. Invertebr. Path. 7(3):378-383.
M. domestica is the first species of Diptera to show protective immunity following inoculation with a sub-lethal dose of antigen. ABSTRACTS: BA 47:8559(99748), Oct. 15, 1966; RAE-B 54:24, Jan., 1966.

3000. Bettini, S., Boccacci, M. and **Natalizi, G. 1960** Ricerche sugli effetti tossici di alcuni bromo-e chloroacetati in ceppi sensibili e resistenti di *Musca domestica* L. [Studies of the toxic effect of some bromo- and chloroacetates to susceptible and resistance strains of *Musca domestic* L.] Rend. Ist. Sup. San. (Rome) 23:634-651.
Speed of action greater for iodo- and bromoacetic acids than for chloroacetates. Some strains suceptible to myristyl and cetyl compounds. ABSTRACTS: BA 36:5884(62110), Sept. 15, 1961; RAE-B 51:120, June, 1963; TDB 58:389, Mar., 1961.

3001. **Bettini, S. Boccacci, M.** and **Natalizi, G. 1960** Cumulative Effect of Sublethal Doses of Myristyl Chloroacetate on a Highly DDT-Resistant Strain of *Musca domestica* L. J. Econ. Ent. 53(1):99-101.
Accumulation of toxic effects correlated with the time intervals involved. ABSTRACTS: BA 35:3455(39608), July 15, 1960; RAE-B 49:81, Apr., 1961.

Bettini, S., joint author. *See* Gattone, F., et al., 1963; Vicari, G., et al., 1965.

Beye, E., joint author. *See* Reiff, M., et al., 1960.

3002. **Bhalla, S. C. 1964** Some Measures of Fitness in Different Genotypes of the Housefly. J. Kans. Ent. Soc. 37:202-207.
Concerns two variables: (1) per cent adult emergence, (2) mean dry weights of the emerged adults. Females were consistently heavier than males. ABSTRACT: BA 46-1066(13467), Feb. 1, 1965.

3003. **Bhalla, S. C.** and **Sokal, R. R. 1964** Competition among Genotypes in the Housefly at Varying Densities and Proportions: The Green Strain. Evolution. 18(2):312-330.
Studied 4 genotypes. Fitness of a genotype in mixed culture is not predictable from its fitness in pure culture. Mean dry weight decreases upon crowding before emergence does. ABSTRACT: BA 45:8104(102863), Dec. 15, 1964.

3004. **Bhaskaran, G.** and **Deoras, P. J. 1963** Secretory Activity of the Corpus Cardiacum during the Pupal Development of *Musca nebulo. In* 16th Int. Cong. Zool., Proc. Int. Cong. Zool. 16(2):125. (Abstract only.)
Describes histological evidence indicating a definite secretory function for the chromophile cells of this organ. ABSTRACT: BA 45:4742(57838), July 1, 1964.

3005. **Bhaskaran, G.** and **Deoras, P. J. 1967** Studies on the Neuroendocrine System in the Indian House Fly, *Musca nebulo* Fabr. II. Histological Changes in the Neuroendocrine System during Metamorphosis. J. Univ. Bombay 35:41-58.
Histological detail of the neurosecretory system and the ring gland during metamorphosis. Probable significance. For other papers of this series: *See* Deoras.

3006. **Bhaskaran, G.** and **Nair, K. K. 1965** Inhibition of Nuclear RNA Synthesis in Allatectomised Female Housefly. Proc. Seminar Int. Cell Biol. pp. 64-66. (Paper 12.)
Autoradiographic techniques used to establish the role of corpus allatum hormone in the regulation of RNA synthesis in fat body and ovary.

3007. **Bhaskaran, G.** and **Nair, K. K. 1967** Effect of Allatectomy on RNA and Protein Synthesis in Tissues of the Female House Fly. J. Animal Morph. Physiol. 14(1):141.
A continuation of earlier studies. Fat body and ovaries are the tissues concerned.

3008. **Bhaskaran, G., Ramakrishnan, V.** and **Adeesan, C. 1970** Effects of Benzamide on Embryonic Development of the Housefly. Develop., Growth and Different. 11(4):265-276.
Substance is a chromosomal RNA inhibitor. Eggs exposed when first laid developed only to blastema stage. Synthesis of new RNA perhaps required for further growth. Treatment after blastema stage reversible; before this, not.

3009. **Bhaskaran, G.** and **Sivasubramanian, P. 1969** Metamorphosis of Imaginal Disks of the Housefly: Evagination of Transplanted Disks. J. Exp. Zool. 171(4):385-395.
New technique for transplanting disks. Results suggest that hormones

may induce or inhibit morphogenetic cell movement in imaginal disks. ABSTRACT: BA 51:3372(34879), Mar. 15, 1970.

3010. **Bhaskaran, G.** and **Sivasubramanian, P. 1969** Development of Transplanted Imaginal Disks in X-irradiated House Fly Pupae. Nature (London) 222:786-787.
Two types of indirect effect from radiation: (1) manifested in the whole body of the pupa, (2) localized in the irradiated epidermal cells and transmitted through cell contact.

Bhaskaran, G., joint author. *See* Deoras, P. J., et al., 1962, 1963, 1967; Nair, K. K., et al., 1967; Sivasubramanian, P., et al., 1970.

3011. **Bhatnagar, P. L., Rockstein, M.** and **Dauer, M. 1965** X-ray Treatment of Housefly Pupae. Exp. Gerontol. 1:149.
Such exposure, if non-lethal, extends the life span of adult male flies.

Bhatnagar, P. L., joint author. *See* Dauer, M., et al., 1965; Rockstein, M., et al., 1965, 1966, 1967; Simon, J., et al., 1968.

Bhatnagar, V. N., joint author. *See* Raghavan, N. G. S., et al., 1967.

Bianche, G., joint author. *See* Burgerjon, A., et al., 1967.

3012. **Bianchi, U.** and **Laudani, U. 1964** Attività del complesso succinossidasico durante il ciclo vitale di *Musca domestica* L. [Activity of complex succinoxidase during the life cycle of *Musca domestica* L.] Atti. Acc. Naz. It. Ent. A 11:148-154. (English summary.)
Activity not constant. Increases from egg to 3 day larva, then decreases (lowest point) until pupa is in third day. Increases from fourth day on, and continues in adult.

3013. **Bianchini, C. A. 1969** [Some breeding places of flies.] Arch. Ital. Sci. Med. Trop. 50:197-218. (In Spanish.)

3014. **Bieber, L. L. 1968** Incorporation of Trimethylamino-ethylphosphonic Acid and Dimethyl-aminoethylphosphonic Acid into Lipids of Housefly Larvae. Biochim. Biophys. Acta 152(4):778-780.
Both TMAP and DMAP incorporated into housefly larvae lipids. Larvae have little or no ability for methylating DMAP or the lipid which contains it. ABSTRACT: BA 51:2334(28183), Feb. 15, 1970.

3015. **Bieber, L. L.** and **Monroe, R. E. 1969** The Relation of Carnitine to the Formation of Phosphatidyl-Beta-methylcholine by *Tenebrio molitor* L. Larvae. Lipids 4(4):293-298.
Concerns lipid metabolism. *Tenebrio* does not metabolize carnitine to *beta* methylcholine as housefly and blowfly larvae do. ABSTRACT: BA 51:562(5646), Jan. 1, 1970.

3016. **Bieber, L. L., O'Connor, J. D.** and **Sweeley, C. C. 1969** The Occurrence of Tetradecasphing-4-enine and Hexadecasphing-4-enine as the Principal Sphingosines of *Musca domestica* Larvae and Adults. Biochim. Biophys. Acta 187:157-159. (Research paper No. 4636, Mich. State Agr. Exp. Sta.)
Concerns sphingolipids obtained from larvae reared aseptically on a *beta* methylcholine-containing diet, and from nonaseptically reared flies.

3017. **Bieber, L. L., Sellers, L. G.** and **Kumar, S. S. 1968** Studies on the Biosynthesis and Degradation of Phosphatidyl Beta Methyl Choline by House Fly Larvae. Fed. Proc. 27(2, pt. 1):457(1354).
Housefly larvae can convert several quarternary bases into PMC.

3018. **Bieber, L. L., Sellers, L. G.** and **Kumar, S. S. 1969** Studies on the Formation and Degradation of Phosphatidyl-Beta-methylcholine, Beta-methylcholine derivatives, and Carnitine by Housefly Larvae. [*Musca domestica.*] J. Biol. Chem. 244(3):630-636.
Larvae of *M. domestica* synthesize and degrade phosphatidyl-*beta* methyl-

choline in a manner analogous to the synthesis and degradation of phosphatidylcholine. ABSTRACT: BA 50:8096(84270), Aug. 1, 1969.

Bieber, L. L., joint author. *See* Hildenbrant, G. R., et al., 1969; Kumar, S. S., et al., 1970.

Biedebach, M. C., joint author. *See* McCann, G. D., et al., 1966.

3019. Bier, K. 1963 Synthese, interzellulärer Transport, und Abbau von Ribonukleinsäure im Ovar der Stubenfliege *Musca domestica*. [Synthesis, intercellular transport and destruction of RNA in the ovary of the housefly *Musca domestica*.] J. Cell. Biol. 16(2):436-440.

Observed processes of RNA activity are similar to what could be expected for messenger RNA. ABSTRACT: BA 43:360(3934), July 1, 1963.

Bier, K., joint author. *See* Engels, W., et al., 1967; Petzelt, C., et al., 1970.

3020. Bigley, W. S. 1966 Inhibition of Cholinesterase and Ali-Esterase in Parathion and Paraoxon Poisoning in the House Fly. J. Econ. Ent. 59:60-65.

Inhibition of body cholinesterase correlated with the effects of the insecticides; inhibition of head cholinesterase appeared to be a function of the dosage. ABSTRACTS: BA 47:6400(74888), Aug. 1, 1966; RAE-B 54:97, May, 1966.

3021. Bigley, W. S. and **Plapp, F. W., Jr. 1960** Cholinesterase and Aliesterase Activity in Organophosphorus-susceptible and Resistant House Flies. Ann. Ent. Soc. Am. 53(3):360-364.

Appears to be relationship between lower aliphatic esterase activity and resistance. Cholinesterase activity levels nearly identical in susceptible and resistant flies. ABSTRACTS: BA 35:4674(54307), Oct. 1, 1960; RAE-B 49:191, Sept., 1961; TDB 57:1229, Nov., 1960.

3022. Bigley, W. S. and **Plapp, F. W., Jr. 1961** Esterase Activity and Susceptibility to Parathion at Different Stages in the Life Cycle of Organophosphorus-resistant and Susceptible House Flies. J. Econ. Ent. 54(5):904-907.

Levels of aliphatic esterase activity less in immature stages of resistant flies. Cholinesterase activity levels similar for all fly strains at stages tested. ABSTRACT: RAE-B 50:173, Aug., 1962.

Bigley, W. S., joint author. *See* Plapp, W. F., et al., 1961, 1962, 1963, 1964.

3023. Binning, A., Darby, F. J., Hennan, M. P. and **Smith, J. N. 1967** The Conjugation of Phenols with Phosphate in Grass Grubs and Flies. Biochem. J. 103(1):42-48.

After dosing with phenols, bodies and/or excreta of flies contain substantial amounts of monoaryl phosphates. An unidentified metabolite of [^{14}C] 1-naphthol was present in extracts of flies dosed with that substance. ABSTRACT: BA 48:880(98945), Oct. 1, 1967.

Biralo, T. I., joint author. *See* Sukhova, M. N., et al., 1965.

3024. Bishop, L. G. 1968 Spectral Response of Single Neurons Recorded in the Optic Lobes of the House Fly and Blow Fly. Nature (London) 219(5161):1372-1373.

3025. Bishop, L. G. 1969 A Search for Color Encoding in the Responses of a Class of Fly Interneurons. Zeit. Vergl. Physiol. 64(4):355-371.

Interneurons were found to respond to the direction of movement of objects. Do not encode color. Suggests that optomotor response is determined by the activities of retinula cells 1-6. ABSTRACT: BA 51:3372 (34876), Mar. 15, 1970.

3026. **Bishop, L. G.** and **Keehn, D. G. 1966** Two Types of Neurones Sensitive to Motion in the Optic Lobe of the Fly. (*Eucalliphora lilaea, Musca domestica, Calliphora phaenicia.*) Nature 212(5068):1374-1376.
Believed to be a part of the optomotor system. *Musca domestica* has a large ommatidial angle, enabling a clear reversal in the direction indicated by the directionally selective units. (Indicated by torque measurements.) ABSTRACT: BA 48:6970(78007), Aug. 1, 1967.

3027. **Bishop, L. G.** and **Keehn, D. C. 1967** Neural Correlates of the Optomotor Response in the Fly. Kybernetik 3(6):288-295.
Gives data from extracellular micro electrode recordings from the optic lobes of houseflies and blowflies. Concerns the optomotor response (tendency to turn in response to a moving pattern).

3028. **Bishop, L. G., Keehn, D. G.** and **McCann, G. D. 1968** Motion Detection by Interneurons of Optic Lobes of the Flies *Calliphora phaenicia* and *Musca domestica*. J. Neurophysiol. 31(4):509-525.
Correlates the retinal geometry with the properties of form and motion detection, as recorded extracellularly from certain neurons in the fly's brain and optic lobe. ABSTRACT: BA 49:10285(113622), Nov. 1, 1968.

Bishop, P. W., joint author. *See* Bailey, D. L., et al., 1967, 1968.

Black, R. J., joint author. *See* Hart, S. A., et al., 1960; Smith, A. C., et al., 1960; Campbell, E., et al., 1960.

3029. **Blair, E. H., Kauer, K. C.** and **Kenaga, E. E. 1963** Synthesis and Insecticidal Activity of 0-methyl 0-(2, 4, 5-trichlorophenyl) Phosphoramidothioates and Related Compounds. J. Agric. Fd. Chem. 11(3):237-240.
Results of tests show a wide spectrum of toxicity for these compounds. Are residually toxic to houseflies when the latter are exposed to treated panels. ABSTRACT: RAE-B 51:217, Oct., 1963.

3030. **Blair, E. H., Kauer, K. C.** and **Kenaga, E. E. 1966** Synthesis and Insecticidal Activity of 0-Alkyl 0-2, 4, 5-Trichlorophenyl Phosphoramidothioates. Agric. Fd. Chem. 14:298-301.
Tested on several insect species, including houseflies, these compounds show a wide spectrum of insecticidal activity. They are phytoxic, but can be used against household insects and grain pests.

3031. **Blair, E. H., Wasco, J. L.** and **Kenaga, E. E. 1965** Synthesis and Insecticide Activity of Methyl 2, 4, 5-Trichlorophenyl Phosphoroamidates. Agric. Fd. Chem. 13(4):383-385.
These compounds have a wide spectrum of insecticidal activity, but are toxic to plants and have too short a residual toxicity to be practical. Tests were run on several species, including houseflies.

3032. **Blakitnaya, L. P. 1962** [On hibernation and certain other features specific to the biology and ecology of synanthropic flies in Northern Kirghizia.] Med. Parazit. i Parazitarn. (Bolezni) 31(4):424-429. (English summary.)
Housefly overwinters both as adults (active and inactive) and as larvae of the III instar and as pupae. Overwintering flies appear active at end of Feb. and beginning of March. ABSTRACT: BA 41:297(3956), Jan. 1, 1963.

3033. **Bland, R. G. 1966** Toxicity of Ronnel, Dimethoate, and Malathion to Little House Fly Adults. J. Econ. Ent. 59:1435-1437.
Ronnel and dimethoate produced rapid knock-down of *Fannia canicularis* (L.). Mortality from residual effect was high for 2 weeks. Malathion toxicity was low. ABSTRACTS: BA 48:3230(35579), Apr. 1, 1967; RAE-B 55:67(248), Apr., 1967; TDB 64:438, Apr., 1967.

3034. **Blazquez, V. J.** and **Gabaldon, A. 1965** Efecto repelente del DDT en *Musca domestica* [Repellent effect of DDT on *Musca domestica.*] Boln. inf. Dir. Malar. 5(4):193-195. (Summary in English.)

3035. **Blickle, R. L. 1961** Parasites of the Face Fly (*Musca autumnalis*) in New Hampshire. J. Econ. Ent. 54(5):802.
A survey of the parasites of the face fly indicated that three species of Hymenoptera, *Aphaereta pillipes* (Say), *Xyalophora quinquelineata* (Say), and *Eucoila* sp. parasitize the puparia. *Aphaereta* was made to parasitize *Musca domestica* but attempts with the other two failed. ABSTRACT: RAE-B 50:153, July, 1962.

3036. **Blum, M. S.** and **Pratt, J. J., Jr. 1960** Relationships Between Structure and Insecticidal Activity of Some Organotin Compounds. J. Econ. Ent. 53(3):445-448.
The nature of the anion influences the toxicity of the compound. Replacement of one of the anionic groups with an aryl or alkyl radical to form diphenyl and di-n-butyl compounds increased toxicity of some organotin compounds. ABSTRACT: RAE-B 49:120, June, 1961.

Blum, M. S., joint author. *See* Sannasi, A., et al., 1969.

Boccacci, M., joint author. *See* Grasso, A., et al., 1961, 1962; Bettini, S., et al., 1960.

Böcker, E., joint author. *See* Behrenz, W., et al., 1965.

3036a. **Bodenstein, O. F., Fales, J. H.** and **Walker, R. L. 1970** Laboratory Evaluations of Materials as Repellents for the Face Fly. J. Econ. Ent. 63(6):1752-1755.
Repellent activity found with certain sulfides and sulfoxides and a variety of N, N-disubstituted amides.

Bodenstein, O. F., joint author. *See* Fales, J. H., et al., 1960, 1961, 1962, 1964, 1965, 1966, 1968, 1970; Anthony, R. W., et al., 1961.

3037. **Bodnaryk, R. P. 1968** Dietary Amino Acid and Nucleic Acid Imbalance Leading to Altered Tumour Frequency in the Adult Housefly, *Musca domestica.* J. Insect Physiol. 14:223-242.
Internal melanotic tumors, also melanotic sclerosis, were produced in both sexes by elevating the L-phenylalanine content of fully chemically defined diet beyond 1 per cent. ABSTRACT: BA 49:6289(70356), July 1, 1968.

3038. **Bodnaryk, R. P. 1970** Biosynthesis of *gamma*-1-glutamyl-1-*phenylalanine* by the Larva of the Housefly. J. Insect Physiol. 16(5):919-929.
This bipeptide is the major biosynthetic product of the larvae of the housefly and other species of the genus *Musca.* Plays a key role in puparium formation. ABSTRACT: BA 51:11472(116508), Oct. 15, 1970.

3039. **Bodnaryk, R. P.** and **Morrison, P. E. 1966** The Relationship between Nutrition, Haemolymph Proteins, and Ovarian Development in *Musca domestica* L. J. Insect Physiol. 12(8):963-976.
Suggests that adequate protein diet elevates total protein content (or content of 1 or more specific fractions) in the haemolymph to the point where (directly or indirectly) it initiates and sustains ovarian development. ABSTRACTS: BA 47:9493(109892), Nov. 15, 1966; RAE-B 55: 233(799), Dec., 1967.

3040. **Bodnaryk, R. P.** and **Morrison, P. E. 1968** Asynchronous Ovarian Development in the Housefly *Musca domestica* L. (Diptera: Muscidae) Caused by Dietary gamma-aminobutyric acid (Gaba). Comp. Biochem. Physiol. 25:573-579.
Gaba, besides being toxic to flies, caused a higher rate of failure in

ovarian development among survivors, than other isomers and related compounds. ABSTRACT: BA 50:1051(11016), Jan. 15, 1969.

3041. **Bodnaryk, R. P.** and **Morrison, P. E. 1968** The Effect of Essential Amino Acid Deficiency on the Sex-specific Accumulation of a Blood Protein in Female House-flies (*Musca domestica* L.). Comp. Biochem. Physiol. 27(1):365-369.
Concerns mainly the blood protein fractions 3 and 4 of female house-fles during the first cycle of ovarian development. ABSTRACT: BA 50: 4806(50544), May 1, 1969.

3042. **Bodnaryk, R. P.** and **Morrison, P. E. 1968** Immunochemical Analysis of the Origin of a Sex-specific Accumulated Blood Protein in Female Houseflies. J. Insect Physiol. 14:1141-1146.
Fraction 3 originates in larval fat body and becomes depleted in other adults. Fraction 4 arises *de novo* in adult fat body to meet adult female requirements. ABSTRACT: BA 49:10813(119082), Nov. 15, 1968.

3043. **Bojanowska, A., et al. 1966** [Evaluation of the susceptibility of *Musca domestica* L. to gamma-HCH. Studies on the stability of insecticidal bases containing gamma-HCH. 1.] Przegl. Epidem. 20:193-198. (In Polish.)

3044. **Bojanowska, A.** and **Domicz-styczyńska, B. 1963** Fluctuations of Resistance to DDT in the Housefly (*Musca domestica* L.) in Poland. J. Hyg. Epidemiol., Microbiol., Immunol. 7(1):90-96.
Mean temperature of summer months affects resistance. Warmer conditions make for faster detoxication of DDT, also favors shorter life cycles, which increases the number of generations undergoing selection. Resistance rises. ABSTRACTS: BA 45:1758(21924), Mar. 1, 1964; TDB 61:108, Jan., 1964.

3045. **Bojanowska, A., Goszczyńska, K.** and **Mańkowska, H. 1966** Evaluation of Sensitivity of Flies (*Musca domestica* L.) to gamma-HCH. I. Studies on the Stability of Insecticidal Deposits of gamma-HCH. Epidem. Rev. 20(1-2):95-99.
Filter paper impregnated with acetone solutions of gamma-HCH maintained constant toxicity for at least 7 months; i.e. for the duration of the fly season.

3046. **Bojanowska, A., Goszczyńska, K.** and **Mańkowska, H. 1966** Terenowa metoda oceny wrażliwości much na gamma HCH. [Elaboration of a field study for the evaluation of sensitivity of flies (*Musca domestica* L.) to gamma-HCH.] Przegl. Epidem. 20(4):435-442. (In Polish, with Russian and English summaries).
Five methods for assaying resistance in domestic flies were compared. Under field conditions "Petri-plate" method was preferred. ABSTRACTS: TDB 64:1272, Nov., 1967; RAE-B 58:97(384), Apr., 1970.

3047. **Bojanowska, A.** and **Wojciak, Z. 1960** Stan oparnosci na DDT much (*Musca domestica* L.) w polsce w latach 1953-1958. [Resistance to DDT of flies (*Musca domestica* L.) in Poland in 1953-1958.] Przegl. Epidem. 14(1):67-81. Warsaw. (English summary.)
Highest resistance level occurred in 1955. Since 1957, some regression was observed, probably caused by use of gamma-HCH in mixture with DDT, in emulsions. ABSTRACT: TDB 57:1121-1122, Oct., 1960.

3048. **Bojanowska, A.** and **Styczysńka, B. 1966** Efektywność insektycydów fosforoorganicznych trichlorfonu, malationu i dichlorwosu w odniesieniu do larw much *Musca domestica* L. [Effectiveness of phosphoorganic insecticides (trichlorphon, malathion and dichlorovos) against larvae of the fly *Musca domestica* L.] Roczn. Pánstw. Zakt. Hig. 17:513-516. (In Polish, with Russian and English summaries.)

Inscticides were added to milk medium. Trilaphon was the most effective.

3049. **Bolchi-Serini, G. 1961** Sul sistema stomatogastrico di *Musca domestica*. [About the stomatogastric system of *Musca domestica*.] Atti Soc. Ital. Sci. Nat. Mus. Civico Storia Nat. Milano 100(1-2):205-207.

This portion of the sympathetic nervous system, in larval and adult *Musca domestica*, is described. ABSTRACT: BA 39:316(3963), July 1, 1962.

3050. **Bolotova, T. A. 1961** [Experience with chlorophos insecticide paper in the control of houseflies in Mystischi during 1959.] Zhur. Mikrobiol. 32:123-124.

3051. **Bolotova, T. A. 1961** [The use of chlorophos insecticide paper against houseflies in Mytishchi in 1959.] Zhur. Mikrobiol. [Eng.] 32:931-932.

Bolotova, T. A., joint author. *See* Sukhova, M. N., et al., 1965.

3052. **Bolton, H. T.** and **Hansens, E. J. 1970** Ability of the House Fly, *Musca domestica*, to Ingest and Transmit Viable Spores of Selected Fungi. Ann. Ent. Soc. Am. 63(1):98-100.

House flies can ingest and pass spores of Deuteromycetes not exceeding a diameter of 0·050 mm. Spores, unharmed, were all able to germinate. ABSTRACT: BA 51:5665(58133), May 15, 1970.

3053. **Boose, R. B.** and **Terriere, L. C. 1967** Quantitative Aspects of the Detoxication of Naphthalene by Resistant and Susceptible House Flies. J. Econ. Ent. 60:580-586.

R and S strains were compared, both quantitatively and qualitatively for naphthol metabolism. No differences were found, either in rate of metabolic production or in relative levels of metabolites. ABSTRACTS: BA 48:6970(78008), Aug. 1, 1967; RAE-B 55:147(517), Aug., 1967.

Boose, R. B., joint author. *See* Terriere, L. C., et al., 1960, 1961.

3054. **Booth, G. M.** and **Metcalf, R. L. 1970** Histochemical Evidence for Localized Inhibition of Cholinesterase in the House Fly. Ann. Ent. Soc. Am. 63(1):197-204.

Peripheral regions of thoracic ganglion are primary area of ChE inhibition. Fly brain is a secondary site of action. ABSTRACTS: BA 51:5670 (58182), May 15, 1970; TDB 67:577(1144), May, 1970.

Borck, K., joint author. *See* Metcalf, R. L., et al., 1966.

Boreham, M. M., joint author. *See* Kliewer, J. W., et al., 1964.

3055. **Borgatti, A. L. 1961** The Toxicology of Commercial Formulation of *Bacillus thuringiensis* Berliner to Japanese Quail and House Fly Larvae. Diss. Absts. 22(6):2110-2111.

ABSTRACT: BA 39:959(12190), Aug. 1, 1962.

3056. **Borgatti, A. L.** and **Guyer, G. E. 1962** Formulations of *Bacillus thuringiensis* Berliner Found to be Contaminated with Chlorinated Hydrocarbon Insecticides. J. Econ. Ent. 55:1015-1016.

Has to do with toxicity to warm-blooded organisms (Japanese quail). Uncontaminated, the bacillus may be safely ingested by domestic or wild animals.

3057. **Borgatti, A. L.** and **Guyer, G. E. 1963** The Effectiveness of Commercial Formulations of *Bacillus thuringiensis* Berliner on House-fly Larvae. J. Insect Pathol. 5(3):377-384.

House flies breeding in quail droppings were effectively reduced by the addition of selected amounts of spore preparations to the birds' diet. ABSTRACTS: BA 45:663(8250), Jan. 15, 1964; RAE-B 52:64, Apr., 1964; TDB 61:225, Feb., 1964.

3058. Bořkovec, A. B. 1962 Sexual Sterilization of Insects by Chemicals. Science 137(3535):1034-1037.
Data indicate a relationship between chemosterilants and anti-tumor compounds used in cancer therapy. (Rapidly-dividing cells of the reproductive system in some respects resemble tumor cells.)

3059. Bořkovec, A. B. 1966 New Chemical Approaches to Insect Control. Proc. 3rd Brit. Insect. & Fung. Confer. November 8-11, 1965, Brighton, Sussex. (Publ. 1966) pp. 3-17.
ABSTRACT: RAE-B 55:201(697), Nov., 1967.

3060. Bořkovec, A. B., Chang, S. C. and Limburg, A. M. 1964 Effect of pH on Sterilizing Activity of Tepa and Metepa in Male House Flies. J. Econ. Ent. 57(6):815-817.
No significant variations in relation to pH. Concludes that in partially degraded solutions sterilizing activity is proportional to contents of intact tepa (or metepa) rather than to total contents of aziridine function. ABSTRACTS: BA 46:2538(31671), Apr. 1, 1965; RAE-B 53:56, Mar., 1965; TDB 62:476, May, 1965.

3061. Bořkovec, A. B. and DeMilo, A. B. 1967 Insect Chemosterilants. V. Derivatives of Melamine. J. Med. Chem. 10(3):457-461.
Structures of 64 derivatives of melamine were correlated with chemosterilant activity for *M. domestica*. Size and spatial requirements of substituents exert more influence on sterilizing activity than do electronic characteristics. ABSTRACT: BA 48:10149(114012), Nov. 15, 1967.

3062. Bořkovec, A. B., Fye, R. L. and LaBrecque, G. C. 1968 Aziridinyl Chemosterilants for House Flies. U.S. Dept. Agric., Agric. Res. Serv. ARS 33-129. 60 pp. (Mimeographed document.)
Lists all aziridinyl compounds tested on houseflies, with summary of screening data, chemical structure, and commercial or other source.

3063. Bořkovec, A. B., LaBrecque, G. C. and DeMilo, A. B. 1967 S-Triazine Herbicides as Chemosterilants of House Flies. J. Econ. Ent. 60: 893-894.
No compound of this group proved highly effective or potentially useful. ABSTRACT: BA 48:11211(124521), Dec. 15, 1967.

3064. Bořkovec, A. B., Settepani, J. A., LaBrecque, G. C. and Fye, R. L. 1969 Boron Compounds as Chemosterilants for House Flies. J. Econ. Ent. 62(6):1472-1480.
Sixty-nine organic and inorganic compounds containing boron inhibited reproduction in *M. domestica*. Sterilizing activity greater when compounds yielding boric or boronic acids on hydrolysis were administered in diet, than when such acids were administered as such. ABSTRACTS: TDB 67: 462(920), Apr., 1970; BA 51:7937(81044), July 15, 1970; RAE-B 58: 146(610), May, 1970.

3065. Bořkovec, A. B. and Woods, C. W. 1963 Aziridine Chemosterilants. Sulfur-containing Aziridines. Advanc. Chem. Ser. 41:47-55.
Most important group of chemically-related insect chemosterilants are aziridine derivatives. Substitution of methyl on the aziridine ring carbon lowers activity of the compound. ABSTRACT: RAE-B 53:15, Jan., 1965.

3066. Bořkovec, A. B. and Woods, C. W. 1965 Insect Chemosterilants. II. N-carbamoylaziridines. J. Med. Chem. 8:545-547.
Twenty N-carbamoyl aziridines were synthesized, prepared as candidate chemosterilants and screened, using houseflies. Replacement of an acyl with a carbamoyl group in the diaziridinyl group is significant.

3067. **Bořkovec, A. B., Woods, C. W.** and **Brown, R. T. 1966** Insect Chemosterilants. III. 1-aziridinylphosphine oxides. J. Med. Chem. 9:522-526.
Tested sterilizing activity of 40 variously substituted oxides. This was correlated with the degree of substitution on the aziridinyl ring and with the type of nonaziridinyl moieties attached to the phosphinylidyne group.

Bořkovec, A. B., joint author. *See* Chang, S. C., et al., 1964, 1966, 1967, 1968, 1969; Terry, P. H., et al., 1964, 1967; Smith, C. N., et al., 1964; Akov, S., et al., 1968; Ascher, K. R. S., et al., 1967.

3068. **Borlund, H. P. 1966** Afprøvning af fluemidler på gårde. [Testing of insecticides against flies on farms.] Ann. Rept. (1965) Govern. Pest Infestation Laboratory. pp. 32-36. Lyngby, Denmark. (In Danish, with English translation.)
Residual sprays with dimethoate and/or formothion were effective. For summer months, fly strips with dimethoate and residual fumigants with dichlorvos gave good control.

3069. **Borlund, H. P. 1967** Afprøvning af fluemidler på gårde. [Testing of insecticides against flies on farms.] Ann. Rept. (1966) Govern. Pest Infest. Lab., pp. 28-30, 32-35. Denmark.
Bromophos, dimethoate, and trichlorfon tested as residual sprays. Only dimethoate was satisfactory. Fention and dichlorvos fly strips not effective in ventilated areas.

3070. **Borlund, H. P. 1968** Afprøvning af fluemidler på gårde. [Testing of insecticides against flies on farms.] Ann. Rept. (1967) Govern. Pest Infest. Lab., pp. 31-33, 35-37. Denmark.
Of three residual type sprays dimethoate had the most persistent effect (though less so than previously). Paint-on type baits gave decreasingly satisfactory control.

3071. **Borlund, H. P. 1969** Afprøvning af fluemidler på gårde. [Testing of insecticides against flies on farms.] Ann. Rept. (1968) Govern. Pest Infest. Lab., pp. 29, 35. Denmark.
Trials of housefly control with sprays, powders, and baits. Trichlorfon formulations not effective. Dimethoate excelled in bait formulations.

3072. **Borlund, H. P.** and **Mourier, H. 1969** Apparat til fangst af gedehamse og spyfluer. [A trap for hornets and blow-flies.] Ann. Rept. (1968) Govern. Pest Infest. Lab., pp. 24-25 and 28-29. Denmark.
An improvement over the 1967 model. Especially effective against blow-flies in a fish shop.

3073. **Bornemissza, G. F. 1968** Studies on the Histerid Beetle *Pachylister chinensis* in Fiji, and its Possible Value in the Control of Buffalo-fly in Australia. Australian J. Zool. 16(4):673-688.
A predaceous beetle which feeds on fly larvae living in the dung of herbivores. Introduced from Java in 1938 against houseflies. Because of its irregular distribution, is not likely to prove dependable against the buffalo fly. Abstract: BA 50:2648(27590), Mar. 1, 1969.

3073a. **Bornemissza, G. F. 1970** Insectary Studies on the Control of Dung Breeding Flies by the Activity of the Dung Beetle, *Onthopagus gazella* F. (Coleoptera: Scarabaeinae). J. Australian Ent. Soc. 9(1): 31-41.
The rapid burial of cattle dung by the Afro-Asian dung beetle causes 80-100 per cent reduction in the numbers of *Musca vetustissima* Walker emerging; any surviving flies are small with low or nil reproductivity. Abstract: BA 51:13685(139245), Dec. 15, 1970.

Bornstein, A. A., joint author. *See* Greenberg, B., et al., 1963, 1964.

Borst, P., joint author. *See* Van Bruggen, E. F. J., et al., 1968.

3074. Boschek, C. B. 1970 [On the structure and synaptic organization of the first optic ganglion in the fly.] Zeit. Naturforsch. (B) 25:560.

Bose, B. N., joint author. *See* Pradham, S., et al., 1960.

3075. Bouillant, A., Lee, V. H. and **Hanson, R. P. 1965** *Epizootiology* on Mink Enteritis. II. *Musca domestica* L. as a Possible Vector of Virus. Canad. J. Comp. Med. Vet. Sci. 29:148-152. (French summary.)

Viruses are transmitted from infected to susceptible mink by flies having access to both. Disease may also be induced by susceptible animals eating flies. ABSTRACT: BA 47:3736(43993), May 1, 1966.

Bowen, W. R., joint author. *See* Bell, D. D., et al., 1965; Deal, A. S., et al., 1965; Georghiou, G. P., et al., 1965, 1966; Loomis, E. C., et al., 1968.

Bowers, W. S., joint author. *See* Fales, J. H., et al., 1970.

3076. Bowman, M. C., Beroza, M., Gordon, C. H., Miller, R. W. and **Morgan, N. O. 1968** A Method of Analyzing the Milk and Feces of Cows for Coumaphos and its Oxygen Analog after Feeding Coumaphos for Control of House Fly Larvae. J. Econ. Ent. 61(2):358-362.

Chromatography technique. Coumaphos was recovered only from animals fed this compound while the oxygen analog was undetected. Degree of larval control related to level of coumaphos fed. ABSTRACTS: RAE-B 56:185(671), Sept., 1968; BA 49:9742(107859), Oct. 15, 1968.

Bowman, M. C., joint author. *See* Morgan, P. B., et al., 1968; Miller, R. W., et al., 1970.

Bowness, J. M., joint author. *See* Wolken, J. J., et al., 1960.

3077. Boyer, A. C. 1967 Vinyl Phosphate Insecticide Sorption to Proteins and Its Effect on Cholinesterase I_{50} Values. J. Agric. Fd. Chem. 15(2):282-286.

A study of the effect which protein-insecticide reaction has on the inhibition of cholinesterase as contained in preparations of mammalian blood plasma and homogenates of housefly heads. ABSTRACT: BA 49: 715(7931), Jan. 15, 1968.

3078. Boyes, J. W. 1967 The Cytology of Muscoid Flies. Chapter 10 *in* Genetics of Insect Vectors of Disease. (J. W. Wright & R. Pal, Editors.) Amsterdam. Elsevier Publ. Co. pp. 371-384.

Contains much detailed information on *M. domestica.* Treats of mitosis, meiosis, karyotypes, polytene chromosomes, sex determination, radiation effects, etc. ABSTRACT: RAE-B 56:157(552), Aug., 1968.

3079. Boyes, J. W., Corey, M. J. and **Paterson, H. E. 1964** Somatic Chromosomes of Higher Diptera. IX. Karyotypes of some Muscid Species. Canad. J. Zool. 42:1025-1036.

Concerns several species and subspecies of *Musca.* Suggests that species having 10 chromosome karyotypes may have originated independently in different subfamilies and genera of the Muscidae. ABSTRACTS: BA 46:2190(27317), Mar. 15, 1965; TDB 62:365, Apr., 1965.

3080. Boyes, J. W. and **Naylor, A. F. 1962** Somatic Chromosomes of Higher Diptera. VI. Allosome-autosome Length Relations in *Musca domestica* L. Canad. J. Zool. 40(5):777-784.

Length of the two heterochromatic X-chromosomes of pair I is compared with the total length of the five autosomal pairs in 50 mitotic chromosome complements of varying length in cells without colchicine or other treatment prior to fixation. ABSTRACT: BA 41:1330(17001), Mar. 1, 1963.

Bracha, P., joint author. *See* Gratz, N. G., et al., 1963.

3081. **Bradbury, F. R.** (?) Biophysical Aspect of Insecticide Resistance. S. C. I. Monograph 29:47-53. (Reprint, not dated.)
Reports narcosis experiments with 12 compounds, showing a difference in physical properties between 2 strains of *M. domestica*, one susceptible to γBHC, the other resistant.

3082. **Bradbury, F. R.** and **O'Carrol, F. M. 1966** Measurement of Narcotic Potency using Houseflies. Ann. Appl. Biol. 57:15-31.
Narcotic potency measurements pose mathematical problems different from toxicity determinations. Paper sets forth a suitable method and gives data obtained by its use. ABSTRACTS: BA 48:1686(18256), Feb. 15, 1967; TDB 63:905, Aug., 1966.

3083. **Bradbury, F. R.** and **Standen, H. 1960** Mechanisms of Insect Resistance to the Chlorohydrocarbon Insecticides. J. Sci. Fd. Agric. 11(2):92-100. London.

3084. **Brady, U. E. 1965** Studies on *in vivo* Inhibition and Recovery of Cholinesterase Activity in Organophosphate Treated Insects (Orthoptera, Diptera). Diss. Absts. 26(1):38.

3085. **Brady, U. E. 1966** A Technique of Continuous Exposure for Determining Resistance of House Flies to Insecticides. J. Econ. Ent. 59(3):764-765.
Gives exposure in minutes required for LT_{50} and LT_{90} with 5 strains of flies exposed to known concentrations of diazinon and malathion. Equates with knockdown, since flies did not recover.

3086. **Brady, U. E., Jr.** and **Arthur, B. W. 1962** Absorption and Metabolism of Ruelene by Arthropods. J. Econ. Ent. 55:833-836.
M. domestica one of 16 arthropod species tested. Used P^{32} technique. ABSTRACT: RAE-B 51:56, Mar., 1963.

3087. **Brady, U. E., Jr.** and **Arthur, B. W. 1963** Biological and Chemical Properties of Dimethoate and Related Derivatives. J. Econ. Ent. 56:477-482.
Of 17 dimethoate derivatives tested those containing dimethoxy groups were more effective against *M. domestica* than the diethoxy compounds. The most toxic materials were those having the C=O band between 5·90 and 5·95 microns. ABSTRACT: RAE-B 51:237, Nov., 1963.

3088. **Brady, U. E., Jr., Dorough, H. W.** and **Arthur, B. W. 1960** Selective Toxicity and Animal Systemic Effectiveness of Several Organophosphates. J. Econ. Ent. 53(1):6-8.
Used rats, rabbits, houseflies, and other arthropods. Of 17 compounds tested, dimethoate excelled for toxicity, Bayer 29493 for duration of residual effect. ABSTRACT: RAE-B 49:79, Apr., 1961.

3089. **Brady, U. E.** and **LaBrecque, G. C. 1966** Larvicides for the Control of House Flies in Poultry Houses. J. Econ. Ent. 59:1521.
Of 16 compounds tested on larvae in poultry droppings, most were inferior to dimethoate. Those equal to or slightly better, were Shell compounds. None were effective for longer than 1 week. ABSTRACTS: BA 48:3230(35580), Apr. 1, 1967; RAE-B 55:69(255), Apr., 1967; TDB 64:438, Apr., 1967.

3090. **Brady, U. E., Jr., Meifert, D. W.** and **LaBrecque, G. C. 1966** Residual Sprays for the Control of House-flies in Field Tests. J. Econ. Ent. 59:1522-1523.
Of the sprays tested dimethoate was most effective. Gave satisfactory control for about 2 weeks. ABSTRACTS: BA 48:3230(35581), Apr. 1, 1967; RAE-B 55:70(256), Apr., 1967; TDB 64:809, July, 1967.

3091. **Brady, U. E., Jr.** and **Sternburg, J. 1967** Studies on *in-vivo* Cholinesterase Inhibition and Poisoning Symptoms in House Flies. J. Insect. Physiol. 13(3):369-379.
For each of 6 organophosphate anticholinesterases, a specific level of inhibition was required for knock-down. ABSTRACTS: BA 48:5112(57178), June 1, 1967; RAE-B 56:35(111), Feb., 1968; TDB 64:686, June, 1967.

3092. **Brady, U. E., Jr.** and **Sternburg, J. 1966** Recovery of Cholinesterase Activity in Organophosphate Treated Insects. J. Insect Physiol. 12(9):1171-1185.
A study of *in vivo* and *in vitro* aging of dimethyl phosphorylated cholinesterase. Very slow aging *in vivo*; also slow recovery. Believed a result of cholinesterase synthesis; not a reactivation of inhibited cholinesterase. ABSTRACT: TDB 63:1402, Dec., 1966.

3093. **Braitenberg, V. 1966** Unsymmetrische Projekton der Retinulazellen auf die Lamina ganglionaris bei der Fliege *Musca domestica.* [Asymmetrical projection of the retinula cells on the lamina ganglionaris in the fly, *Musca domestica.*] Zeit. Vergl. Physiol. 52(2):212-214.
ABSTRACT: BA 48:1437(15577), Feb. 1, 1967.

3094. **Braitenberg, V. 1966** Patterns of Projection in the Visual System of the Fly: I. Retina-lamina Projections. MS copy: To be published in Exp. Brain Res. (*See* Braitenberg, 1967.)

3095. **Braitenberg, V. 1967** Patterns of Projection in the Visual System of the Fly: I. Retina-lamina Projections (*Musca domestica*). Exp. Brain Res. 3(3):271-298.
Sets forth the principle that all fibers carrying information from the same point of the optical environment are united into one synaptic site of the lamina ganglionaris. ABSTRACT: BA 48:9249(103997), Oct. 15, 1967.

3096. **Braitenberg, V.** and **Ferretti, C. T. 1966** Landing Reaction of *Musca domestica* Induced by Visual Stimuli. Naturwissenschaften 53(6): 155-156.
Used a rotating disk, facing the insect, on which a black (arithmetical) spiral was painted. Rotation gave impression of expansion or contraction, i.e. of approach or recession. Landing reaction occurred only when rotation suggested expansion. ABSTRACT: BA 48:3264(35996), Apr. 1, 1967.

Braitenberg, V., joint author. *See* Reichardt, W., et al., 1968.

Bramhall, E. L., joint author. *See* Loomis, E. C., et al., 1968.

Brandt, K. F., joint author. *See* Rockstein, M., et al., 1962, 1963.

Bratanov, H., joint author. *See* Sarbova, S., et al., 1968.

3097. **Bratkowski, T. A.** and **Knowles, C. O. 1968** Properties of an Acetylcholinesterase from the Face Fly, *Musca autumnalis.* Ann. Ent. Soc. Am. 61(2):397-402.
Is similar to corresponding enzyme of the housefly with respect to substrate specificity, pH specificity, centrifugal localization and sensitivity to certain organophosphate and carbamate compounds. ABSTRACTS: BA 49:6775(75561), July 15, 1968; RAE-B 56:205(743), Oct., 1968.

Bravo-Becherelle, M. A., joint author. *See* Varela, G., et al., 1964.

3098. **Brebbia, D. R., et al., 1962** The effects of Sodium, Potassium, and Calcium Ions on the Heartbeat and the Electrocardiogram of the Housefly, *Musca domestica* Linnaeus. Diss. Absts. 23(1):278-279.

3099. **Brebbia, D. R.** and **Ludwig, D. 1962** Saline Solutions for Maintaining the Isolated Heart of the House Fly, *Musca domestica* L. Ann. Ent. Soc. Am. 55(1):131-135.
The values for pH, osmotic pressures, sodium and potassium ion con-

centrations; also dextrose and phosphate buffer are given for the maintaining of adult, larval and pupal housefly hearts. ABSTRACT: BA 38:941 (12267), May 1, 1962.

Breeland, S. G., joint author. *See* Smith, G. E., et al., 1965.

3100. Bremer, H. G. 1963 Insektizide Reaktionsaerosole. Zeit. Pfl. Krankh. 70(3):147-157. Stuttgart. (English summary.)
Relates to German Patent No. 1058895. Aerosol-forming reaction yields no noxious bi-products. No equipment necessary; safe to handle; economical. ABSTRACT: RAE-B 52:92, May, 1964.

3101. Bremer, H. G. 1964 Pyrethrum Reaction Aerosol. J. Econ. Ent. 57: 62-67.
The yield of pyrethrins from pyrethrum flowers in burning is not less than by extraction. Insecticidal effect of aerosols depends on the pyrethrin content of the flowers used. ABSTRACTS: RAE-B 52:96, May, 1964; TDB 61:973, Sept., 1964.

Brezhneva, I. M., joint author. *See* Lineva, V. A., et al., 1964.

3102. Bridges, R. G., Crone, H. D. and **Beard, J. R. 1962** A Study of the Phospholipids of Dieldrin-resistant and Susceptible Houseflies, with Particular Reference to Those of the Thoracic Ganglion. *In* Internat. Atomic Energy Agency. Radioisotopes and Radiation in Entomology. Proc. of Symp. Int. Publ., Inc., New York. 22:145-153.
Four major phospholipid fractions can be separated from the thoracic ganglion (as from whole flies). Distribution of labelled phosphate shows only slight difference of turnover, in the major fractions from resistant and susceptible flies. ABSTRACTS: BA 42:321(3887), Apr. 1, 1963; RAE-B 51:213, Oct., 1963.

3103. Bridges, R. G. and **Holden, J. S. 1969** The Incorporation of Aminoalcohols into the Phospholipids of the Adult Housefly, *Musca domestica*. J. Insect Physiol. 15(5):779-788.
Amino-alcohols replace naturally-occurring phosphatidylethanolamine and phosphatidylcholine. Compete for enzymes involved in metabolic pathways. ABSTRACT: BA 50:11406(118007), Nov. 1, 1969.

3104. Bridges, R. G. and **Price, G. M. 1970** The Phospholipid Composition of Various Organs from Larvae of the Housefly, *M. domestica* fed on Normal Diets and on Diets containing 2-amino butan-1-ol. Comp. Biochem. Physiol. 34(1):47-60.
Phosphatidylethanolamine the major phospholipid found in all organs and subcellular fractions examined. Larval diet affects findings. ABSTRACT: BA 51:11471(116493), Oct. 15, 1970.

3105. Bridges, R. G. and **Ricketts, J. 1966** Formation of a Phosphonolipid by Larvae of the Housefly, *Musca domestica.* Nature 211(5045):199-200.
Larvae replace lipid-bound choline and ethanolamine with other amino and N-methylated amino-alcohols when these are included in the larval diet. ABSTRACT: BA 47:9494(109895), Nov. 15, 1966.

3106. Bridges, R. G. and **Ricketts, J. 1967** The Incorporation of 2-amino Butanols into the Phospholipids of Housefly (*Musca domestica*) Larvae. Abstract. Biochem. J. 102(1):4P.
Greatest effect is from 2-amino butan-1-ol and 2-amino-2-methylpropan-1-ol. Cause death of larvae before sixth day after hatching. They replace, partially, the choline and enthanolamine of the larval phospholipids.

3107. Bridges, R. G. and **Ricketts, J. 1967** The Incorporation, *in vivo*, of Aminoalcohols into the Phospholipids of the Larva of the Housefly, *Musca domestica.* J. Insect Physiol. 13(6):835-850.
The nearer the structure of the amino alcohol is to that of ethanolamine,

the larger will be the phosphatidyl derivative found in larvae when expressed as percentage of total phospholipid. ABSTRACT: BA 48:8333 (93634), Sept. 15, 1967.

3108. **Bridges, R. G.** and **Ricketts, J. 1968** The Effect of 2-aminobutan-1-ols on the Growth of the Housefly (*Musca domestica*). Comp. Biochem. Physiol. 25:383-400.
Larval growth was retarded by a number of N-alkyl derivatives, 2-amino butan-1-ol being the most active. Choline and related compounds reversed the retardation. ABSTRACT: BA 49:9250(102722), Oct. 1, 1968.

3109. **Bridges, R. G.** and **Ricketts, J. 1970** The Incorporation of Analogues of Choline into the Phospholipids of the Larvae of the Housefly, *Musca domestica*. J. Insect Physiol. 16(4):579-593.
Describes the three general types of choline analogues incorporated into larval phospholipids. ABSTRACTS: BA 51:10892(110654), Oct. 1, 1970; TDB 67:1292(2508), Oct., 1970; RAE-B 59:170(723), May, 1971.

3110. **Bridges, R. G., Ricketts, J.** and **Cox, J. T. 1965** The Replacement of Lipid-bound Choline by other Bases in the Phospholipids of the Housefly, *Musca domestica*. J. Insect Physiol. 11(3):225-236.
Larvae incorporate beta-methylcholine (or carnitine) into phospholipids in place of choline, when these replace choline in the diet. *Adults* cannot use carnitine, and use beta-methylcholine but slightly. They are unable to decarboxylate the carnitine. ABSTRACT: BA 46:4800(59435), July 1, 1965.

3110a. **Bridges, R. G., et al. 1970** The Phospholipid Composition of Various Organs from Larvae of the Housefly, *Musca domestica*, Fed on Normal Diets and on Diets Containing 2-aminobutan-1-ol. Comp. Biochem. Physiol. 34:47-60.

Bridges, R. G., joint author. *See* Crone, H. D., et al., 1963.

3111. **Briggs, John D. 1960** Reduction of Adult House-fly Emergence by the Effects of *Bacillus* spp. on the Development of Immature Forms. J. Insect Pathol. 2(4):418-432.
A filterable principle produced by these bacteria interferes with the development of *M. domestica*. As an additive in chicken feed, it reduced fly emergence from hen manure 99 per cent. ABSTRACTS: BA 36:2096 (25541), Apr. 15, 1961; RAE-B 49:122, June, 1961; TDB 58:646, May, 1961.

Brikman, L. I., joint author. *See* Vashkov, V. I., et al., 1962.

Bronskill, J. F., joint author. *See* Lange, R., et al., 1964.

Brooker, P. J., joint author. *See* Geering, Q. A., et al., 1965.

3112. **Brooks, G. T. 1960** Mechanisms of Resistance of the Adult Housefly (*Musca domestica*) to "Cyclodiene" Insecticides (Correspondence). Nature 186:96-98.
Isodrin and endrin, labelled with C^{14} in the terminal unchlorinated ring, and with aldrin and dieldrin in the chlorinated ring. Suggests that resistance of R flies to dieldrin is *not* due to cuticular penetration, differences in metabolism, or excretion of insecticides. ABSTRACTS: BA 35:3739 (42991), Aug. 1, 1960; RAE-B 50:102, May, 1962; TDB 57:856, Aug., 1960.

3113. **Brooks, G. T. 1967** Biochemical Implications of Resistance in Diptera to the Cyclodiene Insecticides. Abstract. Biochem. J. 102(1):3P.
Appear to act as central nerve poisons. Commercial cyclodienes appear to be remarkably stable, both in normal and resistant flies. Inheritance of resistance to cyclodienes relatively simple (and similar) in a number of species.

3114. **Brooks, G. T. 1968** Mechanisms of Resistance of Chlorohydrocarbon Insecticides. Wld. Rev. Pest Control 7(3):127-134.
Treats of detoxication mechanisms, such as dehydrochlorinations and hydroxylation. Alternative methods discussed. ABSTRACTS: BA 50:13064 (134737), Dec. 15, 1969; RAE-B 57:40(150), Feb., 1969.

3115. **Brooks, G. T.** and **Harrison, A. 1963** Relations between Structure, Metabolism and Toxicity of the "Cyclodiene" Insecticides. Nature 198(4886):1169-1171.
A discussion of known facts, with suggestions as to possible interpretations. ABSTRACT: RAE-B 52:113, June, 1964.

3116. **Brooks, G. T.** and **Harrison, A. 1963** Metabolism and Toxicity of the "Cyclodiene" Insecticides. Proc. Biochem. Soc. 87(1):5P-6P.
The most toxic of three cyclodiene compounds of low toxicity formed more epoxide than the least, which formed none at all as the compounds were converted to polar substances. The low toxicity of the epoxides when used as insecticides is due to instability *in vivo*. ABSTRACT: RAE-B 52:114, June, 1964.

3117. **Brooks, G. T.** and **Harrison, A. 1964** The Metabolism of Some Cyclodiene Insecticides in Relation to Dieldrin Resistance in the Adult Housefly, *Musca domestica* L. J. Insect Physiol. 10(4):633-641.
Dieldrin-resistant flies show complete cross resistance to a number of related cyclodiene compounds. Resistance to these compounds *not* dependent on their metabolism *in vivo*. Sesamex does not synergize dieldrin against R flies. ABSTRACTS: RAE-B 53:40, Feb., 1965; BA 46: 354, Jan. 1, 1965.

3118. **Brooks, G. T.** and **Harrison, A. 1964** The Effect of Pyrethrin Synergists especially Sesamex on the Insecticidal Potency of Hexachlorocyclopentadiene Derivatives ("cyclodiene" insecticides) in the Adult Housefly, *Musca domestica* L. Biochem. Pharmacol. 13:827-840.
Only 7 of 51 compounds were noticeably active against resistant strains of flies. Almost none were synergized by sesamex. Data on cross-resistance. ABSTRACTS: BA 45:8401(106718), Dec. 15, 1964; RAE-B 53:40, Feb., 1965.

3119. **Brooks, G. T.** and **Harrison, A. 1965** Structure-activity Relationships among Insecticidal Compounds Derived from Chlordene. Nature 205(4975):1031-1032.
Different toxicities of the heptachlor epoxides. May be a consequence of their stereochemical differences. Suggests reasons for low toxicity of chlordene. ABSTRACT: RAE-B 54:94, May, 1966.

3120. **Brooks, G. T.** and **Harrison, A. 1966** Metabolism of Aldrin and Dihydroaldrin by Houseflies (*M. domestica* L.) *in vivo* and by Housefly and Pig Liver Microsomes. Life Sci. 5(24):2315-2320. Oxford.
Conversion of aldrin to its insecticidal epoxide dieldrin, is perhaps the only metabolic conversion of this compound in *M. domestica*. Dihydroaldrin is hydroxylated *in vitro* to exo- and endo-6-hydroxydihydroaldrin by microsomes in the presence of NADPH. ABSTRACTS: BA 48:10300 (115739), Dec. 1, 1967; RAE-B 56:107(372), May, 1968.

3121. **Brooks, G. T.** and **Harrison, A. 1967** The Toxicity of Alpha-dihydroheptachlor and Related Compounds to the Housefly (*M. domestica* L.) and their Metabolism by Housefly and Pig Liver Microsomes. Life Sci. 6:1439-1448.
Microsomal metabolic products from *alpha*, *beta* and *gamma* dihydroheptachlor are qualitatively more or less similar for mammalian and insect systems. Conversions are inhibited by sesamex. ABSTRACTS: BA: 49:2153(24171), Mar. 1, 1968; RAE-B 55:217(751), Dec., 1967.

3122. Brooks, G. T. and **Harrison, A. 1967** The Metabolism of Dihydrochlordene and Related Compounds by Housefly (*M. domestica* L.) and Pig Liver Microsomes. Life Sci. 6(7):681-689. Oxford.
Hydroxylation of dihydrochlordene into chemically indentified products, provides evidence for structures assigned to primary metabolites. Sesamex inhibits hydroxylation by microsome enzymes. Hydrochlordane has a low intrinsic toxicity. ABSTRACTS: RAE-B 56:107(373), May, 1968; BA 49:3110(34794), Apr., 1968.

3123. Brooks, G. T. and **Harrison, A. 1969** The Oxidative Metabolism of Aldrin and Dihydroaldrin by Houseflies (*Musca domestica*: Dipt. Muscidae) Housefly Microsomes and Pig Liver Microsomes and the Effect of Inhibitors. Biochem. Pharmacol. 18:557-568.

3124. Brooks, G. T., Harrison, A. and **Cox, J. T. 1963** Significance of the Epoxidation of the Isomeric Insecticides Aldrin and Isodrin by the Adult Housefly *in vivo*. (Correspondence.) Nature 197:311-312.
Aldrin and Isodrin have lower intrinsic toxicities than corresponding epoxides. Insecticidal properties probably not due entirely to epoxidation. (Related compounds are somewhat toxic without undergoing epoxidation *in vivo*.) ABSTRACTS: BA 44:643(8611), Oct. 15, 1963; RAE-B 52:64, Apr., 1964; TDB 60:598, June, 1963.

3125. Brooks, G. T., Harrison, A. and **Lewis, S. E. 1970** Cyclodiene Epoxide Ring Hydration by Microsomes from Mammalian Liver and Houseflies. Biochem. Pharmacology. 19:255-273.
Microsomes hydrate enzymically the epoxide rings of several cyclodiene type insecticides. When $NADPH_2$ and oxygen at present, oxidation products are formed in some cases, by action of mixed function oxidases. ABSTRACT: BA 51:8309(84783), Aug. 1, 1970.

Brooks, I. C., joint author. *See* Price, R. W., et al., 1962.

3126. Brown, A. W. A. 1960 Mechanisms of Resistance Against Insecticides. Ann. Rev. Ent. 5:301-326.
Notes 6 types of resistance, the most important being DDT resistance, dieldrin resistance, and organophosphorus (OP) resistance.

3127. Brown, A. W. A. 1960 The Resistance Problem, Vector Control and W.H.O. Misc. Publ. Ent. Soc. Am. 2(1):59-67.
A review of measures undertaken by the World Health Organization.

3128. Brown, A. W. A. 1961 Negatively-Correlated Insecticides. A Possible Countermeasure for Insecticide Resistance. Pest Control 29(9): 24-26.
An excellent review of research reports relating to negatively-correlated insecticides.

3129. Brown, A. W. A. 1961 The Challenge of Insecticide Resistance. Bull. Ent. Soc. Am. 7(1):6-19.
An informative presentation of the genetic and physiological mechanisms of resistance in insects. ABSTRACT: BA 36:4076(50463), Aug. 1, 1961.

3130. Brown, A. W. A. 1962 Insecticides and Human Health. Wld. Rev. Pest Control 1(3):6-17.
Reviews the effectiveness of various methods of application of DDT and other insecticides against the major insects causing human disease and misery throughout the world.

3131. Brown, A. W. A. 1963 Insecticides and World Health. I. Pest Control. 31(4):18-20, 22, 25. Map.
A review of the role of insects in sickness and death in human populations. Examples are cited of the control of insects and the development

of resistance among others. The importance of DDT, lindane, chlordane, parathion, heptachlor, malathion, diazinon, trichlorfon, fention, and other synthetic insecticides is discussed. ABSTRACT: BA 43:1336(16564), Aug. 15, 1963.

3132. Brown, A. W. A. 1963 Meeting the Resistance Problem. Vector Control Prospects in the Light of Present Knowledge of Insecticidal Resistance. Bull. W. H. O. 29:41-50. (Suppl.)
Treats of many insecticides, cross-resistance between them, most effective methods of application, biochemical mechanisms of resistance and attempts to inhibit them.

3133. Brown, A. W. A. 1963-1964 Insect Resistance. I. Nature and Prevalence of Resistance. Farm Chemicals. 126(10):21-25; II. Mechanisms of Resistance. III. Development and Inheritance of Resistance. T.c. (11):24-28; IV. Countermeasures for Resistance. 127(1):58-60.
Treats broadly of the entire subject. Houseflies mentioned chiefly in connection with DDT.

3134. Brown, A. W. A. 1967 Insecticide Resistance, Genetic Implications and Applications. Wld. Rev. Pest Control 6(4):104-114.
Includes discussion of "marker-mutants" in *M. domestica* and other species, to show the position of resistance genes or chromosomes. Chromosome maps. ABSTRACT: TDB 66:876(1745), Aug., 1969.

3135. Brown, A. W. A. 1967 Genetics of Insecticide Resistance in Insect Vectors. *In* Genetics of Insect Vectors of Disease. (J. W. Wright and R. Pal, Editors) Amsterdam. Elsevier Publ. Co., pp. 505-552.
ABSTRACT: RAE-B 56:158(559), Aug., 1968.

3136. Brown, A. W. A. 1967 Genetic Implications and Applications of Resistance to Insecticides. Abstr. Int. Pflanzenschutz-Kong. [Int. Congr. Plant Prot.] 30:28-29.

3137. Brown, A. W. A. 1968 Insecticide Resistance Comes of Age. Bull. Ent. Soc. Am. 14(1):3-9.
A detailed, tabulated review of the subject, with emphasis on the need to meet the problem with new kinds of chemicals and standardized methods for testing resistance levels.

3138. Brown, A. W. A., West, A. S. and **Lockley, A. S. 1961** Chemical Attractants for the Adult House Fly. J. Econ. Ent. 54(5):670-674.
Many compounds tested in an olfactometer. Most attractive retested in baited traps in closed rooms and best of these then tested in the field. First choice: a mixture (aqueous solution) of malt extract 5 per cent ethyl alcohol 0·5 per cent, skatole 0·02 per cent, acetal 1 per cent, made into a paste with finely ground peat and alfalfa meal. ABSTRACT: RAE-B 50:147, July, 1962.

Brown, A. W. A., joint author. *See* Knipling, E. F., et al., 1968.

3139. Brown, J. F. 1968 A Systematic Insecticide for Control of the Face Fly, *Musca autumnalis*. Diss. Absts. 28(12B):5061-B.

3140. Brown, N. C., Chadwick, P. R. and **Wickham, J. C. 1967** The Role of Synergists in the Formulation of Insecticides. Int. Pest Contr., Nov.-Dec., 4 pp.
Greatest use is with pyrethrins and DDT. Will become more important as resistant strains become more common. Present synergists do *not* increase mammalian toxicity of insecticides.

3141. Brown, P., Wong, W. and **Jelenfy, I. 1970** A Survey of the Fly Production from Household Refuse Containers in the City of Salinas, California. Calif. Vector Views 17(4):19-23.
Concerns the green blow fly (*Phaenicia*). Single containers produced

0 to 30,150 larvae in 1 week. Only 4 of 30 containers produced no flies during the 6-week study.

Brown, R. T., joint author. *See* Bořkovec, A. B., et al., 1966.

3142. **Browne, L. B.** and **Kerr, R. W. 1967** The Response of the Labellar Taste Receptors of DDT-resistant and Non-resistant Houseflies (*Musca domestica*) to DDT. Ent. Exp. Appl. 10(3-4):337-346.
Concludes that resistance of strain HR flies is fully expressed at the level of the chemoreceptor hairs on their labella, and is mainly due to the ability to recover from DDT poisoning. ABSTRACTS: BA 49:5302(59496), June 1, 1968; TDB 65:1396(3126), Nov., 1968.

Brthour, J. R., joint author. *See* Harvey, T. L., et al., 1960.

3143. **Bruce, W. G. 1964** The Face Fly in North Carolina. N. C. Agr. Ext. Serv. Folder 227:no page number.

3144. **Bruce, W. N., Moore, S., III** and **Decker, G. C. 1960** Face Fly Control. J. Econ. Ent. 53:450-451.
In Illinois, treatment with compounds R-326 and Tabatrex gave practical control. Face applications of dimethoate and DDVP in syrup bait eliminated the flies. ABSTRACT: RAE-B 49:120, June, 1961.

3145. **Bryant, E. H.** and **Sokal, R. R. 1967** The Fate of Immature House Fly Populations at Low and High Densities. Res. Pop. Ecol. 9(1): 19:44. (Japanese summary.)
Development takes longer at higher density, but inception of developmental stages is not postponed. Main differences in outcome produced by crowding result from different responses of third instar larvae. ABSTRACT: BA 50:7537(78576), July 15, 1969.

3146. **Bryant, E. H.** and **Sokal, R. R. 1968** Genetic Differences Affecting the Fates of Immature Housefly Populations at Two Densities. Res. Pop. Ecol. 10:140-170.
Eggs of *ge* and *bwb* strains were reared under 2 densities and the fate of immature forms analyzed. Treats of how genetic differences are reflected in components of the life history process.

3147. **Bryant, E. H. 1969** The Fates of Immatures in Mixtures of Two Housefly Strains. Ecology 50(6):1049-1069.
An analysis was made of ecological factors determining numbers of adults produced in pure and mixed larval populations of two housefly strains at low and high densities. ABSTRACT: BA 51:7464(76275), July 15, 1970.

Bryant, E. H., joint author. *See* Sokal, R. R., et al., 1967, 1968, 1970.

3148. **Brydon, H. W. 1967** Response of Larval *Fannia femoralis* (Diptera: Anthomyiidae) to Light. Ann. Ent. Soc. Am. 60(2):478-480.
Tests conducted at approximately 75°F. and 45 per cent RH. Larvae responded rapidly and negatively to light when freshly exposed. Were less responsive following exercise. ABSTRACT: BA 48:5579(62457), June 15, 1967.

Brydon, H. W., joint author. *See* Eastwood, R. E., et al., 1967; Legner, E. G., et al., 1966; Stone, R. S., et al., 1965.

Buckner, A. J., joint author. *See* Perry, A. S., et al., 1964, 1965.

3149. **Budyl'nikova, M. A., Kashina, D. A., Surchakov, A. V.** and **Shvartsshtein, E. I. 1962** [Tests of the effectiveness of vat residues of dichloroethane on the larvae of flies in liquid substrates. Annotation.] Med. Parazit. i Parazitarn. (Bolezni) 31(6):748.
ABSTRACT: BA 42:1906(24061), June 15, 1963.

3150. Buèi, K. 1963 [Biological Differences between Resistant and Susceptible Strains of the House Fly.] Botyu-Kagaku (Bull. Inst. Insect Contr.) 28(4):98-103. (In Japanese, with English summary.)
Resistant and susceptible strains of subspecies *domestica* and subspecies *vicina* compared as to their egg and larval periods, pupal period, per cent pupation, per cent emergence, sex ratio and weight of pupae. ABSTRACTS: BA 46:7779(96193), Nov. 1, 1965; RAE-B 55:38(143), Feb., 1967.

3151. Buèi, K. 1967 [The relationship between the body weights of pupae of house fly and number of matured eggs in the ovaries.] Jap. Jour. Sanit. Zool. 18(1):18-20. (In Japanese, with English summary.)
Flies having a heavier pupal weight developed more eggs in their ovaries. No significant difference between wild strain and laboratory strain. (*M. domestica vicina.*) ABSTRACTS: RAE-B 58:72(287), March, 1970; BA 51: 9698(98692), Sept. 1, 1970.

3152. Buèi, K. and **Fukuhara, Y. 1961** [Field test of fly control using Nankov (0,0-dimethyl 0-2, 4, 5-trichlorophenyl phosphorothioate).] Jap. Jour. Sanit. Zool. 12(3):183-186. (English summary.)
Spraying in barns and houses reduced fly population for 7 days. Application of emulsion in latrines arrested larval production for 11 days.

3153. Buèi, K. and **Fukuhara, Y. 1964** [The relation between the length of the larval period and adult susceptibilities to insecticides in the housefly (Diptera).] Botyu-Kagaku 29(1):9-14. (English summary.)
DDT-resistant and susceptible strains of *M. domestica vicina* were selected for 15 generations on basis of early or late pupation. Measurable differences in susceptibility demonstrated. LD_{50} of each colony did not change thereafter (35 to 50 generations). ABSTRACT: RAE-B 55:38(144), Feb., 1967.

3154. Buèi, K., Asada, S., Hasuo, A., Omori, M. and **Fujimoto, N. 1966** [Synergistic effect of synthetic synergists on phthalthrin against adults of the common house fly, *Musca domestica vicina* Macq.: Studies on the biological assay of pyrethroids. Rept. 3.] Botyu Kagaku (Bull. Inst. Insect. Contr.) 31(2):86-90. (In Japanese, with English summary.)
Used topical applications and settling mists. Synergism of phthalthrin plus safroxan inferior to that of pyrethrins plus piperonyl butoxide. Synergists did not increase knockdown effect of phthalthrin alone. ABSTRACTS: BA 48:3703(41094), Apr. 15, 1967; RAE-B 56:153(536), Aug., 1968.

3155. Buèi, K., Asada, S. and **Kodama, M. 1963** [Synergistic effect of synthetic synergists on pyrethrins and allethrin against adults on the common housefly, *Musca domestica vicina* (Diptera).] Botyu-Kagaku 28(3):47-55. (English summary.)
Pyrethrins and allethrin were mixed with one of 8 synergists respectively in kerosene solution at ratios of 1:5 and 1:10. Knockdown and lethal effects were evaluated by tests on *M. domestica vicina.* Used Nagasawa's settling mist apparatus. ABSTRACT: RAE-B 55:38(142), Feb., 1967.

3156. Buèi, K., Asada, S., Kodama, M., Hasuo, A. and **Omori, M. 1965** [Evaluation of pyrethroid in kerosene and deobase against adults of the common housefly, *Musca domestica vicina* (Diptera) by settling mist method. Studies on the biological assay of pyrethroids 2.] Botyu-Kagaku (Bull. Inst. Insect Contr.) 30(1):37-44. (English summary.)
Five pyrethroids evaluated. *Knockdown* was best with phthalthrin; pyre-

thrins and allethrin following. Barthrin and dimethrin were less effective. *Lethal* effect of dimethrin was excellent, that of barthrin and phthalthrin about equal. Allethrin and pyrethrins were less effective. ABSTRACT: RAE-B 55:175(608), Sept., 1967.

3157. Buèi, K., Asada, S. and **Sotani, I. 1968** On Some Fluctuating Factors by Using the Settling Mist Apparatus on the Biological Assay. Contributions to the Study of the Laboratory Test Methods of Insecticides. 3:46-49.
Factors include: (1) Mechanics of the spraying device; (2) Time the fly is exposed to spray; (3) Gauze over fly containers (= interference with settling).

Buèi, K., joint author. *See* Fugito, S., et al., 1962, 1963, 1966.

3158. Bull, D. L., Lindquist, D. A. and **Grabbe, R. R. 1967** Comparative Fate of the Geometric Isomers of Phosphamidon in Plants and Animals. J. Econ. Ent. 60(21):332-341.
Trans isomers of phosphamidon and its N-diethyl derivative are more toxic to house flies than the *cis* isomers. Are more potent inhibitors of acetyl cholinesterase. ABSTRACT: BA 48:6927(77463), Aug. 1, 1967.

Burall, R. K., joint author. *See* Fales, J. H., et al., 1960.

Burdick, D. J., joint author. *See* Smith, T. A., et al., 1966.

3159. Burger, J. F. and **Anderson, J. R. 1970** Association of the Face Fly, *Musca autumnalis*, with Bison in Western North America. Ann. Ent. Soc. Am. 63(3):635-639.
Has become a pest, breeding in bison droppings. Bison's eyes are affected (discoloration, eruption, blindness). First record of such pathology. Fear danger to other large game animals.

Burger, J. F., joint author. *See* Poorbaugh, J. H., et al., 1968.

3160. Burgerjon, A. and **Bianche, G. 1967** Divers effets spéciaux et symptômes tératologiques de la toxine thermostable de *B. thuringiensis* en fonction de l'âge physiologique des insectes. [Several special effects and teratological symptoms of the thermostable toxin of *B. thuringiensis* in terms of the physiological age of insects.] Ann. Soc. Ent. Fr. (N.S.) 3(4):929-952. (English summary.)
May cause death of larva in pre-moulting stage, or cause cessation of feeding after moulting, with death following in a few days. Tests run on *Pieris brassicae* and *Musca domestica.*

3161. Burgerjon, A. and **Bianche, G. 1967** Effets teratologiques chez les nymphes et les adultes d'insectes, dont les larves ont ingéré des doses subléthales de toxine thermostables de *Bacillus thuringiensis* Berliner. [Teratological effects in nymphs and adult insects, the larvae of which had ingested sublethal doses of thermostable toxin of *Bacillus thuringiensis* Berliner.] Compt. Rend. Hebdom. Acad. Sci., Ser. D. Sci. Natur. (Paris) 264(20):2423-2425.
Adult *M. domestica* shows atrophy of the buccal apparatus; particularly a marked reduction in the length of the labrum-epipharynx. ABSTRACT: BA 50:6965(72759), July 1, 1969.

3162. Burgerjon, A. and **DeBarjac, H. 1967** Another Serotype (4:4a, 4c) of *Bacillus thuringiensis* which Produces Thermostable Toxin. [*Pieris brassicae, Barathra brassicae, Musica domestica, Drosophila melangoaster*]. J. Invertebr. Path. 9(4):574-577.

3163. Burgerjon, A. and **Galichet, P. F. 1965** The Effectiveness of the Heat-stable Toxin of *Bacillus thuringiensis* var. *thuringiensis* Berliner on

Larvae of *Musca domestica* Linnaeus. J. Invertebr. Path. 7(2):263-264.
Table shows mortality of *M. domestica* larvae as a result of the presence of heat stable toxin in the larval nutritive medium. Number of viable spores in 1 gram of medium important. ABSTRACTS: BA 46:8152(100736), Nov. 15, 1965; RAE-B 53:248, Dec., 1965.

3164. **Burges, H. D. 1966-1967** The Standardization of *Bacillus thuringiensis*: Tests on Three Candidate Reference Materials. *In*: Insect Pathology and Microbiol. Control. Proc. Internat. Colloqium, 5-10 September. Wageningen, Neth. North-Holland Publ. Co., Amsterdam, Netherlands, pp. 314-319.
M. domestica one of six insect species considered best for routine bioassay. ABSTRACT: BA 50:3707(38904), Apr. 1, 1969.

Burgess, J. B., joint author. *See* Deal, A. S., et al., 1965.

Burk, J., joint author. *See* Metcalf, R. L., et al., 1966.

3165. **Burkhard, R. K.** and **Klaassen, D. H. 1966** A Comparative Study of the Amino Acid Composition of Insoluble Mitochondrial Proteins. Comp. Biochem. Physiol. 18(1):231-235.
Includes data on proteins from the large mitochondria of housefly flight muscles: ABSTRACT: BA 47:8240(95979), Oct. 15, 1966.

Burkman, A. M., joint author. *See* Greenberg, B., et al., 1963.

3166. **Burns, E. C.** and **Nipper, W. A. 1960** House Fly Control in Pig Parlors. J. Econ. Ent. 53(4):539.
Cotton cords soaked in emulsifiable concentrates of ronnel, diazinon, and Bayer 29493 were allowed to dry and hung from rafters. Arranged to provide 28·5 feet of vertical cord per 100 square feet of floor space. ABSTRACTS: BA 35:5511(64856), Nov., 1960; RAE-B 49:149, July, 1961.

3167. **Burns, E. C., Wilson, B. H.** and **Tower, B. A. 1961** Effect of Feeding *Bacillus thuringiensis* to Caged Layers for Fly Control. J. Econ. Ent. 54(5):913-915.
Commercially-prepared spore powders, fed to hens, gave variable results against *M. domestica.* Above a certain dose level, physical condition of layers decreased. ABSTRACT: RAE-B 50:174, Aug., 1962.

Burns, E. C., joint author. *See* Wilson, B. H., et al., 1968.

3168. **Burton, R. P.** and **Turner, E. C., Jr. 1968** Laboratory Propagation of *Muscidifurax raptor* on Face Fly Pupae. J. Econ. Ent. 61(5):1380-1383.
Less than 5 per cent of adult parasites were able to emerge, unaided, from host puparia. ABSTRACT: BA 50:4219(44396), Apr. 15, 1969.

3169. **Burton, R. P.** and **Turner, E. C., Jr. 1970** Mortality in Field Populations of Face Fly Larvae and Pupae. J. Econ. Ent. 63(5):1592-1594.
Larval mortality averaged 49 per cent in 1966; 63 per cent in 1967. Adult flies emerged from only 47·4 per cent of exposed pupae. Indicates that exposure to the biotic environment reduces number of adult flies 80·8 per cent.

Burton, R. P., joint author. *See* Turner, E. C., Jr., et al., 1965.

3170. **Burton, V. E., Anderson, J. R.** and **Stanger, W. 1965** Fly Control Costs on Northern California Poultry Ranches. J. Econ. Ent. 58(2): 306-309.
The average cost was found to be 7.4 cents per bird year. Figure ran higher on the smaller ranches.

3171. **Busvine, J. R. 1961** Insecticide-Resistance Strains of Insects in England by 1961. Sanitarian, for December 1961:192-193.
Definite resistance confined essentially to houseflies (usually at refuse tips) and German cockroaches (at ports).

3172. **Busvine, J. R. 1962** The Problem of Insecticide Resistance. *In* Symposium on the Problems of Epidemiologically Important Arthropods and their Resistance to Insecticides. J. Hyg. Epidemiol., Microbiol. and Immunol. (Prague) 6(3):243-255. (Summaries in French, German and Italian.)
Treats of genetic studies, toxicological studies, and biochemical research. Suggests 5 "policies" for control of resistance. ABSTRACT: BA 42:1249 (15801), May 15, 1963.

3173. **Busvine, J. R. 1964** The Insecticidal Potency of γ-BHC and the Chlorinated Cyclodiene Compounds and the Significance of Resistance to Them. Bull. Ent. Res. 55(2):271-288.
Musca domestica and *Lucilia cuprina* were used to test 36 compounds of the chlorinated terpene and cyclodiene group, including chlorinated adamantane and *gamma*-BHC for insecticide potency and cross resistance patterns. ABSTRACT: TDB 62:69, Jan., 1965.

3174. **Busvine, J. R. 1965** Wire Mesh Screening for the Exclusion of Houseflies. J. Hygiene (Camb.) 63:305-309.
Where smaller insects are no problem, use aperture 2·17 mm. square. Allows for greater light admittance and for ventilation. ABSTRACTS: BA 47:7250(84562), Sept. 1, 1966; TDB 63:102, Jan., 1966.

3175. **Busvine, J. R. 1968** Resistance to Organo-phosphorus Insecticides in Insects. Meded. Rijksfac Landbouwwetensch Gent. 33(3):605-620.
M. domestica and other species. Treats of enzymes (cholinesterase, aliesterase) also of dieldrin and synergism.

3176. **Busvine, J. R. 1968** Detection and Measurement of Insecticide Resistance in Arthropods of Agriculture or Veterinary Importance. Wld. Rev. Pest Control 7(1):27-41.
Discusses test methods and their standardization. ABSTRACTS: RAE-B 56:194(701), Oct., 1968; BA 50:4781(50257), May 1, 1969.

3177. **Busvine, J. R.** and **Haupt, A. 1968** House Fly Size as Affected by Larval Food Competition. Trans. Roy. Soc. Trop. Med. Hyg. 62(1): 12.
Overcrowding (more than 800 eggs per rearing jar) results in reduced size of adults. Degree can be measured by dry weight of flies or by wing dimensions. Wild-caught flies rarely have small wing measurements.

3178. **Busvine, J. R.** and **Pal, R. 1969** The Impact of Insecticide-resistance on Control of Vectors and Vector-borne Diseases. Bull. W.H.O. 40(5):731-744.
Data on housefly voluminous. Hungary, India and Canada report insecticide failure associated with prevalence of fly-borne disease. ABSTRACT: TDB 67:90(196), Jan., 1970.

3179. **Busvine, J. R.** and **Townsend, M. G. 1963** The Significance of BHC Degradation in Resistant House-flies. Bull. Ent. Res. 53(4):763-768.
M. domestica shows greater tolerance of BHC than of endrin and isodrin. Degradation believed enzymatic, but enhanced in some manner. ABSTRACTS: RAE-B 51:50, Mar., 1963; TDB 60:597, June, 1963.

Busvine, J. R., joint author. *See* Bell, J. C., et al., 1967; Gil, L., et al., 1968; Guneidy, A. M., et al., 1964; Haupt, A., et al., 1969; Khattat, F. H., et al., 1965; Nguy, V. D., et al., 1960.

3180. **Buttram, J. R.** and **Arthur, B. W. 1961** Absorption and Metabolism of Bayer 22408 by Dairy Cows and Residues in Milk. J. Econ. Ent. 54(3):446-451.
Detectable quantities of the intact insecticides were isolated from the

milk the first 6 days after treatment. Metabolites passed in the feces were toxic to *Stomoxys calcitrans* (L) but not to *Musca domestica* L. ABSTRACTS: BA 36:5562(58140), Sept. 1, 1961; RAE-B 50:127, June, 1962.

Bykhovskaia, A. M., joint author. *See* Magdiev, R. R., et al., 1960.

3181. Calciu-Dumitreasa, A., Curcaneanu, N. and **Popescu-Pretor, I. 1968** Cercetări experimentalé de resensibilizare cu solutii insecticide DDT + AR pé *Musca domestica* DDT-resistenză. [Experimental studies on the resensitization of DDT-resistant *Musca domestica* with DDT + AR insecticide solutions.] Igiena 17(4):217-222. (English summary.)
Investigations were conducted on the efficiency of the DDT + AR associated product in housefly destruction. Maximum mortality results if the ratio of five parts DDT to 1 part AR is used. ABSTRACT: TDB 65:1187 (2717), Sept., 1968.

3182. Calciu-Dumitreasa, A. and **Gafiteanu, L. 1969** Contributii la studiul fenomenului de insecticifo-resistentă la *Musca domestica* L. [Contributions to study of the phenomenon of resistance to insecticides in *Musca domestica* L.] Igiena, 18(9):519-524.
Sensitivity curve shows that mortality is not always influenced by the insecticide used. Resistance in Rumania (1968) not high enough to jeopardize control measures. ABSTRACT: TDB 67:366(767, Mar., 1970.

3183. Callahan, R. F. 1962 Effects of Parental Age on the Life Cycle of the House Fly *Musca domestica* Linnaeus. (Diptera:Muscidae.) J. New York Ent. Soc. 70(3):150-158.
Adult flies from young parents survive unfavorable conditions longer than those from older parents. ABSTRACT: BA 45:2860(35480), Apr. 15, 1964.

3184. Callahan, R. F. 1962 Effects of Parental Age on the Life Cycle of the House Fly, *Musca domestica* Linnaeus. Diss. Absts. 23(3):802-803. ABSTRACT: BA 42:1287(16257), June 1, 1963.

3185. Calvert, C. C., Martin, R. D. and **Morgan, N. O. 1969** House Fly Pupae as Food for Poultry. J. Econ. Ent. 62(4):938-939.
Dried pupae meal a possible substitute for soy bean meal. Use of larvae to convert human waste a possible device for reducing pollution. ABSTRACT: RAE-B 58:27(99), Jan., 1970.

3186. Calvert, C. C., Martin, R. D. and **Morgan, N. O. 1969** Dual Roles for House Flies in Poultry Manure Disposal. Poultry Sci. 48(5): 1793.

3186a. Calvert, C. C., Morgan, N. O. and **Martin, R. D. 1970** House Fly Larvae: Biodegradation of Hen Excreta to Useful Products. Poultry Sci. 49(2):588-589.
Newly laid house fly eggs were seeded at the rate of 3 eggs per gram of excreta, which was reduced in 8 days to friable, nearly odorless material. Can be used as a fertilizer. Dried pupae used to replace soybean meal in chick diet.

Cambiesco, I., joint author. *See* Duport, M., et al., 1963.

3187. Camp, H. B. and **Arthur, B. W. 1967** Absorption and Metabolism of Carbaryl by Several Insect Species. J. Econ. Ent. 60(3):803-807.
Major metabolite in feces of house flies a highly polar material. ABSTRACTS: BA 49:1418(15900), Feb. 1, 1968; RAE-B 55:190(656), Oct., 1967.

3188. **Campbell, E.** and **Black, R. J. 1960** The Problem of Migration of Mature Fly Larve from Refuse Containers and its Implication on the Frequency of Refuse Collection. Calif. Vector Views 7(2):9-15.
Many larvae complete development and migrate from container before the weekly collection. Wrapping of garbage, indoor disposal, twice a week collections, are possible solutions.

3189. **Campos, F. 1963** [The fly.] Rev. Ecuat. Hig. 20:7-8.

3190. **Cantwell, G. E., Heimpel, A. M.** and **Thompson, M. J. 1964** The Production of an Exotoxin by Various Crystal-forming Bacteria Related to *Bacillus thuringiensis* var. *thuringiensis* Berliner. J. Insect Pathol. 6(4):466-480.
Twenty-eight varieties tested. Only those isolates previously described as *B. thuringiensis* var. *thuringiensis* (and not all of these) produced the exotoxin. ABSTRACT: RAE-B 53:176, Sept., 1965.

3191. **Cantwell, G. E.** and **Laird, M. 1966** The World Health Organization Kit for the Collection and Shipment of Pathogens and Parasites of Diseased Vectors. J. Invertebr. Path. 8(4):442-451.
For preserving and consigning for diagnosis, all types of pathogens and parasites. ABSTRACT: TDB 64:571, May, 1967.

3192. **Cantwell, G. E., Shortino, T. J.** and **Robbins, W. E. 1966** The Histopathological Effects of Certain Carcinogenic 2-fluorenamine Derivatives on Larvae of the House Fly. J. Invertebr. Path. 8:167-174.
Affect growth, pupation and adult emergence. Also produce focal lesions in the hypodermis, midgut and fat bodies. ABSTRACTS: BA 47:8559 (99751), Oct. 15, 1966; RAE-B 55:15(42), Jan., 1967.

Cantwell, G. E., joint author. *See* Loy, V. A., et al., 1966; West, J. A., et al., 1968; Shortino, J. T., et al., 1963.

3193. Duplicate reference. Withdrawn.

3194. **Caprotti, M.** and **Rubini, P. G. 1966** Osservazioni sulla interazioni fra gas ambientali e radiosensibilita di pupe di *Musca domestica* L. [Observations on the interaction between atmospheric gases and radiosensitivity of the pupae of *Musca domestica* L.] Symp. Genet. et Biol. Ital. 13:379-386.
Under helium, oxygen or normal atmosphere, exposure to X-rays does not affect the oxygen consumption of the pupae. Helium exerts a conspicuous radio protection.

Caprotti, M., joint author. *See* Rubini, P. G., et al., 1966.

Carbanza, N., joint author. *See* Quevedo, F., et al., 1966.

3195. **Cardaras, P.** and **Vocino, G. 1962** Il ruolo di *Musca domestica* L. nella diffusione del virus aftoso. [The role of *Musca domestica* L. in the diffusion of foot-and-mouth disease virus.] Igiene San. Pubbl. 18 (11/12):566-576.
There is a high incidence of fly contamination. Mechanical vectorship considered probable.

Carey, W. F., joint author. *See* Khan, M. A. Q., et al., 1970.

Carmichael, A., joint author. *See* Gratz, N. G., et al., 1963.

3196. **Carney, G. C. 1965** Swelling and Shrinkage Properties of Housefly Sarcosomes after *in vivo* Exposure to X-rays. Radiat. Res. 25(4): 637-645.
Swelling was rapid and in proportion to dose. Reversal occurred with the addition of ATP. ABSTRACT: BA 47:437(5394), Jan. 15, 1966.

3197. **Carney, G. C. 1965** The Effects of Ionizing Radiation on the Sarcosomes of Housefly Flight Muscle. Diss. Absts. 25(9):4917.

3198. Carney, G. C. 1966 The Effects of Different Isolation Media on the Respiration and Morphology of Housefly Sarcosomes. J. Insect Physiol. 12(9):1093-1103.
Concerns structural effects, oxygen uptake, respiratory response to ADP, and oxidation of alpha-glycerophosphate as studied in a KCl medium and a sucrose medium. ABSTRACTS: BA 47:10360(119783), Dec. 15, 1966; TDB 63:1400, Dec., 1966.

3199. Carney, G. C. 1969 The Utilization of ^{14}C-labelled Adenosine Diphosphate During the *in vitro* Respiration of House Fly Sarcosomes. Life Sci. 8:453-464.
Concerns flight-muscle mitochrondria. Added ADP ends up in an equilibrium mixture containing about 45 per cent of original ADP, regardless of experimental conditions or rate of oxygen uptake.

3200. Carney, W. P. 1964 A Pseudo Natural Biological Control of Insects. Biologist 47(1/2):17-21.
Concerns the "sterile male" method of insect control, citing screw worm eradication as the prime example. Mentions the alkylating agents used as chemosterilants with the house fly. ABSTRACT: BA 46:8879(109484), Dec. 15, 1965.

Carpenter, P. D., joint author. *See* Greenberg, B., et al., 1960.

3201. Carson, N. A. and **Martinez, E. F. 1967** Fly Identification by the Morphology of the Head and Head Appendages. J. Ass. Offic. Anal. Chem. 50(5):1146-1193.
Detailed descriptions and numerous drawings presented for identifying fly heads, antennal segments, mouth parts, and other fragments, as found in foods. Ten species identified by special key. ABSTRACT: BA 49:2960 (32019), Mar. 15, 1968.

3201a. Casagrande, R. A. and **Hansens, E. J. 1969** The Effects of Temperature and Humidity on Longevity of Insecticide Resistant and Susceptible *Musca domestica* Linnaeus. J. New York Ent. Soc. 77(4):257-263.
No difference in longevity between the insecticide resistant and susceptible strains tested. Desiccation factor appears to be primary mechanism involved in determining longevity with temperature determining the rate. ABSTRACT: BA 51:13730(139781), Dec. 15, 1970.

3202. Casida, J. E. 1968 Activity of House Fly Microsome NADPH-2 Enzyme System and its Relation to Synergism and Resistance Mechanisms for Insecticide Chemicals. Acta Vitaminol. et Enzymol. 22(1/2):39.

3203. Casida, J. E., Augustinsson, K.-B. and **Johnsson, G. 1960** Stability, Toxicity and Reaction Mechanism with Esterases of Certain Carbamate Insecticides. J. Econ. Ent. 53(2):205-212.
Results indicate that competition rather than carbamoylation is the primary mechanism of cholinesterase inhibition and presumably toxicity for the N-alkyl and N, N-dialkylcarbamates studied. ABSTRACT: BA 35: 3739(42993), Aug. 1, 1960.

3204. Casida, J. E., Engel, J. L., Esaac, E. G. (as **Essac, E. G.**), **Kamienski, F. X.** and **Kuwatsuka, S. 1966** Methylene-C^{14}-dioxyphenyl Compounds: Metabolism in Relation to their Synergistic Action. Science 153(3740):1130-1133.
Certain compounds containing the dioxyphenyl group increase the duration of action and toxicity of many insecticide chemicals. Believed to serve as alternate substrates for an enzymatic hydroxylation system of microsomes. ABSTRACT: BA 48:4855(54267), June 1, 1967.

3205. **Casida, J. E., Engel, J. L.** and **Nishizawa, Y. 1966** 3', 5'-Diesters of 5-fluoro-2'-deoxyuridine and Thymidine. Hydrolysis by Esterases in Human, Mouse and Insect Tissue. Biochem. Pharmacol. 15(5): 627:644.
House fly abdomens were used in work primarily concerned with tumor tissue from mice and human beings. ABSTRACT: BA 47:7797(90912), Oct. 1, 1966.

3206. **Casida, J. E., Shrivastava, S. P.** and **Esaac, E. G. 1968** Selective Recovery of Volatile Products from House Flies Treated with Radioactive Insecticide Chemicals and Synergists. J. Econ. Ent. 61(5):1339-1344.
Besides giving off expired C^{14} O_2, flies give off various radio-labelled, volatile, organic compounds. (Earlier workers considered C^{14} O_2 the only material released from insects so treated.) ABSTRACT: BA 50:4250 (44801), Apr. 15, 1969.

3207. **Casida, J. E.** and **Yamamota, I. 1967** Metabolism of Pyrethroids Insecticides. Abstr. Int. Pflanzenschutz-Kong. [Abstr. Int. Plant Prot. Congr.]. 30:233-234.

Casida, J. E., joint author. *See* Agarwal, H. C., et al., 1960; Berteau, P. E., et al., 1969; Dorough, H. W., et al., 1964; Esaac, E. G., et al., 1968, 1969; Fukami, J.-I., et al., 1967, 1969; Krueger, H. R., et al., 1961; Mengle, D. C., et al., 1960; Peiper, G. R., et al., 1965; Plapp, F. W., Jr., et al., 1969, 1970; Shrivastava, S. P., et al., 1969; Slade, M., et al., 1970; Tsukamoto, M., et al., 1967, 1968; Yamamoto, I., et al., 1969; Zubari, M. Y., et al., 1965.

3208. Duplicate reference. Withdrawn.

3209. **Cassidy, J. D., Smith, E.** and **Hodgson, E. 1969** An Ultra Structural Analysis of Microsomal Preparations from *Musca domestica* and *Prodenia eridania*. J. Insect. Physiol. 15(9):1573-1578.
Electron microscope findings from gut, abdomen, fat body and microsomal preparations. Used adult *M. domestica* and larva of *Prodenia*. ABSTRACT: BA 51:2230(23137), Feb. 15, 1970.

3210. **Castle, R. E.** and **Ristich, S. S. 1966** Structure-activity Relationships in Apholate Analogs. J. Agric. Fd. Chem. 14:301-303.
Number of aziridinyl substitutions required for chemosterilant activity is related to the water-solubility of the compound. ABSTRACTS: BA 48: 10606(119008), Dec. 1, 1967; RAE-B 55:57(211), Mar., 1967.

3211. **Cavazzini, G. 1965** [On the attractiveness of colors to some insects.] Riv. Ital. Igiene 25:216-226.

Cawley, B., joint author. *See* Sullivan, W. N., et al., 1970.

Cecere, T., joint author. *See* Russo-Caia, et al., 1961.

3212. **Celedova, V., Prokesova, A., Havlik, B.** and **Müller, O. 1963** Demonstration of Migration of *Musca domestica* from Pig Sheds into Human Dwellings. J. Hyg. Epidem. (Praha) 7:360-370. (French, German and Spanish summaries.)
Described as a "mass migration". *M. domestica* made up one-third of the synanthropic flies collected. ABSTRACT: BA 46:5147(63901), July 15, 1965.

Celedova, V., joint author. *See* Havlik, B., et al., 1962.

Ceplea, A. G., joint author. *See* Combiesco, I., et al., 1967.

3213. **Cernov, Yu. I. 1965** Kompleks sinantropnykh dvukrylykh (Diptera) v tundrovoi zone SSSR. [Complex of synanthropic Diptera in the tundra zone of the USSR.] Ent. Obozrenie 44(1):73-83. (English summary.)
As man cultivates and settles in the tundra zone, synanthropic forms

follow northward. Native forms do not easily become synanthropic. ABSTRACT: BA 47:2931(34685), Apr. 1, 1966.

Cerny, V., joint author. *See* Rezabova, B., et al., 1968.

3213a. Chabora, P. C. 1970 Studies of Parasite—Host Interaction II. Reproductive and Developmental Response of the Parasite *Nasonia vitripennis* (Hymenoptera; Pteromalidae) to Strains of the House Fly Host, *Musca domestica.* Ann. Ent. Soc. Am. 63(6):1632-1636.

Parasite larvae in fly pupae (New York strain) showed 20 per cent greater mortality than those in pupae of flies from Florida. Genetics of local host population should be considered in biological control work.

3213b. Chabora, P. C. 1970 Studies in Parasite—Host Interaction III. Host Race Effect on the Life Table and Population Growth Statistics of the Parasite *Nasonia vitripennis.* Ann. Ent. Soc. Am. 63(6):1637-1642.

Parasites reared from New York flies showed a life expectancy of 3·94 days for males, 7·82 for females; those from Florida hosts 3·38 (males) and 11·79 (females). Mean number of offspring was 35·2 (with New York flies); 298·5 (with Florida hosts).

3214. Chabora, P. C. and **Pimentel, D. 1966** Effect of Host *Musca domestica* Linnaeus Age on the Pteromalid Parasite *Nasonia vitripennis* (Walker). Canad. Ent. 98(11):1226-1231.

Most younger hosts had high percentage yielding neither fly nor parasite. Actual ranking for successful parasitism by age of pupa: 2, 3, 1, 4, 5 days old. ABSTRACT: RAE-B 56:242(897), Dec., 1968.

Chadha, D. B., joint author. *See* Agarwal, P. N., et al., 1963; Perti, S. L., et al., 1966.

Chadroff, S., joint author. *See* Tracy, R. L., et al., 1960.

3215. Chadwick, L. E. 1964 Inhibition of *Musca domestica* Fly-head Cholinesterase *in vitro* by Pilocarpine and Atropine. J. Insect Physiol. 10: 573-585.

Both are rather weak, competitive, reversible inhibitors of fly head ChE. Do not interfere with each other. Inhibitory potency not related to degree of ionization. ABSTRACT: BA 46:354(4458), Jan. 1, 1965.

Chadwick, L. E., joint author. *See* Krysan, J. L., et al., 1962, 1963, 1966; Snyder, F. M., et al., 1964.

3216. Chadwick, P. R. 1961 A Comparison of Safroxan and Piperonyl Butoxide as Pyrethrum Synergists. Pyrethrum Post 6(2):30-37.

Effect of pyrethrins plus safroxan on knockdown of houseflies was less than with P.B. Direct toxic action limited; safroxan found to be a true pyrethrins synergist. ABSTRACT: RAE-B 51:159, Aug., 1963.

3217. Chadwick, P. R. 1963 A Comparison of MGK 264 and Piperonyl Butoxide as Pyrethrum Synergists. Pyrethrum Post 7(1):11-15, 48.

Effect of pyrethrins with MGK 264 on the knockdown of houseflies was less than with P.B. Mixture of P.B. and MGK 264 was less effective as a synergist than P.B. alone at the same concentration. ABSTRACT: RAE-B 53:122, June, 1965.

3218. Chadwick, P. R. 1963 The Use of Pyrethrum Synergists. A Discussion for Formulators. Pyrethrum Post 7(1):25-32.

Shows how cost of using pyrethrins can be reduced by using the appropriate synergist at the most economical ratio.

3219. Chadwick, P. R. and **Jones, G. D. G. 1960** The Relative Toxicities of Barthrin and Pyrethrum. Pyrethrum Post 5(3):14-16.

Barthrin much lower in toxicity than pyrethrum. Tests with houseflies indicate that piperonyl butoxide is not an effective synergist for barthrin. ABSTRACT: RAE-B 50:3, Jan., 1962.

Chadwick, P. R., joint author. *See* Jones, G. D. G., et al., 1962; Krysan, J. L., et al., 1970; Brown, N. C., et al., 1967.

Chai, K., joint author. *See* Kung, K., et al., 1963, 1965, 1968.

3220. Chakraborty, J., Sissons, C. H. and **Smith, J. N. 1967** Inhibitors of Microsomal Oxidations in Insect Homogenates. Biochem. J. 102: 492-497.
In homogenates of whole flies and of fly heads or thoraces, inhibitory power was high. Thought to involve both irreversible inactivation of the enzyme and the removal of essential cofactors.

3221. Chakraborty, J. and **Smith, J. N. 1967** Enzymic Oxidation of some Alkylbenzenes in Insects and Vertebrates. Biochem. J. 102:498-503.
Oxidation of alkylbenzenes was greater in the fly than in the locust; about equal to that found in a range of vertebrate species. Piperonyl butoxide and similar synergists inhibited the oxidation.

Chalkley, J., joint author. *See* Lipke, H., et al., 1962.

Chamberlain, W. F., joint author. *See* Weidhaas, D. E., et al., 1962.

Chambers, D. L., joint author. *See* Louloudes, S. J., et al., 1962; Monroe, R. E., et al., 1963.

3221a. Chan, S. K. 1970 Phospholipid Composition in the Mitochrondria of the Housefly *Musca domestica*: a Re-examination. J. Insect Physiol. 16(8):1575-1577.
Phosphatidylethanolamine content is high but little phosphatidylcholine is present. The phospholipid composition with respect to the content of cardiolipin is similar to that of other aerobic organisms. ABSTRACT: TDB 67:1292(2507), Oct., 1970.

3222. Chang, C. P., Pai, C. H., Sun, Y. J., Li, H. Y., Li, S. M. and **Su, T. K. 1965** [Studies of oögenesis in housefly: I. Stages of oögenesis in housefly.] Acta Ent. Sinica 14(6):523-533. (In Chinese, with English summary.)
Eleven stages in the oögenesis of the housefly *Musca domestica vicina* Macq. were distinguished and described. Cytological changes of the oöcyte, nurse cell and follicle were described for each stage and their significance discussed. ABSTRACT: BA 50:1044(10942), Jan. 15, 1969.

Chang, C. P., joint author. *See* Meng, L., et al., 1965; Ts'ao, T.-F., et al., 1966.

Chang, J. L., joint author. *See* Khan, M. A. Q., et al., 1970.

3223. Chang, J. T. and **Chiang, Y. C. 1964** [Studies on insect chemosterilants. III. The sterilizing effect of thio-tepa on the common house fly, *Musca domestica vicina* Macq.] Acta Ent. Sinica 13(5):679-688. (In Chinese, with English summary.)
A feeding technique proved to be the best method of sterilizing flies. Complete sterility could be induced. Topical application was also effective. ABSTRACT: RAE-B 53:171, Aug., 1965.

3224. Chang, J. T., Tsao T. P. and **Chiang, Y. C. 1963** [Studies on insect chemosterilants. I. Screen tests of 35 chemicals as insect chemosterilants.] Acta Ent. Sinica 12(4):394-401. (In Chinese, with English summary.)
Of the chemical tested only thiotepa, added in solution to milk powder, proved to be an effective sterilant. ABSTRACT: RAE-B 52:137, Aug., 1964.

Chang, M. T. Y., joint author. *See* Sherman, M., et al., 1963, 1965, 1967, 1969.

3225. Chang, S. C. 1961 Insecticide Assay. Chromatographic Separation of Active Components of Natural Pyrethrins and their Characterizations. Agr. and Food Chem. 9(5):390-394.

Toxicity tests were made with CSMA flies, 2 to 3 days old. Techniques summarized.

3226. Chang, S. C. 1965 Chemosterilization and Mating Behavior of Male House Flies. J. Econ. Ent. 58:669-672.

Includes age of sexual maturity, time after emergence for mating, duration of copulation. Males of any age, injected with tepa were fully sterile in about 3½ hours and remained so for about a week. ABSTRACT: BA 46:8879(109485), Dec. 15, 1965; RAE-B 53:220, Nov., 1965; TDB 63:102, Jan., 1966.

3227. Chang, S. C. and **Bořkovec, A. B. 1964** Quantitative Effects of Tepa, Metepa, and Apholate on Sterilization of Male House Flies. J. Econ. Ent. 57(4):488-490.

Males injected with aqueous solutions of graded concentration. Sterile eggs from matings counted. Tepa proved 4 times as effective as apholate, 12·5 times as effective as metepa. ABSTRACTS: BA 45:8062(102261), Dec. 1, 1964; RAE-B 52:203, Nov., 1964; TDB 62:157, Feb., 1965.

3228. Chang, S. C. and **Bořkovec, A. B. 1966** Structure-activity Relationship in Analogs of Tepa and Hempa. J. Econ. Ent. 59(6):1359-1362.

Toxicity of several compounds tested on male houseflies. No definite trend disclosed. Toxicity was entirely unrelated to the sterilizing effect. ABSTRACTS: BA 48:3234(35621), Apr. 1, 1967; RAE-B 55:64(239), Apr., 1967.

3229. Chang, S. C. and **Bořkovec, A. B. 1969** Comparative Metabolism of ^{14}C-Labeled Hempa by House Flies Susceptible or Resistant to Isolan. J. Econ. Ent. 62(6):1417-1421.

Substantial differences occur in amount of unchanged hempa recovered, radioactivity of excreta, metabolites produced. Values were changed by treatment with the synergist tropital. ABSTRACT;: BA 51:7978(81529), July 15, 1970; RAE-B 58:145(606), May, 1970.

3230. Chang, S. C., Bořkovec, A. B. and **Woods, C. W. 1966** Fate of Tepa Uniformly Labeled with C^{14} in Male House Flies. J. Econ. Ent. 59(4):937-944.

Radioactivity present in treated flies as tepa or aziridinyl metabolites; radioactive metabolites in excreta did not contain aziridinyl groups. Radioactivity was transferred to female flies by copulation. ABSTRACTS: BA 47:10323(119315), Dec. 15, 1966; RAE-B 54:209, Nov., 1966; TDB 64:218, Feb., 1967.

3231. Chang, S. C. DeMilo, A. B., Woods, C. W. and **Bořkovec, A. B. 1968** Metabolism of C^{14}-Labeled Hemel in Male House Flies. J. Econ. Ent. 61(5):1357-1365.

Hemel is metabolized to lower methylmelamines and to other (possibly acyclic) products. Lists several methylmelamines isolated from treated flies and from their excreta. ABSTRACT: BA 50:4250(44802), Apr. 15, 1969.

3232. Chang, S. C. and **Kearns, C. W. 1962** Effect of Sesamex on Toxicities of Individual Pyrethrins. J. Econ. Ent. 55(6):919-922.

Individual pyrethrins, separated by column chromatography, were applied topically to flies. In each case, addition of sesamex resulted in a substantial increase in toxicity. ABSTRACT: RAE-B 51:58, Mar., 1963.

3233. Chang, S. C. and **Kearns, C. W. 1964** Metabolism *in vivo* of C^{14}-Labelled Pyrethrin I and Cinerin I by House Flies with Special

Reference to the Synergistic Mechanism. J. Econ. Ent. 57:397-404. Metabolisms of PI and CI follow a similar pattern; differences being quantitative rather than qualitative. Potency difference is due to the rate of detoxification. There is no difference in the absorption rate. ABSTRACTS: BA 45:8420(106978), Dec. 15, 1964; RAE-B 52:166, Sept., 1964; TDB 61:1260, Dec., 1964.

3234. **Chang, S. C., Terry, P.** and **Bořkovec, A. B. 1964** Insect Chemosterilants with Low Toxicity for Mammals. Science 144:57-58. Hexamethylphosphoramide and hexamethylmelamine are effective as male housefly chemosterilants.

3235. **Chang, S. C., Terry, P. H., Woods, C. W.** and **Bořkovec, A. B. 1967** Metabolism of Hempa Uniformly Labeled with C^{14} in Male Houseflies. J. Econ. Ent. 60(6):1623-1631. Total recovery of radioactivity from excreta was about 90 per cent. Only major metabolite of hempa found in flies and excreta was pentamethylphosphoric triamide. Sterility was 50 per cent in 4 hours, 45 min.; complete after 7 hours. ABSTRACTS: BA 49:3838(43157), Apr. 15, 1968; RAE-B 56:101(348), May, 1968; TDB 65:1306(2954), Oct., 1968.

3236. **Chang, S. C., Woods, C. W.** and **Bořkovec, A. B. 1970** Metabolism of ^{14}C-Labelled N^2, N^2, N^4, N^4-Tetramethylmelamine in Male House Flies. J. Econ. Ent. 63(5):1510-1513. This substance (=TMM) was metabolized to N^3, N^2, N^4 trimethylmelamine and N^2, N^4-dimethylmelamine. Main metabolic pathway was a demethylation, and did not involve ring cleavage.

3236a. **Chang, S. C., Woods, C. W.** and **Bořkovec, A. B. 1970** Sterilizing Activity of Bis (1-aziridinyl) phosphine Oxides and Sulfides in Male House Flies. J. Econ. Ent. 63(6):1744-1746. Alkylamino-substituted compounds were the most effective sterilants. Only minor differences between analogous phosphine oxides and sulfides.

Chang, S. C., joint author. *See* Bořkovec, A. B., et al., 1964.

Chang, Y., joint author. *See* Meng, L., et al., 1965.

Chao, Y. C., joint author. *See* Hsieh, C.-Y., et al., 1964; Shieh, T. R., et at., 1964.

Chapman, G. A., joint author. *See* Plapp, F. W., Jr., et al., 1962, 1963, 1964.

Chatenay, G., joint author. *See* Floch, H., et al., 1966.

Cheema, P. S., joint author. *See* Dixit, R. S., et al., 1962.

3237. **Chefurka, W. 1963** Comparative Study of the Dinitrophenol-induced ATP'ase Activity in Relation to Mitochondrial Aging. Life Sci. 6: 399-406. Utilized sarcosomes of thoracic flight muscles of *M. domestica.* A close relation exists between DNP-induced ATP'ase activity and fatty acid production during sarcosomal aging.

Chefurka, W., joint author. *See* Ilivicky, J., et al., 1967; Tsiapalis, C. M., et al., 1967; Ishaaya, I., et al., 1968; McAllan, J. W., et al., 1961.

Chen, D., joint author. *See* Shieh, T. R., et al., 1964, 1968.

Chen, L., joint author. *See* Leng, H., et al., 1965.

Cheng, H.-M., joint author. *See* Eto, M., et al., 1968.

3238. **Cheng, T.-H. 1967** Frequency of Pinkeye Incidence in Cattle in Relation to Face Fly Abundance. J. Econ. Ent. 60:598-599. Tests in Pennsylvania during a year of heavy fly infestation supported the view that pink eye is carried by these flies. ABSTRACTS: BA 48:6838 (76386), Aug. 1, 1967; RAE-B 55:148(520), Aug., 1967.

3239. **Cheng, T.-H., Frear, D. E. H.** and **Enos, H. F., Jr. 1961** Fly Control in Dairy Barns Sprayed with Dimethoate and the Determination of Dimethoate Residues in Milk. J. Econ. Ent. 54(4):740-743.
Single applications of emulsion sprays containing 0·5 to 1 per cent dimethoate were highly effective for five weeks against *Musca domestica* L. and for more than nine weeks against *Siphona* (*Haematobia*) *irritans* (L). A second application gave even longer control. ABSTRACT: RAE-B 50: 150, July, 1962.

3240. **Cheng, T.-H., Frear, D. E. H.** and **Enos, H. F., Jr. 1962** The Use of Spray and Aerosol Formulations Containing R-1207 and Dimethoate for Fly Control on Cattle and the Determination of Dimethoate Residues in Milk. J. Econ. Ent. 55(1):39-43.
Excellent control was obtained for the horn fly, moderate control for the stable fly and only limited control of the house and face fly. Only one milk sample showed dimethoate, possibly from contamination.

3241. **Cheng, T.-H., Hower, A. A.** and **Sprenkel, R. K. 1965** Oil-based and Water-based Ciodrin Sprays for Fly Control on Dairy Cattle. J. Econ. Ent. 58(5):910-913.
Oil-based spray showed greater initial toxic effect than water-based spray, but the latter was more lasting against *M. autumnalis*. ABSTRACTS: BA 47:2046(24556), Mar. 1, 1966; RAE-B 54:14, Jan., 1966.

3242. **Cheng, T.-H.** and **Kesler, E. M. 1961** A Three Year Study on the Effect of Fly Control on Milk Production by Selected and Randomized Dairy Herds. J. Econ. Ent. 54(4):751-757.
Over a three year period fly control in carefully selected, well managed, well fed dairy herds did not have any significant effect on milk production. Of herds randomly selected and managed poorly one responded very well to fly control. ABSTRACT: RAE-B 50:151, July, 1962.

Cheng, T. H., joint author. *See* Hower, A. A., Jr., et al., 1968; Tung, S.-C., et al., 1969.

Cheung, L., joint author. *See* O'Brien, R. D., et al., 1965.

Chiang, Y. C., joint author. *See* Chang, J. T., et al., 1963, 1964.

3243. **Chikamoto, T. 1965** The Toxicity of Chrysanthemyl-allethronyl and Piperonyl Piperonylates and the Higher Homologues. Agric. Biol. Chem. 28(9):633-638.
Three of the esters formed showed a slight toxicity. Further elongation of the ester linkage by methylene groups reduced the biological activity. ABSTRACT: BA 47:3356(39592), Apr. 15, 1966.

Childress, C. C., joint author. *See* Sacktor, B., et al., 1965.

3244. **Chillcott, J. G. 1960** A Revision of the Nearctic Species of Fanniinae (Diptera:Muscidae). Canad. Ent. 92(14):1-295.
Concerns historical data, diagnostic characters, evolutionary trends, biology and speciation. Gives keys to genus and species of adults and mature larvae. There are 46 new species of *Fannia* and one each of *Euroyomma* and *Piezura*. ABSTRACT: BA 36:4088(50600), Aug. 1, 1961.

3245. **Chiu, Y. C., Hassan, A., Guthrie, F. E.** and **Dauterman, W. C. 1968** Studies on a Series of Branched-chain Analogs of Diethyl Malathion and Malaxon with Regard to Toxicity and *in vitro* Enzymatic Reactions. Toxicol. Appl. Pharmacol. 12(2):219-228.
Housefly LD_{50} data showed beta-glutarate analogs to be the least toxic of the groups studied. Using mice, alpha-glutarate was the most toxic. No correlation between toxicity and *in vitro* enzymatic activity. ABSTRACT: BA 50:5631(59019), June 1, 1969.

3246. **Chmurzýnski, J. A. 1967** On the Orientation of the Housefly (*Musca domestica* L.) towards White Light of Various Intensities. Bull. Acad. Pol. Sci. Sér. Sci. Biol. 15(7):415-422. (Russian summary).
Finds a correlation between number of flies which select one of two white lights and relative intensity of the light. Discussion is mathematical. ABSTRACT: BA 49:8300(92678). Sept. 15, 1968.

3246a. **Chmurzýnski, J. A. 1969** Orientation of Blowflies (*Calliphoridae*) towards White Light of Various Intensities. Bull. Acad. Pol. Sci. Cl. II. 17(5):321-324.
Proportion of insects (here blowflies) seeking lighted area of a maze depends not on light density, but on the structure of the maze. Treatment is mathematical.

Chmurzýnski, J. A., joint author. *See* Lipinska-Skawinska, B., et al., 1968.

3247. **Chou, T.-W. 1963** [A preliminary study on the species composition and seasonal prevalence of the fly population in Tsinan.] Acta Ent. Sinica 12(2):233-242. (In Chinese, with English summaryl)
From May 1960 to April 1961 a study of the species composition and seasonal distribution was carried out in the urban and suburb districts of Tsinan using traps and baits. Forty-two species were captured. Peaks and percentages of total catch are given for the species taken. ABSTRACTS: BA 45:2838(35203), Apr. 15, 1964; RAE-B 52:74, Apr., 1964.

Choudhury, D. S., joint author. *See* Raghavan, N. G. S., et al., 1967.

Christenson, L. D., joint author. *See* Steiner, L. T., et al., 1961.

3248. **Chubareva, L. A. 1968** [Homologous series of chromosome polymorphism in natural populations of flies.] Tsitologiia 10:1248-1256.

Cima, L., joint author. *See* Grigolo, A., et al., 1969.

3249. **Clark, A. G., et al. 1969** The Metabolism of Hexachlorocyclohexenes and Pentachlorocyclohexenes in Flies and Grass Grubs. Biochem. J. 113:89-96.

3250. **Clark, A. G., Hitchcock, M.** and **Smith, J. N. 1966** Metabolism of Gammexane in Flies, Ticks and Locusts [Correspondence]. Nature 209:103.
Major metabolic product in flies dosed with gammexane, or in enzyme extracts fortified with glutathione, is an aromatized molecule which appears identical with S-(2, 4-dichlorophenyl) glutathione. ABSTRACTS: RAE-B 55:137(492), Aug., 1967; TDB 63:493, April, 1966.

Clark, A. G., joint author. *See* Balabaskaran, S., et al., 1968.

Clifford, C. R., joint author. *See* Krause, G. T., et al. 1961.

3251. **Cline, R. E. 1968** Evaluation of Chemosterilant Damage to the Testes of the Housefly *Musca domestica,* by Microscopic Observation and by Measurement of the Uptake of ^{14}C-compounds. J. Insect Physiol. 14(7):945-953.
Various chemosterilants, added to the food of male flies for 3 days following emergence led to progressive decrease in content or synthesis of various constituents of the testis, e.g. microscopic examination on fourth day revealed a shortage of spermatogonia. ABSTRACTS: BA 49: 9744(107878), Oct. 15, 1968; TDB 65:1397(3127), Nov. 1968.

3252. **Cline, R. E.** and **Pearce, G. W. 1966** Similar Effects of DDT and Convulsive Hydrazides on Housefly Metabolism. J. Insect Physiol. 12:153-162.
Numerous hydrazides and thiourea derivatives tested for toxicity. The most toxic, thiocarbohydrazide (TCH) was as toxic to a DDT-resistant

strain as to a susceptible strain. ABSTRACTS: BA 47:5101(59843), June 15, 1966; RAE-B 54:246, Dec., 1966; TDB 63:712, June, 1966.

3253. Cline, R. E. and **Pearce, G. W. 1963** Unique Effects of DDT and Other Chlorinated Hydrocarbons on the Metabolism of Formate and Proline in the Housefly. Biochemistry 2(4):657-662.
Of injected compounds, DDT interfered most with the metabolism of formate, glycine and proline. Metabolism of formate not affected by pyrethrum, organic phosphates, phosphonates, and a carbamate. ABSTRACTS: BA 45:323, Jan. 1, 1964; TDB 61:727, July, 1964.

Cochran, D. G., joint author. *See* Guerra, A. A., et al., 1970.

3254. Coffey, M. D. 1966 Studies on the Association of Flies (Diptera) with Dung in Southeastern Washington. Ann. Ent. Soc. Am. 59:207-218.
Treats of the relative attractiveness of 8 types of dung to the fly population. Concerns dung preferences, seasonal and geographic distribution, climatic influence and other bionomic matters. ABSTRACT: BA 47:3797 (44697), May 1, 1966.

3255. Cohen, C. F. and **Barker, R. J. 1962** The Relation of Vitamin A to Vision in *Musca domestica* L. *In* 59th Annual Meeting of the American Society of Zoologists, Philadelphia, 1962. Am. Zool. 2(4):514 (Abstract only), No. 179.
Flies were photosensitive regardless of the diet, but response differed in some spectral regions. ABSTRACT: BA 46:8913(109939), Dec. 15, 1965.

3256. Cohen, C. F. and **Barker, R. J. 1963** Vitamin A Content and Spectral Response of House Flies Reared on Diets with and without a Vitamin A Source. J. Cell. Comp. Physiol. 62:43-47.
Study of vitamin A-depleted flies suggests that the retinene system of rod vision is not solely essential for insect vision. From eggs with less than 10^{-6} micrograms of vitamin A each, flies can be reared through ten generations with no vitamin A in the diet. ABSTRACT: BA 45:3614 (44503), May 15, 1964.

Cohen, C. F., joint author. *See* Goldsmith, T. H., et al., 1964; Mitlin, N., et al., 1961.

3257. Coker, W. Z., Frizzi, G., Kitzmiller, J. B., Laven, H., Mason, G. F., Milani, R., Pal, R., Sharma, G. P. and **Wright, J. W. 1966** Standardized Strains of Insects of Public Health Importance. [Memoranda.] Bull. W. H. O. 34(3):437-460. (French summary.)
Includes: (1) definition of strains, (2) necessity for standardized strains, (3) methods for development of standardized strains, (4) existing and needed standardized strains, (5) rearing, maintenance and dispatch of strains, (6) international collaboration. ABSTRACT: BA 63:1392, Dec., 1966.

3258. Coleman, V. R. and **Roberts, J. E. 1962** The Face Fly. 1st Year in Georgia. Georgia Agric. Res. 3(4):12. Portrait.
M. autumnalis first collected in Georgia in 1961. Species identification and control methods. ABSTRACT: BA 40:1916(25186), Dec. 15, 1962.

3259. Collard, R. V. and **Smythe, V. R. 1967** The Efficiency of "Dimetilan" for Fly Control in Dairy Sheds (*Musca domestica*). Australian J. Dairy Tech. 22(2):71-74.
Applied as a bait spray, at minimum recommended rates, this proprietary insecticide effectively controlled fly populations for 4-5 weeks. ABSTRACT: BA 48:11207(124480), Dec. 15, 1967.

3260. Collier, B. D. 1967 Some Experiments on the Evolution of Emigratory Behavior using *Musca domestica* in a Laboratory Model. Diss. Absts. 28(1B):51-B-52-B.

Collins, C., joint author. *See* Metcalf, R. L., et al., 1966.

3261. Collins, W. J. 1965 A Study of House Fly Carboxyl-esterase with Particular Reference to Organophosphate Tolerance Mechanisms. Diss. Absts. 26(1):456.

3262. Collins, W. J. and **Forgash, A. J. 1968** Acrylamide Gel Electrophoresis of Housefly Esterases. J. Insect Physiol. 14(10):1515-1523.
Esterase patterns vary considerably between strains and within strains, but no clear-cut, consistent differences were observed between resistant and susceptible flies. ABSTRACT: BA 50:7538(78582), July 15, 1969.

3263. Collins, W. J. and **Forgash, A. J. 1970** Mechanisms of Insecticide Resistance in *Musca domestica*: Carboxylesterase and Degradative Enzymes. J. Econ. Ent. 63(2):394-400.
Total diazinon resistance is the product of several protective mechanisms that act in an integrated manner. ABSTRACTS: BA 51:8567(87361), Aug. 1, 1970; RAE-B 58:301(1235), Sept., 1970; TDB 68:140(273), Jan., 1971.

Collins, J. M., joint author. *See* Miller, S., et al., 1970.

Collotti, C., joint author. *See* Vicari, G., et al., 1965.

3264. Colombo, G. and **Pinamonti, S. 1967** Analisi biochimica di alcuni mutanti di *Musca domestica* L. [Biochemical analysis of some mutants of *Musca domestica*.] Atti. Ass. Genet. Ital. 12:341-344. (English summary.)
The free metabolites of tryptophan to ommochromes were determined, by chromatographic methods, in pupae of four mutants: *yellow eyes* (*ye*), *white*^{3} (*w*^{3}), *ocra* (*ocra*) and *green eyes* (*ge*). ABSTRACT: BA 49:8275 (92389), Sept. 15, 1968.

Colwell, W. T., joint author. *See* Skinner, W. A., et al., 1966, 1967.

3265. Combiesco, I., Duport, M. and **Enesco A. 1967** Sterilization d'l'espèce *Musca domestica* L. par traitemont avec divers produit chimiques. Note I. Recherches de laboratoire. [Sterilization of the species *Musca domestica* L. by treatment with different chemical products. Note. I. Laboratory investigations.] Arch. Roum. Path. Exp. Microbiol. 26(1):205-214.
ABSTRACT: TDB 65:341(785), Mar., 1968.

3266. Combiesco, I. and **Enesco, A. 1968** La stérilisation de l'espèce *Musca domestica* L. par du thiotepa au stade de nymphe. [Sterilization with thiotepa of the species *Musca domestica* in the pupal stage.] Arch. Roum. Path. Exp. Microbiol. 27(4):715-720.
At various concentrations, and for all periods of treatment, the percentage rate of sterility in adults was high. Oviposition was not inhibited, but hatching was. ABSTRACT: TDB 66:757(1505), July, 1969.

3267. Combiesco, I., Enesco, A., Ciplea, A. G. and **Tîcu, V. 1967** Etude de l'action des chémostérilants-thiotepa et apholate sur le développement des ovaires de l'espèce *Musca domestica* L. [Study of the action of chemosterilants-thiotepa and apholate on the development of the ovaries in the species *Musca domestica* L.] Note II. Arch. Roum. Path. Exp. Microbiol. 26(1):215-227.
ABSTRACT: TDB 65:341(785), Mar., 1968.

Combiesco, I., joint author. *See* Lupasco, G., et al., 1963, 1966; Duport, M., et al., 1968.

3268. Comes, R., Frada, I. and **Macri, D. 1962** Ricerche sulla DDT-deidroclorurasi in ceppi di *Musca domestica* a vario comportamenti nei confronti del DDT e di altri insetticide di sintesi. [Studies on DDT dehydrochlorinase in strains of *M. domestica* varying in suscepti-

bility to DDT and other synthetic insecticides.] Riv. Parassit. 23(2): 151-160. (English summary.)
A correlation was shown between resistance and dehydrochlorination in *Musca domestica*, except in a pyrethrum-resistant strain. ABSTRACTS: TDB 60:276, March, 1963; RAE-B 52:175, Sept., 1964.

3269. Conner, R. M. 1966 Effects of Valine, Leucine and Isoleucine on Fly Toxin Production by *Bacillus thuringiensis* and Related Varieties. Diss. Absts. 27(4B):1028-B-1029-B.

3270. Conner, R. M. and **Hansen, P. A. 1967** Effect of Valine, Leucine, and Isoleucine on the Production of Fly Toxin by *Bacillus thuringirnsis* and Related Organisms. J. Invertebr. Path. 9:114-125.
Three varieties of crystalliferous bacteria were inoculated into media containing all possible combinations of the three substances, each at three concentrations. Potency of heat-stable exotoxin was evaluated by bioassay against *M. domestica* larvae. ABSTRACTS: BA 48:7416(82963), Aug. 15, 1967; RAE-B 56:81(267), Apr., 1968.

3271. Conner, R. M. and **Hansen, P. A. 1967** Effects of Valine, Leucine, and Isoleucine on the Growth of *Bacillus thuringiensis* and Related Bacteria. J. Invertebr. 9:12-18.
Generally, valine and leucine enhanced the growth of all three varieties cultured. Isoleucine partially inhibited the growth of var. *thuringiensis*; stimulated the growth of var. *sotto* and var. *entomocidus*.

3272. Cook, B. J., De la Cuesta, M. and **Pomonis, J. G. 1969** The Distribution of Factor S in the Cockroach *Periplaneta americana* and its Role in Stress Paralysis. J. Insect Physiol. 15(6):963-975.
Factor S, a pharmacologically active substance, is released under stress, in houseflies, as well as in cockroaches. Chemical composition unknown. ABSTRACT: BA 50:11406(118012), Nov. 1, 1969.

Cook, B. J., joint author. *See* Forgash, A. J., et al., 1962.

Corey, M. J., joint author. *See* Boyes, J. W., et al., 1964.

3273. Corey, R. A., Moye, W. C. and **Hall, W. E. 1965** Laboratory and Field Evaluation of SD9129 as an Insecticide. J. Econ. Ent. 58:658-660.
This compound, synthesized after discovery of a highly toxic metabolite of BIDRIN in *M. domestica*, proved promising for the control of agricultural pests.

Cornesky, R. A., joint author. *See* Marak, G. E., et al., 1969.

3274. Cornwell, P. B. Insect Electrocutors. Int. Pest Contr. 11(6):44.
Includes ultraviolet attractancy with reference to *M. domestica* and *Fannia canicularis*.

Corona, A. O., joint author. *See* Loaeza, R. M., et al., 1965.

Cory, M., joint author. *See* Skinner, W. A., et al., 1967.

3275. Cova García, P. 1964 Moscas de Venezuela. 18 pp., 30 pls. Maracay: Division de Endemias Rurales, Dirrección de Malariologia y Saneamiento Ambiental, Ministerio de Sanidad y Asistencia Social, Aragua, Venezuela.
A list, with plates, in color. No descriptive detail. ABSTRACT: TDB 62: 693, July, 1965.

3276. Cova Garcia. P., et al. 1960+ (A series of mimeographed documents issued by the Division de Malariologia, Dirección de Salud Publica, Ministerio de Sanidad y Asistencia Social): (a) Estudio de las moscas en relacion con el aseo domiciliario y el uso de insecticidas fosforados. By P. Cova-Garcia and E. O. Sitil. 24 pp; (b) Identification de algunas moscas de Venezuela. By P. Cova-García and

O. M. Suarez. 17 pp; (c) Estudio sobre las moscas en relation con los metodos de disposicion de Basuras y Desechos. By P. Cova-García and E. S. Oramas. 16 pp; (d) Moscas mas comunes en Venezuela. By P. Cova-García, O. M. Suarez and J. A. Rausseo. 12pp.
Titles explain contents.

3277. **Cowan, B. D.** and **Rogoff, W. M. 1968** Variation and Heritability of Responsiveness of Individual Male House Flies, *Musca domestica*, to the Female Sex Pheromone. Ann. Ent. Soc. Am. 61(5):1215-1218.
Certain males respond to models (pseudoflies) treated with benzene extract of females flies. This tendency appeared to be inherited. ABSTRACT: BA 50:1090(11505), Feb. 1, 1969.

3278. **Cowgill, U. M. 1966** *Perodicticus potto* and some insects (*Galleria melonella, Sarcophaga bullata, Musca domestica, Periplaneta americana.*) J. Mammalogy 47(1):156-157.
Postulates that a secretion from the apocrine glands, located in the skin of the scrotum and the vulva of these nocturnal prosimians from Kenya, acts as an attractant for insects. *P. potto* is an insect feeder. ABSTRACT: BA 48:1014(11006), Feb. 1, 1967.

Cox, J. T., joint author. *See* Brooks, G. T., et al., 1963; Bridges, R. G., et al., 1965.

3279. **Crabtree, D. H.** and **Kranzler, G. A. 1968** Response of Houseflies to Recorded Self-sound. Virginia J. Sci. 19(3):166. (Abstract.)

Craig, G. B., joint author. *See* Knipling, E. F., et al., 1968.

Craig, R., joint author. *See* Enan, O., et al., 1964; Menzel, D. B., et al., 1963.

Cranston, F., joint author. *See* Pimentel, D., et al., 1960.

3280. **Crone, H. D. 1964** The Reaction of Diazomethane with Synthetic Phosphatides and with Insect (*Musca domestica, Paraplaneta americana*) Lipid Extracts. Biochim. Biophys. Acta 84(6):665-680.
Three synthetic phosphatides reacted similarly (at different rates), forming dimethyl esters of phosphatidic acid and a nitrogenous compound. With insect extracts reaction resembled that of diazomethane with natural phosphatides. ABSTRACT: BA 46:3324(41189), May 1, 1965.

3281. **Crone, H. D. 1964** Phospholipid Composition of Flight Muscle Sarcosomes from the Housefly *Musca domestica* L. J. Insect Physiol. 10(3):499-507.
Compounds found, expressed as phosphorus per cent of total lipid phosphorus: Phosphatidylethanolamine, 69 per cent; phosphtidylcholine, 9 per cent; polyglycerolphosphatide, 6 per cent; alkali-stable phospholipids, 3 per cent. ABSTRACT: BA 45:6739(84662), Oct. 1, 1964.

3282. **Crone, H. D. 1967** The Relationship between Phosphatide Serine and Ethanolamine in Larvae of the Housefly, *Musca domestica*. J. Insect Physiol. 13(1):81-90.
There was a rapid incorporation of labelled serine and ethanolamine into the phospholipids of whole fat bodies of housefly larvae. ABSTRACT: BA 48:3265(36015), Apr. 1, 1967.

3283. **Crone, H. D. 1967** The Calcium-stimulated Incorporation of Isotopic Serine and Ethanolamine into the Phospholipids of House Fly (*Musca domestica*) Larvae. Biochem. J. 102(1):4P-5P. (Abstract.)
Is interpreted as a direct exchange with existing bases in the phospholipids. Incorporation is dependent on the calcium ion concentration.

3284. Crone, H. D. 1967 The Calcium-stimulated Incorporation of Ethanolamine and Serine into Phospholipids of the Housefly *Musca domestica*. Biochem. J. 104:695-704.
Ethanolamine and serine act as competitive inhibitors with one another. The buffer was also a competitor with these compounds. Several other amino alcohols were inhibitory.

3285. Crone, H. D. and **Bridges, R. G. 1963** The Phospholipids of the Housefly *Musca domestica*. Biochem. J. 89(1):11-21.
Phosphatidylethanolamine is the predominant phospholipid in houseflies. (65 per cent of total lipid phosphorus.) Lists four main differences between housefly and mammalian phospholipids. ABSTRACTS: BA 46: 8147(100679), Nov. 15, 1965; TDB 61:222, Feb., 1964.

Crone, H. D., joint author. *See* Bridges, R. G., et al., 1962.

Crosby, D. G., joint author. *See* Henderson, G. L., et al., 1967.

Crouch, G. W., joint author. *See* Dorough, H. W., et al., 1966.

Crowe, P. A., joint author. *See* Ludwig, D., et al., 1964.

Cundell, A., joint author. *See* Balabaskaran, S., et al., 1968.

Cunningham, C. J., joint author. *See* Dorsey, C. K., et al., 1962‘ 1966.

Cunningham, R. T., joint author. *See* Steiner, L. F., et al., 1970.

Curcăneanu, N., joint author. *See* Calciu-Dumitreasă, A., et al., 1968.

3286. Curtin, T. J. 1962 Status of Mosquito and Fly Insecticide Susceptibility in Turkey. Mosquito News 22(2):142-146.
In 1960 tests were carried out in four places representing the major geographical regions of Turkey to determine the susceptibility to various insecticides of four species of medically important insects, *Culex pipiens* L, *C. vagans* Wied. adult *Musca domestica* L. and *Anopheles sacharovi* Favr. Results are given in LD 50's and parts per million. ABSTRACTS: RAE-B 51:191, Sept., 1963; TDB 59:1211, Dec., 1962.

Cutkomp, L. K., joint author. *See* Patel, N. G., et al., 1961, 1968; Soliman, S. A., et al., 1963.

Daehnert, R. H., joint author. *See* Bell, D., et al., 1962.

3287. Dahm, P. A., Gurland, J., Lee, I. and **Berlin, J. 1961** A Comparison of Some House Fly Bioassay Methods. J. Econ. Ent. 54(2):343-347.
Most satisfactory procedure resulted from use of 4-day-old female flies, topical application, and a 4 $\times$ 4 design composed of four 50-fly replicates for each of four graded doses of DDT. ABSTRACT: RAE-B 50:109, May, 1962.

3288. Dahm, P. A., Kopecky, B. E. and **Walker, C. B. 1962** Activation of Organophosphorus Insecticides by Rat Liver Microsomes. Toxicol. and Appl. Pharmacol. 4:683-696.
Activation of insecticides compared by aerobically incubating them with male rat liver microsomes, $NADH_2$, nicotinamide and magnesium ions, and manometrically assaying the products with rat brain and fly head cholinesterase preparations.

Dahm, P. A., joint author. *See* Ishida, M., et al., 1965; Johnsen, R. E., et al., 1966; Nakatsugawa, T., et al., 1962, 1965, 1968, 1969; Gurland, J., et al., 1960.

van Dam, K., joint author. *See* Tulp, A., et al., 1969.

3289. Dame, D. A. and **Fye, R. L. 1964** Studies on Feeding Behavior of House Flies. J. Econ. Ent. 57(5):776-777.
Flies fed on liquid bait by the 5th hour, and on dry baits by the 12th hour. If mating occurs around the 16th hour, chemosterilsants in liquid baits should be effective. ABSTRACTS: BA 46:702(8799), Jan. 15, 1965; RAE-B 53:14, Jan., 1965.

3290. **Dame, D. A.** and **Schmidt, C. H. 1964** Uptake of Metepa and Its Effect on Two Species of Mosquitoes (*Anopheles quadrimaculatus, Aedes aegypti*) and House Flies (*Musca domestica*). J. Econ. Ent. 57:77-81.
A high sterility rate (99 per cent) was induced in houseflies by their absorption of labeled metepa from treated surfaces. Male vigor was not reduced. ABSTRACTS: BA 45:5068(62054), July 15, 1964; TDB 61:848, Aug., 1964.

Dame, D. A., joint author. *See* Schmidt, C. H., et al., 1964.

3291. **Damodar, P., Ausat, A., Koshi, T.** and **Perti, S. L. 1962** The Relative Efficiency of Insecticides as Fly and Mosquito Larvicides. Indian J. Malariol. 16:243-247.
Concerns *Musca nebulo* L. and *Culex fatigans* Weid. Most satisfactory synthetic contact insecticides were aldrin, dieldrin, diazinon and endrin. ABSTRACTS: RAE-B 53:42, Feb., 1965; TDB 69:691, July, 1963.

Damodar, P., joint author. *See* Wal, Y. C., et al., 1962.

3291a. **Damodar, P., Perti, S. L.** and **Agarwal, P. N. 1964** The Toxicity of Solvent Extract of the Fungus *Macrosporium* sp. to Flies and Mosquitoes. Indian J. Ent. 24(1):110-112.
Concerns *M. nebulo*. An admixture of fungal extract with DDT, at percentages used, effected a kill of 84 per cent. Extract alone was more effective against mosquitoes than houseflies.

3292. **Dann, O., Bamberg, K. J.** and **Sucker, H. 1968** 3, 4-Diphenylthiophendicarbonsaeure-(2,5)-bis-[(Bidiaethyl-amino)-aethylestermethojodid], ein curareartiger, ruhepotentialstabilisierender Ester: XII. Acetylcholin. [3, 4-Diphenylthiophendicarboxylic acid-(2, 6)-bis-[(B-diethyl-amino)-ethylestermethiodide], a curiform, resting potential stabilizing ester: XII. Acetylcholine.] Pharmazie 23(3):135-145.
Study makes use of aromatic carboxyl acids, which inhibit the formation of acetylcholine (Ach) in housefly extracts. ABSTRACT: BA 50:783(8139), Jan. 15, 1969.

3293. **Darabantu, C. 1967** Contributii la cuno asterea unor specii de Muscini si Stomoxydini (Diptera) din fauna Romaniei. [Contributions to the knowledge of Muscini and Stomoxydini (Diptera) species from Romanian fauna.] Stud. Univ. Babes Bolyaiser Biol. 12(2):103-105. (In Russian, with English summary.)
Musca osiris Wd., *Dasyphora cyanella* Mg. and *Haematobia etripalpis* Bzz. are recorded for first time from Romania. The male genital system of *M. osiris* is described. ABSTRACT: BA 49:10262(113310), Nov. 1, 1968.

3294. **Darby, F. J., Heenan, M. P.** and **Smith, J. N. 1966** The Absence of Glucuronide Conjugates from 1-naphthol Dosed Flies and Grass Grubs: Detection of 1-naphthylphosphate. Life Sci. 5(16):1499-1502.
The biosynthesis of glucosid uronates in flies has been reported during the detoxification of phenols, but the metabolism of 1-naphthol in *M. domestica* does not yield a glucosid uronate. (Substantial amounts of 1-naphthyl-dihydrogen phosphate were produced.) ABSTRACT: BA 48: 2361(25822), Mar. 1, 1967.

Darby, F. J., joint author. *See* Binning, A., et al., 1967.

Darrow, D. I., joint author. *See* Plapp, W. F., et al., 1961.

3295. **Datta, S. K. 1960** Need for Disinsectisation of Aircrafts on International Flights. Bull. Nat. Soc. India for Malaria and Other Mosquito-Borne Diseases. 8(5/6): 153-157.
Houseflies found in 112 aircraft of 6,115 searched at Bombay airport

over 4½ year period. Results of an aerosol spray for killing *Culex* in aircraft reported. ABSTRACT: TDB 59:92, Jan., 1962.

Daturi, S., joint author. *See* Laudani, U., et al., 1965.

3296. **Dauer, M., Bhatnagar, P. L.** and **Rockstein, M. 1965** X-irradiation of Pupae of the House Fly, *Musca domestica* L., and Male Survival. J. Geront. 20:219-223.
All male flies emerging from irradiated 3-4 day old pupae (X-rayed at 10,000-30,000 rads) were apparently normal. Average life span was increased by exposure to 10,000-15,000 rads; remained unchanged at 20,000; decreased after exposure to 30,000. ABSTRACT: BA 46:7040 (87414), Oct. 15, 1965.

Dauer, M., joint author. *See* Rockstein, M., et al., 1965, 1967, 1969.

3297. **Dauterman, W. C.** and **Mehrotra, K. N. 1963** The N-alkyl Group Specificity of Cholinesterase from the House Fly, *Musca domestica* L. and the Two-spotted Spider Mite, *Tetranychus telarius* L. J. Insect Physiol. 9(2):257-263.
Housefly enzyme hydrolyzed the dimethyl and diethyl alkyl analogues rapidly; whereas activity to the dipropyl and dibutyl alkyl ester was low. ABSTRACTS: BA 43:1353(16812), Aug. 15, 1963; RAE-B 52:158, Sept., 1964.

Dauterman, W. C., joint author. *See* Chiu, Y. C., et al., 1968; Hassan, A., et al., 1968; Krueger, H. R., et al., 1960; Matsumura, F., et al., 1964; Mehrotra, K. N., et al., 1963.

3298. **Davé, K. H.** and **Wallis, R. C. 1965** Survival of Type 1 and Type 3 Polio Vaccine Virus in Blowflies (*Phaenicia sericata*) at 40 degrees C. Proc. Soc. Exp. Biol. and Med. 119:121-124.
No evidence of multiplication in either type 1 or type 3 attenuated vaccine strains. No shift in genetic character toward a pattern of greater virulence. Virulent strains grow in elevated temperatures, but attenuated ones do not.

3299. **Davidson, A. 1962** Trapping Houseflies in the Rural Areas of Israel. Riv. di Parassit. (Rome) 23(1):61-70.
Author gives much practical advice on trapping. Types of bait used: chicken entrails and yeast infusion. ABSTRACTS: RAE-B 52:174, Sept., 1964; TDB 60:76, Jan., 1963.

3300. **Davies, D. M.** and **Anderson, E. L. 1967** Egg Production of House-fly Adults Fed Human Blood Containing Various Supplements. Proc. Ent. Soc. Ont. 97:111-114.
Adults fed on blood produced fewer eggs, but most of those laid were viable. Experimental work indicates that phosphate may be the nutritional factor needed to provide maximum production of eggs. ABSTRACT: BA 49:2890(32442), Mar. 15, 1968.

3301. **Davies, D. M., Morrison, P. E., Taylor, G. C.** and **Goodman, T. 1965** Larval Nutrient Stores, Adult Diets and Reproduction in Diptera. *Repr. from*: Proc. XII Int. Congr. Ent. London 1964(1965). p. 174 only.
M. domestica normally shows little or no autogeny, but certain larval diets made it possible for flies to produce eggs on minimum diet. Oviposition soon ceased however, while milk-fed flies continued to lay eggs.

Davies, D. M., joint author. *See* Goodman, T., et al., 1968; Morrison, P. E., et al., 1964.

Davies (=Lloyd-Davies), T. A., joint author. *See* Hale, J. H., et al., 1960.

Davis, D. H. S. (Editor). *See* Paterson, H. E., et al., 1964.

Davis, D. L., joint author. *See* Thompson, R. K., et al., 1953.

3302. **Davis, H. G.** and **Eddy, G. W. 1966** Some Effects of Chemosterilants on the Little House Fly. J. Econ. Ent. 59(4):993-996.
Tests on tepa, metepa, apholate, hempa and hemel, showed tepa to be the most effective. Males were more susceptible. Only a small margin between doses causing sterility and doses causing mortality. ABSTRACTS: BA 47:10324(119316), Dec. 15, 1966; RAE-B 54:210, Nov., 1966; TDB 64:219, Feb., 1967.

De, S. P., joint author. *See* Sinha, R., et al., 1967.

3303. **Deal, A. S. 1968** The Effect of Temperature and Moisture on the Development of *Fannia canicularis* and *Fannia femoralis* (Diptera, Muscidae). Diss. Absts. 28(9B):3737-B-3738-B.

3304. **Deal, A. S., Loomis, E. C., Burgess, J. B.** and **Bowen, W. R. 1965** Fly (*Musca domestica*) Control in Cattle Feedlots with Residual Sprays. Calif Agric. 19(9):6-7.
Spraying rigs had booms to convey spray hoses over fences. Diazinon was the most effective insecticide used. ABSTRACT: BA 47:1198(14487), Feb. 1, 1966.

3305. **Deal, A. S., Loomis, E. C., Pelissier, C. L.** and **Anderson, J. R. 1968** Fly Control on the Dairy. Univ. Calif. Agr. Ext. Serv. AXT-198 Revised.
M. domestica one of 8 species (or groups) concerned. Gives various formulations for dusts, sprays and baits.

Deal, A. S., joint author. *See* Loomis, E. C., et al., 1967, 1968; Anderson, J. R., et al., 1968.

3306. **Deay, H. O.** and **Taylor, J. G. 1962** Response of the House Fly, *Musca domestica* L., to Electric Lamps. Proc. Indiana Acad. Sci. 72:161-166.
Used lamps of different intensities, wave lengths, types of spectra, and electromagnetic energy. Sample observation: Time of day flies were most attracted to a 15 watt, fluorescent 360 BL blacklight lamp was at night. ABSTRACTS: BA 46:1057(13346), Feb. 1, 1965; RAE-B 53:131, July, 1965.

Deb, B. C., joint author. *See* Sinha, R., et al., 1967.

DeBarjac, H., joint author. *See* Burgerjon, A., et al., 1967.

3307. **DeCapito, T. 1963** Isolation of *Salmonella* from Flies. Amer. J. Trop. Med. Hyg. 12(6):892.
Thirteen isolations of 9 serological types were obtained during a study of 3,263 pools of 65,273 flies. ABSTRACT: BA 45:2395(29683), Apr. 1, 1964.

DeCapito, T. M., joint author. *See* Richards, C. S., et al., 1961.

Decker, G. C., joint author. *See* Bruce, W. N., et al., 1960.

3308. **DeFoliart, G. R. 1963** Preventive Spraying Schedules for Dairy Farm Fly Control. J. Econ. Ent. 56:649-654.
Operations begun before flies were numerous and repeated at reduced intervals to prevent buildup proved most effective. Dimethoate or ronnel were the insecticides of choice. ABSTRACTS: BA 45:1758(21927), Mar. 1, 1964; RAE-B 52:20, January, 1964.

3309. **DeFoliart, G. R.** and **Eschle, J. L. 1961** Barn Fogging as a Fly Control Method. J. Econ. Ent. 54(5):862-865.
Fogging gives adequate control in dairy barns, but is 4 times more costly than residual spraying, and provides no pasture control on cattle. ABSTRACT: RAE-B 50:172, Aug., 1962.

DeGraw, J. I., joint author. *See* Skinner, W. A., et al., 1967.

Degrugillier, M., joint author. *See* LaChance, L. E., et al., 1969, 1970.

De la Cuesta, M., joint author. *See* Cook, B. J., et al., 1969.

Del Rivero, J. M., joint author. *See* Planes-Garcia, S., et al., 1963.

De Marquez, M., joint author. *See* Nocerino, F., et al., 1966.

3310. Demaure, J. C. 1968 *In vitro* Culture of Larval Organs of Diptera. Compt. Rend. Seances Soc. Biol. Filiales 162(1):224-227.

DeMeo, G. M., joint author. *See* Sacca, G., et al., 1966.

DeMilo, A. B., joint author. *See* Bořkovec, A. B., et al., 1967; Chang, S. C., et al., 1968.

3311. Deoras, P. J. and **Bhaskaran, G. 1962** Observations on the Endocrine Glands of *Musca nebulo*. Current Sci. 31(8):336-337.
A detailed study of the ring gland in larvae and in adult flies, from a structural standpoint. Discussion of functional relations. ABSTRACT: BA 43:1973(25494), Sept. 15, 1963.

3312. Deoras, P. J. and **Bhaskaran, G. 1963** Hormonal Control of Ovary Development in the House-fly, *Musca nebulo*. Proc. Intern. Cong. Zool. 16(2):126 (abstract only).

3313. Deoras, P. J. and **Bhaskaran, G. 1967** Studies on the Neuroendocrine System in the Indian House Fly, *Musca nebulo*, Fabr. I. Larval Organs. J. Univ. Bombay 35:28-40.
Descriptions accompanied by micrographs, showing the histology of the ring gland and neurosecretory cells of the central nervous system. (For Paper II of this series, *see* Bhaskaran and Deoras, 1967.)

3314. Deoras, P. J. and **Bhaskaran, G. 1967** Studies on the Neuroendocrine System in the Indian House Fly, *Musca nebulo*, Fabr. III. Adult Organs. J. Univ. Bombay 35:59-72.
Completes the anatomical studies of the endocrine structures as found in the adult fly. (The *corpus cardiacum allatum* complex.)

3315. Deoras, P. J. and **Bhaskaran, G. 1967** Studies on [the] Neuroendocrine System in the Housefly *Musca nebulo* (Fabr.). IV. Hormonal Control of Ovary Development. J. Univ. Bombay 35:73-87.
Allatectomy inhibited ovary development, but yolk deposition could be initiated in such flies by transplanting corpora allata from mature adults of either sex. Median neurosecretory cells also, were found necessary for normal egg development.

3316. Deoras, P. J. and **Bhaskaran, G. 1967** Studies on [the] Neuroendocrine System in the Housefly (*Musca nebulo* Fabr.). V. The Probable Origin and Development of the Ring Gland. Indian J. Ent. 29(4): 385-388.
Traces the embryological origins of the corpus cardiacum, hypocerebral ganglion, sympathetic ganglia, corpus allatum, and neurosecretory cells. Origin of peritracheal glands obscure. ABSTRACT: BA 50:4805(50535), May 1, 1969.

3317. Deoras, P. J. and **Karnikar, K. 1967** Studies of House Flies. Current Science 36(18):491-493.
Concerns *M. sorbens* and *M. nebulo*. Morphological differences, external and internal. Rearing techniques. ABSTRACT: BA 49:9755(108025), Oct. 15, 1968.

Deoras, P. J., joint author. *See* Bhaskaran, G., et al., 1963, 1967.

3318. DePietre-Tonelli, P. and **Vian, I. 1966** The Effect of Formulation on Residual Activity of Dimethoate to the House Fly. J. Econ. Ent. 59:4-9.
Residual activity far higher with wettable powder formulations than with emulsion concentrates. In certain tests the former had residual effect for several months, the latter for only a few days. ABSTRACTS: BA 47:6404 (74926), Aug. 1, 1966; RAE-B 54:96, May, 1966; TDB 63:907, Aug., 1966.

3319. **Depner, K. R. 1969** Distribution of the Face Fly, *Musca autumnalis* (Diptera:Muscidae) in Western Canada and the Relation Between its Environment and Population Density. Canad. Ent. 101(1):97-100.
As of this date (1969) species had been recorded from all provinces. Moisture and shade, in pastures, mean larger numbers of these flies. ABSTRACT: BA 50:6977(72886), July 1, 1969.

3320. **Derbeneva-Ukhova, V. P. 1966** [On controlling flies under rural conditions.] Gig. I Sanit. 31:79-82. (In Russian.)

3321. **Derbeneva-Ukhova, V. P. 1967** Effect of Climatic Factors on Significance of Flies as Potential Vectors. Wiad Parazytol. 13(4/5):591-594. (Polish summary.)
Includes a review of Russian literature on synanthropic flies. ABSTRACT: BA 50:5763(60338), 1969.

3322. **Derbeneva-Ukhova, V. P., Lineva, V. A.** and **Drobozina, V. P. 1966** The Development of DDT-resistance in *Musca domestica* and *Protophormia terraenovae*. Bull. W.H.O. 34(6):939-952.
Changes in susceptibility to DDT are accomplished by changes in the percentage of females with disturbance of oögenesis, but further observations under natural conditions are needed to validate this. ABSTRACTS: BA 48:1893(20539), Feb. 15, 1967; RAE-B 56:200(727), Oct., 1968; TDB 64:217, Feb., 1967.

Derbeneva-Ukhova, V. P., joint author. *See* Balashov, Yu. S., et al., 1968.

Derse, P. H., joint author. *See* Schmolesky, G. E., et al., 1963.

3323. **Desmoras, J., Fournel, J., Koenig, F. H.** and **Métivier, J. 1964** Étude au laboratoire et en serre d'un nouvel ester phosphorique: la phosalone ou 11.974 R.P. [A study in the laboratory and greenhouse of a new phosphoric ester: phosalone or R.P. 11974.] Phytiat.-Phytopharm. 13(1):33-43. Paris.
Insecticidal properties tested against *M. domestica.* LC_{90} in dry deposit from sprays on glass was obtained at concentration of 30 µg. per ml. ABSTRACT: RAE-B 53:249, Dec., 1965.

3324. **Dethier, V. G. 1962** To Know a Fly. pp. viii + 119 illus. Holden-Day, 728 Montgomery St., San Francisco 11, Calif.
An amusing as well as informative volume intended chiefly for the non-technical reader. Of only 119 pages, it is illustrated in a manner to enhance the story. The book is introduced through a foreword by N. Tinbergen. ABSTRACT: BA 45:2525(31296), Apr. 15, 1963.

3325. **Detinova, T. S. 1962** Age Grouping Methods in Diptera of Medical Importance with Special Reference to some Vectors of Malaria. W. H. O. Monograph Series No. 47. 216 pp., illus.
Sublethal doses of DDT associated with abortion of eggs in *M. domestica.* ABSTRACT: TDB 59:1113, Nov., 1962.

3326. **Detroux, L.** and **Seutin, E. 1965** Méthode de détermination de l'action, insecticide des bombes aérosol. [Method for determining the insecticidal action of the aerosol bomb.] Parasitica 21(2):40-63. Gembloux.
Exposure of adults of *M. domestica* in chambers with a capacity of 0·3 cu. m.

Devries, D. M., joint author. *See* San Jean, J., et al., 1961.

3327. **Dewey, J. E.** and **Parker, B. L. 1965** Increase in Toxicity of *Drosophila melanogaster* of Phorate-treated Soils. J. Econ. Ent. 58(3):491-497.
These soils proved highly toxic to phosphate-resistant *M. domestica.* ABSTRACT: BA 47:1601(19381), Feb. 15, 1966.

3328. **Dhennin, L'eone, Heim de Balsac, H., Verge, J.** and **Dhennin, Louis. 1961** Du rôle des parasites dans la transmission naturelle et expérimentale du virus de la fièvre aphteuse. [The role of parasites in the natural and experimental transmission of the virus of foot-and-mouth disease.] Rec. Méd. Vét. 137:95-104. Paris. (English and Spanish summaries.)
M. domestica and *Lucilia sericata* can be contaminated by the virus and when homogenated, can infect guinea pigs. ABSTRACT: RAE-B 51:273, Dec., 1963.

3329. **Díaz-Ungría, C. 1965** Transmission del *Trypanosoma cruzi* en los vertebrados. Rev. Ibér. Parasit. 25(3/4):312-356.
Flies fed on feces of *Rhodinus prolixus* were then fed to puppies. Check by xenodiagnosis, 25-28 days later, showed infection present. ABSTRACT: BA 48:422(4386), Jan. 1, 1967.

3330. **Díaz-Ungría, C. 1966** La *Musca domestica* como transmissora del *Trypanosoma cruzi.* [*Musca domestica* as vector of *Trypanosoma cruzi.*] Mimeo. Document. 15 pp., September. Escuela de Ciencias Veterinaires Centro Experimental de Estudios Superiores Barquisimeto-Venezuela.
Under experimental conditions *M. domestica* can infect a dog with *T. cruzi* orally, during at least 12 hours.

3331. **Díaz-Ungría, C. 1966** Transmissión du *Trypanosoma cruzi* chez les Mammifères. Ann. Parasit. Hum. Comp. 41(6):449-471.
Flies infected in nature by feeding on feces of *Rhodnius* containing *T. cruzi* may contaminate milk, cheese etc. with their feces, thus setting the stage for oral infection of the human host.

3332. **Díaz-Ungría, C. 1968** La contaminacion por via bucogastrica y ocular en los trypanosomas. [Contamination with trypanosomes by the oral-gastric and ocular route.] Revista Univ. de Zulia No. 41, pp. 45-75.
Includes role of *M. domestica* in the transmission of *Trypanosoma evansi.*

3333. **Díaz-Ungría, C. 1969** Papel del veterinario en la lucha contra la enfermedad de Chagas. [Role of the veterinarian in the war against Chagas' disease.] Bol. Oficina Sanitaria Panamericana. 72(6):497-506 (año 48).
Summarizes experimental procedure involving *Rhodnius* feces, and *M. domestica.* Fly abdomens fed to dogs confirmed infection.

3334. **Díaz-Ungría, C., Yepez, M. S.** and **Artigas, R. T. 1966** Transmissión bucal con *Trypanosoma cruzi.* [Oral transmission of *Trypanosoma cruzi.*] Trabajo presentado al Coloquio de la Academia de Ciencias Fisicas, Matimaticas y Naturales de Caracas. November. 13 pp., mimeo.
T. cruzi, transported by *M. domestica*, can positively infect dogs and may well infect man (by way of the mouth).

3335. **Dietrick, E. J. 1970** Biological Control of Flies. Quarter Horse Journal 22(10):90, 94, 104.
Cites work at Univ. of Calif. (Riverside). Histerid and Staphylinid beetles known to destroy 95 per cent of the immature stages of manure-breeding flies.

3336. **DiJeso, F.** and **Gaetani, M. T. 1966** Sur les phosphagènes isolés à partir d'oeufs d'Insectes. [[On phosphagens isolated from insect eggs.] Compt. Rend. Soc. Biol. (Paris) 160(7):1408-1410.
M. domestica one of four species studied. ABSTRACT: BA 48:9249(104012), Oct. 15, 1967.

3337. **DiJeso, F., Malcovati, M., Gaetani, M. T.** and **Speranza, M. L. 1967** The Identification and Determination of Phosphagen in Insects and their Eggs. Comp. Biochem. Physiol. 20:607-618.
M. domestica among species studied. Phosphagen content is correlated more closely with type of tissue than with species.

Dill, J. C., joint author. *See* McCann, G. D., et al., 1969.

3338. **Dinamarca, M. L., Saavedra, I.** and **Valdes, E. 1969** DDT-dehydrochlorinase. I. Purification and Characterization. Comp. Biochem. Physiol. 31:269-282.
Enzyme from *M. domestica* appears to have a molecular weight of 120,000; formed by four monomers of 30,000 molecular weight. Activity proceeds only with aggregates of a minimum molecular weight of 90,000. ABSTRACT: BA 51:3942(40710), Apr. 1, 1970.

Dinamarca, M. L., joint author. *See* Gil, L., et al., 1968.

Dinther, J. B. M. *See* Van Dinther, J. B. M.

Disney, R. W., joint author. *See* Winteringham, F. P. W., et al., 1964.

3339. **Dixit, R. S.** and **Perti, S. L. 1962** The Efficiency of Longifolyl Thiocyanoacetate against Flies and Mosquitoes. Indian J. Ent. 24(4): 239-243. New Delhi.
This compound nearly as effective as cyclohexylthiocyanoacetate, but less toxic than DDT or pyrethrins. ABSTRACT: RAE-B 52:24, Jan., 1964.

3340. **Dixit, R. S.** and **Perti, S. L. 1962** The Relative Efficiency of Insecticides as Space Sprays. Indian J. Malariol. 16:231-235.
Several sprays tested. Most satisfactory were pyrethrins, diazinon, lindane, and TEPP. ABSTRACTS: RAE-B 53:42, Feb., 1965; TDB 60:796, Aug., 1963.

3341. **Dixit, R. S., Perti, S. L., Cheema, P. S.** and **Ranganathan, S. K. 1962** Mode of Action of DDT on the Housefly *Musca nebulo* Linn. II. Evidence for the Dissolution of DDT in the Cuticular Lipoids of the Insect. Indian J. Malariol. 16:237-242.
As a contact poison, DDT first becomes dissolved in the cuticular lipoids, then penetrates into the tissues of the body. ABSTRACTS: RAE-B 53:48, Feb., 1965; TDB 60:903, Sept., 1963.

3342. **Dixit, R. S., Verma, R. N.** and **Somaya, C. I. 1968** A Study of the Fumigant Property of Baygon. Labdev J. Sci. Technol. 6B(4):234-237.
Baygon, used as a fumigant, is effective against numerous insects, including the housefly, *M. nebulo* Linn. ABSTRACT: BA 50:8072(84006), Aug. 1, 1969.

Dixit, R. S., joint author. *See* Agarwal, P. N., et al., 1963; Paul, C. F., et al., 1963; Perti, S. L., et al., 1965, 1966; Saxena, B. N. et al., 1965; Srivastava, A. P., et al., 1965.

Dobrolyubova, V. P., joint author. *See* Fomicheva, V. S., et al., 1967.

3343. **Dobson, R. C.** and **Huber, D. A. 1961** Control of Face Flies (*Musca autumnalis*) on Beef Cattle in Indiana. J. Econ. Ent. 54(3):434-436.
Organic insecticides applied by cable-rubber devices and successful in reducing fly populations were methoxychlor, toxaphene, DDT and ronnel. ABSTRACTS: BA 36:5560(58116), Sept. 1, 1961; RAE-B 50:126, June, 1962.

3344. **Dobson, R. C.** and **Kutz, F. W. 1970** Control of House Flies in Swine-Finishing Units by Improved Methods of Waste Disposal. J. Econ. Ent. 63(1):171-174.
Two year study indicates that fly production can be greatly reduced by any of three improved waste disposal systems. Each unit was completely screened. No insecticides were used. ABSTRACT: BA 51:9098(92599), Aug. 15, 1970.

3345. Dobson, R. C. and **Matthew, D. L. 1961** Field Observations of Face Fly *Musca autumnalis* (DeGeer) in Indiana, 1960. Proc. Indiana Acad. Sci. 70:152-153.

Females normally rest on cattle, males more commonly on gates, posts, and foliage. Only fresh manure is used for oviposition. These flies invade homes and attics in cool weather. ABSTRACT: BA 38:567(7711), Apr. 1, 1962.

3346. Dobson, R. C. and **Sanders, D. P. 1965** Low-Volume, High-Concentration Spraying for Horn Fly and Face Fly Control on Beef Cattle. J. Econ. Ent. 58(2):379.

Technical malathion applied on an area basis gave excellent control and continued effective for several days. ABSTRACTS: BA 46:5142(63862), July 15, 1965; RAE-B 53:130, July, 1965.

Dobson, R. C., joint author. *See* Sanders, D. P., et al., 1966.

3347. Dobzhansky, T. 1966 Of Flies and Men. Am. Psychol. 21(7):618.

3348. Dodge, H. R. 1960 An Effective, Economical Fly Trap. J. Econ. Ent. 53(6):1131-1132.

Utilizes cardboard carton of 9 in. diameter, a screen cone, mesh cloth and transparent plastic. Inner walls may be coated with insecticide. ABSTRACT: RAE-B 49:223, Oct., 1961.

3349. Döhring, E. 1964 Zur Wirksamkeit von Phosphorsäureestern und Carbamaten in Sprühmitteln gegenüber Stubenfliegen verschieden empfindlicher Stämme. [On the effectiveness of phosphoric esters and carbamates in spray preparations against houseflies of different susceptible strains.] Zeit. Angew. Zool. 51(3):311-334. (English summary.)

Tests run for fumigant as well as contact effect. Strains susceptible to DDT + gamma-BHC most susceptible to other chemicals. Resistant strains least susceptible. ABSTRACT: RAE-B 54:57, Mar., 1966.

3350. Döhring, E. 1965 Zur Wirksamkeit von einigen phosphorsäureestem und carbamaten auf Stubenfliegen (*Musca domestica*) verschieden empfindlicher stämme. [On the efficacy of several organophosphorus compounds and carbamates on house flies (*Musca domestica*) of differently susceptible strains.] Proc. XIIth International Congress of Entomology. Section 8. Insecticides and Toxicology. pp. 491-493. (See preceding reference.)

3351. Döhring, E. 1966 Zur Wirksamkeit und Anwendungstechnik von Fliegenstreifen. [On the effectiveness and application technique of employment of fly strips.] Zeit. Angew. Zool. 53(1):95-121. (English summary.)

Tested against strains of *M. domestica*. Gives results of five types of commercial fly strips impregnated with different insecticides. Other types of Diptera likely to be controlled by strips are reviewed. ABSTRACT: RAE-B 55:149(524), Aug., 1967.

Döhring, E., joint author. *See* Privora, M., et al., 1969.

Dojmi di Delupis, G., joint author. *See* Filliponi, A., et al., 1963, 1964.

Domicz-Styczynska, B., joint author. *See* Bojanowska, A., et al., 1963.

3352. Donnellan, J. F. and **Beechey, R. B. 1969** Factors Affecting the Oxidation of Glycerol-1-phosphate by Insect Flight Muscle Mitochondria. *Sarcophaga barbata, Pieris brassicae*; *Apis millifera, Schistocerca gregaria, Musca domestica.* J. Insect Physiol. 15(3): 367-372.

Low levels of calcium and strontium ions act as heterotropic allosteric effectors of this oxidation by both mitochondria and sub-mitochondrial particles. In the absence of such ions, L-Glycerol-1-phosphate also acts as a homotropic effector. ABSTRACT: BA 50-8655(89898), Aug. 15, 1969.

3353. Dorough, H. W. 1968 Metabolism of Furadan (NIA-10242) in Rats and Houseflies. J. Agric. Fd. Chem. 16(2):319-325.
Both hydrolytic and oxidative mechanisms are involved. Four groups of intact conjugates were resolved by thin layer chromatography. ABSTRACT: BA 49:10048(111044), Nov. 1, 1968.

3354. Dorough, H. W. and **Arthur, B. W. 1961** Toxicity of Several Organophosphates Administered in the Diet of Broilers to House Fly Larvae in the Feces. J. Econ. Ent. 54(6):1117-1121.
Ronnel, butonate, Ruelene, and several Bayer formulations were tested at different concentrations as to consumption by the birds, weight gain (or loss), lethal effect on birds, toxicity to fly larvae. ABSTRACT: RAE-B 50:185, Sept., 1962.

3355. Dorough, H. W. and **Casida, J. E. 1964** Nature of Certain Carbamate Metabolites of the Insecticide Sevin. J. Agric. Fd. Chem. 12(4): 294-304. Easton, Pa.
Concerns 8 metabolites, 5 of which were carbamates. Bioassays on these indicated reduced biological activity compared with original insecticides. Insects used were *M. domestica* and *Periplaneta americana*. ABSTRACT: RAE-B 52:217, Dec., 1964.

3356. Dorough, H. W. and **Crouch, G. W. 1966** Persistence and Residual Effectiveness of Various Formulations of Baygon (Bayer 39007) Against the House Fly. J. Econ. Ent. 59:1188-1190.
Lack of residual effectiveness (half-life 15 days) thought to be caused by loss of the formulating materials necessary for transfer of toxicant from treated surface to the fly. ABSTRACTS: BA 48:1417(15346), Feb. 1, 1967; RAE-B 55:13, Jan., 1967; TDB 64:324, Mar., 1967.

Dorough, H. W., joint author. *See* Andrawes, N. R., et al., 1967; Brady, U. E., et al., 1960; Green, L. R., et al., 1968.

3357. Dorsey, C. K. 1966 Face Fly Control Experiments on Quarter Horses 1962-64. J. Econ. Ent. 59(1):86-89.
Dichlorvos-plastic strand halters, when correctly installed, adjusted and maintained, provided good to excellent control of face flies without observable toxicity to the horses. ABSTRACTS: BA 47:6373(74548), Aug. 1, 1966; RAE-B 54:98, May, 1966.

3358. Dorsey, C. K. 1967 Experimental Use of Apholate to Control Face Flies in Pastures and House Flies in Barns. West Virginia Agr. Exp. Sta. Bull. 555T:1-16.
This chemosterilant successfully reduced face fly reproduction in pastures, also housefly reproduction in and about barns. Several types of dry and liquid apholate baits were used.

3359. Dorsey, C. K. 1968 Field Experiments with Attractants for the Face Fly. J. Econ. Ent. 61(6):1695-1696.
Attractants were more effective when applied on the animals than when placed in petri dishes. Gut slime, placenta powder, blood hydrolysate, and bone phosphate were among the substances tested. ABSTRACT: BA 50:4777(50213), 1969.

3360. Dorsey, C. K., Heishmann, J. O. and **Cunningham, C. J. 1966** Face Fly and Horn Fly Control on Cattle, 1962-1964. J. Econ. Ent. 59: 726-732.
Used mainly self-treatment devices, such as 3-cable back rubbers and dust bags. A large number of insecticides were tested. ABSTRACTS: BA 47:8521(99265), Oct. 15, 1966; RAE-B 54:181, Sept., 1966.

3361. **Dorsey, C. K., Kidder, H. E.** and **Cunningham, C. J. 1962** Face Fly Control Studies in West Virginia in 1960 and 1961. J. Econ. Ent. 55(3):369-374.
Formulations were applied as sprays, dusts and smears. Diazinon and dimethoate preparations gave fair protection up to 14 days. ABSTRACT: RAE-B 50:272, Dec., 1962.

3362. **Dotsenko, T. K., et al. 1962** Opyt primeneniya novykh insektitsidov v bor'be s sinantropnymi mukhami v neizolirovannykh uchastkakh. [Experiments with the use of new insecticides in controlling synanthropic flies in non-isolated areas.] Med. Parazit. (Moskva) 31:355-358.
The use of polychlorpinene as a larvicide around drains and cess pools, the use of a DDT emulsion on interior surfaces of food processing plants, and the use of a flypaper coated with a mixture of sugar, an insecticidal hydrocarbon and an organophosphate are several methods resulting in successful fly control. ABSTRACT: BA 40:1924(25267), Dec. 15, 1962.

Doty, A. E., joint author. *See* Kenaga, E. E., et al., 1962, 1965; Whitney, W. K., et al., 1969.

Doty, R. E., joint author. *See* Strickland, W. B., et al., 1970.

3363. **Downe, A. E. R. 1962** Serology of Insect Proteins. I. Preliminary Studies on the Proteins of Insect Cuticle. Canad. J. Zool. 40(6): 957-9 67.
Precipitin tests on extracts of larval cuticle of *M. domestica* showed presence of several antigenic protein components. One such component from third stage larva possessed phenol-oxidase activity. ABSTRACT: BA 41:1311(16785), Feb. 15, 1963.

3364. **Downey, T. W. 1963** Polioviruses and Flies: Studies on the Epidemiology of Enteroviruses in an Urban Area. Yale J. Biol. and Med. 35(4):341-352.
States that in areas of good sanitation practices, a high level of infection (i.e. an epidemic) must be present before flies will become contaminated. ABSTRACT: BA 43:1904(24606), Sept. 15, 1963.

3365. **Draber-Monko, A. 1966** Materialy do znajomosci Muscinae (Diptera) Polski. [Materials to the knowledge of Polish Muscinae (Diptera).] Fragmenta Faunistica 12(18):309-331. (German and Russian summaries.)
M. larvipara Portsch. listed as new to the fauna of Poland. New biological data obtained concerning 3 species of *Muscina*. ABSTRACT: BA 48:2821(31013), Mar. 15, 1967.

3366. **Drea, J. J. 1966** Studies of *Aleochara tristis* (Coleoptera: Staphylinidae) a Natural Enemy of the Face Fly. J. Econ. Ent. 59:1368-1373.
Summarizes life history studies made in France (1962-1964) prior to the introduction of *Aleochara tristis* into the United States. ABSTRACTS: BA 48:3233(35608), Apr. 1, 1967; RAE-B 55:64(240), Apr., 1967.

3367. **Dremova, V. P. 1962** Organization of the Fight against Flies in Hot Climates. J. Hyg. Epidem. (Praha) 6:278-285. (Summaries in French, German and Italian.)
Concerns southeastern USSR. *M. domestica vicina* is epidemiologically the most significant species in this region. ABSTRACT: BA 42:1250 (15802), May 15, 1963.

3368. **Dremova, V. P.** and **Grozdeva, I. V. 1961** [Determination of the sensitivity of the domestic fly to chlorophos by the method of individual feeding.] Med. Parazit. (Moskva) 30:739-741. (In Russian, with English summary.)
For trichlorphon, LD_{50}'s obtained for laboratory and wild strains were

0·27 and 0·3 μg per fly, respectively, when fed to flies in sugar; 0·73 and 0·77 μg. when applied topically. ABSTRACTS: BA 40:1298(17051), Nov. 15, 1962 (Indicates "Bolezni"); RAE-B 53:206, Oct., 1965.

3369. **Dremova, V. P.** and **Troshin, I. S. 1961** [Preservation of residual toxicity of chlorophos on various surfaces and its use in fly control.] Med. Parazit. (Moskva) 30:223-225. (In Russian, with English summary.)
Carried out in Soviet Union in which various building materials were sprayed with aqueous solution of chlorofos and tested for toxicity to adults of *M. domestica.* ABSTRACT: RAE-B 53:50, Mar., 1965.

Dremova, V. P., joint author. *See* Magdiev, R. R., et al., 1960.

3370. **Dresden, D. 1965** Enzymes and Mutations in Insect Resistance. Meded. Landb. Hogesch. Opzoek Stns. Gent. 30(3):1382-1389. (Summaries in Flemish, French, and German.)
Resistance to organophosphorus compounds in *M. domestica* depends on action of specific breakdown enzymes capable of hydrolysing compounds to which a particular strain is resistant. Enzyme produced under influence of special mutated allele of gene (a+) that produces aliesterase of unknown function. ABSTRACTS: RAE-B 54:212, Nov., 1966; RAE-A 54: 574, Nov., 1966.

3371. **Drobozina, V. P. 1963** Soderzhanie askorbinovoi kisloty i glyutationa v organizme *Musca domestica* L. na raznykh stadiyakh ovogeneza. [Asorbic acid and glutathione level in the organism of *Musca domestica* at the different stages of ovogenesis.] Nauch. Dokl. Vyssh. Shkoly Biol. Nauk. 4:39-42.
Ascorbic acid and glutathione level decrease after pupa hatches. A restoration of glutathione is connected to protein intake, while that of ascorbic acid is connected with protein synthesis in the flies and with its presence in the food. (A good translation of the paper may be found in REFZH-Biol., 1964, No. 19E 377.) ABSTRACT: BA 47:2504(29932), Mar. 15, 1966.

3372. **Drobozina, V. P. 1966** O razvitii rezistent nosti k DDT u *Protophormia terrae-novae.* [Development of resistance to DDT in *Porotophormia terrae-novae* R.-D.] Med. Parazit. i Parazitarn. (Bolez.) 35(5):525-531.
Resistance to DDT is developed much slower in *P. terrae-novae* than in *Musca domestica.* Fluctuations in the development of resistance in a population are described. ABSTRACT: BA 49:1420(15918), Feb. 1, 1968.

3373. **Drobozina, V. P. 1968** Povedenie *Musca domestica* L. i *Phormia terrae -novae* R.-D. pri Kontakta s DDT [Behavior of *Musca domestica* L. and *Phormia terrae-novae* R.-D. during contact with DDT.] Med. Parazit. i Parazitarn. (Bolez.) 37(1):36-41. (English summary.)
M. domestica tends to avoid DDT-treated surfaces by moving about. *P. terrae-novae*, a crawler, has more contact with the treated surface (but develops resistance more slowly, even so).

Drobozina, V. P., joint author. *See* Derbeneva-Ukhova, V. P., et al., 1966.

Drozdova, O. I., joint author. *See* Fomicheva, V. S., et al., 1967.

3374. **Drummond, R. O. 1963** Toxicity to House Flies and Horn Flies of Manure from Insecticide-fed Cattle. J. Econ. Ent. 56:344-347.
Percentage effectiveness of insecticides tested determined according to formula: Number of adults emerged $\times 100 \div$ number of eggs (or larvae and eggs). ABSTRACTS: BA 44:630(8444), Oct. 15, 1963; RAE-B 51:202, Sept., 1963; TDB 60:1084, Nov., 1963.

3375. **Drummond, R. O. 1967** Further Evaluation of Animal Systemic Insecticides, 1966. J. Econ. Ent. 60(3):733-737.
Data presented for 14 insecticides. Two, fed to cattle, caused manure to be toxic to *M. domestica* and *Haematobia irritans*. ABSTRACT: RAE-B 55:187(646), Oct., 1967.

3376. **Drummond, R. O.** and **Gladney, W. J. 1969** Further Evaluation of Animal Systemic Insecticides. J. Econ. Ent. 62(4):934-936.
Paper included because it is one of a series on the subject. *M. domestica* mentioned very briefly. ABSTRACT: RAE-B 58:27(98), Jan., 1970.

3377. **Drummond, R. O., Whetstone, T. M.** and **Ernst, S. E. 1967** Control of Larvae of the House-fly and Horn Fly in Manure of Insecticide-fed Cattle. J. Econ. Ent. 60:1306-1308.
Several insecticides effective. Face fly larvae were less susceptible than horn fly larvae; more susceptible than house fly larvae. ABSTRACTS: RAE-B 56:64(210), Mar., 1968; BA 49:1395(15627), Feb. 1, 1968.

Duckles, C. K., joint author. *See* Roberts, R. B., et al., 1969.

3378. **Dunaeva, I. D. 1964** Ostatochnoe deistvie khlorofosa v substrate no predimaginal'nye fazy komnatnoi mukhi. [Residual action of chlorophos in the substrate on the preimaginal phase of house flies.] Med. Parazit. (Moskva) 33:13-15.
Residual action is dependent upon the dose of the formulation ABSTRACTS: BA 45:7738(97732), Nov. 15, 1964; RAE-B 55:198(690), Nov., 1967.

Duncan, J., joint author. *See* Hadaway, A. B., et al., 1963.

3379. **Dun, P. H. 1960** Control of House Flies in Bovine Feces by a Feed Additive containing *Bacillus thuringiensis* var. *thuringiensis* Berliner. J. Insect. Pathol. 2(1):13-16.
Accomplished 92 per cent prevention of imago development. Mortality occurred chiefly in the pupal stage. A toxin, rather than a septicemia, is probably responsible. ABSTRACTS: BA 35:3453(39554), July 15, 1960; RAE-B 49:59-60, Mar., 1961.

Dunning, L. L., joint author. *See* Loomis, E. C., et al., 1968.

3380. **Duport, M., Cambiesco, I., Atanasiu, A.** and **Ianco, L. 1963** Contribution à l'étude de la variation de la sensibilité au DDT de l'espèce *M. domestica* L. par rapport à differente facteurs. [Contribution to the study of the variation in susceptibility to DDT of the species *M. domestica* in regard to diverse factors.] Arch. Roum. Path. Exp. Microbiol. 22(3):757-762.
In an assessment of the fly resistance problem in Rumania in 1955 it was noted that resistance declined in the autumn, F_1 progeny were more susceptible than field specimens, female flies were less susceptible to DDT than males, and 4 or 5 day old flies the most resistant age group. ABSTRACTS: TDB 61:532, May, 1964; RAE-B 53:237, Dec., 1965.

3381. **Duport, M., et al. 1965** [Current problems in the use of insecticides.] Microbiologia (Bucur) 10:551-562. (In Rumanian.)

3382. **Duport, M., Combiesco, I.** and **Enesco, A. 1968** Traitement expérimental de l'espèce *Musca domestica* L. par du thiotépa, sur le terrain. [Treatment of the species *Musca domestica* with thiotepa in field trials.] Arch. Roum. Path. Exp. Microbiol. 27(3):707-714.
Solutions sprayed or placed about as sugar baits. Fly densities decreased, and a high rate of sterility prevailed in eggs from captured flies. Question of toxic and mutagenic risks to man and animals. ABSTRACT: TDB 66: 1339(2535), Dec., 1969.

Duport, M., joint author. *See* Combiesco, I., et al., 1967; Lupasco, G., et al., 1963, 1966.

3383. **Durden, J. A., Jr.** and **Weiden, M. H. 1969** Insecticidal 2-(methylcarbamoyloxyphenyl) 1,3-dioxolanes-oxathiolanes, and -dithiolanes. J. Agric. Fd. Chem. 17(1):94-100.
Compounds were effective inhibitors of fly head acetyl cholinesterase. Showed activity against several insects, including *M. domestica.* ABSTRACT: BA 50:10829(112103), Oct. 15, 1969.

3384. **Dutky, R. C., Robbins, W. E., Kaplanis, J. N.** and **Shortino, D. J. 1963** The Sterol Esters of Housefly Eggs. Comp. Biochem. and Physiol. 9(3):251-255.
Free sterols and sterol esters were separated from eggs and from adult females by column chromatography. Sterol ester fraction was 41 per cent of total sterol from eggs; 8·4 per cent of total sterol from flies. ABSTRACT: BA 45:323, Jan. 1, 1964.

3385. **Dutky, R. C., Robbins, W. E., Shortino, T. J., Kaplanis, J. N.** and **Vroman, H. E. 1967** The Conversion of Cholestanone to Cholestanol by the House Fly, *Musca domestica* L. J. Insect Physiol. 13(10): 1501-1510.
Both males and females can make this conversion, which seems of considerable importance in larval life. This is the first example of the reduction of a ketosteroid to its corresponding alcohol by an insect. ABSTRACTS: BA 49:948(10689), Jan. 15, 1968; RAE-B 56:59(197), Mar., 1968.

Dutky, R. C., joint author. *See* Kaplanis, J. N., et al., 1961.

Dutt, S. C., joint author. *See* Srivastava, H. D., et al., 1963.

3386. **Dutton, G. J.** and **Ko, V. 1964** The Apparent Absence of Uridine Diphosphate Glucuronyl-transferase for Detoxication in *Musca domestica*. Comp. Biochem. Physiol. 11(3):269-272.
M. domestica has a uridine diphosphate glucose glycosyltransferase, but lacks the glucuronate glucuronyl-transferase. It may be that this species is unlike other insects in possessing a mechanism for detoxication by glucosiduronate formation. ABSTRACT: BA 45:5434(66798), Aug. 1, 1964.

3387. **Earle, N. W. 1963** The Fate of Cyclodiene Insecticides Administered to Susceptible and Resistant Houseflies. J. Agric. Fd. Chem. 11: 281-285. Easton, Pa.
Metabolism of cyclodiene insecticides is a relatively unimportant resistance mechanism in houseflies. ABSTRACT: RAE-B 52:56, Mar., 1964.

Earle, N. W., joint author. *See* Sun, Y.-P., et al., 1963.

Earp, V. F., joint author. *See* Kranzler, G. A., et al., 1966, 1968.

3388. **Eastwood, R. E., Kada, J. M.** and **Schoenburg, R. B. 1966** Plastic Tarpaulins for Controlling Flies in Stockpiled Poultry Manure Fertilizer. J. Econ. Ent. 59(6):1507-1511.
Polyethylene tarps gave best overall fly control; black polyethylene best for retaining fertilizer elements. Counts and identification of larvae were made and recorded. ABSTRACTS: BA 48:3231(35590), Apr. 1, 1967; RAE-B 55:68(253), Apr., 1967; TDB 64:438, Apr., 1967.

3389. **Eastwood, R. E., Kada, J. M., Shoenburg, R. B.** and **Brydon, H. W. 1967** Investigations on Fly Control by Composting Poultry Manures. J. Econ. Ent. 60(1):88-98.
Windrow composting of poultry manure reduced fly larvae numbers. Once composted, manure was not suitable as an oviposition site for fly larvae development. ABSTRACTS: BA 48:6008(67152), July 1, 1967; RAE-B 55:104(371), June, 1967; TDB 64:810, July, 1967.

Ebina, R., joint author. *See* Miyamoto, K., et al., 1965.

3390. **Ecke, D. H. 1966** A Comparison of Fly Larval Production from 6 Refuse Systems. J. Environ. Health 29(2):156-160.

3391. Eddy, G. W. 1961 Laboratory Tests of Residues of Organophosphorus Compounds against House Flys. J. Econ. Ent. 54(2):386-389.
Flies were exposed to plywood panels sprayed with acetone solutions of 22 organophosphorus compounds to test for toxicity, knockdown and duration of effectiveness. ABSTRACT: RAE-B 50:110, May, 1962.

3392. Eddy, G. W. and **Roth, A. R. 1961** Toxicity to Fly Larvae of the Feces of Insecticide-Fed Cattle. J. Econ. Ent. 54(3):408-411.
Comparative data presented on susceptibility of the housefly, stable fly and horn fly to several insecticides which, when fed to cattle, appeared in manure. ABSTRACTS: BA 36:5570(58203), Sept. 1, 1961; RAE-B 50: 125, June, 1962.

3393. Eddy, G. W., Roth, A. R. and **Plapp, F. W. 1962** Studies on the Flight Habits of Some Marked Insects. J. Econ. Ent. 55(5):603-607.
M. domestica among several species tested. ABSTRACTS: BA 41:955(12420), Feb. 1, 1963; RAE-B 51:36, Feb., 1963.

Eddy, G. W., joint author. *See* Davis, H. G., et al., 1966; Plapp, W. F. Jr., et al., 1961, 1962, 1963.

3394. Edelstein, A. and **Taylor, R. B. 1961** Automation and Pyrethrum Control Flies Around the Clock. Pyrethrum Post 6(2):43-48. Nakuru.
A technique using a dispenser emitting a fixed dose automatically at regular intervals gives satisfactory control in spaces up to 10,000 cu. ft. ABSTRACT: RAE-B 51:159, Aug., 1963.

Edmunds, L. N. Jr., joint author. *See* Barker, R. J., et al., 1963.

Edwards, J. G., joint author. *See* Ecke, D. H., et al., 1965.

Edwards, T. D., joint author. *See* Axtell, R. C., et al., 1970.

3395. Eichenbaum, D. and **Goldsmith, T. H. 1967** Physiological Properties of Intact Isolated Photo Receptor Cells. J. Gen. Physiol. 50(10): 2479-2480.
Imaginal disc of eye from 3rd instar larvae, transplanted into host larvae, resulted in normal adult with abdominal 3rd eye having retinular cells, pigment cells and cornea but devoid of all other nervous elements.

3396. Eichenbaum, D. M. and **Goldsmith, T. H. 1968** Properties of Intact Photoreceptor Cells Lacking Synapses. J. Exp. Zool. 169(1):15-32.
Development of synapses of the lamina ganglionaris not necessary for normal physiological development of the receptors; responses of photoreceptors of normal eyes can be generated without significant feedback control from optic lobe. ABSTRACT: BA 50:5347(56153, May 15, 1969.

Eichner, J. T., joint author. *See* Kay, R. E., et al., 1967.

3397. Eide, P. E. and **Reinecke, J. P. 1970** A Physiological Saline Solution for Sperm of the House Fly and the Black Blow Fly. J. Econ. Ent. 63:1006.
A pH of 6·8 and osmolarities of 350-450 are optimal for motility. Gives formula devised by authors.

Eisenberg, B. C., joint author. *See* MacLaren, W. R., et al., 1960.

3398. El-Aziz, S. A. E.-S. 1968 Insect Enz Oxidases and Detoxication of Carbamate Insecticide. Diss. Absts. 28(108):4164-B.

3399. El-Aziz, S. A., Metcalf, R. L. and **Fukuto, T. R. 1969** Physiological Factors Influencing the Toxicity of Carbamate Insecticides to Insects. J. Econ. Ent. 62(2):318-324.
Topical LD_{50} values of the carbamates were markedly affected by age and sex in the housefly. Variations in toxicity in the carbamates were well correlated with the titre of soluble phenolases in the various life stages, sexes and adults of different ages.

El-Aziz, A., joint author. *See* Metcalf, R. L., et al., 1966.

3400. ElBashir, S. 1967 Causes of Resistance to DDT in a Diazinon-selected and a DDT-selected Strain of House Flies. Ent. Exp. Appl. 10(1): 111-126. (German summary.)
There seemed to be considerable differences in the mechanism of resistance in the two strains studied. DDE was the only metabolite identified. ABSTRACTS: RAE-B 55:176(610), Sept., 1967; TDB 64:1361, Dec., 1967.

3401. ElBashir, S. and **Lord, K. A. 1965** DDT Tolerance in Diazinon-selected and DDT-selected Strains of Houseflies. Chem. and Ind. 37:1598-1599.
The causes of resistance to DDT in strains selected with diazinon and with DDT, differ. Rate of penetration, degradation, and changing to metabolites, vary for the strains. ABSTRACT: RAE-B 56:47(157), Feb., 1968.

3402. ElBashir, S. and **Oppenoorth, F. J. 1969** Microsomal Oxidations of Organophosphate Insecticides in Some Resistant Strains of Houseflies. Nature 223(5202):210-211.
Measured, *in vitro*, microsomal reactions in 2 OP-resistant strains and some susceptible strains. Low specificity of the microsomal mechanism is remarkable. ABSTRACT: TBD 67:724(1405), June, 1970.

3403. El-Deeb, A. L., Hammad, S. M. and **Gaaboob, I. A. 1969** Laboratory Tests to Determine Susceptibility of Adult Females of the House Fly to Malathion. (Diptera: Muscidae.) Bull. Ent. Soc. Egypt. Econ. Ser. 3:179-185.
Used both topical application and baits. There was found to be more homogeneity in flies from laboratory culture than in flies from the field. ABSTRACT: BA 51:10266(104378), Sept. 15, 1970.

3404. Eldefrawi, M. E. and **Hoskins, W. M. 1961** Relation of the Rate of Penetration and Metabolism to the Toxicity of Sevin to Three Insect Species. J. Econ. Ent. 54:401-405.
Sevin absorbed rapidly into and metabolized by housefly; resistant flies have faster metabolism. Inhibition of metabolism occurs with synergists such as sesamex. ABSTRACTS: BA 36:5562(58143), Sept., 1961; RAE-B 50:124, June, 1962.

3405. Eldefrawi, M. E., Miskus, R. and **Sutcher, V. 1960** Methylenedioxyphenyl Derivatives as Synergists for Carbamate Insecticides on Susceptible, DDT- and Parathion-resistant House Flies. J. Econ. Ent. 53(2):231-234.
Found compounds (with the exception of one) to be active synergists. Carbamates studied were less active inhibitors for fly head cholinesterase than for human plasma cholinesterase. ABSTRACTS: BA 35:3739(42995), Aug. 1, 1960; RAE-B 49:102, May, 1961.

Eldefrawi, M. E., joint author. *See* Gordon, H. T., et al., 1960.

3406. Elliott, M., Farnham, A. W., Janes, N. F., Needham, P. H. and **Pearson, B. C. 1967** 5-Benzyl-3-furymethyl Chrysanthemate: A New Potent Insecticide. Nature 213(5075):493-494.
Presents the chemical composition and discusses the effectiveness of a new non-ketonic ester of chrysanthemic acid extremely toxic to the housefly and other insects. ABSTRACT: BA 48:7866(88162), Sept. 1, 1967.

3407. Elliott, M., Farnham, A. W., Janes, N. F., Needham, P. H., Pearson, B. C. and **Stevenson, J. H. 1965** New Synthetic Insecticidal Compounds Related to the Pyrethrins. Proc. 3rd Brit. Insect. and Fung. Conf. pp. 437-443.
Paper seeks to show that compounds containing only carbon, hydrogen and oxygen can equal and surpass, in toxicity to insects, compounds from other classes.

3408. **Elliott, M., Janes, N. F., Jeffs, K. A., Needham, P. H., Sawicki, R. M.** and **Stevenson, J. H. 1965** New Insecticidal Esters of Chrysanthemic Acid. Proc. 3rd Brit. Insect. and Fung. Conf. (November 8-11, Brighton, Sussex.) pp. 432-439.

Certain esters of cis-trans-chrysanthemic acid were highly toxic to insects. 4-allyl-2, b-dimethylbenzyl chrysanthemate and 4-allylbenzyl chrysanthemate were more than twice as toxic as allethin and pyrethrin to *Musca domestica.* ABSTRACT: RAE-B 55:202(699), Nov., 1967.

Elliott, M., joint author. *See* Sawicki, R. M., et al., 1962, 1965.

El-Moursy, A. A., joint author. *See* Hafez, M., et al., 1969.

3409. **El'Piner, L. E. 1966** Novoe v biofizike ul'trazvukovykh voln. [Advances in the biophysics of ultrasonic waves.] USP Sovrem Biol. 61(2): 212-229.

A review with 90 references. *M. domestica* included, with various other plant and animal spp. ABSTRACT: BA 48:5203(58209), June 15, 1967.

3410. **El'Piner, L. E., Faikin, I. M.** and **Basurmanova, O. K. 1965** O vnutrikletochnykh mikropotokakh, vyzyvaemykh ul'trazvukovymi volnami. [On intracellular microcurrents evoked by ultrasonic waves.] Biofizika 10:805-812. (In Russian.)

Worked with first visual ganglion of *Musca domestica.* Experimental possibility of inducing spatial interrelationships between intracellular microscopic and submicroscopic structures was demonstrated for the first time. ABSTRACT: BA 49:3359(37565), Apr. 1, 1968.

3411. **El-Sebae, A. H., Metcalf, R. L.** and **Fukuto, T. R. 1964** Carbamate Insecticides: Synergism by Organothiocyanates. J. Econ. Ent. 57: 478-482.

Synergism with Thanite involves detoxification inhibition of the carbamates within the fly body. Differing degress of synergism were shown by Thanite and piperonyl butoxide. ABSTRACT: RAE-B 52:203, Nov., 1964.

El-Sherif, A. F., joint author. *See* Salem, H. H., et al., 1960, 1961.

Ely, D. G., joint author. *See* Harvey, T. L., et al., 1969.

El-Ziady, S., joint author. *See* Hafez, M., et al., 1969.

3412. **Enan, O. H. 1965** Laboratory Studies on the Effect of Crowding on Percent Emergence, Size of Adult, and Period of Development of the Housefly *Musca domestica.* J. Egyptian Pub. Health Ass. 40: 177:183.

With 40-160 eggs per cup, per cent of adult emergence varied little. With higher density, emergence dropped markedly. Size of adult fairly normal. Period of development = 17-19 days, (longer than reported by most workers.) ABSTRACT: TDB 62:1272, Dec., 1965.

3413. **Enan, O.** and **Gordon, H. T. 1965** Temperature Effects on Toxicity of Synergized Carbamate Insecticides on House Flies. J. Econ. Ent. 58(3):513-516.

In the absence of synergists there is a positive temp. coefficient (more toxicity with increase in temp.). With the addition of synergists toxicity is increased but temp. coefficient becomes zero or negative. ABSTRACTS: BA 47:1601(19382), Feb. 15, 1966; RAE-B 53:180, Sept., 1965; TDB 63:238, Feb., 1966.

3414. **Enan, O., Miskus, R.** and **Craig, R. 1964** A Comparison of the Sterols in Resistant and Susceptible House Flies, *Musca domestica.* J. Econ. Ent. 57(3):364-366.

Relative amounts of several sterols isolated by gas chromatography vary with the housefly strain. Cholesterol is not the major constituent of the isolated sterols. ABSTRACTS: BA 45:8420(106980), Dec. 15, 1964; RAE-B 52:166, Sept., 1964; TDB 61:1260, Dec., 1964.

Endo, T., joint author. *See* Fujito, S., et al., 1962.

Enesco, A., joint author. *See* Combiesco, I., et al., 1967, 1968; Lupasco, G., et al., 1966; Duport, M., et al., 1968.

Engel, J. L., joint author. *See* Casida, J. E., et al., 1966.

3415. Engels, W. 1968 Anerobioseversuche mit *Musca domestica.* Alters-und geschlechtsunterschieda von Überlebensrate und Erholfähigkeit. [Anaerobiasis experiments with *Musca domestica.* Age- and sex differences from survival rate and recovery ability.] J. Insect. Physiol. 14(2):253-260.

Adults kept in atmosphere of pure N_2 for 1-25 hours. Discusses mortality, survival and recovery rates from the viewpoint of carbohydrate stores available for anaerobic metabolism. ABSTRACTS: BA 49:4802(53990), May 15, 1968; TDB 65:726(1566), May, 1968.

3416. Engels, W. 1968 Anaerobioseversuche mit *Musca domestica.* Verwertung normaler und experimentell erzeugter Kohlenhydratreserven. [Achievement of natural and experimental rearing on carbohydrates.] J. Insect Physiol. 14(6):869-879. (English summary.)

Female flies may mobilize for themselves glycogen of the oöcyte (full grown; covered with a chorion). Utilization depends upon the maternal metabolic situation. ABSTRACT: BA 49:10287(113643), Nov., 1968.

3417. Engels, W. 1969 Zur Wirkung und Lokalisation von H^3-Actinomycin D in Eifollikeln von *Musca domestica* nach in vivo-Applikation. [Effects and localization of H^3-actinomycin D in egg follicles of *Musca domestica* after *in vivo* application.] Histochemie 19(3):225-234.

In short time experiments actinomycin was always found in connection with nuclear DNA. Prolonged treatment brought on premature degeneration. ABSTRACT: BA 51:5666(58145), May 15, 1970.

3417a. Engels, W. 1970 [Cold as trigger of glycogen storage during oogenesis of *Musca domestica.*] Experientia 26:884-885. (In German.)

3418. Engels, W. and **Bier, K. 1967** Zur Glykogenspeicherung während der oogenese und ihrer vorzeitigen auslösung durch blockierung der RNS-Versorgung. (Untersuchungen an *Musca domestica* L.) [On glycogen development during oogenesis, and its premature release during blockage of the RNS supply. (Researches on *Musca domestica* L.).] Roux' Archiv für Entwicklungsmechanik. 158:64-88.

Glycogen development can be induced at an earlier stage than normal, indicating that the inactive glycogen synthetase is present. At a later stage, this process is activated by the lack of RNA supply or by some subsequent process. Discusses the validity of an antagonism between RNA depending protein metabolism and glycogen synthesis.

3419. Engels, W. and **Ribbert, D. 1969** Nucleic Acid, Amino Acid and Carbohydrate Metabolism of Nurse Cell Nucleoli in *Musca domestica.* Experientia 25:805-807.

Besides showing high RNA metabolism, the multiple nucleoli of *Musca* NCN have been proved to be sites of rapid ^{3}H-amino acid and ^{3}H-glucose incorporation.

Enos, H. T., joint author. *See* Cheng, T. H., et al., 1961, 1962.

3420. Ensing, K. J. 1968 Resistance Patterns and Fecundity of House Flies, *Musca domestica* Selected with Diazinon Insectic Dimetilan Insectic and Combinations of the Two Insecticides. Diss. Absts. 28(108): 4165-B-4166-B.

Erakey, A. S., joint author. *See* Hafez, M., et al., 1969, 1970.

3421. **Ermokhina, T. M., Mekhanik, M. L., Zaitseva, G. N.** and **Belozerskii, A. N. 1966** Izuchenie fenilalanit-RNK-simtetaz i fenilalaninovykh SRNK w drozhzhei i nasekomykh. [A study of phenylalanyl-RNA-synthetase and phenylalanine SRNA's in yeasts and insects.] Dokl. Akad. Nauk. SSSR 170(4):974-977.
Two phenylalanyl-RNA synthases were isolated from fly larvae the specific activities of which were nearly identical but one of which was present in twice the concentration of the other. Further experiments with the enzymes are reported. ABSTRACT: BA 48:10181(114401), Nov. 15, 1967.

Ernst, S. E., joint author. *See* Drummond, R. O., et al., 1967.

3422. **Erofeeva, T. V. 1967** Raspredelenie lichinok otdelnykh vidov sinantrophykh mukh v zavisimosti ot reaktsii sredy pitatelnego substrata. [The effect of pH occurring in the feeding substrate of the environment on the distribution of the larvae of some synanthropic fly species.] Wiad. Parazytol. 13(4/5):609-613. (Polish summary.)
Optimum pH for *Musca domestica* between 7 and 8. ABSTRACT: BA 51: 565(5684), Jan. 1, 1970.

3423. **Esaac, E. G.** and **Casida, J. E. 1968** Piperonylic Acid Conjugates with Alanine, Glutamate, Glutamine, Glycine and Serine in Living Houseflies. J. Insect Physiol. 14(7):913-925.
Following their injection into adult female houseflies, each of piperonal, piperonyl alcohol, safrole, and tropital synergists oxidizes to piperonylic acid, which converts into 5 N-piperonyl conjugates involving the following amino acids: alanine, glutamate, glutamine, glycine and serine. The ratio of the conjugates formed and excreted by flies depends upon the administered precursor for piperonylic acid. ABSTRACT: BA 49:9772 (108232), Oct. 15, 1968.

3424. **Esaac, E. G.** and **Casida, J. E. 1969** Metabolism in Relation to Mode of Action of Methylenedioxyphenyl Synergists in Houseflies. J. Agric. Fd. Chem. 17(3):539-550.
Ten C^{14}-labelled MDP compounds studied give evidence for the hypothesis that the synergistic activity of MDP compounds results from competitive inhibition of the microsomal mixed-function oxidases responsible for insecticide detoxification. ABSTRACT: BA 51:572(5751), Jan. 1, 1970.

Esaac, E. G., joint author. *See* Casida, J. E., et al., 1966, 1968.

Eschle, J. L., joint author. *See* DeFoliart, G. R., et al., 1961.

Eskina, G. V., joint author. *See* Tukhmanyants, A. A., et al., 1963.

3425. **Esther, H. 1962** Einige Versuchsergebnisse zur Selektion eines gegen Trichlorphon resistenten Laborstammes von *Musca domestica* L. [Some experimental results on the selection of a laboratory strain of *Musca domestica* L., resistant to trichlorphon.] J. Hyg., Epidemiol., Microbiol., Immunol. (Praha) 6:286-295. (Summaries in English, French and Italian.)
Investigates the possibility, rate and speed of the development of housefly resistance to Trichlorphon. Concludes:a decrease of efficacy must be anticipated in practical fly control. ABSTRACTS: BA 42:1241(15715), May 15, 1963; RAE-B 53:18, Jan., 1965.

3426. **Eto, M. 1969** Specificity and Mechanism in the Action of Saligenin Cyclic Phosphorus Esters. Residue Reviews 25:187-200.
Discusses the mechanism of the metabolic formation, and the specificity in the biological activity, of the saligenin cyclic phosphorus esters. Data from rats, chickens and houseflies.

3427. Eto, M., Kishimoto, K., Matsumura, K., Ohshita, N. and **Oshima, Y. 1966** Studies on Saligenin Cyclic Phosphorus Esters with Insecticidal Activity. IX. Derivatives of Phosphonic and Phosphonothionic Acids. Agric. Biol. Chem. (Tokyo) 30(2):181-185.
Compounds named in title proved less stable than corresponding phosphate esters. 2-ethyl-4H-1, 3, 2-benzodioxaphosphorin-2-sulfide was the most effective against houseflies. ABSTRACT: TDB 63:1037, Sept., 1966.

3428. Eto, M., Kobayashi, K., Sasamoto, T., Cheng, H.-M., Aikawa, T., Kume, T. and **Oshima, Y. 1968** Studies on Saligenin Cyclic Phosphorus Esters with Insecticidal Activity. XII. Insecticidal Activity of Ring-substituted Derivatives. Botyu-Kagaku 33(3):73-77. (Japanese summary.)
Appeared to be no relation between insecticidal activity and the electron-withdrawing or electron-releasing character of substituents at the para-position of phenolic ester linkage. About 50 esters were tested. ABSTRACT: RAE-B 57:213(773), Oct., 1969.

3429. Eto, M., Matsuo, S. and **Oshima, Y. 1963** Metabolic Formation of Saligenin Cyclic Phosphates from 0-Tolyl Phosphates in House Flies, *Musca domestica*. Agric. Biol. Chem. (Tokyo) 27(12):870-875.
Examined effects on flies of several tri-esters of phosphoric acid. Certain esters converted to biologically active metabolites. These were confirmed by chromatography to be saligenin cyclic phosphates. ABSTRACTS: BA 45:7415(93630), Nov. 1, 1964; TDB 61:613, June, 1964.

3430. Eto, M., Oshima, Y., Kitakata, S., Tanaka, F. and **Kojima, K. 1966** Studies on Saligenin Cyclic Phosphorus Esters with Insecticidal Activity. X. Synergism of Malathion against Susceptible and Resistant Insects. Botyu-Kagaku. (Bull. Inst. Insect Contr.) 31(1):33-38.
Aryl derivatives of cyclic esters are not insecticidal but are synergistic with malathion and show high activity against resistant strains. Alkyl derivatives showed low synergism against resistant strains. ABSTRACT: RAE-B 56:128(450), July, 1968.

Eto, M., joint author. *See* Ohkawa, H., et al., 1968; Kobayashi, K., et al., 1969.

3431. Evans, B. R. and **Porter, J. E. 1965** The Incidence, Importance and Control of Insects Found in Stored Food and Food-handling Areas of Ships. J. Econ. Ent. 48:479-481.
Lists insects collected from food-handling and dining areas of ships from foreign ports arriving in Miami (1957-1961) and New Orleans (1960-1963). *Musca domestica* L. included. ABSTRACT: RAE-B 53:179, Sept., 1965.

3432. Evans, E. S., Jr. and **Hansens, E. J. 1968** Effect of Post-Treatment Temperature on Toxicity of Insecticides to Resistant and Susceptible House Flies. J. Econ. Ent. 61(2):543-546.
The level of resistance in a strain of houseflies influenced the toxicity of diazinon and lindane differently at different temperatures. Also, the same temperature influenced one insecticide more than another. ABSTRACTS: BA 49:7252(80827), Aug. 1, 1968; RAE-B 56:191(689), Sept., 1968; TDB 66:486(932), May, 1969.

Evans, E. S., Jr., joint author. *See* Hansens, E. J., et al., 1967, 1968.

3433. Eversole, J. W., Lilly, J. H. and **Shaw, F. R. 1965** Comparative Effectiveness and Persistence of Certain Insecticides in Poultry Droppings against Larvae of the Little House Fly. J. Econ. Ent. 58:704-709.
Of materials compared for effectiveness and persistence against larvae of *Fannia canicularis* (L), dimethoate was substantially more effective

than the others. Coumaphos and diazinon were next in effectiveness as larvicides. ABSTRACTS: BA 46:8886(109572), Dec. 15, 1965; RAE-B 53: 222, Nov., 1965.

3434. **Eversole, J. W., Lilly, J. H.** and **Shaw, F. R. 1965** Toxicity of Droppings from Coumaphos-Fed Hens to Little House Fly Larvae. J. Econ. Ent. 58:709-710.
Coumaphos provided an effective degree of control of larvae in droppings when fed to poultry, but approximately a 70-fold decrease in effectiveness occurred during passage of the insecticide through the bird. ABSTRACTS: BA 46:8886(109573), Dec. 15, 1965; RAE-B 53:223, Nov., 1965.

Evtyushina, T. M., joint author. *See* Rostavtseva, S. A., et al., 1969.

3435. **Fahmy, M. A. H.** and **Gordon, H. T. 1965** Selective Synergism of Carbamate Insecticides on House Flies by Aryloxyalkylamines. J. Econ. Ent. 58:451-455.
Experimental results suggest two or more major carbamate detoxication mechanisms. Some of the compounds tested are stronger carbamate synergists than sesamex. ABSTRACTS: BA 47:1601(19384), Feb. 15, 1966; RAE-B 53:179, Sept., 1965.

3436. **Fahmy, M. A. H., Metcalf, R. L., Fukuto, T. R.** and **Hennessy, D. J. 1966** Effects of Deuteration, Fluorination, and Other Structural Modifications of the Carbamyl Moiety Upon the Anticholinesterase and Insecticidal Activities of Phenyl N-methylcarbamates. J. Agric. Fd. Chem. 14:79-83.
Alterations of the $NHCH_3$ carbamate insecticide resulted in decreased anticholinesterase and insecticidal activity to housefly and mosquito larvae. The $NHCF_3$ and NH_2 carbamates showed substantial synergism when evaluated in admixture with piperonyl butoxide. ABSTRACT: RAE-B 54:160, Aug., 1966.

Fahmy, M. A. H., joint author. *See* Fukuto, T. R., et al., 1967; Metcalf, R. L., et al., 1966.

Faikin, I. M., joint author. *See* El'Piner, L. E., et al., 1965.

3437. **Fales, J. H.** and **Bodenstein, O. F. 1961** Promising Synergist for DDT. Soap and Chem. Spec. 37(11):77-80, 188.
WAPF antiresistant shows a high degree of synergism when combined with DDT against resistant house flies. ABSTRACTS: RAE-B 51:63, Mar., 1963; TDB 59:621, June, 1962.

3438. **Fales, J. H.** and **Bodenstein, O. F. 1964** DDT Aerosols Containing WARF Antiresistant. Soap and Chem. Spec. 40(4):96-97.
States that commercially available formulas could be improved by the addition of this preparation. ABSTRACT: RAE-B 53:217, Nov., 1965.

3439. **Fales, J. H.** and **Bodenstein, O. F. 1965** New Household Insecticide. Soap and Chem. Spec. 41(11):91-94, 123-124.
Refers to Bromophos (=CELA S-1942), a promising insecticide, in the experimental stage. ABSTRACT: RAE-B 55:23(63), Feb., 1967.

3440. **Fales, J. H.** and **Bodenstein, O. F. 1966** A Promising New Insecticide. Soap and Chem. Spec. 42(6):80-81, 84, 86, 88; (7):66-68, 104-106.
Neopynamin, a Japanese product, found superior to pyrethrins and allethrin in fly knock-down. Is of the same order of effectiveness as these two, in the matter of kill. ABSTRACT: RAE-B 55:200(696), Nov., 1967.

3441. **Fales, J. H., Bodenstein, O. F.** and **Bowers, W. S. 1970** Seven Juvenile Hormone Analogues as Synergists for Pyrethrins Against House Flies. J. Econ. Ent. 63(4):1379-1380.
Three of the seven proved to be good synergists, since they increased

both knock-down and kill. But no combination was so effective as piperonyl butoxide.

3442. **Fales, J. H., Bodenstein, O. F.** and **Burall, R. K. 1960** Effectiveness of Pyrethrum Sprays against Mosquitoes and House Flies in Darkness and Light. Mosquito News 20(2):1 page only.
Darkness reduces the effectiveness of aerosols against house flies, but not against mosquitoes.

3443. **Fales, J. H., Bodenstein, O. F.** and **Fulton, R. A. 1964** DDVP in Aerosol Insecticides.
Is dichlorvos. To control resistant flies, a higher concentration would have to be used than is currently permitted in a space spray aerosol.

3444. **Fales, J. H., Bodenstein, O. F.** and **Keller, J. C. 1962** Face Fly Investigations in Maryland in 1961. Soap and Chem. Spec. 38(2):85, 87, 89, 91, 109, 197.
Tested 9 oil-based spray formulations of pyrethrins, barthrin or dimethrin with piperonyl butoxides and repellents on cattle from low pressure aerosols. Procedure was generally unsatisfactory in reducing fairly dense populations of face fiies. ABSTRACT: RAE-B 51:106, June, 1963.

3445. **Fales, J. H., Bodenstein, O. F., Mills, G. D., Jr.** and **Fields, E. S. 1968** Evaluation of Aerosols, Sprays, and other Formulations of SBP-1382. Proc. 55th Ann. Meeting of Chem. Spec. Manuf. Assoc. Inc., pp. 152-163.
Manifests a high degree of activity as an insecticide. Can be used in aerosols, space sprays, and probably for residues, in baits and in larvicides. Not registered for use in U.S. as of this date.

3446. **Fales, J. H., Bodenstein, O. F., Mills, G. D., Jr.** and **Fields, E. S. 1968** Tropital-New Insecticide Synergist. Aerosol Age 13(11):57, 59, 62, 64; (12):57, 59, 62, 65.
Also known as G-1572. Is recommended as an alternative to piperonyl butoxide and sulfoxide. Has a pleasant odor and produces a minimum of irritation.

3447. **Fales, J. H., Bodenstein, O. F., Mills, G. D., Jr.** and **Wessel, L. H. 1964** Preliminary Studies on Face Fly Dispersion. Ann. Ent. Soc. Am. 57:135-137.
Flies marked in various ways were released among 4 herds of cattle. Some went directly to the eyes of cows, but others scattered. Marked flies were recaptured up to three-quarters of a mile from points of release. ABSTRACTS: RAE-B 52:193, Oct., 1964; BA 45:4313(53185), June 15, 1964.

3448. **Fales, J. H., Fulton, R. A.** and **Bodenstein, O. F. 1960** Aerosol Insecticides for Aircraft. Soap and Chem. Spec. 36(4):137, 140, 185-187.
Two, medium pressure (55 p.s.i.) aerosols, G-1029 and G-1152, devised to replace older, high pressure (70 p.s.i.) formulas.

3449. **Fales, J. H., Keller, J. C.** and **Bodenstein, O. F. 1961** Experiments on Control of the Face Fly. J. Econ. Ent. 54(6):1147-1151.
Effectiveness of various insecticides tested by painting and spraying poison baits on the heads and backs of cattle. ABSTRACT: RAE-B 50: 186, Sept., 1962.

Fales, J. H., joint author. *See* Piquette, P. G., et al., 1966; Bodenstein, O. F., et al., 1970.

3450. **Falter, J.** and **Spencer, M. 1967** Fly Control for Dairy Farms. N.C. Agr. Ext. Folder 241 (not paged).

3451. **Fanatsu, M., Hayashi, K., Inaba, T., Namihira, G.** and **Kariya, M. 1968** The Nature of Latent Enzyme Phenol Oxidase in the Prepupae of House Fly. Acta Vitaminol. et Enzymol. 22(1/2):30.

3452. **Farish, D. J.** and **Axtell, R. C. 1966** Sensory Functions of the Palps and First Tarsi of *Macrocheles muscaedomesticae* (Acarina: Macrochelidae) a Predator of the House Fly. Ann. Ent. Soc. Am. 59:165-170.
Olfactory receptors are located on Tarsi I; contact receptors on palpi. ABSTRACTS: BA 47:4252(49910), May 15, 1966; RAE-B 54:106, June, 1966.

3453. **Farnham, A. W., Gregory, G. E.** and **Sawicki, R. M. 1966** Bioassay and Histological Studies of the Poisoning and Recovery of House Flies (*Musca domestica* L.), Treated with Diazinon and Diazoxon. Bull. Ent. Res. 57(1):107-118.
A close correlation exists between inhibition of cholinesterase activity in the thoracic ganglia and external symptoms of poisoning. ABSTRACTS: BA 48:3277(36145), Apr. 1, 1967; TDB 64:1039, Sept., 1967.

3454. **Farnham, A. W., Lord, K. A.** and **Sawicki, R. M. 1965** Study of Some of the Mechanisms Connected with Resistance to Diazinon and Diazoxon in a Diazinon-resistant Strain of Houseflies. J. Insect Physiol. 11:1475-1488.
Both compounds more toxic by injection than by topical application. Difference in susceptibility, thus demonstrated, was greater with resistant than with susceptible flies. Resistance of both strains increased with age. ABSTRACTS: BA 47:3380(39886), Apr. 15, 1966; TDB 63:487, Apr., 1966; RAE-B 54:218, Nov., 1966.

Farnham, A. W., joint author. *See* Elliott, M., et al., 1965, 1967; Sawicki, R. M., et al., 1967, 1968, 1969.

Fatland, C. L., joint author. *See* Adams, T. S., et al., 1970.

3455. **Fay, R. W., Kilpatrick, J. W.** and **Baker, J. T. 1963** Rearing and Isotopic Labelling of *Fannia canicularis*. J. Econ. Ent. 56(1):69-71.
Culture methods. Used milk containing 2·5 millicuries of P^{32} per liter as the only food for 24 hours, to give adult flies with net counts of 675 to 1,375 counts per minute at 8 to 10 days. Was satisfactory for dispersal studies. ABSTRACTS: BA 42:1907(24065), June 15, 1963; RAE-B 51:92, May, 1963.

3456. **Fedorov, V. G. 1962** O roli komnatnoi mukhi (*Musca domestica* L.) v rasprastranenii zimoi yaits gel'mintov i tsist kishechynky. [On the role of houseflies (*Musca domestica* L.) in the distribution of helminth eggs and intestinal protozoan cysts during the winter.] Med. Parazit. (Moskva) 31:618-620.
The numbers of helminth ova found in and on flies suggests greater attention must be given to the elimination of winter flies in private homes and public buildings. ABSTRACT: BA 42:1487(18771), June 1, 1963.

3457. **Feigin, J. M. 1963** Exposure of the House Fly to Selection by *Bacillus thuringiensis*. Ann. Ent. Soc. Am. 56:878-879.
There was no noticeable decrease in mortality over 27 generations, but about 34 per cent of the surviving adults emerged with wings that failed to expand. These flies had a reduced longevity. ABSTRACTS: BA 45:2477 (30754), Apr. 1, 1964; RAE-B 52:145, Aug., 1964; TDB 61:448, Apr., 1964.

3458. **Feinberg, E. H. 1963** The Evolutionary Adjustment of Two Dipteran Species, *Phaenicia* (*Lucilia*) *sericata* (Calliphoridae) and *Musca domestica* (Muscidae), Competing for the Same Niche in a Laboratory Population Cage. Diss. Absts. 23(5):1493-1949.

3459. Feinberg, E. H. and **Pimentel, D. 1966** Evolution of Increased "Female Sex Ratio" in the Blow Fly (*Phaenicia sericata*) Under Laboratory Competition with the Housefly (*Musca domestica*). Am. Nat. 100 (912): 235-244.
Musca went to extinction between the 64th and 65th weekly census. Preponderance of *Phaenicia* female adults observed after one-half year. This continued after cessation of competition. ABSTRACT: BA 48:3738 (41578), Apr. 15, 1967.

Feinberg, E. H., joint author. *See* Pimentel, D., et al., 1965.

3460. Fellig. J., Barnes, J. R., Rachlin, A. I., O'Brien, J. P. and **Focella, A. 1970** Substituted Phenyl 2-Propynyl Esters as Carbamate Synergists. J. Agric. Fd. Chem. 18:78-80.
Substitution with more than one nitro group or more than three halogens decreased activity, as did the introduction of an alkyl, alkoxy, or aryl group, or the formation of a thio-ether. ABSTRACT: BA51:6220(63634), June 1, 1970.

Fellig, J., joint author. *See* Barnes, J. R., et al., 1969.

3461. Felton, J. C. 1968 Insecticidal Activity of Some Oxime Carbamates (from Symposium on Pesticidal Carbamates). J. Sci. Fd. Agric. Suppl. 19:32-37.
Within the aldoxine carbamates, compounds of the oxy- series are particularly active against *M. domestica.* Study based on over-all insecticidal activity and ability to inhibit head acetylcholinesterase of houseflies. ABSTRACT: RAE-B 58:35(144), Feb., 1970.

3462. Feng, Y. S. and **Tsao, C. C. 1960** [A Preliminary Study of the Species Composition and Seasonal Prevalence of the Fly Population in Tsingtao.] Acta Ent. Sinica 10(1):79-85. (In Chinese, with English summary.)
Baited traps indicated flies could be collected from March to December with the peak in September. *Lucilia* spp. were the most abundant; *Musca* spp. made up only 1·35 per cent. ABSTRACT: BA 35:4003(46102), Aug. 15, 1960.

3463. Fermi, G. and **Reichardt, W. 1963** Optomotorische Reaktionen der Fliege *Musca domestica.* Abhängigkeit der Reaktion von der Wellenlänge, der Geschwindigkeit, dem Kontrast und der mittleren Leuchtdichte bewegter periodischer Muster. [Optomotor Reactions of the Fly *Musca domestica.* Dependence of the Reaction on Wave Length, Velocity, Contrast and Median Brightness of Periodically Moved Stimulus Patterns.] Kybernetik 2:15-28. (In German, with English summary.)
The torque exerted by the housefly during fixed flight was used as a measure of the optomotor reaction of the insect, elicited by the rotation of cylindrical patterns with periodic distributions of surface brightness.

3464. Fernandez, A. T. 1967 Effects of Various Photoperiods on the Life Cycle and Susceptibility of the House Fly, *Musca domestica* to Insecticide Residues. Diss. Absts. 28(2B):726-B.

3465. Fernandez, A. T. and **Randolph, N. M. 1966** The Susceptibility of House Flies Reared under Various Photoperiods to Insecticide Residues. J. Econ. Ent. 59:37-39.
Seven photoperiods utilized. Flies reared under Light-Dark 14:10 photoperiod were most consistent in susceptibility to film residues of 3 insecticides tested. Those reared under LD 15:9 were least susceptible to DDT and dieldrin. (Third insecticide was endrin.) ABSTRACTS: BA 47:6401 (74894), Aug. 1, 1966; RAE-B 54:97, May, 1966; TDB 63:906, Aug., 1966.

3466. **Fernandez, A. T.** and **Randolph, N. M. 1967** A Photoperiodic Effect on the Daily Susceptibility of the House Fly to Trichlorfon. J. Econ. Ent. 60(6):1633-1636.

Experiments carried on at 80 ± 2°F and 50-70 per cent R.H. Flies displayed maximum sensitivity at or near "Lights-on"; minimum sensitivity at "Lights-off". (Equiv. to *dawn* and *dusk*.) ABSTRACTS: BA 49:3811 (42812), Apr. 15, 1968; RAE-B 56:101(349), May, 1968; TDB 65:1304 (2951), Oct., 1968.

Fernandez, H. R., joint author. *See* Goldsmith, T. H., et al., 1966.

3467. **Ferrer, F. R., Grosch, D. S.** and **Guthrie, F. E. 1968** Effect of Chemical Protectants on the Action of Apholate to the Housefly. J. Econ. Ent. 61(3):719-724.

Five protectants reduced the effect of apholate. Is consistent with the hypothesis of free radical reaction at the nucleic acid level. Thiourea became *less* effective in protecting fecundity, as treatment age increased from 1 to 3 days. ABSTRACTS: BA 49:10247(113107), Nov. 1, 1968; RAE-B 56:219(799), Nov., 1968; TDB 66:756(1506), July, 1969.

Ferretti, C. T., joint author. *See* Braitenberg, V., et al., 1966.

Fields, E. S., joint author. *See* Fales, J. H., et al., 1968.

3468. **Filipponi, A. 1964** Experimental Taxonomy Applied to the Macrochelidae (Acari:Mesostigmata). Acarologia, fasc. h.s. 1964 [C.R. Ier Congrès Int. d'Acarologie, Fort Collins, Col., U.S.A. 1963.]

Emphasis on methods of breeding and crossing. Group includes *Macrocheles muscaedomesticae*, which attaches to adult flies, and attacks housefly eggs and larvae, as do certain other species.

3469. **Filipponi, A. 1965** Facultative Viviparity in Macrochelidae (Acari: Mesostigmata). Proc. XII Int. Congr. Ent. London, 1964 (publ. 1965), pp. 309-310.

All of 12 species studied proved capable of depositing either eggs or larvae according to ecological conditions.

3470. **Filipponi, A.** and **Dojmi di Delupis, G. 1963** Sul regime dietetico di alcuni macrochelidi (Acari:Mesostigmata), associati in natura a muscidi di interesse sanitario. [On the food habits of some Macrochelids (Acari:Mesostigmata) associated in the field with synanthropic flies.] Riv. di Parassit. 24(4):277-288. (English summary.)

Six species studied all proved to be active predators. Highest rate of increase resulted from feeding either on house fly eggs or on nematodes. Food preferred is not always the same as that which ensures greatest reproductive rate. ABSTRACT: RAE-B 53:175, Sept., 1965.

3471. **Filipponi, A.** and **Dojmi di Delupis, G. 1964** Sulla biologia e capacitá riproduttiva di *Macrcheles peniculatus* Berlese (Acari:Mesostigmata) in condizioni sperimentali di laboratorio. [On the biology and reproductive capacity of *Macrocheles peniculatus* Berlese (Acari:Mesostigmata) in the Laboratory.] Riv. di Parassit. 25(2): 93:111.

Studied more than 50 strains, all reproducing by obligate thelytoky. Reproductive capacity indicates that *M. peniculatus* is able to out number its probable prey, *M. domestica*.

3472. **Filipponi, A.** and **Francaviglia, G. 1964** Larviparità facoltativa in alcuni macrochelidi (Acari:Mesostigmata) associati a muscidae di interesse sanitario. [Facultative larviparity in certain macrochelids associated with Muscidae of sanitary interest.] Parassitologia 6(1-2):99-113.

Plentiful food favors oviparity, scanty food, larviparity, in all 5 species studied. Other factors (thelytokous or arrhenotokous; and if the latter, whether eggs are fertilized) influence different species in different ways. Ecological requirements of each species more or less unique.

3473. **Filipponi, A.** and **Mosna, B. 1968** Influenza di fattori ecologici e genetici sulla natalità e mortalità di *Macrocheles robustulus* (Berlese 1904). [The influence of ecological and genetic factors on birth-rate and death-rate of *Macrocheles robustulus* (Berlese 1904).] Ann. Ist. Sup. Sanità 4:551-571.

Temperature range for reproduction, 13-33°C (optimum for fecundity, 20-28°C; for highest reproductive rate, 26-28°C). Other effects of temperature cited. Strain differences, when temperature is constant, illustrate importance of genetic factors.

3474. **Filipponi, A.** and **Passariello, S. 1969** Influenza della temperatura sulla fecondità, longevità, e capacità moltiplicativa di *Macrocheles peniculatus* Berlese (Acari, Mesostigmata). [The influence of temperature on fecundity, longevity and capacity for reproduction in *Macrocheles peniculatus* Berlese (Acari, Mesostigmata).] Riv. di Parassit. 30(4):295-310.

A comparison with *M. domestica* on which *Macrocheles* is predaceous. Each of the statistics related to the speed of multiplication has its own temperature range. The influence of each component on the *rm* and *Rm* curves is characteristic for each species.

3475. **Filipponi, A.** and **Petrelli, M. 1966** L'innata capacità di incremento numerico di *Musca domestica* L. [The innate capacity for increase in numbers of *Musca domestica* L.] Riv. di Parassit. 27(4):235-260. (Detailed summary in English.)

The corresponding curves of the two strains studied were not parallel for all parameters, which showed the effect of genetic components that reacted differently at different temperatures. Optimum temperature not the same for different components and for capacity for increase. Abstracts: BA 49:3360(37572), Apr. 1, 1968; RAE-B 57:24(104), Jan., 1969; TDB 64:909, Aug., 1967.

3476. **Filipponi, A.** and **Petrelli, M. G. 1967** Autoecologia e capacità moltiplicata di *Macrcheles muscaedomesticae* (Scopoli) (Acari:Mesostigmata). [Autoecology and capacity for increase in numbers of *Macrocheles muscaedomesticae* (Scopoli) (Acari:Mesostigmata.] Riv. di Parassit. 28(2):129-156.

Multiplicative capacity of *Macrocheles muscaedomesticae* shown to be greater than that of *M. peniculatus*, which, in turn is greater than that of *Musca domestica*. Abstract: RAE-B 57:219(796), Nov., 1969.

3477. **Filipponi, A.** and **Petrelli, M. G. 1969** Sulla capacità moltiplicativa di *Macrocheles perglaber* (Acarina, Mesostigmata) in condizioni di laboratorio. [On the multiplicative capacity of *Macrocheles perglaber* —under laboratory conditions.] Parassitologia 11(1-2):Estratto, 4 pp.

Presents comparisons with 2 other species of *Macrocheles* and with *Musca domestica*.

Filonova, I. A., joint author. *See* Vashkov, V. I., et al., 1968.

3478. **Fine, B. C. 1961** Pattern of Pyrethrin-resistance in Houseflies. Nature (London) 191:884-885.

One strain subjected to selection with pyrethrum, the other with DDT. Both became more resistant to pyrethrum. Is there a single mechanism of resistance which acts with equal facility against either insecticide? (Both strains proved also more or less resistant to dieldrin and malathion.) Abstracts: BA 37:1584(15817), Feb. 15, 1962; TDB 59:212, Feb., 1962; RAE-B 50:253, Nov., 1962.

3479. **Fine, B. C. 1963** The Present Status of Resistance to Pyrethroid Insecticides. Pyrethrum Post 7(2):18-21, 27.
A progress report. Three strains of *M. domestica* tested, along with several other Arthropods. At least 6 Arthropod species have the potential to develop pyrethroid resistance.

3480. **Fine, B. C., Godin, P. J.** and **Thain, E. M. 1963** Penetration of Pyrethrin I Labelled with Carbon-14 into Susceptible and Pyrethroid Resistant Houseflies. [Correspondence.] Nature 199:927-928.
Penetration into the resistance strain was significantly slower than into the other strains. A definite correlation exists between rates of absorption of ^{14}C-pyrethrin I and pyrethroid resistance, in the strains studied. ABSTRACTS: BA 45:3222(39901), May 1, 1964; RAE-B 52:119, July, 1964; TDB 61:107, Jan., 1964.

3481. **Fine, B. C., Godin, P. J., Thain, E. M.** and **Marks, T. B. 1967** Resistance to Pyrethrins and DDT in a Strain of Houseflies *Musca domestica* L. I. The Sorption of a Synthetic ^{14}C Pyrethrin I. J. Sci. Fd. Agric. 18(5):220:224.
The resistance of the flies to pyrethrin was due in part to reduced penetration when applied topically in acetone. ABSTRACT: RAE-B 56:111(388), June, 1968.

Fine, B. C., joint author. *See* Agosin, M., et al., 1966; Gill, L., et al., 1968; Srivastava, S. C., et al., 1967.

3482. **Firtel, R. A. 1967** Characteristics of Phenol Oxidases in *Mormoniella vitripennis* (Walker). J. Insect Physiol. 13(8):1197-1206.
The activation of phenol oxidase activity in *M. vitripennis* (Walker) is similar to that in *Drosophila* and *Musca*. There is an initial lag period followed by an autocatalytical increase. ABSTRACT: BA 48:11244(124891), Dec. 15, 1967.

Fisher, E. H., joint author. *See* Berge, O. I., et al., 1964.

3483. **Flanders, S. E.** and **Bay, E. C. 1964** Standardization of Mass Rearing Procedures for Entomophaga. Bull. W. H. O. 31(4):505-507.
Covers nutritional factors, standardization of media, maintenance of proper sex ratios, control of mating, etc., Houseflies listed among ideal hosts for laboratory work. ABSTRACT: RAE-B 55:26(92), Feb., 1967.

3484. **Fleming, M. M. 1967** Epizootiology of Muscido Trypanosomids. Diss. Absts. 27(98):3335B.
Musca domestica linked with cattle host.

3485. **Flitters, N. E. 1968** Insect Body Temperatures Determined with Microthermoelectric Thermometry. Ann. Ent. Soc. Am. 61(1):36-38.
M. domestica among 5 species studied. Peak body temperature for houseflies, 112°F. ABSTRACT: BA 49:4802(53993), May 15, 1968.

3486. **Floch, H., Kramer, R.** and **Chatenay, G. 1966** Le malathion est doué de propriétés insecticides rémanentes utilisables dans la lutte contre divers moustiques. [Malathion has residual insecticidal properties usable in the control of various mosquitoes.] Bull. Soc. Path. Exot. 59(5):874-880.
Includes report on *M. domestica*. Malathion on wood kills 95 per cent of exposed houseflies after 18 weeks. ABSTRACT: BA 50:5865(61443), 1969.

3487. **Flowers, A. I. 1964** Interspecies Transmission of *Salmonella*. Proc. Salmonella Seminar. Agric. Res. Serv. USDA, Hyattsville, Md.
Not known whether the occurrence of *Salmonella* in flies is merely a reflection of their environment or not.

Flury, V. P., joint author. *See* Jensen, J. A., et al., 1965.

3488. **Flynn, A. D.** and **Schoof, H. F. 1966** Effect of Surface on Residual Activity of Selected Compounds. J. Econ. Ent. 59:678-681.
Twenty-two compounds were tested against *M. domestica* and *Blatella germanica* on painted and unpainted galvanized metal, tempered masonite, and asphalt tile. ABSTRACTS: BA 47:8522(99277), Oct. 15, 1966; RAE-B 54:179, Sept., 1966; TDB 63:1253, Nov., 1966.

Flynn, A. D., joint author. *See* Mathis, W., et al., 1967.

Foa, S., joint author. *See* Laudani, U., et al., 1966, 1969.

Focella, A., joint author. *See* Fellig, J., et al., 1970.

3489. **Fomicheva, V. S., Geshvind, G. N., Drozdova, O. I., Krupina, E. P., Aganina, V. M., Urakova, A. K., Orlova, V. P.** and **Dobrolyubova, V. P. 1967** Kubovye ostatki-piropolimery (otkhod proizvodstva zadova "Neftegaz") kak sredstvo bor'by s priemaginal' nymi fazami sinantropnykh mukh. [Still tails polymers (wastes of oil-gas plant production) as a means of control of preimaginal stages of synanthropic flies.] Med. Parazit. i Parazitarn. (Bolez.) 36(5):608-611 (English summary.)
The use of still tail wastes in the form of 30 per cent aqueous emulsion weekly or monthly in cesspools resulted in very economical control of fly larvae. ABSTRACT: BA 50:2099(22018), Feb. 15, 1969.

3490. **Forgash, A. J. 1964** Laboratory Studies with WARF Anti-resistant on House Flies. J. Econ. Ent. 57:644-645.
DDT-resistant flies at first showed considerable susceptibility to DDT + WARF; became resistant after three generations of selection. ABSTRACTS: BA 46:698(8747), Jan. 15, 1965; RAE-B 53:10, Jan., 1965.

3491. **Forgash, A. J. 1967** Joint Action of SKF525-A and Sesamex with Insecticides in Susceptible and Resistant House Flies. J. Econ. Ent. 60(6):1596-1600.
This combination enhanced the action of many insecticidal compounds against both susceptible and polyresistant flies. ABSTRACTS: BA 49:3811 (42815), Apr. 15, 1968; RAE-B 56:99(345), May, 1968; TDB 65:1305 (2952), Oct., 1968.

3492. **Forgash, A. J. 1967** WARF Anti-Resistant: DDT Selection and House Fly Cross-Resistance. J. Econ. Ent. 60(6):1750-1751.
Repeated exposure to DDT + WARF resulted in resistance to the combination, and an intensification of resistance to DDT. ABSTRACTS: BA 49:3811(42816), Apr. 15, 1968; RAE-B 56:103(358), May, 1968; TDB 65:1305(2953), Oct., 1968.

3493. **Forgash, A. J., Cook, B. J.** and **Riley, R. C. 1962** Mechanisms of Resistance in Diazinon-Selected Multi-Resistant *Musca domestica.* J. Econ. Ent. 55:544-551.
Reduction in permeability of the cuticle an important factor in resistance to Diazinon. (Not the only factor, however.) ABSTRACTS: BA 40:1634 (21318), Dec. 1, 1962; RAE-B 51:5, Jan., 1963; TDB 60:182, Feb., 1963.

3494. **Forgash, A. J.** and **Hansens, E. J. 1960** Further Studies on the Toxicity of Insecticides to Diazinon-Resistant *Musca domestica.* J. Econ. Ent. 53(5):741-745.
Resistant flies showed increased tolerance toward a very large number of insecticides. So much cross-resistant requires a new approach in research ABSTRACTS: BA 36:757(8997), Feb. 1, 1961; RAE-B 49:178, Aug., 1961.

3495. **Forgash, A. J.** and **Hansens, E. J. 1962** Effect of Selection on Cross Resistance in Diazinon-Resistant *Musca domestica.* J. Econ. Ent.

55(5):679-682.
Gives rate of increase in resistance under selection and rate of regression when selection is terminated. ABSTRACTS: BA 41:949(12349), Feb. 1, 1963; RAE-B 51:39, Feb., 1963.

3496. Forgash, A. J. and **Hansens, E. J. 1967** Resistance Levels in Diazinon-Pressured and Non-pressured Polyresistant House Flies. J. Econ. Ent. 60:1241-1247.
In many instances diazinon-treated wild populations reached a resistance peak which declined after several generations, then stabilized at a lower level. ABSTRACTS: BA 49:1907(21454), Feb. 15, 1968; RAE-B 56:61(204), Mar., 1968; TDB 64:540(1177), Apr., 1968.

Forgash, A. J., joint author. *See* Granett, P., et al., 1962; Reed, W. T., et al., 1968, 1969, 1970; Collins, W. J., et al., 1968, 1970.

Foster, F. J., Jr., joint author. *See* Post, F. J., et al., 1965.

3497. Foster, W. 1967 Hormone-mediated Nutritional Control of Sexual Behavior in Male Dung Flies. Science 158:1596-1597.
Male *Scatophaga stercoraria* must prey on other Diptera before displaying sexual behavior, developing accessory cells of the ejaculatory ducts and achieving full length of the testes (*Musca* used as food in this research).

Fournel, J., joint author. *See* Desmoras, J., et al., 1964.

Fowler, K. S., joint author. *See* Lewis, S. E., et al., 1960.

Frada, I., joint author. *See* Comes, R., et al., 1962.

3498. Fraenkel, G. 1962 Tanning of the Adult Fly: a New Hormone Action. *In* 59th Ann. Meeting of the Amer. Soc. Zool. Philadelphia. (Abstract only.)
ABSTRACT: BA 46:8914(109952), Dec. 15, 1965.

3499. Fraenkel, G. and **Hsiao, C. 1962** Hormonal and Nervous Control of Tanning in the Fly. Science 138(3536):27-29.
Darkening of the adult fly involves a neurosecretion and the action of a hormone other than ecdyson. (Studied *Phormia*, *Calliphora* and *Sarcophaga*.)

3500. Fraenkel, G. and **Hsiao, C. 1963** Tanning in the Adult Fly: a New Function of Neurosecretion in the Brain. Science 141(3585):1057-1058.
Concerns a hormone found in the *pars intercerebralis* of the brain and in the compound ganglion of the thorax, where the concentration is 6 times as great. This hormone appears to be different from both the prothoracotropic and gonadotropic hormones. ABSTRACT: BA 45:3980 (48949), June 1, 1964.

3501. Fraenkel, G. and **Hsiao, C. 1965** Bursicon, a Hormone Which Mediates Tanning of the Cuticle in the Adult Fly and Other Insects. J. Insect Physiol. 11(5):513-556.
This hormone, produced in the *pars intercerebralis* of the brain, occurs in the blood of newly emerged flies. It acts directly on the cuticle. ABSTRACT: BA 46:5896(73251), Aug. 15, 1965.

3502. Fraenkel, G. and **Hsiao, C. 1967** Calcification, Tanning, and the Role of Ecdyson in the Formation of the Puparium of the Facefly, *Musca autumnalis*. J. Insect Physiol. 13(9):1387-1394.
Puparium of the face fly is brittle, opaque, and white. Calcification has taken the place of tanning. Like tanning in other flies, this is controlled by ecdyson. Amount of $CaCO_3$ in puparium is 65 to 80 per cent. ABSTRACT: BA 49:1421(15927), Feb. 1, 1968.

3503. **Fraenkel, G., Hsiao, C.** and **Seligmann, M. 1966** Properties of Bursicon: An Insect Protein Hormone that Controls Cuticular Tanning. Science 151(3706):91-93.
Molecular weight of bursicon is about 40,000. Its properties, in the Diptera and in the Orthoptera are similar but not identical.

Fraenkel, G., joint author. *See* Galun, R., et al., 1961; Berreur, P., et al., 1969.

3504. **Franco, M. G. 1961** Osservazioni preliminari sulla ereditarietà della resistenza al DDT di un ceppo di mosche selezionate con diazinone. [Preliminary observations on the inheritance of resistance to DDT in a strain of flies selected with diazinon.] Genetica Agraria 14: 252-266.
Multiresistant strain IPR, obtained by selection with diazinon, varied greatly as to knockdown when exposed by tarsal contact with DDT. Selection for early or delayed knockdown gets a marked response by the 3rd generation.

3505. **Franco, M. G. 1962** Il V° gruppo di concatenazione in *Musca domestica* L. [The fifth linkage group in *Musca domestica* L.] Boll. Zoologia 29:821-830.
Linkage between the housefly mutant character *aristapedia* (*ar*) and *carmine* (*cm*) has been confirmed. Rate of cross-over is 17·18 per cent.

3506. **Franco, M. G.** and **Lanna, T. M. 1962** Misura della concatenazione tra i mutanti *ocra* e *curly* di *Musca domestica* L. [Measurement of the linkage between the mutants *ocra* and *curly* in *Musca domestica* L.] Symposia Genetica et Biologica Italica 13:341-358.
Factors affecting sex-ratio and inheritance of the *bwb* linkage group do not interfere with inheritance of *curly* and *ocra*; they may, however, cause differences in fertility, homogeneity between cultures segregating for these mutants, and penetrance of *curly*.

3507. **Franco, M. G., Lanna, T. M.** and **Milani, R. 1961** Eredita legata al sesso nella *Musca domestica* L. [Sex-linked inheritance in *Musca domestica* L.] *In*: Atti della VII Riunione scientifica della Associazione Genetica Italiana, Parma, 9-10 Ottobre. Atti Ass. Genet. Ital. 7:198-212. (English summary.)
Sex-linked criss-cross inheritance has never been observed in the housefly. Holandric inheritance has. Present authors noted holandric inheritance of normal alleles only in strains from Orlando. Is introduced by either sex. ABSTRACT: BA 47:844(10300), Feb. 1, 1966.

3508. **Franco, M. G.** and **Oppenoorth, F. J. 1962** Genetical Experiments on the Gene for Low Aliesterase Activity and Organophosphate Resistance in *Musca domestica* L. Ent. Exp. Appl. 5(2):119-123. (German summary.)
Gene *a* was located on chromosome 5, after crosses of a dizon-resistant strain with several strains carrying recessive marker genes. Resistance was found to segregate in opposition to *ar* amd *cm*, indicating that *a*, *ar*, and *cm* are located on the same (5th) chromosome. ABSTRACTS: BA 40: 990(13177), Nov. 15, 1962; RAE-B 51:119, June, 1963; TDB 59:1120, Nov., 1962.

3509. **Franco, M. G.** and **Rubini, P. G. 1966** Genetics of *Musca domestica* L. Mutants, Inheritance and Cytology. Symposia Genetica et Biologica Italica 13:393-453.
A richly informative paper, superbly illustrated. Treats of distribution and variability of the species; mutants (their phenotypic effects, fre-

quency, usefulness as markers); linkage, holandric inheritance, sex abnormalities; evidence for chromosomal polymorphism.

Franco, M. G., joint author. *See* Grigolo, A., et al., 1965; Lanna, T. M., et al., 1961; Milani, R., et al., 1959, 1960, 1961, 1965, 1967; Rubini, P. G., et al., 1963, 1965, 1966, 1968; Sawicki, R. M., et al., 1966.

Frank, D. E., joint author. *See* MacLaren, W. R., et al., 1960.

3510. Frantisek, G. 1961 Klic k určovani synantropnich dvoukridlych pro praktickow potrebu zdravotniků. [Tables for the identification of Central European synanthropic Diptera for medical workers.] Zool. Listy 24(3):193-202. (German summary.)

Includes a key for identification of most important central European species, genera, and families. Designed for the practical medical worker. ABSTRACT: BA 37:293(3885), Jan. 1, 1962.

3511. Fraser, A. 1960 Humoral Control of Metamorphosis and Diapause in the Larvae of Certain Calliphoridae (Diptera:Cyclorrhapha). Proc. Roy Soc. Edinb. (B) 67 (pt. 2, No. 8):127-140.

Thoracic gland cells produce a hormone which promotes several processes relating to pupation. Thoracic gland activity is stimulated by the brain, through nerves to the corpus cardiacum, which liberates a humoral material into the blood. ABSTRACT: RAE-B 51:151, July, 1963.

Fratantoni, J., joint author. *See* Hennesy, D. J., et al., 1961.

3512. Frear, D. E. H. and **Dills, L. E. 1967** Mechanism of the Insecticidal Action of Mercury and Mercury Salts. J. Econ. Ent. 60(4):970-974.

The ovicidal action of mercury salts applied to soil is due to metallic mercury in the gaseous state. Soil organic matter, moisture, soil pH and temperature affect the reduction of the salts to metallic mercury.

Frear, D. E. H., joint author. *See* Cheng, T.-H., et al., 1961, 1962.

Frederickson, M., joint author. *See* Metcalf, R. L., et al., 1964, 1965; Fukuto, T. R., et al., 1964.

3513. French, A. and **Hoopingarner, R. 1965** Gametogenesis in the House Fly, *Musca domestica.* Ann. Ent. Soc. Am. 58:650-657.

Spermatogonial divisions occur in third stage larvae and newly-formed pupae; oogonial divisions during early pupation. Sperm maturation essentially complete by emergence. First egg chambers formed before emergence. ABSTRACTS: BA 47:835(10189), Feb. 1, 1966; TDB 63:339, Mar., 1966.

Friedman, S., joint author. *See* Benson, R. L., et al., 1970.

3514. Frishman, A. M. and **Matthysse, J. G. 1966** Olfactory Responses of the Face Fly *Musca autumnalis* De Geer and the Housefly *Musca domestica* Linn. Mem. Cornell Univ. Agric. Exp. Stn. 394. 89 pp.

Tested 425 materials for value as attractants or repellents in the laboratory with face flies; 370 with house flies. Field tests (with face flies only) involved 200 test materials. In general, field and laboratory results were similar, except that highly volatile substances were not effective in the field. ABSTRACT: RAE-B 57:154(557), Aug., 1969.

3515. Fromme, H. G. 1968 The Description of Animal and Plant Structures in the Surface Screen Electron Microscope. Natur. Mus. 98(12): 482-490.

Includes observations on the *Culex* mosquito and on *M. domestica.*

Frontali, N., joint author. *See* Vicari, G., et al., 1965.

3516. Frudden, L. and **Wellso, S. G. 1968** Daily Susceptibility of House Flies to Malathion. J. Econ. Ent. 61(6):1692-1694.

Flies did *not* exhibit a daily susceptibility rhythm to malathion, but those reared under a 14:10 light:dark (LD) cycle, or under continuous light, were 21 per cent less suceptible to malathion than flies reared under LD 10:14. ABSTRACTS: BA 50:4777(50215), May 1, 1969; RAE-B 57:89 (309); TDB 66:634(1254), June, 1969.

Fuertes-Polo, C., joint author. *See* Metcalf, R. L., et al., 1963.

Fuhremann, T. W., joint author. *See* Lichtenstein, E. P., et al., 1969.

Fujimoto, N., joint author. *See* Buei, K., et al., 1966.

3517. Fujit, S. and **Ashiba, M. 1963** The Influence of Soil Fumigant Against the Pupae of *Musca domestica vicina* in the Soil. Jap. J. Sanit. Zool. 14(2):92-94.
Concerns tests conducted in 1960 with soil fumigants D-D and DBCP. At appropriate concentrations emergence from puparia was wholly prevented with D-D, 90 per cent prevented with DBCP.

3518. Fujito, S., Buei, K., Saito, T., Niihara, T., Tsujimoto, S. and **Nakatani, S. 1963** Field Test of Fly and Mosquito Control Using Sumithion, 0,0-dimethyl-0-(3-methyl-4-nitrophenyl) Phosphorothioate. Jap. J. Sanit. Zool. 14(1):53-57.
Spraying and dusting in a rural district markedly decreased the fly population, particularly in barns.

3519. Fujito, S., Buei, K., Saito, T. and **Taniguchi, M. 1966** A Field Experiment on Fly Control With Residual Treatments of Baytex and Dimethoate in a Poultry Ranch. Jap. J. Sanit. Zool. 17(3):218-222.
Flesh fly populations on the ranch and in neighbouring houses were markedly reduced. Effect lasted one month. *M. domestica* was reduced also, but not to the same degree. ABSTRACT: RAE-B 57:82(289), May, 1969.

3520. Fujito, S., Endo, T., Buei, K., Hayama, S., Fukuhara, Y. and **Ohno, T. 1962** Field Test of Fly Control Using Dibrom, 1,2-dibromo-2,2-dichloroethyl Dimethyl Phosphate. Jap. J. Sanit. Zool. 13(3):224-228.
Spraying or suspending emulsion-soaked cords gradually reduced the fly population in 3 rural districts of Osaka Prefecture. Effect appeared to be limited to 7 days.

3520a. Fukami, J.-I., Shishido, T., Fukunaga, K. and **Casida, J. E. 1969** Oxidative Metabolism of Rotenone in Mammals, Fish, and Insects and Its Relation to Selective Toxicity. Agric. Food. Chem. 17(6): 1217-1226.
The results of *in vivo* and *in vitro* studies on rotenone detoxification indicate that the effects of components in the soluble fraction possibly are related to the selective toxicity of rotenone to mammals, fish, and insects. *M. domestica* among the species cited.

3521. Fukami, J.-I., Yamamoto, I. and **Casida, J. E. 1967** Metabolism of Rotenone *in vitro* by Tissue Homogenates from Mammals and Insects. Science 155(3763):713-716.
Products have been identified for the hydroxylation of rotenone in vitro in the enzyme system composed of microsomes and reduced nicotinamide-odenine dinucleotide phosphate, in living mice and in houseflies. ABSTRACT: BA 48:6266(69877), July 15, 1967.

Fukami, J. I., joint author. *See* Mitsui, T., et al., 1969.

3522. Fukuda, M. 1960 [Survey of fly pupae hibernating near various breeding places in a farm village.] Endemic Dis. Bull. Nagasaki Univ. 2(2):141-153. (In Japanese, with English summary.)
From a survey in 1954 of pupae, no adult *Musca domestica* or *Stomoxys*

calcitrans were obtained in the major breeding sources. Reasons for this are discussed. ABSTRACTS: BA 36:5570(58204), Sept. 1, 1961; RAE-B 51:267, Dec., 1963.

3523. Fukuda, M. 1960 [On the effect of physical condition of setting place upon the number of flies collected by fish-baited trap. (Studies on the ecology and control of flies, 2.)] Endem. Dis. Bull. Nagasaki Univ. 2(3):222-228. (In Japanese, with English summary.)
For a population census of flies, traps must be set in different kinds of places. In general, the numbers of individuals, species, genus and family of flies trapped are most numerous under the shade of a tree, next in the forest, and least in a sunny yard. Seasonal changes in dominancy of 7 dominant fly species are illustrated. ABSTRACT: BA 39:669(8302), July 15, 1962.

3524. Fukuda, M. 1960 [Seasonal prevalence of pupae found about the breeding places of flies in a farm village in 1955. (Studies on the ecology and control of flies, 3.)] Endemic Dis. Bull. Nagasaki Univ. 2(4):281-286. (In Japanese, with English summary.)
The prevalence of various species is reported. Seasonal prevalences of the soil-collected pupae almost agree with those of wild caught adults ABSTRACT: BA 36:6483(70178), Oct. 15, 1961.

3525. Fukuda, M. 1961 [Fly control by screening barns.] End. Dis. bull. Nagasaki Univ. 3(1):68-74.
In a village the control measures resulted in a substantial reduction of the numbers of flies of several species in 1956, including *Musca domestica.* Flies breeding away from barns in other sources remained as prevalent. ABSTRACT: BA 36:6483(70179), Oct. 15, 1961.

Fukuda, M., joint author. *See* Suenaga, O., et al., 1963, 1964.

Fukuhara, Y., joint author. *See* Buei, K., et al., 1961, 1964; Fugito, S., et al., 1962.

Fukunaga, K., joint author. *See* Mitsui, T., et al., 1969; Kazano, H., et al., 1968; Fukami, J.-I., et al., 1969.

3526. Fukushi, Y. 1967 Genetic and Biochemical Studies on Amino Acid Compositions and Color Manifestation in Pupal Sheaths of Insects. Jap. J. Genet. 42(17):11-21.
Content of β-alanine in pupal sheath of *Musca* strains whose adult body colors are pale, is higher than in the darker, wild-type strains. Various experiments established a clear correlation between grade of color and content of β-alanine in pupal sheaths. ABSTRACT: BA 49:508(5670), Jan. 15, 1968.

3527. Fukushi, Y. and **Seki, T. 1965** Differences in Amino Acid Compositions of Pupal Sheaths between Wild and Black Pupa Strains in Some Species of Insects. Jap. J. Genet. 40(3):203-208.
Worked with *Bombyx*, *Drosophila* and *Musca.* In the pupal sheath of a hybrid (bp/+) *M. domestica*, whose pupal color was indistinguishable from that of a wild one, β-alanine was detected in as great an amount as in the pupal sheath of the wild type (+/+). ABSTRACT: BA 47:10361 (119796), Dec. 15, 1966.

3528. Fukuto, T. R., Fahmy, M. A. H. and **Metcalf, R. L. 1967** Alkaline Hydrolysis, Anti-cholinesterase, and Insecticidal Properties of Some Nitro-substituted Phenyl Carbamates. J. Agric. Fd. Chem. 15(2):273-281.
Practically all these compounds, used alone, were poor in housefly toxicity, but several were strongly synergized with piperonyl butoxide. ABSTRACT: RAE-B 55:218(756), Dec., 1967.

3529. **Fukuto, T. R., Metcalf, R. L., Frederickson, M.** and **Winton, M. Y. 1964** Insecticidal Properties of Some Diethyl Nitronaphythyl Phosphates. J. Agric. Fd. Chem. 12(3):288-231.
Compounds were examined for hydrolysis rates, fly-brain cholinesterase inhibition and toxicity to house flies. Diethyl 6-nitro-2 napthyl phosphate showed highest toxicity. ABSTRACT: RAE-B 52:127, July, 1964.

3530. **Fukuto, T. R., Metcalf, R. L.** and **Winton, M. Y. 1961** The Insecticidal Properties of Esters of Phosphoric and Phosphinic Acids. J. Econ. Ent. 54(5):955-962.
Alkyl p-nitrophenyl ethyl phosphates maintained high insecticidal activity over a wide variety of alkyl groups. Phosphinate esters were less toxic to houseflies than analogous phosphonate esters. ABSTRACTS: BA 37: 1119 (11647), Feb. 1, 1962; RAE-B 50:174, Aug., 1962.

3531. **Fukuto, T. R., Metcalf, R. L.** and **Winton, M. Y. 1964** Carbamate Insecticides: Insecticidal Properties of Some Optically Active Substituted Phenyl N-Methyl-Carbamates. J. Econ. Ent. 57:10-12.
Includes a discussion of the mechanics of competitive inhibition of cholinesterase. ABSTRACTS: BA 45:4728(57650), July 1, 1964; RAE-B 52:95, May, 1964; TDB 61:615, June, 1964.

3532. **Fukuto, T. R., Metcalf, R. L., Winton, M. Y.** and **March, R. B. 1963** Structure and Insecticidal Activity of Alkyl 2, 4, 5-Trichlorophenyl N-Alkylphosphoramidates. J. Econ. Ent. 56:808-810.
Compounds examined for anticholinesterase activity and toxicity to insects. Evidence indicated that toxicity may be dependent on *in vivo* formation of an activated intermediate.

3533. **Fukuto, T. R., Metcalf, R. L., Winton, M. Y.** and **Roberts, P. A. 1962** The Synergism of Substituted Phenyl N-methylcarbamates by Piperonyl Butoxide. J. Econ. Ent. 55:341-345.
Forty compounds re-evaluated with P.B. as a synergist. Most carbamates had their toxicity increased to levels which parallel anticholinesterase activity in the fly brain. Variability in toxicity of carbamates alone due probably to different detoxification rates in the insects. ABSTRACT: RAE-B 50:270, Dec., 1962.

Fukuto, T. R., joint author. *See* El-Aziz, S. A., et al., 1969; El-Sebae, A. H., et al., 1964; Bakry, N., et al., 1968; Fahmy, M. A. H., et al., 1966; Hollingworth, R. M., et al., 1967; Jones, R. L., et al., 1969; Metcalf, R. L., et al., 1960, 1961, 1962, 1963, 1964, 1965, 1966, 1967, 1968; Wilkinson, C., et al., 1966; Sacher, R. M., et al., 1968, 1969; Quistad, G. B., et al., 1970.

Fukuyoshi, S., joint author. *See* Kono, I., et al., 1967.

Fuller, R. G., joint author. *See* Brydon, H. W., et al., 1966.

3534. **Fulton, R. A., Yeomans, A. H.** and **Sullivan, W. N. 1963** Ethylene Oxide as a Fumigant Against Insects. J. Econ. Ent. 56:906.
M. domestica among the test insects used. Housefly eggs were more susceptible to the fumigant than adult flies. ABSTRACTS: RAE-B 52:55, Mar., 1964; TDB 62:816, Aug., 1965.

Fulton, R. A., joint author. *See* Knipling, G. D., et al., 1961; Fales, J. H., et al., 1964, 1960.

3535. **Funatsu, M. 1966** [Variations of enzyme activities during metamorphosis of insects. On phenol oxidase system in the fly.] Protein, Nucleic Acid and Enzymes (Tokyo) 11:220-227. March. (In Japanese.)

3536. Funatsu, M., Hayashi, K. and **Namihira, G. 1967** The Protein Components of a Latent Phenoloxidase System in the Prepupae of Housefly (*Musca vicina*). Agric. Biol. Chem. 31(11):1379-1380.
Three types of protein participate in the organization of an inactive (latent or dormant) phenoloxidase system in housefly prepupae. ABSTRACT: BA 49:5303(59508), June 1, 1968.

3537. Funatsu, M., Hayashi, K., Inaba, T., Namihira, G. and **Kariva, M. 1967** The Nature of Latent Phenoloxidase in the Prepupae of Housefly. 7th Internat. Cong. Biochem. Tokyo. August 19-25. Reprinted from Abstracts, G-34.
The phenoloxidase surviving as a proenzyme in the prepupa, can be artifically activated; either by treating the homogenate of prepupae with anionic detergents (plus an activator from aged pupae), or by dialyzing it with water (which lowers the ionic strength.)

3538. Funatsu, M. and **Inaba, T. 1962** Studies on Tyrosinase in Housefly. I. Protyrosinase in the Pupae of Housefly and its Activation. Agric. Biol. Chem. 26(8):535-540.
No appreciable tyrosinase activity in prepupae. Enzyme exists as inactive protyrosinase. Aged pupae contain an activator for this. ABSTRACT: BA 42:322(3897), Apr. 1, 1963.

Funatsu, M., joint author. *See* Inaba, T., et al., 1963, 1964.

3539. Funder, J. V. 1964 Fliegenversuche auf dänischen Höfen mit R Baytex 40-Spritzpulver. [Trial controls of flies on Danish farms with Baytex 40 wettable powder.] Pfl. Schutz-Nachr. Bayer 17(2):68-88. (Leverkusen.)
Baytex 40 contains 40 per cent fenthion. One or two applications suppressed both *Stomoxys* and *M. domestica* (including diazinon-resistant flies). Some resistance developed during the third year of use. ABSTRACT: RAE-B 54-85, Apr., 1966.

3540. Funder, J. V. 1966 Fluebekaempelse på pelsdyrfarme. [Fly control on fur farms.] Ann. Rept. (1965) of the Govern. Pest Infest. Lab., pp. 25-28. Lyngby, Denmark. (In Danish, with complete English translation.)
Concerns *Fannia canicularis*. Dimethoate was the most effective larvicide; trichlorfon in sugar bait and dimethoate-treated gauze were most effective against adults.

3541. Funder, J. V. 1966 Fluebekaempelse; spattegulusstald. [Fly control in a piggery with "slotted-floor".] Ann. Rept. (1965) Govt. Pest Infest. Lab., pp. 39-40. Lyngby, Denmark.
Dimethoate the insecticide used.

3542. Funder, J. V. 1966 Afprøving af insektmidler-laboratorieforsøg Fluemidler. [Testing of insecticides in the laboratory. Insecticides for fly control.] Ann. Rept. (1965) Govt. Pest Infest. Lab., pp. 28-31. Lyngby, Denmark. (In Danish with complete English translation.)
Reports effectiveness of new commercial products; also the results of repeat tests of various insecticides for durability.

3543. Funder, J. V. 1966 Fluebekaempelsen. [Control of Houseflies.] St. Skadedyrlab., 3 pp. (In Danish.)
Covers types of control procedure, e.g. fly baits, fly strips, etc.

3544. Funder, J. V. 196? Fluer på minkfarme. [Flies on Minkfarms.] St. Skadedyrlab., pp. 286, 289-290. (In Danish.)
Discusses types of fly control.

3545. Funder, J. V. 1967 Fluebekaempelsen 1963-1966. [Control of houseflies 1963-1966.] Ann. Rept. (1966) Govt. Pest. Infest. Lab., pp. 16-18. Lyngby, Denmark.
Covers both normal and subnormal years as to fly population. Residual sprays increased in effectiveness; fly strips decreased.

3546. Funder, J. V. 1967 Fluebekaempelse på pelsdyrfarme. [Fly control on furfarms.] Ann. Rept. (1966) Govt. Pest Infest. Lab., pp. 18-20. Lyngby, Denmark.
Fly strips impregnated with parathion were best against *Fannia*; paint-on bait best against *Muscina.*

3547. Funder, J. V. 1967 Afprøvning af insektmidler-laboratorieforsøg. [Testing of insecticides in the laboratory.] Ann. Rept. (1966) Govt. Pest Infest. Lab., pp. 20-27. Lyngby, Denmark.
Concerns Bromophos-ethyl, methylcarbamate, and formothion. New formulations, in general, proved inferior to standard ones.

3548. Funder, J. V. 1967 Fluebekaempelse; spaltegulvsstald. [Fly control in piggery with "slotted floor".] Ann. Rept. (1966) Govt. Pest Infest. Lab., pp. 32 and 37. Lyngby, Denmark.
Reports continuing program, with flies showing moderate resistance to dimethoate.

3549. Funder, J. V. 1968 Fluebekaempelsen; 1967. [Control of houseflies 1967.] Ann. Rev. (1967) Govt. Pest Infest. Lab. pp. 23-24. Lyngby, Denmark.
Emphasis on residual dimethoate sprays.

3550. Funder, J. V. 1968 Afprøvning af insektmidler-laboratorieforsøg Fluemidler. [Testing of insecticides in the laboratory. Insecticides for fly control.] Ann. Rept. (1967) Govt. Pest Infest. Lab., pp. 24-26; 28-29. Lyngby, Denmark.
Aerosols based on pyrethrins superior. Paint-on trichlorfon bait showed good results on second testing.

3551. Funder, J. V. 1969 Fluebekaempelsen; 1968. (Control of houseflies, 1968.] Ann. Rept. (1968) Govt. Pest Infest. Lab., pp. 19 and 20. Lyngby, Denmark.
Impregnated strips and synergized pyrethrum preparations used to supplement dimethoate residual sprays.

3552. Funder, J. V. 1969 Afprøvning af insektmidler. Fluemidler. Laboratorieforsøg. [Testing of insecticides in the laboratory. Insecticides for fly control.] Ann. Rept. (1968) Govt. Pest Infest. Lab., pp. 20, 21, 22, 25, 26. Lyngby, Denmark.
Concerns dual-purpose aerosols and residual spraying with wettable powders. Of several ingredients, fenthion remained on a porous surface the longest.

3553. Funder, J. V. and **Mourier, H. 1965** Investigations of flies on Danish fur farms. Arsberetn. St. Skadedyrlab. 1964, pp. 17-18. Lyngby, Denmark.
Lists species and gives control measures. The use of a fly index (number of flies caught per 10 m) is discussed.
ABSTRACT:RAE-B 54:47, Feb., 1966.

3554. Funder, J. V. and **Mourier, H. 1965** (?) Fluerne på pelsdyrfarmen. [Flies on fur farms.] Statens Skadedyrlab., pp. 250-251.
Lists families, and gives information on important species concerned.

Furuno, A., joint author. *See* Ogata, K., et al., 1961.

3555. Fye, R. L. 1967 Screening of Chemosterilants against House Flies. J. Econ. Ent. 60:605-607.
A report on 31 compounds, administered dietetically. Lists minimum concentrations which will give desired effects. ABSTRACTS: BA 48:6927 (77468), Aug. 1, 1967; RAE-B 55:148(521), Aug., 1967; TDB 64:909, Aug., 1967.

3556. Fye, R. L., Gouck, H. K. and **LaBrecque, G. C. 1965** Compounds Causing Sterility in Adult House Flies. J. Econ. Ent. 58:446-448.
Of 173 compounds tested, 23 were effective in causing sterility. ABSTRACTS: BA 47:371(4492), Jan. 1, 1966; RAE-B 53:178, Sept., 1965; TDB 62:1274, Dec., 1965.

3557. Fye, R. L. and **LaBrecque, G. C. 1966** Sexual Acceptability of Laboratory Strains of Male House Flies in Competition with Wild Strains. J. Econ. Ent. 59:538-540.
Females of all strains mated more readily with males of their own strain, whether or not the males were sterilized. ABSTRACTS: BA 47:8554(99693), Oct. 15, 1966; RAE-B 54:175, Sept., 1966.

3558. Fye, R. L. and **LaBrecque, G. C. 1967** Sterility in House Flies Offered a Choice of Untreated Diets and Diets Treated With Chemosterilants. J. Econ. Ent. 60:1284-1286.
Chemosterilants were somewhat repellent alone, but if combined with an acceptable bait, will be consumed in sufficient amount to sterilize. ABSTRACTS: BA 49:1399(15666), Feb., 1968; RAE-B 56:63(208), Mar., 1968; TDB 65:541(1180), Apr., 1968.

3559. Fye, R. L. and **LaBrecque, G. C.** Further Developments on the Acetosterilization of House Flies. I. Evaluation of Newer Chemosterilants. In Press.
Two new compounds discovered as effective as tepa.

3560. Fye, R. L., LaBrecque, G. C., Borkovec, A. B. and **Morgan, J., Jr. 1969** Compounds Affecting Fertility of Adult House Flies. J. Econ. Ent. 62(2):522-524.
Reports on 25 compounds administered in the diet of which 3 proved highly effective. Highest rated was 3, 5-bis(dimethylamine)-1, 2, 4-dithiazolium chloride. ABSTRACTS: BA 50:10829(112104), Oct. 15, 1969; RAE-B 57:187(667), Sept., 1969.

3561. Fye, R. L., LaBrecque, G. C. and **Gouck, H. K. 1966** Screening Tests of Chemicals for Sterilization of Adult House Flies. J. Econ. Ent. 59:485-487.
Reports on 26 experimental chemicals tested for effect on fertility. Table gives comparison with *tepa* (a standard chemosterilant). ABSTRACTS: BA 47:7234(84394), Sept. 1, 1966; RAE-B 54:134, July, 1966; TDB 63:1156, Oct., 1966.

Fye, R. L., joint author. *See* Dame, D. A., et al., 1964; Meifert, D. W., et al., 1963; Murvosh, C. M., et al., 1964; Borkovec, A. B., et al., 1968, 1969.

Gaaboub, I. A., joint author. *See* El-Deeb, A. L., et al., 1969.

Gabaldon, A., joint author. *See* Blazquez, V. J., et al., 1965.

Gabriyanik, I. A., joint author. *See* Yaguzhinskaya, L. V., et al., 1966, 1967, 1969.

3562. Gadallah, A. I., Kilgore, W. W. and **Painter, R. R. 1970** Isolation and Identification of Egg Ribosomes of the Housefly, *Musca domestica*. J. Insect Physiol. 16:1245-1248.
Ribosomes exist in monomeric, dimaric, and polymeric forms. Are

sensitive to RN ase but not to DN ase. ABSTRACT: BA 51:13729(139770), Dec. 15, 1970.

3562a. Gadallah, A. I., Kilgore, W. W. and **Painter, R. R. 1970** Metabolism of Nucleic Acids and Proteins of Normal and Chemosterilized House Flies during Oögenesis and Embryogenesis. J. Econ. Ent. 63(6):1777-1783.

Purine-pyramidine ratio of the RNA nucleotides of treated ovaries was consistently higher than that from normal ovaries. This ratio also *became* higher in RNA from chemosterilized eggs, after 6 hours of incubation at 37°C.

3563. Gadzhei, E. F. 1963 [Fly migrations in urban areas.] Med. Parazit. i Parazitarn. (Bolezni) 32(4):465-468. (English summary.)

Most flies were found within 30 m., but some went as far as 250 m. on the first day. These distances should be considered in undertaking epidemiological and entomological control during disease outbreaks. ABSTRACTS: BA 45:3204(39651), May 1, 1964; TDB 61:726, July, 1964.

Gadzhei, E. F., joint author. *See* Arskii, V. G., et al., 1961; Il'Yashenko, L. Y., et al., 1962.

Gaetani, M. T., joint author. *See* DiJeso, F., et al., 1966, 1967.

Gafiteanu, L., joint author. *See* Calciu-Dumitreasă, A., et al., 1969.

Gahan, J. B., joint author. *See* Gouck, H. K., et al., 1963; Wilson, H. G., et al., 1961, 1967.

Gaines, T. B., joint author. *See* Kimbrough, R., et al., 1966; Mattson, A. M., et al., 1960.

3564. Galichet, P. F. 1966 Administration aux animaux domestiques d'une toxine thermostable sécrétée par *Bacillus thuringiensis* Berliner, en vue d'empêcher la multiplication de *Musca domestica* Linnaeus and les fèces. [Administration to domestic animals of a thermostable toxin secreted by *Bacillus thuringiensis* Berliner with a view to impeding the multiplication of *Musca domestica* Linnaeus in the feces.] Ann. Zootech. (Paris) 15(2):135-145. (English summary.]

Greater amounts needed to prevent insect from developing in pig feces than calf feces. Death most frequently delayed until nymphal stage. Evidence does not indicate that mortality is due to proliferation of the micro-organism.

3565. Galichet, P. F. 1967 Sensitivity to the Soluble Heat-stable Toxin of *Bacillus thuringiensis* of Strains of *Musca domestica* Tolerant to Chemical Insecticides. J. Invertebr. Path. 9(2):261-262.

Experiments show there is no relation between the characters conferring resistance to one or several groups of chemical insecticides, and to the bacterial substance. ABSTRACTS: BA 48:10187(114471), Nov. 15, 1967; RAE-B 56:144(506), July, 1968.

Galichet, P. F., joint author. *See* Burgerjon, A., et al., 1965.

3566. Galley, D. J. 1967 The Effect of Gas Velocity on the Fumigant Action of Nicotine, Dichlorvos and Hydrogen Cyanide. J. Stored Prod. Res. 3:17-27.

Toxicities of nicotine and dichlorvos as fumigants were enhanced as gas velocity increased; hydrogen cyanide was barely affected. Increased kills were not obtained by velocity changes.

3567. Galley, D. J. 1967 The Effect of Gas Velocity on the Absorption of Fumigant Nicotine by *Musca domestica* L. J. Stored Prod. Res. 3(3):213-222.

Fumigant air movement enhanced absorption, the effect increasing with velocity until maximum limiting rate was reached. Epicuticular wax affects uptake and rate of uptake. ABSTRACT: BA 49:1907(21455), Feb. 15, 1968.

Gallick, C. J., joint author. *See* Marak, G. E., et al., 1968, 1969.

3568. Galun, R. and **Fraenkel, G. 1961** The Effect of Low Atmospheric Pressure on Adult *Aedes aegypti* and on Housefly Pupae. J. Insect Physiol. 7(3/4):161-176.
Mortality can be due to desiccation, lack of oxygen or low pressure per se. Authors designed apparatus for measuring respiration at very low atmospheric pressures. ABSTRACTS: BA 37:2531(25072), Mar. 15, 1962; RAE-B 51:70, Apr., 1963.

3569. Gamal-Eddin, F. M. 1964 Further Studies on the Ecology of *Musca sorbens* Wied. in Egypt (Diptera-Muscidae). J. Egyptian Pub. Health Ass. 39(1):53-59.

Gamal-Eddin, F. M., joint author. *See* Hafez, M., et al., 1961, 1963, 1966.

Ganapathipillai, A., joint author. *See* Wharton, R. H., et al., 1962.

Gandolfa, D., joint author. *See* Saccà, G., et al., 1968, 1969.

Ganla, V. G., joint author. *See* Narasimhan, M. J., et al., 1967.

3570. Garanrvolgyi, N. 1966 Polarizing Microscopic Studies on Insect Flight Muscle Fibrils. Acta Biochim. Biophys. Acad. Sci. Hung. 1(3):336.

Garber, M. J., joint author. *See* Georghiou, G. P., et al., 1965.

Gard, I., joint author. *See* Rogoff, M. H., et al., 1969.

3571. Garrett, D. A. 1965 The Sensory Responses of the House Fly, *Musca domestica* Linn., to Attractants. (Ph.D. Thesis, Oklahoma State University.) Diss.Absts. 27(1B):16-B, 17-B.
Several tests, conducted on sensory response to attractants, included: anesthetization; aspiration; attraction to other flies, food and visual objects; aggregation and tracking patterns.

3572. Gattone, F. and Bettini, S. 1963 Infection of Houseflies (*Musca domestica* L.) through Contamination by *Serratia* sp. during Routine Toxicological Tests by Injection. Riv. di Parassit. (Rome) 24(3): 213-214.
Contamination of a stock solution of *Serratia* sp. caused unusually high mortality of flies. Flies showed a reddish tinge about 24 hours later. ABSTRACTS: RAE-B 53:175, Sept., 1965; TDB 61:851, Aug., 1964.

3573. Geering, Q. A., Brooker, P. J. and **Parsons, J. H. 1965** The Chemosterilant Activity of Some Substituted Phenyl Esters of Aziridine-1-carboxylic Acid. J. Econ. Ent. 58:574-575.
Two compounds were shown to have significant sterilant activity. ABSTRACTS: BA 47:801(9739), Jan. 15, 1966; RAE-B 53:183, Sept., 1965; TDB 63:102, Jan., 1966.

3574. Geisthardt, G. 1967 The Influence of Stable Wall Surfaces on the Residual Activity of House Fly Sprays Demonstrated with Bromophos Insecticide. Abstr. Int. Pflanzenschutz-Kong. (Int. Congr. Plant. Prot.) 3:240-241.

Gelvin, D. E., joint author. *See* Kay, R. E., et al., 1967.

3575. Genis, D. E. 1962 Opyt primeneniya khlorofosa v bor̕bes mukhami. [Experience with the application of chlorophos in fly control.] Med. Parazit. (Moskva) 31:620-622. (English summary.)
Carefully placed aqueous solutions of 0·2 per cent chlorophos, applied

as a fly agaric, greatly reduced or eradicated fly population. More economical than 1-2 per cent chlorophos solution. ABSTRACT: BA 41: 1646(20707), Mar. 1, 1963.

3576. Genrich, E. G. II. 1967 Metabolic and Enzymatic Degradation of Several Aromatic Carbamate Insecticides. Diss. Absts. 27(7B): 2245.
Concerns *Musca domestica* and *Blaberus giganteus*.

Gentry, T., joint author. *See* Childo, D. P., et al., 1964.

3577. Georghiou, G. P. 1962 Carbamate Insecticides: The Cross-Resistance Spectra of Four Carbamate-Resistant Strains of the House Fly After Protracted Selection Pressure. J. Econ. Ent. 55:494-497.
Spectra resemble those of organophosphorus-selected more than those of chlorinated hydrocarbon-selected strains. ABSTRACTS: BA 40:1634 (21320), Dec. 1, 1962; RAE-B 51:3, Jan., 1963; TDB 60:183, Feb., 1963.

3578. Georghiou, G. P. 1962 Carbamate Insecticides: Toxic Action of Synergized Carbamates against Twelve Resistant Strains of the House Fly. J. Econ. Ent. 55(5):768-772.
Carbamate resistance due in part to components insensitive to action of piperonyl butoxide. ABSTRACTS: BA 41:949(12350), Feb. 1, 1963; RAE-B 51:42, Feb., 1963.

3579. Georghiou, G. P. 1964 The Stability of Resistance to Carbamate Insecticides in the Housefly after Cessation of Selection Pressure. Bull. W. H. O. 30:85-90. (French summary.)
The Isolan-selected strain developed a more stable resistant phenotype; resistance remained for 41 generations. The m-isopropylphenyl methylcarbamate strain developed an unstable phenotype; resistance regressed rapidly. ABSTRACTS: BA 46:5496(68401), Aug. 1, 1965; RAE-B 54:78, Apr., 1966; TDB 61:1180, Nov., 1964.

3580. Georghiou, G. P. 1964(1965) Physiological and Genetical Bases of Resistance to Carbamate Insecticides in House Flies and Mosquitoes. Proc. XII Int. Congr. Ent. London 1964. (Publ. 1965), p. 494.
Two different expressions of the character of resistance are recognized. Pronounced intensification of detoxification mechanisms characterize the physiological basis; the presence of a semi-dominant, autosomal gene characterizes the genetic basis.

3581. Georghiou, G. P. 1965 Effects of Carbamates on House Fly Fecundity, Longevity, and Food Intake. J. Econ. Ent. 58:58-62.
Topical application of Isolan reduced egg laying but did not affect mating longevity or egg fertility. Cholinesterase inhibition is probably not directly responsible for suppression of egg production. ABSTRACTS: BA 46:4069(50335), June 1, 1965; RAE-B 53:96, May, 1965; TDB 62:811, Aug., 1965.

3582. Georghiou, G. P. 1965 Genetic Studies on Insecticide Resistance. Advances in Pest Contr. Res. 6:171-230. (263 ref.). A Review.
Includes: inheritance of resistance to insecticides; determination of genotype; negatively correlated insecticides; cross resistance and multiple resistance; vigor tolerance.

3583. Georghiou, G. P. 1966 Distribution of Insecticide-resistant House Flies on Neighboring Farms. J. Econ. Ent. 59:341-346.
Only a few resistant flies dispersed to other farms constitute nucleus for rapid development of resistance upon treatment with the insecticide concerned. ABSTRACTS: BA 47:7250(84563), Sept. 1, 1966; RAE-B 54: 129, July, 1966; TDB 63:907, Aug., 1966.

3584. Georghiou, G. P. 1967 Differential Susceptibility and Resistance to Insecticides of Coexisting Populations of *Musca domestica, Fannia canicularis, F. femoralis,* and *Ophyra leucostoma.* J. Econ. Ent. 60: 1338-1344.
No evidence of a pronounced natural tolerance towards any of the test compounds. Housefly most resistant to a majority of the compounds. ABSTRACTS: BA 49:1395(15628), Feb. 1, 1968; RAE-B 56:65(214), Mar., 1968.

3584a. Georghiou, G. P. 1969 Genetics of Resistance to Insecticides in Houseflies and Mosquitoes. Exptl. Parasitol. 26(2):224-255.
Reviews a number of advanced contributions to insecticide research. Argues that rational explanation of cross-resistance is greatly helped by deeper knowledge of genetics. Practical application in choice of alternate chemicals. ABSTRACT: TDB 67:1056(2053), Aug., 1970.

3585. Georghiou, G. P. and **Bowen, W. R. 1966** An Analysis of House Fly Resistance to Insecticides in California. J. Econ. Ent. 59:204-214.
Pattern of insecticide resistance reflects the differing chemical control practices in each area. ABSTRACTS: BA 47:6376(74577), Aug. 1, 1966; RAE-B 54:100, May, 1966.

3586. Georghiou, G. P., Bowen, W. R., Loomis, E. C. and **Deal, A. S. 1965** The Present Status of Housefly Resistance to Insecticides in California. Calif. Agric. 19(10):8-10.
Studies show degree of resistance to insecticides; new insecticides effective but resistant strains rapidly develop resistance to them. Sanitation practices still an effective means of reducing flies. ABSTRACT: BA 47: 2060(24708), Mar. 1, 1966.

3587. Georghiou, G. P. and **Garber, M. J. 1965** Studies on the Inheritance of Carbamate-resistance in the Housefly (*Musca domestica* L.) Bull. W. H. O. 32:181-196.
Resistance to Isolan is inherited as a partially dominant major single factor, without sex linkage or appreciable cytoplasmic influence. ABSTRACTS: BA 47:3847(45306), May 15, 1966; RAE-B 55:87(308), May, 1967; TDB 62:1273, Dec., 1965.

3588. Georghiou, G. P., Hawley, M. K. and **Loomis, E. C. 1967** A Progress Report on Insecticide Resistance in the Fly Complex of California Poultry Ranches. Calif. Agric. 21(4):8-11.
Objective to study: to determine the changing pattern of resistance following a changeover in chemical control practices; investigate presence and extent of resistance; propose means of delaying resistance. ABSTRACT: BA 48:8309(93263), Sept. 15, 1967.

3589. Georghiou, G. P., March, R. B. and **Printy, G. E. 1963** Induced Regression of Dieldrin-Resistance in the Housefly (*Musca domestica* L.). Bull. W. H. O. (Geneva) 29(2):167-176.
Found cyclodiene resistance of unknown origin. Involves entire cyclodiene group and lindane; is inherited as monofactorial character transmitted by either male or female. ABSTRACTS: BA 45:2155(26953), Apr. 1, 1964; RAE-B 53:199, Oct., 1965; TDB 61:222, Feb., 1964

3590. Georghiou, G. P., March, R. B. and **Printy, G. E. 1963** A Study of the Genetics of Dieldrin-Resistance in the Housefly (*Musca domestica* L.). Bull. W. H. O. (Geneva) 29(2):155-165.
Appears that inheritance of resistance is due to a single major factor, without sex linkage or cytoplasmic effects. ABSTRACTS: BA 45:2155 (26953), Apr. 1, 1964; RAE-B 53:199, Oct., 1965; TDB 61:109, Jan., 1964.

3591. Georghiou, G. P. and **Metcalf, R. L. 1961** Synergism of Carbamate Insecticides with Octachlorodipropyl Ether. J. Econ. Ent. 54(1):

150-152.
Synergistic effect obtained with ratios as low as 1:20. ABSTRACTS: BA 36:2889(35439), June, 1961; RAE-B 49:257, Nov., 1961.

3592. Georghiou, G. P. and **Metcalf, R. L. 1961** The Absorption and Metabolism of 3-Isopropylphenyl N-Methylcarbamate by Susceptible and Carbamate-selected Strains of House Flies. J. Econ. Ent. 54(2): 231-233.
Used manometric cholinesterase inhibition method to determine rate of absorption. ABSTRACT: RAE-B 50:107, May, 1962.

3593. Georghiou, G. P. and **Metcalf, R. L. 1962** Carbamate Insecticides: Comparative Insect Toxicity of Sevin, Zectran, and Other New Materials. J. Econ. Ent. 55:125-127.
All have pronounced insecticidal effectiveness. ABSTRACT: RAE-B 50: 225, Oct., 1962.

3594. Georghiou G. P., Metcalf, R. L. and **March, R. B. 1961** The Development and Characterization of Resistance to Carbamate Insecticides in the House Fly, *Musca domestica.* J. Econ. Ent. 54(1):132-140.
A selection experiment involving two subcolonies and many compounds. ABSTRACT: BA 36:2889(35440), June, 1961.

3595. Georghiou, G. P., Metcalf, R. L. and **VonZboray, E. P. 1965** Toxicity of Certain New Compounds to Insecticide-Resistant Houseflies. Bull. W. H. O. 33(4):479-484. (French summary.)
Contact and oral toxicity of new compounds tested; 5 found to be as toxic to 3 resistant strains as to a susceptible strain showing steep low-dosage/probit mortality lines. ABSTRACTS: BA 47:5531(64855), July 1, 1966; RAE-B 55:209(725), Nov., 1967; TDB 63:711, June, 1966.

Georghiou, G. P., joint author. *See* March, R. B., et al., 1964; Metcalf, R. L., et al., 1962.

Gerhardt, R. R., joint author. *See* Turner, E. C., Jr., et al., 1968.

3596. Gerling, D. J. 1967 The Eggs of the Pupal Parasites of *Musca domestica* L. Israel J. Ent. 2:11-13.
Ova of 5 species of Pteromalidae figured and described. ABSTRACT: BA 50:5323(55889), May 15, 1969.

3597. Gerling, D. and **Legner, E. F. 1968** Developmental History and Reproduction of *Spalangia cameroni,* Parasite of Synanthropic Flies. Ann. Ent. Soc. Am. 61(6):1436-1443.
Describes in detail developmental history, with observations on the physiology of ovum formation and sperm translocation thru the male tract. ABSTRACTS: BA 50:3733(39199), Apr. 1, 1969; RAE-B 57:152(549). Aug., 1969.

Gerling, D. J., joint author. *See* Legner, E. F., et al., 1967.

3598. Gerolt, P. 1961 Investigation into the Problem of Insecticide Sorption by Soils. Bull. W. H. O. 24(4-5):577-591. (French summary.)
Studies using DDT labelled with ^{14}C showed that effect of humidity is the only factor influencing toxicity. Movement of insecticide in soil is blocked at both very high and very low humidity. Migration of soil water causes insecticide to move in same direction. ABSTRACT: RAE-B 51:115, June, 1963.

3599. Gerolt, P. 1963 Influence of Relative Humidity on the Uptake of Insecticides from Residual Films. Nature 197(4868):721.
The greater effectiveness of insecticides under humid conditions is due to a greater availability of toxicant and a greater takeup by insect. Humidity should be considered in bioassay tests.

3600. Gerolt, P. 1965 The Fate of Dieldrin in Insects. J. Econ. Ent. 58:849-857.
Rates of penetration, translocation of sublethal amounts, concentration in body tissues, and conversion, appear similar in resistant and susceptible flies under normal conditions. Excitation increases rate of penetration in susceptible flies. ABSTRACTS: RAE-B 54:12, Jan., 1966; TDB 63:486, Apr., 1966.

3601. Gerolt, P. 1969 Mode of Entry of Contact Insecticides. J. Insect Physiol. 15(4):563-580.
Route of entry of insecticides to an action site is via the integument of the tracheal system. In houseflies, insecticides do *not* penetrate integument and enter haemolymph. ABSTRACTS: BA 50:9217(95522), Sept. 1, 1969; RAE-B 58:111(449), Apr., 1970.

3602. Gersdorff, W. A. and **Piquett, P. G. 1961** The Relative Effectiveness of Two Synthetic Pyrethroids more Toxic to House Flies than Pyrethrins in Kerosene Sprays. J. Econ. Ent. 54(5):1250-1252.
One analogue of allethrin, with allyl group replaced, was 64 per cent as toxic as allethrin and 195 per cent as toxic as pyrethrins. Another, with part of acid component replaced, was 41 per cent as toxic as allethrin and 120 per cent as toxic as pyrethrins. ABSTRACTS: RAE-B 50:189, Sept., 1962; TDB 59:498, May, 1962.

3603. Gersdorff, W. A., Piquett, P. G., Gertler, S. I. and **Alexander, B. H. 1960** Comparative Value of Some Synthetic Compounds as Synergists with Barthrin in House Fly Sprays. J. Econ. Ent. 53(2):282-285.
Compounds tested by Campbell turntable method. Most effective were sesamex, piperonyl butoxide, a new synthetic compound, and its ethoxy analog. They raised the toxicity of the mixtures about 3 times that expected. ABSTRACTS: BA 35:3740(42997), Aug. 1, 1960; RAE-B 49:103, May, 1961.

3604. Gersdorff, W. A., Piquett, P. G. and **Mitlin, N. 1961** Stability of Allethrin and Kerosene Solutions of it in Storage as Demonstrated by Bioassay. J. Econ. Ent. 54(4):731-733.
During a 10 year period, tests showed the toxicity to be lessened greatly by light but only after 6 years by room temperature. Hydroquinone was a good inhibitor and kerosene a good diluent.

3605. Gersdorff, W. A., Piquett, P. G., Mitlin, N. and **Green, N. 1961** Reproductibility of the Toxicity Ratio of Allethrin to Pyrethrins Applied to House Flies by the Turntable Method. J. Econ. Ent. 54:580-583.
Summarization of 52 comparisons made in 10 years of the ratio of toxicity at the 50 per cent mortality level; values ranged from 2·23 to 3·61. ABSTRACTS: BA 36:5563(58145), Sept. 1, 1961; RAE-B 50:131, June, 1962.

Gersdorff, W. A., joint author. *See* Piquett, P. G., et al., 1960.

Gertler, S. I., joint author. *See* Gersdorff, W. A., et al., 1960.

Geshvind, G. N., joint author. *See* Fomicheva, V. S., et al., 1967.

3606. Ghiasuddin, S. M. 1967(?) Sex Susceptibility of the House Fly, *Musca domestica nebulo*, to Different Insecticides. Symp. Pestic. A., pp. 17-18.

Gibson, A., joint author. *See* Hudson, L. D., et al., 1960.

Gibson, J. B., joint author. *See* Thoday, J. M., et al., 1970.

3607. Gil, L., Fine, B. C., Dinamarca, M. L., Balazs, I., Busvine, J. R. and **Agosin, M. 1968** Biochemical Studies on Insecticide Resistance in *Musca domestica*. Ent. Exp. Appl. 11(1):15-29. (German summary.)
Microsomal enzymes are DDT-induced in at least 2 of the 76 strains

used. The increase in enzyme activity is paralleled by an increased rate of incorporation of labelled precursors into total RNA. ABSTRACTS: BA 49:9251(102733), Oct. 1, 1968; RAE-B 56:245(905), Dec., 1968; TDB 66:165(299), Feb., 1969.

3608. Gilgan, M. W. and **Retallack, N. E. 1967** Insect Molt-hormone Activity of an Extract of the Japanese Yew (*Taxus cuspidata*). J. Fish. Res. Board Can. (Halifax Lab., Halifax, N.S., Canada) 24(11): 2497-2499.

Housefly bioassay. ABSTRACT: BA 49:3835(43128), Apr. 15, 1968.

Gillham, E. M., joint author. *See* Ward, J., et al., 1960.

3609. Gingrich, R. E. 1965 *Bacillus thuringiensis* as a Feed Additive to Control Dipterous Pests of Cattle. J. Econ. Ent. 58(2):363-364.

Horn fly, *Haematobia irritans*, most susceptible; *Stomoxys calcitrans* the most resistant; *Musca domestica* intermediate. ABSTRACTS: BA 46:5143 (63868), July 15, 1965; RAE-B 53:129, July, 1965; TDB 62:944, Sept., 1965.

3610. Gip, L. and **Svensson, S. A. 1968** Can Flies Cause the Spread of Dermatophytosis? Acta Derm.-Venereol. (Stockholm) 48:26-29.

Sixty-nine flies were caught, classified and examined; 59 were *M. domestica. Trichophytor mentagrophytes* var. *granulosum* was isolated from 3 *M. domestica.* ABSTRACT: BA 49:6651(74171), July 15, 1968.

3611. Giulio, L. 1963 Elektroretinographische Beweisführing dichroitischer Eigenschaften des Komplexauges bei Zweiflüglern. [Electroretinographic demonstration of the analytical capabilities of the compound eyes in Diptera.] Zeit. Vergl. Physiol. 46:491-495. (English summary.)

Experiments indicate presence of an analyser for polarized light in the rhabdomeres.

3612. Giulio, L. 1966 The Role of Sulphydryl Groups in the Visual Excitation of the Compound Eye in Diptera. *In* Proc. Internat. Sympos.-Stockholm, Sweden. October, 1965. Symp. Pub. Div., Pergamon Press: London and New York. 7:151-162.

Expository paper. Finds sulphydryl groups of considerable importance. Discussion includes *M. domestica.* ABSTRACT: BA 48:9250(104017), Oct. 15, 1967.

Giulio, L., joint author. *See* Ercolini, A., et al., 1967.

Gladney, W. J., joint author. *See* Drummond, R. O., et al., 1969.

Glazunova, A. Y., joint author. *See* Shura-Bura, B. L., et al., 1963.

3613. Glofke, E. 1966 Untersuchungen von Milchproben auf DDT-Rückstände. [Examination of milk samples for DDT residues.] Pflanzen. 33(11-12):163-169. (English summary.)

Only 2 of 121 milk samples assayed contained DDT. Was traced to recent periodic spraying of the cowshed. DDT content decreased with cessation of spraying. ABSTRACT: RAE-B 54:247, Dec., 1966.

3614. Godin, P. J., Stevenson, J. H. and **Sawicki, R. M. 1965** The Insecticidal Activity of Jasmolin II and its Isolation from Pyrethrum (*Chrysanthemum cinerariaefolium* Vis.). J. Econ. Ent. 58:548-551.

Jasmolin II nearly as toxic as cinerin II against *M. domestica.* ABSTRACTS: BA 47:801(9739), Jan. 15, 1966 (Godin only); RAE-B 53:181, Sept., 1965.

Godin, P. J., joint author. *See* Fine, B. C., et al., 1963, 1967.

3615. Gohda, M., Sakai, S., Miura, H., Koizumi, H. and **Nakagoshi, S. 1966** [A case of control of sanitary insect pests by Bell 47-G2 helicopter dusting in Japan.] Botyu-Kagaku 31(1):38-47. (In Japanese, with

English summary.)
Recorded: direct toxicity of dusts; relation between effectiveness against flies and mosquitoes and amount of dust settling; population fluctuations after aerial applications; distribution pattern of deposited dusts; and meteorological factors. ABSTRACT: RAE-B 56:129(451), July, 1968.

3616. Gojmerac, W. L. 1967 Wall Whitening Agents, their Effect on Residual Toxicity of Ronnel to House Flies. J. Econ. Ent. 60:872-873.
Talc-based agents prolonged residual activity; lime reduced it. ABSTRACTS: BA 48:11208(124486), Dec. 15, 1967; TDB 65:82(184), Jan., 1968.

3617. Goldberg, A. A., Head, S. and **Johnson, P. 1965** Distillation of Pyrethrum Extract in a Wiped-wall, Falling-film, Short-path Still: Separation of 'Pyrethrin I' and 'Pyrethrin II' and Determination of their Relative Biological Activities. J. Sci. Fd. Agric. (London) 16(2):104-116.
Pyrethrum oleo resin was distilled. Biological assays showed the activities of the distilled and undistilled extracts to be identical. Knockdown effect of pyrethrum I and II reported. ABSTRACT: RAE-B 54:44, Feb., 1966.

3618. Goldberg, A. A., Head, S. and **Johnson, P. 1965** Action of Heat on Pyrethrum Extract: the Isomerisation of Pyrethrins to Isopyrethrins. J. Sci. Fd. Agric. (London) 16(1):43-51.
Isomerization is a first order reaction. Study concerns the biological activities of normal, partly isomerised and completely isomerised extracts on flies and weevils. The fully isomerized extract (to isopyrethrins) has half the power of the normal. ABSTRACT: RAE-B 54:44, Feb., 1966.

3619. Golden, C. A. 1968 Studies for the Development of Procedures and Standards for Fly Free Composting Techniques. Compost Sci. 9(2):21.
Concerns chemical and mechanical control.

Goldina, G. S., joint author. *See* Neselovskaya, V. K., et al., 1968; Sukhova, M. N., et al., 1965.

3620. Goldsmith, T. H. 1964 Photoreceptor Sensitivity of Carotenoid-deficient House Flies (Diptera). Am. Zool. 4(4):382. (Abstract.)
Under conditions of the experiment, photoreceptor sensitivity fell to more than 4 log units below normal by the third generation. No apparent growth requirement for vitamin A or carotenoids.

3621. Goldsmith, T. H. 1965 Do Flies Have a Red Receptor? J. Gen. Physiol. 49(2):265-287.
Summarizes experimental results on sensitivity of flies to red and green light. An attempt is made to resolve the problem of the reported presence of a non-existent red receptor. ABSTRACT: BA 47:2497(34907), Apr. 1, 1966.

3622. Goldsmith, T. H., Barker, R. J. and **Cohen, C. F. 1964** Sensitivity of Visual Receptors of Carotenoid-depleted Flies: A Vitamin A deficiency in an Invertebrate. Science 146(3640):65-67.
Carotenoid stored in eggs prevents complete blindness in adults. Microorganisms of environment supply small amounts also. ABSTRACT: BA 46:1459(18229), Feb. 15, 1965.

3623. Goldsmith, T. H. and **Fernandez, H. R. 1966** Some Photochemical and Physiological Aspects of Visual Excitation in Compound Eyes. *In* Proc. Intern. Sympos. —— Stockholm, Sweden. October, 1965. Symp. Pub. Div., Pergamon Press: London and New York 7:125-143.
In houseflies, the absence of retinal, retinol and carotenoids from the

diet can induce a condition analogous to "nutritional night blindness". Evidence interpreted as signifying different kinds of receptors in the fly may really indicate only that different numbers are stimulated. ABSTRACT: BA 48:10649(119590), Dec. 1, 1967.

3624. Goldsmith, T. H. and **Fernandez, H. R. 1966** Sensitivity of Compound Eyes to Ultraviolet Light. Am. Zool. 6(4):538.
A mutant white eye housefly was used in work to extend the measurements of spectral sensitivity further into the ultraviolet than previously attempted.

3625. Goldsmith, T. H. and **Fernandez, H. R. 1968** Comparative Studies of Crustacean Spectral Sensitivity. Zeit. Vergl. Physiol. 60:156-175.
Spectral sensitivities of the lateral eyes of several crustaceans have been measured. *Musca domestica*, having ultraviolet receptors, is referred to in comparisons.

3626. Goldsmith, T. H. and **Fernandez, H. R. 1968** The Sensitivity of Housefly Photoreceptors in the Mid-ultraviolet and the Limits of the Visible Spectrum. J. Exp. Biol. 49(3):669-677.
Short wavelength limit of housefly's visible spectrum, determined by UV light availability, is about 300, in nature. The long wavelength is set by the falling absorption of visual pigment in the red. ABSTRACT: BA 50:9765(101151), Sept. 15, 1969.

Goldsmith, T. H., joint author. *See* Eichenbaum, D., et al., 1967; Hays, D., et al., 1969; Post, C. T., Jr., et al., 1969; Waterman, T. H., et al., 1969.

Gomes, F., joint author. *See* Machado, C., et al., 1967.

3626a. Goodchild, B., et al. 1970 The Separation of Multiple Forms of Housefly 1, 1, 1-Trichloro-2, 2-bis-(p-chlorophenyl) Ethane (DDT) Dehydrogenase by Electrofocusing and Electrophoresis. Biochem. J. 117:1005-1009.
Separation occurred in a sucrose gradient at pH values between 5 and 8. There are two major fractions. ABSTRACT: BA 51:12050(122379), Nov. 1, 1970.

3627. Goodhue, L. D. 1965 How it all Began. Aerosol Age. May. 5 pp.
Resume of early days of modern aerosol industry.

3628. Goodhue, L. D. and **Howell, D. E. 1960** Repellents and Attractants in Pest Control Operations. Pest Control Magazine. August. 4 pp.
Eight repellents are described; several of use against the housefly.

Goodhue, L. D., joint author. *See* Stansbury, R. E., et al., 1960.

Goodland, H., joint author. *See* Pedler, C., et al., 1965.

Goodman, R. M., joint author. *See* Terner, J. Y., et al., 1965.

3629. Goodman, T., Morrison, P. E. and **Davies, D. M. 1968** Cytological Changes in the Developing Ovary of the House Fly Fed Milk and Other Diets. Canad. J. Zool. 46(3):409-421.
Names assigned to developing ovarian stages in *Drosophila melanogaster* are used for the 14 stages in *Musca domestica*. ABSTRACTS: BA 49:9251 (102735), Oct. 1, 1968; RAE-B 57:267(1012), Dec., 1969.

3630. Gorbacheva, E. A., Semenikhina, A. D. and **Sadykova, V. R. 1967** [Information on the mechanical transport of helminth eggs by flies.] Med. Zhur. Uzbekistana 2:70. (Abstract only.)
ABSTRACT: BA 49:814(9036), Jan. 15, 1968.

Gordon, C. H., joint author. *See* Bowman, M. C., et al., 1968; Miller, R. W., et al., 1970.

3631. Gordon, H. T. and **Eldefrawi, M. E. 1960** Analog-synergism of Several Carbamate Insecticides. J. Econ. Ent. 53(6):1004-1009.

Combinations of 2 carbamate insecticides (Pyrolan and Dimetilan) applied jointly increased mortality beyond the expected additive effect when applied to houseflies and cockroaches. ABSTRACT: RAE-B 49:218, Oct., 1961.

Gordon, H. T., joint author. *See* Enan, O., et al., 1965; Fahmy, M. A. H., et al., 1965; Jao, L. T., et al., 1969.

3632. Gorecki, K. and **Swiech, J. 1966** [The repellency of some compounds tested on laboratory-reared strains of *Musca domestica* L. (Diptera) and *Aedes aegypti* L. (Diptera).] Pol. Pismo Ent. Ser. B. Ent. Stosowana 3/4(43/44):179-190. (In Polish, with English and Russian summaries.)

Four repellents were tested; indalone and dimethyl phthalate were highly effective against both insects. N, N-diethyl p-toluamide was strongest of the four, and very effective. ABSTRACTS: BA 48:5549(62033), June 15, 1967; RAE-B 57:31(127), Feb., 1969.

3633. Gorham, J. R. 1969 Hospital Pests and Hospital Pathogens. A Missing Link. Sanscript 3(1):17.

Mentions fly as a vector of diseases in hospitals. Concerns epidemic of human *Salmonella*.

3634. Goriacheva, T. A. 1966 On the Biology of *Scopeuma stercoparium* L. in the Leningrad Region. Med. Parasit. and Parasitic. Dis. 8(4): 1 page. (In Russian.)

Reference included here because *Scopeuma stercoparium* is classified in the Muscidae.

3635. Gostick, K. G. and **Hewlett, P. S. 1960** Killing House-flies (*Musca domestica* L.) by Means of Hanging Drops of Insecticide. Bull. Ent. Res. (London) 51(3):523-532.

Oil drops at lower end of vertical wires; used Shell Risella Oil 17(R17). Works in daylight, as flies attempt to light on rod. Economical in amount of insecticide required but further development needed before practical application. ABSTRACT: RAE-B 49)1):20, Jan., 1961; TDB 58:390, Mar. 1961.

3636. Goszczyńska, K. 1966 Analiza skojarzonego dzialania niektórych substancji owadobójczych w mieszaninach dwuskladnikowych. Cześć I. Mieszaniny weglowodorów chlorowanych. [Analysis of joint action of selected insecticides in two componential mixtures. Part I. Mixtures of chlorinated hydrocarbons.] Roczn. Pánstw. Zakl. Hig. 19:103-111. (In Polish, with Russian and English summaries.)

Nine two-compound mixtures of insecticides of the chlorinated hydrocarbons group were tested. Seven mixtures showed additive action. Two mixtures showed increased toxicity: DMDT with dieldrine or toxaphen.

3637. Goszczyńska, K. 1966 Analiza skojarzonego dzialania niektórych substancji owadobójczych w mieszaninach dwuskladnikowych. Cześć II. Mieszaniny zwiazków fosforoorganicznych. [Analysis of joint action of selected insecticides in two componential mixtures. Part II. Mixtures of organic phosphorus compounds.] Roczn. Pánstw. Zakl. Hig. 17:215-220. (In Polish, with Russian and English summaries.)

Four mixtures showed toxicity close to sum toxicity of both components. Five mixtures showed increased toxicity. Only dipterex with malathion showed synergism.

3638. Goszczyńska, K. 1966 Analiza skojarzonego dzialania niektórych substancji owadobójczych w mieszaninach dwuskladnikowych. Cześć III. Mieszaniny weglowodorów chlorowanych ze zwiazkami fosforoorganicznymi. [Analysis of the joint action of selected insecticides in two componential mixtures. Part III. Mixtures of chlorated hydrocarbons with phosphoorganic compounds.] Roczn. Pánstw. Zakl. Hig. 17:409-415. (In Polish, with Russian and English summaries.)
Studied 16 two-compound mixtures. Majority showed toxicity similar to sum of constituent toxicities; 5 showed increased toxicity and 2, lowered toxicity.

3639. Goszczyńska, K. 1967 Ocena skojarzonego dzialania na owady malationu i trichlorfonu w mieszaninach o róznym skladzie. [Evaluation of the joint action on insects of malathion and trichlorphon, applied in mixtures of various compounds.] Roczn. Pánstw. Zakl. Hig. 18:689-692. (In Polish, with Russian and English summaries.)
The mixtures showed increased toxicity in relation to the sum of the individual components. Most active was 0·75-4·5 parts trichlorphon to 1 part malathion (w/w).

3640. Goszczyńska, K. and **Mańkowska, H. 1967** Metoda oceny dzialania fumigacyjnego lindanu na muche domowa *Musca domestica* L. [A method for evaluation of the action of fumigating with lindane on the fly, *Musca domestica* L.] Roczn. Pánstw. Zakl. Hig. 18:471-476. (In Polish, with Russian and English summaries.)
Quantitative determination is possible according to a reference curve. Lindane was dissolved in acetone or in ethylketone.

3641. Goszczyńska, K. and **Mańkowska, H. 1968** Badania nad skojarzonym dzialaniem DDT, metoksychloru i lindanu w preparatach "Tritox". [A study on the joint action of DDT, methoxychlor and lindane applied as the preparation: Tritox.] Roczn. Pánstw. Zakl. Hig. 19:125-129. (In Polish, with Russian and English summaries.)
Mixtures, tested on *M. domestica*, showed no synergistic action; only Tritox emulsifiable concentrate showed increased toxicity.

3642. Goszczyńska, K. and **Mańkowska, H. 1969** Obserwacja zmian wrazliwósci na DDT u much *M. domestica* L. w hodowli dóswiadczalnej. [Observation of changes in the susceptibility to DDT of a laboratory strain of the housefly, *M. domestica* L.] Roczn. Pánstw. Zakl. Hig. 20:563-567. (In Polish, with Russian and English summaries.)
Topical application used for determining the increasing LD_{50} values of laboratory reared flies. Assumed that the milk fed to flies contained DDT and caused them to lose their susceptibility.

3643. Goszczyńska, K. and **Mańkowska, H. 1970** Skutecznosc preparatów Tritox (Zawierajacych mieszanine DDT + DMDT + γHCH) na muchy *M. domestica* L. oporne na DDT. [The effectiveness of Tritox preparations (containing a mixture of DDT-DMDT-gamma-HCH) against houseflies, *M. domestica* L. resistant to DDT.] Roczn. Pánstw. Zakl. Hig. 21:83-86. (In Polish, with Russian and English summaries.)
Only effective ingredient of the mixture was gamma-HCH; DDT and DMDT were completely ineffective against the resistant flies.

3644. Goszczyńska, K., Mańkowska, H., Styczyńska, B., Krzemińska, A. and **Bojanowska, A. 1968** Ocena stanu opornosci much na gamma-HCH na terenie Polski. [Evaluation of the resistance of flies to

BHC in Poland.] Roczn. Pánstw. Zakl. Hig. 19:337-342. (In Polish, with Russian and English summaries.)
A survey of flies from 7 regions in Poland showed resistance to lindane in 4 areas and susceptibility in three. The flies were most resistant around Warsaw.

3645. Goszczyńska, K. and **Styczyńska, B. 1968** Badanie hamowania aktywności esterazy cholinowej u much *M. domestica* L. pod wplywem zwiazkow fosforoorganicznych. [Inhibition of housefly head cholinesterase by organophosphorus compounds.] Roczn. Pánstw. Zakl. Hig. 19:491-497. (In Polish, with Russian and English summaries.)
Studied inhibition by malathion and diazinon; found to be more active *in vivo* than *in vitro*. ChE inhibition in flies (in state of knockdown) proved to be independent of the dose applied.

3646. Goszczyńska, K., Styczyńska, B., Mańkowska, H. and **Krzemińska, A. 1967** Skuteczność fenchlorfosu produkcji krajowej w stosunku do muchy *Musca domestica* L. [Effectiveness of the preparation fenchlorfos (Ronnel) produced in Poland against the housefly, *Musca domestica* L.] Roczn. Pánstw. Zakl. Hig. 18:519-525. (In Polish, with Russian and English summaries.)
Fenchlorfos was shown to have equal effectiveness and stability (on impregnated surfaces) as a concentrate, or in the form of a technical product used to prepare aqueous emulsions.

Goszczyńska, J., joint author. *See* Bojanowska, A., et al., 1966; Styczyńska, B., et al., 1969.

3647. Gouck, H. K. 1964 Chemosterilization of House Flies by Treatment in the Pupal Stage. J. Econ. Ent. 57(2):239-241.
Dipping 2 day old pupae in 2·5 per cent and 5 per cent apholate and metepa gave consistent sterility; dipping 1 day old pupae in tepa resulted in sterility. ABSTRACTS: BA 45:6395(79959), Sept. 15, 1964; RAE-B 52:124, July, 1964; TDB 61:1103, Oct., 1964.

3648. Gouck, H. K. and **LaBrecque, G. C. 1964** Chemicals Affecting Fertility in Adult House Flies. J. Econ. Ent. 57(5):663-664.
Ethyl bis (1-aziridinyl) phosphinylcarbamate, in fly food and sugar, induced sterility over the broadest concentration range. ABSTRACTS: BA 46:1083 (13670), Feb. 1, 1965; RAE-B 53:10, Jan., 1965; TDB 62:158, Feb., 1965.

3649. Gouck, H. K., Meifert, D. W. and **Gahan, J. B. 1963** A Field Experiment with Apholate as a Chemosterilant for the Control of House Flies. J. Econ. Ent. 56:445-446.
Cornmeal bait containing 0·75 per cent of apholate when applied on a dump 1-5 times per week decreased the fly population as shown by grid counts and egg hatch. Male fertility decreased little with weekly application, but averaged low. ABSTRACTS: BA 45:1410(17527), Feb. 15, 1964; RAE-B 51:234, Nov., 1963; TDB 61:224, Feb., 1964.

Gouck, H. K., joint author. *See* LaBrecque, G. C., et al., 1963; Fye, R. L., et al., 1965, 1966.

Govind, R., joint author. *See* Mukerjea, T. D., et al., 1960.

Gower, J. C., joint author. *See* Sawicki, R. M., et al., 1962.

Grabbe, R. R., joint author. *See* Bull, D. L., et al., 1967.

3650. Gradidge, J. M. G. 1963 The Control of Flies Breeding in Poultry Houses. Sanitarian (London) 71(8):400-403.
Reviews bionomics of *Fannia canicularis* frequently found in poultry houses of Britain. Gives directions for use of fenchlorphos and diazinon on fly strings. Proved to be most effective method of control. ABSTRACT: RAE-B 54:9, Jan., 1966.

3651. Gradidge, J. M. G. 1965 Now is the Time to Deal with Flies in Poultry Houses. Reprinted from "Agriculture" (Mar., 1965), pp. 138-141.
Deals chiefly with *Fannia canicularis*, its life history and control. Emphasizes use of fly strings and sprays.

Graham, O. H., joint author. *See* Hoffman, R. A., et al., 1965.

3652. Granett, P. and **Hansens, E. J. 1961** Tests against Face Flies on Cattle in New Jersey During 1960. J. Econ. Ent. 54(3):562-566.
No treatment considered outstanding. Some relief from annoyance attributed to use of synergized pyrethrins plus repellents. ABSTRACTS: BA 36(17):5560(58117), Sept. 1, 1961; RAE-B 50:129, June, 1962.

3653. Granett, P., Hansens, E. J. and **Forgash, A. J. 1962** Tests Against Face Flies on Cattle in New Jersey During 1961. J. Econ. Ent. 55(5):655-659.
None of the methods of application gave more than a few hours relief, but stabilized dust of DDVP was most effective. ABSTRACTS: BA 41:946 (12330), Feb. 1, 1963; RAE-B 51:38, Feb., 1963.

Granett, P., joint author. *See* Hansens, E. J., et al., 1963, 1965.

3654. Grasso, A., Boccacci, M. and **Quintiliani, M. 1962** Lethal Effects of X-rays on the Housefly, *Musca domestica* L. Sci. Rept. Ist. Super. Sanita 2(3):292-300. (Rec. 1963.)
Adults highly resistant (LD_{50} = 60,000r). Preimaginal stages more radiosensitive (pupae LD_{50} = 800r; 3rd stage larvae LD_{50} = 700r). Iodoacetic acid administered before irradiation enhances the effect. ABSTRACT: BA 45:2559(31786), Apr. 15, 1964.

3655. Grasso, A., Boccacci, M. and **Quintiliani, M. 1963** Lethal Effects of X-rays on *Musca domestica* L. Int. J. Radiat. Biol. 6(4):383. (Abstract only.)
Radiosensitivity shows no relation to resistance to insecticides; varies with the biological cycle, the preimaginal stages being more radiosensitive. Effects of radiation under other situations also given. ABSTRACT: BA 45:35, Jan. 1, 1964.

3656. Grasso, A. and **Majori, G. 1969** Toxicity of Ouabain (G-strophanthin) to Adult House Flies. J. Econ. Ent. 62(4):944-945.
Very toxic if injected; less so when imbibed; least toxic when applied topically. Three strains of flies used.

3657. Gratz, N. G. 1961 Insecticide-Resistance to Bed Bugs and Flies in Zanzibar. Bull. W. H. O. 24:668-670.
Houseflies from several rural areas of the island showed considerable resistance to dieldrin and DDT. ABSTRACT: RAE-B 51:117, June, 1963.

3658. Gratz, N. G. 1966 The Effect of the Development of Dieldrin Resistance on the Biotic Potential of House Flies in Liberia. Acta Tropica 23(2):108-136. (German and French summaries.)
In the strains examined, there is a significant increase in the number of eggs laid by 1st generation progeny of flies selected for dieldrin resistance. ABSTRACTS: BA 47:9909(114592), Dec. 1, 1966; RAE-B 56:153(537), Aug., 1968; TDB 63:1401, Dec., 1966.

3659. Gratz, N. G., Saccà, G. and **Keiding, J. 1964** Dichlorvos for the Control of Houseflies. Riv. di Parassit. (Rome) 25(4):269-278.
In Italy and Denmark good results were achieved by using a formulation consisting of plastic resin strips or cords, impregnated with 20 per cent dichlorvos. ABSTRACT: TDB 62:1061, Oct., 1965.

Greathead, D. J., joint author. *See* Legner, E. F., et al., 1969.

Green G., joint author. *See* Sawicki, R. M., et al., 1965.

3660. **Green, L. R.** and **Dorough, H. W. 1968** House Fly Age as a Factor in Their Response to Certain Carbamates. J. Econ. Ent. 61:88-90.
Adult house flies treated topically with Baygon®, Banol® and carbaryl were most resistant to these chemicals when 5 days old. Susceptibility was greatest in adults 1 and 15 days old. ABSTRACTS: BA 49:5303(59511), June 1, 1968; RAE-B 56:138(483), July, 1968; TDB 65:1062(2412), Aug., 1968.

Green, N., joint author. *See* Gersdorff, W. A., et al., 1961.

3661. **Greenberg, B. 1960** House Fly Nutrition. II. Comparative Survival Values of Sucrose and Water. Ann. Ent. Soc. Am. 53(1):125-128.
Sucrose takes precedence over water in fly survival. Fly's position at edge of food is a resultant of the two forces of hunger drive and an avoidance reaction. ABSTRACT: BA 35:2918(33183), June 15, 1960.

3662. **Greenberg, B. 1960** Hostcontaminant Biology of Muscoid Flies: I. Bacterial Survival in the Pre-adult Stages and Adults of Four Species of Blow Flies. J. Insect Pathol. 2(1):44-54.
Blow fly prepupae less variable than house fly in their bacterial counts. Other comparisons. House fly has many sterile or low-count individuals. ABSTRACT: BA 35:3997(46046), Aug. 15, 1960.

3663. **Greenberg, B. 1961** Mite Orientation and Survival on Flies. Nature 190(4770):107-108.
Describes the behavior of *Myianoetus muscarum* (Anoetidae) on adult *Musca stabulans.*

3664. **Greenberg, B. 1962** Host-contaminant Biology of Muscoid Flies. II. Bacterial Survival in the Stable Fly, False Stable Fly, and the Little House Fly. J. Insect Pathol. 4(2):216-223.
A similarity exists between the bacterial survival patterns in the false stable fly and the house fly. Results indicate that sterilization of the digestive tract is a usual consequence of metamorphosis in this group. ABSTRACT: BA 40:323(4180), Oct. 1, 1962.

3665. **Greenberg, B. 1962** Host-contaminant Biology of Muscoid Flies. III. Effect of Hibernation, Diapause, and Larval Bactericides on Normal Flora of Blow-fly Prepupae. J. Insect Pathol. 4:415-428.
Mentions demonstration of acid middle portion of *Musca domestica* intestinal tract, in comparison with that of blow flies.

3666. **Greenberg, B. 1962** Autosomal Recessive White-eyed Mutation in *Muscina stabulans* (Fall). Nature 195(4845):1026.
Refers to sex-linked and sex-limited white-eyed mutation described by Paterson (1958). This (autosomal) is believed to be the first such case reported in muscoid Diptera.

3667. **Greenberg, B. 1964** Experimental Transmission of *Salmonella typhimurium* by Houseflies to Man. Am. J. Hyg. 80(2):149-156.
Infection was from a dog to human volunteers by means of flies and drink. Discusses dosage delivery of flies in relation to human infective doses and fecal titers in salmonellosis, shigellosis and poliomyelitis. ABSTRACT: BA 46:1728(21586), Mar. 1, 1965.

3668. **Greenberg, B. 1965** Flies and Disease. Sci. Amer. 213:92-99. July.
Discusses the symbiosis of flies and microorganisms and its use in better understanding infectious diseases. Reviews author's work with *Salmonella* and *Proteus* bacteria in the fly's digestive tract.

3669. **Greenberg, B. 1966** Bacterial Interactions in Gnotobiotic Flies. IX. International Congress for Microbiology. Moscow. Symp. D II Gnotobiology pp. 371-380.
Investigations dealing with the biotic community within the fly and

with its population dynamics. Description given of typical bacterial interactions occurring in each stage of the life cycle of *Calliphora vicina* and in successive regions of the larva's gut.

3670. Greenberg, B. 1968 Model for Destruction of Bacteria in the Midgut of Blow Fly Maggots. J. Med. Ent. 5(1):31-38.
Three different models were devised employing various physical and biotic factors known to be present in the maggot gut.

3671. Greenberg, B. 1968 Micro-Potentiometric pH Determinations of Muscoid Maggot Digestive Tracts. Ann. Ent. Soc. Am. 61(2):365-368.
Crop pH is unbuffered and fluctuates with food pH which is influenced by microbes. Three mid-gut regions are buffered; pH there is microbe-independent. Gives pH for several divisions of the digestive tract of *Musca domestica*. ABSTRACT: BA 49:6777(75583), July 15, 1968.

3672. Greenberg, B. 1969 *Salmonella* Suppression by Known Populations of Bacteria in Flies. J. Bact. 99:629-635.
Used *Calliphora vicina*. In dibiotic interaction, *Salmonella typhimurium* dominated *Streptococcus faecalis*, was dominated by *Proteus mirabilis*. In tribiotic, which included *Pseudomonas mirabilis*, *Salmonella* tended to be eliminated from the gut.

3673. Greenberg, B. and **Bornstein, A. A. 1964** Fly Dispersion from a Rural Mexican Slaughterhouse. Am. J. Trop. Med. Hyg. 13(6):881-886.
Fluorescein-tagged flies (6 species) were recovered up to 3 miles from their origin; from residential sites, market places, a dairy and a neighboring village. ABSTRACTS: BA 46:2182(27212), Mar. 15, 1965; RAE-B 54:125, July, 1966.

3674. Greenberg, B. and **Burkman, A. M. 1963** Effect of B-vitamins and a Mixed Flora on the Longevity of Germ-free Adult Houseflies, *Musca* domestica L. J. Cell. and Comp. Physiol. 62:17-22.
Three experimental diets were fed to flies. Results suggest that a sufficient B-vitamin reserve is carried over from the larvae to last the life of the fly. ABSTRACT: BA 45:3614(44514), May 15, 1964.

3675. Greenberg, B. and **Carpenter, P. D. 1960** Factors on Phoretic Association of a Mite and Fly. Science 132:738-739.
Pupae produce a volatile substance which is attractive to hypopi (deutonymphs). Mites (*Myianoetus muscarum* L.) attracted as readily in dark as in light. *M. domestica* pupae found attractive.

3676. Greenberg, B. and **Miggiano, V. 1963** Host-Contaminant Biology of Muscoid Flies. IV. Microbial Competition in a Blowfly. J. Infect. Dis. 112:37-46.
Concerns quantitative survival of *Salmonella typhimurium* in *Calliphora*. Observation not previously noted in other flies: bacteria (*Escherichia coli* and *Proteus mirabilis*) usually multiplied in conventional mono- and bi-contaminated early pupae and adults.

3677. Greenberg, B., Varela, G., Bornstein, A. and **Hernandez, H. 1963** Salmonellae from Flies in a Mexican Slaughter House. Am. J. Hyg. 77:177-183.
Twelve salmonella types isolated; 5 from *Musca domestica*. *Salmonella derby* was most prevalent. This is the first report of salmonella isolation from flies in Mexico. ABSTRACTS: BA 45:6638(83447), Oct. 1, 1964; RAE-B 52:86, May, 1964.

3678. Greenwood, E. S. 1963 *Bacillus thuringiensis* (Eubacteriales) in the Control of Some Diptera. Symposium. Proc. Ent. Soc. New Zealand (May 14-16). Page 1.

3679. **Greenwood, E. S. 1964** *Bacillus thuringiensis* in the Control of *Lucilia sericata* and *Musca domestica*. New Zeal. J. Sci. 7(2):221-226.
Suspensions of *B. thuringiensis* in larval rearing medium killed many larvae, resulted in few normal adults, and those surviving produced fewer and mostly infertile eggs. ABSTRACTS: BA 46:2178(27166), Mar. 15, 1965; RAE-B 53:188, Sept., 1965.

3680. **Greenwood, E. S. 1964** Bacterium Kills Insect Pests. New Zealand J. Agric. 2 pp. April.
A review of the possible use of *Bacillus thuringiensis* in fly control.

3681. **Gregg, C. T., et al. 1964** Inhibition of Oxidative Phosphorylation and Related Reactions in Insect Mitochondria. Biochim. Biophys. Acta 82:343-349.

3682. **Gregg, C. T., Heisler, C. R.** and **Remmert, L. F. 1960** Oxidative Phosphorylation and Respiratory Control in Housefly Mitochondria. Biochim. Biophys. Acta 45:561-570.

3683. **Gregor, F.** and **Povolony, D. 1964** Eine Ausbeute von Synanthropen Fliegen aus Tirol. [On Synanthropic flies in Tyrol.] Zool. Listy 13(3):229-248.
Does not treat of *M. domestica*. Many other species listed with excellent illustrations of genitalia.

Gregory, G. E., joint author. *See* Farnham, A. W., et al., 1966; Sawicki, R. M., et al., 1966.

3684. **Grigolo, A. 1967** Determinazione quantitativa di alcuni metaboliti del triptofano in ceppi normali e mutanti per il colore degli occhi di *Musca domestica* L. [Quantitative determination of certain metabolites of tryptophane in normal strains and in eye-color-mutant strains of *Musca domestica*.] Boll. di Zool. 34:125.
Studies were made on pupae aged 1, 2, 3, 4, and 5 days, and on adult flies 3 days emerged, without distinction as to sex.

3685. **Grigolo, A., et al. 1969** [Research on the activity of succinate dehydrogenase in mitochondria preparations of *Musca domestica* L.] Riv. di Parassit. 30:319-327. (In Italian.)

3686. **Grigolo, A. 1969** Variazioni quantitative del triptofano, dell acido chinurenico e della 3-OH chinurenina in ceppi di *Musca domestica* L. Estratta da Redia 51:179-185. (English summary.)
Determined variations of the amount of tryptophan, kynurenic acid and 3-0H kynurenine during pupal period and for 3 day old adults in 4 normal and 6 eye-color-mutant strains of *M. domestica*.

3687. **Grigolo, A. 1969** Attività della chinurenina-formamidasi (formilasi) in adulti di ceppi normali e mutanti per il colore dell'occhio di *Musca domestica* L. Estratto da Redia 51:169-178.
The activity of the enzyme kynurenine-formamidase has been determined by the spectrophotometric method for normal strains and eye-color-mutant strains of *M. domestica*.

3688. **Grigolo, A.** and **Cima, L. 1969** Variazioni quantitative degli ommocromi in vari ceppi di *Musca domestica* L. [Quantitative variations in ommochromes in some strains of *Musca domestica* L.] Riv. di Parassit. 30(3):243-248.
Determined the quantitative variation in ommochromes for males and females of normal strains and for eye color mutants of *M. domestica*. Normal strains have greater amounts of ommochromes, the males having more than the females, in all strains. ABSTRACT: BA 51:12641(128547), Dec. 1, 1970.

3689. **Grigolo, A.** and **Laudani, U. 1966** Osservazioni sul consumo di ossigeno durante il ciclo vitale di *Musca domestica* L. [Observations on the consumption of oxygen during the life cycle of *Musca domestica* L.] Riv. di Parassit (Rome) 27(1):39-46. (English summary.)
Oxygen consumption, during entire life cycle of *M. domestica*, has been referred to the wet weight and to the single fly. During preimaginal development, 3-day larvae have greatest uptake. Adult values depend on sex. ABSTRACT: BA 49:3835(43130), Apr. 15, 1968.

3690. **Grigolo, A.** and **Laudani, U. 1969** Richerche sull' attivita' dell' enzima succino-deidrogenasi in preparazioni mitocondriali di *Musca domestica* L. [Research on the succinic dehydrogenase activity in mitochondrial preparations of *Musca domestica* L.] Riv. di Parassit. 30(4):319-327.
Studied mitochrondria in thoracic muscles. Males show greater activity; reaches maximum 4th day after emergence. Females show gradual increase with peak on the 12th day.

3691. **Grigolo, A., Laudani, U.** and **Franco, M. G. 1965** Resistance to DDT and dehydrochlorinase activity in the strain *bwb kdr* of *Musca domestica* L. Genetica Agraria 19(4):314-326.
Determined mortality rates at different hours following topical applications of DDT. The DDT recovered from flies used to determine their ability to metabolize DDT to DDE. ABSTRACT: BA 48:8786(98705), Oct. 1, 1967.

3692. **Grigolo, A.** and **Oppenoorth, F. J. 1966** The importance of DDT-dehydrochlorinase for the effect of the resistance gene *kdr* in the housefly *Musca domestica* L. Genetica 37(2):159-170.
Alleles producing DDT-dehydrochlorinases, when combined with the *kdr* factor in the heterozygous condition, act as dominance-modifiers. ABSTRACTS: BA 48:2864(31557), Apr. 1, 1967; RAE-B 57:115(408), July, 1969.

Grigolo, A., joint author. *See* Laudani, U., et al., 1965, 1966, 1968, 1969; Visona, L., et al., 1965.

3693. **Gromov, A. I., 1968** [On the significance of flies in the epidemiology of dysentery on Sakhalin.] Zhur. Mikrobiol. 45:104-105. (In Russian.)

Grosch, D. S., joint author. *See* Ferrer, F. R., et al., 1968.

3694. **Gross, B. 1960** Emergency Vector Control Following a Tsunami. Calif. Vector Views. 7(11):65-67.
Vector control program very successful; demonstrated importance of speedy mobilization of existing resources and early initiation of control measures.

Grozdeva, I. V., joint author. *See* Dremova, V. P., et al., 1961; Sukhova, M. N., et al., 1965.

Grudtsina, M. V., joint author. *See* Sychevskaya, V. I., et al., 1960.

3695. **Grzywacz, M. 1967** [Toxic effect of new diacylhydrazines and arylhydrazines of caroboxylic acids on *Musca domestica* L.] Wiad. Parazytol. 13:399-404. (In Polish.)

Guarniera, D., joint author. *See* Saccà, G., et al., 1967.

3696. **Gudnadóttir, M. G. 1961** Studies of the Fate of Type 1 Polioviruses in Flies. J. Exp. Med. 113:159-176.
Phormia regina selected for most experiments; *M. domestica* used in one. Flies and excreta remained infective for 11 days at room temperature and for 3 months in hibernation.

3697. Gudnadóttir, M. G. and **Paul, J. R. 1960** Studies of the Fate of Type 1 Polioviruses in Flies. Bact. Proc. 60:104. (Abstract only.)

3698. Guenthner, A. W. and **Ware, G. W. 1967** Effects of X-irradiation on Toxicity of Malathion, Heptachlor, and Temick to the House Fly. J. Econ. Ent. 60:369-373.

The varied effect of irradiation with different toxicants may, in some cases, be attributed to altered ratios or increased levels of esterases involved in detoxication. ABSTRACTS: BA 48:6927(77469), Aug. 1, 1967; RAE-B 55:141(503, Aug., 1967; TDB 64:1040, Sept., 1967.

3699. Guerra, A. A. 1966 A Comparative Study of the Structure and Biochemical Activity of Flight Muscles for Several Insect Species. Diss. Absts. 26(10):6247.

Includes *Musca autumnalis.*

3700. Guerra, A. A. and **Cochran, D. G. 1970** Respiration During the Life Cycle of the Face Fly. J. Econ. Ent. 63(3):918-921.

Using a Warburg constant volume respirometer, rapid respiration was demonstrated in embryonic, larval, prepupal and adult stages. Results interpreted in terms of O_2 consumption as an indicator of metabolic activity. ABSTRACT: BA 51:13731(139796), Dec. 15, 1970.

3701. Gunderson, H. 1964 Controlling Household Pests! Pamphlet 143(Rev). Iowa State Univ. Science Tech. Cooperative Extension Service. 11 pp.

A pamphlet concerning common household pests which may carry disease, be annoying, or attack fabrics, foods and structures. Control methods are presented, using trade names of effective pesticides.

3702. Güneidy, A. M. and **Busvine, J. R. 1964** Genetical Studies on Dieldrin-Resistance in *Musca domestica* L. and *Lucilia cuprina* (Wied.). Bull. Ent. Res. 55(3):499-507.

Mode of inheritance investigated by crossing experiments. A monofactorial inheritance is indicated. ABSTRACTS: BA 47:6431(75278), Aug. 15, 1966; RAE-B 53:31, Feb., 1965; TDB 62:601, June, 1965.

3703. Günther, S. 1964 Bedeutung und Vorkommen synanthroper Dipteren auf Müllhalden. [Significance and occurrence of synanthropic Diptera on dumping grounds.] Zeit. Ges. Hyg. 10:601-607. (Russian and English summaries.)

A survey of 5 dumping grounds shows that a strong influence is exerted upon neighboring biozoenoses insofar as the number of flies leaving exceeds those coming to the dumps from elsewhere. Regular covering with sand prevents fly population build-up.

3704. Günther, S. 1965 Synanthrope Fliegen und ihre medizinische Bedeutung. [Synanthropic flies and their medical importance.] Angew. Parasit. 5 Beih. 1:87-89.

Relative frequency of certain domestic flies of importance to public health was investigated in East Germany. Information is compared with that reported by other authors in Europe. ABSTRACT: RAE-B 55:158(555), Sept., 1967.

3705. Gurland, J., Ilbok, L. and **Dahm, P. A. 1960** Polychotomous Quantal Response in Biological Assay. Biometrics 16(3):382-398.

House fly used in biological assay to estimate the amount of guthion residue present in a test preparation prepared from guthion sprayed alfalfa. Made note of flies affected 17 hours after exposure.

Gurland, J., joint author. *See* Dahm, P. A., et al., 1961.

3706. Gürtunca, S. 1965 Acetylcholin ve cholinesteraz. [Acetycholine and cholinesterase.] Ankara Univ. Vet. Fak. Dergisi 12(4):337-344.

(English summary.)
Presence of active cholinesterase in housefly head demonstrated by color reaction based upon formation of hydroxamic acid which reacts with ferric chloride solution. Discusses possible mechanisms underlying the effect of organic phosphate insecticides. ABSTRACT: BA 51:5085(52375), May, 1970.

Gutfreund, D. E., joint author. *See* Rockstein, M., et al., 1961.

3707. **Guthrie, F. E.** and **Hodgson, E. 1968** Adaptation of Insects to Nornicotine. Ann. Ent. Soc. Am. 61(2):545-547.
Nornicotine 2nd most important toxic constituent of commercial tobacco leaves. Metabolic processes of treated insects (including housefly) followed by chromatographic procedures. Metabolic end product not identified but characteristics of configuration established. ABSTRACT: BA 49:6777(75586), July 15, 1968.

Guthrie, F. E., joint author. *See* Chiu, Y. C., et al., 1968; Ferrer, F. R., et al., 1968; Meksongsee, B., et al., 1967.

Gutsevich, A. V., joint author. *See* Balashov, Y. S., et al., 1968.

Guyer, G. E., joint author. *See* Borgatti, A. L., et al., 1962, 1963; Hoopingarner, R. A., et al., 1966.

3708. **Gvozdeva, I. V. 1969** [Synthesis and investigation of synergistic properties in bicarpolate.] Med. Parazit. (Moskva) 38:57-62. (In Russian.)

3709. **Gvozdeva, I. V. 1962** [Insecticidal properties of aqueous solutions of chlorophos with ammonium carbonate and its use for destruction of houseflies in Central Asia.] Med. Zhur. Uzbekistana 3:15-19. (In Russian.)

3710. **Gvozdeva, I. V. 1962** [Use of a new insecticide—trichlorometa-phosphate 3—— in controlling fly larvae in cities of Tashkent and Iagiiula.] Med. Zhur. Uzbekistana:79-80. (In Russian.)

3711. **Gwiazda, M.** and **Bakuniak, E. 1967** [Studies on the control of houseflies using 0,0-dimethyl-0-2,4,5-trichlorphenyl phosphorothioate.] Wiad. Parazytol. 13:431-435. (In Polish.)

3712. **Gwiazda, M.** and **Lord, K. A. 1967** Factors Affecting the Toxicity of Diazinon to *Musca domestica* L. Ann. Appl. Biol. 59(2):221-232.
Evidence shows that the slower penetration of diazinon through the integument of resistant flies and their ability to decompose and detoxify diazinon seem to be important causes of resistance. ABSTRACTS: BA 48: 10182(114408), Nov. 15, 1967; RAE-B 56:47(158), Feb., 1968; TDB 64:1039, Sept., 1967.

Habutsu, Y., joint author. *See* Kano, R., et al., 1965.

3713. **Hadaway, A. B., Barlow, F.** and **Duncan, J. 1963** Effects of Piperonyl Butoxide on Insecticidal Potency. Bull. Ent. Res. 53(4):769-778.
High degree of synergism obtained with natural pyrethrins and with carbamate against house flies. P.B. was less effective with synthetic pyrethroids (allethrin and dimethrin) and was found to antagonize the action of malathion. ABSTRACT: RAE-B 51:51, Mar., 1963.

Haenlein, G. F. W., joint author. *See* MacCreary, D., et al., 1962.

3714. **Hafez, M.** and **Gamal-Eddin, F. M. 1961** The Behaviour of the Stable Fly Larvae, *Stomoxys calcitrans* L. Towards Some Environmental Factors (Diptera:Muscidae). Bull. Soc. Ent. Egypt. 45:341-367.
Compares behavior of *Stomoxys* with that of *M. domestica* and *M. sorbens* in relation to humidity, temperature, odor and normal environment. ABSTRACT: RAE-B 52:221, Dec., 1964.

3715. Hafez, M. and **Gamal-Eddin, F. M. 1966** On the Bionomics of *Musca crassirostris* Stein, in Egypt (Diptera:Muscidae). Bull. Soc. Ent. Egypt. 50:25-40.
Reports similarities and differences between *M. crassirostris* and other muscoid species, previously studied. ABSTRACT: BA 51:12573(127832), Nov. 15, 1970.

3716. Hafez, M., Osman, F. M., El-Ziady, S., El-Moursy, A. A. and **Erakey, A. S. 1969** Studies on Control of House Flies in Egypt by Chemosterilants. I. Laboratory Studies on *Musca domestica vicina*. J. Econ. Ent. 62(2):324-329.
Tested tepa, metepa, apholate and hempa, administered in adult food, by exposure to residues on glass and by treatment of larvae. ABSTRACTS: RAE-B 57:183(654), Sept., 1969; TDB 66:1339(2536), Dec., 1969.

3717. Hafez, M., Osman, M. F. and **Erakey, M. A. S. 1970** Studies on Control of House Flies in Egypt by Chemosterilants. II. Effect of Metepa on *Musca sorbens*. J. Econ. Ent. 63(1):213-214.
Metepa successfully sterilized adult flies when used as a food additive, also a residual film. Females proved more susceptible than males to this chemosterilant. ABSTRACTS: BA 51:10164(103300), Sept. 15, 1970; RAE-B 58:231(948), July, 1970.

3718. Hafez, M., Osman, M. F. and **Erakey, M. A. S. 1970** Studies on Control of House Flies in Egypt by Chemosterilants. III. Sterilization of *Musca sorbens* by Apholate, Tepa and Hempa. J. Econ. Ent. 63(4):1167-1169.
Concentration of 0·4-1·6 per cent apholate and 0·4-0·8 per cent tepa inhibited oviposition in females and induced partial sterility in males. With hempa, minimum effective concentration for females was 0·2 per cent; for males 1·2 per cent.

Hafis, S., joint author. *See* Zuberi, R. I., et al., 1969.

Hagino, K., joint author. *See* Miyamoto, K., et al., 1965.

3719. Hair, J. A. and **Adkins, T. R., Jr. 1964** Sterilization of the Face Fly, *Musca autumnalis*, with Apholate and Tepa. J. Econ. Ent. 57(4): 586-589.
Pupae were dipped in a solution of apholate, and adults received either apholate or tepa in their diet. Complete or partial sterility achieved, according to concentration of chemo-sterilant and/or duration of diet. ABSTRACTS: BA 45:8063(102266), Dec. 1, 1964; RAE-B 52:206, Nov., 1964; TDB 61:1261, Dec., 1964.

3720. Hair, J. A. and **Adkins, T. R., Jr. 1965** Dusting Stations and Cable Backrubbers as Self-Applicatory Devices for Control of the Face Fly. J. Econ. Ent. 58:39-41.
Good to excellent resuιts obtained with various insecticides by both cable rubbers and dusting bags. ABSTRACT: BA 46:4042(50018), June 1, 1965.

3721. Hair, J. A. and **Turner, E. C., Jr. 1965** Attempted Propagation of *Nasonia vitripennis* on the Face Fly. J. Econ. Ent. 58:159-160.
Was unsuccessful; thousands of face fly puparia subjected to adult parasites failed to produce parasite offspring. ABSTRACTS: BA 46:4043 (50024), June 1, 1965; RAE-B 53:98, May, 1965.

3722. Hair, J. A. and **Turner, E. C., Jr. 1966** Susceptibility of Mature and Newly Emerged Face Flies to Chemosterilization with Apholate. J. Econ. Ent. 59:452-454.
The two groups equally susceptible to topical application, but mature females were more easily rendered sterile by apholate in the diet. Explained by the fact that mature females consume twice as much food

as newly emerged females. ABSTRACTS: BA 47:7234(84397), Sept. 1, 1966; RAE-B 54:133, July, 1966.

Hair, J. A., joint author. *See* Turner, E. C., Jr., et al., 1966, 1967.

Halberg, F., joint author. *See* Sullivan, W. N., et al., 1970.

3723. **Hale, J. H., Davies (Lloyd-Davies), T. A.** and **NG Cheng Hin, W. K. 1960** Flies in Aeroplanes as Vectors of Faecal-borne Disease. Trans. Roy. Soc. Trop. Med. Hyg. 54:261-262.
Faecal organisms were recovered from 9 of 34 flies taken from 9 planes. Flies from 32 planes were cultured for cholera, dysentery and salmonella, but no pathogens were recovered. Authors advocate rigid fly control. ABSTRACT: BA 36:2707(25656), April 15, 1961.

Halevy, A. H., joint author. *See* Tahori, A. S., et al., 1965.

Hall, M. S., joint author. *See* Ballard, R. C., et al., 1969.

3724. **Hall, W. E.** and **Sun, Y.-P. 1965** Mechanism of Detoxication and Synergism of Bidrin Insecticide in House Flies and Soil. J. Econ. Ent. 58:845-849.
Effect of synergist is not on the rate of penetration or transport to active site. Probably 2 metabolic pathways for Bidrin: (1) with hydrolysis, (2) with oxidative dealkylation of the —$N(CH_3)_2$ group. Sesamex affects both. ABSTRACTS: BA 47:2949(34930), Apr. 1, 1966; TDB 63:239, Feb., 1966.

Hall, W. E., joint author. *See* Corey, R. A., et al., 1965.

Hamada, M., joint author. *See* Ohno, M., et al., 1960.

Hammad, S. M., joint author. *See* El-Deeb, A. L., et al., 1969.

3725. **ben Hannine, S.** and **Keiding, J. 1967** Strategisk bekaempelse af fluer i stalde. [Strategic fly control in piggeries and cowsheds.] Ann. Rept. 1966 Govern. Pest Infest. Lab. pp. 30-32, 36-37.
Strategic spraying with dimethoate gave satisfactory results even with 10 per cent and 30 per cent coverage of the area, if the areas sprayed are well chosen.

3726. **ben Hannine, S.** and **Keiding, J. 1968** Strategisk bekaempelse af fluer i stalde. [Strategic fly control in piggeries and cowsheds.] Ann. Rept. 1967 Govern. Pest Infest. Lab. Denmark. pp. 33-35, 37-41.
Results presented as graphs. Strategically selected surfaces were either sprayed or painted.

ben Hannine, S., joint author. *See* Keiding, J., et al., 1964, 1966, 1967; Mourier, H., et al., 1969.

Hansen, P. A., joint author. *See* Conner, R. M., et al., 1967.

3727. **Hansens, E. J. 1960** Field Studies of House Fly Resistance to Organophosphorus Insecticides. J. Econ. Ent. 53(2):313-317.
Cross resistance to Diazinon, ronnel, lindane and DDT existed throughout the area studied, with highest resistance to DDT. Lindane, DDT and related materials will not give control where organophosphorus resistance exists. ABSTRACT: BA 35:3740(42999), Aug. 1, 1960.

3728. **Hansens, E. J. 1961** Face Fly Promises to Become No. 1 Cattle Problem in N. E. New Jersey Agric. 43(3):11-12.
The situation of the face fly problem in the Northeastern U.S. is discussed. ABSTRACT: BA 36:6871(74118), Nov. 1, 1961.

3729. **Hansens, E. J. 1963** Fly Populations in Dairy Barns. J. Econ. Ent. 56:842-844.
Variations in abundance of species in New Jersey. *M. domestica* increases greatly late in the season. *Fannia* spp. are abundant early but disappear with increasing warmth. *Stomoxys* fluctuates with local breeding conditions. ABSTRACTS: BA 45:3600(44350), May 15, 1964; RAE-B 52:52, Mar., 1964.

3730. Hansens, E. J. 1963 Area Control of *Fannia canicularis*. J. Econ. Ent. 56(4):541.
A mist blower used to apply a spray made up of 8 gal. DDT25E and 2 qts. malathion 57E per 100 gal. Used along roadsides and in yards of the community, it greatly reduced populations of the little housefly near a poultry farm in New Jersey. ABSTRACT: BA 45:1413(17562), Feb. 15, 1964.

3731. Hansens, E. J. 1965 How Far Will a House Fly (Diptera) Fly? New Jersey Agric. 47(2):11.

3732. Hansens, E. J. 1965 Apholate Reduces Fertility of Flies. New Jersey Agric. 47(4):11-12.

3733. Hansens, E. J. 1965 Effects of Apholate on Restricted Populations of Insecticide-resistant House Flies, *Musca domestica*. J. Econ. Ent. 58:944-946.
Caged populations of flies were treated with 2 per cent apholate sugar bait, apholate-impregnated cords or trichlorfon bait in various combinations. In all cases, the fly populations were significantly reduced. ABSTRACTS: BA 47:2926(34638), Apr. 1, 1966; RAE-B 54:15, Jan., 1966; TDB 63:490, Apr., 1966.

3733a. Hansens, E. J. and **Anderson, W. F. 1970** House Fly Control and Insecticidal Resistance in New Jersey. J. Econ. Ent. 63(6):1924-1926.
Several compounds used. Fly resistance increased during the season with all insecticides. Tests carried out by topical application to F_1 generation of flies collected from barns prior to treatment dates.

3734. Hansens, E. J., Benezet, H. J. and **Evans, E. S., Jr. 1967** House Fly Control and Insecticide Resistance with Continued use of Diazinon, Ronnel and Dimethoate. J. Econ. Ent. 60(4):1057-1064.
Degree of resistance varied with type of exposure, surface sprayed, sanitation, and period of fly season. Resistance to diazinon, ronnel, malathion, dimethoate, lindane and DDT was greatest in barns sprayed with diazinon. Use of ronnel did not produce high ronnel resistance. ABSTRACTS: BA 48:11208(124489), Dec. 15, 1967; RAE-B 56:6(19), Jan., 1968; TDB 65:540(1178), Apr., 1968.

3735. Hansens, E. J., Evans, E. S., Jr. and **Shibles, D. B. 1968** Tests of Bromophos for House Fly Control in New Jersey. J. Econ. Ent. 61(3):833-836.
At first gave control when used as 0·5 and 1·0 per cent residual sprays. Resistance developed to 6-14 fold (LD50 level), as compared with susceptible strains. ABSTRACTS: BA 49:10245(113084), Nov. 1, 1968; RAE-B 56:222(809), Nov., 1968.

3736. Hansens, E. J. and **Granett, P. 1963** Tests of Ciodrin and Other Materials Against Face Fly, *Musca autumnalis*. J. Econ. Ent. 56: 24-29.
Ciodrin proved considerably superior to other materials tested, and reduced face flies to a tolerable level for animals. ABSTRACTS: BA 42: 1899(23983), June 15, 1963; RAE-B 51:90, May, 1963.

3737. Hansens, E. J. and **Granett, P. 1965** Effects of Apholate on a Restricted Population of House Flies. J. Econ. Ent. 58:157-158.
Females mating with males fed on apholate did not produce fewer eggs, but those laid were less viable. Study indicates that even with best techniques, control cannot be achieved with chemosterilants alone. ABSTRACTS: BA 4044(50038), June 1, 1965; RAE-B 53:98, May, 1965; TDB 62:943, Sept., 1965.

3738. **Hansens, E. J.** and **Morris, A. P. 1962** Field Studies of House Fly Resistance to Diazinon, Ronnel, and Other Insecticides. J. Econ. Ent. 55(5):702-708.
A 2-year study. Barns were sprayed with one of the insecticides. Resistance patterns were correlated with breeding potential, sanitation, weather conditions and fly migration. Resistance to lindane and DDT were at high level throughout the area. ABSTRACTS: BA 41:955(12421), Feb. 1, 1963; RAE-B 51:41, Feb., 1963.

3739. **Hansens, E. J.** and **Valiela, I. 1967** Activity of the Face Fly in New Jersey, J. Econ. Ent. 60:26-28.
Activity influenced chiefly by incident light (if temperature is above 60-65°F.). Flies may move more than 1500 m. in 72 hours after release. Are influenced by color and by activity of the cows. ABSTRACTS: BA 48:6008(67155), July 1, 1967; RAE-B 55:103(367), June, 1967.

Hansens, E. J., joint author. *See* Benezet, H. J., et al., 1967; Abasa, R. D., et al., 1969; Evans, E. S., Jr., et al., 1968; Forgash, A. J., et al., 1960, 1962, 1967; Granett, P., et al., 1961, 1962; Morris, A. P., et al., 1966; Casagrande, R. A., et al., 1969.

Hanson, R. P., joint author. *See* Bouillant, A., et al., 1965.

Hardy, J. L., joint author. *See* Kenaga, E. E., et al., 1962, 1965; Whitney, W. K., et al., 1969.

3740. **Harris, C. R.** and **Lichtenstein, E. P. 1961** Factors Affecting the Volatilization of Insecticidal Residue From Soils. J. Econ. Ent. 54(5):1038-1045.
From 16-38 per cent of applied aldrin volatilizes from the soil, rapidly at first, then at a more constant rate. Factors involved: concentration of insecticide, soil moisture, soil temperature, RH of air above soil, rate of air movement over soil. ABSTRACT: BA 37:1119(11647), Feb. 1 1962.

Harris, E. J., joint author. *See* Steiner, L. F., et al., 1970.

3741. **Harris, R. L., Weorden, S.** and **Poan, C. C. 1961** Preliminary Study of the Genetics of House Fly (*Musca domestica*) Resistance to Malathion. J. Econ. Ent. 54(1):40-45.
Resistant strain showed a longer period from egg to adult, also a lower fecundity than the susceptible. Males of susceptible strain emerged earlier than the females. No sex-linkage in malathion resistance.

Harrison, A., joint author. *See* Brooks, G. T., et al., 1963, 1964, 1965, 1966, 1967, 1969, 1970; Winteringham, F. P. W., et al., 1961.

3742. **Harrison, R. A. 1964** How to Recognize Nuisance Flies. New Zealand J. Agric. 109(5):407, 409, 410.
Treats of *M. domestica*, *Fannia canicularis*, blowflies, mosquitoes, midges, sandflies and others.

3743. **Hart, R. J. 1962** Houseflies Resistant to Organic Phosphate in Australia. [Correspondence.] Nature 195(4846):1123-1124.
Flies resistant to DDT, dieldrin, and diazinon were collected and genetic studies begun. ABSTRACTS: BA 42:1901(23998), June 15, 1963; RAE-B 51:230, Oct., 1963; TDB 60:78, Jan., 1963.

3744. **Hart, R. J. 1963** The Inheritance of Diazinon Resistance in an Australian Strain of *Musca domestica* L. Bull. Ent. Res. 54:461-465.
Presents strong evidence for a fully dominant factor in the inheritance of diazinon resistance (Perth strain). Uses symbol OP for "resistance to organic phosphate". ABSTRACTS: RAE-B 52:4, Jan., 1964; TDB 61: 850, Aug., 1964.

3745. Hart, S. A., Black, R. J. and **Smith, A. C. 1960** Dairy Manure Sanitation Study. Calif. Vector Views 7(7):44-47.
Successful manure disposal must be measured in terms of the maintainence of general farm sanitation and the elimination of fly production.

Hart, S. A., joint author. *See* Smith, A. C., et al., 1960.

Hart, W. G., joint author. *See* Steiner, L. F., et al., 1970.

3746. Hartigan, J. J. 1964 The Synthesis and Toxicity to House Flies of Ortho Substituted Analogs of DDT. Diss. Absts. 24(10):3981-3982.

Hartigan, J., joint author. *See* Hennesy, D. J., et al., 1961.

Hartle, R. J., joint author. *See* Thayer, H. I., et al., 1965.

Hartsock, J. G., joint author. *See* Killough, R. A., et al., 1965; Pickens, L. G., et al., 1967.

3747. Harvey, T. L. 1964 House Fly Resistance to *Bacillus thuringiensis* Berliner, a Microbial Insecticide. Diss. Absts. 25(4):2463.
A resistance to *B. thuringiensis* was developed by selection. This did not induce cross resistance to DDT or malathion, or vice versa. ABSTRACT: BA 46:4777(59121), July 1, 1965.

3748. Harvey, T. L. and **Brthour, J. R. 1960** Feed Additives for Control of House Fly Larvae in Livestock Feces. J. Econ. Ent. 53(5):774-776.
Polybor 3, an effective larvicide when applied to manure, did not reduce fly numbers when mixed with cattle feed. *Bacillus thuringiensis* spores, fed daily to animals, did reduce flies from cow manure and hen droppings. ABSTRACTS: BA 36:763(9076), Feb. 1, 1961; TDB 58:533-534.

3749. Harvey, T. L. and **Howell, D. E. 1963** The Effect of *Bacillus thuringiensis* Berliner on *Drosophila melanogaster* Meigen. J. Insect Pathol. 5(4):495-497.
Experiments to test commercial product (Bakthane L 69). Known susceptibility of *M. domestica* larvae and pupae compared with findings on *Drosophila*. See following reference.

3750. Harvey, T. L. and **Howell, D. E. 1965** Resistance of the House Fly to *Bacillus thuringiensis* Berliner. J. Invertebr. Path. 7:92-100.
Both age and size of house fly larvae appear to be factors in their susceptibility to Bakthane. (*B. thuringiensis* spore powder.) ABSTRACTS: BA 47:2508(29968), Mar. 15, 1966; RAE-B 53:215, Nov., 1965; TDB 62:814, Aug., 1965.

Hashimoto, M., joint author. *See* Shimizu, F., et al., 1965.

3751. Hassan, A. and **Dauterman, W. C. 1968** Studies on the Optically Active Isomers of 0,0-diethyl Malathion and 0,0-diethyl Malaoxon. Biochem. Pharmacol. 17(7):1431-1439.
The d-isomers of malathion and malaoxon proved more toxic to houseflies than the l-isomers corresponding to them. ABSTRACT: BA 50:3483 (36527), Apr. 1, 1969.

Hassan, A., joint author. *See* Chiu, Y. C., et al., 1968.

Hassemer, M. M., joint author. *See* Ludwig, D., et al., 1964.

Hasuo, A., joint author. *See* Buei, K., et al., 1966.

Hatanaka, M., joint author. *See* Takagi, U., et al., 1965.

Hatsukade, M., joint author. *See* Hayashi, A., et al., 1967, 1968, 1969.

3752. Haupt, A. and **Busvine, J. R. 1969** The Effect of Overcrowding on the Size of Houseflies (*Musca domestica* L.). Trans. R. Ent. Soc. Lond. 120(15):297-311.
Measurements taken daily showed a rapid, then a slow fall in pupal weight. Linear measurements of wing and the ratio of frons to head

width were also taken. ABSTRACT: RAE-B 58:200(831), June, 1970.

Haupt, A., joint author. *See* Busvine, J. R., et al., 1968.

Haus, J. B., joint author. *See* Price, R. W., et al., 1962.

3753. Havlík, B. 1964 [The sanitary problem of synanthropic flies (Diptera) of Greater Prague.] Wiad Parazytol. 10:588-589. (In Polish, with English summary.)
A 4-year study. Treats of qualitative and quantitative incidence of flies in communities and in homes, alimentary presence of Enterobacteriaceae and migration of flies from animal areas to human dwellings. ABSTRACT: BA 47:4663(54622), June 1, 1966.

3754. Havlík, B. and **Batova, B. 1961** A Study of the Most Abundant Synanthropic Flies Occurring in Prague. Casopis Ceskoslov. Spolecnosti Entomol. 58(1):1-11. (Czech summary.)
Flies captured from 4 different localities showed Calliphoridae leading in numbers. *M. domestica* among the leading 9 species. Seasonal dynamics shown by graphs. ABSTRACT: BA 45:2486(30838), Apr. 1, 1964.

3755. Havlík, B. and **Celedova, V. 1962** K Hygienickému Významu Synantropńich much (Diptera) Pražských Bytu. [On the sanitary significance of synantropic flies (Diptera) on the Prague flats.] Ceskoslov. Hyg. 8:468-474. (Russian and English summaries.)
Flies captured between 1955-1959 represented 8 families (23 types), 27 per cent of which were *M. domestica.* Discusses sanitary aspects of most frequently encountered species, and recommends methods for their eradication.

Havlík, B., joint author. *See* Celedova, V., et al., 1963.

Hawkins, W. B., joint author. *See* Rockstein, M., et al., 1965.

3756. Hawley, M. K. and **Georghiou, G. P. 1970** Isolation and Description of Twelve Morphological Mutants of *Fannia canicularis.* (Diptera: Muscidae). Ann. Ent. Soc. Am. 63(4):980-985.
Mutants obtained by inbreeding of field-collected flies. Several were established as true-breeding strains. Brown body (*bwb*) is inherited as a single, recessive gene.

Hawley, M. K., joint author. *See* Georghiou, G. P., et al., 1967.

Hayama, S., joint author. *See* Fujiito, S., et al., 1962.

3757. Hayashi, A. 1966 [Inhibition of Development of the Housefly by Synergists.] Botyu Kagaku (Bull. Inst. Insect Contr.) 31(3):135-136 (In Japanese, with English summary.)
Sulfoxide and safroxan inhibited development if added to culture medium; caused a decrease in pupation and adult emergence if topically applied to first stage larvae. ABSTRACTS: BA 48:7888(88440), Sept. 1, 1967; RAE-B 57:22(94), Jan., 1969.

3758. Hayashi, A. 1966 [Studies on the increment of the efficacy of insecticides. II. On the effects of combining two synergists.] Jap. J. Sanit. Zool. 17(3):205-208. (In Japanese, with English summary.)
As tested on *M. domestica vicina*, a combination of 2 synergists with allethrin was ineffective. (Two kinds of synergists operate antagonistically.) ABSTRACT: RAE-B 57:81(286), May, 1969.

3759. Hayashi, A. 1967 [Studies on the increment of the efficacy of insecticides. III. On the toxic properties of S-421 against the house fly, *Musca domestica vicina* Macq.] Jap. J. Sanit. Zool. 18(1):35-38. (In Japanese, with English summary.)
S-421, if sufficiently concentrated, caused excellent knockdown, and its

lethal effect is greater than that of other known synergists. Synergistic effect increased during activity of the housefly. ABSTRACT: BA 51:6804 (69499), June 15, 1970.

3760. Hayashi, A. 1969 [Effet attractif des excréments de la mouche domestique sur celle-ci.] Jap. J. Sanit. Zool. 20(3):215-217. (In Japanese, with French summary.)

Extract of a paper filter contaminated by housefly excrement and vomitus was strongly attractive. There is believed to exist, besides various amino acids, an attractive unknown substance apart from the components of such acids.

3761. Hayashi, A. 1970 [Joint action of pyrethroids against houseflies.] Jap. J. Sanit. Zool. 20(4):261-263. (In Japanese, with English summary.)

Calculates whether effect is one of (1) similar joint action, (2) independent action, (3) synergism, (4) antagonism. Chrythron plus allethrin, also chrythron plus phthalthrin, show high synergistic action.

3762. Hayashi, A., Aoki, H. and **Saito, T. 1968** [Studies on the increment of the efficacy of insecticides: IX. On the synergistic action of pyrethrosin with pyrethroid.] Botyu Kagaku 33(4):130-134. (In Japanese, with English summary.)

With kerosene spray, no synergistic effects of pure pyrethrosin with pyrethroids were observed. Knock-down was studied by the dusting method. Combination of pyrethrosins with allethrin showed lower order of synergism than a similar combination with pyrethrins. ABSTRACT: BA 50:7510(78289), July 15, 1969.

3763. Hayashi, A. and **Hatsukade, M. 1967** [Studies on the increment of the efficacy of insecticides. VI. On the quantitative analysis of allethrin and synergists in the house fly *Musca domestica vicina* Macq.] Botyu Kagaku (Bull. Inst. Insect Contr.) 32(1):1-5. (In Japanese, with English summary.)

Quantities of allethrin and its synergists penetrating through the cuticle determined by gas chromatography. Most rapid penetration of allethrin occurred with the addition of S-421. ABSTRACTS: BA 49:1400(15669), Feb. 1, 1968; RAE-B 57:213(772), Oct., 1969.

3764. Hayashi, A. and **Hatsukade, M. 1967** Sur la sensibilité au pyréthroide chez la mouche domestique (*Musca domestica* L.) des quatre colonies. Botyu-Kagaku 32:61-63.

Concerns pyrethrins, allethrin, and phthalthrin. Colonies differed as to knock-down and mortality. Flies of European origin showed greater resistance than Japanese flies. ABSTRACT: BA 49:7757(86603), Aug. 15, 1968.

3765. Hayashi, A. and **Hatsukade, M. 1968** [Studies on the increment of the efficacy of insecticides. VII. On the effect of synergized pyrethroid on various strains of house flies.] Botyu Kagaku 33(2):39-41. (In Japanese, with English summary.)

Seven strains of flies were treated topically with pyrethroids, with or without synergists. Pyrethroids synergized with S-421 or piperonyl butoxide were most effective. ABSTRACT: BA 50:9170(95051), Sept. 1, 1969.

3766. Hayashi, A. and **Hatsukade, M. 1969** [On the knock-down effect of pyrethroids on various strains of house flies.] Jap. J. Sanit. Zool. 20(2):105-107. (In Japanese, with English summary.)

Several strains tested to determine the relative effectiveness of pyrethrins, allethrin and phthalthrin. Order of resistance levels, based on KT-50 values, given for each insecticide. ABSTRACT: RAE-B 59:456(1897), Dec., 1971.

3767. **Hayashi, A.** and **Ikeno, N. 1968** [The length of larval stage in pyrethroid-resistant and -susceptible strains of the house fly, *Musca domestica* L.] Jap. J. Sanit. Zool. 19(3):204-206. (In Japanese, with English summary.)
Two strains of *M. domestica* (SMA, susceptible; and 213-ab, resistant); one strain of *M. domestica vicina* (Takatsuki; no insecticidal pressure in laboratory). Duration of egg-larval stage longest in Takatsuki strain; pupal body weight also the greatest. ABSTRACT: BA 51:2234(23180), Feb. 15, 1970.

3768. **Hayashi, A.** and **Itoga, A. 1966** [Studies on the testing methods for larvicides. I. Difference in evaluating the effectiveness of larvicides by various testing methods.] Botyu-Kagaku. (Bull. Inst. Insect Contr.) 31(1):25-29. (In Japanese, with English summary.)
Three larvicides evaluated by 5 methods. Dipping in emulsions for 2 hours and counting percentage emergence of pupae considered the method of choice for any larvicide. ABSTRACT: RAE-B 56:128(488), July, 1968.

3769. **Hayashi, A., Saito, T.** and **Iyatomi, K. 1968** [Studies on the increment of the efficacy of insecticides. VIII. Metabolism of ^{3}H-pyrethroids in the adult housefly, *Musca domestica vicina* Macq.] Botyu-Kagaku 33(3):90-95.
Synergists appeared to reduce the permeability of the integument to pyrethroids, and to depress the decomposition of pyrethroids in the body. S-421 most effective synergist for allethrin; piperonyl butoxide for phthalthrin. ABSTRACT: RAE-B 57:213(772), Oct., 1969.

Hayashi, K., joint author. *See* Funatsu, M., et al., 1967, 1968.

Hayashi, Y., joint author. *See* Tsiapalis, C. M., et al., 1967.

Hayes, D. K., joint author. *See* Sullivan, W. N., et al., 1970.

Hayes, J. T., joint author. *See* Pimentel, D., et al., 1965.

Hays, R. L., joint author. *See* Hays, S. B., et al., 1969.

3770. **Hays, S. B. 1965** Some Effects of Reserpine, a Tranquilizer, on the House Fly. J. Econ. Ent. 58:782-783.
Continuously treated flies flew but little, made only feeble attempts at mating, produced no eggs, had shorter life span. Flies treated for 8 days only appeared normal 1 week later. ABSTRACTS: BA 47:2505(29935), Mar. 15, 1966; RAE-B 53:225, Nov., 1965; TDB 63:237, Feb., 1966.

3771. **Hays, S. B. 1968** Chemosterilizing Activity and Toxicity of P, P-Bis (1-aziridinyl)-N-Methylphosphonic Amide and P,P-Bis(1-aziridinyl)-N-(3-Methoxypropyl) phosphinothioic Amide Against the House Fly. J. Econ. Ent. 61(3):800-802.
Achieved sterility without excessive mortality. Males more affected than females. Permanence of sterility not complete, but egg hatch remained low weeks later. ABSTRACTS: BA 49:10247(113111), Nov. 1, 1968; RAE-B 56:221(805), Nov., 1968, TDB 66:757(1507), July, 1969.

3772. **Hays, S. B. 1968** Reproduction Inhibition in House Flies with Triphenyl Tin Acetate and Triphenyl Tin Chloride Alone and in Combination with other Compounds. J. Econ. Ent. 61(5):1154-1157.
Both compounds compared favorably with tepa in reducing hatch when offered to sexually mature females. Tepa and triphenyl tin chloride offered as a free choice diet completely inhibited reproduction. Population was completely eliminated in 35 days. ABSTRACTS: BA 50:4221 (44420), Apr. 15, 1969; RAE-B 57:49(175), Mar., 1969; TDB 66:1186 (2310), Nov., 1969.

3773. **Hays, S. B.** and **Amerson, G. M. 1967** Reproductive Control in the House Fly with Reserpine. J. Econ. Ent. 60:781-783.
Will control reproduction at rates of 1·8 mg. reserpine per gram of diet. The longer the feeding period, the greater the control. Flies fed treated diet in early life produced fewer eggs than those fed treatment later. ABSTRACTS: BA 48:11208(124490), Dec. 15, 1967; RAE-B 55:189(653), Oct., 1967; TDB 65:83(186), Jan., 1968.

3774. **Hays, S. B., Hays, R. L.** and **Mims, I. S. 1969** Comparative Effects of Reserpine and Serotonin Creatine Sulfate on Oviposition in the House Fly. Ann. Ent. Soc. Am. 62(3):663-664.
Reserpine may release serotonin and/or similar substances from storage sites in fly tissues. (There is an increase in extractable materials of this type in reserpine-treated flies.) Tranquilizing effect may be due to such release from sites in nervous tissue. ABSTRACT: RAE-B 58:47(193), Feb., 1970.

Hays, S. B., joint author. *See* Kissam, J. B., et al., 1966, 1967; Wicht, M. C., Jr., et al., 1967; Wilson, J. A., et al., 1969.

Head, S., joint author. *See* Goldberg, A. A., et al., 1965.

3775. **Hecht, O. 1963** On the Visual Orientation of House-Flies in their Search of Resting Sites. Ent. Exp. Appl. (Amsterdam) 6(2):107- 113.
Flies showed a preference for black over gray and for gray over white. ABSTRACTS: BA 45:4448(54165), July 1, 1964; RAE-B 51:283, Dec., 1963; TDB 60:1168, Dec., 1963.

3776. **Hecht, O. 1964** Aspectos etologicos y fisiologicos de la percepcion de colores en los insectos. [Ecological and physiological aspects of the perception of color by insects.] Rev. Sociedad Mexicana de Historia Natural. 25:127-148.
Examples drawn from several Orders. Diptera illustrated by *Aedes*, *Stomoxys* and *M. domestica.* Paper is primarily a review of the work of many authors.

3777. **Hecht, O. 1970** Ecologia y comportamiento de las moscas domesticas. [Ecology and behavior of domestic Flies. Part I: *Musca domestica.*] L. Contr. Instituto Politecnico National al Año Internat. de la Educ. Mexico, D. F. xii + 113 pp., illus.
An up-to-date compilation based on a carefully selected bibliography plus many of the author's personal observations. A useful work.

3778. **Hecht, O. 1970** Light and Color Reactions of *Musca domestica* under Different Conditions. Bull. Ent. Soc. Am. 16(2):94-98.
Discusses and interprets discrepancies in the findings of various authors, including his own. Laboratory tests and tests out-of-doors do not yield the same results; nor do studies on different species and subspecies (e.g. *M. domestica, M. d. vicina, M. sorbens*).

3779. **Hecht, O., Muñiz, R.** and **MacGregor, R. 1964** On the Selection of Resting Sites by *Stomoxys calcitrans* and *Musca domestica.* Folia Ent. Mex. (7/8):72. Abstract.
Stomoxys sought illuminated spaces or lighter surfaces. *M. domestica* preferred darker surfaces or darkened areas. Possible influence of surface temperature under study.

3780. **Hecht, O., Muñiz, R.** and **Nava, A. 1968** Contrary Responses of *Musca domestica* Concerning their Selection of Different Shades and Hues. Ent. Exp. Appl. 11:1-14.
Outdoor tests showed preference for yellow (first), white (second). This was in contrast with indoor tests where black was favored. ABSTRACTS: BA 49:8722(97071), Sept. 15, 1968; RAE-B 56:245(904), Dec., 1968; TDB 66:164(298), Feb., 1969.

3781. **Hecht, O., Quijano, E., Muñiz, R.** and **Landgrave, G. 1969** Ensayo sobre el aprovechamiento de la atraccion de *Musca domestica* L., a colores con propositos de combate. [Experiment on utilization of the attraction of *Musca domestica* to colors, with (for) the purpose of combat (i.e. control).] Acta Zool. Mex. 9(5):9-24.
Flies gathered in large numbers on a certain yellow. Suggestions: (1) Treat surface with chemosterilants. (2) Use surface to expose wild flies to pathogens. ABSTRACT: BA 51:7349(75054), July 1, 1970.

Hecht, O., joint author. *See* Muñiz, R., et al., 1968.

3782. **Heenan, M. P.** and **Smith, J. N. 1967** Conjugation of 1-naphthol and p-nitrophenol in Flies: Formation of Glucoside-6-phosphates. Life Sci. 6(16 pt. 2):1753-1757.
Acidic conjugates detected as metabolites of these substances appear to be the 6-phosphate esters of the corresponding phenolic glucosides.

Heenan, M. P., joint author. *See* Darby, F. J., et al., 1966.

3783. **Hegdekar, B. M.** and **Smallman, B. N. 1967** Lysosomal Acid Phosphatase During Metamorphosis of *Musca domestica* (Linn.). Can. J. Biochem. 45(7):1202-1206.
Records marked changes in the proportion of bound and free enzymes during larval and pupal stages. ABSTRACT: BA 48:11245(124900), Dec. 15, 1967.

3784. **Hegdekar, B. M.** and **Smallman, B. N. 1969** Intracellular Distribution of β-glucuronidase Activity During Metamorphosis of the Housefly, *Musca domestica* L. Canad. J. Zool. 47(1):45-49.
Beta-glucuronidase activity, found chiefly in lysosomes and microsomes, increases from *none* in early third stage larvae, to *ten times* its first manifestation, in the pupal stage. Has a pH optimum of 4. ABSTRACTS: BA 50:10317(106750), Oct. 1, 1969; TDB 66:486(931), May, 1969.

3785. **Heimpel, A. M. 1967** A Critical Review of *Bacillus thuringiensis* var. *thuringiensis* Berliner and other Crystalliferous Bacteria. Ann. Rev. Ent. 12:287-322.
Treats of taxonomy, pathogenicity, standardization, and insect control. A valuable review, based on 137 references.

Heimpel, A. M., joint author. *See* Cantwell, G. E., et al., 1964.

Heinc de Balsac, H., joint author. *See* Dhennin, L., et al., 1961.

Heishmann, J. O., joint author. *See* Dorsey, C. K., et al., 1966.

Heisler, C. R., joint author. *See* Gregg, C. T., et al., 1960.

3786. **Hellenbrand, K. 1967** Inhibition of Housefly Acetylcholinesterase by Carbamates. J. Agric. Fd. Chem. 15(5):825-829.
A study of housefly cholinesterase inhibition by two methyl- and two dimethyl- carbamates. Fly-head cholinesterase appears more sensitive to these carbamates than bovine erythrocyte cholinesterase. Rate constants for carbamylation and decarbamylation were determined. ABSTRACTS: BA 49:1422(15939), Feb. 1, 1968; RAE-B 56:143(497), July, 1968.

Hellyer, G. C., joint author. *See* Winteringham, F. P., et al., 1960.

3787. **Heming, W. E. 1968** Cluster and Other Flies that Overwinter in Homes. Ont. Dept. Agr. Food Publ. 18, pp. 1-5.
Treats of prevention and use of insecticides.

3788. **Henderson, G. L.** and **Crosby, D. G. 1967** Photodecomposition of Dieldrin and Aldrin. J. Agric. Fd. Chem. 15(5):888-893.
Both insecticides dechlorinated by ultraviolet irradiation. Product is less toxic to flies. Wave lengths required are lower than those found in sunlight; products not detected under field conditions. ABSTRACTS: BA 49:1400(15670), Feb. 1, 1968; RAE-A 56:376(1463), July, 1968.

3789. Hennesy, D. J., Fratantoni, J. and **Hartigan, J. 1961** Toxicity of 2-(2-hologen-4-chlorophenyl)-2-(4-chlorophenyl)-1, 1, 1-trichloroethanes to Normal and DDT-resistant Houseflies. Nature 190(4773):341.
These are DDT compounds with a halogen substituted in the ortho position. In the susceptible strain studied, a direct correlation was shown between size of the halogen molecule and the LD50 value of the compound.

Hennessy, D. J., joint author. *See* Fahmy, M. A. H., et al., 1966; Perry, A. S., et al., 1967.

3790. Henson, J. L. 1966 The Effects of Apholate on the Reproductive Organs of the Face Fly, *Musca autumnalis*. Diss. Absts. 27(5B), Page 1655-B.

Hernandez, H., joint author. *See* Greenberg, B., et al., 1963.

Herrick, R., joint author. *See* Sherman, M., et al., 1965, 1967, 1968, 1969.

3791. Heslop, J. P. 1964 The Estimation of Adenosine Triphosphate and Related Compounds in Insect Tissue. Biochem. J. 91:183-187.
Used freezing and extraction techniques to determine ATP in tissue from thorax of *M. domestica*. Believed applicable to other tissues. ABSTRACT: BA 46:1080(13632), Feb. 1, 1965.

3792. Heslop, J. P., Price, G. M. and **Ray, J. W. 1963** Anaerobic Metabolism in the Housefly, *Musca domestica* L. Biochem. J. 87:35-38.
A study of the effects in adult male flies, of various periods of anoxia on tissue concentrations of glycerol 1-phosphate, orthophosphate, adenosine triphosphate, arginine phosphate, acid-soluble phosphorus, phospholipid phosphorus, residual phosphorus and glycerol. ABSTRACTS: BA 43:1973(25498), Sept. 15, 1963; TDB 60:690, July, 1963.

3793. Heslop, J. P. and **Ray, J. W. 1963** The Metabolism of Glycerol 1-Phosphate in Resistant and Susceptible Houseflies (*Musca domestica* L.) and the Effect of Dieldrin. Biochem. J. 87:31-34.
Dieldrin poisoning progressively lowered the amount of thoracic glycerol 1-phosphate in resistant and susceptible houseflies, whether or not conditions were anaerobic. Not associated with normal differences between strains, or with presence of dieldrin alone. ABSTRACTS: BA 43:1973(25499), Sept. 15, 1963; RAE-B 52:115, June, 1964; TDB 60:691, July, 1963.

3794. Heslop, J. P. and **Ray, J. W. 1964** Glycerol 1-Phosphate Metabolism in the Housefly (*Musca domestica* L.) and the Effects of Poisons. Biochem. J. 91(1):187-195.
Concentrations of glycerol 1-phosphate vary widely from one generation to another. Some insecticides affect the thoracic glycerol 1-phosphate and ATP concentrations in aerobic flies. ABSTRACTS: BA 46:1083(13671), Feb. 1, 1965; TDB 61:1103, Oct., 1964.

Heslop, J. P., joint author. *See* Ray, J. W., et al., 1963.

3795. Hewitt, P. H. 1965 The Degradation of Four N-Methyl Carbamates and the Reactivation of Cholinesterase in the House Fly, *Musca domestica* L. Diss. Absts. 25(11):6516.

3796. Hewlett, P. S. 1969 The Potentiation Between Thanite and Arprocarb in their Action on Houseflies (*Musca domestica*). Ann. Appl. Biol. 63(3):477-481.
Thanite probably potentiates arprocarb, but not vice versa. Believed to potentiate isolan or m-iso-propylphenyl N-methylcarbamate more than it does arprocarb. (Used *M. domestica* females of susceptible strain.) ABSTRACTS: BA 50:11362(117510), Nov. 1, 1969; RAE-B 58:61(238), Feb., 1970.

3797. **Hewlett, P. S. 1969** Measurement of the Potencies of Drug Mixtures. Biometrics. (The Biometric Society) 25(3):477-457. (French summary.)
Refers to synergism of DDT and piperonyl cyclonene, as used against houseflies. Diagrams, calculations.

3798. **Hewlett, P. S.** and **Wilkinson, C. F. 1967** Quantitative Aspects of the Synergism Between Carbaryl and Some 1, 3-benzodioxole (methy lene-dioxyphenyl) Compounds in Houseflies. J. Sci. Fd. Agric. 18(7):279-282.
Graphical analysis showed a common pattern in the results for 18 synergists. There was a 700-fold difference between the most and the least active. Doses of carbaryl and synergist, in different combinations that gave 50 per cent mortality (LD50), were estimated. ABSTRACT: RAE-B 57:22(96), Jan., 1969.

Hewlett, P. S., joint author. *See* Gostick, K. G., et al., 1960; Bates, A. N., et al., 1965; Winteringham, F. P. W., et al., 1964.

3799. **Hijikuru, S. 1968** Studies on the Biosynthetic Pathways of Beta-alanine in Wild and Black Puparium Strains of the Housefly. Jap. J. Genet. 43:419. (Address in Japanese.) (Abstract.) No English translation available.

3800. **Hildenbrant, G. R., Kumar, S. S., Peterson, D.** and **Bieber, L. L. 1969** Formation and Degradation of Glycerophosphorylcholine, Glycerophosphoryl-beta-methylcholine and Glycerophosphorylethanolamine by Enzyme Preparations Derived from *Musca domestica* Larvae. Abstract. Fed. Proc. 28(2 pt. 1):860(3367).
Housefly larvae microsomes deacylate three compounds. Formation of the glycerophosphoryl bases is stimulated by lauryl sulfate. Phosphodiesterase in prepupating larvae can cleave several glycerophosphoryl compounds.

Hin. W. K., joint author. *See* Hale, J. H., et al., 1960.

3801. **Hinshaw, W. R. 1964** Epidemiology of Salmonellosis. Proc. Salmonella Seminar. Agric. Res. Serv. USDA, Hyattsville, Md.
Reports isolation of *S. typhimurium* from houseflies in an infected turkey ranch.

3802. **Hinton, H. E. 1967** The Respiratory System of the Egg-shell of the Common Housefly. J. Insect Physiol. 13(5):647-651.
Egg shell examined with transmission and stereoscan electron microscopes. A new, thin, outer layer discovered. Other species of *Musca* compared with *M. domestica.* ABSTRACT: BA 48:7435(83224), Aug. 15, 1967.

Hintz, A. M., joint author. *See* Adams, T. S., et al., 1968.

3803. **Hiraga, S. 1964** Tryptophan Metabolism in Eye-color Mutants of the Housefly. Jap. J. Genet. 39(4):240-253.
Ten eye-color mutants studied. Mutants differ as to the step at which tryptophan metabolism is "blocked". It is likely that *ge* and *ocra* mutants of *M. domestica* are homologous to the *v* and *cn* mutants of *Drosophila.* ABSTRACT: BA 46:4839(59951), July 15, 1965.

Hiraga, S., joint author. *See* Tsukamoto, M., et al., 1961.

3804. **Hirakoso, S. 1960** [Studies on the behavior of housefly in relation to the control effect of the insecticide impregnated tapes.] Jap. J. Sanit. Zool. 11:42-53. (In Japanese, with English summary.)
Of the chemicals used, 5 per cent diazinon was the most effective. ABSTRACT: RAE-B 51:271, Dec., 1963.

3805. **Hirakoso, S. 1962** The Effects of Water Contents in Culture on the Development and Insecticide Susceptibility of the House Fly. Jap. J. Exptl. Med. 32:387-397.
Too low or too high water concentrations, in the larval media, hindered larval development and left the insects more susceptible to DDVP and/or Lindane, when they became adult flies. ABSTRACTS: BA 43:361(3937), July 1, 1963; TDB 60:902, 1963.

3806. **Hirakoso, S. 1965** [Some factors affecting the residual effects of the insecticides to the adult housefly and the mosquito.] Jap. J. Sanit. Zool. 16(3):231-238. (In Japanese, with English summary.)
The absorption of solutions, containing toxicants, on or into the treated surface, the release of toxicants into the air by evaporation and the translocation of toxicants from treated to untreated surfaces were the factors affecting differences in effects of insecticidal deposits on glass and paper. ABSTRACT: RAE-B 56:131(458), July, 1968.

Hirakoso, S., joint author. *See* Suzuki, T., et al., 1961.

Hirano, T., joint author. *See* Kobayashi, K., et al., 1969.

Hiroo, A., joint author. *See* Hayashi, A., et al., 1968.

3807. **Hiroyashi, T. 1960** Some New Mutants and Linkage Groups of the Housefly. J. Econ. Ent. 53(6):985-990.
Reports 4 eye-color mutants; also 1 for eye shape, 1 for wing shape, 1 for wing venation. Five linkage groups demonstrated. No sex-linked mutant detected. *See also* Japanese Journal "The Heredity", 14(12): 41-46. (Plus 4 plates.)

3808. **Hiroyoshi, T. 1961** The Linkage Map of the House Fly, *Musca domestica* L. Genetics 46(10):1373-1380.
Map is based on recombination values in the female. Experiments to detect crossing over in males gave negative results. Males found with sex-limited inheritance (Chromosome II).

3809. **Hiroyoshi, T. 1964** Sex-limited Inheritance and Abnormal Sex Ratio in Strains of the Housefly. Genetics 50:373-385.
Genetical and cytological analysis indicated these two phenomena as due to translocation of a part of the Y chromosome, on which the male determinant (*s*) and the viability factor were located, to chromosome II. ABSTRACTS: BA 46:744(9386), Feb. 1, 1965; TDB 62:68, Jan., 1965.

Hiroyoshi, T., joint author. *See* Sullivan, R. L., et al., 1960.

Hirsch, I., joint author. *See* Ascher, K. R. S., et al., 1961, 1963, 1965.

3810. **Hirsh, J.** and **Zaiman, H. 1965** Vectors and Victims: Being a Collection of Essays About Flies Without Zippers and Other Nuisances of Man. x + 70 pp. Charles C. Thomas, Publisher: Springfield, Ill.
A book providing an informal and witty comment on animals which are carriers of disease, directly and indirectly to man. The house fly is reported to carry at least 10 bacteria, 1 or 2 viruses and 2 protozoans. ABSTRACT: BA 46-7291(90613), Oct. 15, 1965.

Hitchcock, M., joint author. *See* Clark, A. G., et al., 1966.

Hkashi, K. P., joint author. *See* Qadri, S. S., et al., 1969.

3811. **Hocking, B. 1960** An Insect-proof Doorway. Bull. Ent. Res. 51:135-144.
A special design based on study of the factors influencing passage of flying or crawling insects into doorways. Tested extensively with *Apis* and *Musca*. ABSTRACT: RAE-B 48:110, June, 1960.

3812. **Hocking, B. 1960** Smell in Insects: A Bibliography with Abstracts. EP Technical Report No. 8. Defence Research Board. Dept. National Defence. 266 pp. Ottawa.
Includes many references to Muscoid Diptera, *M. domestica* among them.

3813. **Hocking, B. 1961** Further Consideration Regarding the Repellency of Spray Components. Bull. Ent. Res. 52(1):1-5.
Diptera used in studies of olfactory repellency designed to identify responsible ingredients. Consideration given to possibility of finding substitutes with less repellent or even attractive properties. ABSTRACT: RAE-B 49:125, June, 1961.

3814. **Hodgson, E. S. 1968** Taste Receptors of Arthropods. Symp. Zool. Soc. Lond. 23:269-277.
Because of the non-availability of an economic non-repellent material, it is suggested that: (1) repellency of spray components might be counteracted by the addition of specific attractants to spray formulae; (2) a useful space repellent spray, containing no toxicant, might be formulated from the lighter fractions of solvent materials.

3815. **Hodgson, E. S.** and **Steinhardt, R. A. 1967** Hydrocarbon Inhibition of Primary Chemoreceptor Cells. Olfaction and Taste II. Proc. Sec. Int. Symp. pp. 737-748.
Worked on *Phormia regina*, studying the effects, on chemoreceptor cells, of straight chain hydrocarbons. Lower alcohols and long-chain amines act in two stages: (1) a reversible inhibition; (2) injury of salt, water, and sugar receptors in labellar sensilla of the flies. Included in this bibliography as a contribution to muscoid biology. (Hodgson and associates have published several papers in this field, but none on *M. domestica*).

Hodgson, E., joint author. *See* Guthrie, F. E., et al., 1968; Khan, M. A. Q., et al., 1967; Mathews, H. B., et al., 1966; Cassidy, J. D., et al., 1969; Steinhardt, R. A., et al., 1966.

3816. **Hoffman, R. A., Berry, I. L.** and **Graham, O. H. 1965** Control of Flies on Cattle by Frequent, Low-Volume Mist Spray Applications of Ciodrin. J. Econ. Ent. 58:815-817.
System devised to control horn flies, was also effective against house flies, stable flies, horse flies and mosquitoes, at least temporarily. ABSTRACTS: BA 47:2047(24558), Mar. 1, 1966; RAE-B 54:11, Jan., 1966.

Hoffman, R. A., joint author. *See* Smith, C. N., et al., 1960.

3817. **Hoffman, Von G., Köhler, G.** and **Janitschke, K. 1967** Ein Beitrag zur Kenntniss der veterinärmedizinish wichtigen Dipteren Syriens. [A Contribution to the Knowledge of Syrian Diptera of Veterinary and Medical Importance.] Zeit. Angew. Zool. 54(2):297-301.
A three year study. Includes species of *Musca*, *Stomoxys*, *Lucilia*, *Calliphora*, *Sarcophaga* and other related genera; also certain Pupipara and Nematocera. ABSTRACT: RAE-B 58:45(188), Feb., 1970.

Hogendijk, C. J., joint author. *See* Matsumura, F., et al., 1964.

Holbrook, D. V., joint author. *See* Sawicki, R. M., et al., 1961.

Holcomb, B., joint author. *See* Ballard, R. C., et al., 1965.

Holden, J. S., joint author. *See* Bridges, R. G., et al., 1969.

3818. **Holdsworth, R. P. 1962** Control of Face Flies Attacking Commercial Dairy Herds. J. Econ. Ent. 55(1):146-147.
Conscientious daily application of DDVP sirup bait will greatly alleviate face fly annoyance. Irregular use is not satisfactory. ABSTRACTS: BA 38:1580(20339), June 1, 1962; RAE-B 50:226, Oct., 1962.

3819. **Holling, C. S. 1966** The Functional Response of Invertebrate Predators to Prey Density. Memoirs of the Entomological Society of Canada. No. 48, 3 + 86 pp. Canada Dept. Forestry, Ent. and Pathol. Branch, Ottawa, Canada.
Concerns chiefly the praying mantis, *Hierodula crassa*. All instars beyond

the third were fed *Musca domestica.* An elaborate study, contributing to the development of a realistic model of the predation system. Includes "hunger" as an important component of attack. ABSTRACT: BA 47: 8691(100657), Nov. 1, 1966.

3820. Hollingworth, R. M. 1966 Biochemical Factors Determining Selective Toxicity of the Insecticide Sumithion and Its Analogs. Diss. Absts. 27(2B):340-B-341-B.

3821. Hollingworth, R. M., Fukuto, T. R. and **Metcalf, R. L. 1967** Selectivity of Sumithion Compared with Methyl Parathion: Influence of Structure on Anticholinesterase Activity. J. Agric. Fd. Chem. 15(2):235-241.

In general, these compounds were uniformly toxic to susceptible flies and cockroaches, but varied widely in toxicity to mice. Difference in inhibitory effect believed due to variation in the distance between the anionic and esteratic sites in enzyme from insect and mammalian sources. ABSTRACT: BA 48:9532(107170), Nov. 1, 1967; RAE-B 55:217(753), Dec., 1967.

3822. Hollingworth, R. M., Metcalf, R. L. and **Fukuto, T. R. 1967** The Selectivity of Sumithion Compared with Methyl Parathion: Metabolism in Susceptible and Resistant Houseflies. J. Agric. Fd. Chem. 15(2): 250-255.

The two compounds were identical with respect to penetration, activation, and degradation in the two strains. Greater phosphatase activity in resistant strain lowered the level of active toxicants. Slower penetration into resistant flies may also contribute. ABSTRACTS: BA 48:9532 (107172), Nov. 1, 1967; RAE-B 55:218(755), Dec., 1967.

3823. Homan, E. R. 1963 The Influence of Some Environmental Factors on the Appraisal of Experimental Housefly Control in Dairy Barns in New York State. Diss. Absts. 23(9):3056.

Hooper, G. H. S., joint author. *See* Wan, T. K., et al., 1967, 1969.

3824. Hoopingarner, R. A., Guyer, G. E. and **Krause, D. H. 1966** The Mackinac Island Fly Problem. I. History of Insecticide Use and Characterization of Resistance. Mich. Agr. Expt. Sta. Quart. Bull. 48(4): 559-564.

A natural strain from the island possessed a high level of resistance to other organophosphate insecticides. Sub-strains pressured with an organophosphate and a carbamate indicate a rapid acquisition of organophosphate resistance. ABSTRACTS: BA 47:9039(104517), Nov. 1, 1966; RAE-B 56:106(368), May, 1968.

3825. Hoopingarner, R. A. and **Krause, D. H. 1968** The Mackinac Island Fly Problem. II. Induced Insecticide Resistance. Mich. Agr. Exp. Sta. Quart. Bull. 50(3):281-284.

The induced resistance to carbamate and organophosphorus insecticides increased or maintained resistance to DDT and malathion. Major factors for resistance to Baygon, Dursban, DDT and malathion are located on chromosome V. ABSTRACT: BA 49:6314(70696), July 15, 1968.

Hoopingarner, R., joint author. *See* French, A., et al., 1965.

Hooven, N. W., joint author. *See* Anthony, R. W., et al., 1961.

3826. Hopkins, L. O. and **Maciver, D. R. 1965** Tropital — A New Synergist for Pyrethrins. Pyrethrum Post 8(2):3-5.

Discusses the properties of tropital and tests in which it was used with pyrethrins in comparison with piperonyl butoxide against insects. It is comparable with the latter against *M. domestica.* ABSTRACT: RAE-B 55:139(496), Aug., 1967.

3827. Hopkins, T. L. 1962 Radioisotope Techniques and Recent Research on Metabolism of Insecticides in Insects. *In* Proc. Symp. Radioisotopes and Radiation in Entomology. —— Bombay, December 5-9. pp. 101-109.
Studied quantitative fate and metabolism of P^{32} labelled Dipterex in normal and Dipterex-resistant houseflies. Resistant strain detoxified the insecticide and excreted the water soluble metabolites more rapidly than normal flies. Metabolites identified by paper chromatography. ABSTRACT: RAE-B 51:213, Oct., 1963.

Hopkins, T. L., joint author. *See* Valder, S. M., et al., 1968; Monroe, R. E., et al., 1967; Murdock, L. L., et al., 1968; Pitts, C. W., et al., 1964.

Hora, J., joint author. *See* Rezabova, B., et al., 1968.

3828. Horie, Y. and **Chefurka, W. 1966** The Distribution and Properties of Transhydrogenases in Insect Tissues. Comp. Biochem. Physiol. 17: 1-18.
Concerns transhydrogenase in the gut and thoracic muscle mitochondria of the cockroach, *Periplaneta americana* and of the housefly, *Musca domestica.* Techniques described.

Horstman, D. M., joint author. *See* Riordan, J. T., et al., 1961.

3829. Hoshishima, K. and **Izumi, K. 1967** Study on the Participation of Metals in the Olfactory Response of Housefly. *In* Proc. Second Int. Symp. on Olfaction and Taste. II. September. Tokyo. Pergamon Press: London and New York. Wenner-Gren Center Int. Symp. Ser. 8:779-785.
Dmonstrates relationship between the cobalt ion and olfactory response in *M. domestica.* ABSTRACT: BA 49:7758(86605), Aug. 15, 1968.

3830. Hoskins, W. M. and **Nagasawa, S. 1961** Cross-resistance in Sevin-selected House Flies and a Summary of Resistance Among the Carbamate Insecticides. Botyu-Kagaku 26(4):115-125. (Summary in Japanese.)
Concerns the susceptibility to DDT, parathion, and four carbamates of a Sevin-resistant strain. Covers factors affecting the behavior of carbamates, and their cross-resistance. ABSTRACT: RAE-B 52:72, Apr., 1964.

Hoskins, W. M., joint author. *See* Menzel, D. B., et al., 1963.

Hosoi, T., joint author. *See* Kobayashi, K., et al., 1969.

3831. Houser, E. C. and **Wingo, C. W. 1967** *Aphaereta pallipes* as a Parasite of the Face Fly in Missouri, with Notes on Laboratory Culture and Biology. J. Econ. Ent. 60(3):731-733.
M. autumnalis was third preferred host. Highest parasitism recorded, 78 per cent. *Aphaereta* a possible agent for the biological control of face flies. ABSTRACTS: BA 48:11210(124505), Dec. 15, 1967; RAE-B 55:186 (645), Oct., 1967.

3832. Houser, E. C. and **Wingo, C. W. 1967** Laboratory Culture and Biology of *Orthellia caesarion* with Notes on Determining Larval Instars. J. Econ. Ent. 60(5):1355-1358.
A thorough study of the life cycle under controlled conditions. This species used to support the laboratory culture of *Aphaereta pallipes*, for experiments on the biological control of *Musca.* ABSTRACT: BA 49:1422 (15943), Feb. 1, 1968.

Houx, N. W. H., joint author. *See* Oppenoorth, F. J., et al., 1968.

Howden, G. F., joint author. *See* Rothschild, J., et al., 1961.

3833. **Howell, D. E. 1961** Fly Baits. Pest Control 29(4):9-11.
Baits a useful means of control, the choice of bait depending on the situation served. Most effective insecticides for use in baits were Dibrom and DDVP (Vapona). Tan colored baits killed the most flies, followed by yellow, brown and pink. ABSTRACT: BA 36:5571(58208), Sept. 1, 1961.

Howell, D. E., joint author. *See* Goodhue, L. D., et al., 1960; Harvey, T. L., et al., 1963, 1965.

3834. **Hower, A. A., Jr. 1967** Field Studies on the Biological Control of the Face Fly, *Musca autumnalis* with Special Reference to Ecological Factors and Feeding Behavior. Diss. Absts. 28(6B):2665-B-2666-B.
Includes role of *Bacillus thuringiensis* in limiting fly development.

3835. **Hower, A. A., Jr.** and **Cheng, T.-H. 1968** Inhibitive Effect of *Bacillus thuringiensis* on the Development of the Face Fly in Cow Manure. J. Econ. Ent. 61:26-31.
Spores of *B. thuringiensis* included in cattle feed. No deterrent effect on ovipositing females, but with sufficient dosage to animals, manure yielded less than 1 per cent of the usual number of flies. Cows remained healthy. ABSTRACTS: BA 49:6245(69844), July 1, 1968; RAE-B 56:137 (477), July, 1968.

3836. **Hower, A. A., Jr.** and **Cheng, T.-H. 1968** Oviposition Behavior of the Face Fly in Caged Cow Manure Pats. J. Econ. Ent. 61(3):701-702.
Studied oviposition on caged cowpats under field conditions. *M. autumnalis* tended to deposit eggs chiefly in the central portion. Number of ova steadily declined toward the edge. ABSTRACTS: BA 49:9742(107861), Oct. 15, 1968; RAE-B 56:218(797), Nov., 1968.

Hower, A. A., joint author. *See* Cheng, T.-H., et al., 1965.

3837. **Hoyer, R. F. 1966** Some New Mutants of the House Fly, *Musca domestica*, with Notations of Related Phenomena. J. Econ. Ent. 59: 133-137.
Several mutants were isolated from normal and gamma-irradiated strains. Had to do chiefly with wing form, wing positioning, or the pattern of wing venation. ABSTRACTS: BA 47:6402(74903), Aug. 1, 1966; RAE-B 54:99, May, 1966.

3838. **Hoyer, R. F.** and **Plapp, F. W., Jr. 1966** A Gross Genetic Analysis of Two DDT-Resistant House Fly Strains. J. Econ. Ent. 59:495-501.
In both strains, a fifth chromosome dominant confers moderate resistance to *ortho*-chloro DDT. The Orlando strain possesses a second chromosome recessive, conferring moderate resistance to DDT and high tolerance to *o*-chloro DDT (1000-fold). Is probably the same as Milani's factor *kdr-o*. ABSTRACTS: BA 48:39(309), Jan. 1, 1967; RAE-B 54:175, Sept., 1966; TDB 63:1254, Nov., 1966.

3839. **Hoyer, R. F.** and **Plapp, F. W., Jr. 1968** Insecticide Resistance in the House Fly: Identification of a Gene that Confers Resistance to Organotin Insecticides and Acts as an Intensifier of Parathion Resistance. J. Econ. Ent. 61(5):1269-1276.
A strain of *M. domestica* (Cal. P-R) is *extremely* resistant to parathion and moderately resistant to organotin toxicants (e.g. TBTC). High tolerance of parathion believed to be due to the interaction of 2, independent, major genes. ABSTRACTS: BA 50:6469(67745), July 1, 1969; RAE-B 57:51(181), Mar., 1969; TDB 67:240(513), Feb., 1970.

3840. **Hoyer, R. F., Plapp, F. W., Jr.** and **Orchard, R. D. 1965** Linkage Relationships of Several Insecticide Resistance Factors in the House Fly (*Musca domestica* L.). Ent. Exp. Appl. 8:65-73.
Strains resistant to parathion, malathion, Isolan, and DDT were crossed

with a recessive strain (stubby wing, *stw*), then backcrossed. Major factors for resistance shown to be *all* located on the same chromosome. Factors for DDT and malathion resistance, though linked, are genetically distinct. ABSTRACTS: BA 47:3415(40325), May 1, 1966; RAE-B 53:225, Nov., 1965; TDB 62:813, Aug., 1965.

Hoyer, R. F., joint author. *See* Plapp, F. W., Jr., et al., 1967, 1968.

Hsiao, C., joint author. *See* Fraenkel, G., et al., 1962, 1963, 1965, 1966, 1967.

Hsiao, T. H., joint author. *See* Saunders, R. C., et al., 1970.

3841. Hsieh, C.-Y., Len, H. F. and **Chao, Y. C. 1964** [Esterase inhibition of housefly poisoning by organophosphorus insecticides.] Acta Ent. Sinica 13(4):503-509.

Treats of the correlation between esterase inhibition and housefly poisoning by Rogor, Dipterex and TOCP. Cholinesterase inhibition seems to account for the major toxic action of organophosphorus insecticides, but aliesterase inhibition may also be involved. ABSTRACT: BA 49:1422 (15944), Feb. 1, 1968.

3842. Hsieh, H.-M. and **Shi, C. Y. 1965** [Preliminary observation on the diurnal and nocturnal activity of common flies in Tienchin, North China.] Acta Ent. Sinica 14(5):511-514. (In Chinese.)

Huber, D. A., joint author. *See* Dobson, R. C., et al., 1961.

3843. Hudson, Le Verne, D. and **Gibson, A. 1960** Fly Control for 1960 at the Illinois State Fair. Pest Control 28(4):21-26.

A plan is outlined stressing the need for sanitation, the use of bait sprays, insecticidal mists and the elimination of fly breeding. ABSTRACT: BA 35:3745(43045), Aug. 1, 1960.

3844. Hughes, R. D. and **Nicholas, W. L. 1969** *Heterotylenchus* spp. Parasitising the Australian Bush Fly (*Musca vetustissima* Wlk.) in Australia. Additional Information on the Origin of the Parasite of the Face Fly (*M. autumnalis* Deg.) in the United States. J. Econ. Ent. 62(2):520-521.

Australian parasite very similar to parasite of *M. autumnalis* in Europe and North America. Could be same species which may not be very host-specific. Appears to parasitize other muscoid flies in Australia. ABSTRACTS: BA 50:11379(117684), Nov. 1, 1969; RAE-B 57:187(666), Sept., 1969.

Hughes, R. D., joint author. *See* Nicholas, W. L., et al., 1970; Tyndale-Biscoe, M., et al., 1969.

3845. Hunter, P. E. 1964 Observations on the Biology of *Laelaspis vitzthumi* (Acarina:Laelaptidae). J. Kans. Ent. Soc. 37(4):289-292.

Was successfully reared on crushed pieces of freshly killed or frozen *M. domestica*. ABSTRACT: RAE-B 54:108, June, 1966.

3846. Hurkova, J. 1962 [On the occurrence of DDT-lindane resistant houseflies, *Musca domestica* L., in Czechoslovakia.] Zool. Listy 11(2): 125-130. (English summary.)

Strain originated in Senec, Westslovakian region; LD50 values determined by topical application of insecticides on female flies. ABSTRACTS: BA 41:298(3959), Jan. 1, 1963; RAE-B 51:186, Sept., 1963.

3847. Hutchison, J. A. 1963 The Genus *Entomophthora* in the Western Hemisphere. Trans. Kans. Acad. Sci. 66(2):237-254.

Lists 40 species in this genus, which extend from Canada to Argentina. Gives distribution maps, keys for identification, list of hosts.

3848. **Hutchison, J. A. 1962** Studies on a New *Entomophthora* Attacking Calyptrate Flies. Mycologia 54(3):258-271.
This phycomycete, *E. kansana*, readily infects houseflies. The disease cycle of the fungus required approximately 96 hours in the laboratory. ABSTRACT: BA 43:943(11361), Aug. 1, 1963.

3849. **Hyzer, W. G. 1962** Flight Behavior of a Fly Alighting on a Ceiling. Science 137(3530):609-610.
Salient features of this maneuver were recorded in photographs exposed in a continuous-writing, high-speed framing camera. ABSTRACT: BA 41:1301(16662), Feb. 15, 1963.

Iaguzhinskaia, L. V., et al. *See* Yaguzhinskaya, L. V., et al.

Iasinskii, A. V., joint author. *See* Arskii, V. G., et al., 1961.

3850. **Ibragimov, S. Yu. 1964** [Species composition and vertical-zone distribution of synantropic flies in Dagestan.] SB Aspirantskikh Rabot Dagestanskii Univ. Estest Fiz Mat Nauk. 36-51. From: Ref. Zh. Biol. 1966, No. 1K402.
Specimens were collected, March to November, from 1958 to 1961 in 137 populated areas. More than 400,000 specimens of flies from 86 species were gathered. Most widely distributed flies were those from the genera *Musca* and *Fannia*. ABSTRACT: BA 47:8530(99359), Oct. 15, 1966.

3851. **Ibusuki, O. 1960** [A fundamental study on the effect of several insecticides against *Musca domestica vicina*.] J. Osaka City Med. Center 9(7):2237-2251. (In Japanese, with English summary.)
A fundamental study was made of the insecticidal effect and synergistic actions of 14 insecticides topically applied singly or in combination to *M. domestica vicina*. ABSTRACT: BA 36:2642(32449), May 15, 1961.

Ignoffo, C. M., joint author. *See* Rogoff, M. H., et al., 1969.

Iirkovskii, G. G., joint author. *See* Palii, V. F., et al., 1962.

Ikeda, J., joint author. *See* Sherman, M., et al., 1967.

3852. **Ikeme, M. M. 1967** Kerato-conjunctivitis in Cattle in the Plateau Area of Northern Nigeria: A Study of *Thelazia rhodesi* as a Possible Aetiological Agent. Bull. Epizoot. Dis. Afr. 15(4):363-367. (French summary.)
Thelazia rhodesi larvae isolated from *Musca* sp. ABSTRACT: BA 50:1015 (10563), Jan. 15, 1969.

3853. **Ikemoto, H. 1962** [Development of DDT-resistance in the so-called "Takatsuki" strain of the house-fly, *Musca domestica vicina*.] Botyu-Kagaku 27(3):76-78. (In Japanese, with English summary.)
The second of two related papers. *See also* S. Nagasawa, 1962. This paper gives a description of the development of resistance to DDT in flies of the Takatsuki strain that had previously been reared for many years without intentional exposure to insecticides. ABSTRACT: RAE-B 52:139, Aug., 1964.

3854. **Ikemoto, H. 1964** [Cross-resistance in the "Takatsuki" strain of the House Fly, *Musca domestica vicina* selected with DDT. Studies on insect resistance to insecticides. 1.] Botyu-Kagaku 29(3):59-60. (English summary.)
Results are tabulated in the form of LD50's for the original and resistant strains; considered to show cross-resistance similar to that found in European and American strains of *M. domestica domestica* L. ABSTRACT: RAE-B 55:89(316), May, 1967.

3855. **Ikemoto, H. 1964** [Some notes on Sevin-resistance in the house fly, *Musca domestica* and *Musca domestica vicina*.] Botyu-Kagaku

29(4):68-73. (In Japanese, with English summary.)
Carbamate insecticides appear to fall into two or more groups. ABSTRACTS: BA 47:8150(94922), Oct. 1, 1966; RAE-B 55:173(601), Sept., 1967.

3856. Ikemoto, H. 1965 [Studies on susceptibility for DDT of "Takatsuki" strain of the house fly, *Musca domestica vicina*. Studies on insect resistance to insecticides. 3.] Botyu-Kagaku (Bull. Inst. Insect Contr.) 30(1):1-8.
Concerns experimental work to check the findings of other investigators that the "Takatsuki" strain is resistant to DDT. Females of the least DDT-resistant of the "Takatsuki" populations were 30 times as resistant as those of the susceptible strain; males were 80 times as resistant. ABSTRACT: RAE-B 55:174(604), Sept., 1967.

Ikeno, N., joint author. *See* Hayashi, A., et al., 1968.

3857. Ikeshoji, T. 1960 Observations on the Retarded Lethal Effect of Lindane and Dieldrin on the Housefly. Jap. J. Exptl. Med. 30:83-87.
Larval, pupal and adult stages of *M. vicina* Macq. treated with 0·05 per cent lindane or dieldrin emulsion. Results (per cent of chemicals present per initially absorbed): lindane — 46-49 per cent with 2 day larvae and pupae, 35 per cent with adults; dieldrin — 83-97 per cent with 2 day larvae and pupae, 47 per cent with adults. ABSTRACT: TDB 57:1228, Nov., 1960.

3858. Ikeshoji, T. Umino, T. and **Namiki, T. 1967** [An automatic spraying device for test of aerosol insecticides.] Jap. J. Sanit. Zool. 18(2-3): 118-121.
Standard deviations for knock-down were smaller than with hand-operated aerosol sprays. Doses of labelled insecticide picked up by flies showed proportional increase in accordance with increased spraying periods. ABSTRACTS: BA 51:9099(92612), Aug. 15, 1970; RAE-B 58:278 (1150), Aug., 1970.

Ikeshoji, T., joint author. *See* Patel, N. G., et al., 1968.

Ikeuchi, M., joint author. *See* Oshio, Y., et al., 1960, 1964, 1965.

Ilbok, L., joint author. *See* Gurland, J., et al., 1960.

3859. Ilivicky, J., Chefurka, W. and **Casida, J. E. 1967** Oxidative Phosphorylation and Sensitivity to Uncouplers of House Fly Mitochondria; Influence of Isolation Medium. J. Econ. Ent. 60:1404-1407.
Gives formulae for isolation and suspension media. ABSTRACTS: BA 49: 1422(15945), Feb. 1, 1968 (Chefurn, W.); RAE-B 56:69(220), Mar., 1968; TDB 65:343(789), Mar., 1968.

Ilivicky, J., joint author. *See* Agosin, M., et al., 1966.

Ilnytsky, S., joint author. *See* Morrison, F. O., et al., 1968.

3860. Il'Yashenko, L. Ya. 1963 [Hibernation of *Musca domestica vicina* in the Hissar Valley in Tadzhik SSR.] Med. Parazit. i Parazitarn. (Bolezni) 32(5):565-567. (In Russian, with English summary.)
Unsteady winter with continuously rising and falling temperatures and the absence of fat body accumulation result in destruction of adult flies. These flies hibernate at the pupa stage; but freezing and excessive humidity of the soil and substrates result in a low survival rate. ABSTRACT: BA 45:3224(39930), May 1, 1964.

3861. Il'Yashenko, L. Ya. 1964 [Migrations of flies (*Musca domestica vicina*) in rural areas.] Med. Parazit. i Parazitarn. (Bolezni) 33(1):9-13. (In Russian.)
P^{32}-labelled flies proved the presence of fly migrations in rural areas

from animal-breeding farms to residential areas and vice versa. Migrations are determined by physiological condition of the insects. ABSTRACT: BA 45:7738(97733), Nov. 15, 1964.

3862. Il'Yashenko, L. Ya. and **Gadzhei, E. F. 1962** [A method to determine the state of pupae in *Musca domestica vicina* Macq.] Med. Parazit. i Parazitarn. (Bolezni) (Moskva) 31:617-618.
Convenient method for determining the state of pupae to study problem of overwintering and in taking account of the efficiency of sanitary measures directed towards disinsectation of various substrates. ABSTRACT: BA 41:1646(20709), Mar. 1, 1963.

3863. Inaba, T. and **Funatsu, M. 1964** Studies on Tyrosinase in the Housefly. III. Activation of Protyrosinase by Natural Activator. Agric. Biol. Chem. 28(4):206-215.
Partial purification of a natural activator occurring in the aged pupae of houseflies. Natural activator is an enzyme and protyrosinase is its substrate. ABSTRACT: BA 45:8420(106983), Dec. 15, 1964.

3864. Inaba, T., Suetake, Y. and **Funatsu, M. 1963** Studies on Tyrosinase in the Housefly. II. Activation of Protyrosinase by Sodium Dodecyl Sulfate. Agric. Biol. Chem. 27(5):332-339.
Studied occurrence of tyrosinase over wide ranges of sodium dodecyl sulfate (SDS) concentrations, pH values and temperatures. A certain range of ratios between the concentration of SDS and that of protyrosinase is effective for the activation. ABSTRACT: BA 45:3222(39905), May 1, 1964.

Inaba, T., joint author. *See* Funatsu, M., et al., 1962, 1967.

3865. Incho, H. H. 1970 New Insecticide for Aerosols. Soap and Chem. Spec. February. pp. 37-40, 68-69, 72.
"Neo-pynamin" combined with NIA 17370 provides a new ingredient for oil-based aerosols and water-based pressurized formulations. The first gives rapid knockdown, the latter, a high killing power. Mammalian toxicity low.

3866. Incho, H. H. and **Odeneal, J. F. 1962** A Study of Two New Synergists for Pyrethrins, Allethrin. Soap and Chem. Spec. 38(8):69-72, 173-175.
Safroxane (analog of piperonyl butoxide) and S421 were generally more effective than MGK-264 when combined with pyrethrins and allethrin in low-pressure aerosol formulations, but were not considered Grade A aerosols except at high concentrations. They were not as effective as pyrethrins or allethrin with piperonyl butoxide. ABSTRACT: RAE-B 51: 209, Oct., 1963.

3867. Incho, H. H. and **Odeneal, J. F. 1963** A Study of Two New Synergists for Pyrethrins, Allethrin. Pyrethrum Post 7(2):37-43.
This is the same paper as the preceding one, reprinted in the Pyrethrum Post.

3868. Inoue, Y. 1964 Studies on Biological Assay of Insecticide Residue (IV) Effect of the Moisture Content of the Treated Surface on the Blooming of Insecticide Residue. Jap. J. Sanit. Zool. 15(1):41-45. (In Japanese, with English summary.)
Concludes: efflorescence increases with an increase in humidity. Knockdown effect greater as moisture content increased. ABSTRACT: RAE-B 54:198, Oct., 1966.

Inoue, M., joint author. *See* Ogushi, K., et al., 1964.

Ioffe, I. D., joint author. *See* Yagushinskaya, L. V., et al., 1967,

1969; Tamarina, N. A., et al., 1960.

Ionica, M., joint author. *See* Lupasco, G., et al., 1966.

Ionova, A. I., joint author. *See* Lineva, V. A., et al., 1960.

3869. Ishaaya, I. and **Chefurka, W. 1968** Effect of DDT on Microsomal RNA and Protein Biosynthesis in Susceptible and DDT-resistant Houseflies. Riv. di Parassit 29:289-296. (Italian summary.)
In resistant flies DDT stimulates RNA biosynthesis in microsomal fraction more than in susceptible flies. Protein synthesis enhanced in resistant flies. Author suggests DDT might act as other inducers. ABSTRACTS: BA 51:570(5733), Jan. 1, 1970; TDB 67:723(1404), June, 1970.

3870. Ishida, M. 1968 Comparative Studies on BHC Metabolizing Enzymes, DDT Dehydrochlorinase and Glutathione s-Transferases. Agr. Biol. Chem. 32(8):947-955.
It was demonstrated by gel filtration that the enzymes from house flies, for the metabolism of 8 compounds, were indistinguishable from one another in their molecular sizes which were estimated at 36,000 to 38,000 by using reference proteins. ABSTRACT: BA 50:5633(59033), June, 1969.

3871. Ishida, M. and **Dahm, P. A. 1965** Metabolism of Benezene Hexachloride Isomers and Related Compounds In Vitro. I. Properties and Distribution of the Enzyme. J. Econ. Ent. 58:383-392.
Housefly is outstanding in its capability to metabolize BHC and PCCH isomers. Gas-liquid chromatography used to estimate metabolism. ABSTRACTS: BA 47:2505(29937), Mar. 15, 1966; RAE-B 53:177, Sept., 1965.

3872. Ishida, M. and **Dahm, P. A. 1965** Metabolism of Benzene Hexachloride Isomers and Related Compounds In Vitro. II. Purification and Stereospecificity of House Fly Enzymes. J. Econ. Ent. 58:602-607.
Analyses of effluents from ion exchange columns indicated at least 3 enzyme systems metabolizing both alpha-BHC and gamma-PCCH with different ratios of metabolizing activity for the 2 substrates. ABSTRACTS: BA 47:3381(39896), Apr. 15, 1966; RAE-B 53:219, Nov., 1965.

Ishida, M., joint author. *See* Takagi, U., et al., 1965; Nakatsugawa, T., et al., 1965.

3873. Ishijima, H. 1967 Revision of the Third Stage Larvae of Synanthropic Flies of Japan (Diptera:Anthomyiidae, Muscidae, Calliphoridae and Sarcophagidae). Jap. J. Sanit. Zool. 18(2-3):47-100.
Describes third stage larvae of 70 species in 33 genera of 4 families; also gives keys to families and species. Includes photographs of internal and external structures. Six species of Muscidae and one of Sarcophagidae, for which the larvae were not seen, are also cited and included in the keys.

Ishikawa, K., joint author. *See* Yasutomi, K., et al., 1966.

Islam, A. A., joint author. *See* Raghuwanshi, O. P., et al., 1968.

Isokoski, M., joint author. *See* Nuorteva, P., et al., 1967.

3874. Ito, H. and **Matsubara, H. 1962** [On the synergistic action of natural and synethetic synergists with Diazinon toward resistant houseflies. Studies on the control of organo-phosphorus insecticides-resistant insects I.] Botyu-Kagaku (Bull. Inst. Insect Control) 27(2):43-48. (In Japanese, with English summary.)
Concludes: diazinon-resistant house-flies cannot be controlled by the addition of either natural or synthetic synergists to diazinon. ABSTRACTS: BA 40:1917(25201), Dec. 15, 1962; RAE-B 52:128, July, 1964.

Ito, H., joint author. *See* Matsubara, H., et al., 1964.

Itoga, A., joint author. *See* Hayashi, A., et al., 1962, 1966.

Ivanova, G. B., joint author. *See* Roslavtseva, S. A., et al., 1970.

3875. Ivashkin, V. M., et al. 1966 [*Musca vitripennis*—an intermediary host of *Thelazia gulosa.*] Veterinariya 43:51-52. (In Russian.)

Iwata, T., joint author. *See* Ogushi, K., et al., 1966, 1967, 1968.

Iyatomi, K., joint author. *See* Hayashi, A., et al., 1968.

Jabaratnam, G., joint author. *See* Wharton, R. H., et al., 1962.

3876. Jackson, H. and **Schneider, H. 1968** Pharmacology of Reproduction and Fertility. *In* Ann. Rev. Pharmacology 8:467-490.
Treats of many animal species, including housefly. ABSTRACT: BA 49: 8013(89375), Sept. 1, 1968.

3877. Jacobson, M. 1967 The Structure of Echinacein, the Insecticidal Component of American Coneflower Roots. J. Org. Chem. 32(5):1646-1647.
Echinacein is insecticidal to mosquito larvae and housefly adults. Chemical name given; identical to neoherculin and alpha-sanshool, which are derived from Rutaceae. ABSTRACT: RAE-B 57:41(153), Mar., 1969.

Jackson, W. B., joint author. *See* Richards, C. S., et al., 1961 (Abstract).

Jagushinskaya, *see* Yagushinskaya.

Jakob, W. L., joint author. *See* Schoof, H. F., et al., 1964.

3878. Jalil, M. and **Rodriguez, J. G. 1970** Biology of and Odor Perception by *Fuscuropoda vegetans* (Acarina:Uropodiae), a Predator of the House Fly. Ann. Ent. Soc. Am. 63(4):935-938.
Feeds on *M. domestica* eggs and first stage larvae. However, all immature stages and the adult mite prefer nematodes to housefly eggs or larvae.

3879. Jalil, M. and **Rodriguez, J. G. 1970** Studies of Behavior of *Macrocheles muscae-domesticae* (Acarina:Macrochelidae) with Emphasis on its Attraction to the Housefly. Ann. Ent. Soc. Am. 63:738-744.
Electron micrographs show setae on tips of tarsi (*M. muscaedomesticae*) apparently responsible for odor perception. Due to a chemical attraction, the phoresy is influenced by temperature, mite stage of development, chemical stimuli, and density of mites and flies. Mites prefer odor of adult flies to eggs. ABSTRACT: BA 51:10855(110195), Oct. 1, 1970.

3880. James, M. 1969 A Study in the Origin of Parasitism. Bull. Ent. Soc. Am. 15(3):251-253.
In the genus *Musca*, the blood sucking habit is believed to have arisen with the lapping of fluid from wounds made by other species. This culminates in the species *M. crassirostris*, which makes its own wound. ABSTRACT: BA 51:9702(98734), Sept. 1, 1970.

3881. Jampoler, I. and **Palut, D. 1964** Poszukiwanie elementow roznicujacych wrazlieve i oporne na DDT muchy *Musca domestica* L. [Differences in body composition of houseflies sensitive and resistant to DDT.] Roczn. Pánstw. Zakl. Hig. 25:223-232. (In Polish, with English and Russian summaries.)
Role of lipoids in mechanism of housefly's resistance to DDT studied by extraction with solvents of body fractions of sensitive and resistant flies. Observations confirm view that body lipoids block DDT.

Janes, N. F., joint author. *See* Elliot, M., et al., 1965, 1967.

Janitschke, K., joint author. *See* Hoffman, Von G., et al., 1967.

3882. Jao, L. T. and **Gordon, H. T. 1969** Toxicity of Certain Pyrethroids and Carbamates to the CS Strain of *Oncopeltus fasciatus.* J. Econ. Ent. 62:612-616.
Compares detoxication rate of large milkweed bug with earlier findings for *M. domestica.*

3883. Jarczyk, H. J. 1965 Nachweis und charakterisierung von Phosphor und Thionophosphorsäureister der E 605-Reike spaltenden Hydrolasen aus Mitteldärmen von Lepidopteren-Raupen. [Identification and characterization of Phosphoric and Thionophosphoric acid esters of E605 series decomposing hydrolyses from the mid-intestine of Lepidopterous worms (caterpillars).] Zeit. Naturforsch. 20b(3):257-259.

Mentions hydrolysis of paraoxon and related compounds by homogenates of *M. domestica*. Enzymes hydrolytically splitting paraoxon, respiration parathion type, are called "A-esterases", "phosphotases", and "paroxonases". ABSTRACT: BA 47:7284(84989), Sept. 1, 1966.

3884. Jarczyk, H. J. 1966 [The influences of esterases in insects on the degradation of organophosphates of the ® E605 series.] Pfl. Schutz-Nachr. Bayer 19/1966, 1. 34 pp.

Worked with Lepidoptera. Connection with *M. domestica* through well selected bibliography of 88 titles. ABSTRACT: BA 48:6509(72682), July 15, 1967.

3885. Jedynak-Mankowska, H. 1963 [Sensitivity of male and female houseflies (*Musca domestica* L.) to gamma-hexachlorocyclohexane in various environmental temperatures.] Wiad. Parazytol. 9:155-160 (In Polish, with English summary.)

Males are more susceptible than females to gamma-HCH. There is more of a marked increase in susceptibility to gamma-HCH in females with the rise in temperature of the surroundings. ABSTRACTS: BA 45:2107 (26364), Mar. 15, 1964; RAE-B 52:198, Nov., 1964.

Jeffs, K. A., joint author. *See* Elliott, M., et al., 1965.

Jelenfy, I., joint author. *See* Brown, P., et al., 1970.

Jelsova, M., joint author. *See* Lavtschiev, V., et al., 1968.

3886. Jenkins, C. F. H. 1967 The House Fly, *Musca domestica*. J. Agr. West Australia 8(4th series) (12):503-506. (=Bull. No. 3545.)

Life history, control, disease vector.

3887. Jenkins, D. W. 1964 Pathogens, Parasites and Predators of Medically Important Arthropods. Annotated List and Bibliography. Bull. W. H. O. Geneva 30, suppl., 150 pp. (numerous refs).

Text is planned in 15 sections, each devoted to one of the major groups of medically important insects or acarins. In each section are reported the parasites (or pathogens) and predators affecting the type of vector under consideration. Six pages on houseflies. ABSTRACT: TDB 62:602, June, 1965.

Jensen, E., joint author. *See* Yamamoto, R. T., et al., 1967.

3888. Jensen, J. A., Flury, V. P. and **Schoof, H. R. 1965** Dichlorvos Vapour Disinsection of Aircraft. Bull. W. H. O. 32(2):175-180.

Automatic aircraft disinsection by compressed air shows biological efficiency. ABSTRACT: RAE-B 55:87(307), May, 1967.

3889. Jeu, M.-H. 1962 [The ectoparasitic mites of *Anopheles hyrcanus* var. *sinensis* Wied. and *Musca domestica vicina* Macq. and their effect on the reproduction of the hosts.] Acta Ent. Sinica 11(2):135-137.

Arrhenurus madaraszi Daday is reported from Szechuan for the first time; *Macrocheles muscaedomesticae* Scopoli a new record in China. Neither are natural enemies of the adult fly. ABSTRACT: BA 48:10644 (119518), Dec. 1, 1967.

Johansen, C. A., joint author. *See* Vinopal, J. H., et al., 1967.

3890. Johnsen, R. E. and **Dahm, P. A. 1966** Activation and Degradation Efficiencies of Liver Microsomes from Eight Vertebrate Species, Using Organophosphates as Substrates. J. Econ. Ent. 59(6):1437-1442.
Housefly head cholinesterase used in manometric anticholinesterase assays. Differences in activation and degradation of organophosphates found between species and sexes. Houseflies were insecticide-susceptible, approximately 4 days old, and laboratory reared. ABSTRACT: BA 48: 9952(111879), Nov. 15, 1967.

Johnson, E. R., joint author. *See* Sun, Y.-P., et al., 1960, 1963, 1967, 1969.

Johnson, J. D., joint author. *See* Adams, T. S., et al., 1970.

Johnson, J. O., joint author. *See* Rogoff, W. M., et al., 1964.

3891. Johnson, O. and **Wagoner, D. E. 1968** The House Fly and Hybrid Sterility. Proc. N. Centr. Branch Ent. Soc. Am. 23(1):27.
Results of crosses between laboratory-reared wild stock *M. domestica* from U.S., Italy, and wild strains from other parts of world presented to determine if any natural genetic differences have evolved in wild populations which might result in hybrid sterility or cytoplasmic incompatability when interbreeding occurs. ABSTRACT: RAE-B 58:151(634), May, 1970.

Johnson, O. A., joint author. *See* Wagoner, D. E., et al., 1965, 1968, 1969.

Johnson, P., joint author. *See* Goldberg, A. A., et al., 1965.

Johnsson, G., joint author. *See* Casida, J. E., et al., 1960.

3892. Joiner, R. L. and **Lambremont, E. N. 1969** Hydrocarbon Metabolism in Insects: Oxidation of Hexadecane-1-^{14}C in the Boll Weevil and the House Fly. Ann. Ent. Soc. Am. 62(4):891-894.
Metabolism of hexadecane yielded two metabolic products, $^{14}CO_2$ and fatty acids. Direct conversion of hexadecane to fatty acids by terminal oxidation. Fatty acids distributed to triglycerides, phospholipids, stereolesters, and other fatty-acid containing lipid fractions. ABSTRACTS: BA 51:2232(23158), Feb. 15, 1970; RAE-B 58:114(463), Apr., 1970.

3893. Jones, C. M. 1967 *Aleochara tristis*, a Natural Enemy of Face Fly I. Introduction and Laboratory Rearing. J. Econ. Ent. 60:816-817.
Young larvae of beetle enter puparia of *M. autumnalis* and feed on pupae. ABSTRACTS: BA 48:11210(124508), Dec. 15, 1967; RAE-B 55:190(657), Oct., 1967.

3894. Jones, C. M. 1969 Biology of the Face Fly: Migration of Larvae. J. Econ. Ent. 62(1):255-256.
Observation in Nebraska indicates that the migration of larvae of *M. autumnalis* from hatching site should be considered when natural enemies of the pupae (or larvae of *Aleochara tristis* Gravenhorst) are studied. ABSTRACTS: BA 51:551(5531), Jan. 1, 1970; RAE-B 57:121(431), July, 1969.

3895. Jones, C. M. and **Medley, J. G. 1963** Control of the Face Fly on Cattle with Co-Ral in Grain and on Pasture. J. Econ. Ent. 56:214-215.
Feces of cattle sufficiently toxic to completely inhibit larval development. ABSTRACTS: BA 43:1327(16455), Aug. 15, 1963; RAE-B 51:148, July, 1963.

3896. Jones, C. M. and **Perdue, J. M. 1967** *Heterotylenchus autumnalis*, a Parasite of the Face Fly. J. Econ. Ent. 60:1393-1395.
About 30 per cent of flies studied were infected by this nematode.

ABSTRACTS: BA 49:1397(15651), Feb. 1, 1968; RAE-B 56:68(218), Mar., 1968.

Jones, C. M., joint author. *See* Smith, C. N., et al., 1960.

3897. Jones, G. D. G. and **Chadwick, P. R. 1960** A Comparison of Four Pyrethrum Synergists. Pyrethrum Post 5(3):22-30.
As space spray against houseflies, sulfoxide, piperonyl butoxide, and bucarpolate had similar knockdown rate. S.421 had reduced effect. Also tested flour beetles and grain weevils. ABSTRACT: RAE-B 50:3, Jan., 1962.

3898. Jones, G. D. G. and **Chadwick, P. R. 1962** Observations on the Toxicity of Some Petroleum Oils to Houseflies. Pyrethrum Post 6(3): 27-31. Nakuru.
Toxicity of a range of petroleum based solvents to houseflies determined using measured drop and space spray technique. Odourless kerosene and high aromatic solvents most toxic. No simple relationship appeared between either boiling point or viscosity and toxicity. ABSTRACT: RAE-B 51:275, Dec., 1963.

Jones, G. D. G., joint author. *See* Chadwick, P. R., et al., 1960, 1962.

3899. Jones, G. W. 1962 The Significance of the Plague Fly. An Historical Note. Virginia Med. Monthly 89:87-89, February.

3900. Jones, R. L., Metcalf, R. L. and **Fukutu, T. R. 1969** Use of the Multiple Regression Equation in the Prediction of the Insecticidal Activity of Anticholinesterase Insecticides. J. Econ. Ent. 62(4):801-808.
LD_{50} values affected by both cholinesterase inhibition and lipophylicity with *para*-substituted compounds; by cholinesterase inhibition only with *meta*-substituted compounds. ABSTRACT: RAE-B 58:25(193), Jan., 1970.

3901. Jordan, T. W., Smith, J. N. and **Whitehead, N. 1968** Preparation of house fly oxidation enzymes. Australian J. Pharm. 49(No. 584, suppl. 66):S66-S68.
Object: to prepare housefly oxidizing enzyme which might minimize liberation of natural inhibitors.

Joshi, G. C., joint author. *See* Raghavan, N. G. S., et al., 1967.

Judd, W. W., joint author. *See* Singh, S. B., et al., 1966.

Kada, J. M., joint author. *See* Eastwood, R. E., et al., 1966, 1967.

3902. Kaddou, I. K. 1966 Effects of X-irradiation of *Musca domestica* Pupae on Adult Emergence and Longevity. Bull. Biol. Res. Cent. (Baghdad) 2:36-42. (Summary in Arabic.)
Female emergence and longevity were reduced at 2000 R; more so at higher dosages. Effects on males were more severe. ABSTRACT: BA 51: 9197(93657), Sept. 1, 1970.

3903. *Kahn, N. H., Ansari, J. A., Rehman, J. and **Ahmad, D. 1964** Relative Abundance of Houseflies in India and Their Susceptibility to DDT, BHC, and Dieldrin. J. Bombay Nat. Hist. Soc. 61(3):712-716.
Survey of the relative occurrence of the more important forms of *M. domestica nebulo*, *M. domestica vicina*, *M. sorbens* in different states of India and the susceptibility of *M. domestica nebulo* to DDT, BHC and dieldrin. ABSTRACT: RAE-B 55:177, Oct., 1967.
*Believed identical with N. H. Khan, which see.

Kahn, N. H., joint author. *See* Rahman, S. J., et al., 1964, 1965.

Kai, K., joint author. *See* Matsubara, H., et al., 1964.

Kakizawa, H., joint author. *See* Shimizu, F., et al., 1965.

Kalle, G. P., joint author. *See* Amonkar, S. V., et al., 1967.

3904. Kalmus, H. 1961 The Attenuation of Optomotor Responses in White-eyed Mutants of *Musca domestica* and *Coelopa frigida*. Vision Res. 1(1-2):192-197.
Optomotor reactions, over a wide range of conditions, less than one-twelfth of the normal. ABSTRACT: BA 39:1015(12987), Aug. 15, 1962.

Kalra, R. L., joint author. *See* Sharma, M. I., et al., 1962; Ramakrishnan, S. P., et al., 1963.

Kamakahi, D. C., joint author. *See* Steiner, L. F., et al., 1970.

Kamal, A., joint author. *See* Khan, M. A. Q., et al., 1970.

Kamat, D. N., joint author. *See* Mustafa, M., et al., 1970.

Kamienski, F. X., joint author. *See* Casida, J. E., et al., 1966.

Kamimura, H., joint author. *See* Yamamoto, I., et al., 1968.

3905. Kamyszek, F. 1965 Mucha domowa (*Musca domestica*) jako przenosiciel grzybic. [*Musca domestica* as a carrier of fungal infections.] Med. Weter 21(10):622-624.
Pathogenic fungi can be carried passively by flies into artificial media, the number of carrier flies being directly proportional to the length of their stay on the fungal-infected media. ABSTRACT: BA 48:3162(34783), Apr. 1, 1967.

3906. Kamyszek, F. 1965 [Effect of certain factors on the passive transmission of pathogenic fungi by domestic flies.] Wiad. Parazytol. 11(6):567-572. (In Polish, with English summary.)
Fungus studied: *Trichophyton crateriforme*. Transmission higher with increased humidity; occurs in summer. ABSTRACT: BA 48:7778(87036), Sept. 1, 1967.

Kanamori, S. I., joint author. *See* Matsubara, H., et al., 1964.

Kaneko, K., joint author. *See* Kano, R., et al., 1964, 1965, 1968; Miyamoto, K., et al., 1965; Matsuzawa, H., et al., 1965; Shimizu, F., et al., 1965.

3907. Kano, R. 1964 On the Medically Important Flies Collected by Gentaro Imadaté in Southeast Asia from 1961 to 1962. Nature and Life in Southeast Asia. Vol. III. Publ. by Fauna and Flora Research Society, Kyoto. March. pp. 435-436.
Lists *M. conducens*, *M. convexifrons*, *M. domestica vicina* and *M. sorbens*.

3908. Kano, R., Aniya, K., Kaneko, K., Shinonaga, S. and **Kiuna, H. 1964** Notes on Flies of Medical Importance in Japan. Part XX. Seasonal Occurrence of Medically Important Flies on the Ishigaki Island Ryukyu. Jap. J. Sanit. Zool. 15(1):1-6.
Lists *M. domestica*, *M. sorbens* and *M. ventrosa*.

3909. Kano, R., Kaneko, K., Miyamoto, K., Shinonaga, S., Kiuna, H., Okazaki, T. and **Habutsu, Y. 1965** Notes on Flies of Medical Importance in Japan. Part XXIII. Seasonal Fluctuation of Flies in the Imperial Palace Grounds, Tokyo. J. Med. Ent. (Honolulu) 1(4):387-394.
M. hervei common in pasteurs; *M. pabulorum* in mountain paths and forests. ABSTRACTS: BA 46:4783(59214), July 1, 1965; TDB 62:693, July, 1965.

3910. Kano, R., Kaneko, K. and **Shinonaga, S. 1964** Notes on the Medically Important Flies Collected on the Amami Oshima Islands and Tokara Islands. Kontyu 32(1):129-135.
Lists *M. conducens*, *M. domestica*, *M. hervei*, *M. sorbens*, and *M. ventrosa*.

3911. Kano, R., Kaneko, K. and **Shinonaga, S. 1968** Synantropic Flies in Hong Kong [Anthomyiidae, Muscidae, Calliphoridae, Sarco-

phagidae, Distribution]. Kontyu 36(1):75-87.
In April 1964, a survey of medically important flies resulted in the collection of 4 families, 25 genera and 44 species. Data on these flies, of which 6 are *Musca*, are presented in the paper. ABSTRACT: BA 50:11945 (123493), Nov. 15, 1969.

3912. Kano, R. and **Shinonaga, S. 1964** Notes on Medically Important Flies in the Tohoku District, Japan. Tohoku Konchu Kenkyu 1(1):5-8.
Lists *M. convexifrons*, *M. domestica* and *M. hervei.*

3913. Kano, R. and **Shinonaga, S. 1967** Notes on Flies of Medical Importance in Japan. XXVII. Revisional Notes on the Metallic Muscoid Flies. Jap. J. Sanit. Zool. 18(4):195-212. (Japanese summary.)
Seven metallic muscoid flies described. One species of genus *Orthellia* is new record for Japan. ABSTRACT: BA 51:2214(22948), Feb. 15, 1970.

Kano, R., joint author. *See* Miyamoto, K., et al., 1965, 1967; Suzuki, T., et al., 1962; Shimizu, F., et al., 1965; Shinaga, S., et al., 1966.

3914. Kantack, B. H., Berndt, W. L. and **Balsbaugh, E. V., Jr. 1967** Horn Fly and Face Fly Control on Range Cattle with Aerial Applications of Ultra-Low-Volume Malathion Sprays. J. Econ. Ent. 60:1766-1767.
Excellent control on rangeland beef herds; number of applications (4-6) differs with years. ABSTRACTS: BA 49:3807(42775), Apr. 15, 1968; RAE-B 56:104(360), May, 1968.

3915. Kao, C. M. and **Wei, P. H. 1960** [The fly species of Hopei Province.] Acta Ent. Sinica 10(1):75-78. (In Chinese, with English summary.)
Survey gave 30 species in 14 genera. Most numerous were *M. domestica vicina*, *Muscina stabulans*, *Lucilia sericata*, *Chrysomyia megacephala*, *Calliphora grahami*, *Fannia scalaris*, and 4 species of *Sarcophaga.* ABSTRACTS: BA 35:4003(46103), Aug. 15, 1960; RAE-B 50:247, Nov., 1962.

Kao, G., joint author. *See* Kung, K., et al., 1963, 1965, 1968.

Kaplan, A. M., joint author. *See* Shambaugh, G. F. et al. 1968.

3916. Kaplanis, J. N., Dutky, R. C. and **Robbins, W. E. 1961** The Incorporation of 2-C^{14} Mevalonate into House Fly Lipids. Ann. Ent. Soc. Am. 54(1):114-116.
About the same amount incorporated into saponifiable and unsaponifiable lipids after 18 hours.

3917. Kaplanis, J. N., Monroe, R. E., Robbins, W. E. and **Louloudes, S. J. 1963** The Fate of Dietary H^3-β-Sitosterol in the Adult House Fly. Ann. Ent. Soc. Am. 56:198-201.
Fly utilizes this compound directly without detectable conversion to cholesterol. ABSTRACT: BA 43:1039(12501), Aug. 1, 1963.

3918. Kaplanis, J. N., Robbins, W. E., Monroe, R. E., Shortino, T. J. and **Thompson, M. J. 1965** The Utilization and Fate of β-Sitosterol in the Larva of the Housefly, *Musca domestica* L. J. Insect Physiol. 11(3):251-258.
Larvae of the housefly do not dealkylate ß-sitosterol to form cholesterol. On ß-sitosterol, only 1·4 per cent of the individuals emerged as adults and these laid no viable eggs. This sterol will not fulfil in entirety the sterol requirements of *M. domestica.* ABSTRACT: BA 46:4802(59462), July 1, 1965.

3919. Kaplanis, J. N., Robbins, W. E. and **Tabor, L. A. 1960** The Utilization and Metabolism of 4-C-Cholesterol by the Adult House Fly. Ann. Ent. Soc. Am. 53(2):260-264.
Injected cholesterol efficiently utilized in egg production.

3920. Kaplanis, J. N., Tabor, L. A., Thompson, M. J., Robbins, W. E. and **Shortino, T. J. 1966** Assay for Ecdysone (Molting Hormone) Activity Using the House Fly, *Musca domestica* L. Steroids 8(5): 625-631.
Quantity required for 60 per cent pupation in *M. domestica* about one-quarter to one-third that need by *Calliphora*. ABSTRACT: BA 48:3269 (36052), Apr. 1, 1967.

Kaplanis, J. N., joint author. *See* Dutky, R. C., et al., 1963, 1967; Monroe, R. E., et al., 1961; Vroman, H. E., et al., 1966; Robbins, W. E., et al., 1960, 1970.

3921. Karandikar, K. R. and **Ranade, D. R. 1965** Studies on the Pupation of *Musca domestica nebulo* Farr. (Diptera, Cyclorrhapha, Muscidae). Proc. Indiana Acad. Sci., Sec. B. 61(4):204-213.
Excellent, detailed account of prepupal and pupal characters. ABSTRACT: BA 47:2950(34933), Apr. 1, 1966.

Kariva, M., joint author. *See* Funatsu, M., et al., 1967.

Karnikar, K., joint author. *See* Deoras, P. J., et al., 1967.

3922. Kasai, T. 1965 Genetical and Biochemical Studies on Joint Action of Insecticides. Botyu-Kagaku (Bull. Inst. Insect Contr.) 30(3):73-91. (Japanese summary.)
Work on effect of genetic heterogeneity of insect populations on the joint action of insecticides is discussed and a new scheme for the various types of joint action is proposed. ABSTRACT: RAE-B 56:52(172), Mar. 1968.

3923. Kasai, T. and **Ogita, Z.-I. 1965** A Genetic Study on Sevin-Resistance and Joint Toxic Action of Sevin with Gamma-BHC Against House Flies. Botyu-Kagaku (Bull. Inst. Insect Contr.) 30(1):12-17. (Japanese summary.)
Increase in insecticidal action, associated with the use of carbyl and gamma-BHC together, may be due to a combination of joint action on different sections of a population genetically heterogeneous with regard to resistance and dissimilar joint action. ABSTRACT: RAE-B 55:174(605), Sept., 1967.

Kasai, T., joint author. *See* Ogita, Z. I., et al., 1965, 1966.

Kascheff, A. H., joint author. *See* Soliman, A. A., et al., 1966, 1967.

3924. Kaselis, P. V. 1968 Study of Resistance of Houseflies to Various Insecticides. Bull. New Jersey Acad. Sci. 13(1):111.

Kashi, K. P., joint author. *See* Qadri, S. S. H., et al., 1970.

Kashina, D. A., joint author. *See* Budyl'nikova, M. A., et al., 1962.

Kato, M., joint author. *See* Ogata, K., et al., 1960.

Kato, S., joint author. *See* Ohtaki, T., et al., 1964.

3925. Kato, Y. 1960 [Flies in Nansatu district of Kagosima prefecture.] Igaku Kenkyu 30(11):2996-3018.
Flies belonging to 17 genera and 41 species were collected. In almost all species collected, the female was predominant, except for *Sarcophaga*. ABSTRACT: BA 36:4009(49606), Aug. 1, 1961.

3926. Katsuda, Y. and **Ogami, H. 1966** [Studies on the substituted benzyl esters of chrysanthemic acid.] Botyu-Kagaku (Bull. Inst. Insect Contr.) 31(1):30-33. (In Japanese, with English summary.)
Most toxic to *M. domestica* was 4-allylbenzyl chrysanthemate (benathrin). ABSTRACT: RAE-B 56:128(449), July, 1968.

3927. Katsuda, Y., Ogami, H., Kunishige, T. and **Sugii, Y. 1967** Novel Active Chrysanthemic Esters. Agric. Biol. Chem. 31(2):259-260.

Insecticidal activity on *Musca domestica vicina*. ABSTRACT: BA 48:6012 (67200), July 1, 1967.

3928. **Katsuda, Y., Ogami, H., Kunishige, T.** and **Togashi, E. 1966** [Studies on chrysanthemic esters of substituted-benzyl alcohols, -phenols and -cyclohexanols.] Botyu-Kagaku 31(2):82-86. (In Japanese, with English summary.)
Benathrin (4-allylbenzyl chrysanthemate) more toxic to adult *M. domestica* than 14 other esters. ABSTRACT: RAE-B 56:153(535), Aug., 1968.

Kauer, K. C., joint author. *See* Blair, E. H., et al., 1963, 1966.

3929. **Kaur, D.** and **Steve, P. C. 1969** Induced Sterility in Face Fly—Initial and Sustained Effects of Hempa and Metepa. J. Econ. Ent. 62(6): 1462-1464.
Sterility was sustained for 28 days in both sexes. Longevity of males was affected but at 1·0 per cent level females were not affected. Mortality of females achieved at 2·0 per cent level. ABSTRACTS: BA 51:7937(81043), July 15, 1970; RAE-B 58:146(608), May, 1970.

Kawaguchi, K., joint author. *See* Ohtaki, T., et al., 1964.

3930. **Kawai, S.** and **Suenaga, O. 1960** Studies on the Methods of Collecting Flies. III. On the Effect of Putrefaction of Baits (Fish). Endemic Dis. Bull. Nagasaki Univ. 2(1):61-66.
One day old fish bait was the most effective for catching flies while the freshest bait was least effective. ABSTRACT: BA 39:670(8305), July 15, 1962.

Kawai, S., joint author. *See* Suenaga, O., et al., 1964.

Kawasaki, M., joint author. *See* Matsubara, H., et al., 1964.

3931. **Kay, R. E., Eichner, J. T.** and **Gelvin, D. E. 1967** Quantitative Studies on the Olfactory Potentials of *Lucilia sericata*. Am. J. Physiol. 213(1):1-10.
L. sericata's olfactory potential threshold about 10 times lower than human subjective threshold; considerably lower than behavioral threshhold of *M. domestica* and *Phormia regina*. ABSTRACT: BA 48:11246 (124909), Dec. 15, 1967.

Kazakova, T. I., joint author. *See* Kovalenok, A. V., et al., 1967.

3932. **Kazano, H., Asakawa, M., Tanaka, T.** and **Fukunaga, K. 1968** [Studies on carbamate insecticides. I. Insecticidal activities of substituted phenyl-N-methyl and N, N-dimethylcarbamates to several species of insects.] Jap. J. Appl. Ent. Zool. 12(4):202-210.
Toxicity of these compounds to *M. domestica* is related to the chemical structure of each. Findings agree with published reports of R. L. Metcalf. ABSTRACT: BA 51:9099(92615), Aug. 15, 1970.

3933. **Kazhdan, V. B. 1960** Dikhloridifenilkarbinol kak sinergist DDT. [Dichlorodiphenylmethylcarbinol as a synergist of DDT.] Med. Parazit. i Parazitarn. (Bolezni) 29(2):223-226. (English summary.)
Various types of experiments show DMC to be a very effective synergist with DDT. In tests using DDT alone, resulting in 30 per cent mortality for susceptible and 7 per cent for resistant flies, the use of DDT and DMC obtained a 72-99 per cent for susceptible and 77-100 per cent for resistant flies. ABSTRACTS: BA 36:6196(66268), Oct. 1, 1961; RAE-B 50:207, Sept., 1962.

Kearns, C. W., joint author. *See* Chang, S. C., et al., 1962, 1964.

Keehn, D. G., joint author. *See* Bishop, L. G., et al., 1966, 1967, 1968.

3934. Keiding, J. 1962 The Use of Thiourea Against Insecticide-resistant Houseflies (*Musca domestica* L.). XI. Internat. Kong. für Ent. Wien Bd. II., pp. 618-621.
The mechanism of the pupation inhibition of thiourea is not known but could be hormonal. There is neither a negative or positive correlation between tolerance to thiourea and resistance to the contact insecticides nor is there an increase of thiourea resistance in comparison with a population that had not received thiourea or with a laboratory strain.

3935. Keiding, J. 1963 Possible Reversal of Resistance. *In* Bull. W.H.O. Geneva. Vol. 29. Suppl. pp. 51-62. WHO Seminar Papers, Geneva, 5-9 November.
Prospects of resuming use of an insecticide to which resistance has developed are *not* good unless it is withdrawn at an *early* stage and unless *hybrids* are amenable to control. Aim should be to reserve use of cheap and quick-acting insecticides for emergencies. Other methods of control should be kept in mind (e.g. sanitation, biological control, sterilization, etc.). ABSTRACT: TDB 61:441, Apr., 1964.

3936. Keiding, J. 1965 Observations on the Behaviour of the Housefly in Relation to its Control. Riv. di Parassit. Rome. 26(1):45-60.
Special attention given to characteristic, aggregated, night resting sites. ABSTRACT: RAE-B 55:73(266), Apr., 1967; TDB 62:1061, Oct., 1965.

3937. Keiding, J. 1965 Development of Resistance in Field Populations of House-flies Exposed to Residual Treatments with Organo-phosphorus Compounds. Meded. Landbouwhogesch. Opzoek. Stns. Gent. 30(3): 1362-1381. Ghent. (Summaries in Flemish, French and German.)
Summarizes investigations with nine OP compounds tested on farms during 1953-1964. ABSTRACTS: BA 48:2362(25837), Mar. 1, 1967; RAE-B 54:212, Nov., 1966.

3938. Keiding, J. 1965 Investigations on the Development of Tolerance in House-fly Populations Exposed to Dichlorvos Vapour in the Field and in the Laboratory. W.H.O./Vector Control/160.65, 10 pp., 3 tables.
A strain under selective pressure with dichlorvos vapour for 15 generations showed only doubling of the dichlorvos tolerance with topical applications but a 4 to 8 times increase in vapour tests. This indicates a specific resistant to the vapour effect.

3939. Keiding, J. 1966 Stuefluens resistens mod insektmidler. [Insecticide resistance in houseflies.] Ann. Rept. (1965) Govern. Pest Infest. Lab. Lyngby, Denmark. pp. 40-48.
Tests for resistance to organophosphorus insecticides were carried out on 41 Danish farms mainly to detect resistance development at early stages.

3940. Keiding, J. 1966 Laboratoriestammer. [Laboratory strains.] Ann. Rept. (1965) Govern. Pest Infest. Lab. Lyngby, Denmark. pp. 48-51.
A list of housefly strains kept in the laboratory during 1965.

3941. Keiding, J. 1967 Persistence of Resistant Population After the Removal of the Selection Pressure. Abstract Int. Pflanzenschutz-Kong. [Int. Congr. Plant Prot.] 30:44-46.

3942. Keiding, J. 1967 Development of Housefly Resistance to Fenitrothion (OMS-43) and Bromophos (OMS-658) in the Laboratory and in the Field. W.H.O./V.B.C./67.2.
Development of resistance to OMS-43 and OMS-658 followed through 16-17 generations. Resistance developed to OMS-43 and OMS-658 did not interfere with fly control. Resistance to OMS-658 regressed when selection pressure was withdrawn but that to OMS-43 persisted.

3943. Keiding, J. 1967 Persistence of Resistant Populations After the Relaxation of the Selection Pressure. Wld. Rev. Pest Control 6(4):115-130.
Possibilities of restoring susceptibility to a pesticide to which resistance has once developed, are small. ABSTRACT: RAE-B 56:111(390), June, 1968.

3944. Keiding, J. 1967 Stuefluens resistens mod insektmidler. [Insecticide Resistance in Houseflies.] Ann. Rept. (1966) Govern. Pest Infest. Lab. Lyngby, Denmark. pp. 38-48.
Resistance to organophosphorus compounds investigated on 50 Danish farms with results summarized in table form. Discusses insecticides tested as well as strains of flies involved.

3945. Keiding, J. 1968 Stuefluens resistens mod insektmidler. Resistensundersøgelser i forbindelse med fluebekaempelse ellerforsøg pa garde. [Insecticide resistance in houseflies. Fly populations on farms.] Ann. Rept. (1967) Govein. Pest Infest. Lab. Lyngby, Denmark. pp. 41-45.
Resistance to one or more organophosphorus compounds and other insecticides carried out on 52 Danish farms. Gives table of strains kept in laboratory and names of those exported for study.

3946. Keiding, J. 1968 Selective housefly control on farms based on the resting habits of the flies. Int. Congr. Ent. (Moscow). Abstracts, p. 120.
Studied diurnal distribution and resting habits of houseflies; results used for selecting surfaces for concentrated insecticide treatment. Treated areas comprised only 10-30 per cent of the sprayable surfaces. Discusses the effect of selective treatments.

3947. Keiding, J. 1969 Stuefluens resistens mod insektmidler. Resistensundersøgelser; forbindelse med fluebekaempelse eller forsøg på gårde. [Insecticide resistance in houseflies. Fly populations on farms.] Ann. Rept. (1968) Govern. Pest Infest. Lab. Lyngby, Denmark. pp. 35-37, 39-44.
Investigations concerning resistance to organophosphorus compounds and other insecticides (including newly developed ones), were carried out on 61 farms. Studied: (1) effect of treatment on susceptibility of fly population; (2) resistance acquired from treatment; and (3) previously acquired resistance. Results summarized in a table.

3948. Keiding, J. 1969 Forsøg med stuefluer fra Malaysia. [Tests with Malaysian houseflies.] Ann. Rept. (1968) Govern. Pest Infest. Lab. Lyngby, Denmark. pp. 38, 39, 47.
Flies collected from the Cameron Highlands (which had been treated for several years with OP-compounds), show a high resistance to chlorinated hydrocarbons and OP-compounds.

3949. Keiding, J. 1969 Laboratoriestammer. [Laboratory strains.] Ann. Rept. (1968) Govern. Infest. Lab. Lyngby, Denmark. pp. 39, 47-50.
A table lists the housefly strains kept in the laboratory. They are the same used in 1967 except that 2 recent Danish strains were discarded and the Malaysian strain *Musca domestica nebulo* added.

3950. Keiding, J. 1969 Consultantship for World Health Organization on a Fly Control Problem in Malaysia, September 1968. Ann. Rept. (1968) Govern. Pest Infest. Lab. Lyngby, Denmark.
The breeding of the tropical housefly, *Musca domestica nebulo*, in organic fertilizer used on crops in the Cameron Highlands created a problem. Personal investigations on breeding habits, biology, dispersion, distribution and night habits of flies, and insect resistance were carried out; also small scale experiments on use of impregnated cards, toxic baits and insecticidal treatment of manure for fly control.

3951. Keiding, J. and **ben Hannine, S. 1964** Field Trials of Dichlorvos Resin Strands for Fly Control on Farms in Denmark in 1963. W.H.O./ Vector Control/85. 10pp.

Resin strands, when hung in stables, gave excellent knockdown and control of houseflies and stable flies for 3-4 months. There were no indications of resistance-development to dichlorvos during the season.

3952. Keiding, J. and **ben Hannine, S. 1966** Strategisk fluebekaempelse. [Strategic Fly Control.] Ann. Rept. (1965) Govern. Pest Infest. Lab. Lyngby, Denmark. pp. 38-39.

"Strategic fly control", or treatment of limited, strategically placed areas selected for fly preference, proved effective in controlling flies in pig and cattle barns.

3953. Keiding, J. and **ben Hannine, S. 1966** Forsøg med fluemidler pa garde for Verdenssundhedsorganisationen (W.H.O.) [Small scale field trials of insecticides for fly control on farms for World Health Organization.] Ann. Rept. (1965) Govern. Pest Infest. Lab. Lyngby, Denmark. pp. 36-38. (In Danish, with complete English translation.)

Recommends 1·0 g dimethoate per m^2 for season long fly control.

3954. Keiding, J. and **ben Hannine, S. 1966** Small Scale Field Trials of Certain Dimetholate Formulations for Fly Control on Danish Farms in 1965. W.H.O./Vector Control/66.208. 10 pp.

Concentrations are given for dimethoate formulations tested as residual sprays and as strategic spot treatments; both types of applications gave excellent results in fly control. Tolerance development to dimethoate was moderate.

3955. Keiding, J. and **ben Hannine, S. 1966** Metoder til mäling af resistens. [Method of testing resistance.] Ann. Rept. (1965) Govern. Pest Infest. Lab. Lyngby, Denmark. pp. 43 and 51-52.

Resistance to fumigant action of dichlorvos was tested aiming at LD_{50} between 40-100 minutes.

3956. Keiding, J. and **ben Hannine, S. 1967** Stuefluens biologi og adfaerd. Fluernes adfaerd og fordeling i stalde. [Housefly biology and behavior. Fly behavior and distribution in piggeries and cowsheds.] Ann. Rept. (1966) Govern. Pest Infest. Lab. Lyngby, Denmark. pp. 50-51, 53-54.

Determined distribution of flies to various sites from time of hatching. Flies moved from hatching sites to lower resting sites and animals by day; at night seeking out ceilings and high beams, not being on animals at all.

3957. Keiding, J. and **Yasutomi, K. 1969** Metoder til mäling af resistens hos fluer. [Methods of testing housefly resistance to insecticides.] Ann. Rept. (1968) Govern. Pest Infest. Lab., Lyngby, Denmark. pp. 38, 46, 47.

Discusses several methods used and proposed as standard WHO procedures including the "treated vial" exposure method, topical application and use of anaesthesia. The topical application method using 0·35 ml applied by a micro-capillary was acceptable with volatile and nonvolatile substances.

3958. Keiding, J. and **Yasutomi, K. 1969** Nogle Kombinationer af resistens modfosformidler i danske flaestammer. [Some combinations of resistance to organosphorus compounds in Danish fly strains.]

Ann. Rept. (1968) Govern. Pest Infest. Lab. Lyngby, Denmark. pp. 37, 44.
A summary of tests; results to be published elsewhere. Main trends in field populations show positive correlation between dimethoate-, fenthion- and trichlorfron-resistance, but not so clearly between dimethoate and diazinon or fenitrothion.

Keiding, J., joint author. *See* Gratz, N. G., et al., 1964; ben Hannine, S., et al., 1967, 1968.

Keller, J. C., joint author. *See* Fales, J. H., et al., 1961, 1962.

3959. Kenaga, E. E. 1965 Triphenyl Tin Compounds as Insect Reproduction Inhibitors. J. Econ. Ent. 58:4-8.
Triphenyl tin compounds sterilize adult flies well below a lethal concentration, females being sterilized at lower concentration than males. For some compounds the reproduction control is reversible. Other species of insects and related arthropods are also affected. ABSTRACT: RAE-B 53:94, May, 1965.

3960. Kenaga, E. E. 1969 Some Hydroxynitrosamino Aliphatic Acid Derivatives as Insect Reproduction Inhibitors. J. Econ. Ent. 62(5):1006-1008.
The propionic or butyric acid derivatives having the hydroxynitrosamino group in the *beta* or *gamma* position were most effective. These inhibitors are more effective on the female than on the male housefly. ABSTRACTS: TDB 67:366(768), Mar., 1970; BA 51:9095(92580), Aug. 15, 1970.

3961. Kenaga, E. E. 1969 Insect Reproduction Inhibition Effects of Some Dinitronaphthalene Benezenesulfonamides, Bis (Haloethyl) Phenyl Amines and Hydroxynitrosamino Aliphatic Acids. Congrès International des Antiparasitaires. Milan. 6-8 October. 5 pp. (No. 19). (Summaries in Italian, French, German and Spanish.)
The effects of several compounds are given; most compare favorably with tepa, a standard of comparison for reproduction inhibition. All were effective for time of treatment and were more effective for males than females.

3962. Kenaga, E. E., Doty, A. E. and **Hardy, J. L. 1962** Laboratory Insecticidal Tests with 4-dimethylamino-3,5-xylyl Methylcarbamate. J. Econ. Ent. 55:466-469.
Zectran (trademark) shows a wide spectrum of insecticidal activity, particularly as a residual application because it is relatively non-volatile. Alkaline conditions and sunlight or UV light deteriorate its insecticidal properties. ABSTRACT: RAE-B 51:3, Jan., 1963.

3963. Kenaga, E. E., Whitney, W. K., Hardy, J. L. and **Doty, A. E. 1965** Laboratory Tests with Dursban Insecticide. J. Econ. Ent. 58:1043-1050.
Dursban shows good insecticidal properties being moderately residual on plant surfaces and quite residual on inert surfaces. It is volatile enough to form insecticidal residues on nearby untreated objects and is stable except under rigorous conditions of alkalinity and acidity. ABSTRACT: RAE-B 54:58, Mar., 1966.

Kenaga, E. E., joint author. *See* Blair, E. H., et al., 1963, 1965; Whitney, W. K., et al., 1969; Rigterink, R. H., et al., 1966.

3964. Kerr, R. W. 1960 Sex Limited DDT-resistance in Houseflies. Nature 185(4716):868.
Strain E Y. All males specifically DDT resistant; all females non-resistant. Very different from strains previously studied, which are neither sex-linked nor sex-limited for this trait. ABSTRACT: BA 35:3745 (43046), Aug. 1, 1960.

3965. Kerr, R. W. 1961 Inheritance of DDT-Resistance Involving the Y-Chromosome in the Housefly. (*Musca domestica* L.) Australian J. Biol. Sci. 14(4):605-619.
Demonstrated a genetically new type of resistance, not transmitted through females. While females remained susceptible, males showed eight-fold resistance to DDT. ABSTRACTS: BA 39:1700(21580), Sept. 15, 1962; RAE-B 50:284, Dec., 1962; TDB 59:398, Apr., 1962.

3965a. Kerr, R. W. 1970 Inheritance of DDT Resistance in a Laboratory Colony of the Housefly, *Musca domestica.* Australian J. Biol. Sci. 23(2):377-400.
Flies from the Canberra laboratory colony (established 1939) were examined genetically and cytologically. DDT-resistance was found to be determined by incompletely dominant allele of a gene in chromosome II, that confers the ability to metabolize DDT to DDE. ABSTRACT: RAE-B 58:274(1130), Aug., 1970.

Kerr, R. W., joint author. *See* Barton-Browne, L., et al., 1967; Browne, L. B., et al., 1967.

Kesler, E. M., joint author. *See* Cheng, T. H., et al., 1961.

Kessler, J., joint author. *See* MacLaren, W. R., et al., 1960.

Khalsa, H. G., joint author. *See* Saxena, B. N., et al., 1965.

3966. Khan, M. A. Q. 1969 Some Biochemical Characteristics of the Microsomal Cyclodiene Epoxidase System and Its Inheritance in the House Fly. J. Econ. Ent. 62(2):388-392.
Location of gene for microsomal epoxidase was determined. *Musca domestica* L. microsomal NADP.H oxidizing system can epoxidise aldrin, isodrin and heptachlor at identical rates. ABSTRACT: RAE-B 57: 185(660), Sept., 1969.

3967. Khan, M. A. Q. 1969 DDT-dehydrochlorinase and Aldrin-epoxidase Activity in Corn Earworm and Polyphemus Moth Larvae, and House Fly Adults. J. Econ. Ent. 62:723-725.
In vivo epoxidation of aldrin shown to be higher in the abdomen of houseflies than in any other body regions. Epoxidation by microsomes from head, thorax and abdomen of 3 insecticide-resistant strains showed 80 per cent of total epoxidase activity to be present in abdomen. ABSTRACT: RAE-B 57:239(904), Nov., 1969.

3968. Khan, M. A. Q. 1970 Genetic and Biochemical Characteristics of Cyclodiene Epoxidase in the House Fly. Biochem. Pharmacol. 19: 903:910.
Epoxidase activity in resistant strains increases after emergence of adult flies. Male flies have about half the activity of females. ABSTRACT: BA 51:11474(116525), Oct. 15, 1970.

3968a. Khan, M. A. Q., Chang, J. L., Sutherland, D. J., Rosen, J. D. and **Kamal, A. 1970** House Fly Microsomal Oxidation of Some Foreign Compounds. J. Econ. Ent. 63(6)-1807-1813.
All such reactions are inhibited by very low concentrations of sesamex. The gene *Ox*, which controls activity of the mixed function oxidase, is located on chromosome 2 in the strains studied.

3969. Khan, M. A. Q. and **Hodgson, E. 1967** Phospholipase Activity in *Musca domestica* L. Comp. Biochem. Physiol. 23(3):899-910.
Phospholipase A activity depends on calcium ion concentration in *M. domestica.* Phospholipase B activity, highest in mitochondria, can also be detected in microsomal fraction. ABSTRACT: BA 49:5303(59515), June 1, 1968.

3970. Khan, M. A. Q. and **Hodgson, E. 1967** Phospholipids of Subcellular Fractions from the Housefly, *Musca domestica* L. J. Insect Physiol. 13(5):653-664.
The microsomal fraction contains only 36 per cent phosphatidylethanolamine whereas other fractions contain 50 to 59 per cent. ABSTRACT: BA 48:7436(83231), Aug. 15, 1967.

3971. Khan, M. A. Q., Rosen, J. D. and **Sutherland, D. L. 1969** Insect Metabolism of Photoaldrin and Photodieldrin. Science 164:318-319.
Flies and mosquito larvae metabolize these photoisomers (sunlight products of aldrin and dieldrin) to more toxic materials. Suggests that this conversion is the cause for the enhanced toxicities of the photoproducts.

3972. Khan, M. A. Q., Sutherland, D. J., Rosen, J. D. and **Carey, W. F. 1970** Effects of Sesamex on the Toxicity and Metabolism of Cyclodienes and Their Photoisomers in the House Fly. J. Econ. Ent. 63(2):470-475.
To a susceptible strain of flies, all photoisomers (except photoisodrin) are more toxic and act more rapidly than parent compounds. In resistant strain, sesamex inhibited metabolism of aldrin, photodieldrin and photoaldrin. ABSTRACTS: BA 51:8569(87381), Aug. 1, 1970; RAE-B 58:304 (1242), Sept., 1970.

3973. Khan, M. A. Q. and **Terriere, L. C. 1968** DDT-Dehydrochlorinase Activity in House Fly Strains Resistant to Various Groups of Insecticides. J. Econ. Ent. 61(3):732-736.
DDT-ase activity, of soluble subcell fraction from resistant *M. domestica*, about 10 times that of susceptible strains. Relates to second chromosomal resistance factors. ABSTRACTS: BA 49:10332(114111), Nov. 15, 1968; RAE-B 56:219(800), Nov., 1968; TDB 66:757(1509), July, 1969.

Khan, M. A. Q., joint author. *See* Schonbrod, R. D., et al., 1968; Ashrafi, S. H., et al., 1963; Rosen, J. D., et al., 1969.

3974. Khan, N. H. and **Ahmad, D. 1964** Inheritance of BHC-Resistance in the Housefly, *Musca domestica nebulo* Fabr. Jap. J. Genet. 38:367-373.
From reciprocal crosses between individuals, as well as groups, of a BHC resistant and susceptible fly strain of the subspecies, it was concluded that BHC resistance is controlled by a multiple-gene factor. ABSTRACT: BA 46:1432(17895), Feb. 15, 1965.

3975. Khan, N. H. and **Ahmad, D. 1965** Development of BHC-Resistance in *Musca domestica nebulo* and *Musca domestica vicina*. Angew. Parasit. (Jena) 6:150-156. (Summaries in German and Russian.)
Selectively bred 30 generations of *nebulo* surviving BHC exposure, resulting in 36-fold tolerance when compound applied in acetone, and 19-fold in risella. *M. d. vicina* gave 59-fold and 22-fold tolerance, respectively after 22nd generation. ABSTRACTS: RAE-B 56:48(159), Feb., 1968; TDB 63:597, May, 1966.

3976. Khan, H. N. and **Ansari, J. A. 1964** DDT-resistance Hazard in the Indian House Flies, *Musca domestica nebulo* and *Musca domestica vicina*. Botyu-Kagaku 29(2):15-18.
Resistant strain of *M. d. vicina* developed and found to be 10 times as resistant, after 12 generations, as *M. d. nebulo* was after 15 generations. ABSTRACT: RAE-B 55:87(309), May, 1967.

3977. Khan, H. N. and **Ansari, J. A.** Loss and Reversion of DDT Resistance in *Musca domestica nebulo*. Symp. Pestic. A., p. 41.

3978. Khan, N. H. and **Rehman, S. J. 1964** Inheritance of Thoracic Stripes in the Indian Form of Houseflies. Jap. J. Genet. 38:374-377.
The thoracic stripes in the two fly forms were controlled by a single allele which was dominant in 1 genotype and recessive in the other. ABSTRACT: BA 46:1108(13995), Feb. 15, 1964.

Khan, N. H., joint author. *See* Rahman, S. J., et al., 1965, 1968; Raghuwanshi, O. P., et al., 1968.

3979. Kharlamov, V. P. 1964 Nekotorye fiziko-khimicheskie pokazateli gazoobmena u komnatnykh mukh *Musca domestica* L. i nuckeotid-nogo sostava DNK ikh lichinok (L_{111}) v pervom potomstve ot vnutrennego B-obllucheniya fosfora P^{32}. [Some physico-chemical indices of gas exchange in the housefly *Musca domestica* L. and of nucleotide DNA composition of their larvae (L-3) in the first generation following internal beta-irradiation with P^{32}.] Radio-biologiia 4:893-895.
Treats of changes in metabolism and respiratory quotient of *Musca domestica*, when exposed to B-irradiation in the imago stage. ABSTRACT: BA 46:8590(106029), Dec. 15, 1965.

3980. Khattat, F. H. and **Busvine, J. R. 1965** A Modified Test Method for Measuring Resistance to Dichlorvos Vapour. Bull. W. H. O. 32(4): 551-556.
M. domestica and several other insect species can be tested satisfactorily by use of this method; more compact for field work. Tested under various conditions of temperature and humidity. ABSTRACT: RAE-B 55:129(464), July, 1967.

Kholodova, G. K., joint author. *See* Sukhova, M. N., et al., 1965.

Khromova, L. A., joint author. *See* Tamarina, N. A., et al., 1960.

Khudadov, G. D., joint author. *See* Vashkov, V. I., et al., 1965.

Kidder, H. E., joint author. *See* Dorsey, C. K., et al., 1962.

3981. Kikodze, S. L., et al. 1968 [Experience of organization of planned domestic trash removal from household in Tbilisi.] Med. Parasit. (Moskva) 37:424-428. (In Russian.)

3982. Kilgore, W. W. and **Painter, R. R. 1962** The Effect of 5-Fluorouracil on the Viability of House Fly Eggs. J. Econ. Ent. 55(5):710-712.
Radioactive material fed to adults rendered viability of eggs low for first four days of oviposition. ABSTRACTS: BA 41:1664(20909), Mar. 1, 1963; RAE-B 51:41, Feb., 1963.

3983. Kilgore, W. W. and **Painter, R. R. 1964** Effect of the Chemosterilant Apholate on the Synthesis of Cellular Components in Developing Housefly Eggs. Biochem. J. 92(2):353-357.
Non-viable eggs deposited by insects given a diet treated with apholate do not synthesize any significant quantities of DNA during incubation; this is accompanied by the loss of ability to form lactate dehydrogenase. ABSTRACTS: BA 47:3356(39606), Apr. 15, 1966; TDB 61:1261, Dec., 1964.

3984. Kilgore, W. W. and **Painter, R. R. 1966** Insect Chemosterilants: Incor-poration of 5-Fluorouracil into House Fly Eggs. J. Econ. Ent. 59(3):746-747.
Indication that sterilization may be caused by chemosterilant replacing uracil in RNA. ABSTRACTS: BA 47:8559(99759), Oct. 15, 1966; RAE-B 54:182, Sept., 1966; TDB 63:1402, Dec., 1966.

Kilgore, W. W., joint author. *See* Painter, R. R., et al., 1964, 1965, 1967; Gadallah, A. I., et al., 1970.

3985. Killough, R. A. 1965 Effect of Different Levels of Illumination on the Life Cycle of the Face Fly. J. Econ. Ent. 58(2):368-369.
Life cycle was not affected by 3 different illumination levels of fluorescent lamps. ABSTRACTS: BA 46:5142(63864), July 15, 1965; RAE-B 53:144, July, 1965.

3986. Killough, R. A., Hartsock, J. G., Wolf, W. W. and **Smith, J. W. 1965** Face Fly Dispersal, Nocturnal Resting Places and Activity During Sunset as Observed in 1963. J. Econ. Ent. 58:711-715.
Marked flies released near cattle at sunset. Flies left cattle while natural light level was still high. Spent night in trees and tall grasses rather than barns. Flies moved 2 miles from release site in 24 hours; one 4 miles away in 5 days. ABSTRACTS: BA 46:8877(109462), Dec. 15, 1965; RAE-B 53:223, Nov., 1965.

3987. Killough, R. A. and **McClellan, E. S. 1965** Face Fly Oviposition Studies. J. Econ. Ent. 58:716-719.
Eggs from caged, isolated females averaged about 20 per batch (maximum 31), with 2-8 days between batches. 78·2 per cent eggs developed into pupae. Eggs deposited as soon as they became full size in ovaries. ABSTRACTS: BA 46:8877, Dec. 15, 1965; RAE-B 53:223, Nov., 1965.

3988. Killough, R. A. and **McClellan, E. S. 1969** Laboratory Studies of the Mating Habits of the Face Fly. J. Econ. Ent. 62(3):551-555.
Most females mated only once, remained in copulation 65 minutes and only mated a second time if no sperm was received. An 8:1 ratio of sterilized to unsterilized males reduced pupal production 94 per cent. ABSTRACTS: BA 50:11387(117785), Nov. 1, 1969; RAE-B 57:235(892), Nov., 1969.

3989. Kilpatrick, J. W., Maddock, D. R. and **Miles, J. W. 1962** Modification of a Semiautomatic Liquid-Poison Bait Dispenser for House Fly Control. J. Econ. Ent. 55:951-953.
Liquid or solid formulations of DDVP used in modified poultry waterers, placed at rate of one dispenser per 1,500 sq. feet, efficiently controlled flies on dairy and chicken ranches. ABSTRACTS: BA 42:307(3707), Apr. 1, 1963; RAE-B 51:59, Mar., 1963.

3990. Kilpatrick, J. W., Miles, J. W. and **Schoof, H. F. 1962** New Poison Bait for Housefly Control. Pest Control 30(10):13, 14, 18.
Solid formulations of DDVP used as bait in adapted chicken watering devices gave satisfactory results at concentrations of 0·002-0·01 per cent; comparable to impregnated cord baits. ABSTRACTS: BA 42:972(12259), Apr. 15, 1963; RAE-B 52:59, Mar., 1964.

3991. Kilpatrick, J. W. and **Schoof, H. F. 1963** Effectiveness of Seven Organophosphorus Compounds as Space Applications against *Musca domestica*. J. Econ. Ent. 56:560-563.
Mist applications at various concentrations of compounds are evaluated as to effective dosage and distance of mortality. ABSTRACTS: BA 45:1759 (21934), Mar. 1, 1964; RAE-B 52:17, Jan., 1964; TDB 61:223, Feb., 1964.

3992. Kilpatrick, J. W. and **Schoof, H. F. 1963** Adult House Fly Control with Residual Treatments of Six Organophosphorus Compounds. J. Econ. Ent. 56(1):79-81.
Results are given for dairy treatments using 6 residual-treatment compounds. Dimethoate proves most effective against *M. domestica*, yet has low mammalian toxicity. ABSTRACTS: BA 42:1907(24067), Jan. 15, 1963; RAE-B 51:93, May, 1963; TDB 60:795, Aug., 1963.

Kilpatrick, J. W., joint author. *See* Fay, R. W., et al., 1963; Weinburg,

H. B., et al., 1961.

Kimmel, E. C., joint author. *See* O'Brien, R. D., et al., 1965; Yamamoto, I., et al., 1969.

Kimura, S., joint author. *See* Yasutomi, K., et al., 1966.

Kindler, S. H., joint author. *See* Zahavi, M., et al., 1964.

King, W. E., joint author. *See* Singh, P., et al., 1966.

3993. Kinn, D. N. 1966 Predation by the Mite *Macrocheles muscaedomesticae* (Acarina:Macrochelidae) on Three Species of Flies. J. Med. Ent. 3(2):155-158.

Mite destroys eggs and first stage larvae of *M. domestica.* ABSTRACT: BA 48:1855(20044), Feb. 15, 1967.

3994. Kirchberg, E. 1961 Zur Kenntnis fäkalgebundener Fliegen auf Samos (Greichenland). [Toward knowledge of fecal-associated flies on Samos (Greece).] Zentralbl. Bakt. Parasit. Infekt. 182(2):267-275. (English summary.)

Sarcophaga, Muscina, Fannia, Paregle and *Anthomyia* stressed most as vectors of disease. ABSTRACT: BA 41:1296(16598), Feb. 15, 1963.

3995. Kirchberg, E. 1969 Über den Aussagewert der Fangergebnisse von Insektenfallen: Zubleich eine Stelle ungnahme zu den Arbeiten von H. Peters, Heidelberg, über synanthrope Fleigen. [On the value of articles concerning the catching results of insect traps: Also an opinion on the works of H. Peters, Heidelberg, on synanthropic flies.] Deut. Ent. Zeit. 16(1-3): 131-139.

Only flies reacting to specific stimuli are caught in traps. Fecal and cadaver flies not taken in same traps with *Fannia* and *M. domestica.* Differing opinions on the role of synanthropic flies in spreading human disease. ABSTRACT: BA 51:10854(110192), Oct. 1, 1970.

3996. Kirchberg, E. and **Schulz, L. 1962** Beiträge zur Fliegenfauna Berlins und seiner Umgebung IV. (Diptera). [Contributions to the knowledge of the flies of Berlin and its surroundings. (Diptera).] Mitt. Deutsch Ent. Gesell. 21(3):40-43.

Several records of new and rediscovered Diptera are reported, including eleven species of adult Tachinidae and Sarcophagidae. ABSTRACT: BA 45:7744(97804), Nov. 15, 1964.

3997. Kirschfeld, K. 1965 Das anatomische und das physiologische Sehfeld der Ommatidien im Komplexauge von *Musca.* [The anatomic and physiological visual field of ommatidia in the complex eye of *Musca.*] Kybernetic 2:249-257.

Intracellular micropipette recording used to investigate compound eye. Visual fields of adjacent ommatidia strongly overlap. Discusses consequences of the overlap of visual fields of adjacent ommatidia for perception of motions and patterns.

3998. Kirschfeld, K. 1966 Discrete and Graded Receptor Potentials in the Compound Eye of the Fly (*Musca*). Proc. Int. Symp. on the functional organization of the compound eye, 25-27 October, 1965. Stockholm, Sweden. Symp. Publ. Div. Pergamon Press, London and New York. 7:291-307.

High quantal efficiency (more than 60 per cent) is discussed on the basis of the "unfused" rhabdoeric structure of the dipteran ommatidium. ABSTRACT: BA 48:9251(104026), Oct. 15, 1967.

3999. Kirschfeld, K. 1967 Die Projection der optischen Umwelt auf das Raster der Rhabdomere im Komplexauge von *Musca.* Exp. Brain Res. 3(3):248-270. (English summary.)

The *Musca* compound eye can be regarded as a "neural superposition eye". The dioptrics of the *M. domestica* ommatidium act as an inverting lens system. There exists a one to one correspondence between a lattice fo points in the environment and the lattice of "cartridges" in the lamina.

4000. Kirschfeld, K. and **Franceschini, N. 1969** Ein Mechanismus zur Steuerung des Lichtflusses in den Rhabdomeren des Komplexauges von *Musca*. [A mechanism for the control of the light flow in the rhabdomeres of the complex eye of *Musca*.] Kybernetik 6:13-22. (English summary.)
Changes within a few seconds following illumination in the ommatidia of *Musca* are interpreted as a consequence of interactions between pigment granules in the sense cells and electromagnetic fields induced outside the rhabdomeres by light traveling on the inside.

4001. Kirschfeld, K., et al. 1970 [Optomotor experiments on *Musca* with linearly polarized light.] Zeit. Naturforsch. (B) 25:288.

Kishimoto, K., joint author. *See* Eto, M., et al., 1966.

Kishino, M., joint author. *See* Nagasawa, S., et al., 1963, 1965.

4002. Kissam, J. B. 1967 Mortality and Fertility Response of *Musca domestica* to Certain Known Mutagenic or Anti-tumor Agents. Diss. Absts. 27(12B):4436-B-4437-B.
See reference 4003, Kissam, et al., 1966.

4003. Kissam, J. B. and **Hays, S. B. 1966** Mortality and Fertility Response of *Musca domestica* Adults to Certain Known Mutagenic or Anti-Tumor Agents. J. Econ. Ent. 59(3):748-749.
Materials were screened to determine dosage rate producing sterility without causing significant mortality. ABSTRACTS: BA 47:8523(99285), Oct. 15, 1966; RAE-B 54:182, Sept., 1966.

4004. Kissam, J. B., Wilson, J. A. and **Hays, S. B. 1967** Selective Effects of Certain Anti-fertility Compounds on the House Fly as Shown by Reciprocal Crosses and Histological Sectioning. J. Econ. Ent. 60(4):1130-1135.
"HU" destroys female egg cells; with "MMS" and "TEM", embryos die in the eggs. ABSTRACTS: BA 49:1423(15953), Feb. 1, 1968; RAE-B 56:7(22), Jan., 1968.

4005. Kitahara, K. and **Miyazaki, T. 1966** On Flies of Medical Importance in South Kyushu. III. Fly Fauna in Nase City. Acta Med. Univ. Kagoshimaensis. 8(1):13-34.

Kitahara, T., joint author. *See* Matsui, M., et al., 1967.

Kitakata, S., joint author. *See* Eto, M., et al., 1966.

Kitzmiller, J. B., joint author. *See* Spielman, A., et al., 1967; Knipling, E. F., et al., 1968.

Kiuna, H., joint author. *See* Kano, R., et al., 1964, 1965.

Kivatra, S., joint author. *See* Sharma, G. P., et al., 1966.

Kjordev, N., joint author. *See* Sarbova, S., et al., 1968.

Klaassen, D. H., joint author. *See* Burkhard, R. K., et al., 1966.

Klas-Bertil, A., joint author. *See* Casida, J. E., et al., 1960.

4006. Klein, W., et al. 1969 [Contributions on biological chemistry. XIX. Metabolism of photodieldrin-C^{14} in warm blooded animals, insects and plants.] Tetrahedron Lett. 37:3197-3199.

4007. Kliewer, J. W. and **Boreham, M. M. 1964** Oviposition Studies of the Little House Fly, *Fannia canicularis* (Diptera:Muscidae). Calif. Vector Views 11(4):23-26.
Study to determine the moisture content in fowl droppings preferred for oviposition. Droppings of 33·3 per cent or less unattractive. Optimum range 42·8 to 60 per cent. ABSTRACT: RAE-B 54:215, Nov., 1966.

4008. Knapp, F. W. 1962 Horn Fly and Face Fly Control Studies with Dow M-1816. J. Econ. Ent. 55(5):816-817.
Daily applications not necessary, nor does each animal in herd need to be treated to obtain adequate fly control. ABSTRACTS: BA 41:947(12331), Feb. 1, 1963; RAE-B 51:45, Feb., 1963.

4009. Knapp, F. W. 1965 Free Choice Feeding of Ronnel Mineral Block and Granules for Face Fly, Horn Fly and Cattle Grub Control. J. Econ. Ent. 58(5):836-838.
Fly reduction greatest on cattle having access to face rubber, which was treated with 1 per cent ronnel in No. 2 fuel oil. ABSTRACTS: BA 47: 2047(24559), Mar. 1, 1966; RAE-B 54:12, Jan., 1966.

4010. Knapp, F. W. 1966 Aerial Application of Trichlorfon for Horn Fly and Face Fly Control on Cattle. J. Econ. Ent. 59:468.
Remarkably good control for small amount of chemical used. ABSTRACTS: BA 47:7231(84360), Sept. 1, 1966; RAE-B 54:134, July, 1966.

4011. Knapp, F. W. 1967 Ultra-low-volume Aerial Application of Trichlorfon for Control of Adult Mosquitoes, Face Flies, and Horn Flies. J. Econ. Ent. 60(4):1193.
Control of face flies believed somewhat nullified by migration from untreated herd nearby. ABSTRACTS: BA 49:458(5096), Jan. 1, 1968; RAE-B 56:9(28), Jan., 1968.

4012. Knapp, F. W. 1967 Face Fly Studies. Kentucky Agr. Exp. Sta. Progr. Rep. p. 56.
Reports specific data on reproduction, nutrition and the minimum amount of bovine feces required for adult emergence.

4013. Knapp, F. W. 1968 New Developments in Fly Control for Livestock. Farm Technology 24(4):16, 18. (Summer 1968.)
Describes an automatic spray system used for fly control around swine barns. Operates on the principle of water being forced through pellets of insecticide formulation and forcing it out through spray nozzles.

4014. Knapp, F. W. 1968 Fly Control Tests. *In* Results of Research in 1967. 80th Ann. Rept. Kentucky Agr. Exp. Sta. pp. 55-56.
Reports on self-dusting and water forced over Vapona resin pellets through spray nozzles for controlling horn fly and face fly on beef and dairy cattle.

Knapp, F. W., joint author. *See* Thurston, R., et al., 1966; Rodriguez, J. G., et al., 1968.

4015. Knight, Sister M. R. 1962 Rhythmic Activities of the Alimentary Canal of the Black Blow Fly, *Phormia regina* (Diptera-Calliphoridae). Ann. Ent. Soc. Am. 55:380-382.
Included as a contribution to Muscoid biology.

Knight, S. G., joint author. *See* Matsumura, F., et al., 1967.

4016. Knipling, E. F. 1960 Views on Future Research Contributing to a Solution of the Resistance Problem. Misc. Publ. Ent. Soc. Am. 2(1):157-162.
Musca domestica cited. Recommendations for the future.

4017. Knipling, E. F. 1962 The Use and Limitations of Isotopes and Radiation Sterility in Meeting Insect Problems. J. Appl. Rad. and Isotopes. 13:417-426.
M. domestica briefly cited.

4018. Knipling, E. F. 1962 Potentialities and Progress in the Development of Chemosterilants for Insect Control. J. Econ. Ent. 55:782-786.
Discusses alkylating agents (successful with *M. domestica*) versus gamma radiation as a means of effecting sterilization. ABSTRACT: RAE-B 51:43, Feb., 1963.

4019. **Knipling, E. F. 1964** The Potential Role of the Sterility Method for Insect Population Control with Special Reference to Combining this Method with Conventional Methods. ARS-33-98. iv + 54 pp. Washington, D.C., Dept. Agr.
Includes incidental references to houseflies and face flies. ABSTRACT: RAE-B 55:128(461), July, 1967.

4020. **Knipling, E. F., Laven, H., Craig, G. B., Pal, R., Kitzmiller, J. B., Smith, C. N.** and **Brown, A. W. A. 1968** Genetic Control of Insects of Public Health Importance. Bull. W. H. O. 38(3):421-438.
Cytoplasmic incompatability may be used successfully without use of radiation or chemosterilants. Houseflies discussed briefly.

4021. **Knipling, G. D., Sullivan, W. N.** and **Fulton, R. A. 1961** The Survival of Several Species of Insects in a Nitrogen Atmosphere. J. Econ. Ent. 54(5):1054-1055.
M. domestica had a short survival time (3·5 hours to kill 50 per cent). Results indicate that nitrogen may be an excellent substitute for carbon dioxide in anaesthetizing insects for laboratory handling. ABSTRACT: RAE-B 50:178, Aug., 1962.

4022. **Knowles, C. O.** and **Arthur, B. W. 1966** Metabolism of and Residues Associated with Dermal and Intramuscular Application of Radiolabeled Fenthion to Dairy Cows. J. Econ. Ent. 59(6):1346-1352.
Mortality of house flies and face flies, caged on cows treated dermally with fenthion, was 100 per cent at 1 day after treatment. ABSTRACT: BA 48:3946(43923), May 1, 1967.

4023. **Knowles, C. O.** and **Arthur, B. W. 1967** Biological Activity of N-Methylcarbamate and Dimethyl Phosphorothionate Esters of Various Phenols and Monoterpenoids. J. Econ. Ent. 60(5):1417-1420.
Esters of the substituted phenols were more insecticidally active than those of the terpenoids. Both carbamate and the phosphate esters of the substituted phenols were synergized by piperonyl butoxide. ABSTRACTS: BA 49:1400(15672), Feb. 1, 1968; RAE-B 56:69(222), Mar., 1968; TDB 65:345(799), Mar., 1968.

4024. **Knowles, C. O.** and **Arurkar, S. K. 1969** Acetylcholinesterase Polymorphism in the Face Fly (Diptera:Muscidae). J. Kans. Ent. Soc. 42(1):39-45.
Face fly head acetylcholinesterase resolved electrophoretically in acrylamide gel; three soluble zones were present. ABSTRACT: BA 50:9766 (101155), Sept. 15, 1969.

Knowles, C. O., joint author. *See* Bratkowski, T. A., et al., 1968.

4025. **Knudsen, G. J. 1969** House fly, *Musca domestica.* Wisc. Conserv. Bull. 34(1):30-31.
Reports on reproductive capacity and human disease relations.

Ko, V., joint author. *See* Dutton, G. J., et al., 1964.

4026. **Kobara, R. 1967** [Sexual difference in haemolymph protein of several insects.] Jap. J. Appl. Ent. Zool. 11(2):71-75. (English summary.)
Studies were made by means of an improved discontinuous electrophoresis on acrylamide gels. Most information given concerns the silkworm but similar results are stated for the housefly; e.g. sexual differences were detected at the middle stage of the 5th larval instar and in the pupal stage. ABSTRACT: BA 51:566, Jan. 1, 1970.

4027. **Kobayashi, K., Eto, M., Oshima, Y., Hirano, T., Hosoi, T.** and **Wakamori, S. 1969** [Synthesis and biological activities as insecticides and fungicides of saligenin cyclic phosphorothiolates.] Botyu

Kagaku 34:165-170. (In Japanese, with English summary.)
Several compounds synthesized and evaluated. The smaller the S-alkyl group, the higher the insecticidal activity. (S-methyl thiolate is most active). *M. domestica vicina* used for evaluations (4-5 day old females). Topical application of acetone solutions.

4028. **Kobayashi, K., Hirano, T. Wakamori, S., Eto, M.** and **Oshimo, Y. 1969** [Synthesis and insecticidal activities of 2-β-substituted ethoxy-4H-1, 3, 2-benzodioxaphosphorin-2-oxides and -sulfides.] Botyu-Kagaku 34(2):66-69. (In Japanese, with English summary.)
A halogen, alkoxyl or phenoxyl group, introduced at the β-position of the ethyl ester, decreased insecticidal activity. No compound superior in insecticidal activity to the unsubstituted ethyl ester or methyl esters was found. ABSTRACT: BA 51:5061(52113), May 1, 1970.

Kobayashi, K., joint author. *See* Eto, M., et al., 1968.

4029. **Kobayashi, M.** and **Akai, H. 1967** Action of Ecdysone on some Metabolism During Larval-pupal Transformation of the House Fly *Musca domestica* L. (Diptera:Muscidae). Appl. Ent. Zool. 2(4): 223-224.
Deals largely with nucleic acid synthesis, amino acid metabolism and glucose metabolism.

4030. **Kobayashi, M., Nakanishi, K.** and **Koreeda, M. 1967** The Moulting Hormone Activity of Ponasterones on *Musca domestica* (Diptera) and *Bombyx mori* (Lepidoptera). Steroids 9(5):529-536.
Ponasterones, of plant origin, showed high activity on both species. ABSTRACT: BA 48:7436(83234), Aug. 15, 1967.

4031. **Kobayashi, M., Takemoto, T., Ogawa, S.** and **Nishimoto, N. 1967** The Moulting Hormone Activity of Ecdysterone and Inokosterone Isolated from *Achyranthis radix*. J. Insect Physiol. 13(9):1395-1399.
These chemicals show high activity as the hormone, although pupal-adult mixtures were induced, indirectly by injecting large doses. Tested on *Musca domestica* and *Bombyx mori*. ABSTRACT: BA 49:1423(15954), Feb. 1, 1968.

4032. **Koch, H. A. 1964** [Flies as transmitters of dermatophytes.] Hautarzt 15:365-366.

Kodama, M., joint author. *See* Buei, K., et al., 1963.

Koenig, F. H., joint author. *See* Desmoras, J., et al., 1964.

4032a. **Kogan, M.** and **Legner, E. F. 1970** A Biosystematic Revision of the Genus *Muscidifurax* (Hymenoptera:Pteromalidae) with Descriptions of Four New Species. Canad. Ent. 102(10):1268-1290.
Specimens used in morphological studies were obtained from stocks reared on puparia of wild Riverside and NAIDN—house flies. *M. domestica* was thought to have been the "original" host.

Köhler, G., joint author. *See* Hoffman, Von G., et al., 1967.

4033. **Kohls, R. E., Lemin, A. J.** and **O'Connell, P. W. 1966** New Chemosterilants Against the House Fly. J. Econ. Ent. 59(3):745-746.
Of several compounds administered orally to house flies, pactamycin was by far the most active. At the 1 per cent level, it resulted in complete inhibition of egg-laying. ABSTRACTS: BA 47:8523(99287), Oct. 15, 1966; RAE-B 54:181, Sept., 1966; TDB 63:1402, Dec., 1966.

Koizumi, H., joint author. *See* Gohda, M., et al., 1966.

Kojima, K., joint author. *See* Eto, M., et al., 1966.

Komatsu, G. H., joint author. *See* Sherman, M., et al., 1962, 1963, 1967.

Komatsuzaki, I., joint author. *See* Takagi, U., et al., 1965.

Kondrashova, M. L., joint author. *See* Zaidenov, A. M., et al., 1965.

4034. Kono, I. and **Fukuyoshi, S. 1967** [Leukoderma of the muzzle of cattle induced by a new species of *Stephanofilaria*.] Jap. J. Vet. Sci. 29(6): 301-313. (In Japanese, with English summary.)
M. conducens (which gathers on the muzzles of cattle) was found to contain larvae of a nematode resembling known larvae of *Stephanofilaria*. These larvae, used to infect healthy cattle, caused histological changes typical of leucoderma. ABSTRACT: BA 51:11441(116118), Oct. 15, 1970.

4035. Kontar', V. A. and **Uspenskii, I. V. 1966** O povedenii nasekomykh vblizi diffundiruyushchikh istochnikov razdrazheniya. [Behavior of insects (flies) near diffusing stimulation sources.] Zh. Obshch. Biol. 27(1):32-39. (In Russian, with English summary.)
An attempt was made to apply mathematical methods to a description of insect behavior within the closed space, with a stimulations source. ABSTRACT: BA 48:11246(124914), Dec. 15, 1967.

Kopecky, B. E., joint author. *See* Dahm, P. A., et al., 1962.

4036. Kopelovich, L. 1964 Investigation of Enzymatic Activities in Relation to (Insecticide) Resistance in the Housefly, *Musca domestica* Linnaeus (to Hydrocarbon and Organophorphorus Insecticides). Diss. Absts. 24(9):3894.

Koreeda, M., joint author. *See* Kobayashi, M., et al., 1967.

Koutz, F. R., joint author. *See* Lodha, K. R., et al., 1970.

4037. Kovacs, E. 1966 Fliegenbekämpfung in Rinderställen mit besonderer Berücksichtigung der Resistenz. [Control of flies in cattle-sheds with special regard to resistance (to insecticides).] Mh. Veterinaermed. 21:462-466. (Summaries in Russian and English.)
The author discusses the major species of flies occurring in cowsheds, their mode of life and the possibility of control. An outline of the major insecticides and repellents is given as well as some new means of control. The resistance problem is discussed.

4038. Kovalenok, A. V. and **Kazakova, T. I. 1967** Izmenenie urovnya aktivnosti suktsinodegridazy u komnatnoi mukhi pri otravlenii khlorofosom i gamma-geksakhloranom. [Changes in the level of succinate dehydrase activity in the house fly poisoned by Dipterex and gamma-hexachlorane.] Izv. Sib. Otd. Akad. Nauk. SSSR Ser. Biol. Med. Nauk. 2:118-122.
Poisoning with Dipterex and gamma-hexachlorane is accompanied by an increase in succinate dehydrogenase activity in the indirect flying muscles of the thorax and by the inhibition of activity of this enzyme in the nervous system. ABSTRACTS: BA 50:1052(11032), Jan. 15, 1969; RAE-B 57:153(555), Aug., 1969.

4039. Kraft, K. A. 1960 Sezonnyi khod chislennosti i sutochnyi khod aktivnosti komnatnoi mukhi *Musca domestica* L. v usloviakh Akmolinsk. [Seasonal variations of the population and daily variations of the activity of *Musca domestica* L. in Akmolinsk.] Med. Parazit. (Moskva) 29:726-730. (English summary.) Referat. Zhur., Biol., 1961, No. 10E185.

4040. Kramer, J. P. 1961 *Thelohania thomsoni* n. sp., a Microsporidian Parasite of *Muscina assimilis* (Fallén) (Diptera, Muscidae). J. Insect Pathol. 3:259-265.
A potential parasite of *Musca*. The Microsporidia attack the epithelia of the chyle stomach and the proximal intestine. ABSTRACT: RAE-B 50:122, June, 1962.

4041. Kramer, J. P. 1961 *Herpetomonas muscarum* (Leidy) in the Haemocoele of Larval *Musca domestica* L. Ent. News 72(6):165-166.
Third instar larvae from insecticide-free chicken feces contained large numbers of this parasite in the haemolymph. Could be seen through the integument, with microscope. Normally considered a benign parasite in alimentary tract of adult muscoids. ABSTRACT: BA 36:6879(74252), Nov. 1, 1961.

4042. Kramer, J. P. 1962 The Fate of Spores of *Nosema apis* Zander, Ingested by Muscoid Flies. Entomophaga, Memoire Hors Série Numéro 2. (Colloque Internat. sur la Pathologie des Insectes et la Lutte Microbiologique.) Paris, 16-24 October. pp. 95-99.
Spores of *N. apis* do not parasitize the flies. Remain viable after passing through fly's alimentary tract.

4043. Kramer, J. P. 1964 The Microsporidian *Octosporea muscaedomesticae* Flu, a Parasite of Calypterate Muscoid Flies in Illinois. J. Insect Pathol. 6(3):331-342.
This parasite was found in field-collected *M. domestica* and 4 other muscoid species. Developmental stages described. Never found in reproductive tract or in fat bodies. ABSTRACT: BA 46:4072(50366), June 1, 1965.

4044. Kramer, J. P. 1964 *Nosema kingi* sp. n., a Microsporidian from *Drosophila willistoni* Sturtevant, and Its Infectivity for Other Muscoids. J. Insect Pathol. 6(4):491-499.
A monosporablastic species. Life-history stages described. *Lucilia cuprina*, *Phormia regina* and *M. domestica* are susceptible hosts.

4045. Kramer, J. P. 1965 Generation Time of the Microsporidian *Octosporea muscaedomesticae* Flu in Adult *Phormia regina* (Meigen) (Diptera, Calliphoridae). Zeit. Parasitenk. 25:309-313.
Studied smears and sections of adult *Phormia* intestine. Gives time required for schizogony, sporgony, etc. Newly formed spores present in fecal specks, and recovered therefrom. ABSTRACT: RAE-B 56:82(269), Apr., 1968.

4046. Kramer, J. P. 1965 Effects of an Octosporeosis on the Locomotor Activity of Adult *Phormia regina* (Meigen) (Diptera:Calliphoridae). Entomophaga 10(4):339-342. (German summary.)
For at least two weeks, infected flies were fully active and served as mobile spore disseminators. Disease did, however, eventually depress their locomotor activity (17th to 19th day). Uninfected flies showed no diminution in activity over the whole 19 day period.

4047. Kramer, J. P. 1966 On the Octosporeosis of Muscoid Flies Caused by *Octosporea muscaedomesticae* Flu (Microsporidia). Am. Midland Nat. 75(1):214-220.
Discusses progressive changes in the organs of adult *Phormia*, leading to death usually by the 16th day. Lethality of the disease in *M. domestica* also considered. ABSTRACT: BA 47:5104(59883), June 15, 1966.

4048. Kramer, J. P. 1968 An Octosporeosis of the Black Blowfly, *Phormia regina*: Incidence Rates of Host and Parasite. Zeit. Parasitenk. 30:33-39.
Monthly parasite incidence rate increased steadily as the season progressed, while the fly population declined accordingly (1963).

4049. Kramer, J. P. 1968 An Octosporeosis of the Black Blowfly, *Phormia regina*: Effect of Temperature on the Longevity of Diseased Adults. Texas Repts. Biol. and Med. 26(2):199-204.
Flies held at 12, 21, 27 and 32 ± 2°C were fed spores of the parasite.

After 24 (and 72) hours, infection was established only in flies held at the 3 higher temperatures. Longevity of *all* flies decreased with increasing temperature; that of infected flies being about half of the non-infected at any temperature level.

4050. **Krampitz, H. E.** and **Persoons, C. 1967** Ecto Parasitic Mites on Tsetse Flies. East African Trypanosomiasis Res. Organ Rept. 1966. p. 55.
Included because of similar relation of mites to *M. domestica.*

4051. **Kranzler, G. A. 1969** Response of House Flies to Recorded Flight Sounds. Paper No. 69-306. Am. Soc. Agric. Engineers. Annual Meeting.
Findings from mating communication experiments inconclusive. No significant effect of flight sounds on gregarious or congregational behavior.

4052. **Kranzler, G. A.** and **Earp, U. F. 1968** Response of Face Flies and House Flies to Sonic Energy. Trans. Am. Soc. Agric. Eng. 11(5): 691-693.
Steady and pulse-modulated pure tones produced no significant effects. Propose to investigate amplitude and frequency-modulated signals, superimposed tones, recordings of fly's own sounds. Acoustic fly control may prove impractical. ABSTRACT: BA 50:11925(123261), Nov. 15, 1969.

4053. **Kranzler, G. A., Earp, U. F.** and **Swink, E. T. 1966** Preliminary Studies on the Behavioral Effect of Sonic Energy on Flies. Virginia J. Sci. 17(4):244.
Describes research facility which has been established for studying the response of insects (including face flies and house flies) to sonic energy.

Kranzler, G. A., joint author. *See* Crabtree, D. H., et al., 1968.

Kratochvil, I., joint author. *See* Vostal, Z., et al., 1963.

Krause, D. H., joint author. *See* Hoopingarner, et al., 1966, 1968.

4054. **Krause, G. F.** and **Roan, C. C. 1961** Fluctuation in Response of Male Kun Strain House Flies to Topically Applied Malathion. J. Kans. Ent. Soc. 34(2):101-104.
There was significant "among batch" variation. Lexis series was assumed, after a Chi-square test, to describe the response of the house fly, when several batches were studied. ABSTRACTS: BA 36:6873(74144), Nov. 1, 1961; RAE-B 50:231, Nov., 1962.

Krause, G. F., joint author. *See* Rai, L., et al., 1964.

Krishnamurthy, B. S., joint author. *See* Ramakrishnan, S. P., et al., 1963.

Krishna Rao, J. K., joint author. *See* Qadri, S. H., et al., 1967.

Krishnan, K. S., joint author. *See* Raghavan, N. G. S., et al., 1967.

Kristhurie-Thirumalachar, M. J., joint author. *See* Narasimhan, M. J., Jr., et al., 1967.

Kroon, A. M., joint author. *See* VanBruggen, E. F. J., et al., 1968.

4055. **Krueger, H. R., O'Brien, R. D.** and **Dauterman, W. C. 1960** Relationship between Metabolism and Differential Toxicity in Insects and Mice of Diazinon, Dimethoate, Parathion and Acethion. J. Econ. Ent. 53(1):25-31.
Selectivity of Diazinon attributed to high levels of oxygen analog in susceptible species; of methoate and acethion to a persistence of unaltered parent compound in the whole body. Small difference found in Diazinon absorption and metabolism by normal and resistant houseflies.

4056. **Krueger, H. S.** and **Ballard, R. C. 1965** DNA in the Heart of Male and Female Houseflies of Different Ages. Comp. Biochem. Physiol. 16(1):13-20.
DNA analyses were performed by a microbiological method using

Lactobacillus acidophilus ATCC 11506 and by the Burton Modification of the Dische diphenylamine reaction. Quantity of DNA in hearts of males and females of the same or different age groups is similar. ABSTRACT: BA 46:8915(109965), Dec. 15, 1965.

Krupina, E. P., joint author. *See* Fomicheva, V. S., et al., 1967.

4057. Krysan, J. L. 1965 Studies on Soluble and Particulate Cholinesterase from the House Fly. Diss Absts. 26(5):2838.

4058. Krysan, J. L. and **Chadwick, L. E. 1962** Bimolecular Rate Constants for Organophosphorus Inhibitors of Fly Head Cholinesterase. Ent. Exptl. et Appl. 5(3):179-188. (German summary).
Rate constant for reaction of SD-3562 with fly head cholinesterase determined as $2{\cdot}024 \times 10^4$ liters mole^{-1} min.$^{-1}$ and for TEPP as $1{\cdot}022 \times 10^8$ liters mole^{-1} min.$^{-1}$. Rate constants for 11 other inhibitors cited from the literature. ABSTRACTS: BA 41:1664(20911), Mar. 1, 1963; RAE-B 51:144, July, 1963.

4059. Krysan, J. L. and **Chadwick, L. E. 1963** The Effect of Choline on Measurement of the Activity of Fly Head Cholinesterase. Ent. Exp. Appl. 6(3):199-206. (German summary.)
Choline found to be an inhibitor of fly head cholinesterase (ChE). Can affect ChE activity determinations made at low substrate concentrations. Methods for correcting such determinations. ABSTRACTS: BA 46:3704 (45766), May 15, 1965; RAE-B 52:91, May, 1964.

4060. Krysan, J. L. and **Chadwick, L. W. 1966** The Molecular Weight of Cholinesterase from the House Fly, *Musca domestica* L. J. Insect Physiol. 12(7):781-787.
Experiments and calculations indicated a molecular weight near 160,000. Existence of ChE in other molecular forms was not ruled out. ABSTRACT: BA 47:9496(109920), Nov. 15, 1966.

4061. Krysan, J. L. and **Chadwick, L. E. 1970** The Solubility of Cholinesterase from the Housefly, *Musca domestica.* J. Insect Physiol. 16: 75-82.
Centrifugation of housefly head homogenates at 105,000 *g* for 1 hour, sediments 73 per cent of the cholinesterase (ChE) at pH values of 7, 8 and 9. Was not altered by dialysis or the addition of salts. At pH 5, all the enzyme precipitated. ABSTRACT: BA 51:6250(63963), June 1, 1970.

4062. Krzemińska, A. 1969 Wplyw metepa na reprodukcje muchy domowej *Musca domestica* L. [The effect of metepa on the reproduction of the housefly (*Musca domestica* L.).] Roczn. Pánstw. Zakl. Hig. 20:727-734. (In Polish, with Russian and English summaries.)
Compound administered 3 days to both sexes, with food, at 4 concentrations. Flies feeding on 0·1 per cent metepa laid 5 times less eggs than controls; on 0·5 per cent, 9 times less. Viability of eggs laid became less as dosage was increased.

4063. Krzemińska, A. and **Styczyńska, B. 1968** Badania rozwoju larw muchy domowej *Musca domestica* L. na róznych pozywkach. [A comparative study of the development of larvae of the housefly *Musca domestica* L. on different media.] Roczn. Pánstw. Zakl. Hig. 19: 331-336. (In Polish, with Russian and English summaries.)
Quickest development (10-11 days) was on the Laboratory's Formula "B". On YMA, flies required 11-14 days; on CSMA, 12-16 days. Pupae from medium B weighed more than those from other media.

Krzemińska, A., joint author. *See* Styczyńska, B., et al., 1966, 1967, 1969.

Kudova, J., joint author. *See* Privora, M., et al., 1969.

4064. Kühlhorn, F. 1963 Über die klimatischen Verhältnisse in Viehställen im Hinblick auf den Einflug von Dipteren und deren Verteilung in Stalbraum. [On the climatic situation in cattle stalls in regard to the entrance of Diptera and their distribution in the stable area.] Abhandlungen der Braunschweigischen Wissenshaftlichen Gesellschaft. Band XV, pp. 166-199.

More than 300 species of flies found in this ecology. Temperature and relative humidity studied in different vertical and horizontal regions. Relates to the distribution of important species within the area.

4065. Kühlhorn, F. 1963 Gehöfttyp und Vorkommen von Dipteren in Ställen und Wohnräumen. [Flies in stables and living rooms on farms.] Arch. f. Hyg. u. Bakt. 147:41-57. (English and French summaries.)

Examples given showing how surroundings, both near and distant, together with climatic factors, determine the occurrence of flies in stables and farmhouses. Dung heaps important, also the lay-out of the farmstead. ABSTRACTS: BA 45:229, Jan. 1, 1964; TDB 60:993, Oct. 1963.

4066. Kühlhorn, F. 1964 Über die Dipterenfauna des Stallbiotops. [On the dipterous fauna of the stable ecology.] Beitr. Ent. 14(1/2):85-118. (English and Russian summaries.)

Fauna largely determined by milieu conditions. Lists, to date, 330 species, with data as to time of appearance in the room, and relations to it. Includes many species of hygienic and/or economic importance. ABSTRACT: BA 47:2047(24560), Mar. 1, 1966.

4067. Kühlhorn, Von F. 1965 Untersuchungen über die Beziehungen zurischen den Luftbewegungen und dem Verteilungsverhalten von Dipteren in Stallraum. [Researches into the relation between air currents and the distribution of Diptera in stable area.] Gesundheitswesen und Desinfektion for 1965 (1):8 pp.

Discusses ventilation, natural and forced. Diagrams relate to strategic placing of windows, resulting air currents and effects upon flies within the buildings.

4068. Kühlhorn, F. 1965 Über die mögliche Bedeutung einiger im Lebensbereich des Menschen und seiner Nutztiere vorkommender heinischer Dipterenarten als Gesundheits-schädlinge. [On the possible importance of several domestic dipterous forms (which enter the living quarters of man and his domestic animals) as a detriment to health.] Gesundheitswesen und Desinfektion for 1965 (6):5 pp.

Includes information on species of *Musca* and *Fannia*.

4069. Kühlhorn, F. 1968 Über Substratkontakte und Baumbeziehungen einiger heimischer Dipterenarten von medizinisch-hygienischer Bedeutung. [On substratum contact and the significance of several domestic dipterous forms of medical and hygienic importance.] Zeit. Angew. Zool. 55:257-293. (English summary.)

Concerns the possibilities for transfer of infection by fly's contact with food, feed, excrements, corpse, contagious material, etc. Treats of localities (stables, latrines, silage depots, dwellings, etc.) within which these contacts may occur.

4070. Kühlhorn, F. 1968 Gehöfttyp, Stallanlageform und -ausgestaltung, Aufstallungsweise und Substratlagerstätten in der Sicht des Dipterologen. [Farmstead type, stable planning and construction, manner of stabling and substratum locations in the light of Dipterology.] Sonderdruck aus Abhandlungen der Braunschweigischen Wissen-

shaftlichen Gesellschaft Band XX:43-95.
By logical design and arrangement of the features of man's environment (dwelling, stables, animals, etc.), the fly population may be adequately controlled.

4071. **Kuhr, R. J. 1969** [Possible role of tyrosinase and cytochrome P-450 in the metabolism of 1-naphthyl methylcarbamate (carbaryl) and phenyl methylcarbomate by houseflies.] J. Agric. Food Chem. 17(1):112-115.
Metabolism of both carbamates by housefly microsomes is inhibited by carbon monoxide, but inhibition is partially reversed by light. A soluble tyrosinase, prepared from adult houseflies, does not degrade carbaryl, Baygon, or phenyl methylcarbamate. ABSTRACT: BA 50:10829 (112108), Oct. 15, 1969.

4072. **Kuiper, J. W. 1965** On the Image Formation in a Single Ommatidium of the Compound Eye in Diptera. *In* Proc. Inst. Sympos. October 25-27. Stockholm, Sweden. Symp. Publ. Div. Pergamon Press 7:35-50. Illus.

4073. **Kulshrestha, S. K. 1969** Observations on the Ovulation and Oviposition with Reference to Corpus Luteum Formation in *Musca domestica nebulo* Fabr. (Muscidae; Diptera). J. Nat. Hist. 3(4): 561-570.
Gives time relation of copulation and ovulation. Describes anatomy and histology of the corpus luteum. ABSTRACT: BA 51:9148(93171), Aug. 15, 1970.

4074. **Kulshrestha, S. K. 1970** Morpho-histological Changes in the Ovarioles of *Musca domestica nebulo* Fabr. (Muscidae; Diptera) during Preoviposition Period. J. Nat. Hist. 4(1):137-144.
A description of normal conditions. ABSTRACT: BA 51:9143(93129), Aug. 15, 1970.

Kumakura, M., joint author. *See* Ohtaki, T., et al., 1964.

4075. **Kumar, P. 1967** Field Observations on the Breeding of Muscoid Flies in Organic Wastes in Villages. Indian J. Publ. Health 11(4):199-201.
Concerns *M. domestica*, *Stomoxys calcitrans* and certain Calliphoridae; their use of cattle shed litter, garbage and human excreta as breeding media. *M. domestica* breeds in all three.

Kumar, P., joint author. *See* Sehgal, B. S., et al., 1966.

4075a. **Kumar, S. S., Millay, R. H.** and **Bieber, L. L. 1970** Deacylation of Phospholipids and Acylation and Deacylation of Lysophospholipids Containing Ethanolamine, Choline, and β-methylcholine by Microsomes from Housefly Larvae. Biochemistry 9(4):754-759.
A microsomal preparation from housefly larvae converts exogenous phosphatidylcholine, phosphatidyl-ethanolamine, and phosphatidyl-3-methylcholine into the respective glycerophosphoryl bases. ABSTRACT: 51:12051(122387), Nov. 1, 1970.

Kumar, S. S., joint author. *See* Hildenbrant, G. R., et al., 1969; Bieber, L. L., et al., 1968, 1969.

Kume, T., joint author. *See* Eto, M., et al., 1968.

4076. **Kung, K., Kao, C. Y.** and **Chai, K. T. 1963** [Some biological and inherited characters of a mutant white-eyed strain of housefly and its use in bioassay.] Acta Ent. Sinica 12(3):262-267. (In Chinese, with English summary.)
The white-eyed strain had greater body weight and reproductive power than normal *Musca domestica*. The character of the white eye color was proved to be a simple recessive Mendelian factor and was not sex linked. ABSTRACT: BA 48:7470(83661), Sept. 1, 1967.

4077. Kung, K., Kao, G. and **Chai, K. 1965** [Studies on the insecticide resistance of the housefly. II. The influence of selection of houseflies on gamma-BHC resistance.] Acta Ent. Sinica 14(2):107-117. (In Chinese, with English summary.)
In one experiment, selection was carried out without insecticide pressure, in another gamma-BHC was used. Evidence from the study favorably substantiates the pre-adaption view for housefly resistance. See reference 4078, Kung, et al., 1968. ABSTRACTS: BA 49:3365(37620), Apr. 1, 1968; RAE-B 53:230, Dec., 1965.

4078. Kung, K.-Y., Kao, G.-Y. and **Chai, K.-Y. 1968** [Studies on the insecticide resistance of the house fly (*Musca domestica* L.). II. The selective influence of house-flies on gamma-BHC resistance.] Acta Ent. Sinica (English translation). 1965(1):102-114.
Eggs of most resistant and most susceptible flies, selected from a normal population, were bred and process continued through a series of generations. At end of experiment KD_{50} of tolerant line much higher and KD_{50} of susceptible line much lower than original strain. Second experiment described carried out from survivors of gamma-BHC treatment. See reference 4077, Kung, et al., 1965.

Kunishige, T., joint author. *See* Katsuda, Y., et al., 1966, 1967.

4079. Kunkee, R. E. and **Zweig, G. 1965** Inactivation and Reactivation Rates of Fly and Bee Cholinesterases Inhibited by Sevin. Biochem. Pharmacol. 14(6):1011-1017.
Inactivation rate constant of honeybee cholinesterase, inhibited by Sevin, was 5 times larger than that of housefly cholinesterase. Reactivation rate constants of the inhibited enzymes were about the same for both species. ABSTRACT: BA 46:8149(100699), Nov. 15, 1965.

4080. Kupriyanova, E. S. 1962 Aktivnost' kholinesterazy u samok komnatnoi mukhi (*Musca domestica* L.) na raznykh stadiyakh svogeneza. [Activity of cholinesterase in female house flies (*Musca domestica* L.) in various stages of ovogenesis.] Nauchn. Dokl. Vyssh. Shkoly. Biol. Nauk. 4:55-58. Referat Zhur., Biol. No. 9E163.
The activity of cholinesterase increases from the time of hatching and remains at a moderate level. Changes noted in the activity were related to its effect on secretions of the median protocerebral cells and adjoining bodies regulating oögenesis. ABSTRACT: BA 45:685(8527), Jan. 15, 1964.

4081. Kupriyanova, E. S. 1963 Kizucheniyu fiziologii sinantropnykh mukh i ikh chuvstvitel' nosti k insektitsidam (zhirovye reservy *Musca domestica* L. na raznykh stadiakh ovogeneza). [Studies on the physiology of synanthropic flies and their sensitivity to insecticides (fat reserves of *Musca domestica* L. at various stages of oogenesis).] Vestn. Mosk. Univ. Ser. Biol., Pochvoved. 2:22-29.
The role of the fat reserves of *Musca domestica* are described including a review of the literature. The accumulated fat serves to detoxify insecticides, thus increasing resistance and explaining resistance variations with season, age and sex of flies. ABSTRACT: BA 45:685(8528), Jan. 15, 1963.

4082. Kurchatov, V. I., et al. 1964 [Experience with fly control in Krasnodar.] Gig. Sanit. 29:80-81. (In Russian.)

Kutz, F. W., joint author. *See* Dobson, R. C., et al., 1970.

Kuwatsuka, S., joint author. *See* Casida, J. E., et al., 1966.

4083. Labadan, R. M. 1967 Comparative Effect of Selection in the Larval and Adult Stages on Development of Organophosphorus Insecticide Resistance in the House Fly. Diss. Absts. 28(1B):Page 222-B.

Labadan, R. M., joint author. *See* Travis, B. V., et al., 1967, 1968, 1969.

4084. LaBrecque, G. C. 1961 Studies with Three Alkylating Agents as House Fly Sterilants. J. Econ. Ent. 54:684-689.
Aphoxide, aphomide, and apholate, at concentrations of 1 per cent to 0·5 per cent in adult food, caused sterility in both males and females. In room tests, flies given untreated food produced 40,088 pupae; those given food treated with 1 per cent apholate, 86 to 226. ABSTRACT: RAE-B 50:148, July, 1962.

4085. LaBrecque, G. C. 1963 Chemosterilants for the Control of Houseflies. Advanc. Chem. Ser. No. 41:42-46. Amer. Chem. Soc. Symp., Atlantic City, N.J., September.
Over 40 chemicals, chiefly alkylating agents and antimetabolites, can induce sterility in houseflies. This is usually irreversible. General vigor and mating aggressiveness are unimpaired. ABSTRACTS: BA 45:8401 (106724), Dec. 15, 1964; RAE-B 53:15, 1965.

4086. LaBrecque, G. C., Adcock, P. H. and **Smith, C. N. 1960** Tests with Compounds Affecting House Fly Metabolism. J. Econ. Ent. 53(5): 802-805.
Of 200 compounds tested, 79 had some deleterious effect in larval medium; 10 affected development when combined with adult food. Most active compound, Amethopterin, caused sterility in females but not in males. ABSTRACTS: BA 36:757(9003), Feb. 1, 1961; RAE-B 49:180, Aug., 1961; TDB 58:389, Mar., 1961.

4087. LaBrecque, G. C., Evers, M. C. and **Meifert, D. W. 1966** Control of House Flies in Outdoor Privies with Larvicides. J. Econ. Ent. 59: 245.
Eighteen insecticides tested as larvicides in Grand Turk Island, British West Indies. The most effective, diazinon, gave complete control for 7-13 days; dimethoate, for 3-7 days. ABSTRACTS: BA 47:5951(69650), July 15, 1966; RAE-B 54:101, May, 1966; TDB 63:909, Aug., 1966.

4088. LaBrecque, G. C., Fye, R. L., DeMilo, A. B. and **Bořkovec, A. B. 1968** Substituted Melamines as Chemosterilants of House Flies. J. Econ. Ent. 61(6):1621-1632.
Of 110 compounds tested, 55 effectively inhibited hatch or pupation. Compounds affecting hatch when both sexes were treated, were usually also effective when only males were treated. Inhibitors of pupation were not effective in sterilizing males. ABSTRACT: RAE-B 57:87(302), May, 1969; TDB 66:1340(2537), Dec., 1969.

4089. LaBrecque, G. C. and **Gouck, H. K. 1963** Compounds Affecting Fertility in Adult House Flies. J. Econ. Ent. 56:476.
Of 1,100 compounds tested, 20 caused sterility in adult flies, when given in the food. Three induced sterility without apparent toxic effect over the broadest range of concentrations (5 per cent to 0·25 per cent). ABSTRACTS: BA 45:3595(44292), May 15, 1964; RAE-B 51:237, Nov., 1963; TDB 60:1169, Dec., 1963.

4090. LaBrecque, G. C. and **Meifert, D. W. 1966** Control of House Flies (Diptera:Muscidae) in Poultry Houses with Chemosterilants. J. Med. Ent. 3(3/4):323-326.
Syrup baits containing metepa (1 per cent) and hempa (2·5 per cent) applied to manure under caged hens completely eliminated infestations of *M. domestica* in poultry houses. Apholate (1 per cent) gave near control in one house, but not in another, subject to rapid reinfestation. ABSTRACTS: BA 48:5081(56765), June 1, 1967; RAE-B 56:238(875), Dec., 1968; TDB 64:809, July, 1967.

4091. LaBrecque, G. C. and **Meifert, D. W. 1970** Sterility in Adult House Flies Exposed to Residual Deposits of Chemicals. J. Econ. Ent.

63(5):1716-1717.
An effective, quick, and inexpensive method of causing sterility. Does not depend on the degree of feeding. Degree of sterility determined by percentage of progeny reaching the pupal stage.

4092. **LaBrecque, G. C., Meifert, D. W.** and **Fye, R. L. 1963** A Field Study on Control of House Flies with Chemosterilant Techniques. J. Econ. Ent. 56:150-152.
Baits containing 0·5 per cent metepa were applied to droppings in a poultry house, at weekly intervals for 9 weeks, then twice weekly. Viability of eggs from female flies taken in the area was below 10 per cent during most of the test period. ABSTRACTS: BA 43:1328(16472), Aug. 15, 1963; RAE-B 51:146, July, 1963; TDB 60:904, Sept., 1963.

4093. **LaBrecque, G. C., Meifert, D. W.** and **Gouck, H. K. 1963** Effectiveness of Three 2-methyl-aziridine Derivatives as House Fly Chemosterilants. Florida Ent. 46:7-10.
Metepa induced complete sterility in females at a concentration of 1 per cent, and complete sterility in males at 0·5 per cent. Methiotepa sterilized males, but not females at 1 per cent. Phenyl metepa did not cause complete sterility in either sex at 5 per cent. ABSTRACT: RAE-B 53:69, Apr., 1965.

4094. **LaBrecque, G. C., Meifert, D. W.** and **Smith, C. N. 1962** Mating Competitiveness of Chemosterilized and Normal Male House Flies. Science 136(3514):388-389.
Males, sterilized by a diet containing 1 per cent apholate, were as successful as normal males in competition for mates. ABSTRACT: BA 39:2017 (25442), Sept. 15, 1962.

4095. **LaBrecque, G. C., Morgan, P. B., Meifert, D. W.** and **Fye, R. L. 1966** Effectiveness of Hempa as a House Fly Chemosterilant. J. Med. Ent. (Honolulu) 3(1):40-43.
Hempa (hexamethylphosphoramide) is proving to be a very promising chemosterilant. Is less effective against females than males. Does not impair mating competitiveness in males, or motility of sperm. ABSTRACTS: BA 47:9043(104554), Nov. 1, 1966; RAE-B 55:210(730), Nov., 1967; TDB 63:1255, Nov., 1966.

4096. **LaBrecque, G. C.** and **Smith, C. N. 1960** Tests with Young Poultry for the Control of House Fly Larvae Under Caged Laying Hens. J. Econ. Ent. 53(4):696.
Used one chick per 10 layers. All larvae were eliminated at the end of 1 week and control remained complete for 5 weeks, when test was terminated. Pullets cause less excitement among laying flocks than cockerels. ABSTRACT: RAE-B 49:155, July, 1961.

4097. **LaBrecque, G. C.** and **Smith, C. N. 1968** Principles of Insect Chemosterilization. viii + 354 pp. Illus. Appleton-Century-Crofts: New York.
A compilation of research reports over approximately a ten year period. A large amount of the material has to do with houseflies. The list of investigators is impressive. ABSTRACTS: BA 49:4330(48817), May 15, 1968; TDB 67:594-595, May, 1970.

4098. **LaBrecque, G. C., Smith, C. N.** and **Meifert, D. W. 1962** A Field Experiment in the Control of House Flies with Chemosterilant Baits. J. Econ. Ent. 55:449-451.
Cornmeal baits containing 0·5 per cent aphoxide applied weekly on refuse dump in the Florida Keys. *M. domestica* populations were reduced from 47 per grid to 0 within 4 weeks. Per cent hatch among all eggs laid was reduced to 1 per cent within 5 weeks. ABSTRACTS: BA 41:608 (8151), Jan. 15, 1963; RAE-B 51:2, Jan., 1963; TDB 60:76, Jan., 1963.

4099. LaBrecque, G. C. and **Weidhaas, D. E. 1970** Advantages of Integrating Sterile-Male Releases with Other Methods of Control Against House Flies. J. Econ. Ent. 63(2):379-382.
Use of insecticides with sterilant baits would (1) reduce number of treatments of either sort that would be needed; (2) reduce the number of insects needed for releases; (3) reduce the time required to reach (theoretical) eradication. ABSTRACTS: BA 51:8531(86942), Aug. 1, 1970; RAE-B 58:301(1234), Sept., 1970.

4100. LaBrecque, G. C. and **Wilson, H. G. 1960** Effect of DDT Resistance on the Development of Malathion Resistance in House Flies. J. Econ. Ent. 53(2):320-321.
Resistance to malathion developed much more rapidly in a strain previously selected with DDT than in the regular strain. ABSTRACTS: BA 35:3740(43004), Aug. 1, 1960; TDB 57:960, Sept., 1960.

4101. LaBrecque, G. C. and **Wilson, H. G. 1961** Development of Insecticide Resistance in Three Field Strains of House Flies. J. Econ. Ent. 54(6):1257-1258.
Malathion and Dipterex the chief insecticides used in baits for fly control in Florida dairy barns. High resistance to malathion, which may be attributable to behavioral as well as physiological factors, has caused a shift to the use of DDVP. ABSTRACTS: RAE-B 50:190, Sept., 1962; TDB 59:621, June, 1962.

4102. LaBrecque, G. C., Wilson, H. G., Brady, U. E. and **Gahan, J. B. 1967** Screening Tests of Contact Sprays for Control of Adult House Flies. J. Econ. Ent. 60(3):760-762.
Of 119 insecticides tested in wind tunnel against susceptible and resistant house flies, 40 were more effective against both strains based on LC_{50} levels, than the malathion or ronnel standards. ABSTRACTS: BA 48:11208(124491), Dec. 15, 1967; RAE-B 55:188(650), Oct., 1967; TDB 65:83(185), Jan., 1968.

4103. LaBrecque, G. C., Wilson, H. G. and **Gahan, J. B. 1965** Residual Effectiveness of some Insecticides Against Adult House Flies. U.S. Dept. Agr., A.R.S. 33-103. 12 pp. mimeo.
Of a number of candidate materials (mostly organophosphorus hydrocarbons and carbamates), 16 remained effective for the 24 week test period. Results with the malathion standard were variable.

4104. LaBrecque, G. C. Wilson, H. G., Gahan, J. B. and **Willis, N. L. 1968** Residual Toxicity of some Insecticides to Adult House Flies. Florida Ent. 51(4):217-218.
Of 75 chemicals tested, 11 caused 90 per cent or greater mortality, throughout the 24 week test period, against susceptible flies. Against a resistant strain, Bayer 78389 was effective for 16 weeks; CELA K-37, for 8. ABSTRACT: BA 50:4777(50221), May 1, 1969.

LaBrecque, G. C., joint author. *See* Bailey, D. L., et al., 1967, 1968, 1970; Bořkovec, A. B., et al., 1967, 1968; Boroza, M., et al., 1967; Brady, U. E., et al., 1966; Fye, R. L., et al., 1965, 1966, 1967, 1969; Gouck, H. K., et al., 1964; Meifeıt, D. W., et al., 1963, 1967, 1969; Morgan, P. B., et al., 1962, 1964, 1966, 1967, 1968; Murvosh, C. M., et al., 1963, 1964, 1965; Wilson, H. G., et al., 1960, 1961, 1967; Smith, C. N., et al., 1960, 1964.

4105. LaChance, L. E., Degrugillier, M. and **Leverich, A. P. 1969** Comparative Effects of Chemosterilants on Spermatogenic Stages in the House Fly: I. Induction of Dominant Lethal Mutations in Mature

Sperm and Gonial Cell Death. Mutat. Res. 7(1):63-74.
Aziridines and hempa killed all gonial cells. Sulfonates killed none. Spermatogenic activity remaining normal, though dominant lethals were induced in sperm. Such flies have a chance for recovery from the sterile condition. ABSTRACTS: BA 50:7510(78292), July 15, 1969; TDB 66:879 (1754), Aug., 1969.

4106. LaChance, L. E., Degrugillier, M. and **Leverich, A. P. 1970** Comparative Effects of Chemosterilants on Spermatic Stages in the House Fly. II. Recovery of Fertility and Sperm Transfer in Successive Matings after Sterilization with 1, 3-Propanediol Dimethanesulfonate or X-Rays. Ann. Ent. Soc. Am. 63(2):422-428.
Sperm production never completely interrupted. Flies returned to normal in 11 to 13 days. With X-rays all gonial cells were killed and spermatogenic cycles ceased. ABSTRACT: BA 51:7979(81530), July 15, 1970.

4107. LaChance, L. E. and **Leopold, R. A. 1969** Cytogenetic Effect of Chemosterilants in House Fly Sperm: Incidence of Polyspermy and Expression of Dominant Lethal Mutations in Eaıly Cleavage Divisions. Can. J. Genetics & Cytology 11(3):648-659.
Treatment of males with sterilizing doses of tepa, hempa, or intermediate analogs reduced but little the number of eggs fertilized. Chromosomes were affected however, and embryonic anomalies caused cessation of development.

4108. Lachinova, R. I. 1966 Vliyanie otraleniya DDT na kolichestvo i nabukhanie mitokhondrii muskulatury kryla vydelennykh iz chuvstvitel' nykh û ustoichwykh mukh *Musca domestica* L. [Effect of poisoning with DDT upon the quantity and swelling of mitochrondria of the wing musculature isolated from susceptible and resistant flies (*Musca domestica*).] Med. Parazit. i Parazitarn. (Bolez.) 35(5):519-525. (In Russian, with English summary.)
Susceptible flies showed a swelling or thickening of the sarcosomes depending upon dosage. Resistant flies did not. The significance of the evidence is discussed. ABSTRACT: BA 48:9533(107176), Nov. 1, 1967.

4109. Lachinova, R. I. 1967 Izmenenie aktivnosti tsitokhromoksidazy grudnoi muskulatury mukh na raznykh stadiyakh ovogeneza. [Changes in cytochrome oxidase activity of the thoracic musculature of flies at different stages of oogenesis.] Biol. Nauk. 10(1):41-43.
Cytochrome oxidase activity increases sharply until stage IIa of oogenesis; levels off and decreases sharply after stages III-IV. By using it to detect females in stage II, a uniform sample could be achieved. ABSTRACT: BA 49:6779(75601), July 15, 1968.

4110. Lachinova, R. I. 1967 Biokhimicheskaya geterogennost' mitokhondrii muskulatury kryla mukhi *Musca domestica.* [Biochemical heterogeneity of mitochondria from wing muscles of the fly, *Musca domestica.*] Zh. Evol. Biokhim. Fiziol. 3(4):295-299. (English summary.)
Three mitochrondrial subfractions were obtained, all exhibiting high respiratory activity. Small variations in precipitation technique may affect results calculated per protein unit. ABSTRACT: BA 49:10813 (119084), Nov. 15, 1968.

4111. Lackany, A. 1963 Flies as Carriers of Typhoid Bacilli in Alexandria. Alexandria Med. J. 9(3):186-193.
The peak of seasonal typhoid infection corresponds with the peak of fly population, but no convicing evidence was found to support view of flies as carriers of typhoid bacilli. Other more probable sources were sought. ABSTRACT: BA 45:2751(34149), Apr. 15, 1964.

4112. **Lai-Fook, J. E. I-L. 1967** The Repair of Wounds to the Integument of Insects. Diss. Absts. 27(108):3405-B-3406-B.
Concerns *M. autumnalis.*

Laiho, K., joint author. *See* Nuorteva, P., et al., 1967.

Laird, M., joint author. *See* Cantwell, G. E., et al., 1966.

Lamb, N. J., joint author. *See* Monroe, R. E., et al., 1968.

Lambert, J., joint author. *See* Lhoste, J., et al., 1967, 1968.

Lambremont, E. N., joint author. *See* Joiner, R. L., et al., 1969.

4113. **Lancaster, G. A.** and **Sourkes, T. L. 1969** Effect of Alpha-methyl-DL-tryptophan on Tryptophan Metabolism of *Musca domestica* L. Comp. Biochem. Physiol. 28:1435-1441.
This analogue inhibits pyrrolase by a competitive mechanism. Inhibition insufficient to affect concentration of xanthommatin in the eyes. Tryptophan pyrrolase of whole-body *M. domestica* has a K_M of $4{\cdot}6 \times 10^{-3}$M. ABSTRACT: BA 50:13068(134780), Dec. 15, 1969.

4114. **Lancaster, J. L., Jr.** and **Simco, J. S. 1969** House Fly Control by Chemical Sterilization with Apholate. Arkansas Agr. Exp. Sta. Bull. 737:1-11.
Oral intake most effective route. Apholate baits controlled flies in barns where dairy was the only enterprise; failed if poultry was an additional activity. ABSTRACT: BA 50:10829(112109), Oct. 15, 1969.

Lancaster, J. L., Jr., joint author. *See* Simco, J. S., et al., 1966.

Landa, V., joint author. *See* Rezabova, B., et al., 1967, 1968.

Landers, P. L., joint author. *See* Berberet, R. C., et al., 1969.

Landgrave, G., joint author. *See* Hecht, O., et al., 1969.

Lange, J., joint author. *See* Skinner, W. A., et al., 1967.

4115. **Lange, R. 1964** Der Einfluss der Nährung von *Musca domestica* L. auf die Mortalität des Entoparasiten *Aphaereta pallipes* (Say) (Hym. Braconidae). [The influence of the food of *M. domestica* on mortality of the parasite *A. pallipes.*] Zeit. Angew. Ent. 54(1-2):213-218. (English summary.)
Diet had some effect on suitability of *M. domestica* as alternative host but many larvae were permanently injured by the act of parasite oviposition. Concluded that *M. domestica* is not a suitable host. ABSTRACT: RAE-B 54:200, Oct., 1966.

4116. **Lange, R.** and **Bronskill, J. F. 1964** Reactions of *Musca domestica* L. (Diptera:Muscidae) to Parasitism by *Aphaereta pallipes* (Say) (Hymenoptera:Braconidae), with Special Reference to Host Diet and Parasitoid Toxin. Zeit. Parasitenk. 25:193-210.
Varying host reaction correlated chiefly with diet of host. A considerable increase of parasitoid survival occurred when *M. domestica* larvae were reared on a chemically defined diet. Temperature range an important limiting factor. ABSTRACTS: BA 48:1903(20642), Feb. 15, 1967; RAE-B 56:82(268), Apr., 1968; TDB 62:366, Apr., 1965.

4117. **Lanna, T. M. 1962** Eredità parzialmente legata al sesso in *Musca domestica* L. [Partially sex-linked inheritance in *Musca domestica* L.] Boll. di Zool. 29:831-844.

4118. **Lanna, T. M. 1963** The Inheritance of Insecticide-Resistance. (Demonstration.) Proc. XI Int. Congr. Genet. I:254.

4119. **Lanna, T. M.** and **Franco, M. G. 1961** Modalità di ricombinazione di alcuni mutanti di *Musca domestica* L. [Manner of recombination of some mutations in *Musca domestica* L.] Genetica Agraria 14:297-306.

Lanna, T. M., joint author. *See* Franco, M. G., et al., 1961, 1966.

4120. **Lanzavecchia, G.** and **Stazio, M. 1961** Variazioni della colinesterasi in mosche trattate per applicazione topica con la metilamide dell' acido 0,0-dimetilditiofosforilacetico e con il parathion. [Variation of cholinesterase level in flies submitted to local treatment with 0,0-dimethyldithiophosphoril-acetic acid and with parathion.] Riv. di Parassit. (Rome) 22(4):297-313. (English summary.)
Activity of 4 strains of *M. domestica* was lowered after doses approaching LD_{50} of parathion and 0,0-dimeth. . . . but in all strains there was recovery after 24 hours with parathion. With 0,0-dimeth. . . . second application produced high mortality. ABSTRACTS: RAE-B 51:121, June, 1963; TDB 59:1120, Nov., 1962.

4121. **Larsen, J. R., Pfadt, R. E.** and **Peterson, L. G. 1966** Olfactory and Oviposition Responses of the House Fly to Domestic Manures, with Notes on an Autogenous Strain. J. Econ. Ent. 59:610-615.
In olfactometer test, cow manure was the most attractive to flies of 8 manures provided, but pig manure was the most favored site for oviposition, followed by human, chicken, dog, calf, horse, sheep, and cow. ABSTRACTS: BA 47:8555(99709), Oct. 15, 1966; RAE-B 54:177, Sept., 1966; TDB 63:1400, Dec., 1966.

4122. **Laudani, U. 1964** Attività citocromo-c ossidasica in adulti del ceppo SRS/1 di *Musca domestica* L. [Cytochrome-c oxidase activity in adults of the SRS/1 strain of *Musca domestica* L.] Boll. di Zool. 31:695-696.

4123. **Laudani, U. 1965** Alcune osservazioni sul metabolismo ossidativo di *Musca domestica* L. [Some observations on the oxygen metabolism of *Musca domestica* L.) Boll. di Zool. 32:751-758.
Oxygen uptake of eggs, larvae, and pupae was studied at different temperatures. Succininoxidase and cytochromeoxidase activity was tested over entire life cycle.

4124. **Laudani, U. 1966** Attività citrocromo-c ossidasica in adulti del ceppo SRS/1 di *Musca domestica* L. [Cytochrome-c oxidase activity in adults of the SRS/1 strain of *Musca domestica* L.] Symp. Gen. et Biol. It. 13:359-368.
In males, activity rises from emergence for 4 days, then decreases slowly to emergence level. In females, maximum activity is on the eighth day, followed by a decrease to much below emergence level at end of life cycle.

4125. **Laudani, U.** and **Grigolo, A. 1965** Determinazione della citocromo-c ossidasi in uova, larve e pupe del ceppo SRS di *Musca domestica* L. [Determination of cytochrome-c oxidase in eggs, larvae and pupae of the SRS strain of *Musca domestica* L.] Riv. di Parassit. 26:187-192. (English summary.)
Enzymatic activity highest in eggs, diminishing to a minimum in 48 hour pupae. Rises from this point until emergence.

4126. **Laudani, U.** and **Grigolo, A. 1968** Sulla presenza di alcuni metaboliti del triptofano, pterine ed altre sostanze U.V. fluorescenti in individui di *Musca domestica* L., normali e mutanti per il colore degli occhi. [Tryptophan metabolites, pterines, and other ultraviolet fluorescent substances in normal and mutant adult house fly, *Musca domestica*.] Atti Ass. Genet. Ital. 13:180-182.
Several ommochrome precursors in adult flies were detected by applying cellulose thin-layer chromotography to full-body and head extracts from flies of four normal and six eye-color mutant strains.

4127. Laudani, U. and **Grigolo, A. 1969** Ommochrome Precursors and U. V. Fluorescent Substances in Eye Colour Mutants of *Musca domestica* L. Monitore Zoo. Ital. (n.s.) 3:99-104.
Trytophan, kynurenine, 3-hydroxykynurenine, xanthurenic and kynurenic acids were found in normal flies and in the mutants *cm*, *pink*, *ye* and w^3. In *ge* flies only tryptophan; in *ocra* flies, tryptophan, kynurenine and kynurenic acid. ABSTRACT: BA 51:10343(105164), Oct., 1970.

4127a. Laudani, U. and **Grigolo, A. 1969** Ricerche sull'attivita' dell-enzima succino-deidrogenasi in preparazioni mitocondriali di *Musca domestica* L. [Research on the succinic dehydrogenase activity in mitochondrial preparations of *Musca domestica* L.] Riv. di Parassit. 30(4):319-327.
Activity was tested in mitochondria from thoracic muscles. Males show more activity which is greatest on the 4th day after emergence. Female activity increases slowly until 12th day. ABSTRACT: BA 51:12052(122396), Nov. 1, 1970.

4128. Laudani, U., Grigolo, A. and **Daturi, S. 1965** Respirazione delle uova di *Musca domestica* L., ceppo SRS. [The Respiration of Eggs of *Musca domestica* L. SRS strain.] Riv. di Parassit. 26:133-142. (English summary.)
Oxygen consumption varies with temperature. Is considerably more when incubation is at 38°C than when carried out at 27°C. Respiration in homogenates of bacteriologically sterile eggs also studied.

4129. Laudani, U., Grigolo, A. and **Foa, S. 1966** Studio di fosfatasi acide in muscoli toracici di *Musca domestica* L. [Study of acid phosphatase in the thoracic muscles of *Musca domestica* L.] Boll. di Zool. 33(1):140.
Used a modification of the technique of Torriani with *Escherichia coli*.

4130. Laudani, U., Grigolo, A. and **Foa, S. 1969** Study of Acid Phosphatases in Thoracic Muscles of *Musca domestica* L. Monitore Zool. Ital. (n.s.) 3:105-115.
Tests were at pH 4·8 and 38°C, with p-nitrophenyl phosphate (NPP) as a substrate. Activity decreases with time, being greater in females until the 9th day. After this, rises in males, to a maximum on the 15th day.

4131. Laudani, U., Grigolo, A. and **Visona, L. 1965** Influenza delle variazioni di temperatura sulla respirazione di pupe di *Musca domestica* L. [Influence of variations in temperature on respiration in the pupa of *Musca domestica* L.] Genetica Agraria 19(4):351-358. (English summary.)
Oxygen consumption tested at 14°, 18°, 22°, 26°, 30°, 34°, 38°C. Reduced most at 14°; least at 30°C. Is correlated with temperature and with age. ABSTRACT: BA 48:9251(104031), Oct. 15, 1967.

Laudani, U., joint author. *See* Bernardini, P. M., et al., 1965; Bianchi, U., et al., 1964; Grigolo, A., et al., 1965, 1966, 1969; Mosconi-Bernardini, P., et al., 1965.

Laven, H., joint author. *See* Knipling, E. F., et al., 1968.

4132. Lavtschiev, V. and **Jelsova, M. 1968** Experimentelle Untersuchungen über die morphologische Veränderlichkeit von *Musca domestica domestica* L. (Dipt. Muscidae). [Experimental investigations of morphological plasticity in *Musca domestica domestica* L. (Dipt. Muscidae).] Zool. Anz. 181(5/6):411-421.
Morphology of houseflies reared in the laboratory varies according to

temperature. Flies answer description of *M. domestica vicina* at 30°C; of *M. domestica domestica* (s. str.) at 18°C. ABSTRACT: BA 51:3371 (34869), Mar. 15, 1970.

4133. Lawson, F. R. 1967 Theory of Control of Insect Populations by Sexually Sterile Males. Ann. Ent. Soc. Am. 60:713-722. Errata T.c. No. 5, p. 1076.
Theory examined by establishing basic premises and deriving algebraic equations. Many factors considered, including expense. Is an example of how mathematics can make biological judgement more reliable. ABSTRACTS: RAE-B 56:2(3), Jan., 1968; TDB 65:531(1152), Apr., 1968.

4134. Lebedeva, A. P. and **Tolstoshei, O. N. 1963** Khimicheskom i biologicheskam opredelenii geksakhorana v pochve. [Chemical and biological determination of BHC in the soil.] Gosmedizdat USSR (Kiev) 4:226-230; Referat. Zhur., Biol., 1964, No. 12E42.
Chemical methods for determining amount of BHC in soil are described. A biological method is based on a comparison of the number of flies (*Musca domestica*) dying on soil samples as compared to a previously developed standard using various doses of BHC. ABSTRACT: BA 46: 1433(17898), Feb. 15, 1965.

Lechtenstein, E. P., joint author. *See* Harris, C. R., et al., 1961.

4135. LeClercq, M. 1968 *Stomoxyinae* et *Muscinae* hématophages de la République Démocratique du Congo (Diptera Muscidae). Bull. Recherches Agron. Gembloux. N.S. 3(4):697-702.
List includes *Musca* (*Philaematomyia*) *crassirostris* Stein. ABSTRACT: BA 51:13145(133775), Dec. 1, 1970.

4136. LeClercq, M. 1969 Entomologie en Gerechtelijke Geneeskunde. [Entomology in Legal Medicine.] Entomoligische Berichten, Deel 29, 1. VI. p. 104.
Necrophilic arthropod fauna attacks a corpse in successive waves. *Calliphora*, *Musca*, and *Muscina* larvae appear in first wave. Author lists 7 other groups, in order.

Lee, H. H., joint author. *See* Travis, B. V., et al., 1969.

Lee, I., joint author. *See* Dahm, P. A., et al., 1961.

4137. Lee, J. E. 1967 Aerosol Insecticide Formulations up to Date. Soap and Chem. Spec. 43(4):62, 64, 66, 70, 72, 193.
A review of the development of aerosols for the control of insect pests in houses, chiefly flies and mosquitoes. ABSTRACT: RAE-B 56:194(699), Oct., 1968.

4138. Lee, T. and **Li, V. 1964** [Studies on ovarian development of the adult housefly with regard to the effects of different diets and amino acid requirements.] Acta Ent. Sinica 13(6):803-814. Peking. (In Chinese, with English summary.)
Hen's egg solution and synthetically prepared hen's egg solution were satisfactory diets while sucrose alone was not. Essential amino acid requirements determined. Histidine and methionine not essential. ABSTRACT: RAE-B 53:197, Oct., 1965.

Lee, V. H., joint author. *See* Bouillant, A., et al., 1965.

4139. Lee, W.-Y., Wang, Y.-H. and **Lin, F.-J. 1967** The Effect of Apholate on Oxygen Consumption of the Eggs of House Flies, *Musca domestica* L. Bull. Inst. Zool. Acad. Sinica 6(2):87-91.
Flies were fed graded concentration of apholate. The higher the concentration fed to adults, the less oxygen the ova consumed. Oxygen con-

sumption by ova was measured by a conventional Warburg respirometer at 25°C. ABSTRACT: BA 50:10284(106370), Oct. 1, 1969.

Lee, W.-Y., joint author. *See* Tung, A. S.-C., et al., 1968.

Legath, V., joint author. *See* Vostál, Z., et al., 1963.

4140. Legner, E. F. 1965 Natural and Induced Control of House Flies with Parasites in Southern California. The Calif. Golden Egg 4(6):18. December.
Parasitism of *Musca* and *Fannia* by Hymenoptera of three families can be a significant force in spring, summer and fall. (Activity is very low in winter.)

4141. Legner, E. F. 1966 Parasites of the House Fly and Other Filth-breeding Dipteria in Southern California. J. Econ. Ent. 59(4):999-1001.
Lists 8 species of parasitic Hymenoptera (falling in 5 families) as attacking larval and pupal stages. Possibility of importing others. ABSTRACTS: BA 47:10322(119303), Dec. 15, 1966; RAE-B 54:211, Nov., 1966; TDB 64:216, Feb., 1967.

4142. Legner, E. F. 1967 Behavior Changes the Reproduction of *Spalangia cameroni*, *S. endius*, *Muscidifurax raptor*, and *Nasonia vitripennis* (Hymenoptera:Pteromalidae) at Increasing Fly Host Densities. Ann. Ent. Soc. Am. 60:819-826.
Progeny production always increased with an increased host density limited by specific reproductive capacity. *Spalangia* species showed a greater searching capacity than *Muscidifurax* or *Nasonia* and would be favored for biological control. ABSTRACTS: BA 48:10183(114427), Nov. 15, 1967; RAE-B 56:2(5), Jan., 1968.

4143. Legner, E. F. 1967 The Status of *Nasonia vitripennis* as a Natural Parasite of the House Fly, *Musca domestica*. Canad. Ent. 99(3): 308-309.
Does not, as a rule, parasitize *M. domestica* in nature. Prefers *Fannia*, *Lucilia*, or *Calliphora*. Author lists several species known to attack *M. domestica* in the Western Hemisphere. ABSTRACTS: BA 48:7864(88149), Sept. 1, 1967; RAE-B 57:31(126), Feb., 1969.

4144. Legner, E. F. 1969 Reproductive Isolation and Size Variation in the *Muscidifurax raptor* Complex. Ann. Ent. Soc. Am. 62(2):382-385.
There are 7 biologically distinct sibling forms of this species, separately situated at diverse collection sites. Each must be considered a separate entity for colonization in biological control of noxious flies. ABSTRACT: RAE-B 57:268(1016), Dec., 1969.

4144a. Legner, E. F. 1970 Contemporary Considerations of the Biological Suppression of Noxious Brachycerous Diptera that Breed in Accumulated Animal Wastes. Proc. Pap. 38th Ann. Conf. Calif. Mosquito Control Assoc.; Jan. 26-29, 1970. pp. 88-89.
Cites practical demonstrations of integrated fly control in Orange and San Bernardino Counties, California. Requires proper management of manure and use of residual adulticides at one or two critical times per year.

4145. Legner, E. F. and **Bay, E. C. 1970** The Introduction of Natural Enemies in California for the Biological Control of Noxious Flies and Gnats. Proc. & Papers 37th Ann. Conf. Calif. Mosquito Control Assoc. January 27-29, 1969. Publ. January 22, 1970. pp. 126-129.
Gives information on 14 colonizations of natural enemies for the purpose of controlling species of Muscidae, Calliphoridae and Sarcophagidae. Possibility of importing additional species.

4146. Legner, E. F., Bay, E. C., Brydon, H. W. and **McCoy, C. W. 1966** Research with Parasites for Biological Control of House Flies in Southern California. Calif. Agric. 20(4):10-12.
Suggests introduction of exotic species of larval and pupal parasites to supplement American forms. Existing parasites are not as effective, however, against *M. domestica* as against *Fannia* and others. ABSTRACT: BA 48:6925(77447), Aug. 1, 1967.

4147. Legner, E. F., Bay, E. C. and **McCoy, C. W. 1965** Parasitic Natural Regulatory Agents Attacking *Musca domestica* L. in Puerto Rico. J. Agric. Univ. Puerto Rico 49(3):368-376. (Spanish summary.)
Six hymenopterous parasites discovered; one species of Diapriidae, five Pteromalidae. Parasitization rate ranged from 1·4 to 25 per cent of viable housefly pupae. ABSTRACT: RAE-B 56:108(375), May, 1968.

4148. Legner, E. F., Bay, E. C. and **White, E. B. 1967** Activity of Parasites from Diptera: *Musca domestica, Stomoxys calcitrans, Fannia canicularis,* and *F. femoralis,* at Sites in the Western Hemisphere. Ann. Ent. Soc. Am. 60(2):462-468.
Concerns 14 larval and pupal parasites. Gives information for each species on: (1) percent parasitization; (2) number of sites at which parasite was active; (3) mean host pupal density sampled at those sites. ABSTRACTS: BA 48:5550(62044), June 15, 1967; RAE-B 55:205(710), Nov., 1967; TDB 64:910, Aug., 1967.

4149. Legner, E. F. and **Brydon, H. W. 1966** Suppression of Dung-Inhabiting Fly Populations by Pupal Parasites. Ann. Ent. Soc. Am. 59(4): 638-651.
Muscidifurax raptor and *Spalangia endius* most important parasites of *Fannia femoralis* and *Ophyra leucostoma. M. raptor* is prominent in June, July, and August; *S. endius*, in September, October, and November. Discussion of biological control. ABSTRACTS: BA 47:9913(114628), Dec. 1, 1966; RAE-B 54:240, Dec., 1966.

4150. Legner, E. F. and **Gerling, D. 1967** Host-Feeding and Oviposition on *Musca domestica* by *Spalangia cameroni, Nasonia vitripennis* and *Muscidifurax raptor*. (Hymenoptera:Pteromalidae) Influences their Longevity and Fecundity. Ann. Ent. Soc. Am. 60:678-691.
Adults of *S. cameroni* and *N. vitripennis* require hosts when very young for maximum longevity and oviposition. *M. raptor* performed best when deprived of hosts for 4 days or more after emergence. ABSTRACTS: BA 48:7864(88150), Sept. 1, 1967; RAE-B 55:220(762), Dec., 1967.

4151. Legner, E. F. and **Greathead, D. J. 1969** Parasitism of Pupae in East African Populations of *Musca domestica* and *Stomoxys calcitrans*. Ann. Ent. Soc. Am. 62(1):128-133.
Combined effect of 6 hymenopterous species perhaps responsible for the annual scarcity of these flies as pests in Uganda. Evolution of parasites discussed. ABSTRACTS: BA 50:6966(72768), July 1, 1969; RAE-B 57:200 (724), Oct., 1969; TDB 67:241(515), Feb., 1970.

4152. Legner, E. F. and **McCoy, C. W. 1966** The Housefly, *Musca domestica* Linnaeus as an Exotic Species in the Western Hemisphere Incites Biological Control Studies. Canad. Ent. 98(3):243-248.
Lists the principal larval and pupal parasites found attacking *M. domestica* in North, Central, and South America. This is compared with a list of those recorded from the Eastern Hemisphere. ABSTRACT: RAE-B 55:193(670), Oct., 1967.

4153. Legner, E. F. and **Olton, G. S. 1968** The Biological Method and Integrated Control of House and Stable Flies in California. Calif. Agric. 22(6):2-4.
The use of residual poisons to control flies did not interfere with natural enemy complexes in the tests reported. ABSTRACT: BA 49:10784(118705), Nov. 15, 1968.

4154. Legner, E. F. and **Olton, G. S. 1968** Activity of Parasites from Diptera: *Musca domestica*, *Stomoxys calcitrans*, and Species of *Fannia*, *Muscina* and *Ophyra*. II. At Sites in the Eastern Hemisphere and Pacific Area. Ann. Ent. Soc. Am. 61(5):1306-1314.
Parasitism recorded at collection sites in Palaearctic, Ethiopian and Australian regions and in the Pacific. Comparison with data from Western Hemisphere. Importance of parasites (parasitoids) as biological control agents. ABSTRACTS: BA 50:490(5026), Jan. 1, 1969; RAE-B 57:97(344), June, 1969; TDB 66:96(150), Jan., 1969.

4155. Legner, E. F. and **Olton, G. S. 1970** Worldwide Survey and Comparison of Adult Predator and Scavenger Insect Populations Associated with Domestic Animal Manure Where Livestock is Artificially Congregated. Hilgardia 40(9):225-266.
Concerns predation on larvae or pupae of *M. domestica* and other Diptera breeding in animal manures. An extensive study, well reported. Coleoptera of dominant importance.

4156. Legner, E. F. and **Olton, G. S.** Distribution and Relative Abundance of Diptera and their Parasitoids Inhabiting Animal Manure Accumulations in the Southwestern United States. (MS of paper to be published in Hilgardia, 1970.)

Legner, E. F., joint author. *See* Gerling, D., et al., 1968; White, E. B., et al., 1966; Anderson, J. R., et al., 1968; Kogan, M., et al., 1970.

Lemin, A. J., joint author. *See* Kohls, R. E., et al., 1966.

Len, H. F., joint author. *See* Hsieh, C.-Y., et al., 1964.

4157. Leng, H.-F. and **Chen, D.-L. 1965** True and Pseudo-ChE Inhibition of House-fly Poisoning by Organic Phosphorus Insecticides. Acta. Ent. Sinica 14(1):10-14. (In Chinese, with English summary.)

4158. Leng, H.-F. and **Chen, D.-L. 1968** The Inhibitory Effect upon True- and Pseudo-ChE in House Flies [*Musca domestica* L.] Produced by Several Types of Organic Phosphorus Insecticides, After Poisoning. Acta. Ent. Sinica (English translation). 1:11-16.
Results in *M. domestica vicina* showed close correlation between enzyme inhibition and symptoms of poisoning, and true cholinesterase inhibited to a greater degree than pseudo-ChE. ABSTRACTS: BA 49:3365(37626), Apr. 1, 1968; RAE-B 53:229, Dec., 1965.

Leng, H., joint author. *See* Shieh, T. R., et al., 1964.

4159. Lentz, E. C. 1964 Survival and Waste Disposal. Military Med. 129(3): 247-252.
Prisoners of war in Carbanatuan, P.I., following the fall of Bataan, evolved sanitary engineering methods as needed, for control of disease and insect vectors. ABSTRACT: BA 46:1730(21629), Mar. 1, 1965.

4160. Leopold, R. A. 1970 Cytological and Cytochemical Studies on the Ejaculatory Duct and Accessory Secretion in *Musca domestica* L. MS of paper to be published in J. Insect Physiol.
Accessory secretion found to be a protienaceous substance relatively resistant to pepsin hydrolysis, but digestible by pronase and trypsin. Renewed synthesis follows depletion by repeated matings.

Leopold, R. A., joint author. *See* LaChance, L. E., et al., 1969.

4161. Leski, R. A. and **Cutkomp, L. K. 1962** The Influence of Parathion and Para-oxon on Sensory Hairs of Flies. J. Econ. Ent. 55:281-285.
Describes results of responses by labellar hairs of *M. domestica* and *Phormia regina*. ABSTRACTS: BA 40:323(4184), Oct. 1, 1962; RAE-B 50:268, Dec., 1962.

4162. Leverich, A. P. and **LaChance, L. E. 1968** Cytological Basis of Sterility Induced by Alkylating and Non-alkylating Chemosterilants in House Flies. Proc. N. Centr. Branch Am. Ass. Econ. Ent. 23:28-29. (Abstract.)
Male flies were treated with tepa (alkylating) or hempa (non-alkylating). It is likely that these two chemicals have a different mode of action at the molecular level, but that the expression of dominant lethal mutations induced in sperm during embryonic development is similar for both chemosterilants. ABSTRACT: RAE-B 58:151(637), May, 1970.

Leverich, A. P., joint author. *See* La Chance, L. E., et al., 1969, 1970.

Levi, V., joint author. *See* Sarbova, S., et al., 1968.

4163. Leviev, P. I. A. 1961 [On the possibility and development of resistance in houseflies to chlorophos.] Med. Parazit. (Moskva):30:742-743. (In Russian, with English summary.)
Continuous exposure of adults of each generation to sublethal deposits on glass resulted in reduced susceptibility. ABSTRACTS: BA 40:1285 (16882), Nov. 15, 1962 (indicates Bolezni); RAE-B 53:207, Oct., 1965.

4164. Leviev, P. Ya. 1963 Toksichnost' preparator DDT i GKhTsG dlya mukh v zavisimosti ot nekotorykh faktorov vneshnei sredy. [Toxicity of DDT and hexachloro-cyclohexane preparations to flies, depending on some environmental factors.] Med. Parazit. i Parazitarn. (Bolezni) 27(6):732. Referat. Zhur. Biol. 1959, No. 87622 (Translation).
Under similar environmental conditions, the effect of DDT and hexachlorocyclohexane (HCCH) depends upon the nature of the treated surface, glass surfaces yielding the greatest mortality. ABSTRACT: BA 44:951(12820), Nov. 1, 1963.

4165. Levinson, Z. H. 1960 Food of Housefly Larvae. Nature 188(4748): 427-428.
Nutritional value of *Escherichia coli* cells very high. Is complete food for larvae, if supplemented by cholesterol and water. ABSTRACTS: RAE-B 50:143, July, 1962; BA 36:2364(28942), May, 1961.

4166. Levinson, Z. H. 1962 The Function of Dietary Sterols in Phytophagous Insects. J. Insect Physiol. 8:191-198.
Larvae of both *Musca* and *Calliphora* show a very *low* rate of conversion of C_{28-29} sterols to cholesterol.

4167. Levinson, Z. H. 1963 Production of Metabolic Deficiencies as a Possible Approach to Insect Control. Bull. Res. Counc. of Israel. Sect. E Experimental Medicine 10E(3-4):116-123.
Treats principally of *Dermestes*, but refers to *M. domestica* and *M. vicina*, from literature.

4168. Levinson, Z. H. 1963-1964 Some Comments on Rachel Carson's "Silent Spring" and Several Research Ideas Stimulated by It. Israel J. Exp. Med. 11:190-193.
Advocates research on sterilizing techniques, also on microbial pathogens to serve as "biological insecticides". States that exposure of experimental animals to insecticides should last at least for the same number of generations as it takes for a strain of houseflies to become resistant.

4169. Levinson, Z. H. 1969 The Nutritional Requirements of the Levant Housefly *Musca vicina* (Macq.). Ph.D. Thesis, Univ. of Jerusalem (1958). (Chapters 1-5 in Hebrew.)
Body composition of the fly at various developmental stages. New rearing methods. Artificial and synthetic diets.

Levinson, Z. H., joint author. *See* Bergmann, E. D., et al., 1966; Kodiceck, E., et al., 1960; Shaaya, E., et al., 1966.

Levkovich, V. G., joint author. *See* Sergeeva, Z. D., et al., 1961.

4170. Lewis, C. T. 1963 Some Applications of Radioisotopes to the Study of the Contamination of Insects by Insecticide Solutions. Reprinted from Radiation and Radioisotopes Applied to Insects of Agricultural Importance. Int. Atomic Energy Agency, Vienna. pp. 135-145.
Work done with *Tribolium castaneum*, with comparative reference to *Phormia terraenovae*, *Aedes aegypti* and *Musca domestica.*

4171. Lewis, C. T. 1965 Influence of Cuticle Structure and Hypodermal Cells on DDT Absorption by *Phormia terraenovae* R-D. J. Insect Physiol. 11:683-694.
Solution of DDT in lanoline applied to structurally different parts of integument. After a short initial period, rate of diffusion through cuticle appears to be governed by the hypodermal cells.

4172. Lewis, J. B. 1969 Detoxification of Diazinon by Subcellular Fractions of Diazinon-resistant and Susceptible Houseflies. Nature (London) 224:917-918.
Identifies 3 routes by which the degradation process may proceed. ABSTRACT: TDB 67:723(1406), June, 1970.

4173. Lewis, S. E. 1967 Effect of Carbon Monoxide on Metabolism of Insecticides *in Vivo.* Nature (London) 215:1408-1409.
A component sensitive to carbon monoxide is believed to be involved in the metabolism of some insecticides *in vivo*; is tentatively identified as the microsomal cytochrome P-450. ABSTRACTS: BA 49:950(10707), Jan. 15, 1968; RAE-B 56:246(910), Dec., 1968; TDB 65:342(787), Mar., 1968.

4174. Lewis, S. E., Wilkinson, C. F. and Ray, J. W. 1967 The Relationship Between Microsomal Epoxidation and Lipid Peroxidation in Houseflies and Pig-liver and the Inhibitory Effect of Derivatives of 1,3-Benzodioxole (methylenedioxybenzene). Biochem. Pharmacol. 16 (7):1195-1210.
Epoxidation system of houseflies markedly susceptible to inhibition by these derivatives. Housefly preparations contained an endogenous inhibitor of lipid peroxidation. ABSTRACTS: BA 48:10428(117058), Dec. 1, 1967; RAE-B 56:226(823), Nov., 1968.

Lewis, S. E., joint author. *See* Brooks, G. T., et al., 1970.

4175. Lhoste, J., Lambert, J. and Rauch, F. 1967 Etude de l'action insecticide du d. trans-chrysanthemate de d.1. alléthrolone sur *Musca domestica* L. [Study of the insecticide action of the D transchrysanthemate of D 1 allethrolone insectic on *Musca domestica* L.] Compt. Rend. Seances Acad. Agr. Fr. 53(9):686-691.
An addition to the pyrethrinoid group of insecticides. Can be synergized by piperonyl butoxide and by S421, the same as pyrethrins, and with comparable effect. ABSTRACT: RAE-B 58:139(583), May, 1970.

4176. Lhoste, J., Martel, J. and Rauch, F. 1969 A New Insecticide 5-benzyl 3-furylmethyl d-trans ethanochrysanthemate. Proc. 5th Br. Insectic.

Fungic. Conf. pp. 554-557.
RU-11.679. Has no knockdown effect, but lethal action is great. Synergized with piperonyl butoxide, proved more toxic to *M. domestica* than to several other species tested.

4177. Lhoste, J. and **Rauch, F. 1969** Sur les proprietes insecticides du d-trans-"ethanochrysanthemate" de benzyl-5 furylmethyle-3. [Insecticide properties of d-trans-"ethanochrysanthemate" of benzyl-5 furylmethyl-3.] Compt. Rend. Herd Seances Acad. Sci. Ser. D. Sci. Natur. (Paris) 268(26):3218-3220.
Tested on *M. domestica*, *Blatella germanica*, and *Sitophilus granarius*; showed insecticidal properties superior to any products of this series available to date. Abstract: BA 51:1096(11300), Jan. 15, 1970.

4178. Lhoste, J. and **Rauch, F. 1969** Remarques sur quelques chrysanthémates insecticides de synthèse. [Some observations on synthetic insecticide chrysanthemates.] Rev. Zool. Agric. et Appl. for 1969 (4-6):53-66. (English summary.)
Discusses two types of synthetic insecticide chrysanthemates, one having rapid knockdown with no after-effects, the other having no noticeable knockdown effect but a more powerful lethal effect due to penetration.

4179. Lhoste, J., Rauch, F. and **Lambert, J. 1968** Contribution à l'étude de l'effect insecticide de quelques synergistes associés au *d. trans.* chrysanthémate de *de.* 1 alléthrolone. [Contribution to a study of the insecticidal effect of certain synergists associated with *d. trans.* chrysanthemate of *d.* 1 allethrolone.] XIII Congrès International d'Entomologie de Moscow. Section 9(2).
Gives D.L50 values for *M. domestica* with 11 different synergists.

4180. Lhoste, J., Rauch, F. and **Lambert, J. 1968** Action du d-trans-chrysanthémate de dl-alléthrolone sur quelques insectes. [Action of d-trans-chrysanthemate of dl-allethrolone on certain insects.] Phytiat.-Phytopharm. 2:143-150.
Action against a number of other insect species compared with known lethal effect against *M. domestica.*

Li, H. Y., joint author, *See* Chang, C. P., et al., 1965.

Li, S. M., joint author, *See* Chang, C. P., et al., 1965.

Li, V., joint author, *See* Lee, T., et al., 1964.

Liang, T. T., joint author, *See* Lichtenstein, E. P., et al., 1969.

4181. Lichtenstein, E. P. 1966 Increase of Persistence and Toxicity of Parathion and Diazinon in Soils with Detergents. J. Econ. Ent. 59(4): 985-993.
Addition of detergents ABS and LAS to insecticide-treated soils increased the persistence of organophosphorus residues, especially parathion and diazinon. Abstract: BA 47:10324(119326), Dec. 15, 1966.

4182. Lichtenstein, E. P. 1967 Effects of Detergents on the Movement, Persistence and Toxicity of some Insecticides. Abstr. Int. Pflanzenschutz-Kong. [Int. Congr. Plant Prot.] 30:502-503.

4183. Lichstenstein, E. P., Schulz, K. R., Fuhremann, T. W. and **Liang, T. T. 1969** Biological Interaction Between Plasticizers and Insecticides. J. Econ. Ent. 62(4):761-765.
Many polychlorinated diphenyl compounds proved toxic to *M. domestica*, but not so toxic as dieldrin, or DDT. Toxicity increases with decrease in chlorine content. Abstracts: BA 51:4482(46180), Apr. 15, 1970; TDB 67:1057(2055), Aug., 1970.

4184. Lichtwardt, E. T. 1964 A Mutant linked to the DDT-Resistance of an Illinois Strain of House Flies. Ent. Exp. Appl. (Amsterdam) 7:296-309.
Refers to mutant cye-color *carnation*, located on chromosome V, as is dominant gene R (*resistant*). ABSTRACTS: BA 47:3415(40329), May 1, 1966; RAE-B 53:175, Sept., 1965; TDB 62:365, Apr., 1965.

Lidror, R., joint author. *See* Barkai, A., et al., 1966.

Lilly, J. H., joint author. *See* Eversole, J. W., et al., 1965; Steve, P. C., et al., 1965.

Limburg, A. M., joint author. *See* Borkovec, A. B., et al., 1964.

Lin, F.-J., joint author. *See* Lee, W.-Y., et al., 1967.

Lindquist, D. A., joint author. *See* Bull, D. L., et al., 1967; Andrawes, N. R., et al., 1967.

4185. Lineva, V. A. 1961 [Procedures used to determine resistance of housefly to insecticides. Communication 1.] Med. Parasit. and Parasitic. Dis. (Moscow) 30(4):465-470.
Instructions given for rearing and maintaining colony, and for making and calibrating microloops for application of insecticide. Concerns DDT, BHC and chlorophos. ABSTRACT: TDB 59:99, Jan., 1962.

4186. Lineva, V. A. 1962 [Alterations in the ovogenesis of the housefly (*Musca domestica* L.) under the influence of insecticides.] XI. Internat. Kongr. für Ent. Wien, 17 bio 25 August. 1960. Band II. Sektion X. Medizinische und Vet. Ent. Verh. XI, 2:448-450.
Account given of pathological irregularities observed in the laboratory and in nature. Abnormalities decreased fertility and sometimes caused sterility. ABSTRACT: RAE-B 51:246, Nov., 1963.

4187. Lineva, V. A. 1962 Change in the Susceptibility of the Housefly (*Musca domestica* L.) to Chlorophose over a Period of Five Seasons. J. Hyg. Epidemiol., Microbiol., Immunol. (Prague) 6(3):271-277. (Summary in French, German and Italian).
Use of chlorophose against DDT-resistant fly population increased its resistance to DDT. Produces disturbances of ovogenesis in females. ABSTRACTS: BA 42:1241(15720), May 15, 1963; TDB 60:384, Apr., 1963.

4188. Lineva, V. A. 1963 [Changes in the sensitivity of the domestic fly *Musca domestica* L. to chlorphos, applied for 5 seasons under field conditions.] Med. Parazit. (Moskva) 32:92-95. (In Russian.)
See preceding reference. ABSTRACTS: BA 45:3204(39655), May 1, 1964; RAE-B 54:172, Sept., 1966; TDB 60:598, June, 1963.

4189. Lineva, V. A. 1964 [Experience with fly control by adding chlorophos to drinking water destined for chicken.] Med. Parasit. and Parasitic. Dis. (Moscow) 33(1):15-19. (In Russian, with English summary.)
Mentions *M. domestica*, *Phormia terraenovae*, *Fannia* and *Hydrotaea*. Droppings of treated birds found to be toxic to larval instars of *M. domestica*. Toxicity decreased with age. No effect or trace in birds. ABSTRACTS: BA 45:7400(93438), Nov. 1, 1964; RAE-B 55:198(691), Nov., 1967; TDB 61:613, June, 1964.

4190. Lineva, V. A. 1965 Sezonnyi khod chislennosti komnatnoi mukhi *Musca domestica vicina* Macq. v uslovujakh Lenkorani. [Seasonal incidence of the house fly *Musca domestica* in Lenkoran.] Med. Parazit. i Parazitarn. (Bolezni) 34(3):334-336.
A dip in the seasonal peak of flies in July corresponds with increasing dryness of the air, resulting in a redistribution of flies from indoors to outdoors. ABSTRACT: BA 47:2931(34688), Apr. 1, 1966.

4191. **Lineva, V. A. 1966** [The pattern of the effect of chlorophos upon the ovogensis of house flies (*Musca domestica* L.) in its long application under natural conditions.] Med. Parasit. and Parasitic. Dis. (Moscow) 35(2):217-223. (In Russian, with English summary.)
Disorders in egg production were due to abnormalities in the mechanism of oviposition and disorders in the formation of follicles. ABSTRACTS: RAE-B 56:232(848), Dec., 1968; TDB 63:815, July, 1966.

4192. **Lineva, V. A. 1967** [On metatoxical action of insecticides on *Musca domestica* L.] Wiad. Parazytol. 13:405-411.
Author defines metatoxical action as any action by toxic substances which causes no death of the insect, but hinders the normal development of its generations. This paper deals chiefly with the "disordering of oogenesis".

4193. **Lineva, V. A. 1970** [The effect of prolonged action of chlorofos on natural population of house flies (*Musca domestica* L.).] Med. Parazit. 48(1):73-77. (In Russian.)
Kept population at low levels despite the fact that flies develop tolerance to chlorophos. Believed to cause biological disorders. ABSTRACT: TDB 67:723(1403), June, 1970.

4194. **Lineva, V. A., Abezgauz, I.** and **Ionova, A. 1960** Primenenie sukhogo "mikhomora" s deistvuyushchim veshchestvom khlorofos v bor'be s mukhami. [Use of dry fly-paper "mukhomor" with chlorphos as the active substance in fly control.] Med. Parazit. (Mosk) 29:330-334. (In Russian.)
Papers impregnated with 50 mg khlorofos per 100 cm^2 were very effective killing nearly all flies released in test area in 7-8 days. An excess of khlorofos reduced the papers attractiveness to flies. ABSTRACTS: BA 36: 6197(66269), Oct. 1, 1961; RAE-B 50:249, Nov., 1962.

4195. **Lineva, V. A.** and **Brezhneva, I. M. 1964** Perspektivy primeneniya repellentov protiv mukh. [Prospects of using repellents against flies.] Med. Parazit. i Parazitarn. (Bolezni) 33(5):532-536. (English summary.)
Types of tests used to determine the effectiveness of several formulations of repellents included those in an olfactometer to determine repelling power, in glass tubes to determine distant action and with screens of gauze to determine food protection capacity. ABSTRACT: BA 46:6622 (82214), Sept. 15, 1965.

4196. **Lineva, V. A., Osipova, L. S.** and **Tamarina, N. A. 1961** Metodika opredeleniya ustoichivosti komnatnoi mukhi *Musca domestica* L. k insektitsidam (Soo. bshch. I.). [Method for determining the resistance of the house-fly, *Musca domestica* L. to insecticides. (First report).] Med. Parazit. i Parazitarn. (Bolezni) 30(4):465-470.
Test involves treating the mesotergite of a narcotized fly with solutions of the insecticide in the form of a 10 per cent DDT solution in acetone diluted to various concentrations. Flies are kept in clean breeding areas, fed, and the LD_{50} determined. ABSTRACTS: BA 39:661(8225), July 15, 1962; RAE-B 53:169, Aug., 1965.

4197. **Lineva, V. A., Osipova, L. S.** and **Tamarina, N. A. 1961** [Method for determining the resistance of the housefly, *Musca domestica* L. to insecticides. (2nd report).] Med. Parazit. i Parazitarn. (Bolezni) 30(5):603-608. Referat. Zhur. Biol. No. 10K119. (English summary.)
Describes conventional procedure in probit analysis plus modifications by Litchfield and Wilcoxon (1949) and of Roth (1959). Uses conversion tables of Bayesain (1957). ABSTRACTS: BA 40:1276(16802), Nov. 15, 1962; RAE-B 53:205, Oct., 1965.

Lineva, V. A., joint author. *See* Derbeneva-Ukhova, V. P., et al., 1966.

4198. Linsdale, D. D. 1969 Domestic Flies: Their Recognition, Biology, and Control. *In* J. E. Brooks and T. D. Peck (ed.) Community Pest and Related Vector Control. Pest Control Operators of Calif., Inc. Los Angeles. pp. 82-96.

Linsdale, D. D., joint author. *See* Ecke, D. H., et al., 1965, 1967; Walsh, J. D., et al., 1968; Smith, T. A., et al., 1966, 1967, 1968, 1969, 1970; Strickland, W. B., et al., 1970.

4199. Lipínska-Skawínska, B. and **Chmurzynski, J. A. 1968** Phototactic Choice between Two White Light Sources of Various Intensity in Blowfly, *Calliphora erythrocephala* Meig. Experientia 24:283-284.
Is sequel to 1967 paper by junior author, which see.

4200. Lipke, H. and **Chalkley, J. 1962** Glutathione, Oxidized and Reduced, in some Dipterans Treated with 1, 1, 1-trichloro-2, 2-di-(p-chlorophenyl) ethane. Biochem. J. 85:104-109.
In *M. domestica*, the content of reduced plus oxidized glutathione is about 2μ equiv./g. of tissue, with a GSH/GSSG ratio approximating to 7:1. GSH is the principal free thiol present in the tissues. ABSTRACT: BA 41:1664(20914), Mar. 1, 1963.

Lisk, D., joint author. *See* Matthysse, J. G., et al., 1968.

4201. Litvak, S. 1967 Protein Synthesis in Cell-free Preparations from *Musca domestica*. Arch. Biol. Med. Exp. 4(1/2):216. Abstract.
See reference 4203. Litvak, et al., 1967.

4202. Litvak, S. and **Agosin, M. 1968** Protein Synthesis in Polysomes from Houseflies and the Effect of 2, 2-bis(p-chlorophenyl)-1, 1, 1-trichloromethane. Biochemistry 7(4):1560-1567.
The role of DDT as an enzyme inducer is mediated through the deoxyribonucleic acid dependent syntheses of ribonucleic acid species, presumably messenger RNA. A translational mechanism at the polysomal level must still be considered. ABSTRACT: BA 49:9773(108240), Oct. 15, 1968.

4203. Litvak, S., Mancilla, J. and **Agosin, M. 1967** Protein Synthesis in Cell-free Preparations from *Musca domestica* L. Comp. Biochem. Physiol. 21(3):441-448.
Protein synthesis occurs by the pathway involving amino acyladinylates and amino acyl-soluble ribonuclei acid as intermediates. System requires ATP-generating system divalent cation. ABSTRACT: BA 48:10184(114428), Nov. 15, 1967.

4204. Litvinova, N. F. 1965 Sinantro pnye mukhi Blagoveshchenska Osobennosti i ekologii mukh semeistva Muscidae. [Synanthropic flies found in Blagoveshchensk. Specific ecological features of flies from the family Muscidae.] Med. Parazit. i Parazitarn. (Bolezni) 34(5): 529-533. (In Russian, with English summary.)
In rainy seasons *Muscina stabulans* becomes the most prevalent fly species and is numerous in dwellings. *Musca domestica* occurs as a facultative hematophage in pastures. It was found that *F. incisurata* overwinters as a larva. ABSTRACT: BA 47:8115(94492), Oct. 1, 1966.

4205. Lin Hsuen, Leng (H.-F.) and **Yang, P.-N. 1965** [Studies on selective organophosphorus insecticides: chemical structure and biological activity of acethion analogs.] Acta Ent. Sinica 14(4):339-346. Peking. (In Chinese, with English summary.)
Studies included an examination for anticholinesterase activities on house fly cholinesterase and their toxicities to house flies. ABSTRACTS: BA 50:1329(13908), Feb. 1, 1969; RAE-B 54:82, Apr., 1966.

4206. Liu, S.-Y. 1962 Observations on the Bionomics of Filariasis Vector and Flies of Public Health Importance on Hsi-Yü Island of Peng-hu Archipelago, Taiwan. J. Formosan Med. Ass. 61:391-402.
Predominant housefly is *M. d. vicina*. Breeds in pig excreta and litter, compost and garbage, cow, human and canine excreta, in that order. *M. sorbens* and flesh flies favor human and canine excreta. ABSTRACT: TDB 60:902, Sept., 1963.

Llorente, V., joint author. *See* Peris, S. V., et al., 1963.

Lloyd, C. J., joint author. *See* Bates, A. N., et al., 1965; Hewlett, P. S., et al., 1961.

4207. Lloyd, J. E. and **Matthysse, J. G. 1966** Polymer Insecticide Systems for use as Livestock Feed Additives. J. Econ. Ent. 59(2):363-367.
Polyvinyl chloride most promising polymer for minimum loss in the animal, and maximum release in manure. ABSTRACT: BA 47:7231(84361), Sept. 1, 1966.

4208. Lloyd, J. E. and **Matthysse, J. G. 1970** Polyvinyl Chloride—Insecticide Pellets Fed to Cattle to Control Face Fly Larvae in Manure. J. Econ. Ent. 63(4):1271-1281.
Several preparations tested. Diazinon and dichlorvos in PVC the most promising. Results affected by size of pellet, concentration of insecticide, specific gravity and frequency of feeding.

Lloyd-Davies, T.A., joint author, *See* Hale, J.H., et al., 1960.

4209. Loaeza, R. M. and **Corona, A. O. 1965** Esterilizacion de la mosca domestica con apholate. [Sterilization of the domestic fly with apholate.] Folia Ent. Mex. Núm. 10, Septiembre 1965. 15 pp.
Reports on a series of experiments to estimate action and persistence of apholate, variously applied.

4210. Lobanov, A. M. 1960 [Certain observations on the attraction of various baits for exophilic types of synanthropic flies.] Med. Parazit. (Moskva) 29:720-726.

4211. Lobanov, A. M. 1960 On the Problem of the Role of Flies in the Epidemiology of Intestinal Infections. Zhur. Mikrobiol. 31:116-121.
The participation of flies in the spread of intestinal infections cannot be determined by numbers of the species occurring but by the degree of its contact with feces, which in turn is determined by meteorological factors, chiefly atmospheric humidity. ABSTRACT: BA 36:978(11623), Feb. 15, 1961.

4212. Lobanov, A. M. 1966 Kizucheniyu sinantropnykh mukh v zakrytykh statsiyakh. [A study of synanthropic flies in closed stations.] Med. Parazitol. i Parazitarn. (Bolezni) 35(1):55-60. (In Russian, with English summary.)
Over thirty species of flies known to enter dwellings can be divided into several groups according to the extent and character of their association with premises. The extent of endophylly in various landscapes and climatic conditions varies and merits study. ABSTRACT: BA 47:10322 (119294), Dec. 15, 1966.

Lockley, A. S., joint author. *See* Brown, A. W. A., et al., 1961.

4213. Lodha, K. R., Treece, R. E. and **Koutz, F. R. 1970** Oviposition and Hatchability of Eggs of the Face Fly. J. Econ. Ent. 63(2):446-449.
As fly age increased, there was a reduction in rate of oviposition and hatchability of eggs. The latter could be induced by removing citrated bovine blood from standard diet. One mating sustains egg production through at least 3 gonadotrophic cycles. ABSTRACTS: BA 51:8566(87348), Aug. 1, 1970; RAE-B 58:303(1240), Sept., 1970.

4214. Lodha, K. R., Treece, R. E. and **Koutz, F. R. 1970** Studies on the Mating Behavior of the Face Fly. J. Econ. Ent. 63(1):207-212.
Treats of pre-copulatory behavior, seizure, copulation. Visual, olfactory and tactile senses are all involved (Copulation does not always follow seizure.) ABSTRACT: RAE-B 58:231(947), July, 1970.

4215. Logan, J. C. P. and **Walker, M. 1964** A Case of Endemic Cutaneous Myiasis. Brit. J. Derm. 76:218-222.
Hundreds of third stage larvae of *M. domestica* were recovered from sores on buttocks and nearby tissues of a doubly incontinent patient. ABSTRACT: BA 46:3838(47517), June 1, 1965.

4216. Loginovskii, G. E. 1963 [Seasonal changes in the quantity of house-flies in Kurgan.] Med. Parazit. (Moskva) 32:563-565.

4217. Long, C. L. and **Silverman, P. H. 1965** Serological Characterization of Five Strains of Insecticide-Resistant and Susceptible House Flies. J. Econ. Ent. 58:1070-1074.
Characterization by means of the agar-gel double-diffusion technique and the production of antisera in rabbits showed definite difference between resistant and susceptible strains. Precipitation bands (representing DDT-dehydrochlorinase in the case of the DDT resistant strain) were not found in the susceptible strains. ABSTRACTS: RAE-B 54:59, Mar., 1966; TDB 63:905, Aug., 1966.

4218. Loomis, E. C. 1964 Agricultural Sanitation and the Domestic Fly Problem. (Diptera) Proc. Pap. Ann. Conf. Calif. Mosq. Contr. Assoc., Inc. 32:34-36.

4219. Loomis, E. C. 1969 Domestic Fly Control by Application of DOW Liquid Fly Larvicide to Animal Manures. Down to Earth 25(1): 27-31.
Also known as DLFL. Was effective when broadcast over animal manures on a weekly or biweekly schedule. Better results against *Musca* than *Fannia*. ABSTRACT: BA 50:11923(123240), Nov. 15, 1969.

4220. Loomis, E. C., Anderson, J. R. and **Deal, A. S. 1967** Identification of Common Flies Associated with Livestock and Poultry. Agr. Sanit. Series. Univ. Calif. June. 5M. AXT-236, 12 pp.
Contains excellent illustrations, several in color. Gives both larval and adult characters.

4221. Loomis, E. C., Bowen, W. R. and **Dunning, L. L. 1968** Hymenopterous Parasitism in the Little House Fly. J. Econ. Ent. 61(4):1105-1107.
Puparia of *Fannia canicularis* parasitized by *Muscidifurax raptor* and by *Stilpnus* sp. ABSTRACTS: BA 50:3709(38930), Apr. 1, 1969; RAE-B 57: 5(11), Jan., 1969.

4222. Loomis, E. C., Bramhall, E. L. and **Dunning, L. L. 1968** A Preliminary Report —— Zytron as a Larvicide for Fly Control. Calif. Agric. 22(6):6-7. June.
Reports experimental use as an emulsifiable concentrate, to be mixed with water and sprayed on manure piles. Showed promise. (Had not yet been registered or recommended by the University of California.) ABSTRACT: BA 49:10784(118701), Nov. 15, 1968.

4223. Loomis, E. C., Deal, A. S. and **Bowen, W. R. 1968** The Relative Effectiveness of Coumaphos as a Poultry Feed Additive to Control Synanthropic Fly Larvae in Manure. J. Econ. Ent. 61(4):904-908.
Achieved 43 per cent control of *M. domestica*; very poor control of *Fannia canicularis* and *Musca stabulans*. Succeeded best with *F. femoralis* (50-84 per cent) and with *Ophyra leucostoma* (84-99 per cent). ABSTRACTS:

BA 50:3174(33189), Mar. 15, 1969; RAE-B 57:2(2), Jan., 1969.

Loomis, E. C., joint author. *See* Deal, A. S., et al., 1965, 1968; Georghiou, G. P., et al., 1967; Anderson, J. R., et al., 1968.

4224. Lord, K. A. 1965 Causes of Resistance in Houseflies and some Practical Implications. Proc. 3rd Brit. Insect. and Fung. Conf. Nov. 8-11. Brighton, Sussex. pp. 61-64.
Reviews work showing resistance to diazinon; discusses reasons and suggests criteria for governing choice of an insecticide to overcome resistance. ABSTRACT: RAE-B 55:201(698), Nov., 1967.

4225. Lord, K. A., Molloy, F. M. and **Potter, C. 1963** Penetration of Diazoxon and Acetyl Choline into the Thoracic Ganglia in Susceptible and Resistant House-flies and the Effects of Fixatives. Bull. Ent. Res. 54:189-197.
Nerve sheath of thoracic ganglion found permeable to diazoxon but impermeable to acetyl choline and acetyl thiocholine. Holds for both susceptible and resistant strains. Inhibition of cholinesterase in the thoracic ganglion probably not the cause of death. ABSTRACTS: RAE-B 51:223, Oct., 1963: TDB 60:1169, Dec., 1963.

4226. Lord, K. A. and **Solly, S. R. B. 1961** Changes in the Amino-acid Content of Houseflies after Organophosphorus Poisoning. Chem. and Ind. pp. 1359-1360.
M. domestica shows a definite fall in amino acid content after exposure to diazinon.

4227. Lord, K. A. and **Solly, S. R. B. 1964** Effect of Insecticides, Especially Diazinon, on the Amino Acids of Adult Houseflies *Musca domestica.* Biochem. Pharmacol. 13:1341-1349.
Organophosphorus poisons may affect the metabolism of fats or carbohydrates in addition to inhibiting cholinesterase of the nervous system. Amounts of free amino acids change normally with age and nutritional condition. ABSTRACTS: BA 46:1083(13673), Feb. 1, 1965; RAE-B 54:217, Nov., 1966.

Lord, K. A., joint author. *See* Farnham, A. W., et al., 1965; El Basheir, El S., et al., 1965; Gwiazda, M., et al., 1967.

Lord, T. H., joint author. *See* Zaharis, J. L., et al., 1961.

4228. Loughran, J. X., III and **Ballard, R. C. 1970** Evidence for a Neural Influence of the House Fly Heart. Ann. Ent. Soc. Am. 63(5): 1460-1461.
Surface potentials were recorded by suction electrodes placed on two segments of the heart. Simultaneous recordings suggest an anterior influence on the heart beat.

4229. Louloudes, S. J., Chambers, D. L., Moyer, D. B. and **Starkey, J. H., III. 1962** The Hydrocarbons of Adult House Flies. Ann. Ent. Soc. Am. 55:442-448.
Gives technique of isolation. Saturated, unsaturated, cyclic, and straight-chain hydrocarbon types were present. Carbon numbers ranged from C-16 to C-35 with odd-numbered compounds predominant. ABSTRACT: BA 40:642(8515), Oct. 15, 1962.

Louloudes, S. J., joint author. *See* Kaplanis, J. N., et al., 1963; Thompson, M. J., et al., 1962, 1963; Robbins, W. E., et al., 1960.

4230. Lovell, J. B. 1963 The Relationship of Anticholinesterase Activity, Penetration, and Insect and Mammalian Toxicity of Certain Organophorus Insecticides. J. Econ. Ent. 56(3):310-317.
All compounds tested had some ability to inhibit *M. domestica* head

cholinesterase *in vitro*. But best inhibitors were not always the most active against living organisms, and conversely. ABSTRACT: BA 45:889 (11022), Feb. 1, 1964.

Lovell, J. B., joint author. *See* Snyder, F. M., et al., 1964.

4231. Loy, V. A. and **Cantwell, G. E. 1966** Mortality of House Flies Exposed to Dichlorvos Resin Strips in Permanent Field Latrines. J. Econ. Ent. 59:1527.

Author considers this an excellent means of fly control. Mortality is highest if strips are located where little or no air movement occurs. ABSTRACTS: BA 48:3232(35596), Apr. 1, 1967; TDB 64:809, July, 1967; RAE-B 55:70(258), Apr., 1967.

4232. Ludwig, D., Crowe, P. A. and **Hassemer, M. M. 1964** Free Fat and Glycogen during Metamorphosis of *Musca domestica* L. J. New York Ent. Soc. 72(1):23-28.

Both are used as direct energy sources. Report gives percentage gains and losses of each during the period of metamorphosis, with the temperature held at 25°C. ABSTRACT: BA 45:5761(71268), Aug. 15, 1964.

Ludwig, D., joint author. *See* Brebbia, D., et al., 1962.

4233. Lupasco, G. Duport, M., Combiesco, I. and **Atanasiu, A. 1963** Sensibilité au DDT de l'espece *Musca domestica* L. testée dans differéntes localités de Roumanie. [Susceptibility to DDT in the species *Musca domestica* L., tested in different localities in Rumania.] Arch. Roum. Path. Exp. Microbiol. 22(3):749-756. (Summaries in Russian, German and English.)

Resistance tended to be lower in autumn, (the period when they are entering hibernation,) than in summer. ABSTRACTS: RAE-B 53:237, Dec., 1965; TDB 61:532, May, 1964.

4234. Lupasco, G., Duport, M., Combiesco, I., Enesco, A. and **Ionică, M. 1966** Contributions à l'etude de la résistance de *Musca domestica* L. aux différents insecticides rémanents (en Roumanie). [Contributions to the study of resistance in *M. domestica* to various lasting insecticides (in Roumania).] Arch. Roum. Path. Exp. Microbiol. 25(1):205-210.

Flies of various localities where insecticides had been used for some years were found to be very resistant to DDT, γ-BHC, and slightly to trichlorphon and endrin. ABSTRACTS: RAE-B 56:226(824), Nov., 1968; TDB 63:1254, Nov., 1966.

Lurik, B. B., joint author. *See* Misnik, Y. N., et al., 1967.

4235. Lysenko, O. and **Povolný, D. 1961** The Microflora of Synanthropic Flies in Czechoslovakia. Folia Microbiol. 6(1):27-32. (Russian summary.)

Some species of flies (reasons not clear) can act as hosts to more species of bacteria than other flies from the same environment. Study of a hygienic-epidemiological situation should include examination of the flies' microflora. ABSTRACT: BA 38:1182(15283), May 15, 1962.

4236. McAllan, J. W. and **Chefurka, W. 1961** Some Physiological Aspects of Glutamate-asparate Transamination in Insects. Comp. Biochem. Physiol. 2:290-299.

The main site for interconversion of amino acids during housefly development is muscle tissue. Enzymatic activity rises during larval growth, falls during the pupal stage and rises during emergence, creating a U-shaped curve related to the curve for protein synthesis during morphogenesis.

4237. McCann, G. D. 1965 Nervous System Research in Man and *Musca domestica* with Computers. Science 148(3677):1565-1571.
Aim is to make mathematically precise correlation of each type of response with stimulus and with other responses. ABSTRACT: BA 47: 158(1923), Jan. 1, 1966.

4238. McCann, G. D. 1966 Dimensions of the Information Sciences. (*Musca domestica*, nervous system.) Yale Sci. 40(4):6-8.

4239. McCann, G. D. and **Dill, J. C. 1969** Fundamental Properties of Intensity, Form, and Motion Perception in the Visual Nervous Systems of *Calliphora phaenicia* and *Musca domestica*. J. Gen. Physiol. 53(4):385-413.
Response of both form and motion detection units is independent of the direction of pattern intensity gradation. ABSTRACT: BA 50:9766 (101157), Sept. 15, 1969.

4240. McCann, G. D. and **MacGinitie, G. F. 1965** Optomotor Response Studies of Insect Vision. Proc. Roy. Soc. Ser. B. (Biol.) 163:369-401.
Precise methods are given for the study of the optomotor response in correlation with microprobing and histology. ABSTRACT: BA 47:2950 (34936), Apr. 1, 1966.

4241. McCann, G. D., Sasaki, Y. and **Biedebach, M. C. 1966** Correlated Studies of Insect Visual Nervous Systems. *In* Proc. Internat. Symp. on . . . , Stockholm, Sweden, 1965. Publ. Div. Pergamon Press: 7:559-583.
Initial information transduced by photoreceptors can be determined primarily from total light flux entering the receptor. ABSTRACT: BA 48: 9252(104038), Oct. 15, 1967.

4242. McCann, G. D., et al. 1969 Fundamental Properties of Intensity, Form and Motion Perception in the Visual Nervous System of *Calliphora, Phaenicia* and *Musca domestica*. J. Gen. Physiol. 53:385-413.
Functional and histological relationship have been established among interneurons in the medulla, lobula-lobular plate and central brain regions of *Musca domestica* and accurately positioned visual stimuli.

McCauley, T. R., joint author. *See* Sullivan, W. N., et al., 1960.

McClellan, E. S., joint author. *See* Killough, R. A., et al., 1965, 1969.

4243. McCord, D. J. 1968 Fly Control Success. Pest Control 36(4):8.
Concerns chiefly insecticide resin strips.

4244. McCoy, C. W. 1965 Biological Control Studies of *Musca domestica* and *Fannia* species on Southern California Poultry Ranches. Proc. Pap. 33rd Ann. Conf. Calif. Mosq. Control Assoc. Inc. 33:40-42. Visalia, Calif.
Concerns effectiveness of the pupal parasite, *Muscidifurax raptor* Gir. and Sand., against flies breeding in droppings under poultry cages. ABSTRACT: RAE-B 56:203(735), Oct., 1968.

McCoy, C. W., joint author. *See* Legner, E. F., et al., 1965, 1966.

4245. MacCreary, D. and **Haenlein, G. F. W. 1962** House Fly Breeding in Oak Sawdust and Peanut Hulls Used as Bedding in Calf Pens. J. Econ. Ent. 55:419.
There is a much lower fly population in sawdust. It packs more firmly, is less moist, does not heat up as do peanut hulls or straw bedding. ABSTRACTS: BA 40:307(3970), Oct. 1, 1962; RAE-B 50:274, Dec., 1962; TDB 59:1118, Nov., 1962.

4246. **McDonald, I. C. 1969** Detection and Partial Characterization of a Crossover Suppressor on Chromosome V of the Housefly. (Abstract.) Genetics 61(2) suppl. 2: S39-S40.
The mating procedure to obtain the crossover suppressor in the female fly is described. Evidence indicates that a translocation is responsible.

4247. **McDonald, I. C. 1970** Population Cage Studies with Wild-Type and Mutant Strains of the House Fly. Ann. Ent. Soc. Am. 63(1):187-191.
A rearing cycle procedure was devised to test the ability of 11 recessive mutant strains of the housefly, *Musca domestica*, to survive and reproduce in population cages. ABSTRACT: RAE-B 58:329(1353), Oct., 1970.

4248. **McDuffie, W. C. 1960** Current Status of Insecticide Resistance in Livestock Pests. *In*: Symposium on research progress in insecticide resistance. Misc. Publ. Ent. Soc. Am. 2(1):49-54.
Resistance of many insect pests is increasing yearly, yet it is not the acute problem in the U.S. that it is in some countries. Paper includes an informative discussion of the problem of housefly resistance to organophosphates. ABSTRACT: BA 36:1020(12268), Feb. 15, 1960.

MacGinitie, G. F., joint author. *See* McCann, G. D., et al., 1965.

4249. **MacGregor, R. 1964** Experiments with the Chemosterilant Apholate on *Musca domestica*. Folia Ent. Mex. (7/8):67. (Abstract.)

4250. **MacGregor-Loaeza, R.** and **Ortega-Corona, A. 1965** Esterilización de la mosca domestica con Apholate. [Sterilization of the housefly with Apholate.] Folia Ent. Mex. 10:1-15 (+5 pp.). English summary.
Apholate gave good results when applied daily over deposited poultry manure but doses were too high to be recommended for commercial use. Apholate-impregnated threads gave good results only in closed areas. ABSTRACTS: BA 47:2927(34641), Apr. 1, 1966; RAE-B 56:107(374), May, 1968.

MacGregor, R., joint author. *See* Hecht, O., et al., 1964.

4251. **McGuire, J. U., Jr., Sullivan, W. N.** and **Starkweather, R. 1966** Analysis of the Effectiveness of Aircraft-disinfection Aerosols Against Resistant Houseflies and Mosquitoes. Bull. W. H. O. 34:151-154.
Musca domestica was exposed to the test aerosol for 30 minutes and mortality counts made after 24 hours. ABSTRACTS: BA 48:916(9783), Jan. 15, 1967; TDB 63:1035, Sept., 1966.

MacIver, D. R., joint author. *See* Hopkins, L. O., et al., 1965.

McKay, M. A., joint author. *See* Winteringham, F. P. W., et al., 1960.

4252. **McLintock, J. R. 1964** Puparium Formation in Diptera (*Sarcophaga*, *Musca*, *Cuterebra*, *Hypoderma*). Nature 201(4925):1245.
Musca domestica, *M. autumnalis*, *Hypoderma bovis* and *Cuterebra* sp. are compared. ABSTRACT: BA 45:8420(106975), Dec. 15, 1964.

4253. **McMillan, J. W. 1963** An Invasion of Maggots. J. Christ. Med. Ass. India. 38:588-589.

4254. **MacNay, G. G. 1965 and 1966** Highlights of the Occurrence of Insects and other Arthropods in Canada, 1965. Canad. Insect Pest Rev. 43(4):93-96. Ottawa. *Also in* Co-op. Econ. Insect Rep. 16(4):56-58. Hyattsville, Md. 1966.
A survey in 1965 indicated that *M. autumnalis* was established in southern areas of Manitoba (cf. RAE-B 53:135) and Saskatchewan and was spreading westward, rapidly. ABSTRACT: RAE-B 54:70, Apr., 1966.

4255. **Machado, C., Silva, S.** and **Gomes, F. 1967** Transmission of Acquired Toxoplasmosis. Hospital (Rio de Janiero) 71(1):137-149.
Musca domestica is mentioned as a probable carrier, along with certain

Hemiptera and Acarina.

Maciver, D. R., joint author. *See* MacIver, D. R., above.

Macri, D., joint author. *See* Comes, R., et al., 1962.

4255a. Madden, J. L. 1964 Ecological Studies of the Parasite *Nasonia vitripennis* (Walk.) and the Housefly Host, *Musca domestica* Linn. Diss. Absts. 24(9):3508-3509.

4256. Madden, J. L. and **Pimentel, D. 1965** Density and Spacial Relationship Between a Wasp Parasite and its Housefly Host. Canad. Ent. 97(10): 1031-1037.

Parasitism of pupae of *M. domestica* by *Nasonia vitripennis* found to decrease as spatial area increased. Density relationship discussed. ABSTRACT: RAE-B 55:72(264), Apr., 1967.

Madden, J. L., joint author. *See* Pimentel, D., et al., 1963.

4257. Maddock, D. R., Sedlak, V. A. and **Schoof, H. F. 1961** Dosage-mortality Response of *Musca domestica* Exposed to DDVP Vapour. Bull. W. H. O. 24:643-644.

Concentrations are given which obtain a 95-100 per cent mortality for control of insects in aircraft. ABSTRACT: RAE-B 51:116, June, 1963.

4258. Maddock, D. R., Sedlak, V. A. and **Schoof, H. F. 1961** Preliminary Tests with DDVP Vapor for Aircraft Disinsection. Publ. Health Repts. (Washington) 76(9):777-780.

Vapor residues are more persistent on vinyl plastic surfaces than on upholstery, carpet or curtains. Vapor method considered preferable to aerosol method then in use. ABSTRACT: TDB 59:399, Apr., 1962.

Maddock, D. R., joint author. *See* Kilpatrick, J. W., et al., 1962.

4259. Maeda, O. 1962 [Field studies on the evaluating method of insecticidal effect. 1. Experiments on the maggots of flies in the privy.] Endemic Dis. Bull. Nagasaki Univ. 4(2):135-140. June 23. (In Japanese, with English summary.)

Effectiveness of insecticides applied during 1960 and 1961 was compared by the mean number per treatment per time of survival of larvae including newly hatched ones 2 days following treatment and of all larvae a week following or just before the next treatment. Results are summarized. Dinzinon, Nankor, and DDVP proved the most effective insecticides. ABSTRACTS: BA 43:1016(12203), Aug. 1, 1963; TDB 60:384, Apr., 1963.

4260. Maeda, O. 1963 [Field studies on the evaluating method of insecticidal effect. 4. Expeiiments for the maggots of flies in the privy (2).] Endemic Dis. Bull. Nagasaki Univ. 5:67-71. (In Japanese, with English summary.)

Tests during 1962 using diazinon, Nankor and DDVP as privy insecticides showed a difference in residual effect of the three. Nankor and diazinon resulted in fewer living larvae a week following treating. ABSTRACTS: BA 44:1258(17009), Nov. 15, 1963; RAE-B 53:107, June, 1965; TDB 60:994, Oct., 1963.

Maeda, S., joint author. *See* Oshio, Y., et al., 1965.

4261. Magaudda, P. L., Saccà, G. and **Guarniera, D. 1969** Sterile Male Method Integrated by Insecticides for the Control of *Musca domestica* L. in the Island of Vulcano, Italy. Ann. Ist. Sup. Sanita (5) 29-38.

In the eradication of flies from the island the fly population was first reduced with insecticides, then sterilized male flies were released. A new fly population in competition with sterile males still remained in slow reproduction. ABSTRACT: BA 51:7938(81049), July 15, 1970.

Magaudda, P. L., joint author. *See* Saccà, G., et al., 1967.

4262. Magdiev, R. R., Dremova, V. P., Bykhovskaia, A. M. and **Zueva, E. V. 1960** [Control of synanthropic flies in the city of Kattcuburgan UzSSR by means of the prevention of their breeding in manure.] Med. Parazit. (Mosk.) 29:72-76.

Mager, J., joint author. *See* Zahavi, M., et al., 1968, 1969.

Magrone, R., joint author. *See* Saccà, G., et al., 1964, 1965.

4263. Mailloux, M. 1960 Some Means of Fight Against Flies. Maroc. Med. 39:786-787.

Majori, G., joint author. *See* Grasso, A., et al., 1969.

Majumder, S. K., joint author. *See* Qadri, S. H., et al., 1967, 1968.

Malcovati, M., joint author. *See* DiJeso, F., et al., 1967.

4264. Maldonado Sampedro, M. 1962 La lucha contra las moscas y su resistencia a los insecticidas. [The fight against flies and their resistance to insecticides.] Rev. Sanidad e Hig. Publ. 36(4/6):180-207.

Before attributing resistance to any insects, possible deficiencies of procedure or carelessness should be checked. In one area, fly resistance was reduced significantly by carefully applied sprayings. ABSTRACT: BA 40:1534(20138), Dec. 1, 1962.

4265. Mallis, A. and **Miller, A. C. 1964** Prolonged Resistance in the House Fly and Bed Bug. J. Econ. Ent. (57)4:608-609.

The Renshaw strain of housefly was still resistant to DDT after an isolation of almost 5 years from it and other insecticides. ABSTRACTS: BA 46:1433(17900), Feb. 15, 1965; RAE-B 52:207, Nov., 1964; TDB 61:1261, Dec., 1964.

Mallis, A., joint author. *See* Thayer, H. I., et al., 1965.

Mancilla, J., joint author. *See* Litvak, S., et al., 1967.

Mandel'Baum, Ya A., joint author. *See* Roslavtseva, S. A., et al., 1963.

4266. Mangum, C. L. 1962 Fly Baiting in Honduras. Pest Control 30(4):20.

Sugar DDVP bait applied around dwellings by loose scattering was effective in controlling large fly populations from banana refuse heaps. ABSTRACT: BA 39:968(12275), Aug. 1, 1962.

4267. Mánkowska, H. and **Goszczynska, K. 1969** Efektywność dichlorfosu jako trwalego fumiganta. [Studies on effectiveness of dichlorvos as a residual fumigant insecticide.] Roczn. Panst. Zakl. Hig. 20: 319-328. (In Polish, with Russian and English summaries.)

Vapona strip formulation, containing 20 per cent by weight dichlorvos, was extremely effective under field conditions against houseflies and bed bugs for several months.

Mánkowska, H., joint author. *See* Bojanowska, A., et al., 1966; Goszczynska, K., et al., 1967, 1968, 1970.

4267a. Marak, G. E., Jr., Cornesky, R. A., Gallik, G. J. and **Wolken, J. J. 1969** Insect Visual Pigments. Insect Ophthalmol. 8(4):462.

Treats of houseflies, cockroaches, honey-bees and fireflies.

4268. Marak, G. E., Gallik, C. J. and **Wolken, J. J. 1968** Photosensitive Pigments in House Fly Heads. *Musca domestica.* J. Cell. Biol. 39(2 pt. 2):84A-85A (204).

Pigments isolated from housefly heads appeared to be pteridines or ommochromes; some had spectral properties similar to those of vertebrate rhodopsins.

4269. March, R. B. 1960 Biochemical Aspects of Organophosphorus Resistance. *In*: Symposium on research progress on insecticide resistance. Misc. Publ. Ent. Soc. Am. 2(1):139-144.

Contains many points of interest, specific and general, on the mechanism of organophosphorous resistance. Evidence suggests that it is biochemically limited and relatively unstable. ABSTRACT: BA 36:1020(12267), Feb. 15, 1961.

4270. **March, R. B., Georghiou, G. P., Metcalf, R. L.** and **Printy, G. E. 1964** The Comparative Toxicity of Phosphoramidothionates and Phosphoramidates to Susceptible and Insecticide-resistant Houseflies and Mosquitoes. Bull. W. H. O. 30:71-84. Geneva. (Summary in French.)

2, 4, 5-trichlorophenyl series is the most active; 4-chlorophenyl series the least active, of the groups tested. ABSTRACTS: BA 46:5497(68410), Aug. 1, 1965; RAE-B 54:77, Apr., 1966.

March, R. B., joint author. *See* Georghiou, G. P., et al., 1961, 1963; Fukuto, T. R., et al., 1963.

4271. **Markar'iants, L. A. 1968** [Combination of synanthropic flies with eggs of helminths in Western Pamir.] Med. Parazit. (Moskva) 37:102-103. (In Russian.)

Marks, T. B., joint author. *See* Fine, B. C., et al., 1963, 1967.

4272. **Marsh, P. M. 1969** Two New Species of *Aphaereta* with Notes on Other Nearctic Species (Hymenoptera:Braconidae:Alysiinae). Proc. Ent. Soc. Wash. 71(3):416-420.

Two new species of reared *Aphaereta* are described. One, *A. muesebecki* is from *Fannia canicularis* and *Musca domestica.* ABSTRACTS: BA 51: 2217(22978), Feb. 15, 1970; RAE-B 58:51(215), Feb., 1970.

Martel, J., joint author. *See* Lhoste, J., et al., 1969.

Martin, R. D., joint author. *See* Calvert, C. C., et al., 1969, 1970.

Martinez, E. F., joint author. *See* Carson, N. A., et al., 1967.

Mastrilli, M. L., joint author. *See* Saccà, G., et al., 1966, 1969.

7273. **Mathew, D. L.** and **Dobson, R. C. 1960** *Musca autumnalis* (DeGeer), A New Livestock Pest in Indiana. *In* Proc. Indiana Acad. Sci. (for 1959) 69:165:166.

Taxonomic characters are compared with those of *Musca domestica.* Notes on habits and behavior.

4274. **Mathews, H. B.** and **Hodgson, E. 1966** Naturally Occurring Inhibitor(s) of Microsomal Oxidations from the House Fly. J. Econ. Ent. 59: 1286-1288.

Housefly microsomes are inactive unless dialyzed extensively before use. ABSTRACTS: BA 48:3271(36074), Apr. 1, 1967; RAE-B 55:15(37), Jan., 1967.

4275. **Mathis, W., Flynn, A. D.** and **Schoof, H. F. 1967** Parathion Resistance in House Fly Populations in the Savannah, Georgia Area. J. Econ. Ent. 60:1407-1409.

Exposure of field collected flies showed a loss of susceptibility to parathion and diazinon on farms where none of these had been used. Discusses how it may have been acquired. ABSTRACTS: BA 49:1396(15634), Feb. 1, 1968; RAE-B 56:69(221), Mar., 1968.

4276. **Mathis, W.** and **Schoof, H. F. 1964** Field Test of Dichlorvos, General Chemical 4072, Hooker Compound 1422 and Synergized DDT Against *Musca domestica.* J. Econ. Ent. 57(2):256-258.

Field test results were favorable. Dichlorvos in a resin base retards dissipation in wall deposits. ABSTRACTS: BA 45:6401(80046), Sept. 15, 1964; RAE-B 52:125, July, 1964; TDB 61:850, Aug., 1964.

4277. Mathis, W. and **Schoof, H. F. 1965** Studies on House Fly Control. J. Econ. Ent. 58(2):291-293.
Miscellaneous results are given in several situations using various new chemical preparations. Neither apholate nor dichlorvos-resin strands gave good results. ABSTRACTS: BA 46:5147(63903), July 15, 1965; RAE-B 53:129, July, 1965; TDB 62:942, Sept., 1965.

4278. Mathis, W. and **Schoof, H. F. 1968** Chemical Control of House Flies in Dairy Barns and Chicken Ranches. J. Econ. Ent. 61(4):1071-1073.
Several residual insecticides, cords and dry baits were tested for effectiveness (Anthio, Mobam, Bomyl). ABSTRACTS: BA 50:3706(38901), Apr. 1, 1969; RAE-B 57:5(9), Jan., 1969.

4279. Mathis, W., Schoof, H. F. and **Mullenix, T. L. 1969** Fly Production in Relation to Refuse Disposal in Recreational Areas. J. Econ. Ent 62(6):1288-1291.
Flies were trapped and control measures worked out for the species involved. Dichlorvos dispensers were the most effective means of control, giving 95 per cent or more reduction of all flies.

4280. Mathis, W., Smith, E. A. and **Schoof, H. F. 1970** Use of Air Barriers to Prevent Entrance of House Flies. J. Econ. Ent. 63(1):29-31.
An air barrier with velocities in a range of 1500-2200 feet per minute and a discharge opening of 1, 3, or 5 inches can exclude over 90 per cent of all flies.

Mathis, W., joint author. *See* Schoof, H. F., et al., 1964.

4281. Matolin, S. 1969 The Effect of Chemosterilants on the Embryonic Development of *Musca domestica.* Acta Ent. Bohemoslav 66(2): 65-69. (Summary in Czechoslovakian.)
Apholate, tepa and hempa inhibit embryonic development at the stage of a few cleavage nuclei. The compound, 6-azauridine, stops embryonic development only at advanced organogeny. ABSTRACTS: BA 50:11365 (117536), Nov. 1, 1969; RAE-B 58:340(1397), Oct., 1970.

4282. Matsubara, H. 1961 [On the synergistic effect of natural and synthetic synergists on barthrin.] Botyu-Kagaku 26(4):125-132. (In Japanese, with English summary.)
Tests made with the housefly, *Musca domestica vicina* using a settling dust apparatus, showed that a combination of barthrin with natural and synthetic synergists, results in a lower order of synergism than the similar combination of pyrethrins or allethrin. ABSTRACT: RAE-B 52:73, Apr., 1964.

4283. Matsubara, H., Ito, H., Kawasaki, M., Kai, K. and **Kanamori, S.-I. 1964** [On the synergism of barthrin, dimethrin and naphthyl N-methylcarbamate with pyrethrum synergists against diazinon-resistant and susceptible houseflies (Diptera). Studies on the control of organo phosphorus insecticides-resistant insects. II., Botyu-Kagaku 29(1):1-9. (In Japanese, with English summary.)
Combinations of dimethrin with pyrethrum synergists show a higher order of synergism than similar combinations of barthrin. Sevin alone, or in synergistic combination was very effective against houseflies. ABSTRACT: RAE-B 55:39(145), Feb., 1967.

4284. Matsubara, H. and **Tanimura, R. 1966** [On the utilization of constituents of pepper as an insecticide and pyrethrins or allethrin synergist. Studies on synergist for insecticides. XXIV., Botyu-Kagaku 31(4):162-167. (In Japanese, with English summary.)
Three constituents of *Pipa nigrum* L. have considerable synergistic effect

for pyrethrins, as tested on houseflies. ABSTRACTS: BA 48:7866(88165), Sept. 1, 1967; RAE-B 57:22(95), Jan., 1969.

Matsubara, H., joint author. *See* Ito, H., et al., 1962.

4285. Matsui, M. and **Kitahara, T. 1967** Studies on chrysanthemic acid. XVIII. A new biologically active acid component related to chrysanthemic acid. Agric. Biol. Chem. 31(10):1143-1150.
Tests were made with houseflies using 17 rethronyl esters. ABSTRACTS: BA 49:4282(48189), May 1, 1968; TDB 65:550(1021), Apr., 1968.

4286. Matsumura, F. and **Dauterman, W. C. 1964** Effect of Malathion Analogues on a Malathion-resistant Housefly Strain which Possesses a Detoxication Enzyme, Carboxyesterase. Nature (London) 202: 1356-1358.
Experimental data discussed and interpreted. ABSTRACTS: BA 45:8421 (106986), Dec., 1964; RAE-B 53:139, July, 1965; TDB 61:1103, Oct., 1964.

4287. Matsumura, F. and **Hogendijk, C. J. 1964** The Enzymatic Degradation of Malathion in Organophosphate Resistant and Susceptible Strains of *Musca domestica*. Ent. Exp. Appl. 7(2):179-193.
Resistant strains have a superior ability to degrade malathion to its monocarboxylic acid derivative. ABSTRACTS: BA 46:7363(91595), Oct. 15, 1965; RAE-B 53:139, July, 1965; TDB 61:1260, Dec., 1964.

4288. Matsumura, F. and **Hogendijk, C. J. 1964** The Enzymatic Degradation of Parathion in Organophosphate-susceptible and resistant Houseflies. J. Agric. Fd. Chem. 12:447-453.
Resistant strains have a superior ability to degrade parathion to diethyl phosphorothionate. ABSTRACT: RAE-B 53:149, Aug., 1965.

4289. Matsumura, F. and **Knight, S. G. 1967** Toxicity and Chemosterilizing Activity of Aflatoxin Against Insects. J. Econ. Ent. 60(3):871-872.
This product of *Asperigillus flavus* Link reduced the number of eggs laid and percentage of hatch for *Aedes*, *Musca* and *Drosophila*. ABSTRACTS: BA 48:11213(124541), Dec. 15, 1967; RAE-B 55:192(662), Oct., 1967; TDB 65:80(181), Jan., 1968.

Matsumura, K., joint author. *See* Eto, M., et al., 1966.

Matsunaga, H., joint author. *See* Suzuki, T., et al., 1961; Mizutani, K., et al., 1968.

4290. Matsuo, K. 1962 [Effects of water contents of medium and population density on the larval development of *Musca domestica vicina* and *Sarcophaga peregrina*.] Endemic Dis. Bull. Nagasaki Univ. 4(1): 74-81. (In Japanese, with English summary.)
S. peregrina is less affected by changes in water content and population densities of culture medium than *Musca domestica vicina*. An increase in larval density and water content prolongs the larval plus pupal period for the latter. ABSTRACTS: BA 45:1759(21939), Mar., 1964; RAE-B 52:235, Dec., 1964.

4290a. Matsuo, K. and **Tamura, T. 1967** [Seasonal abundance of flies collected by trap with poison bait.] Jap. J. Sanit. Zool. 18(1):21-26.
Fannia canicularis collected from late February to November; *Muscina stabulans* from early March to early Sept. ABSTRACT: BA 51:10267 (104381), Sept. 15, 1970.

Matsuo, S., joint author. *See* Eto, M., et al., 1963.

4291. Matsuzawa, H. 1961 [On the seasonal prevalence of some important insects from the sanitary view in the house at the paddy field district in Kagawa prefecture.] Tech. Bull. Fac. Agric. Kagawa

Univ. 12(1960)No. 2:255-259. (In Japanese, with English summary.)
The *Musca-Fannia* group were present from spring to autumn with peak in July and August. ABSTRACT: RAE-B 52:23, Jan., 1964.

4292. **Matsuzawa, H. 1962** [Seasonal prevalence of some species of flies appearing in the lavatory (Ecological studies on some important insects from the sanitary view in Shikoku. III).] Tech. Bull. Fac. Agric. Kagawa Univ. 13(2):159-162. (In Japanese, with English summary.)
Musca is less abundant than *Calliphora*, *Sarcophaga* and *Fannia*.

4293. **Matsuzawa, H. 1963** [On the crowd of flies gathering to the body of the fish during the period from spring to summer in Kagawa. (Ecological studies on some important insects from the sanitary view in Shikoku. VI).] Tech. Bull. Fac. Agric. Kagawa Univ. 14(2):141-144. (In Japanese, with English summary.)
Calliphora, *Sarcophaga* and *Lucilia* are dominant. *Musca* and *Fannia* are of lesser importance. ABSTRACT: BA 45:4319(53259), June 15, 1964.

4294. **Matsuzawa, H. 1967** [On the insecticidal effect of the emulsion of BHC, Sevin, and of their mixture (1:1, v/v) on the housefly, *Musca domestica vicina*.] Jap. J. Sanit. Zool. 18(2/3):129-131. (In Japanese, with English summary.)
No synergistic action was shown—only independent joint action. The BHC-emulsion showed a greater effect on adult houseflies than that of the Sevin-emulsion. ABSTRACTS: BA 51:2758(28557), Mar. 1, 1970; RAE-B 58:279(1153), Aug., 1970.

4295. **Matsuzawa, H.** and **Kaneoka, K. 1965** [Resistability to Lindane and Diazionon of the adult of the common house fly, *Musca domestica vicina* (Diptera) collected from various stations in Kagawa-Prefecture (Japan). (Ecological studies on some important insects in Shikoku from sanitary point of view. IX).] Kagawa Daigaku Nogakubu Gakuzyutu Hokoku 16(2):119-126. (In Japanese, with English summary.)
A variation is reported in resistance to lindane and diazinon. The greatest resistance was shown to lindane. ABSTRACT: RAE-B 57:133(474), Aug., 1969.

4296. **Matsuzawa, H.** and **Nakada, H. 1967** [Susceptibility to several kinds of insecticides of the adult house flies reared on different culture media.] Kagawa Digaku Nógakubu Gakuzyutu Hôkoku 18(2): 103-109. (In Japanese, with English summary.)
No difference was shown in the susceptibility to Sumithion, Malathion and Diazinon. In all cases male adults were more susceptible to the five kinds of insecticides than were female adults.

4297. **Matsuzawa, H.** and **Shiozaki, S. 1962** [On the seasonal prevalence of the two species of flies, *Musca domestica vicina* Macquart and *Fannia canicularis* L. in Kagawa Prefecture.] Tech. Bull. Fac. Agric. Kagawa Univ. 14:32-39. (In Japanese, with English summary.)
Adult *Musca domestica vicina* were found from March to November; *Fannia canicularis* were found from April through June. ABSTRACTS: BA 45:4319(53258), June 15, 1964; RAE-B 51:270, Dec., 1963.

4298. **Matsuzawa, H.** and **Yukihiko, F. 1968** [On the sterilizing effect of Hempa on the common housefly, *Musca domestica vicina* (II). A field test in Sei Island.] Jap. J. Sanit. Zool. 19(3):210-212. (In Japanese, with English summary.)
Good control of the housefly in autumn was observed in the treated area. Some problems in technique of application remain to be solved. ABSTRACT: BA 51:2761(28586), Mar. 1, 1970.

4299. Matthew, D. L. and **Dobson, R. C. 1960** *Musca autumnalis* (DeGeer), a New Livestock Pest in Indiana. Proc. Indiana Acad. Sci. (for 1959) 69:165-166.
Description and life history of *Musca autumnalis* (DeGeer), first appearing on this continent in 1952 and at this reporting in a north-central state. ABSTRACT: BA 36:1560(19797), Mar. 15, 1961.

Matthew, D. L., joint author. *See* Dobson, R. C., et al., 1960.

Matthysse, J. G., joint author. *See* Lloyd, J. E., et al., 1966, 1970; Ode, P. E., et al., 1964, 1967; Stoffolano, J. G., et al., 1967; Frishman, A. M., et al., 1966.

4300. Mattson, A. M., Fray, R. W., Gaines, T. B. and **Pearce, G. W. 1960** Preparation and Biological Activity of a Series of Halogenated Ethyl and Vinyl Dimethyl Phosphate Esters. J. Agric. Fd. Chem. 8(3):196-198.
Tests showed this series of compounds to be effective against houseflies and of *low* toxicity to white rats. ABSTRACT: RAE-B 49:2, Jan., 1961.

4301. Mauri, R. A. 1963 Sobre dos ácaros entomofagos. [Notes on two entomophagous mites.] Physis Rev. Assoc. Argent. Cienc. Natur. 24(67): 159-160.
Two species of mites associated with insects are recorded. One is *Macrocheles muscaedomesticae*, which parasitizes flies. ABSTRACT: BA 50:1555 (16342), Feb. 1, 1969.

4302. Mayer, M. S. and **Thaggard, C. W. 1966** Investigations of an Olfactory Attractant Specific for Males of the Housefly, *Musca domestica.* J. Insect Physiol. 12(8):891-897.
An attractant was found associated with live virgin females and contaminated holding cylinders, also dead and mated females. ABSTRACTS: BA 47:9497(109927), Nov. 15, 1966; RAE-B 55:232(797), Dec., 17;96 TDB 64:216, Feb., 1967.

Mazel, P., joint author. *See* Shargel, L., et al., 1969.

4303. Mazokhin-Porshnyakov, G. A. and **Vishnevskaya, T. M. 1966** "Krasnyi" svetopriemnik mukh i tsvetovoe *Drosophila melanogaster.* [Red light receiver of flies and the color vision of *Drosophila melanogaster.*] Biofizika 11(6):1034-1041.
Musca domestica and *Calliphora* also gave evidence of true color vision. Two light receptors were found in the three species. The "red light" receptor may be of a filtering nature. ABSTRACT: BA 48:9252(104037), Oct. 15, 1967.

4304. Mazzotti, L. 1967 Casos humanos de maisis intestinal. [A case of human intestinal myiasis.] Ciencia (Mex.) 25(5):167-168. (English summary.)
The patient vomited food with larvae of *Musca domestica.* ABSTRACT: BA 50:496(5098), Jan. 1, 1969.

Medley, J. G., joint author. *See* Jones, C. M., et al., 1963.

4305. Mehrotra, K. N. 1961 Properties of Choline Acetylase from the House Fly *Musca domestica* L. J. Insect Physiol. 6(3):215-221.
The preparation of choline acetylase from the heads of houseflies is described. Various chemical properties are summarized. ABSTRACT: BA 36:75736(80951), Dec. 1, 1961.

4306. Mehrotra, K. N. and **Dauterman, W. C. 1963** The N-alkyl Group Specificity of Choline Acetylase from the Housefly *Musca domestica* L. and the Two-spotted Spider Mite, *Tetranychus telarius* L. J. Insect. Physiol. 9:293-298.
Twenty-one analogues were tested as substrates for choline acetylase.

Mehrotra, K. N., joint author. *See* Dauterman, W. C., et al., 1963.

4307. **Meifert, D. W., Fye, R. L.** and **LaBrecque, G. C. 1963** Effect on House Flies of Exposure to Residual Applications of Chemosterilants. Florida Ent. 46:161-168.
Tarsal contact was used to sterilize both sexes of *Musca domestica.* ABSTRACTS: BA 45:1754(21878), Mar. 1, 1964; RAE-B 53:113, June, 1965.

4308. **Meifert, D. W., LaBrecque, G. C.** and **Rye, J. R., Jr. 1969** House Fly, *Musca domestica*, Control with Chemosterilants and Insecticides. Florida Ent. 52(2):55-60.
Control of flies in poultry-caged-layer operation was achieved by several baits and a dropping treatment. In a pig-rearing establishment, the weekly release of sterile flies and larvicide treatments reduced the fly population at least 5-fold. ABSTRACT: BA 50:11362(117512), Nov. 1, 1969.

4309. **Meifert, D. W., LaBrecque, G. C., Smith, C. N.** and **Morgan, P. B. 1967** Control of House Flies on Some West Indies Islands with Metepa, Apholate, and Trichlorfon Baits. J. Econ. Ent. 60:480-485.
Flies did not develop resistance to chemosterilants during two years of exposure to them. Abundance reduced 50-90 per cent, depending upon local conditions. ABSTRACTS: RAE-B 55:144(511), Aug., 1967; TDB 64:1273, Nov., 1967.

4310. **Meifert, D. W., Morgan, P. B.** and **LaBrecque, G. C. 1967** Infertility Induced in Male House Flies by Sterilant-Bearing Females. J. Econ. Ent. 60:1336-1338.
The method was not considered practical. ABSTRACTS: BA 49:1400 (15676), Feb. 1, 1968; RAE-B 56:65(213), Mar., 1968; TDB 65:542 (1181), Apr., 1968.

Meifert, D. W., joint author. *See* Bailey, D. L., et al., 1968; Brady, U. E., et al., 1966; Morgan, P. B., et al., 1967; LaBrecque, G. C., et al., 1970.

Meisner, J., joint author. *See* Ascher, K. R. S., et al., 1967, 1968.

4311. **Meister, G. 1962** Biologische Beobachtungen bei der Laboratoriumszucht von *Musca domestica* L. [Biological observations in the laboratory cultivation of *Musca domestica* L. (A contribution to the problem of the standardization of cultivation methods).] Zeit. Tropenmed. Parasit. 13:102-133. (In German, with English summary.)
Housefly strains exhibiting similar characteristics cannot be produced within a reasonably short time from field specimens. ABSTRACTS: BA 39:1323 (16735), Aug. 15, 1962; RAE-B 52:46, Mar., 1964; TDB 59:1118, Nov., 1962.

4312. **Meister, G. 1962** Beobachtungen bei der Testung von Zuchtfliegen, *Musca domestica* L., mit Insektiziden (Ein Beitrag zur Frage der Standardisierung von Testmethoden). [Observations on the testing of *Musca domestica* L. (in the breeding period) with insecticides. (Contribution to the problem of standardization of test methods).] Zeit. Tropenmed. Parasit. 13(2):220-238. (In German, with English summary.)
The resistance of the housefly to insecticides with special reference to "Dipterex" was examined. A modified glass tube test was used and is recommended as standard procedure. ABSTRACTS: BA 41:610(8165), Jan. 15, 1963; RAE-B 52:47, Mar., 1964; TDB 60:77, Jan., 1963.

Mekhanik, M. L., joint author. *See* Ermokhina, T. M., et al., 1966.

4313. Meksongsee, B., Yang, R. S. and Guthrie, F. E. 1967 Effect of Inhibitors and Inducers of Microsomal Enzymes on the Toxicity of Carbamate Insecticides to Mice and Insects. J. Econ. Ent. 60(5):1469-1471.
Treating adult *M. domestica* (DDT-resistant strain) with DDT or phenobarbital before treatment with Zectran, dimetilan or VC20047A did not decrease the toxicity of the carbamates. Other experiments reported. ABSTRACTS: RAE-B 56:70(227), Mar., 1968.

4314. Mello, D., Mello, E. J. R. and Pigatti, A. 1961 Estudos sõbre uma colônia de moscas domésticas multiplo-resistentes a insecticidas no municepio des cosmôpolis, São Paulo. [Laboratory studies on an insecticide multi-resistant strain of houseflies from Cosmopolis.] Arq. Inst. Biol. 28:63-70.
Flies proved to be resistant to DDT, *gamma* BHC, Dieldrin and Sevin, susceptible to Diazinon, Malathion and Dipterex. ABSTRACTS: BA 40: 1609(21038), Dec. 1, 1962; RAE-B 52:85, May, 1964.

Mello, D., joint author. *See* Mello, E. J. R., et al., 1961, 1962; Queiroz, J. C., et al., 1962.

4315. Mello, E. J. R., Mello, D., Pigatti, A. and Queiroz, J. C. 1961 Tolerância nas condições de Laboratõrio, das môscas domésticas do estado de São Paulo aos insecticidas orgânicos (1961). [Survey of house fly (*Musca domestica* L.) tolerance to several organic insecticides, in the state of São Paulo (Brazil) 1961., I. House Flies from Itapetininga, Guararema, Nova Odessa, Campinas, Braganca Paulista, and Aguai.] Arq. Inst. Biol. (São Paulo) 28:119-125. (English summary.)
Flies generally were tolerant to DDT and Lindane, and susceptible to organophosphorous compounds. ABSTRACTS: BA 40:1608(21039), Dec., 1962; RAE-B 52:86, May, 1964.

4316. Mello, E. J. R., Mello, D. and Queiroz, J. C. 1962 Ação do ronnel sôbre môscas domésticas resistentes ao DDT. [Effectiveness of ronnel against a DDT-resistant strain of houseflies.] Arq. Inst. Biol. (São Paulo) 29:109-115. (English summary.)
Flies are not as susceptible to ronnel as to diazinon. ABSTRACTS: BA 45:1413(17565), Feb. 15, 1964; RAE-B 52:85, May, 1964.

4317. Mello, E. J. R. and Pigatti, A. 1961 Resistencia da *Musca domestica* (L.) e das larvas do *Culex pipiens fatigans* (Wied.) ao DDT e ao isômero *gama* do BHC, em São Paulo. [Resistance of *M. domestica* L. and the larvae of *Culex pipiens fatigans* (Wied.) to DDT and the *gamma* isomer of BHC in São Paulo.] Arq. Inst. Biol. 28:25-34. (English summary.)
The authors' data are compared with those of LePage, et al. (1945). Resistance to *gamma* BHC estimated at 40-fold; to DDT, at 20-fold, over 1945 findings. ABSTRACT: RAE-B 52:84, May, 1964; BA 40:1609 (21040), Dec. 1, 1962. (Gives epagination as 23-24.)

4318. Duplicate reference. Withdrawn.

Mello, E. J. R., joint author. *See* Mello, D., et al., 1961; Pigatti, A., et al., 1961; Queiroz, J. C., et al., 1962.

4319. Meltzer, J. and Welle, H. B. A. 1967 Insecticidal Activity of Phenyl Carbamates. Abstr. Int. Pflanzenschutz-Kong. Int. Plant Prot. Congr. 30:188-189.
Musca domestica and four other species were used in the tests. Variations in size, number and position of ring substituents were found to be important.

4320. Meltzer, J. and **Welle, H. B. A. 1969** Insecticidal Activity of Substituted Phenyl N-methylcarbamates. Ent. Exp. Appl. 12(2):169-182.
Responses of the test insects indicate that various carbamates exhibit selective properties. For the housefly, meta-isomers were the more active. ABSTRACT: BA 51:5642(57896), May 15, 1970.

Mendoza, C. E., joint author. *See* Travis, B. V., et al., 1967.

Menez, C. F., joint author. *See* Tacal, J. V., Jr., et al., 1967.

4321. Meng, L., Chang, Y. and **Chang, C. P. 1965** [Studies on the embryonic development of the common housefly, *Musca domestica vicina*, Macq.] Acta Zool. Sinica 17(2):157-163. (In Chinese, with English summary.)
Early embryonic development of the housefly is described from the formation of the polar body and fusion of pronuclei to the foundation of the blastoderm. ABSTRACT: BA 49:8744(97354), Sept. 15, 1968.

4322. Mengle, D. C. and **Casida, J. E. 1960** Biochemical Factors in the Acquired Resistance of Houseflies to Organophosphate Insecticides. J. Agric. Fd. Chem. 8(6):431-437. Easton, Pa.
Metabolic fate of radiolabeled Diazinon, malathion and methyl parathion was studied in three organophosphate-resistant strains. Evidence indicated that the cholinesterase inhibition rate was reduced in the abdomen and thorax. This contributes to resistance. ABSTRACT: RAE-B 50:75, Apr., 1962.

4323. Mengle, D. C. and **O'Brian, R. D. 1960** The Spontaneous and Induced Recovery of Fly-Brain Cholinesterase after Inhibition by Organophosphates. Biochem. J. 75(1):201-2071.
In vivo and *in vitro* studies were made with six organophosphates inhibiting fly brain cholinesterase. Recovery of flies and reactivation rate depends upon the alkoxy substituents of the organophosphate. ABSTRACT: TDB 57:762, July, 1960.

4324. Menn, J. J. and **Szabo, K. 1965** The Synthesis and Biological Properties of new O-alkyl S-aryl Alkylphonodithioates. J. Econ. Ent. 58(4):734-739; Errata (5):1037.
Toxicity apparently is related to the carbon-to-phosphorus bond of a given alkyl group. Two strains of multiresistant house flies were used in the tests. ABSTRACTS: BA 47:2508(29970), Mar. 16, 1966; RAE-B 53:224, Nov., 1965.

Menn, J. J., joint author. *See* Sherman, M., et al., 1965, 1967; Szabo, K., et al., 1969.

4325. Menon, M. 1965 Endocrine Influences of Yolk Deposition in Insects. J. Animal Morph. Physiol. 12(1):76-80.
Refers to work of Bhaskaran (1962) on *Musca* in which allatectomy resulted in a rise in fatty acid. Ovariectomy did not interfere with fat depletion (Menon worked with cockroaches removing the corpora allata.) ABSTRACT: BA 48:4191(46714), May 1, 1967.

Mensink, F. T., joint author. *See* VanDinther, J. B. M., et al., 1965.

Mento, G., joint author. *See* Saccà, G., et al., 1968.

4326. Menzel, D. B., Craig, R. and **Hoskins, W. M. 1963** Electrophoretic Properties of Esterases from Susceptible and Resistant Strains of the Housefly, *Musca domestica* L. J. Insect Physiol. 9(4):479-493.
Detailed studies tend to support the theory that resistance to organophosphorus insecticides (e.g. malathion) is related to higher phosphatase activity. ABSTRACT: RAE-B 52:179, Oct., 1964.

4327. Mer, G. G. and **Paz, M. 1960** Control of House-flies in Rural Areas by Means of Traps. Riv. di Parassit. (Rome) 21(2):143-150. (Italian

summary.)
Used in sufficient numbers and well placed, traps reduced fly populations to a considerable extent. Form chiefly concerned was *M. domestica vicina*. ABSTRACTS: RAE-B 51:103, May, 1963; TDB 58:143, Jan., 1961.

Messina, F., joint author. *See* Ercolini, A., et al., 1967.

Metcalf, E. R., joint author. *See* Metcalf, R. L., et al., 1966.

4328. Metcalf, R.L. 1967 Mode of Action of Insecticide Synergists. Ann. Rev. Ent. 12:229-256.
Includes sections on pyrethroids, carbamates, organophosphates and DDT. Many references to work carried out with houseflies.

4329. Metcalf, R. L. and **Frederickson, M. 1965** Selective Insecticidal Action of Isopropyl Parathion and Analogues. J. Econ. Ent. 58:143-147.
Isopropyl parathion was far less toxic to the bee than to the fly, although the dimethyl, diethyl and dipropyl analogues were nearly equitoxic to both species. Reasons are given. ABSTRACT: BA 46:4778(59145), July 1, 1965.

4330. Metcalf, R. L., Fuertes-Polo, C. and **Fukuto, T. R. 1963** Carbamate Insecticides: Multisubstituted Chlor and Methyl-phenyl N-methylcarbamates. J. Econ. Ent. 56:862-864.
Chloro-substituted compounds had reduced activity. With dimethylphenyl methylcarbamates, maximum biological activity was associated with substitution in meta position on phenyl ring, most active being 2,3-, 3,4- and 3.5-dimethylphenyl compounds. Detoxification patterns discussed. *M. domestica* one of 3 species studied. ABSTRACT: RAE-B 52:54, Mar., 1964.

4331. Metcalf, R. L. and **Fukuto, T. R. 1960**. 0-Ethyl S-2-(ethylthio)-ethyl Alkylphosphonothioates as Systemic Insecticides. J. Econ. Ent. 53(1):127-130.
Compounds tested topically against *M. domestica* and systemically against insects and mites on leaves. Contact toxicity of 0-ethyl S-2-(ethylthio)ethyl. . . . found to decrease as alkyl chain increased from methyl to iso-pentyl, varying with polar substituent constant. ABSTRACT: RAE-B 49:83, Apr., 1961.

4332. Metcalf, R. L. and **Fukuto, T. R. 1961** Toxicity of Diisopropyl 1, 2, 2, 2-Tetrachloroethyl Phosphate and its Vinyl Analogue to Resistant Houseflies. Bull. W. H. O. 24:670-672.
A toxic impurity was believed present, but due to the over-all low degree of toxicity shown to *Musca domestica*, it was not practical for it to be determined. ABSTRACT: RAE-B 51:118, June, 1963.

4333. Metcalf, R. L. and **Fukuto, T. R. 1962** Meta-sulfurpentafluorophenyl diethyl Phosphate and Meta-sulfurpentafluorophenyl N-methylcarbamate as Insecticides and Anticholinesterases. J. Econ. Ent. 55:340-341.
Tested by topical application to adult females of *M. domestica*. Activity of compounds in excellent agreement with electron withdrawing ability of their substituents as expressed by Hammett's *sigma* constant. Diethyl m-sulphurpentafluorophenyl phosphate a good toxic and anticholinesterase agent. ABSTRACT: RAE-B 50:270, Dec., 1962.

4334. Metcalf, R. L. and **Fukuto, T. R. 1965** Silicon-containing Carbamate Insecticides. J. Econ. Ent. 58:1151.
Of the *o*-, *m*- and *p*-isomers of trimethylsilylphenyl methylcarbamate, the *m*- is the most active inhibitor of fly head cholinesterase. Against topically treated houseflies the *o*-isomer is most toxic when applied alone. ABSTRACT: RAE-B 54:60, Mar., 1966.

4335. **Metcalf, R. L.** and **Fukuto, T. R. 1965** Effects of Chemical Structure on Intoxication and Detoxication of Phenyl N-Methylcarbamates in Insects. J. Agric. Fd. Chem. 13:220-231.
A review of fifteen years of research by the authors on this group of insecticides.

4336. **Metcalf, R. L.** and **Fukuto, T. R. 1967** Some Effects of Molecular Structure upon Anticholinesterase and Insecticidal Activity of Substituted Phenyl-N-methyl-carbamates. J. Agric. Fd. Chem. 15(6):1022-1029.
The results of investigations are discussed in terms of the competition of the carbamate with acetylcholine for interaction with the esteratic and anionic sites of cholinesterase, and their detoxication by phenolase enzymes. (This action can be inhibited by piperonyl butoxide.) ABSTRACT: RAE-B 56:216(789), Nov., 1968.

4337. **Metcalf, R. L.** and **Fukuto, T. R. 1968** The Comparative Toxicity of DDT and Analogues to Susceptible and Resistant Houseflies and Mosquitoes. Bull. W. H. O. 38(4):633-647. (French summary.)
120 DDT analogues, were tested. Correlation of structure, DDT-like activity, and resistant ratios gives information on the DDT mode of action and the nature of resistance to it. ABSTRACTS: BA 51:571(5745), Jan. 1, 1970; RAE-B 58:142(593), May, 1970; TDB 66:94(143), Jan., 1969.

4338. **Metcalf, R. L., Fukuto, T. R., Collins, C., Borck, K., Burk, J., Reynolds, H. T.** and **Osman, M. F. 1966** Metabolism of 2-methyl-2-(Methylthio)-Propionaldehyde 0-(Methylcarbamoyl)-Oxime in Plant and Insect. J. Agric. Fd. Chem. 14(6):579-584.
Temik (registered tradename for the above mentioned carbamate insecticide) is readily and completely oxidized to its sulfoxide which is the active metabolite and a cholinesterase inhibitor. Metabolism in the housefly followed a pattern similar to that in cotton plants. ABSTRACT: BA 49:6539(72996), July 15, 1968.

4339. **Metcalf, R. L., Fukuto, T. R.** and **Frederickson, M. 1964** Para-substituted Metaxylenyl Diethyl Phosphates and N-methylcarbamates as Anti-cholinesterases and Insecticides. J. Agric. Fd. Chem. 12(3): 231-236. Easton, Pa.
The activities of the compounds varied, depending upon the mesomeric effects of the para substituent and the steric interference of the 3,5-methyl groups with the resonance of the para substituent. Biological behavior was compared with that of para-substituted phenyl esters and the results are discussed in terms of physical-organic theory. ABSTRACT: RAE-B 52:128, July, 1964.

4340. **Metcalf, R. L., Fukuto, T. R., Frederickson, M.** and **Peak, L. 1965** Insecticidal Activity of Alkylthiophenyl N-methylcarbamates. J. Agric. Fd. Chem. 13(5):473-477. Easton, Pa.
Thirty substituted compounds were compared for anti-cholinesterase activity and toxicity to insects, including *Musca domestica*. High toxicity was shown by some of the thioether carbamates. ABSTRACT: RAE-B 54:128, July, 1966.

4341. **Metcalf, R. L., Fukuto, T. R., Wilkinson, C., Fahmy, M. H., El-Aziz, A.** and **Metcalf, E. R. 1966** Mode of Action of Carbamate Synergists. J. Agric. Fd. Chem. 14(6):555-562.
The insecticidal carbamates are synergized by a wide variety of methylenedioxyphenyl compounds which act as inhibitors of phenolase enzymes. Tyrosinase, abundant in the housefly, serves as a model enzyme for the study of the kinetics of this interaction. ABSTRACT: RAE-B 55:90(319), May, 1967.

4342. **Metcalf, R. L., Fukuto, T. R.** and **Winton, M. Y. 1960** Alkoxyphenyl N-methylcarbamates as Insecticides. J. Econ. Ent. 53(5):828-832.
Table gives toxic properties of 24 compounds. However, the 2-isopropxy-5-methoxyphenyl N-methylcarbamate, more recently prepared, was found to be much more toxic to *M. domestica* than the 3,5-dimethoxyphenyl compound, the most effective of those dealt with in this paper. ABSTRACT: RAE-B 49:182, Aug., 1961.

4343. **Metcalf, R. L., Fukuto, T. R.** and **Winton, M. Y. 1962** Insecticidal Carbamates: Position Isomerism in Relation to Activity of Substituted Phenyl N-methylcarbamates. J. Econ. Ent. 55(6):889-894.
Results of the testing of 49 compounds for anticholinesterase and insecticidal activity. Tables and discussion. Maximum activity seen when structures exhibited closest complementariness to the active site of cholinesterase. ABSTRACT: RAE-B 51:58, Mar., 1963.

4344. **Metcalf, R. L., Fukuto, T. R.** and **Winton, M. Y. 1962** Insecticidal Carbamates: Comparison of the Activities of N-methyl- and N,N-dimethyl-carbamates of Various Phenols. J. Econ. Ent. 55:345-347. June.
Direct comparison of the biological activity of methylcarbamates and dimethylcarbamates of eight active aryl groups showed that the former were 4-40 times as effective as the latter in inhibiting fly-brain cholinesterase, and were mostly more toxic than the latter to adult females of *M. domestica*. ABSTRACTS: RAE-B 50:271, Dec., 1962; TDB 59:1213, Dec., 1962.

4345. **Metcalf, R. L., Fukuto, T. R.** and **Winton, M. Y. 1963** Chemical and Biological Behaviour of Fenthion Residues. Bull. W. H. O. 29(2): 219-226. (French summary.)
The residues are oxidized to sulfoxide and sulfone derivatives which are insecticidal to flies as very active anticholinesterases. ABSTRACTS: BA 45:2481(30793), Apr. 1, 1964; RAE-B 53:201, Oct., 1965.

4346. **Metcalf, R. L.** and **Georghiou, G. P. 1962** Cross Tolerances of Dieldrin-resistant Flies and Mosquitoes to Various Cyclodiene Insecticides. Bull. W. H. O. 27:251-256.
Nineteen cyclodiene insecticides were studied, including heptaclor, telodrin and others. Telodrin showed the least cross resistance and was quite effective against dieldrin-resistant house flies.

4347. **Metcalf, R. L., Osman, M. F.** and **Fukuto, T. R. 1967** Metabolism of C^{14} Labeled Carbamate Insecticides to C^{14} 0_2 in the House Fly. J. Econ. Ent. 60:445-450.
Four processes were shown to be concerned with detoxication. ABSTRACTS: BA 48:6512(72716), July 15, 1967; RAE-B 55:143(509), Aug., 1967.

Metcalf, R. L., joint author. *See* Bakry, N., et al., 1968; El-Aziz, S. A., et al., 1969; El-Sebae, A. H., et al., 1964; Fahmy, M. A. H., et al., 1966; Fukuto, T. R., et al., 1962, 1963, 1964, 1967; Georghiou, G. P., et al., 1961, 1962, 1965; Hollingworth, R. M., et al., 1967; Jones, R. L., et al., 1969; March, R. B., et al., 1964; Quistad, G. B., et al., 1970; Sacher, R. M., et al., 1969; Wilkinson, C., et al., 1966; Williamson, R. L., et al., 1967.

Metivier, J., joint author. *See* Desmoras, J., et al., 1964.

4348. **Mialo, Il. 1961** [Polychloropinene in fly control.] Med. Parazit. (Moskva) 30:616-617. (In Russian.)

Mibutani, K., joint author. *See* Suzuki, T., et al., 1968.

4349. Michaeli, D. 1963 Biochemical Factors in Cross-resistance to DDT in Houseflies Resistant to Other Insecticides. Diss. Absts. 24:2253.

4350. Michelsen, A. 1960 Experiments on the Period of Maturation of the Male House-fly, *Musca domestica* L. Oikos 11(2):250-254.
At 28°C, male reaches sexual maturity 18-27 hours after leaving puparium. Over a temperature range of 34° to 15° time for maturation varies from 24 hours to 60 hours. High temperatures for more than 24 hours tend to sterilize.

Miggiano, V., joint author. *See* Greenberg, B., et al., 1963.

4351. Mihályi, F. 1965 Rearing Flies from Faeces of Meat, Infected under Natural Condition. Acta Zool. Hung. 11(1/2):153-164.
No *Musca domestica* L. issued from bait set in traps for feeding and egg laying. ABSTRACT: RAE-B 55:21, Feb., 1967.

4352. Mihályi, F. 1966 Flies Visiting Fruit and Meat in an Open Air Market in Budapest. Acta Zool. Acad. Sci. Hung. 12(3/4):331-337.
Musca domestica made up only 0·3 per cent of flies. The summary claims that this species rarely visits feces! ABSTRACT: BA 48:4167(46411), Mar. 1, 1967.

4353. Mihályi, F. 1967 Seasonal Distribution of the Synanthropic Flies in Hungary. Ann. Hist.-Nat. Mus. Natn. Hung. 59:327-344.
Most species and specimens caught in traps were from the Muscidae. *Musca domestica* was greatly diminished in numbers due to insecticides, season and climate of the collecting period.

4354. Mihályi, F. 1967 Separating Rural and Urban Synanthropic Fly Faunas. Acta Zool. Acad. Sci. Hung. 13(3/4):379-383. (Russian summary.)
Flies were separated on the basis of breeding in feces of man or of animals. Presence and development of *Musca domestica* around dung practically nil. ABSTRACT: BA 49:3808(42782), Apr. 15, 1968.

4355. Mihályi, F. 1967 The Danger-Index of the Synanthropic Flies. Acta Zool. Acad. Sci. Hung. 12(3/4):373-377.
Using several species, including *Musca domestica*, a mathematical equation was proposed that should allow for the calculation of the grade of "dangerousness", or the probable share in the transmission of enteric disease.

4356. Milani, R. 1961 Results of Genetics Research on *Musca domestica* L. Atti Ass. Genet. Ital. 6:427-438.
Many genes identified and illustrated. It is believed that some recessive genes are present in the heterozygous condition with high frequencies, over the greater part of the geographic distribution of the species.

4357. Milani, R. 1962 Observations on Intra-specific Differentiation, Genetic Variability, Sex-limited Inheritance, DDT-resistance and Aspects of Sexual Behaviour in *Musca domestica* L. Symp. Genet. et Biol. Ital. 9:312-327.

4358. Milani, R. 1963 Genetical Aspects of Insecticide Resistance. *In* Bull. W. H. O., 29(suppl.), pp. 77-87.
Chiefly a summary. 70 strains (of 22 species) have been studied genetically from three lines of approach: (1) inheritance of particular forms of resistance; (2) genetical aspects of cross-resistance or multiple resistance and, (3) genetics of biochemical properties causing or related to resistance. ABSTRACT: TDB 61:442, Apr., 1964.

4359. Milani, R. 1964 Citologia della mosca domestica (*Musca domestica* L.) [Cytology of the house fly, *Musca domestica* L.] Relazione Scientifica del C. N. R. sull'atività 1962-1963. La Ricerca Scientifica, (Quaderni), 25:109-116.

Studies relate primarily to chromosomes.

4360. Milani, R. 1964 Recenti sviluppi delle ricerche genetiche sulla resistenza agli insetticidi. [Recent developments in the genetic research on resistance to insecticides.] Genetica Agraria 18(3):555-556.

4361. Milani, R. 1965 The Genetics of the Housefly. Proc. XII Int. Congr. Ent. London. pp. 276-277.

A brief summing up of recent work by the author and others. Contains a bibliography of 16 titles, the earliest having been published in 1954.

4362. Milani, R. 1967 The Genetics of *Musca domestica* and of other Muscoid flies. *In* Genetics of Insect Vectors of Disease. Amsterdam. Elsevier Publ. Co. pp. 315-369.

Reviews in detail genetics of *M. domestica* L. (in addition to *M. d. domestica*, *M. d. nebulo* F., *M. d. vicina* Macq., *M. d. curviforceps* Saccà and Rivosecchi and *M. d. calleva* Wlk.). Includes geographical and ecological differentiation, chromosome complement, and described mutants. ABSTRACT: RAE-B 56:157(551), Aug., 1968.

4363. Milani, R. and **Franco, M. G. 1960** Modalità di ricombinazione del mutante *ac* di *Musca domestica* L. con tre marcatori e osservazioni sulle segregazioni mendeliane. Symposia Genetica et Biologica Italica. 7:59-74.

4364. Milani, R. and **Franco, M. G. 1960** Morphometric Observations on a Field Sample of *Musca domestica* L. Symp. Genet. et Biol. Ital. 7:201-216.

The field sample was from a remote Canadian locality. Some morphological and color pattern differences were noted after 23 generations of selection with DDT. Sexually abnormal flies were extremely rare.

4365. Milani, R. and **Franco, M. G. 1961** Il terzo gruppo di concatenazione di *Musca domestica*. [The third linkage group of *Musca domestica*.] Genetica Agraria 14(3-4):288-296.

The test-cross of males heterozygous for the two markers *ac* and *ctc* with double homozygous females gives progeny of only two phenotypes—(those of the original P_1). Test-cross of heterozygous females gives 4 phenotypic classes, with some 20 per cent recombination.

4366. Milani, R. and **Franco, M. G. 1965** Holandric Inheritance in *Musca domestica* L. Proc. Symp. on the Mutational Process. Prague. August 9-11. pp. 85-88.

Cites Kerr (1960, 1961). A strain in which males are approximately 8 times as resistant to DDT as females shows true holandric inheritance, the resistance being transmitted only by males to all their male progeny. Discussion of many genes and their inheritance.

4367. Milani, R., Rubini, P. G. and **Franco, M. G. 1967** Osservazioni sulla determinazione del sesso in *Musca domestica* L. [Observations on sex determination in *Musca domestica* L.] Atti Ass. Genet. Ital. 12: 372-373.

In some atypical strains of *Musca* of American, African or Asiatic origin, both sexes have two heterochromatic chromosomes similar to X. A strain in which both sexes are XY, and one in which heterochromosomes have been increased to 6X in females and 5X1Y in males, have been established. ABSTRACT: BA 49:11383(124935), Dec. 15, 1968.

4368. **Milani, R., Rubini, P. G.** and **Franco, M. G. 1967** Determinazione del sesso nella mosca domestica. [Sex determination in the housefly., Genetica Agraria 21:385-411.
Factors of standard and atypical housefly strains for determining sex, were isolated and variously recombined. Four types of strains, 2 natural and 2 synthesized, were kept in stock.

Milani, R., joint author. *See* Franco, M. G., et al., 1966; Sawicki, R. M., et al., 1966.

Milburn, N. S., joint author. *See* Simon, J., et al., 1968.

4369. **Miles, J. W., Pearce, G. W.** and **Woehst, J. E. 1962** Stable Formulations for Sustained Release of DDVP. J. Agric. Fd. Chem. 10(3): 240-244.
DDVP, formulated with a number of anhydrous, hydrophobic materials to produce a series of solid products containing up to 60 per cent of the toxicant, is effective in controlling mosquitoes and houseflies. ABSTRACT: RAE-B 50:239, Nov., 1962.

Miles, J. W., joint author. *See* Kilpatrick, J. W., et al., 1962; Miles, J. W., et al., 1967.

4370. **Millar, E. S. 1965** *Bacillus thuringiensis* in the Control of Flies Breeding in the Droppings of Caged Hens. New Zealand J. Agr. Res. 8(3):721-722.
Bacteria seemed more effective when applied to droppings than when included in food. ABSTRACT: BA 47:1197(14473), Feb. 1, 1966.

4370a. **Millar, L. A.** and **Hooper, C. H. S. 1969** Resistance to Insecticides of three Strains of *Musca domestica* L. from Brisbane, Queensland. J. Australian Ent. Soc. 8:113-115.
Three strains, established from field collected adults, displayed high resistance to DDT and dieldrin; lower levels of resistance to Malathion and Lindane. Were susceptible to diazinon.

Millay, R. H., joint author. *See* Kumar, S. S., et al., 1970.

Miller, A. C., joint author. *See* Mallis, A., et al., 1964.

4371. **Miller, B. F.** and **Shaw, J. H. 1969** Digestion of Poultry Manure by Diptera. Poultry Sci. 48(5):1844-1845.
Concerns *M. domestica*; photo-negative response.

Miller, E. M., joint author. *See* Yendol, W. G., et al., 1967, 1968.

4371a. **Miller, R. W., Gordon, C. H., Bowman, M. C., Beroza, M.** and **Morgan, N. O. 1970** Gardona as a Feed Additive for Control of Fly Larvae in Cow Manure. J. Econ. Ent. 63(5):1420-1423.
In a series of 3 trials, this substance, at concentrations fed, killed 94 per cent or more larvae of *M. domestica*, seeded on to feces. At 2 lower concentrations no Gardona appeared in the milk. ABSTRACT: BA 52: 2000(19095), Feb. 15, 1971.

4372. **Miller, R. W., Gordon, C. H., Morgan, N. O., Bowman, M. C.** and **Beroza, M. 1970** Coumaphos as a Feed Additive for the Control of House Fly Larvae in Cow Manure. J. Econ. Ent. 63(3):853-855.
Increasing the concentrations of coumaphous in rations of cattle increased the mortality of the 1st stage housefly larvae seeded into manure of these cattle. Residues did not appear in milk nor did milk production or feed intake become affected.

Miller, R. W., joint author. *See* Bowman, M. C., et al., 1968.

4373. **Miller, S. 1969** One-carbon Metabolism in the Developing Ovary of the Housefly *Musca domestica*. J. Insect Physiol. 15(12):2263-2271.
Adult female houseflies incorporate 1-carbon units from radioactive

formate, glycine and serine into purines of ovarian RNA. ABSTRACT: BA 51:3944(40720), Apr. 1, 1970.

4374. **Miller, S. 1969** Tetrahydrofolate Cofactors in the Housefly *Musca domestica*. Comp. Biochem. Physiol. 30(5):955-963.
Three tetrahydrofolate cofactors were separated and measured by microbial assays from a chromatographed, dialyzed aqueous extract of houseflies. The major cofactor was N^{10}-formyl-THFA (tetrahydrofolic acid). ABSTRACT: BA 51:5082(52349), May 1, 1970.

4374a. **Miller, S., and Collins, J. M. 1970** DNA Synthesis in the Developing Ovary of the House Fly. Comp. Biochem. Physiol. 36:559-567.
A method is described for the isolation of DNA from housefly ovaries where small quantities of the nucleic acid are present. ABSTRACT: BA 52:2137(20445), Feb. 15, 1971.

4375. **Miller, S.** and **Perry, A. S. 1965** Isolation, Purification and Characterization of N^5-Formyltetra-hydrofolic Acid (Folinic acid) from the Housefly. Life Sci. 4:1573-1580.
Procedural techniques are given, together with considerable explanatory information. Was first research study undertaken for this purpose; hence listed here, rather than in Section VII. ABSTRACT: BA 47:6403 (74920), Aug. 1, 1966.

4376. **Miller, S.** and **Perry, A. S. 1968** RNA Synthesis in the Developing Ovary of the Housefly: Incorporation of 14^c-formate. J. Insect Physiol. 14:581-589.
Following the administration of C^{14}-labelled formate the C^{14} is extensively incorporated into the adenine and gaunine moieties of RNA in the early stages of ovarian development. ABSTRACT: BA 49:8241(91946), Sept. 1, 1968.

Miller, S., joint author. *See* Perry, A. S., et al., 1965.

4377. **Miller, T. A. 1967** Some Relationships of *Musca autumnalis* (Diptera: Muscidae) Feeding, Ovarian Development and Incidence on Dairy Cattle. Diss. Absts. 28(6B):2467-B.

4378. **Miller, T. A.** and **Treece, R. E. 1968** Some Relationships of Face Fly Feeding, Ovarian Development, and Incidence on Dairy Cattle. J. Econ. Ent. 61:250-257.
Incidence on cattle varied in cyclic patterns with age, the highest being in females 3·8 days old. ABSTRACTS: BA 49:5304(59521), June 1, 1968; RAE-B 56:141(491), July, 1968.

4379. **Miller, T. A.** and **Treece, R. E. 1968** Gonadotrophic Cycles in the Face Fly, *Musca autumnalis*. Ann. Ent. Soc. Am. 61(3):690-696.
Two gonadotrophic cycles were completed in most females within a twelve day period, the first by 6-7 days and the second by 10-11 days. ABSTRACTS: BA 49:8241(91947), Sept. 1, 1968; RAE-B 56:236(866), Dec., 1968.

Miller, T. A., joint author. *See* Treece, R. E., et al., 1966, 1968.

Mills, G. D., Jr., joint author. *See* Fales, J. H., et al., 1964, 1968.

Mims, I. S., joint author. *See* Hays, S. B., et al., 1969.

4380. **Minkin, J. L.** and **Scott, H. G. 1960** House Fly Pupation under Baseboards. J. Econ. Ent. 53:479-480.
The unusual site of house fly pupation illustrates the tremendous survival potential of the house fly and reemphasizes the difficulty of control. ABSTRACTS: BA 35:4678(54357), Oct. 1, 1960; RAE-B 49:122, June, 1961.

Miskus, R. P., joint author. *See* Enan, O., et al., 1964; Eldefrawi, M. E., et al., 1960; Roberts, R. B., et al., 1969.

4381. Misnik, Yu. N. 1965 Perekrestnaya tolerantnost' k piretrinam u kul'tur *Musca domestica* L., vysokorezistentnykh k deistviyu khlorirovannykh uglevodorodov. [Cross tolerance of pyrethrins in *Musca domestica* cultures that are highly resistant to the action of chlorinated hydrocarbons., Zhur. Mikrobiol. Epidemiol. i Immunobiol. 42(8):14-17.

Gives results of a study using many strains resistant to a variety of insecticides. Topical application and aerosol sprays were used. ABSTRACT: BA 47:5973(69894), July 16, 1966.

4382. Misnik, Y. N., Lurik, B. B., Sukhova, M. N. and **Starkov, A. V. 1967** Sravnitel'noe izuchenie effektivnosti nekotorykh sinergistov piretrinov na *M. domestica.* [Comparative study of the effectiveness of some synergists of pyrethrins in *Musca domestica.*] Med. Parazit. 36(1):48-54. (English summary.)

The relative effectiveness of the synergists under study was: piperonylbutoxide—1·0; bicarpolate—0·9; octachlordipropyl ether—0·5; nitro methylenedioxybenzole—0·3; nitropiperonal—0·2; nitroethylenedioxybenzole—0·1. All but the last are synergists of pyrethrins. ABSTRACTS: BA 49:461(5126), Jan. 1, 1968; TDB 64:686, June, 1967.

4383. Misnik, Yu. N., Sukhova, M. N., Tsetlin, V. M., Zhuk, E. B., Starkov, A. V. and **Lupik, B. B. 1967** Deistvie nekotorykh insectitsidov i ikh smesei soktakhlordipropilovym efirom v forme aerozolei na komnatnuyu mukhu, (*Musca domestica* L.). [Effect of certain insecticides and their mixtures with octachlordipropyl ether in the form of aerosol on room flies (*Musca domestica* L.).] Tr. Tsent Nauchnoizsled Dezinfek Inst. 18(2):60-64. From: Referat. Zhur., Biol. 1968, 7K217 (Translation).

Room flies resistant to individual insecticides, applied topically or as aerosols, could be controlled with a mixture containing pyrethrin, BHC and ODE (octachlordipropyl ether) in a 1:6 ratio, applied as an aerosol. ABSTRACT: BA 50:12457(128637), Dec. 1, 1969.

Misnik, Y. N., joint author. *See* Sukhova, M. N., et al., 1965.

4384. Misra, B. and **Nayak, B. C. 1966** Infectious Kerato-conjunctivitis Pink Eye in Cattle in Orissa. Ind. Vet. J. 43(7):652-656.

Musca vicina was found to be involved in its transmission.

4385. Misra, J. N., Perti, S. L. and **Paul, R. K. 1963** A Note on Residual Effectiveness of Insecticides Applied on Mud Surfaces. Ind. J. Malariol. 17:107-111. Delhi.

Using *Musca nebulo* and *Culex fatigans*, the most effective insecticides were gamma BHC, dieldrin, DDT and chlordane. Malathion was least satisfactory. ABSTRACT: RAE-B 54:40, Feb., 1966.

4386. Mitlin, N. 1962 The Composition of Ribonucleic Acid in Several Holometabolous Insects. Ann. Ent. Soc. Am. 55(1):104-105.

Insect RNA's are not of one type and may differ in the larvae and adult of a species. Flies have bases different from other insects, with a change in the base ratios occurring at metamorphosis.

4387. Mitlin, N. and **Cohen, C. F. 1961** The Composition of Ribonucleic Acid in the Developing House Fly Ovary. J. Econ. Ent. 54:651-653.

Adenine and uracil predominate. Nucleic acid composition did not differ with age or strain. ABSTRACT: RAE-B 50:146, July, 1962.

4388. Mitsui, T., Fukami, J.-I., Fukanaga, K., Sagawa, T., Takahashi, N. and **Tamura, S. 1969** Studies on Piercidin. I. Effects of Piercidin

A and B on Mitochondrial Electron Transport in Insect Muscle Comparing with Rotenone. (Inst. Insect Contr., Kyoto Univ.) Botyu-Kagaku 34(3):126.

Work done with cockroaches, but included here as part of a research program involving *M. domestica vicina.*

4389. **Mitsui, T., Sagawa, T., Fukami, J., Fukunaga, K., Takahashi, N.** and **Tamura, S. 1969** Studies on Piericidin. II. Insecticidal Effects and Respiratory Inhibition of Piericidin A-related Compounds. Botyu-Kagaku (Bull. Inst. Insect Control) 34(3):135-139.

A homologous series of Piericidin A-related compounds was tested against several insect species including *Musca domestica vicina* Macq. Acetylation and hydrogenation of the parent structure decreased toxicity. None of the compounds tested proved superior to Piericidin A. ABSTRACT: BA 51:6839(69896), June 15, 1970.

Mittler, T. E., joint author. *See* Smith, R. F., et al., 1970.

Miura, H., joint author. *See* Gohda, M., et al., 1966.

4390. **Miyamoto, K. 1965** [Studies on intermediate hosts of bovine thelazia. II. Investigation on Miyake Island, Tokyo.] Jap. J. Sanit. Zool. 16(4):270-273. (In Japanese, with English summary.)

The female *Musca convexifrons* was determined to be the vector of *Thelazia rhodesi* and was suspected as vector of *T. skrjabini* also. ABSTRACT: RAE-B 56:133(463), July, 1968.

4391. **Miyamoto, K., Kano, R., Kaneko, K., Shimizu, N., Akamatsu, T., Nagashima, A., Nagaoka, S., Ebina, R., Hagino, K.** and **Morita, K. 1965** [Studies on intermediate hosts of bovine thelazia. I. Investigation in Niikappu Pasture, Hokkaido.] Jap. J. Sanit. Zool. 16(3): 194-200. (In Japanese, with English summary.)

Of several species of flies found on cattle, *Musca convexifrons* only, harbored thelazian larvae. These were most abundant at the time the flies' ovaries were fully developed. ABSTRACT: RAE-B 56:130(454), July, 1968.

4392. **Miyamoto, K., Tanaka, H.** and **Kano, R. 1967** [Morphological studies on larvae of bovine *Thelazia*.] Jap. J. Parasitol. 16(6):458-463. (In Japanese, with English summary.)

Thelazia rhodesi larvae were collected from females of *Musca convexifrons* and *M. hervei*. The latter also contained a second species believed to be *T. skrjabini*.

4393. **Miyamoto, K., Tanaka, H.** and **Kano, R. 1967** [Studies on intermediate hosts of bovine Thelazia. III. Surveys in Hokkaido and Kanto districts.] Jap. J. Sanit. Zool. 18(4):255-259. (In Japanese, with English summary.)

Larvae of *Thelazia* develop differently in the two hosts, female *Musca convexifrons* and *Musca hervei*. ABSTRACT: BA 51:2206(22841), Feb. 15, 1970.

Miyamoto, K., joint author. *See* Shimizu, F., et al., 1965.

Miyazaki, T., joint author. *See* Kitahara, K., et al., 1966.

Mizogucki, K., joint author. *See* Muto, T., et al., 1968.

4394. **Mizutani, K., Matsunaga, H.** and **Suzuki, T. 1968** Studies on Test Methods of Stomach Poison Insecticides to the Housefly and the German Cockroach and the Comparative Effectiveness of Several Insecticides. Contributions to the Study of the Laboratory Test Methods of Insecticides. Publ. 1968. (3):20-24.

Two test methods of the oral effect of insecticides were designed. Dimethoate, dichlorvos, malathion and trichlorfon were more effective by oral administration to houseflies than they were by contact treatment.

Mizutani, K., joint author. *See* Seki, M., et al., 1968.

Moen, D. J., joint author. *See* Riemann, J. G., et al., 1967.

Mohamed, M. A., joint author. *See* Toppozada, A., et al., 1967.

4395. Moherek, E. A. 1965 A study of the Labellar Contact Chemoreceptors of the Face Fly *Musca autumnalis* DeGeer. Diss. Absts. 25(11): 6129.

Mohiuddin, S., joint author. *See* Ashrafi, S. H., et al., 1969.

4396. Molev, E. V. 1962 Unichtozhenie mukh s pomoshch'yu khlorofosa. [Extermination of flies with the aid of Dipterex.] Zhivotnovudstvo 1:54.
Dipterex successfully controlled flies in areas of human habitation, offices, kitchens and dining areas, hotels, etc. and is recommended because of its lack of odor, and low toxicity to animals and man. ABSTRACT: BA 39:311(3882), July 1, 1962.

4397. Molloy, F. M. 1961 The Histochemistry of the Cholinesterases in the Central Nervous System of Susceptible and Resistant Strains of the House-fly *Musca domestica* L. in Relation to Diazinon Poisoning. Bull. Ent. Res. 52(4):667-681.
Data provides strong evidence that if death is caused by inhibition of cholinesterase of the nervous system it is due to local, not generalized inhibition. Good diagrams of the CNS and photos of frozen sections are presented. ABSTRACTS: BA 39:1336(16890), Aug. 15, 1962; RAE-B 50:29, Feb. 15, 1962. TDB 59:498, May, 1962.

Molloy, F. M., joint author. *See* Lord, K. A., et al., 1963.

4398. Monro, J. and **Bailey, P. T. 1965** Influence of Radiation on Ovarian Maturation and Histolysis of Pupal Fat Body in Diptera. Nature 207(4995):437-438.
The histolysis of the pupal fat body is partly dependent upon ovarian maturation. Radiation does not act hormonally but damages the nurse cells of the ovary directly.

4399. Monroe, R. E. 1960 Effect of Dietary Cholesterol on House Fly Reproduction. Ann. Ent. Soc. Am. 53(6):821-824.
Cholesterol deficiency caused nearly 80 per cent reduction in egg hatch. ABSTRACT: BA 36:2016(24579), Apr. 15, 1961.

4400. Monroe, R. E. 1964 Metabolism and Utilization of Cholesterol-4-C^{14} for Growth and Reproduction of Aseptically-reared House-flies, *Musca domestica* L. Diss. Absts. 25(4):2673-2674.

4401. Monroe, R. E., Hopkins, T. L. and **Valder, S. A. 1967** Metabolism and Utilization of Cholesterol-4-C^{14} for Growth and Reproduction of Aseptically Reared Houseflies, *Musca domestica* L. J. Insect Physiol. 13(2):219-233.
Sterol is stored in the larval stage, lost to eggs by ovipositing adults. ABSTRACTS: BA 48:4191(46716), May 1, 1967; RAE-B 56:35(111), Feb., 1968.

4402. Monroe, R. E., Kaplanis, J. N. and **Robbins, W. E. 1961** Sterol Storage and Reproduction in the Housefly. Ann. Ent. Soc. Am. 54:537-539.
A qualitative rather than a quantitative change in sterol content of larval medium found responsible for an increase in viable egg production. ABSTRACT: RAE-B 50:227, Oct., 1962.

4403. Monroe, R. E. and **Lamb, N. J. 1968** Effect of Commercial Proteins on House Fly Reproduction. Ann. Ent. Soc. Am. 61(2):456-459.
Egg albumen, yeast hydrolysate and defatted milk, sustained good ovarian development. Casein sources had little or no effect. ABSTRACTS: BA 49:6780(75614), July 15, 1968; RAE-B 56:205(744), Oct., 1968.

4404. Monroe, R. E., Polityka, C. S. and **Lamb, N. J. 1968** Utilization of Larval Cholesterol-4-C^{14} for Reproduction in House Flies Fed Unlabelled Cholesterol in the Adult Diet. Ann. Ent. Soc. Am. 61(2):292-296.
Sterols, initially incorporated into eggs at 0·53 microgram per milligram (μg/mg) rapidly decreased with later oviposition. At or below 0·15 μg/mg hatchability is affected. ABSTRACTS: BA 49:6780(75615), July 15, 1968; RAE-B 56:205(741), Oct., 1968.

4405. Monroe, R. E., Robbins, W. E., Chambers, D. L. and **Tabor, L. A. 1963** Sterol Antagonists and House Fly Reproduction. Ann. Ent. Soc. Am. 56:124-125.
Several compounds are compared as to their ability to influence egg production and percentage hatch. ABSTRACTS: BA 42:983(12396), Apr. 15, 1963; RAE-B 51:199, Sept., 1963.

Monroe, R. E., joint author. *See* Kaplanis, J. N., et al., 1963, 1965; Robbins, W. E., et al., 1960; Bauer, J. W., et al., 1969; Bieber, L. L., et al., 1969.

Mookherjee, P. B., joint author. *See* Pradham, S., et al., 1960.

Mooney, J. W., joint author. *See* Childo, D. P., et al., 1964.

Moore, S., III., joint author. *See* Bruce, W. N., et al., 1960.

4406. Moorefield, H. H. and **Tefft, E. R. 1960** Evaluation of 2-(3,5 dichlor-2-biphenylyloxy) Triethylamine as an Insecticide Adjuvant. Contr. Boyce Thompson Inst. 20(4):293-298. Yonkers, N.Y.
Compound offers no promise in regard to DDT resistant strains of houseflies.

4407. Moorefield, H. H. and **Weiden, M. H. J. 1964** Influence of the Methylenedioxyphenyl Structure in Synergism of a Carbamate Insecticide for Houseflies. Contrib. Boyce Thompson Inst. Plant. Res. 22(8): 425-433.
The integrity of the 1,2-methylenedioxy structure is essential for maximum potentiation of house fly toxicity of the carbamates, paralleling the case for pyrethrins synergism. ABSTRACTS: BA 46:2179(27179), Mar. 15, 1965; RAE-B 53:151, Aug., 1965.

Moorefield, H. H., joint author. *See* Wieden, M. H. J., et al., 1964, 1965.

4408. Morallo, B. D. and **Sherman, M. 1967** Toxicity and Anticholinesterase Activity of Four Organophorous Insecticides to Four Species of Flies. J. Econ. Ent. 60(2):509-515.
High insecticidal activity was shown by all compounds. There appeared to be little or no correlation between toxicity and *in vitro* or *in vivo* inhibition of cholinesterase. ABSTRACTS: BA 48:6928(77475), Aug. 1, 1967; RAE-B 55:145(512), Aug., 1967.

4409. Morello, A., Spencer, E. Y. and **Vardanis, A. 1967** Biochemical Mechanisms in the Toxicity of the Geometrical Isomers of Two Vinyl Organophosphates. Biochem. Pharmacol. 16(9):1703-1710.
Fly head cholinesterase is more susceptible to inhibition than the mouse brain or bovine red cell enzyme. ABSTRACT: BA 49:1188(13293), Feb. 1, 1968.

4410. Morello, A., Vardanis, A. and **Spencer, E. Y. 1968** Comparative Metabolism of Two Vinyl Phosphorothianate Isomers (thiono Phosdrin) by the Mouse and the Fly. Biochem. Pharmacol. 17(9):1795-1802.
An attempt is made to account for different toxicity of thiono phosdrin isomers to the fly and mouse. ABSTRACT: BA 50:3485(36542), Apr. 1, 1969.

Morgan, J. W., joint author. *See* Plapp, F. W., Jr., et al., 1965.

4411. Morgan, N. O. 1968 House Fly Response to Light Traps. Folia Ent. Mex. (for) 1968 (18-19):79-80.
Responses of *M. domestica* in relation to colored lights differ, depending on temperature, etc. Male response greater at cool temperatures; female, at warmer. ABSTRACT: RAE-B 58:414(1728), Dec., 1970.

4412. Morgan, N. O. and **Pickens, L. G. 1967** Cold Tolerance of Adults and Pupae of the Face Fly. J. Econ. Ent. 60:1464-1466.
Flies can endure refrigeration for one or more weeks either in pupal or adult stages. ABSTRACTS: BA 49:1424(15967), Feb. 1, 1968; RAE-B 56:70(226), Mar., 1968.

4413. Morgan, N. O. and **Pickens, L. G. 1968** Influence of Air Temperature on the Attractiveness of Electric Lamps to House Flies. J. Econ. Ent. 61(5):1257-1259. Errata T.c. No. 6, p. 1491.
Males are most responsive (expecially to green and orange) at low temperatures (19°C). Females responded best at 32°C to green, blue and ultraviolet. ABSTRACTS: BA 50:3736(39223), Apr. 1, 1969; RAE-B 57: 50(179), Mar., 1969; TDB 66:1185(2308), Nov., 1969.

4414. Morgan, N. O., Pickens, L. G. and **Thimijan, R. W. 1970** House Flies and Stable Flies Captured by Two Types of Traps. J. Econ. Ent. 63(2):672-673.
One trap, an electrocutor grid, caught about three times as many female stable flies as the other trap which had a removable screen cage. Flies may have been attracted to the stench of burning insects in the first cage, as the second did not kill them. ABSTRACT: RAE-B 58:307(1251), Sept., 1970.

Morgan, N. O., joint author. *See* Bowman, M. C., et al., 1968; Calvert, C. C., et al., 1969, 1970; Pickens, L. G., et al., 1967, 1969; Thimijan, R. W., et al., 1970; Miller, R. W., et al., 1970.

4415. Morgan, P. B. 1967 Booby-Trapped Female House Flies as Sterilant Carriers. J. Econ. Ent. 60:612-613.
None of the eggs laid by booby-trapped females hatched. ABSTRACTS: BA 48:6928(77476), Aug. 1, 1967; RAE-B 55:149(522), Aug., 1967; TDB 64:1273, Nov., 1967.

4416. Morgan, P. B. 1967 Effects of Hempa on the Ovarian Development of the House Fly, *Musca domestica* (Diptera:Muscidae). Ann. Ent. Soc. Am. 60:812-818.
Effects are dependent upon hours of exposure and percentage concentration of chemosterilant. ABSTRACTS: BA 48:10184(114436), Nov. 15, 1967; RAE-B 56:2(4), Jan., 1968; TDB 65:341(786), May, 1968.

4417. Morgan, P. B. 1967 Inhibition by 5-Fluoroorotic Acid of the Ovarian Development of House Flies, *Musca domestica*. Ann. Ent. Soc. Am. 60:1158-1161.
Acid administered in food of adult females caused vacuolation of nurse cells and egg size to be two-thirds normal. ABSTRACTS: BA 49:3367 (37638), Apr. 1, 1968; RAE-B 56:116(403), June, 1968; TDB 65:726 (1569), May, 1968.

4418. **Morgan, P. B., Bowman, M. C.** and **LaBrecque, G. C. 1968** Uptake and Persistence of Metepa and Hempa in the House Fly. J. Econ. Ent. 61(3):805-808.
Gas chromotography was used to measure chemosterilant uptake and persistence in flies. The possibility of its use in determining the sterilizing dose is proposed. ABSTRACTS: BA 49:10247(113112), Nov. 1, 1968; RAE-B 56:221(806), Nov., 1968; TDB 66:880(1755), Aug., 1969.

4419. **Morgan, P. B.** and **LaBrecque, G. C. 1962** The Effect of Apholate on the Ovarian Development of House Flies. J. Econ. Ent. 55(5): 626-628.
Exerts the greatest influence on the second egg chamber and the germarium. ABSTRACTS: BA 41:1661(20877), Mar. 1, 1963; RAE-B 51:37, Feb., 1963.

4420. **Morgan, P. B.** and **LaBrecque, G. C. 1964** Effect of Tepa and Metepa on Ovarian Development of House Flies. J. Econ. Ent. 57:896-899.
Oocytes in the first egg chambers of tepa- and metepa-treated ovaries were inhibited and began degenerating at 48 and 72 hours, respectively. ABSTRACTS: BA 46:2565(31991), Apr. 1, 1965; TDB 62:599, June, 1965.

4421. **Morgan, P. B., LaBrecque, G. C.** and **Wilson, H. G. 1966** Tests with Larvicides for the Control of House Flies, *Musca domestica* (Diptera:Muscidae) in Poultry Houses. Florida Ent. 49(2):91-93.
Results were highly variable, with Dimethoate and Bayer 39007 being somewhat promising. ABSTRACTS: BA 47:9040(104523), Nov. 1, 1966; RAE-B 56:110(387), June, 1968.

4422. **Morgan, P. B., LaBrecque, G. C., Smith, C. W., Meifert, C. M.** and **Murvosh, C. M. 1967** Cumulative Effects of Substerilizing Doses of Apholate and Metepa on Laboratory Populations of the House Fly. J. Econ. Ent. 60(4):1064-1067.
A diet containing 0·05 per cent apholate caused extinction of a colony in the F_7 generation. Metepa, at 0·2 per cent did so in the F_{10}. ABSTRACTS: BA 48:11213(124544), Dec. 15, 1967; RAE-B 56:7(20), Jan., 1968; TDB 64:1363, Dec., 1967.

Morgan, P. B., joint author. *See* Meifert, D. W., et al., 1967.

4423. **Morikawa, O.** and **Saito, T. 1966** Degradations of Vamidothion and Dimethoate in Plants, Insects and Mammals. Botyu-Kagaku (Bull. Inst. Insect Contr.) 31(3):130-135. (Japanese summary.)
The optimum pH for degradation of these substances for insect homogenates ranges from 7·0 to 7·4. ABSTRACT: BA 49:8532(95071), Sept. 15, 1968.

Morita, H., joint author. *See* Steinhardt, R. A., et al., 1966.

Morita, K., joint author. *See* Miyamoto, K., et al., 1965.

4424. **Morris, A. P. 1964** Studies of the Dispersion of Insecticide Resistant Populations of the House Fly *Musca domestica* L. Diss. Absts. 25(4):2463-2464.

4425. **Morris, A. P.** and **Hansens, E. J. 1966** Dispersion of Insecticide-resistant Populations of the House Fly, *Musca domestica* L. J. Econ. Ent. 59:45-50.
Fluorescent dye-marked flies collected at night indicated movement of a mile a day in all directions. Population measurements at intervals indicated size of total population and numbers migratory into a given area. ABSTRACTS: BA 47:6376(74579), Aug. 1, 1966; RAE-B 54:97, May, 1966; TDB 63:814, July, 1966.

Morris, A. P., joint author. *See* Hansens, E. J., et al., 1962.

4426. **Morrison, F. O.** and **Ilyntsky, S. 1968** Biological Indicators of Soil Insecticide Residues. Rech. Agron. 13:22.

Concerns aldrin, dieldrin, heptachlor and lindane. Housefly the test insect.

4427. **Morrison, F. O.** and **Ozburn, G. W. 1965** The Pattern and Extent of DDT-Resistance Development in House Fly (*Musca domestica* L.) Strains as Related to the Sex and Stage Exposed to the Selective Pressure. Proc. XII Int. Congr. Ent., London, 1964 (Publ. 1965). pp. 497-498.

All colonies under pressure developed resistance by the fifth generation.

Morrison, F. O., joint author. *See* Chan, K., et al., 1965; Ozburn, G. W., et al., 1963, 1965; Shaikh, M. U., et al., 1966.

4428. **Morrison, P. E.** and **Davies, D. M. 1964** Feeding of Dry Chemically Defined Diets and Egg Production in the Adult House Fly. Nature 201(4914):104-105.

Two dry, synthetic diets compared with fresh milk, for effectiveness in egg production. Abstract: TDB 61:531, May, 1964.

4429. **Morrison, P. E.** and **Davies, D. M. 1964** Repeated Ovarian Cycles with Ribonucleic Acid in the Diet of Adult House-flies. (Correspondence.) Nature 201:948-949.

Full chemical formulation is given for two types of liquid diet. Abstract: TDB 61:726, July, 1964.

Morrison, P. E., joint author. *See* Bodnaryk, R. P., et al., 1966, 1968; Goodman, T., et al., 1968; Davies, D. M., et al., 1965.

4430. **Mosconi-Bernardini, P., Vecchi, M. L.** and **Laudani, U. 1965** Localizzazione istochimica della citocromo-c ossidase nelle pupe di *Musca domestica* L. [Histochemical localization of cytochrome-c oxidase in the pupa of *Musca domestica*.] Boll. di Zool. 32:759-764.

Moscowitz, J., joint author. *See* Ascher, K. R. S., et al., 1967, 1968.

Mosettig, E., joint author. *See* Thompson, M. J., et al., 1962, 1963.

4431. **Moskalets, N. D. 1960** [A study of synanthropic flies in Uzhgorod in the transcarpathian region of the Ukraine., Med. Parazit. (Mosk) 29:575-578. Referat. Zhur., Biol., 1961, No. 10E179.

Mosna, B., joint author. *See* Filipponi, A., et al., 1968.

4432. **Mount, G. A. 1964** The Use of Factorial Experiments to Determine the Sensory Perception of Various Bait Factors by *Musca domestica* L. Diss. Absts. 25:709.

4433. **Mourier, H. 1964** Circling Food-searching Behaviour of the House Fly (*Musca domestica* L.) Vidensk. Medd. Dansk. Naturh. Foren. (Kobenhavn) 127:181-194.

Gyration movements are the same as those demonstrated for *Phormia regina* (Dethier, 1957). Abstract: BA 47:5973(69895), July 15, 1966.

4434. **Mourier, H. 1965** The Behaviour of House Flies (*Musca domestica* L.) towards New Objects. Vidensk. Medd Dansk. Naturh. Foren. 128:221-231.

New objects are highly attractive at first but attraction declines to "normal" after about 20 minutes.

4435. **Mourier, H.** and **Ben Hannine, S. 1969** Activity of Pupal Parasites from *Musca domestica* (Diptera) in Denmark. Vidensk. Medd. Dansk. Naturh. Foren. 132:211-216.

Three parasitic Hymenoptera from pupae on dung heaps (parasitizing an average of 12·4 per cent but in one locality 59 per cent) were *Spalangia cameroni, Muscidifurax raptor, Phygadeuon sp.* Abstract: BA 52:414 (13357), Feb. 1, 1971.

4436. **Mourier, H.** and **Ben Hannine, S. 1969** Undersøgelse af de naturlige fjenders indflydelse på stue fluepopulationer under danske forhold. [Survey of the natural enemies of houseflies in Denmark.] 1968 Ann. Rept. Govern. Pest Infest. Lab. (Lyngby). pp. 50 and 51.
Samples from dung heaps showed mites play the greatest role quantitatively, but beetles of the families *Staphylinidae Hydrophilidae*, *Ptiliidae*, and the earwig *Labia minor*, are also common. Laboratory colonies maintained for study.

Mourier, H., joint author. *See* Borlund, H. P., et al., 1969; Funder, J. V., et al,. 1965.

Moye, W. C., joint author. *See* Corey, R. A., et al., 1965.

Moyer, D. B., joint author. *See* Louloudes, S. J., et al., 1962.

4437. **Mozhaev, E. A. 1961** [Extermination of flies in chambers with the use of a red light.] Med. Parazitol. i Parazitarn. (Bolezni) 30(4):479, 509. (In Russian, with English summary.)
ABSTRACT: BA 39:670(8308), July 15, 1962.

Mue, K., joint author. *See* Takemoto, T., et al., 1967.

4438. **Muesebeck, C. F. W. 1961** A new Japanese *Trichopria* Parasitic on the House Fly (Hymenoptera:Diapriidae). Mushi 35(1):1-2. Fukuoka.
The species is being mass-produced in Japan for liberation in a fly-control program. Full description of male and female is given. Type in U.S. Natural History Museum. No. 65, 398. ABSTRACT: RAE-B 51:70 Apr., 1963.

4439. **Mukerjea, T. D.** and **Govind, R. 1960** Studies of Indigenous Insecticidal Plants: III. *Acorus calamus* Linn. J. Sci. and Indust. Res. (India) 19(5):112-119.
The ether extract of *A. calamus* was found to be 17 times less toxic than DDT when applied as a contact insecticide. It also exhibits ovicidal activity but has no residual toxicity. ABSTRACT: BA 37:1584(15825), Feb. 15, 1962.

Muller, O., joint author. *See* Celedova, V., et al., 1963.

4440. **Müller, P. 1966** Untersuchungen zur Wirkung einiger substituierter Phenyl-N-methyl Carbamate auf Hygieneschädlinge. [The effects of substituted phenyl-N-methyl-carbamates on pests of public health importance.] Zeit. Ges. Hyg. u. ihre Grenzgebiete. (Berlin) 12(1):1-20. (In German, with English summary.)
Adults of *M. domestica* tested. Literature review included on effect of substituted carbamate compounds on pests and some data on toxicity of tested compounds on mammals. ABSTRACT: TDB 63:911, Aug., 1966.

4441. **Müller, P. 1967** Der Einfluss von Piperonylbutoxid auf die Wirkung einiger substituierter Phenyl-N-methyl-carbamate. [Influence of piperonylbutoxide on the effect of some substituted phenyl-n-methyl-carbamates.] Angew. Parasit. 8(2):101-114.
The degree of influence depends upon the proportion of components in the mixture. In *Musca* the piperonyl butoxide had an inhibitory effect on the degradation of carbamate, thus reducing recovery. ABSTRACT: BA 50:11925(123262), Nov. 15, 1969.

4441a. **Muñiz (Vélez), R. 1968** Preferencias de *Musca domestica* L. en la selección de superficies de color. Ensayos en el campo. [Preferences of *M. domestica* in the selection of colored surfaces. Field Tests.] Folia Ent. Méx. No. 18-19:101. VIth National Congress of Entomology organized by the Mexican Society of Entomology, 23rd-26th Oct., 1967.

Colored surfaces attractive to *M. domestica* in the laboratory are not the same as in the field. Yellow is preferred color in field. ABSTRACT: RAE-B 58(12):415(1731), Dec., 1970.

4442. **Muñiz, R.** and **Hecht, O. 1968** Observaciones sobre ea distribución de *Stomoxys calcitrans* en un predio ganadero y ensayos sobre la selección de superficies de colores al aire libre. [Observations on the distribution of *Stomoxys calcitrans* on a cattle farm and an experiment concerning their selection of colored surfaces in the open air.] An. Esc. Nac. Cien. Biol., Méx. 17:225-243.

Stomoxys preferred black, yellow, red, green, and brown, over gray or white. Blue was selected *less* often than black, gray, or white. Reduced to formula: Red (65) — brown (47) — green (46) — black (34) — yellow (33) — gray (24) — white (12) — blue (0). Reactions *not the same* as with *Musca domestica*. (See earlier publications.)

Muñiz, R., joint author. *See* Hecht, O., et al., 1964, 1968, 1969.

4443. **Munt, R. H. 1964** Fly Prevention on the Poultry Farm. Queensland Agric. J. 90(7):394-397. Reprinted as Division of Animal Industry Advisory Leaflet No. 429. 5 pp.

Necessity of using baits and sprays will diminish if breeding ground for flies is removed. Remarks are made on "flame thrower" as a useful device. ABSTRACT: BA 47:9361(108315), Nov. 15, 1966.

Murata, M., joint author. *See* Ogata, K., et al., 1961.

4444. **Murdock, L. L.** and **Hopkins, T. L. 1968** Insecticidal, Anticholinesterase, and Hydrolytic Properties of 0,0-dialkyl-S-aryl Phosphorothiolates in Relation to Structure. J. Agric. Fd. Chem. 16(6): 954-958.

A positive correlation exists between *in vitro* inhibition of fly head cholinesterase and toxicity; similar to other organophosphorous anticholinesterases. ABSTRACT: BA 50:4221(44421), Apr. 15, 1969.

4445. **Murray, W. D. 1968** Blow Fly Control in Residential Areas. Proc. and Papers Ann. Conf. Calif. Mosquito Contr. Ass. 36:72-75. October 8.

Murtuza, S. S. M., joint author. *See* Ashrafi, S. H., et al., 1963, 1966, 1969.

4446. **Murvosh, C. M., Fye, R. L.** and **LaBrecque, G. C. 1964** Studies on the Mating Behavior of the House Fly, *Musca domestica* L. Ohio J. Sci. 64(4):264-271.

An imperfect sex recognition exists in the male. The female exudes a sex attractant of a low order. ABSTRACT: BA 46:2930(36476), Apr. 15, 1965.

4447. **Murvosh, C. M., LaBrecque, G. C.** and **Smith, C. N. 1963** Sex Attraction Studies with the House Fly. Am. Zool. 3:506. (An abstract.) Complete publication in: Ohio Journal of Science 65(2):68-71. (Ref. No. 4449.)

ABSTRACT: BA 46:1446(18070), Feb. 15, 1965.

4448. **Murvosh, C. M., LaBrecque, G. C.** and **Smith, C. N. 1964** Effect of Three Chemosterilants on House Fly Longevity and Sterility. J. Econ. Ent. 57:89-93.

Tepa sterilizes at lower concentrations than metepa and apholate. ABSTRACTS: BA 45:4729(57659), July 1, 1964; RAE-B 52:98, May, 1964; TDB 61:849, Aug., 1964.

4449. Murvosh, C. M., LaBrecque, G. C. and Smith, C. N. 1965 Sex Attraction in the House Fly, *Musca domestica* L. Ohio J. Sci. 65(2):68-71.
There is an attraction to virgin females of both virgin males and females. This is not due to the effect of moisture, motion, or sound. ABSTRACTS: BA 47:49(586), Jan. 1, 1966; RAE-B 54:247, Dec., 1966.

4450. Murvosh, C. M. and Thaggard, C. W. 1966 Ecological Studies of the House Fly. Ann. Ent. Soc. Am. 59:533-547.
A high correlation exists between a grill count and a total kitchen population. Males were found to predominate. ABSTRACTS: BA 47:9045 (104583), Nov. 1, 1966; RAE-B 54:197, Oct., 1966; TDB 64:120, Jan., 1967.

Murvosh, C. M., joint author. *See* Morgan, P. B., et al., 1967.

Musonov, V. B., joint author. *See* Sukhova, M. N., et al., 1965.

4451. Mustafa, M. and Kamat, D. N. 1970 Mucopolysaccharide Histochemistry of *Musca domestica.* 1. A Report on the Occurrence of a New Type of KOH-labile Alcianophilia. Histochemie 21:54-63.

4451a. Mustafa, M. and Kamat, D. N. 1970 Mucopolysaccharide Histochemistry of *Musca domestica.* II. The Peritrophic Membrane. Folia Histochem. Cytochem. (Krakow) 8:77-83. (Russian summary.)
Studied *M. d. nebulo* (Fabr.). Concerns PAS (positiveness of membrane) and possible methods of formation. ABSTRACT: BA 51:12607(128239), Nov. 15, 1970.

Mutchmor, J. A., joint author. *See* Thiessen, C. I., et al., 1967.

4452. Muto, T. and Sugawara, R. 1970 1, 3-Diolein. A House Fly Attractant in the Mushroom *Amanita muscaria. In*: Wood, D. L., R. M. Silverstein, M. Nakajima (ed.) Control of Insect Behavior by Natural Products. Symposium. Acad. Press. Inc.: New York, N.Y. pp. 189-208.

4453. Muto, T., Sugawara, R. and Mizoguchi, K. 1968 The Housefly Attractants in Mushrooms. II. Identification of the Fraction D_3 and the Activities of Some Related Compounds. *Amanita muscaria.* Agric. Biol. Chem. 32(5):624-627.
The attracting substance, referred to as D_3, was identified with 1, 3-diolein. Related compounds of known formula were found to have an activity 10 times that of 1, 3-diolein. ABSTRACT: BA 49:11856 (129926), Dec. 15, 1968.

4454. Muus, J. 1968 Gel Electrophoresis of Some Dehydrogenases from Flight Muscles of Tsetse Flies (*Glossina morsitans*) and Houseflies (*Musca domestica*). Comp. Biochem. Physiol. 24:527-536.
Some interesting differences are reported between the two species. ABSTRACT: BA 50:2132(22410), Feb. 15, 1969.

4455. Myalo, I. I. 1961 Polikhlorpinen dlya bor'by s mukhami. [Polychlorpinene for fly control.] Med. Parazitol. i Parazitarn. (Bolezni) 30(5):616-617. Referat. Zhur., Biol. 1962, No. 4 K142.
An emulsion of a polychlorpinene paralyzed and killed flies in the laboratory and in a hog barn when applied to walls. The effective duration of a treatment was about 2 weeks. ABSTRACT: BA 40:1924(25274), Dec. 15, 1962.

4456. Nachtigall, W. 1967 Getrennte Systeme in Fliegenthorax für die Erzeugung und Verteilung Aerodynamischer Kräfte. [Separate systems in the fly thorax for the production and dispersion of aeroclynamic power.] Naturwissenschaften 54. Jg. Heft. 12, p. 323.
Worked with Calliphoridae. Study concerns antagonistic (and other) muscle groups; their complexity, as contrasted with the musculature of more primitive insects, e.g. Orthoptera.

4457. Nadzhafov, I. G. 1967 O roli raxlichnykh vidov sinantropnykh mukh v rasprostranemii onkosfer *Taeniarhynchus saginatus.* [Role of different species of synanthropic flies in dissemination of onchospheres of *Taeniarhynchus saginatus.*] Med. Parazitol. i Parazitarn. (Bolez.) 36(2):144-149. (English summary.)
Synanthropic flies play a definite role in dissemination of beef tapeworm oncospheres, therefore, elimination of flies must be included into measures of taeniarhynchosis control. ABSTRACT: BA 49:814(9040), Jan. 15, 1968.

Nagai, N., joint author. *See* Ogata, K., et al., 1960.

4458. Nagasawa, S. 1960 The Toxic Action of a Mixture of Sevin and pp'-DDT to the House Fly. J. Econ. Ent. 53(5):709-711.
Sevin and p, p'-DDT (2:1) applied topically in acetone to females 3-5 days old. Results indicate that insecticides have different types of toxic action. Effectiveness of mixture depends on total concentration of insecticide. ABSTRACTS: BA 36:758(9013), Feb. 1, 1961; RAE-B 49:178, Aug., 1961.

4459. Nagasawa, S. 1962 [DDT-resistance in the so-called "Takatsuki" strain of the common house fly, *Musca domestica vicina.*] Botyu-Kagaku 27(3):67-76. (In Japanese, with English summary.) The first of 2 related papers. *See also* Ikemoto, H., 1962.
The strain appears resistant when tested by topical application method, but LD_{50} could not be determined. Author discusses strain used for biological assay of insecticides in Japan and the effect of contamination on research. Suggestions for keeping strain pure. ABSTRACT: RAE-B 52:139, Aug., 1964.

4460. Nagasawa, S. 1962 [Lindane-resistance in the so-called "Takatsuki" strain of the common house fly, *Musca domestica vicina.*] Botyu-Kagaku 27(4):108-112. Kyoto. (In Japanese, with English summary.)
Strains made DDT-resistant by DDT pressure show no significant cross-resistance to *gamma*-BHC. Tests made on 13 populations, divided from original strain, and bred separately for 2 years. ABSTRACTS: BA 46:8515(105200), Dec. 1, 1965; RAE-B 52:173, Sept., 1964.

4461. Nagasawa, S. and Hoskins, W. M. 1962 [The relation between temperature and lethal action of P,P-DDT on adults of the housefly, *Musca domestica* L.] Ent. Exp. Appl. 5(2):139-146. (In Japanese, with German summary.)
Effect of temperature upon toxicity of DDT to both a sensitive and a resistant strain of houseflies was determined over a range of 5° to 35°C. Maximum toxicity at approximately 10°C. ABSTRACTS: BA 40:948(12631), Nov. 1, 1962; RAE-B 51:119, June, 1963; TDB 59:1213, Dec., 1962.

4462. Nagasawa, S. and Hoskins, W. M. 1964 Toxicity of p, p' DDT in Acetone and in Acetone with Oil to the House Flies of a DDT-resistant Strain. Jap. J. Appl. Ent. Zool. 8(4):295-299. Tokyo. (Japanese summary.)
Dosages above 6·2 mg DDT found more toxic in acetone with oil than in acetone alone. At lower dosages, DDT more toxic in acetone alone. ABSTRACT: RAE-B 53:196, Oct., 1965.

4463. Nagasawa, S. and Hoskins, W. M. 1968 The Relation Between Temperature and Lethal Action of Carbaryl on Adults of the Housefly and the Oriental Housefly (Diptera:Muscidae). Appl. Ent. Zool. 3(3):139-147.
Temperature coefficient of lethal action of carbaryl to housefly was

negative in temperature range 5-25°C. and positive between 25-35°C. ABSTRACTS: BA 50:10284(106373), Oct. 1, 1969; RAE-B 57:153(556), Aug., 1969.

4464. Nagasawa, S. and **Kishino, M. 1963** [On the relation between the diameter of rearing container and the duration from oviposition to emergence of the common housefly *Musca domestica vicina* when the powdered biscuit for experimental animals was used as the culture medium. (Problems on the breeding of insects for biological assay of insecticides. XXXI).] Botyu-Kagaku 28:4-8. (In Japanese, with English summary.)
Time from oviposition to emergence decreased with the increase of diameter. Relation between the two variables could be expressed by a linear equation. ABSTRACT: RAE-B 54:199, Oct., 1966.

4465. Nagasawa, S. and **Kishino, M. 1965** Application of Pradhan's Formula to the Pupal Development of the Common House Fly, *Musca domestica vicina* Macquart. Jap. J. Appl. Ent. Zool. 9(2):94-98. Tokyo. (Japanese summary.)
Formula substantially applicable to the present data. Pupal development of housefly checked at 24 different temperatures between 18·7 and 39·0°C.

4466. Nagasawa, S. and **Shiba, M. 1964** Joint Toxic Action of Mixtures Between Lindane and Hercules 5727 Against the Common House Fly. Botyu-Kagaku 29(4):73-76.
Joint action was synergistic. Maximum mortality from a ratio of lindane to Hercules 5727 of L3·7:H2·3. ABSTRACTS: BA 47:8154(94959), Oct. 1, 1966; RAE-B 55:173(602), Sept., 1967.

4467. Nagasawa, S. and **Shiba, M. 1964** [Effect of humidity and solvent on the toxicity of dimethoate against the common house fly.] Jap. J. Sanit. Zool. 15(3):182-186. (In Japanese, with English summary.)
Effect of atmospheric humidity after treatment with dimethoate highest among factors concerned with absorption and toxicity. Effect of sesame oil not significant. ABSTRACT: RAE-B 54:244, Dec., 1966.

4468. Nagasawa, S. and **Shiba, M. 1964** [Effect of temperature and solvent on the toxicity of DDT against the common house fly.] Jap. J. Appl. Ent. Zool. 8(3):203-209. (In Japanese, with English summary.)
Toxicity of DDT at 20°C greater than at 30°C. ABSTRACTS: BA 47:3816 (44914), May 1, 1966; RAE-B 53:55, Mar., 1965.

4469. Nagasawa, S. and **Shiba, M. 1964** [Effect of temperature and solvent on the toxicity of naled against the common house fly. Studies on the biological assay of insecticides—XLIV.] Botyu-Kagaku 29(2): 31-36. (In Japanese, with English summary.)
Adjuvant effect of soybean oil was negligible. Toxicity of naled greater at lower temperatures and seems to be negative in range from 20°-30°C. ABSTRACT: RAE-B 55:88(313), May, 1967.

4470. Nagasawa, S. and **Shiba, M. 1964** [Toxicity of malathion to the common house fly, evaluated by the impregnated filter paper method. Studies of the biological assay of insecticides—XLVII.] Botyu-Kagaku 29(3):46-51. (In Japanese, with English summary.)
Under applied experimental conditions, deposit more important factor than exposure time in determining mortality. Data obtained are expressed in 2 probit regression planes and analysed by Finney's method. Equations given. ABSTRACT: RAE-B 55:88(314), May, 1967.

4471. **Nagasawa, S.** and **Shiba, M. 1965** [Toxicity of B-1946 to the common housefly evaluated by the impregnated filter paper method. Studies on the biological assay of insecticides—L.] Botyu-Kagaku. (Bull. Inst. Insect Contr.) 30(1):30-33. (In Japanese, with English summary.)
Probit analysis of data shows that toxicity of B-1946 was slightly decreased when sesame oil was added to acetone, as solvent. Toxicant deposit almost as important as exposure time in determining mortality. ABSTRACT: RAE-B 55:175(607), Sept., 1967.

4472. **Nagasawa, S.** and **Shiba, M. 1965** [Comparison of synergistic action of anti-resistant DDT and DMC with DDT against the common house fly, evaluated by the impregnated filter paper method. Studies on the biological assay of Insecticides—LIII.] Botyu-Kagaku (Bull. Inst. Insect Contr.) 30(2):58-61. (In Japanese, with English summary.)
In the case of anti-resistant DDT, amount of deposit is as important as exposure time in determining mortality. With DMC, exposure time is more important. Importance of deposit in determining mortality the same in both synergists. ABSTRACT: RAE-B 56:52(170), Mar., 1968.

4473. **Nagasawa, S., Shiba, M.** and **Fushimi, S. 1964** [Difference in susceptibility to the letal effect of malathion between adults of the common house fly, *Musca domestica vicina* reared on the culture medium prepared with "Okara" and those on agar. Problems on the breeding of insects for biological assay of insecticides—XXXVI.] Botyu-Kagaku 29(2):25-30. (In Japanese, with English summary.)
Flies reared on agar were 2·06 times as susceptible as those reared on "Okara", to several insecticides tested. Difference believed due to nutritional differences between culture media. ABSTRACT: RAE-B 55:88(312), May, 1967.

4474. **Nagasawa, S., Shiba, M.** and **Fushimi, S. 1965** [Difference in susceptibility to the lethal effect of malathion between adults of the common house fly, *Musca domestica vicina* Macq. reared on the culture medium prepared with okara and those on the CSMA culture medium. Problems on the breeding of insects for biological assay of insecticides—XL.] Botyu-Kagaku (Bull. Inst. Insect Contr.) 30(2):61-66. (In Japanese, with English summary.)
No significant difference recognized between malathion-susceptibilities of houseflies reared on CSMA and "Okara" media. Data obtained adjusted for mortality rate among controls, and analyzed by probit method of Finney. ABSTRACT: RAE-B 56:52(171), Mar., 1968.

4475. **Nagasawa, S.** and **Tsuruoka, Y. 1962** [Joint toxic action of mixtures of malathion, dimethoate and Dibrom to adults of the common house fly, *Musca domestica vicina*.] Botyu-Kagaku 27(3):78-81. (In Japanese, with English summary.)
Similar joint toxic action obtained between malathion and dimethoate, at three different ratios. Malathion and Dibrom joint toxic action increased with decrease in relative quantity of malathion to Dibrom. Synergism in 57:43 mixture of malathion and Dibrom statistically significant. ABSTRACT: RAE-B 52:139, Aug., 1964.

Nagasawa, S., joint author. *See* Asano, S., et al., 1963; Hoskins, W. M., et al., 1961; Nakayama, I., et al., 1969.

Nagashima, A., joint author. *See* Miyamoto, K., et al., 1965.

Nagata, C., joint author. *See* Ban, T., et al., 1966.

4476. **Nagel, W. P. 1963** The Population Dynamics of an Experimental Parasite-host System, with *Nasonia vitripennis* (Walk.) (Pteromalidae Hymenoptera) as the Parasite and *Musca domestica* L. (Muscidae, Diptera) as the Host. Diss. Absts. 23(9):3563-3564.

4477. **Nagel, W. P.** and **Pimentel, D. 1963** Some Ecological Attributes of a Pteromalid Parasite and its Housefly Host. Canad. Ent. 95(2):208-213.
Many aspects studied, including dispersal ratio for host and parasite, host parasite ratios, superparasitism. ABSTRACTS: BA 43:1944(25138), Sept. 15, 1963; RAE-B 51:272, Dec., 1963.

4478. **Nagel, W. P.** and **Pimentel, D. 1964** The Intrinsic Rate of Natural Increase of the Pteromalid Parasite *Nasonia vitripennis* (Walk.) on its Muscoid Host *Musca domestica* L. Ecology 45(3):658-660.
Finite rate of increase calculated to be 1·3 per day. ABSTRACT: BA 46:1430(17868), Feb. 15, 1965.

Nagel, W. P., joint author. *See* Pimentel, D., et al., 1963.

Nagooka, S., joint author. *See* Miyamoto, K., et al., 1965.

Naidu, M. B., joint author. *See* Pershad, S. B., et al., 1966; Qadri, S. H., et al., 1967.

4479. **Nair, K. K. 1962** Preliminary Studies on the Effects of *Gamma*-radiation on Housefly Pupae with Special Reference to the Critical Periods in Relation to the Mechanism of Emergence. *In* Internat. Atomic Energy Agency. Radioisotopes and Radiation in Entomology. Proc. Symp. Internat. Publ. Inc., New York 22:207-211. (Russian, French and Spanish summaries.)
Dose levels from 500 to 10,000r. Author considers early stages of development most sensitive to radiation. ABSTRACTS: BA 42:324(3921), Apr. 1, 1963; RAE-B 51:213, Oct., 1963.

4480. **Nair, K. K., Bhaskaran, G.** and **Sivasubramanian, P. 1967** Effect on Adult Emergence of Whole and Partial X-irradiation of Pupae of the Housefly, *Musca domestica nebulo*. Canad. Ent. 99(6):597-598.
Target organ, damage to which prevents emergence, lies in anterior 7 segments. ABSTRACTS: BA 48:11248(124933), Dec. 15, 1967; RAE-B 57(3):47(170), March, 1969.

Nair, K. K., joint author. *See* Amonkar, S. V., et al., 1965, 1967 Bhaskaran, G., et al., 1965, 1967; Sivasubramanian, P., et al., 1970.

Nakada, H., joint author. *See* Matsuzawa, H., et al., 1967.

Nakagoshi, S., joint author. *See* Gohda, M., et al., 1966.

Nakanishi, K., joint author. *See* Kobayashi, M., et al., 1967.

4481. **Nakasuji, F. 1965** [On the predaceous habits of a histerid, *Saprinus speciosus* Erichson, feeding on maggots of medically important flies.] Kontyú 33(4):434-440. (In Japanese, with English summary.)
Both larvae and adults predaceous. Causes decrease in fly population, as number of *Saprinus* rises.

4482. **Nakasuji, F. 1965** [Difference of the activity of hymenopterous parasites with special reference to the depth of the pupation site of fly larvae.] Kontyú 33(4):441-445. (In Japanese, with English summary.)
Sparangia nigra able to parasitize fly pupae at 4-6 cm. below surface, but searching is more efficient near surface. (Believed to be a misprint of *Spalangia nigra*).

Nakatani, S., joint author. *See* Fujito, S., et al., 1963.

4483. Nakatsugawa, T. and **Dahm, P. A. 1962** Activation of Guthion by Tissue Preparations from the American Cockroach. J. Econ. Ent. 55(5):594-599.

Manometric determinations made by inhibition of housefly cholinesterase.

4484. Nakatsugawa, T., Ishida, M. and **Dahm, P. A. 1965** Microsomal Epoxidation of Cyclodiene Insecticides. Biochem. Pharmacol. 14: 1853-1865.

CSMA strain of *M. domestica* and cockroach used as sources of insect enzymes. Epoxidase activities found in rabbit and rat liver microsomes and in insect homogenates.

4485. Nakatsugawa, T., Tolman, N. M. and **Dahm, P. A. 1968** Degradation and Activation of Parathion Analogs by Microsomal Enzymes. Biochem. Pharmacol. 17(8):1517-1528.

Used rat and rabbit livers and housefly abdomens for *in vitro* studies. Fly abdomen preparations used because they were more active with microsomal oxidations. ABSTRACT: BA 49:1138(122360), Dec. 1, 1968.

4486. Nakatsugawa, T., Tolman, N. M. and **Dahm, P. A. 1969** Metabolism of S^{35}-Parathion in the House Fly. J. Econ. Ent. 62(2):408-411.

Products of metabolism indicate oxidative enzymes involved in degradation and metabolism of parathion in *M. domestica.* ABSTRACT: RAE-B 57:185(661), Sept., 1969.

4487. Nakayama, I., Nagasawa, S. and **Shimizu, H. 1969** [Sterilizing effect of hempa on *Drosophila melanogaster* Meigen. Studies on the chemosterilants of insects. XV.] Botyu-Kagaku (Bull. Inst. Insect Contr.) 34(1):6-12. (In Japanese, with English summary.)

Drosophila melanogaster compared to *M. domestica vicina* in chemosterilant studies. Both species showed same rate of susceptibility to hempa. Main effect, inhibition of hatchability. ABSTRACTS: BA 51:2196 (22739), Feb. 15, 1970; RAE-B 58:276(1141), Aug., 1970.

Namihira, G., joint author. *See* Funatsu, M., et al., 1967.

Namiki, T., joint author. *See* Ikeshoji, T., et al., 1967.

4488. Nappi, A. J. and **Stoffolano, J. G., Jr. 1970** Hemocytic Reactions of *Musca domestica* L. Larvae Against the Nematode Parasite *Heterotylenchus autumnalis* Nickle. Second Intern. Congr. Parasitol., Proc. 56(4):246-247. (Abstract.)

Initial reaction of fly larva is lysis of oenocytoids, and release of material into haemolymph and/or cuticle of the parasite. A marked increase in the total number of non-fused hemocytes, while others aggregate (and fuse). Syncytia not limited to close vicinity of parasite.

4489. Narahashi, T. 1964 Insecticide Resistance and Nerve Sensitivity. *In* Symp. on Insecticide Resistance of Pest Insects, Tokyo, December 10, 1963. Jap. J. Med. Sci. Biol. 17(1):46-53.

Sensitivity of nerve lower in resistant strains than in those suceptible to DDT, BHC and Dieldrin. ABSTRACT: BA 47:3382(39906), Apr. 15, 1966.

Narahashi, T., joint author. *See* Yamasaki, T., et al., 1962; Tsukamoto, M., et al., 1965.

4490. Narasimham, M. J., Jr., Kristhuri-Thirumalachar, M. J. and **Ganla, V. G. 1967** Colour Vision and Colour Preference in the Tropical Eye-fly. Experientia 23(10):818-819. (German summary.)

Attraction toward eye appears to be due to the shine of the reflected light in the eye. One reference made to main receptors which are present in the housefly eye.

Nasrat, G. E., joint author. *See* Oppenoorth, F. J., et al., 1966.

Natalizi, G., joint author. *See* Bettini, S., et al., 1960.

4490a. Nauimin, I. V. 1960 [Reaction of houseflies *Musca domestica* L. to hexachlorane.] Med. Parazit. (Moskva) 29:731-733. (In Russian.)

4490b. Nava, P. A. 1968 Preferencias de *Musca domestica* L. en la selección de superficies de color. Ensayos en el laboratorio. [Preferences of *Musca domestica* L. in the selection of colored surfaces. Laboratory tests.] Folia Ent. Mex. for 1968, No. 18-19:100-101. Mexico, D. F.
In cage tests, using two color choices, black was always more attractive to flies than other colors. ABSTRACT: RAE-B 58:414(1730), Dec., 1970.

Nava, A., joint author. *See* Hecht, O., et al., 1968.

Nayak, B. C., joint author. *See* Misra, B., et al., 1966.

4491. Nayar, J. K. 1963 Effect of Synthetic "Queen Substance" (9-Oxodec-trans-2-Enoic Acid) on Ovary Development of the House-fly, *Musca domestica* L. (Correspondence.) Nature 197(4870):923-924.
Continuous supply of queen substance necessary for inhibition of ovary development. ABSTRACTS: BA 43:1355(16846), Aug. 15, 1963; RAE-B 52:65, Apr., 1964; TDB 60:690, July, 1963.

Naylor, A. F., joint author. *See* Boyes, J. W., et al., 1962.

Needham, P. H., joint author. *See* Elliott, M., et al., 1965, 1967.

Neil, W., joint author. *See* Tarshis, I., et al., 1970.

Nelms, N. M., joint author. *See* Wingo, C. W., et al., 1967.

4492. Nelson, D. R., Adams, T. S. and **Pomonis, J. G. 1969** Initial Studies on the Extraction of the Active Substance Inducing Monocoitic Behavior in House Flies, Black Blow Flies and Screw-worm Flies. J. Econ. Ent. 62(3):634-639.
Injected into virgin females, prevented mating. Is neither a lipid nor a protein. ABSTRACTS: BA 50:11409(118039), Nov. 1, 1969; RAE-B 57: 237(897), Nov., 1969.

Nelson, D. R., joint author. *See* Adams, T. S., et al., 1968.

Nelson, E. L., joint author. *See* O'Donnell, A. E., et al., 1967.

4493. Nelson, M. 1964 Some Properties of Uricase from the Housefly *Musca domestica.* Comp. Biochem. Physiol. 12(1):37-46.
Unlike mammalian uricase, can be extracted directly by aqueous solvents. ABSTRACT: BA 45:6741(84674), Oct. 1, 1964.

Neri, I., joint author. *See* Ascher, K. R. S., et al., 1961.

4494. Neselovskaya, V. K., Goldina, G. S. and **Ryazantsev, V. F. 1968** Primenenie khlorofosnykh mukhomorov v sochetanii s pishche-vymi primankami dlya bor'by s sinantropnymi mukhami. [Use of chlorophose fly-poisoners in combination with food baits for control of synanthropic flies.] Med. Parazitol. i Parazitarn. (Bolez.) 37(2):230-232. (English summary.)
When chlorophose is used in attractive food baits such as soft cheese (the most promising), liver or blood, fly kill is greatly increased. ABSTRACT: BA 50:1004(10438), Jan. 15, 1969.

Neselovskaya, V. K., joint author. *See* Sukhova, M. N., et al., 1965.

4495. Neuimin, L. V. 1960 Reaktsiya komatnykh mukh (*Musca domestica* L.) na geksakhloran. [The reaction of house flies, (*Musca domestica* L.) to hexachlorocyclohexane.] Med. Parazitol. i. Parazitarn. (Bolezni) 29(6):731-733. (English summary.)
Flies resistant to hexachlorocyclohexane (HCH) were bred from a natural population not previously in contact with the preparation. Reactions of resistant and sensitive flies in cages partially or entirely treated with HCH are reported. ABSTRACT: BA 38:1589(20446), June 1, 1962.

4496. Neuimin, L. V. 1961 Deistvie DDT i geksakhlorana na lichinok komnatnykh mukh (*Musca domestica* L.). [Effect of DDT and hexachlorane on housefly, *Musca domestica* L. larvae.] Med. Parazit. (Moskva) 30:214-218, 253. Referat. Zhur., Biol. 1961, No. 20E 102.
Percentage of flies hatched, pupal weights, life span of flies and sensitivity to the insecticide are reported for flies raised from substrates for breeding the larvae, which contained known amounts of DDT or HCH. ABSTRACT: BA 38(1):571(7753), Apr. 1, 1962.

4497. Nguy, V. D. and **Busvine, J. R. 1960** Studies on the Genetics of Resistance to Parathion and Malathion in the Housefly. Bull. W. H. O. (Geneva) 22(5):531-542.
Resistant strains crossed with normal. Both types of resistance inherited through single dominant gene pair. Two genes are either alleles or closely linked. Good description of rearing and crossing techniques. ABSTRACTS: RAE-B 50:80, Apr., 1962; TDB 58:1085, Sept., 1961.

4498. Nicholas, W. L. and **Hughes, R. D. 1970** *Heterotylenchus* sp. (Nematoda:Sphaerulariidae), a Parasite of the Australian Bush Fly, *Musca vetustissima*. J. Parasit. 56(1):116-122.
Closely related to *Heterotylenchus autumnalis*, parasite of face fly in N. America. Reports experimental infection of lab-bred flies. Life cycle and seasonal distribution under Australian conditions discussed.

Nicholas, W. L., joint author. *See* Hughes, R. D., et al., 1969.

4499. Nickel, G. A. and **Wagoner, D. E. 1968** A test for Meiotic-Drive Factors in the House Fly. Proc. N. Centr. Branch Ent. Soc. Am. 23(1):28.
F_2 progeny from matings scored for ratio distortion. Characteristics those for which 1:1 ratio was expected, but where 3:1 appeared, were retested to determine if a meiotic-drive factor (ratio distorter) had been induced. ABSTRACT: RAE-B 58:151(636), May, 1970.

4500. Nickel, C. A. and **Wagoner, D. E. 1970** Some New Mutants of House Flies and their Linkage Groups and Map Positions. J. Econ. Ent. 63(5):1385-1390.
Map positions determined for masked eyes (*Mk*), pointed wing (*pw*), stubby wings (*stw*), irregular veins (*Iv*). Other information.

Nickel, C. A., joint author. *See* Wagoner, D. E., et al., 1965, 1969.

4501. Nickle, W. R. 1967 *Heterotylenchus autumnalis* sp. n. (Nematoda: Sphaerulariidae), a Parasite of the Face Fly, *Musca autumnalis* De Geer. J. Parasit. 53:398-401.
Description and illustration of nematode parasite causing sterility in female *M. autumnalis*. ABSTRACTS: BA 48:6940(77613), Aug. 1, 1967; RAE-B 5:92(324), May, 1969.

Nickle, W. R., joint author. *See* Stoffolano, J. G., Jr., et al., 1966.

Niihara, T., joint author. *See* Fujito, S., et al., 1963.

Nipper, W. A., joint author. *See* Burns, E. C., et al., 1960.

Nishimoto, N., joint author. *See* Kobayashi, M., et al., 1967; Takemoto, T., et al., 1967.

Nishizawa, Y., joint author. *See* Casida, J. E., et al., 1966.

Nissim, S., joint author. *See* Ascher, K. R. S., et al., 1964.

4502. Nocerino, F. 1963 Nuevos jaulones para pruebas con insectos en condiciones similinaturales. [New cages for experiments with insects under essentially natural conditions.] Bol. Inform. Dirección Malariol. y San. Ambient. (Maracay) 3(2):59-65.
Adapted for various species including *M. domestica*. Mortality tables. ABSTRACTS: RAE-B 53:212, Nov., 1965; TDB 61:449, Apr., 1964.

4503. Nocerino, F. 1965 La Humedad Relativa y la Mortalidad de Insectos Expuestos a la Accion de Insecticidas. [Relative humidity and mortality of insects exposed to the action of insecticides.] Bol. Inform. Dirrección Malariol. y San. Ambient. 5(1):7-10.

Studied *M. domestica*, *Culex fatigans* and *Rhodnius prolixus*. Results of tests with dieldrin and cereclor presented in tabular form.

4504. Nocerino, F. 1967 Algunas consideraciones sobre la eficacia del dieldrin y del DDT antiresistente en relación con la naturaleza del substrato. [Some considerations on the effectiveness of dieldrin and of DDT with WARF Antiresistant in relation to the nature of the substrate.] Bol. Inform. Dirrection Malariol. y San. Ambient. 7(5): 257-267.

Insecticide tests primarily to control the vectors of Chagas' disease, indicated effectiveness influenced by wall coatings. Rubber-based paints reduced action of dieldrin and DDT. *M. domestica* and mosquito species also tested. Abstract: RAE-B 58:33(135), Feb. 1970.

4505. Nocerino, F. and **De Marquez, M. 1966** Comparación de las DL_{90} y DL_{50} en *Musca domestica* Cepas RR(Resistant) y SS(Susceptibles) Tratadas con una Solución Acetónica de WARF-DDT(1-5) por el Método March-Metcalf. [Comparison of RR(Resistant) and SS (Susceptible) strains of *Musca domestica* for DL_{90} and DL_{50} treated with a WARF-DDT (1-5) in acetone solution according to the method of March-Metcalf.] Bol. Inform. Dirección Malariol. y San. Ambient. 6(2):74-77.

Authors conclude that antiresistance represents a synergistic action which augments the insecticidal action of DDT, but is not capable of cancelling the condition of resistance acquired by the insects.

4506. Norris, K. R 1966 Notes on the Ecology of the Bushfly, *Musca vetustissima* Walk. (Diptera:Muscidae) in the Canberra District. Australian J. Zool. 14(6):1139-1156.

Has been bred from faeces of man, cattle, sheep, dog. Abstract: BA 48:3702(41081), Apr. 15, 1967.

Norris, K. R., joint author. *See* Waterhouse, D. F., et al., 1966; Paterson, H. E., et al., 1970.

4507. North, D. T. 1966 Sperm Storage in the House Fly. Effect on the Number of Dominant Lethal Mutations Recovered after Treatment with Several Alkylating Agents or their Nonalkylating Analogs. Genetics 54(1, pt. 2):S55, 352.

Storage reduces frequency of recovered dominant lethal mutations by more than 50 per cent. Phenomenon is dose and treatment dependent. Agents prevent its occurrence. Alkylating ability not directly responsible for this; may be due to carrier portion of molecule.

4508. North, D. T. 1967 Sperm Storage: Modification of Recovered Dominant Lethal Mutations Induced by Tretamine and Analogs. Mutat. Res. 4:225-228.

Effect of several alkylating agents on mature housefly sperm reported. Alkylating agents possess equal number of aziridinyl groups and a non-alkylating analog of these compounds.

North, D. T., joint author. *See* Holt, G. G., et al., 1970; Pomonis, J. G., et al., MS.

4509. Nuorteva, P. 1960 Studies on the Significance of Flies in the Transmission of Poliomyelitis. VI. On the Influence of the Icosaeonic Climatic Fluctuation on the Incidence of Poliomyelitis and on the

Occurrence of *Lucilia* Species in Finland. Ann. Ent. Fenn. 26(4): 273-280.
Dominance of *Lucilia sericata* coincides with increase in incidence of paralytic poliomyelitis, and rise in summer temperatures. ABSTRACT: BA 43:1956(25273), Sept. 15, 1963.

4510. **Nuorteva, P. 1961** Studies on the Significance of Flies in the Transmission of Poliomyelitis. VII. Attraction of Bowflies to a House by the Honeydew of the Aphid *Phorodon humuli* (Schrk.). Ann. Ent. Fenn. 27:51-53.
In this way aphids may cause an increase in the degree of contamination of fruits and vegetables with fly faeces.

4511. **Nuorteva, P. 1962** Lier-Ja Raatokärpästen esiintyminen urjalan havaintoasemalla vuonna 1961. [The occurrence of pollenine and calliphorine flies in Urjala during the year 1961.] Ylipainos Lounais-Harneen Luonto. 13:3-7. (In Finnish, with English summary.)
Decrease of dominance of genus *Lucilia* followed by practically complete absence of poliomyelitis.

4512. **Nuorteva, P. 1963** Die Rolle der Fliegen in der Epidemiologie der Poliomyelitis. [The role of flies in the epidemiology of poliomyelitis.] Anz. Schaedlingsk. 36:149-155. (In German, with English summary.)
Excellent summary of existing literature on the subject. ABSTRACT: BA 46:2105(26248), Mar. 15, 1965.

4513. **Nuorteva, P., Isokoski, M.** and **Laiho, K. 1967** Studies on the Possibilities of Using Blowflies (Dipt.) as Medicolegal Indicators in Finland. Ann. Ent. Fenn. 33(4):217-225.
Indoor cadavers infested chiefly by *Calliphora erythrocephala*. Reference included here because of subject matter interest in which *M. domestica* is sometimes concerned.

4514. **Nuorteva, P.** and **Skaren, U. 1960** Studies on the Significance of Flies in the Transmission of Poliomyelitis. V. Observations on the Attraction of Blowflies to the Carcasses of Micro-mammals in the Commune of Kuhmo, East Finland. Ann. Ent. Fenn. 26(3):221-226.
Lucilia richardsi believed well suited to transmit viruses of poliomyelitis. ABSTRACT: BA 43:1956(25274), Sept. 15, 1963.

O'Brien, J. P., joint author. *See* Fellig, J., et al., 1970.

4515. **O'Brien, R. D. 1961** Selective Toxicity of Insecticides. Advances in Pest Contr. Res. 4:75-116.
Over 100 references cited. Comparisons of insect toxicity include *M. domestica.*

4516. **O'Brien, R. D. 1961** Esterase Inhibition in Organophosphorus Poisoning of House Flies. J. Econ. Ent. 54(6):1161-1164.
Difference between enzyme levels of killed and surviving flies at any time was greatest for cholinesterase. Conclusion: Death due to cholinesterase inhibition. ABSTRACT: RAE-B 50:187, Sept., 1962.

4517. **O'Brien, R. D., Cheung, L.** and **Kimmel, E. C. 1965** Inhibition of the *a*-Glycerophosphate Shuttle in House Fly Flight Muscle. J. Insect Physiol. 11(9):1241-1246.
Shuttle permits oxidation of exogenous $NADH_2$. Neither the 6 analogs tested, nor the effective inhibitors of the shuttle (ring-substituted cinnamic acids) had any toxic effect. ABSTRACT: BA 47:1227(14899), Feb. 1, 1966.

O'Brien, R. D., joint author. *See* Krueger, H. R., et al., 1960; Mengle, D. C., et al., 1960; Uchida, T., et al., 1965; Toppozada Eldefrawi,

A., et al., 1970.

O'Carroll, F. M., joint author. *See* Bradbury, F. R., et al., 1966.

O'Connell, P. W., joint author. *See* Kohls, R. E., et al., 1966.

O'Conner, J. D., joint author. *See* Bieber, L. L., et al., 1969.

4518. Oda, T. 1966 [Studies on the dispersal of the housefly *Musca domestica vicina* by mark-and-release method.] Endemic Dis. Bull. Nagasaki Univ. 8(3):136-144. (In Japanese, with English summary.)
Flies marked with ^{32}P showed a short random flight and longer dispersal flight, depending upon proximity to dwellings. ABSTRACT: RAE-B 57: 113(401), July, 1969.

Oda, T., joint author. *See* Wada, Y., et al., 1963.

4519. Ode, P. E. 1966 Fluctuations of Face Fly, *Musca autumnalis*, Populations on Pasturing Cattle and Dispersal among Herds of Pasturing Cattle. Diss. Absts. 26(11):6927-6928.

4520. Ode, P. E. and **Matthysse, J. G. 1964** Face Fly Control Experiments. J. Econ. Ent. 57(5):631-636.
Dichlorvos and naled syrup baits, applied to cows were effective for 1 day, sometimes two. ABSTRACTS: BA 46:695(8723), Jan. 15, 1965; RAE-B 53:9, Jan., 1965.

4521. Ode, P. E. and **Matthysse, J. G. 1964** Feed Additive Larviciding to Control Face Fly. J. Econ. Ent. 57(5):637-640.
Several insecticides found effective as tested against first stage larvae. ABSTRACTS: BA 46:695(8724), Jan. 15, 1965; RAE-B 53:9, Jan., 1965.

4521a. Ode, P. E. and **Matthysse, J. G. 1967** Bionomics of the Face Fly *Musca autumnalis* DeGeer. Mem. Cornell Univ. Agric. Exp. Sta. No. 402. 91 pages.
Concerns population peaks (both sexes), mating, oviposition, age and color of most frequented cattle, day and night resting places, effects of temperature and weather. ABSTRACT: RAE-B 58:339(1393), Oct., 1970.

Odeneal, J. F., joint author. *See* Incho, H. H., et al., 1962, 1963.

4522. Odintsov, V. S. 1970 [Study of the central nervous system esterases of housefly larvae (*Musca domestica* L.) by ultramicroelectrophoresis on polyacrylamide gel.] Dokl. Akad. Nauk, SSSR 193: 1185-1188. (In Russian.)

4522a. Odintsov, V. S. and **Petrenko, V. S. 1969** Esterazy v ontogeneze komnatnykh mukh *Musca domestica*. [Esterases in the development of the house fly *Musca domestica*.] Zool. Zhur. 48(1):146-148. (English summary.)
Activity of esterases are studied and found to increase or decrease in developmental stages of the egg and pupae of *M. domestica*. Speculation concerning roles played by these enzymes. ABSTRACT: BA 50:9766 (101158), Sept. 15, 1969.

4523. Odintsov, V. S., et al. 1968 [Acetylcholinesterase activity in metamorphosis of house fly (Diptera: *Musca domestica* L.).] Ukr. Biokhim. Zhur. 40:608-610. (In Ukranian.)
See also Odyntsov, V. S., et al.

4524. O'Donnell, A. E. and **Axtell, R. C. 1965** Predation of *Fuscuropoda vegetans* (Acarina:Uropodidae) on the House Fly (*Musca domestica*). Ann. Ent. Soc. Am. 58:403-404.
Macrocheles muscaedomesticae preferred fly eggs to larvae; *F. vegetans* the reverse. ABSTRACTS: BA 46:6987(86780), Oct. 1, 1965; RAE-B 53:195, Oct., 1965; TDB 63:713, June, 1966.

4525. O'Donnell, A. E. and **Nelson, E. L. 1967** Predation of *Fuscuropoda vegetans* (Acarina:Uropodidae) and *Macrocheles muscaedomesticae* (Acarina:Macrochelidae) on the Eggs of the Little House Fly, *Fannia canicularis*. J. Kans. Ent. Soc. 40(3):441-443.
Average consumption for adult female *M. muscaedomesticae* greatly exceeds that of adult male and/or female *F. vegetans*. ABSTRACTS: BA 48:11232(124753), Dec. 15, 1967; RAE-B 58:91(359), March, 1970.

4526. Odyntsov, V. S. and **Petrenko, V. S. 1968** Aktyvnist'atsetylkholinesterazy pry metamorfozi kimnatnykh mukh (Diptera:*Musca domestica* L.). [Acetylcholine esterase activity in the metamorphosis of Diptera: *Musca domestica* L.] Ukr. Biokhim. Zhur. 40(6):608-610. (Russian and English summaries.)
By means of the Hestrin colorimeter method it was established that during metamorphosis the activity of acetylcholine esterase (AchE) increases in larvae and pupae of *M. domestica*. ABSTRACT: BA 50:13070 (134795), Dec. 15, 1969.

4527. Oftedal, P. 1961 Studies with Radioactive Yttrium in Flies. I. Retention and Distribution in *Drosophila* after Injection. Int. J. Radiat. Biol. 3:211-221. (Abstracts in French and German.)
After injection in male flies ^{91}Y citrate is completely retained. Distribution depends on injection site. Microscopically, 2 alternative patterns of distribution occur. Mechanisms of patterns discussed. (Included here as introductory to references which follow.)

4528. Duplicate reference. Withdrawn.

4529. Oftedal, P. 1961 Studies with Radioactive Yttrium in Flies. II. Retention and Distribution in *Drosophila* and *Musca* after Ingestion. Int. J. Radiat. Biol. 3:222-230. (Abstracts in French and German.)
90-99 per cent ^{91}Y citrate ingested is excreted within 2-3 days in *Drosophila*, within one week in *Musca*. That retained in *Drosophila* found in narrow sections of midgut; in *Musca* in a wide band in posterior region of midgut. Heart and pericardium also contain radioactivity.

Ogami, H., joint author. *See* Katsuda, Y., et al., 1966, 1967.

4530. Ogata, K. 1960 [Seasonal change of the fly fauna in a dwelling house in Kawasaki City.] Jap. J. Sanit. Zool. 11:17-21. Tokyo. (In Japanese, with English summary.)
Used fly catch ribbon. 20 spp. from living room, 14 from lavatory. Majority: *Fannia canicularis* (65·9 per cent) and *M. d. vicina* (31·4 per cent). Appears that both species do not occur abundantly the same season, at the same time. ABSTRACT: RAE-B 51:271, Dec., 1963.

4531. Ogata, K. 1961 [Observation on the entrance and exit of house flies in a Residence.] Jap. J. Sanit. Zool. 12(4):247-249. (In Japanese, with English summary.)
Flies were collected from ceilings at night, marked on scutum with color paints and released. Subsequent counts showed percentage remaining in house, plus new invaders.

4532. Ogata, K., Murata, M., Furuno, A., Uchida, S. and **Sasa, M. 1961** [Detection of the poliomyelitis viruses from flies in an epidemic area in Hokkaido, Japan.] Jap. J. Sanit. Zool. 12:165-168. Tokyo. (In Japanese, with English summary.)
Flies collected just after peak of epidemic by netting and bait trapping. Kept in deep freeze. Of 14 test samples, 3 positive for polio virus (type I), were *Phaenicia sericata*. ABSTRACT: RAE-B 51:272, Dec., 1963.

4533. Ogata, K., Nagai, N., Koshimizu, N., Kato, M. and **Wada, A. 1960** [Release studies on the dispersion of the house flies and blow flies in the suburban area of Kawasaki city, Japan.] J. Jap. Sanit. Zool. 11:181-188. Tokyo. (In Japanese, with English summary.)
Wild *M. d. vicina* and *Phaenicia* spp. trapped, marked by feeding with radioactive phosphorus. Released at night and recovered by catch ribbons and cage traps. Most moved within 100 m. of release point. *One* fly went 700 m. Topography influenced direction. ABSTRACT: RAE-B 51:272, Dec., 1963.

4534. Ogata, K. and **Suzuki, T. 1960** [Release studies on the dispersion of the lesser housefly, *Fannia canicularis*, in the residential area of Bibai, Hokkaido.] Botyu-Kagaku 25(2):51-57. Kyoto. (In Japanese, with English summary.)
F. canicularis reared from field-collected pupae, marked, released at evening in privy. Trapping in houses within 480 yards of release point for 6 days revealed less than 10 per cent recovered and all within 115 yards. Dispersal appeared to be random. ABSTRACT: RAE-B 52:27, Feb., 1964.

4535. Ogata, K. and **Tanaka, I. 1965** [Effectiveness of several compounds as house fly chemosterilants (Abstracts).] Jap. J. Exptl. Med. 35:24pp. (Proc. Monthly Meeting).
Metepa, hempa and apholate tested as possible sterilants for houseflies in lab. and field experiments. Results given.

4536. Ogata, K. and **Tanaka, I. 1967** [Notes on the mating competitiveness of hempa-treated and normal male houseflies.] Jap. J. Sanit. Zool. 18(1):32-34. (In Japanese, with English summary.)
Treated males found to be sexually as competitive as, or slightly less vigorous than, normal males. ABSTRACT: BA 51:2759(28573), Mar. 1, 1970.

4537. Ogata, K., Tanaka, I. and **Suzuki, T. 1966** [Observations on the chemosterilization of two alkylating agents, metepa and hempa on house flies.] Jap. J. Sanit. Zool. 17(3):201-204. (In Japanese, with English summary.)
Flies treated by topical application and bait methods. Results given for treatment of male or female of mating pair and amount of chemical required for effective sterilization. ABSTRACT: RAE-B 57:81(285), May, 1969.

Ogawa, S., joint author. *See* Kobayashi, M., et al., 1967; Takemoto, T., et al., 1967.

4538. Ogita, Z. 1962 Genetico-biochemical Analysis of the Enzyme-activities in the House Fly by Agar Gel Electrophoresis. Jap. J. Genet. 37(6):518-521.
Description of technique with zymograms of esterase and amylase in various phenotypes. ABSTRACT: BA 43:1388(17277), Sept. 1, 1963.

4539. Ogita, Z. 1968 [Genetico-biochemical study of amylase-isozymes in the housefly.] Jap. J. Genet. 43:434. (In Japanese.)

4540. Ogita, Z. I. and **Kasai, T. 1965** A Genetic Analysis of Synergistic Action on Sulfonamide Derivatives with DDT Against House Flies. Botyu-Kagaku (Bull. Inst. Insect Contr.) 30(4):119-128. (Japanese summary.)
Structural similarities between synergistic sulfonamides and hypoglycaemic sulphonylureas suggested investigations which showed that chemicals causing hypoglycaemia in mice were synergistic to DDT and *vice versa.* Genetic analysis also given. ABSTRACT: RAE-B 56:53(175), Mar., 1968.

4541. Ogita, Z. I. and **Kasai, T. 1965** Genetic Control of Multiple Molecular Forms of the Acid Phosphomonoesterases in the House Fly *Musca domestica.* Jap. J. Genet. 40(3):185-197.
Inheritance of electrophoretic variants controlled by a pair of co-dominant alleles of *Phos*-locus on 5th chromosome. ABSTRACT: BA 47:9970(115354), Dec. 15, 1966.

4542. Ogita, Z. I. and **Kasai, T. 1965** Genetic Control of Multiple Esterases in *Musca domestica* Jap. J. Genet. 40(1):1-14.
Esterase bands which are active upon naphthylacetate esters not correlated with the diazinon-resistance gene. ABSTRACT: BA 47:9970(115352), Dec. 15, 1966.

4543. Ogita, Z. I. and **Kasai, T. 1965** Genetico-biochemical Analysis of Specific Esterases in *Musca domestica.* Jap. J. Genet. 40(3):173-184.
Suggests that 4 types of esterases may be distinguished in housefly homogenates. ABSTRACT: BA 47-9970(115353), Dec. 15, 1966.

4544. Ogita, Z. I. and **Kasai, T. 1966** A Genetic Study of Resistance to Nicotine Sulfate in House Flies. Botyu-Kagaku. (Bull. Inst. Insect Contr.) 31(1):14-18. (Japanese summary.)
A strain, after selection for resistance to sulfate through 18-22 generations was found to have become 8 times as resistant to diazinon as parent susceptible strain; 1-3 times to DDT, γ-BHC and Sevin. Genetic analysis shows resistance to nicotine sulfate to be controlled by more than 1 factor. ABSTRACT: RAE-B 56:128(447), July, 1968.

Ogita, Z. I., joint author. *See* Kasai, T., et al., 1965.

4545. Ogushi, K. and **Tokumitsu, I. 1964** [Influences of relative humidity and temperature on the effect of several insecticides.] Jap. J. Sanit. Zool. 15(1):28-32. (In Japanese, with English summary.)
Effect of R.H. on knockdown by deposits was greatest with dichlorvos, fairly great with naled and trichlorphon, smaller with diazinon, fenchlorphos and fenitrothion, slight with fenthion. Temperature effects also reported. ABSTRACT: RAE-B 54:197, Oct., 1966.

4546. Ogushi, K. and **Tokumitsu, I. 1968** Studies on Some Factors in Laboratory Breeding Affecting the Suceptibility to Insecticides in the Housefly. Contributions to the Study of the Laboratory Test Methods of Insecticides. Sanit. Insecticide Assoc., Japan (for) 1968: 1-6.
Time from emergence to test on fly had great influence on insecticide susceptibility. Median lethal doses of insecticide should be demonstrated by the unit of microgramme per fly—*not* per gramme body weight of fly.

4547. Ogushi, K. and **Tokumitsu, I. 1969** Laboratory Evaluation of Dichlorvos Vapor Against Housefly, Common Mosquito and German Cockroach. Jap. J. Sanit. Zool. 20(3):208-211.
Effectiveness of 90 per cent mortality could be equated to concentration, (c), in air and exposure time, (t), by the following equation $c^n t = k$. Relationship between number of days, (t^1), after opening seal of dichlorvos resin dispenser and dose, (d), of released insecticide vapor expressed by, $d^{1 \cdot 351} + t^1 = 977 \cdot 24$. ABSTRACT: BA 51:5643(57899), May 15, 1970.

4548. Ogushi, K., Tokumitsu, I. and **Inoue, M. 1964** [Effect of several synergists on the mixture of pyrethrins and lindane to houseflies.] Jap. J. Sanit. Zool. 15(3):179-181. (In Japanese, with English summary.)
Evaluated by topical application; ratio of toxicants to synergists 1:5. LD_{50}'s used to report relative effectiveness of piperonyl butoxide, sulfoxide, S-421 and cynethrin. ABSTRACT: RAE-B 54:244, Dec., 1966.

4549. Ogushi, K., Tokumitsu, I. and **Iwata, T. 1966** [Effect of insecticides for house flies and German cockroaches by dipping test.] Jap. J. Sanit. Zool. 17(3):209-213. (In Japanese, with English summary.)
Effectiveness of 11 chlorinated-hydrocarbon or organophosphorus insecticides tested by immersion of insects for 3 minutes in aqueous dilutions of acetone or ethanol solutions of insecticides. Mortality recorded after 24 hours. ABSTRACT: RAE-B 57:82(287), May, 1969.

4550. Ogushi, K., Tokumitsu, I. and **Iwata, T. 1967** [Studies on laboratory evaluation of several insecticides against two colonies of houseflies from Kajiki, Kagoshima Prefecture.] Jap. J. Sanit. Zool. 18(4): 294-303. (In Japanese, with English summary.)
Concerns effects of 11 insecticides by contact to residue, topical application, spray and dipping on the two colonies of mixed susceptible and resistant populations. ABSTRACTS: BA 51:2194(221712), Feb. 15, 1970; RAE-B 58:323(1328), Oct., 1970.

4550a. Ogushi, K., Tokumitsu, I. and **Iwata, T. 1967** [Observation on the influences of solvents in the effect of insecticide emulsion to house fly adults.] Jap. J. Sanit. Zool. 18(4):306-308. (In Japanese, with English summary.)
Housefly kills were highest from papers impregnated with diazinon from an emulsion concentrate in which the solvent was kerosene. Xylene was less satisfactory. ABSTRACT: RAE-B 58:324(1331), Oct., 1970.

4551. Ogushi, K., Tokumitsu, I. and **Iwata, T. 1968** Study on Evaluation Methods of Effectiveness of Insecticides to Larvae of the Housefly and the Flesh Fly. Contributions to the Study of the Laboratory Test Methods of Insecticides. Sanit. Insecticide Assoc., Japan (for) 1968:6-15.
Worked with DDT, lindane, dieldrin, DDVP, and Dibrom. Emphasizes importance of solvents in larvicides.

4552. Ohgushi, P.-I. 1960 Studies on the Host Selection by *Nasonia vitripennis* (Walk.) Parasitic on Housefly Pupae. Physiol. Ecol. (Japan) 9(1):19-31.
This small pteromalid wasp was reared on pupae of *Musca, Phaenicia* and *Calliphora*. Studies were concerned with host preference, physiological and reproductive differences and longevity. ABSTRACT: BA 35: 4673(54301), Oct. 1, 1960.

4553. Ohkawa, H., Eto, M., Oshima, Y., Tanaka, F. and **Umeda, K. 1968** Two Types of Carboxyesterase Degrading Malathion in Resistant Houseflies and their Inhibition by Synergists. Botyu-Kagaku 33(4): 139-145.
Esterases inhibiting organophosphorus esters have higher activity in resistant strains. Synergistic activity to malathion is shown by saligenin cyclic phenyl phosphate (K-2) which is thought to inhibit the carboxy esterase degrading malathion. ABSTRACT: BA 50:13070(134796), Dec. 15, 1969.

4554. Ohnishi, K. 1966 Studies on Cytochrome *b*. I. Isolation, Purification and Some Properties of Cytochrome *b* from Beef Heart Muscle. J. Biochem. 59:1-8.
Basic to subsequent studies involving *M. domestica*.

4555. Ohnishi, K. 1966 Studies on Cytochrome *b*. II. Crystallization and Some Properties of Cytochrome *b* from Larvae of the Housefly, *Musca domestica* L. J. Biochem. 59:9-16.
Crystalline cytochrome *b* obtained from larvae by ammonium sulfate fractionation, etc. ABSTRACT: BA 48:466(4960), Jan. 1, 1967.

4556. Ohnishi, K. 1966 Studies on Cytochrome *b*. III. Comparison of Cytochrome *b's* from Beef Heart Muscle and Larvae of the Housefly. J. Biochem. 59:17-23.

Beef heart cytochrome *b* made up of aggregates; larval material monomeric. Other differences. ABSTRACT: BA 47:9563(110723), Dec. 1, 1966.

4557. Ohnishi, K. and **Okuniiki, K. 1965** Crystallization of Cytochrome *b* from Larvae of the Housefly, *Musca domestica* L. Biochim. Biophys. Acta 99(3):575-578.

Paper describes the further purification of the water-soluble cytochrome *b* by chromatography on a hydroxylapatite column, its crystallization from ammonium sulfate solution and some properties of this cytochrome. ABSTRACT: BA 47:1640(19862), Feb. 15, 1966.

4558. Ohno, M., Hamada, M. and **Takahara, H. 1960** [Synthesis of 3, 4-methylenedioxyphenyl ethers and their synergistic activities on pyrethrins and allethrin.] Botyu-Kagaku 25(2):74-77. (In Japanese, with English summary.)

Among ethers tested, 2-(3, 4-methylenedioxyphenyl)-tetrahydrofuryl ether was best in synergistic action (1·4 times as effective as piperonyl butoxide). ABSTRACT: RAE-B 52:27, Feb., 1964.

Ohno, T., joint author. *See* Fujito, S., et al., 1962.

Ohshita, N., joint author. *See* Eto, M., et al., 1966.

4559. Ohtaki, T. 1965 [The resistant level of the housefly to several synthetic insecticides in Saitama Prefecture, Japan.] Jap. J. Sanit. Zool. 16(3):253-254. (In Japanese.)

Of 9 colonies tested, all were resistant to technical DDT or p, p' DDT, 3 to γ-BHC and dieldrin. (Dieldrin had not been used locally for fly control.) LD_{50}'s of 8 locally used insecticides were determined. ABSTRACT: RAE-B 56:132(461), July, 1968.

4560. Ohtaki, T., Kato, S., Kumakura, M. and **Kawaguchi, K. 1964** [Field tests with dimetilan baits for control of house flies.] Jap. J. Sanit. Zool. 15(1):33-35. (In Japanese, with English summary.)

Effectiveness of baits (100 mg. dimetilan) investigated in rural districts in dining rooms or kitchens, estimating numbers of flies present by means of ribbon traps. Most promising results obtained by combination of larvicidal treatment with use of 3 baits per house. ABSTRACT: RAE-B 54:198, Oct., 1966.

4561. Ohtaki, T., et al., 1964 [Control tests of non-biting stable flies in poultryfarm.] Jap. J. Sanit. Zool. 15(3):193-198. (In Japanese, with English summary.)

Fenchlorphos, fenitrothion and trichlorphon applied to poultry droppings all gave good control of *Muscina stabulans* but populations of *M. domestica,* which did not breed in treated droppings, rose markedly with decline of *Muscina*. ABSTRACT: RAE-B 54:245, Dec., 1966.

4562. Okada, Y. and **Okunuki, K. 1969** Studies on Cytochrome *b*-555 from Larvae of the Housefly, *Musca domestica* L. I. Purification and Properties of Cytochrome *b*-555. J. Biochem. (Tokyo) 65(4):581-596.

Concerns absorption spectra, prosthetic groups, sedimentation rate, molecular weight, and other properties. ABSTRACT: BA 51:571(5744), Jan. 1, 1970.

4562a. Okada, Y., et al., 1970 Studies of Cytochrome *b*-555 from Larvae of the Housefly, *Musca domestica* L. II. Isoelectric Point, Amino Acid Composition and Circular Dichroic Spectra. J. Biochem. (Tokyo) 67:487-496.

Concerns 2 or 3 components of a crude larval cytochrome *b*-555 preparation, separated by isoelectric focusing. ABSTRACT: BA 52:13089 (131185), Dec. 1, 1971.

Okazaki, T., joint author. *See* Kano, R., et al., 1965.

Okinata, K., joint author. *See* Steiner, L. F., et al., 1970.

4563. Okulov, V. P. 1961 Sluchai letnei depressii chislennosti u komnatnoi mukhi *Musca domestica* L. v Feodosii. [Cases of a summer depression in the number of house flies (*Musca domestica* L.) in Feodosiya.] Med. Parazitol. i Parazitarn. (Bolezni) 30(4):477-478.
A dry summer with a great deficiency in moisture resulted in a peak curve during 1957 with a dip during July and August. ABSTRACT: BA 39:670(8310), July 15, 1962.

4564. Okulov, V. P. 1962 Bor'ba s mukhami v feodosii n 1959 i 1960 gg. [Fly control in Feodosia in 1959 and 1960.] Med. Parazit. (Moskva) 31:351-355.
Improved measures of sanitary disposal and the use of chlorophos in 1960 cut down the frequency of sanitary treatment for flies. ABSTRACT: BA 40:1925(25276), Dec. 15, 1962.

4565. Okulov, V. P. 1969 [On the study of synanthropic flies in the eastern Crimea (Feodosia and its surroundings).] Med. Parazit. (Moskva) 38:744-746. (In Russian.)

Okunuki, K., joint author. *See* Ohnishi, K., et al., 1965; Okada, Y., et al., 1969.

4566. Oldroyd, H. 1964 The Natural History of Flies. London, Weidenfeld and Nicolson. xiv + 324 pp.
Discusses evolution of the Order Diptera, peculiarities of habit and anatomical structure of Families, and those Diptera having impact on man as carriers of disease. ABSTRACTS: BA 47:2498(29855), Mar. 15, 1966; RAE-A 52:235, May, 1964; RAE-B 52:94, May, 1964.

Oliver, J. E., joint author. *See* Akov, S., et al., 1968.

Olstad, G., joint author. *See* Adams, T. S., et al., 1970.

Olton, G. S., joint author. *See* Legner, E. F., et al., 1968, 1970.

4567. Omori, N., Suenaga, O., Ori, S. and **Shimogama, M. 1962** [A method of destroying the maggots of housefly and stablefly in animal manure by airtight vinyl cover.] Endemic Dis. Bull. Nagasaki Univ. 4(1):52-56. (In Japanese, with English summary.)
A description, and instructions for the use of the cover, are given. ABSTRACT: BA 45:1759(21936), Mar. 1, 1964.

4568. Omori, N. and **Tanikawa, T. 1960** [Invasion of flies and mosquitoes into the privy through the ventilator pipe.] Endemic Dis. Bull. Nagasaki Univ. 2(1):53-60. (In Japanese, with English summary.)
Investigations to find the numbers and species of flies showed *Sarcophaga peregrina* the most abundant of 23 species collected. *Lucilia sericata* and *M. domestica vicina* (in small numbers) taken in traps not shaded. ABSTRACTS: BA 36:6484(70192), Oct. 15, 1961; RAE-B 51:266, Dec., 1963.

Omori, N., joint author. *See* Buei, K., et al., 1965, 1966.

Oota, W., joint author. *See* Shimizu, F., et al., 1965.

4569. Oppenoorth, F. J. 1964 Some Cases of Resistance Caused by the Alteration of Enzymes. Proc. XII Int. Cong. Ent. London. Two pages MS.
Studies of recent results on DDT-dehydrochlorinases described. Three strains of *M. domestica* were involved in these studies. DDT-ase levels were determined in these strains.

4570. Oppenoorth, F. J. 1965 Biochemical Genetics of Insecticide Resistance (Diptera, Acarina). Ann. Rev. Ent. 10:185-206.
Five pages devoted to organophosphorus resistance and two to DDT resistance in the housefly. A review to March, 1964. ABSTRACTS: BA 46:8881(109505), Dec. 15, 1965; TDB 64:690, June, 1967.

4571. Oppenoorth, F. J. 1965 DDT-resistance in the Housefly Dependent on Different Mechanisms and the Action of Synergists. Meded. Landb. Hogesch. Opzoek Stns. Gent. 30(3):1390-1396. (Summaries in Flemish, French and German.)
DDT-ases produced by alleles of a gene on chromosome 5. Many-fold basis for resistance to DDT-synergist combinations. ABSTRACTS: BA 48:2363(25853), March 1, 1967; RAE-B 54:212, Nov., 1966.

4572. Oppenoorth, F. J. 1967 Two Types of Sesamex-suppressible Resistance in the Housefly. Ent. Exp. Appl. 10(1):75-86. (German summary.)
One controlled by 3rd chromosome, the other by 5th. ABSTRACTS: BA 48:10212(114764), Dec. 1, 1967; TDB 64:1362, Dec., 1967; RAE-B 55: 175(609), Sept., 1967.

4573. Oppenoorth, F. J. 1967 Biochemical Mechanisms of Insect Resistance to Anti-cholinesterases. (Abstract.) Biochem. J. 102(1):2P-3P.
Persistence essentially the result of selection of biochemically abberant individuals pre-existing in the original population. Oligogenes responsible for important part of resistance. Housefly relies on detoxication mechanisms in resistance to anti-cholinesterase.

4574. Oppenoorth, F. J. and **Houx, N. W. H. 1968** DDT Resistance in the Housefly Caused by Microsomal Degradation. Ent. Exp. Appl. 11(1):81-93. (German summary.)
Responsible gene (on third chromosome) designated DDT-*md* (microsomal detoxication). ABSTRACTS: BA 49:9253(102752), Oct. 1, 1968; RAE-B 56:246(907), Dec., 1968; TDB 66:165(300), Feb., 1969.

4575. Oppenoorth, F. J. and **Houx, N. W. H. 1968** Resistance in the Housefly Caused by Microsomal Oxidation. Overdruk Vit: Meded. Rijksfak. Landbouwwetensch. Gent. 33(3):641-646. (Summaries in Flemish, French and German.)
Genetic, toxicological and biochemical evidence given that oxidative microsomal degradation of DDT forms resistance mechanism in DDT and diazinon strain of housefly. ABSTRACT: RAE-B 58:340(1396), Oct., 1969.

4576. Oppenoorth, F. J. and **Nasrat, G. E. 1966** Genetics of Dieldrin and γ-BHC Resistance in the Housefly. Ent. Exp. Appl. 9(2):223-231. Amsterdam. (German summary.)
Several factors on different chromosomes believed responsible for total resistance. ABSTRACTS: BA 47:9970(115355), Dec. 15, 1966; RAE-B 54: 206, Nov., 1966; TDB 64:120, Jan., 1967.

4577. Oppenoorth, F. J. and **Van Asperen, K. 1960** Allelic Genes in the Housefly Producing Modified Enzymes that Cause Organophosphate Resistance. Science 132:298-299.
A mutant gene is present in resistant strains which produces an altered ali-esterase. Modified enzymes no longer irreversibly inhibited by oxygen analogs of insecticides to which strains are resistant, but can slowly convert them. In five of 6 strains the resistance was caused by this gene only. ABSTRACT: TDB 57:1319, Dec., 1960.

4578. Oppenoorth, F. J. and **Van Asperen, K. 1961** The Detoxication Enzymes Causing Organophosphate Resistance in the Housefly; Properties, Inhibitions and the Action of Inhibitors as Synergists.

Ent. Exp. Appl. 4(4):311-333.
Comparison of breakdown enzymes in OP-resistant houseflies and oli-esterol of susceptible flies shows differences as well as similarities. In general, both are rapidly phosphorylated by OP-compounds, but only in case of breakdown enzymes does dephosphorylation occur. ABSTRACTS: BA 38:597(8185), Apr. 15, 1962; RAE-B 50:195, Sept., 1962.

Oppenoorth, F. J., joint author. *See* Grigolo, A., et al., 1966; Franco, M. G., et al., 1962; ElBashir, S., et al., 1969; Van Asperen, K., et al., 1960.

Orchard, R. D., joint author. *See* Hoyer, R. F., et al., 1965; Plapp, F. W., Jr., et al., 1965.

4579. Ori, S. and **Shimogama, M. 1960** [Studies on the resting place of *Musca domestica vicina* in houses in semifarm villages near the city of Nagasaki.] Endemic Dis. Bull. Nagasaki Univ. 2(2):154-159. (In Japanese.)
Resting places for the housefly varied in the different types of houses and barns examined. Ceilings were the most preferred places, except in cattle barns. Daytime and nighttime resting places were different. ABSTRACT: BA 36:5571(58211), Sept. 1, 1961.

4580. Ori, S., Shimogama, M. and **Takatsuki, Y. 1960** [Studies of the methods of collecting flies. IV. On the effect of colored cage trap.] Endemic Dis. Bull. Nagasaki Univ. 2(3):229-235. (In Japanese, with English summary.)
The differences in radiated light from colored traps in hot or cold weather affects the numbers and species of flies drawn to them. ABSTRACT: BA 39:670(8311), July 15, 1962.

Ori, S., joint author. *See* Omori, N., et al., 1962; Suenaga, O., et al., 1962.

4581. Osborn, A. W., Shipp, E. and **Rodger, J. C. 1970** House Fly Fecundity in Relation to Density. J. Econ. Ent. 63(3):1020-1021.
Fecundity not reduced by adult densities up to 4·77 flies/inch, however, greater number of eggs were laid on water containers, etc. Adult flies can be maintained more densely than currently recommended. ABSTRACT: RAE-B 58:376(1548), Nov., 1970.

Osborn, A. W., joint author. *See* Shipp, E., et al., 1967.

Oshima, Y., joint author. *See* Ohkawa, H., et al., 1968; Eto, M., et al., 1963, 1966, 1968; Kobayashi, K., et al., 1969.

4582. Oshio, Y. and **Ikeuchi, M. 1960** On the Cattle, Horse and Swine Feces as Breeding Sources of House-fly (*Musca domestica vicina*). Jap. J. Vet. Sci. 22(suppl.):555.

4583. Oshio, Y. and **Ikeuchi, M. 1964** [Studies on the relation between live-stock keeping and occurrence of flies (4). On the observations of fly fauna in the four farms of Tokyo keeping a large number of swine, surveyed in May and September, 1963.] Jap. J. Sanit. Zool. 15(3):199-304. (In Japanese, with English summary.)
Of 48,254 flies collected on ribbon traps, over 75 per cent belonged to small species such as Drosophilids and Psychodids, 11·8 per cent were *M. domestica*, 7·9 per cent *Fannia canicularis*, 2 per cent *Muscina stabulans*. *Musca* occurred both inside and outside of dwellings. ABSTRACT: RAE-B 54:245, Dec., 1966.

4584. Oshio, Y., Ikeuchi, M. and **Malda, S. 1965** [Studies on the relation between keeping livestock and breeding flies. (5). On the observations of fly dispersal in the farm keeping several kinds of livestock in the limited area.] Jap. J. Sanit. Zool. 16(1):80-85. (In Japanese,

with English summary.)
M. domestica most numerous from places where it bred, i.e. pig housing. ABSTRACT: RAE-B 55:136(489), Aug., 1967.

Oshio, Y., joint author. *See* Ikeuchi, M., et al., 1965.

Osipova, L. S., joint author. *See* Lineva, V. A., et al., 1961.

Osman, F. M., joint author. *See* Hafez, M., et al., 1969, 1970; Metcalf, R. L., et al., 1966, 1967, 1969.

4585. Ozburn, G. W. and **Morrison, F. O. 1963** The Effect of Diluting a Colony of DDT Resistant Houseflies with Non-resistant Houseflies. Phytoprotection 44(1):32-36. (French summary.)
Dilution each generation caused loss of all resistance by generation 4. ABSTRACTS: BA 45:3596(44298), May 15, 1964; RAE-B 52:127, July, 1964; RAE-B 53:98, May, 1965.

4586. Ozburn, G. W. and **Morrison, F. O. 1965** The Pattern and Extent of DDT-resistance Development in House Fly (*Musca domestica*) Strains as Related to the Sex and Stage Exposed to Selective Pressure. Ann. Ent. Soc. Quebec 10(1):26-32. (French summary.)
Neither intensity of, nor stage subjected to DDT pressure is a major factor in assuring selection for resistance. Gene involved is *not* sex-linked. ABSTRACT: RAE-B 55:71(263), Apr., 1967.

Ozburn, G. W., joint author. *See* Morrison, F. O., et al., 1965.

4587. Pagano, S. 1964 Sanitary Landfill Operations in New York State. Publ. Health Repts. 79(6):543-548.
Included as an example of good engineering practice in the combined handling of nuisance, rodent and fly problems with reference to the disposal of urban waste.

Pai, C. H., joint author. *See* Chang, C. P., et al., 1965.

4588. Paik, Y. H. 1960 [Studies on insecticide resistance of medical insects in Korea. A preliminary report.] Botyu-Kagaku 25(1):1:5. Kyoto. (In Japanese, with English summary.)
Author records investigations demonstrating resistance to DDT in strains of *Pediculus humanus humanus* L. (*P. h. corporis* Deg.) and *M. domestica* L., from Seoul, Korea. ABSTRACT: RAE-B 52:27, Feb., 1964.

4589. Paik, Y. H. 1960 [On the sexual difference of susceptibility of the Korean housefly, *Musca domestica* L. to p,p'-DDT, γ-BHC and malathion.] Botyu-Kagaku 25(1):5-10. Kyoto. (In Japanese, with English summary.)
Insecticide deposited at 25 mg per sq. ft. in glass jars. Males show greater susceptibility. ABSTRACT: RAE-B 52:27, Feb., 1964.

4590. Paik, Y. H. 1960 [On the resistance of the Korean Housefly, *Musca domestica* L. to p, p'-DDT, γ-BHC and malathion.] Botyu-Kagaku 25(1):14-16. (In Japanese, with English summary.)
Tests on *M. domestica* indicated that females of Seoul strain were about 97 times as resistant to p, p' DDT, about 143 times to γ-BHC and about 4·6 times to malathion, as a susceptible laboratory strain. ABSTRACT: RAE-B 52:27, Feb., 1964.

4591. Paikin, D. M. and **Sazonova, I. N. 1961** Perspektivy primeneniya khlorofosa. [Prospects for the application of Dipterex.] Zashchita Rast. Vreditelei i Boleznei 7:36-37. Referat. Zhur., Biol. 1962, No. 10 Zh. 459. (Translation.)
Dipterex has been found to be very effective in controlling many insects, including *Musca domestica*. In an alkaline medium it is converted to DDVP. ABSTRACT: BA 41:610(8168), Jan. 15, 1963.

4592. Paim, G. V., et al., 1963 Capacidade da *Musca domestica* para albergar o *Toxoplasma gondii.* [Ability of *Musca domestica* to host *Toxoplasma gondii.*] Arq. Hig. Saúde Pub. (S. Paulo) 28:213-216.
Survival of the toxoplasms in domestic flies is of short duration. Probably does not multiply in arthropod carrier.

4593. Painter, R. R. and **Kilgore, W. W. 1964** Temporary and Permanent Sterilization of House Flies with Chemosterilants. J. Econ. Ent. 57:154-157.
Sixteen compounds were fed to recently emerged adults for 48 hours. At levels given, 9 were negative; 2 induced sterility with no oviposition; 3 were only temporary sterilants; 2 gave permanent sterility with oviposition. ABSTRACTS: BA 45:4729(57660), July 1, 1964; RAE-B 52:99, May, 1964; TDB 61:850, Aug., 1964.

4594. Painter, R. R. and **Kilgore, W. W. 1965** Chemosterilant Effect of 5-Fluoroorotic Acid on House Flies. J. Econ. Ent. 58:888-891.
Feeding at a concentration of 1·0 per cent for 48 hours after emergence produced permanent sterilization, eliminating oviposition. ABSTRACTS: BA 47:2051(24609), Mar. 1, 1966; RAE-B 54:13, Jan., 1966; TDB 63: 491, Apr., 1966.

4594a. Painter, R. R. and **Kilgore, W. W. 1967** Some Physical and Chemical Characteristics of Normal Eggs, Larvae, and Chorions of the House Fly *Musca domestica.* Ann. Ent. Soc. Am. 60(6):1163-1166.
Average weight for ova, 55 mg.; for chorions 5 mg.; for newly hatched larvae 47 mg. Moisture content of eggs just laid, 75 per cent; protein content, 10 per cent. ABSTRACTS: BA 49:3367(37644), Apr. 1, 1968; TDB 65:725(1565), May, 1968.

4595. Painter, R. R. and **Kilgore, W. W. 1967** The Effect of Apholate and Thiotepa on Nucleic Acid Synthesis and Nucleotide Ratios in Housefly Eggs. J. Insect. Physiol. 13(7):1105-1118.
Normal embryonic development produces a many-fold increase in DNA. No increase occurs in chemo-sterilized eggs. Apholate-sterilized eggs were lower in adenylic acid than normal fly-egg RNA. RNA of thiotepa-sterilized ova was slightly lower in guanylic acid, and contained a compound not in normal egg RNA. ABSTRACTS: BA 48:10184(11440), Nov. 15, 1967; RAE-B 56:59(196), Mar., 1968; TDB 64:1139, Oct., 1967.

Painter, R. R., joint author. *See* Kilgore, W. W., et al., 1962, 1964, 1966; Gadallah, A. I., et al., 1970.

4596. Pal, R. 1967 Genetics of Insect Vectors of Disease. W. H. O. Chronicle 21(8):343-350.
The article discusses recent advances in vector genetics (insecticide resistance, genetic systems, vector ability, standardized strains), the World Health Organization Program, and the Vector Genetics Information Service. *M. domestica* one of many species considered. ABSTRACT: BA 49:8277(92406), Sept. 15, 1968.

Pal, R., joint author. *See* Knipling, E. F., et al., 1968.

Palaveyeva, M., joint author. *See* Srbova, S., et al., 1962.

Palenzona, D., joint author. *See* Rubini, P. G., et al., 1967.

4597. Palii, V. F. and **Iirkovskii, G. G. 1962** Issledovaniya vliyaniya ioniziruyushchei radiatsii na razvitie nasekomykh. [Studies on the effect of ionizing radiation on the development of insects.] Sb. Ent. Rabot. Akad. Nauk Kirghizsk. SSR, Kirghizsk. Otd. Vses. Ent. Obshch. 1:70-79; Referat. Zhur., Biol. 1963, No. 14 E 12.
Report of a three year study of the effects of radiation on the housefly

and the white cabbage butterfly. Literature on radiation effect on animals, as related to fauna size, is reviewed. ABSTRACT: BA 45:3225(39936), May 1, 1964.

Palmas, G., joint author. *See* Bergamini, P. G., et al., 1970.

Palut, D., joint author. *See* Jampoler, I., et al., 1964.

4598. Panda, N. 1966 Effect of Larval Nutrition on the Tolerance of Adult *Musca nebulo* F. to Insecticides. Indian J. Ent. 28(1):94-97.
Eight compounds added to diet, made flies more tolerant to DDT. Four made them more susceptible. Report also concerns nicotine sulphate. ABSTRACT: RAE-B 55:202(701), Nov., 1967.

Pankaskie, J. E., joint author. *See* Sun, Y.-P., et al., 1963.

4599. Pant, R. and **Varma, R. 1968** Variation of Citrate in the Tissues of *Musca domestica* L. During Metamorphosis. Indian J. Biochem. 5(4):187-188.
No significant citrate titre change observed during pupal development; nor was there indication of relationship between citrate accumulation and oxygen uptake. ABSTRACT: TDB 66:633(1251), June, 1969.

4600. Parish, J. C. and **Arthur, B. W. 1965** Chemosterilization of House Flies Fed Certain Ethylenimine Derivatives. J. Econ. Ent. 58:699-702.
Compounds fed daily to flies in condensed milk and water. Data on concentrations producing various effects. Of chemosterilants tested, those most *toxic* to flies were: tepa, metepa, thiotepa, and 1-methanesulfonyl-aziridine. ABSTRACTS: BA 46:8881(109506), Dec. 15, 1965; RAE-B 53: 222, Nov., 1965; TDB 63:239, Feb., 1966.

4601. Parish, J. C. and **Arthur, B. W. 1965** Mammalian and Insect Metabolism of the Chemosterilant Thiotepa. J. Econ. Ent. 58(5):976-979.
P^{32}-labelled thiotepa was administered topically to 4 insect species, including *M. domestica*. Maximum absorption had occurred by 4 hours after treatment. Tepa (oxygen analog of thiotepa) was the only chloroform-soluble metabolite recovered from insects. ABSTRACTS: BA 47:2927 (34645), Apr. 1, 1966; RAE-B 54:16, Jan., 1966.

4602. Parker, A. H. 1962 Studies on the Diurnal Rhythms of the Housefly, *Musca domestica* L. in a Dry Tropical Environment. Acta Tropica. (Basle) 19(2):97-119, 13 figs., 15 refs.
Endogenous rhythms tested by various changes in light conditions: (1) constant dim light; (2) bright daytime illumination; (3) reversed illumination; (4) continuous bright light. Effect of low temperatures discussed. ABSTRACTS: BA 40:1618(21137), Dec., 1962; RAE-B 52:195, Oct., 1964; TDB 59:1119, Nov., 1962.

Parker, B. L., joint author. *See* Dewey, J. E., et al., 1965.

Parshad, R., joint author. *See* Sharma, G. P., et al., 1966.

Parsons, J. H., joint author. *See* Geering, Q. A., et al., 1965.

4603. Pasquali-Ronchetti, I. 1969 The Organization of the Sarcoplasmic Reticulum and T System in the Femoral Muscle of the Housefly, *Musca domestica*. J. Cell. Biol. 40(1):269-273.
Presents a three-dimensional reconstruction of the structural relationships of the T system and reticulum. A dyad in the femoral muscle is believed to be a variation of the typical triadic structure. ABSTRACT: BA 50:12502 (129151), Dec. 1, 1969.

4603a. Pasquali-Ronchetti, I. 1970 The Ultrastructural Organization of Femoral Muscles in *Musca domestica* (Diptera). Tissue Cell 2:339-354.
The muscles are organized as typical synchronous muscle. Description

is given of the microscopic components. ABSTRACT: BA 52:1554(14785), 15, 1971.

Pass, B. C., joint author. *See* Thurston, R., et al., 1966.

Passariello, S., joint author. *See* Filipponi, A., et al., 1969.

4604. Pastrana, A. N. 1967 Preferencias de *Musca domestica* L. en la seleccion de superficies de color. [Preferences of *Musca domestica* in the selection of colored surfaces.] Tesis Profesional. Instituto Politecnico Nacional. México, D. F. 33 pp., plus 7 tab. Also figs. and graphs.

An informative study. At most temperatures, flies show a strong preference for black over white or gray.

4605. Patel, N. G. 1961 Toxicity Studies of Some Proteolytic Enzymes to the House Fly (*Musca domestica* L.). Diss. Absts. 21(Pt. 4, No. 10): 2841-2842.

Used three enzymes from infected honey-bee larvae. Oral toxicity not evident with housefly, but toxicity by injection was shown in 6 species, *M. domestica* being one. Variation in toxicity to houseflies ascribed to sex, age, strain, and/or temperature coefficient. ABSTRACT: BA 37:2060 (20652), Mar., 1962.

4606. Patel, N. G. and **Cutkomp, L. K. 1961** The Toxicity of Enzyme Fractions of *Bacillus larvae*. J. Econ. Ent. 54:773-777.

Three enzyme fractions separated from crude extract. Fraction III studied intensively. Toxicity associated with proteolytic activity; was greater at higher temperatures. Toxicity scarcely affected by strain, age, or sex. ABSTRACT: RAE-B 50:152, July, 1962.

4607. Patel, N. G., Cutkomp, L. K. and **Ikeshoji, R. 1968** Ovarian Measurements in DDT-susceptible and DDT-resistant House Flies. J. Econ. Ent. 61(4):1079-1081.

Higher reproductive potential of DDT-susceptible strain (NAIDM) over DDT-resistant strains is positively correlated with quantitative measurements of ovaries. NAIDM had a lower LD_{50}, greater dry weight of ovaries, higher value of ovarian nitrogen. ABSTRACTS: BA 50:3736 (39229), Apr. 1, 1969; RAE-B 57:5(10), Jan., 1969.

4608. Paterson, H. E. 1963 On the Naming of the Indigenous Houseflies of the Ethiopian Region (Diptera:Muscidae). J. Ent. Soc. S. Africa 26:226-227.

Presents evidence for 2 subspecies of *M. domestica*; *M. d. cuthbertsoni* and *M. d. curviforceps*. ABSTRACTS: BA 45:2123(26539), Mar. 15, 1964; RAE-B 52:75, Apr., 1964; TDB 61:319, Mar., 1964.

4609. Paterson, H. E. 1964 Population Genetic Studies in Areas of Overlap of Two Subspecies of *Musca domestica* L. *In* Ecological Studies in Southern Africa. The Hague, Dr. W. Junk. (D. H. S. Davis, Editor.) pp. 244-254.

In spite of free exchange of genes of *M. d. calleva* wlk. and *M. d. curviforceps* Saccà and Rivossecchi, stable domestic and feral populations coexist in areas in which secondary intergradation occurs. Probable reasons given. ABSTRACT: RAE-B 53:184, Sept., 1965.

4610. Paterson, H. E. and **Norris, K. R. 1970** The *Musca sorbens* Complex: The Relative Status of the Australian and Two African Populations. Australian J. Zool. 18:231-245.

The two forms found in Africa, and the single type found in Australia, are probably three distinct species. Conclusion drawn from both morphometric and genetic data. ABSTRACT: BA 52:933(8581), Jan. 15, 1971.

Paterson, H. E., joint author. *See* Boyes, J. W., et al., 1964.

4611. Patnaik, B. and **Roy, S. P. 1966** On the Life Cycle of the Filariid *Stephanofilaria assamensis* Pande, 1936, in the Arthropod Vector *Musca conducens* Walter, 1859. Indian J. Anim. Hlth. 5(2):91-101.
Musca conducens is shown to be an intermediate host of *S. assamiensis*. ABSTRACTS: RAE-B 56:48(161), Feb., 1968; Helminthol. Abstr. 36(4), No. 3152, 1967.

4612. Patnaik, B. and **Roy, S. P. 1966** Experimental Infection of Laboratory Bred *Musca conducens* with Microfilaria of *Stephanofilaria assamensis*. Proc. Ind. Sci. Congr. 53(3):352-353.
A parasite of cattle, with *Musca* the alternate host.

Patterson, H. E., joint author. *See* Boyes, J. W., et al., 1964.

Patterson, N. A., joint author. *See* Pickett, A. D., et al., 1963.

4613. Paul, C. F., Dixit, R. S. and **Agarwal, P. N. 1963** Evaluation of the Insecticidal Properties of the Seed Oil and Leaf Extract of the Common Indian Neem, *Azadirachta indica* Linn. Sci. Cult. 29(8): 412-413.
Insecticidal activity against such insects as *Musca nebulo* was not exhibited by *Azadirachta indica* Linn. ABSTRACT: BA 47:1602(19391), Feb. 15, 1966.

Paul, C. F., joint author. *See* Agarwal, P. N., et al., 1963.

Paul, J. R., joint author. *See* Gudnadottir, M., et al., 1960.

4613a. Paul, R. K. and **Perti, S. L. 1962** Studies on Contact Toxicity Part V. —Residual Activity of DDT applied in Admixture with Glue Dichromate on Mud Surface. Defence Sci. J. 12(3):298-308.
This combination, both in water dispersion and in emulsion, enhanced residual activity against *Culex fatigans* and *Musca nebulo*.

Paul, R. K., joint author. *See* Misra, J. N., et al., 1963.

Paul, Y. I., Jr., joint author. *See* Riordan, J. T., et al., 1961.

4614. Pausch, R. D. 1969 A Laboratory Evaluation of Baits and Chemosterilants on the Little House Fly. J. Econ. Ent. 62(1):25-28.
Of 18 chemicals tested, 6 effected sterility at nontoxic concentrations. Hempa, metepa, and tepa were the most satisfactory. ABSTRACTS: BA 51:539(5387), Jan. 1, 1970; TDB 66:634(1256), June, 1969; RAE-B 57: 116(414), July, 1969.

Payne, L. K., joint author. *See* Weiden, M. H. J., et al., 1965.

Paz, M., joint author. *See* Mer, G. G., et al., 1960.

Peak, L., joint author. *See* Metcalf, R. L., et al., 1965.

4615. Pearce, G. W., Schoof, H. F. and **Quarterman, K. D. 1961** Insecticidal Vapours for Aircraft Disinsection. Bull. W. H. O. 24(4-5):611-616. (Summary in French.)
Describes apparatus used in screening insecticides for vapor toxicity. Reports remarkable vapor toxicity of DDVP (0, 0-dimethyl-2, 2-dichlorovinyl phosphate). ABSTRACT: RAE-B 51:116, June, 1963.

Pearce, G. W., joint author. *See* Mattson, A. M., et al., 1960; Cline, R. E., et al., 1963, 1966; Miles, J. W., et al., 1962; Perry, A. S., et al., 1964.

4616. Pearincott, J. V. 1960 Changes in the Lipid Content During Growth and Metamorphosis of the House-fly (*Musca domestica* Linnaeus). J. Cell. and Comp. Physiol. 55:167-174.
Cholesterol is probably a tissue constituent rather than an energy source. Fatty acid an important source of energy during metamorphosis. Phos-

pholipids and fatty acids increase during larval growth, and decrease in the pupal stage, the fatty acids much more rapidly.

Pearson, B. C., joint author. *See* Elliott, M., et al., 1965, 1967.

4617. Peck, J. H. 1969 Arthropod Predators of Immature Diptera Developing in Poultry Droppings in Northern California. II. Laboratory Studies on Feeding Behavior and Predation Potential of Selected Species. J. Med. Ent. 6(2):168-171.

Macrocheles muscaedomesticae, predator on immature *Musca domestica*, inflicts mortality chiefly on eggs and first stage larvae. *Ophyra leucostoma* larvae kill larvae of *M. domestica*, *Muscina stabulans*, *Fannia canicularis* and *Aldrichina grahami*. ABSTRACT: BA 51:6803(69495), June 15, 1970.

4618. Peck, J. H. and **Anderson, J. R. 1969** Arthropod Predators of Immature Diptera Developing in Poultry Droppings in Northern California. I. Determination, Seasonal Abundance and Natural Cohabitation with Prey. J. Med. Ent. 6(2):163-167.

Predators identified and studied. Seasonal abundance and association with prey determined for 10 species. ABSTRACT: BA 51:6803(69494), June 15, 1970.

4619. Peck, J. H. and **Anderson, J. R. 1970** Influence of Poultry-manure-removal Schedules on Various Diptera Larvae and Selected Arthropod Predators. J. Econ. Ent. 63(1):82-90.

Abstension from manure removal favored predators; weekly or biweekly removal favored dipterous larvae. ABSTRACT: BA 51:9093(92558), Aug. 15, 1970.

4620. Pedler, C. and **Goodland, H. 1965** The Compound Eye and First Optic Ganglion of the Fly. A Light and Electron Microscopic Study. J. Roy. Micro. Soc. 84:161-179.

Unlikely that rhabdomes of a particular ommatidium can be differentially stimulated. A new variety of synapse is reported in the first optic ganglion. ABSTRACT: BA 47:1227(14899), Feb. 1, 1966.

4621. Peffly, R. L. 1961 Practical Fly Control. Indian and Trop. Hlth. 4:11 pp.

Concerns districts in which Arabian American Oil Co. operates. A report of conditions and control measures in use. Stresses need for health education and practice of basic sanitation.

4622. Pegg, E. J. 1970 *Toxacara canis* Transmitted by the Common House-fly to the Canine Host. Vet. Rec. 86(17):522 (April 25).

Perdue, J. M., joint author. *See* Jones, C. M., et al., 1967.

4623. Perez de Talens, A. F., et al. 1970 Landing Reaction of *Musca domestica*: Dependence on Dimensions and Position of the Stimulus. J. Exp. Biol. 52:233-256.

4623a. Peris, S. V. 1968 Una nueva especie de *Fannia* de los Pirineos centrales (Diptera, Muscidae). [A new *Fannia* species from the central Pyrenees (Diptera, Muscidae).] Bol. R. Soc. Española Hist. Natur. SECC Biol. 66:5-10. (English summary.)

Fannia liduskae, from the Spanish Central Pyrenees, is described. ABSTRACT: BA 52:1536(14545), Feb. 1, 1971.

4624. Peris, S. V. and **Llorente, V. 1963** Notas sobre Muscini paleárticos y revisión de las especies españolas (Diptera, Muscidae). [Notes on palaearctic Muscini and a revision of the Spanish species.] Bol. Soc. Esp. Hist. Nat. (Madrid) B61:209-269.

Perlman, D., joint author. *See* Ristich, S. S., et al., 1965.

4625. Perlstein, J. M. 1965 Pictorial Keys to Common Domestic Flies in California. Calif. Vector Views 12(5): no page numbers.

4626. Perry, A. S. 1960 Metabolism of Insecticides by Various Insect Species. J. Agric. Fd. Chem. 8(4):252-272. (Symposium on the Mechanism of Action of Pesticide Chemicals.)
Several metabolic pathways, depending on the chemical structure of the compound, and the species against which it is used; e.g. DDT is metabolized by houseflies, body lice, certain mosquitoes, and other insects, but process follows 4 or 5 pathways.

4627. Perry, A. S. 1964 The Physiology of Insecticide Resistance by Insects. Chapter 6(pp. 285-378) of "Physiology of Insecta" Vol. 3. Academic Press Inc., New York.
An excellent discussion of the subject, with full bibliography. ABSTRACT: BA 46:5516(68630), Aug. 1, 1965.

4628. Perry, A. S. 1970 Studies on Microsomal Cytochrome P-450 in Resistant and Susceptible Houseflies. Life Sci. 9:335-350.

4629. Perry, A. S., Hennessy, D. J. and **Miles, J. W. 1967** Comparative Toxicity and Metabolism of p, p'-DDT and Various Substituted DDT-Derivatives by Susceptible and Resistant House Flies. J. Econ. Ent. 60:568-573.
Compound most toxic to DDT-resistant flies was *o*-CI-DDT. (Reduces enzymatic dehydrochlorination.) Deuterated-DDT was only slightly more toxic than DDT. ABSTRACTS: BA 48:6928(77479), Aug. 1, 1967; RAE-B 55:147(517), Aug., 1967; TDB 64:1038, Sept., 1967.

4630. Perry, A. S. and **Miller, S.** (assisted by **A. J. Buckner). 1965** The Essential Role of Folic Acid and the Effect of Antimetabolites on Growth and Metamorphosis of Housefly Larvae, *Musca domestica* L. J. Insect Physiol. London 11(9):1277-1287.
Larvae depend completely on dietary folic acid (PGA) for normal growth and metamorphosis. They cannot synthesize this from available precursors. ABSTRACTS: BA 47:1227(14900), Feb. 1, 1966; RAE-B 54: 218, Nov., 1966; TDB 63:101, Jan., 1966.

4631. Perry, A. S., Pearce, G. W. and **Buckner, A. J. 1964** The Absorption, Distribution and Fate of C^{14}-aldrin and C^{14}-dieldrin by Susceptible and Resistant House Flies. J. Econ. Ent. 57(6):867-872.
Aldrin is epoxidized to dieldrin by both susceptible and dieldrin-resistant flies. No difference was found in the distribution of either substance in the head, thorax and abdomen of susceptible and resistant house flies. ABSTRACTS: BA 46:2539(31683), Apr. 1, 1965; RAE-B 53:57, Mar., 1965; TDB 62:476, May, 1965.

Perry, A. S., joint author. *See* Miller, S., et al., 1965, 1968.

4632. Pershad, S. B. and **Naidu, M. B. 1966** Sexual Sterility Induced in the House Fly by Contact Exposure to Apholate. J. Econ. Ent. 59(4): 948-950.
A 12 hour exposure, given in intermittent doses of 2 hours per day for 6 consecutive days, produced almost complete sterility with but low mortality. ABSTRACTS: BA 47:10325(119331), Dec. 15, 1966; RAE-B 54:210, Nov., 1966.

Persoons, C., joint author. *See* Krampitz, H. E., et al., 1967.

4633. Perti, S. L. and **Dixit, R. S. 1965** Control of Insects by Thermal Vaporization of Insecticides. Labdev. J. Sci. Technol. 3(1):62-63. Kanpur.
Concerns tests run on *M. domestica nebulo,* with an electric vaporiser dispensing γ-BHC. Knockdown and kill were high in a closed room. ABSTRACT: RAE-B 54:83, Apr., 1966.

4634. Perti, S. L., Dixit, R. S. and **Chadha, D. B. 1966** Selection of DDT-resistant Strains of Houseflies in the Laboratory. Labdev. J. Sci. Technol. 4:69-70. Kanpur, U.P.

M. d. nebulo surviving DDT were bred with larvae being reared on medium containing DDT. After 72-85 generations no resistance developed. Concluded that specific genes for resistance were not present in strains investigated. ABSTRACT: RAE-B 54:142, Aug., 1966.

Perti, S. L., joint author. *See* Agarwal, P. N., et al., 1963; Damodar, P., et al., 1962, 1964; Dixit, R. S., et al., 1962, 1963; Misra, J. N., et al., 1963; Paul, R. K., et al., 1962; Wal, Y. C., et al., 1962; Rajak, R. L., et al., 1967, 1968.

4635. Pesson, P. and **Ramade, F. 1960** Etude d'une souche de *Musca domestica* L. résistante au lindane. Influence de divers facteurs sur cette résistance. [Study on a strain of *Musca domestica* L. resistant to lindane. Influence of various factors on resistance.] Ann. Inst. Nat. Agron. 46:15-36.

After 75 generations, treated adults were up to 50,000 times as resistant at LD_{50} level as susceptible strain with increased resistance to aldrin, dieldrin and endrin, and tolerance to DDT. ABSTRACT: RAE-B 53:112, June, 1965.

4636. Pesson, P. and **Ramade, F. (1964) 1965** Modalités de l'acquisition et de la perte de résistence au lindane dans une population de *Musca domestica.* [Methods of acquisition and loss of resistance to lindane in a population of *M. domestica.*] Phytiat.-Phytopharm. 13(4):203-209. Paris.

Results of tests show that the build up and loss of resistance were progressive and continuous. It is suggested that the physiological changes undergone by adults may be cumulative, though underlying genetic factors intervene. ABSTRACT: RAE-B 54:142, Aug., 1966.

4637. Peters, H. 1965 The House Fly as a Disease Carrier. Pest Control 33(5):16-20.

A discussion of a fly survey of 21 months in Stuttgart, Germany, with reference to the species found and their relationship to humans and establishments used to prepare human food. ABSTRACT: BA 46:7292 (90618), Oct. 15, 1965.

4638. Peters, R. F. 1963 Urbanizations's Impact on the Poultry Industry. Nuisances and Co-existence. Calif. Vector Views 10(11):69-72.

Stresses *M. domestica, Fannia canicularis* and *Phaenicia* sp. Notes that fly problem is worsening, and urges Poultry Industry to move toward neutralizing public criticism.

4639. Peters, R. F. 1968 The Three Rs of Vector Control. Proc. Pap. Ann. Conf. Calif. Mosq. Contr. Assoc. Inc. 36:70-71.

Concerns refuse, rats, and reduction of noxious insects, particularly mosquitoes and synanthropic flies.

Peterson, D., joint author. *See* Hildenbrant, G. R., et al., 1969.

Peterson, L. G., joint author. *See* Larsen, J. R., et al., 1966.

Petrelli, M. G., joint author. *See* Filipponi, A., et al., 1966, 1967, 1969.

Petrenko, V. S., joint author. *See* Odintsov (Odyntsov), V. S., et al., 1968, 1969.

4640. Petrov, V. I. 1967 [Sanitary-hygienic assessment of the method of planned refuse disposal of apartments.] Gig. I Sanit. 32:96-98. (In Russian.)

4641. Petrov, V. P. 1964 O sezonnosti zabolevaemosti dizenteriei v Dnepropetrovskoi oblasti za poslednie gody. [Seasonal incidence of dysentery in Dnepropetrovsk Oblast in recent years.] Zhur. Mikrobiol. Epidemiol. i Immunobiol. 41(12):19-21. (English summary.)
Observations in three regions correlate the seasonal incidence of dysentery having a peak in August-October with the so-called "fly factor". ABSTRACT: BA 46:6907(85783), Oct. 1, 1965.

4642. Petrova, A. D. 1960 [Some data on mites of the Family Macrochelidae Vitz., spread by synanthropic flies.] Med. Parazit. (Mosk) 29:211-213.

4643. Petrova, A. D. 1964 Rol' sinantropnykh mukh v rasprostranenii kleshchei semeistva Macrochelidae Vitzt., 1930. [Role of synanthropic flies in the spread of Macrochelidae mites.] Med. Parazit. i Parazitarn. (Bolezni) 33(5):553-557. (English summary.)
Macrocheles muscaedomesticae was most frequently found on flies, its preference being with the family Muscidae. Several other species were found. Numbers of mites per fly increased in the autumn. ABSTRACT: BA 46:6621(82207), Sept. 15, 1965.

4644. Petrova, B. K. 1966 [Some data on the ecology and behaviour of pasture flies (Diptera) in the southern part of the Maritime Territory, (USSR).] Ent. Obozrenie 45(1):76-82. In Russian. English Translation in Ent. Rev. 45(1):42-45.
Includes 3 species of *Musca* (*M. amica*, *M. convexifrons*, *M. tempestiva*). ABSTRACTS: BA 48:1418(15352), Feb. 1, 1967; RAE-B 55:158(553), Sept., 1967.

4645. Petzelt, C. and **Bier, K. 1970** Hemmung und Induktion von Proteinsynthesen durch Actinomycin in den wachsenden Oocyten von *Musca domestica*. [Inhibition and induction of protein synthesis by actinomycin in the growing oocytes of *Musca domestica*.] Wilhelm Roux'Arch. Entwicklungsmech. Organismen 164(4):341-358. (English summary.)
The very high RNA synthesis in nurse cells and follicle epithelium is completely blocked by actinomycin. Protein synthesis in the euplasm of the growing oocyte is high and remains so, despite absence of RNA. ABSTRACT: BA 51:10895(110680), Oct. 1, 1970.

4646. Petzelt, C. and **Bier, K. 1970** Synthese der Haemolymphprotiene und die Aufnahme der Dotterfraktion in die Oocyte unter Actinomycin-Einfluss. (Untersuchungen an *Musca domestica*.) [Synthesis of haemolymph protein and the uptake of the yolk fraction in the oocyte during actinomycin treatment. (Studies on *Musca domestica*).] Wilhelm Roux' Arch. Entwicklungsmech. Organismen. 164(4):359-366. (English summary.)
Used autoradiographic and electrophoretic techniques. Yolk protein uptake is inhibited by actinomycin, but synthesis continues as normal.

4647. Petzsch, H. 1960 Zur Toxizität des Fliegenpilzes (*Amanita muscaria* L.) für Diptera, inskesondere die Grosse Stubenfliege (*Musca domestica* L.). [The toxicity of the fly agaric *Amanita muscaria* to Diptera, especially the large house fly *Musca domestica* L.] Beitr. Ent. 10(3/4):405-409. (English and Russian summaries.)
Suggests investigations to isolate the toxic principle in the fungus, and perhaps put it to use in overcoming the resistance problem. ABSTRACTS: BA 36:6818(73354), Nov. 1, 1961; RAE-B 50:86, Apr., 1962.

4648. Petzsch, H. 1962 Inwiefern könnten Fliegen eine Mittlerolle bei der Tollwut (Lyssa)-Ausbreitung unter Wild-und Haussaugetieren spielen? Ein Diskussionsbeitrag. [To what extent may flies play an intermediate role in the transmission of rabies among wild and domestic animals? A discussion.] Anz. Schädlingsk. 35(8):123-125. (English, French and Russian summaries.)

The question of rabies transmission by necrophagous Diptera is discussed. Three possible routes of transmission are presented. ABSTRACT: BA 44: 952(12824), Nov. 1, 1963.

Pfadt, R. E., joint author. *See* Larsen, J. R., et al., 1966.

4649. Philleo, W. W., Schonbrod, R. D. and **Terriere, L. C. 1965** Methylene-dioxyphenyl Compounds as Inhibitors of the Hydroxylation of Naphthalene in Houseflies. J. Agric. Fd. Chem. 13(2):113-115. Easton, Pa.

Five commercial and nine non-commercial compounds containing this structure proved inhibitory. Of the commercial synergists piperonyl cyclonene is the most potent inhibitor of microsomal hydroxylation. Several commercial synergists proved effective with naphthalene. ABSTRACT: RAE-B 53:196, Oct., 1965.

Philleo, W. W., joint author. *See* Schonbrod, R. D., et al., 1965.

Piantelli, F., joint author. *See* Bergamini, P. G., et al., 1970.

Pickard, E., joint author. *See* Smith, G. E., et al., 1965; Breeland, S. G., et al., 1965.

4650. Pickens, L. G., Morgan, N. O., Hartsock, J. G. and **Smith, J. W. 1967** Dispersal Patterns and Populations of the House Fly Affected by Sanitation and Weather in Rural Maryland. J. Econ. Ent. 60:1250-1255.

Rate of dispersal increased when temperatures were above 53°F and suitable breeding materials were scarce at emergence sites. Age and sex did not influence rate or direction of dispersal. Flies tend to move against a 2-7 mph wind. ABSTRACTS: BA 49:1396(15637), Feb. 1, 1968; RAE-B 56:62(205), Mar., 1968; TDB 64:538(1175), Apr., 1968.

4651. Pickens, L. G., Morgan, N. O. and **Thimijan, R. W. 1969** House Fly Response to Fluorescent Lamps: Influenced by Fly Age and Nutrition, Air Temperature, and Position of Lamps. J. Econ. Ent. 62(3): 536-539.

Used blacklight (BL) and blacklight blue (BLB) at elevations of 0·6 m. and 2 m. Responses differed according to age, sex, and hunger of the flies. BL lamps most attractive when placed near east end of barn. ABSTRACTS: BA 50:11363(117517), Nov. 1, 1969; RAE-B 57:235(892), Nov., 1969; TDB 66:1052(2045), Oct., 1969.

Pickens, L. G., joint author. *See* Morgan, N. O., et al., 1967, 1968, 1970; Thimijan, R. W., et al., 1970.

4652. Pickett, A. D. and **Patterson, N. A. 1963** Arsenates: Effect on Fecundity in Some Diptera. Science 140(3566):493-494.

M. domestica one of the four species studied. Feeding of sublethal doses to young adults reduces egg production. Calcium and lead arsenate were used in tests. ABSTRACT: BA 45:2869(35574), Apr. 15, 1964.

4653. Pieper, G. R. 1965 House Fly (Diptera) ATP-ase and its Inhibition by Insecticidal Organotin Compounds (Abstract). Diss. Absts. 26(4):1897-1898.

4654. Pieper, G. R. and **Casida, J. E. 1965** House Fly Adenosine Triphosphatases and their Inhibition by Insecticidal Organotin Compounds. J. Econ. Ent. 58:392-400.

Trisubstituted organotins proved very active as ATP-ase inhibitors, also as toxicants. Insecticidal action of trialkyltins and oligomycin may result from interference with ATP-ase activity or related processes associated with oxidative phosphorylation. ABSTRACTS: BA 47:2506(29953), Mar. 15, 1966; RAE-B 53:177, Sept., 1965.

Pier, A. C., joint author. *See* Richard, J. L., et al., 1966.

4655. Pigatti, A. and **Mello, E. J. R. 1961** Acão dos insecticidas organicos sôbre larvas do mosquito *Culex pipiens fatigans* Wied. e sôbre môscas domésticas (*Musca domestica* L.) do municipio de São Paulo. [The action of organic insecticides on larvae of *Culex pipiens fatigans* (Wied.) and *Musca domestica* L. in São Paulo.] Arq. Inst. Biol. (São Paulo) 28:101-112. Illus. (English summary.)

Houseflies were exposed on glass surfaces to residues of several insecticides and LD_{50}'s obtained for DDT, BHC, dieldrin, malathion, Diazinon, and Dipterex. ABSTRACTS: BA 40:1610(21043), Dec., 1962; RAE-B 52:84, May, 1964.

Pigatti, A., joint author. *See* Mello, D., et al., 1961; Mello, E. J. R., et al., 1961; Queiroz, J. C., et al., 1962.

Pigatti, P., joint author. *See* Queiroz, J. C., et al., 1962.

Pillai, M. K. K., joint author. *See* Pradhan, S., et al., 1960.

4656. Pimentel, D. 1966 Wasp Parasite (*Nasonia vitripennis*) Survival on its House Fly Host (*Musca domestica*) Reared on Various Foods. Ann. Ent. Soc. Am. 59(6):1031-1038.

Wide variety in diet provided. Results indicate that host nutrition may cause changes up to 4 times the normal level in parasite reproduction. ABSTRACTS: BA 48:3273(36100), Apr. 1, 1967; TDB 64:326, Mar., 1967.

4657. Pimentel, D. and **Al-Hafidh, R. 1965** Ecological Control of a Parasite Population by Genetic Evolution in the Parasite-host System. Ann. Ent. Soc. Am. 58:1-6.

Concerns *Nasonia vitripennis* and *M. domestica*. During a study lasting 1004 days the host and parasite evolved toward ecological homeostasis. Reproductive capacity of the parasite declined from approximately 135 to 39 progeny per female. ABSTRACTS: RAE-B 53:123, June, 1965; TDB 62:603, June, 1965.

4658. Pimentel, D. and **Cranston, F. 1960** The House Cricket, *Acheta domestica* and the House Fly, *Musca domestica*, as a Model Predator-Prey System. J. Econ. Ent. 53(1):171-172.

A balanced population of 165 crickets (all stages) consumed about 100 house fly pupae per day. Smaller pupae are preferred. Crickets find pupae by contact, using antennae and maxillary palps; cannot locate pupae covered with sand $\frac{1}{32}$ inch deep.

4659. Pimentel, D., Feinberg, E. H., Wood, P. W. and **Hayes, J. T. 1965** Selection, Spatial Distribution, and Coexistence of Competing Fly Species. Am. Nat. 99(905):97-109.

Houseflies and blowflies reared together. At first the houseflies dominated; later the blowflies. (When a competing species becomes dominant, it is at an evolutionary disadvantage, as there is no longer interspecific selection pressure.) ABSTRACT: BA 46:7019(87171), Oct. 15, 1965.

4660. **Pimentel, D., Nagel, W. P.** and **Madden, J. L. 1963** Space-time Structure of the Environment and the Survival of Parasite-Host Systems. Am. Nat. 97(894):141-167.
A study of *Nasonia vitripennis*, parasite on *M. domestica* and *Phaenicia sericata*. (Qualitative changes must be considered in studies of population dynanics.)

4661. **Pimentel, D., Smith, G. J. C.** and **Soans, J. 1967** A Population Model of Sympatric Speciation. Am. Nat. 101(922):493-504.
Used *M. domestica* for "gene-flow" experiments, involving 5 to 30 per cent migration between subpopulations. ABSTRACT: BA 49:10798(118892), Nov. 15, 1968.

4662. **Pimentel, D.** and **Stone, F. A. 1968** Evolution and Population Ecology of Parasite-host Systems. Canad. Ent. 100(6):655-662.
Nasonia vitripennis, parasite of *M. domestica*, showed less population fluctuation, and a lower mean number, in an association that had become relatively homeostatic. Control, in a newly associated parasite-host system, fluctuated with great intensity. ABSTRACT: BA 49:9791(108463), Nov. 1, 1968.

4663. **Pimentel, D.** and **Uhler, L. 1969** Ants and the Control of House Flies in the Philippines. J. Econ. Ent. 62(1):248.
Ants of the *Pheidologelon affinis* group fed voraciously on maggots in various situations, and obviously exerted high predaceous pressure. Probably effected 80-90 per cent control. ABSTRACTS: BA 51:538(5370), Jan. 1, 1970; RAE-B 57:120(430), July, 1969.

Pimentel, D., joint author. *See* Feinberg, E. H., et al., 1966; Chabora, P. C., et al., 1966; Madden, J. L., et al., 1965; Nagel, W. P., et al., 1963, 1964; Pimentel, D., et al., 1966, 1967; Smith, G. J. C., et al., 1969.

4664. **Pin, T. 1968** [A preliminary observation of the mechanism of sterilization of the house-fly, *Musca vicina* Macquart. treated with thiotepa.] Acta Ent. Sinica (Transl.) 2:90-98.
Ovarian follicles become opaque and number of oogonia decrease, followed by complete disappearance of oogonia and atrophy of the ovary. Effects proportioned to dosage and duration of treatment. ABSTRACT: Ent. Abstr. 1(1):76. Item E261.

Pinomonti, S., joint author. *See* Colombo, G., et al., 1967.

4665. **Piquett, P. G.** and **Fales, J. H. 1966** Evaluation of Several Carbamates in Space Sprays Alone and with Piperonyl Butoxide. J. Econ. Ent. 59(4):1020-1022.
Tables given, showing effectiveness of carbaryl, dimetilan, dimetan and nine proprietary compounds; alone and in combination with PB. ABSTRACTS: BA 47:10325(119332), Dec. 15, 1966; RAE-B 54:211, Nov., 1966; TDB 64:218, Feb., 1967.

4666. **Piquett, P. G., Gersdorff, W. A.** and **Alexander, B. H. 1960** Screening Tests with Synthetic Compounds as Synergists for House Fly Sprays. J. Econ. Ent. 53(2):299-301.
Tested 56 synthetic compounds. Most effective synergists were the butyl 2 methyl-3-(3,4-methylenedioxyphenyl) propyl acetal of acetaldehyde and its isobutyl analog. ABSTRACT: BA 35:3741(43007), Aug. 1, 1960.

Piquett, P. G., joint author. *See* Gersdorff, W. A., et al., 1960, 1961.

4667. **Pitts, C. W. 1965** Lipid Composition of Hibernating and Reproducing Face Flies, *Musca autumnalis* Degeer. Diss. Absts. 26(1):561.

4668. Pitts, C. W. and **Hopkins, T. L. 1964** Toxicological Studies on Dichlorvos Feed-additive Formulations to Control House Flies and Face Flies in Cattle Feces. J. Econ. Ent. 57(6):881-884.

Face fly larvae twice as susceptible to dichlorvos as housefly larvae (LD_{50}). Polyvinyl chloride resin formulations should control both species at dosage levels tested. ABSTRACTS: BA 46:2542(31724), Apr. 1, 1965; RAE-B 53:57, Mar., 1965; TDB 62:365, Apr., 1965.

Pitts, C. W., joint author. *See* Bay, D. E., et al., 1968, 1969; Skaptason, J. S., et al., 1962.

4669. Planes-Garcia, S. and **Del Rivero, J. M. 1963** Ensayos de pulverización cebo contra la mosca de los frutos *Ceratitis capitata* y mosca comun. [Experiments with bait sprays against the fruit fly *Ceratitis capitata* and the common fly.] Bol. Patol. Vegetal. Ent. Agr. 26: 291-297.

The carbamate compounds Bayer 5006 and 5024, wettable dusts, are promising in housefly and olive tree fly control. Sugar was a better bait for flies, except for the olive tree fly, which preferred protein. ABSTRACT: BA 46:3675(45397), May 15, 1965.

4669a. Planning Research and Action Institute. 1967 Report of an Action Research Study "Developing Effective Fly Control Programme for Rural Areas" (December 1964 to November 1967). ii + 21 pages. Published by the Institute, Uttar Pradesh, Lucknow.

Concerns the effect of fly control measures on the epidemiology of diarrhea in children. Treats of bore hole pits, fly baiting, sanitary composting and health education.

4670. Plapp, F. W., Jr. 1970 Inheritance of Dominant Factors for Resistance to Carbamate Insecticides in the House Fly. J. Econ. Ent. 63(1): 138-141.

Resistance to carbamates relates to genes on chromosomes 1, 3, and 5. Total resistance of strains tested apparently due to interaction of several genes, acting together. ABSTRACTS: BA 51:8601(87679), Aug. 15, 1970; RAE-B 58:229(942), July, 1970.

4671. Plapp, F. W., Jr. 1970 Changes in Glucose Metabolism Associated with Resistance to DDT and Dieldrin in the House Fly. J. Econ. Ent. 63(6):1768-1772.

Treatment stimulated glucose metabolism, with effect greater on metabolism via the pentose pathway. May be a factor in resistance to chlorinated insecticides.

4672. Plapp, F. W., Jr. 1970 On the Molecular Biology of Insecticide Resistance. *In* Biochemical Toxicology of Insecticides. Ed. R. D. O'Brien and I. Yamamoto. (Symposiom.) New York and London. Acad. Press, Inc., pp. 179-192.

Concerns mutations in *M. domestica.*

4673. Plapp, F. W. and **Bigley, W. S. 1961** Inhibition of House Fly Ali-Esterase and Cholinesterase Under *in vivo* Conditions by Parathion and Malathion. J. Econ. Ent. 54(1):103-108.

Patterns of inhibition were similar with both insecticides and did not seem to be related to toxic action. The cholinesterase activity was more slowly and less reversibly inhibited. ABSTRACT: BA 36:2890(35455), June, 1961.

4674. Plapp, F. W., Jr. and **Bigley, W. S. 1961** Carbamate Insecticides and Ali-esterase Activity in Insects. J. Econ. Ent. 54(4):793-796.

Aliphatic esterase activity, like acetylcholinesterase, is susceptible to inhibition both *in vivo* and *in vitro* by insecticidal carbamates (Serin and Isolan). ABSTRACTS: RAE-B 50:152, July, 1962; TDB 59:99, Jan., 1962.

4675. Plapp, F. W., Jr., Bigley, W. S., Chapman, G. A. and **Eddy, G. W. 1962** Metabolism of Methaphoxide in Mosquitoes, House Flies, and Mice. J. Econ. Ent. 55(5):607-613.
Adult house flies degraded 50 per cent of large dosages within 2 hours. Rates of degradation were similar in a susceptible strain and two organophosphate-resistant strains. ABSTRACTS: BA 41:1666(20928), Mar. 1, 1962; RAE-B 51:37, Feb., 1963; TDB 60:184, Feb., 1963.

4676. Plapp, F. W., Jr., Bigley, W. S., Chapman, G. A. and **Eddy, G. W. 1963** Synergism of Malathion Against Resistant House Flies and Mosquitoes. J. Econ. Ent. 56:643-649.
Eighteen tri-substituted derivatives of phosphoric acid were evaluated as synergists. Many were effective against M-resistant strains, but caused little or no increase in the toxicity of malathion to susceptible insects. ABSTRACTS: BA 45:1754(21879), Mar. 1, 1964; RAE-B 52:19, Jan., 1964; TDB 61:224, Feb., 1964.

4677. Plapp, F. W., Bigley, W. S., Darrow, D. I. and **Eddy, G. W. 1961** Studies on Parathion Metabolism in Normal and Parathion-resistant House Flies. J. Econ. Ent. 54(2):389-392.
No differences in rate of absorption of toxicants between susceptible and resistant flies, but detoxication was more rapid in the resistant strain, which also excreted para-oxon and/or its metabolites more quickly. ABSTRACTS: RAE-B 50:110, May, 1962; TDB 58:962, Aug., 1961.

4678. Plapp, F. W., Jr. and **Casida, J. E. 1969** Genetic Control of House Fly NADPH-Dependent Oxidases: Relation to Insecticide Chemical Metabolism and Resistance. J. Econ. Ent. 62(5):1174-1179.
Studies with 9 insecticide chemicals showed that gene(s) on autosome 2 (*M. domestica* resistant strain R-Baygon) and on autosome 5 (strain R-FE) control the level of activity of reduced NADPH-dependent oxidases. ABSTRACTS: RAE-B 58:82(326), Mar., 1970; BA 51:8601 (87681), Aug. 15, 1970.

4679. Plapp, F. W., Jr. and **Casida, J. 1970** Induction by DDT and Dieldrin of Insecticide Metabolism by House Fly Enzymes. J. Econ. Ent. 63:1091-1092.
Homogenates from abdomens of flies fed high dietary levels of DDT or dieldrin showed higher ability to metabolize 5 insecticide substrates than similar homogenates from untreated flies. Increases ranged from 20 per cent to 150 per cent. ABSTRACT: BA 52:314(2818), Jan. 1, 1971.

4680. Plapp, F. W., Jr., Chapman, G. A. and **Bigley, W. S. 1964** A Mechanism of Resistance to Isolan in the House Fly. J. Econ. Ent. 57(5):692-695.
Resistance to Isolan, and low aliphatic esterase activity, believed due to the same gene, or to factors that are genetically linked. ABSTRACTS: BA 46:3729(46101), June 1, 1965; RAE-B 53:11, Jan., 1965.

4681. Plapp, F. W. and **Eddy, G. W. 1961** Synergism of Malathion against Resistant Insects. Science 134:2043-2044.
Several tri-substituted derivatives of phosphoric acid synergized the toxicity of malathion to resistant houseflies and mosquitoes. ABSTRACT: TDB 59:496, May, 1962.

4682. Plapp, F. W., Jr. and **Hoyer, R. F. 1967** Insecticide Resistance in the House Fly: Resistance Spectra and Preliminary Genetics of Resistance in Eight Strains. J. Econ. Ent. 60(3):768-774.
Eight strains tested with thirteen insecticides. Resistance to organophos-

phates and carbamates associated primarily with semidominant genes of chromosome 5. Chlorinated hydrocarbon resistance traced to semidominant genes of chromosome 5 and recessive genes of chromosome 2. ABSTRACTS: BA 49:1469(16554), Feb. 15, 1968; RAE-B 55:189(651), Oct., 1967; TDB 65:81(183), Jan., 1968.

4683. Plapp, F. W., Jr. and **Hoyer, R. F. 1968** Insecticide Resistance in the House Fly: Decreased Rate of Absorption as the Mechanism of Action of a Gene that Acts as an Intensifier of Resistance. J. Econ. Ent. 61(5):1298-1303.

A gene named *organotin-R*, with the mutant symbol *tin*, confers resistance to organotin insecticides in *M. domestica.* Also acts as an intensifier of resistance in flies resistant to other insecticides. Both functions probably achieved through a decreased rate of absorption. ABSTRACTS: BA 50: 6470(67756), July 1, 1969; RAE-B 57:51(182), Mar., 1969; TDB 67:240 (514), Feb., 1970.

4684. Plapp, F. W., Jr. and **Hoyer, R. F. 1968** Possible Pleiotropism of a Gene Conferring Resistance to DDT, DDT Analogs, and Pyrethrins in the House Fly and *Culex tarsalis.* J. Econ. Ent. 61(3):761-765.

The third chromosome gene, *kdr*-0, imparts resistance not only to DDT, but also to several DDT analogs, to pyrethrins, and to pyrethrins-piperonyl butoxide mixtures. Crossing experiments support this. ABSTRACTS: BA 49:10333(114122), Nov. 15, 1968; RAE-B 56:220(803), Nov., 1968; TDB 66:879(1753), Aug., 1969.

4685. Plapp, F. W., Jr., Orchard, R. D. and **Morgan, J. W. 1965** Analogs of Parathion and Malathion as Substitute Insecticides for the Control of Resistant House Flies and the Mosquito *Culex tarsalis.* J. Econ. Ent. 58:953-956.

Diisopropyl analogs were outstanding for toxicity against houseflies either alone, or in combination with parathion or malathion. ABSTRACTS: BA 47:2927(34646), Apr. 1, 1966; RAE-B 54:15, Jan., 1966; TDB 63: 596, May, 1966.

4686. Plapp, F. W., Jr. and **Tong, H. H. C. 1966** Synergism of Malathion and Parathion Against Resistant Insects: Phosphorus Esters with Synergistic Properties. J. Econ. Ent. 59(1):11-15.

Many phosphorus esters tested with flies and mosquitoes. Against *M. domestica*, esters containing isopropyl, propyl, or butyl groups proved strongly synergistic. ABSTRACTS: BA 47:7236(84420), Sept. 1, 1966; RAE-B 54:96, May, 1966; TDB 63:907, Aug., 1966.

4687. Plapp, F. W., Jr. and **Valega, T. M. 1967** Synergism of Carbamate and Organophosphate Insecticides by Non-insecticidal Carbamates. J. Econ. Ent. 60(4):1094-1102.

Nearly 200 materials were evaluated as synergists against *M. domestica.* With malathion the most effective reduced resistance from 300-fold to 5-fold or less. Experiments support the hypothesis that resistance to carbamates and to organophosphates are conferred by different alleles of the same gene. ABSTRACTS: BA 48:11214(124549), Dec. 15, 1967; RAE-B 56:7(21), Jan., 1968; TDB 65:549(1199), Apr., 1968.

Plapp, F. W., Jr., joint author. *See* Hoyer, R. F., et al., 1965, 1966, 1968; Bigley, W. S., et al., 1961; Eddy, G. W., et al., 1962; Rogoff, W. M., et al., 1964; Schonbrod, R. D., et al., 1968.

Poan, C. C., joint author. *See* Harris, R. L., et al., 1961.

4688. Poindexter, C. E. and **Adkins, T. R., Jr. 1970** Control of the Face Fly and the Horn Fly with Self-Applicatory Dust Bags. J. Econ. Ent. 63(3):946-948.
Various formulations of insecticides reduced horn-fly populations significantly. Several gave good to excellent control of face fly (*M. autumnalis*). Cattle had free access to dustbags. ABSTRACTS: BA 51:13684(139242), Dec. 15, 1970; RAE-B 58:375(1544), Nov., 1970.

4689. Poleshchuk, V. D. 1965 [A device for using liquid insecticide baits against flies.] Med. Parazit. (Moskva) 34:358-359. (In Russian.)

4690. Poleshchuk, V. D. 1967 [A trap for cockroaches and winged flies.] Med. Parazit. (Moskva) 36:240. (In Russian.)

Poleshchuk, V. D., joint author. *See* Vashkov, V. I., et al., 1966.

4691. Politov, A. K. 1961 [Fly control measures in the campaign for municipal cleanliness.] Gig. i Sanit. 26:64-65. (In Russian.)

4692. Politov, A. K. 1962 [Combined measures for control of flies in market places.] Gig. i Sanit. 27:74-76. (In Russian.)

4693. Politov, A. K. 1962 [Experience in control of flies on pig farms.] Gig. i Sanit. 27:80-81. (In Russian.)

4694. Politov, A. K. 1965 Sinantropnye mukhi g Groznogo. [Synanthropic flies in Grozny, (USSR).] Med. Parazit. i Parazitarn. (Bolezni) 34(2):210-215. (English summary.)
Among flies of the town of Grozny, 81 species are classified as synantropic forms; their activity occurring for 215-225 days a year. *Musca domestica* is listed as one of the more epidemiologically dangerous species. Several species were reported for the first time in some areas, among them *Musca domestica vicina.* ABSTRACT: BA 47:3372(39788), Apr. 15, 1966.

Polityka, C. S., joint author. *See* Monroe, R. E., et al., 1968.

Polyakova, V. K., joint author. *See* Roslavtseva, S. A., et al., 1970.

4695. Pomonis, J. G., North, D. T. and **Zaylskie, R. G. (19 ?)** Modification of Recovered Dominant Lethal Mutations Induced by Heteroaromatic Aziridines in Stored House Fly Sperm. MS for publication in J. Med. Chem.
Compounds fed to males, to determine the effect upon their sperm, when stored in the body of the female. Number of recovered dominant lethal mutations decreased when sperm, treated with the appropriate chemical, were stored in the female for 7 days.

Pomonis, J. G., joint author. *See* Nelson, D. R., et al., 1969; Cook, B. J., et al., 1969.

Poonawalla, Z. T., joint author. *See* Quarishi, M. S., et al., 1969.

4696. Poorbaugh, J. H. 1969 Laboratory Colonization of the Canyon Fly, *Fannia benjamini* Malloch, and Speculation on the Larval Habitat. Calif. Vector Views 16(3):21-24.
A successful colony was established from wild-caught female flies, using rodent and rabbit fecal pellets. ABSTRACT: BA 51:13728(139763), Dec. 15, 1970.

4697. Poorbaugh, J. H., Anderson, J. R. and **Burger, J. F. 1968** The Insect Inhabitants of Undisturbed Cattle Droppings in Northern California. Calif. Vector Views. 15(3):17-36.
M. domestica and *M. autumnalis* among 151 insect species attracted to and reared from cowpats during a 3 year study. Includes an annotated key to California species. ABSTRACT: RAE-B 58:383(1580), Nov., 1970.

4697a. Poorbaugh, J. H. and **Smith, C. R. 1968** The Face Fly (*Musca autumnalis* deGeer) found in California. Calif. Vector Views 15(5):54.
An infestation was found on cattle near the Oregon-California border and was believed to have overwintered in that area from the previous summer. All flies captured were fertilized females. ABSTRACT: RAE-B 59(2):63(189), Feb., 1971.

Poorbaugh, J. H., joint author. *See* Anderson, J. R., et al., 1964, 1965, 1968.

Popescu-Pretor, I., joint author. *See* Calciu-Dumitreasa, A., et al., 1968.

Popov, P. V., joint author. *See* Roslavtseva, S. A., et al., 1969.

Porter, J. E., joint author. *See* Evans, B. R., et al., 1965.

4698. Pospíšil, J. 1961 Influence of the Cultivation Media of Larvae on the Chemotaxis of Adult Flies. [Vliv Chovného Prostředí Larev Na Chemotaxi Dospělých Much.] Zool. Listy. Folia Zoologica. Ročník X (XXIV):219-221. (In English, with Czech summary.)
Both positive and negative chemotaxis of adult flies depends, among other factors, on the food of the larvae in preceding generations.

4699. Pospíšil, J. 1962 On visual orientation of the housefly (*Musca domestica*) to colors. [Zraková orientace mouchy domácí (*Musca domestica*) k barvám.] Casopis Ceskoslov. Společnosti Entomol. 59(1): 1-8. (In English, with Czech summary.)
M. domestica more led by wave length than by intensity. Red, interpreted as "dark" attracts more flies than yellow, green, or blue. ABSTRACTS: BA 45:3616(44532), May 15, 1964; BA 45:3616(44532), May 15, 1964.

4700. Post, F. J. and **Foster, F. J., Jr. 1965** Distribution and Characterization of Fecal Streptococci in Muscoid Flies. J. Invertebr. Path. 7(1):22-28.
Fecal streptococci were recovered from 78 per cent of 1174 muscoid flies representing 18 species. *Enterococcus* and enterococcus biotype groups mostly from 249 individuals. Flies assumed to have had frequent contact with feces, refuse, plants, or soil. ABSTRACTS: BA 47:2508(29971), Mar. 15, 1966; RAE-B 53:215, Nov., 1965.

Potter, C., joint author. *See* Lord, K. A., et al., 1962; Ward, J., et al., 1960.

4701. Potworowski, A. 1965 Review of the House Fly. Trav. Jeunes Sci. 2(1):11.

4702. Povolný, D. and **P˘ívora, M. 1961** Kritische Bewertung mikrobiologischer Befunde bei synanthropen Fliegen in Mitteleuropa. [A critical study of microbiological findings in synanthropic flies in central Europe.] Angew. Parasit. (Jena) 2(3):66-74. (English summary.)
Besides non-pathogens, several pathogenic forms were isolated, including pneumococci and pyogenic staphylococci. ABSTRACT: TDB 60:275, Mar., 1963.

Povolný, D., joint author. *See* Lysenko, O., et al., 1961; Gregor, F., et al., 1964.

4703. Pradhan, S., Mookherjee, P. B., Pillai, M. K. K. and **Bose, B. N. 1960** Chemical Control of Fly-breeding in Compost. Indian J. Ent. 22(3):214-225 (publ. 1961). New Delhi.
Seventeen types of treatment studied and compared, in seven experiments. Among insecticides, aldrin, endrin, and *gamma*-BHC gave

satisfactory results; DDT and malathion not. ABSTRACTS: BA 41:298 (3693), Jan. 1, 1963; RAE-B 50:220, Oct., 1962.

Prakash, O. M., joint author. *See* Raghuwanshi, O. P., et al., 1968.

Pratt, J. J., joint author. *See* Blum, M. S., et al., 1960; Shambaugh, G. F., et al., 1968.

4704. Price, G. M. 1961 Some Aspects of Amino Acid Metabolism in the Housefly, *Musca domestica*. Proc. Biochem. Soc. November 18, 1960. Biochem. J. 78(2):21-22.

Protein hydrolysate from "labelled" flies contained *alpha*-alanine, aspartic acid, glutamic acid and proline. Glutamic-alanine transaminase and glutamic-aspartic acid transaminase were demonstrated in flight muscle sarcosomes, but were absent from haemolymph. ABSTRACT: TDB 58:962, Aug., 1961.

4705. Price, G. M. 1961 Some Aspects of Amino Acid Metabolism in the Adult Housefly (*Musca domestica*). Biochem. J. 80:420-428.

Lists amino acids identified in the extracts of adult flies. Aspartic and glutamic acids are concentrated in the tissues; hardly detectable in the haemolymph. *Gamma*-aminobutyric acid is detectable only in head extracts. ABSTRACT: BA 36:7991(85028), Dec. 15, 1961.

4706. Price, G. M. 1963 The Effects of Anoxia on Metabolism in the Adult Housefly, *Musca domestica*. Biochem. J. 86(2):372-378.

Marked chemical differences reported for aerobic and anaerobic (anoxic) flies. (Latter had been placed in oxygen-free nitrogen for 1-2 hours.) Specific activity of protein amino acids increased in aerobic flies; decreased or remained constant in anoxic flies. ABSTRACTS: BA 43:1358 (16883), Aug. 15, 1963; TDB 60:501, May, 1963.

Price, G. M., joint author. *See* Heslop, J. P., et al., 1963; Bridges, R. G., et al., 1970.

4707. Price, R. W. 1960 Sulfoxide—a Pyrethrum Synergist. Pyrethrum Post 5(3):5-11.

Author describes physical properties and manufacture of sulfoxide, cites literature with results of comparative tests in which it was used with pyrethrins against *M. domestica* and mosquitos. Data on mammalian toxicity reviewed. ABSTRACT: RAE-B 50:2, January, 1962.

4708. Price, R. W., Brooks, I. C. and **Haus, J. B. 1962** Evaluation of a New Antiresistant DDT Insecticide, "Pramex", a New Combination of DDT plus Pyrethrins, and Sulfoxide found to be an Effective Insecticide for DDT Resistant Houseflies, Chlordane-Resistant Roaches. Soap and Chem. Spec. (New York) 38(9):82-86.

From test results it was concluded that this combination will control DDT-resistant houseflies and chlordane-resistant cockroaches. ABSTRACTS: RAE-B 51:275, Dec., 1963; TDB 60:797, Aug., 1963.

Prieto, A. P., joint author. *See* Rogoff, M. H., et al., 1969.

Printy, G. E., joint author. *See* Georghiou, G. P., et al., 1963; March, R. B., et al., 1964.

4709. Prívora, M. and **Döhring, E. 1969** Zur Wirkung von Dichlorvos- Lindan- und Diazinon-Dämpfen auf Stubenfliegen (*Musca domestica* L.). [Concerning the action of dichlorvos, lindane and diazinon vapors on house flies (*Musca domestica* L.).,] Zeit. Angew. Zool. 56(2):221-227. (English summary.)

Four strains of flies tested against these 3 insecticides. Differing results interpreted as due in part to repeated selection, in part to strain characteristics. ABSTRACT: TDB 58:277(1145), Aug., 1970.

4710. Prívora, M. and **Radova, E. 1962** Study on Insecticide Resistance of Flies in Czechoslovakia. J. Hyg. Epidemiol., Microbiol., Immunol. (Praha) 6:265-270. (Summaries in French, German and Italian.)
Resistance to DDT appeared in Slovakia in 1955. Resistance to BHC much less common. Resistance to phosphorus compounds not yet reported. Discontinuance of chlorinated insecticides for a number of years caused resistance to disappear. ABSTRACTS: BA 42:1250(15806), May 15, 1963; TDB 60:503, May, 1963.

4711. Prívora, M., Vranova, J. and **Kudova, J. 1969** Kotazce mechanickeho prenosu mikrobu mouchami: 1. Prezivanl mekterych mikrobu no tarzech *M. domestica* L. [To the problem of mechanical transmission of micro-organisms on tarsi of the fly *Musca domestica* L.] Cesk. Epidemiol., Mikrobiol., Immunol. 18(5/6):353-359. (English summary.)
Laboratory experiments performed to determine survival of *Serratia marcescens*, *Corynebacterium* sp. and *Staphylococcus aureus* on fly tarsi. *Serratia* disappeared quickly; the other two persisted substantially longer. ABSTRACT: BA 51:5554(56945), May 15, 1970.

4712. Prívora, M. and **Zemanova, J. 1968** Über die Wirkung der Dampfphase einiger Insektizide auf Stubenfliegen (*Musca domestica* L.). [On the effect of the vapour phase of some insecticides on *M. domestica*.] Zeit. Angew. Zool. 55(1):101-105.
A DDT-resistant and susceptible strain were exposed to several insecticide vapors. The toxicities of DDT, malathion and fenitrothon were negligible, those of Imidan and diazinon were at most moderate and short lived, that of trichlorphon was high at first, then fell quickly. The action of *gamma*-BHC and of fenitrothon mixed with dichlorvos lasted longest. ABSTRACT: RAE-B 58:182(757), June, 1970.

Prívora, M., joint author. *See* Povolný, D., et al., 1961.

Prokesova, A., joint author. *See* Celedova, V., et al., 1963.

4713. Przyborowski, T. and **Tyrakowski, M. 1963** Fly Control with Organo-Phosphorous Preparations in Closed Rooms. Bull. Inst. Mar. Trop. Med. Gdánsk. 14:293-297.
Good control of *M. domestica* achieved by colored cards, saturated with Diazionon and DDT. A safe method for use in bakeries, confectionaries, and restaurants.

Puscasu, E., joint author. *See* Ungureanu, E. M., et al., 1963.

Puttler, B., joint author. *See* Thomas, G. D., et al., 1970.

4714. Qadri, S. H. 196? Toxicity of Cyclodiene Compounds at Intermittent and Continuous Exposures to Housefly *Musca domestica nebulo*. Symp. Pestic A, p. 17.

4715. Qadri, S. H. 1968 Modes of Exposure and the Toxicity of Cyclodiene Compounds to Housefly. Bull. Indian Soc. Mal. Com. Dis. 5(1/2): 35-44.
Intermittent exposure affected more flies than continuous exposure. Difference in toxicity is attributed to the quantity of insecticide entering the insect's body. ABSTRACTS: BA 51:3371(34868), Mar. 15, 1970; TDB 67:577(1145), May, 1970.

4716. Qadri, S. H. 1969 *Megaselia scalaris* (L.W.) (Dipt., Phoridae) for the Bioassay of Organophosphorus Insecticides. Bull. Ent. Res. 59 (p. 3):389-392.
A procedure of mass-rearing *M. scalaris* is described and its suitability as 1-day-old adults for the bioassay of organophosphorus insecticides

is compared with that of *Drosophila* and *Musca domestica nebulo* F. Mixed population bioassay would be useful, greater sensitivity being achieved with the use of males. ABSTRACT: BA 51:7938(81053), July 15, 1970.

4716a. **Qadri, S. S. H.** and **Kashi, K. P. 1970** Estimation of Malathion Residues on Jute Bags by Bioassay Using Houseflies (*Musca domestica nebulo*). J. Food Sci. Technol. 7(1):8-10.
Estimation assessed in terms of mortality rate. About 50 per cent of the initial deposit remained after 4 weeks, on treated sacks. ABSTRACT: BA 52:2008(19184): Feb. 15, 1971.

4717. **Qadri, S. H., Krishna Rao, J. K.** and **Majumder, S. K. 1967** Evaluation of Repellent Cream Formulations Against Housefly and Mosquito. Int. Pest Contr. 9(2):25-26, 29.
Creams containing various proportions of citronella, clove oil and powdered rhizome of *Kaempferia galanga*, were tested against *M. domestica nebulo* and *Culex pipiens fatigans*. ABSTRACT: RAE-B 58:243(987), July, 1970.

4718. **Qadri, S. H.** and **Majumder, S. K. 1968** Repellency of Arrowroot-*M. maranta*-Arundinaceae-M. to Some Important Insect Pests. Pesticides (Bombay) 2(1).
Houseflies among species tested.

4719. **Qadri, S. H.** and **Naidu, M. B. 1967** Pick-up of Insecticide through Different Pairs of House Fly Tarsi. Ent. Exp. Appl. 10(3-4):476-484. (German summary.)
A technique for controlled exposure of active house flies to insecticide deposits is described. The second pair tarsi were found most important in pick-up of dry insecticide, followed by the third and first pairs respectively. ABSTRACTS: BA 49:5283(59280), June 1, 1968; RAE-B 56:105 (366), May, 1968; TDB 66:486(933), May, 1969.

4720. **Qadri, S. S.** and **Hkashi, K. P. 1969** Effect of Combining Plant Tubers with Chlorinated Insecticides against House Flies. Int. Pest Contr. 11(5):19-21.
Concerns arrowroot, combined with DDT or lindane.

4721. **Quarterman, K. D. 1960** Test Methods for Establishing Levels of Susceptibility and Detecting the Development of Resistance in Insects of Public Health Importance. *In* Symp. on research progress in insecticide resistance. Misc. Publ. Ent. Soc. Am. 2(1):95-102.
Test methods are given for various insects. For the housefly, there is no standard WHO test, but a wide variety of usable tests and a standard Armed Forces test, are available. ABSTRACT: BA 36:1021(12272), Feb. 15, 1961.

Quarterman, K. D., joint author. *See* Pearce, G. W., et al., 1961.

4722. **Queiroz, J. C., Pigatti, P., Mello, D., Pigatti, A.** and **Mello, E. J. R. 1962** Tolerância nas condições de laboratorio, das môscas domésticas do estado de São Paulo aos insecticidas orgânicos. [Survey of housefly (*Musca domestica* L.) tolerance to several organic insecticides, in the state of São Paulo. . . .] Arq. Inst. Biol. (São Paulo) 29:139-144. (English summary.)
Flies from all localities were resistant to DDT and *gamma*-BHC. An initial resistance to dieldrin and malathion was noted. ABSTRACTS: BA 45:1413(17567), Feb. 15, 1964; RAE-B 52:86, May, 1964.

Queiroz, J. C., joint author. *See* Mello, E. J. R., et al., 1961, 1962; Paim, G. V., et al., 1963.

4723. Quevedo, F. and **Carbanza, N. 1966** Le role des mouches dans le contaminations des aliments au Perou. [The role of flies in food contamination in Peru.] Ann. Inst. Pasteur, Lille. 17:199-202.
In samples of flies from places of epidemiological interest, *Staphylococcus aureus* was detected in 8 out of 10. *Proteus* and *Escherichia* were found most frequently. All samples showed the sulfite reducing *Clostridia*. ABSTRACT: BA 49:814(904), Jan. 15, 1968.

Quijano, E., joint author. *See* Hecht, O., et al., 1969.

Quintiliani, M., joint author. *See* Grasso, A., et al., 1962, 1963.

4724. Quistad, G. B., Fukuto, T. R. and **Metcalf, R. L. 1970** Insecticidal, Anticholinesterase, and Hydrolytic Properties of Phosphoramidothiolates. J. Agric. Fd. Chem. 18(2):189-194.
Emphasis on structure-activity relationships. Toxicity of compounds to houseflies relates directly to the rate of alkaline hydrolysis and the inhibition of cholinesterase. ABSTRACT: BA 51:9098(92605), Aug. 15, 1970.

4725. Quraishi, M. S. and **Poonawalla, Z. T. 1969** Radioautographic Study of the Diffusion of Topically Applied DDT-C^{14} into the House Fly and Its Distribution in Internal Organs. J. Econ. Ent. 62(5):988-994.
Insecticide penetrated into the body cavity through membraneous areas, rather than through the cuticle. The site of maximum concentration was the alimentary canal. ABSTRACTS: TDB 67:366(770), Mar., 1970; BA 51:8569(87378), Aug. 1, 1970.

4726. Quraishi, M. S. and **Thorsteinson, A. J. 1965** Toxicity of Some Straight Chain Saturated Fatty Acids to House Fly Larvae. J. Econ. Ent. 58(3):400-402.
Fatty acids containing 8-11 carbon atoms were found toxic. Dodecanoic acid was toxic only if larvae were wet. ABSTRACTS: BA 47:1600(19365), Feb. 15, 1966; RAE-B 53:177, Sept., 1965; TDB 62:1274, Dec., 1965.

Rachlin, A. I., joint author. *See* Fellig, J., et al., 1970.

Radova, E., joint author. *See* Privora, M., et al., 1962.

4727. Radvan, R. 1960 Persistence of Bacteria During Development in Flies. I. Basic Possibilities of Survival. Folia Microbiol. 5(1):50-56.
Concerns 12 species of bacteria borne by *M. domestica* and *Protophormia terrae-novae*. Microorganisms transmitted on the surface of ova (not within them). Survival of pathogens such as *Salmonella typhi* considered rare under natural conditions. ABSTRACT: BA 37:1480(14526), Feb. 15, 1962.

4728. Radvan, R. 1960 Persistence of Bacteria During Metamorphosis in Flies. II. The Number of Surviving Bacteria. Folia Microbiol. 5: 85-91.
About 60 per cent of adult flies contain from 1 to 100,000 bacteria. Only 10 per cent contain from 100,000 to 10,000,000. Ten to fifty times as many bacteria remain in puparium as are found on fly. Perhaps 30 per cent of the imagos are bacteriologically sterile.

4729. Radvan, R. 1960 Persistence of Bacteria During Development in Flies. III. Localization of the Bacteria and Transmission After Emergence of the Fly. Folia Microbiol. 5(3):149-156. (Russian summary.)
Concerns *Salmonella schottmülleri*. Organisms surviving pupation are localized in space between developing fly and pupal exuviae. Fly's alimentary tract usually sterile. Body surface can carry and spread bacteria not more than 24 hours. ABSTRACT: BA 37:711(7771), Jan. 15, 1962.

4730. **Radvan, R. 1960** Epidemiological Importance of the Spread and Transmission of Bacteria Surviving During the Development of Flies. Cesk. Epidem. 9:497-500.
(See annotations for three preceding references.)

4731. **Raghavan, N. G. S., Wattal, B. L., Bhatnagar, V. N., Choudhury, D. S., Joshi, G. C.** and **Krishnan, K. S. 1967** Present Status of Susceptibility of Arthropods of Public Health Importance to Insecticides in India. Bull. Indian. Soc. Mal. Com. Dis. 4(3):209-245.
Houseflies reported as resistant to organochlorine insecticides, but not to organophophosrus insecticides. ABSTRACT: TDB 66:621(1219), June, 1969.

4732. **Raghuwanshi, O. P., Islam, A.** and **Khan, N. H. 1968** Effects of Apholate on the Bionomics of *Musca domestica nebulo* Fabr. Botyu-Kagaku 33(4):119-122.
Apholate used as a chemosterilant even at very low concentrations, reduced oviposition and viability of eggs laid. Adult life span was shortened in both sexes. ABSTRACTS: BA 50:9731(100722), Sept. 15, 1969; RAE-B 58:88(349), Mar., 1970.

4733. **Rahman, S. J. 1965** Bionomics of Dieldrin-Resistant and Normal Strains of Indian Housefly, *Musca domestica nebulo* Fabr. Bull. Indian Soc. Mal. Com. Dis. 2(2):150-153.
Female flies resistant to dieldrin laid half the normal number of eggs, had a shorter oviposition period and shorter adult life. Differences in larval, pupal and incubation period in normal and resistant strains not significant. ABSTRACT: RAE-B 56:211(764), Oct., 1968.

4734. **Rahman, S. J.** and **Kahn, N. H. 1964** Inheritance of Dieldrin-resistance in *Musca domestica nebulo*. Botyu-Kagaku 29(2):19-21. (Japanese summary.)
Dieldrin resistance appeared to be governed by a multiple-gene factor. ABSTRACT: RAE-B 55:87(310), May, 1967.

4735. **Rahman, S. J.** and **Khan, N. H. 1968** The Inheritance of DDT-resistance in the House-fly, *Musca domestica nebulo* Fabr. Indian J. Ent. 30(1):69-76.
Crossing of strains, production of F_2, and back cross of F_1 hybrids to susceptible parents indicate DDT-resistance as due to a single dominant, autosomal gene. ABSTRACTS: BA 51:2270(23551), Mar. 1, 1970; RAE-B 58:35(147), Feb., 1970.

4736. **Rahman, S. J., Riaz, N.** and **Kahn, N. H. 1964** The Sensitivity of *Musca domestica nebulo* F. and *M. domestica vicina* Macq. to DDT, BHC and Dieldrin, as Shown by Dosage-Mortality Curves. Bull. Ent. Res. 55(2):355-365.
Insecticides proved slightly less toxic in acetone, and more toxic in ethanol, than in Risella oil. Males of both species were more susceptible than females to all 3 insecticides. BHC was found the most toxic, DDT the least, to both subspecies. ABSTRACTS: RAE-B 52:187, Oct., 1964; TDB 62:68, Jan., 1965; BA 47:4254(49930), May 15, 1966.

Rahmati, H. S., joint author. *See* Uchida, T., et al., 1965.

4737. **Rai, L., Krause, G. F.** and **Roan, C. C. 1964** The Effects of Incremental Doses of Malathion Applied Topically to Adult Houseflies. J. Kans. Ent. Soc. 37(2):106-112.
Findings suggest that the proportions of insecticide going into intoxication and detoxication pathways are of the same order within the range of total dosages used. Greater than additive effect with highest dosages tested and shortest application interval. ABSTRACTS: BA 46: 337(4244), Jan. 1, 1965; RAE-B 54:107, June, 1966.

4738. **Rai, L.** and **Roan, C. C. 1963** Effects of Route of Administration to Houseflies on the Interactions of Malathion and Piperonyl Butoxide. J. Kans. Ent. Soc. 36(3):161-165.

Flies which receive a topical application of piperonyl butoxide and are then exposed to milk containing malathion show an antagonistic effect. This disappears by the 72 hour feeding period.

4738a. **Rajak, R. L.** and **Perti, S. L. 1967** Insecticidal Activation. Labdev. J. Sci. Technol. 5(3):262-263. Kanpur, U. P.

Mixtures of sublethal quantities of DDT and toxaphene exhibited synergism, and had a marked lethal effect. Mixtures of DDT and diazinon showed similar effect, but less pronounced. ABSTRACT: RAE-B 58:340 (1394), Oct., 1970.

4739. **Rajak, R. L.** and **Perti, S. L. 1968** Toxicity of Bacterial Spores to Fly and Mosquito Larvae. Pesticides (Bombay) 1(9):45.

Treats of the relation of *Bacillus cereus*, var. *thuringiensis* to *M. nebulo*, *M. domestica* and *Culex fatigans*.

4740. **Ramade, F. 1962** Étude au développement post-embryonnaire du testicule et de la spermatogenèse chez l'asticot de *Musca domestica* L. [Study of the post embryonic development of the testis and of spermatogenesis in the larva of *Musca domestica* L.] Thesis. University of Paris. 63 pp., 9 plates.

Larvae are distinguishable as to sex. Author presents new knowledge regarding histology and cytology of the testis during spermatogenesis and spermiogenesis, as observed in third stage larva. Chromosomes described and photographed.

4741. **Ramade, F. 1963** Données sur le spectre de résistance aux insecticides d'une souche de *Musca domestica* L. [Data on the spectrum of resistance to insecticides of a strain of *Musca domestica* L.] Compt. Rend. Seances Acad. Agr. Fr. 49(12):1062-1066. Paris.

Table gives coefficient of resistance $\left(\frac{\text{DL50R}}{\text{DL50S}}\right)$ for 10 insecticides, including DDT, parathion and Dipterex. ABSTRACT: RAE-B 53:250, Dec., 1965.

4742. **Ramade, F. 1965** L'action anticholinestérasique de quelques insecticides organophosphorés sur le système nerveux central de *Musca domestica*. [Anticholinesterase action of certain organophosphorus insecticides on the central nervous system of *Musca domestica*.] Ann. Soc. Ent. Fr. (N.S.) 1(3):549-566. (English summary.)

Decrease in cholinesterase activity causes uncoordinated excitation in the fly. Total inhibition leads to knockdown, then paralysis. With phosphoric esters, cerebral acetylcholinesterase is involved. In paralysis caused by Dipterex, however, no brain change occurs. ABSTRACT: RAE-B 55:39(146), Feb., 1967.

4743. **Ramade, F. 1966** Sur l'ultrastructure de la pars intercerebralis chez *Musca domestica* L. [On the microstructure of the pars intercerebralis in *Musca domestica*.] Compt. Rend. Acad. Sci. (Paris) 263: 271-274. Série D.

Electron microscope discloses 3 kinds of neurosecretory cells, differing as to shape, dimension of elementary grains of neurosecretin, and the appearance of their nuclei. Cytoplasms include an *ergastoplasm* and a well-developed Golgi apparatus, also many dense bodies of lysosome type.

4744. **Ramade, F. 1966** Sur la presence d'altérations ultrastructurales dans le cerveau de *Musca domestica* L. apres intoxication au lindane. [On the presence of microstructural alterations in the brain of *Musca domestica* L. after poisoning with lindane.] Compt. Rend. Acad. Sci. (Paris) 263:371-373. Série D.
Numerous microstructural lesions. The number of multivesicular bodies (releated to lysosomes) increases in the perinuclear portion of the neurones. The ergastoplasm (reticular endoplasm) and Golgi apparatus are distorted.

4745. **Ramade, F. 1967** Ultrastrukturelle Veränderungen im Gehirn mit Lindan vergifteter *Musca domestica* und deren physiologische Bedeutung. [On the presence of ultrastructural alterations in the brain of *Musca domestica* poisoned by lindane insecticide, and their physiological meaning.] Abstr. Int. Pflanzenschutz-Kong. [Int. Plant Prot. Congr.] 30:595-597. (In German, English and French.)
Females flies treated with LD_{50} (or stronger) doses of lindane. Caused vacuolation of axonic content and swelling of neurotubules. Paralyzed flies showed not a few multivesicular bodies (replacing many lysosome type bodies) also a number of pre-mortem alterations.

4746. **Ramade, F. 1968** Sur la presence de cellules vacuolaries dans le cerveau de *Musca domestica* L. [The presence of vacuolar cells in the brain of *Musca domestica* L.] Compt. Rend. Acad. Sci. (Paris) 266(26): 2437-2440. Série D. (Sci. Natur.)
Certain cells in the cerebral cortex of adult flies have large, cytoplasmic vacuoles, the size and number of which vary during the imaginal period. Cells grow while their number diminishes, as the insect grows older. ABSTRACT: BA 50:12502(129152), Dec. 1, 1969.

4747. **Ramade, F. 1969** Données histologiques, histochimiques et ultrastructurales sur la *pars intercerebralis* de *Musca domestica* L. [Histological, histochemical and ultra-structural data on the *pars intercerebralis* of *Musca domestica* L.] Mem. du Mus. Nat. d'Hist. Nat. 58(2):113-142. Série A.
Treats of materials and techniques, microscopic anatomy, neurosecretory cells (their location, structure, function), vacuolate cells and various ultrastructures. ABSTRACT: BA 51:13158(133950), Dec. 1, 1970.

4748. **Ramade, F. 1969** Mise en évidence de cellules neurosécrétrices Gomori négatives dans la *pars intercerebralis* de *Musca domestica* L. par une étude comparative en microscopie ordinaire et électronique. [Presentation in elucidation of Gomori-negative neurosecretory cells in the *pars intercerebralis* of *Musca domestica* by a comparative study with ordinary and electronic microscopy.] Compt. Rend. Acad. Sci. (Paris) 268:1945-1947. Série D.
Recognizes *two* distinct categories of cells: (1) Large macrogranular neurones; (2) Small, microgranular neurones. Reconciles this with earlier histological classification of neurosecretory cells *three* ways into types A, B and C.

4749. **Ramade, F. 1969** Comparative Study of the Pars Intercerebralis of *Musca domestica* in the Ordinary and Electron Microscope. Evidence of Negative Gomori Neuro-secretory cells. Bull. Soc. Zool. Fr. 94(2):303-304.
A demonstration of ultrastructure.

4750. **Ramade, F.** and **Rivière, J. L. 1970** Recherches histochimiques sur les proteines associées aux neurosécrétions protocérébrales de *Musca domestica* L. [Histochemical researches on the proteins associated with the protocerebral secretions of *Musca domestica* L.] Compt. Rend. Acad. Sci. (Paris) 270:1803-1806.

All neurosecretory cells of the protocerebrum elaborate a material of protein nature. This contains arginine and tryptophane in histochemically detectable amounts, but as a rule amino acids are most in evidence.

Ramade, F., joint author. *See* Pesson, P., et al., 1960, 1965.

4751. **Ramakrishnan, S. P., Krishnamurthy, B. S.** and **Kalra, R. L. 1963** A Note on the Toxicity of Dichlorvos to Insects of Public Health Importance. Indian J. Malariol. 17:115-118.

Compares toxicity of this insecticide for *Cimex*, *Xenopsylla*, *Periplaneta*, *Musca nebulo* and *M. vicina.* ABSTRACT: RAE-B 54:91, May, 1966.

Ramirez-Genel, M., joint author. *See* Vazquez-Gonzalez, J., et al., 1962, 1963.

4752. **Rampazzo, G. 1962** Lotta contro la mosca (*Musca domestica* L.) in una stazione di cura e soggiorno. [Effort against the fly (*Musca domestica* L.) at a health resort.] Riv. di Parassit. 33(4):299-308.

Reports a housefly control trial with Dithion 18 in Abano Terme (Padua) during 1961. Residual activity satisfactory for approximately 100 days. ABSTRACTS: RAE-B 52:200, Nov., 1964; TDB 60:795, Aug., 1963.

4753. **Rampazzo, G. 1968** Sperimentazione di lotta contro la mosca (*Musca domestica* L.) con Vapona striscia, in Abano Terme. [Experiment in the control of the house fly *Musca domestica* in Abano Terme using Vapona strips.] Riv. di Parassit. 29(2):143-149.

Reports tests made in 1966. Plastic strips, impregnated with dichlorvos (DDVP) used to reduce fly densities in kitchens and stables. Residual effect lasted 90 days. ABSTRACT: BA 50:5221(55879), 1969.

4754. **Ranade, D. R. 1960(1961)** Comparative Notes on the Female Terminalia of Some Common Indian Species of *Musca* L. Indian J. Ent. 22:152-159.

A study of 8 forms in 7 species. *M. fletcheri* and *M. pattoni* are structurally similar. The other six resemble one another and presumably constitute a natural group. ABSTRACT: BA 41:306(4071), Jan. 1, 1963.

4755. **Ranade, D. R. 1964** Observations on the Ring Gland of the Larva of *Musca domestica nebulo* Fabr. (Diptera:Cyclorrhapha, Muscidae). Sci. Cult. 30(4):205-206.

Good histological description. Brief mention of hormone function. ABSTRACT: BA 47:398(4873), Jan. 1, 1966.

4756. **Ranade, D. R. 1964** The Muscular System of the Third Instar Larva of *Musca domestica nebulo* Fabr. (Diptera:Cyclorrhapha-Muscidae). Univ. Poona J. Sci. Technol. 28:67-76.

A full description of 3 muscle groups: cephalopharyngeal, body wall, and anal. ABSTRACT: BA 49:3354(37523), Apr. 1, 1968.

4757. **Ranade, D. R. 1964** Developmental Anatomy of *Musca domestica nebulo* Fabr. Part I. Life History of the Imago and External Morphology of the Larval Instars. J. Animal Morph. Physiol. 11(2): 245-246.

Emphasizes cephalopharyngeal skeleton and posterior spiracles. ABSTRACT: BA 48:3273(36105), Apr. 1, 1967.

4758. **Ranade, D. R. 1965** The Anatomy of the Tracheal System of the Larva of *Musca domestica nebulo* Fabr. (Diptera:Muscidae). Indian J.

Ent. 27(2):172-181.
A complete morphological description, well illustrated. As in all larvae of this group, system is amphipneustic.

4759. **Ranade, D. R. 1967** The Circulatory System of the Third Instar Larva of *Musca domestica nebulo* Fabr. (Diptera:Muscidae). Bull. Ent. 8(1):13-16.
Covers 3-chambered heart, aorta, dorsal diaphragm (3 pairs of alary muscles), pericardial sinus, haemolymph, visceral sinus and blood circulation. ABSTRACT: BA 50:527(5518), Jan. 1, 1969.

4760. **Ranade, D. R. 1967** The Digestive System of the Third Instar Larva of *Musca domestica nebulo* Fabr. (Diptera:Muscidae). Bull. Ent. 8(1):29-35.
Covers oral structures, stomodaeum, ventriculus, proctodaeum, malpighian tubes, salivary glands. Well illustrated. ABSTRACT: BA 50:1587 (16774), Feb. 1, 1969.

4761. **Ranade, D. R. 1967** The Anatomy of the Nervous System of the Third Instar Larva of *Musca domestica nebulo* Fabr. (Diptera-Cyclorrhapha-Muscidae). Proc. Indian Acad. Sci. (Sect. B) 65(1):34-41.
Covers cerebral ganglia, ventral ganglionic mass, cerebral and related nerves, thoracic and abdominal nerves, median nerves, accessory nerves, and stomadaeal nervous system. ABSTRACT: BA 48:6037(67480), July 1, 1967.

Ranade, D. R., joint author. *See* Karandikar, K. R., et al., 1965.

Randolph, N. M., joint author. *See* Fernandez, A. T., et al., 1966, 1967.

Ranganathan, S. K., joint author. *See* Dixit, R. S., et al., 1962.

4762. **Rao, N. S. 1968** Household Pest Control. Pesticides (Bombay) 1(11): 101-104.

4763. **Rasheed, S.** (?) Role of Food on the Toxicity of Insecticides to House Fly, *Musca domestica nebulo*. Symp. Pestic. A., p. 42.

4764. **Ratcliffe, R. H.** and **Ristich, S. S. 1965** Insect Sterilant Experiments in Outdoor Cages with Apholate, Metepa, and four Bifunctional Aziridine Chemicals Against the House Fly. J. Econ. Ent. 58: 1079-1082.
Performance of baits affected by method of placement, fly age, and environmental conditions. Apholate preparations ranked as follows: (1) liquid and granular baits; (2) ribbon treatments; (3) string treatments; (4) residual treatments on plywood. ABSTRACTS: RAE-B 54:59, Mar., 1966; TDB 63:908, Aug., 1966.

Ratcliffe, R. H., joint author. *See* Ristich, S. S., et al., 1965.

Rauch, F., joint author. *See* Lhoste, J., et al., 1967, 1968, 1969.

4765. **Ray, J. W. 1967** The Epoxidation of Aldrin by Housefly Microsomes and Its Inhibition by Carbon Monoxide. Biochem. Pharmacol. 16(1):99-107.
M. vicina microsomes contain a system which converts aldrin to dieldrin and which requires the presence of NADPH and O_2. Optimum pH is 8·2. Reaction is inhibited by EDTA, sesamex and CO. ABSTRACTS: BA 48:5113(57192), June 1, 1967; RAE-B 56:225(822), Nov., 1968.

4766. **Ray, J. W.** and **Heslop, J. P. 1963** Phosphorus Metabolism of the House-fly (*Musca domestica* L.) During Recovery from Anoxia. Biochem. J. 83:39-42.
During anoxia, the ATP is replaced, in the thorax, by AMP and IMP. During recovery, AMP is rapidly phosphorylated, but IMP remains

unchanged for at least 2 hours. Return of glycerol 1-phosphate concentration to normal is at first very rapid; later, very slow. ABSTRACTS: BA 43:1974(25506), Sept. 15, 1963; TDB 60:690, July, 1963.

Ray, J. W., joint author. *See* Heslop, J. P., et al., 1963, 1964; Lewis, S. E., et al., 1967.

4767. Raybould, J. N. 1966 Further Studies on Techniques for Sampling the Density of African House Fly Populations. I. A Field Comparison of the Use of the Scudder Grill and the Sticky-Flytrap Method for Sampling the Indoor Density of African House Flies. J. Econ. Ent. 59:639-644.
A one year study made at Muheza, near Tanga. Sticky flytrap gave results more in agreement with obvious fly densities than the Scudder grill. Also gave more information regarding species and sex. ABSTRACTS: BA 47:8530(99361), Oct. 15, 1966; RAE-B 54:178, Sept., 1966; TDB 64:323, Mar., 1967.

4768. Raybould, J. N. 1966 Further Studies on Techniques for Sampling the Density of African House Fly Populations. 2. A Field Comparison of the Scudder Grill and the Sticky-Flytrap Method for Sampling Outdoor Density of African House Flies. J. Econ. Ent. 59:644-648.
Sticky flytrap, designed to be resistant to natural weather conditions, enabled investigator to make separate counts of *M. d. calleva*, *M. d. curviforceps*, and *M. sorbens*. Scudder grill yielded data on fly movements and resting sites. Both methods useful. ABSTRACTS: BA 47:8530 (99362), Oct. 15, 1966; RAE-B 54:178, Sept., 1966; TDB 64:323, Mar., 1967.

4769. Reed, W. T. and **Forgash, A. J. 1968** Lindane: Metabolism to a New Isomer of Pentachlorocyclohexene. Science 160(3833):1232.
Two isomers identified by gas-liquid chromatography and mass spectroscopy. One identical with the previously known compound; the second is here reported for the first time. ABSTRACT: BA 50:5324(55902), May 15, 1969.

4770. Reed, W. T. and **Forgash, A. J. 1969** Metabolism of Lindane to Tetrachlorobenzene. J. Agric. Fd. Chem. 17(4):896-897.
Is the *third* isomer of pentachlorocyclohexene to be reported as a metabolite of lindane in susceptible and resistant strains of house flies.

4771. Reed, W. T. and **Forgash, A. J. 1970** Metabolism of Lindane to Organic-soluble Products by Houseflies. J. Agric. Fd. Chem. 18(3): 475-481.
This process a minor resistance mechanism. Authors compared susceptible, moderately resistant, and highly resistant flies. ABSTRACT: BA 51: 12051(122384), Nov. 1, 1970.

Rehman, J., joint author. *See* Kahn, N. H., et al., 1965.

4772. Rehman, S. J. 1966 Studies on the Development of Dieldrin-Resistance in *Musca domestica nebulo* and *Musca domestica vicina*. Angew. Parasit. (Jena) 7(2):115-119. (Summaries in German and Russian.)
M. d. nebulo and *M. d. vicina* surviving treatment of dieldrin allowed to breed and selection continued 37 and 20 generations, respectively. *M. d. nebulo* only 8·4 times as tolerant as normal strain, while *M. d. vicina* was 22·5. ABSTRACTS: RAE-B 56:193(694), Oct., 1968; TDB 63: 1034, Sept., 1966.

Rehman, S. J., joint author. *See* Khan, N. H., et al., 1964.

4773. **Reichardt, H. 1962** Praktische Erfahrungen bei der Answendung von Trichlorphon zur Bekämpfung von Hygiene-schädlingen. [Practical results in the use of Trichlorphon for control of sanitary pests.] *In* Symp. on problems of control of epidemiologically important arthropods and their resistance to insecticides, 1961. J. Hyg., Epidemiol., Microbiol., Immunol. (Prague) 6(3):328-333. (Summaries in English, French and Spanish.)

Flibol-E, containing 50 per cent trichlorfon has given good control where 2 chlorinated hydrocarbons have failed. Used in sugar bait on fly papers, in fly plates against *M. domestica*, and in fly balls against *Fannia canicularis*. Trichlorfon involves no risk to man. ABSTRACTS: BA 42:1250(15807), May 15, 1963; RAE-B 53:19, Jan., 1965.

4774. **Reichardt, W. E. 1965** Quantum Sensitivity of Light Receptors in the Compound Eye of the Fly *Musca*. Cold Spring Harbor Symp. Quant. Biol. 30:505-515.

Coincidences of two or more quanta not observed in the eye of the fly for time intervals up to one-twentieth of a second. ABSTRACT: BA 47:6404(74930), Aug. 1, 1966.

4775. **Reichardt, W. E. 1966** Detection of Single Quanta by the Compound Eye of the Fly, *Musca*. *In* Proc. Internat. Symp. on . . . 25-27 October, 1965. Stockholm, Sweden. Symp. Publ. Div. Pergamon Press. 7:267-289.

It is believed that in the optomotor reaction system of *Musca*, there are no temporal and/or spatial coincidences before perception of motion is carried out by the central nervous system. ABSTRACT: BA 48:9253 (104048), Oct. 15, 1967.

4776. **Reichardt, W. E. 1966** Optomotor Reactions of the Beetle *Chlorphaus*, the Housefly *Musca* and the Fruit Fly *Prosophila*. *In*: Neurosciences Research Symp. Summ., Mass. Inst. Technol. Press. 1:157-163. (Available in book form only—not as reprints.)

4777. **Reichardt, W., Braitenberg, V. and Weidel, G. 1968** Auslösung von Elementarprozessen durch einzelne Lichtquanten im Fliegenauge. Verhaltensexperimente an der Stubenfliege *Musca*. [Release of elementary processes through single light quanta in fly eyes. Behavior experiments with the housefly *Musca*.] Kybernetik 5: 148-169. (English summary.)

Supports the hypothesis that one single quantum of light is sufficient to trigger an elementary photochemical reaction, and that in turn, one single photochemical event can elicit a miniature receptor potential.

4778. **Reichardt, W. and Wenking, H. 1969** Optical Detection and Fixation of Objects by Fixed Flying Flies. Naturwissenschaften 56(8):424-425.

A servo-system was developed for studying the fly's behavior under "closed-loop" conditions: a fixed flying individual, by its own torque response, controls the angular velocity of its optical "surround".

Reichardt, W., joint author. *See* Fermi, G., et al., 1963.

4779. **Reid, E. T. 1960** Insecticide Resistance. Centr. Afr. J. Med. 6:528-534. Salisbury, S. Rhod.

Concerns resistance of *M. domestica* to BHC in Southern Rhodesia. Laboratory experiments included work with *M. d. cuthbertsoni*, *M. d. curviforceps* and hybrids of the two. ABSTRACT: RAE-B 51:260, Dec., 1963.

4780. Reiff, M. and **Beye, F. 1960** Stoffwechselvorgänge bei sensiblen und resistenten Fliegen unter Einfluss der DDT-Substanz. [Metabolic processes in sensitive and resistant flies under DDT control.] Acta Tropica 17:1-47. (Summary in French and English.)
Concerns physiological aspects of resistance in houseflies, which have "highly active protective mechanisms and the ability to counter-balance toxic actions". Special terminology. ABSTRACTS: BA 41:1666(20932), Mar. 1, 1963; RAE-B 50:65, Apr., 1962; TDB 57:1122, Oct., 1960.

Reinecke, J. P., joint author. *See* Eide, P. E., et al., 1970.

Remmert, L., Jr., joint author. *See* Gregg, C. T., et al., 1960.

Retallack, N. E., joint author. *See* Gilgan, M. W., et al., 1967.

Reynolds, H. T., joint author. *See* Metcalf, R. L., et al., 1966.

4781. Rezabova, B. 1968 Changes in the Metabolism of Nucleic Acids in the Ovaries of the House Fly, *Musca domestica,* After Application of Chemosterilants. Acta Ent. Bohemoslov. 65(5):331-340.
After application of (5 different) chemosterilants, an increase in the nucleic acid content occurs. Using labelled precursors, one may follow divisions of the nuclei in the follicular epithelium, and the formation of a tumor. ABSTRACT: BA 50:2679(27954), Mar. 1, 1969.

4782. Rezábová, B., Hora, J., Landa, V., Cerny, V. and **Sorm, F. 1968** On Steroids. CXIII. Sterilizing Effect of Some 6-ketosteroids on Housefly (*Musca domestica* L.). Steroids 11:475-496.
Effect of 47 compounds on growth and development of the ovaries. Most of these showed a sterilizing activity. Describes synthesis of new 6-keto-steroids. ABSTRACT: BA 49:8724(97113), Sept. 15, 1968.

4783. Rezábová, B. and **Landa, V. 1967** Effect of 6-azauridine on the Development of the Ovaries in the House Fly *Musca domestica* L. (Diptera). Acta Ent. Bohemoslov. 64(5):344-351.
Is a cytostatic agent. A concentration of 0·01 per cent in adult food completely inhibits ovarian growth. There is a proliferation of the follicular epithelium of the egg chamber. ABSTRACT: BA 49:3368(37654), Apr. 1, 1968.

4784. Rezábová, B. and **Landa, V. 1968** Effects of Thalidomide on the Development of the House Fly *Musca domestica* L. Acta Ent. Bohemoslov. 65(3):212-215.
Application in concentration of 0·1 per cent during embryonal and larval development blocks ovarian development completely. Applied during larval stages only, permits third egg chamber to develop. ABSTRACT: BA 49:10813(119079), Nov. 15, 1968; RAE-B 57:281(1065), Dec., 1969.

Riaz, N., joint author. *See* Ansari, J. A., et al., 1965; Rahman, S. J., et al., 1964, 1965.

Ribbert, D., joint author. *See* Engels, W., et al., 1969.

4785. Richard, J. L. and **Pier, A. C. 1966** Transmission of *Dermatophilus congolensis* by *Stomoxys calcitrans* and *Musca domestica.* Am. J. Vet. Res. 27(117):419-423.
Both species transmitted infection from diseased to normal rabbits. Moistening of feeding sites enhanced transmission, but mechanical disruption of host's skin was not necessary. ABSTRACTS: BA 47:5868(68667), July 15, 1966; RAE-B 57:73(257), May, 1969.

4786. Richards, C. S., Jackson, W. B., DeCapito, T. M. and **Maier, P. P. 1961** Studies on Rates of Recovery of *Shigella* from Domestic Flies and from Humans in Southwestern United States. Am. J.

Trop. Med. Hyg. 10(1):44-48.
A total of 65,273 flies taken outdoors yielded 69 isolations of *Shigella*. Indoor collection of 5,664 flies yielded 12. Rectal swabs from humans gave indices generally parallel to indices from flies. Eleven types of *Shigella* concerned. *M. domestica* the most abundant fly. ABSTRACT: BA 36:2592(31829), May 15, 1961.

Ricketts, J., joint author. *See* Bridges, R. G., et al., 1965, 1966, 1967, 1968, 1970.

Riehl, L. A., joint author. *See* Rodriguez, J. L., et al., 1962.

4787. Riemann, J. G. 1969 Competitive Mating Tests Between Normal and Aspermic Male House Flies. J. Econ. Ent. 62(1):276-277.
Female flies were caged (1) with normal males; (2) with males which had been surgically castrated. At the end of 1 hour, 80 per cent of the first group, 67 per cent of the second, had copulated. ABSTRACTS: BA 51:539(5389), Jan. 1, 1970; RAE-B 57:122(436), July, 1969; TDB 67: 462(921), Apr., 1970.

4788. Riemann, J. G., Moen, D. J. and **Thorson, B. J. 1967** Female Monogamy and Its Control in Houseflies. J. Insect Physiol. 13(3):407-418.
Loss of sexual receptivity by females after mating, caused by male seminal fluid, not by mechanical stimulation, or the presence of sperm. After oviposition for 20 days, 25 per cent recovered. ABSTRACTS: BA 48:5113(57194), June 1, 1967; TDB 64:686, June, 1967.

4789. Riemann, J. G. and **Thorson, B. J. 1968** Radiation Levels Required to Kill the Spermatogonia in Three Species of Diptera. Proc. N. Centr. Branch Am. Ass. Econ. Ent. 23:24 (Abstract).
Dosage of X-rays required to destroy all spermatogonia in *Musca domestica* is approximately 3,000 R. ABSTRACT: RAE-B 58:151(633), May, 1970.

4790. Riemann, J. G. and **Thorson, B. J. 1969** Effect of Male Accessory Material on Oviposition and Mating by Female House Flies. Ann. Ent. Soc. Am. 62(4):828-834.
Lesser amounts of material can induce oviposition without loss of receptivity to males. Females mating early recover receptivity more often than females mated after reaching full matury. ABSTRACTS: BA 51:1666(17413), Feb. 1, 1970; RAE-B 58:113(459), Apr., 1970; Ent. Abstr. 1:75(E259), Nov., 1969; TDB 67:903(1732), July, 1970.

4791. Riemann, J. G. and **Thorson, B. J. 1969** Comparison of Effects of Irradiation on the Primary Spermatogonia and Mature Sperm of Three Species of Diptera. Ann. Ent. Soc. Am. 62(3):613-617.
Doses of 2500, 5500, and 8500 roentgens were required to stop sperm production permanently in *M. domestica*, *Phormia regina*, and *Cochliomyia macellaria*, respectively. No close correlation with doses required to sterilize males by inducing dominant lethal mutations. ABSTRACT: BA 50:9173(95083), Sept. 1, 1969.

Rigato, M., joint author. *See* Bergamini, P. G., et al., 1970.

4792. Rigterink, R. H. and **Kenaga, E. E. 1966** Synthesis and Insecticidal Activity of some 0, 0-Dialkyl 0-3, 5, 6-Trihalo-2-pyridyl Phosphate and Phosphorothiotes. Agric. Food Chem. 14:304-306.
Tests showed that these compounds have a wide spectrum of insect toxicity; also that they are effective in several different solvent systems. *M. domestica* important among test insects used.

Riley, R. C., joint author. *See* Forgash, A. J., et al., 1962.

4793. Riordan, J. T., Paul, A., Yoshioka, I. and **Horstman, D. M. 1961** The Detection of Poliovirus and Other Enteric Viruses in Flies. Results of a Test Carried Out During an Oral Poliovirus Vaccine Trial. Am. J. Hyg. 74:123-136.
In southern Arizona, enteric viruses from flies, from rectal swabs of children, and from privies proved to be much the same. Fly believed to play a part both in *infecting* and in *immunizing* human populations. ABSTRACT: BA 37:624(6717), Jan. 15, 1962.

4794. Ristich, S. S., Ratcliffe, R. H. and **Perlman, D. 1965** Chemosterilant Properties, Cytotoxicity, and Mammalian Toxicity of Apholate and Other P-N Ring Chemicals. J. Econ. Ent. 58(5):929-932.
A positive correlation between per cent ethylene imine in the molecule, water solubility, and degree to which sterilization is successful. (*M. domestica.*) ABSTRACTS: BA 47:2928(34649), Apr. 1, 1966. (Gives pages as 927-932); RAE-B 54:15, Jan., 1966; TDB 63:341, Mar., 1966.

Ristich, S. S., joint author. *See* Castle, R. E., et al., 1966; Ratcliffe, R. H., et al., 1965.

Rivière, J. L., joint author. *See* Ramade, F., et al., 1970.

4795. Rivnay, E. 1968 Biological Control of Pests in Israel. Israel J. Ent. 3(1):1-156.
A review (1905-1965) of insect pests and the insecticides and other control measures used against them. Appropriate emphasis on houseflies.

4796. Rivosecchi, L. 1960 Osservacioni teratologiche su alcuni esemplari di *Musca domestica* L. raccolti in provincia di Latina. [Teratological observations on some specimens of *Musca domestica* L. collected in the province of Latina.] Rend. Ist. Sup. Sanit. 23(5):449-481.
Classifies 85 sexually abnormal specimens into 8 categories. Discusses sexual behavior and histological structure. ABSTRACTS: BA 36:2364, May 1, 1961; TDB 58:142, Jan., 1961.

4797. Rivosecchi, L. 1962 Un esperimento sul campo con maschi irradiate di mosca domestica in una zona rurale della provincia de Latina. [A field experiment with irradiated male houseflies in a rural area of Latina province.] Riv. di Parassit. 23(1):71-74. (English summary.)
Sterile males set free in March. Sterile females appeared in April. Female population low for 2 months. Program discontinued in July. Fertile females appeared in August, predominated in September. ABSTRACTS: RAE-B 52:174, Sept., 1964; TDB 59:1212, Dec., 1962.

Roan, C. C., joint author. *See* Rai, L., et al., 1963, 1964; Zaharis, J. L., et al., 1961.

4798. Robbins, W. E. 1963 Studies on the Utilization, Metabolism and Function of Sterols in the House-fly, *Musca domestica*. Proc. Symp. (on) Radiation and Radioisotopes Applied to Insects of Agricultural Importance. Athens, April 22-26. Publ. Vienna, Int. Atomic Energy Agency. Int. Publ. Inc., New York 16:269-280.
Reports radiotracer investigations showing sterols in diet required for sustained production of viable eggs; cholesterol also involved in mobilization and utilization of food reserves associated with ovarian maturation. ABSTRACTS: RAE-B 53:16, Jan., 1965; BA 46:7008(87034), Oct. 1, 1965.

4799. Robbins, W. E., Kaplanis, J. N., Louloudes, S. J. and **Monroe, R. E. 1960** Utilization of 1-C^{14}-Acetate in Lipid Synthesis by Adult House Flies. Ann. Ent. Soc. Am. 53(1):128-129.
A large percentage of acetate is incorporated into the unsaponifiable

fraction of adult housefly homogenate. Indicates a high turnover rate for this fraction.

4799a. Robbins, W. E., Kaplanis, J. N., Thompson, M. J. and **Shortino, T. J. 1970** Ecdysomes and Synthetic Analogs: Molting Hormone Activity and Inhibitive Effects on Insect Growth, Metamorphosis and Reproduction. Steroids 16(1):105-125.
House fly assay utilized to assess 5 *beta* steroids for molting hormone. Authors discuss relation of structure to molting hormone and to inhibitive effects of compounds tested. ABSTRACT: BA 52:962(8914), Jan. 15, 1971.

4800. Robbins, W. E. and **Shortino, T. J. 1962** Effect of Cholesterol in the Larval Diet on Ovarian Development in the Adult House-fly. Nature (London) 194(4827):502-503.
Flies reared on standard larval medium plus cholesterol developed mature ovaries when held on an adult diet of sucrose and water. ABSTRACTS: BA 40:1298(17042), Nov. 15, 1962; RAE-B 51:166, Aug., 1963.

4801. Robbins, W. E., Thompson, M. J., Yamamoto, R. T. and **Shortino, T. J. 1965** Feeding Stimulants for the Female House Fly, *Musca domestica* Linnaeus. Science 147:628-630.
Casein hydrolysate (contains several amino acids) and yeast hydrolysate (contains guanosine monophosphate) greatly stimulate feeding. Solution in phosphate buffer is necessary to obtain maximum activity. ABSTRACTS: BA 46:4805(59484), July 1, 1965; TDB 62:694, July, 1965.

Robbins, W. E., joint author. *See* Kaplanis, J. N., et al., 1960, 1961, 1963, 1965, 1966; Cantwell, G. E., et al., 1966; Dutky, R. C., et al., 1963, 1967; Monroe, R. E., et al., 1961, 1963; Vroman, H. E., et al., 1966; Thompson, M. J., et al., 1962, 1963; Shortino, J. T., et al., 1963.

4802. Roberts, J. E. 1963 Control of Flies on Beef and Dairy Cattle. Georgia Agr. Exp. Sta. Mimeo. Ser. 186:1-11. December.
Face fly populations, in north Georgia, were not appreciably reduced by rubbing devices. Horn fly populations were. ABSTRACT: BA 46:4412 (54592), June 15, 1965.

4803. Roberts, J. E. 1965 Evaluation of Impregnated Bands and Residual Sprays for Control of House Flies in Barns. Georgia Agr. Exp. Sta. Leaflet (N.S.) 46:1-7.
With few exceptions, control in 3 dairy barns was excellent throughout test period. Dichlorvos and Dimetlan used for bands; Ronnel for residual sprays.

4804. Roberts, J. E. 1965 Evaluation of Insecticidal Formulations and Application Methods for Control of Flies on Dairy Cattle. Georgia Agr. Expt. Sta. Leaflet (N.S.) 47:1-4.
Ciodrin applied by 3 methods. A 2 per cent preparation, applied as a mist spray, averaged 96 and 99 per cent control with hornflies, but was only 55 per cent effective against face flies. (Described as a "degree of relief".)

4805. Roberts, J. E. 1965 Rubbing Devices Treated with Insecticides for Horn Fly and Face Fly Control on Beef Cattle. Georgia Agr. Expt. Sta. Mimeo. Series N.S. 227:1-7.
Back rubbers consisted of burlap-wrapped cable, suspended between posts, and treated with insecticides. Dust bag rubbing devices, placed near salt blocks, were also used.

Roberts, J. E., joint author. *See* Coleman, V. R., et al., 1962.

Roberts, P. A., joint author. *See* Fukuto, T. R., et al., 1962.

4806. Roberts, R. B., Miskus, R. P., Duckles, C. K. and **Sakai, T. T. 1969** *In vivo* fate of the insecticide Zectran in spruce budworm, tobacco budworm and housefly larvae. (*Choristoneura occidentalis, Heliothis virescens, Musca domestica.*) J. Agric. Fd. Chem. 17(1):107-111.
Zectran not strongly toxic to *M. domestica*; was metabolized to 10 metabolites in houseflies. Rate of cuticular penetration was much the same for all 3 insects until the third hour. ABSTRACT: BA 50:10860(112419), Oct. 15, 1969.

4807. Robertson, R. L. 1969 Simple, Safe, Effective Fly Control for Dairy Farms. North Carolina Agr. Extension Service. Ext. Folder 241 (Revised).
Gives advice on which insecticides not to use, and methods of application for effective ones.

4808. Rockstein, M. 1960 Aging of Insects. *Repr. from* The Biology of Aging. Am. Inst. Biol. Sci. Symp. 6:243-245.
Seeks sound biochemical criteria for aging, in terms of enzyme systems and related substrates. House fly much used in this research.

4809. Rockstein, M. 1964 Some Physical-Biochemical Correlates of Aging of Motor Ability. Duke Univ. Council of Gerontology. Proc. Seminars 1961-1965. Seminar Feb. 4. pp. 240-251.
Alpha-glycerophosphate dehydrogenase activity, which normally fails in houseflies 36 to 48 hours before wing loss, remains at original high level, if wings are removed early.

4810. Rockstein, M. 1966 Biology of Aging in Insects. *In*: Symp. on topics in the biology of aging, 4-6 November 1965. San Diego, Calif. Interscience Publishers, John Wiley & Sons: New York and London. pp. 43-61. Illus.
Concerns flight muscles of *M. domestica* and *Apis mellifera.* Stresses need to study the mechanism as well as the parameters of the aging process. ABSTRACT: BA 48:7889(88458), Sept. 1, 1967.

4811. Rockstein, M. 1967 Wing Beat Frequency in the Aging Housefly. J. Gen. Physiol. 50(4):1082.

4812. Rockstein, M. and **Bhatnagar, P. L. 1965** Age Changes in Size and Number of Giant Mitochondria in the Flight Muscle of the Common House Fly (*Musca domestica* L.). J. Insect Physiol. 11(4): 481-491.
Variations in mitochondrial size and number are both sex-related and age-dependent. (Giant mitochondria are also known as "sarcosomes".) ABSTRACT: BA 46:5189(64446), Aug. 1, 1965.

4813. Rockstein, M. and **Bhatnagar, P. L. 1966** Duration and Frequency of Wing Beat in the Aging House Fly, *Musca domestica* L. Biol. Bull. 131(3):479-486.
Female attains maximum frequency by 5th day of adult life. Decline in frequency is gradual. Male reaches peak frequency on 4th day. Decline is abrupt. ABSTRACT: BA 48:3740(41603), Apr. 15, 1967.

4814. Rockstein, M. and **Bhatnagar, P. L. 1966** X-irradiation and Wing Retention in the Common Fly, *Musca domestica* L. Naturwissenschaften 53(24):702-703.
High doses of X-rays (10,000 to 30,000 rads) to house fly pupae increase adult longevity, and delay the usual loss of wings early in adult life. (Especially true of normally aging males.)

4815. Rockstein, M. and **Brandt, K. 1962** The Biochemical Basis for Aging of Flight Ability in the Male Housefly. *In*: 59th Ann. Meet. Amer.

Soc. Zool., Philadelphia. Am. Zool. 2(4):552. (Abstract only.)
Data suggest a programmed chemical senescence in the form of a genetically scheduled failure of a particular enzyme system. ABSTRACT: BA 46:8916(109980), Dec. 15, 1965.

4816. Rockstein, M. and **Brandt, K. F. 1963** Enzyme Changes in Flight Muscle Correlated with Aging and Flight Ability in the Male Housefly. Science 139:1049-1051.
Preceding loss of wings, there is a decline in the activity of an *alpha* glycerophosphate dehydrogenase, located in the extra-mitochondrial fraction. ABSTRACTS: BA 44:643(8617), Oct. 15, 1963; TDB 60:690, July, 1963.

4817. Rockstein, M. and **Dauer, M. 1969** Effects of X-irradiation on Longevity and Rate of Aging in the Common Male Housefly. Radiation Res. 39(2):467. (Abstract.)

4818. Rockstein, M., Dauer, M. and **Bhatnagar, P. L. 1965** Adult Emergence of the House Fly, *Musca domestica*, from X-Irradiated Pupae. Ann. Ent. Soc. Am. 58:375-379.
The number of flies dying in the pupal stage increases with an increase in radiation dosage, e.g., a large majority survive to emerge at 10,000-15,000 rads; 42 per cent at 20,000; 19·5 per cent at 30,000. ABSTRACTS: BA 46:6676(82914), Oct. 1, 1965; RAE-B 53:195, Oct., 1965; TDB 63: 340, Mar., 1966.

4819. Rockstein, M., Dauer, M. and **Bhatnager, L. 1967** Further Studies on the Effect of X-irradiation on the House Fly, *Musca domestica* L. Radiat. Res. 31(4):840-845.
Exposure of pupae to low levels of X-irradiation (2000-8000 rads) had little or no effect on percentage of emergence or normal appearance of adults. ABSTRACT: BA 48:10806(120178), Dec. 15, 1967.

4820. Rockstein, M. and **Gutfreund, D. E. 1961** Age Changes in Adenine Nucleotides in Flight Muscle of Male House Fly. Science 133(3463): 1476-1477.
Content of adenosine triphosphate increases five-fold, while content of adenosine monophosphate diminishes accordingly.

4821. Rockstein, M. and **Hawkins, W. B. 1965** Thiamine in Aging House Flies. Physiologist 8(3):261.
Changes in thiamine content parallel closely the chemical changes in Mg-activated adenosine triphosphate (ATPase) and alpha-glycerophosphate dehydrogenase activity in relation to senescence (particularly loss of flight ability in male flies with advancing age).

4822. Rockstein, M. and **Srivastava, P. N. 1967** Trehalose in the Flight Muscle of the House Fly, *Musca domestica* L. in Relation to Age. Experientia 23:636-637. (German summary.)
Age-related changes in the quantity of trehalose in flight muscles of the male fly believed to be correlated with wing activity and (eventually) loss of wings. ABSTRACT: BA 49:4314(48610), May 1, 1968.

Rockstein, M., joint author. *See* Dauer, M., et al., 1965; Srivastava, P. N., et al., 1969.

Rodger, J. C., joint author. *See* Osborn, A. W., et al., 1970.

4823. Rodriguez, J. G. 1969 The Manure Acarine Complex. Kentucky Agric. Exp. Sta. Res. for 1968. Page 67.
Concerns *Fuscuropoda vegetans*, *M. domestica* and *Fannia canicularis*.

4824. **Rodriguez, J. G.** and **Knapp, F. W. 1968** Integrated Control of Muscids in Poultry Houses. Kentucky Agr. Exp. Sta. Prog. Rep. for 1967. p. 59.
Treats of *M. domestica* and *Fannia canicularis*. Control by natural enemies (*Macrocheles muscae-domesticae*) and by insecticides.

4825. **Rodriguez, J. G., Singh, P.** and **Taylor, B. 1970** Manure Mites and Their Role in Fly Control. J. Med. Ent. 7(3):335-341.
Acarine control of house flies in a poultry house (semi-field conditions) ranged from 86 to 99 per cent, depending on the mites involved. Five insecticides (of 14 tested) found relatively toxic to fly maggots and relatively non-toxic to *Macrocheles muscaedomesticae*. ABSTRACT: BA 51: 13685(139254), Dec. 15, 1970.

4826. **Rodriguez, J. G.** and **Taylor, B. 1969** Integrated Control of Flies in Poultry House. Kentucky Agric. Exp. Sta. Res. for 1968. Page 68.
Concerns discreet use of four insecticides without elimination of *Macrocheles muscaedomesticae*, in the control of *Fannia* and *Musca*.

4827. **Rodriguez, J. G.** and **Wade, C. F. 1961** The Nutrition of *Macrocheles muscae-domesticae* (Acarina:Macrochelidae) in Relation to its Predatory Action on the House Fly Egg. Ann. Ent. Soc. Am. 54(6):782-788.
Biological control tests showed that amount of predation varied with the type of substrate and with periods between feedings. A mean of 19·8 eggs per mite was destroyed when ova and mites were in contact on filter paper or in conditioned CSMA larva media. ABSTRACT: RAE-B 50:241, Nov., 1962.

4828. **Rodriguez, J. G.** and **Wade, C. F. 1962** Preliminary Studies on Biological Control of Housefly Eggs using Macrochelid Mites. XI. Internationaler Kongress für Entomologie Wien. 1960. Sonderdruck aus den Verhandlungen Bd. III. 1 p.
Concerns *Macrocheles muscae-domesticae* and *M. plumiventris*. Technique of rearing mites in laboratory. Observations relative to fly control.

4829. **Rodriguez, J. G., Wade, C. F.** and **Wells, C. N. 1962** Nematodes as a Natural Food for *Macrocheles muscae-domesticae* (Acarina: Macrochelidae) a Predator of the House Fly Egg. Ann. Ent. Soc. Am. 55:507-511.
A two-year ecological study. Adult mites preferred housefly eggs over nematodes, while proto- and deuto-nymphes preferred the nematodes, providing environmental conditions were the same. ABSTRACTS: BA 41: 1312(16805), Feb., 1963; RAE-B 51:135, June, 1963; TDB 60:78, Jan., 1963.

Rodriguez, J. G., joint author. *See* Wade, C. F., et al., 1961; Wallwork, J. H., et al., 1963; Singh, P., et al., 1966; Jalil, M., et al., 1970; Taylor, B., et al., 1968, 1969.

4830. **Rodriguez, J. L.** and **Riehl, L. A. 1962** Control of Flies in Manure of Chickens and Rabbits by Cockerels in Southern California. J. Econ. Ent. 55:473-477.
The use of one cockerel on the ground for every 20-100 hens (or 5 rabbits) in the cages overhead reduced larvae and pupae of *M. domestica* to a count of zero. ABSTRACTS: BA 40:1608(21024), Dec. 1, 1962; RAE-B 51:16, Jan., 1963.

Rogers, M. R., joint author. *See* Shambaugh, G. F., et al., 1968.

4831. **Rogoff, M. H., Ignoffo, C. M., Singer, S., Gard, I.** and **Prieto, A. P. 1969** Insecticidal Activity of Thirty-one Strains of *Bacillus* Against

Five Insect Species. J. Invertebr. Path. 14(2):122-129.
Used animal-protein-base medium or plant-protein-base medium supplemented with casein. Three strains active against the house fly were also active against the house mosquito. *Bacillus thuringiensis* var. *anagastae*-30 active against all insects. ABSTRACT: BA 51:6253(64002), June 1, 1970.

Rogoff, M. H., joint author. *See* Shieh, T. R., et al., 1968.

4832. Rogoff, W. M. 1961 Chemical Control of Insect Pests of Domestic Animals. Advances in Pest Contr. Res. 4:153-181.
Includes an 8 page list of 39 common entomocides in commercial or experimental use. Common names only, used for insects. Housefly given appropriate attention. Bibliography contains 212 references.

4833. Rogoff, W. M. 1965 Mating of the House Fly, *Musca domestica* L. in Monitored Darkness. J. Med. Ent. 2:54-56.
Light not a necessary condition for successful coitus. House flies can mate in the absence of both ultraviolet and visible light. ABSTRACTS: BA 46:7466(92124), Nov. 1, 1965; TDB 63:1253, Nov., 1966.

4834. Rogoff, W. M., Beltz, A. D., Johnsen, J. O. and **Plapp, F. W. 1964** A Sex Pheromone in the Housefly, *Musca domestica* L. J. Insect Physiol. (London) 10(2):239-246.
Used on olfactometer and simulated, fly-treated models. Female flies produce 1 or more volatile chemicals which attract the males and/or excite mating behavior patterns. Substance is species-related. Extracts of *Stomoxys* or *M. autumnalis* will not attract *M. domestica*. ABSTRACTS: BA 45:5761(71274), Aug. 15, 1964; RAE-B 52:221, Dec., 1964; TDB 61:726, July, 1964.

Rogoff, W. M., joint author. *See* Cowan, B. D., et al., 1968.

4835. Rohe, D. L. 1964 A Trap for Fly Larvae Migrating from Garbage Cans. Calif. Vector Views 11(1):4-6. Berkeley, Calif.
Details given for construction of trap that simplifies collection and counting of Dipterous larvae migrating from dust pans. ABSTRACT: RAE-B 54:47, Feb., 1966.

Rohwer, G. G., joint author. *See* Steiner, L. T., et al., 1961.

Rones, G., joint author. *See* Ascher, K. R. S., et al., 1965.

4836. Rose, G. J. 1963 Crop Protection. 2nd Ed. xvi + 490 pp. London, L. Hill (Books), Ltd.
Includes notes on the control of *M. domestica*. ABSTRACT: RAE-B 51: 119, June, 1963.

4837. Rosen, J. D., Sutherland, D. J. and **Khan, M. A. Q. 1969** Properties of Photoisomers of Heptachlor and Isodrin. Agric. Food Chem. 17(2):404-405.
Photoisomer of heptachlor is more toxic to *M. domestica* than heptachlor itself; but the photoisomer of isodrin is less toxic than isodrin. (Due, it is believed, to the more rapid metabolism of this isomer.)

Rosen, J. D., joint author. *See* Khan, M. A. Q., et al., 1969, 1970.

4838. Rosenberg, P. 1964 Effect of Flies (Diptera) on Man and their Control. J. Environ. Hlth. 26(6):415.

Rosenthal, J., joint author. *See* Sullivan, W. N., et al., 1970.

4839. Roslavtseva, S. A. 1963 Ingibirovanie kholinesterazy i aliesterazy fosfororganicheskimi i soedineniyami. [Inhibition of cholinesterase and aliesterase by phosphororganic compounds.] Khim. Sel'skom. Khoz. 1:30-31. Referat. Zhur., Biol. 1964, No. 11E430 (Translation).
Of seven compounds the thiolic and thionic isomers of Meta-Systox,

Dipterex and methylacetophos suppressed cholinesterase more, but acetone and DDVP had a greater effect on aliesterase. Acetophos inhibited both to approximately the same degree. Mixtures gave a greater effect than single components. ABSTRACT: BA 47:3357(39614), Apr. 15, 1966. (Author given as "Rostlavtseva".)

4840. **Roslavtseva, S. A. 1968** Studies of Esterase Specificity in Organophosphorus-resistant House Flies. Abstracts of Papers XIIIth International Congress of Entomology. Moscow. Page 220.

4841. **Roslavtseva, S. A. 1970** Cross-resistance in Three Organophosphorus-resistant Strains of House Flies. Chim. Selskom. Choz. 8(7):30.

4842. **Roslavtseva, S. A., Evtyushina, T. M.** and **Popov, P. V. 1969** Aktivnost' nekotorykh esteraz i priobretennaya resistentnost' komnatnykh mukh. [Activity of some esterases and acquired resistance in house flies.] Med. Parazit. i Parazitarn. (Bolez.) 38(1):35-38. (English summary.)

Values of resistance and activity of enzymes determined for resistant and normal strains by the colorimetric method of Hestrin. ABSTRACT: BA 51:2235(23185), Feb. 15, 1970.

4843. **Roslavtseva, S. A.** and **Mandel'Baum, Ya. A. 1963** Perspektivnye insektitsidy iz grupp fosfororganicheskikh soedinenii. [Possible insecticides from the group of organophosphorous compounds.] Med. Parazit. i Parazitarn. (Bolezni) 32(3):338-340.

Acetaphos and methylacetaphos are insecticides with a broad range of action, are very stable on glass surfaces, and have a low toxicity to warm-blooded animals and man. ABSTRACTS: BA 45:3601(44356), May 15, 1964; RAE-B 54:235, Dec., 1966.

4844. **Roslavtseva, S. A., Polyvakova, V. K.** and **Ivanova, G. B. 1970** Izmenenie resistentnosti mukh k metiletiltioforu i ftalofosu v laboratornykh opytakh. [Changes in the resistance of flies to methylethylthiophos and phthalophos in laboratory experiments.] Med. Parazit. i Parazitarn. (Bolez.) 39(3):345-349. (English summary.)

After 26 selections with methylethylthiophos the level of resistance was reduced. Treatment with phthalophos over 7 generations increased resistance three-fold. ABSTRACT: BA 52:1411(13320), Feb. 1, 1970.

4844a. **Ross, E.** and **Sherman, M. 1960** Toxicity to House Fly Larvae of Droppings from Chickens Fed Insecticide-Treated Rations. J. Econ. Ent. 53:429-434.

Co-Ral, at 89 ppm., gave over 90 per cent larval mortality. Similar results with Diazinon at 154 ppm., Dipterex at 89-132 ppm., ronnel at 176-220 ppm., Dow ET-15 at 89 ppm.

Ross, E. S., joint author. *See* Sherman, M., et al., 1960, 1961, 1962, 1963, 1968.

4845. **Ross Institute Information and Advisory Service. 1969** The Housefly and its Control. London School of Hygiene and Tropical Medicine. Bull. No. 5, 24 pp. (Originally issued November 1950. Revisions: 1953, 1956, 1962, 1965.)

An informative and educational publication. Well illustrated. ABSTRACT: TDB 62:942, Sept., 1965.

Roth, A. R., joint author. *See* Eddy, G. W., et al., 1961, 1962.

4846. **Rothschild, J.** and **Howden, G. F. 1961** Effect of Chlorinated Hydrocarbon Insecticides on Insect Choline Acetylase, Condensing Enzyme and Acetylkinase. Nature 192(4799):283-284.

Tissue homogenates of *M. domestica* heads or whole flies were added to mixtures containing coenzyme A, a cholinesterase inhibitor, choline chloride and other compounds, with sodium acetate or sodium citrate as acetyl donor. Effect of insecticide tested by adding it in acetone. Agrees with work of Lewis (1953). ABSTRACT: RAE-B 50:280, Dec., 1962.

Roubal, W. T., joint author. *See* Terriere, L. C., et al., 1960, 1961.

4847. Round, M. C. 1961 Observations on the Possible Role of Filth Flies in the Epizootiology of Bovine Cysticercosis in Kenya. J. Hygiene (London) 59:503-513.

Concerns primarily *Chrysomyia albiceps*, *C. chloropyga* and a species of *Sarcophaga*. Refers to *M. domestica* as a known distributor of helminth ova.

4848. Rousell, P. G. 1965 Comparative Insecticidal Susceptibility of Field-Collected and Laboratory-Reared Face Flies, *Musca autumnalis*. J. Econ. Ent. 58:674-676.

Both strains proved susceptible to commonly used insecticides; the field-collected flies slightly more so. ABSTRACTS: BA 46:8878(109469), Dec. 15, 1965; RAE-B 53:221, Nov., 1965.

4849. Rousell, P. G. 1967 Activities of Respiratory Enzymes During the Metamorphosis of the Face Fly, *Musca autumnalis* DeGeer. J. New York Ent. Soc. 75(3):119-125.

Activities of ten enzymes determined. Greatest activity was shown by malic dehydrogenase. Information presented in comparative tables. ABSTRACT: BA 49:952(10728), Jan. 15, 1968.

Roy, S. P., joint author. *See* Patnaik, B., et al., 1966.

4850. Rozov, A. A. 1966 Vyzhivaemost' virusa yashchura na poverkhnosti tela i v organizme komnatnykh mukh. [Survival of the foot and mouth disease virus on the surface and inside the body of house flies.] Tr. Vses Nauchn-Issled. Inst. Vet. Sanit. 25:104-109. (German summary.) Referat Zhur., Biol. 1967. No. 7B164.

4851. Rubini, P. G. 1964 Polimorfismo cromosomico in *Musca domestica* L. [Chromosomal polymorphism in *Musca domestica* L.] Boll. di Zool. 31(2):679-694. (Read at XXXIII Convegno dell' U.Z.I., Roma, October 5-10, 1964.)

In two T(YII) strains the chromosomal complement in females is usually normal, but males have one, two or three large heterochromosomes; the Y is practically indistinguishable from the X; when three are present, one is thicker.

4852. Rubini, P. G. 1966 Osservazioni citologiche preliminari su *Musca domestica* L. [Preliminary cytological observations on *Musca domestica* L.] Symp. Genet. et Biol. Ital. 13:332-340.

In strain SRS-1 both X and Y chromosomes are heterochromatic and metacentric; X is comparable to largest autosome in length; Y approaches one arm of X. Variations in other strains described.

4853. Rubini, P. G. 1967 Ulteriori osservazioni sui determinanti sessuali di *Musca domestica* L. [Recent observations on sex determining factors in *Musca domestica* L.] Genetica Agraria 21:363-384.

New combinations of heterochromosomes and of autosomal sex factors with their linked genes can be obtained by crosses between standard and atypical strains. Present information lacking on possible functional differences. ABSTRACT: BA 51:2274(23595), Mar. 1, 1970.

4854. Rubini, P. G. and **Caprotti, M. 1966** Variazioni della sensibilità ai raggi X durante il periodo pupale di *Musca domestica* L. [Variation in sensitivity to X-rays during the pupal period of *Musca*

domestica L.] Symp. Genet. et Biol. Ital. 13:369-378.
Emergence rates show that sensitivity of housefly pupae declines along with pupal aging. There is a conspicuous lowering during the second day in the pupal state.

4855. Rubini, P. G. and **Caprotti, M. 1966** Osservazioni sull'-effetto letale di raggi X di diversa durezza su pupe di *Musca domestica* L. [Observations concerning the lethal effect of X-rays of different hardness on the pupae of *Musca domestica* L.] Symp. Genet. et Biol. Ital. 13:387-392.
Lethal effect greater from harder X-rays (230 Kv filtered 5 mm. Al) than from 80 Kv unfiltered, softer rays. With equal doses of constant voltage (80 Kv), using different degrees of filtration, the mortality consistently increased with hardness.

4856. Rubini, P. G. and **Franco, M. G. 1963** Formal Genetics of the Housefly. (Demonstration.) Proc. XI Int. Congr. Genet. I:254.
Summarizes best known facts concerning mutations, their dominance, etc. States that close inbreeding leads quickly to sterility.

4857. Rubini, P. G. and **Franco, M. G. 1965** Aneuploidy of Heterochromosomes and Considerations on Sex Determination in *Musca domestica* L. Paper read at XXXIV Convegno Zoologica Italiana (Pallanza, October, 1965).
Loss or gain of one heterochromosome has been observed in both sexes, in almost every strain studied. In one, selected line, the majority of the flies has 6X's or 5X's and 1Y, according to sex.

4858. Rubini, P. G. and **Franco, M. G. 1965** Osservazioni sugli sterocromosomi e considerazioni sulla determinazione del sesso in *Musca domestica* L. (Nota preliminare). [Observations on the heterochromosomes and consideration of sex determination in *Musca domestica* L.] Boll. di Zool. 32:823-827.
Males of standard strains have 2 heterochromosomes, very different in size; males of atypical strains, 2 large heterochromosomes, as in females. Suggests different systems of sex balance in standard and atypical strains.

4859. Rubini, P. G. and **Franco, M. G. 1965** Polymorphism and Aneuploydy of the Heterochromosomes in *Musca domestica* L. Paper read at Symposium on the mutational process. Praga. Carbon of typed copy, with illus., in L. S. West collection of separates. pp. 89-92.
Males (normally XY) may have XXY (6 per cent) or OY (3 per cent). Females (normally XX) may have XXX (1 per cent) or XO (4 per cent).

4860. Rubini, P. G. and **Franco, M. G. 1966** Formazione ed analisi genetica preliminare di un ceppo YY di *Musca domestica* L. [Production and preliminary genetical analysis of a YY strain of *Musca domestica* L.] Estratto dal Boll. di Zool. 33(1):141.
Brief note. Speculates as to breeding behavior.

4861. Rubini, P. G. and **Franco, M. G. 1968** Osservazioni sui determinanti sessuali presenti in un ceppo YY di *Musca domestica* L. [Observations on sex determining factors present in a YY strain of *Musca domestica* L.] Estratto dal Boll. di Zool. 35:437.
Concerns chromosome behavior in certain crosses, and resulting sexual differentiation.

4862. Rubini, P. G. and **Franco, M. G. 1968** Additional Information on Sex Determination in *Musca domestica*. Atti. Ass. Genet. Ital. 13: 114-116.
Reports experiments which show that the YY *m m* strain has undergone

a natural selection, leading to its present heterogeneity for sex factors and to new equilibria departing from those artificially introduced.

4863. Rubini, P. G. and **Palenzona, D. 1967** Response to Selection for High Number of Heterochromosomes in *Musca domestica* L. Genetica Agraria 21(1/2/3):101-110. (Italian summary.)
In several strains, individuals trisomic for X-chromosomes were observed. The aneuploids were phenotypically like normal disomics. Either the supernumary chromosomes are inactive, or some compensatory mechanism exists. Selection for high number of heterochromosomes was successful. ABSTRACT: BA 50:7579(79013), Aug. 1, 1969.

Rubini, P. G., joint author. *See* Caprotti, M., et al., 1966; Franco, M. G., et al., 1966; Milani, R., et al., 1967.

4864. Rumiantsev, I. N., Bagirov, K. S. and **Askerov, G. A. 1961** [On the use of oil of turpentine in control of flies.] Voennomed Zhur. 7: 76-77.

Runner, C. M., joint author. *See* Van Bruggen, E. F. J., et al., 1968.

4865. Ruprah, N. S. 1967 Effects of Bovine Diet on Oviposition, Larval Development and Pupation of the Face Fly, *Musca autumnalis*. Diss. Absts. 28(3B):751-B.
See next reference, *also* Treece, 1966.

4866. Ruprah, N. S. and **Treece, R. E. 1968** Further Studies on the Effect of Bovine Diet on Face Fly Development. J. Econ. Ent. 61(5):1147-1150.
M. autumnalis was exposed to feces from cows fed 18 different diets. Flies showed preference for manure from grass- or alfalfa-fed animals, over manure from diets containing corn silage. Larval survival not appreciably affected by cow's diet. ABSTRACTS: BA 50:4218(44392), Apr. 15, 1969; RAE-B 57:49(174), Mar., 1969.

4867. Russo-Caia, S. 1960 [Biochemical aspects of the metamorphosis of insects. The metabolism of nitrogenous substances during the larval development and metamorphosis of *Musca domestica* L.] Riv. Biol. 53:409-430.

4868. Russo-Caia, S. 1961 Aspetti biochimici della metamorfosi degli insetti. Gli enzimi proteolitici nell'accrescimento larvale e nella metamorfosi di *Musca domestica* L. [Biochemical aspects of insect metamorphosis. Proteolytic enzymes during the larval growth and the metamorphosis of *Musca domestica* L., Acta. Embryol. et Morph. Expt. 3(2):131-145.
Each active enzyme has its own peculiar behavior. Study included alanylglycine-dipeptidase; acid and alkaline proteinases. ABSTRACT: BA 36: 7991(85031), Dec. 15, 1961.

4869. Russo-Caia, S. 1963 Ricerche sul corpo grasso degli Insetti. Osservazioni autoradiografiche sulla incorporazione di adenina-C^{14} e acido orotico C^{14} nei trofociti larvali di *Musca domestica.* [Research on the fat body of insects. Autoradiographic observations on the incorporation of adenine-C^{14} and orotic acid-C^{14} in growing larvae of *Musca domestica.*] Rend. Ist. Sci. Univ. Camerino 4(3):216-228. (English summary.)
ABSTRACT: BA 47:5975(69917), July 15, 1966.

4870. Russo-Caia, S. and **Cecere, T. 1961** Le variazioni del peso e del contenuto di acqua durante l'accrescimento larvale e la metamorfosi di *Musca domestica* L. [The variations of weight and water content during the larval growth metamorphosis of *Musca domestica* L.] Rend. Ist. Sci. Univ. Camerino 1(2):126-135.

A weight increase occurs in the growth phase of larval development and a weight loss in the premetamorphosis phase. The percentage of water loss decreases during the entire life cycle. ABSTRACT: BA 36:7992 (85033), Dec. 15, 1961.

Ruttenberg, G. J. C. M., joint author. *See* Van Bruggen, E. F. J., et al., 1968.

Ryazantsev, V. F., joint author. *See* Neselovskaya, V. K., et al., 1968.

Rye, J. R., Jr., joint author. *See* Meifert, D. W., et al., 1969.

Saavedra, I., joint author. *See* Dinamaria, M. L., et al., 1969.

4871. Sabrosky, C. W. 1961 Our first decade with the Face Fly, *Musca autumnalis*. J. Econ. Ent. 54(4):761-763.

A year by year documentation of the history of the face fly in North America, from 1951. Believed to be an immigrant species. ABSTRACT: RAE-B 50:152, July, 1962.

4872. Saccà, G. 1960 Esperienze con mosche domestiche, sterilizzate con raggi X. [Experiments with domestic flies, sterilized by X-rays.] Atti Acad. Naz. Ital. Ent. Rendic. Anno VIII. pp. 91-98.

Sterile males compete effectively with normal males for mates. Females of such matings, usually monogamous, are therefore sterile for life.

4873. Saccà, G. 1961 Esperienze con mosche domestiche sterilizzate con raggi X. [Experiments with house flies, sterilized by X-rays.] Rend. Ist. Sup. Sanit. 24:5-12. Rome. (English summary.)

Pupae are irradiated with 3000 rads, 2-3 days before emergence. Except for their sterility, flies exhibit normal viability, longevity, and sexual behavior. ABSTRACTS: BA 36:6484(70194), Oct. 15, 1961; RAE-B 51:274, Dec., 1963; TDB 58:1192, Oct., 1961.

4874. Saccà, G. 1964 Nota sulla presenza in Europa di *Ophyra aenescens* Wied. (Diptera:Muscidae). Riv. di Parassit. 25(4):295-296.

Discussion compares this species with *M. domestica* in regard to ecology, life history, and distribution.

4875. Saccà, G. 1964 Comparative Bionomics in the Genus *Musca*. Ann. Rev. Ent. 9:341-358.

Brings out relation to several diseases. Cites 79 references.

4876. Saccà, G. 1965 Considerazioni sull'origine del "complesso" *domestica*. (Diptera, Muscidae, gen. *Musca*). [A consideration of the origin of the *domestica* complex. (Diptera, Muscidae, gen. *Musca*).] Boll. di Zool. 32(2):789-799.

Suggests that Ethiopian forms, *calleva* and *curviforceps*, may, through hybridization and geographical transfer, have given rise to the *domestica-vicina-nebulo* complex.

4877. Saccà, G. 1964 Comparative Bionomics in the Genus *Musca*. Ann. Rev. Ent. 9:341-358.

A concise review, with special mention of *M. domestica* and *M. sorbens*. Lists 79 references published prior to March, 1963. ABSTRACTS: BA 45: 8409(106841), Dec. 15, 1964; TDB 64:686, June, 1967.

4878. Saccà, G. 1967 Speciation in *Musca*. *In* Genetics of Insect Vectors of Disease. (J. W. Wright and R. Pal, Editors.) Amsterdam. Elsevier Publ. Co. Chapter 11, pp. 385-399.

A well illustrated exposition of hybrid types and their evolutionary significance. ABSTRACT: RAE-B 56:157(553), Aug., 1968.

4879. Saccà, G. 1969 Ricerche sperimentali sulla chemosterilizzazione in *Musca domestica* (1963-1968). [Experimental research on chemosterilization in *Musca domestica.*] Congrès. Internat. des Anti parasitaires. Milan, October 6-8. 7 pp. (Several foreign summaries.)
Covers 6 years of work on chemosterilants in liquid sugar baits, with emphasis on tepa and hempa. A strain resistant to metepa was selected, and an experiment run on "integrated control".

4880. Saccà, G. and **Benetti, M. P. 1960** Ricerche sperimentali sulla maturità e sul comportamento sessuale di *Musca domestica* L. (Diptera, Muscidae). [Experimental research on the maturity and the sexual behaviour of *Musca domestica* L.] Rend. Ist. Sup. Sanita 23(5): 423-432.
Feeding flies on an incomplete diet of sugar and water does not inhibit ripening of the testes, mating, or insemination. It does, however, prevent the ovaries from ripening. While females usually mate but once, males may fertilize numerous females. ABSTRACTS: BA 36:2017(24587), Apr. 15, 1961; RAE-B 50:165, Aug., 1962; TDB 58:142, Jan., 1961.

4881. Saccà, G., Gandalfo, D. and **Mastrilli, M. L. 1969** La sterilizzazione di *Musca domestica* L. con i raggi gamma: Importanza dell'età delle pupe irradiate. [Sterilization of *Musca domestica* with gamma rays: Importance of radiating pupae.] Parassitol. 11:3-7.
Doses of 2000 and 3000 rads inhibit emergence of flies when applied before the third day of pupal life in a temperature of 25-26°C. Later application permits emergence, but flies are sterile.

4882. Saccà, G., Gandolfo, D. and **Stella, E. 1968** Ricerche sulla sterilizzazione della mosche mediante raggi gamma. [Research on the sterilization of flies by means of gamma rays.] Parassitol. 10(2-3): 185-194.
Males irradiated with 2000 rads have very low fertility while 3000 rads sterilize completely. Sterile males behave normally in competing for females.

4883. Saccà, G., Magaudda, P. L. and **Guarniera, D. 1967** Un esperimento di lotta integrata (chimica e biologica) contro *Musca domestica* L. alle Isole Lipari—(Nota preliminaire). [An experiment with integrated (chemical and biological) control of *Musca domestica* L. in the Lipari Islands—(Preliminary Note).] Riv. di Parassit. 28(4): 295-307. (English summary.)
Expresses the belief that reasonably good control may be achieved by discontinuing chemical insecticides, and increasing the release of sterile flies. ABSTRACTS: BA 49:8724(97114), Sept. 15, 1968; TDB 65:952(2141), July, 1968; RAE-B 58:84(335), March, 1970.

4884. Saccà, G., Magrone, R. and **Scirocchi, A. 1965** Sulla repellanza esercitata da alcuni chemosterilanti verso *Musca domestica* L. [Repellency exercised by some chemosterilants toward *Musca domestica.*] Riv. di Parassit. (Rome) 26(1):61-66. (English summary.)
Tests run with sugar baits. While Apholate exercises a moderate degree of repellency, Metepa and Hempa are highly repellent. Repellency may be lessened (especially for metepa) by adding malt extract to the baits. ABSTRACTS: RAE-B 55:73(267), Apr., 1967; TDB 62:1063, Oct., 1965.

4885. Saccà, G. and **Scirocchi, A. 1965** Ricerche sui chemosterilanti: un ceppo di *Musca domestica* L. Resistente all'azione sterilizzante del metepa. [Research on chemosterilization: a strain of *Musca domes-*

tica L. resistant to the sterilizing action of metepa.] Estratto dagli Atti del VI Congress. Nazionale Italiano di Entomolgia, Padova, Settembre 11-14. pp. 75-77.

Tables show relative fertility for ten generations of treated, untreated and control populations of *M. domestica*, demonstrating resistance to metepa in one strain.

4886. Saccà, G., Scirocchi, A., De Meo, G. M. and **Mastrilli, M. L. 1966** Una prova di campo con il chemosterilante hempa (Esametilfosforammide) contro *M. domestica* L. [A field test with the chemosterilant hempa (Esametilphosphorammide) against *M. domestica* L.] Atti Soc. Peloritana Sci. Fis. Mat. Nat. 12(1/2):457-464.

Garbage dump treated with liquid formulations of various concentrations. That containing 33 per cent sugar and 2·5 per cent hempa, with 2 per cent malt extract was the most effective. After second treatment hatch fell to 2·6 per cent. ABSTRACT: RAE-B 57:195(701), Oct., 1969.

4887. Saccà, G., Scirocchi, A. and **Stella, E. 1966** Un esperimento di laboratorio sulla efficacia e sulla persistenza di un' esca liquida a base di "Hempa" (Esametilfosforammide) per il controllo di *M. domestica* L. [A laboratory experiment dealing with the efficiency and persistence of a liquid bait based on "Hempa" (Esametilphosphorammide) for the control of *M. domestica* L.] Atti. Soc. Peloritana Sci. Fis. Mat. Nat. 12(1/2):465-468.

A syrup bait, containing 2·5 per cent chemosterilant, 33 per cent sugar, 2 per cent malt extract, of pH 7·2 was highly effective 7 months after set-up, though pH had been lowered to 4. ABSTRACTS: TDB 64:910, Aug., 1967; RAE-B 57:195(702), Oct., 1969.

4888. Saccà, G., Scirocchi, A., Stella, E., Mastrilli, M. L. and **De Meo, G. M. 1966** Studio sperimentale di un ceppo di *M. domestica* L., selezionato con il chemosterilante metepa. [Experimental study of a strain of *M. domestica* L. selected with the chemosterilant metepa.] Atti Soc. Peloritana Sci. Fis. Mat. Nat. 12(1/2):447-456.

In 22 generations there was no significant change in the Hempa-selected strain. With Metepa, the medium effective dosage showed some increase, but this resistance disappeared as soon as selection pressure was withdrawn. ABSTRACTS: TDB 64:910, Aug., 1967; RAE-B 57:195(700), Oct., 1969.

4889. Saccà, G. and **Stella, E. 1964** Una prova di campo per il controllo di *Musca domestica* L. mediante esche liquide a base del chemosterilante tepa (=afoxide). [A field trial against *Musca domestica* with liquid baits of tepa.] Riv. di Parassit. (Rome) 25(4):279-294. (English summary.)

Utilized garbage dumps. A high degree of relative and absolute sterility was found in flies collected on the garbage 1 hour after spraying a 0·0625 per cent solution with 1 per cent malt extract as an attractant. ABSTRACT: TDB 62:1062, Oct., 1965.

4890. Saccà, G., Stella, E. and **Gandolfo, D. 1968** Studi sull'azione residua su superfici solide del DDVP erogato sotto forma Gassosa. [Studies of residual action on solid surfaces of DDVP applied in vapor form.] Parassitol. 10(2-3):1-7.

DDVP vapors deposited toxic residues on walls and other surfaces, which then remained effective in killing flies for 23 days. In a room, from which vapor containers had been removed, flies were killed up to 5 weeks later (exposure time 24 hours).

4891. Saccà, G., Stella, E., Gandolfo, D. and **Mastrilli, M. L. 1969** Prime osservazioni sui livelli di resistenza in *Musca domestica* L. nella Provincia di Latina, a 24 anni dall' introduzione degli insetticidi di contatto. [First observations on the level of resistance in *Musca domestica* L. in the Province of Latina, 24 years after the introdiction of contact insecticides.] Parassitol. 11(2):3-8.

LD_{50} values determined by topical application of acetone solution. Resistance to chlorinated insecticides still very high. LD_{50} values rise in the autumn (as compared with spring), probably due to selection pressure from summer use of insecticides.

4892. Saccà, G., Stella, E., Gandolfo, D. and **Mastrilli, M. 1969** Tossicometria comparata di alcuni insetticidi sperimentati populazioni di *Musca domestica* L. in provincia di Latina, a 24 anni dalla introduzione degli insetticidi murali. [Comparative toxicity of some experimental insecticides to populations of *Musca domestica* in the province of Latina, 24 years after the introduction of surface insecticides.] Cong. Internat. des Antiparasitaires, Milan, October 6-8. pp. 1-10.

Experiments assessing resistance levels disclosed a consistently high tolerance of chlorinated insecticides, while resistance to organophosphorus compounds and to carbamates was of a lower order.

4893. Saccà, G., Stella, E. and **Magrone, R. 1964** Ricerche di laboratorio sull' efficacia sterilizzante del tepa (afoxide) e dell' afolato in *Musca domestica.* [Laboratory studies on the sterilizing effectiveness of Tepa and Apholate on *Musca domestica.*] Riv. di Parassit. (Rome) 25(3):207-216. (English summary.)

With tepa the average (SD_{50}) dosage for absolute sterility is 0·7-1·1 mg. in males; 7 mg. in females. SD_{50} for relative sterility (assessed from percentage of sterile eggs) is 0·2-0·4 mg. in males; 2·6 in females. If both sexes are treated, 3 mg. of tepa prevents the hatching of 99·5 per cent of the eggs. ABSTRACTS: RAE-B 54:101, May, 1966; TDB 62:1062, Oct., 1965.

4894. Saccà, G., Stella, E. and **Magrone, R. 1964** Ricerche preliminari sull' azione dell'Afoxide (ossido di tris-(1-aziridinil) fosfato) su *Musca domestica* L. [Preliminary study of the action of Afoxide (——) on *Musca domestica* L.] Parassitol. 6(1-2):229-234.

Tables give percentage sterility achieved with male and female flies, using various dosages and means of application.

4895. Saccà, G., Stella, E., Mastrilli, M. L. and **Gandolfo, D. 1969** Sul sinergismo tra Hempa e Tepa, usati per la chemosterilizzazione di *Musca domestica* L. [On synergism between Hempa and Tepa, used in the chemosterilization of *Musca domestica* L.] Parassitol. 10(3):271-275.

Hempa, Tepa, and mixtures of the two were administered orally to male house flies. The two active ingredients exercise a strong synergistic action on each other. Activity of Tepa increases some 20 times; that of Hempa, 5 times. Dosages in liquid sugar baits can be reduced accordingly. ABSTRACT: TDB 67:1293(2510), Oct., 1970.

4896. Saccà, G., Stella, E. and **Mento, G. 1968** La dispersione delle mosche domestiche ricerche sperimentali nella provincia di Latina. [The dispersion of domestic flies studied experimentally in the Province of Latina.] Estratto da: Arch. Atti Soc. Medico-Chirurgica di Messina. Anno XII fasciolo III. pp. 1-8.

Marked flies were released from 4 different environmental locations. Results: (1) flies have the ability to migrate unusual distances; (2) they rarely do so, preferring to remain sedentary, when the environment is suitable. They strike out only when "compulsed" to seek their domestic biotype.

Saccà, G., joint author. *See* Gratz, N. G., et al., 1964.

4897. Sacher, R. M., Metcalf, R. L. and **Fukuto, T. R. 1969** Selectivity of Carbaryl-2, 3-methylene-dioxynaphthalene Combination. Metabolism of the Synergist in Two Strains of Houseflies and in Mice. J. Agric. Fd. Chem. 17(3):551-557.

Although efficient and rapid degradation of 2, 3-methylenedioxynaphthalene took place in mice, this substance proved extremely stable in flies, where it required a full 24 hours for deactivation. (Used in combination with carbaryl-2 in both cases.) No significant *qualitative* differences in the metabolites produced. ABSTRACT: BA 51:886(9003), Jan. 15, 1970.

4898. Sacher, R. M., Metcalf, R. L. and **Fukuto, T. R. 1968** Propynyl Naphthyl Ethers as Selective Carbamate Synergists. J. Agric. Fd. Chem. 16(5):779-786.

Thirty derivatives of propynyl naphthyl ether were evaluated as synergists for carbaryl. Most active compound was 1-naphthyl 3-butynyl ether, with a synergistic ratio of 176·5 against houseflies. Was *not* synergistic against mice. ABSTRACT: BA 50:2650(27601), Mar. 1, 1969.

4899. Sacktor, B., Childress, C. C. and **Shavity, E. W. 1965** A Special Role for Proline Metabolism in Flight Muscle (Diptera). Fed. Proc. 24(2) pt. 1:471(1894).

Proline penetrates the mitochondrial membrane, forming intramitochondrial precursors of oxaloacetate, enabling the synthesis of citrate, and effecting the complete oxidation of pyruvate via the Krebs cycle. (Flight muscle lacks endogenous mitochondrial precursors of oxaloacetate.)

Sadykova, V. R., joint author. *See* Gorbacheva, E. A., et al., 1967.

Sagawa, T., joint author. *See* Mitsui, T., et al., 1969.

Saito, T., joint author. *See* Morikawa, O., et al., 1966; Fujito, S., et al., 1963, 1966; Hayashi, A., et al., 1968.

4900. Sakai, M. 1964 Studies on the Insecticidal Action of Nereistoxin, 4-N, N-dimethylamino-1, 2-dithiolane. I. Insecticidal Properties. Jap. J. Appl. Ent. Zool. 8(4):323-324. Tokyo. (Japanese summary.)

On preliminary test appeared inferior to DDT and to parathion, but did have a rapid knockdown. Its synthetic hydrogen oxalate proved an effective bait for *M. domestica*, mortality increasing with exposure. Knockdown was quite often reversible. ABSTRACT: RAE-B 53:197, Oct., 1965.

4901. Sakai, M. 1966 Studies on the Insecticidal Action of Nereistoxin 4-N, N-dimethylamino-1, 2-dithiolane. IV. Role of the Anti-cholinesterase Activity in the Insecticidal Action to Housefly, *Musca domestica* L. (Diptera:Muscidae). Appl. Ent. Zool. 1(2):73-82.

Results indicated that the anticholinesterase activity of the compound did not play an important role in its insecticidal action. ABSTRACT: RAE-B 55:89(318), May, 1967.

4902. Sakai, M. 1966 Studies on the Insecticidal Action of Nereistoxin, 4-N, N-dimethylamino-1, 2-dithiolane. II. Symptomatology. Botyu-Kagaku 31(2):53-61.

Symptoms of poisoning in *Blatella germanica* (L) and *Musca domestica* L. were investigated by direct observation of treated insects, by mechanical recording of leg-jerk and by recording nerve action potentials. ABSTRACT: RAE-B 56:152(533), Aug., 1968.

4903. Sakai, M. 1967 Hydrolysis of Acetyl Thiocholine and Butyryl Thiocholine by Enz Choline Esterases of Insects and a Mite. Appl. Ent. Zool. 2(2).
M. domestica among the species studied.

Sakai, S., joint author. *See* Ghoda, M., et al., 1966.

Sakai, T. T., joint author. *See* Roberts, R. B., et al., 1969.

4904. Saldarriaga, A. V. 1966 Control de la cria de la mosca doméstica en los gallineros. [Control of the breeding of house flies on poultry ranches.] Agric. Trop. 22(2):83-86.

4905. Salem, H. H. and **El-Sharif, A. F. 1960** A New Muscoid Fly from Gebel Elba (Diptera: Muscidae). Bull. Soc. Ent. d'Egypte 44:171-174.
Concerns *Musca efflatouni efflatouni.* ABSTRACT: BA 38:930(12034), May 1, 1962.

4906. Salem, H. and **El-Sherif, A. T. 1960** A Study of the Female Terminalia of *Musca vitripennis* Meig. and Description of the Third Larval Stage. Bull. Soc. Ent. d'Egypte 44:175-178.

4907. Salsbury, D. L. 1969 Dusting Bags Control Flies. Veterinary Medicine 64:719.

Samsonova, A. M., joint author. *See* Sukhova, M. N., et al., 1965.

4908. Sanchez, F. F. 1965 Factors Affecting the Latent Toxicity of Aldrin, DDT, and Heptachlor to Resistant and Susceptible Strains of the Housefly. Diss. Absts. 26(6):2945-2946.

4909. Sanchez, F. F. and **Sherman, M. 1966** Penetration and Metabolism of DDT in Resistant and Susceptible House Flies and the Effect on Latent Toxicity. J. Econ. Ent. 59(2):272-277.
Cuticular penetration of DDT was more rapid in the larvae of the susceptible than in the larvae of the resistant strain. The latter rapidly converted DDT to DDE. There was no carry over of DDT to the adult state in the resistant strain, indicating complete detoxification. ABSTRACTS: BA 47:7286(85010), Sept. 1, 1966; RAE-B 54:128, July, 1966; TDB 63:1253, Nov., 1966.

Sanchez, F. F., joint author. *See* Sherman, M., et al., 1963, 1964.

4910. Sanders, D. P. and **Dobson, R. C. 1966** The Insect Complex Associated with Bovine Manure in Indiana. Ann. Ent. Soc. Am. 59:955-959.
Samples from vicinity of Lafayette yielded 20 species of Diptera, 15 of Coleoptera, 3 of parasitic Hymenoptera and 2 species of mites. List of flies included *M. autumnalis.*

Sanders, D. P., joint author. *See* Bodson, R. C., et al., 1965.

Sanjean, J., joint author. *See* Sun, P.-Y., et al., 1961.

4911. Sannasi, A. and **Blum, M. S. 1969** Pathological Effects of Fire Ant Venom on the Integument and Blood of Housefly Larvae. J. Georgia Ent. Soc. 4(3):103-110.
Larvae of *M. domestica* treated with venom of *Solenopsis saevissima* became black, due to melanosis of cuticle and blood. Melanosis may be inhibited by certain chemical substances, but toxicity remains undiminished, and the larva dies. ABSTRACT: BA 50:13075(134842), Dec. 15, 1969.

4912. Sarbova, S., Avramova, S., Levi, V., Stoicheva, R., Atanasova, N., Bratanov, H. and **Kjordev, N. 1968** Iz uchenie stepeni rezistentnost: K ddt, lindanu i neguvonu populyatsii *Musca domestica* v nekotorykh paionakh strany. [Studies on the degree of resistance to DDT, lindane and neguvone of *Musca domestica* populations in some areas of the country.] Sci. Works. Res. Inst. Epidemiol. Microbiol. (Sofia) 13:253-259. (English summary.)

Points out dependency of LD_{50} of *M. domestica* on lipoid content of the tarsus. Lipoid content is higher in resistant strains. ABSTRACT: 51:10266 (104379), Sept. 15, 1970.

Sasa, M., joint author. *See* Ogata, K., et al., 1961.

Sasaki, Y., joint author. *See* McCann, G. D., et al., 1965.

Sasamoto, T., joint author. *See* Eto, M., et al., 1968.

4913. Satija, R. C. and **Sharma, M. L. 1967** Differentiation of Eye Ganglia in *Musca domestica* (Diptera). Res. Bull. Punjab Univ. 18(3/4): 343-349.

A histological description of eye ganglion development. Ganglion of 4-day pupa closely resembles the adult condition. ABSTRACT: BA 50: 8095(84262), Aug. 1, 1969.

4914. Satija, R. C. and **Sharma, M. L. 1968** Postembryonic Development of the Brain of *Musca domestica.* Res. Bull. Panjab Univ. (N.S.) 19(1/2):71-80.

The 3 neuromeres are differentiated in the first instar. In 2-3 day pupa, lobes of the protocerebrum and corpora ventralia can be recognized. Suboesophageal ganglion fuses completely with the brain in 3-day pupa, and adult structures are complete by the fourth day of pupal life. ABSTRACT: BA 50:12502(129155), Dec. 1, 1969.

4915. Sato, Y. 1968 Insecticidal Action of Phytoecdysones. Appl. Ent. Zool. 3(4):155-162.

Ponasterone, ecdysterone, inokosterone, and cyasterone were tested against various species. *M. domestica* larvae proved tolerant of these substances at most concentrations used. Dead individuals were in the form of abnormal pupae. ABSTRACTS: BA 50:9173(95085), Sept. 1, 1969; RAE-B 58:109(444), Apr., 1970.

4916. Saunders, D. S. 1965 Larval Diapause of Maternal Origin: Induction of Diapause in *Nasonia vitripennis* (Walker) (Hymenoptera:Pteromalidae). J. Exp. Biol. 42:495-509.

A parasite of *M. domestica.* See next reference.

4917. Saunders, D. S. 1966 Larval Diapause of Maternal Origin. II.. The Effect of Photoperiod and Temperature on *Nasonia vitripennis.* J. Insect Physiol. 12(5):569-581.

Species is a long-day insect with a temperature-modified photoperiodic response. If female wasps are exposed to both low temperature and low photo period, they will produce *mostly* diapause larvae. If both are high, *few* diapause larvae will result. Is a parasite of *M. domestica.* ABSTRACT: RAE-B 55:212(735), Nov., 1967.

4918. Saunders, D. S. 1966 Larval Diapause of Maternal Origin. III. The Effect of Host Shortage on *Nasonia vitripennis.* J. Insect Physiol. 12(8):899-908.

Host deprivation affects longevity and fecundity but little. It does increase the proportion and number of diapause larvae, since it delays the start of oviposition, resulting in an "adding up" of short day cycles. ABSTRACT: RAE-B 55:232(798), Dec., 1967.

4919. Saunders, R. C. and **Hsiao, T. H. 1970** Biology and Laboratory Propagation of *Amblymerus bruchophagi* (Hymenoptera:Pteromalidae), a Parasite of the Alfalfa Seed Chalcid. Ann. Ent. Soc. Am. 63(3): 744:749.

Reports successful rearing of this parasite on five unnatural hosts, of which *M. domestica* was one. ABSTRACT: BA 51:10854(110194), Oct. 1, 1970.

4920. Sawicki, R. M. 1961 The Effect of Safroxan on the Knock-down and the 24 hr. Toxicity of Commercial Pyrethrum Extract Against Houseflies (*Musca domestica* L.). Pyrethrum Post 6(2):38-42.
Did not increase the effectiveness of pyrethrum in the first 15 minutes (when knockdown takes place), but did increase the toxicity of the extract 24 hours after treatment. Is considered a poor synergist. ABSTRACT: RAE-B 51:159, Aug., 1963.

4921. Sawicki, R. M. 1962 Insecticidal Activity of Pyrethrum Extract and its Four Insecticidal Constituents Against House Flies. II. Synergistic Activity of Piperonyl Butoxide with the Four Constituents. J. Sci. Fd. Agric. 13(4):260-264.
Order and magnitude of the toxicities of synergized constituents vary with the amount of synergist present. At 1·8 (pyrethroid/piperonyl butoxide) relative toxicites were: pyrethrum extract 0·1; pyrethrin I, 1·31; cinerin I, 1·28; pyrethrin II, 0·90; cinerin II, 0·58. ABSTRACT: RAE-B 51:109, June, 1963.

4922. Sawicki, R. M. 1962 Insecticidal Activity of Pyrethrum Extract and its Four Insecticidal Constituents Against House Flies. III. Knockdown and Recovery of Flies Treated with Pyrethrum Extract With and Without Piperonyl Butoxide. J. Sci. Fd. Agric. 13(5):283-292.
Deals with technique for immobilizing and treating flies. Defines terms used in author's series of related papers. For comments on knockdown, see Sawicki and Thain, 1962. ABSTRACT: RAE-B 51:109, June, 1963.

4923. Sawicki, R. M. 1962 Insecticidal Activity of Pyrethrum Extract and its Four Insecticidal Constituents Against House Flies. V. Knockdown Activity of the Four Constituents with Piperonyl Butoxide. J. Sci. Fd. Agric. 13(11):591-598. London.
Knockdown was faster, lasted longer, and the knockdown-end point was reached later, when constituents were applied in the presence of piperonyl butoxide, which presumably inhibits the recovery mechanism of the flies. Effect of synergist is least during the first few minutes after treatment; increases most rapidly during first 2 hours. ABSTRACT: RAE-B 51:281, Dec., 1963.

4924. Sawicki, R. M. 1965 Similarity in Response to Diazinon, Dieldrin and Pyrethrum Extract by Allatectomized and Normal Houseflies (*Musca domestica* L.). Bull Ent. Res. 55(4):727-732.
Allatectomy of newly emerged flies, which prevents maturation of the ovaries, does not affect resistance to insecticides. Describes apparatus and procedure for removing corpus allatum and corpora cardiaca. ABSTRACTS: BA 47:4254(49933), May 15, 1966; RAE-B 53:112, June, 1965; TDB 62:943, Sept., 1965 (as Sawicki and Green).

4925. Sawicki, R. M., Elliott, M., Gower, J. C., Snarey, M. and **Thain, E. M. 1962** Insecticidal Activity of Pyrethrum Extract and Its Four Insecticidal Constituents Against House Flies. I. Preparation and Relative Toxicity of the Pure Constituents; Statistical Analysis of the Action of Mixtures of These Components. J. Sci. Fd. Agric. 13(3):172-185. London.
Relative toxicity to *M. domestica* at 20°C: Pyrethrum extract, 1·0; pyrethrin II, 1·3-1·5; pyrethrin I, 0·9-1·0; cinerin II, 0·5-0·6; cinerin I, 0·4-0·5. The four esters prepared from natural material are biologically identical with those prepared by reconstitution from other substances. (Chemical and physical properties alike also.) (For Parts II and III, *See* Sawicki, R. M., 1962; For Part IV, *See* Sawicki, R. M. and Thain, E. M., 1962.) ABSTRACT: RAE-B 51:109, June, 1963.

4926. Sawicki, R. M. and **Elliott, M. 1965** Insecticidal Activity of Pyrethrum Extract and Its Four Insecticidal Constituents Against House Flies. VI. Relative Toxicity of Pyrethrin I and Pyrethrin II Against Four Strains of House Flies. J. Sci. Fd. Agric. 16(2):85-89. London.

Pyrethrin II was 1·21-1·50 times more toxic than pyrethrin I, 24 hours after treatment; 1·09-1·54 times more, 48 hours after. Two strains resistant to organophosphorus insecticides were strongly resistant to knockdown, but not to kill, by pyrethrins. ABSTRACT: RAE-B 54:44, Feb., 1966.

4927. Sawicki, R. M. and **Farnham, A. W. 1967** Genetics of Resistance to Insecticides of the SKA Strain of *Musca domestica*. I. Location of the Main Factors Responsible for the Maintenance of High DDT-resistance in the Diazinon-selected SKA Flies. Ent. Exp. Appl. 10(2):253-262. (German summary.)

A resistance factor (R_3) inhibited by sesamex, most probably common to DDT and diazinon is on the III linkage group. Another is recessive, segregates independently, and is unaffected by synergists. When heterozygous, R_3 gives only moderate resistance to DDT. ABSTRACTS: BA 49: 3409(38125), Apr. 15, 1968; TDB 65:342(788), Mar., 1968; RAE-B 56:10(31), Jan., 1968.

4928. Sawicki, R. M. and **Farnham, A. W. 1967** Genetics of Resistance to Insecticides of the SKA Strain of *Musca domestica*. II. Isolation of the Dominent Factors of Resistance to Diazinon. Ent. Exp. Appl. 10(3-4):363-376. (German summary.)

Reports isolation, in homozygous condition, of R_3 on III linkage group, and gene *a* (for low ali-esterase), on V linkage group. Mutants of the *ar* type carry only R_3, and *ocra* mutants only gene *a*. Sesamex synergized diazinon against flies with R_3 only, but *increased* resistance to diazinon by $\times 2$ in flies with gene *a* only. ABSTRACTS: BA 49:5335(59884), June 15, 1968; TDB 65:884(1876), June, 1968; RAE-B 56:105(364), May, 1968.

4929. Sawicki, R. M. and **Farnham, A. W. 1968** Genetics of Resistance to Insecticides of the SKA Strain of *Musca domestica*. III. Location and Isolation of the Factors of Resistance to Dieldrin. Ent. Exp. Appl. 11(2):133-142. (German summary.)

Major factor designated DR_4, on the IV linkage group. Minor factor designated R_2, on II linkage group. Presence of DR_4 in SKA flies probably not the result of selection with diazinon. Is almost certainly inherited from the chlordane-resistant parents of the SKA strain. ABSTRACTS: TDB 66:238(476), Mar., 1969; BA 50:5384(56544), June 1, 1969; RAE-B 57:215(779), Oct., 1969.

4930. Sawicki, R. M. and **Farnham, A. W. 1968** The Use of Visible Mutant Markers in the Study of Resistance of House Flies to Insecticides. Proc. 4th Br. Insectic. Fungic. Conf. 1967. I. pp. 355-363. Croydon, Surrey.

Recessive visible mutants on known linkage groups can be used to locate and isolate single resistance factors in strains where resistance to one or more insecticides is controlled by more than one gene. In the work here reported a diazinon-selected SKA strain was used. ABSTRACT: RAE-B 57:24(102), Jan., 1969.

4931. Sawicki, R. M. and **Farnham, A. W. 1969** Examination of the Isolated Autosomes of the SKA Strains of House Flies (*Musca domestica* L.), for Resistance to Several Insecticides With and Without Pre-

treatment with Sesamex and TBTP. Bull. Ent. Res. 59(1968) pt. 3: 409:421.
TBTP, an ali-esterase inhibitor, greatly synergized organophosphates only against flies with gene *a*. Pretreatment with sesamex, an inhibitor of microsomal activity, gave 2 types of response: (1) antagonism with most of the thioates and (2) synergism with the phosphates.

4932. Sawicki, R. M., Franco, M. G. and **Milani, R. 1966** Genetic Analysis of Non-recessive Factors of Resistance to Diazinon in the SKA Strain of the Housefly (*Musca domestica* L.). W. H. O. Bull. 35(6): 893-903.
Crosses between SKA flies (diazinon-resistant) and four susceptible recessive marker strains, followed by test-crosses with recessive markers and by bioassays of each cross, show that non-recessive factors for resistance to diazinon are present on IV and V linkage groups. ABSTRACTS: BA 48:7928(88959), Sept. 15, 1967; RAE-B 57:108(382), 1969.

4933. Sawicki, R. M. and **Green, G. 1965** Changes in the Susceptibility of Normal and Resistant Houseflies (*Musca domestica* L.) to Diazinon with Age. Bull. Ent. Res. 55(4):715-725.
Increasing age augments resistance to diazinon in both normal and resistant flies. To obtain reproducible values, the degree of resistance to diazinon should be measured not earlier than three days after emergence. ABSTRACTS: BA 47:4254(49934), May 15, 1966; RAE-B 53:111, June, 1965; TDB 62:943, Sept., 1965.

4934. Sawicki, R. M. and **Thain, E. M. 1961** Chemical and Biological Examination of Commercial Pyrethrum Extracts for Insecticidal Constituents. J. Sci. Food Agric. 12(2):137-145.
Insecticidal activity is restricted to fractions identified chemically as cinerin I, pyrethrin I, cinerin II, and pyrethrin II (the most active). Fractions were obtained from commercial pyrethrum extract by extraction with nitromethane and separated by displacement chromatography.

4935. Sawicki, R. M. and **Thain, E. M. 1962** Insecticidal Activity of Pyrethrum Extract—— IV. Knock-down Activities of the Four Constituents. J. Sci. Fd. Agric. 13(5):292-297.
Used measured drop technique on unanaesthetized flies held by suction during dosage. Inspection at 5 hour intervals showed differences in relative toxicities due to time. Toxicities of pyrethrin II and cinerin II decreased because flies recovered sooner from these two than from pyrethrin I and cinerin I. ABSTRACT: RAE-B 51:109, June, 1963.

Sawicki, R. M., joint author. *See* Farnham, A. W., et al., 1965, 1966; Elliott, M., et al., 1965; Godin, P. J., et al., 1965.

4936. Saxena, B. N., Dixit, R. S. and **Khalsa, H. G. 1965** Insecticidal Properties of Turmeric. Labdev. J. Sci. Technol. 3(3):212-213. Kanpur.
Reports tests made in India with petroleum-ether extracts of the rhizomes of this plant. Is toxic to *M. domestica nebulo*, and more so to *Culex*. ABSTRACT: RAE-B 54:83, Apr., 1966.

Sazonov, A. P., joint author. *See* Shapiro, I. D., et al., 1967.

Sazonova, I. N., joint author. *See* Paikin, D. M., et al., 1961.

4937. Schaefer, C. H. 1967 Studies on the Mode of Action of the Chemosterilants 2-imidazolidinone and 4-imidazolin-2-one in the House Fly and in the Large Milkweek Bug. Life Sci. 6:2677-2683.
These substances are temporary sterilants in house flies; permanent, in milkweed bugs. Flies can eliminate large amounts of both compounds through excretion and through metabolism. ABSTRACT: BA 49:7253 (80873), Aug. 1, 1968.

4938. Schaefer, C. H. and **Sun, Y. P. 1967** A Study of Dieldrin in the House Fly Central Nervous System in Relation to Dieldrin Resistance. J. Econ. Ent. 60(6):1580-1583.
No evidence that binding of dieldrin to tissues within the central nervous system is a significant factor in resistance. Dieldrin resistance may be due largely to insensitivity at the receptor site. ABSTRACTS: BA 49:3837 (43152), Apr. 15, 1968; RAE-B 56:98(341), May, 1968; TDB 65:726 (1568), May, 1968.

4939. Schaefer, C. H. and **Tieman, C. H. 1967** 4-imidazolin-2-one: an Insect Growth-inhibitor and Chemosterilant. J. Econ. Ent. 60(2):542-546.
Tested against *M. domestica* and *Anopheles albimanus*. Is a potent inhibitor of reproduction and immature development in several species. This substance, and its saturated analog, have low acute toxicity to birds and mammals. ABSTRACTS: RAE-B 55:146(515), Aug., 1967; TDB 64:910, Aug., 1967.

Schaefer, C. H., joint author. *See* Sun, Y. P., et al., 1967.

4940. Schafer, J. A. and **Terriere, L. C. 1970** Enzymatic and Physical Factors in House Fly Resistance to Naphthalene. J. Econ. Ent. 63:787-792.
Three factors involved: (1) on chromosome II, a gene which steps up oxidase activity; (2) on chromosome III, a gene which reduces the penetration rate of the cuticle; (3) on chromosome V, a factor which augments the effectiveness of the first two 6 to 10-fold, depending on age of flies. ABSTRACT: RAE-B 58:372(1534), Nov., 1970.

4941. Schechter, M. S., Sullivan, W. N. and **Yeomans, A. H. 1960** Studies on Fine-particle, Coarse-particle, and Concentrated Aerosols. J. Econ. Ent. 53(5):908-914.
Size of particle makes little difference, the important factor being the concentration of insecticide dispersed per unit volume of air. Concentrated aerosols containing up to 10 per cent allethrin, with no adjuvants, were highly effective against both resistant and non-resistant flies. ABSTRACT: RAE-B 49:185, Aug., 1961.

Schechter, M. S., joint author. *See* Sullivan, W. N., et al., 1960; Yeomans, A. H., et al., 1964.

Scheer, I. J., joint author. *See* Wolken, J. J., et al., 1960.

4942. Scheibner, R. A. and **Knapp, F. W. 1968** Insecticide Recommendations for Lactating Dairy Animals. Univ. Kentucky Coop. Ext. Serv. Misc. Publ. 256 E.
Gives proven, conventional procedures for the control of mosquitoes, flies, and lice.

4943. Schellen, O. 1968 Trachea in Wings of Diptera. Ent. Ber. (Berlin). Pages 109-112.
Includes microphotographs of *Calliphora*, *Muscina*, *Chrysozona* and *Fannia* (*canicularis*). Cited here as a contribution to muscoid biology.

4944. Schenone, H. 1962 Medidas de profilaxis de las moscas. [Preventive measures against flies.] Bol. Chileno Parasit. 17(1):23-25.
Not a research report. Objective of publication: to be educational and stimulative. ABSTRACT: BA 39:671(8315), July 15, 1962.

4945. Schlegel, P. 1969 Investigations of the Mechano Receptors of the Fly Antennae. Zool. Anz. Suppl. (32).
Falls in the area of electrophysiology. Concerns sensory cells.

4946. Schmidt, C. H., Dame, D. A. and **Weidhaas, D. E. 1964** Radiosterilization vs. Chemosterilization in House Flies and Mosquitoes. J. Econ. Ent. 57:753-756.
Criterion for comparison was reduction in number of viable egg batches from females confined with both treated and untreated males. Results

from chemosterilization equalled or surpassed those from radiosterilization. ABSTRACTS: BA 46:700(8769), Jan. 15, 1965; RAE-B 53:13, Jan., 1965; TDB 62:365, Apr., 1965.

Schmidt, C. H., joint author. *See* Weidhaas, D. E., et al., 1962; Dame, D. A., et al., 1964.

4947. Schmidt, G. H. 1966 Insektizidresistenz bei Dipteren. [Insecticide resistance in Diptera.] Anz. Schädlingsk. 39(3):33-39. (English summary.) Deals chiefly with mosquitoes and *M. domestica*. Recognizes insecticide resistance as a genetically interrelated characteristic. Insecticides act solely in the sense of a positive selection factor. ABSTRACT: RAE-B 56: 248(914), Dec., 1968.

4948. Schmolesky, G. E. and **Derse, P. H. 1963** Antiresistant/DDT Continues to Offer Superior Fly Control. Agric. Chem. 18(3):24-26, 111. Reports a series of tests with the butyl analog of Antiresistant/DDT, applied as an emulsion. Gave 80-100 per cent mortality through 10 weeks. (Better than wettable powder or oil formulation.) ABSTRACT: RAE-B 52:215, Dec., 1964.

Schneider, H., joint author. *See* Jackson, H., et al., 1968.

Schoenburg, R., joint author. *See* Eastwood, R., et al., 1966, 1967.

4949. Scholes, J., et al. 1969 The Quantal Content of Optomotor Stimuli and the Electrical Responses of Receptors in the Compound Eye of the Fly *Musca*. Kybernetik 6:74-79.

4950. Schonbrod, R. D., Khan, M. A. Q., Terriere, L. C. and **Plapp, F. W., Jr. 1968** Microsomal Oxidases in the Housefly: A Survey of Fourteen Strains. Life Sci. Physiol. Pharmacol. 7(13) pt. 1:681-688. Two oxidases studied with naphthalene, carbamate insecticides, DDT and alkylbenzenes as substrates. No simple relationship between insecticide resistance and levels of microsome oxidase. Perhaps more than 1 enzyme. ABSTRACT: BA 50:8097(84285), Aug. 1, 1969.

4951. Schonbrod, R. D., Philleo, W. W. and **Terriere, L. C. 1965** Hydroxylation as a Factor in Resistance in House Flies and Blow Flies. J. Econ. Ent. 58(1):74-77. Studied hydroxylation of naphthalene by microsomes prepared from *M. domestica* and *Phormia regina*. Older flies have greater hydroxylating capacity than do younger. Flies highly resistant to DDT and dieldrin show twice the activity *in vitro* as those moderately resistant. ABSTRACTS: BA 46:4070(50344), June 1, 1965; RAE-B 53:96, May, 1965; TDB 62: 813, Aug., 1965.

Schonbrod, R. D., joint author. *See* Philleo, W. W., et al., 1965; Terriere, L. C., et al., 1965.

4952. Schoof, H. F., Jakob, W. L. and **Mathis, W. 1964** Response of *Musca domestica* and Several Species of Mosquitoes to Various Insecticides. Proc. 51st Ann. Meeting New Jersey Mosquito Exterm. Ass. 51:176-180. March 10-12. Report concerns toxicity of several insecticides applied as a vapor, as a larvicide, and as a residual deposit. Results show that toxicity to one species cannot be used to predict effect on others. ABSTRACTS: RAE-B 55:50(186), Mar., 1967; TDB 62:1059, Oct., 1965.

Schoof, H. F., joint author. *See* Flynn, A. D., et al., 1966; Kilpatrick, J. W., et al., 1962, 1963; Jensen, J. A., et al., 1965; Maddock, D. R., et al., 1961; Mathis, W., et al., 1964, 1965, 1967, 1968, 1970; Weinburg, H. B., et al., 1961; Pearce, G. W., et al., 1961.

Schulz, K. R., joint author. *See* Lichtenstein, E. P., et al., 1969.

Schulz, L., joint author. *See* Kirchberg, E., et al., 1962.

4953. Schumann, H. 1961 Die Bedeutung symboviner Fliegen als Verbreiter von Mastitis-Erregern. [The importance of cattle-frequenting flies as distributors of mastitis infection.] Mh. Veterinaer. Med. 16(16): 624-626. Leipzig.

Streptococcus agalactiae, which causes bovine mastitis, when fed to flies in broth or milk, was found in all parts of the alimentary tract of infected flies and could be isolated, with no lessening of their viability, up to five days after the feeding. Transmission by this route seems possible. ABSTRACT: RAE-B 51:210, Oct., 1963.

4954. Schumann, H. 1963 Zur Larvalsystematik der Muscinae nebst. Beschreibung einiger Musciden- und Anthomyidenlarven. [On larval classification in the Muscinae, along with the description of several muscid and anthomyiid larvae.] Deut. Ent. Zeit. (Berlin) (N.F.) 10:134-163.

Schuurmans Stekhoven, F. M. A. H., joint author. *See* Van Bruggen, E. F. J., et al., 1968.

Scirocchi, A., joint author. *See* Saccà, G., et al., 1965, 1966.

4955. Scott, H. G. 1964 Human Myiasis in North America. Florida Ent. 47(4):255-261.

Lists 40 species, including *M. domestica.*

Scott, H. G., joint author. *See* Minkin, J. L., et al., 1960.

Scott, W. L., joint author. *See* Wallis, R. C., et al., 1970.

4956. Seawright, J. A. and **Adkins, T. R. 1968** Dust Stations for Control of the Face Fly in South Carolina. J. Econ. Ent. 61(2):504-505.

Self applicatory, insecticide-filled dust bags were installed in 24 pastures in 2 counties. Numbers of *M. autumnalis* per head were significantly reduced within one week (from 9·1 to 1·5-4·2). Remained low for duration of experiment. ABSTRACTS: BA 49:7253(80837), Aug. 1, 1968; RAE-B 56:189(683), Sept., 1968.

Seawright, J. A., joint author. *See* Adkins, T. R., Jr., et al., 1967.

Sedlak, V. A., joint author. *See* Maddock, D. R., et al., 1961.

4957. Sehgal, B. S. and **Kumar, P. 1966** A Study of the Seasonal Fluctuations in Fly Populations in Two Villages near Lucknow. Indian J. Med. Res. 54(12):1175-1181.

Cattle litter the most important larval substrate. Garbage important during monsoon season and human excreta during the spring months. Highest incidence of adult flies in March and April. Second highest peak in November. ABSTRACTS: RAE-B 56:213(775), Nov., 1968; TDB 64: 808, July, 1967.

4958. Seki, M., Mizutani, K. and **Suzuki, T. 1968** Studies on Aerosol-type Insecticides. II. Effect of Synergized Pyrethroids to the Housefly. Cont. Study of Laboratory Test Methods of Insecticides 1968 (3):57-59.

Very little difference in effectiveness between oil-based and water-based formulations. Synergized pyrethroids were generally more effective than corresponding synergized allethrin .

Seki, M., joint author. *See* Shimada, A., et al., 1969; Suzuki, T., et al., 1968.

Seki, T., joint author. *See* Fukushi, Y., et al., 1965.

Seligmann, W., joint author. *See* Fraenkel, G., et al., 1966.

Sellers, L. G., joint author. *See* Bieber, L. L., et al., 1968, 1969.

4959. Semenikhina, A. D. 1960 [On the origin of the development of DDT and hexachlorane-resistant flies and methods for its prevention.] Med. Zhur. Uzbekistana 10:17-22. (In Russian.)

Semenikhina, A. D., joint author. *See* Gorbacheva, E. A., et al., 1967.

4960. Sendo, T. 1961 [Fly control experiments by improved treatment of animal manure.] Endemic Dis. Bull. Nagasaki Univ. 3(2):139-144. (In Japanese, with English summary.)

The "closed tank" method of manure treatment was found the most effective for controlling flies, at the same time preserving desirable ingredients of the manure. ABSTRACT: BA 40:1619(21142), Dec. 1, 1962.

4961. Sento, T. 1965 [Fly (Diptera) control experiment by the large 4-roomed closed tank for animal manure.] Endemic Dis. Bull. Nagasaki Univ. 7(1):38-63. (In Japanese, with English abstract.)

Manure was stored in compartmentized brick and concrete containers for a fixed number of days, from which larvae could fall into 4 inches of water and drown when seeking pupation sites. With improved sanitation and the use of diazinon for adults, fly grid counts of *Musca domestica vicina* Macq. and *Stomoxys calcitrans* L. remained very low. ABSTRACT: RAE-B 55:172(600), 1967.

4962. Sento, T. 1966 [Fly control experiment by the large 4-roomed closed tank for animal manure. 2. The results obtained in 1964 and 1965.] Endemic Dis. Bull. Nagasaki Univ. 8(2):107-119. (In Japanese, with English abstract.)

The use of trichlorphon bait along with the storage of manure as previously reported further reduced numbers of *Musca domestica vicina* Macq. and *Stomoxys calcitrans* if invasion occurred from neighboring piggeries. ABSTRACT: RAE-B 57:113(400), 1969.

Seow, C. L., joint author. *See* Wharton, R. H., et al., 1962.

Serafimova, A. M., joint author. *See* Sukhova, M. N., et al., 1965.

4963. Serbova, S. and **Avramova, S. 1966** Studies on the Lipoid Content of *M. domestica* L. Strains Susceptible and Resistant to DDT and Lindane. Arch. Roum. Path. Exp. Microbiol. 25(4):853-858. (Summaries in Russian, French and German.)

Lipoid content of tarsi in strains resistant to DDT and lindane is higher, and is related to LD_{50} level, regardless of the manner of selection (thoracic or tarsal application). Tends to confirm work of Reiff (1956). ABSTRACT: RAE-B 57:93(330), June, 1969.

4964. Serebrovsky, A. S. 1969 On the Possibility of a New Method for the Control of Insect Pests. *In* Sterile-Male Technique for Eradication or Control of Harmful Insects. International Atomic Energy Agency. Vienna, Austria. Pages 123-137.

Concerns translocation of the heterozygous chromosome and its relation to hybrid sterility.

4965. Sergeeva, Z. D. and **Levkovich, V. G. 1961** [The structure of the sexual apparatus in some synanthropic flies.] Zool. Zhur. 40(5): 719-724. (English summary.)

Gives both group and species characteristics. Among the latter are: (1) structure of the chitinous capsule of the testis, (2) diameter of the ductus ejaculatorius in its various portions and (3) structure of the receptaculum seminis. ABSTRACT: BA 39:975(12370), Aug. 1, 1962.

4965a. Serini, G. B. 1961 Sul sistema stomatogastrico di *Musca domestica* L. [On the stomatogastric system of *Musca domestica* L.] Att Soc. Ital. Sci. Nat. (Milano) 100:205-207.
Observations parallel findings of previous workers on *Calliphora erythrocephala*. Differences and resemblances recorded and discussed.

4966. Setlepani, J. A., Stokes, J. B. and **Bořkovec, A. B. 1970** Insect Chemosterilants: VIII. Boron Compounds. J. Med. Chem. 13(1):128-131.
Benzeneboronic acid and seven homologs containing electron-withdrawing substituents proved to be moderately effective chemosterilants of *M. domestica*. ABSTRACT: BA 51:11435(116050), Oct. 15, 1970.

Seutin, E., joint author. *See* Detroux, L., et al., 1965.

4967. Shaaya, E. and **Levinson, H. Z. 1966** Hormonal Balance in the Blood of Blowfly Larvae. Riv. di Parassit. 27(3):211-215.
Hormonal balance evaluated by blood transfusions between larvae of various ages. (Reference included here because of presumed relevance of principles and techniques to the study of *Musca* and other genera.)

Shaaya, E., joint author. *See* Levinson, Z. H., et al., 1966.

Shafiq, S. A., joint author. *See* Ahmad, M., et al., 1965.

Shaidurov, V. S., joint author. *See* Sychevskaya, V. I., et al., 1965.

4968. Shaikh, M. U. and **Morrison, F. O. 1966** Susceptibility of Nine Insect Species to Infection by *Bacillus thuringiensis* var. *thuringiensis*. J. Invertebr. Path. 8(3):347-350.
Organism is a natural enemy of *M. domestica*.

Shakhurina, E. A., joint author. *See* Tukhmanyants, A. A., et al., 1963.

4969. Shambaugh, G. F., Pratt, J. J., Jr., Kaplan, A. M. and **Rogers, M. R. 1968** Repellency of Some Phenylphenols and Related Compounds to House Flies. J. Econ. Ent. 61(6):1485-1487.
Biphenyl and 4-chloro-2-phenylphenol were the most active of single compounds tested. A mixture of the two was more repellent than any single component. ABSTRACTS: BA 50:4777(50225), 1969; RAE-B 57:83(293), May, 1969; TDB 66:634(1253), June, 1969.

4970. Shanahan, G. J. 1965 A Review of the Flystrike Problem of Sheep in Australia. J. Australian Inst. Agr. Sci. 31(1):11-24.
M. domestica is reported as a tertiary species in the flystrike problem. ABSTRACT: BA 47:2047(24564), Mar. 1, 1966.

4971. Shapiro, I. D., Vilkova, N. A. and **Sazonov, A. P. 1967** [Adaptive features of the structure of digestive tracts of larvae of some fly species (Diptera, Brachycera, Cyclorrapha).] Zool. Zhur. 46(4):540-550. (In Russian, with English summary.)
All larvae carry on some extra-intestinal digestion. *Musca* and *Calliphora* show a well-developed intra-intestinal digestion. Pyloric processes as secretory organs. ABSTRACT: BA 49:11350(124547), Dec. 1, 1968.

4972. Shargel, L., Akov, S. and **Mazel, P. 1969** The Reduction of Nitro and Azo Compounds by House Fly Microsomes. Toxicol. and Appl. Pharmacol. 14(3):645.
Concerns microsomal enzymes in *M. domestica* and in rat liver.

4973. Sharma, G. P., Parshad, R. and **Kwatra, S. 1966** Studies on the Chromosomes of the Housefly, *Musca domestica nebulo*. Proc. Indian Sci. Congr. 53(3):318-319.
Salivary gland studies revealed four chromosomes, A, B, C, D, and seven free ends. Diploid number for this species is 12. Sex chromosomes,

which are heterochromatic, lie near the poles at anaphase I and lag behind at anaphase II.

4974. Sharma, J. C. 1967 Biology of *Muscidifurax raptor* (Pteromalidae: Hymenoptera) a Pupal Parasite of Common House Fly, *Musca domestica*. Proc. Indian Sci. Congr. 54(3):466.

4975. Sharma, M. I. D. and **Kalra, R. L. 1962** A Note on the Relative Toxicity of Organo-phosphorus Insecticides Against Houseflies. Indian J. Malariol. 16:75-80.

Reports studies with *M. domestica nebulo*, to determine relative toxicities of baytex, diazinon, malathion and trithion. ABSTRACTS: RAE-B 52:81, May, 1964; TDB 60:503, May, 1963.

Sharma, M. I. D., joint author. *See* Satija, R. C., et al., 1967, 1968.

4976. Sharma, V. N. 1967 Histochemical Studies of Nucleolus and Nucleolar Extrusions in Dipteran Oogenesis. Cytologia 32(3/4):524-531.

Studied 5 species of Diptera, including *M. domestica*. In the genus *Musca*, the oocyte nucleolus lacks RNA. DNA appears in the oocyte during its growth period. ABSTRACT: BA 50:11960(123676), Nov. 15, 1969.

4977. Sharma, V. N. 1969 Effect of Starvation on the Trophocyte Nucleus of *Musca domestica*. Indian J. Exp. Biol. 7:182-183.

4978. Sharp, R. D. 1963 Butonate. A New Safer Phosphate Household and Industrial Insecticide. Agric. Vet. Chem. 4(5):135-136. Sevenoaks, Kent.

Butonate is odorless, non-staining, soluble in most organic solvents. Is toxic to white rats and very effective against houseflies. ABSTRACT: RAE-B 53:216, Nov., 1965.

4979. Shatoury, H. H. 1963 Effect of N-Butanol on Esterase Activity in the Housefly (*Musca domestica* L.). J. Insect Physiol. (London) 9(2):165-176.

N-Butanol found to increase hydrolytic activity of housefly heads to phenylacetate and most choline esters. In the thorax the hydrolysis of phenylacetate was inhibited. N-Butanol perhaps inhibits both cholinesterase and aliesterase by denaturation. ABSTRACTS: BA 43:1356(16857), Aug. 15, 1963; RAE-B 52:158, Sept., 1964; TDB 60:794, Aug., 1963.

4980. Shatoury, H. H. 1963 *In Vitro* Effect of Lowering Surface Energy on Esterase Activity of Musca Homogenates. (Correspondence.) Nature 199:1192.

With each of several compounds, the level of inhibition was greater at the concentration which gave lower surface tension value. "Inhibition" relates to interference with the action of the enzyme cholinesterase (obtained from fly heads). ABSTRACTS: BA 45:3223(39919), May 1, 1964; RAE-B 52:159, Sept., 1964; TDB 61:107, Jan., 1964.

4981. Shatoury, H. H. 1965 Effect of Solvated Ions on the Activity of Esterases in the Housefly. (Correspondence.) Nature 205:501-502.

The order of the effect of a number of electrolytes on cholin- and aliesterase of the housefly corresponds to the order of their energy of dehydration. ABSTRACTS: BA 46:7364(91611), Oct. 15, 1965; TDB 62:694, July, 1965.

4982. Shatoury, H. H. 1968 The Use of Ion Exchange Resins in Biochemical Studies of Insect Blood. Laboratory Practice 17(11):1233-1235.

Houseflies are prompt in acting to maintain a normal level of blood sodium, which maintains a normal osmotic balance. Increases in the concentration of H ions are buffered by blood proteins and the movement into the cells of H ions, resulting in the movement outside of K ions.

4983. Shatoury, H. H. 1969 Retention of Intra-cellular Potassium by the Housefly. Ent. Exp. Appl. 12(1):19-22. (German summary.)
Protein-starved and metal-free sugar-fed flies excrete Na and K in a ratio different from the ratio within the cells. K excretion is negligible, compared with Na, due to loss of blood-Na and retention of intracellular K. ABSTRACT: BA 51:563(5666), Jan. 1, 1970.

4984. Shatoury, H. H. 1969 Intracellular Concentration of Electrolytes in The Housefly. Nature (London) 221:282.

Shavity, E. W., joint author. *See* Sacktor, B., et al., 1965.

Shaw, F. R., joint author. *See* Eversole, J. W., et al., 1965; Wasti, S. S., et al., 1970.

Shaw, J. H., joint author. *See* Miller, B. F., et al., 1970.

Shehata, M. N., joint author. *See* Soliman, A. A., et al., 1966, 1967.

Shellenberger, T. E., joint author. *See* Skinner, W. A., et al., 1966, 1967.

4985. Sheremet'ev, N. N. 1964 Dinamika vydeleniya vaktsinnykh shtammov virusa poliomielita ot mukh posle provedeniya vaktsinatsii. [Dynamics of the isolation of poliomyelitis vaccinal virus strains from flies after vaccination.] Zh. Mikrobiol., Epidemiol., i Immunobiol. 41(10):102-106. (English summary.)
Data from studies during 1961-1962 show that synantropic flies play a definite role not only in the mechanism of polio infection transmission, but also in the spread of vaccinal strains among the population. ABSTRACT: BA 46:5419(67409), Aug. 1, 1965.

4986. Sherman, M. 1965 The Effectiveness of Insecticides Administered Orally to the Fowl as a Deterrent to the Breeding of Flies in Droppings. Proc. Hawaiian Ent. Soc. 19(1):111-117.
Research aims 4-fold: To determine (1) toxicity of such droppings to larvae; (2) toxicity of insecticides to poultry; (3) quantitative deposition of insecticides in poultry tissue and in eggs; (4) metabolic fate of certain organophosphorus insecticides in poultry. ABSTRACTS: BA 46:8516 (105202), Dec. 1, 1965; RAE-B 55:109(391), June, 1967.

4987. Sherman, M., Chang, M. T. Y. and **Herrick, R. B. 1969** Fly Control, Chronic Toxicity and Residues from Feeding Propyl Thiopyrophosphate to Laying Hens. J. Econ. Ent. 62(6):1494-1499.
Treatment resulted in excellent control of *Musca domestica* L. and *Calliphora,* but poor control of *Parasarcophaga* and *Fannia.* No mortality among the hens and but few physiological changes. ABSTRACTS: BA 51: 7937(81038), July 15, 1970; RAE-B 58:147(614), May, 1970.

4988. Sherman, M. and **Herrick, R. B. 1970** Acute Toxicity of Five Insect Chemosterilants, Hemel, Hempa, Tepa, Metepa, and Methotrexate, for Cockerels. Toxicol. and Appl. Pharmacol. 16:100-107.

4989. Sherman, M., Herrick, R. and **Chang, M. T. Y. 1965** Toxicity of Imidan (Cholinesterase Inhibitor) and Five Analogs to the Chick, Rat and House Fly. *In* Fourth Ann. Meet. Soc. Toxicol., Williamsburg, Va. March 8-10. Toxicol. and Appl. Pharmacol. 7(3):497 (Abstract only).
Of the 6 compounds tested, imidan was the most toxic to the housefly and the least toxic to the chick and rat. No clear correlation between cholinesterase inhibition and the toxicity of this series of compounds.

4990. Sherman, M., Herrick, R. B., Chang, M. T. Y. and **Menn, J. J. 1967** Comparative Toxicity of Imidan and Homologs Containing Asymmetrical Esters to the Chick, Rat, and Housefly. J. Med. Ent. 4(4):

451-455.
See preceding reference. ABSTRACTS: BA 49:4546(51117), May 15, 1968; TDB 65:954(2149), July, 1968.

4991. Sherman, M. and **Komatsu, G. H. 1963** Maggot Development in Droppings from Chicks Fed Organophosphorus Insecticide-treated Rations. J. Econ. Ent. 56(6):847-850.
At concentrations in poultry feed of 50-100 p.p.m. ASP-51, Bayer 18779, and Methyl Trithion were highly toxic to *M. domestica*, *Chrysomya megacephala* and *Parasarcophaga argyrostoma*. *Fannia pusio* proved more tolerant of these insecticides than the other species. ABSTRACTS: BA 45:3601(44357), May 15, 1964; RAE-B 52:53, Mar., 1964.

4992. Sherman, M. and **Komatsu, G. H. 1965** Toxicity to Fly Larvae of Droppings from Chicks Reared on Insecticide-treated Feed. J. Econ. Ent. 58(2):203-206.
N-2404 was *more* toxic after passage through the chick than when inoculated directly into droppings. Ciodrin was *less* toxic to larvae after passage through the chick. ABSTRACTS: BA 46:5147(63905), July 15, 1965; RAE-B 53:127, July, 1965; TDB 62:816, Aug., 1965.

4993. Sherman, M., Komatsu, G. H. and **Ikeda, J. 1967** Larvicidal Activity to Flies of Manure from Chicks Administered Insecticide-treated Feed. J. Econ. Ent. 60(5):1395-1403.
Forty-four compounds were tested. A very large number were effective in controlling at least three species of the four studied (*M. domestica*, *Fannia pusio*, *Chrysomya megacephala*, *Parasarcephaga argyrostoma*). The last named was the most tolerant to insecticide-containing manures. ABSTRACTS: BA 49:1396(15638), Feb. 1, 1968; RAE-B 56:68(219), Mar., 1968; TDB 65:199(445), Feb., 1968.

4994. Sherman, M. and **Ross, E. 1960** Toxicity to House Fly Larvae of Droppings from Chickens Fed Insecticide-treated Rations. J. Econ. Ent. 53(3):429-432.
Several insecticides tested. Greater than 90 per cent larval mortality achieved by feeding Co-Ral, 89 p.p.m.; Diazinon, 154 p.p.m.; Dipterex, 89-132 p.p.m.; ronnel, 176-220 p.p.m.; Dow ET-15, 89 p.p.m. Malathion- and phenothiazine-treated rations caused relatively low mortality among larvae in hen manure. ABSTRACTS: BA 35:4679(54359), Oct. 1, 1960; RAE-B 49:119, June, 1961.

4995. Sherman, M. and **Ross, E. 1960** Toxicity to House Fly Larvae of Droppings from Chicks given Dipterex-treated Water. J. Econ. Ent. 53(6):1066-1070.
Dipterex at 30 p.p.m. in water at pH levels below 7·0 caused high larval mortality in droppings. Water solutions at pH 8·0 *decreased* the toxicity greatly. (Different metabolic pathway, in chicks, converts Dipterex to DDVP with a different end product, less toxic to larvae.) ABSTRACTS: BA 36:1827(22141), Apr. 1, 1961; RAE-B 49:220, Oct., 1961.

4996. Sherman, M. and **Ross, E. 1961** Toxicity to House Fly Larvae of Droppings from Chicks Administered Insecticides in Feed, Water, and as Single, Oral Dosages. J. Econ. Ent. 54(3):573-578.
Twenty-three insecticides tested. Examples: Dimethoate and phosphamidon, administered to chicks in drinking water at 22 and 220 p.p.m. respectively, resulted in greater than 90 per cent mortality in larvae placed in droppings. ABSTRACT: RAE-B 50:130, June, 1962.

4997. **Sherman, M., Ross, E.** and **Komatsu, G. H. 1962** Differential Susceptibility of Maggots of Several Species to Droppings from Chicken Fed Insecticide-treated Rations. J. Econ. Ent. 55(6):990-993.
Twenty-two insecticides tested. Muscid larvae proved more tolerant of insecticide-containing manure than sarcophagids or calliphorids. American Cyanamid 12008, Bayer 22408, and dimethoate proved highly toxic. ABSTRACTS: BA 42:308(3712), Apr. 1, 1963; RAE-B 51:60, Mar., 1963.

4998. **Sherman, M., Ross, E., Sanchez, F. F.** and **Chang, M. T. Y. 1963** Chronic Toxicity of Dimethoate to Hens. J. Econ. Ent. 56(1): 10-15.
Emulsifiable dimethoate added to drinking water of hens gave excellent larval control in droppings (*M. domestica* and others). Weight gain in hens slowed somewhat, food intake was less, egg production remained the same. ABSTRACTS: BA 42:1901(24003), June 15, 1963; RAE-B 51:89, May, 1963.

4999. **Sherman, M.** and **Sanchez, F. F. 1964** Latent Toxicity of Insecticides to Resistant and Susceptible Strains of the House Fly. J. Econ. Ent. 57(6):842-845.
In the susceptible SCR-60 strain, latent toxicity was shown by aldrin, chlordane, DDT, dieldrin, and heptachlor. Aldrin and heptachlor showed very low levels of latent toxicity to the resistant (Hawaiian) strain. Both strains were equally susceptible to diazinon. ABSTRACTS: BA 46:2539 (31685), Apr. 1, 1965; RAE-B 53:56, Mar., 1965; TDB 62:475, May, 1965.

5000. **Sherman, M., Takei, G. H., Herrick, R. B.** and **Ross, E. 1968** Chronic Toxicity to Laying Hens and Degradation of Bayer 18779. Poultry Sci. 47(2):648-654.
Bayer 18779, applied in the feed of laying hens, proved too unstable to effect a high level of protection against the breeding of maggots in the droppings from these hens. ABSTRACT: BA 50:6177(64633), June 15, 1969.

5001. **Duplicate reference.** Withdrawn.

Sherman, M., joint author. *See* Morallo, B. D., et al., 1967; Ross, E., et al., 1960; Sanchez, F. F., et al., 1966; Yates, J. R. III, et al., 1970.

Shi, C. Y., joint author. *See* Hsieh, H.-M., et al., 1965.

Shiba, M., joint author. *See* Nagasawa, S., et al., 1964, 1965.

Shibles, D. B., joint author. *See* Hansens, E. J., et al., 1968.

5002. **Shieh, T., Leng, H., Chen, D.** and **Chao, Y. 1964** [Esterase inhibition of house-fly poisoning by organophosphorus insecticides.] Acta Ent. Sinica 13(4):503-509. (In Chinese, with English summary.)
Cholinesterase inhibition is the major toxic action of the organophosphates. Insecticides used in this study were: dimethoate (Rogor), trichlorphon (Dipterex) and tri-o-cresyl phosphate (TOCP). ABSTRACT: RAE-B 53:55, Mar., 1965.

5003. **Shieh, T. R., Anderson, R. F.** and **Rogoff, M. H. 1968** Regulation of Exo-Toxin Production in *Bacillus thuringiensis*. Bact. Proc. for 1968, page 6. (Abstract.)
Concerns housefly larvae. Involves RNA synthesis.

5004. **Shimada, A., Seki, M.** and **Suzuki, T. 1969** [Studies on aerosol type insecticides. III. The fluctuating factors on evaluating the effect of the Peet-Grady tests.] Jap. J. Sanit. Zool. 20(3):195-201. (In Japanese, with English summary.)
Worked with 5 colonies of *M. domestica vicina*. Older flies tend to rest

on floor or lower side walls; younger flies on ceiling or upper sides. Ceiling more used by males than females. Percentage knockdown different with oil-base and water-base aerosols. Other problems. ABSTRACT: BA 51:5642(57893), May 15, 1970.

5005. Shimizu, F., Hashimoto, M., Taniguchi, H., Oota, W., Kakizawa, H., Takada, R., Kano, R., Targe, H., Kaneko, K., Shinonaga, S. and **Miyamoto, K. 1965** Epidemiological Studies on Fly-borne Epidemics. Report I. Significant Role of Flies in Relation to Intestinal Disorders. Jap. J. Sanit. Zool. 16(3):201-211. (Japanese summary.)
Concerns bacterial organisms found on and in the bodies of flies. Strong emphasis on *Morganella* spp., some of which cause summer diarrhea.

Shimizu, N., joint author. *See* Miyamoto, K., et al., 1965; Nakayama, I., et al., 1969.

5006. Shimogama, M. 1960 [Field experiments of controlling house-fly by residual sprays and insecticide impregnated cords in semi-farm villages in Nagasaki City.] Endemic Dis. Bull. Nagasaki Univ. 2(4):287-295. (English summary.)
Sprays and cords, using DDT and diazinon proved effective in farming communities against *M. domestica vicina.* ABSTRACTS: RAE-B 51:268, Dec., 1963; BA 39:671(8316), July 15, 1962.

Shimogama, M., joint author. *See* Omori, N., et al., 1962; Ori, S., et al., 1960; Suenaga, O., et al., 1962, 1964.

Shinonaga, S., joint author. *See* Kano, R., et al., 1964, 1965, 1967, 1968; Shimizu, F. et al., 1965.

Shiozaki, S., joint author. *See* Matsuzawa, H., et al., 1962.

Shipitsyna, N. K., joint author. *See* Balashov, Yu. S., et al., 1968.

5007. Shipp, E. and **Osborn, A. W. 1967** The effect of Protein Sources and of the Frequency of Egg Collection on Egg Production by the Housefly (*Musca domestica* L.). Bull. W. H. O. 37(2):331-335.
Higher fecundities achieved with yeast pastes than with milk-based diets. Also involves less labor. Egg collection on alternate days gave optimal results. ABSTRACTS: BA 49:4314(48618), May 1, 1968; TDB 65:843(1873), June, 1968; RAE-B 58:14(48), Jan., 1970.

Shipp, E., joint author. *See* Osborn, A. W., et al., 1970.

Shishido, T., joint author. *See* Fukami, J.-I., et al., 1969.

Shivistava, S. P., joint author. *See* Tsukamoto, M., et al., 1968.

Shlyaposhnikov, M. S., joint author. *See* Trofimov, G. K., et al., 1961.

Shnaider, E. V., joint author. *See* Vashkov, V. I., et al., 1961, 1962.

5008. Shortino, J. T., Cantwell, G. E. and **Robbins, W. E. 1963** Effect of Certain Carcinogenic 2-fluorenamine Derivatives on Larvae of the Housefly, *Musca domestica* Linnaeus. J. Insect Pathol. 5(4):489-492.
All the derivatives tested profoundly affected larval development. In certain cases malformed larvae and pupae contained growths similar to hereditary tumors in *Drosophila.* ABSTRACTS: BA 45:4339(53530), June 15, 1964; RAE-B 52:86, May, 1964.

Shortino, J. T., joint author. *See* Cantwell, G. E., et al., 1966; Dutky, R. C., et al., 1963, 1967; Kaplanis, J. N., et al., 1965, 1966; West, J. A., et al., 1968; Robbins, W. E., et al., 1962, 1965, 1970.

5009. Shrivastava, S. P. 1968 Metabolism of *o*-Isopropoxy-phenyl Methylcarbamate (Baygon) in House Flies. Diss. Absts. 28(78):2888B.
Compounds were separately labelled with C_{14}. Findings suggest that detoxification by hydroxylation mechanisms may be a major limiting factor in the susceptibility of house flies to carbamate toxicants.

5010. Shrivastava, S. P., Tsukamoto, M. and **Casida, J. E. 1969** Oxidative Metabolism of C^{14}-Labeled Baygon by Living House Flies and by House Fly Enzyme Preparations. J. Econ. Ent. 62(2):483-498.
Supports the hypothesis of detoxication by hydroxylation mechanisms. ABSTRACT: RAE-B 57:186(664), Sept., 1969.

Shrivastava, S. P., joint author. *See* Casida, J. E., et al., 1968.

Shugal, N. F., joint author. *See* Vashkov, V. I., et al., 1966, 1968.

5011. Shulman, S. 1967 Allergic Responses to Insects. Ann. Rev. Ent. 12: 323-346.
Problems of hay-fever and asthma attributed to insects, including the housefly. Refers to comparative skin tests.

5012. Shura-Bura, B. L. and **Glazunova, A. Y. 1963** Insektitsidnye svoistu khlorofosnoi bumagi "mukhomor". [The insecticidal properties of chlorophos "Fly Paper".] Med. Parazit. (Moskva) 32:554-558.
Chlorophos paper containing 26 mg. of dimethyl-2,2,2-trichloro-1-ethyl phosphate per sheet attracts flies better than that of higher concentrations, and very effectively can kill up to 70 per cent of flies up to a distance of 100 cm. from the bait. ABSTRACTS: BA 45:2840(35213), Apr. 15, 1964; TDB 61:109, Jan., 1964.

Sidorova, M. V., joint author. *See* Vashkov, V. I., et al., 1966, 1967.

Silva, S., joint author. *See* Machado, C., et al., 1967.

Silverman, P. H., joint author. *See* Long, C. L., et al., 1965.

5013. Simco, J. S. and **Lancaster, J. L., Jr. 1966** Field Test to Determine the Effectiveness of Coumaphos as a Feed Additive to Control House Fly Larvae Under Caged Layers. J. Econ. Ent. 59(3):671-672.
Coumaphos found effective in rendering hen manure larvicidal. ABSTRACTS: BA 47:8521(99268), Oct. 15, 1966; RAE-B 54:179, Sept., 1966.

Simco, J. S., joint author. *See* Lancaster, J. L., et al., 1969.

5014. Simkover, H. G. 1964 2-Imidazolidinone as an Insect Growth Inhibitor and Chemosterilant. J. Econ. Ent. 57:574-579.
Modifies growth and development of the immature stages of *M. domestica* and other insects. Sterilizes males and females, and serves as a larvicide in manure, when consumed by poultry. ABSTRACTS: RAE-B 52:206, Nov., 1964; TDB 62:158, Feb., 1965.

5015. Simon, J., Bhatnagar, P. L. and **Milburn, N. S. 1969** An Electron Microscope Study of Changes in Mitochondria of Flight Muscle of Aging Houseflies, *Musca domestica*. J. Insect Physiol. 15(1): 135-140.
Two kinds of mitochondria recognized, A and B, distinguished by their cristae and density of their matrix. Type B increases in number about the 12th day after emergence. Each type has its own pattern of degeneration. ABSTRACTS: BA 50:7536(78560), July 1, 1969; TDB 66:360(707), Apr., 1969.

Singer, S., joint author. *See* Rogoff, M. H., et al., 1969.

5016. Singh, P., King, W. E. and **Rodriguez, J. G. 1966** Biological Control of Muscids as influenced by Host Preference of *Macrocheles muscaedomesticae* (Acarina:Macrochelidae). J. Med. Ent. (Honolulu) 3(1):78-81.
This predatory mite attacks the ova of *M. domestica, M. autumnalis* and *Fannia canicularis*. Shows marked preference for the first two. ABSTRACTS: BA 47:9041(104540), Nov. 1, 1966; TDB 63:1256, Nov., 1966; RAE-B 55:211(733), Nov., 1967.

Singh, P., joint author. *See* Rodriguez, J. G., et al., 1970.

5017. **Singh, S. B.** and **Judd, W. W. (1965) 1966** A Comparative Study of the Alimentary Canal of Adult Calyptrate Diptera. Proc. Ent. Soc. Ont. 96:29-80.
Studied 12 species, belonging to 10 families. Comparative data taken on: (1) alimentary canal indices (ratio of body length to total length of food canal, (2) coiling of the canal, (3) relative position of the rectal valve.

5018. **Sinha, R., Deb, B. C., De, S. P., Abou-Gareeb, A. H.** and **Shrivastava, D. L. 1967** Cholera Carrier Studies in Calcutta in 1966-1967. Bull. W. H. O. 37(1):89-100.
Ninety batches of fly samples were collected from fifty-five suspected houses. Four were found positive for *Vibrio cholerae* and three for non-agglutinable vibrios. ABSTRACT: BA 49:4663(52366), May 15, 1968.

Sissons, C. H., joint author. *See* Chakraborty, J., et al., 1967.

5019. **Sivasubramanian, P., Bhaskaran, G.** and **Nair, K. K. 1970** Effects of X-rays on Morphogenesis in the Housefly. J. Insect Physiol. 16: 89-97.
Concentrated on radiosensitivity of the imaginal discs. Morphogenetic movements are not affected by X-rays up to 10,000 rads, but differentiation is. Critical phase ends after the evagination of the head (when the gene activation process is probably complete.) ABSTRACT: BA 51:6303 (64494), June 15, 1970.

5020. **Sivasubramanian, P., Bhaskaran, G.** and **Nair, K. K. 1970** Differentiation of the Imaginal Muscles in X-Irradiated House Fly Pupae. Ann. Ent. Soc. Am. 63(4):1019-1022.
Radiation less than 20 hours after puparium formation induces drastic dystrophic changes in the structure of both tubular and fibrillar muscles. If radiation is delayed, damage is less.

Sivasubramanian, P., joint author. *See* Nair, K. K., et al., 1967; Bhaskaran, G., et al., 1969.

5021. **Skaptason, J. S.** and **Pitts, C. W. 1962** Fly Control in Feces from Cattle Fed Co-ral. J. Econ. Ent. 55:404-405.
Co-ral in the diet of feeder cattle significantly reduced the development of *Musca domestica* in the fecal material. ABSTRACTS: BA 40:307(3974), Oct. 1, 1962; RAE-B 50:274, Dec., 1962; TDB 59:1119, Nov., 1962.

Skarén, U., joint author. *See* Nuorteva, P., et al., 1960.

5022. **Skelton, T. E.** and **Hunter, P. E. 1970** Flight and Respiration Responses in House Flies Following Topical Application of Sterilization Levels of Tepa. Ann. Ent. Soc. Am. 63(3):770-773.
No significant changes in 7 flight characteristics of males. A significant ($P<0{\cdot}05$) reduction in O_2 consumption by sterilized males. Tepa-sterilization had no significant effect on O_2 consumption in females. ABSTRACT: BA 51:10854(110185), Oct. 1, 1970.

5023. **Skinner, W. A., Cory, M.** and **DeGraw, J. I. 1967** Effect of Organic Compounds on Reproductive Processes. VII. Bis-N, N'-carbamoyl-aziridines (House Fly Chemosterilant). J. Med. Chem. 10(6):1186-1188.
Studies on some effective chemosterilants for houseflies emphasize importance of spacing between two alkylating functions of the compounds. ABSTRACT: BA 49:3813(42829), Apr. 15, 1968.

5024. **Skinner, W. A., Cory, M., Shellenberger, T. E.** and **DeGraw, J. I. 1967** Effect of Organic Compounds on Reproductive Processes. V. Alkylating Agents Derived from Aryl-, Aralkyl-, and Cyclohexyl-methylenediamines. J. Med. Chem. 10(1):120-121.
The influence of the carrier moiety on the chemosterilant activity of

bis(aziridineacetyldiamines) is further explored. ABSTRACT: BA 49:1190 (13313), Feb. 1, 1968.

5025. Skinner, W. A., Hayford, J., Shellenberger, T. E. and **Colwell, W. T. 1966** Effect of Organic Compounds on Reproductive Processes [in *Musca domestica*]: II. Alkylating Agents Derived from Various α, *w*-Alkylenediols. J. Med. Chem. 9(4):605-607.
In an evaluation as inhibitors of reproduction, none of the compounds tested had a significant effect on egg production or fertility. ABSTRACT: BA 47:10364(119828), Dec. 15, 1966.

5026. Skinner, W. A., Lange, J., Shellenberger, T. E. and **Colwell, W. T. 1967** Effect of Organic Compounds on Reproductive Processes. VI. Alkylating Agents Derived from Various Diamines. J. Med. Chem. 10(5):949-950.
Of compounds formed and studied, results seem to further define the requirement that the alkylating group must be a to the carbonyl in this series.

5027. Slade, M. and **Casida, J. E. 1970** Metabolic Fate of 3, 4, 5- and 2, 3, 5-Trimethylphenyl Methylcarbamates, the Major Constituents in Landrin Insecticide. J. Agric. Fd. Chem. 18(3):467-474.
Included mixed-function oxidase systems of housefly abdomens, among several media employed to determine the fate of these compounds. ABSTRACT: BA 51:12870(130887), Dec. 1, 1970.

Slater, E. C., joint author. *See* Bergh, S. G., et al., 1960, 1962; Veldsema-Currie, R. D., et al., 1968.

5028. Sleeper, D. A. 1964 Paper Electrophoretic Separation of Hemolymph Proteins as a Taxonomic Character in Certain Muscoid Diptera. Diss. Absts. 24(10):4326.

Smallman, B. N., joint author. *See* Hegdekar, B. M., et al., 1967, 1969.

5029. Smith, A. C., Black, R. J. and **Hart, S. A. 1960** Dairy Manure Sanitation Study. Part II: Entomological Aspects. Calif. Vector Views 7(12):69-75.
Ramp storage method unsatisfactory. Good fly control accomplished by initial liquefaction followed by thin spreading and rapid drying on topsoil, sawdust (or shavings) and previously dried manure.

Smith, A. E., joint author. *See* Hart, S. A., et al., 1960.

5030. Smith, A. G. 1961 An Indication of the Present Resistance to Insecticides of the Housefly, *Musca domestica* L. in New Zealand. New Zealand J. Sci. 4(2):288-291.
Studies revealed houseflies resistant to dieldrin and lindane, with high level of resistance to DDT. No evidence of resistance to organophorphorus compounds. ABSTRACTS: BA 36:8114(86688), Dec. 15, 1961; RAE-B 50:246, Nov., 1962.

5031. Smith, C. N., LaBrecque, G. C. and **Bořkovec, A. B. 1964** Insect Chemosterilants. Ann. Rev. Ent. 9:269-284.
A review up to March 1963. Includes table of 43 chemosterilants listing insects affected and references for each. ABSTRACT: TDB 64:690, June, 1967.

5032. Smith, C. N., LaBrecque, G. C., Wilson, H. G., Hoffman, R. A., Jones, C. M. and **Warren, J. W. 1960** Dimetilan Baits, Fly Ribbons and Cords for the Control of House Flies. J. Econ. Ent. 53(5):898-902.
Results are presented following the use of dimetilan in various forms of application in dairies, poultry and pig barns. It effectively controlled

Musca domestica resistant to other insecticides. ABSTRACTS: BA 36:764 (9085), Feb. 1, 1961; RAE-B 49:184, Aug., 1961; TDB 58:389, Mar., 1961.

Smith, C. N., joint author. *See* Murvosh, C. M., et al., 1963, 1964, 1965; Sullivan, W. N., et al., 1960; Knipling, E. F., et al., 1968.

Smith, C. R., joint author. *See* Poorbaugh, J. H., et al., 1968.

Smith, C. T., joint author. *See* Wasti, S. S., et al., 1970.

Smith, C. W., joint author. *See* Morgan, P. B., et al., 1967.

5033. Smith, D. L. 1966 The Face Fly, *Musca autumnalis* (DeGeer) in Manitoba. Proc. Ent. Soc. Manitoba 22:30-31.
Species well established in Manitoba from Morden and Elm Creek west to Saskatchewan boundary and from U.S. border north to Riding Mt. Nat. Park. ABSTRACTS: BA 49:939(10567), Jan. 15, 1968; RAE-B 58: 272(1124), Aug., 1970.

Smith, E., joint author. *See* Cassidy, J. D., et al., 1969.

5034. Smith, G. E., Breeland, S. G. and **Pickard, E. 1965** The Malaise Trap —a Survey Tool in Medical Entomology. Mosquito News 25(4): 398-400.
Collections using the Malaise trap in 1964 contained muscoid flies, a significant number being *Stomoxys calcitrans*, reflecting a local outbreak of this species. ABSTRACT: RAE-B 55:163(568), Sept., 1967.

5035. Smith, G. J. C. 1968 The Efficiency of *Nasonia vitripennis* (Hymenoptera:Pteromalidae) as a Parasite of 2 Host Species, *Musca domestica* and *Phaenicia sericata*. Diss Absts. 28(78):2889B.

5036. Smith, G. J. C. 1969 Host Selection and Oviposition Behavior of *Nasonia vitripennis* (Hymenoptera:Pteromalidae) on Two Host Species. Canad. Ent. 101(5):533-538.
At all densities of *Nasonia* eggs, more wasps matured on blowfly pupae than on housefly pupae of the same size. ABSTRACT: RAE-B 58:308 (1257), Sept., 1970.

5037. Smith, G. J. C. and **Pimentel, D. 1969** The Effect of Two Host Species on the Longevity and Fertility of *Nasonia vitripennis*. Ann. Ent. Soc. Am. 62(2):305-308.
The adult female parasitoid has shorter longevity and produces less progeny when feeding on the housefly as compared to the blowfly. ABSTRACT: BA 50:8652(89872), Aug. 15, 1969.

Smith, G. J. C., joint author. *See* Pimentel, D., et al., 1967.

5038. Smith, J. N. and **Turbert, H. B. 1964** Comparative Detoxication. II. Conjugations of 1-Naphthol and some other Phenols in Houseflies and Locusts. Biochem. J. 92(1):127-131.
Glucosides and ethereal sulphates of *m*-aminophenol, 1-naphthol and 4-methylumbelliferone detected in excreta of flies dosed with these phenols.

Smith, J. N., joint author. *See* Chakraborty, J., et al., 1967; Clark, A. G., et al., 1966; Heenan, M. P., et al., 1967; Jordan, T. W., et al., 1968; Darby, F. J., et al., 1966; Balabaskaran, S., et al., 1968.

Smith, J. W., Jr., joint author. *See* Killough, R. A., et al., 1965 Pickens, L. G., et al., 1967.

Smith, M., joint author. *See* Tarshis, I. B., et al., 1961.

5039. Smith, R. F. and **Mittler, T. E. 1970** Annual Review of Entomology. Vol. 15, 502 pp. Palo Alto, Calif. Annual Reviews, Inc.
Reviews in this volume include: Lofgren, C. S., "Ultralow volume applications of concentrated insecticides in medical and veterinary entomology," pp. 321-342, in which sprays against Diptera are reviewed. ABSTRACT: RAE-B 58:98(398), Apr., 1970.

5040. Smith, T. A. 1968 Comparison of Known Age with Physiological Aging in the Adult Female House Fly, *Musca domestica* L. J. Med. Ent. 5:1-4.
Ovarian dissection techniques used. It is not possible to determine the exact number of ovipositions of parous flies by an inspection of the sex organs. Only slightly more than half of the females of older age groups were parous. ABSTRACTS: BA 49:6783(75641), July 15, 1968; TDB 65:1061(2410), Aug., 1968; RAE-B 58:115(465), Apr., 1970.

5041. Smith, T. A. 1969 The Maturation of Fly Larvae Following Removal from the Larval Medium. Calif. Vector Views 16(8):73-78.
M. domestica larvae were able to emerge if desiccated when they were 2 days old but with very reduced proportions; emergence can be prevented if manure is removed every 4 days and rapidly dried. ABSTRACT: BA 51:6220(63639), June, 1970.

5042. Smith, T. A. and **Linsdale, D. D. 1967** First Supplement to an Annotated Bibliography of the Face Fly, *Musca autumnalis* DeGeer, in North America. Calif. Vector Views 14(11):74-76. 50 refs.
Lists publications not included in first list (1952-1965). CITED: RAE-B 58:248(1015), July, 1970.

5043. Smith, T. A. and **Linsdale, D. D. 1968** Second Supplement to an Annotated Bibliography of the Face Fly, *Musca autumnalis* DeGeer, in North America. Calif. Vector Views 15(11):119-120.
Lists some publications of earlier dates not included in 2 previous lists. ABSTRACT: BA 50:5914(61970), June 15, 1969.

5044. Smith, T. A. and **Linsdale, D. D. 1969** Third Supplement to an Annotated Bibliography of the Face Fly, *Musca autumnalis* DeGeer, in North America. Calif Vector Views 16(12):119-123.
Gives 98 references in addition to those listed in Cal. Vect. Views, 13(6); 14(11); 15(11). See reference No. 5045 for original paper.

5045. Smith, T. A., Linsdale, D. D. and **Burdick, D. J. 1966** An Annotated Bibliography of the Face Fly, *Musca autumnalis* DeGeer, in North America. Calif. Vector Views 13(6):43-53.
Introductory paragraphs cover geographical distribution, vectorship, life history, etc. Bibliography lists literature from 1952 through 1965. ABSTRACT: RAE-B 57:98(350), June, 1969.

5046. Smith, T. A., Walsh, J. D. and **Linsdale, D. D. 1970** Fly Occurrence in Waste Fruit in Fresno County Orchards. Calif. Vector Views 17(4):24-28.
Refers to fallen peaches, plums, nectarines. Larval infestation occurred from mid-July to November; *Drosophila* most abundant, *Muscina assimilis* second.

5047. Smith, W. W. 1962 Laboratory Tests of Some Soil Fumigants as Housefly Larvicides. J. Econ. Ent. 55(2):265.
Nine compounds tested. Most toxic were 1,2-dibromo-3-chloropropane and ethylene dibromide. ABSTRACTS: BA 39-961(12213), Aug. 1, 1962; RAE-B 50:239, Nov., 1962; TDB 59:1017, Oct., 1962.

5048. Smith, W. W. 1962 Field Tests of Some Soil Fumigants as House Fly Larvicides. J. Econ. Ent. 55:1001-1003.
Certain soil fumigants very effective as fly larvicides, but lack residual toxicity. Must be applied twice weekly. ABSTRACTS: BA 42:308(3714), Apr. 1, 1963; RAE-B 51:61, Mar., 1963; TDB 60:599, June, 1963.

5049. Smith, W. W. and **Yearian, W. C. 1964** Studies of Behavioristic Insecticide Resistance in Houseflies. J. Kans. Ent. Soc. 37:63-77.
Behavioristic resistance tends to lag behind physiological resistance in

its changes, and probably cannot change as rapidly. Concept of 2 different types of resistance supported by this work. ABSTRACTS: BA 46:703(8817), Jan. 15, 1965; RAE-B 54:57, Mar., 1966.

5050. **Smittle, B. J. 1967** Effect of Aeration on *Gamma* Irradiation of House Fly Pupae. J. Econ. Ent. 60(6):1594-1596.
Exposure of 3 day old pupae to O_2, CO_2 and air during *gamma* irradiation = no effect on eclosion; but eclosion was reduced up to 10 per cent by exposure to 6000 R. Percentage sterility produced by *gamma* rays in males and females was reduced by CO_2. ABSTRACT: RAE-B 56:99(344), May, 1968.

Smythe, V. R., joint author. *See* Collard, R. V., et al., 1967.

Snarey, M., joint author. *See* Sawicki, R. M., et al., 1962.

5051. **Snyder, F. M.** and **Chadwick, L. E. 1964** The Course of Poisoning of Normal and Pyrethrins-Resistant House Flies by Pyrethrins—Piperonyl Butoxide Residues: A Kinetic Analysis. Ent. Exp. Appl. (Amsterdam) 7(3):229-240.
Supports hypothesis that flies possess a mechanism for detoxifying Py, and that PB exerts synergistic action by interfering with the detoxification process. ABSTRACTS: RAE-B 53:174, Sept., 1965; TDB 62:245, Mar., 1965.

5052. **Snyder, F. M., Lovell, J. B.** and **Chadwick, L. E. 1964** The Dosage-Mortality Relationship for Two Strains of House Flies Exposed to Vapor of Di-isopropyl-phosphorofluoridate. Ent. Exp. Appl. (Amsterdam) 7(4):277-286.
Males of either strain 1·5x as sensitive as respective females. Strains differ by a factor of about 10. Believed that an appreciable portion of entering DFP is diverted by nervous system cholinesterase to nontoxic pathways. ABSTRACTS: BA 47:1198(14499), Feb. 1, 1966; RAE-B 53:174, Sept., 1965; TDB 62:597, June, 1965.

Soans, J., joint author. *See* Pimentel, D., et al., 1967.

5053. **Soeda, Y.** and **Yamamoto, I. 1968** Studies on Nicotinoids as an Insecticide. V. Inhibition of House Fly Head Cholinesterase by Nicotine. Agric. Biol. Chem. 32(5):568-573.
Indicates that the cationic head of the nicotinium ion interacts with the anionic site in the active center of cholinesterase. ABSTRACT: BA 49: 11857(129935), Dec. 15, 1968.

5054. **Soeda, Y.** and **Yamamoto, I. 1968** Studies on Nicotinoids as an Insecticide. VI. Relation of Structure to Toxicity of Pyridylalkylamines. Agric. Biol. Chem. 32(6):747-752.
Highest toxicities against housefly were seen in the N-substituted 3-pyridylmethylamine series which were provided with the highly basic nitrogen. ABSTRACT: BA 50:8628(89595), Aug. 15, 1969.

5055. **Soeda, Y.** and **Yamamoto, I. 1969** [Studies on nicotinoids as an insecticide. VIII. Physiological activities of the optical isomers of nicotinoids., Botyu-Kagaku 34(2):57-62. (In Japanese, with English summary.)
Relative toxicities of optical isomers given: physiological activities reviewed. All available data on physiological activities of the isomers suggests a similar site of action but different processes of reaching the site. ABSTRACT: BA 51:5087(52394), May 1, 1970.

Soeda, Y., joint author. *See* Yamamoto, I., et al., 1968.

5055a. Sohal, R. S. and **Allison, V. F. 1970** Fine Structure of the Fibrillar Flight Muscles in the Housefly, *Musca domestica* (Diptera). J. Grad. Res. Cent., S. Methodist Univ. 38(2/4):27-36.

Describes finer structures observed by means of the electron microscope. ABSTRACT: BA 52:1555(14788), Feb. 15, 1971.

5056. Sokal, R. R. and **Bryant, E. H. 1967** Computing a Population Budget from Sequentially Sacrificed, Replicated Cultures, *Musca domestica*. Res. Pop. Ecol. 9(1):10-18. (Japanese summary.)

It was found possible to obtain a population budget by a randomization treatment of census results obtained at two densities. Computations and graphing were carried out on a digital computer. ABSTRACT: BA 50:6496 (68011), July 1, 1969.

5057. Sokal, R. R. and **Bryant, E. H. 1968** Genetic Differences Affecting the Fates of Immature Housefly Populations at Two Densities. Res. Pop. Ecol. 10:140-170.

5057a. Sokal, R. R., Bryant, E. H. and **Wool, D. 1970** Selection for Changes in Genetic Facilitation: Negative Results in *Tribolium* and *Musca*. Heredity 25(2):299-306.

Studied *T. castaneum* and *M. domestica*. Two strains of each placed in competition at low and high densities. Adult survivors permitted to mate with their own kind only.

5058. Sokal, R. R. and **Sonleiter, F. J. 1965** Components of Selection in *Tribolium* (Coleoptera) and Houseflies. Cont. No. 1253 from Dept. of Entomology, Univ. of Kansas, Lawrence, Kans. U.S.A. Also in Proc. XII Int. Congr. Ent. 1964(1965):274-275. London.

With increasing density, adult weight is reduced before survival declines. Housefly adults can sustain a weight reduction up to 40 per cent with almost no effect on fecundity or innate capacity for increase.

5059. Sokal, R. R. and **Sullivan, R. L. 1963** Competition Between Mutant and Wild-type House Fly Strains at Varying Densities. Ecology 44(2):314-322.

Wild type has slightly higher emergence, is heavier, and has a shorter developmental period in mixed than in pure culture. Mutant type, the opposite, in all respects (at least at higher densities). ABSTRACT: BA 45: 389(4903), Jan. 15, 1964.

Sokal, R. R., joint author. *See* Bhalla, S. C., et al., 1964; Bryant, E. H., et al., 1967, 1968; Sullivan, R. L., et al., 1963, 1965.

5060. Soliman, A. A., Kascheff, A. H. and **Shehata, M. N. 1966/1967** Mating Behaviour in the House Fly *Musca domestica vicina* Macq. and Its Response to Insecticide Treatments. Proc. Egypt. Acad. Sci. 20: 45-58.

Detailed descriptions. Topically applied methoxychlor or dipterex in sublethal doses resulted in almost complete inhibition of copulation. ABSTRACT: BA 49:11853(129900), Dec. 15, 1968.

5061. Soliman, A. A., Kascheff, A. H. and **Shehata, M. N. 1967** Effect of Methoxychlor and Dipterex on the Chemoreceptors of the House Fly, *Musca domestica vicina* Macq. and the Whole Fly Response. Zeit. Angew. Ent. 59(4):443-447.

Trichlorphon (Dipterex), although more toxic than methoxychlor (methoxy-DDT), has no direct effect on the chemoreceptor nerves; chemoreceptors are effected by much lower doses of methoxy-DDT than is the general nervous system. Sensory peripheral nervous system primary site of action of this chemical. ABSTRACT: RAE-B 57:190(683), Oct., 1969.

5062. Soliman, S. A. and **Cutkomp, L. K. 1963** A Comparison of Chemoreceptor and Whole-Fly Responses to DDT and Parathion. J. Econ. Ent. 56:492-494.

Sensitivity of labellar hairs to sucrose increased 6× by topical application of DDT, one and one half minutes after application (whole-fly response took 4 minutes). Parathion did not affect labellar sensitivity prior to hyperactivity, or later. ABSTRACTS: BA 45:2866(35547), Apr. 15, 1964; RAE-B 51:238, Nov., 1963; TDB 60:1165, Dec., 1963.

Solly, S. R. B., joint author. *See* Lord, K. A., et al., 1961, 1964.

5063. Soloway, S. B. 1965 Correlation between Biological Activity and Molecular Structure of the Cyclodiene Insecticides. Advances in Pest Contr. Res. 6:85-126.

Entomological data presented for 6 species of insects, including *M. domestica*. Considerable information given.

Somaya, C. I., joint author. *See* Dixit, R. S., et al., 1968.

5064. Sømme, L. 1960 Resistance to Organophosphorus Insecticides in Norwegian House Flies. Norsk. Ent. Tidsskr. 11(3-4):150-159. Oslo.

No definite patterns of resistance were found; appears that small levels of resistance develop to several organophosphorus compounds as a result of selection to one of them. ABSTRACTS: BA 38:1882(24116), June 15, 1962; RAE-B 49:242, Nov., 1961.

5065. Sømme, L. 1961 On the Overwintering of House Flies (*Musca domestica* L.) and Stable Flies) *Stomoxys calcitrans* L.) in Norway. Norsk. Ent. Tidsskr. 11(5-6):191-223. Oslo.

From literature and a survey of about 80 farms it was concluded that house and stable flies pass the winter by continuous breeding in warm animal houses. ABSTRACTS: BA 41:956(12428), Feb. 1, 1963; RAE-B 50:263, Nov., 1962.

Sonleiter, F. J., joint author. *See* Sokal, R. R., et al., 1965.

Sorm, F., joint author. *See* Rezabova, B., et al., 1968.

Sotani, I., joint author. *See* Buei, K., et al., 1968.

Sourkes, T. L., joint author. *See* Lancaster, G. A., et al., 1969.

Speck, U., joint author. *See* Becker, G., et al., 1964.

Spencer, E. Y., joint author. *See* Morello, A., et al., 1967, 1968.

Spencer, M., joint author. *See* Falter, J., et al., 1967.

Speranza, M. L., joint author. *See* DiJeso, F., et al., 1967.

5066. Spielman, A. and **Kitzmiller, J. B. 1967** Genetics of Populations of Medically-important Arthropods. *In* Genetics of Insect Vectors of Disease. (J. W. Wright and R. Pal, Editors.) Amsterdam. Elsevier Publ. Co. pp. 459-485.

Includes *Aedes aegypti*, *Culex pipiens*, *Anopheles gambiae*, *Musca domestica* and others. Outlines processes by which vector populations maintain individuality and their adaption to a changing environment. ABSTRACT: RAE-B 56:157(557), Aug., 1968.

5067. Spiller, D. 1961 A Digest of Available Information on the Insecticide Malathion. Advances in Pest. Contr. Res. 4:249-335. 658 refs.

Musca domestica among the species concerned.

5068. Spiller, D. 1963 Insecticide Resistance: Effects of WARF Antiresistant on Toxicity of DDT to Adult Houseflies. Science 142(3592):585-586.

N,N-dibutyl-p-chlorobenzene sulphonamide, in amount equal to one-fifth dose of DDT, increased toxicity thereof to DDT-resistant group, but not to susceptible portion of population. ABSTRACTS: BA 45:2832 (35138), Apr. 15, 1964; RAE-B 53:250, Dec., 1965; TDB 61:533, May, 1964.

5069. Spiller, D. 1964 Nutrition and Diet of Muscoid Flies. Bull. W. H. O. 31:551-554.
Laboratory populations are continuously exposed to natural selection for maximum fitness on the medium supplied; hence optimum diet for one colony will not be the same as that for another. ABSTRACTS: BA 47: 5529(64832), July 1, 1966; RAE-B 55:27(102), Feb., 1967; TDB 62:932, Sept., 1965.

5070. Spiller, D. 1964 Bacterial Status of a Housefly Rearing Medium. New Zealand Ent. 3(3):33-38.
Bacteria comprised *Proteus vulgaris*, *P. rettgeri*, *Escherichia freundii*, *Aerobacter aerogenes*, *Streptococcus faecalis* and two unidentified species. Discusses larval and adult flora in relation to media. ABSTRACT: RAE-B 55:72(265), Apr., 1967.

5071. Spiller, D. 1966 House Flies. *In* Insect Colonization and Mass Production. Edited by C. N. Smith, Academic Press. New York and London. pp. 203-225. 13 figs, 1½ pp. refs.
Concerned with establishment of laboratory colonies and large scale production of species. ABSTRACT: RAE-B 56:1(1), Jan., 1968.

Sprenkel, R. K., joint author. *See* Cheng, T. H., et al., 1965.

5072. Spotarenko, S. S., et al. 1969 [Evaluation of the role of flies in the epidemiology of dysentery and infectious hepatitis.] Zhur. Mikrobiol. 46:43-48.

5073. Srbova, S. and **Palaveyeva, M. 1962** Study of the Insecticidal Effect of Some Plants. J. Hyg., Epidemiol., Microbiol., Immunol. (Prague) 6(4):498-502. (Summaries in French, German and Italian.)
Among 23 species of plants tested, 13 had an insecticidal effect on a laboratory strain of *M. domestica*. Relatively good results were obtained with: *Fumaria schleicheri*, *Veratrum lobelianum*, *Delphinium consolida*, *D. orientale*, *Anthemis tinctoria*, *Lysimachia punctata*, *Verbascum tapsiformae* and *Euphorbia cyparissias*. ABSTRACT: TDB 60:797, Aug., 1963.

5074. Srivastava, A. P., Dixit, R. S. and **Agarwal, P. N. 1965** Efficiency of a Newer Insecticide DDVP as a Space Spray. Defence Sci. J. 15(3): 221-224.
Work done with *M. nebulo* and *Culex fatigans*. DDVP found much superior to pyrethrins in killing effect. Knockdown, however, is poor.

5075. Srivastava, A. S. 1964 Presence of Acetylcholinesterase Activity in *Musca domestica nebulo* Fabricius and Its Kinetics. Zeit. Angew. Ent. 55:186-189. (German summary.)

5076. Srivastava, A. S. 1965 Studies on the Mechanism of Action of Insecticides. II. Role of Epicuticular Wax in the Mechanism of Entry of p, p'DDT in Flies. Zeit Angew. Ent. 56(4):326-329.
Dissolved DDT lowers the melting point of the cuticular wax inducing a semi-solid condition which facilitates the entry of the insecticide. ABSTRACT: RAE-B 56:92(307), Apr., 1968.

5077. Srivastava, A. S. 1966 Studies on the Mechanism of Action of Insecticides, Intoxication, and Detoxication of Methoxychlor in Flies. Labdev. J. Sci. Tech. 4(3):162-164.

5078. Srivastava, A. S. 1969 Studies on the Mechanism of Action of Insecticides in Insects. III. Inhibition of the Acetylcholinesterase Activity. Labdev. J. Sci. Technol. 7B(1):53-55.
DDT inhibition of cholinesterase activity occurs in head and thorax of

houseflies; in abdomen there is practically none. Gradual inhibition occurs in whole fly. ABSTRACT: BA 51:571(5746), Jan. 1, 1970.

5079. Srivastava, H. D. and **Dutt, S. C. 1963** Stephanofilariasis. Agric. Res. (India) 3(3):236.
Concerns *Musca conducens.*

5080. Srivastava, H. D. and **Dutt, S. C. 1963** Studies on the Life History of *Stephanofilaria assamensis*, the Causative Parasite of "Humpsore" of Indian Cattle. Indian J. Vet. Sci. 33(4):173-177.
Musca conducens was found to contain the filarial larvae parasite; more abundant in males. ABSTRACT: RAE-B 54:22, Jan., 1966.

5080a. Srivastava, J. B. 1970 Insecticide and Larvicide Activity in the Extract of *Piper peepuloides* Royle (Piperaceae). Indian J. Exp. Biol. 8(3):224-225.
Petroleum ether extract of *P. peepuloides* fruit at 0·125 and 0·625 per cent concentration, exhibited insecticidal and larvicidal properties with adults of *M. domestica nebulo* and both larvae and adults of *Aedes aegypti.* No fumigant activity was shown. ABSTRACT: 52:799(7312), Jan. 15, 1971.

5081. Srivastava, P. N. and **Rockstein, M. 1969** The Utilization of Trehalose During Flight by the Housefly, *Musca domestica* L. J. Insect Physiol. 15(7):1181-1186.
In male houseflies, rate of utilization is most rapid during first few minutes of flight, then decreases. To restore and maintain the thoracic trehalose level for energizing of flight, data show that a complex metabolic route is involved. ABSTRACT: BA 50:13072(134819), Dec. 15, 1969.

Srivastava, P. N., joint author. *See* Rockstein, M., et al., 1967.

5082. Srivastava, S. C. and **Fine, B. C. 1967** Effects on Resistance of Exposure of House Flies to Sub-lethal and Partially Lethal Doses of DDT. Indian J. Ent. 29(2):164-169.
Sublethal doses have no effect on mating ability of flies. Oogenesis was disturbed and there existed a slight reduction in reproductive capacity and egg hatchability, but this was only maintained for 7 generations. Doses causing mortality gave a high level of resistance. ABSTRACTS: RAE-B 57:48(172), Mar., 1969; BA 50:6435(67386), June 15, 1969.

Stalker, J. M., joint author. *See* Batth, S. S., et al., 1970.

Standen, H., joint author. *See* Bradbury, F. R., et al., 1960.

5083. Stansbury, R. E. and **Goodhue, L. D. 1960** New Fly Repellents in Dairy Sprays. Agric. Chem. 4 pp. (October issue).
Mixture of 1207 and MGK better than synergized pyrethrum against *Stomoxys* and *Siphona.* Presumably capable of repelling non-bloodsuckers as well.

Starkey, J. H. III, joint author. *See* Louloudes, S. J., et al., 1962.

Starkov, A. V., joint author. *See* Vashkov, V. I., et al., 1966, 1967; Misnik, Y. N., et al., 1967.

5084. Starkweather, R. J. and **Sullivan, W. N. 1964** Insect Tolerance to Increased Atmospheric Pressures. J. Econ. Ent. 57:766-767.
Used *M. domestica* and other species. Results indicate that pressure above atmospheric could be used in control of some insects.

Starkweather, R. J., joint author. *See* Yeomans, A. H., et al., 1964.

5085. Statens Skadedyrlaboratorium. Arsberetning Government Pest Infestation Laboratory (Stored Products and Household Pests). Lyngby, Denmark, 1948-1968.
Annual Reports of research on the control of flies and other pests in

Denmark. Published in both Danish and English texts. Treats of such topics as insecticide resistance in houseflies, the testing of new insecticides, laboratory methods and field control.

Stazio, M., joint author. *See* Lanzavecchia, G., et al., 1961.

Steele, J. A., joint author. *See* Thompson, M. J., et al., 1962, 1963.

5086. Steele, W. F. 1966 Oxidative Phosphorylation and Related Reactions in Particulate Fractions from Insects. Diss. Absts. 27(1B):89B.
Used *M. domestica* and *Phormia regina.*

5087. Stegwee, D. 1960 The Role of Esterase Inhibition in Tetraethylpyrophosphate Poisoning in the Housefly *Musca domestica.* Canad. J. Biochem. 38:1417-1430.
Organophosphate poisoning results in hyperactivity followed by paralysis. ABSTRACTS: RAE-B 51:232, Oct., 1963; TDB 58:534, April, 1961.

Stella, E., joint author. *See* Saccà, G., et al., 1964, 1966, 1968, 1969.

5088. Duplicate reference. Withdrawn.

5089. Stepanova, M. M. 1969 K metodike opredeleniya perspektionosti nekotarykh khemosterilizuyushchikh preparatov iz gruppy etileniminoproizvodnykh po otnosheniyu k otdel' nym vidam sinantropnykh mukh. [Technique for determining the prospects of using some chemosterilizing preparations from the group of ethylenimine derivatives against some species of synantrophic flies.] Med. Parazit. i Parazitarn. (Bolez.) 38(1):39-46. (English summary.)
Phosphamide a good sterilizer. Thiophosphamide more powerful, but repels insects. Neither is toxic to flies. Tested 5 species: *M. domestica, Muscina stabulans, Calliphora vicina, Phormia terraenovae* and *Lucilia sericata.* An accurate method of dosed feeding was developed. ABSTRACTS: TDB 66:634(1255), June, 1969; BA 51:9100(92627), Aug. 15, 1970.

Sternburg, J., joint author. *See* Brady, U. E., et al., 1966, 1967.

5090. Steve, P. C. and **Lilly, J. H. 1965** Investigations on Transmissability of *Moraxella bovis* by the Face Fly. J. Econ. Ent. 58:444-446.
Findings tend to incriminate *Musca autumnalis* as mechanical vector of bovine keratitis. ABSTRACTS: BA 47:370(4472), Jan. 1, 1966; RAE-B 53:178, Sept., 1965.

Steve, P. C., joint author. *See* Kaur, D., et al., 1969.

Stevenson, J. H., joint author. *See* Godin, P. J., et al., 1965; Elliott, M., et al., 1965.

5091. Stoffolano, J. G., Jr. 1967 The Synchronization of the Life Cycle of Diapausing Face Flies, *Musca autumnalis,* and of the Nematode *Heterotylenchus autumnalis.* J. Invertebr. Path. 9(3):395-397.
Nondiapausing flies contained immature nematodes; dispausing flies did not. ABSTRACTS: BA 49:2375(26712), Mar. 1, 1968; RAE-B 56:198(722), Oct., 1968.

5092. Stoffolano, J. G., Jr. 1968 Distribution of the Nematode *Heterotylenchus autumnalis,* a Parasite of the Face Fly, in New England with Notes on its Origin. J. Econ. Ent. 61(3):861-863.
Found from coast to coast and probably throughout the present range of the fly in North America. ABSTRACTS: BA 50:503(5180), Jan. 1, 1969; RAE-B 56:223(813), Nov., 1968.

5093. Stoffolano, J. G., Jr. 1968 The Effect of Diapause and Age on the Tarsal Acceptance Threshhold of the Fly, *Musca autumnalis.* J. Insect Physiol. 14(9):1205-1214
Proboscis extension used as index of acceptance. There was exhibited an elevation of the tarsal acceptance threshold for dispausing and gravid

adults. ABSTRACTS: BA 50:2678(27943), Mar. 1, 1969; RAE-B 57:270 (1024), Dec., 1969.

5094. **Stoffolano, J. G., Jr. 1969** Nematode parasites of the Face Fly and Onion Maggot in France and Denmark. J. Econ. Ent. 62(4):792-795.
Heterotylenchus autumnalis Nickle parasitizes face flies in France, but at very low rates. Findings of study support the hypothesis that *H. autumnalis* is of Palaearctic origin and probably was carried to North America by *M. autumnalis*. ABSTRACTS: RAE-B 58:25(92), Jan., 1970; BA 51:8538(87012), Aug. 1, 1970.

5095. **Stoffolano, J. G., Jr. 1970** Experimental Parasitization of *Musca domestica* L., *Orthellia caesarion* Meig., *Ravinia therminieri* R.-D., and *Musca autumnalis* DeGeer, by the Nematode *Heterotylenchus autumnalis* Nickle, with Special Reference to Host Reactions, Hemocytic Involvement, Anal Organ and the Hemocytopoietic Organ. Doctoral dissertation. Univ. of Connecticut. Very well illustrated. *See also*: Second Intern. Congr. Parasit. Proc. 56(4): 331 (Abstract).
Host larvae manifest protective reactions to nematode larvae, such as: encapsulation, melanization, parasite expulsion, tracheolar proliferation, cephalic caps, cuticular spotting (relates to healing of wound) and girdling.

5096. **Stoffolano, J. G., Jr. 1970** The Anal Organ of Larvae of *Musca autumnalis*, *M. domestica* and *Orthellia caesarion* (Diptera-Muscidae). Ann. Ent. Soc. Am. 63(6):1647-1654.
Structure made visible by immersion in acetocarmine or dilute silver nitrate. It surrounds the anus, and differs in size according to species. Believed important as a taxonomic character. ABSTRACT: 52:2738(26401), Mar. 1, 1971.

5097. **Stoffolano, J. G., Jr. 1970** Nematodes Associated with the Genus *Musca* (Diptera:Muscidae). Bull Ent. Soc. Am. 16(4):194-203.
A compilation of the literature with excellent discussion.

5097a. **Stoffolano, J. G., Jr. 1970** Parasitism of *Heterotylenchus autumnalis* Nickle (Nematoda:Sphaerulariidae) to the Face Fly, *Musca autumnalis* De Geer (Diptera:Muscidae). J. Nematology 2(4):324-329.
Principal damage to host is castration of the female. Nematode larva does not leave host before female fly is 11 days old. ABSTRACT: BA 52: 2613(25157), Mar. 1, 1971.

5098. **Stoffolano, J. G., Jr.** and **Matthysse, J. G. 1967** Influence of Photoperiod and Temperature on Diapause in the Face Fly, *Musca autumnalis* (Diptera:Muscidae). Ann. Ent. Soc. Am. 60:1242:1246.
Continuous light and high temperature (83°F) prevented diapause. Continuous dark at 65°F caused nearly all flies to diapause; they failed to feed or mate, became photonegative and sought hibernation sites. ABSTRACTS: BA 49:3370(37678), Apr. 1, 1968; RAE-B 56:117(406), June, 1968.

5099. **Stoffolano, J. G., Jr.** and **Nickle, W. R. 1966** Nematode Parasite (*Heterotylenchus* sp.) of Face Fly in New York State. J. Econ. Ent. 59:221-222.
Question as to whether parasite entered U.S. with the face fly, or whether it is an endemic species, adapting to a new host. ABSTRACTS: BA 47: 6829(79814), Aug. 15, 1966; BA 47:5944(69555), July 15, 1966.

Stoffolano, J. G., joint author. *See* Nappi, A. J., et al., 1970.

Stoicheva, R., joint author. *See* Sarbova, S., et al., 1968.

Stokes, J. B., joint author. *See* Setlepani, J. A., et al., 1970.

Stone, F. A., joint author. *See* Pimentel, D., et al., 1968.

5100. Stone, R. S. and **Brydon, H. W. 1965** The Effectiveness of Three Methods for the Control of Immature *Fannia* species in Poultry Manure. J. Med. Ent. (Honolulu) 2(2):145:149. *Also* in MS form with English (rather than metric) units.
The only satisfactory control (99·1 per cent) consisted of stirring manure, plus application of diazinon and gypsum dust. ABSTRACTS: BA 46:8516 (105203), Dec. 1, 1965; RAE-B 55:36(137), Feb., 1967.

Storozheva, E. M., joint author. *See* Sukhova, M. N., et al., 1965.

Stranger, W., joint author. *See* Burton, V. E., et al., 1965.

5101. Strickland, W. B., Linsdale, D. D. and **Doty, R. E. 1970** Hibernation of Face Flies in Buildings in Humboldt County, California. Calif. Vector Views. 17(8):79-83.
M. autumnalis has become a severe pest in buildings in this locality. Dichlorvos strips a satisfactory means of control in attics.

5102. Stringfellow, T. L. 1966 Meth. Bioassay of Residues of Insecticides Endrin, Diazinon, Carbaryl and Naled on Sorghum-M Grain. Diss. Absts. 26(12, pt. 1):6943).
M. domestica used in testing.

5103. Strother, G. K. 1966 Absorption of *Musca domestica* Screening Pigment. J. Gen. Physiol. 49(5):1087-1088.
Confirms Goldsmith's work. Red receptor absent.

Stuht, J. N., joint author. *See* Tarshis, I. B., et al., 1970.

5104. Sturckow, B. 1963 Electrophysiological Studies of a Single Taste Hair of the Fly During Stimulation by a Flowing System. *In* 16th Int. Congr. Zool., Proc. Int. Congr. Zool. 16(3):102-104. (Abstract only).

5105. Styczyńska, B. and **Krzemińsak, A. 1966** Wybór pozywki dla larw muchy domowej *M. domestica* L. w hodowli laboratoryjnej. [The choice of medium for the laboratory cultivation of larvae of the domestic house-fly (*Musca domestica* L.).] Przegl. Epidem. 20(3): 325-330. (Russian and English summaries.)
Milk alone was inadequate for the development of larvae. Supplementation with other nutrients indicated. Addition of cholesterol and Biovitana had beneficial effect on development of larvae, their metamorphosis and fertility of the females. ABSTRACT: TDB 64:1272, Nov., 1967; RAE-B 58:97(383), Apr., 1970.

5106. Styczyńska, B. and **Krzemińska, A. 1967** Zniekształcenie bobowek *Musca domestica* L. pod wplywem srodkow fosforo-organicznych. [Deformation of puparia results when house fly (*Musca domestica* L.) larvae are subjected to treatment with organo-phosphorus compounds.] Wiad. Parazytol. 13:413-419. (Russian and English summaries.)
Three-day-old larvae treated with malathion, dichlorvos, and trichlorfon. Pupation was delayed 24 hours (compared to controls). 46·4 per cent of 807 puparia showed morphological deformations. Deformed pupae weighed considerably less than normal ones.

5107. Styczyńska, B., Krzemińska, A. and **Goszczyńska, K. 1969** Wplyw niektórych insektycydów fosforoorganicznych na plodnosc muchy domowej *Musca domestica* L. [Effects of some organo-phosphorus insecticides on fertility of *Musca domestica* L.] Roczn. Pánstw. Zakl. Hig. 20:249-253. (Russian and English summaries.)
Doses of 0·40 μg of "trichlorfos" and 0·11 μg of "fenchlorfos" per fly, applied topically, caused reduction of fertility in treated flies. Findings reported as mean no. of eggs produced per female flv.

5108. Styczyńska, B., et al. 1967 [Deformation of puparia of the housefly *Musca domestica* L. by the effect of organophorus insecticides.] Roczn. Pánstw. Zakl. Hig. 18:337-343. (In Polish.)

Styczyńska, B., joint author. *See* Bojanowska, A., et al., 1960; Goszeznska, K., et al., 1967, 1968; Krysan, J. L., et al., 1970.

Su, T. K., joint author. *See* Chang, C. P., et al., 1965.

5109. Sucker, H. 1965 [Acetylcholine. IX. Acetokinase and phosphotransacetylase of the housefly.] Liebig Ann. Chem. 683:225-238. (In German.)

Sucker, H., joint author. *See* Dann, O., et al., 1968.

5110. Suenaga, O. 1962 [Ecological studies of flies. 6. Peculiar breeding sources of housefly in some fishing villages.] Endemic Dis. Bull. Nagasaki Univ. 4(1):60-63. (In Japanese, with English summary.)
In fishing villages piles of fish scales removed in the washing and boiling of sardines, and sandy soil soaked in stock containing waste from fish manure preparation, were major breeding areas for flies. ABSTRACT: BA 45:1759(21941), Mar. 1, 1964.

5111. Suenaga, Osamu. 1962 [On the flies of medical importance in the Ryukyus.] Endemic Dis. Bull. Nagasaki Univ. 4(3):206-208.
Flies collected from fish baited traps. Predominant species were *M. domestica vicina* Macq. and *Chrysomyia megacephala*. Also recorded are *M. sorbens* and *M. ventrosa*. A first time recording of *Lucilia sericata*. ABSTRACTS: RAE-B 52:237, Dec., 1964; BA 44:1195(16182), Nov. 15, 1963.

5112. Suenaga, O. 1963 [Ecological studies of flies. 8. The diurnal activities of flies attracted to the fish baited trap.] Endemic Dis. Bull. Nagasaki Univ. 5(2):136-144. (In Japanese, with English summary.)

5113. Suenaga, O. 1969 [Age-grouping method by ovariole changes following oviposition in females of *Musca domestica vicina*, and its application to field populations.] Trop. Med. 11(2):76-90. (In Japanese, with English summary.)
Table of diagrammatic calendar ages of female houseflies devised. Attempt was made to evaluate the effect of an insecticide by using the calendar for determining age distribution of female wild populations before and after insecticide application. ABSTRACT: TDB 67:461(919) Apr., 1970.

5114. Suenaga, O. and **Fukuda, M. 1963** [Ecological studies of flies. 7. The species and season prevalence of flies breeding out from a privy and a urinary pit in a farm village.] Endemic Dis. Bull. Nagasaki Univ. 5:72-80. (In Japanese, with English summary.)
Survey revealed: length of time of fly breeding, the most prevalent times and overwintering habits. *M. domestica vicina* was not found to breed in privies. *Fannia scalaris* F. may do so. ABSTRACTS: BA 46:1058(13352), Feb. 1, 1965; RAE-B 53:107, June, 1965; TDB 60:993, Oct., 1963.

5115. Suenaga, O., Ori, S., Shimogama, M. and **Takatsuki, Y. 1962** [On the flies breeding out from steepers containing spoiled rice-bran, rotten vegetable, or wastes of melon in pickle shops at Nagasaki City.] Endemic Dis. Bull. Nagasaki Univ. 4(1):57-59. (In Japanese, with English summary.)
The flies found utilizing the pickle wastes as breeding sites were: *Fannia canicularis*, *Muscina stabulans*, *Musca domestica vicina* and a few others. ABSTRACT: BA 45:1760(21942), Mar. 1, 1964.

5116. Suenaga, O., Shimogama, M., Kawai, S., Fukuda, M. and **Tanikawa, T. 1964** [Ecological studies of flies. 9. Some notes on the flies trapped by fish baited traps in different places in middle and southwestern Japan. (Diptera).] Endemic Dis. Bull. Nagasaki Univ. 6(1):34.

Suenaga, O., joint author. *See* Kawai, S., et al., 1960; Omori, N., et al., 1962.

Suetake, Y., joint author. *See* Inaba, T., et al., 1963.

Sugawara, R., joint author. *See* Muto, T., et al., 1968.

Sugii, Y., joint author. *See* Katsuda, Y., et al., 1967.

Sugiyama, C., joint author. *See* Tsuruoka, Y., et al., 1963.

5117. Sukhova, M. N. 1962 [Control of flies and its organization in Uzbekistan.] Med. Zhur. Uzbekistana 3:3-14. (In Russian.)

5118. Sukhova, M. N. 1962 [On methods for the application of insecticides for the eradication of synanthropic flies.] Zhur. Mikrobiol. 33: 15-19. (In Russian.)

5119. Sukhova, M. N. 1967 O klassifikatsii osnovnykh vidov semeistv Muscidae, Calliphoridae i Sarcophagidae v zavisimosti ot stepeni ikh sinantropizma i sviazei s domashnimi zhivotnymi. [On the classification of main fly species of the families Muscidae, Calliphoridae and Sarcophagidae, depending on their synanthropism and connections with domestic animals.] Wiad. Parazytol. 13(4/5):595-601. (Polish summary.)

The fly species are distinguished according to food requirements and hydrothermic properties of their breeding places. *Musca domestica* and *M. sorbens* are included with the obligatory synanthropic flies. ABSTRACT: BA 50:6965(72750), July 1, 1969.

5120. Sukhova, M. N., Grozdeva, I. V., Misnik, Yu. N., Teterovskaya, T. O., Bolotova, T. A., Kholodova, G. K., Storozheva, E. M., Samsonova, A. M., Musonov, V. B., Neselovskaya, V. K., Gol'dina, G. S., Serafimova, A. M., Biralo, T. I. and **Vasilenko, L. N. 1965** Chuvsvitel'nost'k khlorofosu, trikhlormetafosu-3, DDT, GkhtsG, polikhlorpinenu u populyatsii komnatnykh mukh posle mnogoletnego promeneniya etikh insektisidov. [Sensitivity of housefly (*Musca domestica*) populations to Dipterex (0, 0-Dimethy l-2, 2, 2-trichloro-1-hydroxy-ethylphosphonate), trichlorometaphos-3, DDT, BHC hexachlorocyclohexane, and polychloropinene after these insecticides have been used for many years.] Zhur. Microbiol., Epidemiol., Immunobiol. 42(8)-71-74.

In the U.S.S.R. the development of tolerance did not occur in some areas, but resistance to BHC and DDT did increase, due to the use of polychloropinene in those areas. A moderate tolerance of DDT was noted. ABSTRACT: BA 47:5977(69934), July 15, 1966.

Sukhova, M. N., joint author. *See* Misnik, Y. N., et al., 1967.

5121. Sullivan, R. L. 1961 Abnormal Sex Ratios in the House Fly. Abstract; Proc. N. Centr. Branch Ent. Soc. Am. 16:20.

Several field collections showed better than 60 per cent males. High male ratio is inherited; can be raised up to 100 per cent by selection. Is carried by male, as shown by reciprocal crossing. High male strains appear to compete successfully with normal flies in nature.

5122. **Sullivan, R. L. 1961** Linkage and Sex Limitation of Several Loci in the House Fly. J. Heredity 52(6):282-286.
Crossover tests, female carrying the markers in repulsion phase, showed brown body (*bwb*) to be 20·4 crossover units from white eye (*w*), 28·2 from green eye (*ge*). Between *ge* and *w*, 39·1 units. Crossover in males 0·1 per cent.

5123. **Sullivan, R. L.** and **Hiroyoshi, T. 1960** A Preliminary Report on Mutations in the House Fly [*Musca domestica* L.]. J. Econ. Ent. 53(2):213-215.
Eleven mutants studied. Behave as simple recessives. ABSTRACT: RAE-B 49:108, May, 1961.

5124. **Sullivan, R. L.** and **Sokal, R. R. 1963** The Effects of Larval Density on Several Strains of the Housefly. Ecology 44:120-130.
A density of 320 eggs per 36g of CSMA medium caused reduction in dry weights of adults. At density 1280, adult emergence decreased in all strains. ABSTRACT: BA 44:684(9234), Nov. 1, 1963.

5125. **Sullivan, R. L.** and **Sokal, R. R. 1965** Further Experiments on Competition Between Strains of House Flies. Ecology 46(1/2):172-182.
Relative fitness of wild phenotype over *bwb* mutant increases markedly with density. ABSTRACT: BA 46:4092(50602), June 15, 1965.

Sullivan, R. L., joint author *See* Sokal, R. R., et al., 1963.

5126. **Sullivan, W. N., Cawley, B., Hayes, D. K., Rosenthal, J.** and **Halberg, F. 1970** Circadian Rhythm in Susceptibility of House Flies and Madeira Cockroaches to Pyrethrum. J. Econ. Ent. 63(1):159-163.
The crest of susceptibility in both species occurred during the last quarter of the daily light span, about midafternoon. ABSTRACTS: BA 51:9098(92606), Aug. 15, 1970; RAE-B 58:230(944), July, 1970.

5127. **Sullivan, W. N.** and **McCauley, T. R. 1960** Effect of Acceleration Force on Insect Mortality. J. Econ. Ent. 53(4):691-692.
Insects tested (including *M. domestica*) can tolerate any acceleration forces involved in escaping or returning to heavenly bodies.

5128. **Sullivan, W. N.** and **Smith, C. N. 1960** Exposure of House Flies and Oriental Rat Fleas on a High-Altitude Balloon Flight. J. Econ. Ent. 53(2):247-248.
Sixteen hour exposure to altitude of 78,000-82,000 ft. failed to affect reproduction or physical structure in succeeding generations. ABSTRACTS: BA 35:3236(36805), July 15, 1960; RAE-B 49:102, May, 1961.

5129. **Sullivan, W. N., Yeomans, A. H.** and **Schechter, M. S. 1960** The Effectiveness of Liquified-gas-propelled Concentrated Allethrin Aerosols and Air Atomized Dibrom Aerosols Against Normal and Resistant House Flies. J. Econ. Ent. 53(5):956.
Allethrin and Dibrom are very effective against normal flies but Dibrom is superior against resistant flies. Usefulness in field areas obvious, provided toxicological clearance is obtained.

Sullivan, W. N., joint author *See* McGuire, J. U., Jr., et al., 1966; Knipling, G. D., et al., 1961; Fulton, R. A., et al., 1963; Thornton, B. C., et al., 1964; Schechter, M. S., et al., 1960; Starkweather, R. J., et al., 1964; Yeomans, A. H., et al., 1964.

Sun, J.-Y. Tung, joint author. *See* Sun, Y.-P., et al., 1963.

Sun. Y.-J., joint author. *See* Chang, C. P., et al., 1965.

5130. **Sun, Y.-P. 1963** Bioassay-Insects. *In* Analytical Methods for Pesticides and Food Additives. 1(15):399-423.
M. domestica among the insect species used for assay purposes. *Droso-*

phila sometimes preferred. Houseflies are more resistant to CO_2 anesthesia, can stand prolonged starvation, and are less sensitive to minor contamination than *Drosophila*.

5131. Sun, Y.-P. 1968 Dynamics of Insect Toxicology—a Mathematical and Graphical Evaluation of the Relationship Between Insect Toxicity and Rates of Penetration and Detoxication of Insecticides. J. Econ. Ent. 61:949-955.
Relationship between toxicity and rates of penetration (and detoxication) expressed mathematically and graphically. It is conceivable that a make-up of resistant factors (one reducing penetration and one increasing detoxication) could render an insect immune to an insecticide. ABSTRACTS: RAE-B 57:3(5), Jan., 1969; TDB 65:1484(3294), Dec., 1968; BA 50: 2650(27602), Mar. 1, 1969.

5132. Duplicate reference. Withdrawn.

5133. Sun, Y.-P. and **Johnson, E. R. 1960** Analysis of Joint Action of Insecticides Against House Flies. J. Econ. Ent. 53(5):887-892.
Toxicity indices of components and their mixture are determined by dosage-mortality curves. ABSTRACTS: BA 36:759(9017), Feb. 1, 1961; RAE-B 49:184, Aug., 1961.

5134. Sun, Y.-P. and **Johnson, E. R. 1960** Synergistic and Antagonistic Actions of Insecticide-synergist Combinations and Their Mode of Action. (Symposium on the Mechanism of Action of Pesticide Chemicals). J. Agric. Fd. Chem. 8(4):261-266.
Toxicity results, colorimetric and enzymic analyses indicate that synergistic or antagonistic action by pyrethrin synergists may be due to inhibition of certain biological oxidations either activating or detoxifying the compounds. ABSTRACT: RAE-B 50:35-36, Feb., 1962.

5135. Sun, Y.-P. and **Johnson, E. R. 1965** Integration of Physico-Chemical and Biological Techniques in Specific Bioassay with Special Reference to Bidrin Insecticide. J. Econ. Ent. 58(5):838-844.
Gives table indicating relative toxicity to houseflies of 29 organophosphorus insecticides; 3 carbamate insecticides; 17 chlorinated insecticides with or without synergist. Also their interference on bioassay of Bidrin (with or without synergist).

5136. Sun, Y.-P. and **Johnson, E. R. 1969** Relationship Between Structure of Several Azodrin^R Insecticide Homologues and Their Toxicities to House Flies, Tested by Injection, Infusion, Topical Application and Spray Methods with and without Synergist. J. Econ. Ent. 62(5):1130-1135.
Toxicities obtained generally decreased from low to high homologues. By injecting toxicants into flies pretreated with sesamex, the potential toxicity of compounds can be more fully indicated. ABSTRACTS: TDB 67:366(769), Mar., 1970; BA 51:9006(92584), Aug. 15, 1970.

5137. Sun, Y.-P., Johnson, E. R., Pankaskie, J. E., Earle, N. W. and **Sun, J.-Y. 1963** Factors Affecting Residue-Film Bioassay of Insecticide Residues. J. Ass. Offic. Agric. Chem. 46(3):530-542.
More accurate results may be obtained if (a) all test jars contain equal amounts of the same extractives; (b) a corresponding standard curve is prepared from an uncontaminated check for each series of tests; (c) insects used for each series of tests are properly sampled from a mixture of several cultures.

5138. Sun, Y.-P., Johnson, E. R. and **Ward, F. L., Jr. 1967** Evaluation of Synergistic Mixtures Containing Sesamex and Organophosphorus

or Chlorinated Insecticides Tested against House Flies. J. Econ. Ent. 60:828-835.
Sesamex increased activity of low toxic compounds more than high ones. ABSTRACTS: BA 48:11214(124554), Dec. 15, 1967; RAE-B 55:191(658), Oct., 1967.

1539. Sun, Y.-P. and **Sanjean, J. 1961** Specificity of Bioassay of Insecticide Residues with Special Reference to Phosdrin. J. Econ. Ent. 54(5): 841-846.
A relatively specific bioassay method developed for assaying Phosdrin residues in presence of any one or a mixture of 2-4 insecticides.

5140. Sun, Y.-P., Schaefer, C. H. and **Johnson, E. R. 1967** Effects of Application Methods on the Toxicity and Distribution of Dieldrin in House Flies. J. Econ. Ent. 60(4):1033-1037.
Dieldrin, *infused* into house flies is much less toxic than when *injected* into them. ABSTRACTS: BA 49:1426(15995), Feb. 1, 1968; RAE-B 56:6 (17), Jan., 1968; TDB 65:539(1176), Apr., 1968.

Sun, Y.-P., joint author. *See* Hall, W. E , et al , 1965; Schaefer, C. H., et al., 1967; Sanjean, J., et al., 1961.

Sutcher, V., joint author. *See* Eldefrawi, M. E., et al., 1960.

Sutherland, D. J., joint author. *See* Khan, M. A. Q., et al., 1969, 1970; Rosen J. D. et al., 1969.

Suzecki, A., joint author. *See* Takahashi, N., et al., 1967.

Suzuki, I., joint author. *See* Takagi, U., et al., 1965.

5141. Suzuki, K. 1962 Polyphenolic Derivatives in Rhabdomeres of the Compound Eye (*Musca domestica*). Nature 196(4858): 994.
Regarded as proof of the presence of phenolic derivatives. Worked with *Musca vicina* Macq. Rhabdomeres proved positive for Millon's reaction, potassium iodate, potassium bichromate; and ammoniacal silver; negative for x-amino radical (tested by two methods). ABSTRACT: BA 43:362 (3953), July 1, 1963.

Suzuki, R., joint author. *See* Tsukamoto, M., et al., 1964, 1966.

5142. Suzuki, T. 1960 A Method for Estimating the Effect of Insecticides in a Field Test. Jap. J. Exptl. Med. 30:67-74.
Method based on rapidity of knockdown effect of the houseflies exposed to the insecticide residues. Easier to conduct than conventional methods and is able to detect a minute amount of residues just available for the insects. ABSTRACT: RAE-B 51:157, Aug., 1963.

5143. Suzuki, T. 1963 Insecticide Resistance in Flies, Mosquitoes and Cockroaches in Japan Evaluated by Topical Application Tests, with Special Reference to the Susceptibility Levels of the Insects. Jap. J. Exptl. Med. 33(1):69-83.
Resistance of 9 species of insects to 7 chlorinated hydrocarbon and organophosphorus insecticides evaluated by topical application method. Susceptibility levels also reported by summarizing LD-50's as determined by author and other domestic and foreign workers. Eleven colonies of *M. domestica vicina* used. ABSTRACTS: BA 44:1580, Dec. 1, 1963; TDB 61:322, Mar., 1964.

5144. Suzuki, T., Hirakoso, S. and **Matsunaga, H. 1961** Diazinon Resistance in House Flies of Japan. Jap. J. Exptl. Med. 31(5):351-364.
In the fall of 1960, a failure of house fly control by diazinon spray occurred in Japan where diazinon had been applied for 1-3 years. Flies were collected, reared in laboratories and resistance to insecticides examined. Resistance to diazinon confirmed. ABSTRACTS: RAE-B 51:158, Aug., 1963; TDB 59:942, Sept., 1962.

5145. Suzuki, T. and **Kano, R. 1962** Susceptibility or Resistance to Insecticides in Lesser Housefly, Sarcophagid Fly and Two Species of Blowfly in Japan. Jap. J. Exptl. Med. 32:309-313.
Bibai colony of *Fannia caniculuris* resistant to DDT; partly so to lindane. ABSTRACTS: BA 41:1639(20619), Mar. 1, 1963; TDB 60:275, Mar., 1963.

5146. Suzuki, T., Seki, M. and **Mibutani, K. 1968** Studies on Aerosol-type Insecticides. I. Comparison of the Effectiveness Evaluated by Various Test Methods. Contributions to the Study of the Laboratory Test Methods of Insecticides. Sanitary Insecticide Assoc. Japan. 1968(3):53-56.
The test method using a Peet-Grady chamber seemed to be preferable for the evaluation of the effect of aerosol-type insecticides to the housefly.

Suzuki, T., joint author. *See* Mizutani, K., et al., 1968; Seki, M., et al., 1968; Ogata, K., et al., 1960, 1965, 1966; Shimada, A., et al., 1969.

Svensson, S. A., joint author. *See* Gip, L., et al., 1967, 1968.

Swanson, M. H., joint author. *See* Anderson, J. R., et al., 1968.

Sweeley, C. C., joint author. *See* Bieber, L. L., et al., 1969.

Swiech, J., joint author. *See* Gorecki, K., et al., 1966.

5147. Swift, H. 1962 Nucleic Acids and Cell Morphology in Dipteran Salivary Glands. *In* The Molecular Control of Cellular Activity. A Symposium. J. M. Allen, Editor. McGraw-Hill Book Co., New York. Illustrated. pp. 72-125.

Swink, E. T., joint author. *See* Kranzler, G. A., et al., 1966.

5148. Sychevskaya, V. I. 1960 O fenologii sinantropnykh mukh Uzbekistana. [On the phenology of synanthropic flies in Uzbekistan.] Med. Parazit. (Mosk) 29:66-72. (In Russian, with English summary.)
Fly control measures should be based on meterological conditions of current and preceding years since these determine phenology of flies. ABSTRACT: BA 36:7262(78317), Nov. 15, 1961.

5149. Sychevskaya, V. I. 1960 [Methodological and practical aspects of phenologic observations of synanthropic flies (according to observations in Uzbekistan).] Med. Parazit. (Moskva) 29:712-720. (In Russian, with English summary.)
For the 20-25 species of synanthropic flies, it is suggested that methods be devised for observation of mature flies, their larvae and pupae and that control methods be based on more phenologically grounded dates.

5150. Sychevskaya, V. I. 1960 [On the morphology and biology of synanthropic species of the genus *Fannia* R. D. (Diptera:Muscidae).] Ent. Obozrenie 39(2):225-232. (In Russian, with English summary.) *Also in* Ent. Rev. 39(2):349-360, 1960.
Observations on three species, *Fannia canicularis* L., *F. scalaris* F. and *F. leucosticta* Mg. were made with regard to egg laying and genital structures. ABSTRACT: BA 36:2357(28836), May 1, 1960.

5151. Sychevskaya, V. I. 1962 [Seasonal changes in the sensitivity to DDT of some species of synanthropic flies in Samarkand.] Zool. Zhur. 41(10):1509-1515. (In Russian, with English summary.)
Longevity for 5 species studied, changed during a season in the form of a 2-peak curve, peaks occurring in spring and autumn. This depends on the physiological condition of the flies, also on the air temperature at which poisoned insects are kept. ABSTRACTS: BA 41:2009(24917), Mar. 15, 1963; RAE-B 52:105, June, 1964.

5152. Sychevskaya, V. I. 1962 [On changes in the daily dynamics of the specific composition of flies associated with man in the course of

the season.] Rev. Ent. URSS 41(3):545-553. (In Russian, with English summary.)
The fauna of synanthropic flies inhabiting a definite biotope undergoes seasonal and daily changes in relation to shifting of temperature ranges of their activity, within a season and in 24 hours. ABSTRACT: RAE-B 52:194, Oct., 1964.

5153. Sychevskaya, V. I. 1962 Pereponchatokrylye vyvedennye iz pupariev sinantropnykh mukh v Srednei Azii. [Hymenoptera hatched from puparia of synanthropic flies in Central Asia.] *In* Voprosy ekologii. Kiev. Univ., Kiev. 8:116. Referat. Zhur., Biol. 1963, No. 2E81 (Translation).
Data is given on biology (mainly parasitism), of three species of the thirteen parasitic Hymenoptera hatched. ABSTRACT: BA 43:1945(25147), Sept. 15, 1963.

5154. Sychevskaya, V. I. 1963 Nekotorye Chalcidoidae, vyvedennye iz pupariev sinantropnykh mukh v Uzbekistane. [Some Chalcidoidea reared from puparia of synanthropic flies in Uzbekistan.] Zool. Zhur. 42(6):858-864. (English summary.)
Seven chalcid species were reared from the puparia of eleven species of synanthropic flies collected. *Spalangia nigra* Latr., and *S. stomoxysiae* Girault are the most abundant and effective parasitoids. Are capable of decreasing the fly population of some species by 82-95 per cent. ABSTRACT: BA 45:2107(26358), Mar. 15, 1964.

5155. Sychevskaia, V. I. 1964 O Kleshchokh, obnaruz hennykh na synantropnykh mukhakh v Uzbekistane. Sychevskaia, V. I. 1964. [On ticks found on synanthropic flies in Uzbekistan.] Med. Parazit. i Parazitarn. (Bolezni) (Moskva) 33(5):557-560. (English summary.)
Species of Acarina found on 23 species of synantropic flies include four species of Macrochelidae, one species of Parasitidae and four species of gamasoid ticks. ABSTRACT: BA 46:8142(100625), Nov. 15, 1965.

5156. Sychevskaya, V. I. 1964 [Hymenopterous parasites of flies associated with man in Central Asia.] Ent. Obozrenie 43(2):391-404. (In Russian with English summary.) Engl. Transl. in Ent. Rev. 43(2): 199-206.
Over 18 species of parasitic Hymenoptera were reared from larvae and puparia of flies associated with man. ABSTRACT: RAE-B 54:239, Dec., 1966.

5157. Sychevskaya, V. I. 1965 Biology and Ecology of *Calliphora vicina* R.D. in Central Asia. Zool. J. (Zhur.) 44(4):552-560.
Increasing temperatures inhibit the ovaries of *Calliphora vicina*; egg laying and population peaks occur after late August.

5158. Sychevskaya, V. I. 1966 [On synantrophic flies of Pamir.] Zool. Zhur. 45(3):390-399.
Not only phenology but breeding sites change in alpian and subalpian belts. Darker colors of higher-dwelling flies, as compared to the same species in lower areas, aid in their rapid warming up and also protects them from ultra-violet rays.

5159. Sychevskaya, V. I. and **Shaidurov, V. S. 1965** O temperature tela nekotorykh sinantropnykh mukh na Vostochmom Pamire. [Body temperature in some synanthropic flies in the eastern Pamirs.] Zool. Zhur. 44(5):779-783.
Maximal fly activity coincides with hours of maximal radiation as observed with *Phormia terrae-novae*, R. D. and *Calliphora uralensis* Vill. ABSTRACT: BA 47:4688(54922), June 1, 1966.

5160. Syrowatka, T.1965 Wplyw temperatury na zuzycie thenu przez muchy (*Musca domestica* L.) poddane dzialaniu γ-HCH. [Effect of temperature on the oxygen consumption of flies (*Musca domestica* L.) exposed to γ-HCH.] Wiad. Parazytol. 11:185-190. (Russian and English summaries.)

As with control insects, oxygen consumption of insects poisoned by γ-HCH is not linear, but exhibits a characteristic peak. ABSTRACT: BA 49:1426(15996), Feb. 1, 1968.

5161. Syrowatka, T. 1967 Wplyw szésciochlorocykloheksanu na aktywnośc oksydazy cytochromowej u much (*Musca domestica* L.). [Effect of hexachlorocyclohexane on the activity of cytochrome oxydase in flies (*Musca domestica* L.).] Wiad. Parazytol. 13:247-256. (Russian and English summaries.)

Studied the rate at which cytochrome oxidase disappears at various temperatures and the influence gamma-HCH has on this. Inhibition occured at certain concentrations. (Temperature was of little importance.) ABSTRACT: BA 50:1590(16800), Feb. 1, 1969.

5162. Syrowatka, T. 1967 Wplyn zatrucia 0, 0-dwumetylofosforanem 0-(2, 2-dwuchlorowinylu) (dichlorfos) na zawartosc wolnych nukleotydow adeninowych w tkankach much *Musca domestica* L. [Effect of poisoning with 0, 0-dimethyl-0-(2, 2-dichloro vinyl) phosphate (dichlorfos) on content of free adenine nucleotides in tissues of the fly, *Musca domestica* L.] Roczn. Pánstw. Zakl. Hig. 18:463-469. (Russian and English summaries.)

The content of ATP significantly increased in *Musca domestica* poisoned with dichlorfos, as compared with normal flies.

5163. Syrowatka, T. 1968 Wplyw zatruca dichlorfosem na zawartosc koenzyman flawinowych w tkankach muchy, *Musca domestica* L. [The effect of dichlorphos poisoning on the content of flavin coenzymes in the tissues of *Musca domestica* L.]. Wiad. Parazytol. 14:275-281. (Russian and English summaries.)

Dichlorphos poisoning exerts no influence on the FAD level or the total content of flavins. ABSTRACT: BA 50:8657(89925), Aug. 15, 1969.

5164. Szabo, K. and **Menn, J. J. 1969** Synthesis and Biological Properties of Insecticidal N-(Mercaptomethyl) Phthalimide-S-(0-alkyl)- Alkylphosphonodithioates and Thiolates. Agric. Food Chem. 17(4): 863-868.

Examined for anticholinesterase activity, toxicity to rats, and insecticidal and miticidal properties, the isobutyl ester is the most preferred member of the series.

Szabo, K., joint author. *See* Menn, J. J., et al., 1965.

5165. Sztankay-Gulyas, M. and **Zoltai, N. 1962** Investigation of the DDT Resistance of House-flies and the Newer Methods of Control in Hungary. J. Hyg. Epidem. (Praha) 6:296-302. (French, German and Spanish summaries.)

Resistance to DDT existed since 1960. Diazinon was recommended as effective without danger to the inhabitants. ABSTRACTS: BA 43:1016 (12210), Aug. 1, 1963; TDB 60:502, May, 1963.

5166. Sztankay-Gulyas, M. and **Barsy, G. 1962** Routinemethoden zur Wertbestimmung von Kontaktinsektiziden. [Routine methods for the evaluation of contact insecticides.] J. Hyg. Epidemiol. Microbiol.,

Immunol. 6:322-327.
The Petri dish method is the most simple. Extensive examination involves timed exposures and use of probit analysis.

Tabor, L. A., joint author. *See* Kaplanis, J. N., et al., 1960, 1966; Monroe, R. E., et al., 1963.

5167. Tacal, J. V., Jr. and **Menez, C. F. 1967** Salmonella Studies in the Philippines: VII. The Isolation of *Salmonella derby* from Abattoir Flies and Chicken Ascarids. Philipp. J. Vet. Med. 6(1/2):106-111.
M. domestica among flies examined for *salmonella.* ABSTRACT: BA 50: 5764(60346), June 1, 1969.

5168. Taha, A. M. 1963 Notes on House Fly Dispersion in Egypt. J. Egyptian Pub. Health Ass. 38:143-151.
M. domestica vicina trapped, marked and released. Results show flies cannot be prevented from returning to dwellings merely by dumping refuse at a distance. Flies appear to follow vehicles along connecting roads. ABSTRACT: TDB 60:1168, Dec., 1963.

5169. Tahori, A. S. 1960 The Larvicidal Effect of Phenylthiourea Against Resistant and Susceptible Housefly Strains. Bull. W. H. O. 22: 584-585.
Tested 5 strains of *M. domestica.* Only one (malathion-resistant) gave larval kill at all concentrations tried. Doses required for larval kill too high to be good larvicide. PTU does depress growth. ABSTRACTS: RAE-B 50:81, Apr., 1962; TDB 57:1319-1320, Dec., 1960.

5170. Tahori, A. S. 1961 A Method for Selection for DDT-susceptibility. J. Econ. Ent. 54:611.
Selection of *M. domestica* for greater susceptibility to DDT by technique based on the negative temperature coefficient of the compound. ABSTRACT: RAE-B 50:132, June, 1962.

5171. Tahori, A. S. 1963 Selection for a Fluoroacetate Resistant Strain of House Flies and Investigation of Its Resistance Pattern. J. Econ. Ent. 56(1):67-69.
After 25 generations a 7-fold resistance developed; lasted 8 generations after selection with fluoroacetate was discontinued. Increased resistance to starvation and lack of water also developed. ABSTRACTS: BA 42:1901 (24004), June 15, 1963; RAE-B 51:91, May, 1963; TDB 60:903, Sept., 1963.

5172. Tahori, A. S. 1966 Resistance Pattern of Fluoracetate-resistant Fly Strain. Israel J. Ent. 1:179-182.
Reports results of selection of a susceptible strain with fluoroacetate for 32 additional generations. ABSTRACTS: RAE-B 54:216, Nov., 1966; BA 50:6435(67389), June 15, 1969.

5173. Tahori, A. S. 1966 Changes in the Resistance Pattern of a Fluoroacetate-resistant Fly Strain. J. Econ. Ent. 59(2):462-464.
A change in the resistance pattern to other insecticides was investigated along with the increase in resistance to fluoroacetate. ABSTRACTS: BA 47:7286(85017), Sept. 1, 1966; RAE-B 54:133, July, 1966; TDB 63:1254, Nov., 1966.

5174. Tahori, A. S., Zeidler, G. and **Halevy, A. H. 1965** The Effect of Phosphon (2, 4-dichlorobenzyltributyl phosphonium chloride) as a Housefly Sterilant. Naturwissenschaften. 52(13):400.
No effect on female houseflies by feeding or topical application.

Tahori, A. S., joint author. *See* Zahavi, M., et al., 1969.

Takada, R., joint author. *See* Shimizu, F., et al., 1965.

5175. Takagi, U., Suzuki, I., Ishida, M., Hatanaka, M. and **Komatsuzaki, I. 1965** [Experiments on the residual effect of pesticide-fog. I. Effects of 0.3 per cent DDVP oil solution fog against housefly.] Jap. J. Sanit. Zool. 16(4):303-306. (In Japanese, with English summary.)
Time for 50 per cent knockdown about equal on painted and plain veneer surfaces, and progressively greater on fibreboard, textured board, etc. Method of application also important. ABSTRACT: RAE-B 56:134(468), July, 1968.

Takahara, H., joint author. *See* Ohno, M., et al., 1960.

5176. Takahashi, F. and **Pimentel, D. 1966** pH Changes in Various Housefly Media. Canad. Ent. 98(6):633-635.
As larvae grew and agitated the medium, pH increased. ABSTRACT: RAE-B 55:221(764), Dec., 1967.

5177. Takahashi, F. and **Pimentel, D. 1967** Wasp Preference for Black-, Brown-, and Hybrid-type Pupae of the House Fly. Ann. Ent. Soc. Am. 60(3):623-625.
Parasite wasp, *Nasonia vitripennis*, preferred the black pupae of a strain of *M. domestica*; in a controlled population flies producing black pupae rapidly declined. ABSTRACTS: BA 48:8332(93625), Sept. 15, 1967; RAE-B 55:220(761), Dec., 1967.

5178. Takahashi, N., Suzuki, A. and **Tamura, S. 1967** The New Natural Insecticides, Piericidin-A Insecticide and Piericidin-B Insecticide. Abstr. Int. Pflanzenschutz-Kong. [Int. Congr. Plant Prot.] 30:228-229.

Takahashi, N., joint author. *See* Mitsui, T., et al., 1969.

Takatsuki, Y., joint author. *See* Ori, S., et al., 1960; Suenaga, O., et al., 1962.

Takei, G. H., joint author. *See* Sherman, M., et al., 1968.

5179. Takei, M. 1961 [Observation on houseflies packed in canned food.] Bull. Tokai Reg. Fish. Res. Lab. 30:111-116. (In Japanese, with English summary.)
In an attempt to secure a criterion to distinguish houseflies found in canned food from others it was shown that black spots developed on the wings of houseflies recovered from canned food. Most fly characteristics were lost. ABSTRACT: BA 36:7686(82604), Dec. 1, 1961.

5180. Takei, M. 1962 [Study of the protection from harm by fly on marine products. III. Smell attracting fly (2).] Bull. Tokai Reg. Fish Res. Lab. 35:15-22. (In Japanese, with English summary.)
The relation between smell from fish meat and flies attracted to it was studied. The acid fraction of ethanol extract attracted the most flies, the basic fraction of ethanol extract, solution and residium of water extract being also effective. Combination of the extracts was more effective than single components. ABSTRACT: BA 44:1580(21284), Dec. 1, 1963.

5181. Takei, M. 1966 [Study on the protection from harm by fly on marine products. VII. Fly repellent odours setting with meat (2).] Bull. Tokai Reg. Fish Res. Lab. 46:61-68. (In Japanese, with English summary.)
Among perfumes and smoke flavors tested for repellent effect on meat against flies, peppermint oil, clove oil, d-limonen, geraniol, n-octyl alcohol, anisale, and anetole were most effective. A mixture of anise oil and citronella oil or turpentine was also effective as were some smoke flavors. ABSTRACT: BA 48:6922(77423), Aug. 1, 1967.

5182. **Takei, S. 1962** [Studies on the synthesis of the pyrethrin analogues and their biological activities.] Botyu-Kagaku 27(3):51-65. (In Japanese, with English summary.)
Esters of 6 substituted phenyl cyclopropancarboxylic acids showed some toxicity against *M. domestica vicina*. ABSTRACT: RAE-B 52:138, Aug., 1964.

5183. **Takemoto, T., Ogawa, S., Nishimoto, N.** and **Mue, K. 1967** [Studies on the constituents of *Achyranthes radix*: V. Insect hormone activity of ecdysterone and inokosterone on the flies.] Yakugaku Zasshi 87(12):1481-1483. (In Japanese, with English summary.)
Ecdysterone and inokosterone showed marked hormonal activity when assayed by the use of free abdomen from various flies, including *Musca domestica*. ABSTRACT: BA 49:9254(102765), Oct. 1, 1968.

Takemoto, T., joint author. *See* Kobayashi, M., et al., 1967.

5184. **Tamarina, N. A. 1964** Ukladka dlya opredeleniya ustoichivosti sinantropnykh mukh k insektitsidam v laboratosiyakh i polevykh usloviyakh. [Materials for the determination of the resistance of synantropic flies to insecticides in laboratories and under field conditions.] Med. Parazit. i Parazitarn. (Bolezni) 33(2):197-201.
A set of specially designed equipment and materials is proposed for fly collecting, obtaining new generations and staging toxicological tests. Entirely portable, the kit includes a thermostat which can operate on car and a.c. mains. ABSTRACT: BA 45(22):7739(97739), Nov. 15, 1964.

5185. **Tamarina, N. A., Khromova, L. A.** and **Ioffe, I. D. 1960** Vliyanie temperatury na chuvstvitel' nost' k DDT nekotorykh vidov sinantropnykh mukh. [The influence of temperature on the sensitivity of certain species of synanthropic flies to DDT.] Med. Parazitol. i Parazitarn. (Bolezni) 29(6):733-739. (English summary.)
In tests, in which *Musca domestica* was included, it was found that the action of DDT with temperature was composed of 2 movements: the speed of indications of poisoning and the rapidity with which poisoned individuals die. (With a temperature increase, DDT poisoning symptoms appeared later, but dying occurred sooner.) ABSTRACTS: BA 38:1584 (20374), June 1, 1962; RAE-B 51:21, Feb., 1963.

5186. **Tamarina, N. A.** and **Zhelezova, V. F. 1962** [Temperature optimum of activities of some synanthropic flies.] Med. Parazit. (Moskva) 31:593-599. (In Russian).

Tamarina, N. A., joint author. *See* Lineva, V. A., et al., 1961.

Tamura, S., joint author. *See* Mitsui, T., et al., 1969; Takahashi, N., et al., 1967; Matsuo, K., et al., 1967.

Tanaka, F., joint author. *See* Eto, M., et al., 1966; Ohkawa, H., et al., 1968.

Tanaka, H., joint author. *See* Miyamoto, K., et al., 1967; Ohkawa, H., et al., 1968.

Tanaka, I., joint author. *See* Ogata, K., et al., 1965, 1966, 1967.

Tanaka, T., joint author. *See* Kazano, H., et al., 1968.

Tanco, L., joint author. *See* Duport, M., et al., 1963.

Taniguchi, H., joint author. *See* Shimizu, F., et al., 1965.

Taniguchi, M., joint author. *See* Fugito, S., et al., 1966.

5187. **Tanikawa, T. 1962** [An experiment of controlling fly maggots in the privy by dusting BHC powder after each bowel movement.] Endemic Dis. Bull. Nagasaki Univ. 4(1):46-51. (In Japanese, with English

summary.)
The number of full grown maggots was low from privies where dusting with 3 per cent BHC powder occurred after each bowel movement. Overall fly population in the village was also lowered. ABSTRACTS: BA 45:1760(21943), Mar. 1, 1964; RAE-B 52:235, Dec., 1964.

Tanikawa, T., joint author. *See* Omori, N., et al., 1960; Suenaga, O., et al., 1964.

Tanimura, R., joint author. *See* Matsubara, H., et al., 1966.

Targe, H., joint author. *See* Shimuzu, F., et al., 1965.

5188. Tarry, D. W. 1968 The Control of *Fannia canicularis* in a Poultry House Using a Black-light Technique. Brit. Poult. Sci. 9:323-328.
Traps consisting of 2UV bars (20w) mounted before a high voltage grid greatly reduced the number of *Fannia* in an area with free fly access. ABSTRACT: BA 50:13009(134129), Dec. 15, 1969.

5189. Tarshis, I. B. and **Smith, M. 1961** Field Tests for Effective Control of Flies in Farm and Ranch Buildings. Calif. Agric. 15(7):11-13.
Dibrom in sprays or in sugar baits gave successful fly control on a cattle ranch, poultry ranch, and in a dairy. ABSTRACT: BA 41:2009 (24918), Mar. 15, 1963.

5190. Tashmukhamedov, B. 1962 [Features of the abdominal receptors of stretching in flies.] Zh. Obshch. Biol. 23:76-80.

Tateno, K., joint author. *See* Yasutomi, K., et al., 1966.

5191. Tauber, M. J. 1967 The Biology and Behaviour of 2 Species of *Fannia* (Diptera:Muscidae). Diss. Absts. 28(4B):1564-B-1565-B.
See following reference.

5192. Tauber, M. J. 1968 Biology, Behavior and Emergence Rhythm of Two Species of *Fannia* (Diptera:Muscidae). Univ. Calif. Publs. Ent. 50:1-77.
F. femoralis and *F. canicularis*. A yellow-eyed *F. fem.* mutant was discovered; is due to an autosomal recessive gene. Reproductive behavior and innate capacity for increase in the 2 species discussed.

Tawfik, M. F. S., joint author. *See* Azab, A. K., et al., 1964.

5193. Taylor, B. and **Rodriguez, J. G. 1969** Biological and Behavioral Studies on *Fuscuropoda vegetans*. Kentucky Agric. Exp. Sta. Results Res. 1968-1969. Page 68.
This species is predatory on muscoid larvae.

Taylor, B., joint author. *See* Rodriguez, J. G., et al., 1968, 1969, 1970.

Taylor, H. O., joint author. *See* Deay, H. O., et al., 1962.

Taylor, R. B., joint author. *See* Edelstein, A., et al., 1961.

Taylor, W. C., joint author. *See* Whitten, M. J., et al., 1970.

Têcer, V., joint author. *See* Combiesco, I., et al., 1967.

Tefft, E. R., joint author. *See* Moorefield, H. H., et al., 1960.

5194. Telford, H. S. 1968 Face Fly Control. Washington Agr. Exp. Sta. Bull. 707:34.
Stresses use of malathion and *Heterotylenchus autumnalis*.

Telippo, C., joint author. *See* Russo-Caia, et al., 1961.

5195. T'eng, P. 1968 A preliminary observation of the mechanism of sterilization of the housefly (*Musca* [*domestica*] *vicina* Macquart) treated with thio-tepa. Acta Ent. Sinica (English translation) 1965(2): 90-98. Washington, D.C.
See Tung Bing (1965) for original reference. ABSTRACT: RAE-B 58:107 (433), Apr., 1970 (Title only); CITES: RAE-B 54:38, Feb., 1966.

Teodorescu, C., joint author. *See* Ungureanu, E. M., et al., 1963.

Terranova, A. C., joint author. *See* Ware, G. W., et al., 1960.

5196. Terriere, L. C., Boose, R. B. and **Roubal, W. T. 1960** The Metabolism of Naphthalene and 1-Naphthol by Houseflies and Rats. Biochem. J. 79:620-623.

Animals fed these 2 substances in radioactive form. Houseflies metabolize naphthalene in a manner similar to rats, but 1-naphthol follows different pathways in the two species.

5197. Terriere, L. C. and **Schonbrod, R. D. 1965** Relationship of *In Vitro* Hydroxylation to Resistance in *Musca domestica* and *Phormia regina.* Proc. 12th Int. Congr. Ent. (London) 1964 (Publ. 1965).

Studied microsomes (technique of preparation given). No significant sex difference noted in hydroxylation, but age differences were great. Total results support the idea that hydroxylation is an important biochemical reaction in the detoxication of DDT, dieldrin and similar insecticides.

Terriere, L. C., joint author. *See* Arias, R. O., et al., 1962; Boose, R. B., et al., 1967; Schonbrod, R. D., et al., 1965, 1968; Khan, M. A. Q., et al., 1968; Philleo, W. W., et al., 1965; Schafer, J. A., et al., 1970; Walker, C. R., et al., 1970.

5198. Terry, P. H. and **Bořkovec, A. B. 1967** Insect Chemosterilants. IV. Phosphoramides. J. Med. Chem. 10(1):118-119.

Fifty compounds tested. Only hexamethyl thiophosphoric triamide sterilized houseflies as efficiently as HEMPA (hexamethyl phosphoric triamide).

Terry, P. H., joint author. *See* Chang, S. C., et al., 1967.

5199. Teschner, D. 1960 Zum Verhalten an Fäkalien und zur Ernährung der Art *Muscina stabulans* (Fall.) 1823 (Muscidae, Diptera). [On the behavior toward faeces and on the feeding habits of the species *M. stabulans.*] Zeit. Angew. Ent. 46(2):221-227.

Females landed quickly, fed and then oviposited on faeces; males did not usually land or feed, but were attracted by the presence of the females. Abstract: RAE-B 50:49, Mar., 1962.

5200. Teschner, D. 1961 Zur Dipterénfauna an Kinderkot. [On the Dipterous fauna of infant feces.] Deut. Ent. Zeit. Neue Folger Band 8(1-2): 63-72.

Over 75 per cent of collected species were muscoid. Six species of *Fannia* and two of *Muscina* were among those listed, with females of *F. canicularis* far exceeding all others.

5201. Teschner, D. 1961 Beiträge zur Kenntnis der Fauna eines Müllplatzes in Hamburg. 6. Die Fliegen eines Hamburger Müllplatzes. [Contributions toward a knowledge of the fauna of a refuse site in Hamburg. 6. The flies of a Hamburg refuse site.] Ent. Mitt. (Hamburg). 35:189-214.

Listed as abundant were *Calliphora erythrocephala, Muscina stabulans, Fannia canicularis, Lucilia caesar, Protophormia terrae-novae* and *Lucilia sericata. M. domestica* females were present.

5202. Teschner, D. 1962 Fliegen einer Hamburger Wohnung und in Hamburg neuaufgefundene Fliegen Arten (Diptera). [Flies of a Hamburg dwelling, and newly discovered dipterous forms in Hamburg.] Ent. Mitt. (Hamburg) 37(2):221-232.

Forty-nine species taken at windows facing south. Paper also includes completion of author's 1961 list, and pertinent information on 43 species newly found in Hamburg. (*Fannia ornata* and *Calliphora loewi* were among the latter.)

5203. Teschner, D. 1964 Die Bedeutung der Nester verwilderter Tauben in Grosstädten. [The significance of the nests of wild pigeons in large cities.] Anz. Schädlingsk. 37(3):40-43.
Nests were sprayed with DDT and gathered for zoological study. *Fannia canicularis* shared primary dominance with the bird mite (*Dermanyssus gallinae*).

5204. Teskey, H. J. 1960 A Review of the Life-history and Habits of *Musca autumnalis* De Geer (Diptera:Muscidae). Canad. Ent. 92(5):360-367.
Adults facultative blood suckers, feeding otherwise on nectar, animal secretions, dung fluids. Annoying to cattle. Eggs deposited in fresh cattle dung, in pasture. Pupate near edge of pat. Adults enter buildings only to hibernate. ABSTRACT: RAE-B 49:193, Sept., 1961.

5205. Teskey, H. J. 1969 On the Behavior and Ecology of the Face Fly, *Musca autumnalis* (Diptera:Muscidae). Canad. Ent. 101(6):561-576.
Peak populations near Aug. 1 in Guelph, Ontario. Feeding, mating and oviposition occur in livestock pastures. Thorough observations on life cycle, behavior, ecology and parasites of the species. ABSTRACT: BA 50:11389(117807), Nov. 1, 1969.

Teterovskaia, T. D., joint author. *See* Sukhova, M. N., et al., 1965.

Tetsuo, S., joint author. *See* Hayashi, A., et al., 1968.

Thaggard, C. W., joint author. *See* Mayer, M. S., et al., 1966; Murvosh, C. M., et al., 1966.

Thain, E. M., joint author. *See* Fine, B. C., et al., 1963, 1967; Sawicki, R. M., et al., 1961, 1962.

Tharumarajah, K., joint author. *See* Thevasagayam, E. S., et al., 1962.

5206. Thayer, H. I., Hartle, R. J. and **Mallis, A. 1965** N-alkyl Carbamates as Insecticides and Pyrethrins Synergists. J. Agric. Fd. Chem. 13(1):43-48. Easton, Pa.
Of a series of N-alkyl carbamates 3 were found to show appreciable insecticidal activity against houseflies. Others (not toxic) proved effective synergists for pyrethrins and allethrin. ABSTRACT: RAE-B 53:196, Oct., 1965.

5207. Thevasagayam, E. S. and **Tharumarajah, K. 1962** Fly-breeding in Night-soil Trenches, and its Reduction by the Use of Green Vegetation. Ann. Trop. Med. Parasit. 56:127-129.
Fly emergence can be reduced by as much as two-thirds if a 3-inch layer of green vegetation is placed on top of the night soil before the trench is filled with excavated earth. ABSTRACT: RAE-B 51:141, July, 1963.

5208. Thiessen, C. I. and **Mutchmor, J. A. 1967** Some Effects of Thermal Acclimation on Muscle Apyrase Activity and Mitochondrial Number in *Periplaneta americana* and *Musca domestica*. J. Insect Physiol. 13(12):1837-1842.
Apyrase activity rates were higher and temperature coefficients lower in cold-acclimated insects. Mitochondria were greater in cold. ABSTRACT: BA 49:3371(37689), Apr. 1, 1968.

5209. Thimijan, R. W., Pickens, L. G. and **Morgan, N. O. 1970** A Trap for House Flies. J. Econ. Ent. 63(3):1030-1031.
A cage type trap for use with attractant lamps designed and described. Results of comparisons with similar traps reported.

Thimijan, R. W., joint author. *See* Pickens, L. G., et al., 1969; Morgan, N. O., et al., 1970.

5210. Thoday, J. M. and **Gibson, J. B. 1970** Environmental and Genetical Contributions to Class Differences: A Model Experiment. Science 167(3920):990-992.

In a fly-raising experiment, after 9 generations, the genetic component of the intergroup difference was 42 per cent; portion of intragroup variance that was genetic was 13 per cent. ABSTRACT: BA 51:6288(64341), June 15, 1970.

5211. Thomas, G. D. IV. 1968 Natural Enemies of the Face Fly *Musca autumnalis* in Missouri, U.S.A. Diss. Absts. 28(78):2890-B-2891-B.

5212. Thomas, G. D. and **Puttler, B. 1970** Seasonal Parasitism of the Face Fly by the Nematode *Heterotylenchus autumnalis* in Central Missouri, 1968. J. Econ. Ent. 63(6):1922-1923.

Parasitism, ranging from 4 to 84 per cent, occurred in all collected samples but one. High rate of parasitism associated with low face fly densities.

5213. Thomas, G. D. and **Wingo, C. W. 1968** Parasites of the Face Fly and Two Other Species of Dung-Inhabiting Flies in Missouri. J. Econ. Ent. 61(1):147-152.

Of face fly parasites in Missouri, *Eucoila impatiens* is the most common, *Aphaereta pallipes*, 2nd most common and *Aleochara bimaculata*, least common. ABSTRACTS: BA 49:5803(64976), June 15, 1968; RAE-B 56:140 (487), July, 1968.

Thomas, G. D., joint author. *See* Wingo, C. W., et al., 1967.

5214. Thomas, V. H. 1966 Selection of Larvae of a Laboratory Colony of *Culex pipiens fatigans* for DDT resistance. Med. J. Malaya 20(3): 221-229.

Also concerns houseflies.

5215. Thompson, M. J., Louloudes, S. J., Robbins, W. E., Waters, J. A., Steele, J. A. and **Mosettig, E. 1962** Identity of the "House Fly Sterol". Biochem. and Biophys. Res. Commun. 9(2):113-119.

Housefly sterol (previously designated as Muscasterol) was identified as campesterol and is suggested to be derived from the CSMA media. ABSTRACT: BA 43:362(3954), July 1, 1963.

5216. Thompson, M. J., Louloudes, S. J., Robbins, W. E., Waters, J. A., Steele, J. A. and **Mosettig, E. 1963** The Identity of the Major Sterol from Houseflies Reared by the CSMA Procedure. J. Insect Physiol. (London) 9:615-622.

Muscasterol is a mixture of at least two sterols, campesterol and b-sitosterol, both originating from the CSMA medium. ABSTRACTS: BA 45: 2867(35551), Apr. 15, 1964; RAE-B 52:212, Nov., 1964; TDB 61:222, Feb., 1964.

Thompson, M. J., joint author. *See* Cantwell, G. E., et al., 1964; Kaplanis, J. N., et al., 1965, 1966; Robbins, W. E., et al., 1965, 1970.

5217. Thornton, B. C. and **Sullivan, W. N. 1964** Effects of a High Vacuum on Insect Mortality. J. Econ. Ent. 57(6):852-854.

Fly mortality under various physical stresses measured. Under pressures of 0·05 to 0·03 mm Hg. for periods of 2-64 minutes, greatest mortality occurred at 64 minutes. ABSTRACT: BA 46:2565(31996), Apr. 1, 1965.

Thorson, B. J., joint author. *See* Riemann, J. G., et al., 1967, 1969.

Thorsteinson, A. J., joint author. *See* Quraishi, M. S., et al., 1965.

5218. **Thurston, R., Knapp, F. W.** and **Pass, B. C. 1966** Toxicity of a Resin Formulation of Dichlorvos to Flies and Aphids at Various Temperatures and Humidities. J. Econ. Ent. 59(5):1297-1298.
Controls best at lower humidity and higher temperature. ABSTRACTS: BA 48:1857(20064), Feb. 15, 1967; RAE-B 55:15(37), Jan., 1967; RAE-A 55:46, Jan., 1967.

Ticû, V., joint author. *See* Combiesco, I., etal, 1967.

Tieman, C. H., joint author. *See* Schaefer, C. H., et al., 1967.

5219. **Tiwari, B. K., Bajpai, V. N.** and **Agarwal, P. N. 1966** Evaluation of Insecticidal Fumigant and Repellent Properties of Lemon Grass Oil. Indian J. Exp. Biol. 4(2):128-129.
Shows toxicity against *M. nebulo* at dosage of 7·5 μg. per insect where it produces 90 per cent mortality. Very effective fumigant (evaporating from a cotton ball). Repellent action poor. ABSTRACTS: BA 47:9040 (104528), Nov. 1, 1966; RAE-B 57:41(152), Mar., 1969; TDB 64:571, May, 1967.

Togashi, E., joint author. *See* Katsuda, Y., et al., 1966.

Tokuchi, S., joint author. *See* Yasutomi, K., et al., 1966.

Tokumitsu, I., joint author. *See* Ogushi, K., et al., 1964, 1966, 1967, 1968, 1969.

Tölg, G., joint author. *See* Ballschmiter, K., et al., 1966.

Tolman, N. M., joint author. *See* Nakatsugawa, T., et al., 1968, 1969.

Tolstoshei, O. N., joint author. *See* Lebedeva, A. P., et al., 1963.

Tomiko, I., joint author. *See* Ogushi, K., et al., 1967.

Tong, H. H. C., joint author. *See* Plapp, F. W., Jr., et al., 1966.

5220. **Tonkonozhenko, A. P. 1967** Toksichnost' ne kotorykh entomopatogennykh bakterii dlya komnatnoi mukhi i deistvie termostabil'nogo ekzotoksina *Bac. thuringienses* na lichinki mukh. [Toxicity of some entomopathogenic bacteria for the house fly and the effect of the thermostable exotoxin of *Bacillus thuringiensis* on fly larvae.] Tr. Vses. Nauch-Issled Inst. Vet. Sanit. 28:332-337. (German summary.)
Vegetative forms of all of the bacteria strains were not toxic to 3rd instar larvae, nor to the imago. For 2-day old flies there occurred slight toxicity and a high mortality of the pupa could be reached using a medium containing $3{\cdot}6 \times 10^9$ spores/g. ABSTRACT: BA 50:4779(50241), May 1, 1969.

5220a. **Tonkonozhenko, A. P. 1967** [The pathogenicity of some bacterial preparations for *Musca domestica*.] Veterinariya 44(4):98-100. (In Russian.)
Preparations contained *Bacillus thuringiensis*. Biotrol was most effective against larvae; Thuricide against adult. Least effective was Entobakterin. ABSTRACT: RAE-B 58:344(1406), Oct., 1970.

5221. **Toppozada, A., Eldefrawi, A.** and **O'Brien, R. D. 1970** Binding of Muscarone by Extracts of Housefly Brain: Relationship to Receptors of Acetycholine. J. Neurochem. 17(8):1287-1293.
Binding, which seems to involve acetylcholine receptors, was reversible. There was strong blockade by 17 cholinergic agents. Interpretation. ABSTRACT: BA 52:961(8904), Jan. 15, 1971.

5221a. **Toppozada, A., Mohamed, M. A.** and **Eldefrawi, M. E. 1967** Resistance of the Housefly (*Musca domestica* L.) to Insecticides in Egypt. Bull. W.H.O. 36(6):937-948.
Beginning with the use of DDT in 1946-1949 and then lindane, mala-

thion and trichlorphon dichlorvos in 1964, the resistance developed was tested and calculated in 1965 in 8 strains, against susceptible strains as standard. ABSTRACTS: BA 49:3808(42786), Apr. 15, 1968; TDB 65:726 (1567), Mar., 1968; RAE-B 57:224(852), Nov., 1969.

Tower, B. A., joint author. *See* Burns, E. C., et al., 1961.

Townsend, M. G., joint author. *See* Busvine, J. R., et al., 1963.

5222. Tracy, R. L., Woodcock, J. G. and **Chodroff, S. 1960** Toxicological Aspects of 2, 2-Dichlorovinyl Dimethyl Phosphate (DDVP) in Cows, Horses, and White Rats. J. Econ. Ent. 53(4):593-601.
The effects of exposing horses to DDVP vapors, of feeding DDVP to cows suckling calves and to rats nursing young are reported. Fresh liver macerates from various animals detoxified dichlorvos according to bioassay tests with *Musca domestica* L. ABSTRACT: RAE-B 49:150, July, 1961.

5223. Travis, B. V. and **Labadan, R. M. 1967** Arthropods of Medical Importance in Asia and the European USSR. Part II. Sec. G. Non-biting flies. Technical Report 67-65-ES. U.S. Army Natick Laboratories. U.S. Army Materiel Command, Earth Sciences Division ES-32. (Processed document) pp. 500-509.
Includes references to *M. domestica*, *sorbens*, *tempestiva*, and *vitripennis*, with brief notations. Similar notes on *Calliphora*, *Lucilia*, *Muscina*, *Phaenicia*, *Sarcophaga* and others.

5223a. Travis, B. V. and **Labadan, R. M. 1967** Arthropods of Medical Importance in Latin America. Part II. Sec. G. Tech. Rept. 68-30-ES. ES-35 pp. 375-383.
Includes references to *Callitroga*, *Fannia*, *Lucilia*, and *Sarcophaga.*

5223b. Travis, B. V., Labadan, R. M. and **Lee, H. H. 1968** Arthropods of Medical Importance in Australia and the Pacific Islands. Sec. G. Non-biting Flies. Tech. Rept. 68-61-ES. ES-36 pp. 193-196.
Includes references to *M. domestica*, *vetustissima*, *sorbens*, *terraeregina*, *ventrosa*, also *Chrysomyia* and *Phaenicia.*

5223c. Travis, B. V., Lee, H. H. and **Labadan, R. M. 1969** Arthropods of Medical Importance in America North of Mexico. Sec. G. Non-biting flies. Tech. Rept. 69-2-ES. ES-47 pp. 225-234.
Includes references to *M. domestica* with brief notations. Similar notes on *Calliphora*, *Fannia*, *Muscina*, *Lucilia*, *Phaenicia*, *Phormia*, and *Sarcophaga.*

5223d. Travis, B. V., Mendoza, C. E. and **Labadan, R. M. 1967** Arthropods of Medical Importance in Africa. Part II. Sec. G. Non-biting Flies. Technical Report 67-55-ES U.S. Army Natick Laboratories. U.S. Army Materiel Command, Earth Sciences Division ES-31. (Processed document) pp. 635-650.
Includes references to *M. conducens*, *crassirostris*, *domestica*, *fasciata*, *nebulo*, *sorbens*, *vicina*, *xanthomelas* and *yerburyi*, with brief annotations. Similar notes on *Calliphora*, *Fannia*, *Muscina*, and other genera.

Tray, P. W., joint author. *See* Mattson, A. M., et al., 1960.

5224. Treece, R. E. 1961 A Comparison of the Susceptibility of the Face Fly, *Musca autumnalis*, and the House Fly, *Musca domestica* to Insecticides in the Laboratory. J. Econ. Ent. 54(4):803-804.
Susceptibility of faceflies is comparable to that of housefly. Greater difficulty in control could be attributed to other factors. ABSTRACT: RAE-B 50:154, July, 1962.

5225. Treece, R. E. 1962 Feed Additives for Control of Face Fly Larvae in Cattle Dung. J. Econ. Ent. 55(5):765-768.
Insecticides tested for control of larvae in cattle dung gave variable results. Perhaps 50 per cent reduction in total fly population as a result of treatment. ABSTRACTS: BA 41:947(12335), Feb. 1, 1963; RAE-B 51:42, Feb., 1963.

5226. Treece, R. E. 1964 Evaluation of Some Chemicals as Feed Additives to Control Face Fly Larvae. J. Econ. Ent. 57(6):962-963.
Additional testing of insecticides as feed additives reported. ABSTRACT: BA 46:2537(31661), Apr. 1, 1965.

5227. Treece, R. E. 1966 Effect of Bovine Diet on Face Fly Development—A Preliminary Report. J. Econ. Ent. 59(1):153-156.
Feces from animals on diets containing alfalfa prove most attractive to face fly oviposition. ABSTRACTS: BA 47:6405(74937), Aug. 1, 1966; RAE-B 54:100, May, 1966.

5228. Treece, R. E. and **Miller, T. A. 1968** Observations on *Heterotylenchus autumnalis* in Relation to the Face Fly. J. Econ. Ent. 61(2):454-456.
Nematode-infested flies usually complete only one gonadotrophic cycle. ABSTRACTS: BA 49:8243(91967), Sept. 1, 1968; RAE-B 56:187(676), Sept., 1968.

5229. Treece, R. E. and **Miller, T. A. 1966** Fly Control in Ohio Barns. Res. Circ. Ohio Agric. Res. Dev. Cent. 148:7 pp. Wooster, Ohio.
Against *Stomoxys calcitrans* L. and *M. domestica*, baited ribbons, strips, and residual sprays were applied. Compounds are listed and their effective ranges and mode of application given. Sprays with 1 per cent dimethoate were effective for 2 or more months. ABSTRACT: RAE-B 57:214 (778), Oct., 1969.

Treece, R. E., joint author. *See* Ruprah, N. S., et al., 1968; Lodha, K. R., et al., 1970.

5230. Trofimov, G. K. 1963 [The bazaar fly *Musca sorbens* Wd. (Diptera, Muscidae) in Azerbaijan.] Ent. Obozrenie 42(4):757-764. (Leningrad.) (In Russian, with English summary.)
Specific geographic locations where *M. sorbens* was found to occur are noted. Maximum numbers occur in August-September. Because it reproduced exclusively in human excrement and was a nuisance in markets and open-air eating places, improved hygiene and treatment of breeding materials with mineral oil are advocated. ABSTRACT: RAE-B 54:70, Apr., 1966.

5231. Trofimov, G. K. 1965 Kratkii obzor fauny sinantropnykh mukh Cem. Muscidae, Calliphoridae i Sarcophagidae (Diptera) Tal'sha. [A short review of synanthropic flies (family Muscidae, Calliphoridae, Sarcophagidae) of the Talysh region of the Caucasus.] Ent. Obozrenie 44(3):605-612.
ABSTRACT: BA 47:9057(104763), Nov. 1, 1966.

5232. Trofimov, G. K., Bagirov, G. A. and **Shlyaposhnikov, M. S. 1961** Primenenie zelenogo masla v bor'be s imaginal'noi stadiei mukh. [The application of green oil for the control of flies in the imago-stage.] Med. Parazit. i Parazitarn. (Bolezni) 30(1):104.
Green oil may be atomized with the apparatus "Avtamacs". Directions, and quantities for spraying are given. ABSTRACT: BA 37:2040(20348), Mar., 1962.

Troshin, I. S., joint author. *See* Dremova, V. P., et al., 1961.

5233. **Trpis, M. 1962** [Activity and seasonal dynamics of flies in the locations of their hiding-places in the vegetation of the Danube Valley forests.] Biologia (Bratisl) 17:263-282.

5234. **Trujillo-Cenóz, O. 1969** Some Aspects of the Structural Organization of the Medulla in Muscoid Flies. J. Ultrastruct. Res. 27(516): 533-553.

An illustrated description using electron micrographs. ABSTRACT: BA 50:11957(123642), Nov. 15, 1969.

Tsao, C. C., joint author. *See* Feng, Y. S., et al., 1960.

5235. **Ts'ao, T.-F.** and **Chang, C. P. 1966** [Studies on insect chemosterilants. IV. Further results on screening of insect chemosterilants.] Acta. Ent. Sinica 15(1):13-27. (In Chinese, with English summary.)

One hundred and two chemicals, most of them newly synthesized, tested on houseflies. ABSTRACT: BA 50:1557(16359), Feb. 1, 1969.

Tsao, T. P., joint author. *See* Chang, J. T., et al., 1963.

Tsetlin, V. M., joint author. *See* Misnik, Y. N., et al., 1967; Bessonova, I. V., et al., 1970.

5236. **Tsiapalis, C. M., Hayashi, Y.** and **Chefurka, W. 1967** Polyribosomes from Houseflies. Nature (London) 214:358-361.

Investigates properties of ribosomes and polyribosomes isolated from adult houseflies and the part they play in protein synthesis. Suggests nascent protein is the important constituent that holds the polyribosome together. ABSTRACT: BA 48:10186(114459), Nov. 15, 1967.

Tsujimoto, S., joint author. *See* Fujito, S., et al., 1963.

5237. **Tsukamoto, M. 1960** [Japanese title; concerns genetics of the fly.] The Heredity 14(12):41-46, plus plates. (In Japanese.)

Excellent plates, illustrating mutant types of houseflies.

5238. **Tsukamoto, M. 1964** [Methods for the linkage-group determination of insecticide-resistance factors in the house fly (Diptera).] Botyu-Kagaku 29(3):51-59.

Multiple visible mutants as chromosome markers. ABSTRACT: RAE-B 55:88(315), May, 1967.

5239. **Tsukamoto, M. 1965** The Estimation of Recombination Values in Backcross Data When Penetrance is Incomplete, with a Special Reference to its Application to Genetic Analysis of Insecticide-resistance. Jap. J. Genet. 40(3):159-171.

Concerns methods of estimating recombination values and the application of such methods to the determination of the locus for an insecticide-resistance gene on a given chromosome. ABSTRACT: BA 47:9971 (115365), Dec. 15, 1966.

5240. **Tsukamoto, M. 1969** Biochemical Genetics of Insecticide Resistance in the Housefly. Residue Reviews 25:289-314. Ed. F. A. Gunther.

In *M. domestica* the major genes for insecticide resistance are mostly linked to the second or fifth chromosome, except for the fourth-chromosome control of dieldrin-resistance. ABSTRACT: RAE-B 57:274(1038), Dec., 1969.

5241. **Tsukamoto, M., Baba, Y.** and **Hiraga, S. 1961** Mutations and Linkage Groups in Japanese Strains of the Housefly. Jap. J. Genet. 36(5/6): 168-174.

Fifteen visible mutants were isolated from Japanese populations and assigned to one or other of five autosomal linkage groups.

5242. **Tsukamoto, M.** and **Casida, J. E. 1967** Metabolism of Methylcarbamate Insecticides by the $NADPH_2$-requiring Enzyme System from Houseflies. Nature (London) 213(5071):49-51.
In certain resistant strains a higher enzyme activity for the oxidation of insecticidal chemicals may contribute to resistance mechanisms. ABSTRACTS: RAE-B 56:111(391), June, 1968; TDB 64:686, June, 1967.

5243. **Tsukamoto, M.** and **Casida, J. E. 1967** Albumin Enhancement of Oxidative Metabolism of Methylcarbamate Insecticide Chemicals by the House Fly Microsome—$NADPH_2$ System. J. Econ. Ent. 60:617-619.
Albumin is effective in enhancing microsomal oxidation of methylcarbamate insecticide chemicals by housefly abdomen enzymes and in stabilizing these preparations. ABSTRACTS: BA 48:6516(72761), July 15, 1967; RAE-B 55:149(523), Aug., 1967.

5244. **Tsukamoto, M., Narahashi, T.** and **Yamasaki, T. 1965** Genetic Control of Low Nerve Sensitivity to DDT in Insecticide-resistance Houseflies. Botyu-Kagaku 30:128-132.
Recessive gene pair on second chromosome controls low nerve sensitivity to DDT, a major factor in resistance. ABSTRACT: RAE-B 56:53(176), Mar., 1968.

5245. **Tsukamoto, M., Shivastava, S. P.** and **Casida, J. E. 1968** Biochemical Genetics of House Fly Resistance to Carbamate Insecticide Chemicals. J. Econ. Ent. 61(1):50-55.
Comparisons showed the importance of the fifth chromosomal effect in conferring both high carbamate resistance and high activity for microsomal oxidases involved in ring hydroxylation and attack on N-methyl and O-alkyl groups. ABSTRACTS: BA 49:6315(70714), July 15, 1968; RAE-B 56:138(480), July, 1968; TDB 65:843(1875), June, 1968.

5246. **Tsukamoto, M.** and **Suzuki, R. 1964** Genetic Analysis of DDT Resistance in Two Strains of the House Fly *Musca domestica* L. Botyu-Kagaku 29(4):76-89. (Japanese summary.)
High resistance believed due largely to a fifth chromosome dominant gene and a second chromosome incomplete recessive. ABSTRACTS: BA 47:7750(90372), Oct. 1, 1966; RAE-B 55:173(603), Sept., 1967.

5247. **Tsukamoto, M.** and **Suzuki, R. 1966** Genetic Analyses of Diazinon Resistance in the House Fly. Botyu-Kagaku Bull. (Inst. Insect Contr.) 31(1):1-14.
Using backcrosses in a Japanese strain, diazinon resistance was traced to a multifactorial system with fifth chromosome exerting major influence and a single locus responsible for its effect on resistance. ABSTRACT: RAE-B 56:127(446), July, 1968.

Tsukamoto, M., joint author. *See* Shrivastava, S. P., et al., 1969.

5248. **Tsuruoka, Y.** and **Sugiyama, C. 1963** [Toxicities of alcohols to the common house fly, *Musca domestica vicina* Macquart.] Jap. J. Appl. Ent. Zool. 7(1):79. (In Japanese.)
Tests by topical application show the toxicities of methyl, ethyl, propyl, butyl, amyl, hexy, and octyl alcohol to *Musca domestica vicina* Macq. Increase is in the order given, the last being 20 times as toxic as methanol and ethanol. ABSTRACT: RAE-B 52:138, Aug., 1964.

Tsuruoka, Y., joint author. *See* Nagasawa, S., et al., 1962.

5249. **Tsutomu, O. 1966** [Studies on the dispersal of the house fly.] Endemic Dis. Bull. Nagasaki Univ. 8(3):136-144.

5250. **Tsutsumi, C. 1966** Studies on the Behavior of the Housefly, *Musca domestica* L. I. The Behavior and Activity Patterns Under Experimental Conditions with Special Reference to the Night Time Resting Habit. Jap. J. Med. Sci. Biol. 19(3):155-164.
Night resting behavior may be coupled with some internal factor which controls circadian rhythms of activity in the fly. ABSTRACTS: BA 48:5178 (57920), June 15, 1967; TDB 64:808, July, 1967.

5251. **Tsutsumi, C. 1968** Studies on the Behavior of the Housefly, *Musca domestica* L. II. Some Environmental Factors Affecting the Night Time Resting Behavior of Flies. Jap. J. Med. Sci. Biol. 21(3): 195-204.
Extreme temperature or lack of food results in deviations from usual pattern in flies. Decreased state of activity due to 2 things: (1) some factors controlling circadian rhythms causing flies to rest on ceiling; (2) food intake, which keeps flies on the floor for at least 3 hours. ABSTRACTS: TDB 66:164(297), Feb., 1969; BA 50:6991(73032), July, 1969.

5252. **Tukhmanyants, A. A., Shakhurina, E. A.** and **Eskina, G. V. 1963** [Contribution to the ecology of *Musca larvipara* (Portsch, 1910) an intermediate host of *Thelasia rhodesi* (Desmarest, 1827) from horned Cattle.] Uzbeksk Biol. Zhur. 7(2):57-62.
Sexually mature parasites found in eyes of cattle when flight of the fly is at its peak. ABSTRACT: BA 45:6719(84393), Oct. 1, 1964.

5253. **Tulp, A.** and **van Dam, K. 1969** The Effect of Mersalyl on the Oxidation of Succinate by Housefly Mitochondria. Biochim. Biophys. Acta 189(3):337-341.
Succinate is oxidized rapidly by housefly flight-muscle mitochondria in presence of rotenone. Phosphate (or arsenate) inhibits the succinate oxidation and is believed to compete with it. ABSTRACT: BA 51:3942 (40700), Apr. 1, 1970.

5254. **Tung, A. S.-C.** and **Lee, W.-Y. 1968** The Comparative Study of the Central Nervous System Between Larvae and Adults of Houseflies, *Musca domestica* L. (Muscidae, Diptera, Hexapoda). Bull. Inst. Zool. Acad. Sinica (Taipei) 7(1):7-25.
Illustrated description. ABSTRACTS: BA 50:12503(129163), Dec. 1, 1969; Ent. Absts. 1(1):41(E159), Nov., 1969.

5255. **Tung Bing. 1965** [A preliminary observation on the mechanism of sterilization of the house-flies (*Musca vicina* Macquart), treated with thio-tepa.] Acta Ent. Sinica 14(3):250-256. Peking. (In Chinese, with English summary.)
Sterilizaton due to degeneration of oogonia which finally disappear. Whole ovary then atrophies. Degree of degeneration varied with dosage of thiotepa and treatment duration. ABSTRACTS: BA 49:4317(48634), May 1, 1968; RAE-B 54:38, Feb., 1966.

Turbert, H. B., joint author. *See* Smith, J N., et al., 1964.

5256. **Turner, E. C., Jr. 1965** Area Control of the Face Fly Using Self-Applicating Devices. J. Econ. Ent. 58(1):103-105.
Several insecticides tested in self-applicating devices on cattle substantially reduced number of faceflies but did not give complete control. ABSTRACTS: BA 46:4043(50023), June 1, 1965; RAE-B 53:97, May, 1965.

5257. Turner, E. C., Jr., Burton, R. P. and **Gerhardt, R. R. 1968** Natural Parasitism of Dung-Breeding Diptera: A Comparison Between Native Hosts and an Introduced Host, the Face Fly. J. Econ. Ent. 61(4):1012-1015.
Gives list of parasites of dung-breeding Diptera. Higher percentage of parasites emerge from puparia in middle of summer than early or late in season, due to low temperatures at two extremes. ABSTRACTS: BA 50:3175(33204), Mar. 15, 1969; RAE-B 57:4(8), Jan., 1969.

5258. Turner, E. C., Jr. and **Hair, J. A. 1966** Effect of Temperature on Certain Life Stages of the Face Fly. J. Econ. Ent. 59:1275-1276.
Six temperatures tested. Optimal emergence from pupae to adults occurred at 85°F. Increase in temperature decreased longevity. Optimal temperature for best progeny production was 80°F. ABSTRACTS: BA 48:2365(25874), Mar. 1, 1967; RAE-B 55:14(35), Jan., 1967.

5259. Turner, E. C. and **Hair, J. A. 1967** Effect of Diet on Longevity and Fecundity of Laboratory-Reared Face Flies. J. Econ. Ent. 60:857-860.
Sugar essential for adult survival and a proteinaceous substance is necessary for reproduction. ABSTRACTS: BA 49:1427(15998), Feb. 1, 1968: RAE-B 55:191(659), Oct., 1967.

5260. Turner, E. C. and **Wang, C. M. 1964** Residual and Topical Toxicity of Certain Insecticides to Laboratory-Reared Face Flies. J. Econ. Ent. 57(5):716-719.
Residual effectiveness of several insecticides tested indicates that face fly is susceptible to wide range of organic insecticides. Poor control in field must be due to factors other than toxicity. ABSTRACTS: BA 46:696 (8725), Jan. 15, 1965; RAE-B 53:11, Jan., 1965.

Turner, E. C., Jr., joint author. *See* Burton, R. P., et al., 1968, 1970; Hair, J. A., et al., 1965, 1966; Wallace, J. B., et al., 1962, 1964.

5261. Tyler, P. S. 1961 Cluster Flies and Swarming Flies: Their Behavior and Control. Sanitarian 69:285-290. London.
Bionomics and insecticidal control of four species of Diptera: *Pollenia rudis* (F.), *Musca autumnalis* Deg., *Dasyphora cyanella* (Mg.) and *Thaumatomyia notata* (Mg.). ABSTRACT: RAE-B 51:274, Dec., 1963.

5262. Tyndale-Biscoe, M. and **Hughes, R. D. 1969** Changes in the Female Reproductive System as Age Indicators in the Bushfly *Musca vetustissima* Wlk. Bull. Ent. Res. 59(1):129-141.
Factors influencing development of female from emergence to fourth ovarian cycle (notably protein feeding) show that valid estimates can only be made of minimum age. (Seventeen cagetories of development recognized.) ABSTRACTS: BA 51:5668(58164), May 15, 1970; RAE-B 57:170(613), Sept., 1969.

Tyrakowski, M., joint author. *See* Przyborowski, T., et al., 1963.

Uchida, S., joint author. *See* Ogata, K., et al., 1961.

5263. Uchida, T., Rahmati, H. S. and **O'Brien, R. D. 1965** The Penetration and Metabolism of H^3-Dimethoate in Insects. J. Econ. Ent. 58: 831-835.
Five insect species were studied including adults of both sexes of *Musca domestica*. Results show no relation between degradation rate of dimethoate or the amount of oxygen analogue present and the toxicity. ABSTRACTS: RAE-B 54:11, Jan., 1966; TDB 63:599, May, 1966.

Uhler, L., joint author. *See* Pimentel, D., et al., 1969.

Umeda, K., joint author. *See* Ohkawa, H., et al., 1968.

Umino, I., joint author. *See* Ikeshoji, T., et al., 1967.

5264. Duplicate Reference. Withdrawn.

5265. Ungureanu, E. M., Teodorescu, C., Ungureanu, S., Puscasu, E. and **Violeta, Z. 1963** Studies on the Resistance of *Musca domestica* L. to Insecticides. J. Hyg. Epidem. (Praha) 7:252-256. (French, German, and Spanish summaries.)
Spraying with diazinon and Dipterex gave satisfactory control of flies in Rumania that had become resistant to DDT, gamma-BHC and also dieldrin, methoxychlor, aldrin and endrin. ABSTRACTS: BA 44:1258 (17013), Nov. 15, 1963; TDB 61:614, June, 1964.

Ungureanu, S., joint author. *See* Ungureanu, E. M., et al., 1963.

Uspenskii, I. V., joint author. *See* Kontar, V. A., et al., 1966.

Valder, S. A., joint author. *See* Monroe, R. E., et al., 1967; Valder, S. M., et al., 1969.

5266. Valder, S. M. 1969 Reproductive System Morphology of the Face Fly, *Musca autumnalis* with Notes on a Gynandromorph. J. Kans. Ent. Soc. 42(2):176-182.
Morphology of the reproductive systems of male and female face flies is described and illustrated, as is that of a gynandromorph face fly from the Kansas State University colony. ABSTRACT: BA 50:10315(106729), Oct. 1, 1969.

5267. Valder, S. M. and **Hopkins, T. L. 1968** Effects of Nutrition on Egg Development and Survival in the Face Fly, *Musca autumnalis.* Ann. Ent. Soc. Am. 61(4):827-834.
The maximum development was obtained with diets of from 40-70 per cent protein. ABSTRACTS: BA 50:2672(27882), Mar. 1, 1969; RAE-B 57:68(235), Apr., 1969.

5268. Valder, S. M. and **Hopkins, T. L. 1968** Artificially Induced Diapause in the Face Fly, *Musca autumnalis* DeGeer. Proc. N. Centr. Branch Ent. Soc. Am. 23(1):57. (Abstract only.)
Flies were sensitive to diapause-inducing conditions only in the first 24-48 hours after leaving puparia. ABSTRACT: RAE-B 58:153(641), May, 1970.

5269. Valder, S. M., Hopkins, T. L. and **Valder, S. A. 1969** Diapause Induction and Changes in Lipid Composition in Diapausing and Reproducing Faceflies, *Musca autumnalis.* J. Insect Physiol. 15:1199-1214.
A simulated autumn environment induced a uniformly high percentage of diapause. ABSTRACT: BA 50:13073(134826), Dec. 15, 1969.

Valdes, E., joint author. *See* Dinamarca, M. L., et al., 1969.

Valega, T. M., joint author. *See* Plapp, F. W., et al., 1967.

5270. Valentry, D. 1969 The Fantastic Fly. Todays Health, pp. 56, 57. (August Issue.)
Cites work of Dr. Vincent G. Dethier on fly brain, using brain wave recorders. Dethier has transplanted a brain from one fly to another.

5271. Valiela, I. 1969 An Experimental Study of the Mortality Factors of Larval *Musca autumnalis* DeGeer. Ecol. Monogr. 39(2):199-225.
Factors include temperature, density of larvae, and dung-inhabiting predators. ABSTRACT: BA 51:10311(104868), Sept. 15, 1970.

Valiela, I., joint author. *See* Hansens, E. J., et al., 1967.

5272. Van Asperen, K. 1960 Mode of Action on Metabolism of Some Organic Phosphorus Insecticides in Houseflies. Proc. IVth Internat. Congr. Crop Prot. Hamburg, September. Section XIII. Insecticides, pp. 1173-1176.
Exposure to dichlorvos, parathion, paraoxon, diazinon and coumaphos resulted in *in vivo* inhibition of cholinesterase at time of knockdown. ABSTRACT: RAE-B 51:107, June, 1963.

5273. Van Asperen, K. 1962 A Study of Housefly Esterases by Means of a Sensitive Colorimetric Method. J. Insect Physiol. (London) 8:401-416.
Cholinesterase is strongly pH dependent; ali-esterase weakly so. Both are more active at a higher pH. Electrophoresis shows the presence of about seven electrophoretically different esterases, occurring in more or less strain-specific patterns. ABSTRACTS: RAE-B 52:10, Jan., 1964; TDB 59:1120, Nov., 1962.

5274. Van Asperen, K. 1964 Biochemistry and Genetics of Esterases in Houseflies (*Musca domestica*) with Special Reference to the Development of Resistance to Organophosphorus Compounds. Ent. Exp. Appl. (Amsterdam) 7(3):205-214.
A review and reconsideration of experimental work on esterases in the housefly. Mode of action of organophosphorus insecticides and mechanism of resistance of these compounds, also possible function of esterases, are discussed. ABSTRACTS: RAE-B 53:173, Sept., 1945; TDB 62:245, Mar., 1965.

5275. Van Asperen, K. and **Oppenoorth, F. J. 1960** The Interaction Between Organophosphorus Insecticides and Esterases in Homogenates of Organophosphate Susceptible and Resistant Houseflies. Ent. Exp. Appl. (Amsterdam) 3(1):68-83.
Cholinesterases and ali-esterases rank first in competing esterases. The result of competition depends upon the amount of the esterases present and the reaction rates. ABSTRACTS: RAE-B 49:76, Apr., 1961; TDB 57:960, Sept., 1966.

5276. Van Asperen, K. and **Van Mazijk, M. E. 1965** Agar Gel-Electrophoretic Esterase Patterns in Houseflies. Nature 205:1291-1292.
Agar gel-electrophoretic patterns were obtained for mass homogenates of eleven strains of flies. ABSTRACTS: BA 46:6270(77856), Sept. 1, 1965; RAE-B 54:95, May, 1966; TDB 62:812, Aug., 1965.

5277. Van Asperen, K., Van Mazijk, M. E. and **Oppenoorth, F. J. 1965** Relation Between Electrophoretic Esterase Patterns and Organophosphate Resistance in *Musca domestica*. Ent. Exp. Appl. 8(3): 163-174. (German summary.)
A gene on the 5th chromosome, causing a difference in the mobility of an esterase electrophoresis, occurs in the Italian OP-resistant strain. ABSTRACTS: BA 47:5977(69941), July 15, 1966; RAE-B 54:68, Apr., 1966.

Van Asperen, K., joint author. *See* Oppenoorth, F. J., et al., 1960, 1961; Velthuis, H. H. W., et al., 1963.

5278. Van Bruggen, E. F. J., Runner, C. M., Borst, P., Ruttenberg, G. J. C. M., Kroon, A. M. and **Schuurmans Stekhoven, F. M. A. H. 1968** Mitochondrial DNA: III. Electron Microscopy of DNA Released from Mitochondria by Osmotic Shock. Biochim. Biophys. Acta 161(2):402-414.
Osmotic shock released predominantly circular DNA from mitochondrial preparations of housefly flight muscle. ABSTRACT: BA 50:2700(28230), Mar. 15, 1969.

5279. Van den Bergh, S. G. 1964 Pyruvate Oxidation and Permeability of Housefly Sarcosomes. Biochem. J. 93(1):128-136.
Oxidation of pyruvate by isolated housefly sarcosomes is by way of the Kreb cycle. A barrier prevents the cyclic products from leaking sarcosomes. ABSTRACT: BA 47:2508(29965), Mar. 15, 1966.

5280. **Van Dinther, J. B. M. 1964** Studies on the Starvation Capacities of Some Carabid Species. Overdruk uit de Mededelingen van de Landbouwhogeschool en de Opzoekingsstations van de Staat te Gent. 29(3):1088-1096.
The 2nd and 3rd instar larvae of the housefly were used as food in these experiments.

5281. **Van Dinther, J. B. M. 1966** Laboratory Experiments on the Consumption Capacities of some Carabidae. Meded. Rijksfak. Landouw-Wetenschappen Gent. 31(3):730-739.
For food, eggs, larvae and pupae of houseflies were used. ABSTRACT: BA 49:2375(26713), Mar. 1, 1968.

5282. **Van Dinther, J. B. M.** and **Mensink, F. T. 1965** Egg Consumption by *Bembidion ustulatum* and *Bembidion lampros* (fam. Carabidae) in Laboratory Prey Density Experiments with House Fly Eggs. Meded. Landbouwhogesch. Opzoekings. Staat Gent. 30(3):1542-1554. (German, French, and Dutch summaries.)
House fly eggs were substituted for cabbage root fly eggs in tests with a predator of the cabbage root fly. ABSTRACT: BA 48:417(4327), Jan. 1, 1967.

Van Mazijk, M. E., joint author. *See* Van Asperen, K., et al., 1965.

Vardanis, A., joint author. *See* Morello, A., et al., 1967, 1968.

5283. **Varela, G., Arroyo-Bornstein, J. A.** and **Bravo-Becherelle, M. A. 1964** Aislamiento de *Salmonella* en moscas de Tlalnepantla, Mexico. [Isolation of *Salmonella* in flies from Tlalnepantla, Mexico.] Rev. Inst. Salubr. Enferm. Trop. (Mexico) 24(1/4):3-6.
Musca domestica was one of the more numerous flies collected at a slaughterhouse. Carrion flies carried amounts of *Salmonella* suggesting their vectorship as they visit human dwellings. ABSTRACT: BA 46:5420 (67413), Aug. 1, 1965.

Varela, G., joint author. *See* Greenberg, B., et al., 1963.

Varma, R., joint author. *See* Pant, R., et al., 1968.

5284. **Vashkov, V. I. 1962** [Sensitivity to DDT of houseflies in various climatic zones of the USSR.] Zh. Mikrobiol. 33:20-24. (In Russian.)

5285. **Vashkov, V. I., Khudadov, G. D.** and **Zakolodkina, V. I. 1965** Skorost' proniknoveniya i nakopleniya khlorofosa, mechennogo p32 v razlichnykh organakh i tkanyakh' komnatnykh mukh. [Rate of the penetration and accumulation of P32-labelled Dipterex (0, 0-dimethyl-2, 2, 2-trichlor-1-hydroxyethyl' phosphonate) in the different organs and tissues of house flies (*Musca domestica*).] Zhur. Mikrobiol., Epidemiol., i Immunobiol. 42(8):3-6. (English summary.)
Dipterex, or its metabolites, first increase upon penetration, then within a matter of hours decrease. The hemolymph and digestive system are areas of greatest concentration. ABSTRACT: BA 47:5978(69942), July 15, 1966.

5286. **Vashkov, V. I.** and **Poleshchuk, V. D. 1966** Vliyanie rentgensoblucheniya na chuvstvitel' nost' nasekomykh k insektitsidam. [Effect of X-irradiation on the sensitivity of insects to insecticides.] Zhur. Mikrobiol., Epidemiol., i Immunobiol. 43(8):9-11. (English summary.)
Sensitivity to DDT, hexachlorane and chlorphos increased following irradiation. ABSTRACT: BA 48:6577(73494), Aug. 1, 1967.

5287. **Vashkov, V. I.** and **Shnaider, E. V. 1961** The Insecticidal Properties of Dimethyl-dichlorovinyl Phosphate (DDVP). J. Microbiol., Epidemiol. and Immunobiol. (London) 32(4):747-754. (Translated from Russian.)
DDVP has a high toxicity for many insects including the housefly, both as a residual spray and as a larvicide. ABSTRACT: TDB 60:79, Jan., 1963.

5288. **Vashkov, V. I.** and **Shnaider, E. V. 1962** [Khlorofos (insecticidal properties and use).] (In Russian.) Moscow, Medgiz. Price 83 kop.
In this booklet on trichlorphon, its properties and mode of action in insects are noted, and directions given for use in controlling houseflies and other insects. Bactericidal properties, use against ectoparasites and pests of plants, also its toxicity to animals, are discussed. ABSTRACT: RAE-B 50:219, Oct., 1962.

5289. **Vashkov, V. I., Vinogradskaya, O. N., Volkov, Yu P., Volkova, A. P., Sidorova, M. V.** and **Starkov, A. V. 1966** Issledovanie v oblast: piretrinov i rodstvenarykh soedinenni. VIII. Primenenie piperonil-d, 1-tsis, trans-khrizantemata s smesi s sinergistami v kachestve insektsida. [Studies in the field of pyrethrins and affiliated compounds. Report VIII. The use of piperonyl-d, 1-cis, trans-crysanthemate in a mixture with synergists as an insecticide.] Zhur. Mikrobiol., Epidemiol., i Immunobiol. 43(12):108-112. (English summary.)
Insecticidal activity is enhanced with the aid of synergists used for increasing activity of natural pyrethrins. These synergists manifested a low animal toxicity. ABSTRACTS: BA 49:461(5129), Jan. 1, 1968; TDB 64:570, May, 1967.

5290. **Vashkov, V. I., Vinogradskaya, O. N., Volkov, Yu P., Volkova, A. P., Sidorova, M. V.** and **Starkov, A. V. 1967** Issledovaniya, v oblasti piretrinov v rodetvennylch soedinenii. VI. Sentez pipetrina, ego insektitsidnye i toksicheskie svoistva. [Studies in the field of pyrethrins and related compounds. Synthesis of pipethrin, its insecticidal and toxic properties.] Zhur. Mikrobiol., Epidemiol., i Immunobiol. 44(4):89-94. (English summary.)
The method of obtaining piperonyl chrysanthemate (pyrethrin) is described. DD_{50} for *Musca domestica* is given. Pipethrin is inferior to chlorophorus and pyrethrins but is active enough for flies, and nontoxic to warm-blooded animals. ABSTRACT: BA 49:3340(37379), Apr. 1, 1968.

5291. **Vashkov, V. I., Vinogradskaya, O. N., Volkov, Yu. P., Zubova, G. M.** and **Shugal, N. F. 1968** Insektitsidnye svoistva Bartrina i Dimetrina. [Insecticidal properties of Bartrin and Dimetrin.] Zhur. Mikrobiol., Epidemiol., i Immunobiol. 45(9):9-13. (English summary.)
LD_{50} was established for domestic flies using the synthetic analogues of pyrethrins. The synergist piperonyl butoxide intensified activity. ABSTRACT: BA 50:9173(95086), Sept. 1, 1969.

5292. **Vashkov, V. I., Volkov, Yu. P., Zubova, G. M.** and **Shugal, N. F. 1966** Issledovanie pir etr inov i rodstvennykh soedinenii. IX. Sintez i insektitsidyne svoistva nekotorykh efirov d, 1-tsis, trans-khrizantemovoi monokarbonovoi kisloty: 6-zameshchennykh piperonilovykh spirtov. [Study of pyrethrins and related compounds. IX. Synthesis and insecticide properties of some esters d, 1-cis, of transchrysanthemic moncarbonic acids and 6-replaced pipe-

ronyl alcohols.] Zhur. Mikrobiol., Epidemiol., i Immunobiol. 43(10):106-109.
Insecticidal properties of synthesized compounds tested on *Musca domestica*, *Cimex lectularius* L. and *Blatella germanica* were not encouraging. Only one compound, brombartrun, showed moderate activity. ABSTRACT: BA 48:7417(82975), Aug. 15, 1967.

5293. Vashkov, V. I., et al. 1963 Chuvstvitel'nost'komatnylch mulch k khlorofosu do ego primeneniya. [Sensitivity of domestic flies to chlorofos prior to its use.] Zhur. Mikrobiol., Epidemiol., i Immunobiol. (Moscow) 40(7):3-7. (English summary.)
A study of flies in the USSR in 1959-61 showed that those from cities, in particular those having an increased DDT resistance, were more sensitive to chlorophos than those caught in rural areas, where insecticides were not previously employed. ABSTRACTS: BA 44:1580, Dec. 1, 1963; TDB 60:1083, Nov., 1963.

5294. Vashkov, V. I. and twenty-one others. 1962 [DDT sensitivity of synanthropic flies in different climatic zones of the U.S.S.R.] Zhur. Mikrobiol., Epidemiol., i Immunobiol. (Moscow) 33(8):20-24. (In Russian.)
DDT sensitivity varied in localities studied. In the majority of cities, flies already had acquired sensitivity, the level of which changed during the fly season. ABSTRACT: TDB 60:275, Mar., 1963.

Vashkov, V. I., joint author. *See* Bessonova, I. V., et al., 1968, 1970.

Vasilenko, L. N., joint author. *See* Sukhova, M. N., et al., 1965.

5294a. Vasilev, L. V. 1970 [Incidence of dysentery and changes in the etiological structure of dysentery in relation to meteorological conditions and number of flies.] Zhur. Mikrobiol., Epidemiol., i Immunobiol. (Moscow) 33(8):20-24. (In Russian.)

5295. Vazquez-Gonzalez, J., Young, W. R. and **Ramirez-Genel, M. 1962/63** Reducción de la poblacion de mosca domestica en gallinaza por la mosca soldado en el tropico. [The use of soldier flies (*Hermetia illucens*) for lowering the *Musca domestica* population in hen dung in the tropics.] Agric. Tec. Mex. 2(2):53-57.
The use of soldier flies is promising, but because the degree of success is dependent upon the moisture content of the dung, water must be added if dryness occurs. ABSTRACT: BA 46:5877(73004), Aug. 15, 1965.

Vecchi, M. L., joint author. *See* Bernardini, P. M., et al., 1965; Mosconi-Bernardini, P., et al., 1965.

5296. Veldsema-Currie, R. D. and **Slater, E. C. 1968** Inhibition by Anions of Dinitrophenol-induced ATPase of Mitochondria. Biochim. Biophys. Acta 162:310-319.
Inhibitory effects are explained by competition with dinitrophenol for penetration of the mitochondrial matrix. ABSTRACT: BA 50:2224(23397), Mar. 1, 1969.

5297. Velthuis, H. H. W. and **Van Asperen, K. 1963** Occurrence and Inheritance of Esterases in *Musca domestica*. Ent. Exp. Appl. (Amsterdam) 6:79-81. (German summary.)
Ten esterases resulting from nine esterase genes occur in a housefly strain. Only strain C of all the houseflies studied was shown to be heterogeneous for the occurrence of a number of electrophoretically shown esterase bands, and for the genes responsible for their occurrence. ABSTRACTS: BA 45:355(4478), Jan. 15, 1964; RAE-B 51:234, Nov., 1963; TDB 60:903, Sept., 1963.

Verma, R. N., joint author. *See* Dixit, R. S., et al., 1968.

5298. Veselkin, G. A. 1964 [The control of flies on livestock farms.] Veterinariya 41(6):106-109. Moscow. (In Russian.)
Studies in the Tyumen region of Siberia in 1961 revealed 112 species of flies from 21 families. *Musca autumnalis* and *M. larvipara* were among those listed. Trikhlormentafos, trichlorphon and polychoropinene were used as insecticides. ABSTRACT: RAE-B 54:183, Sept., 1966.

5299. Veselkin, G. A. 1966 Mukhi (Diptera)-sputniki domashnikh zhivotnykhi chelveka vuyzhnoi chasti tyumenskoi oblasti. [Flies (Diptera) satellites of domestic animals and man, in the southern part of the Tyumen region (USSR).] Ent. Obozrenie 45(4):779-792. Map.

5300. Veselkin, G. A. 1966 Diptera Associates of Domestic Animals and Man in the South of the Tyumen Province. Ent. Rev. 45(4):439-447.
Concerns flies associating with pigs, cattle, and manure.

Vian, I., joint author. *See* De Pietre-Tonelli, P., et al., 1966.

5301. Vicari, G., Bettini, S., Collotti, C. and **Frontali, N. 1965** Action of *Latrodectus mactans tredecimguttatus* Venom and Fractions on Cells Cultivated *In Vitro*. Toxicon 3:101-106.
Two of three fractions obtained electrophoretically from the venom of the black widow spider were toxic to the housefly, one (LV_1) causing "quick" paralysis, the other (LV_2) a "slow" paralysis.

5302. Vickery, D. S. and **Arthur, B. W. 1960** Animal Systemic Activity, Metabolism and Stability of Co-Ral (Bayer 21/199). J. Econ. Ent. 53(6):1037-1043.
Musca domestica was used as one of the test species in studying Coumaphos, its oxygen analogue chloroferron, and Potasan. Labelled P^{32} Coumaphos was also used. ABSTRACT: RAE-B 49:219, Oct., 1961.

5303. Vilagiova, I. 1962 Význam Múch pre vývin očných parazitov—pôvodcov telàziözy hovädzieho dobytka. [The importance of flies for the development of eye parasites—the casual agents of thelaziosis of cattle.] Biologia 17(4):297-299. (Summaries in Russian and German.)
Investigations of pasture flies in 1960 to determine the host of cattle thelaziosis identified *Musca autumnalis* Deg. (70 per cent of total flies collected) and *M. larvipara* Portsch as carriers of the causative bacilli, *Thelazia rhodesi* and *T. gulosa*. ABSTRACT: RAE-B 53:29, Feb., 1965.

5304. Vilagiova, I. 1967 Results of Experimental Studies on the Development of Preinvasive Stages of Worms of the Genus *Thelazia* Bosc., 1819 (Spirurata:Nematoda), Parasitic in the Eye of Cattle. Folia Parasitol. (Praha) 14(3):275-280.
Studies in Czechoslovakia of the life cycle of three species of *Thelazia*. Intermediate host used was *Musca autumnalis*. ABSTRACT: BA 50:8634 (89663), Aug. 15, 1969.

5305. Vilagiova, I. 1967-1968 Ecological Specificity of Intermediate Hosts of *Thelazia* in Cattle and the Relation of their Infective Larvae to Some Unusual Hosts. Helminthologia Vol. VIII-IX, (1-4):656 pp. Illus. Academia Scientificarum Slovaca: Bvatislava, Czechoslovakia. Edited by J. Hovorka.
Concerns the genus *Musca*.

5306. Vilagiova, I. 1968 *Heterotylenchus autumnalis* Nickle (1967): A Parasite of Pasture Flies. Biologia (Bratislava) 23(5):397:400. (Czechoslovakian, German, and Russian summaries).
The first record of *H. autumnalis* in Slovakia. Most were found in *M.*

autumnalis DeGeer but a few single specimens were from other flies. ABSTRACT: BA 50:10297(106498), Oct. 1, 1969.

Vilkova, N. A., joint author. *See* Shapiro, I. D., et al., 1967.

5307. Vinogradskaya, O. N. 1964 Deistvie khlorofosa i temperatury na dykhatel' nyi ritm mukh. [Effect of Dipterex (0, 0-Dimethyl-2, 2, 2-trichlor-1-oxethyl-phosphonate) and temperature on the respiratory rate in flies.) Med. Parazit. i Parazitarn. (Bolezni) 33(5): 527-532. (English summary.)

Dipterex, effective as a fumigant and as a contact insecticide, changes the respiratory rhythm, prolonging the periods of trachael system ventilation. This leads to desiccation, exhaustion and death. ABSTRACT: BA 46:7364(91623), Oct. 15, 1965.

Vinogradskaya, O. N., joint author. *See* Vashkov, V. I., et al., 1966, 1967, 1968.

5308. Vinopal, J. H. and **Johansen, C. A. 1967** Selective Toxicity of Four 0-(methylcarbomyl) Oximes to the House Fly and Honey Bee. J. Econ. Ent. 60(3):794-798.

Four carbamate acaricide-insecticides were tested against *Musca domestica* L. and *Apis mellifera* L. by oral and topical drop methods. ABSTRACTS: BA 48:11215(124559), Dec. 15, 1967; RAE-B 55:190(655), Oct., 1967.

Violeta, Z., joint author. *See* Ungureanu, E. M., et al., 1963.

Vishnevskaya, T. M., joint author. *See* Mazokhin-Porshnyakov, G. A., et al., 1966.

Visona, L., joint author. *See* Laudani, U., et al., 1965.

5309. Vitanovic, R. 1966 [The role of synanthropic flies in the transmission of diarrheal diseases.] Vojnosanit Pregl. 23:474-476.

Vocino, G., joint author. *See* Cardaras, P., et al., 1962.

Volkov, Yu P., joint author. *See* Bessonova, I. V., et al., 1968, 1970; Vashkov, V. I., et al., 1966, 1967, 1968.

Volkova, A. P., joint author. *See* Vashkov, V. I., et al., 1966, 1967.

5310. Von Frisch, K. 1960 Ten Little Housemates. Pergamon Press: New York. Republished in 1964 and 1969.

The book gives accurate features of the life and habits of ten common house pests in interesting, non-technical language. It includes information on getting rid of each. Common housefly is one of the ten. ABSTRACTS: BA 36:3216(39332), July 1, 1961; RAE-B 52:212, Nov., 1964.

Von Zboray, E. P., joint author. *See* Georghiou, G. P., et al., 1965.

5311. Vostál, Z. 1960 Výskt resistence KDDTU mouchy domáci (*Musca domestica* L.) na vychodnin Slovensku. [The DDT resistance of the house-fly (*Musca domestica* L.) in eastern Slovenskia.] Zool. Listy/ Folia Zoologica 23(1):89-93. Prague. (In Czechoslovakian, with German summary.)

Resistance was reported in a locality where DDT had been used for 3-5 years. No resistant flies were found where DDT had not been used, although at a distance of but 5 km. was a site with flies of the highest resistance. ABSTRACTS: RAE-B 51:19, Dec., 1963; BA 35:5233(61151), Nov. 1, 1960.

5312. Vostál, Z. 1964 Odolnost proti malationu u mouchy domácí, *Musca domestica* L. na východním Slovensku. [Malathion resistance in the house fly *Musca domestica* L. in eastern Slovakia.] Folia Zool. 13(2):161-164. (In Slovakian, with German summary.)

Malathion resistance reported for the first time in the domain of CSSR at a place in eastern Slovakia in the year 1960. It was believed due to selection pressure. ABSTRACT: RAE-B 53:249, Dec., 1965.

5313. Vostál, Z., Kratochvil, I. and **Legath, V. 1963** Resistenz probleme bei synanthropen Fliegen in der Ostslowakei. [The problem of resistance in synanthropic flies in eastern Slovakia.] Angew, Parasit. (Jena) 4(3):135-138. (English summary.)
In 1954, DDT and BHC resistant populations were observed. A cessation of spraying with chlorinated hydrocarbons resulted in a reversion to susceptibility in two years. Trichlorphon, malathion and diazinon were effective against the resistant flies. ABSTRACTS: RAE-B 53:67, Apr., 1965; TDB 61:448, Apr., 1964.

5314. Vowles, D. M. 1966 The Receptive Fields of Cells in the Retina of the Housefly (*Musca domestica*). Proc. Roy. Soc. Biol. 164:552-576.
A theory of movement has been proposed to explain the optomotor responses of the housefly. Three properties were studied: the interommatidial angle, the receptive field of retinula cells, and the relationship between light intesity and the magnitude of the generator potential. ABSTRACT: BA 47:8558(99741), Oct. 15, 1966.

Vranchan, Z. E., joint author. *See* Andrew, V. P., et al., 1961.

Vranova, J., joint author. *See* Privora, M., et al., 1969.

5315. Vroman, H. E., Kaplanis, J. N. and **Robbins, W. E. 1966** Cholesterol Turnover in the House Fly. Am. Zool. 6(4):505.
Only male flies used. *M. domestica* appears to have a surplus of cholesterol ester, which decreases rapidly to two-thirds of the initial concentration.

Vroman, H. E., joint author. *See* Dutky, R. C., et al., 1967.

Vtorov, P. P., joint author. *See* Sychevskaya, V. I., et al., 1969.

Vyrvikhovst, L. A., joint author. *See* Sychevskaya, V. I., et al., 1960.

Wada, A., joint author. *See* Ogata, K., et al., 1960.

5316. Wada, Y. and **Oda, T. 1963** [On the range and the number in the dispersal of the housefly, *Musca domestica vicina*, around hog houses.] End. Dis. Bull. Nagasaki Univ. 5(2):116-122. (English summary.)
An estimation of the dispersal of flies around animal buildings was made by marking, releasing, and recapturing flies, noting the sizes of populations at releasing sites. Most flies did not move far. ABSTRACTS: BA 45:7739(97740), Nov. 15, 1964; RAE-B 53:148, Aug., 1965.

5317. Waddell, A. H. 1969 A Survey of *Habronema* spp. and the Identification of Third-stage Larvae of *Habronema megastoma* and *Habronema muscae* in Section. Australian Vet. J. 45(1):20-21.
Heads of infected *Musca domestica* were sectioned and studied. Longitudinal cuticular ridges on *Harbronema muscae* and smooth cuticle of *H. megastroma* were used to identify the two species. ABSTRACT: BA 50: 11374(117631), Nov. 1, 1969.

5318. Wade, C. F. and **Rodriguez, J. G. 1961** Life History of *Macrocheles muscaedomesticae* (Acarina-Macrochelidae), a Predator of the Housefly. Ann. Ent. Soc. Am. 54(6):776-781.
Standardization of rearing techniques and further study of the life cycle. ABSTRACT: RAE-B 50:241, Nov., 1962.

Wade, C. F., joint author. *See* Rodriguez, J. G., et al., 1961, 1962.

Wagner, E. D., joint author. *See* Winkler, L. R., et al., 1961.

5319. Wagner, S. and **Whittmann, W. 1966** Experimentelle Untersuchungen zum Problem der Tollwutübertragung durch Fliegen. [Experimental studies on the problem of rabies transmission by flies.] Arch. Exp.

Veterinaermed. 20:821-824.
Concludes that dead flies do not play an intermediate role in the spread of hydrophobia virus during epidemics in nature.

5320. Wagoner, D. E. 1966 The Linkage Group Karyotype Relationship in *Musca domestica.* Proc. N. Centr. Branch Ent. Soc. Am. 21:104-105.
Five linkage groups found in the house fly are thought to correspond to the five pairs of autosomes known to be present.

5321. Wagoner, D. E. 1967 Translocation Induction and Analysis in the House Fly *Musca domestica.* Genetics 56(3) pt. 2:594.
Over 150 X-ray induced translocations in the housefly have been analyzed for sex ratio of progeny, ratio of translocation to non-translocation bearing progeny and egg hatchability. More promising stocks were selected for laboratory studies on the population-depressing effects of translocations.

5322. Wagoner, D. E. 1967 Linkage Group-Karyotype Correlation in the House Fly Determined by Cytological Analysis of X-ray Induced Translocations. Genetics 57(3):729-739.
A unification of nomenclature and genetic knowledge. Objective of the work was the correlation of the linkage groups to particular chromosomes. ABSTRACT: BA 49:2432(27346), Mar. 15, 1968.

5323. Wagoner, D. E. 1967 Insect Population Control by Genetic Methods Other than the Sterile-Male Technique. A. Introduction of Translocations into Insect Populations—*Musca domestica* L. Proc. N. Centr. Branch Ent. Soc. Am. 22:140. (Abstract and discussion.)
Translocation combinations currently under investigation to determine which are the best for decreasing fertility. Homozygous and heterozygous translocations as well as double and triple exchanges in various combinations were tested.

5324. Wagoner, D. E. 1969 Linkage Group-Karyotype Correlation in the House Fly, *Musca domestica* L., Confirmed by Cytological Analysis of X-ray Induced Y-Autosomal Translocations. Genetics 62:115-121.
Perje's method of numbering chromosomes was used. Males containing XX replaced males with X in subsequent generations. ABSTRACT: BA 51:2832(29390), Mar. 15, 1970.

5325. Wagoner, D. E. 1969 Suppression of Insect Populations by the Introduction of Heterozygous Reciprocal Chromosome Translocations. Bull. Ent. Soc. Am. 15(3):220.
Comparative fecundity, egg hatch and recovery of adult progeny from a known number of eggs showed that males and females bearing a specific translocation reduced the population to a low level in one generation and even more so in the 2nd generation.

5326. Wagoner, D. E. 1969 Presence of Male Determining Factors Found on Three Autosomes in the House Fly *Musca domestica.* Nature 223(5202):187-188.
Three autosomes in the Bowhill fly strain from Australia contained male-determining factors as shown by holandric inheritance. Sex determination in *Musca* appears more complex than previously recognized.

5327. Wagoner, D. E. 1969 Male Determining Factors Found on 3 Autosomes in the House Fly *Musca domestica.* Genetics 61(2) pt. 2 (S61). (Abstract.)
The Bowhill strain of fly was found to contain dominant male determining factors on autosomes II, III and V. Strain also contains XX males and females, and carries a female determining factor epistatic to all of the male determining factors.

5328. Wagoner, D. E. 1969 The Ruby Eye-color Mutant in the House Fly, *Musca domestica* L.; A Case of Duplicate Genes. Genetics 62: 103-113.
This is the first known case in the housefly of a recessive mutant controlled by two loci. The mutant is a phenotypic result of recessive duplicate genes situated on two nonhomologous chromosomes. ABSTRACT: BA 51:2832(29391), Mar. 15, 1970.

5329. Wagoner, D. E. and **Johnson, O. A. 1968** Reproductive Barriers in Various Geographic Strains of House Flies. Genetics 60(1) pt. 2: 233-234.
Matings of flies from various continents showed no classical cases of hybrid sterility or cytoplasmic incompatibility, but did show cases of reduced reproductive potential and some cases of sexual incompatibility.

5330. Wagoner, D. E., Johnson, O. A. and **Nickel, C. A. 1969** A New Modification of the Sex Determination Mechanism in an Australian Strain of the House Fly, *Musca domestica* L. Proc. N. Centr. Branch Ent. Soc. Am. 24(1):47-48.
The second pair of autosomes instead of the third was found to be involved in sex determination, with other pairs having modifying effects.

5331. Wagoner, D. E., Johnson, O. and **Nickel, C. A. 1965** The Linkage Group Karyotype Relationship in the House Fly, *Musca domestica.* Genetics 52(2) pt. 2:482-483.
Translocations and their cytological analysis have been used as a method of determining which linkage groups should be assigned to various chromosome pairs in the housefly karyotype.

5332. Wagoner, D. E., Nickel, C. A. and **Johnson, O. A. 1969** Chromosomal Translocation Heterozygotes in the House Fly. J. Heredity 60(5): 301-304.
Studies were made measuring the effects of irradiation on the fertility of treated flies, the frequency of induced translocations and reduction in fertility, the sex ratios of progeny, and the transmission of the translocation to progeny of heterozygous translocation bearing males. ABSTRACT: BA 51:5708(58532), June 1, 1970.

Wagoner, D. E., joint author. *See* Johnson, O., et al., 1968; Nickel, C. A., et al., 1968, 1969, 1970.

Wakamori, S., joint author. *See* Kobayashi, K., et al., 1969.

5333. Wal, Y. C., Perti, S. L. and **Damodar, P. 1962** Suceptibility of Some Insect Vectors of Diseases to Synthetic Contact Insecticides. Indian J. Malariol. 16(2):129-136.
Concerns the susceptibility of *Musca nebulo*, L., *Culex fatigans* Wied. and *Aedes aegypti* Linn. to modern synthetic contact insecticides such as aldrin, endrin, toxaphene, HETP, TEPP, etc. in relation to DDT, lindane and pyrethrins. ABSTRACTS: RAE-B 52:160, Sept., 1964; TDB 60:499, May, 1963.

Walker, C. B., joint author. *See* Dahm, P. A., et al., 1962.

5333a. Walker, C. R. and **Terriere, L. C. 1970** Induction of Microsomal Oxidases by Dieldrin in *Musca domestica.* Ent. Exp. Appl. 13(3): 260-274.
Dieldrin causes up to 5-fold increases in the activity of naphthalene hydroxylase and heptachlor epoxidase in the housefly. ABSTRACT: BA 52:317(2857), Jan. 1, 1971.

Walker, M., joint author. *See* Logan, J. C. P., et al., 1964.

Walker, R. L., joint author. *See* Bodenstein, O. F., et al., 1970.

5334. Walker, T. F. 1969 Investigations of Possible Sub-lethal Effects of DDT on Oögenesis in Houseflies. Trans. Roy. Soc. Trop. Med. Hyg. 63:430.

5335. Wallace, J. B. 1969 Notes on *Dendrophaonia querceti* with Descriptions of the Larva and Pupa. Diptera:Muscidae. J. Georgia Ent. Soc. 4(1):1-32.

A large number of *Fannia* (sp.) (probably *scalaris*) were found in yellow jacket's nest with *Dendrophaonia*. Identification was from the larvae as no hatching occurred. The report deals mainly with *Dendrophaonia*. ABSTRACT: BA 50:6978(72898), July 1, 1969.

5336. Wallace, J. B. and **Turner, E. C., Jr. 1962** Experiments for Control of the Face Fly in Virginia. J. Econ. Ent. 55(3):415-416.

Methods of control involved a residual insecticide application and feeding of a systemic insecticide. ABSTRACT: RAE-B 50:274, Dec., 1962.

5337. Wallace, J. B. and **Turner, E. C., Jr. 1964** Low-level Feeding of Ronnel in a Mineral Salt Mixture for Area Control of the Face Fly, *Musca autumnalis*. J. Econ. Ent. 57:264-267.

Evaluation of the feeding method: number of larvae are reduced, but not that of adult flies due to reinfestation pressure. ABSTRACTS: BA 45: 6394(79941), Sept. 15, 1964; RAE-B 52:126, July, 1964.

5338. Wallis, D. I. 1961 Response of the Labellar Hairs of the Blowfly, *Phormia regina* Meigen, to Protein. Nature 191(4791):917-918.

Three principal types of chemosensory hairs: (1) large peripheral; (2) intermediate; (3) small central (and scattered). Tracings indicate several specifically reacting nerve fibers in *one* sensory hair.

5339. Wallis, D. I. 1962 The Sense Organs on the Ovipositor of the Blowfly, *Phormia regina* Meigen. J. Insect Physiol. 8:453-467.

Organs include articulated setae, small pegs (cones), and possibly double-walled hairs. Only anal leaflets bear the last two. Most hairs on the ovipositor are of the "slow adapting" type.

5340. Wallis, D. I. 1962 Olfactory Stimuli and Oviposition in the Blowfly, *Phormia regina* Meigen. J. Exp. Biol. 39:603-615. London.

Attempts were made to elucidate sensory mechanisms involved in oviposition. This, and preceding references, included as contributions to muscid biology. ABSTRACT: RAE-B 51:273, Dec., 1963.

5341. Wallis, R. C. 1961 Common Connecticut flies. Conn. Agr. Expt. Sta. Bull. 650:1-23.

Musca domestica L. is now outranked, numerically, by blow flies and flesh flies. ABSTRACT: BA 40:312(4054), Oct. 1, 1962.

5342. Wallis, R. C. and **Scott, W. L. 1970** Primary Monolayer Cell Culture from the House Fly, *Musca domestica*. Ann. Ent. Soc. Am. 63(6): 1788-1790.

Concerns culture media and techniques. Monolayer cell sheets adequate for virus growth studies. Authors hope to obtain a continuous cell line.

5343. Wallwork, J. H. and **Rodriguez, J. G. 1963** The Effect of Ammonia on the Predation Rate of *Macrocheles muscaedomesticae* (Acarina: Macrochelidae) on House Fly Eggs. Advances in Acarology 1:60-69.

Increased biological control by the mites feeding upon housefly eggs in manure was attributed to a biting and puncturing response elicited by a critical level of liberated ammonia in the medium. ABSTRACT: BA 46: 3674(45387), May 15, 1965.

5344. Walsh, J. D. 1964 A Survey of Fly Production in Cattle Feedlots in the San Joaquin Valley. Calif. Vector Views 11(6):33-39.

Musca domestica and *Stomoxys calcitrans* were the most common species, especially where horses were concerned. Many recommendations were made. ABSTRACT: RAE-B 55:21(57), Feb., 1967.

5345. Walsh, J. D., Linsdale, D. D., White, K. E. and **Bergstrom, R. E. 1968** Fly Larval Migration from Residential Rcfuse Containers in the City of Fresno. Calif. Vector Views 15(6):55-62.
Migrating flies were mostly *Phaenicia cuprina* (green blow fly).

Walsh, J. D., joint author. *See* Smith, T. A., et al., 1970.

Walton, G. S., joint author. *See* Beard, R. L., et al., 1965.

5346. Wan, T. K. and **Hooper, G. H. S. 1967** A Possible Role for Aliesterase Enzymes in Insects. J. Australian Ent. Soc. 6(1):20-26.
The aliesterase enzymes are probably concerned with reproduction, especially öogenesis. Greater amounts occur in females than males, also in mated flies of both sexes. A cyclic activity with age, is shown. ABSTRACT: RAE-B 57:23(101), Jan. 1969.

5347. Wan, T. K. and **Hooper, G. H. S. 1969** Aliesterase Activity and Reproduction in *Musca domestica*: Effects of DDT and Metepa. Ent. Exp. Appl. 12(2):221-228.
Oral and topically administered metepa reduced aliesterase activity and hatch of eggs. It also prolonged the pre-ovipositional period. Orally administered DDT reduced egg viability but affected aliesterase activity little. ABSTRACTS: BA 51:4519(46559), Apr. 15, 1970; TDB 67:905(1737), July, 1970; RAE-B 58:270(1114), Aug., 1970.

5348. Wang, C. M. 1964 Laboratory Observations on the Life History and Habits of the Face Fly, *Musca autumnalis* (Diptera:Muscidae). Ann. Ent. Soc. Am. 57(5):563-569.
Stages of the life cycle are described and the effect of environmental factors and artificial mass-rearing techniques discussed. ABSTRACTS: BA 46:354(4454), Jan. 1, 1965; RAE-B 53:74, Apr., 1965; TDB 62:68, Jan., 1965.

Wang, C. M., joint author. *See* Turner, E. C., et al., 1964.

5349. Wang, Y. 1965 [On the overwintering of house-flies (*Musca domestica vicina* Macq.) in Chengtu.] Acta Ent. Sinica 14(2):163-170. (In Chinese, with English summary.)
Breeding of flies continued through the winter in towns but semi-hibernation prevailed in rural areas. Temperatures for survival, different ways of overwintering, hibernation, breeding and functioning of the fat body during hibernation are discussed; also methods of controlling larvae, pupae and adults. ABSTRACTS: BA 49:6745(75224), July 15, 1968; RAE-B 53:230, Dec., 1965.

5350. Wang, Y.-C. 1968 On the Overwintering of Houseflies (*Musca domestica vicina* Macq.) in Chengtu. Acta Ent. Sinica (English Transl.) 1965 (2):1-9. Washington, D.C.

Wang, Y.-H., joint author. *See* Lee, W.-Y., et al., 1967.

Ward, F. L., Jr., joint author. *See* Sun, Y. P., et al., 1967.

Ward, G. L., joint author. *See* Bay, D. E., et al., 1968, 1969, 1970.

5351. Ward, J. L., Gillham, E. M. and **Potter, C. 1960** A Thermal Preference Method of Bioassay of the Toxicity of Insecticidal Films on Houseflies. Bull. Ent. Res. 51(2):379-387.
The floor of the testing chamber was maintained at a higher (preferred) temperature than were the walls and lid. This helped to keep flies on the treated surface during the test. ABSTRACTS: BA 36:6874(74154), Nov. 1, 1961; RAE-B 48:182-183, Oct., 1960; TDB 58:144, Jan., 1961.

5352. **Ware, G. W. 1960** The Penetration of Piperonyl Butoxide as a Synergist and as an Antagonist in *Musca domestica* L. J. Econ. Ent. 53(1):14-16.
The penetration rate of piperonyl butoxide does not appear to determine its role as a synergist or antagonist. ABSTRACTS: BA 35:3456(39628), July 15, 1960; RAE-B 49:80, Apr., 1961.

5353. **Ware, G. W. 1966** Power-mower Flies. J. Econ. Ent. 59(2):477-478.
Muscina, *Fannia*, *Leptocera* and *Stomoxys* larvae and/or pupae were found in green, caked residue. ABSTRACTS: BA 47:7229(84339), Sept. 1, 1966; RAE-B 54:134, July, 1966.

5354. **Ware, G. W.** and **Terranova, A. C. 1960** The Effects on House Flies of Adding Wheat Germ and Corn Oils to Fly Media. J. Econ. Ent. 53(2):325-326.
Considerable variation occurred from generation to generation. However, differences reveal no meaningful biological conclusions or any evidence of carry over effect of diet to subsequent generations. ABSTRACTS: BA 35:3895(44892), Aug. 15, 1960; RAE-B 49:106, May, 1961; TDB 57:1227, Nov., 1960.

5355. **Ware, G. W.** and **Whitacre, G. K. 1970** Similar Effects of X-Irradiation and Piperonyl Butoxide on Toxicity of Malathion and Malaoxon to House Flies. J. Econ. Ent. 63(5):1621-1622.
Application of PB to adults, and X-irradiation of pupae with 10,000 R, caused similar and additive effects on flies treated topically with either of these two insecticides. ABSTRACT: BA 52:2002(19122), Feb. 15, 1971.

Ware, G. W., joint author. *See* Barnes, W. W., et al., 1965; Guenthner, A. W., et al., 1967; Whitacre, G. C., et al., 1970.

Warren, J. W., joint author. *See* Smith, C. N., et al., 1960.

Wasco, J. L., joint author. *See* Blair, E. H., et al., 1965.

5355a. **Wasti, S. S., Shaw, F. R.** and **Smith, C. T. 1970** Detection of Residues of Rabon in Manure of Rhode-Island-Red Hens. J. Econ. Ent. 63(4):1355-1356.
Relates to control of *M. domestica* by systemic insecticides.

5356. **Waterhouse, D. F.** and **Norris, K. R. 1966** Bushfly [*Musca vetustissima* Walk.] Repellents [Human Pest]. Australian J. Sci. 28(9):351.
Repellents containing di-N-propyl isocinchomeronate proved the most effective. ABSTRACT: BA 48:416(4318), Jan. 1, 1967.

Waters, J. A., joint author. *See* Thompson, M. J., et al., 1962, 1963.

Wattal, B. L., joint author. *See* Raghaven, N. G. S., et al., 1967.

Wei, P. H., joint author. *See* Kao, C. M., et al., 1960.

Weidel, G., joint author. *See* Reichardt, W., et al., 1968.

5357. **Weiden, M. H. J. 1968** Insecticidal Carbamoyloximes from Symposium on Pesticidal Carbamates. J. Sci. Fd. Agric. Suppl., 1968, pp. 19-31.
Carbamoyloximes have resemblance to carbamates of cyclic enols and phenols. Methylenedioxyphenyl derivatives synergize many of these carbamates against *Musca domestica* L. ABSTRACT: RAE-B 58:34(143), Feb., 1970.

5358. **Weiden, M. H. J.** and **Moorefield, H. H. 1964** Insecticidal Activity of the Commercial and Experimental Carbamates. Wld. Rev. Pest Control 3(2):102-107.
Current summary of the insecticidal activities of the carbamates. Screening operations over past decade tested 29 compounds against 5 test insects including *Musca domestica* L. ABSTRACTS: BA 48:3705(41109), Apr. 15, 1967; RAE-B 52:201, Nov., 1964.

5359. **Weiden, M. H. J., Moorefield, H. H.** and **Payne, L. K. 1965** 0-(Methylcarbamoyl) Oximes: a New Class of Carbamate Insecticide-acaricides. J. Econ. Ent. 58:154-155.
Five compounds were tested for toxicity to insects including *M. domestica.* Paper also deals with housefly head cholinesterase inhibition. ABSTRACTS: RAE-B 53:97, May, 1965; TDB 62:696, July, 1965.

Weiden, M. H. J., joint author. *See* Moorefield, H. H., et al., 1964; Durden, J. A., Jr., et al., 1969.

5360. **Weidhaas, D. E. 1968** Field Development and Evaluation of Chemosterilants. *In*: Principles of Insect Chemosterilization. Appleton-Century-Crofts. New York. pp. 275-314.
A review of sterility methods by chemicals. ABSTRACT: BA 49:8726 (97119), Sept. 15, 1968.

5361. **Weidhaas, D. E., Schmidt, C. H.** and **Chamberlain, W. F. 1962** Research on Radiation in Insect Control. *In*: Proc. Symp. on Radioisotopes and Radiation in Entomology — Bombay. December 5-9. pp. 257-263.
The technique of releasing x-radiation-sterilized males is discussed; also the lethal effects of irradiation on insects of medical importance, including *Musca domestica.* ABSTRACT: RAE-B 51:213, Oct., 1963.

Weidhaas, D. E., joint author. *See* Schmidt, C. H., et al., 1964; LaBrecque, G. C., et al., 1970.

5362. **Weinburgh, H. B., Kilpatrick, J. W.** and **Schoof, H. F. 1961** Field Studies on the Control of Resistant House Flies. J. Econ. Ent. 54(1):114-116.
Bayer 29493 as a residual insecticide, Dimetilan-impregnated fly bands and Ronnel-impregnated cords gave best results. ABSTRACT: BA 36: 2895(35504), June, 1961.

5363. **Weiser, J. 1965** The Bacterial Pollution of Agricultural Air. Arch. Belg. Med. Soc. Hyg. Med. Trav. Med. Leg. 23(4):274-279.
The influence of air movement, dust, the human factor, flies, and the worker's body posture upon the aerobic aerial bacterial flora was investigated during the performance of different tasks and types of agricultural work. ABSTRACT: BA 47:6776(79204), Aug. 15, 1966.

Welle, H. B. A., joint author. *See* Meltzer, J., et al., 1967, 1969.

Wells, C. N., joint author. *See* Rodriguez, J. G., et al., 1962.

Wellso, S. G., joint author. *See* Frudden, L., et al., 1968.

Wenking, H., joint author. *See* Reichardt, W., et al., 1969.

Weorden, S., joint author. *See* Harris, R. L., et al., 1961.

West, A. S., Jr., joint author. *See* Brown, A. W. A., et al., 1961.

5364. **West, J. A., Cantwell, G. E.** and **Shortino, T. J. 1968** Embryology of the House Fly, *Musca domestica* (Diptera:Muscidae) to the Blastoderm Stage. Ann. Ent. Soc. Am. 61(1):13-17.
The pattern of normal embryonic development was established. Serial sections of dechorionated eggs were used for study. ABSTRACTS: BA 49: 5301(59486), June 1, 1968; TDB 65:843(1872), June, 1968.

5365. **Wharton, R. H., Seow, C. L., Ganapathipillai, A.** and **Jabaratnam, G. 1962** Housefly Populations and Their Dispersion in Malaya with Particular Reference to the Fly Problem in the Cameron Highlands. Med. J. Malaya 17:115-131.
House flies are scarcer in Malaya than in other hot countries, but because *Musca domestica vicina* poses a problem by breeding in fertilizer for vegetable gardens, marking experiments using ^{32}P were made, to study

fly movements. Paper also deals with resistance. ABSTRACTS: BA 45:4320 (53261), June 15, 1964; TDB 60:795, Aug., 1963; RAE-B 53:47, Feb., 1965.

Whetstone, T. M., joint author. *See* Drummond, R. O., et al., 1967.

5366. Whitacre, G. K. and **Ware, G. W. 1970** Effects of Gamma- and X-Irradiation on the Toxicity of Malathion to House Flies. J. Econ. Ent. 63(2):424-426.

An increase in resistance of flies reared from irradiated pupae may be due to an increase in levels of detoxifying enzymes. Gamma-Irradiation had a more pronounced effect than did X-Irradiation. ABSTRACTS: BA 51:8567(87360), Aug. 1, 1970; RAE-B 58:303(1238), Sept., 1970.

Whitacre, G. K., joint author. *See* Ware, G. W., et al., 1970.

5367. White, E. B. and **Legner, E. F. 1966** Notes on the Life History of *Aleochara taeniata*, a Staphylinid Parasite of the House Fly, *Musca domestica*. Ann. Ent. Soc. Am. 59:573-577.

Stages of the life cycle of the parasite were studied and described as a foundation for use of biological control against *Musca domestica*. ABSTRACTS: BA 47:9054(104722), Nov. 1, 1966; RAE-B 54:197, Oct., 1966.

White, E. B., joint author. *See* Legner, E. F., et al., 1967.

White, K. E., joint author. *See* Walsh, J. D., et al., 1968; Ecke, D. H., et al., 1965.

Whitehead, N., joint author. *See* Jordan, T. W., et al., 1968.

Whitfield, T. L., joint author. *See* Bailey, D. L., et al., 1970.

Woods, C. W., joint author. *See* Chang, S. C., et al., 1970.

5368. Whiting, A. R. 1967 The Biology of the Parasitic Wasp *Mormoniella vitripennis* [=*Nasonia brevicornis*]. (Walker.) Quart. Rev. Biol. 42(3):333-406.

The biology, economic importance, physiology, ecology, cytology and genetics of the parasitic wasp are reviewed, and the possibility for research use is considered. ABSTRACT: BA 49:5296(59436), June 1, 1968.

5369. Whitney, W. K. 1967 Laboratory Test with Dursban and Other Insecticides in Soil. J. Econ. Ent. 60(1):68-74.

The effectiveness of soil insecticides was studied on *Musca domestica* and some other (non-soil) insects. Dursban was superior to 31 other commercial insecticides. ABSTRACT: BA 48:6013(67212), July 1, 1967.

5370. Whitney, W. K., Kenaga, E. E., Hardy, J. L. and **Doty, A. E. 1969** Rapid Knockdown Activity of Dowco^R 217, a New Insecticide. J. Econ. Ent. 62(3):567-568.

Dowco^R 217 showed an unusually fast knockdown action against mosquitoes, cockroaches and the housefly, and had a low mammalian toxicity. ABSTRACTS: BA 50:11366(117549), Nov. 1, 1969; RAE-B 57:236 (894), Nov., 1969.

Whitney, W. K., joint author. *See* Kenaga, E. E., et al., 1965.

5371. Whitten, J. M. 1963 Observations on the Cyclorrhaphan Larval Peripheral Nervous System: Muscle and Tracheal Receptor Organs and Independent Peripheral Type II Neurons Associated with the Lateral Segmental Nerves. Ann. Ent. Soc. Am. 56:755-763.

Included for its value in muscoid biology. Work done chiefly on *Sarcophaga bullata*.

5372. Whitten, J. M. 1963 Giant Polytene Chromosome Development in Cells of the Hypodermis and Heart of Fly Pupae. *In* 16th Int. Cong. Zool. Proc. Int. Congr. Zool. 16(2):276. (Abstract only.)

5373. **Whitten, J. M. 1964** Giant Polytene Chromosomes in Hypodermal Cells of Developing Footpads of Dipteran Pupae. Science 143 (3613):1437-1438.
Suggests the use of these conveniently located cells for experimental and genetic studies. ABSTRACT: BA 45:6102(76114), Sept. 15, 1964.

5374. **Whitten, J. M. 1964** Stretch Receptor-like Organs in the Fly Larva. Their Possible Role in Growth Regulation. Science 143(3603): 260-261.
Cells present in larvae of *Sarcophaga bullata* and other flies, which are attached to certain structures, may be concerned with the control of metamorphosis of these structures. ABSTRACT: BA 45:5095(62396), July 15, 1964.

5375. **Whitten, J. M. 1965** Differentiation in the Fly Foot: Coordinated Changes in Hemocytes, Tenent Hair Cells and Giant Foot-pad Cells and Chromosomes. Am. Zool. 5(4):708.

5376. **Whitten, M. J.** and **Taylor, W. C. 1970** A Role for Sterile Females in Insect Control. J. Econ. Ent. 63(1):269-272.
Laboratory tests indicate that the release of sterile females leads to ineffective or noninsemination of normal females due to limited mating capacity of males (usually 10 mates) thus possibly decreasing the population size.

5377. **Wicht, M. C., Jr.** and **Hays, S. B. 1967** Effect of Reserpine on Reproduction of the House Fly. J. Econ. Ent. 60:36-38.
Reserpine reduced mating activity and oviposition in female house flies, although egg development within ovaries remained normal. ABSTRACTS: BA 48:6045(67559), July 1, 1967; RAE-B 55:103(368), June, 1967.

Wickham, J. C., joint author. *See* Brown, N. C., et al., 1967.

5378. **Wickramasinghe, D. N. T. 1965** Observations on the Mode of Action of 2-inidazolidinone, a Female Sterilant of the Adult House Fly, *Musca domestica* L. (Diptera:Muscidae). Diss. Absts. 26(2):680.
ABSTRACT: BA 47:4256(49954), May 15, 1966.

5379. **Wiesmann, R. 1960** Zum Nahrunsproblem der freilebenden Stubenfliegen *Musca domestica* L. [Feeding habits of wild *Musca domestica.*] Zeit. Angew. Zool. 47:159-181.
Crop contents of wild flies were analysed chromatographically and the amino-acids, proteins, lipoids and different kinds of sugars assessed. Comparisons are made with domestic living flies. ABSTRACT: TDB 58: 142, Jan., 1961.

5380. **Wiesmann, R. 1960** Untersuchungen über die Sinnesfunktionen der Antennen von *Musca domestica* L. im Zusammenhang mit dem Köderproblem. [Studies of the sensory function of the antennae of *Musca domestica* in connection with the problem of baits.] Mitt. Schweiz. Ent. Ges. 33(3):121-154.
Morphology and histology of the antennae are described in detail. Studies of baits show smell alone cannot be used to attract flies to bait. (The antennae are sensitive to odor, humidity, heat, shock and air disturbances.) ABSTRACT: RAE-B 50:210, Oct., 1962.

5381. **Wiesmann, R. 1960** Neue Mittel und Methoden zur Fliegenbekampfung im Stall. [New methods of combating flies in stables.] Schweiz. Arch. Tierheilk. 102(3):134-146.
Laboratory experiments in Switzerland with Dimetilan showed it to be effective against DDT resistant strains when applied on cords hanging in barns. ABSTRACTS: BA 35:3741(43013), Aug. 1, 1960; RAE-B 50:49, Mar., 1962.

5382. **Wiesmann, R. 1961** Untersuchungen über Fliegenkot-konzentrationen und Fliegenansammlungen in Viehställen. [Investigations on concentrations of fly excreta and aggregation of flies in cattle sheds.] Mitt. Schweiz. Ent. Ges. 34(2):187-209.
Flies fed on milk, as those in milking barns, flew less and tended to aggregate in places on ceilings where the temperature was 1 or more degrees higher. The lipoids of milk were shown to be a rather unsuitable. food for flies. ABSTRACT: RAE-B 51:194, Sept., 1963.

5383. **Wiesmann, R. 1961** The Control of Polyvalent Resistant Houseflies in Switzerland. Bull. W. H. O. 24:672-674.
Control of *Musca domestica* during 1958-1959 was excellent, using Dimetilane. Stomach poisons were preferable to contact poisons ABSTRACT: RAE-B 51:136, June, 1963.

5384. **Wiesmann, R. 1962** Neue Erkentnisse aus der Biologie von *Musca domestica* L. im Zusammenhang mit der Insektizidresinstenz. [The biology of *Musca domestica* in relation to insecticide resistance.] J. Hyg., Epidemiol., Microbiol., Immunol. (Prague) 6(3):303-321. (Summaries in English, French and Italian.)
In a search for new and effective fly control methods, detailed investigations were made of behavior of flies in natural biotope, physiology of nourishment, sensory function in connection with olfactory orientation and the question of olfactory baits, and reactions to various colors. ABSTRACT: BA 42:1250(15809), May 15, 1963; TDB 60:501, May, 1963.

5385. **Wiesmann, R. 1962** Geruchsorientierung der Stubenfliege *Musca domestica* L. [Olfactory orientation of *Musca domestica*.] Zeit. Angew. Ent. 50:74-81. (English summary.)
Attraction of flies to food being fed upon by other flies is due to the formation of sugar solutions combined with and supplemented by the visual aggregate instinct of flies. ABSTRACT: RAE-B 51:233, Nov., 1963.

5386. **Wiesmann, R. 1962** Unterushcungen über den "Fly-factor" und den Herdentrieb bei der Stubenfliege, *Musca domestica* L. [Investigations on the "Fly Factor" and the gregarious instinct in *Musca domestica*.] Mitt. Schweiz. Ent. Ges. 35(1-2):69-114.
A summary of results on the attraction of flies to certain baits. No unknown attractive substance is produced and deposited by flies, the "fly factor" being only attraction to other flies feeding. ABSTRACT: RAE-B 52:115, June, 1964.

5387. **Wiesmann, R. 1965** The Open Stall, an Entomological Problem. Schweiz. Arch. Tierheilk. 107(1):10-18.

5388. **Wigglesworth, V. B. 1963** A Further Function of the Air Sacs in Some Insects (*Lucillia*, *Calliphora*, *Drosophila*, *Musca*). Nature 198(4875): 106.
Original observations were made on *Drosophilia* but are believed to be the same for *Musca* and *Calliphora*, namely, the restriction of circulating blood volume, thus increasing the speed and efficiency of transport (especially sugar from abdominal fat body to flight muscles). ABSTRACT: BA 43:1671(21245), Sept. 1, 1963.

5389. **Wilkinson, C. F. 1966** The Relationship of Structure and Mode of Action of Synergists for Carbamate Insecticides. Diss. Absts. 26(12) pt. 1:7055-7056.

5390. **Wilkinson, C. F. 1967** Penetration, Metabolism, and Synergistic Activity with Carbaryl and Some Simple Derivatives of 1, 3-benzodioxide in the Housefly. J. Agric. Fd. Chem. 15(1):139-147.
Twenty-three derivatives were evaluated at each of four synergist-

insecticide ratios, against the housefly. Many were superior to sesamex, especially at low synergistic-insecticide ratios, the most marked in this respect being those incorporating nuclear nitro and methoxy groups. ABSTRACTS: BA 48:9958(111945), Nov. 15, 1967; RAE-B 55:128(462), July, 1967.

5391. Wilkinson, C. F., Metcalf, R. L. and **Fukuto, T. R. 1966** Some Structural Requirements of Methylenedioxyphenyl Derivatives as Synergists of Carbamate Insecticides. J. Agric. Fd. Chem. 14:73-79.
Sixty-two compounds were evaluated. Maximum synergistic activity is associated with the planar methylenedioxyphenyl ring system. Replacement of oxygen by sulfur decreases activity slightly. ABSTRACT: RAE-B 54:160, Aug., 1966.

Wilkinson, C. F., joint author. *See* Hewlett, P. S., et al., 1967; Lewis, S. E., et al., 1967; Metcalf, R. L., et al., 1966.

5391a. Williamson, R. L. 1968 Mechanisms of Inhibition and Uncoupling of Oxidative Phosphorylation in House Fly Mitochondria. Diss. Absts. Part B. Sci. Eng. Page 235-B.
Utilized radioactive phosphorus.

5392. Williamson, R. L. and **Metcalf, R. L. 1967** Salicylanilides: A New Group of Active Uncouplers of Oxidative Phosphorylation. Science 158(3809):1694-1695.
The salicylanides are the most effective uncoupling agents reported so far. Housefly mitochondria utilized in study. ABSTRACT: BA 49:3453 (38719), Apr. 15, 1968.

5393. Willis, R. R. and **Axtell, R. C. 1967** Evaluation of *Fuscuropoda vegetans* (Acarina: Uropodidae) as a Predator of the Immature Stages of the Housefly. J. Elisha Mitchell Scientific Society 83(3) (Fall number): 1 page.
F. vegetans is a predator of the 1st instar larvae only; combined with *Macrocheles muscaedomesticae*, the joint predation rate is very high.

5394. Willis, R. R. and **Axtell, R. C. 1968** Mite Predators of the House Fly: A Comparison of *Fuscuropoda vegetans* and *Macrocheles muscaedomesticae*. J. Econ. Ent. 61(6):1669-1674.
Results from laboratory and field observations show that *F. vegetans* cannot penetrate fly eggs but feeds on ruptured ones. It dwells more deeply within the manure and is at maximum number after 5-6 weeks. *M. muscaedomesticae* is found on outer layers and is at a maximum density at 2-3 weeks. ABSTRACTS: BA 50:4779(50244), May 1, 1969; RAE-B 57:88(308), May, 1969; TDB 66:1186(2311), Nov., 1969.

5395. Wilson, B. H. 1962 Effectiveness of WARF-antiresistant as an Additive to DDT for Control of Resistant Houseflies. J. Econ. Ent. 55(5):792-793.
Toxicity values are given for DDT and for DDT plus WARF antiresistant. ABSTRACTS: BA 41:950(12367), Feb. 1, 1963; RAE-B 51:43, Feb., 1963.

5396. Wilson, B. H. and **Burns, E. C. 1968** Induction of Resistance to *Bacillus thuringiensis* in a Laboratory Strain of House Flies. J. Econ. Ent. 6(6):1747-1748.
Experimental exposure over 35 generations produced a 6-fold resistance. ABSTRACTS: BA 50:4780(50245), May 1, 1969; RAE-B 57:89(312), May, 1969.

Wilson, B. H., joint author. *See* Burns, E. C., et al., 1961.

5397. Wilson, D. M. and **Wyman, R. J. 1963** Phasically Unpatterned Nervous Control of Dipteran Flight. J. Insect Physiol. 9(6):859-865.
Flight co-ordination in flies was studied, using *Musca domestica* and

Eucalliphora lilaea, from nature. ABSTRACT: BA 45:3982(48967), June 1, 1964.

5398. Wilson, H. G., Gahan, J. B. and **LaBrecque, G. C. 1967** New Insecticides That Show Residual Toxicity to Adult Houseflies. ARS 33-124, 15 pp. Beltsville, Maryland.
Over a period of three years 168 compounds, mostly organophosphorus compounds and carbamates were tested in the U.S. as contact sprays against a strain of *Musca domestica* kept since 1943 without exposure to insecticides. Ten remained effective for 24 weeks, 2 for 20 weeks and 23 for 4-12 weeks. ABSTRACT: RAE-B 57:214(776), Oct., 1969.

5399. Wilson, H. G. and **LaBrecque, G. C. 1960** Tests with Larvicides for the Control of House Flies in Poultry Houses. Florida Ent. 43(1):19-21.
Eight compounds, chiefly dusts or sprays of pyrophyllite, were tested as larvicides against natural populations of houseflies. Baytex (not a pyrophyllite) was effective at low dosages. ABSTRACT: RAE-B 49:243, Nov., 1961.

5400. Wilson, H. G., LaBrecque, G. C. and **Gahan, J. B. 1961** Laboratory Tests of Selected House Fly Repellents. Florida Ent. 44:123-124.
Twenty-six additional chemicals were tested by the same methods, as vapor or contact repellents. ABSTRACTS: BA 38:284(3869), Apr. 1, 1962; RAE-B 51:89, May, 1963.

Wilson, H. G., joint author. *See* Morgan, P. B., et al., 1966; Smith, C. N., et al., 1964.

5401. Wilson, J. A. and **Hays, S. B. 1969** Histological Changes in the Gonads and Reproductive Behavior of House Flies Following Treatment with Chemosterilants, p, p-bis (1-aziridinyl), N-Methylphosphinic Amide and p, p-bis (1-aziridinyl)-N-(3-Methoxyprophy) Phosphinothioic Amide. J. Econ. Ent. 62(3):690-692.
A reduction in spermatocyte cells, but not aspermia, resulted in males fed the compounds. Females exhibited changes in the primary oöcytes, yolk vacuolization, and atrophy of the ovaries after the 10th day. No changes occurred in mating behavior. ABSTRACTS: BA 50:11366(117550), Nov. 1, 1969; TDB 67:905(1735), July, 1970.

Wilson, J. A., joint author. *See* Kissam, J. B., et al., 1967.

5402. Wilton, D. P. 1961 Refuse Containers as a Source of Flies in Honolulu and Nearby Communities. Proc. Hawaiian Ent. Soc. 17(3):477-481.
The dominant species were *Phaenicia cuprina* Wied. and *Musca domestica.* Flies were collected from 13 per cent of individual dwelling sites and from 42 per cent of the apartment houses. ABSTRACTS: BA 37:1131(11825), Feb. 1, 1962; RAE-B 51:159, Aug., 1963.

5403. Wilton, D. P. 1963 Dog Excrement as a Factor in Community Fly Problems. Proc. Hawaiian Ent. Soc. (1962) 18(2):311-317.
Musca domestica, *Sarcophagula occidua* and *Musca sorbens* account for 99 per cent of the flies reared from nature. Rain and lawn watering prevent drying action. Dogs frequently carry *Salmonella.* ABSTRACT: RAE-B 53:134, July, 1965.

5404. Wingo, C. W. 1970 Laboratory Adaptation of an Indigenous Braconid Parasite to the Face Fly. J. Econ. Ent. 63(3):748-751.
A colony of *Aphaereta pallipes* (Say) was developed by subjecting the abnormal host to oviposition through 27 generations. Highest parasitism occurred during the 13th generation, the highest rate of emergence was in the 18th. The colony declined in ability to parasitize and break out of the puparium. ABSTRACTS: BA 51:13161(133977), Dec. 1, 1970; RAE-B 58:372(1532), Nov., 1970.

5405. Wingo, C. W., Thomas, G. D. and **Nelms, N. M. 1967** Laboratory Evaluation of Two Aleocharine Parasites of the Face Fly. J. Econ. Ent. 60:1514-1517.

A staphylinid from France, *Aleochara tristis*, showed an average of 13·5 per cent parasitism of pupae (indicating low searching capacity) and caused a reduction of approximately 10 per cent of face flies reaching the pupal stage by predation on face fly eggs and larvae. *A. bimaculata*, an American species, caused a reduction of 19·4 per cent by predation. ABSTRACTS: BA 49:3809(42800), Apr. 15, 1968; RAE-B 56:97(338), May, 1968.

Wingo, C. W., joint author. *See* Benson, O. L., et al., 1963; Houser, E. C., et al., 1967; Zapanta, H. M., et al., 1968; Thomas, G. D., et al., 1968.

5406. Winkler, L. R. and **Wagner, E. D. 1961** A Cultured Life Cycle of the Canyon Fly, *Fannia benjamini* Malloch with Observations on the Natural History. Trans. Am. Micr. Soc. 80(2):179-185.

Flies were reared from eggs in the laboratory. Photographs and photomicrographs illustrate the stages of the developing fly. ABSTRACT: BA 36:5898(62276), Sept. 15, 1961.

5407. Winteringham, F. P. W. 1960 Phosphorylated Compounds in the Head and Thoracic Tissues of the Adult Housefly (*Musca domestica* L.) During Flight, Rest, Anoxia and Starvation. Biochem J. 75(1): 38-45.

Several significant observations were made. Example: Starvation to the point of prostration caused a fall in head adenosine triphosphate, which could be reversed by injection of aqueous glucose. ABSTRACT: TDB 57:856, Aug., 1960.

5408. Winteringham, F. P. W. 1966 Metabolism and Significance of Acetylcholine in the Brain of the Adult Housefly, *Musca domestica* L. J. Insect Physiol. 12(8):909-924.

Methods for determination of [^{14}C] acetylcholine are given. Problems of insect neurophysiology are discussed. ABSTRACT: BA 47:9500(109961), Nov. 15, 1966.

5409. Winteringham, F. P. W. and **Disney, R. W. 1964** A Radiometric Study of Cholinesterase and its Inhibition. Biochem. J. 91:506-514.

A simple radiometric method for microestimation of acetylcholinesterase activity over a wide range of substrate concentrations is described. Adult housefly heads (and mammalian blood) were used. Indicates that an underestimation of cholinesterase inhibition by carbamates may exist in earlier studies. ABSTRACT: RAE-B 54:128, July, 1966.

5410. Winteringham, F. P. W. and **Harrison, A. 1961** Incorporation of [2-^{14}C] Acetate into Acetylcholine of the Adult Housefly *in vivo* under Conditions of Rest, Activity and Insecticidal Action. Biochem. J. 78(2) Proc. Biochem. Soc. p. 22.

Continuous cyclopropane anaesthesia lowered the rate of formation of [^{14}C] acetylcholine and the metabolism of acetate. There is no evidence that dieldrin directly affects metabolism of acetycholine *in vivo*. ABSTRACT: TDB 58:962, Aug., 1961.

5411. Winteringham, F. P. W., Hellyer, G. C. and **McKay, M. A. 1960** Effects of the Insecticides DDT and Dieldrin on Phosphorus Metabolism of the Adult Housefly [*Musca domestica* L.]. Biochem. J. 76(3):543-548.

A fall in ATP and respiration rate of flies poisoned by DDT was not due to exhaustion of endogenous reserves or to hypermotor activity induced by DDT. In flies poisoned with both insecticides there occurred a fall in thoracic OC glycerophosphate which could not be reversed by cyclopropane anaesthesia. ABSTRACTS: BA 35:5797(68608), Dec. 1, 1960; RAE-B 51:68, Mar., 1963; TDB 57:1319, Dec., 1960.

5412. **Winteringham, F. P. W.** and **Hewlett, P. S. 1964** Insect Cross Resistance Phenomena: Their Practical and Fundamental Implications. Chem. and Ind. for 1964, pp. 1512-1518.
Many diagrams of molecular structure are included. Discussion indicates that the practice of alternating pesticides is probably advantageous, provided cross-resistance between the two insecticides is negligible.

Winton, M. Y., joint author. *See* Fukuto, T. R., et al., 1961, 1962, 1963, 1964; Metcalf, R. L., et al., 1960, 1962, 1963.

5413. **Witt, J. M. 1966** The Metabolism of Carbon-14-labelled DDT by the Large Milkweed Bug, the Housefly and the American Cockroach. Diss. Absts. 26(12) pt. 1:7020.

Wittmann, W., joint author. *See* Wagner, S., et al., 1966.

Woehst, J. E., joint author. *See* Miles, J. W., et al., 1962.

Woeken, J. J., joint author. *See* Marak, G. E., et al., 1968.

Wojciak, Z., joint author. *See* Bojanowska, A., et al., 1960.

Wolf, W. W., joint author. *See* Killough, R. A., et al., 1965.

5414. **Wolff, H. L., et al. 1969** Houseflies, the Availability of Water, and Diarrhoeal Diseases. Bull. W. H. O 41:952-959.
Concerns *Salmonella, Shigella* and *Vibrio cholerae.*

5415. **Wolken, J. J., Bowness, J. M.** and **Scheer, I. J. 1960** The Visual Complex of the Insect: Retinene in the Housefly. Biochim. Biophys. Acta. 43:531-537.
Retinal reported in *Musca domestica.* Was formerly not found by these authors. (Bowness and Wolken, 1959.)

Wolken, J. J., joint author. *See* Marak, G. E., et al., 1968, 1969.

Wong, W., joint author. *See* Brown, P., et al., 1970.

Wood, P. W., joint author. *See* Pimentel, D., et al., 1965.

Woodcock, J. G., joint author. *See* Tracy, R. L., et al., 1960.

Woods, C. W., joint author. *See* Bořkovec, A. B., et al., 1963, 1965, 1966; Chang, S. C., et al., 1966, 1967, 1968, 1970.

Wool, D., joint author. *See* Sokal, R. L., et al., 1970.

5416. **World Health Organization. 1960** Insecticide Resistance and Vector Control. Tenth Report of the Expert Committee on Insecticides. W. H. O. Tech. Rept. Ser. N. 191. Geneva. 98 pp., illus.
An appropriate portion is devoted to fly control.

5417. **World Health Organization Scientific Group. 1964** Genetics of Vectors and Insecticide Resistance. W. H. O. Tech. Rept. Ser. 268:1-40.
Particular attention is given to *Musca domestica*, and the mosquitoes *A. aegypti* and *C. pipiens.* For *Musca domestica* over 100 visible mutants have been found, thirty examined in some detail. ABSTRACT: RAE-B 53:65, Apr., 1965.

5418. **World Health Organization. 1965** Standard Reference Strain of *Musca domestica* (SRS/*Musca domestica*/1) WHO/Vector Control/113.65, 7(+2) pp. Geneva. [2 pls., multigraph.]

5419. **World Health Organization. 1966** Standardized Strains of Insects of Public Health Importance. Bull. W. H. O. 34:437-460.
Compiled by nine scientists, representing the areas of Entomology concerned. States that an adequate number of standardized strains of *M. domestica* are available, for current research needs.

5420. Wrich, M. J. 1970 Horn Fly and Face Fly Control on Beef Cattle Using Back Rubbers and Dust Bags Containing Coumaphos or Fenthion. J. Econ. Ent. 63(4):1123-1128.
Both chemicals, regardless of concentrations, provided good horn fly control and aided in reducing face fly populations. A tank-type rubber containing 1 per cent fenthion completely controlled *M. autumnalis.*

Wright, A. M., joint author. *See* Hodgson, E. S., et al., 1963.

5421. Wright, G. G. 1964 Flies (Diptera) are Filthy Pests and are Disease Carriers. New Zealand J. Agr. 109(5):445-456.

5421a. Wright, J. W. and **Pal, R. (Editors). 1968** Genetics of Insect Vectors of Disease. W. H. O., Geneva. Publ. by American Elsevier, New York. 813 pp., 50 tables, 312 illus.
Section B deals with Genetics of Muscoid Flies.

5422. Wylie, H. G. 1962 An Effect of Host Age on Female Longevity and Fecundity in *Nasonia vitripennis* (Walk.) (Hymenoptera:Pteromalidae). Canad. Ent. 94:990-993.
The longevity and fecundity of the female parasite is greatest when she feeds on pupae less than 48 hours old. ABSTRACT: RAE-B 51:219, Oct., 1963.

5423. Wylie, H. G. 1963 Some Effects of Host Age on Parasitism by *Nasonia vitripennis* (Walk.) (Hymenoptera:Pteromalidae). Canad. Ent. 95 (8):881-886.
Mortality of immature *Nasonia* was least on young house fly pupae and increased with increasing host age. This was principally manifested by a decrease in female adult progeny.

5424. Wylie, H. G. 1964 Effect of Host Age on Rate of Development of *Nasonia vitripennis* (Walk.) (Hymenoptera:Pteromalidae). Canad. Ent. 96(7):1023-1027.
Development time for the parasite was shorter on young pupae (less than 48 hours old) of *Musca domestica.* Temperature at 24·5 ±0·5°C.

5425. Wylie, H. G. 1965 Effects of Superparasitism on *Nasonia vitripennis* (Walk.) (Hymenoptera:Pteromalidae). Canad. Ent. 97(3):326-331.
Superparasitism creates a food shortage reducing survival and size of *Nasonia* adults and also the percentage of females in adult progeny. ABSTRACT: RAE-B 54:82, Apr., 1966.

5426. Wylie, H. G. 1965 Discrimination Between Parasitized and Unparasitized House Fly Pupae by Females of *Nasonia vitripennis* (Walk.) (Hymenoptera:Pteromalidae). Canad. Ent. 97(3):279-286.
Female *Nasonia* laid fewer eggs on parasitized pupae of *M. domestica.* Chemical and/or physical conditions of the parasitized pupae are detected by the female's ovipositor, resulting in restraint. ABSTRACT: RAE-B 54:82, Apr., 1966.

5427. Wylie, H. G. 1965 Some Factors that Reduce the Reproductive Rate of *Nasonia vitripennis* (Walk.) at High Population Densities. Canad. Ent. 97:970-977.
The reproductive rate decreased with increasing parasite/host (*M. domestica*) ratios. Females laid fewer eggs and an increased larval mortality resulted from superparasitism (simultaneous attacks by 2 females on each fly pupa). ABSTRACT: RAE-B 55:42(159), Mar., 1967.

5428. Wylie, H. G. 1966 Some Mechanisms that Affect the Sex Ratio of *Nasonia vitripennis* (Walk.) (Hymenoptera-Pteromalidae) Reared from Super-parasitized Housefly Pupae. Canad Ent. 98(6):645-653.
Smaller percentage of female progeny due to a smaller number of eggs being fertilized. Males come from unfertilized eggs. More female than

male larvae die on superparasitized hosts. ABSTRACT: RAE-B 55:221 (765), Dec., 1967.

5429. Wylie, H. G. 1966 Some Effects of Female Parasite Size on Reproduction of *Nasonia vitripennis* (Walk.) (Hymenoptera:Pteromalidae). Canad. Ent. 98(2):196-198.
Small females parasitized fewer hosts during their life time due to a shorter life span and a lower egg maturation rate. ABSTRACT: RAE-B 55:92(331), May, 1967.

5430. Wylie, H. G. 1966 Survival and Reproduction of *Nasonia vitripennis* (Walk.) at Different Host Population Densities. Canad. Ent. 98(3): 275-281.
Female *Nasonia* lived longer, found more hosts, and produced more progeny, at high host/population densities because they feed upon the pupae they parasitize, thus obtaining more food when host population is dense. ABSTRACT: RAE-B 55:194(671), Oct., 1967.

5431. Wylie, H. G. 1967 Some Effects of Host Size on *Nasonia vitripennis* and *Muscidifurax raptor* (Hymenoptera:Pteromalidae). Canad. Ent. 99(7):742-748.
Female *Nasonia* discovers large house fly puparia more easily than small ones. Survival of immature *Nasonia* and *Muscidifurax* is greater on large hosts than on small ones. ABSTRACT: BA 49:1428(16009), Feb. 1, 1968.

5432. Wylie, H. G. 1970 Oviposition Restraint of *Nasonia vitripennis* (Hymenoptera:Pteromalidae) on Hosts Parasitized by Other Hymenopterous Species. Canad. Ent. 102(7):886-894.
Females lay fewer eggs on housefly pupae already parasitized by *Muscidifurax raptor* G. and S. or *Spalangia cameroni* Perk. ABSTRACT: BA 51: 13160(133968), Dec. 1, 1970.

5433. Wyman, R. J. 1967 Comparative Study of Motor Output Patterns in Diptera. J. Gen. Physiol. 50(10):2483-2484.
Musca domestica and five other muscoids all exhibited some preferred phase relationships between different units in the same muscle. It is not synchrony nor is it dependent upon frequency.

5434. Wyman, R. J. 1969 Lateral Inhibition in a Motor Output System. I. Reciprocal Inhibition in Dipteran Flight Motor System. J. Neurophysiol. 32(3):297-306.
Concerns patterns of motor neuron activation of indirect flight muscles of several dipteran insects (including *Musca domestica*). ABSTRACT: BA 50:10861(112433), Oct. 15, 1969.

5435. Wyman, R. J. 1969 Lateral Inhibition in a Motor Output System. II. Diverse Forms of Patterning. J. Neurophysiol. 32(3):307-314.
Motor output patterns were generated by the flight motor system of several dipteran flies including *Musca domestica*. The diversity of the patterns is explained. ABSTRACT: BA 50:10861(112434), Oct. 15, 1969.

5435a. Wyman, R. [J.] 1970 Patterns of Frequency Variation in Dipteran Flight Motor Units. Comp. Biochem. Physiol. 35(1):1-16.
Concerns the long-term pattern of motor output to the indirect flight muscles of Diptera. Flies can maintain frequency trends up to one minute. ABSTRACT: BA 52:318(2863), Jan. 1, 1971.

Wyman, R. J., joint author. *See* Wilson, D. M., et al., 1963.

5436. Yaguzhinskaya, L. V. 1963 Some Findings on the Role of Tissue Oxidation Processes in Resistance of the Housefly (*M. domestica*) to DDT. J. Hyg. Epidemiol., Microbiol., Immunol. 7:105-112. Prague. (Summary in French, German, and Italian.)
Results show, that in phases in the life of insects in which an increase

in DDT resistance is observed, an increase may occur in the enzyme activity of the anaerobic link of the system of aerobic oxidation of succinic acid, without an increase in oxidase activity which forms the aerobic link. ABSTRACTS: BA 45:3223(39921), May 1, 1964; TDB 60:903, Sept., 1963.

5437. Yaguzhinskaya, L. V. and **Gabriyanik, I. A. 1966** [On the mechanism of resistance to DDT in insects. Changes of tissue respiration in house flies (*M. domestica*) in the process of increasing resistance to DDT.] Med. Parasit. and Parasitic. Dis. 35(2):212-216. (In Russian, with English summary.)
Increased resistance to DDT in female *Musca domestica* was accompanied by decreases in cytochrome oxidase activity and increases in succinic dehydrogenase. ABSTRACT: RAE-B 56:231(847), Dec., 1968.

5438. Yaguzhinskaya, (as **Jagushinskaya**) **L. V., Ioffe, I. O.** and **Gabriyanik, I. A. 1967** The Role of a Specific Factor (DDT-dehydrochlorination) in Development of DDT-resistance in Laboratory Strains of *Musca domestica* L. Wiad. Parazytol. 13(4/5):393-397.

5439. Yaguzhinskaya, L. V., Ioffe, I. D. and **Gabriyanik, I. A. 1969** Izuchnie aktivnosti nekotorykh fermentov tkanevogo dykhaniya pri razvitii ustoichivosti k DDT u *Musca domestica* pod vozdeistviem subletal'noi dozy. [Study of the activity of some tissue respiration enzymes during the development of resistance to DDT in *Musca domestica* under the effect of a sublethal dose.] Med. Parazit. i Parazitarn. (Bolez.) 38(3):331-337. (English summary.)
Activity of cytochromoxidase showed adaptive changes as well as a compensatory relationship with succinic dehydrogenase although these were masked considerably by background activity and relationships.

5440. Yale, T. H. and **Ballard, R. C. 1966** The Determination of Ribose Nucleic Acid in the Heart and Associated Tissues of House Flies of Various Ages. Comp. Biochem. Physiol. 19(1):29-34.
RNA in houseflies does not vary significantly with respect to sex and age. ABSTRACT: BA 48:468(4971), Jan. 1, 1967.

5441. Yamada, S. 1960 Studies on Body Passage of *Musca vicina* Macquart with Several Protozoa. J. Osaka City Med. Center 9(5):1523-1540. (In Japanese.)
Observations were made of survival periods and biological characters of *E. gingivalis*, *E. histolytica*, *T. tenax* and *T. vaginalis* in crop, intestines, feces and on body surfaces of *Musca vicina* and on foodstuffs. The first three could possibly be transmitted to humans by flies. ABSTRACTS: BA 36:2642(32455), May 15, 1960; TDB 57:1265, Dec., 1960.

5442. Yamamoto, I. 1965 Nicotinids as Insecticides. Advances in Pest Contr. Res. 6:231-260.
For high toxicity it is essential that the molecule contain a highly basic nitrogen. This would be protonated in the insect body.

5443. Yamamoto, I. and **Casida, J. E. 1966** 0-Dimethyl Pyrethrin II Analogs from Oxidation of Pyrethrin I, Allethrin, Dimethrin, and Phthalthrin by a House Fly Enzyme System. J. Econ. Ent. 59:1542-1543.
Used isolated insect enzyme (or enzyme system) for elucidation of pyrethroid metabolism as related to detoxification. ABSTRACTS: BA 48:3741 (41610), Apr. 15, 1967; RAE-B 55:71(262), Apr. 1967; TDB 64:438, Apr., 1967.

5444. Yamamoto, I., Soeda, Y., Kamimura, H. and **Yamamoto, R. 1968** Studies on Nicotinoids as an Insecticide. VII. Cholinesterase Inhibition by Nicotinoids and Pyridylalkylamines—Its Significance to

Mode of Action. Agric. Biol. Chem. 32(11):1341-1348.
Twelve nicotinoids and 26 pyridylalkylamines were studied. The significant correlation between toxicity and inhibition indicate that there are some similarities between the receptor for toxic action and the active center of cholinesterase for combining the molecule.

5445. **Yamamoto, I., Kimmel, E. C.** and **Casida, J. E. 1969** Oxidative Metabolism of Pyrethroids in Houseflies. J. Agric. Fd. Chem. 17(6): 1227-1236.
Allethrin is metabolised in living houseflies (*Musca domestica* L.) and in the housefly mixed-function oxidase system. The method of attack during metabolism is described. ABSTRACT: BA 51:7399(75633), July 1, 1970.

Yamamoto, I., joint author. *See* Fukami, J. I., et al., 1967; Casida, J. E., et al., 1967; Soeda, Y., et al., 1968, 1969.

5446. **Yamamoto, R. T.** and **Jensen, E. 1967** Ingestion of Feeding Stimulants and Protein by the Female Housefly, *Musca domestica* L. J. Insect Physiol. 13(1):91-98.
Protein ingestion is possibly related to the developmental cycle of the ovaries. ABSTRACTS: BA 48:3277(36138), Apr. 1, 1967; RAE-B 56:34 (110), Feb., 1968; TDB 64:438, Apr., 1967.

Yamamoto, R., joint author. *See* Yamamoto, I., et al., 1968; Robbins, W. E., et al., 1965.

5447. **Yamasaki, T.** and **Narahashi, T. 1962** Nerve Sensitivity and Resistance to DDT in Houseflies. Jap. J. Appl. Ent. Zool. 6(4):293-297.
Three strains of houseflies compared. Nerve sensitivity highest in highly susceptible strain; moderate in moderately susceptible strain; lowest in resistant strain. Conclusion: nerve sensibility plays an important role in resistance. ABSTRACT: RAE-B 52:119, July, 1964.

Yamasaki, T., joint author. *See* Tsukamoto, M., et al., 1965.

Yang, P.-N., joint author. *See* Liu Hsuen, Leng (Hsin-fu), et al., 1965.

Yang, R. S., joint author. *See* Meksongsee, B., et al., 1967.

5448. **Yasutomi, K. 1961** [Studies on the insecticide-resistance in Japanese insects of medical importance with special reference to the human lice and housefly.] Jap. J. Sanit. Zool. 12(1):36-76. (In Japanese, with English summary.)
Paper relates laboratory and field work in Japan with various insecticides against *Musca domestica vicina*. ABSTRACT: RAE-B 53:35, Feb., 1965.

5449. **Yasutomi, K. 1964** [Studies on the insect-resistance to insecticides. XII. Control of the DDT-resistant house-fly by DDT containing some synergists.] Jap. J. Sanit. Zool. 15(1):46-49. (In Japanese, with English summary.)
Part of a series in which several synergists and WARF antiresistant were tested, and contributed favorably to knockdown time when flies were exposed to DDT. WARF antiresistant and isothiocyanate were most effective. ABSTRACT: RAE-B 54:198, Oct., 1966.

5450. **Yasutomi, K. 1966** [Insecticide resistance of houseflies outbroken at the dumping site, Yumenoshima-island, Tokyo.] Jap J. Sanit. Zool. 17(1):71-73. (In Japanese, with English summary.)
Adult *Musca domestica* from the dump were highly resistant to DDT, gamma BHC, malathion and diazinon, and slightly resistant to DDVP, fenthion and fenitrothion. The larvae were resistant to malathion, but susceptible to fenthion, fenitrothion and dichlorvos. ABSTRACT: RAE-B 56:136(474), July, 1968.

5451. Yasutomi, K., Fujisaki, Y., Miyamoto, S., Iwahara, T., Tokuchi, S., Kimura, S. and **Tateno, K. 1964** [Field test with mixture of Sumithion and DDVP for the control of diazinon-resistant houseflies.] Jap. J. Sanit. Zool. 15:263-266. (In Japanese, with English summary.)

The fly population was markedly decreased during the experimental period when fenitrothion (Sumithion) and dichlorvos (DDVP) were used as a mixture. DDVP alone was unsatisfactory. ABSTRACT: RAE-B 55:53 (199), Mar., 1967.

5452. Yasutomi, K., Tokuchi, S., Tateno, K. and **Kimura, S. 1965** [Control tests of diazinon-resistant houseflies. (IV).] Jap. J. Sanit. Zool. 16(4):307-310. (In Japanese, with English summary.)

Deposits from sprays of fenitrothion or a mixture of diazinon and a form of diazoxon (referred to as isopropyl diazoxon) reduced the fly population. Similar applications to diazinon were ineffective. ABSTRACT: RAE-B 56:134(468), July, 1968.

5453. Yasutomi, K. and **Keiding, J. 1969** Virkning af synergister på resistens mod insekticider. [Effect of synergists on resistance to insecticides.] Ann. Rept. Govt. Pest Infest. Lab. 1968:37, 45, 46.

Tested the effect of sesamex on resistance to various organophosphorus compounds and chlorinated hydrocrabons. Tests also made with piperonyl butoxide. May be less expensive, but is not as effective.

5454. Yasutomi, K., Tokuchi, S., Ishikawa, K., Tateno, K. and **Kimura, S. 1966** [Field control tests of diazinon-resistant housefly larvae.] Jap. J. Sanit. Zool. 17(4):247-251. (In Japanese, with English summary.)

Applied organophosphorus insecticides to manure heaps. Fenitrothion most effective; fenitrothion and fenthion more effective than diazinon and dichlorvos. ABSTRACT: RAE-B 58:21(74), Jan., 1970.

5455. Yasutomi, K., et al. 1961 [Control tests of diazinon-resistant houseflies.] Jap. J. Sanit. Zool. 12(4):283-288. (In Japanese, with English summary.)

This paper and the one immediately following contain results of tests on diazinon-resistant houseflies, *Musca domestica vicina* Macq., using fenchlorphos (Nankor), trichlorphon (Dipterex) and dichlorvos (DDVP) alone, or in combination. ABSTRACT: RAE-B 53:37, Feb., 1965.

5456. Yasutomi, K., et al. 1962 [Control tests of diazinon-resistant houseflies.] Jap. J. Sanit. Zool. 13(4):290-294. (In Japanese, with English summary.)

See preceding annotation. ABSTRACT: RAE-B 53:37, Feb., 1965.

5457. Yasutomi, K., et al. 1962 [Field test with Sumithion for the control of diazinon-resistant houseflies.] Jap. J. Sanit. Zool. 13(4):295-297. (In Japanese, with English summary.)

Fly density was decreased, particularly when breeding sites were treated. ABSTRACT: RAE-B 53:37, Feb., 1965.

5458. Yasutomi, K., et al. 1964 [Field tests for the control of resistant houseflies.] Jap. J. Sanit. Zool. 15(1):50-52. (In Japanese, with English summary.)

DDT and diazinon-resistant flies were decreased considerably when 2·5 gm DDT/sq. m. was used alone or with a synergist. Fenthion (Baytex) emulsion also gave good results. ABSTRACT: RAE-B 54:199, Oct., 1966.

Yasutomi, K., joint author. *See* Keiding, J., et al., 1969.

5459. Yates, J. R., III, and **Sherman, M. 1970** Latent and Differential Toxicity of Insecticides to Larvae and Adults of Six Fly Species. J. Econ. Ent. 63(1):18-23.

Species investigated included *M. domestica* and *Fannia pusio* (Wiedemann). Median-lethal dosages and dosages causing 90 per cent mortality were calculated by probit analysis.

Yearian, W. C., joint author. *See* Smith, W. W., et al., 1964.

5460. Yendol, W. G. and **Miller, E. M. 1967** Susceptibility of the Face Fly to Commercial Preparations of *Bacillus thuringiensis*. J. Econ. Ent. 60:860-864.

B. thuringiensis toxins prove fatal to face fly; even in manure from cattle fed commercial preparation of toxins. ABSTRACTS: BA 48:11210(124510), Dec. 15, 1967; RAE-B 55:191(660), Oct., 1967.

5461. Yendol, W. G., Miller, E. M. and **Behnke, C. N. 1968** Toxic Substances from Entomophthoraceous Fungi. J. Invertebr. Path. 10(2):313-319.

Reconstituted freeze-dried filtrates were toxic to *M. autumnalis* adults and two species of moth larvae. ABSTRACT: BA 50:1054(11056), Jan. 15, 1969.

Yendol, W. G., joint author. *See* Tung, S. C., et al., 1969.

5462. Yeomans, A. H., Sullivan, W. N., Schechter, M. S. and **Starkweather, R. J. 1964** Aerosol Fly Sprays. Soap and Chem. Spec. October. 2 pp.

Thermally generated aerosols with particle size of two microns m. m. d. gave as good a kill of house flies as liquified gas propelled aerosols with particle sizes of 10 or 20 microns m. m. d., provided exposure was maintained for at least 2 hours.

Yepez, M. S., joint author. *See* Diaz-Ungria, C., et al., 1966.

5462a. Yew, N. G. K. 1970 The Control of Housefly Breeding in Organic Fertilizers at Cameron Heights, Pahang. Malaysian Agric. J. 47(3): 323-332.

Prawn dust protected from fly breeding: (1) by storage in waterproof bags; (2) by adding any of a number of insecticidal dusts to prawn dust soon after treatment (holds up to 30 days). ABSTRACT: BA 52:1409 (13302), Feb. 1, 1971.

5463. Young, D. D. 1961 Dissection of the Common Housefly. Am. Biology Teacher. 23(8):524-525.

Young, R. G., joint author. *See* Berger, R. S., et al., 1962.

Young, W. R., joint author. *See* Vazquez-Gonzalez, J., et al., 1962, 1963.

Yukihiko, F., joint author. *See* Matsuzawa, H., et al., 1968.

5464. Zaharis, J. L. 1960 Pigmented Microorganisms Occurring with House Fly Cultures and Their Relationship to Two House Fly Strains. Diss. Absts. 21(pt. 1, no. 3):712.

Investigated a violet color developing on cellucotton watering pads in rearing cages of insecticide-resistant flies (CAL). An attempt is made to explain its absence in rearing cages of insecticide-susceptible flies (KUN). ABSTRACTS: BA 36:6878(74238), Nov. 1, 1961; BA 36:1482 (17833), Mar. 15, 1961.

5465. Zaharis, J. L., Roan, C. C. and **Lord, T. H. 1961** Pigmented Bacteria Occurring in Fly Cultures and Some Relationships to House Fly Strains. J. Kans. Ent. Soc. 34(2):91-100.

Suggests that the insecticide-susceptible strain of flies (KUN) is able to inhibit either the agent responsible for development of the violet color, or the process of color development on the cellucotton watering pads.

5466. Zahavi, M. and **Tahori, A. S. 1965** Citric Acid Accumulation with Age in Houseflies and Other Diptera. J. Insect. Physiol. 11:811-816.
Declines (in house flies) for 1 day, then increases significantly (but decreases in starved flies). Keto acids increase slightly with age. Flies showing signs of senescence had especially high citric acid concentrations.

5467. Zahavi, M., Tahori, A. S. and **Kindler, S. H. 1964** Studies on the Biochemistry of Fluoroacetate Resistance in House Flies (*Musca domestica*). Proc. XXXIV Meeting of the Israel Chemical Society. Israel J. Chem. 2(5a):320-321.
Since fluoroacetate inhibits respiration of washed mitochondria without causing citrate accumulation, it is inferred that fluoroacetate interferes with pyruvate oxidation at a stage unrelated to that of aconitase. Conclusion supported by results. ABSTRACT: BA 46:7008(87039), Oct. 1, 1965.

5468. Zahavi, M., Tahori, A. S. and **Mager, J. 1968** Studies on the Biochemical Basis of Susceptibility and Resistance of the Housefly to Fluoroacetate. Biochim. Biophys. Acta 153:787-798.
Fluoroacetate administration elevates citrate content of sensitive houseflies. Mechanisms underlying toxicity of fluoroacetate in houseflies are considered. ABSTRACT: BA 49:9255(102774), Oct. 1, 1968.

5469. Zahavi, M., Tahori, A. S. and **Mager, J. 1969** Stimulatory Effects of NAD on Respiration and Oxidative Phosphorylation in Housefly Sarcosomes. Israel J. Chem. Proc. 7(4):142.
Relatively high nicotinamide adenine dinucleotide (NAD) requirement was *not* due to its breakdown in the course of incubation.

5470. Zaidenov, A. M. 1960 [On the study of the dispersal of house-flies (Diptera, Muscidae) by means of the luminescent method of marking in the city of Chita.] Rev. Ent. URSS 39(3):574-584. (In Russian.)
Fluorescein in dilute NaOH mixed with a bait was very good as a fly marker. Exposure for 60-90 minutes resulted in 45-40 per cent flies marked; living flies were *distinguishable* for 72 hours and dead ones for many months. ABSTRACT: RAE-B 50:48, Mar., 1962.

5471. Zaidenov, A. M. 1961 [Study of house fly (Diptera:Muscidae) migrations in Chita by means of luminescent tagging.] Ent. Obozrenie 39(3):406-414.
A luminescent tagging method proposed by B. L. Shura-Bura and B. L. Gagaese is suitable for studying housefly dispersal and allows tagging under natural conditions (as in movement from privies to kitchen and eating areas). ABSTRACT: BA 36:3481(42811), July 1, 1961.

5472. Zaidenov, A. M. and **Kondrashova, M. L. 1965** Nablyudeniya nad komnatoi mukhoi kak perenoschikom kishechnykh infektsii. [Observations on the role of domestic flies as carriers of enteric infections.] Med. Parazit. i Parazitarn. (Bolezni) 34(5):525-528. (In Russian, with English summary.)
By marking the content of privies, flies were tagged. The epidemiological role of populations of domestic flies at different periods of the season differs according to the time of contact of these insects with infected substrates. ABSTRACT: BA 47:8964(103623), Nov. 1, 1966.
For additional paper under this authorship see Zaydenov, A. M. 1961.

Zaiman, H., joint author. *See* Hirsch, J., et al., 1965.

Zaitseva, G. N., joint author. *See* Ermokhina, T. M., et al., 1966.

5473. Zakamyrdin, I. A. 1966 [Possibilities of using polychlorpinene to control house flies (*Musca domestica*) and stable flies (*Stomoxys calcitrans*), in live stock farms.] Med. Parazit. i Parazitarn. (Bolez.) 35(5):618-619.

Compound is recommended for the control of fly larvae in refuse, particularly dung. Also effective against adults when sprayed on walls. For *Stomoxys* it could replace DDT and BHC. ABSTRACTS: BA 49:927 (10400), Jan. 15, 1968; RAE-B 57:259(984), Dec., 1969.

5474. Zakharova, N. F. 1962 Nekotorye dannye po ekologii synantropnykh mukh ashkkabada i ego okrestnostei. [Some data on the ecology of synanthropic flies in Ashkhabad and vicinity.] Med. Parazit. (Moskva) 31:605-607.

Data on various aspects of 12 species of this group: proportion of each trapped at 3 different sites (windows, doors, near privies); relative attraction of different types of baits; favorite oviposition sites; and daily activity. ABSTRACT: BA 42:309(3719), Apr. 1, 1963.

5475. Zakharova, N. F. 1966 Pioski movykh khemosterilyantov. [Search for new chemosterilants.] Med. Parazit. i Parazitarn. (Bolez.) 35(5): 515-519. (English summary.)

Three drugs tested caused sterilization of an insecticide-sensitive strain of housefly, *M. domestica domestica* (5-fluor-uracil, dipine and Thiotef). ABSTRACT: BA 48:10150(114020), Nov. 15, 1967.

5476. Zakharov, S. S. 1964 [Experience with the use of polychlorpinene for control of flies in Saratov.] Med. Parazit. (Moskva.) 33:97-99. (In Russian.)

5477. Zakolodkina, V. I. 1963 Cytochrome Oxidase and Succinodehydrogenase Activity in Tissues of the Housefly (*Musca domestica*) after Exposure to some Insecticides. J. Hyg., Epidemiol., Microbiol., Immunol. (Prague) 7:97-104. (Summary in French, German, and Italian.)

After the application of BHC, chlorophos and carbophos, changes in the activity of cytochrome oxidase and succinodehydrogenase were observed in the epithelium of the mid-gut as well as in the ganglia and muscles. ABSTRACTS: BA 45:3223(39922), May 1, 1964; TDB 60:903, Sept., 1963.

Zakolodkina, V. I., joint author. *See* Vashkov, V. I., et al., 1965.

5478. Zapanta, H. M. and **Wingo, C. W. 1968** Preliminary Evaluation of Heliotrine as a Sterility Agent for Face Flies. J. Econ. Ent. 61(1): 330-331.

Discusses toxicity of various levels of heliotrine to eggs, developing larvae and adult reproduction. ABSTRACTS: BA 48:6247(69866), July 1, 1968; RAE-B 56:142(495), July, 1968.

5479. Zardi, O. 1964 Importanza di "*Musca domestica*" nella transmissione dell'agente del tracoma. [Importance of *Musca domestica* in transmission of the trachoma virus.] Nuovi Ann. Igiene Microb. 15: 587-590.

Fly functions as a "bearer" able to transmit the infection through the feces.

5480. Zardi, O. 1964 Studi epidemiologici sulla toxoplasmosi. Indagini sui vettori. [Epidemiological studies on toxoplasmosis. Findings on vectors.] Nuovi Ann. Igiene Microb. 15:540-544.

Experiments indicate epidmiological importance of *M. domestica*, also certain mosquitoes.

Zatsepen, N. I., joint author. *See* Arskii, V. G., et al., 1961.

5481. Zaydenov, A. M. 1961 [Experience gained from a study of the epidemiological role of synanthropic flies under urban conditions.] Ent. Obozreniye 40(3):554-567. (Translated from Ent. Obozreniye 40(3):299-307.)
Of flies caught mainly in toilet areas, house flies were not numerous, being only 0·19 per cent of larvae in Chita and 1·75 per cent in Stalingrad. Only 5·4 per cent of houseflies were infected with dysentery bacteria in an area near patients. ABSTRACT: BA 40:307(3975), Oct. 1, 1962.

Zaylskie, R. G., joint author. *See* Pomonis, J. G., et al.

Zeidler, G., joint author. *See* Tahori, A. S., et al., 1965.

Zemanova, J., joint author. *See* Privora, M., et al., 1968.

Zhelezova, V. F., joint author. *See* Tamarina, N. A., et al., 1962.

Zhuk, E. B., joint author. *See* Misnik, Y. N., et al., 1967; Bessonova, I. V., et al., 1970.

5482. Zhuzhikov, D. P. 1963 Stroenie peritroficheskoi obolochki dvukrylykh. [The structure of the peritrophic membrane in Diptera.] Vestn. Mosk. Univ., Ser. Biol. 18(1):24-35.
Describes structure and formation of peritrophic membrane. For Muscidae it is a 2 layer membrane. Structure divides the group into Orthorrhapha and Cyclorrhapha. ABSTRACT: BA 43:1036(12476), Aug. 1, 1963.

5483. Zhuzhikov, D. P. 1963 K voprosu o vosmozhnosti perezhivaniya bakteriyami metamorfoza komnatnoi mukhi. [The possibility of bacteria surviving house fly metamorphosis.] Med. Parasit. i Parasitic Dis. (Moscow.) 32:558-562. (English summary.)
Bacteria ingested by larvae survive fly metamorphosis and can be found on the surface of young flies. ABSTRACTS: BA 46:1058(13355), Feb. 1, 1965; TDB 61:107, Jan., 1964.

5484. Zhuzhikov, D. P. 1963 Funktsiya kishechnika komnatnoi mukhi do nachala pytaniya. [Functions of the gut in the house fly (*Musca domestica*) before feeding.] Nauch. Dokl. Vyssh. Shkoly-Biol. Nauk 2:23-27. (From: Referat. Zhur., Biol. no. 23E121.)
Reports changes occurring in the housefly gut, such as the entrance of air up until feeding. Gives description of the peritrophic membrane. ABSTRACT: BA 46:2564(31983), Apr. 1, 1965.

5485. Zhuzhikov, D. P. 1963 Stroenie i funkstii peritroficheskoi obolochki dvukrylykh. [Structure and functions of the peritrophic membrane in Diptera.] *In* Payatee soveshchanie Vsesoyuznogo entomologicheskogo obshchestva 1963. [Fifth meeting of the All-Union Entomological Society, 1963.] Akad. Nauk. SSSR:Moscow-Leningrad, pp. 20-21. Referat. Zhur., Biol., 1964, no. 1E17.

5486. Zhuzhikov, D. P. 1964 Function of the Peritrophic Membrane in *Musca domestica* L. and *Calliphora erythrocephala* Meig. J. Insect Physiol. (London) 10(2):273-278.
Lets through (outwardly) only the end-products of digestion. Each layer (there are 2) permits amylase to pass either way but the whole membrane prevents it from reaching the intestinal epithelium. Whole membrane is permeable both ways to water. ABSTRACTS: BA 45:5762(71283), Aug. 15, 1964; TDB 61:726, July, 1964.

5487. Zhuzhikov, D. P. 1966 Izuchenie peritroficheskoi obolochki nekotorykh dvukrylykh v polyarizovannom svete. [Study of the peritrophic membrane of some Diptera in polarized light.] Vestn. Mosk.

Univ. Ser. VI Biol., Pochvoved 21(1):37-41.
Peritrophic membrane of larvae and imagoes of *M. domestica*, *Calliphora erythrocephala* Meig. and *Aedes aegypti* were found to have a definite internal organization on the submicroscopic level. ABSTRACT: BA 48: 10179(114380), Nov. 15, 1967.

Zoltai, N., joint author. *See* Sztankay-Gulyas, M., et al., 1962.

5488. Zubairi, M. Y. 1966 Detoxication of Dimetilan in Cockroaches and Houseflies. Diss. Absts. 26(7):3620.
See following reference.

5489. Zubairi, M. Y. and **Casida, J. E. 1965** Detoxication of Dimetilan in Cockroaches and House Flies. J. Econ. Ent. 58:403-409.
Four metabolites were detected in *Musca*. Hydroxylation sites are proposed for the major reaction of the initial detoxification. ABSTRACTS: BA 47:3384(39931), Apr. 15, 1966; RAE-B 53:178, Sept., 1965.

5490. Zuberi, R. I., Hafis, S. and **Ashrafi, S. H. 1969** Bacterial and Fungal Isolates from Laboratory-reared *Aedes aegypt.* (Linnaeus), *Musca domestica* L. and *Periplaneta americana* L. Pak. J. Sci. and Ind. Res. 12(1/2):77-82.
Eggs, larvae, pupae and adults of each were plated out on a nutrient media for isolation of normal aerobic bacterial and fungal flora. Found 27 isolates from *M. domestica.* ABSTRACT: BA 51:4511(46474), Apr. 15, 1970.

Zubova, G. M., joint author. *See* Vashkov, V. I., et al., 1966, 1968.

Zueva, E. V., joint author. *See* Magdiev, R. R., et al., 1960.

5491. Zumpt, F. 1963 The Problem of Intestinal Myiasis in Humans. S. Afr. Med. J. 37(23 March):305-307.
Larvae of housefly, if swallowed on human food, are usually killed by gastric secretions. Chitinous skins may pass undamaged in feces. If digestive process is inadequate (due, e. g., to bacterial infection) larvae may be passed alive. Dead or alive, are irritating and may cause nausea, vomiting, diarrhea with abdominal cramps. Theory of paedogenesis *not* accepted.

Zweig, G., joint author. *See* Kunkee, R. E., et al., 1965.

Section VII.

Publications on Research Techniques

References are listed below only when the author places primary emphasis on technical procedure. Because rearing techniques for various muscoid species may usually be adapted for use with other species or groups, certain references are included which deal principally with *Calliphora*, *Lucilia*, *Stomoxys* and related genera. Reports concerned chiefly with experimental findings and which treat but incidentally with new or refined techniques are listed in Sections IV, V and VI.

5492. Adelung, D. and **Karlson, P. 1969** Eine verbesserte, sehr empfindliche Methode zur biologischen Auswertung des Insektenhormones ecdyson. [An improved, very sensitive method for the biological evaluation of the insect hormone *ecdysone*.] J. Insect Physiol. 15(8): 1301-1307.

Authors give method for rearing *Calliphora erythrocephala* and *M. domestica*; also bioassay procedure for determining *ecdysone* in the larvae. ABSTRACTS: BA 51:2796(29019), Mar. 1, 1970; RAE-B 58:409 (1700), Dec., 1970.

5493. Adkins, T. R., Jr. 1968 A Foot-Controlled Device using Carbon Dioxide for the Anesthetization of Insects. J. Econ. Ent. 61(1): 340-341.

A practical aid to laboratory procedure, first used for the study of *M. autumnalis*. ABSTRACT: RAE-B 56:143(496), July, 1968.

5494. Allen, T. C., Dicke, R. J. and **Brooks, J. W. 1943** Rapid Insecticide Testing. Use of the Settling Mist Method for Testing of Vaporized Contact Insecticides against Houseflies. Soap and Sanit. Chem. 19:94-96, 121.

Paper gives several improvements in the method of rearing. Larvae, separated from medium by use of a triple-screened funnel, work their way downward (avoiding light) into dry sand, from which pupae are separated, later, by sifting. ABSTRACT: RAE-B 31:210, 1943.

Amerson, G. M., joint author. *See* Hays, S. B., et al., 1966.

Anderson, C. V., joint author. *See* Doner, M. H., et al., 1947.

5495. Anderson, J. R. and **Poorbaugh, J. H. 1964** A Simplified Technique for the Laboratory Rearing of *Fannia canicularis*. J. Econ. Ent. 57: 254-256.

Larval mortality rises as the moisture content of the medium is increased beyond 35 per cent. ABSTRACTS: RAE-B 52:125, July, 1964; BA 45:6739 (84655), Oct. 1, 1964; TDB 61:851, Aug., 1964.

Archetti, I., joint author. *See* Greenberg, B., et al., 1968.

Arevad, K., joint author. *See* Keiding, J., et al., 1964.

Asano, S., joint author. *See* Nagasawa, S., et al., 1963.

5496. Ascher, K. R. S. 1956 A Novel Method for Isolating Housefly Larvae from Manure. Hasade (The Field) 36:9. (In Hebrew.)

Attia, M. A., joint author. *See* Hafez, M., et al., 1958.

5497. Audemard, H. 1967 L'élevage permanent de la mouche des semis *Phorbia platura* Meigen (*Hylemyia cilicrura* Roudani). (Diptera: Muscidae.) I. Pondoir artificiel. [. . . I. Artificial oviposition site.]

Ann. Epiphyties (Paris) 18(4):551-555.
Used moist sheet of synthetic sponge and three superimposed squares of plastic mesh. Obtained several thousand eggs per day from thirty pairs.

5498. Audemard, H. and **Guennelon, G. 1968** L'Elevage permanent de la mouche des semis *Phorbia platura* Meigen (=*Hylemyia cilicrura* Rondani) (Diptera, Muscidae). II. Milieu artificiel d'elevage des larves. [Continuous breeding of the corn-seed maggot. . . . II. Artificial breeding medium of larvae.] Ann. Epiphyties (Paris) 19(4):713-719. (English summary.)
Breeding medium contains wheat germ, combined with brewer's yeast. Possibly useful in rearing other species of Muscidae. ABSTRACT: BA 50:12499(129121), Dec. 1, 1969.

5499. Bailey, D. L., LaBrecque, G. C. and **Whitfield, T. L. 1970** A Forced-Air Column for Sex Separation of Adult House Flies. J. Econ. Ent. 63(5):1451-1454.
Depends on female: male *ratio* of both weight and volume, which increases steadily with age. Longevity of flies so separated not different from that of flies sexed by hand. ABSTRACT: BA 52:2134(20408), Feb. 15, 1971.

Baker, G. J., joint author. *See* Campau, E. J., et al., 1953.

5500. Basden, E. B. 1947 Breeding the House-fly (*Musca domestica* L.) in the Laboratory. Bull Ent. Res. 37(3):381-387.
Deals with large scale production for testing of fly sprays. Based on Peet-Grady method of 1941, with modifications. ABSTRACT: RAE-B 35:68, 1947.

5501. Beard, R. L. 1958 Laboratory Studies on House Fly Populations. I. A Continuous Rearing System. Bull. Conn. Agric. Exp. Sta. 619 (First portion only, pp. 1-6).
Rearing unit consists of container for larval food attached to a communicating flight chamber, with adult food. Four units connect with one concourse to form a battery. Three, intercommunicating batteries form a series. ABSTRACT: RAE-B 51:120, June, 1963.

5502. Bechtel, R. C. and **Grigarick, A. A. 1954** A Method for the Evaluation and Retreatment of Insecticide-treated Cotton Strings Used for House Fly Control. J. Econ. Ent. 47:369-370.
Trays suspended 3 inches below end of string to catch killed flies. Three foot glass tube, filled with solution, brought up from below to retreat string, then slowly withdrawn. ABSTRACTS: BA 29:207, Jan., 1955; RAE-B 43:27, Feb., 1955; TDB 51:1202, Nov. 1, 1954.

Bell, V. A., joint author. *See* Hill, D. L., et al., 1947.

Benke, R., joint author. *See* Eagleson, C., et al., 1938.

5503. Beran, F. 1953 Ein Beitrag zur Methodik der Insektizidprüfungen. [A Contribution to the Method of Testing Insecticides.] Pflanzenschutzberichte. 11:151-160. Vienna. (English summary.)
M. domestica used as test species. ABSTRACTS: RAE-A 42:273, 1954; TDB 52:103, Jan., 1955; RAE-B 42:123, 1954.

5504. Beroza, M. 1963 Identification of 3,4-Methylenedioxyphenyl Synergists by Thin-Layer Chromatography. Agric. Food Chem. 11(1): 51-54.
Best results obtained on silicic acid plates with acetone in benzene as developing agent. Paper concerned with chemical procedures of chromatography. No mention of housefly but included here because of great importance of synergists in fly control.

5505. Bickoff, E. 1943 House Fly. *In* Laboratory Procedures in Studies of the Chemical Control of Insects, p. 74. Edited by Campbell, F. L. and Moulton, F. R. Publ. Amer. Ass. Adv. Sci. No. 21, viii + 206 pp., 62 figs., 12 pages refs.

Larval food mixed and placed in three-and-a-half-gallon pan. Technician measures about 7,500 ova by calibrated pipette and places them on mixture in pan. Mixture transferred to second pan (with more food) on third day. Pupae separated from dried mash by sieve and air blast. ABSTRACT: RAE-B 32:100, 1944.

5506. Bliss, C. I. 1939 Fly Spray Testing. A Discussion on the Theory of Evaluating Liquid Household Insecticides by the Peet-Grady Method. Soap and Sanit. Chem. 15(4):103, 105, 107, 109, 111.

If half a unit of Official Control Insecticide gives same mortality as 1 unit of the sample, latter is rated 0·5 (50 per cent toxicity rating). All tests must be made on the same culture of flies. ABSTRACT: RAE-B 27:203, 1939.

5507. Block, R. J., LeStrange, R. and **Zweig, G. 1952** Paper Chromatography. Academic Press, Inc. New York. 195 pp.

Included because of increasing use of chromatography in insect techniques.

5508. Boettiger, E. G. and **Furshpan, E. 1952** The Recording of Flight Movements in Insects. Science 116:60-61.

Describes a new method of recording wing movement, which can be used with very small insects. Depends on moving, electrostatically charged bodies, acting as variable condensors. Method gives instantaneous position, direction of movement and velocity during rapid flight.

Bořkovec, A. B., joint author. *See* Gouck, H. K., et al., 1963.

5509. Bovingdon, H. H. S. 1958 An Apparatus for Screening Compounds for Repellency to Flies and Mosquitoes. Ann. Appl. Biol. 46(1): 47-54.

Deals particularly with butoxypolypropylene glycol (MW800). Total number of flies on untreated glass plate divided by total number on treated glass plate gives "repellency quotient". ABSTRACTS: TDB 55:1055, Sept., 1958; Hocking, B., Smell in Insects etc. Tech. Rept. No. 8:129 (3.15), 1960.

Bowman, M. C., joint author. *See* Fye, R. L., et al., 1968.

5510. Brady, U. E. 1966 A Technique of Continuous Exposure for Determining Resistance of House Flies to Insecticides. J. Econ. Ent. 59(3):764-765.

Desired amount of insecticide residue deposited on inner surface of petri dishes by slowly rotating acetone solution of insecticide to dryness. Flies, anesthetized with $C0_2$, transferred to dishes. Controls in dishes treated with acetone only. Lethal time equated with knockdown. ABSTRACTS: RAE-B 54:183, Sept., 1966; BA 47:8522(99273), Oct. 15, 1966.

5511. Briggs, J. D. 1964 Mass Propagation of Bacteria Pathogenic for Insects. Bull. W. H. O. 31(4):495-497.

Treats of organisms, culture media, incubation, recovery, formulation and standardization. ABSTRACT: RAE-B 55:26(89), Feb., 1967.

Brooks, J. W., joint author. *See* Allen, T. C., et al., 1943.

5512. Bruce, W. N. 1953 A New Technique in Control of the House Fly. Illinois Nat. Hist. Surv. Biol. Notes 33:1-8.

Bayer L 13/59 ($1\frac{1}{2}$ grams in 1 pint Karo syrup) applied to window frame of barn with brush. Continued to kill house flies for more than 2 months. ABSTRACTS: BA 28:955, Apr., 1954; Science 120:30, July, 1954.

5513. Bruce, W. N. 1953 Laboratory and Field Evaluation of Factors Affecting the Performance of Fly Repellents. Diss. Absts. 13:1309-1310.
Used gravimetric measurements rather than an optimetric method in tests conducted to evaluate repellent effectiveness against *Stomoxys* and *Tabanus* on cattle. Weights of lactose pellets consumed by *Musca domestica* were used as criteria for measuring percentage repellency of test materials.

5514. Brydon, H. W. 1965 A Sampler for Immature Flies in Poultry Droppings. J. Econ. Ent. 58:697-699.
Gives construction details of a device for extracting an intact 2 inch-wide cross-section segment of coned poultry droppings. Such samples facilitate the gathering of detailed information regarding immature fly inhabitants. ABSTRACT: BA 46:8879(109483), Dec. 15, 1965.

5515. Brydon, H. W. 1965 A Mobile Office-Laboratory. Mosquito News 25(2):160-164.
A 19-foot trailer of complex and unique design. Provides adequate space for three persons, working simultaneously. Approximate cost, $3,600.00. ABSTRACT: BA 47:4957(58172), June 15, 1966.

5516. Brydon, H. W. 1966 A Core Sampler for Immature Flies in Poultry Manure. J. Econ. Ent. 59:1313.
A thin-walled cylinder, 3½ inches in diameter and 8 inches long, with bottom edge sharpened. Plunger plate and 24 inch plunger rod within. Sample released into polyethylene bag by pushing on rod handle. ABSTRACTS: RAE-B 55:15(41), Jan., 1967; BA 48:1854(20032), Feb. 15, 1967.

5517. Brydon, H. W. and **Fuller, R. G. 1966** A Portable Apparatus for Separating Fly Larvae from Poultry Droppings. J. Econ. Ent. 59: 448-452.
Pertains to *Fannia* spp. Employs light and heat from an incandescent bulb in a portable single-funnel unit. Samples spread over galvanized screen. Larvae received below in 70 per cent alcohol in half pint fruit jars. ABSTRACT: BA 47:7227(84317), Sept. 1, 1966.

5518. Buei, K. 1958 [On the single-pair culture of the common house fly, *Musca domestica vicina* Macq. Ecological studies of the flies of medical importance. II.] Botyu-Kagaku 23:177-181. (In Japanese, with English summary.)
Pupae taken from mass cultures. Flies sexed and each pair placed in metal gauze cage. Food supplied by cotton ball soaked in skim milk and honey plus water. Sucrose, 2 per cent, also supplied. Larval medium provided to receive eggs. Rel. humidity 40-50 per cent. Ova incubated at 27°C.

5519. Busvine, J. R. 1962 A Laboratory Technique for Measuring the Susceptibility of Houseflies and Blowflies to Insecticides. Laboratory Practice 11:464-468.
Used a micro-syringe applicator with an attachment to give regular deliveries. Insecticides applied in non-volatile solvents (Risella or dioctyl phthalate) and results from oil carrier compared with acetone solutions. Various strains of flies tested. ABSTRACTS: RAE-B 53:38, Feb., 1965; TDB 59:1017, Oct., 1962.

5520. Buxton, P. A. and **Mellanby, K. 1934** The Measurement and Control of Humidity. Bull Ent. Res. 25(2):171-175.
Gives table for producing desired relative humidity in the atmosphere by use of graduated solutions of sulphuric acid or potassium hydroxide. Latter is recommended in the tropics. ABSTRACT: TDB 31:739, 1934.

5521. Caldwell, A. H. 1948 Small Spray Chamber. II. Additional Notes on the Operation of the Hoskins Caldwell Spray Chamber. Soap and Sanit. Chem. 24(8):133.

Chamber is for use with *M. domestica*, as test species. This paper is supplementary to one by Hoskins and Caldwell, 1947, which see. ABSTRACT: RAE-B 38:183, 1950.

5522. Caldwell, A. H., Jr. 1956 Dry Ice as an Insect Anesthetic. J. Econ. Ent. 49(2):264-265.

Describes apparatus and method. Inexpensive, convenient and not hazardous to operator. Recovery time depends on exposure time; i.e., 30 min. exposure—recover in 3 to 5 min. ABSTRACT: RAE-B 45:80, May, 1957.

Caldwell, A. H., Jr., joint author. *See* Hoskins, W. M., et al., 1947.

5523. Campau, E. J., Baker, G. J. and **Morrison, F. O. 1953** Rearing Stable Fly for Laboratory Tests. J. Econ. Ent. 46:524.

Included here because of many points in common with house fly production for Peet-Grady tests. From a total of 19 cultures, 57 per cent of the eggs produced adults; 86 per cent of the pupae produced flies. ABSTRACT: TDB 50:1096, Nov., 1953.

5524. Campbell, F. L. and **Sullivan, W. N. 1938** Testing Fly Sprays: A Metal Turntable Method for Comparative Tests of Liquid Contact Insecticides. Soap and Sanit. Chem. 14(6):119-125, 149.

Measured amounts of sample sprayed at known pressure into 10 metal cylinders (17 inches high, 8 inch diameter), fitted on top of turntable. Pass in rotation under spray nozzle. Costs less than Peet-Grady chamber. ABSTRACT: RAE-B 26:246, 1938.

Cantrel, K. E., joint author. *See* Goodhue, L. D., et al., 1958.

5525. Carter, C. I. 1965 An Inexpensive Cabinet for Temperature and Humidity Control. Bull. Ent. Res. 56(2):263-268.

Describes basic design of apparatus, easily adjusted to give different combinations of temperature and humidity. Temperatures constant to within 0·5°C over range 5-25°C; relative humidities constant to within 2·5 per cent over range 40-85 per cent. ABSTRACT: TDB 63:493, Apr., 1966.

5526. Cepelák, J. 1955 Usmrcovante a preparácia vyšších much. [Killing and Preparation of Higher Flies.] Biológia (Bratislava) 10(1):84-88.

Procedural directions for collectors and museum workers.

Chadwick, L. E., joint author. *See* Hill, D. L., et al., 1947.

5527. Champlain, R. A., Fisk, F. W. and **Dowdy, A. C. 1954** Some Improvements in Rearing Stable Flies. J. Econ. Ent. 47:940-941.

Modifications relate to Campeau method. Describes apparatus for ovipositing; technique and apparatus described for collecting and seeding eggs. Suggestions given for increasing production. ABSTRACTS: BA 29:2004, Aug., 1955; RAE-B 43:158, Oct., 1955.

5528. Chan, K. and **Morrison, F. O. 1965** A Technique for Comparing the Effectiveness of Insecticidal Paints. Ann. Ent. Soc. Quebec 10(3): 125-127.

Flies anesthetized with CO_2, then transferred by light forceps into plastic petri dish cover with open side held against painted panel. Food provided by cotton wick extending from reservoir of 5 per cent sugar solution through hole in wall of petri dish. ABSTRACT: RAE-B 55:109 (390), June, 1967.

5529. **Chang, S. C. 1965** Improved Bioassay Method for Evaluating the Potency of Chemosterilants against House Flies. J. Econ. Ent. 5:8 796.

A modification of the Chang and Borkovec technique (1964; Section VI). By separating "feeding" and "egg collecting" devices, much labor is saved without affecting reliability of tests. ABSTRACTS: BA 47:371(4489), Jan. 1, 1966; TDB 62:1275, Dec., 1965.

5530. **Chang, T. H.** and **Eide, P. E. 1966** In-vitro Cultivation of House Fly (*Musca domestica*) Embryonic Cells. Can. J. Genet. Cytol. 8(2): 352.

Used 4½-6½ hour old house fly eggs. Trypsinization yielded good cell suspension. Four to six hours after inoculation, cells attach and resume normal shape. Cell groups may contract, like muscle cells. Maintained cultures for more than 30 days.

Chang, T.-H., joint author. *See* Eide, P. E., et al., 1969.

Cheng, Tien-Hsi, joint author. *See* Tung, Sik-Chung, et al., 1969.

5531. **Chiang, H. C. 1963** A Modified Flight Chamber. J. Econ. Ent. 56: 117-118.

Consists of a stand, hood, screen, chamber, control mechanism for light, control mechanisms for air currents, and recording mechanisms. Designed for *Drosophila* and other groups, but believed adaptable for *Musca.* ABSTRACT: BA 42:1914(24152), June 15, 1963.

5532. **Childs, D. P., Mooney, J. W.** and **Gentry, T. 1964** A Mechanical Method of Retaining Flying Insects on a Test Surface. J. Econ. Ent. 57:839-840.

Describes a 2-part unit designed to retain flying insects on a test surface for determining the effectiveness of insecticides. ABSTRACT: TDB 62:479, May, 1965.

5533. **Clark, E. W. 1954** A Simple Technique for the Application of Unknown Material in Paper Chromatography. J. Econ. Ent. 47:934.

Two methods described: (1) (for 1 or 2 chromatograms) hold disk with forceps; pipette at 90° to forceps. (2) paper disks on glass plate; points of pipette (9) rest on disks. Non wetting agent (*Dri Film* or *Desicote*) prevents spread of liquid. Disks sewed to large sheet of Whatman No. 1 filter paper.

Cotty, V. F., joint author. *See* Henry, M. S., et al., 1957.

5534. **Cox, A. J. 1944** Insecticide Testing. A Review of Test Procedures for Evaluating Household Insecticides for Use in the Control of Flies, Clothes Moths, Roaches and Rodents. Soap and Sanit. Chem. 20(6):114-117, 149; (7):123, 125, 129.

Houseflies of uniform age and vitality obtained from larvae reared on moist mixture of coarse, soft, wheat bran, lucerne, malt extract and yeast at 80-85°F. A "large group" modification of the Peet-Grady method. ABSTRACT: RAE-B 33:99, 1945.

Crystal, M. M., joint author. *See* Gouck, H. K., et al., 1963.

5535. **Cummings, E. C. 1959** Preparation of Fly Media. 2 pages. ditto. Stauffer Chemical Co.

Gives formulation used by that laboratory for a number of years. (Has since been considerably modified.)

5536. **Cummings, E. C., Hallet, J. T.** and **Menn, J. J. 1964** A Cylindrical Cage for Fly Rearing. J. Econ. Ent. 57:177.

Considered an improvement over square metal cages, for ease of handling. ABSTRACTS: BA 45:4732(57690), July 1, 1964; TDB 61:727, July, 1964.

5537. Dahm, P. A. and **Pankaskie, J. E. 1949** A Biological Assay Method for Determining Aldrin. J. Econ. Ent. 42:987-988.
Uniformly sampled lots of house flies (approximately 100) exposed to residues of test material in wide-mouth jars, for 48 hours. Mortality per cent recorded at 24 and 48 hours. Moribund flies included with the dead. ABSTRACT: RAE-B 38:164, 1950.

Dahm, P. A., joint author. *See* Hamilton, E. W., et al., 1960.

5538. David, W. A. L. and **Harvey, G. L. 1941** Flies for the Peet-Grady Test. Soap and Sanit. Chem. 17(10):103, 105.
A wind tunnel with converging walls, ends in a wire screen. An endless belt, hand operated, conveys pupae (in rearing medium) past electric fan which has drying effect. Light culture medium blown into a collecting box. Pupae roll down and are collected separately. ABSTRACT: RAE-B 30:92, 1942.

5539. Davidow, B. and **Laug, E. P. 1955** A Surface Aliquot Masking Technique for the Bioassay of Lindane. J. Econ. Ent. 48:659-661.
Describes method for biologically assaying surfaces highly contaminated with lindane. Method is an adaption of a sensitive bioassay method (Laug, 1946) using the housefly. ABSTRACT: RAE-B 45:3, Jan., 1957.

Davies, M., joint author. *See* Goodwin-Bailey, K. F., et al., 1957.

5540. DeBach, P. 1942 A Simple Method of Obtaining Standardized Houseflies. J. Econ. Ent. 35:282-283.
Full-fed larvae strained through 16 mesh wire by stream of water. Or put sieve over battery jar with strong light above. Negative phototropism causes larvae to drop through onto cloth at bottom. Periodic exposure to light causes larvae to crawl away, leaving pupae of similar age. ABSTRACT: RAE-B 31:9, 1943.

5541. Deoras, P. J. 1948 Breeding of *Musca nebulo* F., for Biological Tests of Insecticides. Current Science, 17:301, Oct., 1948.
Describes methods and gives monthly data for breeding *Musca nebulo* F.

5542. Deoras, P. J. 1954 Breeding the Indian Housefly (*Musca domestica nebulo* Fabr.) for Experimental Studies. Parasitology 44:304-309.
Describes technique and apparatus to raise and maintain *M. domestica nebulo* under tropical conditions. Also describes devices for transferring flies from test tube to a covered petri dish or slab of glass. ABSTRACTS: TDB 52:221, Feb., 1955; RAE-B 44:14, Jan., 1956.

Dicke, R. J., joint author. *See* Allen, T. C., et al., 1943.

5543. Dodge, H. R. 1960 An Effective, Economical Flytrap. J. Econ. Ent. 53(6):1131-1132.
Lists materials; gives directions for assembling and for operating. Treats of various baits.

5544. Doner, M. H. and **Anderson, C. V. 1947** Testing Insecticidal Residues. A Micro Method for Biological Testing of Non-fumigant Type Insecticidal Residues on the Housefly (*Musca domestica* L.). Soap and Sanit. Chem. 23(10):124-125, 159.
Describes method and apparatus. Three methods of rating the insecticides are described. Special feature of this method is the use of less than 15 mg. active ingred./sq. ft.; inactivation rapid and results obtained quickly. ABSTRACT: RAE-B 38:41, 1950.

Doner, M. H., joint author. *See* Thomssen, E. G., et al., 1938.

5545. Doner, M. W. 1947 Testing Insecticide Residues. A Review of Methods. Soap and Sanit. Chem. 23(6):139, 141, 143, 193.
Based on a study of techniques practiced in 19 laboratories, plus the

author's own modifications. ABSTRACT: RAE-B 38:10, 1950.

Donnelly, J., joint author. *See* MacLeod, J., et al., 1956, 1957.

5546. Dorman, S. C., Hale, W. C. and **Hoskins, W. M. 1938** The Laboratory Rearing of Flesh Flies and the Relations Between Temperature, Diet and Egg Production. J. Econ. Ent. 31:44-51.

As with *M. domestica*, carbohydrate is essential for continued life of the adult fly, protein for the growth of the ovaries. Only certain proteins favorable. Eggs of old flies more sensitive to sterilizing agents (e.g. lysol) than eggs of young flies.

5547. Dorman, S. C. and **Hall, W. E. 1953** An Improved Spray Tunnel Method for Laboratory Tests. J. Econ. Ent. 46:151-153.

Flies are sprayed in disposable cages, in the spray tunnel. Requires minimum of space and expense. Operating variables do not affect results. ABSTRACT: RAE-B 41:141, Sept., 1953.

5548. Doty, A. E. 1937 Convenient Method of Rearing the Stable Fly. J. Econ. Ent. 30:367-369.

An adaptation of the Grady method for rearing *M. domestica. Stomoxys* is less resistant than *Musca* to Lethane 384, pyrethrum or rotenone. ABSTRACT: RAE-B 25:206, 1937.

Douglas, J. R., joint author. *See* Smith, A. H., et al., 1949.

Dowdy, A. C., joint author. *See* Champlain, R. A., et al., 1954.

Dreiss, J. M., joint author. *See* McGregor, W. S., et al., 1955.

5549. Eagleson, C. 1939 Insect Olfactory Responses. Construction and Use of an Olfactometer for Muscoid Flies and a Discussion of Interpreting Results. Soap and Sanit. Chem. 15(12):123, 125, 127.

Flies are imprisoned in a U tube of adjustable length, one arm of which is perfused with the odor of the attractant (or repellent). Both arms closed by wire screen septa. At regular intervals observer counts number of flies resting on each septum. Reactance "formula" used. ABSTRACT: RAE-B 28:100, 1940.

5550. Eagleson, C. 1940 Livestock sprays. A Rapid Method for Determining their Toxicity. Soap and Sanit. Chem. 16(7):96-99, 117.

Flies placed in cylinders in spray tunnel. After paralysis, are placed in recovery cage. Refers to Peet-Grady method, also to that of Bliss 1939. ABSTRACT: RAE-B 28:240, 1940.

5551. Eagleson, C. 1941 Bioassay of Livestock Spray using Hypnotic Doses Applied in a Spray Tunnel. Soap and Sanit Chem. 17(5):101, 103, 105, 107.

Simulates conditions in well-aired barns, which Peet-Grady Chamber cannot do. Sexes differ little as to "Hypnotic dose". Flies placed in recovery cabinet without removing from cylinder and observed for short period only. (Rules out other causes of mortality.) ABSTRACT: RAE-B 29:185, 1941.

5552. Eagleson, C. and **Benke, R. 1938** A Note on Rearing Houseflies. Soap and Sanit. Chem. 14(11):109, 119.

Larvae caused to pupate on moist crumpled paper towels. Pupae sorted by means of 2 fanning mill screens, one coarse, one finer. Those ending between screens are clean, uniform in size and usually balanced as to sex. ABSTRACT: RAE-B 27:57, 1939.

5553. Eastwood, R. and **Schoenburg, R. 1966** An Evaluation of Two Methods for Extracting Diptera Larvae from Poultry Droppings. J. Econ. Ent. 59:1286.

Hydraulic system, using 2 soil sieves, larger above. Water sprayed from above onto manure, resting on large screen. Material on each screen washed into separate enamel pan and examined for larvae. Flotation method: Salt mixed with water; larvae floated to surface and skimmed off with 50 mesh screen. Hydr. method faster (2×), and more exact for all genera. ABSTRACTS: RAE-B 55:14(36), Jan., 1967; BA 48:2322(25319), Mar. 1, 1967.

Edmonds, E., joint author. *See* Hickman, C., et al., 1966, 1967.

5554. Eide, P. E. and **Chang, T.-H. 1969** Cell Cultures from Dispersed Embryonic House Fly Tissue; Technique, Mitosis and Cell Aggregates. Exptl. Cell. Res. 54(3):302-308.
Used 6 hour old *Musca domestica* eggs. Technique described. Muscle, nerve cells differentiated. ABSTRACT: 50(17):9211(95461), Sept. 1, 1969.

5555. Eide, P. E. and **Reinecke, J. P. 1970** A Physiological Saline Solution for Sperm of the House Fly and the Black Blow Fly. J. Econ. Ent. 63(3):1006.
Fly sperm remain motile for 2-4 hours in a solution known as X-2 medium. Constituents for the medium are given, as well as the optimum requirements for sperm motility.

Eide, P. E., joint author. *See* Chang, T. H., et al., 1966.

Ernst, A. H., joint author. *See* Fay, R. W., et al., 1952.

Farnham, A. W., joint author. *See* Sawicki, R. M., et al., 1964.

5556. Fay, R. W., Stenburg, R. L. and **Ernst, A. H. 1952** A Device for Recording Insecticidal Knockdown of House Flies and Evaluating Residual Deposits. J. Econ. Ent. 45:288-292.
Tape recorder used to secure a continuous record of house flies knocked down by residual deposits of various insecticides. ABSTRACTS: RAE-B 40:129, Aug., 1952; TDB 49:1083, Nov., 1952.

Fay, R. W., joint author. *See* Jensen, J. A., et al., 1951.

5557. Filipponi, A. 1964 The Feasibility of Mass Producing Macrochelid Mites for Field Trials against Houseflies. Bull. W. H. O. 31:499: 501.
Discusses various problems in technical procedure and concludes that mass production would not be excessively difficult. Elucidation of optimum ecological conditions for each species, also selection of prolific strains are important steps. ABSTRACTS: TDB 62:932, Sept., 1965; BA 47:4220(49500), May 15, 1966; RAE-B 55:26(90), Feb., 1967.

5558. Fisher, R. W. and **Jursic, F. 1958** Rearing Houseflies and Roaches for Physiological Research. Canad. Ent. 90:1-7.
Describes current methods for rearing these insects. Important basic feature for *M. domestica* is the use of milk as the major constituent of the larval medium. ABSTRACT: RAE-B 47:98, July, 1959.

5559. Fisher, R. W. and **Morrison, F. O. 1949** Methods of Rearing and Sexing (*Musca domestica* L.). Ann. Rept. Ont. Ent. Soc. 80:41-45.
A basic paper, standardizing the rearing procedure of Hafez (1948), which see. For sorting of sexes a Y-shaped passage with damper gates, one to control speed of approaching fly, the second to direct fly into the proper arm. Mirrors permit operator to view fly's ventral surface. Can sex up to 1800 flies per hour.

Fisk, F. W., joint author. *See* Champlain, R. A., et al., 1954; Lockard, D. H., et al., 1953.

5560. Flynn, A. D. and **Schoof, H. F. 1970** Immersion Technique for Measuring Susceptibility of Houseflies and Cockroaches to Insecticides. J. Econ. Ent. 63(3):883-886.
Large numbers of insects can be exposed to measured doses of toxicants. Serial concentrations of DDT, dieldrin, diazinon and malathion ex-

pressed as mg/ml give responses directly related to the concentration used. Reproducible results obtained. ABSTRACTS: BA 51:13122(133520), Dec. 1, 1970; RAE-B 58:374(1540), Nov., 1970.

5561. Ford, J. H. 1937 A Method for Standardizing Peet-Grady Results. Soap 13(6):116-117, 119.
Inaccuracies in Peet-Grady results believed due to variation in resistance of individual insects. Suggests statistical treatment of data with 50 per cent kill as basis for reporting.

5562. Fowler, K. S. and **Lewis, S. E. 1958** The Extraction of Acetylcholine from Frozen Insect Tissue. J. Physiol. 142(1):165-172. London.
Recommends a method employing ethereal TCA as dispersion medium for the determination of ACh in frozen insect tissue. Essential to ensure complete destruction of enzyme activity before thawing.

5563. Frings, H. 1941 Rearing Blowflies in the Laboratory. J. Econ. Ent. 34:317.
Concerns *Cynomyia cadaverina.* Included here because paper is basic to the three which follow. ABSTRACT: RAE-B 30:44, 1942.

5564. Frings, H. 1947 A Simple Method for Rearing Blowflies without Meat. Science 105(2731):482.
Concerns *Phormia regina.* Based on kibbled dog biscuit kept moist. No disagreeable odors during larval development.

5565. Frings, H. 1948 Rearing Houseflies and Blowflies on Dog Biscuit. Science 107(2789):629-630.
An adaptation of previous technique to serve a larger number of species.

5566. Frings, H. and **Frings, M. 1953** Dog Biscuit as a Larval Medium for *Sarcophaga bullata.* J. Econ. Ent. 46:183.
Refers to previous success with houseflies. ABSTRACT: RAE-B 41:145, Sept., 1953.

Frings, H., joint author. *See* Knipe, F. W., et al., 1952.

Frings, M., joint author. *See* Frings, H., et al., 1953.

Fuller, R. G., joint author. *See* Brydon, H. W., et al., 1966.

Furshpan, E., joint author. *See* Boettiger, E. G., et al., 1952.

5567. Fye, R. L., LaBrecque, G. C., Morgan, P. B. and **Bowman, M. C. 1968** Development of an Autosterilization Technique for the House Fly. J. Econ. Ent. 61:1578-1581.
Both sexes sterilized by passing through expanded polystyrene foam strands which had been immersed in 5 per cent tepa solution, then dried. ABSTRACTS: BA 50:4781(50259), May 1, 1969; RAE-B 57:85(297), May, 1969; TDB 66:1185(2309), Nov., 1969.

Gard, I., joint author. *See* Ignoffo, C. M., et al., 1970.

Gates, J. M., joint author. *See* Hickman, C., et al., 1966, 1967.

5568. General Biological Supply House. 1947 Embedding Specimens in Transparent Plastic. Turtox Service Leaflet No. 33. 4 pp., illus. Chicago.
Included for completeness of coverage. Some workers are interested in this method of preservation and display.

Gentry, T., joint author. *See* Childs, D. P., et al., 1964.

5569. Gerberich, J. B. 1948 Rearing House-flies on Common Bacteriological Media. J. Econ. Ent. 41:125-126.
Of 35 media tested, 16 showed promise. Best was beef lactose agar. Maggots would not grow on sterile media. ABSTRACT: RAE-B 37:173, 1949.

Gerhardt, R. R., joint author. *See* Turner, E. C., et al., 1965.

Giannotti, O., joint author. *See* Lepage, H. S., et al., 1945.

5570. Glaser, R. W. 1924 Rearing Flies for Experimental Purposes with Biological Notes. J. Econ. Ent. 17:486-496.
M. domestica requires sugar or assimilable starch plus a solution of proteins or products of protein hydrolysis. ABSTRACT: RAE-B 12:162, 1924.

5571. Glaser, R. W. 1927 Note on the Continuous Breeding of *Musca domestica.* J. Econ. Ent. 20:432-433.
A suspension of yeast cells in water added to fresh manure for each generation. Medium must be kept moist. ABSTRACT: RAE-B 15:148, 1927.

5572. Glaser, R. W. 1938 A Method for the Sterile Culture of House Flies. J. Parasit. 24:177-179.
Gives method for sterilizing eggs, preparing larval medium and breeding receptacle. Relates to research on origin of bacteriophage. ABSTRACT: RAE-B 26:153, 1938.

5573. Goodhue, L. D. and **Cantrel, K. E. 1958** The Use of Vermiculite in Medium for Stable Fly Larvae. J. Econ. Ent. 51(2):250.
Vermiculite appears to absorb fermentation products, prevents rise in temperature and acts as a fly refuge. (Included because this author has also worked extensively with houseflies.)

5574. Goodhue, L. D. and **Linnard, C. E. 1950** Air Separation Apparatus for Cleaning Fly Pupae. J. Econ. Ent. 43:228.
Pupae separated from media by 5 mesh screen, and dried before a fan. Light material drawn up by air current. Separation 99 per cent. Cleans 5000 pupae per minute. ABSTRACT: RAE-B 38:210, 1950.

5575. Goodhue, L. D. and **Sullivan, W. N. 1941** Insecticidal Smokes. Their Application in the Control of Household Insects. 27th Midyear Meeting Nat. Assoc. of Insecticide and Disinfectant Manufacturers. 3 pp. Chicago.
Various insecticides were vaporized by heat. Methods of stabilizing and increasing insecticidal action are described. Orthodichlorobenzene found most practical synthetic compound against houseflies and other species.

Goodhue, L. D., joint author. *See* Howell, D. E., et al., 1965.

5576. Goodwin-Bailey, K. F., Holborn, J. M. and **Davies, M. 1957** A Technique for the Biological Evaluation of Insecticidal Aerosols. Ann. Appl. Biol. 45:347-360. London.
Gives standard technique for biological testing of aerosols, listing 12 essential items and/or steps. Suggestions and recommendations. ABSTRACTS: BA 32:1791(21281), June, 1958; RAE-B 45:213, Dec., 1957.

Goszczynska, K., joint author. *See* Mankowska, H., et al., 1967.

5577. Gouck, H. K., Crystal, M. M., Bořkovec, A. B. and **Meifert, D. W. 1963** A Comparison of Techniques for Screening Chemosterilants of House Flies and Screw-Worm Flies. J. Econ. Ent. 56:506-509.
Larval dipping not effective. Topical treatment of adults not reliable. Feeding (50 compounds) to adults most effective screening technique. Seven products caused sterility. ABSTRACTS: RAE-B 51:239, Nov., 1963; BA 45:1410(17528), Feb. 15, 1964; TDB 60:1169, Dec., 1963.

5578. Govind, R. 196 A New Design of Insecticide Testing Chamber. Results of Some Preliminary Experiments. Symp. Pestic. A., Mysore, India. Item 2.4, pp. 104-114.
Utilizes preferential behavior of some insects to rest on rough surfaces. Employs treated filter papers. Distinguishes between acutely paralyzed flies and others. Test species, *M. domestica nebulo.*

5579. **Grady, A. G. 1928** Studies in Breeding Insects Throughout the Year for Insecticide Tests. I. House Flies (*Musca domestica*). J. Econ. Ent. 21:598-604.

For larvae: fresh horse manure, water, yeast cells. For adults: milk, bread, sugar, yeast. Constant temperature 86°F. Relative humidity 40 per cent. ABSTRACT: RAE-B 16:254, 1928.

Grady, A. G., joint author. *See* Peet, C. H., et al., 1928.

5580. **Greenberg, B. 1954** A Method for the Sterile Culture of Housefly Larvae, *Musca domestica* L. Canad. Ent. 86:527-528.

Uses a modified CSMA medium. Describes preparation of medium and gives results when sterilized medium was used to raise houseflies. ABSTRACTS: RAE-B 44:113, Aug., 1956; BA 29:2978, Dec., 1955; TDB 53:1493, Dec., 1956.

5581. **Greenberg, B. 1968** Gnotobiotic Insects in Biomedical Research. Reprinted from "Advances in Germfree Research and Gnotobiology". CRC Press, Cleveland, Ohio. pp. 410-416.

A review of methodology in insect gnotogiology with particular emphasis on axenic insects. *M. domestica* one of several species discussed.

5582. **Greenberg, B. 1969** Sterile Culture of *Musca sorbens*. Ann. Ent. Soc. Amer. 62:450.

Concerns the collection and sterilization of eggs of *M. sorbens* and the preparation of sterile culture. First report of the sterile culturing of this species. ABSTRACT: RAE-B 57:269(1018), Dec., 1969.

5583. **Greenberg, B.** and **Archetti, I. 1968** *In Vitro* Cultivation of *Musca domestica* L. and *Musca sorbens* Wiedemann Tissues. Exptl. Cell Res. 54:284-287.

First report of the cultivation of cells from various structures of various stages of these two species. Technique described.

Grigarick, A. A., joint author. *See* Bechtel, R. C., et al., 1954.

5584. **Groat, R. A. 1939** Two New Mounting Media Superior to Canada Balsam and Gum Damar. Anat. Rec. 74:1-6.

Nevillite V and Nevillite no. 1 are cycloparaffin polymers. They are water-white, chemically homogeneous resins. Recommended proportion: 60 per cent of either plus 40 per cent toluene. ABSTRACT: BA 13:962, 1939.

Grose, J. E. H., joint author. *See* Parr, H. C. M., et al., 1961.

Guennelon, G., joint author. *See* Audemard, H., et al., 1968.

5585. **Hafez, M. 1948** A Simple Method for Breeding the House-fly, *Musca domestica* L., in the Laboratory. Bull. Ent. Res. 39(3):385-386.

Three parts milk to 1 part water by volume on pads of cotton wool. Adults both feed and oviposit on pads. Clean pads for larvae. Maintain at 27°C. ABSTRACT: RAE-B 37:57, 1949.

5586. **Hafez, M.** and **Attia, M. A. 1958** Rearing and Culturing *Musca sorbens* Wied. in the Laboratory. Bull. Ent. Res. 49:633-635.

Used artificial media consisting of a mixture of coarse wheat bran and diluted milk. A practical though not perfect substitute for human excrement, which is preferred in nature. ABSTRACTS: BA 35:5512(64867), Nov., 1960; RAE-B 47:21, Feb., 1959.

Haines, T. W., joint author. *See* Lindsay, D. R., et al., 1951.

Hale, W. C., joint author. *See* Dorman, S. C., et al., 1938.

Hall, W. E., joint author. *See* Dorman, S. C., et al., 1953.

Hallett, J. T., joint author. *See* Cummings, E. C., et al., 1964.

5587. Hamilton, E. W. and **Dahm, P. A. 1960** A Versatile Automatic Microapplicator. J. Econ. Ent. 53(5):853-856.
The main use of the microapplicator has been to make topical applications of acetone solutions of insecticides to the mesonotum of house flies, *Musca domestica* L.

5588. Harrison, R. A. 1950 Laboratory Breeding of the Housefly (*Musca domestica* L.). New Zealand J. Sci. (Tech. Ser. B) 30:243-247.
Includes various modifications of Peet-Grady method as adopted in New Zealand, 1946-1947. ABSTRACTS: BA 26:695, Mar., 1952; RAE-B 38:174, 1950.

5589. Harrison, R. A. 1953 The Fly-spray Testing Method in New Zealand. New Zealand J. Sci. (Tech. Ser. B) 34:457-461.
An up-dating of the preceding reference. ABSTRACTS: RAE-B 42:158, Oct., 1954; BA 28:2189, Sept., 1954.

Harrison, R. A., joint author. *See* Smith, A. G., et al., 1951.

Hartsock, J. G., joint author. *See* Wolf, W. W., et al., 1967.

Harvey, G. L., joint author. *See* David, W. A. L., et al., 1941.

Haviland, T. N., joint author. *See* Kampmeier, O. F., et al., 1948.

5590. Hays, S. B. and **Amerson, G. M. 1966** New Devices for Rearing and Handling House Flies in the Laboratory. J. Econ. Ent. 59:1523-1524.
Describes new type cages with devices for handling, maintaining a $C0_2$ atmosphere, feeding of flies, and oviposition. Pertains to studies in chemosterilization. ABSTRACTS: BA 48:10177(114361), Nov. 15, 1967; TDB 64:437, Apr., 1967.

5591. Heinz, H. J. 1949 Methodik zur Untersuchung des Darminhaltes von Fliegen auf pathogene Darmprotozoen des Menschen. [Method of examining the intestinal content of flies for pathogenic intestinal protozoa of man.] Zentralbl. Bakt. I. Abt. Orig. 153(3/5):106-108.
ABSTRACT: BA 25:872, 1951.

5592. Henry, M. S. and **Cotty, V. F. 1957** The Rearing of Aseptic Adult House Flies for Physiological Studies. Contr. Boyce Thompson Inst. 19(2):227-229.
Gives technique. Must know whether compounds formed in the fly are products of fly tissue or of microorganisms. Relates to radioactive tracer experiments. ABSTRACT: BA 32:1791(21282), June, 1958.

5593. Hepburn, G. A. 1943 A Simple Insect Cage-olfactometer. Onderstepoort J. Vet. Sci. 18(1-2):7-12. Pretoria.
Built for work with sheep bot fly, *Lucilia cuprina* Wied., in South Africa. Considered applicable for house flies and other species. ABSTRACT: RAE-B 33:5, 1945.

5594. Hickman, C., Gates, J. and **Edmonds, E. 1966** Fly Tagging, a New Approach. J. Environ. Hlth. 28(4):227-282.
See next reference.

5595. Hickman, C., Gates, J. and **Edmonds, E. 1967** Radioactive Fly Tagging. Pest Control 35(4):20, 22.
Series of lab and field experiments were run with Phosphorus-32 to test possibility of pinpointing the breeding sites of adult house flies.

5596. Hill, D. L., Bell, V. A. and **Chadwick, L. E. 1947** Rearing of the Blowfly, *Phormia regina* Meigen, on a Sterile, Synthetic Diet. Ann. Ent. Soc. Am. 40:213-216.
Advantages over natural medium (putrid meat) are: ease of handling, relative freedom from offensive odor, production of a more uniform and healthier strain of flies for experimental use. ABSTRACT: BA 22: 1500, 1948.

5597. Hockenyos, G. L. 1931 Rearing Houseflies for Testing Contact Insecticides. J. Econ. Ent. 24(3):717-725.
Details of techniques for that period. Used 4 parts pig manure, 3 parts horse manure for larvae; bread, moistened with milk and sugar for adults. Scalding of cages with steam controls parasitic mite. ABSTRACT: RAE-B 19:197, 1931.

Hoffman, R. A., joint author. *See* Roth, A. R., et al., 1952.

Holborn, J. M., joint author. *See* Goodwin-Bailey, K. F., et al., 1957.

Holbrook, D. V., joint author. *See* Sawicki, R. M., et al., 1961.

5598. Hollick, F. S. J. 1940 The Flight of the Dipterous Fly *Muscina stabulans* Fallen. Proc. Roy. Soc. London 129(B):S55-S56.
Special aerodynamic balance and wind tunnel used to study mounted, living insect. Amplitude of wing beat recorded photographically.

5599. Hopkins, T. L. 1962 Radioisotope Techniques and Recent Research on Metabolism of Insecticides in Insects. *In* Proc. Sympos. Radioisotopes and Radiation in Entomology. Bombay, 5-9 December, 1960. pp. 101-109 (Publ. 1962).
Reviews known techniques useful for investigations with radio-labelled insects and insecticides. ABSTRACT: RAE-B 51:213, Oct., 1963.

5600. Hoskins, W. M. and **Caldwell, A. H., Jr. 1947** Development and Use of a Small Spray Chamber. Soap and Sanit. Chem. 23(4):143-145, 161, 163, 165, 167.
A cylindrical device, 40 inches long and 12 inches in diameter. Mounting is nearly horizontal. ABSTRACT: RAE-B 37:201, 1949.

Hoskins, W. M., joint author. *See* Dorman, S. C., et al., 1938.

5601. Howell, D. E. and **Goodhue, L. D. 1965** A Simplified Insect Olfactometer. J. Econ. Ent. 58:1027-1028.
Designed primarily for cockroaches, but serves equally well for flying insects (including *M. domestica*), by the addition of an upper containing screen. Measures both attractants and repellents.

5602. Ignoffo, C. M. and **Gard, I. 1970** Use of an Agar-Base Diet and House Fly Larvae to Assay ß-Exotoxin Activity of *Bacillus thuringiensis*. J. Econ. Ent. 63(6):1987-1989.
Gives diet and techniques used to establish a reproducible assay for characterizing this activity.

5603. Ikeda, Y. 1959 Some Simplified Methods for the Evaluation of the Effectiveness of Fly Repellents in Laboratory and Outdoors. Botyu-Kagaku 24(4):175-181. Kyoto.
Single dose evaluated in laboratory; compared amounts of feeding on lactose pellet placed on repellent-treated filter paper and on untreated paper. Field tests carried out with funnel traps. Results analysed by probit method. ABSTRACT: RAE-B 52:26, Feb., 1964.

5604. Il'inskaya, N. B. 1958 Primenenie metoda vital'noi okraski dlya izucheniya reaktsii tkanei mukh na DDT. [Intra-vitam staining method applied to study the reaction of fly tissues to DDT.] Zool. Inst. Akad. Nauk SSSR. Leningrad. Referat Zhur., Biol., 1959 No. 30 116D.

5605. Ilinskaia, N. B. and **Troshin, A. S. 1954** [The marking of flies and insects by means of radioactive phosphorus.] Zool. Zhur. 33(4): 841-847. (In Russian.)

5606. Incho, H. H. 1954 A Rapid Method for Obtaining Clean House Fly Pupae. J. Econ. Ent. 47:938-939.
Uses layer of vermiculite to which larvae migrate for pupation. Vermi-

culite then removed by vacuum. Loss of pupae very low. ABSTRACTS: BA 29:2005, Aug., 1955; RAE-B 43:158, Oct., 1955.

5607. Janisch, E. 1933 Ueber die Konstanthaltung von Temperatur und Luftfenchtigkeit im biologischen Laboratoriumversuch. [On the maintenance of constant temperature and atmospheric humidity in biological laboratory research.] Abderhaldens Handbuch d. biol. Arbeitsmethoden. Abt. V, Teil 10.

5608. Jensen, J. A. and **Fay, R. W. 1951** Tagging of Adult House Flies and Flesh Flies with Radioactive Phosphorus. J. Trop. Med. 31(4): 523-531.

Insects tagged by incorporating P^{32} into milk fed to flies. Flies reared from fed larvae showed low activity. Tagged insects could be detected in traps. ABSTRACTS: RAE-B 41:187, Nov., 1953; TDB 48:1153, Dec., 1951; BA 26:232, Jan., 1952.

5609. Jepson, J. P. 1909 Notes on Experiments in Colouring Flies for Purposes of Identification. Repts. Local Gov't. Bd. on Publ. Hlth. and Med. Subjects. N.S. No. 16. Further Preliminary Reports on Flies as Carriers of Infection, pp. 4-9.

Powdered chalk, of various colors, preferred. Rice dust on a thin spraying of shellac also satisfactory. (These techniques now chiefly of historical interest.)

Johannsen, O. A., joint author. *See* Kingsbury, B. F., et al., 1935.

Johnson, E. R., joint author. *See* Sun, Yun-Pei, et al., 1965.

Jursie, F., joint author. *See* Fisher, R. W., et al., 1958.

Kalra, N. L., joint author. *See* Wattal, B. L., et al., 1959.

5610. Kampmeier, O. F. and **Haviland, T. N. 1948** On the Mounting of Anatomical Museum Specimens in Transparent Plastics. Anat. Rec. 100:201-231.

Included because of increasing interest in this method of preparing specimens, including insects and arachnids for demonstration or study.

Karlson, P., joint author. *See* Adelung, D., et al., 1969.

5611. Kearns, C. W. and **March, R. B. 1943** Small Chamber Method for Testing Effectiveness of Insecticides against Houseflies. Soap and Sanit. Chem. 19(2):101, 103-4, 128.

Is a horizontal glass cylinder, 26 inches by 12. To be used as a supplement to Peet-Grady test. Results are essentially the same. ABSTRACT: RAE-B 31:166, 1943.

Kearns, C. W., joint author. *See* Roan, C. C., et al., 1948.

5612. Keiding, J. 1951 Simple Equipment and Methods for Use in Insecticidal Tests. Appended (pp. 31-35) to "Statens Skadedyrlaboratorium etc."—which see.

Used inverted lime-washed earthenware pots. Flies introduced thru holes in bottom, later plugged with cotton. Other items of interest. ABSTRACT: RAE-B 40:116, July, 1952.

5613. Keiding, J. and **Arevad, K. 1964** Procedure and Equipment for Rearing a Large Number of Housefly Strains. Bull. W. H. O. 31:527-528.

A short account of methods found most satisfactory by the authors. ABSTRACTS: RAE-B 55:27(98), Feb., 1967; BA 47:5528(64813), July 1, 1966; TDB 62:932, Sept., 1965.

Keller, J. C., joint author. *See* Piquett, P. G., et al., 1962.

5614. **Kessel, E. L. 1962** An Improved Method of Mounting Small and Medium-sized Diptera. Wasmann J. Biol. 20(1):115-127.
Killing should always be done in a cyanide jar. Wings folded upward so they lie dorsal to and at right angles to the body, their surfaces parallel. Thin-cardboard point is drawn across applicator brush saturated with clear nail polish. Immediately, the point is placed with the tip across the right pleural area, a little toward the rear, and the right wing is glued to the point along its whole length.

5615. **Khudadov, G. D. 1959** [The method of labelling insects by introducing radioactive isotopes into their food.] Byull. Mosk. Obshch. Isp. Prir. Otd. Biol. 64(3):35-45. (In Russian, with English summary.)
A review of the literature plus a comprehensive report on investigations in the Soviet Union, to find the most suitable radioactive substances for practical use. ABSTRACT (very full): RAE-B 48:173, Oct., 1960.

Killough, R. A., joint author. *See* Wolf, W. W., et al., 1967.

5616. **Kingsbury, B. F.** and **Johannsen, O. A. 1935** Histological Technique. A guide for Use in a Laboratory Course in Histology. New York. John Wiley & Sons, Inc.
Although one of the older publications, this manual has special value for the entomologist, in that the junior author was a life-long specialist in insect morphology, including special techniques for preparing chitinized tissues.

5617. **Kitzmiller, J. B. 1949** The Use of Dioxane in Insect Microtechnique. Trans. Am. Micr. Soc. 67:227-230.
Is miscible with certain fixatives, with water, and with tissue mat. Calcium oxide may be added to stock bottle to reduce water content.

5618. **Klock, J. W., Pimentel, D.** and **Sternberg, R. L. 1953** A Mechanical Fly-Tagging Device. Science 118:48-49.
One inch length of brightly colored nylon thread is glued to thorax of fly anesthetized in CO_2. One worker can tag 1200 to 1500 flies per day.

5619. **Klots, A. B. 1932** Directions for Collecting and Preserving Insects. Wards Natural Science Establishment, Inc., Rochester, N.Y. 29 pp., illus.
One of many excellent manuals useful to the collector.

5620. **Knipe, F. W.** and **Frings, H. 1952** An Improved Cage for Flies. J. Econ. Ent. 45:1099.
Principal feature is arrangement requiring females to oviposit through screening. ABSTRACT: RAE-B 41:109, July, 1953.

5621. **Knuckles, J. L. 1963** An Apparatus for Studying the Transfer of Pathogenic Bacteria from Fly to Fly. Turtox News 41:284-285.
Based on two 250 ml. pyrex beakers, joined by a migratory tube. ABSTRACT: BA 46:357(4495), Jan. 1, 1965.

5622. **Knuckles, J. L. 1964** A Fly Rack Assembly Useful in Fecal Collection Studies. Turtox News 42(3):82-93.
Eliminates a change in fly position for feeding. Increases seven-fold the number of flies used. Reduces the work of the investigator.

5623. **Kramer, S. 1948** A Staining Procedure for the Study of Insect Musculature. Science 108(2797):141-142.
A seven step procedure, which includes use of Bouin's solution and eosin stain. Muscle differentiated *in toto*.

LaBrecque, G. C., joint author. *See* Fye, R. L., et al., 1968; Morgan, P. B., et al., 1964; Bailey, D. L., et al., 1970.

Lau, S. C., joint author. *See* Sun, Y.-P., et al., 1965.

Laug, E. P., joint author. *See* Davidow, B., et al., 1955.

5624. Leopold, R. A. and **Palmquist, J. 1968** A Method of Studying Early Cleavage in Eggs of House Flies, *Musca domestica*. Ann. Ent. Soc. Am. 61:1624-1626.
Twelve-step staining procedure outlined and illustrated. Puncture of eggs not necessary to obtain fixation. ABSTRACT: BA 59:4804(50526), May 1, 1969.

5625. Lepage, H. S., Giannotti, O. and **Pereira, H. F. 1945** Técnica para o ensáio de inseticidas residuais. [Technique for testing residual insecticides.] Biológico 11(12):320-325. São Paulo.
Laboratory procedure for comparing the toxicity of deposits of DDT and deposits of 666 (benzene hexachloride) to houseflies. ABSTRACT: RAE-B 35:70, 1947.

LeStrange, R., joint author. *See* Block, R. J., et al., 1952.

Lewis, S. E., joint author. *See* Fowler, K. S., et al., 1958.

5626. Lindquist, A. W. and **Madden, A. H. 1946** A Special Chamber for Testing Insecticidal Sprays. U.S. Bur. Entom. and Plant Quarantine. ET-229 (Processed). 5 pp.
Deals with apparatus and techniques of its use.

5627. Lindsay, D. R. and **Haines, T. W. 1951** A Method of Testing the Resistance of House Flies to Residual-type Insecticides. J. Econ. Ent. 44:104-106.
Direct testing of field-collected flies. Eliminates delay of 3 weeks for rearing from eggs. ABSTRACTS: RAE-B 39:138, 1951; TDB 48:1050, Nov., 1951.

Linnard, C. E., joint author. *See* Goodhue, L. D., et al., 1950.

Little, T. M., joint author. *See* Schoenburg, R. B., et al., 1966.

5628. Lockard, D. H. and **Fisk, F. W. 1953** A Turntable Modification of the Peet-Grady Method. J. Econ. Ent. 46:20-24.
Test cages secured to turntable; speed varied according to test. Provides rapid and economical method of testing sprays. This method can be used whether or not insecticides cause knockdown or leave toxic deposit as insects do not come into contact with test chamber walls. ABSTRACTS: RAE-B 41:135, Sept., 1953; TDB 50:865, Sept., 1953.

5629. Louw, B. K. 1964 Physical Aspects of Laboratory Maintenance of Muscoid Fly Colonies. Bull. W. H. O. 31:529-533.
Covers rearing, breeding, testing and shipping. ABSTRACTS: BA 47: 5528(64816), July 1, 1966; RAE-B 55:27(99), Feb., 1967; TDB 62:932, Sept., 1965.

5630. McGregor, W. S. and **Dreiss, J. M. 1955** Rearing Stable Flies in the Laboratory. J. Econ. Ent. 48:327-328.
Medium consists of 1 part (by volume) standard C.S.M.A. fly medium, 5 parts wood shavings and moistened with water. Fill 7 × 9 in. jar half full, mix eggs into medium, cover jar with muslin, place in rearing room at 80°F. Approximately 1000 flies/jar can be reared by this method. ABSTRACT: RAE-B 44:72, May, 1956.

5631. MacLeod, J. and **Donnelly, J. 1956** Methods for the Study of Blowfly Populations. I. Bait Trapping. Significance Limits for Comparative Sampling. Ann. Appl. Biol. 44:80-104.
Worked with 11 species of Calliphorinae. Studied variance of traps sampling blowfly populations from semi-arable land, reclaimed peat bog, and upland sheep walk. ABSTRACT: RAE-B 44:113, Aug., 1956.

5632. MacLeod, J. and **Donnelly, J. 1956** Methods for the Study of Blowfly Populations. II. The Use of Laboratory-bred Material. Ann. Appl.

Biol. 44:643-648.
No rhythm in flies responses to protein bait attractants in first four days. ABSTRACTS: BA 31:1814(18951), June, 1957; RAE-B 45:58, Apr., 1957.

5633. McLeod, J. and **Donnelly, J. 1957** Individual and Group Marking Methods for Fly-population Studies. Bull. Ent. Res. 48(3):585-592.
Describes in detail 4 different methods: (1) individual marking with paints; (2) mass powdering with dyes; (3) radioactive labelling with ^{32}P; (4) combination of the last two. Outlines circumstances affecting choice of method. Worked chiefly with Calliphoridae. ABSTRACTS: RAE-B 45:184, Nov., 1957; TDB 55:109, Jan., 1958.

McLeod, W. S., joint author. *See* Moreland, C. R., et al., 1956, 1957.

Madden, A. H., joint author. *See* Lindquist, A. W., et al., 1946.

5634. Mallison, G. F., Williams, E. R. and **Richards, C. S. 1958** An Apparatus for Rapid Indoor Collection of Adult Flies. Canad. Ent. 90:61-62.
A portable suction-type apparatus is described. ABSTRACT: BA 33:289 (3699), Jan., 1959.

Mammen, M. L., joint author. *See* Wattal, B. L., et al., 1959.

5635. Mánkowska, H. and **Goszczynska, K. 1967** Metodyka oceny efektywności preparatów dezynsekcyjnych z dichlorfosem. [Methods for evaluation of the effectiveness of insecticides containing dichlorfos.] Roczn. Pánstw. Zakl. Hig. 18:513-518. (Russian and English summaries.)
Houseflies exposed to dichlorfos vapors in Petri dishes. Report records time required for lethality of 50 per cent (LT_{50}). Ethymethyl ketone used for dissolving dichlorfos. Method requires neither special apparatus, nor constant temperature.

5636. Duplicate reference. Withdrawn.

March, R. B., joint author. *See* Kearns, C. W., et al., 1943.

5637. Markar'Yants, L. A. 1962 K metodike vyyavleniya yaits gel' mintov na naruzhnykh pokrovakh mukh. [A method for determining the presence of helminth eggs on the external integuments of flies.] Med. Parazit. i Parazitarn. (Bolezni) 31(4):485.
A washing technique, modified from that of Alexander and Dansker. Far superior to earlier methods. ABSTRACT: BA 41:1643(20667), Mar. 1, 1963.

Meifert, D. W., joint author. *See* Gouck, H. K., et al., 1963.

5638. Meijere, J. H. C. de. 1925 [Studies on Dipterous Larvae and Pupae.] Tijdschr. Ent. 68:211.
Kill in hot water or alcohol; clear in strong carbolic acid. Very dark pupae are bleached in diaphanol for several hours.

Mellanby, K., joint author. *See* Buxton, P. A., et al., 1934.

Melnick, J. L., joint author. *See* Penner, L. R., et al., 1952.

Menn, J. J., joint author. *See* Cummings, E. C., et al., 1964.

5639. Miller, T. A. and **Treece, R. E. 1968** A Device for Obtaining Flies of Known Age. Ann. Ent. Soc. Am. 61:548.
Operates automatically on a set timing, ranging from 2 to 24 hours. ABSTRACTS: BA 49:6772(65533), July 15, 1968; RAE-B 56:206(746), Oct., 1968.

5640. Mingo-Peres, E., Soprunov, F. F. and **Riskina, L. P. 1958** [Use of radioisotopes in labelling flies.] Med. Parazit. (Moskva) 27(6): 688-693.

5641. Mitra, R. D. 1952 A Medium for Breeding of House Flies, *Musca nebulo* F. in the Laboratory for the Study of Toxicity of Insecticides and Resistance of Flies to DDT. Sci. Cult. 17(8):341.
Suggests a medium based on meat and sand. ABSTRACT: BA 26:2330, Sept., 1952.

5642. Monroe, R. E. 1962 A Method for Rearing House Fly Larvae Aseptically on a Synthetic Medium. Ann. Ent. Soc. Am. 55:140.
Principal ingredients are micro-pulverized casein and agar. Yields flies of approximately normal size. ABSTRACTS: BA 38:930(12028), May 1, 1962; RAE-B 50:243, Nov., 1962; TDB 59:731, July, 1962.

Mooney, J. W., joint author. *See* Childs, D. P., et al., 1964.

5643. Moorefield, H. H. 1957 Improved Method of Harvesting House Fly Heads for Use in Cholinesterase Studies. Contr. Boyce Thompson Inst. 18(10):463.
Adult flies, frozen at —10°C, are fragmented by agitation. Heads sifted through screen of 100 openings per inch. Permits several thousand heads to be collected in a few minutes. ABSTRACT: RAE-B 46:105, July, 1958.

5644. Moreland, C. R. and **McLeod, W. S. 1956** House Fly Egg-Measuring Techniques. J. Econ. Ent. 49:49-51.
Three methods described: (1) graduated centrifuge tube; (2) graduated pipette tube and (3) pit method. There was no significant difference in puparial yields obtained by the first 2 methods. Centrifuge-tube method preferable for mass rearing because of ease of manipulation. Graduated pipette tube more accurate in the measurement of eggs. ABSTRACTS: BA 30:2414(24212), Aug., 1956; RAE-B 45:37, Mar., 1957; TDB 53:1062, Aug., 1956.

5645. Moreland, C. R. and **McLeod, W. S. 1957** Studies on Rearing the House Fly on a Bran-Alfalfa Medium. J. Econ. Ent. 50:146-150.
Amount of dry ingredients held constant. Changes made in (1) amount of water and (2) yeast and malt. Results given. Reports on the effect of various seeding rates on yield and size of pupae. To prevent lapses in normal testing schedules, eggs oviposited overnight may be used. ABSTRACTS: BA 31:2420(26186), Aug., 1957; RAE-B 46:78, May, 1958; TDB 54:1138, Sept., 1957.

Morgan, N. O., joint author. *See* Pickens, L. G., et al., 1967.

5646. Morgan, P. B. and **LaBrecque, G. C. 1964** Preparation of House Fly Chromosomes. Ann. Ent. Soc. Am. 57:794-795.
A modification of mosquito and *Drosophila* techniques; gave consistently good preparation from adult male flies. ABSTRACTS: BA 46:3725 (46047), June 1, 1965; TDB 62:365, Apr., 1965.

Morgan, P. B., joint author. *See* Fye, R. L., et al., 1968.

Morris, H. M., joint author. *See* Tattersfield, F., et al., 1924.

Morrison, F. O., joint author. *See* Fisher, R. W., et al., 1949; Chan, K., et al., 1965; Campau, E. J., et al., 1953.

5647. Murray, C. A. 1940 A Fundamental Error in the Peet-Grady Method. Soap and Sanit. Chem. 16(6):111, 113, 115, 117, 119, 125.
Comparison with Official Control Insecticide invalid, as flies vary greatly in the amount of insecticide they receive. ABSTRACT: RAE-B 28:238, 1940.

5648. Nagasawa, S. and **Asano, S. 1963** An Inbreeding Method of Rearing the House Fly. Scientific Notes. J. Econ. Ent. 56:714.
A technical alternative to single-pair mating (Buei, 1958). ABSTRACTS: BA 45:3219(39869), May 1, 1964; RAE-B 52:24, Jan., 1964; TDB 61: 222, Feb., 1964.

5649. Nelson, R. H. 1949 A Laboratory Method for Evaluating DDT Residues. J. Econ. Ent. 42:151.
Glass plates are covered with residue to be evaluated. Flies released; mortality counts taken after 24 hours. ABSTRACT: RAE-B 38:56, 1950.

5650. Nocerino, F. 1964 Perfeccionamiento del Jaulón para Pruebas con Insectos en Condiciones Similinaturales. [Perfecting of a cage for experiments with insects under essentially natural conditions.] Bol. Inform. del M.S.A.S. de Venezuela 4(4):255-259.
Diagrams, photos and instructions for use.

5651. Oppenoorth, F. J. and **Voerman, S. 1965** A Method for the Study of Housefly DDT-dehydrochlorinase by Gas-liquid Chromatography. Ent. Exp. Appl. 8(4):293-298. (German summary.)
Describes technique. Is of such sensitivity as to make possible the study of single susceptible houseflies, or measurement of slow reactions with DDT analogues. ABSTRACTS: BA 47:5533(64876), July 1, 1966; RAE-B 54:70, Apr., 1966.

5652. Osborn, A. W. and **Shipp, E. 1965** An Economical Method of Maintaining Adult Diptera. J. Econ. Ent. 58:1023.
A modification of technique originally devised for handling Queensland fruit fly. Adapts easily for use with *M. domestica.*

5653. Ozburn, G. W. 1964 A Simplified Technique for Rearing and Maintaining a Colony of House Flies (*Musca domestica* L.). Papers Mich. Acad. Sci., Arts and Letters 49:203-206.
Modified from the technique of Fisher and Morrison (1949). Uses an 11-day plan to provide a daily supply of flies for experimental use. ABSTRACT: RAE-B 54:204, Oct., 1966.

Palmquist, J., joint author. *See* Leopold, R. A., et al., 1968.

Pankaskie, J. E., joint author. *See* Dahm, P. A., et al., 1949.

5654. Parr, H. C. M. and **Grose, J. E. H. 1961** A Technique for Exposing Individual Flies on Insecticidal Deposits. Nature 192(4801):475.
Used *Stomoxys calcitrans.* Flies do not have to be anaesthetized and are forced to walk on treated surface throughout the period of exposure. Figure and dimensions of apparatus given. Other advantages listed. ABSTRACTS: RAE-B 50:281, Dec., 1962; BA 38:276(3733), Apr. 1, 1962.

5655. Patnaik, B. and **Roy, S. P. 1967** Rearing of *Musca conducens* in the Laboratory. Proc. Indian Sci. Congr. Assoc. 54(3):459.
This species is of importance as the vector of the bovine nematode, *Stephanofilaria assamensis.*

Paulini, E., joint author. *See* Ricciardi, I., et al., 1955.

5656. Peet-Grady Method. 1961 Official Method of the Chemical Specialties Manufacturers Association for Evaluating Liquid Household Insecticides. 1961. Revision. Soap and Sanit. Chem. (Blue Book) pp. 237-239.
Lays down rules for age of flies, food fed to them, etc. Various revisions published from time to time. ABSTRACT: RAE-B 50:1, Jan., 1962.

5657. Peet-Grady Method (and Related Matters). 1966 Soap. Chem. Spec. 42(4a):209-211, 212-214, 223, 274-275.
ABSTRACTS: RAE-B 55:21-22 (Items No. 65, 66, 69), Feb., 1967.

5658. Penner, L. R. and **Melnick, J. L. 1952** Methods for Following the Fate of Infectious Agents Fed to Single Flies. J. Exp. Med. 96: 273-280.
Flies mounted on paraffin blocks and made to ingest virus preparations;

amounts ingested measure by a potometer. Viruses then isolated from flies and their excreta. ABSTRACTS: BA 27:748, Mar., 1953; RAE-B 43:59, Apr., 1955.

Pereira, H. F., joint author. *See* Lepage, H. S., et al., 1945.

5659. Perry, A. S. 1968 An Apparatus for Dispensing House Fly Eggs Mechanically. J. Econ. Ent. 61:1112-1113.

A low cost device which dispenses ova with accuracy and reproducibility of delivery. May be made aseptic by autoclaving. ABSTRACTS: RAE-B 57:6(12), Jan., 1969; BA 50:3210(33605), Mar. 15, 1969.

5660. Pesson, P. and **Ramade, F. 1959** Une methode photographique adaptée à l'étude des tests insecticides sur *Musca domestica.* [A photographic method adapted to the study of test insecticides with *Musca domestica.*] Compt. Rend. Hebdom. Séances Acad. Agric. France 6:278-279.

5661. Pickens, L. G. and **Morgan, N. O. 1967** A Simplified Laboratory Technique for Separating Eggs of the Face Fly from Oviposition Medium. J. Econ. Ent. 60-1479.

Used strips of dung in V-shaped grooves as bait. Dung later mixed with 8 per cent saline, and eggs recovered by flotation. ABSTRACTS: RAE-B 56:71(229), Mar., 1968; BA 49:1417(15880), Feb. 1, 1968.

Pimentel, D., joint author. *See* Klock, J. W., et al., 1953.

5662. Piquett, P. G. and **Keller, J. C. 1962** A Screening Method for Chemosterilants of the House Fly. J. Econ. Ent. 55:261-262.

Batch of 120 fly pupae were dipped into 5 ml. of test solution for 30 minutes. Solution—1 part acetone, 1 part water and various concentrations of chemosterilants. ABSTRACTS: BA 39:989(12570), Aug. 1, 1962; RAE-B 50:238, Nov., 1962.

5663. Poorbaugh, J. H. 1969 Laboratory Colonization of the Canyon Fly, *Fannia benjamini* Malloch, and Speculation on the Larval Habitat. Calif. Vector Views 16(3):21-24.

Colony established from wild-caught females. Used rodent and rabbit fecal pellets for egg deposition and larval development and mammalian blood for adult food. Egg to adult = av. 24·5 days. Males emerge first.

Poorbaugh, J. H., joint author. *See* Anderson, J. R., et al., 1964.

5663a. Quijano, C. A. E. 1968 Un nuevo método de atrapar y combatir moscas en interiores mediante atracciön visual. [A new method of trapping and killing flies indoors by visual attraction.] Folia Ent. Mex. for 1968(18-19):102-103.

Attraction of flies to dishes containing water with a surface agent for drowning was increased by use of black and red paper, over dish. ABSTRACT: RAE-B 58:415(1734), Dec., 1970.

5664. Ramade, F. 1969 Application d'une technique de coloration a la fuchsine paraldehyde sur coupes semifines a l'étude de la *pars intercerebralis* de la mouche domestique. [Application of a technique for dyeing with paraldehyde fuchsin in connection with the semi-final step in the study of the *pars intercerebralis* of the domestic fly.] Ann. Soc. Ent. Fr. (N.S.) 5(3):707-717.

Housefly brains fixed with glutaraldehyde in Cacodylate Buffer and postfixed in a mixture of potassium dichromate and sodium sulfate, in Palade Buffer pH7. Alternate semi-thin (1μ) and ultrathin (500A) sections were cut; the first stained with paraldehyde fuchsin; the others photographed with Electron microscope.

Ramade, F., joint author. *See* Pesson, P., et al., 1959.

5665. **Raybould, J. N. 1964** An Improved Technique for Sampling the Indoor Density of African Housefly Populations. J. Econ. Ent. 57:445-447.

Flies are trapped on sticky, suspended strips and counted later. ABSTRACTS: TDB 62:157, Feb., 1965; BA 45:8404(106756), Dec. 15, 1964; RAE-B 52:202, Nov., 1964.

5666. **Reichardt, W.** and **Wenking, H. 1969** Optical Detection and Fixation of Objects by Fixed Flying Flies. Naturwissenschaften 8:424-425.

A method is described for the analysis of pattern discrimination in insects. Used by authors to investigate behavior of *M. domestica.*

Reinecke, J. P., joint author. *See* Eide, P. E., et al., 1970.

5667. **Rendtorff, R. C. 1953** A Method of Exposing Flies to Infectious Material. J. Parasit. 39:672-673.

Flies were placed in a retaining device, with the heads secured in a notched paper. So held, they were forced to feed on infectious material, which they readily did. Flies were not injured by the immobilization. ABSTRACTS: BA 28:1936(19683), Aug., 1954; RAE-B 43:21, Feb., 1955.

5668. **Ricciardi, I.** and **Paulini, E. 1955** Normas gerais para determinação da densidade de *Musca domestica* em uma localidade. [Procedures for determining the density of *Musca domestica* in a locality.] Rev. Brasil. Malariol. 7(1):93-101.

A modification of the grill method. Very good photographs. ABSTRACT: BA 30:3304(33003), Nov., 1956.

Richards, C. S., joint author. *See* Mallison, G. F., et al., 1958.

5669. **Richardson, H. H. 1932** An Efficient Medium for Rearing House Flies Throughout the Year. Science 76:350-351.

Larval medium utilizes wheat bran, lucerne meal, yeast and diamalt. Adults fed on fresh, diluted milk. ABSTRACT: RAE-B 20:261, 1932.

5670. **Richardson, H. H. 1937** Rearing the House Fly, *Musca domestica,* Throughout the Year. *In* Galtsoff, P. S., et al., Culture Methods for Invertebrate Animals. Ithaca, Comstock Publishing Company, pp. 429-432.

Draws on the experience and publications of several authors.

Riskina, L. P., joint author. *See* Mingo-Penes, E., et al., 1958.

5671. **Roan, C. C.** and **Kearns, C. W. 1948** Testing Insecticide Sprays. Soap and Sanit. Chem. 24:133, 135, 137, 149, 151.

Cage of test insects is placed in a small wind tunnel. Spray is drawn into the latter at controlled velocity. ABSTRACT: RAE-B 38:148, 1950.

5672. **Roth, A. R.** and **Hoffman, R. A. 1952** A New Method of Tagging Insects with P^{32}. J. Econ. Ent. 45:1091.

Worked with both *M. domestica* and *Phormia regina.* Insects immobilized with CO_2 and dipped individually in phosphoric acid solution containing 5μc. of P^{32} per ml., with or without 0·1 per cent of Triton X-100 (Wetting Agent). ABSTRACTS: RAE-B 41:108, July, 1953; TDB 50:573, June, 1953.

5673. **Roy, D. N.** and **Siddons, L. B. 1940** On Continous Breeding of Flies in the Laboratory. Indian J. Med. Res. 28(2):621-624.

Individual, gravid female placed in lamp chimney which sits on egg laying medium in a dish. (Material differs with the species.) Chimney top covered with gauze on which is placed a wet wad of cotton wool. ABSTRACT: RAE-B 29:86, 1941.

Roy, S. P., joint author. *See* Patnaik, B., et al., 1967.

5674. Rummel, R. W. and **Turner, E. C., Jr. 1970** A Refined Technique for Counting Face Fly Eggs. J. Econ. Ent. 63(4):1378-1379.
Female flies were attracted to, and deposited ova on a cheese cloth spread thinly with manure and covering a container of fresh manure. Counting was done by means of a grid, using a low power microscope.

5675. Sawicki, R. M. 1961 A Technique for the Topical Application of Poisons to Non-anaesthetized House-flies for Knockdown Assessments. Bull. Ent. Res. 51(4):715-722.
Flies are immobilized by suction, sexed and the females dosed by a measured-drop apparatus. Procedure is very rapid. ABSTRACTS: RAE-B 49:55-56, Mar., 1961; BA 36:6874(74153), Nov. 1, 1961; TDB 58:757, June, 1961.

5676. Sawicki, R. M. 1964 Some General Considerations on Housefly Rearing Techniques. Bull. W. H. O. 31:535-537.
A universally acceptable medium is highly desirable, but many strains of house flies require special treatment. Discusses merits of CSMA medium, also their own (YMA). Flies from Cooper Technical Bureau could adapt to either. ABSTRACTS: RAE-B 55:27(100), Feb., 1967; BA 47:5529(64827), July 1, 1966; TDB 62:932, Sept., 1965.

5677. Sawicki, R. M. and **Farnham, A. W. 1964** A Dipping Technique for Selecting House-flies *Musca domestica* L., for Resistance to Insecticides. Bull. Ent. Res. 55(3):541-546.
About 2000 flies, chilled until immobile, are tipped into a funnel and fluid added. Formula contains insecticide. Fluid sucked away after 3 minutes. Flies dried and kill recorded after 24 hours. Survivors released into breeding cages. ABSTRACTS: BA 47:4681(54835), June, 1966; RAE-B 53:33, Feb., 1965; TDB 62:598, June, 1965.

5678. Sawicki, R. M. and **Holbrook, D. V. 1961** The Rearing, Handling and Biology of House Flies (*Musca domestica* L.) for Assay of Insecticides by the Application of Measured Drops. Pyrethrum Post 6(2):3-18.
Technique described for breeding flies in pairs, in small numbers and on a large scale. Details of measured drop technique, for which flies are immobilized by chilling. ABSTRACT: RAE-B 51:158, August, 1963.

5679. Schoenburg, R. B. and **Little, T. M. 1966** A Technique for the Statistical Sampling of *Fannia* Larval Densities on Poultry Ranches. J. Econ. Ent. 59:1536-1537.
Describes sequential sampling of under-cage manure for larvae, and statistical treatment of results. *Fannia* larval density determinable within stated confidence levels. ABSTRACTS: BA 48:3232(35601), Apr. 1, 1967; RAE-B 55:71(261), Apr., 1967.

Schoenburg, R., joint author. *See* Eastwood, R., et al., 1966.

5680. Schonbrod, R. D. and **Terriere, L. C. 1966** Improvements in the Methods of Preparation and Storage of House Fly Microsomes. J. Econ. Ent. 59:1411-1413.
Methods of grinding and conditions of incubation and storage are compared, in a further study of the preparation and use of house fly microsomes. ABSTRACTS: BA 48:3737(41565), Apr. 15, 1967; RAE-B 55:66(244), Apr., 1967.

5681. Schoof, H. F. (Not dated) The Detection of Fly Resistance to Chemicals on Community Fly Control Programs. U.S. Dept. H. E. W., Publ. Hlth. Serv., Comm. Dis. Center, Atlanta, Ga. 7 pp.
Apparatus consists of collection cage, test cage and holding cage (all cylindrical and constructed of ice cream cartons), plastic and galvanized screening, and cardboard on plywood squares. Excellent for field use.

5682. **Schoof, H. F. 1952** The Attached Bait Pan Fly Trap. J. Econ. Ent. 45:735-736.
Advantages: Smaller than conventional traps. Can be dissembled, thereby permitting ease of handling in transportation, in killing and removing specimens, and in storage. Can be suspended from trees or fences. Less quantity of bait required per unit. Cost of construction low. ABSTRACT: RAE-B 41:44, Mar., 1953.

5683. **Schoof, H. F. 1964** Laboratory Culture of *Musca, Fannia* and *Stomoxys*. Bull. W. H. O. 31:539-544.
A review of the more dependable methods for rearing muscoid Diptera. ABSTRACTS: RAE-B 55:27(101), Feb., 1967; BA 47:5529(64829), July 1, 1966; TDB 62:932, Sept., 1965.

Schoof, H. F., joint author. *See* Flynn, A. D., et al., 1970.

5684. **Serfling, R. E. 1952** Entomological Survey Methods. Publ. Health Repts. 67(10):1020-1025.
Treats of collecting techniques, types of surveys, and interpretation of results.

5685. **Shamsutdinon, N. K. 1957** [On the methods of extracting larvae of houseflies from substrate.] Med. Zhur. Uzbekistana. (for) 1957(6): 74. Transl. from Referat. Zhur., Biol. 57938, 1958.
Used saturated solution of sodium chloride, poured on substrate. Volume of solution should be $2\frac{1}{2}$ times that of substrate. Larvae and pupae float to surface. ABSTRACT: BA 35:2239(25471), May, 1960.

Shipp, E., joint author. *See* Osborn, A. W., et al., 1965.

Shirai, M., joint author. *See* Suzuki, T., et al., 1960.

Siddons, W. A., joint author. *See* Roy, D. N., et al., 1940.

5686. **Simanton, W. A. 1937** Evaluating Liquid Insecticides. Comments on the 1937 Official Method and Use of the Official Control Insecticide in Grading Liquid Household Sprays. Soap 13(10):103, 105, 107, 115.
Peet-Grady Method adopted in 1932. Official Control Insecticide (a pyrethrum solution) in 1936. Method of reporting worked out in 1937. Not less than 10 tests must be made, in parallel, using flies of the same batch, on the same day. ABSTRACT: RAE-B 27:24, 1939.

5687. **Smith, A. G. 1961** Notes on Breeding Houseflies (*Musca domestica* L.). New Zealand J. Sci. (Tech.) 4(2):292-295.
Technique described, using simple and easily assembled equipment. Cage for adults—cylinder of galvanized wire gauze, closed at ends with tin lids. Plastic bottles at each end for food or used as oviposition site. ABSTRACTS: RAE-B 50:246, Nov., 1962; BA 36:8114(86689), Dec. 15, 1961.

5688. **Smith, A. G.** and **Harrison, R. A. 1951** Notes on Laboratory Breeding of the Housefly (*Musca domestica* L.). New Zealand J. Sci. (Tech.), Sec. B, 33(1):1-4.
Gives 2 modifications of previously described technique: (1) Eggs collected from a piece of hessian instead of cotton wool; (2) Pupae are collected from a petri dish filled with sawdust and buried in this medium before seeding. ABSTRACTS: TDB 50:161, Feb., 1953; BA 26:3222, Dec., 1952.

5689. **Smith, A. H.** and **Douglas, J. R. 1949** An Insect Respirometer. Ann. Ent. Soc. Am. 42:14-18.
For use with insects the size of *M. domestica*. Constant pressure principle adopted as basis for respirometer. Index drop directly measures volume change. Trial data for houseflies cited. ABSTRACT: RAE-B 39:164-165, 1951.

5690. Smith, C. N. 1967 Mass-Breeding Procedures. *In* Genetics of Insect Vectors of Disease. J. W. Wright and R. Pal, Editors. Amsterdam, Elsevier Publ. Co., pp. 653-672.
ABSTRACT: RAE-B 56:158(564), Aug., 1968.

5691. Smith, C. N. (Editor). 1967 Insect Colonization and Mass Production. Academic Press. "Houseflies" by D. Spiller, pp. 203-225. New York and London. 618 pp., 115 figs., refs.
ABSTRACT: RAE-B, 56:1(1) Jan., 1968.

Smith, G. G., joint author. *See* Yule, W. N., et al., 1967.

5692. Solomon, M. E. 1951 Control of Humidity with Potassium Hydroxide, Sulphuric Acid, or other Solutions. Bull. Ent. Res. 42:543-554.
An up-dating of the procedures of Buxton (1932); also Buxton and Mellanby (1934). ABSTRACT: TDB 49:450, Apr., 1952.

Soprunov, F. F., joint author. *See* Mingo-Penes, E., et al., 1958.

5693. Spiller, D. 1963 Procedure for Rearing Houseflies. Nature 199(4891): 405.
Includes several improvements in conventional procedure which save labor, avoid odor, and reduce contamination along with odors resulting therefrom. ABSTRACT: RAE-B 52:118, July, 1964.

5694. Steiner, G. 1942 Eine Zuchtweise für Fleischfliegen. [A rearing method for flesh flies.] Zool. Anz. 138(5-6):97-106.
Used raw, minced meat plus cellulose wadding, to rear *Phormia*, *Lucilia*, *Calliphora* and *Sarcophaga*. Remains of meat in flasks can then be used to rear *M. domestica*. No manure is necessary. ABSTRACT: BA 19:608 (5778), 1945.

Stenburg, R. L., joint author. *See* Fay, R. W., et al., 1952; Klock, J. W., et al., 1953.

Sullivan, W. N., joint author. *See* Goodhue, L. D., et al., 1941; Campbell, F. L., et al., 1938.

5695. Sun, Jung-yi Tung and **Sun, Yun-pei. 1953** Microbioassay of Insecticides in Milk by a Feeding Method. J. Econ. Ent. 46:927-930.
Test consists of feeding contaminated milk to houseflies and interpreting from results. ABSTRACT: RAE-B 42:144, Sept., 1954.

Sun, Jung-yi Tung, joint author. *See* Sun, Yun-pei, et al., 1952.

5696. Sun, Yun-Pei. 1950 Toxicity Index—An Improved Method of Comparing the Relative Toxicity of Insecticides. J. Econ. Ent. 43:45-53.
Equals the inverse ratio of median lethal concentrations multiplied by 100. Houseflies are sprayed with insecticides in a wind tunnel. Technical chlordan used as a standard. ABSTRACTS: RAE-B 38:186, 1950; TDB 47:784, 1950.

5697. Sun, Yun-Pei, Lau, S. C. and **Johnson, E. R. 1965** A Specific Bioassay Method for Determining Bidrin® Insecticide Residues. J. Ass. Offic. Agric. Chem. 48(5):938-942.
Follows the experimental design, procedure and evaluation of bioassay results essentially as in residue-film bioassay.

5698. Sun, Yun-Pei and **Sun, Jung-Yi, T. 1952** Microbioassay of Insecticides, with Special Reference to Aldrin and Dieldrin. J. Econ. Ent. 45: 26-37.
A microbioassay method for determining insecticidal residues on dried deposits. Developed on the principle that deposits containing the same amount of the same toxicant and the same quantity and quality of extracted substances should give the same percentage of mortality. Several modifications of the method are described for testing extracts

and evaluating results for samples containing various amounts of micro-quantities of toxicants. ABSTRACT: TDB 49:1083, Nov., 1952.

Sun, Yun-Pei, joint author. *See* Sun, Jung-yi Tung, et al., 1953.

5699. Suzuki, T. and **Shirai, M. 1960** A New Bioassay Method for Insecticide Residues. Jap. J. Exptl. Med. 30:75-81.
New evaluating method is based on rapidity of knock-down. Time required for 50 per cent knockdown in test section is compared (by means of a mathematical procedure) with time required on standard panels. Does not include insecticide absorbed by substrate, hence not available to insect. Easier to use than chemical assays. ABSTRACTS: TDB 57:1229, Nov., 1960; BA 35:4676(54330), Oct. 1, 1960; RAE-B 51:157, Aug., 1963.

5700. Tamarina, N. A. 1963 [Microsurgical technique for insects.] Zool. Zhur. 42(8):1260-1264. (In Russian, with English summary).
Equipment and procedures for certain microsurgical operations on insects are described and illustrated. Example: larvae of *Calliphora erythrocephala*. ABSTRACT: BA 46:2560(31936), Apr. 1, 1965.

5701. Tate, P. 1948 The Technique for Breeding Pure-line Cultures of the Blow-fly (*Calliphora erythrocephala*). Parasitology 39:102-104.
Describes cage 10 × 10 × 10 in. and various precautions used to prevent contamination by stray flies. Examples: Meat dipped in boiling water and placed in a container surrounded by water to protect it from migrating larvae. Glass top on cage prevents eggs from dropping onto meat. ABSTRACT: RAE-B 39:202, Dec., 1951.

5702. Tattersfield, F. and **Morris, H. M. 1924** An Apparatus for Testing the Toxic Values of Contact Insecticides under Controlled Conditions. Bull. Ent. Res. 14:223-233.
An atomozier, fixed to the lid of a glass jar, throws a fine spray upon insects placed in a dish inside the jar. Apparatus stands on a levelling platform.

Teplykh, V. S., joint author. *See* Yurgenson, I. A., et al., 1967.

Terriere, L. C., joint author. *See* Schonbrod, R. D., et al., 1966.

5703. Tharumarajah, K. and **Thevasagayam, E. S. 1961** A Simple Method for Breeding the House-fly, *Musca domestica vicina* Macquart, in the Laboratory. Bull. Ent. Res. 52(3):457-458.
Makes use of coconut poonac, the cake left over after extraction of oil from coconut pulp. It receives eggs, nourishes larvae, and provides a site for pupation. ABSTRACTS: TDB 59:212, Feb., 1962; BA 39:975 (12373), 1962; RAE-B 49:282, Dec., 1961.

Thevasagayam, E. S., joint author. *See* Tharumarajah, K., et al., 1961.

5704. Thomssen, E. G. and **Doner, M. H. 1938** Breeding Houseflies. A Simplified and More Convenient Method of Rearing and Handling Flies for Peet-Grady Tests. Soap and Sanit. Chem. 14(10):89-90, 101.
Temperature thermostatically controlled at 80°F. Eggs, obtained by placing dish of moist bran in stock cage, are roughly counted, then transferred to 5 gal. drums of Richardson's Medium. Flies emerge into stock cages, 18 inches × 8 inches × 8 inches, screened on 5 sides. ABSTRACT: RAE-B 27:56, 1939.

5705. Tighe, J. F. 1960 Technique of a Housefly Bioassay for Pesticides. J. Ass. Offic. Agric. Chem. 43:82-87.
Describes adaptations of fly bioassay techniques as they have evolved over a span of 15 years. Procedure affords a simple check for possible

presence of pesticides in food and other products. Abstract: BA 36: 7681(82551), Dec. 1, 1961.

Treece, R. E., joint author. *See* Miller, T. A., et al., 1968.

Troshin, A. S., joint author. *See* Ilinskaia, N. B., et al., 1954.

5706. Tung, Sik-Chung, Cheng, Tien-Hsi and **Yendol, W. G. 1969** A Method of Clearing Face Fly Puparia for Sex Determination without Affecting Survival and Development. J. Econ. Ent. 62:1412-1417.
Six day old larvae were satisfactorily cleared by immersion in 0·1 per cent detergent "Joy" for 20 minutes. Sexual dimorphism manifest in compound eyes and adjacent areas. Abstract: BA 51:7971(81451), July 15, 1970.

5707. Turner, E. C., Jr. and **Gerhardt, R. R. 1965** A Material for Rapid Marking of Face Flies for Dispersal Studies. J. Econ. Ent. 58: 584-585.
Large numbers marked rapidly by use of a dust pigment. Very satisfactory. Abstracts: BA 47:1602(19401), Feb. 15, 1966; RAE-B 53:184, Sept., 1965.

Turner, E. C., Jr., joint author. *See* Rummel, R. W., et al., 1970.

5708. Van Asperen, K. 1962 Sensitive Colorimetric Methods for the Estimation of Esterases and Organophosphates. Overdruk vit de Mededelingen Van de Landbouwhogeschool en de Opzoekingsstations van de Staat te Gent. Deel xxvii, No. 3.
Treats of the Method of Kramer and Gramson; also other procedures useful in the study of insect physiology.

Voerman, S., joint author. *See* Oppenoorth, F. J., et al., 1965.

5709. Wattal, B. L., Mammen, M. L. and **Kalra, N. L. 1959** A Simple Medium for Laboratory Rearing of Housefly (*Musca domestica nebulo* Fabricius), with Some Observations on its Biology. Indian J. Malariol. 13(4):175-183.
Water-soaked cotton seed cake for larvae; water-soaked milk powder for adults. Was most effective medium used by Malaria Institute of India. Abstracts: BA 40:1935(25387), Dec., 1962; TDB 58:141, Jan., 1961.

5710. Welling, W. 1968 The Use of a Modified Selector Valve in Recycling Gel Chromography. Sci. Tools. 15(2):24-26.
Modification used chiefly in the purification of an esterase from houseflies. Abstract: BA 50:5430(57034), June 1, 1969.

Wenking, H., joint author. *See* Reichardt, W., et al., 1969.

5710a. West, L. S. 1960 Requirements for Reproduction as Compared with Requirements for Maintenance of Life in Adult Houseflies. *In* High School Biology. Biological Investigations for Secondary School Students. American Institute of Biological Sciences. Biological Sciences Curriculum Study, University of Colorado, Boulder. Photolithographed Document. Pages 369-371.
Prepared for experimental use during the school year 1960-61. Was subsequently revised and expanded for publication in printed form. See Reference 5710b.

5710b. West, L. S. 1963 Requirements for Reproduction as Compared with Requirements for Maintenance of Life in Adult Houseflies. *In* Research Problems in Biology. Investigations for Students. Series 1. Doubleday & Company, Garden City, New York. Project 39, pp.

193-199.
Consists of directions for beginners in research who wish to gain familiarity with methods of rearing, handling, and experimenting with *M. domestica*.

Whitfield, T. L., joint author. *See* Bailey, D. L., et al., 1970.

5711. Wilcox, H. H., III. 1969 An Inexpensive Housefly Pupae Separator. J. Econ. Ent. 62:949-950.
By using an industrial vacuum cleaner equipped with a rheostat, worker can separate 20,000 pupae from vermiculite in about 2 minutes. ABSTRACT: RAE-B 58:28(102), Jan., 1970.

5712. Williams, C. M. 1945 Separatory Funnels as Experimental Chambers in Studies of Insect Physiology. Science 101(2633):622.
Very brief note. Discusses advantages of using separatory funnels in studies where animals are exposed to various gases or vapors. Funnels are durable, inexpensive, transparent, air tight and easily washed.

5713. Williams, C. W. 1946 Continuous Anesthesia for Insects. Science 103(2663):57.
Apparatus provides a continuous flow of CO_2 up through a funnel mounted flush with the table and beneath the dissecting scope.

Williams, E. R., joint author. *See* Mallison, G. F., et al., 1958.

5714. Winteringham, F. P. W. 1962 Radioactive Tracer Techniques in Insect Biochemistry. *In* Internat. Atomic Energy Agency. Radioisotopes and Radiation in Entomology. Proceedings of Symposium. Internat. Publ., Inc., New York. pp. 113-134. Reprinted from Radioisotopes in Tropical Medicine, pp. 283-303.
An excellent review. Suggests use of cytidine pools for studying nucleic acid metabolism in living cells. ABSTRACT: BA 42:323(3911), Apr. 1, 1963.

5715. Wolf, W. W., Killough, R. A. and **Hartsock, J. G.** Small Equipment for Immobilizing Flies with Cool Air. J. Econ. Ent. 60:303-304.
Developed for use with *M. domestica*, *M. autumnalis*, and *Stomoxys calcitrans*. ABSTRACTS: BA 48:6036(67470), July 1, 1967; RAE-B 55:107 (380), June, 1967.

5716. Wolfinson, M. 1953 The Rearing, Marking and Trapping of Houseflies (*Musca domestica vicina*) for Dispersal Studies. Bull. Res. Counc. of Israel 3(3):263-264.
To mark flies, they were fed 2 days before release on a mixture of sugar and ferric oxide; abdomens became red and remained so for several days. Paper also describes rearing and trapping devices. ABSTRACT: RAE-B 43:127, Aug., 1955.

5717. Wollman, E. 1921 La méthode des élevages aseptiques en physiologie. [The method of aseptic rearing through physiology.] Arch. Intern. Physiol. 18:194-199.
Reports that microbe-free cultures of flies can be maintained indefinitely.

Yendol, W. G., joint author. *See* Tung, Sik-Chung, et al., 1969.

5718. Yule, W. N. and **Smith, G. G. 1967** Techniques for Insecticide-bioassay with Houseflies. Canad. Ent. 99(2):219-221.
The use of dry ice (solid CO_2) and of all-glass handling and holding systems simplifies procedure and insures fewer mortalities. ABSTRACT: BA 48:7866(88169), Sept. 1, 1967.

5719. Yurgenson, I. A. and **Teplykh, V. S. 1967** [Use of luminescent microscopy for studying yellow-bodies in the ovaries of insects (*Musca domestica*).] Ent. Obozrenie 46(2):295-298. (In Russian, with English summary.)

First description in literature of this technique. May be employed for revealing regularities in the formation and development of yellow-bodies, thus facilitating the determination of physiological age of insects. ABSTRACT: BA 50:1044(10939), Jan. 15, 1969.

5720. Yurkiewicz, W. J. 1967 A Respirometer Flask for Measuring Oxygen Consumption during Flight on a Turnabout. Ann. Ent. Soc. Am. 60(5):1122-1123.

Developed by work on *Phaenicia* and *Phormia*. Cites earlier papers on *Drosophila* and *Muscina*. Should be useful for any species.

Zweig, G., joint author. *See* Block, R. J., et al., 1952.

Section VIII.
Index

To locate the publications of a given author, the user should consult the four bibliographies here included (Sections IV, V, VI and VII). For all other areas of interest, consult the following alphabetical list of Categories.

After selection of the Category (or Categories) within which pertinent references are likely to occur, turn to that listing and consider the Subdivisions under the Category selected. The reference numbers given for each Subdivision lead directly to the references and their annotations.

It is believed that this relatively simple arrangement, utilizing 15 Categories and 70 Subdivisions, will be of greater benefit to users than would an elaborate system of code numbers and key words. Many reference numbers appear under more than one Category, also in more than one Subdivision, so that the likelihood of overlooking a relevant item is highly remote.

Category 1. Control Measures

This category includes manipulation of the environment as well as the use of dusts, sprays, baits and traps. (One should also consult the following Categories: Insecticides and Toxicology; Military Interests; Parasites and Predators; Sterilization for Control).

In all fifteen Categories, determination of Subdivisions has been arrived at by scrutiny of the literature available, rather than by any theoretical or academic grouping of subject matter. Thus, in Category 1, we have placed the larger number of references in the first, or General Subdivision, with three additional Subdivisions covering special interests. The four groupings are as follows:

Subdivision a. References having to do with space sprays, aerosols, dusts, fumigants, general sanitation, and even a few in the area of X-ray induced translocation of genes.

Subdivision b. References emphasizing the use of poison baits, and directed obviously against adult flies.

Subdivision c. References having to do chiefly with procedures against larval and pupal stages.

Subdivision d. References concerned primarily with fly-traps; their construction, location, baiting, and manner of operation. Electrical devices for killing the entrapped flies are included here also.

Category 1. – *Subdivision a.* (GENERAL PRACTICES).

Reference Numbers: 1, 2, 13, 14, 21, 28, 30, 39, 40, 41, 53, 58, 75, 77, 89, 102, 109, 112, 117, 118, 127, 133, 143, 148, 180, 182, 195, 207, 209, 260, 261, 262, 263, 272, 294, 324, 325, 326, 337, 356, 357, 360, 372, 375, 392, 448, 483, 485, 488,

526, 528, 536, 537, 538, 539, 540, 542, 563, 567, 572, 582, 583, 584, 598, 599, 602, 610, 614, 619, 620, 643, 645, 648, 655, 678, 689, 694, 697, 712, 740, 777, 790, 796, 818, 825, 830, 849, 854, 858, 861, 914, 915, 919, 922, 940, 948, 950, 958, 1000, 1006, 1028,

1031, 1035, 1048, 1060, 1072, 1102, 1110, 1113, 1114, 1137, 1148, 1161, 1174, 1183, 1188, 1218, 1250, 1272, 1279, 1282, 1283, 1293, 1309, 1355, 1358, 1360, 1364, 1365, 1412, 1427, 1435, 1439, 1440, 1442, 1452, 1453, 1469, 1480, 1481, 1515, 1519, 1520, 1522, 1547, 1575, 1603, 1607, 1643, 1656, 1680, 1681, 1684,

1697, 1698, 1702, 1715, 1717, 1718, 1719, 1726, 1729, 1739, 1749, 1775, 1776, 1781, 1791, 1816, 1818, 1819, 1826, 1827, 1828, 1837, 1839, 1845, 1847, 1866, 1910, 1912, 1952, 1963, 1972, 1974, 1975, 1989, 1999, 2000, 2021, 2039, 2058, 2077, 2094, 2126, 2129, 2132, 2133, 2135, 2154, 2159, 2172, 2206, 2243, 2252,

2254, 2268, 2275, 2280, 2287, 2292, 2300, 2333, 2335, 2367, 2373, 2374, 2438, 2439, 2455, 2464, 2465, 2473, 2484, 2488, 2493, 2499, 2514, 2529, 2530, 2533, 2574, 2575, 2588, 2591, 2617, 2637, 2638, 2639, 2645, 2647, 2671, 2681, 2682, 2683, 2700, 2706, 2724, 2732, 2739, 2755, 2761, 2770, 2771, 2788, 2789, 2817,

2844, 2808, 2916, 2918, 2937, 2947, 2961, 2974, 2981, 2985, 3059, 3068, 3069, 3071, 3090, 3100, 3130, 3131, 3144, 3166, 3170, 3174, 3188, 3239, 3240, 3242, 3259, 3295, 3304, 3305, 3308, 3309, 3320, 3343, 3344, 3346, 3357, 3360, 3361, 3362, 3367, 3369, 3388, 3389, 3431, 3450, 3489, 3518, 3519, 3520, 3525, 3539,

3540, 3541, 3542, 3543, 3544, 3545, 3546, 3548, 3549, 3553, 3615, 3619, 3650, 3653, 3659, 3694, 3701, 3720, 3725, 3726, 3730, 3745, 3755, 3804, 3816, 3843, 3888, 3914, 3946, 3952, 4008, 4009, 4010, 4011, 4013, 4020, 4070, 4082, 4167, 4198, 4218, 4231, 4243, 4251, 4257, 4258, 4262, 4263, 4264, 4267, 4276, 4277,

4278, 4280, 4396, 4437, 4443, 4445, 4520, 4561, 4564, 4587, 4621, 4633, 4638, 4639, 4640, 4669, 4669a 4688, 4691, 4692, 4693, 4703, 4713, 4762, 4802, 4803,

Category 1. – *Subdivision b.* (POISON BAITS).

Category 1. – *Subdivision c.* (MEASURES AGAINST IMMATURE FORMS).

Category 1. – *Subdivision d.* (FLY TRAPS).*

*We have received an unpublished description of a jar-type fly trap using pork kidney as bait which was devised and used successfully by Mr. William Kimak, of Wallington, New Jersey. Mr. Kimak's address may be had by communicating directly with the senior author.

Category 2. Disease Relations

Listed here are all references concerned with Public Health as well as those dealing with particular diseases of man, animals, and in a few cases, plants. Older publications dealt very frequently with enteric infections, such as fly-borne typhoid, an interest which abated sharply, with the availability of typhoid vaccine. Fly-borne trachoma, on the other hand is a concern of more modern investigators, and still continues to be so. The Subdivisions of most Categories, in fact, tend to reflect certain historical trends in research; low reference numbers indicating early publications, while contributions published since 1960 bear numbers above 2800. The nine Subdivisions of Category 2 are as follows:

Subdivision a. – References concerned with cholera, typhoid, the dysenteries and other enteric infections.

Subdivision b. – References concerned with possible vectorship of poliomyelitis by filth-feeding and filth-breeding flies.

Subdivision c. – References concerned with the relation of flies to the distribution of tuberculosis.

Subdivision d. – References concerning possible vectorship of yaws, syphilis or pinta (pinto).

Subdivision e. – References having to do with possible transmission of leprosy.

Subdivision f. – References concerned with the invasion of animal tissue or cavities by dipterous larvae (Myiasis), also with the therapeutic use of maggots in the treatment of such conditions as osteomyelitis.

Subdivision g. – References treating of the vectorship of animal infections, also of injury to or significant disturbances of domestic animals by certain species of Muscidae. This area may be summarised by the term "Veterinary Relations".

Subdivision h. – References having to do with the relation of synanthropic flies to the spread of various ophthalmias.

Subdivision i. – References of a miscellaneous or general nature, not easily assigned to a restricted field.

Category 2. – *Subdivision a.* (ENTERIC DISORDERS).

Category 2. – *Subdivision h.* (Ophthalmia).

Category 2. – *Subdivision i.* (Miscellaneous and General).

Category 3. Ecology

Included here are references which deal with environment in the general sense of the term, as well as research reports on carefully controlled experiments pertaining to specific ecological influences. Five Subdivisions are recognized:

Subdivision a. References emphasizing the importance of temperature and relative humidity.

Subdivision b. References pertaining to light and color as ecological factors.

Subdivision c. References treating of altitude and/or barometric pressure as an influence on behavior and life processes.

Subdivision d. References which emphasize availability and quality of food, as a needful factor in the suitability of any environment for fly production.

Subdivision e. Miscellaneous and general references, including occasional data on winds, magnetic fields, pH, ozone concentration, and density of the fly population.

Category 3. – *Subdivision a.* (TEMPERATURE AND HUMIDITY).

Category 3. – *Subdivision b.* (LIGHT AND COLOR).

Category 3. – *Subdivision c.* (BAROMETRIC PRESSURE).

Category 3. – *Subdivision d.* (FOOD RELATIONS).

Category 3. – *Subdivision e.* (MISCELLANEOUS AND GENERAL).

Category 4. Geographical Interests

Assignment to this Category has been highly selective. Listing is confined to those references which convey significant geographical information. Faunal lists, taxonomic revisions, descriptions of new species and appearances of old species in new areas, are obvious inclusions. Also listed here are publications reporting seasonal abundance, peculiarities in the life cycle believed to be territorially limited, areas in which success was first obtained with a newly-devised control technique, and unusual responses to control measures, which seem to be in conflict with results obtained elsewhere. Seven Subdivisions are utilized:

Subdivision a. References concerned primarily with African species or sub-species.

Subdivision b. References concerned primarily with Asian species or sub-species.

Subdivision c. References concerned with species or sub-species found in Australia and/or New Zealand.

Subdivision d. References concerned primarily with European species or subspecies.

Subdivision e. References concerned primarily with North American species or subspecies.

Subdivision f. References concerned primarily with South American species or subspecies.

Subdivision g. References concerning the species or sub-species of various islands of the sea. For *oceanic* islands this is usually the only Subdivision used. In the case of *continental* or *coastal* islands however, the same reference number will be found listed under the appropriate continental Subdivision also.

Category 4. – *Subdivision a.* (AFRICA).

Category 4. – *Subdivision b.* (ASIA).

Category 4. – *Subdivision c.* (AUSTRALIA AND NEW ZEALAND).

Category 4. – *Subdivision d.* (EUROPE).

Category 4. – *Subdivision e.* (NORTH AMERICA).

Category 4. – *Subdivision f.* (South America).

Category 4. – *Subdivision g.* (Islands).

Category 5. Genetics of the House Fly

Not until *Musca domestica* had developed great resistance to DDT and related insecticides, was it realized that our knowledge of hereditary mechanisms in this species was both meager and superficial. Something in the nature of a crash program then came into being, in the attempt to find genetic reasons for resistance phenomena. The vast amount of information on record for the genus *Drosophila*, plus many years of experience in the laboratory rearing of *Musca*, made it possible for workers to push forward rapidly along practical lines, and explain, at least tentatively, the mendelian principles involved in the more common forms of insecticide resistance. Increased knowledge of fly chromosomes, and their behavior, soon led quite naturally to theoretical as well as practical research, with the result that though most of the reliable material on house fly genetics has been published in the last 20 years, the titles are about equally divided between those with a direct bearing on resistance and those of a more academic nature, concerned primarily with basic research. For this reason, we have designated but two Subdivisions:

Subdivision a. References relating chiefly to resistance.

Subdivision b. References of an academic or general nature.

Category 5. – *Subdivision a.* (Resistance).

Category 5. – *Subdivision b.* (BASIC RESEARCH).

Category 6. Histology, Cytology and Microscopy

This is a small Category, by number of titles, but since it is likely to receive increasing attention in the future, separate recognition appears desirable. Zest for purely morphological studies tended to languish, when further improvement of conventional light microscopes became sharply limited, but the advent of the electron microscope opened up new vistas for exploration, in all branches of biology, and we may now look for an expanding body of knowledge in this field. The three Subdivisions listed here may be considered prophetic of research effort yet to come:

Subdivision a. Chiefly older references, concerning tissues studied principally by means of conventional light microscopes, with or without the employment of "phase" microscopy.

Subdivision b. References describing the structure of tissues and cells as observed by means of electron microscopy.

Subdivision c. References concerning the observation of cells in the living state.

Category 6. – *Subdivision a.* (CONVENTIONAL MICROSCOPE STUDIES).

Category 6. – *Subdivision b.* (ELECTRON MICROSCOPE STUDIES).

Category 6. – *Subdivision c.* (TISSUE CULTURE STUDIES).

Category 7. Insecticides, Toxicology and Related Interests

This is obviously a most important Category, one which extends over such a broad field that assignment of references to Subdivision often presents a multiple choice. A synthesized nicotinoid, for example, might properly be associated with "older insecticides" because of the very early use of nicotine; on the other hand, since it is distinctly a modern product, one would be equally justified in assigning such an item to "insecticides contemporary with phosphorus compounds". All points considered, it has seemed better to designate the first three Subdivisions on a chronological basis, rather than by chemical groups. Based on "purpose intended" or "objective sought" the remaining four Subdivisions then fall rather naturally into place, according to the practical aspects concerned. The very great number of "comparative" studies reported, make it necessary that one Subdivision be of a general nature, though many of the reference numbers listed there will also have been assigned to one or more special, preceding groups. The seven Subdivisions are as follows:

Subdivision a. References dealing primarily with older, conventional insecticides, such as derris (rotenone), borax and pyrethrum, together with compounds of later development chemically related to these.

Subdivision b. References treating primarily of DDT and other chlorinated hydrocarbons.

Subdivision c. References reporting on phosphorus insecticides, carbamates, and other insecticidal agents of modern and current production.

Subdivision d. References dealing in one way or another with synergists and adjuvants, e.g. the action of piperonyl butoxide in augmenting the insecticidal value of pyrethrins.

Subdivision e. References reporting on the effectiveness of various attractants (such as may be used in baits) and repellents (valuable in the protective spraying of animals). Some of the numbers assigned here concern substances found in nature which affect the fly's behavior and reproductive activities.

Subdivision f. This Subdivision concerns those insecticidal substances which may be fed to stock or poultry, and which render the feces of such animals toxic to dipterous larvae. This is a field which has received increasing attention in recent years.

Subdivision g. All studies which treat of more than one class or type of insecticide, more especially investigations on their comparative effectiveness, have been assigned here. Also included are many references listed again in Category 10 (Physiology) and Category 12 (Resistance). References concerning fumigation are usually listed here as well.

Category 7. – *Subdivision a.* (OLDER AND CONVENTIONAL INSECTICIDES).

Category 7. – *Subdivision b.* (DDT AND RELATED COMPOUNDS).

Category 7. – *Subdivision c.* (PHOSPHOROUS COMPOUNDS, CARBAMATES AND OTHER MODERN PREPARATIONS).

Category 7. – *Subdivision d.* (SYNERGISTS AND ADJUVANTS).

Category 7. – *Subdivision e.* (REPELLENTS AND ATTRACTANTS).

Category 7. – *Subdivision f.* (SYSTEMIC INSECTICIDES).

Category 7. – *Subdivision g.* (MISCELLANEOUS AND GENERAL).

Category 8. Life History and Related Information

Because modern workers rarely study the *morphology* of the several life stages as an end in itself, we have decided to list all such references under "Life History", in the broadest meaning of that term. Under this approach, observations on *instinctive behavior* may likewise be handled by simply associating each reference with that stage of the life cycle in which the particular behavior or activity takes place. The six Subdivisions listed below are arranged essentially in chronological order, in terms of growth and development. The last three emphasize current trends in research on the biology of adult flies:

Subdivision a. References concerned either with the formation and maturation of gametes, or with developmental changes in the embryo prior to the hatching of the egg.

Subdivision b. References concerning the growth, development, and morphological characteristics of larvae and pupae.

Subdivision c. References emphasizing morphological characteristics and normal activities of adult flies, with special regard to food habits and reproductive behavior.

Subdivision d. References concerned with the aging process as observed (chiefly) in adult flies.

Subdivision e. References concerned with the flight range of normal flies, including numerous studies with marked specimens later recovered at various distances from points of release. There is frequently a cross-connection with Ecology, in as much as direction of flight can be greatly influenced by wind, odors of appropriate food, odors from sites suitable for oviposition, and the location of water bodies large enough to affect local humidity.

Subdivision f. References which do not fit conveniently into one of the foregoing subdivisions, or which convey information in addition to that which determined first assignment. In former times "Hibernation" was a favorite subject for investigation, and had it continued so, a separate Subdivision in this area would have been justified. In as much, however, as interest in this subject was (and is) confined almost wholly to Palaearctic regions, plus the fact that the subject has become of but minor interest in connection with fly control, references concerned either with hibernation or diapause phenomena are here included under Subdivision f.

Category 8. – *Subdivision f.* (Miscellaneous and General).

Category 9. Military Interests

This small category is recognized because of its historical interest. As long as animals, such as horses, mules and, in some parts of the world, camels, were used in large numbers as a necessary part of military operations, the fly control problem was always of great concern. The handling of human waste was also, of course, on a primitive level, both in cantonments and in prison compounds. Not infrequently typhoid and/or dysentery became rampant throughout a military unit and seriously weakened its effectiveness. This was the case with American troops during the Spanish American War, and British units, serving in South Africa, encountered similar problems. World War I still found many animals in use, both for transportation of artillery and as cavalry, especially on the Eastern front. The medical literature of the period continued to reflect grave concern over the relation of fly control to enteric infection under military condtiions.

While World War II saw a tremendous increase in concern for Medical Entomology, the problems of the military became largely identified with those of Public Health in general, and the availability of vaccines, plus proper attention to a pure water supply and sanitary disposal of human waste, together with the mechanization of both cavalry and artillery, removed fly control from its hitherto conspicuous position in military medicine. Scientific work done on and with synanthropic flies by professional entomologists in military service came to be published for the most part in professional journals of non-military type. In this bibliography, we are therefore listing only the earlier publications, in two groups:

Subdivision a. References concerned chiefly with sanitary practices, such as burial or burning of animal manure, and routine disposal of garbage and human waste.

Subdivision b. References dealing with all other aspects of fly control under military conditions, such as the use of larvicides, space sprays, fly traps, screens, fly paper and electrical devices.

Category 9. *Subdivision a.* (SANITARY PRACTICES).

Category 9. – *Subdivision b.* (MISCELLANEOUS AND GENERAL).

Category 10. Physiology and Biochemistry

As with all plant and animal groups, research on synanthropic flies in modern times has turned largely away from taxonomy and morphology,and has become more concerned with the vital processes of the living organism. Increasing knowledge of the chemical nature of hormones, enzymes and protein substances in general has opened a way to a more exact understanding of the mechanisms of nutrition, respiration, elimination and reproduction. For convenience of reference, five Subdivisions are recognized:

Subdivision a. References emphasizing foods, nutrition, and the chemistry of assimilation.

Subdivision b. References concerned with metabolism in the broad sense. Work in this field has been much stimulated by the ability of flies to metabolize insecticides such as DDT, and many of the items listed here deal with metabolic processes involved in the development of resistance.

Subdivision c. Involved with almost every step of a metabolic nature are enzymes, which act, for the most part as catalytic agents. Because of the resistance problem the relation of cholinesterase to the degradation of insecticides has received almost a disproportionate share of attention, and references of this nature tend to dominate the list. We have included here all items in which a particular enzyme, or enzyme group, is mentioned, either in the title, or in the annotation which follows.

Subdivision d. References which concern hormones. This area is chiefly interesting in connection with the initiation of metamorphic change, particularly the function of moulting-fluid glands.

Subdivision e. There is a great array of references which deal with too many aspects of physiologic function to warrant listing in any one of the foregoing Subdivisions. Besides these, there are many which treat of physiology and behaviour in the general sense. Both are assigned to Subdivision e. A recent trend toward intensive investigation of the compound eye and the optical principles involved in visual function will probably establish a special literature of its own in a few years. Until more publications appear, however, it seems best to include the presently available items in the general group.

Category 10. – *Subdivision a.* (DIET AND NUTRITION).

Category 10. – *Subdivision b.* (METABOLISM IN THE BROAD SENSE).

Category 11. Predators, Parasites and Biological Control

The inter-relationships between *Musca domestica* and other organisms which share its environment have been studied since early times. That a certain amount of "natural control" of fly population results from these relationships cannot be questioned. The possibility of so manipulating one or more factors in the complex, with the intention of further checking the fly's natural increase, has led to serious investigations in the area of "biological control". Also, the discrete use of parasites and/or predators, along with the careful employment of insecticides, perhaps on a seasonal basis, in other words "integrated control", shows promise of having considerable future value, especially in view of the fly's ability to develop great resistance to most modern insecticides. The grouping of references into Subdivisions, as listed below, is chiefly along taxonomic lines:

Subdivision a. References concerned with species which act as predators of flies. For the most part these are adults of Coleoptera or Formicidae, which seek the immature stages of the fly as food. The larvae of certain other Diptera have similar habits also.

Subdivision b. References dealing with *insect* species only which either require the housefly as a necessary host (several genera of Hymenoptera) or which achieve transportation by clinging to the fly's exterior. A unique example of the latter is the record of a human louse, believed to be infected with typhus fever, being carried into a hospital ward by a roaming adult fly.

Subdivision c. References dealing with species of Arachnida. Two types of relationship exist. For some mites the fly is simply a carrier or transport host; for others a source of nourishment. Of greatest importance are those species which devour eggs and first stage larvae of the fly, in the latter's breeding medium. Such forms are prime candidates for a place in programs of integrated control.

Subdivision d. We have grouped together here all references having to do with bacteria and fungi, whether the organisms concerned are pathogenic to the fly, normal benign commensals, or etiological agents of human disease. They may be found externally on the fly, or be present in the fly's alimentary tract or within its tissues.

Subdivision e. References dealing in any way with Helminthes found associated with the fly. Of chief importance are several species parasitic in domestic animals, which utilize the fly as an intermediate host. A number of the references listed here will also be found under Category 2 (Disease Relations), Subdivision g (Veterinary Interests).

Subdivision f. References dealing with Protozoa (including Spirochaetes) recorded from the fly's body. Most important here are certain pathogens of man, e.g. the etiological agents of amoebic dysentery and yaws. The same references will be found in many cases under appropriate Subdivisions of Category 2. (Disease Relations).

Subdivision g. References concerning the use of two or more control factors, in a balanced manner, to reduce fly populations. One of these is usually a biological factor, such as predatory mite, but this need not necessarily be the case. A combination of sanitary practice, larvicides and chemosterilization of adults would also be an example of "integrated control". A few references concerning experiments of this type, though not actually appropriate for inclusion in Category 11, have been assigned here nevertheless, in order to bring together related modern ideas on control.

Category 11. – *Subdivision a.* (PRINCIPAL PREDATORS OF FLIES).

Category 11. – *Subdivision b.* (HEXAPOD PARASITES OF FLIES).

Category 11. – *Subdivision c.* (ARACHNIDA PARASITIC OR PREDATORY ON FLIES).

Category 11. – *Subdivision d.* (BACTERIA AND PARASITIC FUNGI).

Category 11. – *Subdivision e.* (HELMINTHES).

Category 11. – *Subdivision f.* (PROTOZOA AND SPIROCHAETES).

Category 11. – *Subdivision g.* (INTEGRATED CONTROL).

Category 12. Resistance

The brief, almost miraculous success of DDT, during and immediately following World War II, was thought at first to have solved problems of fly control for all time to come. When it became apparent that *Musca domestica*, perhaps more than any other insect species had become unbelievably resistant to this insecticide, a very serious situation prevailed. Especially unfortunate was the fact that many sanitary practices, especially in the disposal of animal waste, had been relaxed or even abandoned, because of early success with DDT.

For a time every effort was put forward to find satisfactory substitutes for DDT, and a considerable array of chlorinated hydrocarbon insecticides were brought forward and given practical trial. Synergists and adjuvants were also eagerly sought. Combinations of insecticides were tried, also programs involving alternate use. Other classes of compounds, particularly phosphorus preparations and carbamates have received much serious attention. In spite of this great effort, it is now quite apparent that the perfect insecticide will probably never be found. Flies and many other insects appear genetically capable of adaptation (by mutation and/or selection), and of prevailing ultimately against all preparations of a chemical nature that are in any way practical for insecticidal use. The whole subject of resistance therefore remains tremendously important, in spite of profitable researches in genetics, environmental manipulation, and integrated control. The four Subdivisions of this Category listed below have been arbitrarily chosen as the most expedient way to elucidate the subject:

Subdivision a. References concerning resistance to DDT and chemically related compounds, such as lindane, methoxychlor and toxaphene. To a considerable extent the references assigned here represent the "first historical period" of research on resistance to modern insecticides.

Subdivision b. References concerning resistance to all other insecticides, including phosphorus compounds, carbamates and various synthetic products, some of them chemically related to long-used substances, such as nicotine and the pyrethrins.

Subdivision c. References dealing with those instances where flies which have developed an appreciable resistance to an insecticide in practical use, prove noticeably resistant to a new compound to which they had not previously been exposed. Somewhat related to this phenomenon are instances of negative correlation, in which a strain shows marked resistance to one insecticide, while displaying more than usual susceptibility to another. Certain of the references assigned here represent observations of this type.

Subdivision d. A great many references deal with more than one aspect of the subject, and are best placed in a collective, or general group.

Category 12. – *Subdivision d.* (RESISTANCE. THE GENERAL SUBJECT).

Reference Numbers: 2049, 2050, 2088, 2097, 2100, 2116, 2117, 2127, 2131, 2136, 2152, 2206, 2257, 2299, 2308, 2334, 2359, 2385, 2389, 2406, 2409, 2424, 2459, 2476, 2481, 2486, 2500, 2528, 2585, 2609, 2610, 2662, 2663, 2687, 2717, 2718, 2728, 2747, 2761, 2762, 2775, 2866, 2894, 2920, 2938, 2940, 3081, 3083,

3112, 3113, 3114, 3126, 3127, 3132, 3133, 3137, 3150, 3153, 3171, 3172, 3176, 3178, 3182, 3202, 3263, 3420, 3432, 3465, 3480, 3481, 3490, 3493, 3496, 3583, 3584, 3585, 3586, 3588, 3654, 3655, 3657, 3733a 3738, 3743, 3747, 3923, 3924, 3934, 3935, 3938, 3941, 3943, 4016, 4037, 4081, 4083, 4101, 4197, 4217, 4224,

4248, 4264, 4295, 4314, 4315, 4337, 4370a 4489, 4495, 4559, 4569, 4572, 4588, 4589, 4590, 4598, 4627, 4671, 4672, 4710, 4721, 4741, 4842, 4844, 4885, 4888, 4891, 4940, 4951, 5030, 5049, 5085, 5143, 5145, 5221a 5224, 5265, 5365, 5384, 5416, 5448, 5450.

Category 13. Sterilization and Reduced Fecundity as Approaches to Control

Probably the first successful example of reducing fly populations by the release of previously sterilized males was in relation to the screw-worm fly (*Callitroga*), a long-standing pest of livestock. The possible adaptation of such techniques to *Musca domestica* has been actively investigated, as one of many potential control methods which might augment or replace the use of insecticides to which flies have shown resistance. A wide variety of approaches has been investigated in the laboratory, including administration of the chemosterilant by contact, in the rearing medium, in the food of adult flies and in other ways. Tepa, hempa and apholate are among the chemicals found most promising. Experiments involve females as well as males, and frequently include the histological examination of ovaries, testes, and developing gametes. X-rays have also been studied as a means of sterilization. Although originally stimulated by concern for fly control, a considerable amount of current research can best be termed basic, or academic in character. Because this area of work involves many different techniques, and may not become stabilized for several years, we have not attempted to define Subdivisions along technical lines. Only two groups of references are therefore listed below, one essentially theoretical, the other practical:

Subdivision a. All those references which represent laboratory investigation, without practical testing in the field.

Subdivision b. A considerably smaller group of references concerned with the results of field experiments large or small using one or more sterilizing techniques. Of constant importance in all attempts to make practical use of this approach, is the ability of sterilized individuals to compete effectively with wild flies in finding and serving mates.

Category 13. – *Subdivision a.* (LABORATORY DATA).

Reference Numbers: 2336, 2337, 2340, 2341, 2806, 2834, 2835, 2874, 2878, 2879, 2880, 2881, 2882, 2883, 2885, 2886, 2887, 2889, 2890, 2964, 2980, 2994,

Category 13. – *Subdivison b.* (FIELD DATA, AND TREATMENT OF THE GENERAL SUBJECT).

Category 14. Taxonomy and Group Differentiation

We have brought together here all references which have a significant bearing on the differentiation of species, subspecies and certain larger groups to which these may belong. Most of the earlier, classic work, was based chiefly on external body characters, though W. S. Patton made a serious effort to associate feeding habit with taxonomic position in the various muscoid groups. Refinement and clarification of early concepts came with detailed study of the genitalia, especially the copulatory organs of males. Modern contributions, exemplified by the many excellent papers of Saccà, have been concerned principally with geographical races and subspecies, and with experimental work in the fields of genetics and resistance. It seems best to recognize two Subdivisions, one taxonomic in a limited sense, the other of broader scope:

Subdivision a. Technical Names. This portion lists all species, genera, and taxonomic groups which are cited by name, in titles or in annotations. Because the entire Bibliography concerns *Musca domestica*, either directly or indirectly, no list of reference numbers is given under that particular binomial. Otherwise, the listing which follows is complete for all plant and animal forms concerned. The arrangement is alphabetical.

Subdivision b. Miscellaneous and General. Listed here are faunal lists, taxonomic keys, and certain other papers which discuss the authenticity of species, subspecies, geographical races, and genetically resistant strains. Included also are a few standard works containing information relative to the classification of muscoid groups.

Category 14. – *Subdivision a.* (TECHNICAL NAMES).

Category 14. – *Subdivision b.* (Miscellaneous and General).

Reference Numbers: 33, 43, 44, 45, 46, 99, 100, 135, 137, 188, 231, 264, 353, 363, 487a 488, 508, 561, 557, 578, 595, 607, 638, 676, 809, 838, 864, 865, 1078, 1160, 1187, 1197, 1198, 1204, 1245, 1294, 1295, 1374, 1375, 1390, 1391, 1419, 1458, 1459, 1460, 1461, 1498, 1499, 1500,

1507, 1579, 1581, 1696, 1757, 1833, 1955, 1969, 1983, 2071, 2107, 2108, 2134, 2144, 2204, 2251, 2272, 2429, 2475, 2509, 2538, 2539, 2542, 2546, 2548, 2552, 2554, 2555, 2556, 2558, 2562, 2563, 2564, 2565, 2606, 2613, 2727, 2729, 2736, 2748, 2757, 2794, 3201, 3244, 3275, 3276, 3293, 3365, 3510, 3554, 3683, 3742,

3754, 3755, 3817, 3847, 3873, 3911, 3913, 3915, 3925, 4032a 4132, 4135, 4198, 4220, 4273, 4608, 4610, 4623a 4624, 4625, 4754, 4876, 4905, 4954, 5017, 5028, 5119, 5150, 5192, 5201, 5231, 5298.

Category 15. Techniques and Apparatus

The majority of the references falling in this category are listed in Section VII, and therefore bear numbers above 5000. Many other papers, however, listed in Sections IV, V, and VI have to do at least in a minor way with innovations in technique, and should not be overlooked. Regardless of source, all papers concerned with technical procedure, are here grouped in the five Subdivisions listed below:

Subdivision a. References emphasizing the rearing of laboratory specimens.
Subdivision b. References emphasizing assays and testing procedures.
Subdivision c. References concerned with methods for studying physiological processes.
Subdivision d. References concerned with experiments in the field. This includes tagging techniques.
Subdivision e. Miscellaneous references of a less specialized nature.

Category 15. – *Subdivision a.* (Rearing Procedures).

Reference Numbers: 1131, 1328, 1449, 1477, 1484, 1557, 1564, 1565, 1566, 1658, 1700, 1735, 1751 2011, 2206, 2240, 2646, 2703, 2911, 3618, 3257, 3317, 3416, 3455, 3483, 3845, 3862, 3940, 3949, 4169, 4185, 4247, 4311, 4351, 4464, 4473, 4474, 4502, 4696, 4698, 4716, 4828, 4919, 5007, 5069, 5070, 5071, 5085,

ADDENDUM

The authors realize, that in spite of their best efforts to assemble all significant titles for each decade of research, a considerable number have probably been overlooked. Investigators who do not find their work listed, or who know of publications by others that seem relevant to the scope of this bibliography but have not been included, are invited to communicate with the senior author who may be reached through the Department of Biology at Northern Michigan University. If the present publication is well-received, a Supplementary Bibliography may be considered, to be published at the end of five years. All new references would be listed, along with omissions from the original document.